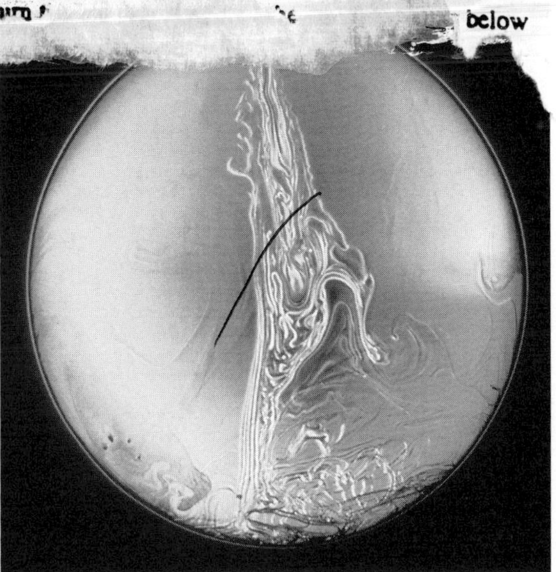

JOHN ANDREW
PAUL RISPOLI

Hodder & Stoughton
A MEMBER OF THE HODDER HEADLINE GROUP

Orders; please contact Bookpoint Ltd, 78 Milton Park, Abingdon, Oxon OX14 4TD.
Telephone: (44) 01235 827720, Fax: (44) 01235 400454. Lines are open from 9.00–6.00,
Monday to Saturday, with a 24 hour message answering service.
Email address: orders@bookpoint.co.uk

British Library Cataloguing in Publication Data
A catalogue record for this title is available from The British Library

ISBN 0 340 65505 4

First published 1999
Impression number 10 9 8 7 6 5 4 3 2 1
Year 2005 2004 2003 2002 2001 2000 1999

Cover photo from Science Photo Library
Typeset by GreenGate Publishing Services, Tonbridge, Kent.
Printed in Great Britain for Hodder and Stoughton Educational, a division of Hodder Headline Plc, 338 Euston Road, London NW1 3BH, by Redwood Books, Trowbridge, Wilts

Contents

Acknowledgements

This second edition of *Chemistry in Focus* has been completely updated and revised to cover the new Advanced Subsidiary and A level syllabus. A lively and accessible text, it retains the successful features of the popular first edition while bringing the subject up to date.

The authors are grateful to the following for their help in the production of this book:

- Our reviewers, especially Nigel Mason and Jeffrey Hancock for constructive criticism and helpful suggestions.
- Friends, colleagues and students who have meticulously worked through sections of the text making many valuable comments.
- Our long-suffering wives and families who cheerfully (for the most part) accepted the long hours spent at the wordprocessor during the gestation of the book.
- All the publishing staff involved at Hodder & Stoughton.
- Many others, too numerous to mention individually, who have contributed in a variety of ways towards the final version.

Of course, any errors or omissions that remain are entirely our own responsibility.

John Andrew and Paul Rispoli

The publishers would like to thank the following individuals, institutions and companies for permission to reproduce photographs in this book. Every effort has been made to trace ownership of copyright. The publishers would be happy to make arrangements with any copyright holder whom it has not been possible to contact:

Action-Plus Photographic (319); Adam Hart-Davis/Science Photo Library (279); Alfred Pasieka/Science Photo Library (195, 434, 505); All-Sport UK Ltd (364); Andrew Lambert (159, 256, 322 top & bottom); Andrew Syred/Science Photo Library (572 top); Andrew Ward (425, 571 top); Astrid & Hans-Frieder Michler/Science Photo Library (585); Barnaby's Picture Library (59 left, 59 top right, 101, 388 bottom); BFI Pictures and Stills (398 bottom); Biology Media/Sience Photo Library (393); Biophoto Associates (329); British Gas/P.Reynolds (197 bottom); Bubbles Photo Library (388 top); Calor Gas Ltd (203); By permission of the Syndics of the Cambridge University Library (4); Charles D. Winters/Science Photo Library (621); Chris Priest/Science Photo Library (398 middle); Claude Nuridsany & Marie Perennov/ Science Photo Library (1); CNRI/Science Photo Library (422); © CORBIS Picture Library (85, 241, 336, 477, 503 left, 569); Crown Paints (522 bottom); CWE Ltd (277); Dave Thompson (397 bottom, 556, 572 bottom); David Parker/Science Photo Library (39); David Nunuk/Science Photo Library (228 bottom right); Department of Clinical Radiology, Salisbury District Hospital/Science Photo Library (440 far right); Elf Atochem UK Ltd. (254 bottom); Emma Lee (269, 398 top, 485, 572 middle, 579 top two, 650); Evostick Information Services (283 top); Geoff Tompkinson/Science Photo Library (411); Graham Burns (600); Greenpeace Communications Ltd (206 left, 206 right); Griffin & George (168); Harrogate Resort Services Department (567); Hays Chemicals/Science Photo Library (45 bottom); Holt Studios International/Nigel Cattlin (564 bottom, 579 bottom); Hulton Getty Picture Collection Ltd (298); Hutchison Library (409); Imperial War Museum (5, 252 left, 252 right); Inga Spence/Holt Studios (560); Jackie Lewin, Royal Free Hospital/Science Photo Library (397 top); J Allan Cash Ltd, Worldwide Photographic Library (565 top left & right)James King-Holmes/Science Photo Library (228 bottom left); James Holmes, Hays Chemicals/Science Photo Library (559); Jerome Yeats/Science Photo Library (619 bottom); Jerry Mason/Science Photo Library (66); John Cox (494); John Mead/Science Photo Library (618); Kingfisher Studios (20, 308 top & bottom, 351 left & right, 440 right); Laboratory for Molecular Biology, Dr Arthur Lesk/Science Photo Library (92); Lawrence Livermore National Laboratory/Science Photo Library (571 bottom); Malcolm Fielding, Johnson Matthey PLC/Science Photo Library (440 far left); Manchester City Council (3); Marie Curie Cancer Care (38); Martin Dohrn/Science Photo Library (45); Mehav Kulyk/Science Photo Library (577); NASA/Science Photo Library (44, 268); National Power, a division of the CEGB (60 right, 197 top, 198); Nursing Standard (254 top, 388 middle); Philippe Plailly/Eurelios/ Science Photo Library (227, 283 bottom); Popperfoto (42 middle, 42 bottom, 101 top, 120 bottom-left column); Quadrant Picture Library (299); Roddy Paine (60 left, 205 bottom, 291); Rosenfeld Images Ltd/Science Photo Library (483); Science Photo Library (86, 237, 564 top); Shell (73 right, 205 top, 208, 219 right); Simon Fraser/Dept. of Neuroradiology, Newcastle General Hospital/Science Photo Library (447); St Bartholomew's Hospital, London, Dept of Medical Illustration (310); St Bartholomew's Hospital, London//Science Photo Library (31); STEAM (162, 353); Telegraph Colour Library (320); Trustees of the Science Museum (30); Tek Image/Science Photo Library (407); U K A E A (43 top, 43 middle, 45 middle); Vandenbergh & Jurgens Ltd (407); West Kent Cold Store (579 top); Will & Deni McIntyre/Science Photo Library (557); Z E F A Picture Library (UK) Ltd (522 top); Zeneca Ltd (351).

We would also like to thank the following examination boards for permission to reproduce questions: AEB, NEAB, UCLES and WJEC.

THEME A

Foundation Chemistry

1. Access to A level Chemistry
2. The structure of the atom
3. An introduction to chemical energetics
4. Chemical bonding and the structure of materials
5. Periodicity
6. Carbon compounds – an introduction to organic chemistry

CHAPTER

1

Access to A level chemistry

Contents

This self-study chapter is designed to help you grasp the essential ideas of chemical quantities, equations and the routine calculations which appear throughout the book. Each section starts with the 'Study Checklist' which tells you exactly what you should be able to do by the end of the section. Then there is an explanation of the topic, usually with worked examples. The section finishes with 'Test yourself ...' questions to check your progress.

1.1 What are atoms?

Study Checklist

At the end of this section you should be able to:

1. State that atoms contain protons, neutrons and electrons and describe the properties of these particles in terms of their relative charge and mass.
2. Explain the term 'isotope'.
3. Explain the terms 'atomic (proton) number' and 'mass (nucleon) number' and use these quantities to work out the numbers of protons, neutrons and electrons in atoms and ions.

You should also have an awareness of the evidence which has led to the nuclear model of atomic structure.

Well over 2000 years ago, the Greek philosopher Democritus proposed that all matter consisted of tiny indivisible particles, which he called **atoms** (a word which means uncuttable). Unfortunately, he had no evidence to support his theory and Plato and Aristotle argued against the existence of the atom saying that if it did explain nature, there would be no need for God. The Ancient Greeks preferred the idea that matter resulted from a mystical combination of 'The Four Elements': fire, earth, water and air. It was not until the 18th century that chemists began to explain their results in terms of atoms.

In this section we shall look at the research and discoveries which have led to the development of the nuclear model of atomic structure.

Dalton's atomic theory

Between 1774 and 1802, chemists proposed a number of laws concerning the ratios by mass of the elements in pure compounds. In 1807 John Dalton, an English schoolteacher, published an **atomic theory** which explained these laws and his own experiments. The main points of the theory are:

- All elements are made of very small particles called **atoms**.

- Atoms cannot be created, destroyed or divided.
- Atoms of the same element are identical but differ in mass from those of other elements.
- Atoms form compounds by combining chemically in simple whole number ratios.

▲ John Dalton (1766–1844) engaged in one of his experiments – collecting marsh gas which is mainly methane. Dalton is regarded as the father of chemical theory. As a devout Quaker, he was very modest about his achievements, even though he gained great acclaim for his work. A teacher from the age of only twelve years, Dalton discovered the law of partial pressure of gases and helped found the British Association for the Advancement of Science. Dalton, who always maintained he was too busy to get married, was an expert on meteorology and was also the first person to describe colour-blindness, being himself colour-blind. This painting, by Ford Madox-Brown, hangs in the Town Hall, Manchester

During the 19th century, the idea of an indivisible atom gradually lost favour and was finally discredited in 1897, when Joseph John Thomson proved the existence of a sub-atomic particle, the **electron**. In 1913, Thomson also discovered that atoms of the same element could have different masses and he called these **isotopes**.

Even with these inaccuracies, Dalton's theory was a landmark in the development of chemical theory and it allows us to define an *atom as the smallest particle of an element which can possess the chemical properties of that element*.

Thomson suggests the existence of sub-atomic particles: electrons and protons

In 1875, William Crookes passed an electrical current through a gas at low pressure and found that invisible radiation was emitted from the cathode (the negatively charged electrode). These **cathode rays** caused a bright glow on striking a fluorescent screen and, like light, they moved in a straight line and cast shadows of objects in their path. In fact, cathode rays are responsible for the same bright glow that we see in neon signs, fluorescent tubes and television screens. By studying their deflections in electric and magnetic fields, Joseph John Thomson, in 1897, concluded that cathode rays were beams of very small, negatively charged particles. He called these particles **electrons** and suggested that they were the substance from which all chemical elements were built. Thomson was awarded the Nobel Prize in 1906 and was further honoured by being buried in Westminster Abbey near the grave of Isaac Newton.

Thomson had been able to measure the charge to mass ratio (e/m) of an electron. Later, in 1909, the American scientist Robert Millikan managed to measure the electronic charge (e) and so the electronic mass could be calculated. The electron was found to be an extremely light particle having a mass (m) of just 9.1×10^{-28} g.

Positive rays were first observed by Goldstein in 1886 using the apparatus shown in Figure 1.1. Using a discharge tube with a perforated cathode, he saw a glow on the fluorescent screens at each end of the tube. The glow at A was due to cathode rays (as expected). However, the radiation at B was attracted towards a negative charge, which meant that it must be positively charged. Thomson followed up this work and found that the charge to mass ratio of these positive rays varied with the gas used in the tube. Since hydrogen gas gave positive rays with the *highest* charge to mass ratio, he concluded that hydrogen produces a positive particle of the *lowest* mass. This particle was accepted as the fundamental positive particle of atomic structure, and was called a **proton**, from the Greek word *proteios* meaning 'the primary one'. Whilst a proton and electron have equal but opposite charges, the proton is about 1835 times heavier than an electron.

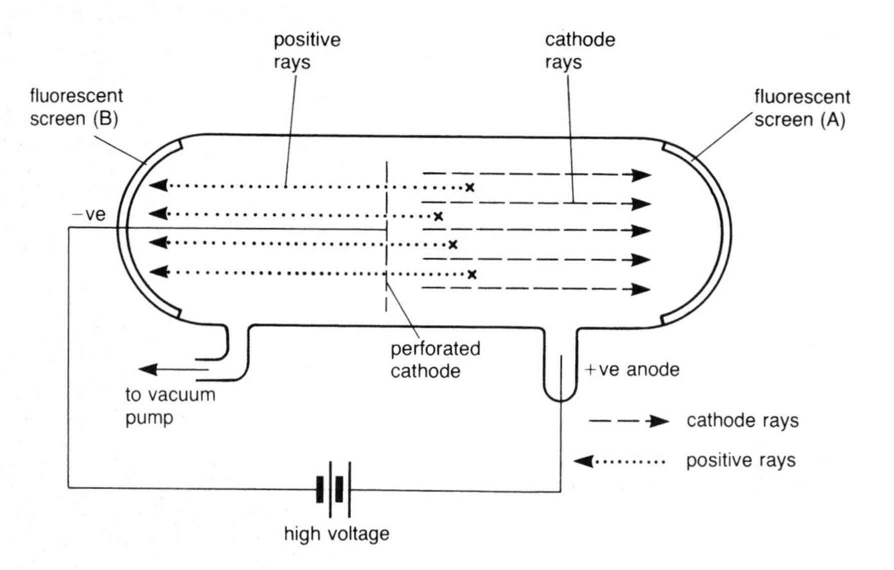

Figure 1.1 ▶
Goldstein's experiment for the detection of positive rays. At points X, a rapidly moving electron strikes a gas molecule or atom to form a positive ion. This moves towards and then through the perforated cathode

Rutherford's idea of the atomic nucleus

Since atoms were known to be electrically neutral, Thomson proposed that they must contain the same number of protons and electrons. Thus, he pictured an atom as a solid sphere of positive charge, made of protons, with electrons embedded in it. Thomson's model was accepted for about ten years and it became known as the 'plum pudding' model of the atom.

A major step forward in the investigation of atomic structure came from work at Manchester University in 1909. Under the supervision of Professor Rutherford, Geiger and Marsden studied the scattering of alpha (α) particles by thin sheets of metal foil. **α-particles** are helium nuclei, He^{2+}, produced by the decay of certain radioactive elements such as radium.

Using the apparatus shown in Figure 1.2 Geiger and Marsden showed that *most of the α-particles passed straight through the foil or were very slightly deflected*, see Figure 1.3. However, to their amazement, *about 1 in 8000 of the*

▲ Ernest Rutherford (1871–1937) received the Nobel Prize for Chemistry in 1907 for his work on radioactivity. He proposed the nuclear model of the atom in 1911

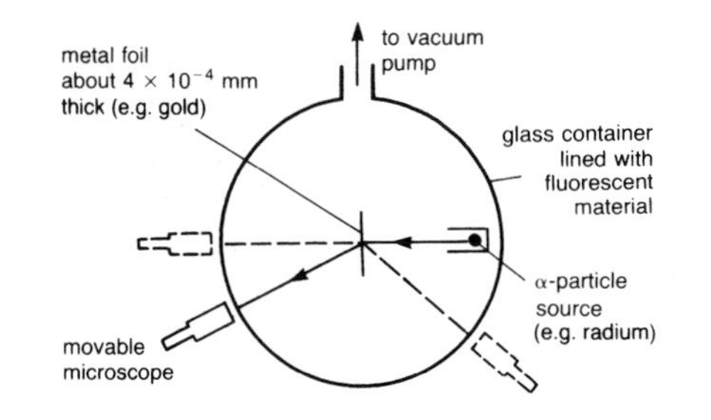

Figure 1.2 ▶
The apparatus used by Geiger and Marsden

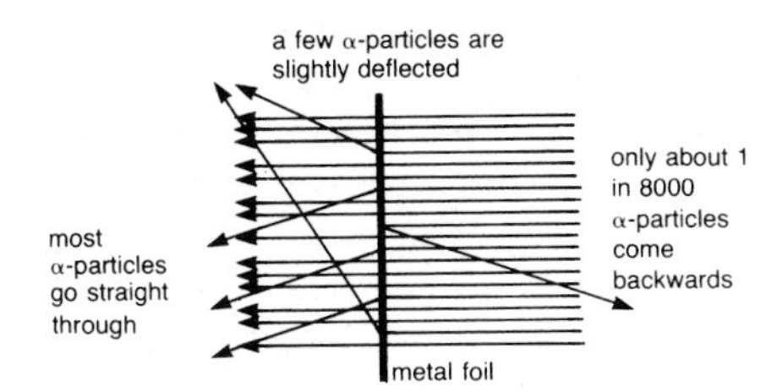

Figure 1.3 ▲
An illustration of Geiger and Marsden's results

α-particles came backwards off the foil. Rutherford's own account of the discovery is worth noting.

> 'I remember two or three days later Geiger coming to me in great excitement and saying, "We have been able to get some α-particles coming backwards ...". It was almost as incredible as if you fired a 15-inch shell at a piece of tissue paper and it came back and hit you.
>
> On consideration, I realised that this scattering backwards must be the result of a single collision ... with a system in which the greater part of the mass of the atom was concentrated in a minute nucleus.'

Rutherford's explanation of Geiger and Marsden's results is described in Figure 1.4.

The fifteen-inch forward guns of HMS Queen Elizabeth photographed circa 1917. These guns had a range of 16 miles ▶

Following these experiments, Rutherford introduced the **nuclear model of the atom** in 1911. *This viewed the atom as a very small, dense nucleus of positive charge with negatively charged electrons arranged around it.* Indeed, modern techniques show that the radius of the nucleus is only about one ten-thousandth of the radius of the atom. Thus, if a nucleus were the size of a football, the atom would be over a mile in diameter. Also, the particularly low mass of the electron means that the nucleus can be taken as the sole contributor to the atomic mass.

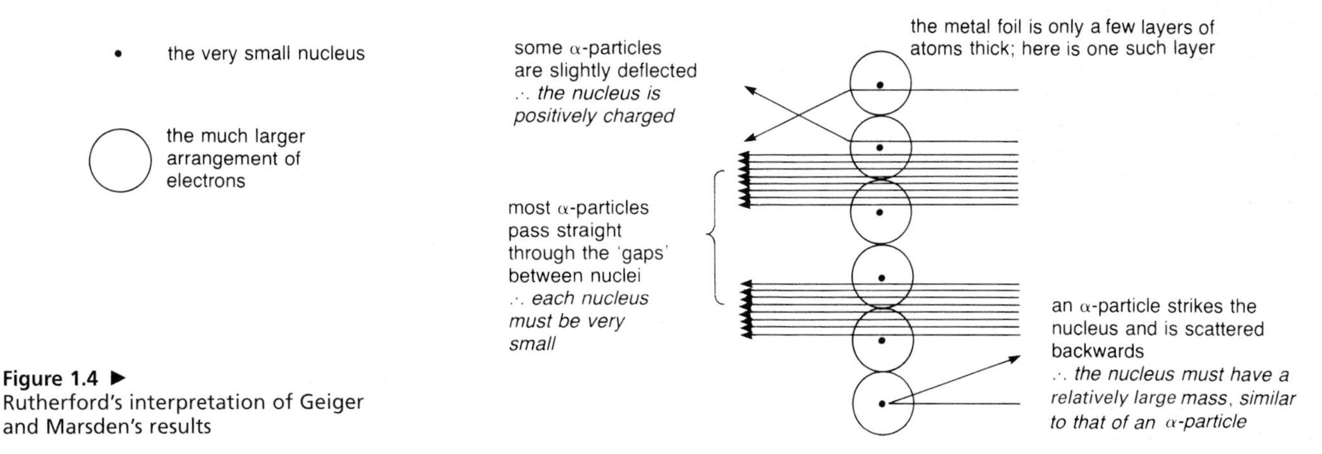

Figure 1.4 ▶
Rutherford's interpretation of Geiger and Marsden's results

At first, Rutherford thought that the positively charged nucleus must be made of protons held together by massive attractive forces. However, further study of the α-particle scattering patterns showed that the mass of the protons in the nucleus was only about half the atomic mass. Consequently, Rutherford suggested that the *nucleus must also contain neutral particles having the same mass as the proton*. These neutral particles were termed **neutrons**.

Moseley investigates the atomic nucleus

In 1869, Dmitri Mendeleev showed that when the elements were arranged in order of increasing atomic mass, elements with similar chemical properties were found at regular intervals. His arrangement became known as the **periodic table of the elements** and is discussed in Chapter 5.

There were, however, some problems. To fit in with the overall pattern, certain elements had to be placed out of 'atomic mass' order. Mendeleev got round this by saying that atomic masses could not be measured accurately enough at that time. When more accurate methods of mass determination were discovered, he expected his order to be proved correct. By the start of the 20th century, however, scientists were seriously questioning whether the periodic table should be based on the order of atomic masses. But what other order might be used?

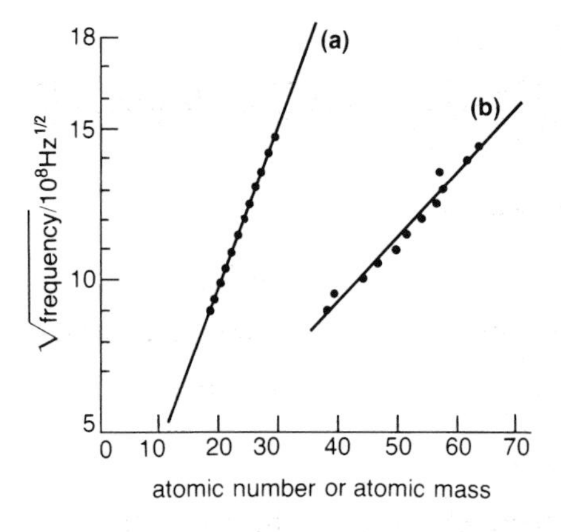

Figure 1.5 ▲
Some of Moseley's results for the X-ray spectra of various metals. (a) An excellent straight line graph for √(X-ray frequency) against atomic number; (b) plotting √(X-ray frequency) against atomic mass gives a poor straight line graph

This debate was resolved in 1913 by Henry Moseley who found that X-rays were released when metals were bombarded with electrons. He measured the frequencies of these X-rays and tried to relate them to (i) the atomic numbers and (ii) the atomic masses of the metals used. After drawing various graphs, he eventually obtained an excellent straight line by plotting √(X-ray frequency) against atomic number, Figure 1.5a. However, a graph of √(X-ray frequency) against atomic mass was not linear, Figure 1.5b. Moseley's impressive evidence confirmed that atomic number, not atomic mass, is the fundamental property upon which we should base our periodic arrangement of the elements.

The **atomic number** *or* **proton number** *of an element is defined as the number of protons in an atomic nucleus*. It also tells us:

- the number of electrons in a neutral atom;
- the numbered position of the element in the periodic table.

When we need to state the atomic number of an element, we write it as a subscript in front of the element symbol. Thus, ${}_{46}Pd$ means that palladium, atomic number = 46, has 46 protons and 46 electrons in a neutral atom. It is the 46th element in the periodic table.

Moseley's research enabled chemists to (i) identify elements from the frequency of the X-rays they emitted and (ii) calculate atomic numbers from experimental results. Moseley's results also acted as a signpost to the discovery of unknown elements such as scandium, ${}_{21}Sc$ and technetium, ${}_{43}Te$.

Chadwick discovers the neutron

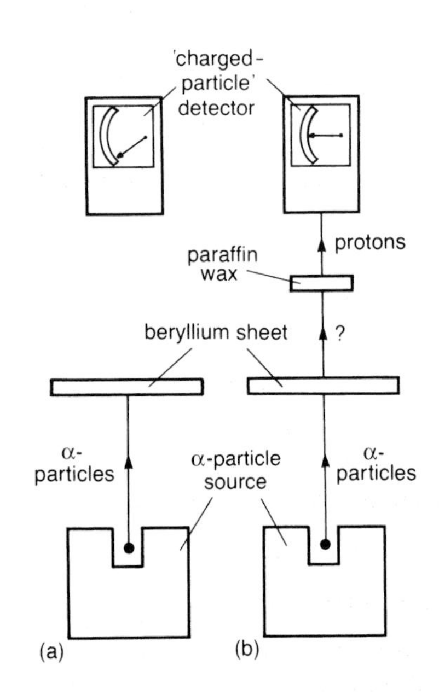

Figure 1.6 ▲
Chadwick's experiment to prove the existence of the neutron

The existence of the neutron was first proved by James Chadwick in 1932. When he bombarded a thin sheet of beryllium metal with α-particles, no charged particles were detected (see Figure 1.6a). However, when a film of paraffin wax was placed between the metal and the detector, there was a reading on the 'charged-particle' detector (see Figure 1.6b). From their charge to mass ratio, these particles were identified as protons. Chadwick proved that these protons must have been dislodged by particles of similar mass. Since the first experiment showed these particles were uncharged, Chadwick had detected neutrons. His work confirmed Rutherford's idea of twenty years earlier, that the atomic nucleus contained neutral particles.

With the exception of hydrogen, which is a single proton, all atomic nuclei contain neutrons. The number of neutrons in the nucleus can be worked out from its **atomic (proton) number** and its **mass number** or **nucleon number**, which is the number of nucleons (protons plus neutrons) in the nucleus. When we need to state the mass (nucleon) number of an atom, we write it as a superscript in front of the element's symbol. Thus, ^{16}O represents an atom of oxygen for which the total of protons plus neutrons is 16. Thus, an ^{16}O atom contains 16 nucleons.

We saw earlier that the atomic number is written as a subscript before the element's symbol. Thus, an atom of $^{19}_{9}F$ has 9 protons, 9 electrons and 10 neutrons; it contains 19 nucleons. The properties of the fundamental sub-atomic particles are summarised in Table 1.1.

Table 1.1 Properties of the fundamental sub-atomic particles

Particle	proton	neutron	electron
charge	+1	0	−1
mass/g	1.67×10^{-24}	1.67×10^{-24}	9.10×10^{-28}
relative mass	1	1	1/1835
location	nucleus	nucleus	around the nucleus
effect of an electric or magnetic field	deflected	no effect	deflected
confirmed by	Rutherford (1911)	Chadwick (1932)	Thomson (1897)

Working out the number of protons, neutrons and electrons in an atom given the atomic and mass numbers

Since the atomic number gives the number of protons in the nucleus and the mass number gives the number of protons plus neutrons, then:

number of neutrons = mass number − atomic number

Atoms are neutral overall, so:

number of electrons = number of protons = atomic number

Example

Give the number of protons, neutrons and electrons in atoms of $^{28}_{14}Si$ (silicon) and $^{56}_{26}Fe$ (iron).

Silicon number of neutrons = 28 − 14
= 14

number of electrons = number of protons = atomic number = 14

So, a silicon atom has 14 protons, 14 neutrons and 14 electrons.

Iron number of neutrons = 56 − 26
= 30

number of electrons = number of protons = atomic number = 26

So, an iron atom has 26 protons, 30 neutrons and 26 electrons.

When atoms lose electrons, **positive ions** are formed, e.g.:

$$\underset{\text{sodium atom}}{Na} \longrightarrow \underset{\text{sodium ion}}{Na^+} + e^-$$

Whilst the sodium ion has the same number of protons and neutrons as its atom, it will have one electron less. The sodium atom has mass and atomic numbers of 23 and 11 respectively, giving it 11 protons, 12 neutrons and 11 electrons. So the sodium ion will have 11 protons, 12 neutrons and 10 electrons.

If atoms gain electrons, e^-, **negative ions** are formed, e.g.:

$$\underset{\text{oxygen atom}}{O} + 2e^- \longrightarrow \underset{\text{oxide ion}}{O^{2-}}$$

Whilst the oxide ion has the same number of protons and neutrons as its atom, it will have two electrons more. An oxygen atom has mass and atomic numbers of 16 and 8 respectively, giving it 8 protons, 8 neutrons and 8 electrons. Thus, the oxide ion will have 8 protons, 8 neutrons and 10 electrons.

As mentioned earlier, Thomson discovered **isotopes** (atoms of the same element which have different masses). Isotopes have the same atomic number but different mass numbers, so they contain different numbers of neutrons. For example, a nickel atom can exist as one of five isotopes with mass numbers of 58, 60, 61, 62 and 64. The atomic number of nickel is 28. Try working out the number of neutrons and electrons in each isotope.

Test yourself

1 What is meant by the term 'atom'?

2 **a)** Name the fundamental particles which make up an atom's structure.
b) How are these particles arranged within the atom?
c) Compare the fundamental particles in terms of charge, mass and deflection in an electric field.

3 What are atomic isotopes?

4 Define the terms 'nucleon, 'atomic (proton) number' and 'mass (nucleon) number' and explain what information we can obtain from them.

5 List the number of protons, electrons and neutrons in the following atoms and ions:

$^{228}_{89}Ac$ $^{133}_{55}Cs$ $^{127}_{53}I$ $^{204}_{82}Pb$ $^{15}_{7}N$ $^{39}_{19}K^{+}$ $^{32}_{16}S^{2-}$ $^{138}_{56}Ba^{2+}$

6 Ruthenium, proton number 44, has seven atomic isotopes with nucleon numbers 96, 98, 99, 100, 101, 102 and 104. List the number of protons, electrons and neutrons in the isotopes.

1.2 Bohr's model of the atom: electrons orbiting the nucleus

Study Checklist

At the end of this section you should be able to:

1. Understand that electrons are arranged around the nucleus in orbits or shells.
2. Write down electronic arrangements for the atoms of atomic number 1–18.
3. Draw dot and cross diagrams to represent the electrons in shells for these atoms.

In 1913, the Danish physicist Niels Bohr suggested a model of the atom in which electrons move round the nucleus in definite orbits similar to the way planets orbit the Sun. He calculated that the electrostatic force between the positive nucleus and each negative electron was balanced by the outward force due to their orbital motion. According to Bohr, the electron was stable in a given orbit and could remain there indefinitely. If the electron accepted a definite amount of energy, called a **quantum** of energy, it could move to a higher energy orbit further from the nucleus – it could not exist between orbits, though. Thus, if the difference in energy between orbits was 100 kJ and 99.9999 kJ supplied, the electron would not jump to the higher energy orbit – if exactly 100 kJ were supplied the transition would occur. We shall return to this idea in section 2.7.

For now, picture the atom as a central nucleus surrounded by electrons arranged in orbits whose energy, and size, increase as they get further from the nucleus. Thus, orbits nearer the nucleus are smaller than those further away and can hold fewer electrons. Each orbit, or **shell**, is identified by its **principal quantum number, n**. We find that the maximum number of electrons which can be held in the first three shells is as follows:

Shell	maximum
$n = 1$	2 electrons
$n = 2$	8 electrons
$n = 3$	18 electrons

Shells increase in energy and size

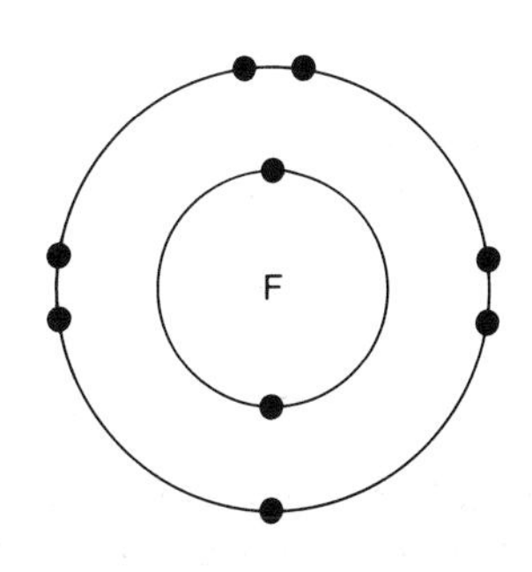

Figure 1.7 ▲
A dot and cross diagram to show the electron arrangement in a fluorine atom. There are 2 electrons in the $n = 1$ shell and 7 electrons in the $n = 2$ shell. The nucleus, which contains 9 protons and 10 neutrons, is not drawn in this type of diagram

Electrons fill the lower energy shells first. Thus, the electron arrangement for fluorine (atomic number 9) is 2 electrons in $n = 1$, 7 electrons in $n = 2$. This arrangement may be written simply as F 2.7. The electron arrangement can be represented by a **dot and cross diagram**, as shown in Figure 1.7.

Example

Write down the electron arrangement of aluminium (atomic number 13). Aluminium will have 2 electrons in $n = 1$, 8 electrons in $n = 2$ and 3 electrons in $n = 3$. The arrangement is:

Al 2.8.3 represented by

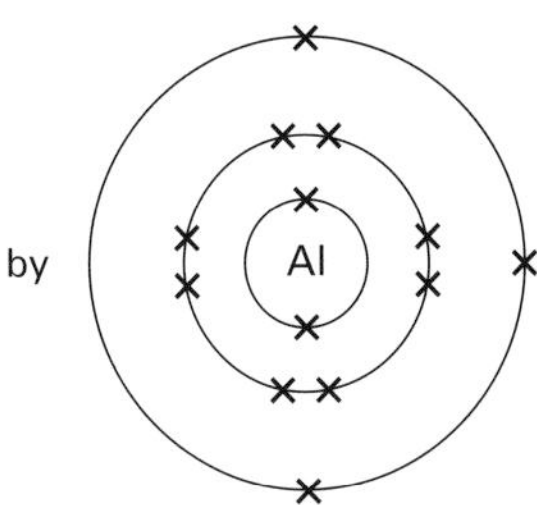

Test yourself

Refer to the periodic table on page 132. You will see that it is arranged into **groups** (vertical) and **periods** (horizontal). The atomic number of each element is given. This question concerns the elements hydrogen to argon, atomic numbers 1–18.

1 Write down the electron arrangement for each atom.

2 Draw a diagram to represent each electron arrangement.

3 For each element identify the relationship between:
a) the number of shells and the period,
b) the number of electrons in the outermost shell and the group.

1.3 Why do atoms react? Ionic and covalent bonding

Study Checklist

At the end of this section you should be able to:

1 Appreciate that when *some* elements react, their atoms achieve the stable electronic arrangement of a noble gas.

2 Explain ionic bonding in terms of the transfer of electrons from an electropositive atom to an electronegative atom.

3 Explain covalent bonding in terms of two atoms sharing one or more pairs of electrons.

4 Explain what is meant by 'coordinate (dative covalent) bonding'.

5 Use dot and cross diagrams to illustrate the electron arrangements in ionic and covalent compounds.

You should also be aware of the different properties of ionic and covalent compounds, though this will be covered in much greater detail in Chapter 4.

When a piece of sodium is placed in fluorine gas, a vigorous reaction occurs and sodium fluoride, NaF is formed:

$$2Na(s) + F_2(g) \longrightarrow 2NaF(g)$$

However neon, the element between sodium and fluorine in the periodic table, is chemically inert. Neon's electron arrangement is 2.8, i.e. it has a full outer shell of electrons. Likewise helium has a full outer $n = 1$ shell and is also inert. The other noble gases are either inert or extremely unreactive – each has eight electrons, an octet, in its outer shell, although this is less than the maximum for that shell. It appears then that a full outer shell, or an octet of outer electrons, is a stable electron arrangement. In many cases other atoms can achieve such an arrangement via chemical bonding. There are two main types of chemical bond: ionic and covalent.

Ionic bonding

When atoms lose or gain electrons, ions are formed. Ionic bonding involves the **transfer** of one or more electrons from one atom to another. Most ionic compounds are formed by a metal donating electrons to a non-metal. For example, in Figure 1.8 the sodium atom donates an electron to the fluorine atom forming a sodium ion and a fluoride ion.

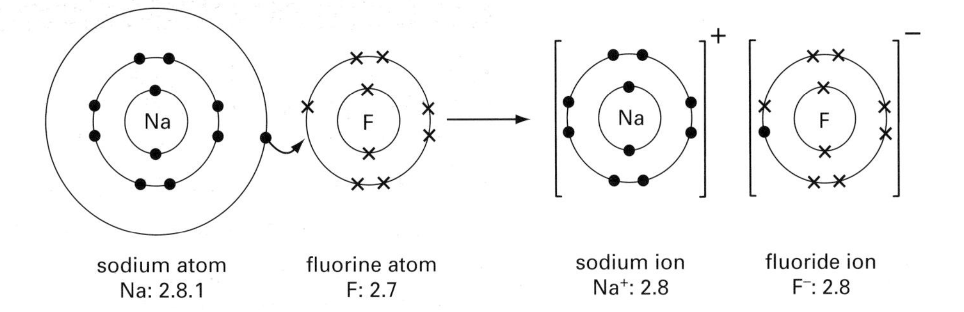

Figure 1.8 ▶
Electron rearrangement during the formation of sodium fluoride

Both ions have full outer $n = 2$ shells of electrons. Notice how the dot and cross diagram helps us to keep an account of the electrons involved in bonding. Remember these diagrams do not represent the actual positions of the electrons – they are simplified 2-D plans. In the above example, the sodium atom is described as being **electropositive**, that is it has a tendency to lose electrons. Fluorine atoms have a tendency to gain electrons – they are **electronegative**.

In lithium oxide, two lithium atoms react with each oxygen atom (see Figure 1.9). Once again, the ions formed have full outer shells of electrons, corresponding to the electron arrangements of the two noble gases: helium (2) and neon (2.8).

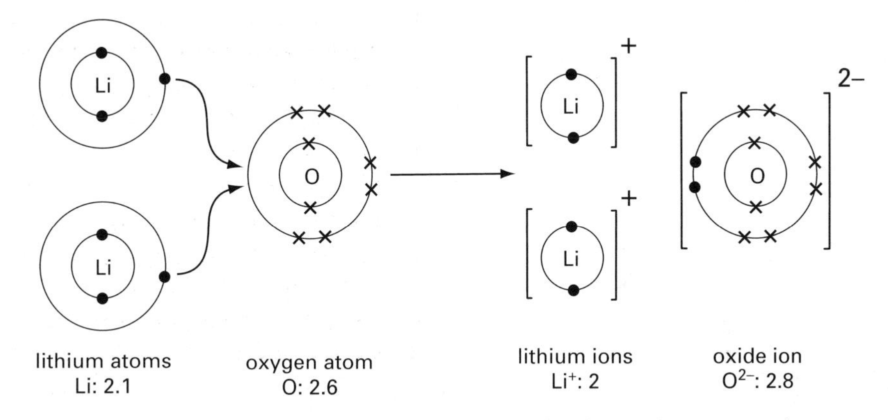

Figure 1.9 ▶
Electron rearrangement during the formation of lithium oxide

Strong electrostatic forces hold the oppositely charged ions together in a 3-D lattice which extends throughout the solid. Since the bonding is strong, ionic solids have high melting points and are hard, brittle substances which do not conduct electricity. However, when molten or dissolved in water, the ions are free to move, and they can conduct electricity.

Covalent bonding

When non-metals react together they usually form covalent bonds. A covalent bond is formed when the two nuclei are mutually attracted to the two bonding electrons – these are said to be **shared** between atoms. An example is the fluorine molecule, F_2 (see Figure 1.10).

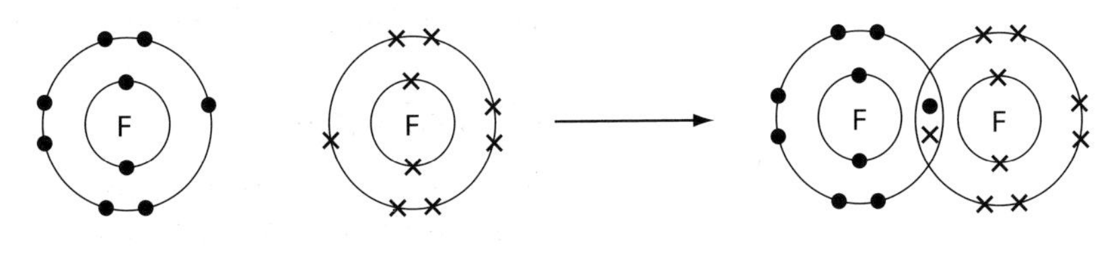

Figure 1.10 ▶
Fluorine atoms form a covalently bonded molecule by sharing a pair of electrons

By sharing two electrons, each fluorine atom has achieved a more stable electron arrangement 2.8, as found in the noble gas neon. In the oxygen molecule, O_2, each atom obtains neon's electron arrangement by forming a double covalent bond (see Figure 1.11).

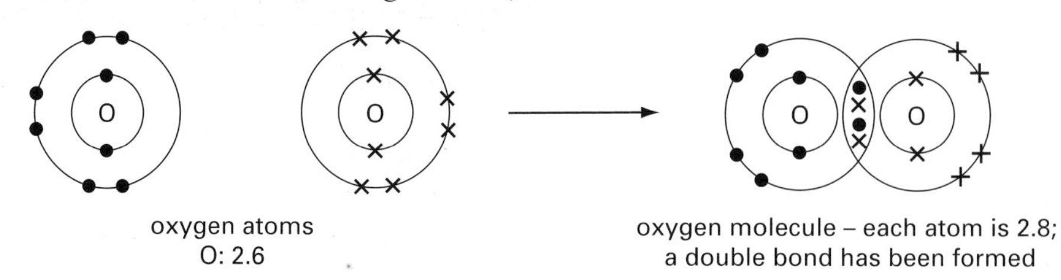

Figure 1.11 ►
The formation of a double covalent bond in the oxygen molecule. Each oxygen atom shares two pairs of bonding electrons

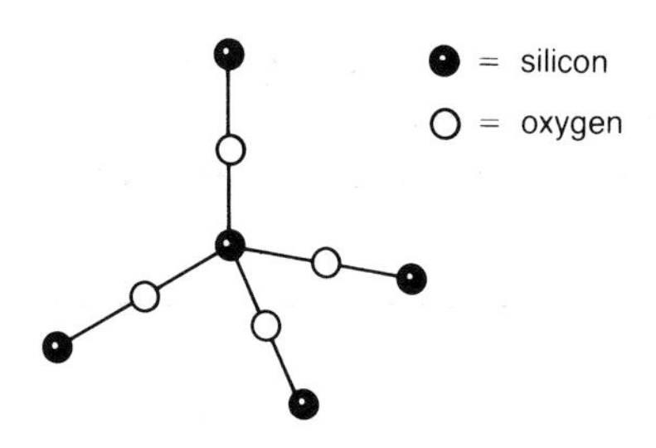

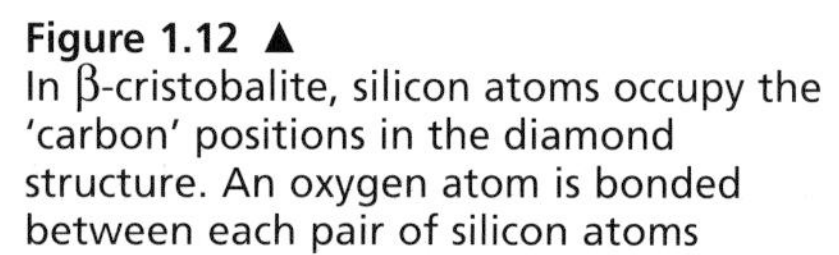
Figure 1.12 ▲
In β-cristobalite, silicon atoms occupy the 'carbon' positions in the diamond structure. An oxygen atom is bonded between each pair of silicon atoms

Covalently bonded substances fall into two categories. In **giant molecular** or **macromolecular structures**, the covalent bonding exists throughout the entire lattice of atoms. An example is sand, which is silicon dioxide, and a small part of the lattice is shown in Figure 1.12. Giant molecular substances are very hard solids with high melting points. Covalent bonds also exist between atoms in **simple molecular** substances such as fluorine and oxygen. Although the covalent bonds within the molecules are strong the attractive forces between the molecules themselves are very weak so they have low melting and boiling points. Most covalent compounds are poor conductors of electricity because they do not contain charged particles.

In some covalent bonds, known as **coordinate** (**dative covalent**) bonds, both bonding electrons are supplied by *one* of the bonding atoms. An example is the molecular addition compound formed when boron trifluoride, BF_3, reacts with ammonia, NH_3, which is shown in Figure 1.13. Coordinate bonding is widespread throughout chemistry as you shall see in later chapters.

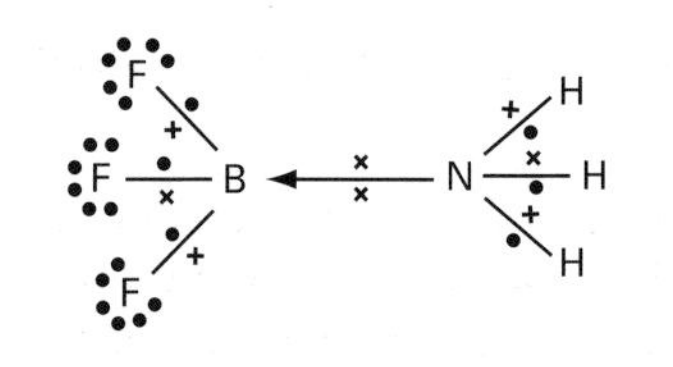

Figure 1.13 ▲
In the molecular addition compound BF_3NH_3, the nitrogen atom donates both electrons to the B—N covalent bond. (Note that the inner shells of the F, B and N atoms have been omitted for clarity)

Test yourself

1 What is the main difference between ionic and covalent bonding?

2 Each pair of elements below forms an ionic compound. Draw dot and cross diagrams for each atom and the ionic bond they form, as shown in Figure 1.8:
a) sodium, chlorine **b)** magnesium, oxygen
c) lithium, sulphur **d)** aluminium, fluorine

3 Use dot and cross diagrams like those in Figures 1.10 and 1.11 to describe the covalent bonding in:
a) hydrogen, H_2 **b)** chlorine, Cl_2 **c)** hydrogen chloride, HCl
d) carbon dioxide, CO_2 **e)** methane, CH_4
f) ethene, C_2H_4 (NOTE that the two carbon atoms are bonded together)

4 By classifying the elements as metals or non-metals, predict which of the following compounds are likely to be ionic. Which are probably covalent?
a) hydrogen fluoride **b)** magnesium sulphide **c)** sodium bromide
d) water, H_2O **e)** carbon monoxide, CO **f)** lithium fluoride
g) calcium oxide **h)** phosphorus trichloride, PCl_3

5 Ammonium nitrate, NH_4NO_3, is a fertiliser. It contains the ammonium ion, NH_4^+, which is formed when an ammonia molecule, NH_3, reacts with a hydrogen ion, H^+.
a) Draw dot and cross diagrams of ammonia and a hydrogen ion.
b) Now combine them in a dot and cross diagram of an ammonium ion.
c) What type of bond is formed?

1.4 Valency and molecular formulae

Study Checklist

At the end of this section you should be able to:

1. Define valency as the number of electrons that the atom donates, receives or shares in order to obtain a more stable electron arrangement.
2. Understand that the valency of an atom describes its combining capacity.
3. Work out the likely valency of atoms from groups 1 to 7 of the periodic table.
4. State the charge on, and give the formulae of, the following ions: sulphate, nitrate, hydroxide, carbonate and ethanoate.
5. Use valencies and ionic charges to write down the molecular formulae of simple compounds, and name them.

Combining capacity of atoms

We have seen that when elements react, they often achieve a full outer shell of electrons. *This is not always the case, for example in the BF_3 molecule, the boron atom has only six electrons in its $n = 2$ shell.* However, the idea of atoms reacting to complete their outer or **valence** shell is useful in helping us to understand the ratios in which atoms combine to form molecules.

Consider the carbon atom with an electron arrangement of 2.4. It could achieve a full valence shell in three ways:

1. Lose four electrons to form a positive ion, C^{4+}, electron arrangement 2. However, a massive amount of energy would be needed to enable the electrons to overcome the nuclear attraction and escape from the atom.
2. Gain four electrons to form a C^{4-} ion, electron arrangement 2.8. Once again, a lot of energy would be needed as the negative electrons are being forced into an increasingly negative ion.
3. Form covalent bonds by sharing four electrons and achieving an electron arrangement of 2.8. This is energetically the most favourable option.

We use the term **valency** to describe the combining capacity of an atom. Think of valency as the number of electrons an atom needs to lose, gain or share to achieve a more stable electron arrangement, such as a full valence shell or an octet of valence electrons. So, the valency of carbon is 4. Not surprisingly, the other elements in group 4 (silicon, germanium, tin and lead) with four electrons in the valence shell also have a valency of four.

Oxygen, a group 6 element with electron arrangement 2.6, could achieve a full valence shell by losing six electrons to form an O^{6+} ion. Energetically, this is a very unfavourable process as the number of protons increasingly outnumbers the electrons present. Oxygen prefers to gain two electrons to form an O^{2-} ion or share two electrons, depending on which atom it reacts with. Thus, with lithium, an ionic oxide would be formed, $(Li^+)_2O^{2-}$. With carbon, a covalently bonded oxide, CO_2, may be obtained in which each oxygen atom shares two electrons with the carbon atom. In both cases, the valency of oxygen is 2 and it obtains a full outer shell ($n = 2$) of electrons. A valency of two is also shown by the other elements in group 6 of the periodic table.

We can use the element's electron arrangement to predict its valency and the possible molecular formulae of its compounds. You should note the following:

- The number of electrons in the valence shell is the same as the element's group number in the periodic table.
- The number of each type of atom in the molecule is written as a subscript after the element's symbol. Thus, a molecule of B_2O_3 contains 2 atoms of boron and 3 atoms of oxygen.
- The names of compounds which are made up of just two elements end in '**-ide**'.

Examples

Predict the formulae of (a) aluminium oxide, (b) silicon hydride, and (c) barium iodide. State whether the compounds are ionic or covalent.

a) Aluminium 2.8.3 might lose 3 or gain 5 electrons, so the likely valency is 3. Oxygen 2.6 might lose 6, gain 2 or share 2 electrons, so the likely valency is 2. Therefore, 2 Al atoms combine with 3 O atoms and the formula is Al_2O_3. Since aluminium is a metal, the bonding will be ionic.

b) Silicon 2.8.4 might lose 4, gain 4 or share 4 electrons, so the likely valency is 4. Hydrogen 1 might lose 1, gain 1 or share 1 electron, so the likely valency is 1. Therefore, 4 H atoms will combine with 1 Si atom and the formula is SiH_4. Silicon and hydrogen are non-metals so the bonding is covalent.

c) Barium, Ba, is in group 2 therefore it has 2 electrons in the valence shell and might lose 2 or gain 6 electrons to gain an octet of valence electrons so the likely valency is 2. Iodine, I, is in group 7, therefore it has 7 electrons in the valence shell and might lose 7, gain 1 or share 1 electron to gain an octet of valence electrons, so the likely valency is 1. So, 1 barium atom, Ba, will combine with 2 iodine atoms and the formula is BaI_2. Like the other elements in group 2, barium is a metal, so the bonding is ionic.

Test yourself

1 Use the method above to predict the formulae of the following compounds and state whether the compounds are ionic or covalent. You will need to refer to the periodic table.

a) calcium oxide **b)** rubidium sulphide **c)** oxygen fluoride
d) strontium bromide **e)** selenium fluoride **f)** carbon chloride
g) lithium iodide **h)** nitrogen chloride **i)** silicon hydride
j) nitrogen hydride (ammonia)

2 Give names to the following formulae:

a) SrO **b)** $AlCl_3$ **c)** BaO **d)** Cs_2S **e)** BF_3
g) $SiCl_4$ **h)** HF **i)** ZnO **j)** Ag_2O **k)** H_2S

Polyatomic ions

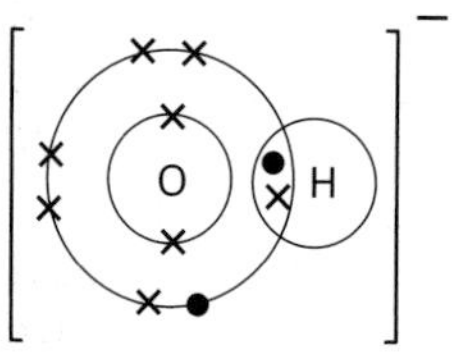

A **polyatomic ion** contains more than one type of atom, for example the hydroxide ion, OH^- (see Figure 1.14). The charge on the ion indicates its combining capacity, so:

sodium hydroxide:
1 sodium ion, Na^+, bonds with 1 OH^- ion $\longrightarrow$ formula NaOH

Figure 1.14 ▲
A dot and cross diagram of the hydroxide ion

magnesium hydroxide:
1 magnesium ion, Mg^{2+}, bonds with 2 OH^- ions ⟶ formula $Mg(OH)_2$

aluminium hydroxide:
1 aluminium ion, Al^{3+}, bonds with 3 OH^- ions ⟶ formula $Al(OH)_3$

Note that the positive and negative charges equate and that brackets are used to show the number of polyatomic ions which are present. When counting atoms, the number after the bracket multiplies the entire contents of that bracket. For example, the formula of aluminium hydroxide is made up of 1 atom of Al and 3 atoms each of O and H. Some other common polyatomic ions are: ethanoate, $CH_3CO_2^-$, nitrate, NO_3^-, sulphate, SO_4^{2-}, carbonate, CO_3^{2-} and ammonium, NH_4^+. You should learn the formulae of these ions.

Examples

Write formulae for sodium carbonate, magnesium nitrate and calcium ethanoate. Count up the atoms in each formula.

Sodium carbonate contains Na^+ (group 1) and CO_3^{2-} ions. The formula is Na_2CO_3. The formula contains 2 Na, 1 C and 3 O atoms.
Magnesium nitrate contains Mg^{2+} (group 2) and NO_3^- ions. The formula is $Mg(NO_3)_2$. The formula contains 1 Mg, 2 N and 6 O atoms.
Calcium ethanoate contains Ca^{2+} (group 2) and $CH_3CO_2^-$ ions. The formula is $(CH_3CO_2)_2Ca$ – by tradition, the ethanoate ion is written before the metal ion. The formula contains 4 C, 6 H, 4 O and 1 Ca atoms.

Variable valencies

As you progress, you will see that *the approach in this section is a simplification of the way in which atoms combine*. In particular, the elements in period 3 upwards may have variable valencies, as shown in Table 1.2, and a more detailed explanation of atomic structure is needed to explain this behaviour. Where variable valency is possible, we identify the valency of an element in a particular compound by putting Roman numerals in the compound's name, e.g.:

iron(II) chloride, $FeCl_2$ — iron(III) sulphate, $Fe_2(SO_4)_3$
tin(II) oxide, SnO — tin(IV) oxide, SnO_2

(Strictly speaking, the Roman numerals indicate the **oxidation state** of the element. This concept will be introduced in section 4.12).

Table 1.2
Some common valencies of selected elements

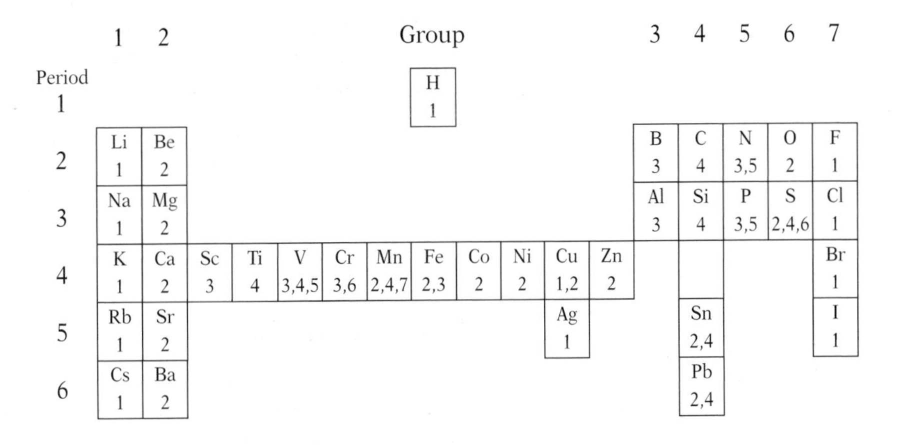

Period \ Group	1	2											3	4	5	6	7
1								H 1									
2	Li 1	Be 2											B 3	C 4	N 3,5	O 2	F 1
3	Na 1	Mg 2											Al 3	Si 4	P 3,5	S 2,4,6	Cl 1
4	K 1	Ca 2	Sc 3	Ti 4	V 3,4,5	Cr 3,6	Mn 2,4,7	Fe 2,3	Co 2	Ni 2	Cu 1,2	Zn 2					Br 1
5	Rb 1	Sr 2									Ag 1			Sn 2,4			I 1
6	Cs 1	Ba 2												Pb 2,4			

Sometimes the name of the compound clarifies the valency, e.g.:

carbon monoxide, CO, valency of C = 2
carbon dioxide, CO_2, valency of C = 4

sulphur difluoride, SF_2, valency of S = 2
sulphur hexafluoride, SF_6, valency of S = 6

Here the prefixes **mon-**, **di-** and **hexa-** mean one, two and six of these atoms respectively. Other prefixes in common use are **tri-** (3), **tetra-** (4) and **penta-** (5). The compounds might also be named using Roman numerals, so: carbon(II) oxide, carbon(IV) oxide, sulphur(II) fluoride and sulphur(VI) fluoride.

Test yourself

You will need to use Table 1.2.

1 Name the following compounds and count up the number of atoms in each formula:
a) $CaCO_3$ **b)** PbO_2 **c)** SF_4 **d)** Rb_2SO_4 **e)** V_2O_5
f) PCl_5 **g)** NO_2 **h)** $Al(NO_3)_3$ **i)** CH_3CO_2Ag **j)** $Fe(OH)_2$

2 Write the formulae for the following compounds and count up the atoms in each:
a) potassium sulphate **b)** nickel(II) nitrate **c)** phosphorus(III) chloride
d) sodium carbonate **e)** nitrogen(V) oxide
f) silicon tetrahydride **g)** manganese(II) ethanoate
h) copper(II) hydroxide **i)** sulphur(VI) oxide **j)** ammonium chloride

3 Classify the compounds in questions 1 and 2 as either ionic or covalent.

1.5 Chemical equations

Study Checklist

At the end of this section you should be able to:

1 Represent chemical reactions using word equations.
2 Represent chemical reactions using balanced chemical equations.

Equations

Have you ever seen a bronze statue in a park or city centre which is covered by a green layer of powdery solid? Bronze is an alloy of copper and tin. The copper reacts very slowly with atmospheric gases and rain water to produce various copper compounds. One of these processes involves two steps. First, the copper reacts with oxygen to form black copper oxide, CuO. We can use a **word equation** to summarise the reaction:

copper + oxygen ⟶ copper oxide

The **chemical equation** for the reaction shows the atomic and molecular formulae:

$$2Cu + O_2 \longrightarrow 2CuO$$

Notice in this equation that the number of atoms is the same on both sides of the arrow and the equation is said to be balanced. Oxygen is a diatomic molecule, O_2 and the 2 in front of the CuO on the right-hand side of the equation multiplies each atom in the molecular formula, giving 2 Cu atoms and 2 O atoms.

In the second step, the copper oxide reacts with rain water, which is a weakly acidic solution of carbon dioxide, CO_2, dissolved in water, to give green copper carbonate:

word equation:	copper oxide	+	carbon dioxide	$\longrightarrow$	copper carbonate
chemical equation:	CuO	+	CO_2	$\longrightarrow$	$CuCO_3$

Once again, the chemical equation is balanced with 1 Cu, 1 C and 3 O atoms on each side of the arrow.

Balancing chemical equations

When you get indigestion, the pain is often caused by a build up of hydrochloric acid, HCl, in the stomach. To reduce the pain, you might take an antacid, in tablet or liquid form, and this reacts with the excess hydrochloric acid. A commonly used antacid is magnesium hydroxide (sold as 'milk of magnesia') and the word equation is:

magnesium hydroxide + hydrochloric acid $\longrightarrow$ magnesium chloride + water

To write the balanced equation, first work out the individual formulae of the compounds involved using ionic charges or valencies:

magnesium hydroxide:
magnesium ion, Mg^{2+}; hydroxide ion OH^-; so the formula is $Mg(OH)_2$

hydrochloric acid (an aqueous solution of hydrogen chloride):
hydrogen, valency 1; chlorine, valency 1; so the formula is HCl

magnesium chloride:
magnesium ion, Mg^{2+}; chloride ion Cl^-; so the formula is $MgCl_2$

Now, write the formulae under the word equation:

magnesium hydroxide	+	hydrochloric acid	$\longrightarrow$	magnesium chloride	+	water
$Mg(OH)_2$	+	HCl	$\longrightarrow$	$MgCl_2$	+	H_2O

Is this balanced? In fact, only the Mg terms are equal on both sides of the equation! To balance the equation we put numbers in front of formulae but remember, **subscripts must not be altered** because the molecular formulae would then be incorrect. Try to work out where the shortages are. Here, for example, we need an extra Cl on the left and an extra O on the right. The balanced equation is:

$$Mg(OH)_2 + 2HCl \longrightarrow MgCl_2 + 2H_2O$$

with 1 Mg, 2 O, 4 H and 2 Cl atoms on each side.

Balanced chemical equations are important in chemistry because they tell us about the **stoichiometry** of the reaction, that is the ratios of the reactant molecules to each other and to the product molecules formed. In the above equation, the stoichiometry is:

$$1\ Mg(OH)_2 : 2\ HCl : 1\ MgCl_2 : 2\ H_2O$$

Example

In the production of iron, iron(III) oxide is heated at about 800°C with carbon monoxide, CO, to produce molten iron and carbon dioxide. Write down the formulae of the reactants and products, the word equation, the chemical equation, and the stoichiometry for the reaction.

iron(III) oxide: iron(III) ion, Fe^{3+}; oxide ion O^{2-}; so the formula is Fe_2O_3; ***carbon monoxide:*** CO; **iron**: Fe; **carbon dioxide**: CO_2.

Word equation:

iron(III) oxide + carbon monoxide $\longrightarrow$ iron + carbon dioxide

Insert formulae:

$$Fe_2O_3 + CO \longrightarrow Fe + CO_2$$

More Fe and O terms are needed on the right, requiring more CO terms on the left. The balanced equation is:

$$Fe_2O_3 \quad + \quad 3\,CO \quad \longrightarrow \quad 2\,Fe \quad + \quad 3\,CO_2$$

Stoichiometry:

$$1\,Fe_2O_3 \quad : \quad 3\,CO \quad : \quad 2\,Fe \quad : \quad 3\,CO_2$$

Test yourself

1 Balance the following equations:
  a) $Al + O_2 \longrightarrow Al_2O_3$
  b) $Na_2CO_3 + HCl \longrightarrow NaCl + CO_2 + H_2O$
  c) $AgNO_3 + BaCl_2 \longrightarrow AgCl + Ba(NO_3)_2$
  d) $Ca(OH)_2 + CH_3COOH \longrightarrow (CH_3COO)_2Ca + H_2O$

2 Write balanced chemical equations from the following word equations:
  a) aluminium + chlorine (a diatomic molecule) ⟶ aluminium chloride
  b) phosphorus + oxygen ⟶ phosphorus(V) oxide
  c) iron(II) sulphate + sodium hydroxide ⟶ iron(II) hydroxide + sodium sulphate

3 For each of the following reactions, write down the word equation, the balanced chemical reaction, and the stoichiometry of the reaction. You may need to look back at the earlier sections to find valencies and ionic charges.
  a) Magnesium burns in oxygen to form magnesium oxide.
  b) Sulphur dioxide reacts with oxygen at 450°C to form sulphur trioxide.
  c) Calcium reacts with sulphuric acid, H_2SO_4, to form calcium sulphate and hydrogen gas.
  d) When a piece of sodium metal is placed in water, it fizzes around on the surface as hydrogen is formed. The other product is sodium hydroxide.
  e) Marble is one form of calcium carbonate. It is attacked by the nitric acid, HNO_3, in acid rain to form calcium nitrate, carbon dioxide and water.
  f) Propane, C_3H_8, is one of the gases in Calor gas. It burns in oxygen to form carbon dioxide and water.
  g) In photosynthesis, carbon dioxide reacts with water in the presence of sunlight to form glucose ($C_6H_{12}O_6$) and oxygen.

1.6 Chemical quantities: relative masses and the mole concept

Study Checklist

At the end of this section you should be able to:

1 Define the terms relative atomic, molecular and formula masses based on the carbon-12 scale.

2 Work out the molecular mass of a compound given the relative atomic masses of its constituent atoms.

3 Calculate the number of moles in a given mass of a substance.

4 Appreciate that the mole is the amount of substance which contains 6.023×10^{23} of a given type of particle and that this number is known as Avogadro's constant.

5 Calculate the number of particles in a given mass of an atomic, molecular or ionic substance.

Relative atomic masses

An analysis of an unknown hydrocarbon shows that it contains 84.0% carbon and 16.0% hydrogen by mass. Which element do you think has the most atoms in the hydrocarbon molecule? Well, to a student new to chemistry an obvious yet incorrect answer would be carbon. In fact, the hydrocarbon is heptane, C_7H_{16}. Although they make up only 16% of the molecular mass, the hydrogen atoms easily outnumber the carbon atoms. The reason for this is that in terms of their **relative atomic masses**, each carbon atom is about twelve times as heavy as a hydrogen atom.

In 1961, it was internationally agreed that the carbon-12 (^{12}C) atom should be the standard against which atomic masses are measured. Thus, for a given element,

$$\textbf{relative atomic mass, } A_r = \frac{\textbf{average mass of one atom of that element}}{\textbf{1/12th of the mass of one atom of carbon-12}}$$

An **atomic mass unit, u**, is taken to be 1/12th of the mass of one atom of carbon-12, which is 1.66×10^{-24} g. So, one atom of carbon-12 weighs exactly 12 u and one atom of hydrogen weighs 1.008 u.

In chemical calculations, relative atomic masses are usually taken as whole numbers. For example, the relative atomic mass of hydrogen, A_r(H), is taken as 1, not 1.008. There are two notable exceptions, chlorine and copper, which are taken to have relative atomic masses of 35.5 and 63.5 respectively. A table of relative atomic masses is given on page 132.

Relative molecular and formula masses

The **relative molecular mass, M_r**, of a compound is the sum of all the relative atomic masses represented by the formula of that compound. To calculate the relative molecular mass of ethanoic acid, CH_3COOH, for example, first list the atoms present and then add up the individual masses, as follows:

CH_3COOH	Atoms present	Substitute A_r values	Add
	2 × C	2 × 12	24
	4 × H	4 × 1	4
	2 × O	2 × 16	32
		$M_r[CH_3COOH]$	**60**

Many compounds do not exist as individual molecules but as a giant lattice of particles held together by electrostatic forces. These particles may be ions, as in potassium chloride, K^+Cl^-, or atoms, as in silicon(IV) oxide, SiO_2. In each case, it is incorrect to say that the compound has a relative 'molecular' mass. Instead, we talk about the **relative formula mass**, which is the mass of the formula showing the simplest whole number ratio of the atoms in the compound. Relative formula masses are obtained using the above method and given the same symbol, M_r, as shown below for calcium nitrate, $Ca(NO_3)_2$:

$Ca(NO_3)_2$	Atoms present	Substitute A_r values	Add
	1 × Ca	1 × 40	40
	2 × N	2 × 14	28
	6 × O	6 × 16	96
		$M_r[Ca(NO_3)_2]$	**164**

Note that there are 2 N atoms and 6 O atoms in each formula mass of $Ca(NO_3)_2$ because the subscript after the bracket multiplies all of the atoms inside the bracket.

Test yourself

You will need to refer to the table of relative atomic masses on page 132.

1 Calculate the relative molecular masses of the following compounds:
a) SF_6 **b)** B_2O_3 **c)** C_4H_{10} **d)** Cl_2O_7 **e)** $C_6H_5NH_2$
f) H_2SO_4 **g)** HNO_3 **h)** $(CH_3CO)_2O$

2 Calculate the relative formula masses of the following compounds:
a) NaCl **b)** Li_2S **c)** SiO_2 **d)** $Pb(NO_3)_2$ **e)** $Mg(ClO_4)_2$
f) K_2CrO_4 **g)** $Fe_2(PO_4)_3$ **h)** $MgSO_4.7H_2O$.

The Avogadro constant and the mole concept

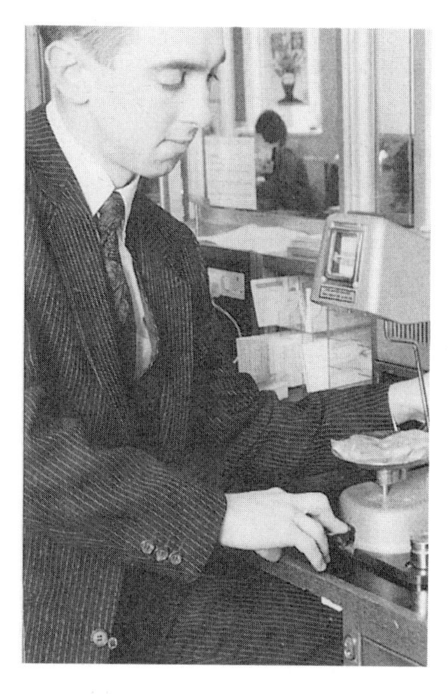

This bank cashier is counting money by weighing coins. He knows, for instance, that £5 in ten pence coins weighs 567 g. In the same way, chemists can work out the number of atoms, molecules or ions in a sample of pure substance by weighing it

Any sample that we weigh out will contain many billions of particles, whether these are atoms, ions or molecules. But how many?

We need to measure them by using a 'standard' number of particles, just as, for example, we buy eggs by the dozen. We define the **mole** as the amount of a substance which contains the same number of particles as there are atoms in 12 g of carbon-12. These particles may be atoms, molecules, ions, neutrons, and so on.

The actual number of particles in a mole, known as the **Avogadro constant, L**, can be measured using X-ray diffraction (see section 4.2). Its value is found to be 6.023×10^{23} mol^{-1}. Looking at Table 1.3, you can see that by weighing out the relative atomic or molecular or ionic mass in grams, we will always be working with one mole of those particles. In chemical calculations, the word 'mole' is shortened to 'mol' (but never to M which has a very different meaning, see section 1.8).

Using a simple ratio, we find that for pure substances:

$$\textbf{number of moles in a sample} = \frac{\textbf{mass of the sample}}{\textbf{mass of one mole of the sample}}$$

For example, 8.05 g of sodium, $A_r = 23$, contains 8.05/23 = 0.350 moles of sodium atoms.

Table 1.3 The relationship between the mass of one particle and the mass of one mole of particles (1 u = 1.66×10^{-24} g)

Particle	boron atom (B)	chlorine molecule (Cl_2)	lithium ion (Li^+)
mass of one particle/u	10.8	70.9	6.9
⇓ $\times 1.66 \times 10^{-24}$	⇓	⇓	⇓
mass of one particle/g	1.79×10^{-23}	11.76×10^{-23}	1.15×10^{-23}
⇓ $\times 6.023 \times 10^{23}$	⇓	⇓	⇓
mass of one mole of particles/g	10.8	70.9	6.9

Examples

1 State the relative atomic masses of sodium, silicon, sulphur and selenium.
2 Calculate the mass of 1 mole of:
a) sulphuric acid, H_2SO_4
b) calcium ethanoate, $(CH_3CO_2)_2Ca$.

3 How many moles are there in:
   a) 38.0 g of sulphuric acid **b)** 7.90 g of calcium ethanoate?
4 Calculate the mass of the following chemical quantities:
   a) 0.75 mol of sulphuric acid **b)** 0.40 mol of calcium ethanoate.
5 Taking the Avogadro constant as 6×10^{23} particles per mole (mol^{-1}), work out the number of particles in the following chemical quantities:
   a) atoms in 28 g of iron, Fe **b)** molecules in 0.88 g of carbon dioxide, CO_2
   c) ions in 117 g of sodium chloride, NaCl.

Solutions

1 $A_r(Na) = 23$, $A_r(Si) = 28$, $A_r(S) = 32$, $A_r(Se) = 79$

2 a)

H_2SO_4	Atoms present	Substitute A_r values	Add
	$2 \times H$	2×1	2
	$1 \times S$	1×32	32
	$4 \times O$	4×16	64
		so mass of 1 mole of H_2SO_4	= 98 g

b)

$(CH_3CO_2)_2Ca$	Atoms present	Substitute A_r values	Add
	$4 \times C$	4×12	48
	$6 \times H$	6×1	6
	$4 \times O$	4×16	64
	$1 \times Ca$	1×40	40
		so mass of 1 mole of $(CH_3CO_2)_2Ca$	=158 g

3 $$\text{number of moles} = \frac{\text{mass of the sample (g)}}{\text{mass of one mole of the sample } (A_r \text{ or } M_r)}$$

a) $\text{mol } H_2SO_4 = \frac{38.0}{98} = 0.39$

b) $\text{mol } (CH_3CO_2)_2Ca = \frac{7.90}{158} = 0.05$

4 From **3** above, it follows that:

mass of the sample (g) = number of moles × mass of 1 mole of the sample (A_r or M_r)

a) Mass of $H_2SO_4 = 0.75 \times 98$
$= 73.5$ g

b) Mass of $(CH_3CO_2)_2Ca = 0.40 \times 158$
$= 63.2$ g

5 a) $A_r(Fe) = 56$; number of moles $= \frac{\text{mass of the sample}}{\text{mass of one mole of the sample}}$

$= \frac{28}{56} = 0.5$ mol

number of particles (Fe atoms) = number of moles $\times 6 \times 10^{23}$
$= 0.5 \times 6 \times 10^{23}$
$= 3 \times 10^{23}$

A similar method is used in parts **b** and **c**.

b) $M_r(CO_2) = 44$; number of moles $= \frac{0.88}{44} = 0.02$ mol number of particles (molecules) = number of moles $\times 6 \times 10^{23}$

$= 0.02 \times 6 \times 10^{23}$

$= 1.2 \times 10^{22}$

c) $M_r(NaCl) = 58.5$ (remember $A_r(Cl) = 35.5$);

number of moles of NaCl $= \frac{117}{58.5} = 2$ mol

But each mole of NaCl contains 2 moles of ions (1 mole Na^+, 1 mole Cl^-), thus 2 moles of NaCl contains $4 \times 6 \times 10^{23} = 2.4 \times 10^{24}$ ions.

Test yourself

Using the table of relative atomic masses on page 132 answer the following questions.

1 List the atomic masses of cerium, iodine, palladium and titanium.

2 What is the mass of one mole of:
a) potassium chloride, KCl
b) iron, **Fe**
c) hydrated sodium thiosulphate, $Na_2S_2O_3.5H_2O$
d) silica, SiO_2?

3 Calculate the mass of the following:
a) 0.50 mol of ammonia molecules, NH_3
b) 0.375 mol of caesium atoms, Cs
c) 1.50 mol of fluorine molecules, F_2
d) 0.275 mol of hydrated sodium carbonate, $Na_2CO_3.10H_2O$.

4 How many moles are there in:
a) 14 g of magnesium atoms, Mg
b) 8 g of carbon dioxide molecules, CO_2
c) 0.50 g of copper(II) chloride, $CuCl_2$
d) 38 g of hydrated copper(II) sulphate, $CuSO_4.5H_2O$?

5 In the following work out the number of:
a) atoms in 2.3 g of sodium
b) molecules in 71 g of chlorine, Cl_2
c) molecules in 5.2 g of silicon fluoride, SiF_4
d) Pb^{2+} and SO_4^{2-} ions in 0.303 g of lead sulphate
e) Al^{3+} and NO_3^- ions in 213 g of aluminium nitrate, $Al(NO_3)_3$

1.7 Chemical equations and reacting masses

Study Checklist

At the end of this section you should be able to:

1 Understand that chemical equations represent molar quantities of reactants and products.

2 Calculate the masses of reactants and products involved in a reaction, given appropriate data and the balanced equation.

3 Deduce the chemical equation for a reaction given the masses of the reactants and products involved.

In section 1.5, we saw that a balanced chemical equation indicates the stoichiometry of the reaction, that is, the ratio of reactant particles to each other and to the product particles formed. The particles may be atoms, molecules or ions, for example:

S	+	O_2	$\longrightarrow$	SO_2
1 sulphur atom	reacts with	1 oxygen molecule	to give	1 molecule of sulphur dioxide
N_2	+	$3H_2$	$\longrightarrow$	$2NH_3$
1 nitrogen molecule	reacts with	3 hydrogen molecules	to give	2 molecules of ammonia
$3Ca^{2+}$	+	$2PO_4^{3-}$	$\longrightarrow$	$Ca_3(PO_4)_2$
3 calcium ions	react with	2 phosphate ions	to give	1 'unit' of calcium phosphate

The last equation is called an **ionic equation**. Notice how the charges on the ions balance.

The stoichiometry of a chemical reaction remains the same regardless of the number of particles involved. For example, when propane camping gas, C_3H_8, is burnt, the equation is:

	C_3H_8	+ $5O_2$	$\longrightarrow$ $3CO_2$	+ $4H_2O$
	1 molecule	5 molecules	3 molecules	4 molecules
or	50 molecules	250 molecules	150 molecules	200 molecules
or	6×10^{23} molecules	30×10^{23} molecules	18×10^{23} molecules	24×10^{23} molecules
i.e.	1 mole	5 moles	3 moles	4 moles

By convention, the amounts in chemical equations are taken as mole quantities and this allows us to work out reacting masses.

Examples

1 Calculate the masses of oxygen used, and carbon dioxide and water formed, when 33 g of propane is completely burnt in oxygen.

	C_3H_8	+ $5O_2$	$\longrightarrow$ $3CO_2$	+ $4H_2O$
Mole ratio	1	5	3	4
Substitute M_r values	1(44)	5(32)	3(44)	4(18)
Gram ratio	44	160	132	72
Divide by 44	1 g	$\frac{160}{44}$ g	$\frac{132}{44}$ g	$\frac{72}{44}$ g
So for 33 g:	33 g	$\frac{33 \times 160}{44}$ g	$\frac{33 \times 132}{44}$ g	$\frac{33 \times 72}{44}$ g
	33 g	120 g	99 g	54 g

Therefore when 33 g of butane is burnt, 120 g of oxygen is needed and 99 g of carbon dioxide and 54 g of water are formed.

2 Calculate the mass of propane which, on combustion, yields 26.4 g of carbon dioxide.

	C_3H_8	+ $5O_2$	$\longrightarrow$ $3CO_2$	+ $4H_2O$
Mole ratio	1	5	3	4
Substitute M_r values	1(44)	5(32)	3(44)	4(18)
Gram ratio	44	160	132	72

Divide by 132	$\frac{44}{132}$ g	$\frac{160}{132}$ g	1 g	$\frac{72}{132}$ g

So for the formation of 26.4 g of carbon dioxide:

	$\frac{26.4 \times 44}{132}$ g	$\frac{26.4 \times 160}{132}$ g	26.4 g	$\frac{26.4 \times 72}{132}$ g
	8.8 g	32 g	26.4 g	14.4 g

The mass of propane burnt is 8.8 g.

Test yourself

1 Ammonia, NH_3, is used to make fertilisers. It is produced in large quantities from nitrogen and hydrogen using the Haber process:

$$N_2 + 3H_2 \longrightarrow 2NH_3$$

a) How many moles of hydrogen molecules are needed to make 40 mol of ammonia?

b) What mass of ammonia could be made from 56 g of nitrogen and 24 g of hydrogen?

c) Which reactant is present in an excess amount? What mass of it will remain unreacted?

2 In winemaking, the glucose in fruit is fermented in the presence of yeast to form carbon dioxide and ethanol, C_2H_5OH:

$$C_6H_{12}O_6 \longrightarrow 2CO_2 + 2C_2H_5OH$$

a) How many moles of carbon dioxide will be formed from 0.7 mol of glucose?

b) What mass of carbon dioxide will be formed from 36 g of glucose?

c) How much glucose would be needed to produce 13.8 g of ethanol?

3 For each of the following processes, calculate the number of moles of each reactant and product and then write a balanced equation.

a) In the laboratory preparation of oxygen gas, 6.8 g of hydrogen peroxide, H_2O_2, decomposed to give 3.6 g of water and 3.2 g of oxygen.

b) When 4.6 g of sodium is added to water, 8 g of sodium hydroxide is formed and 0.2 g of hydrogen gas is evolved.

c) Calcium carbonate, $CaCO_3$, is an antacid. When 1 g of $CaCO_3$ reacts with 0.73 g of excess hydrochloric acid in the stomach, 1.11 g of calcium chloride, 0.18 g of water and 0.44 g of carbon dioxide are formed.

d) Computer chips are made from high purity silicon. 0.28 kg of silicon is produced when 1.70 kg of silicon tetrachloride is heated with 0.48 kg of magnesium metal. The other product is 1.9 kg of magnesium chloride.

1.8 Concentration of solutions

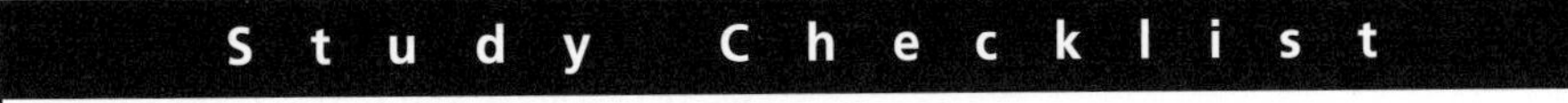

At the end of this section you should be able to:

1. Explain the terms 'solute', 'solvent, 'concentration' and 'molarity'.
2. Calculate the concentration of a substance in aqueous solutions, in mol dm^{-3}, given the mass of solute dissolved and the volume of solution it is dissolved in.
3. Work out the mass of solute that must be dissolved in water to produce a given volume of a standard solution, i.e. one of known concentration.
4. Calculate the number of moles of solute in a solution given its volume and concentration.

Many chemical reactions occur in solution. A solution is formed when a **solute** (a solid, liquid or gas), is dissolved in a **solvent**. Often the solvent is water and an aqueous solution is formed. The **concentration** of the solution is defined as the number of moles of solute which are present in 1 dm^3 of solution (i.e. 1 litre or 1000 cm^3 of solution). So,

$$\text{concentration of a solution} = \frac{\text{moles of solute dissolved}}{\text{volume of solution prepared in dm}^3}$$

For example, if 0.4 mol of sodium hydroxide was dissolved in water and the solution made up to 2 dm^3, the concentration would be 0.2 mol dm^{-3}. The concentration of a solution is sometimes referred to as its **molarity** and given the symbol M.

In the laboratory, we tend to work in cm^3 rather than dm^3. Since 1 dm^3 = 1000 cm^3, the equation above can be written:

$$\text{concentration in mol dm}^{-3} = \frac{\text{mol solute} \times 1000}{\text{volume of solution in cm}^3}$$

Rearranging this equation, we have a formula for calculating the number of moles of solute in a given volume of solution:

$$\textbf{mol solute} = \frac{\textbf{volume (cm}^3\textbf{)} \times \textbf{concentration (mol dm}^{-3}\textbf{)}}{\textbf{1000}}$$

Examples

1 Work out the molar concentrations of these aqueous solutions:
a) 0.5 mol of nitric acid, HNO_3, in 250 cm^3 of solution
b) 0.1 mol of sodium hydroxide in 100 cm^3 of solution.

2 Calculate the number of moles of solute in each of the following solutions:
a) 50 cm^3 of sodium hydroxide of concentration 2 mol dm^{-3}
b) 200 cm^3 of calcium nitrate of concentration 0.1 mol dm^{-3}
c) 24.6 cm^3 of aluminium sulphate of concentration 0.15 mol dm^{-3}.

3 Calculate the molarities of the ions in the following aqueous solutions:
a) Sodium sulphate of concentration 0.5 mol dm^{-3}
b) Aluminium sulphate of concentration 0.6 mol dm^{-3}.

Solutions

1 a) 0.5 mol of HNO_3 in 1000 cm^3 gives a concentration of 0.5 mol dm^{-3}
0.5 mol of HNO_3 in 250 cm^3 gives a concentration of
1000/250 × 0.5 mol dm^{-3}

The concentration of the solution is 2 mol dm^{-3}. Note that dissolving the same number of moles (0.5 mol) in a smaller volume (making up 250 rather than 1000 cm^3) increases the concentration so we multiply by 1000/250.

b) 0.1 mol of NaOH in 1000 cm^3 gives a concentration of 0.1 mol dm^{-3}
0.1 mol of NaOH in 100 cm^3 gives a concentration of
$1000/100 \times 0.1$ mol dm^{-3}
The concentration of the solution is 1 mol dm^{-3}.

2 a) $\text{mol of NaOH} = \frac{\text{volume (cm}^3)}{1000} \times \text{concentration} = \frac{50}{1000} \times 2 = 0.1 \text{ mol}$

b) $\text{mol of Ca(NO}_3)_2 = \frac{\text{volume (cm}^3)}{1000} \times \text{concentration}$

$$= \frac{200}{1000} \times 0.1 = 0.02 \text{ mol}$$

c) $\text{mol of Al}_2(\text{SO}_4)_3 = \frac{\text{volume (cm}^3)}{1000} \times \text{concentration}$

$$= \frac{24.6}{1000} \times 0.15 = 3.69 \times 10^{-3} \text{ mol}$$

3 a) Sodium sulphate dissociates into ions in solution, as shown below:

$$Na_2SO_4 \longrightarrow 2Na^+ + SO_4^{2-}$$

So, from the mole ratios, a 0.5 mol dm^{-3} solution of sodium sulphate will be 1 mol dm^{-3} and 0.5 mol dm^{-3} in Na^+ and SO_4^{2-} ions respectively.

b) Similarly, aluminium sulphate is ionised in acidified aqueous solution:

$$Al_2(SO_4)_3 \longrightarrow 2Al^{3+} + 3SO_4^{2-}$$

So from the mole ratios, a 0.6 mol dm^{-3} solution of aluminium sulphate will be 1.2 mol dm^{-3} and 1.8 mol dm^{-3} in Al^{3+} and SO_4^{2-} ions respectively.

Test yourself

1 Calculate the concentration, in mol dm^{-3}, of the following solutions:
- **a)** 1.17 g of NaCl in 500 cm^3 of solution
- **b)** 3.6 g of K_2SO_4 in 250 cm^3 solution
- **c)** 21.3 g of $Al(NO_3)_3$ in 1 dm^3 of solution
- **d)** 4.74 g of $(CH_3CO_2)_2Ca$ in 200 cm^3 of solution
- **e)** 3.16 g of $KMnO_4$ in 100 cm^3 of solution
- **f)** 28.6 g of $Na_2CO_3.10H_2O$ in 2 dm^3 of solution.

2 What mass of solute must be dissolved to make up the following aqueous solutions?
- **a)** 1 dm^3 of 0.05 mol dm^{-3} KI
- **b)** 2 dm^3 of 0.01 mol dm^{-3} $KHCO_3$
- **c)** 500 cm^3 of 0.05 mol dm^{-3} $FeCl_2$
- **d)** 250 cm^3 of 0.75 mol dm^{-3} $CuSO_4$
- **e)** 100 cm^3 of 1 mol dm^{-3} HNO_3
- **f)** 250 cm^3 of 0.01 mol dm^{-3} $AgNO_3$.

3 Calculate the number of moles of solute in the following volumes of aqueous solution:
- **a)** 25 cm^3 of 0.1 mol dm^{-3} NaOH
- **b)** 50 cm^3 of 2 mol dm^{-3} HCl
- **c)** 23.8 cm^3 of 0.001 mol dm^{-3} $MgCl_2$
- **d)** 18.7 cm^3 of 0.05 mol dm^{-3} $(CH_3CO_2)_2Zn$
- **e)** 20 cm^3 of 0.1 mol dm^{-3} H_2SO_4.

4 Find the total concentration of ions in the volumes of solution in question 3.

5 Name the substances whose chemical formulae are given in questions 1 to 3.

1.9 Volumetric analysis: an introduction

Study Checklist

At the end of this section you should be able to:

1. Explain what is meant by the terms 'volumetric analysis', 'equivalence point', 'titration' and 'indicator'.
2. Describe how volumetric analysis is used to measure the concentration of a solution, giving a simple acid–base reaction as an example.
3. Explain how volumetric analysis can be used to find the mole ratio of the reactants and, thus, to balance equations.
4. Carry out the relevant calculations following from 2 and 3 above.

Volumetric analysis is a common laboratory technique used to determine the number of moles of reactants involved in a reaction. In a **titration**, volumes of two solutions are mixed together until the numbers of moles of each reactant are in the ratio shown by the balanced chemical equation. This point, known as the **equivalence point**, is usually identified using an indicator which changes colour at the instant this mole ratio is achieved. Volumetric analysis can provide valuable information, examples of which follow.

Finding the concentration of a solution

If the concentration of one solution and the chemical equation are known, the concentration of the other solution may be found.

Example

Sulphuric acid and sodium hydroxide react according to the equation:

$$H_2SO_4 + 2NaOH \longrightarrow Na_2SO_4 + 2H_2O$$

If sulphuric acid (concentration 0.100 mol dm^{-3}) is added to 25 cm^3 of NaOH, bromothymol blue indicator changes colour from blue to yellow when 23.5 cm^3 of acid has been added. Calculate the concentration of the sodium hydroxide solution.

The method involves three steps. First, work out the number of moles of reactant provided by the solution of known concentration:

$$\text{mol } H_2SO_4 = \frac{\text{volume (cm}^3\text{)}}{1000} \times \text{concentration} = \frac{23.5}{1000} \times 0.100$$

$$= 2.35 \times 10^{-3}$$

Now, work out the number of moles of the other reactant (unknown concentration) using the mole ratio in the equation:

1 mol H_2SO_4	reacts with	2 mol NaOH
2.35×10^{-3} mol H_2SO_4	reacts with	$2(2.35 \times 10^{-3})$ mol NaOH

So, 4.70×10^{-3} mol of NaOH reacted.
Finally, consider the NaOH solution, of which 25 cm^3 were used:

25 cm^3 of solution contains 4.70×10^{-3} mol of NaOH

$$1\ cm^3 \text{ of solution contains } \frac{4.70 \times 10^{-3}}{25} \text{ mol of NaOH}$$

$$\text{So } 1000\ cm^3\ (1\ dm^3) \text{ of solution contains } \frac{1000 \times 4.70 \times 10^{-3}}{25}$$

$$= 0.118 \text{ mol of NaOH}$$

The concentration of the NaOH solution is 0.188 mol dm^{-3}.

Determining the mole ratio of the reactants

If the concentrations of both solutions are known, it is possible to determine the mole ratios of the reactants and use this information to balance the chemical equation.

Example

When an aqueous solution of silver nitrate, $AgNO_3$, (concentration 0.050 mol dm^{-3}) is titrated with an aqueous solution of a metal chloride, MCl_x, (concentration 0.020 mol dm^{-3}), the equivalence point is reached when 19.8 cm^3 of $AgNO_3$ solution is added to 25.0 cm^3 of MCl_x. The products are silver chloride, AgCl, and the metal nitrate, $M(NO_3)_y$. Find the values of x and y, then write a balanced equation for the reaction.

First, find the moles of each reactant present at the equivalence point:

$$\text{mol } AgNO_3 = \frac{\text{volume } (cm^3)}{1000} \times \text{concentration}$$

$$= \frac{19.8}{1000} \times 0.050 = 9.9 \times 10^{-4}$$

$$\text{mol } MCl_x = \frac{\text{volume } (cm^3)}{1000} \times \text{concentration}$$

$$= \frac{25.0}{1000} \times 0.020 = 5.0 \times 10^{-4}$$

Compare the number of moles used:

5.0×10^{-4} mol MCl_x reacts with 9.9×10^{-4} mol $AgNO_3$

dividing through by 5.0×10^{-4},

1 mol MCl_x reacts with 1.98 mol $AgNO_3$

The molecules react in whole number ratios, so 1 mol MCl_x must react with 2 mol $AgNO_3$.

Write a first attempt at the equation:

$$MCl_x + 2AgNO_3 \longrightarrow M(NO_3)_y + AgCl$$

Then balance it:

2 Ag on the left	$\Longrightarrow$	$MCl_x + 2AgNO_3 \longrightarrow M(NO_3)_y + 2AgCl$
2 Cl on the right	$\Longrightarrow$	$MCl_2 + 2AgNO_3 \longrightarrow M(NO_3)_y + 2AgCl$
2 NO_3 on the right	$\Longrightarrow$	$MCl_2 + 2AgNO_3 \longrightarrow M(NO_3)_2 + 2AgCl$

The equation is balanced and $x = y = 2$.

Test yourself

Answer these questions by using the above methods.

1 Calculate the concentration of the solutions in italics. You will need to write a balanced equation for each reaction.
 a) 20.0 cm³ of 0.10 mol dm⁻³ HCl is neutralised by 25.0 cm³ of *NaOH*.
 b) 15.8 cm³ of 0.050 mol dm⁻³ KOH is neutralised by 20.0 cm³ of HNO_3.
 c) 19.5 cm³ of 0.10 mol dm⁻³ CH_3CO_2H reacts exactly with 25.0 cm³ of *NaOH*.
 d) 25.0 cm³ of H_2SO_4 neutralises 22.3 cm³ of 0.20 mol dm⁻³ *KOH*.

2 Calculate the concentration of the solutions in italics.
 a) When 25.0 cm³ of 0.05 mol dm⁻³ iodine solution reacts exactly with 21.5 cm³ of *sodium thiosulphate solution*, the equation is:

$$I_2 + 2Na_2S_2O_3 \longrightarrow 2NaI + Na_2S_4O_6$$

 b) An equivalence point is obtained when 20.8 cm³ of *potassium iodate*, KIO_3, is added to 25.0 cm³ of 0.02 mol dm⁻³ sodium iodide. The mole ratio of the reactants is 1 KIO_3: 5 NaI.
 c) Aqueous solutions of silver nitrate and *magnesium chloride* react according to the equation:

$$2AgNO_3 + MgCl_2 \longrightarrow 2AgCl + Mg(NO_3)_2$$

 10.0 cm³ of $MgCl_2$ reacts exactly with 22.0 cm³ of 0.01 mol dm⁻³ $AgNO_3$.

3 Find the values of letters *n* to *z* and then balance the equations using the information supplied.
 a) $n\text{HCl} + \text{M(OH)}_p \longrightarrow \text{MCl}_r + s\text{H}_2\text{O}$
 10.0 cm³ of 0.10 mol dm³ HCl reacts with 10.0 cm³ of 0.05 mol dm³ $M(OH)_p$
 b) $t\text{BaCl}_2 + \text{M}_u\text{SO}_4 \longrightarrow \text{BaSO}_4 + \text{MCl}_v$
 25.0 cm³ of 0.10 mol dm³ $BaCl_2$ is equivalent to 10.0 cm³ of 0.25 mol dm³ M_uSO_4.
 c) $w\text{NaOH} + \text{FeCl}_x \longrightarrow \text{Fe(OH)}_y + z\text{NaCl}$
 20.0 cm³ of 0.05 mol dm³ NaOH reacts exactly with 12.5 cm³ of 0.04 mol dm³ $FeCl_x$

4 These questions are harder and use calculations from previous sections.
 a) When 0.985 g of an insoluble metal carbonate of formula MCO_3 is added to 1.0 mol dm³ hydrochloric acid, HCl, exactly 10 cm³ of acid is required to dissolve it. Identify the metal M.
 b) 0.60 g of a metal, N, reacts exactly with 15 cm³ of 1.0 mol dm⁻³ H_2SO_4 to form a sulphate of formula NSO_4. Identify the metal and write a balanced equation for the reaction.
 c) A white crystalline solid has the formula XY, where X is a group 1 atom and Y is a group 7 atom. 0.270 g of XY was dissolved in water and made up to 100 cm³ of solution and all of this was titrated against 0.08 mol dm³ $AgNO_3$, of which 28.1 cm³ was required for complete reaction. What are the possible chemical formulae of XY?

1.10 Measuring atomic masses: the mass spectrometer

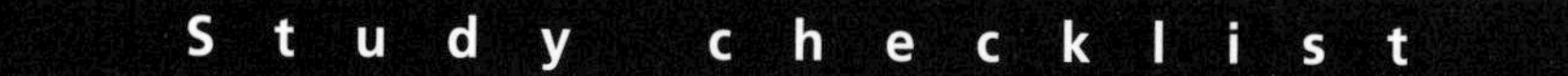

At the end of this section you should be able to:

1. Understand that a mass spectrometer is used to find the relative masses and relative abundance of isotopes in an element.
2. Explain the principles of a simple mass spectrometer in terms of ionisation of atoms, the acceleration and deflection of the resulting positive ions and their detection.
3. Calculate the relative atomic mass of an element given its mass spectrum or relative isotopic abundances.

You should be aware that a mass spectrometer can also be used to determine relative molecular masses and to provide information about the structure of a compound (section 4.2).

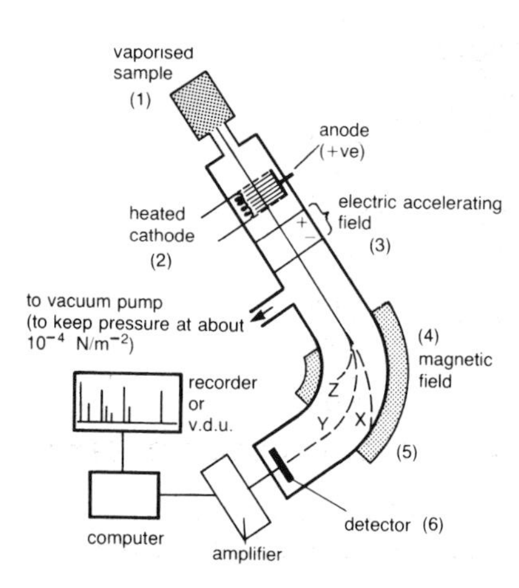

Figure 1.15 ▲
The main components of a mass spectrometer

We saw earlier that atoms of the same element which have different masses are known as **isotopes**. They have the same number of protons but different numbers of neutrons. Chlorine, for example, has two isotopes: ^{35}Cl (17 protons, 18 neutrons) and ^{37}Cl (17 protons, 20 neutrons). The relative isotopic mass is measured on a scale which takes the mass of one atom of carbon-12 to be 12 atomic mass units, or 12 u. (An atomic mass unit, u = 1.66×10^{-27} kg.) Thus, ^{35}Cl has a mass of 35 u and ^{37}Cl has a mass of 37 u. Most elements exist as a mixture of isotopes and the percentage abundance of the isotopes and the relative isotopic masses can be very accurately found using a **mass spectrometer**.

First introduced by Aston in 1919, the mass spectrometer provides the most accurate method of measuring relative atomic masses. The operation of a simple mass spectrometer is described below and the main features are shown in Figure 1.15.

1. The sample is vaporised and flows into the main chamber which is kept at very low pressure.
2. High energy electrons are released by the hot cathode. These strike the gas particles, forming positive ions most of which are singly charged:

e^-	+	X(g)	⟶	e^-	+	e^-	+	$X^+(g)$
'bombarding' electron of high energy				low energy electrons				positive ion

3. An electric field is used to accelerate the positive ions up to the same velocity.
4. Then they enter a magnetic field. While in this field, the ions suffer a deflection, the size of which depends on their charge to mass ratio. *Providing that they have the same charge, we find that ions of lowest mass experience the greatest deflection*. Thus, in Figure 1.15 the mass of the ions shown is $X^+ > Y^+ > Z^+$.
5. By varying the magnetic field strength, then, each of the ions can be deflected so that it strikes the detector.
6. The detector measures the percentage abundance of each ion in the sample. After computer analysis, the results are reproduced on a chart or visual display unit.

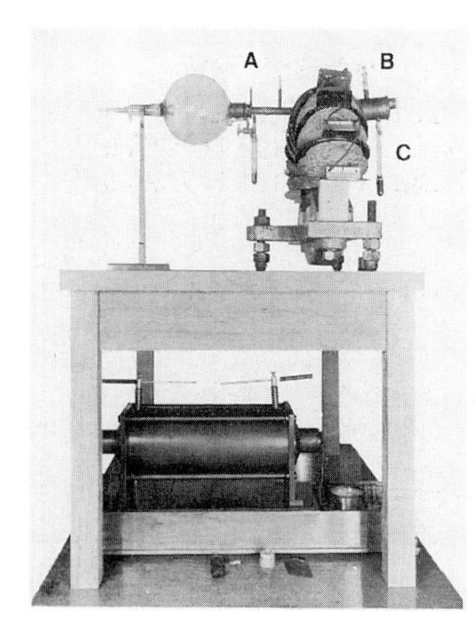

▲ Aston's original mass spectrograph. Gaseous particles pass at high speed from A to B and, in the process, they are deflected by the magnetic field from the electromagnet at C

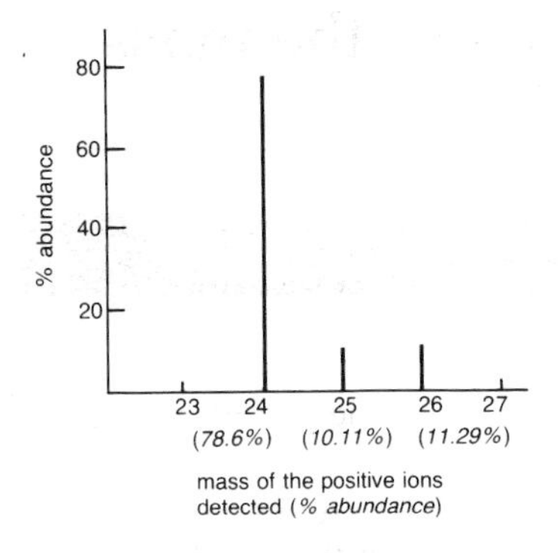

Figure 1.16 ▲
The mass spectrum of naturally occurring magnesium

By carefully varying the electric and magnetic fields, the operator can produce a mass spectrum. Figure 1.16, for example, gives the mass spectrum of magnesium. It shows that a magnesium atom can be one of three isotopes: ^{24}Mg, ^{25}Mg or ^{26}Mg. These have the same number of protons but different numbers of neutrons.

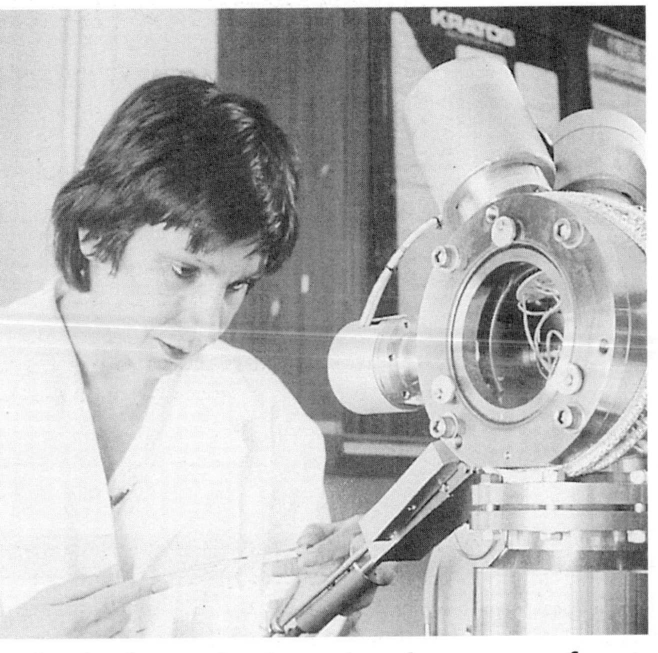

A scientist loading a sample into the mass spectrometer at St Bartholomew's Hospital in London. This instrument is used to identify compounds by exploring their behaviour in a strong magnetic field ▶

Calculating relative atomic masses from mass spectra

The relative atomic mass of an element can be worked out from the percentage abundances of its isotopes. For example, magnesium has the isotopic composition:

isotopic mass/u	% relative abundance
24	78.60
25	10.11
26	11.29

From this data, we can see that 100 atoms of the Mg will contain:

78.60 atoms of mass 24 giving a combined mass of (78.60×24) u
10.11 atoms of mass 25 giving a combined mass of (10.11×25) u
11.29 atoms of mass 26 giving a combined mass of (11.29×26) u

Now,

$$\text{average atomic mass} = \frac{\text{mass of 100 atoms in the isotopic mixture}}{100}$$

$$\text{average mass of 1 Mg atom} = \frac{(78.60 \times 24) + (10.11 \times 25) + (11.29 \times 26)}{100}$$

$$= \frac{1886.4 + 252.75 + 293.54}{100}$$

$$= \frac{2432.69}{100}$$

$$= 24.3 \text{ u}$$

Thus, the average mass of a magnesium atom is 24.3 u and the relative atomic mass of magnesium is 24.3. One mole of magnesium atoms will weigh 24.3 g.

Test yourself

Calculate the relative atomic mass of the following elements given their percentage isotopic abundances:

a) chlorine (75.5% chlorine-35, 24.5% chlorine-37)
b) copper (69.1% copper-63, 30.9% copper-65)
c) silicon (92.2% silicon-28, 4.7% silicon-29, 3.1% silicon-30)
d) nickel (67.7% nickel-58, 26.2% nickel-60, 1.2% nickel-61, 3.7% nickel-62, 1.2% nickel-64).

Questions on Chapter 1

1 The atomic isotopes of bromine, atomic number 35, have mass numbers of 79 and 81, respectively.

a) Explain the difference between these isotopes, stating the atomic symbol in each case.
b) Using a mass spectrometer, the isotopic composition of bromine is found to be 50.52% bromine-79 and 49.48% bromine-81.
i) Explain the main principles underlying the use of a mass spectrometer.
ii) Calculate the relative atomic mass of bromine.

2 Draw dot and cross diagrams to represent the electron arrangements in the following atoms (atomic numbers in brackets): boron (5), potassium (19), chlorine (17), silicon (14) and neon (10).

3 Potassium reacts with chlorine to form potassium chloride, a white solid made up of ions, whereas the chlorides of boron BCl_3, and silicon, $SiCl_4$, contain covalent bonds.

a) Explain the meaning of the word 'ion' and write equations for the formation of potassium and chloride ions from their atoms.
b) What do you notice about the electron arrangements in potassium and chloride ions?
c) What are 'covalent bonds'?
d) Draw dot and cross diagrams to illustrate the covalent bonding in boron and silicon chlorides.

4 Table 1.4 gives some information about three substances. State which one is ionic, which has a simple molecular structure and which is macromolecular.

5 Predict the formulae of the following compounds, indicating whether they are likely to contain ionic or covalent bonds:

a) magnesium chloride
b) calcium hydroxide
c) boron iodide
d) potassium sulphate
e) hydrogen bromide
f) zinc ethanoate
g) carbon(IV) sulphide
h) sulphur(IV) oxide
i) barium carbonate
j) phosphorus(III) oxide
k) carbon monoxide
l) lead(II) oxide
m) sulphur dichloride
n) nitrogen trifluoride
o) manganese(IV) oxide.

6 This question concerns the compounds whose chemical formulae are shown below:

A $C_6H_5NO_2$ B CH_3COOH C $Pb(NO_3)_2$
D $(NH_4)_3PO_4$ E $Al_2(SO_4)_3$ F $NaHCO_3$
G $CaCl_2.2H_2O$ H $Mg(NO_3)_2.6H_2O$

a) Count up the number of each type of atom in the formula of each compound.
b) Calculate the relative molecular mass of each compound.
c) Work out the mass of each of the following molar quantities:
i) 0.05 mol of A
ii) 0.24 mol of D
iii) 2.3×10^{-3} mol of F
iv) 5.6×10^{-4} mol of H
d) How many moles of the compound are found in the following:
i) 23.5 g of B
ii) 1.27 g of C
iii) 13.5 g of F
iv) 7.2 g of G?

Table 1.4 Substances, melting and boiling points, electrical conductivity

Substance	melting point/°C	boiling point/°C	electrical conductivity as a solid	electrical conductivity as a liquid
A	1610	2230	poor	poor
B	750	1392	poor	good
C	−7	59	poor	poor

Questions on Chapter 1 *continued*

7 Calculate the mass of solute required to make up the following aqueous solutions (M stands for mol dm^{-3}):

a) 250 cm^3 of 0.01 M sodium chloride, $NaCl$
b) 500 cm^3 of 2 M sulphuric acid, H_2SO_4
c) 1 dm^3 of 0.1 M lead nitrate, $Pb(NO_3)_2$
d) 300 cm^3 of 0.008 M potassium dichromate, $K_2Cr_2O_7$
e) 250 cm^3 of 0.05 M sodium carbonate decahydrate, $Na_2CO_3.10H_2O$.

8 Which of the following volumes of aqueous solutions contains the greatest number of moles of solute (M stands for mol dm^{-3})?

A 25.2 cm^3 of 0.020 M zinc chloride
B 28.5 cm^3 of 0.015 M calcium nitrate
C 16.5 cm^3 of 0.030 M silver nitrate
D 20.2 cm^3 of 0.018 M potassium hydroxide.

9 Calculate the number of moles of anions (negative ions) present in the volumes of solutions given in question 8.

10 Write balanced chemical equations from the following word equations, checking your answers as you proceed:

a) sodium + bromine ⟶ sodium bromide
b) aluminium + oxygen ⟶ aluminium oxide
c) magnesium + sulphuric acid ⟶ magnesium sulphate + hydrogen
d) potassium hydroxide + nitric acid ⟶ potassium nitrate + water
e) sodium carbonate + sulphuric acid ⟶ sodium sulphate + carbon dioxide + water
f) silver nitrate + sodium chloride ⟶ silver chloride + sodium nitrate

11 Use the balanced equations from question 10 to determine the following quantities:

a) How many grams of sodium bromide will be formed when 0.92 g of sodium is burnt in an excess of bromine?
b) How many grams of aluminium must be burnt in excess oxygen in order to make 30.6 g of aluminium oxide?
c) How many grams of hydrogen will be produced when 2.4 g of magnesium is added to an excess of dilute sulphuric acid?
d) How many grams of silver chloride will be precipitated if we mix aqueous solutions containing 3.40 g of silver nitrate and 2.72 g of sodium chloride, respectively?

12 a) Work out the molar concentrations of the following aqueous solutions:

i) 0.50 mol of sulphuric acid in 1 dm^3
ii) 0.60 mol of sodium hydroxide dissolved in 250 cm^3
iii) 0.25 mol of sodium thiosulphate, $Na_2S_2O_3$ dissolved in 100 cm^3.

b) Calculate the number of moles of solute in each of the following aqueous solutions:

i) 25.0 cm^3 of 0.120 mol dm^{-3} silver nitrate, $AgNO_3$
ii) 22.6 cm^3 of 0.08 mol dm^{-3} potassium manganate(VII), $KMnO_4$
iii) 15.7 cm^3 of 0.5 mol dm^{-3} barium chloride, $BaCl_2$
iv) 11.8 cm^3 of 0.25 mol dm^{-3} copper sulphate, $CuSO_4$.

c) Calculate the total concentration of ions in each of the solutions given in part b.

13 a) Write balanced chemical equations from the following word equations:

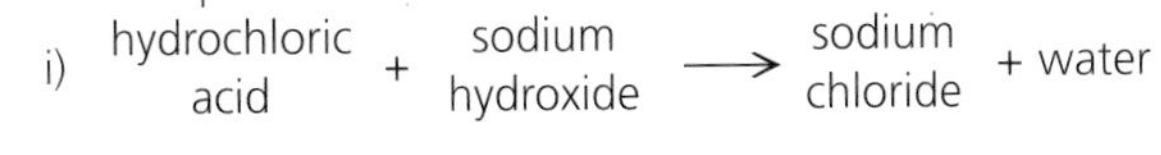

ii) potassium hydroxide + sulphuric acid ⟶ potassium sulphate + water

iii) barium chloride + silver nitrate ⟶ barium nitrate + silver chloride

b) Calculate the concentrations of the aqueous solutions shown in italics from the following titration results:

i) 25 cm^3 of sodium hydroxide, concentration 0.15 mol dm^{-3}, is equivalent to 17.5 cm^3 of *hydrochloric acid*.
ii) 16.4 cm^3 of *barium chloride* is equivalent to 25 cm^3 of 0.01 mol dm^{-3} silver nitrate.
iii) 20 cm^3 of 0.5 mol dm^{-3} potassium hydroxide is equivalent to 12.6 cm^3 of *sulphuric acid*.

14 An experiment was carried out on an antacid tablet to determine the percentage, by mass, of sodium hydrogencarbonate, $NaHCO_3$, which is the active ingredient in the tablet. A tablet was ground up and dissolved in excess water, the mixture left overnight and the undissolved solid was filtered off. Distilled water was added to the filtrate until the volume was exactly 250 cm^3. Then, 25 cm^3 of this solution was titrated against aqueous hydrochloric acid, concentration 0.025 mol dm^{-3}, of which 18.6 cm^3 was required at the equivalence point.

a) Calculate the number of moles of hydrochloric acid in 18.6 cm^3 of 0.025 mol dm^{-3} solution.
b) Write a balanced equation for the reaction between sodium hydrogencarbonate and hydrochloric acid.
c) How many moles of sodium hydrogencarbonate are equivalent to 1 mole of hydrochloric acid?
d) Calculate the number of moles of sodium hydrogencarbonate in 25 cm^3 of the filtrate.
e) How many moles of sodium hydrogencarbonate will be present in all 250 cm^3 of the filtrate?
f) What mass of sodium hydrogencarbonate is present in the antacid tablet?
g) If the mass of the tablet is 0.505 g, calculate the percentage of sodium hydrogencarbonate which is contained in it.

15 Hydrogen peroxide, H_2O_2, is a dangerous atmospheric pollutant, yet in aqueous solution it is used to bleach hair! In an experiment to work out the mass of hydrogen peroxide in an aqueous solution, 10 cm^3 of the solution was diluted to 200 cm^3 with distilled water. Then, 25 cm^3 of this solution was acidified and titrated against 0.11 mol dm^{-3} potassium manganate(VII), $KMnO_4$, of which 21.4 cm^3 was required. The equation for the reaction is:

Questions on Chapter 1 *continued*

$$2KMnO_4(aq) + 5H_2O_2(aq) + 3H_2SO_4(aq) \longrightarrow 2MnSO_4(aq) + K_2SO_4(aq) + 8H_2O(l) + 5O_2(g)$$

a) Calculate the number of moles of potassium manganate(VII) used in the titration.

b) How many moles of hydrogen peroxide reacted with the potassium manganate(VII) in the titration?

c) How many moles of hydrogen peroxide are present in 200 cm^3 of diluted peroxide solution?

d) What mass of hydrogen peroxide is present in 200 cm^3 of diluted peroxide solution?

e) Calculate the mass of hydrogen peroxide which is dissolved in 1 dm^3 of the *concentrated* solution.

Comments on the test yourself sections

Section 1.1

1 An atom is the smallest particle of an element which retains the chemical properties of that element.

2 Refer to Table 1.1.

3 Isotopes of an element are atoms with the same atomic (proton) number but different mass (nucleon) numbers.

4 Protons and neutrons are termed nucleons because they are located in the atomic nucleus. The atomic (proton) number is the number of protons in the nucleus and tells us the number of electrons in an atom and its numbered position in the periodic table. The mass (nucleon) number is the number of protons plus neutrons in the nucleus. Subtracting the atomic number from the mass number gives the number of neutrons in the atom or ion.

5

	Ac	Cs	I	Pb	N	K^+	S^{2-}	Ba^{2+}
protons	89	55	53	82	7	19	16	56
electrons	89	55	53	82	7	18	18	54
neutrons	139	78	74	122	8	20	16	82

6 Each isotope has 44 protons and 44 electrons. The numbers of neutrons are 52, 54, 55, 56, 57, 58 and 60 respectively.

Section 1.2

1

H 1	He 2	Li 2.1
Be 2.2	B 2.3	C 2.4
N 2.5	O 2.6	F 2.7
Ne 2.8	Na 2.8.1	Mg 2.8.2
Al 2.8.3	Si 2.8.4	P 2.8.5
S 2.8.6	Cl 2.8.7	Ar 2.8.8

2 Similar diagrams to Figure 1.7. Three examples are given below:

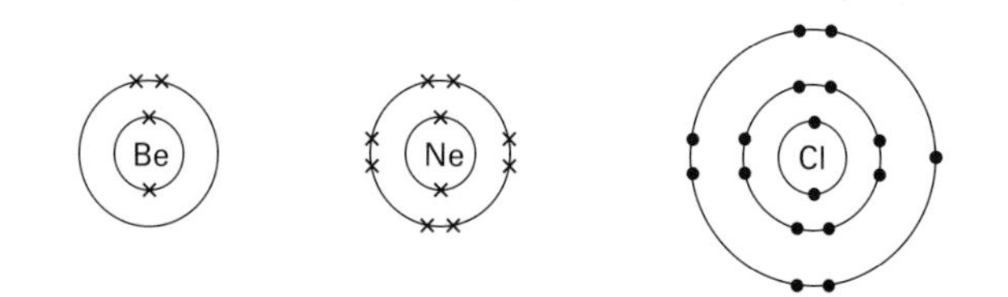

3 **a)** and **b)** The number of shells used = the period (i.e. row).
The number of electrons in the outermost shell = the group number (i.e. column) for groups 1–7. Elements in Group 0 are known as the noble gases, a very unreactive group of elements. We shall find out more about them in the next section.

Section 1.3

1 Ionic bond – electrons are transferred between atoms, giving ions. Covalent bond – the bonded atoms share a pair of electrons.

2

a)

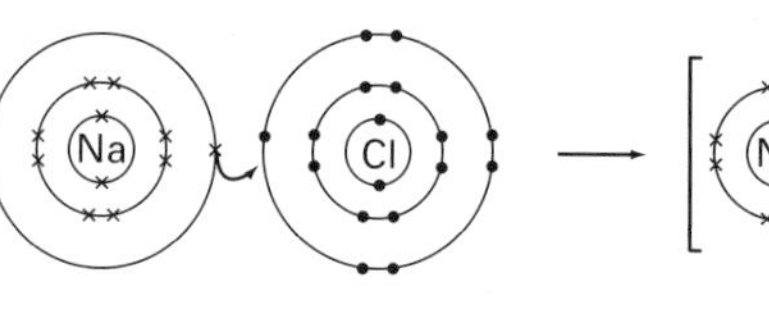

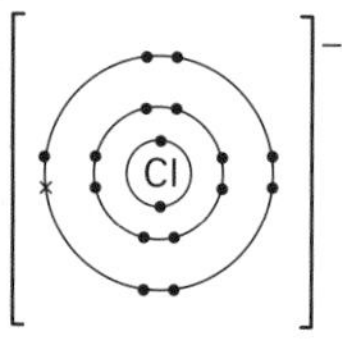

Na: 2.8.1 Cl: 2.8.7 Na^+:2.8 Cl^-: 2.8.8

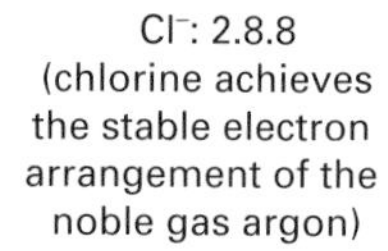

b)

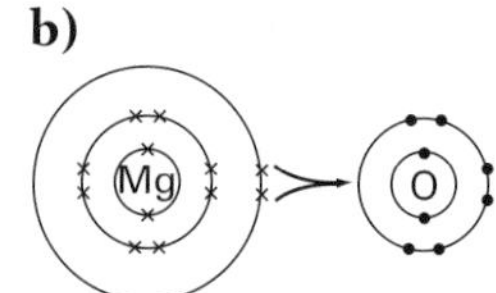

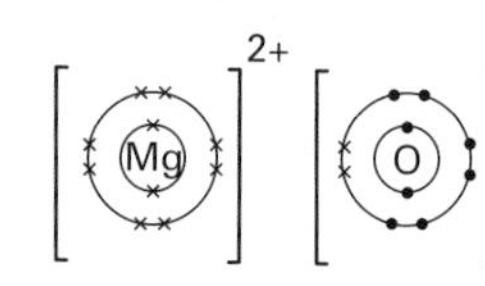

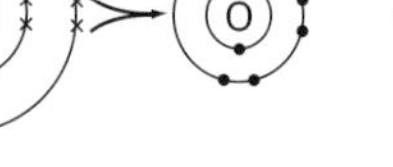

Mg: 2.8.2 O: 2.6 Mg^{2+}:2.8 O^{2-}: 2.8

c)

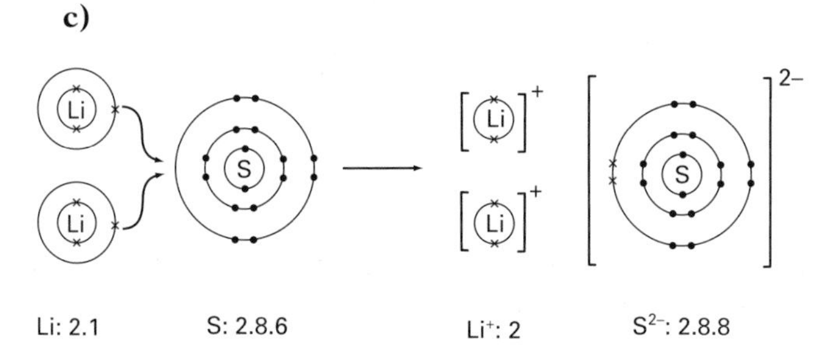

Li: 2.1 S: 2.8.6 Li^+: 2 S^{2-}: 2.8.8

Comments on the test yourself sections *continued*

d)

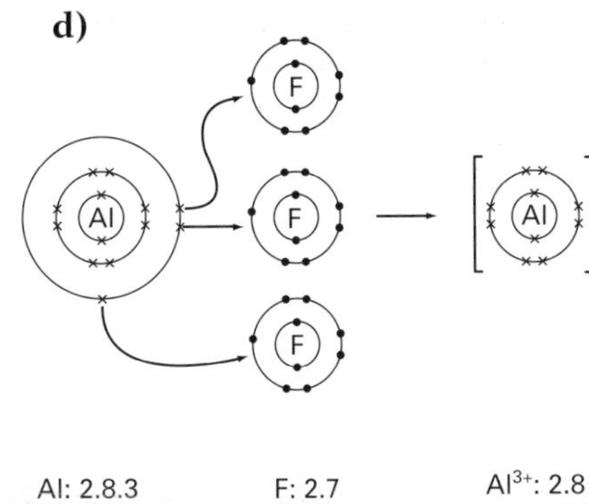

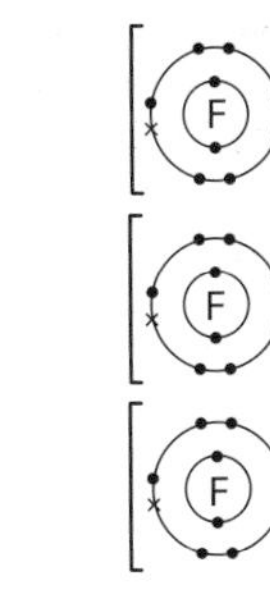

Al: 2.8.3 F: 2.7 Al^{3+}: 2.8 F^-: 2.8

3 a)

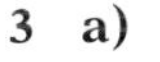

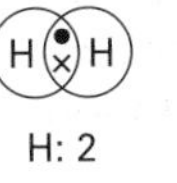

H: 2

b)

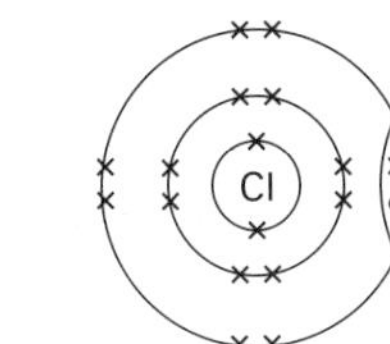

Cl: 2.8.8 (stable octet)

c)

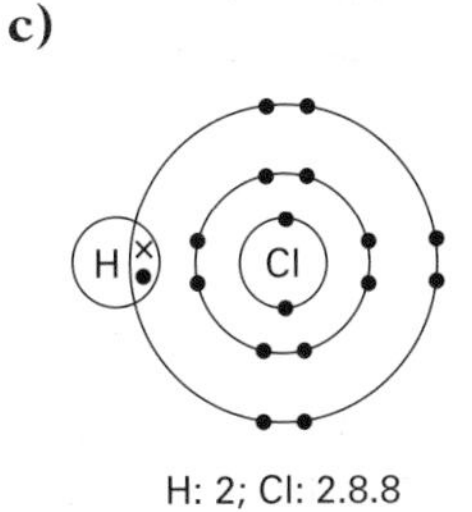

H: 2; Cl: 2.8.8

d)

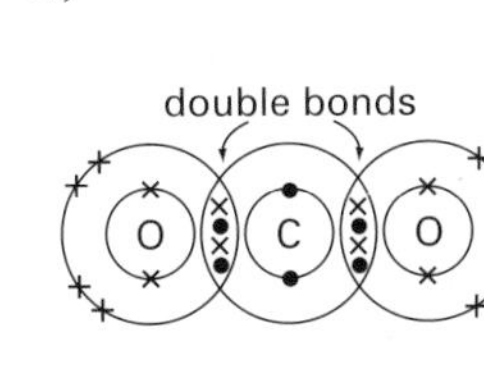

C: 2.8; O: 2.8

e)

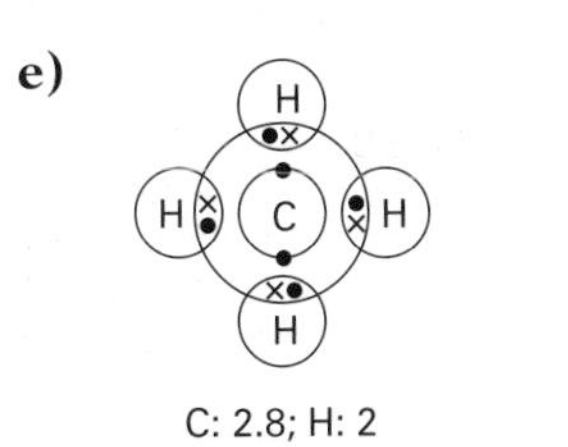

C: 2.8; H: 2

f)

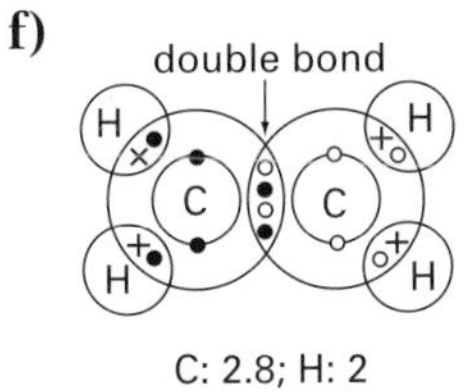

C: 2.8; H: 2

4 Ionic – b, c, f, g. Covalent: a, d, e, h

5 a)

H, N, H, H — H^+ (no electrons)

b)

co-ordinate bond

NH_4^+
N: 2.8; H: 2

c) A coordinate (dative covalent) bond is formed. Nitrogen supplies both electrons.

Section 1.4

Page 14

1 a) CaO; 1Ca, 1O **b)** Rb_2S; 2Rb, 1S
c) F_2O; 2F, 1O **d)** $SrBr_2$; 1Sr, 2Br
e) SeF_2; 1Se, 2F **f)** CCl_4; 1C, 4Cl
g) LiI; 1Li 1I **h)** NCl_3; 1N, 3Cl
i) SiH_4; 1Si, 4H **j)** NH_3; 1N, 3H
Ionic: a), b), d), g). Covalent: c), e), f), h)

2 a) Strontium oxide **b)** Aluminium chloride
c) Barium oxide **d)** Caesium sulphide
e) Boron fluoride **f)** Silicon chloride
g) Hydrogen fluoride **h)** Zinc oxide
i) Silver oxide **j)** Hydrogen sulphide

Page 16

1 a) Calcium carbonate; 1Ca, 1C, 3O
b) Lead(IV) oxide or lead dioxide; 1Pb, 2O
c) Sulphur(IV) fluoride or sulphur tetrafluoride; 1S, 4F
d) Rubidium sulphate; 2Rb, 1S, 4O
e) Vanadium(V) oxide; 2V, 5O
f) Phosphorus(V) chloride or phosphorus pentachloride; 1P, 5Cl
g) Nitrogen(IV) oxide or nitrogen dioxide; 1N, 2O
h) Aluminium nitrate; 1 Al, 3N 9O
i) Silver ethanoate; 1Ag, 2C, 3H, 2O
j) Iron(II) hydroxide; 1Fe, 2O, 2H

2 a) K_2SO_4; 2K, 1S, 4O **b)** $Ni(NO_3)_2$; 1Ni, 2N, 6O
c) PCl_3; 1P, 3Cl **d)** Na_2CO_3; 2Na, 1C, 3O
e) N_2O_5; 2N, 5O **f)** SiH_4; 1 Si, 4H
g) $(CH_3CO_2)_2Mn$; 4C, 6H, 4O, 1Mn
h) $Cu(OH)_2$; 1Cu, 2O, 2H
i) SO_3; 1S, 3O **j)** NH_4Cl; 1N, 4H, 1Cl

3 Ionic: 1a, 1b, 1d, 1e, 1h, 1i, 1j, 2a, 2b, 2d, 2g, 2h, 2j
Covalent: 1c, 1f, 1g, 2c, 2e, 2f, 2i

Section 1.5

1 a) $4Al + 3O_2 \longrightarrow 2Al_2O_3$
b) $Na_2CO_3 + 2HCl \longrightarrow 2NaCl + CO_2 + H_2O$
c) $2AgNO_3 + BaCl_2 \longrightarrow 2AgCl + Ba(NO_3)_2$
d) $Ca(OH)_2 + 2CH_3COOH \longrightarrow (CH_3COO)_2Ca + 2H_2O$

2 a) $2Al + 3Cl_2 \longrightarrow 2AlCl_3$
b) $2P + 5O_2 \longrightarrow P_2O_5$
c) $FeSO_4 + 2NaOH \longrightarrow Fe(OH)_2 + Na_2SO_4$

3 a) $2Mg + O_2 \longrightarrow 2MgO$
b) $2SO_2 + O_2 \longrightarrow 2SO_3$
c) $Ca + H_2SO_4 \longrightarrow CaSO_4 + H_2$
d) $2Na + 2H_2O \longrightarrow 2NaOH + H_2$
e) $CaCO_3 + 2HNO_3 \longrightarrow Ca(NO_3)_2 + CO_2 + H_2O$
f) $C_3H_8 + 5O_2 \longrightarrow 3CO_2 + 4H_2O$
g) $6CO_2 + 6H_2O \longrightarrow C_6H_{12}O_6 + 6O_2$

Comments on the test yourself sections *continued*

Section 1.6

Page 20

1 **a)** 146 **b)** 70 **c)** 58 **d)** 183 **e)** 93 **f)** 98 **g)** 63 **h)** 102

2 **a)**58.5 **b)** 46 **c)** 60 **d)** 331 **e)** 223 **f)** 194 **g)** 397 **h)** 246

Page 22

1 A_r(Ce) = 140; A_r(I) = 127; A_r(Pd) = 106; A_r(Ti) = 48.

2 The molar masses, in grams, are:
a) 74.5; **b)** 56; **c)** 248; **d)** 60.

3 mass of the sample = number of moles × A_r (or M_r)
a) mass of NH_3 molecules = 0.50 × 17 = 8.50 g
b) mass of Cs atoms = 0.375 × 133 = 49.9 g
c) mass of F_2 molecules = 1.50 × 38 = 57.0 g
d) mass of hydrated sodium carbonate = 0.275 × 286 = 78.65 g

4 number of moles $= \dfrac{\text{mass of the sample}}{A_r \text{ (or } M_r)}$

a) number of moles of Mg atoms $= \dfrac{14}{24} = 0.58$

b) number of moles of CO_2 molecules $= \dfrac{8}{44} = 0.18$

c) number of moles of $CuCl_2$ molecules $= \dfrac{0.50}{134.5} = 3.717 \times 10^{-3}$

d) number of moles of $CuSO_4.5H_2O$ molecules $= \dfrac{38}{249.5} = 0.1523$

5 **a)** mol Na atoms $= \dfrac{2.3}{23} = 0.1$

1 mol Na ⟶ 6×10^{23} atoms
thus, 0.1 mol Na ⟶ $0.1 \times 6 \times 10^{23} = 6 \times 10^{24}$ atoms

b) 6×10^{23} molecules
c) 3×10^{22} molecules
d) 6×10^{20} Pb^{2+} ions and 6×10^{20} SO_4^{2-} ions
e) 6×10^{23} Al^{3+} ions but three times as many NO_3^- ions, that is 1.8×10^{24}

Section 1.7

1 **a)** 60 mol **b)** 68 g **c)** hydrogen, 12 g

2 **a)** 1.4 mol **b)** 17.6 g **c)** 27 g

3 **a)** $2H_2O_2 \longrightarrow 2H_2O + O_2$
b) $2Na + 2H_2O \longrightarrow 2NaOH + H_2$
c) $CaCO_3 + 2HCl \longrightarrow CaCl_2 + CO_2 + H_2O$
d) $SiCl_4 + 2Mg \longrightarrow Si + 2MgCl_2$

Section 1.8

1 **a)** 0.04 mol dm^{-3} **b)** 0.083 mol dm^{-3}
c) 0.100 mol dm^{-3} **d)** 0.150 mol dm^{-3}
e) 0.200 mol dm^{-3} **f)** 0.050 mol dm^{-3}

2 **a)** 8.30 g. **b)** 2.00 g **c)** 3.175 g
d) 29.9 g **e)** 6.3 g **f)** 0.425 g

3 **a)** 2.5×10^{-3} mol **b)** 0.1 mol
c) 2.38×10^{-5} mol **d)** 9.35×10^{-4} mol
e) 2.0×10^{-3} mol

4 **a)** 5.0×10^{-3} mol **b)** 0.2 mol
c) 7.14×10^{-5} mol **d)** 2.805×10^{-3} mol
e) 6.0×10^{-3} mol

5
- Sodium chloride, potassium sulphate, aluminium nitrate, calcium ethanoate, potassium manganate(VII), sodium carbonate (hydrated – it contains 10 molecules of water of crystallisation bonded into the solid lattice).
- Potassium iodide, potassium hydrogencarbonate, iron(II) chloride, copper(II) sulphate, nitric acid, silver nitrate.
- Sodium hydroxide, hydrochloric acid, magnesium chloride, zinc ethanoate, sulphuric acid.

Section 1.9

1 **a)** 0.8 mol dm^{-3} **b)** 0.0395 mol dm^{-3}
c) 0.078 mol dm^{-3} **d)** 0.089 mol dm^{-3}

2 **a)** 0.116 mol dm^{-3} **b)** 4.81×10^{-3} mol dm^{-3}
c) 0.011 mol dm^{-3}

3 **a)** $n = p = r = s = 2$;
$2HCl + M(OH)_2 \longrightarrow MCl_2 + 2H_2O$
b) $t = u = 1; v = 2$;
$BaCl_2 + MSO_4 \longrightarrow BaSO_4 + MCl_2$
c) $w = x = y = z = 2$;
$2NaOH + FeCl_2 \longrightarrow Fe(OH)_2 + 2NaCl$

4 **a)** $M_r(MCO_3) = 197$, thus A_r(M) = 137 and M is barium, Ba
b) A_r(N) = 40, thus N is calcium, Ca;
$Ca + H_2SO_4 \longrightarrow CaSO_4 + H_2$
c) Calculated M_r(XY) = 120; XY is *either* potassium bromide, KBr, actual M_r = 119 *or* rubidium chloride, RbCl, actual M_r = 121

Section 1.10

1 A_r(Cl) = 35.5
2 A_r(Cu) = 63.6
3 A_r(Si) = 28.1
4 A_r(Ni) = 58.8

C H A P T E R

2

The structure of the atom

Contents

Study Checklist

After studying this chapter you should be able to:

1. Understand that radioactivity is the spontaneous emission of radiation by an atom and that the rate of radioactive decay is independent of temperature.
2. Distinguish between alpha (α), beta (β) and gamma (γ) radiation in terms of mass, charge, range and relative penetration and be able to balance nuclear equations.
3. Explain the meaning of the term 'half-life', plot a graph of radioactive activity against time and use this to determine the successive half-lives.
4. Appreciate that successive radioactive half-lives are constant for a particular isotope.
5. Write brief notes to explain the difference between nuclear fusion and nuclear fission, describing the main features and mode of operation of a nuclear fission reactor.
6. Briefly describe three uses of radioactivity.
7. Recall the meaning of the following terms: electron energy levels (shells), principal quantum number, first ionisation energy, electron sub-levels (sub-shells) and atomic orbitals.
8. Describe the number and relative energies of the s, p and d orbitals for principal quantum numbers 1, 2 and 3 and also of the 4s and 4p orbitals.
9. Describe the shapes of the s orbitals, p orbitals and the p sub-level, giving a diagram in each case.
10. State the electronic arrangement (configuration) of atoms and ions given their atomic (proton) number and charge.
11. Describe the trends in ionisation energies across a period, and down a group, of the periodic table and relate these to the existence of electron sub-levels.
12. Understand, and use, the relationship between the electronic configurations of elements and their successive ionisation energies.
13. Write down 'electrons-in-boxes' diagrams for elements of atomic number 1–36.

2.1 Radioactivity

In 1896, the French scientist Henri Becquerel noticed the darkening of photographic plates stored in the same drawer as uranium salts. He showed that the salts emitted radiation spontaneously, irrespective of the temperature and chemical nature of the uranium salt. Becquerel used the term **radioactivity** to describe this spontaneous emission of radiation by an atom. Such atoms are said to undergo **radioactive decay** or **radioactive**

disintegration. Following its discovery, scientists questioned whether radioactivity was a unique property of uranium, and this sparked off a search for other radioactive elements.

A major discovery was made by Marie and Pierre Curie in 1898. They found that the intense radioactivity of pitchblende (a uranium ore) could not result only from the uranium in the ore. Using a painstaking technique, they were able to separate the elements **polonium** and **radium** from the uranium ore. These elements are extremely radioactive; in fact radium is over a million times more active than the same amount of uranium.

Over the years many other radioactive elements, of widely varying activity, have been identified, some of which are naturally occurring. An example is the noble gas radon-222, which may rise to harmful levels in houses built over rocks containing heavy radioactive isotopes, such as radium-226:

$$^{226}_{88}Ra \longrightarrow \, ^{222}_{86}Rn + \, ^{4}_{2}He$$

Many other radioactive isotopes are produced by **nuclear transmutations**. In these processes, an atomic nucleus is bombarded with small particles, such as protons and neutrons, which have been accelerated to very high speeds, e.g.:

$$\underset{\text{sodium-23}}{^{23}_{11}Na} + \underset{\text{neutron}}{^{1}_{0}n} \longrightarrow \underset{\text{sodium-24}}{^{24}_{11}Na}$$

Some uses of artificially produced radioactive isotopes (radioisotopes) are described in section 2.5.

Nuclear equations

The equations here are called **nuclear equations**. Notice that the atomic and mass numbers balance on each side of the equation.

▲ Marie and Pierre Curie (1867–1934 and 1859–1906, respectively) at work in their laboratory. Marie Curie, one of Becquerel's students, began studying the radioactive emissions of uranium shortly after their discovery in 1896. She discovered that thorium was radioactive and by the end of 1898, working with her husband, had detected two further radioactive elements, which they named polonium and radium. The Curies were awarded Nobel Prizes in chemistry and physics for their work

2.2 Alpha, beta and gamma radiation

Radioactive isotopes can emit three main types of radiation:

- **alpha (α) particles**
- **beta (β) particles**
- **gamma (γ) rays**

When they pass through a magnetic or an electric field, these behave differently, as shown in Figure 2.1.

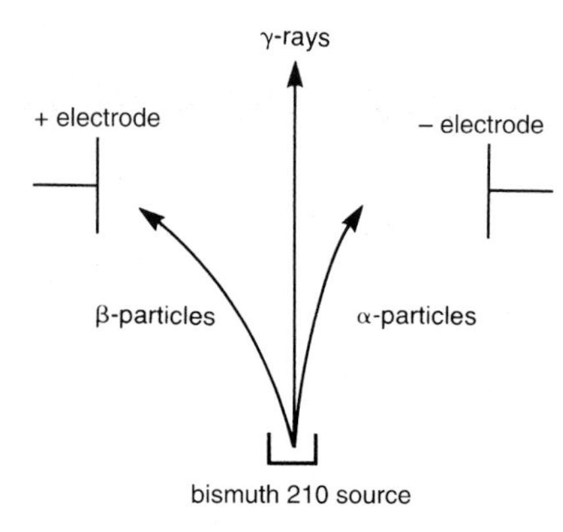

Figure 2.1 ▲
β-particles (electrons) are much lighter than α-particles (helium nuclei) so they experience a much greater deflection in the electric field

α-particles

α-particles are fast-moving helium nuclei, $^{4}_{2}He^{2+}$. However, in nuclear equations an α-particle is given the symbol $^{4}_{2}He$, not $^{4}_{2}He^{2+}$, e.g.:

$$\underset{\text{actinium-225}}{^{225}_{89}Ac} \longrightarrow \underset{\text{francium-221}}{^{221}_{87}Fr} + \underset{\alpha\text{-particle}}{^{4}_{2}He}$$

It is understood that the α-particle is formed without its electrons.

β-particles

β-particles are formed when a neutron disintegrates giving a proton, p, and an electron, e:

$$^{1}_{0}n \longrightarrow \, ^{1}_{1}p + \, ^{0}_{-1}e$$

The proton remains in the nucleus of the atom, so its atomic number increases by one. The electron is rapidly ejected, and can travel up to almost the speed of light. Bismuth-213, for example, decays with β-particle emission to form polonium-213:

$$^{213}_{83}Bi \longrightarrow \, ^{213}_{84}Po + \, ^{0}_{-1}e$$

γ-rays

γ-rays are highly energetic electromagnetic rays, having a wavelength about 100 000 times shorter than that of visible light.

Emission of γ-rays nearly always accompanies the loss of a β-particle. On ejecting an electron an atomic nucleus becomes excited. By releasing this excess energy as γ-radiation, the nucleus is able to return to a more stable energy level.

The properties of α, β and γ radiation are summarised in Table 2.1.

Table 2.1 Properties of the types of radiation emitted by radioactive isotopes

Property	type		
	α-particles	β-particles	γ-rays
nature	helium nuclei, He^{2+}	electrons ejected from the nucleus, moving almost at the speed of light	electromagnetic radiation of very high frequency (like high energy X-rays)
relative mass	4	0.0005; negligible compared to that of the nucleus	none
relative charge	+2	−1	none
range in air	several cm	several m	several km
relative penetration	1	100	10 000
absorbed by ...	paper, air	aluminium sheet	several cm of lead, several m of concrete
deflected by an electric or magnetic field	yes	yes	no

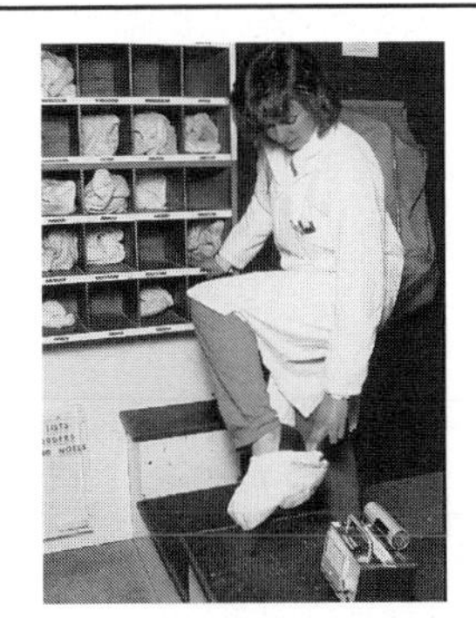

An operator working in the nuclear industry is removing her protective overshoes prior to leaving a laboratory. A Geiger counter is visible in the bottom right of the picture. The Geiger counter works by detecting the ionising effect of radiation. Radiation passes through a mica sheet into a chamber filled with argon gas, causing some of the argon atoms to be ionised.

$$Ar(g) \longrightarrow Ar^{+}(g) + e^{-}$$

The positively charged argon ions move to the cathode (negatively charged) and the electrons move to the anode (positively charged). An electrical pulse is created which is amplified and counted electronically as 'clicks' or displayed as a meter reading. The safety badge she is wearing is specially formulated photographic film which gradually darkens as the amount of radiation increases

Focus 2a

1. **Radioactivity** is the spontaneous emission of radiation by an atom. It results from nuclear disintegration and the **rate of decay** is independent of temperature and the chemical environment of the atom.
2. Naturally occurring radioactive isotopes may emit:
   α-particles (helium nuclei, ${}^{4}_{2}He$) with short range and low penetration;
   β-particles (electrons ${}^{0}_{-1}e$) with medium range and penetration;
   γ-particles (electromagnetic radiation) with long range and high penetration.
3. Some radioactive isotopes can be produced artificially by bombarding a stable nucleus with high-energy particles (e.g. protons, neutrons, or α-particles).

Activity 2.1

During radioactive decay, radiation is emitted and a different atom is formed. If the nucleus of the new atom is also unstable, it will decay further. The actinium radioactive series is shown below. For each nuclear change, fill in the missing atomic and mass numbers or state which type of radiation is emitted.

$${}^{235}_{92}U \xrightarrow{\alpha} Th \longrightarrow {}^{231}_{91}Pa \xrightarrow{\alpha} Ac \longrightarrow {}^{227}_{90}Th \xrightarrow{\alpha} Ra$$

$$\downarrow$$

$$Pb \xleftarrow{\beta} {}^{207}_{81}Tl \longleftarrow Bi \xleftarrow{\beta} {}^{211}_{82}Pb \longleftarrow Po \xleftarrow{\alpha} {}^{219}_{86}Rn$$

2.3 Rate of radioactive decay

Radioactive atoms decay at different rates. For example, iodine-131 emits β-particles about a million times faster than nitrogen-18. The **rate of radioactive decay** may be defined as the number of nuclei disintegrating per second. In activity 2.2 we shall investigate the rates of decay of two nuclei, caesium-130 and xenon-138.

Activity 2.2

The following data were obtained from the decay of caesium-130 and xenon-138.

Time/s	activity of ^{130}Cs /disintegrations s^{-1}	activity of ^{138}Xe /disintegrations s^{-1}
0	200	100
500	165	72
1500	113	36
2500	79	19
3500	54	9
4500	38	4
5500	26	2

The activity of the sample is a measure of the amount of the radioactive isotope that is left, so high activity means many unstable nuclei still remain.

1 For each isotope, plot a graph of activity against time.

2 From your graphs, determine the time taken for the activity of each sample to drop
   a) to one half of its initial value,
   b) from one half to one quarter of its initial value,
   c) from one quarter to one eighth of its initial value.

What do you notice about the values in (a), (b) and (c) for each decay? Compare your answers with those at the end of this chapter.

Table 2.2 Half-lives of some radioactive isotopes

Isotope	half-life
uranium-238	4.51×10^9 years
carbon-14	5730 years
radium-228	6.7 years
phosphorus-32	14.3 days
lead-214	26.8 minutes
radon-220	55.5 seconds
polonium-212	3×10^{-7} seconds

Activity 2.2 shows that:

- *The time it takes for half of the initial number of radioactive atoms to decay, known as the* **half-life** ($t_{1/2}$), is a characteristic property of the isotope. For example, radium-224 and radium-226 both give α-decay but with very different half-lives of 3.64 days and 1622 years, respectively. Radioactive elements, then, can have widely different half-lives, as shown in Table 2.2.
- For a given element, *successive half-lives are constant*, as shown in Figure 2.2. Later on, we shall see that radioactive decay follows first order reaction kinetics.

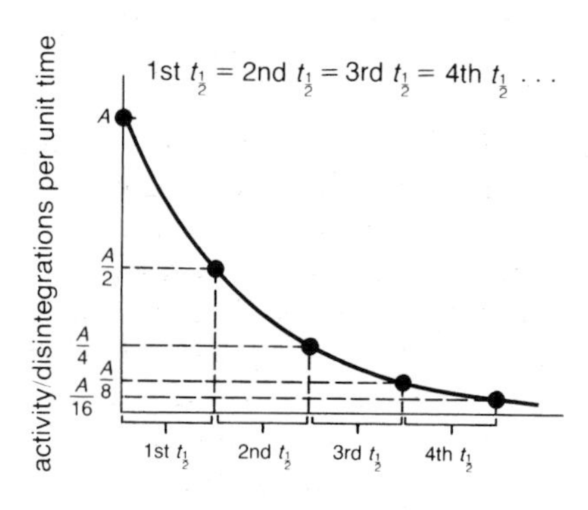

Figure 2.2
Successive half-lives, $t_{1/2}$, are constant during radioactive decay

Unlike chemical reactions, the *rate of radioactive decay is not affected by physical conditions* such as temperature and pressure.

2.4 Nuclear energy

Albert Einstein, in his Theory of Relativity (formed in 1905), proposed that mass, m, can be converted into energy, E, according to the equation $E = mc^2$, where c is the velocity of light. Since c^2 is so large (9×10^{16} m^2 s^{-2}), a very small loss of mass can cause an enormous release in energy and this is what occurs in two nuclear processes, called **fission** and **fusion**.

Nuclear fission

In 1939, Otto Hahn and Fritz Strassman bombarded uranium-235 with low-energy neutrons. They found that, instead of emitting small particles, the uranium-235 nuclei split into large fragments of similar mass. At the same time, two or three rapidly moving (secondary) neutrons were released. This 'splitting' process is called **nuclear fission**.

On slowing down, the secondary neutrons may cause the fission of other uranium-235 nuclei as shown in Figure 2.3a. If a large enough mass of uranium-235 is used, known as the **critical mass**, the neutrons cannot escape without causing more and more fission events and the number of secondary neutrons increases exponentially. The result is a self-sustaining fission process known as a **nuclear chain reaction** as shown in Figure 2.3b.

4 secondary neutrons cause further fission of ^{235}U atoms

Figure 2.3a ▲

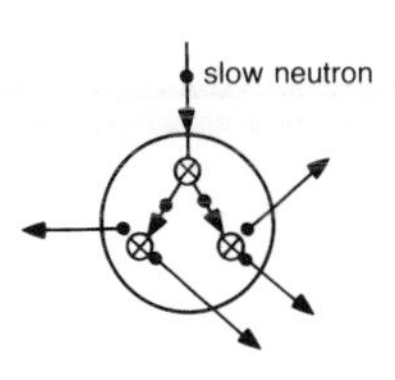

A small piece of ^{235}U is safe because secondary neutrons escape before the chain reaction is set up

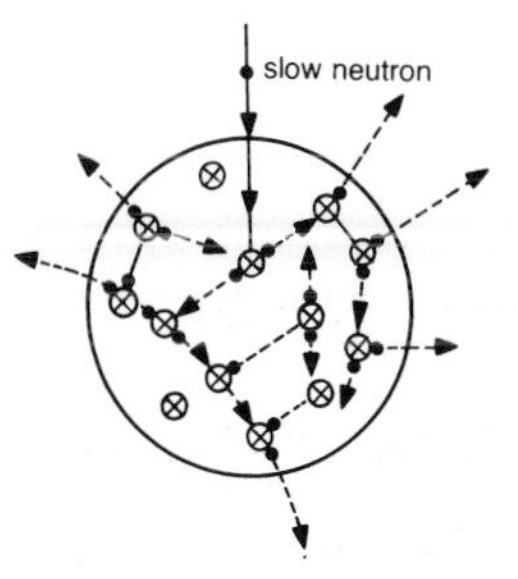

In large samples above the critical mass (about 8 kg) the chain reaction is set up. This may be controlled in a **nuclear reactor** or uncontrolled in an **atomic bomb**

Figure 2.3b ►

A nuclear chain reaction produces a wide variety of fission products, with masses varying from 70 to 160 u. A typical fission reaction is:

$$^{235}_{92}U + \underset{\text{low energy neutron}}{^{1}_{0}n} \longrightarrow ^{94}_{38}Sr + ^{140}_{54}Xe + \underset{\text{secondary neutrons}}{2^{1}_{0}n} + \gamma\text{-rays}$$

Each fission is accompanied by a tiny loss in mass and this is converted into energy. Incredibly, the fission of just 1 g of uranium-235 produces 75 million kJ, a massive release of energy. To obtain the same amount of energy from fossil fuels, we would have to burn about 500 gallons of petrol or 6 tonnes of coal.

A nuclear chain reaction, if uncontrolled, will release an enormous amount of energy in a very short time – in other words, there would be a nuclear explosion! Early in the 1940s, the race was on to find a way of harnessing nuclear energy by controlling the nuclear chain reaction. In 1942, Enrico Fermi and his colleagues at the University of Chicago built the first nuclear reactor, called an **atomic pile**. Fermi found that boron was very effective in absorbing neutrons, so he was able to control the reaction by raising and lowering boron steel rods located between the uranium fuel elements.

At present, there are 37 nuclear reactors in the United Kingdom, generating about 20% of our electricity. The latest to be built is the **pressurised water reactor (PWR)** at Sizewell in Suffolk which was opened in March 1996. The main components of a PWR are shown in Figure 2.4. Uranium-235 will only undergo a nuclear chain reaction if the secondary neutrons are slow-moving, so they are slowed down (**moderated**) by passing pressurised water over the uranium-235 fuel elements. As the chain reaction proceeds, the fuel rods get very hot and the pressurised water carries the heat away from them into a steam generator. The steam drives a turbine and electricity is generated. Cooling water taken from a large natural supply such as a lake or river passes into a condenser and the steam is converted to water which is recirculated. Rods made of boron (a neutron absorber) are used to

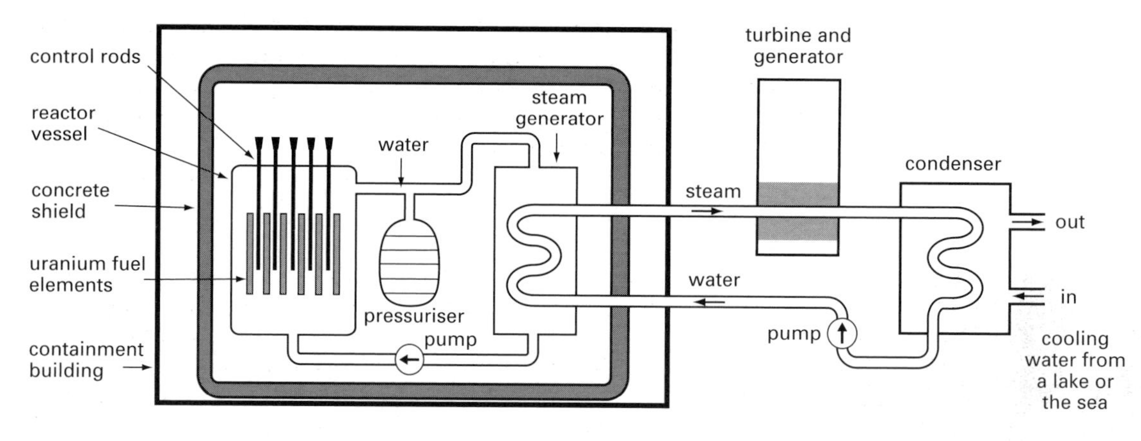

Figure 2.4 ▲
The main features of a pressurised water reactor (PWR)

control the rate of nuclear fission. As they are lowered deeper into the atomic pile, these control rods absorb more neutrons and the reaction gradually slows down. By varying the position of the control rods, the rate of fission, and hence the energy output of the reactor, can be held constant. In an emergency, the control rods would drop automatically into the core and the reactor would be shut down. Nuclear fission produces highly dangerous radioactive waste. To prevent the leakage of any radiation, the core is encased in steel and concrete shields, some 5 m thick. The reactor itself is located in a containment building, so that if any radioactive gas were to be discharged it could not get into the atmosphere.

▲ German fireman washing down a truck after discovering radioactive contamination from the nuclear accident at Chernobyl

After a nuclear reactor has been operating for some time, the nuclear fuel becomes contaminated with large amounts of fission products, many of which are themselves highly radioactive. If these are not removed, they absorb the secondary neutrons, thereby slowing down the nuclear chain reaction. Consequently, at routine intervals, the fuel elements are sent to a reprocessing plant where the fission products are removed. The pure unused fuel is returned to the reactor.

In recent years, there has been much controversy about the safety of nuclear reactors and how best to transport, and store, radioactive waste. The British government caused concern in April 1998 by accepting a consignment of radioactive waste from Georgia for processing at the Dounreay plant sparking fears that Britain may become a dumping ground for other countries' nuclear waste.

At present, the most dangerous radioactive waste is stored in stainless steel tanks which are embedded in thick concrete 'jackets'. However, equipment has recently been developed which converts the waste into very durable glass blocks. Although the waste is stored above ground at present, it is current government policy that it should eventually be buried deep underground.

◀ The world's worst nuclear accident occurred at the Chernobyl power plant, in the Ukraine, on 26 April 1986. One of the four reactors exploded and there was a serious leak of radioactive gases. These gases spread over Northern Europe, with slightly increased levels of radioactivity being detected as far away as the Scottish Highlands. The accident was caused by unauthorised testing of the reactor and the failure to apply several safety regulations. As a result, the top of the reactor was blown off and the core ignited. Unfortunately, the reactor's design had a number of weaknesses, in particular the absence of a containment building which might have prevented radioactive pollutants from entering the atmosphere. Indeed, such a reactor would not be granted a licence for use in the UK. More than 30 people died from radiation exposure and another 300 workers and fire-fighters experienced radiation sickness. The reactor was encased in concrete and about 135 000 people who lived within a 1000 mile radius of the plant were evacuated. Due to the demand for energy in the Ukraine, the remaining three reactors at Chernobyl went back on line in 1988. Now, a decade later, it seems that there are links between the accident and incidence of thyroid cancers in the neighbouring regions

As well as fission products, nuclear reactors also produce plutonium-239, another fissionable material which is used in **fast breeder reactors**. In these reactors, a layer of non-fissile uranium-238 is placed around the uranium-235. Fast electrons from the fission of the uranium-235 strike uranium-238 nuclei and plutonium-239 is produced:

$$^{238}_{92}U + ^{1}_{0}n \longrightarrow ^{239}_{92}U \xrightarrow{-\beta} ^{239}_{93}Np \xrightarrow{-\beta} ^{239}_{94}Pu$$

▲ Nuclear fuel rods being handled. The technician works from behind a thick glass plate window using robotic arms. At present, about 23% of the UK's energy comes from nuclear reactors. Compare this with France, 73%, and the Netherlands, 5%, the highest and lowest in Western Europe, respectively

The Earth's reserves of uranium-235 will be exhausted in about 300 years. However, there is enough uranium-238 to power fast breeder reactors for over 2000 years at our current rate of energy consumption. Although a number of experimental fast reactors have been built, few countries are committed to introducing this technology because of the danger of a leakage of plutonium-239 (an extremely toxic material) and the security risks posed by the theft of plutonium-239, which can be used to make fission bombs.

Nuclear fission bombs work by violently bringing together two pieces of a fissionable isotope, uranium-235 or plutonium-239, to form one piece which is larger than the critical mass (see Figure 2.3b). Uncontrolled fission takes place leading to a violent explosion during which enormously high temperatures and pressure are generated in a fraction of a second. The horrific effects of such conditions were all too apparent after the dropping of the first fission bombs on the Japanese cities of Hiroshima and Nagasaki in 1945. Of the 130 000 people killed by these bombs, some were vaporised while others were buried in the ruins of buildings flattened by the blast. About a fifth of the fatalities were caused by the intense radiation, mainly γ-rays, released on explosion. Many survivors received radiation burns, others were made sterile or, over the years, have developed cancers.

Nuclear fusion

When two nuclei of low mass join together to form a heavier, more stable nucleus, there is a reduction in mass and a lot of energy is released. This process, known as **nuclear fusion**, is responsible for the energy radiated by stars. For example, most of the sun's energy is obtained from the fusion of deuterium nuclei (deuterium is an isotope of hydrogen):

$$^{2}_{1}H + ^{2}_{1}H \longrightarrow ^{3}_{2}He + ^{1}_{0}n$$

▲ Work started on the first, small-scale breeder reactor at Dounreay in Scotland in 1954. The Dounreay Fast Reactor (DFR) became operational in 1956 and continued working until 1976. Such was its success that, in 1966, the Government approved the construction of a much larger, prototype breeder reactor, the Prototype Fast Reactor (PFR), shown in the photograph. The PFR went on-line in 1974 and supplied electricity until, in 1994, the Government decided to put the fast breeder nuclear programme on hold and the reactor was shut down and decommissioned

This reaction liberates about 2.5×10^{13} kJ per mole of deuterium atoms used which is about a million times more energy than is released when the same mass of TNT (trinitrotoluene) explodes. Not only does each nuclear fusion event yield a massive amount of energy but, unlike fission, the products are not radioactive and the fuel, and reactor system, are relatively safe. Moreover, with about 0.1 mole of atoms per dm^3 of sea water, there is an almost limitless supply of deuterium on earth. So, why aren't there any fusion reactors?

As two nuclei approach each other, massive repulsive forces build up between them. During nuclear fusion the nuclei must be forced together and this requires an enormous amount of energy. Thus, nuclear fusion only occurs at very high temperatures of around 100 million °C. At these temperatures hydrogen gas does not exist as atoms but as a **plasma** of unbound nuclei and electrons. Needless to say, a plasma cannot be contained using everyday materials as these could not withstand the temperatures required. Recently, though, it has been found to be possible to contain the plasma in a 'bottle' made of magnetic fields so that it does not touch the vessel walls. One such device, the Tokamak Fusion Test Reactor at Princeton University (USA), has generated an output of 10^7 megawatts of power over a small fraction of a second. Unfortunately, fusion must continue for at least a second before 'ignition' occurs, that is energy goes back into the plasma and the fusion becomes self-sustaining. Even after forty years of research, and many millions of pounds, it appears that controlled nuclear fusion will not make any contribution to our energy needs in the foreseeable future.

Mass defect and binding energy

Consider an atom of deuterium, ${}^{2}_{1}H$. This is an isotope of hydrogen whose nucleus contains 1 proton and 1 neutron. Accurate measurements show:

$$\text{mass of a proton} = 1.67208 \times 10^{-27}\text{ kg}$$
$$\text{mass of a neutron} = 1.67438 \times 10^{-27}\text{ kg}$$

Thus, in theory:

$$\begin{aligned}\text{mass of the } {}^{2}_{1}H \text{ nucleus} &= \text{combined mass of 1 proton and 1 neutron}\\ &= 1.67208 \times 10^{-27} + 1.67438 \times 10^{-27}\text{ kg}\\ &= 3.34646 \times 10^{-27}\text{ kg}\end{aligned}$$

The actual mass of the deuterium nucleus can be measured using a mass spectrometer and is found to be 3.34250×10^{-27} kg. Although the real and theoretical values only disagree by about 0.1%, this difference is too big to be explained by experimental error. In fact, for all nuclei, except ${}^{1}_{1}H$, the nuclear mass is always very slightly less than the combined mass of the protons and neutrons. This loss of mass is known as the **mass defect**.

The mass defect corresponds to the energy released when the protons and neutrons are brought together in the nucleus, and this is known as the **binding energy** for that nucleus. For deuterium then,

$$\begin{aligned}\text{mass defect} &= (3.34646 - 3.34250) \times 10^{-27}\text{ kg}\\ &= 0.00396 \times 10^{-27}\text{ kg}\end{aligned}$$

Substituting the mass defect into Einstein's equation, $E = mc^2$, we obtain the binding energy per ${}^{2}_{1}H$ nucleus:

$$\begin{aligned}\text{binding energy} &= 0.00396 \times 10^{-27} \times (3 \times 10^{8})^2\\ &= 3.56 \times 10^{-13}\text{ J}\end{aligned}$$

Although this is a small amount of energy, it is over 200 000 times greater than the energy liberated when one molecule of methane is burnt! Clearly, chemical bond strengths are minute in comparison to the massive forces which bind the nucleons together in the nucleus.

2.5 Uses of radioactive isotopes

Carbon-14 dating

Carbon-14 is formed in the atmosphere when a nitrogen atom collides with a high-energy neutron:

$$ {}^{14}_{7}N + {}^{1}_{0}n \longrightarrow {}^{14}_{6}C + {}^{1}_{1}H $$

The carbon-14 reacts with oxygen to form carbon-14 dioxide, some of which gets absorbed by plants during photosynthesis. A food chain is set up and eventually all living material reaches a constant level of radioactivity. Of course, when an organism dies it doesn't take in any more carbon-14 and that which is already present in the dead plant or animal starts to decay at a well-known rate, $t_{1/2} = 5700$ years. Thus, by comparing the proportion of carbon-14 in the dead animal with that in the living species, it is possible to estimate the age of the plant or animal remains.

One of the most famous carbon-14 'datings' was of the Shroud of Turin, a piece of cloth believed by many to be the burial shroud in which Jesus Christ was placed after his crucifixion. The cloth, which bears a faint image of a man's face, thought to be Christ, was first exhibited around 1354. Arguments about its authenticity carried on for centuries until, in 1988, a piece of cloth the size of a postage stamp was carbon-14 dated by three different universities. The results dated the shroud at no more than 800 years old, indicating that it could not be Christ's burial shroud and that it might be an elaborate forgery. Even so, no one is quite sure what process was used to get the image on to the cloth!

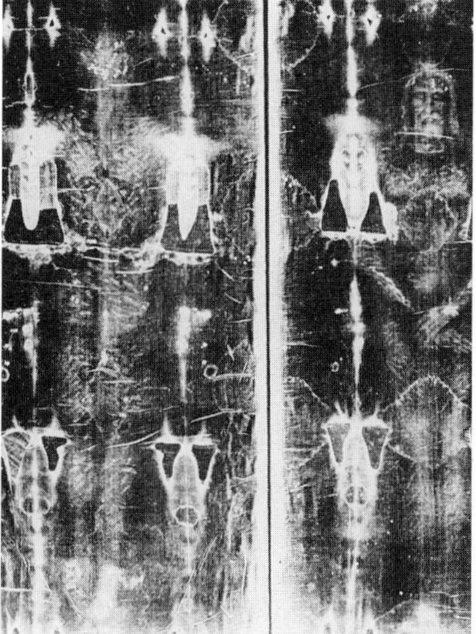

Radio-carbon dating was used to show that the Shroud of Turin, thought by some people to be the burial shroud of Christ, is only about 800 years old

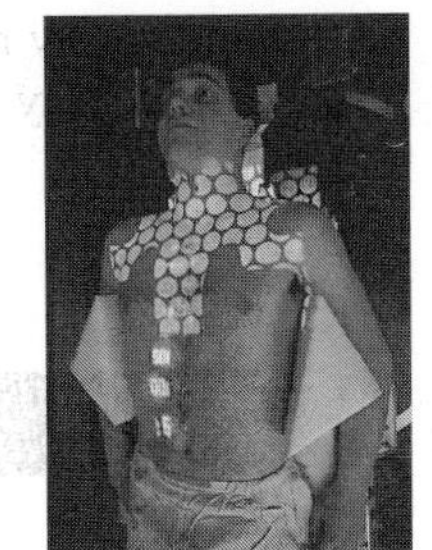

▲ A person undergoing radiotherapy, on a linear accelerator to treat Hogkin's disease, a cancer of the lymphatic system. The illuminated discs over the patient's chest indicate the areas which are to receive radiation. The pattern is defined by an arrangement of lead blocks, suspended below the head of the accelerator, which shields the lungs from excessive irradiation. Linear accelerators are used as a source for high-voltage X-rays, which have the ability to penetrate the skin to a greater depth and deliver a larger X-ray dose to the target area relative to the dose at the skin.

Medical applications

Medical imaging is used in diagnosis to detect disorders of internal organs without invasive surgery. The patient is given a medicine containing a radioisotope attached to a 'carrier' which takes it to a particular location in the body. Its concentration builds up in the tissue and, when the isotope decays, the radiation emitted allows an image of the suspect organ to be recorded. A commonly used radioisotope is technetium-99, which releases γ-rays. The half-life of technetium-99 is 6 hours which is long enough to allow the γ-rays to be detected but not long enough for them to cause a significant amount of tissue damage. A photosensitive image converts γ-radiation intensity into an electrical signal. This is computer analysed and produced as an image on a visual display unit (VDU).

In radiotherapy, a radioisotope is used to treat the patient's condition. The principle is that radiation destroys cancer cells faster than it destroys healthy cells. Thus, γ-rays from a cobalt-60 source are used to destroy inaccessible tumours.

Radioactive tracers

By labelling a molecule with a radioactive atom, we can trace its movement within a given system. This technique has many applications, e.g. monitoring the movement of river waste, finding the rate at which plants absorb phosphorus from the soil and studying the mechanisms of chemical reactions.

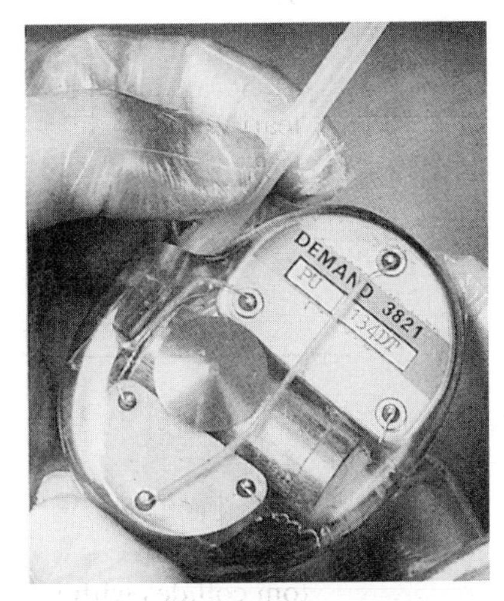

▲ The miniature nuclear battery in this heart pacemaker is about 35 mm long and 15 mm across. The battery has a design life of 20 years. Electricity is generated by the radioactive decay of a small quantity of plutonium-238

Food irradiation

The ability of γ-rays to kill micro-organisms has been extended to food preservation. For example, exposure to γ-radiation slows down the moulding of strawberries and also prevents the sprouting of potatoes in storage. Although food irradiation is widely used throughout Europe, there is some concern that radiation sterilisation might produce carcinogens (chemicals which may cause cancers to develop).

Focus 2b

1. The rate of radioactive decay is directly proportional to the number of radioactive atoms which remain in the sample. The decay follows first order kinetics (section 24.6).
2. The time it takes for half the original number of radioactive atoms to decay is known as the **half-life**. For a given isotope, successive half-lives are constant.
3. **Nuclear fission** occurs when a nucleus splits into two large fragments. An increase in mass defect results in an enormous release of energy. Fission reactors harness the energy released by the fission of uranium-235 or plutonium-239.
4. During **nuclear fusion**, nuclei of low mass join together to form a heavier, more stable, nucleus. In doing so, a massive amount of energy is released.
5. Radioactive isotopes can be used: to date archaeological specimens, measure thickness, study reaction mechanisms, in medical imaging and to destroy cancers.

▲ This worker is checking one of the external radioactive sensors used to monitor the level of liquid chlorine in the tanks

Radon – a deadly gas under the floorboards?

E

Radon, the heaviest noble gas, is colourless and odourless. One of its isotopes, radon-222, is produced by the radioactive decay of uranium-238 which is found in many common rocks, such as granite. Although radon-222 is chemically unreactive, the atom is unstable and decays through a variety of daughters to lead-206, a stable isotope. The decay series is shown below, together with the type of decay and its half-life:

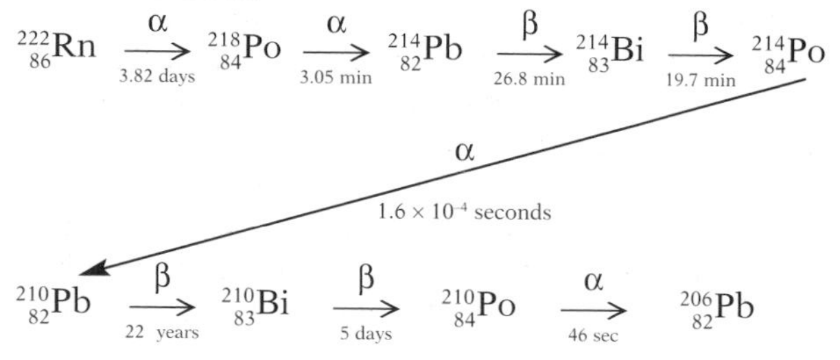

We breathe in radon-222 every day. Whilst most of the gas is exhaled, some is absorbed by the fluid in the lungs and due to radon-222's short half life of 3.82 days, some radioactive decay occurs. Heavy metals (polonium, lead and bismuth) are deposited and react to form ionic compounds which remain in the lungs. More significantly, over a period of time, the radioactive decay of radon-222 produces harmful gamma radiation. This can disrupt the formation of healthy lung cells and cause cancers to develop. The incidence of lung cancer is increased for smokers because radon-222 particles can become absorbed on the particles of dust and tar in a smoker's lungs and so less is exhaled.

There is much evidence to suggest that radon can cause lung cancers. For example, a study of miners who worked at Radium Hill in South Australia showed that they experienced radiation about 30 times above safe limits and this increased their chances of cancer three-fold. Such research prompted concerns that the radon-222 released by rocks might become concentrated under the floorboards of houses and then percolate upwards into living accommodation. By the 1980s, many experts considered radon exposure to be the second most common cause of lung cancer after smoking.

There is a fierce debate about the actual dangers of radon in homes. In Britain, the National Radiological Protection Board estimate that around 2 500 people die each year from radon exposure. However, it is difficult to pinpoint these fatalities to the high radon areas of Britain. Cornwall, for example, has by far the *highest* levels of radon-222 emission but the incidence of lung cancer is well *below* the national average. Indeed, only one of eight major international surveys conducted – from Sweden – has found a statistically significant link between radon in the home and lung cancer.

'There is no question that radon causes lung cancer', says John Boice, the former head of radiation epidemiology at the National Cancer Institute in the US. 'But there is great uncertainty about the effects of low doses among large populations.' There are two difficulties with research methods. Firstly, smoking is known to cause cancer and this might swamp any trend in lung cancer caused by radon. Secondly, only low dosages of radiation are experienced in most homes and, for obvious reasons, it is not possible to increase radon-dosage in homes to test the effects!

Of course, prevention is better than cure. Homes built on a concrete base will be impervious to radon-222 coming from the rocks below. In areas with high radon exposure, homes are designed so as to enhance the natural ventilation in basements and under floor areas and, in some cases, pumps are used to improve ventilation.

(More information about radon in homes can be found in the article 'Undermining our lives?' by Fred Pearce in *The New Scientist* issue 14 March 1998.)

2.6 The arrangement of electrons in atoms

When an unstable atom emits radiation, a nuclear change takes place. For example, in radio-carbon dating, carbon-14 atoms decay to nitrogen-14 and β-particles are emitted:

$$^{14}_{6}C \longrightarrow \, ^{14}_{7}N + \, ^{0}_{-1}e$$

Radioactivity is a property of the atomic nucleus. During a chemical reaction however, the atom's nuclear structure is unaltered. The chemical behaviour of an atom depends solely on the way its electrons are arranged. Carbon-12 and carbon-14, for example, have the same chemical properties because they have identical electron arrangements.

A knowledge of the electron arrangements in atoms and molecules allows us to make accurate predictions about their chemical reactivity. In the following sections, you will learn how the electrons are arranged in atoms and then how the arrangements are shown on paper.

2.7 Atomic emission and electron energy levels

Streetlamps make use of the light emitted when a high voltage is passed through sodium or mercury vapour

Certain atoms emit light when they are supplied with heat or electrical energy. This process, known as **atomic emission**, is responsible for the light given out by fireworks and some street lamps. In 1913, Niels Bohr explained atomic emission by suggesting that electrons move around the nucleus in stable orbits without emitting energy. When an atom absorbs energy, electrons 'jump' from lower to higher energy orbits, where they are less stable and so do not remain. As the electrons return to the lower energy orbits, they release their excess energy as light. The greater the energy difference between the higher and lower energy orbits, or **electron energy levels**, the higher will be the frequency of the light emitted.

In **atomic emission spectroscopy**, the light emitted by an excited atom is split by a prism or diffraction grating into its component frequencies. As different atoms have unique electron energy levels the spectrum produced is characteristic of the element used, so it may be used to identify the presence of that element. Atomic emission spectroscopy is used, for example, to determine the amount of metal ion pollution in river water.

Spectral studies indicate that an atom has an infinite number of electron energy levels (also known as **electron shells**), each of which is represented by a **principal quantum number**, $n = 1$, $n = 2$, $n = 3$, etc. The *greater* the principal quantum number, the *higher* the energy of any electrons in that energy level. In fact, we find that all known atoms, in their lowest energy state, can only hold electrons in energy levels $n = 1$ to $n = 7$.

2.8 How many electrons can occupy an energy level?

On absorbing energy, an atomic electron moves further away from the nucleus, from a lower to a higher electron energy level. If enough energy is absorbed, the electron, e^-, escapes from the atom, M, and a positive ion, M^+ is formed:

$$M(g) \longrightarrow M^+(g) + e^-$$

The **first ionisation energy, $\Delta H_i(1)$** of an element is the minimum energy needed to remove one mole of electrons from one mole of gaseous atoms, to form one mole of gaseous positive ions, each of which has a single charge. For example, when one mole of gaseous magnesium atoms absorb 736 kJ of energy, one mole of electrons will be released:

$$Mg(g) \longrightarrow Mg^+(g) + e^- \qquad \Delta H_i(1) = 736 \text{ kJ}$$

If enough energy is supplied, successive ionisations take place and more highly charged magnesium ions are formed. Thus, the **second ionisation energy, $\Delta H_i(2)$** of an element is the energy needed to remove one mole of electrons from one mole of gaseous unipositive ions:

$$Mg^+(g) \longrightarrow Mg^{2+}(g) + e^- \qquad \Delta H_i(2) = 1448 \text{ kJ}$$

Table 2.3 Successive ionisation energies for magnesium

Number of the electron removed, N	$\Delta H_i(N)$ /kJ mol^{-1}
1	736
2	1448
3	7740
4	10470
5	13490
6	18200
7	21880
8	25700
9	31620
10	35480
11	158300
12	199500

Similar definitions and equations can be written for successive ionisations, for example, for the fifth ionisation energy of magnesium:

$$Mg^{4+}(g) \longrightarrow Mg^{5+}(g) + e^- \qquad \Delta H_i(5) = 13\,490 \text{ kJ}$$

Successive ionisation energies, which can be determined from atomic spectra, provide a lot of information about the arrangement of electrons in atoms.

As electrons are successively removed from an atom, the protons increasingly outnumber the remaining electrons, e.g.:

$$\underset{12p,\ 12e}{Mg(g)} \xrightarrow{-e^-} \underset{12p,\ 11e}{Mg^+(g)} \xrightarrow{-e^-} \underset{12p,\ 10e}{Mg^{2+}(g)} \xrightarrow{-e^-} \underset{12p,\ 9e}{Mg^{3+}(g)} \xrightarrow{-e^-} \underset{12p,\ 8e}{Mg^{4+}(g)} \ldots$$

Consequently, the remaining electrons feel a greater nuclear pull and this makes them more difficult to remove. So, for all atoms, successive ionisation energies increase.

Table 2.3 gives the successive ionisation energies for magnesium ($\Delta H_i(N)$), where N is the number of the electron which is being removed. These can be shown as a graph of $\log_{10} \Delta H_i(N)$ against N, Figure 2.5. As you can see, the $\log_{10} \Delta H_i(N)$ values do increase, but not in a steady way. The graph suggests that the 12 electrons in the magnesium atom are grouped into three energy levels: n = 1, 2 and 3. The lower the energy level, the nearer electrons in it are to the nucleus and the more difficult it is to remove them from the atom, as shown in Figure 2.6.

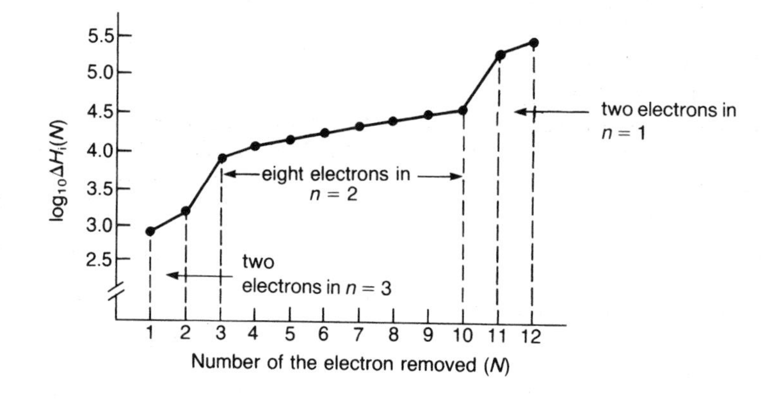

Figure 2.5
A graph of $\log_{10}$ of the successive ionisation energies of magnesium, $\log_{10} \Delta H_i(N)$, plotted against the number of the electron removed (N). As the principal quantum number, n, increases, the mean distance of the electron from the nucleus also increases. The reason why we use $\log_{10} \Delta H_i(N)$ values is that the range of $\Delta H_i(N)$ values is so wide that we would not be able to get a viable scale on the $\Delta H_i(N)$ axis. Using logs merely alters the shape of the graph but does not affect the conclusions that we can draw from it ▶

Patterns in the successive ionisation energies of other atoms confirm this grouping of electrons. Overall, we find that a given energy level, n, can hold a maximum of $2n^2$ electrons.

Energy level	Maximum number of electrons it can hold
n = 1	$2(1)^2$ = 2
n = 2	$2(2)^2$ = 8
n = 3	$2(3)^2$ = 18
n = 4	$2(4)^2$ = 32

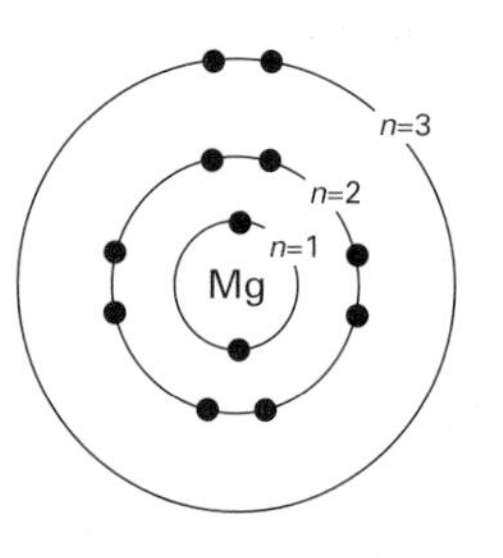

magnesium 2.8.2

Figure 2.6 ▲
A dot diagram for magnesium. Dot and cross diagrams provide a useful way of representing the electrons in energy levels (shells); they are explained in section 1.2

When building up an atom's electron arrangement, electrons first fill the energy levels of lowest energy. For example, in an atom of silicon, atomic number 14, electrons occupy the following energy levels: 2 in n = 1, 8 in n = 2 and 4 in n = 3. This arrangement may be simplified to 2.8.4.

Table 2.3 and Figure 2.5 show some big jumps in the values of the successive ionisation energies of magnesium. Compare, for example, the difference in ionisation energies for the removal of the second (1448 kJ mol^{-1}) and third (7740 kJ mol^{-1}) electrons. The reason for this big jump in values is that the third electron is being taken from a group of electrons (in the n = 2 energy level) which are nearer to the nucleus than the first two to be removed

Activity 2.3

The electron arrangement in magnesium, atomic number 12, may be written as 2.8.2. Write down the corresponding electron arrangements for: beryllium, carbon, argon, phosphorus and sulphur.

and are less screened from it by inner electron shells. Similarly, it is much more difficult to remove the eleventh electron (in $n = 1$) than it is to remove the tenth (in $n = 2$).

By comparing successive ionisation energies, $\Delta H_i(N)$, we can often work out the periodic group in which the element is found (see section 5.1). Consider the first eight ionisation energies of the element X:

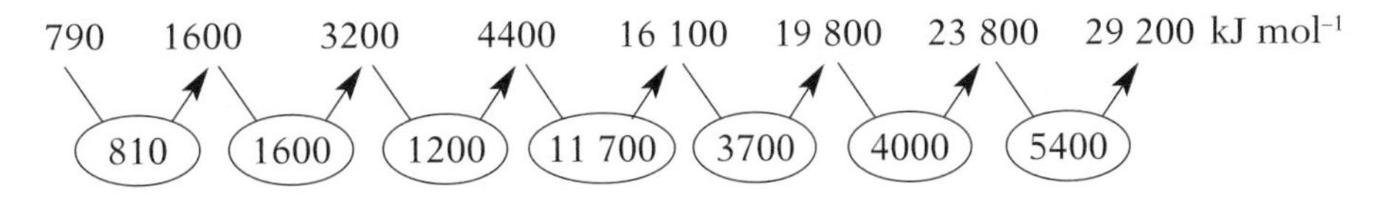

The circled numbers show how much greater an ionisation energy is than the one before. Clearly, four electrons can be removed before the big jump in the $\Delta H_i(N)$ values. Since the big jump occurs when the first electron is removed from an inner energy level, an atom of X must have four electrons in its outermost (or valence) energy level (shell). In section 1.2, we noted that the number of electrons in the outermost electron energy level (electron shell) is the same as the atom's group number in the periodic table. Thus, element X is in group 4 of the periodic table. In fact, the values given for X are actually for silicon, electron arrangement 2.8.4.

Focus 2c

1. The **first ionisation energy**, $\Delta H_i(N)$, of an atom is the minimum energy needed to remove one mole of electrons from one mole of gaseous atoms to form one mole of gaseous positive ions, each of which has a single charge.
2. Trends in successive ionisation energies indicate that there is multiple occupation, by electrons, of **energy levels (shells)**. We find that the '*n*th' energy level can hold a maximum of $2n^2$ electrons.

Activity 2.4

Elements A, B, C, D, E and F have the successive ionisation energies shown below:

	successive ionisation energies/kJ mol⁻¹								
Element	**1st**	**2nd**	**3rd**	**4th**	**5th**	**6th**	**7th**	**8th**	**9th**
A	500	4600	6900	9500	13 400	16 600	20 100		
B	1680	3400	6000	8400	11 000	15 200	17 900	92 000	
C	1520	2700	3900	5800	7200	8800	12 000	13 800	40 800
D	1090	2400	4600	6200	37 800	47 300			
E	1000	2300	3400	4600	7000	8500	27 100	31 700	
F	580	1800	2700	11 600	14 800	18 400	23 300	27 500	

In which periodic group will these elements be found?

2.9 Electron energy sub-levels

Figure 2.7 shows the first ionisation energies of the first 54 elements in the periodic table, plotted against atomic number. You can see that the ionisation energies show a regular pattern, or **periodicity**, with the elements being arranged in groups of 2, 6 or 10. This pattern, together with data from atomic emission spectra, provides evidence for the existence of electron energy **sub-levels** or **sub-shells**. In the three lowest energy levels, the electrons may occupy s, p and d sub-levels (sub-shells), as follows:

the n = 1 level can hold a maximum of 2 electrons,	both in a 1s sub-level
the n = 2 level can hold a maximum of 8 electrons,	2 in a 2s sub-level and 6 in a 2p sub-level
the n = 3 level can hold a maximum of 18 electrons,	2 in a 3s sub-level, 6 in a 3p sub-level and 10 in a 3d sub-level.

Electrons always occupy the lowest energy sub-level which is available, e.g. the 3s before the 3p, 3p before 3d and so on. This is known as the **Aufbau Principle**. Thus, a magnesium atom (atomic number 12) has 2 electrons in the 1s sub-level, 2 electrons in the 2s sub-level, 6 electrons in the 2p sub-level and 2 electrons in the 3s sub-level. More simply, this is written as$1s^2 2s^2 2p^6 3s^2$.

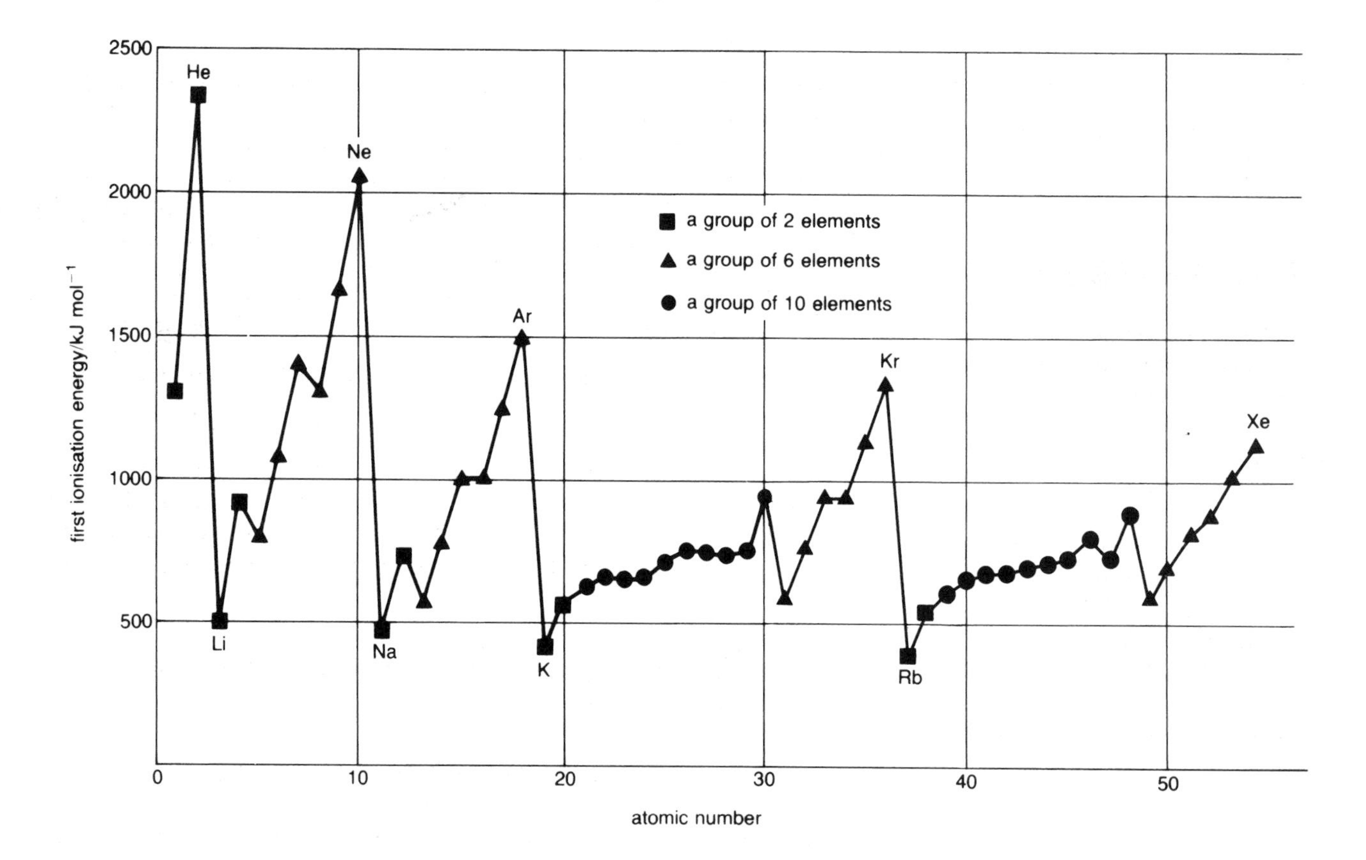

Figure 2.7 ▲
A graph of the first ionisation energies of the elements plotted against atomic number. The graph has a number of interesting features:

- there is a regular pattern, or **periodicity**, with the elements being placed in groups of 2(■), 6(▲) or 10(●).
- the noble gases (He, Ne, Ar, Kr and Xe) have the highest first ionisation energies reflecting the stability of their electronic structures and their very limited chemical reactivity.
- the Group 1 elements, known as the **alkali metals** (Li, Na, K, Rb and Cs), have the lowest first ionisation energies. Each atom has a single electron in its outermost energy level. This electron is relatively easy to remove because it experiences the lowest nuclear charge and consequently has the largest atomic radius in the period.
- elements with atomic numbers 21–30 and 39–48 are found in the **d-block** of the Periodic Table, see page 132. They have fairly constant first ionisation energies indicating that the outermost electrons experience similar nuclear 'pull', and this is explained in section 4.5.

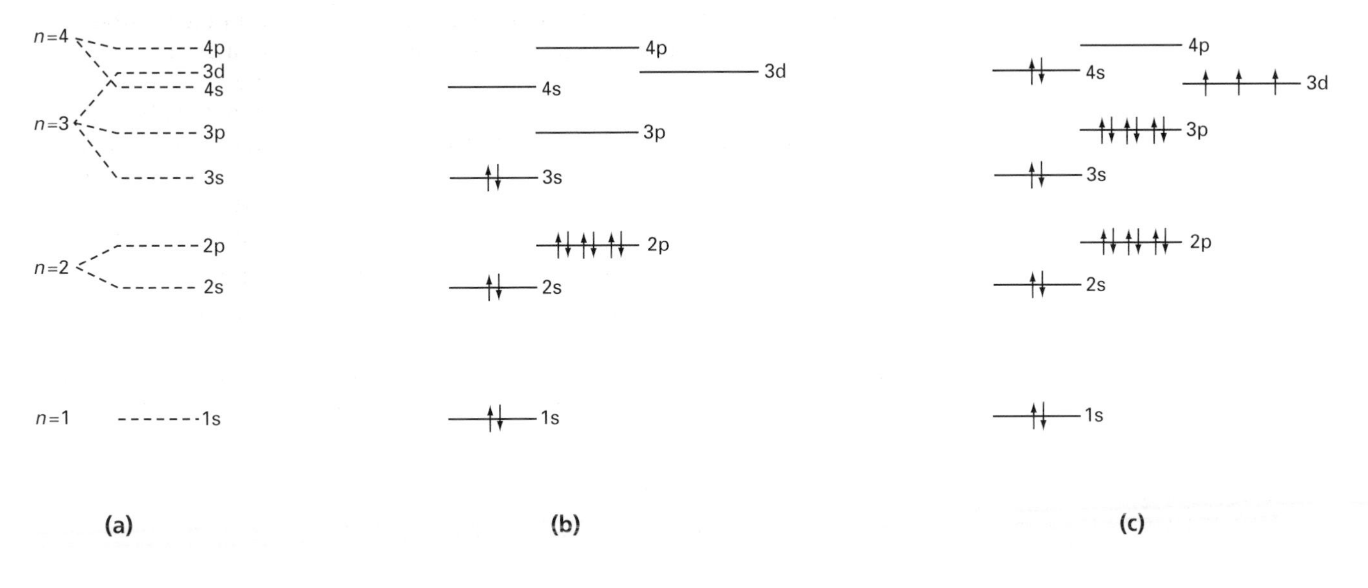

Figure 2.8
(a) Relative energies of the electron sub-levels (sub-shells) in atoms (up to the 4p)
(b) The electronic configuration of magnesium
(c) The electronic configuration of vanadium

Activity 2.5

We often use abbreviations such as [Ne] and [Ar] to represent the electron configurations of the noble gases. Thus, scandium (atomic number 21) could be abbreviated from $1s^2 2s^2 2p^6 3s^2 3p^6 3d^1 4s^2$ to [Ar] $3d^1 4s^2$.

Write similar electron configurations for the following elements (atomic numbers in brackets): H(1), N(7), O(8), Al(13), Ar(18), Ti(22), Cr(24), Fe(26), Cu(29), Ge(32) and Br(35).

When working out electron arrangements, or **electron configurations**, it is important to note that electrons enter the 4s sub-level before the 3d sub-level. Potassium, for example, has the electron configuration $1s^2 2s^2 2p^6 3s^2 3p^6 4s^1$ *not* $1s^2 2s^2 2p^6 3s^2 3p^6 3d^1$. Since it fills first, the 4s sub-level must be at a slightly lower energy than the 3d sub-level, as indicated in Figure 2.8(a). The electron configuration of magnesium, for example, is given in Figure 2.8(b). However, once an electron has entered the 3d sub-level, its energy falls below that of the 4s sub-level. Thus, electron configurations are written with the 3d before the 4s, e.g.:

vanadium, V (atomic number 23): $1s^2 2s^2 2p^6 3s^2 3p^6 3d^3 4s^2$

cobalt, Co (atomic number 27): $1s^2 2s^2 2p^6 3s^2 3p^6 3d^7 4s^2$

The electronic configuration of vanadium is shown in Figure 2.8(c).

On ionisation, the 4s electrons are the first to be lost because they are at higher energy than the 3d electrons. Thus, the electron configurations of the V^{3+} and Co^{2+} ions are $1s^2 2s^2 2p^6 3s^2 3p^6 3d^2$ and $1s^2 2s^2 2p^6 3s^2 3p^6 3d^7$, respectively.

2.10 Orbitals, sub-levels and quantum numbers

If you play a ball game, such as tennis, your success will depend largely on your ability to judge the position and speed of the ball at any time.

These factors can be worked out using the classical laws of physics which govern a particle's motion. In 1913, Bohr used these laws to predict the frequencies of the lines in the emission spectrum of hydrogen (section 1.2). Unfortunately, calculations based on the Bohr model of the atom failed to explain the spectra of atoms with more than one electron.

Compared to a tennis ball, an electron is very small and moves extremely fast. It is impossible to measure the electron's position and speed with certainty because the measurement (no matter how sensitively performed) must interact with the particle. Since the electron has such a low mass, this interaction is bound to affect its position and speed. This is why atomic models based on the classic laws of motion, such as Bohr's, failed to explain the electron arrangement in complex atoms.

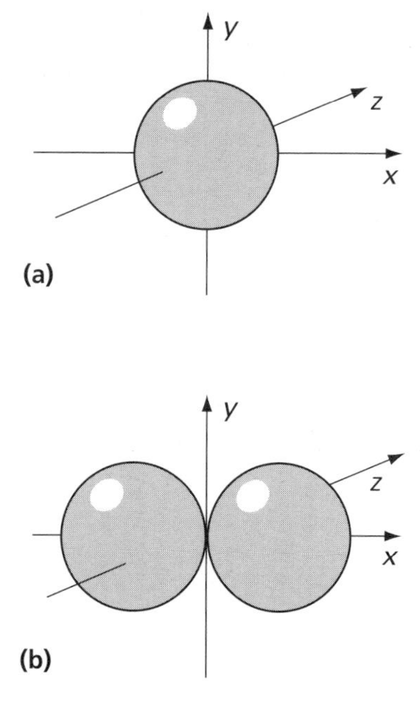

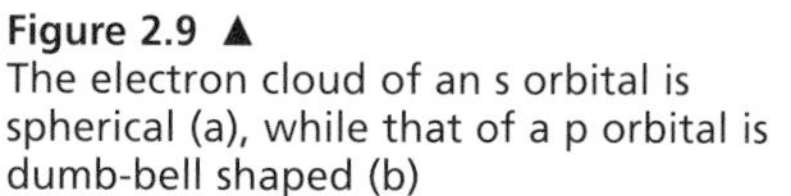

Figure 2.9 ▲
The electron cloud of an s orbital is spherical (a), while that of a p orbital is dumb-bell shaped (b)

In 1928, Erwin Schrodinger solved this problem by adopting a different approach, called **wave mechanics**. He assumed that a given electron may be found anywhere in the space around the nucleus. Then he calculated the **probability** of finding the electron in certain volumes of space. He defined an **atomic orbital** as the volume of space around the nucleus in which the electron spends 90% of its time. Using complex mathematics, Schrodinger worked out the shape of the atomic orbitals, concluding that an s orbital is **spherical** and a p orbital is **'dumb-bell' shaped** (Figure 2.9).

Orbitals can hold either one electron or a 'pair' of electrons, in which case the paired electrons spin in opposite directions. Electron sub-levels (sub-shells) are made up of atomic orbitals, as follows:

- an **s sub-level** can hold up to 2 electrons and has only one spherically shaped orbital
- a **p sub-level** can hold up to 6 electrons in three dumb-bell shaped p orbitals
- a **d sub-level** can hold up to 10 electrons in five d orbitals, the shapes of which are covered later.

Atomic electrons form a cloud of negative charge around the nucleus. Schrodinger used three **quantum numbers** to define the portion of the charge cloud in which a given electron is most likely to be found. A fourth quantum number tells us that when electrons are paired in an atomic orbital, they spin in opposite directions. *Since no two electrons can have the same set of quantum number values* (this is called the **Pauli Exclusion Principle**), each electron will make its own unique contribution to the size, shape and density of this charge cloud. Thus, an electron in either the 1s or 6s sub-level will give rise to a spherical charge cloud but that due to the 6s will be much larger.

2.11 Electrons-in-boxes diagrams

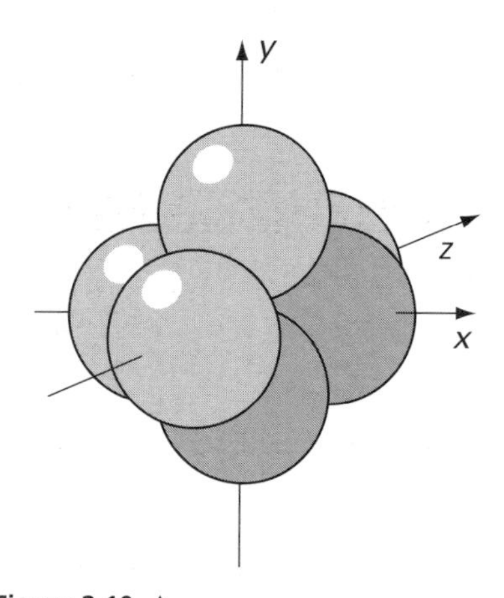

Figure 2.10 ▲
A p subshell is made up of three dumb-bell shaped orbitals at right angles to each other

An atom's electronic configuration can be represented using electrons-in-boxes diagrams. In these diagrams, we use the following symbols:

↑ is one electron,

↑↓ is a pair of electrons, having opposite spins,

□ is an orbital, which can hold one electron or a pair of electrons.

The s sub-level can hold a maximum of two electrons. Therefore, it has only one orbital, i.e. one box. So,

□ means an s sub-level

The p sub-level can hold up to six electrons in three dumb-bell shaped orbitals. So,

□□□ means a p sub-level.

The d sub-level can hold up to ten electrons in five orbitals. So,

□□□□□ means a d sub-level.

We can now draw the sub-levels in order of increasing n value:

Notice that the 3d sub-level is written before the 4s sub-level, even though the latter begins to fill first (section 2.9).

Next, we place the required number of electrons in each sub-level. For example, an atom of sodium (atomic number 11) has the electron configuration $1s^2 2s^2 2p^6 3s^1$. Thus, the electrons-in-boxes diagram for sodium is

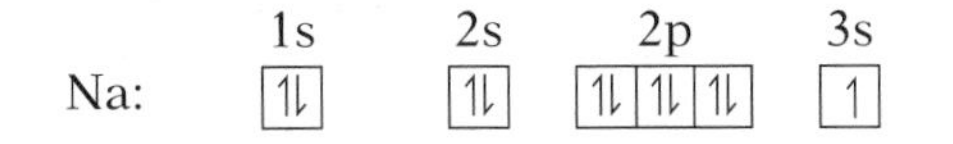

Another example is nitrogen (atomic number 7), electron configuration $1s^2 2s^2 2p^3$:

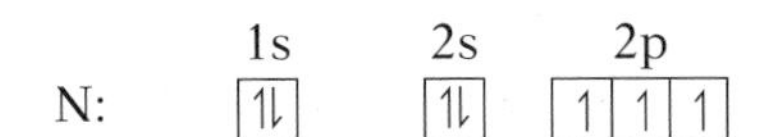

In this example, we have used **Hund's rule**. This says that electrons-in-boxes diagrams must be drawn so that sub-levels contain the maximum number of unpaired electrons. The unpaired electrons tend to have the same spin. Thus, iron (atomic number 26) is written as

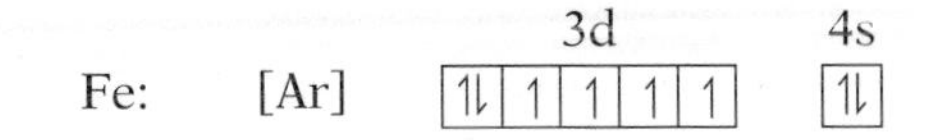

and not

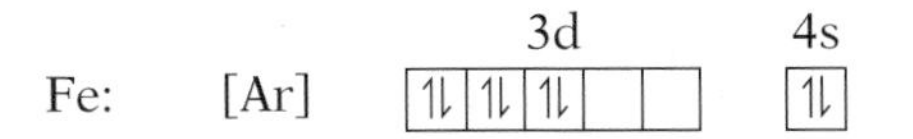

Notice the shorthand use of [noble gas] to represent the configuration of electrons in that atom (i.e. [Ar] = $1s^2 2s^2 2p^6 3s^2 3p^6$).

A final example, then, is sulphur (atomic number 16):

Empty sub-levels need not be drawn but, as we shall see later, they are often useful when explaining chemical bonding.

Activity 2.6

Draw electrons-in-boxes diagrams for the following atoms (atomic numbers in brackets): H(1), C(6), O(8), Ne(10), Al(13), Cl(17), Sc(21), Cr(24), Cu(29), As(33) and Kr(36).

Focus 2d

1 The fine structure in atomic emission spectra and the pattern in the values of $\Delta H_i(N)$ for the elements proves the existence of electron **sub-levels** or **sub-shells**.

2 Since electrons are so small and move very rapidly, we cannot be sure of both their speed and location at any given time. Thus, we talk of the **probability** of finding the electron in a particular volume of space called an **orbital**.

3 s, p and d sub-levels consist of 1, 3 and 5 orbitals respectively. Schrodinger proved that for energy levels n = 1 to 3, sub-levels are available as follows:

	1s	2s	2p	3s	3p	3d	4s	4p	4d
n = 1 can hold a maximum of 2 electrons,	2								
n = 2 can hold a maximum of 8 electrons,		2	6						
n = 3 can hold a maximum of 18 electrons,				2	6	10			

4 Electron sub-levels (up to 4s) increase in energy, thus:

1s 2s 2p 3s 3p 4s 3d

$\longrightarrow$ increasing energy

In stable atoms, electrons first occupy the lowest energy sub-levels (the **Aufbau Principle**).

5 s sub-levels are **spherical**, p sub-levels are made up of three **dumb-bell shaped** orbitals at right angles.

6 **Dot and cross** and **electrons-in-boxes** diagrams provide a simple way of describing the complex 'real-life' nature of atomic electron configurations. As long as we accept their limitations, these diagrams can help us to describe how the electron clouds in atoms move around during a chemical reaction.

Questions on Chapter 2

1 a) What is radioactivity and how can it be detected?

b) α-particles, β-particles and γ-rays are three types of radiation which may be emitted during radioactive decay.

i) Explain the nature of each type of radiation.

ii) Compare them in terms of relative charge and penetration.

iii) What is the effect on the mass (nucleon) number of an atom and the atomic (proton) number when each type of radiation is emitted?

c) Balance the following nuclear equations:

i) ${}^{60}_{27}Co \longrightarrow {}^{59}_{27}Co$

ii) ${}^{220}_{86}Rn \longrightarrow {}^{216}_{84}Po$

iii) ${}^{13}_{7}N \longrightarrow {}^{9}_{5}B$; ${}^{9}_{5}B \longrightarrow {}^{10}_{5}B$

iv) ${}^{235}_{92}U \longrightarrow {}^{236}_{92}U$; ${}^{236}_{92}U \longrightarrow {}^{143}Xe + {}_{38}Sr + 3{}^{1}_{0}n$

2 A plant absorbs phosphorus from the soil via its root system. By labelling plant food with ^{32}P, a beta-particle emitter, scientists can trace the uptake of phosphorus by the plant under various growing conditions. This allows them to grow hybrid strains of plants which absorb phosphorus more quickly and this leads to faster maturing food crops.

a) How can ^{32}P be made from ^{31}P?

b) What are beta-particles?

c) Write a nuclear equation for the decay of ^{32}P.

d) The activity of the ^{32}P decreases with time, as shown in Table 2.4.

i) Plot a graph of activity (vertical axis) against time (horizontal axis)

ii) Define the term 'half-life, $t_{1/2}$' of a radioactive isotope.

iii) Measure the first, second and third half-lives of ^{32}P. What do you notice about the values?

Table 2.4

Time/days	Activity of ^{32}P/disintegrations s^{-1}
0	200
8	130
17	90
23	70
40	30
56	12
71	6

3 The thyroid glands, which are located at the base of the neck, absorb iodine and produce hormones called thyroximes. These control the rate at which cells use oxygen. The iodine uptake by the thyroid gland can be measured by injecting a patient with iodine-132, a radioisotope with a half-life of 13.2 hours. If a dose of 0.0012 g is given, how many grams of iodine-132 will remain in the bloodstream after 39.6 hours?

4 Calculate how long it takes for the various changes in mass of the following radioisotopes:

a) 0.5 g to 0.25 g of cobalt-60, $t_{1/2}$ = 5.27 years; used to destroy malignant tumours.

b) 6.8 g to 1.7 g of caesium-137, $t_{1/2}$ = 30 years; used in food irradiation.

c) 0.6 g to 0.075 g of sulphur-35, $t_{1/2}$ = 87.9 days; used to label, and trace, pesticides.

5 Explain how radioactive isotopes might be used to check that:

a) barrels of oil are being filled to the correct level,

b) there are no flaws or cracks in concrete piping,

c) a wooden axe handle, thought to be about 2000 years old, is a genuine ancient object.

6 a) Explain the meaning of the terms 'nuclear fission' and 'nuclear fusion'. Why do these processes produce energy?

b) Balance the two nuclear equations below and explain which represents 'nuclear fission' and which 'nuclear fusion':

$$^{3}H + {}^{2}H \longrightarrow {}^{4}He + {}^{1}n$$

$$^{235}U + {}^{1}_{0}n \longrightarrow {}^{103}Mo + {}_{50}Sn + 2^{1}_{0}n$$

c) Describe the main features of a nuclear reactor.

d) What factors would you consider when deciding where to locate a nuclear power station?

e) State one environmental advantage of nuclear power stations over coal-burning power stations.

7 a) Define the terms 'first ionisation energy' and 'second ionisation energy'.

b) Write equations which represent the first ionisation energy of sodium and the second ionisation energy of sulphur. Which energy change will be greater? Explain why.

8 a) Sketch graphs for a $\log_{10}$ of successive ionisation energies of

i) sodium

ii) sulphur (refer to Figure 2.5 if you need help)

b) Why do we plot $\log_{10}$ values and not the actual values?

c) Explain the overall slope of the graphs.

d) How do these graphs provide evidence for:

i) the arrangement of electrons in energy levels?

ii) the periodic group in which an element is located?

e) Successive ionisation energies of three elements are shown in the table at the bottom of the page.

i) In which periodic group will they be found?

ii) Which atom needs most energy to form 3^+ ions?

iii) Which atom most easily forms 2^+ ions?

9 a) Sketch a graph of the first ionisation energies of the elements of atomic number 11–30.

b) Explain how this graph provides evidence for the arrangement of electrons in electron sub-levels.

c) Which sub-levels can an electron occupy at energy levels n = 1 to n = 3?

d) The electron configuration of aluminium is [Ne]$3s^2 3p^1$. Give the corresponding configurations for

i) chromium and manganese

ii) copper and zinc.

10 a) Bohr used the classical laws of physics to explain an atom's electronic configuration. Schrodinger used wave mechanics. What is the main difference between the models?

b) What is an 'orbital'?

c) Draw and name the shapes of s and p orbitals.

d) How many p orbitals are there in a p sub-level? How many d orbitals are there in a d sub-level?

11 a) The electron configuration of magnesium is: Mg [Ne] $3s^2$. Using the same convention, write down the electron configurations of boron, neon, sulphur, vanadium, copper and bromine.

b) Electrons-in-boxes diagrams provide a useful way of illustrating the arrangement of electrons in orbitals. The diagram for carbon is:

carbon: [He]

Draw similar diagrams for silicon, phosphorus, iron and selenium.

Table for Question 8e

	Successive ionisation energies kJ mol^{-1}							
	1st	2nd	3rd	4rd	5th	6th	7th	8th
element A	1400	2900	4600	7500	9400	53 300	64 300	
element B	740	1500	7700	10 500	13 600	18 000	21 700	25 700
element C	1310	3400	5300	7500	11 000	13 300	71 300	84 100

Comments on the activities

Activity 2.1

$$^{235}_{92}U \xrightarrow{\alpha} {}^{231}_{90}Th \xrightarrow{\beta} {}^{231}_{91}Pa \xrightarrow{\beta} {}^{231}_{92}Ac \xrightarrow{\alpha} {}^{227}_{90}Th \xrightarrow{\alpha} {}^{223}_{88}Ra$$

$$\downarrow \alpha$$

$$^{207}_{82}Pb \xleftarrow{\beta} {}^{207}_{81}Tl \xleftarrow{\alpha} {}^{211}_{83}Bi \xleftarrow{\beta} {}^{211}_{82}Pb \xleftarrow{\alpha} {}^{215}_{84}Po \xleftarrow{\alpha} {}^{219}_{86}Rn$$

Activity 2.2

1 The graphs are shown in Figure 2.11.

2 Results (in s)

	^{130}Cs	^{138}Xe
a)	1840	1000
b)	1860	1030
c)	1855	1020

d) They are constant within experimental error.

The time taken for half the radioactive substance to decay is known as the half-life. During radioactive decay, successive half-lives are constant.

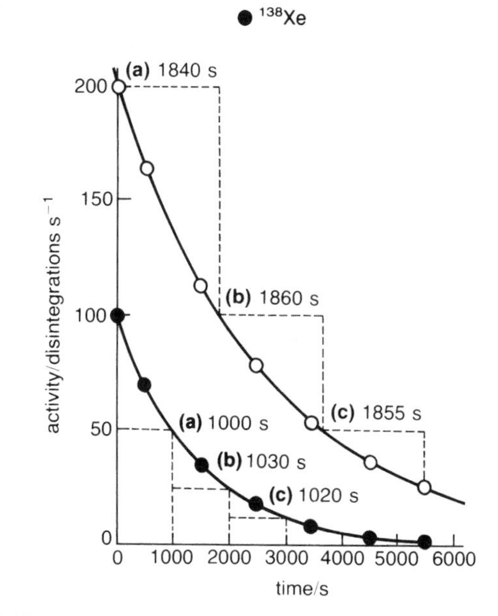

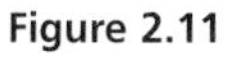

Figure 2.11

Activity 2.3

Be 2.2; C 2.4; Ar 2.8.8; P 2.8.5; S 2.8.6.

Activity 2.4

See the table below.

Element A has the big jump after its 1st ionisation. Therefore, A has one electron in its valence shell and it is in group 1.

Element B has the big jump after its 7th ionisation. Therefore, B has seven electrons in its valence shell and it is in group 7.

Element C has the big jump after its 8th ionisation. Therefore, C has eight electrons in its valence shell and it is in group 0, the noble gases.

Similarly, element D is in group 4, E in group 6 and F in group 3.

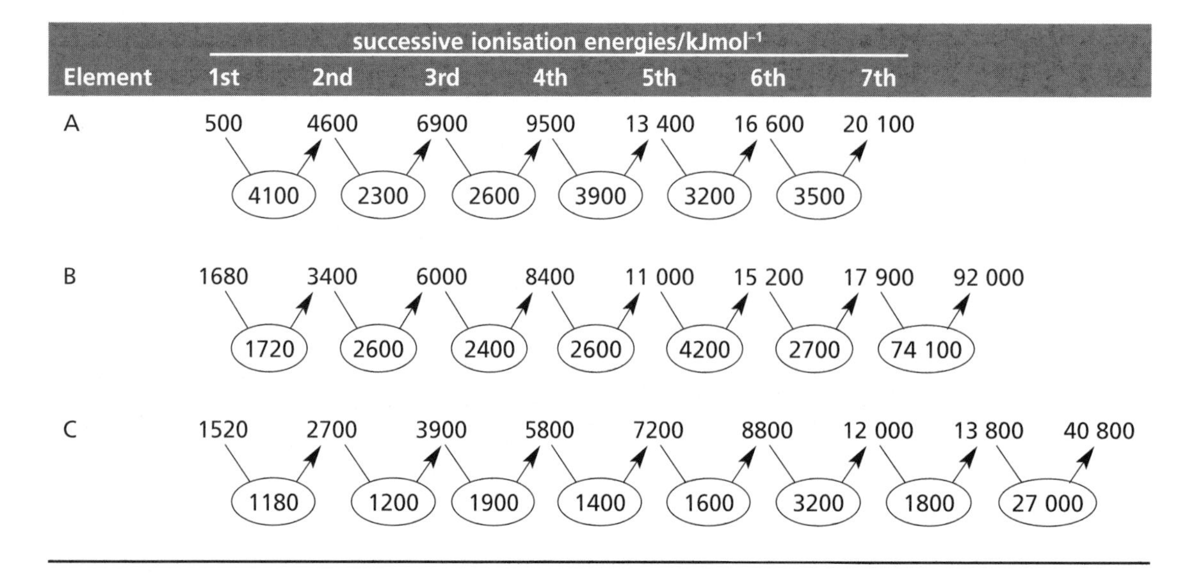

Element	successive ionisation energies/$kJmol^{-1}$								
	1st	2nd	3rd	4th	5th	6th	7th		
A	500	4600	6900	9500	13 400	16 600	20 100		
(differences)	4100	2300	2600	3900	3200	3500			
B	1680	3400	6000	8400	11 000	15 200	17 900	92 000	
(differences)	1720	2600	2400	2600	4200	2700	74 100		
C	1520	2700	3900	5800	7200	8800	12 000	13 800	40 800
(differences)	1180	1200	1900	1400	1600	3200	1800	27 000	

Comments on the activities *continued*

Activity 2.5

H $1s^1$
O $1s^2 2s^2 2p^4$
Ar [Ne] $3s^2 3p^6$
Cr [Ar] $3d^5 4s^1$ (not $3d^4 4s^2$!)
Cu [Ar] $3d^{10} 4s^1$ (not $3d^9 4s^2$!)
Br [Ar]$3d^{10} 4s^2 4p^5$
N $1s^2 2s^2 2p^3$
Al [Ne] $3s^2 3p^1$
Ti [Ar] $3d^2 4s^2$
Fe [Ar] $3d^6 4s^2$
Ge [Ar] $3d^{10} 4s^2 4p^2$

You will probably have incorrect answers (in brackets), for Cr and Cu. In fact, the electron configuration is *stabilised* when the 3d sub-level is *half-full* or *full* (i.e. $3d^5$ or $3d^{10}$). Thus, Cr and Cu adopt the configurations $3d^5 4s^1$ and $3d^{10} 4s^2$, these being more stable than $3d^4 4s^2$ and $3d^9 4s^2$, respectively. Try to remember this anomaly.

Activity 2.6

The electrons-in-boxes diagrams are shown below:

	1s	2s	2p	3s	3p	3d	4s	4p
H:	[↑]							
C:	[↑↓]	[↑↓]	[↑][↑][ ]					
O:	[↑↓]	[↑↓]	[↑↓][↑][↑]					
Ne:	[↑↓]	[↑↓]	[↑↓][↑↓][↑↓]					
Al:		[Ne]		[↑↓]	[↑][ ][ ]			
Cl:		[Ne]		[↑↓]	[↑↓][↑↓][↑]			
Sc:			[Ar]			[↑][ ][ ][ ][ ]	[↑↓]	
Cr:			[Ar]			[↑][↑][↑][↑][↑]	[↑]	
Cu:			[Ar]			[↑↓][↑↓][↑↓][↑↓][↑↓]	[↑]	
As:			[Ar]			[↑↓][↑↓][↑↓][↑↓][↑↓]	[↑↓]	[↑][↑][↑]
Kr:			[Ar]			[↑↓][↑↓][↑↓][↑↓][↑↓]	[↑↓]	[↑↓][↑↓][↑↓]

[Ne] = $1s^2 2s^2 2p^6$; [Ar] $1s^2 2s^2 2p^6 3s^2 3p^6$

C H A P T E R

An introduction to chemical energetics

Contents

Study Checklist

After studying this chapter you should be able to:

1. State that most chemical reactions are accompanied by energy changes and explain the meaning of the term 'enthalpy'.
2. Recall definitions of the following terms: 'standard enthalpy change on reaction', 'standard conditions', 'standard enthalpy of combustion', 'standard enthalpy of formation' and 'standard enthalpy of neutralisation'.
3. Use enthalpy level diagrams to illustrate the difference between exothermic and endothermic reactions.
4. Describe how a polystyrene cup calorimeter may be used to measure a simple enthalpy change in aqueous solution and calculate the enthalpy change from the experimental results.
5. State Hess's law of constant heat summation and use it with appropriate enthalpy cycles to calculate: (a) enthalpy changes which cannot be measured by direct experiment and (b) mean bond enthalpies.
6. Explain what is meant by the term 'activation energy' and draw enthalpy profiles for exothermic and endothermic reactions.

3.1 Energy changes during chemical reactions

Almost all chemical reactions involve energy changes, and a number of these play a vital role in our everyday lives. For example, by absorbing the sun's energy, plants can convert carbon dioxide and water into glucose:

$$6CO_2(g) + 6H_2O(l) \longrightarrow C_6H_{12}O_6(aq) + 6O_2(g)$$

In this process, called **photosynthesis**, light energy is converted into chemical energy, which is stored in the bonds of the glucose molecule. If the plant is now eaten, this chemical energy is converted into

- heat energy, which keeps us warm,
- mechanical energy, which operates our muscles, and
- small impulses of electrical energy, by which messages are passed around our nervous system.

This is how photosynthesis helps us to store up energy from the sun.

The chemical energy in glucose molecules can be converted into heat, mechanical and electrical energy

The environmental effects of burning fossil fuels

The burning of fossil fuels is associated with two environmental problems: the 'enhanced' greenhouse effect and acid rain. The **greenhouse effect** is a natural process by which some of the sun's heat is retained by the earth and its atmosphere. Radiation from the sun reaches the earth by passing through an atmospheric layer of gases, containing carbon dioxide, water vapour, nitrous oxide (N_2O), methane (CH_4) and chlorofluorocarbons (CFCs) which are used as aerosol propellants and refrigerants (section 10.8). On striking the earth's surface some of the radiation is absorbed and some is reflected back towards the gaseous layer. These 'greenhouse' gases act as an absorbing blanket which controls the loss of heat and so keeps the earth at a fairly comfortable temperature. This is the same effect as in a botanical greenhouse, in which the glass transmits the sun's visible light but blocks infrared radiation trying to leave, thereby helping to maintain the temperature.

Without the greenhouse effect, it would be too cold for us to live on earth. It is thought that the proportions of these greenhouse gases in the earth's atmosphere remained fairly constant for thousands of years as there was a balance between the production of greenhouse gases and their destruction by chemical reactions in the atmosphere. With the advent of the Industrial Revolution, about a hundred years ago, this balance has been tipped. Since then, the worldwide use of fossil fuels, such as coal, oil and gas, has steadily increased with a corresponding increase in the emission of greenhouse gases. For example, from 1900 to 1970, the global concentration of carbon dioxide rose by about 7%, from 296 to 318 parts per million (ppm). By 1996, it had reached about 360 ppm, a further 13% rise. Over the last ten years, some developing countries have increased the burden on the ecosystem by cutting down, and burning, trees. This process, known as **deforestation**, not only increases carbon dioxide emissions but it also decreases the number of trees available to photosynthesise the carbon dioxide back into glucose and oxygen. In addition, it erodes the soil-creating desert areas.

The build-up of greenhouse gases has increased the amount of heat absorbed by the earth and its atmosphere. There is concern that this **'enhanced' greenhouse effect**, or **global warming**, if unchecked, might cause an increase of between 1.5 and 4.5°C in the earth's temperature by the end of the next century.

Fossil fuels in use: a stir-fry being prepared using natural gas and coal being delivered to a power station

Such a temperature rise may cause polar icecaps to melt with a consequential rise in sea level. If this happened, our climate would be severely affected and enormous areas of low-lying land would be flooded, with the loss of fertile, food-producing land and the destruction of millions of peoples' homes.

Realising the threat of global warming and other environmental problems, many countries were represented at the Earth Summit in Rio de Janeiro in June 1992. At the summit, it was agreed that governments should aim to reduce CO_2 emissions to 1990 levels by the end of the century, for example by reducing the dependence on fossil fuels and moving to 'renewable' energy sources, such as wind, wave and solar power.

Another problem with fossil fuels is that, on combustion, they release acidic gases such as carbon dioxide, sulphur dioxide and nitrogen dioxide. These gases are carried away by air currents, sometimes travelling hundreds of miles before they dissolve in atmospheric water vapour. The acidified vapour condenses and is precipitated as **acid rain** (or snow). Acid rain causes the erosion of stonework and may kill plants and trees. Some forests, such as Germany's Black Forest, may be irreversibly damaged if acid rain emissions are not diminished. Acid rain also dissolves some poisonous metals, thereby introducing their ions into the water supply and damaging the aquatic life. Norway and Sweden have borne the brunt of the Northern Europe acid rain emissions, with 400 lakes in Norway alone being rendered fishless. To combat the effects of acid rain, European governments are seeking a 30% reduction in sulphur dioxide and nitrogen dioxide emissions by, for example, fitting catalytic converters to cars and incorporating scrubbing devices into industrial chimneys.

Fossil fuels are a 'finite' energy source, that is they will eventually run out and cannot be replaced. It is estimated, at present consumption levels, that oil will run out in 60 years, gas in 70 years and coal in 200 years. In the UK, it is hoped that 10% of our energy needs can soon be obtained from 'renewable' sources, such as wind, wave and solar power. Even so, it is difficult to see how we can meet our long-term energy needs without a significant development of our nuclear energy programme.

Over the years, our lifestyle has become more and more dependent on the interconversion of chemical energy and other forms of energy. For example, imagine what life would be like without the heat, mechanical and electrical energy which is obtained from the chemical energy in fossil fuels.

By measuring the energy change during a reaction, we can get some idea of the relative stabilities of the reactant and product molecules. In industry, chemists often study energy changes because it can help them to optimise reaction conditions and so reduce production costs.

The study of energy changes during chemical reactions is known as **chemical energetics** or **thermodynamics**. In this chapter, we shall:

- explain which energy changes are of most interest to chemists;
- outline some experimental methods for determining these energy changes;
- describe the use of these energy changes in thermochemical calculations.

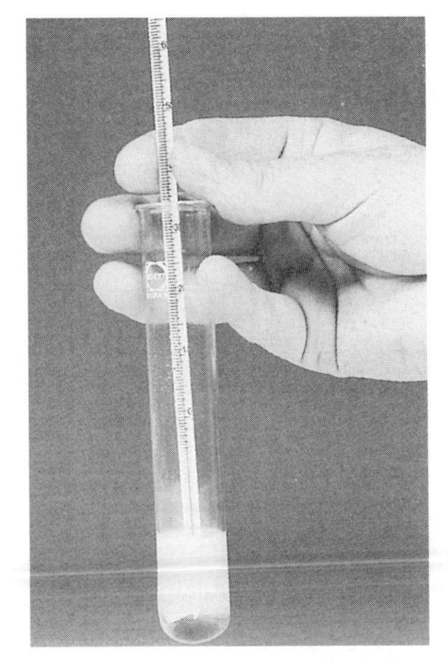

▲ A small piece of magnesium is added to about 2 cm^3 of dilute hydrochloric acid at 20°C. Hydrogen gas is rapidly evolved and the temperature rises to 59°C.

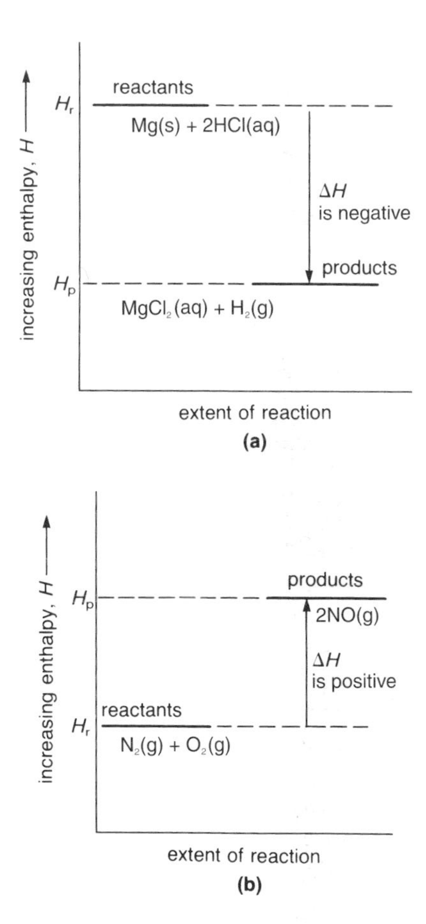

Figure 3.1 ▲
Enthalpy level diagrams for **(a)** an exothermic reaction ($H_p<H_r$) and **(b)** an endothermic reaction ($H_p>H_r$)

3.2 Which energy changes do we measure?

When a small piece of magnesium is placed in dilute hydrochloric acid in a test-tube, a vigorous reaction occurs:

$$\underset{\text{reactants}}{Mg(s) + 2HCl(aq)} \longrightarrow \underset{\text{products}}{MgCl_2(aq) + H_2(g)}$$

Energy is lost by the reactants, causing a rise in the temperature of the reaction mixture. The test-tube itself becomes hot to touch, so some heat must be transferred through it to warm the air around the tube. At the same time, bubbles of hydrogen gas form at the metal surface and then proceed upwards towards the surface of the liquid. To escape into the atmosphere, the gas molecules have to do work against the atmospheric pressure and this uses up some of the energy liberated by the reaction. The overall energy change for this reaction, termed the **enthalpy change** ΔH, is given by the expression:

enthalpy change on reaction	=	heat change when the reactants, Mg(s), HCl(aq), form products, $MgCl_2(aq)$, $H_2(g)$	–	work done by $H_2(g)$ as it leaves the reaction mixture

By definition, the enthalpy change for a reaction is the energy change at constant pressure, so the volume of the system may change (if a gaseous reactant is used up or a gaseous product is formed). Most of the chemical reactions you will do take place in open flasks, or test-tubes, at constant (atmospheric) pressure. Compared to the heat change, the work done by any gases involved in such reactions is very small. Thus, the heat change is almost equal to the enthalpy change and this is often referred to as the **heat change on reaction**.

All substances may be thought of as having an **enthalpy, *H***, and the enthalpy change on reaction ΔH is given by the equation:

$\boldsymbol{\Delta H}$	=	$\boldsymbol{H_p}$ enthalpy of the product species	–	$\boldsymbol{H_r}$ enthalpy of the reactant species

In the reaction of magnesium with hydrochloric acid, the enthalpy of the products is less than that of the reactants, as shown in the **enthalpy level diagram** in Figure 3.1(a). Since H_p is less than H_r, ΔH will be negative and the reaction is said to be **exothermic**, i.e. the reaction releases energy to its surroundings (the water, glass test-tube and air). The **law of conservation of energy** states that energy cannot be created or destroyed but it may be converted from one form to another. So, for an exothermic chemical reaction, the energy *lost* by the reaction equals the energy *gained* by the surroundings.

When petrol is burnt in a car's engine, the high temperature in the cylinders causes a side reaction between the nitrogen and oxygen in air to form nitrogen oxides (notorious atmospheric pollutants):

$$N_2(g) + O_2(g) \longrightarrow 2NO(g) \qquad \text{endothermic, } \Delta H \text{ is positive}$$
$$2NO(g) + O_2(g) \longrightarrow 2NO_2(g) \qquad \text{exothermic, } \Delta H \text{ is negative}$$

The formation of nitrogen monoxide (NO) is an **endothermic** reaction, i.e. energy is absorbed when the reactant molecules form product molecules; H_p is greater than H_r, as shown in Figure 3.1(b). For an endothermic chemical reaction, then, the energy *gained* by the reaction equals the energy *lost* by the surroundings.

Focus 3a

1. Nearly all chemical reactions have an associated energy change.
2. All substances have a heat content or **enthalpy**, H. We are interested in the enthalpy change, ΔH, which accompanies the reaction where

$$\Delta H = H_{products} - H_{reactants}$$

If ΔH is negative, the reaction is **exothermic**. If ΔH is positive, the reaction is **endothermic**.

3.3 Standard enthalpy changes

In the Haber process, ammonia, NH_3, is made by heating nitrogen and hydrogen in the presence of an iron catalyst. The reaction is exothermic, with ΔH equal to –46 kJ per mole of ammonia formed at a temperature of 400 K (127°C). If the temperature is raised to 900 K (627°C), however, the reaction becomes more exothermic with ΔH rising to –59.5 kJ, an increase of more than 30%.

In order to compare enthalpy changes, we need to standardise the amounts of materials used, their physical states and the reaction conditions. The **standard conditions** for thermochemical measurements are as follows:

- The temperature is taken to be 298 K (25°C) and the pressure is taken to be 1 atmosphere (10^5 Pa).
- The substances involved in the reaction must be in their normal physical states at 298 K and 1 atmosphere.
- If a substance exists in more than one structure, we must use the most stable form at 298 K and 1 atmosphere. For example, carbon exists naturally as two allotropes – diamond and graphite but enthalpy changes involving carbon relate to graphite, the more stable allotrope.
- All solutions involved are taken to have a concentration of 1 mole of solute per cubic decimetre of solution (1 mol dm^{-3}).

If an enthalpy change is measured under these standard conditions, it is termed the **standard enthalpy change** for the reaction and given the symbol $\Delta H^\ominus$ or, more precisely, $\Delta H^\ominus_{298}$. (In order to simplify formulae, we shall use the symbol $\Delta H^\ominus$ throughout this chapter.)

Activity 3.1

Which of the following thermochemical equations represent standard enthalpy changes? All ΔH values are in kJ:

1. $C_3H_7OH(l) + \frac{9}{2}O_2(g) \xrightarrow[1\ atm]{298\ K} 3CO_2(g) + 4H_2O(l)$ $\Delta H = -2010$
2. $CH_4(g) + 2O_2(g) \xrightarrow[1\ atm]{298\ K} CO_2(g) + 2H_2O(g)$ $\Delta H = -809$
3. $H_2(g) + CO_2(g) \xrightarrow[10\ atm]{900\ K} H_2O(g) + CO(g)$ $\Delta H = +39$
4. $NH_3(g) + HCl(g) \xrightarrow[1\ atm]{298\ K} NH_4Cl(s)$ $\Delta H = -177$
5. $C_6H_6(l) + O_2(g) \xrightarrow[1\ atm]{298\ K} 6CO_2(g) + 3H_2O(l)$ $\Delta H = -3268$
6. $2C(g) + 6H(g) \xrightarrow[1\ atm]{298\ K} C_2H_6(g)$ $\Delta H = -2844$

Balancing thermochemical equations

Notice in activity 3.1, equation 1, that $\frac{9}{2}O_2(g)$ is needed to balance the equation. This does not mean $\frac{9}{2}$ oxygen molecules but $\frac{9}{2}$ *moles* of molecular oxygen, i.e. $\frac{9}{2} \times 6 \times 10^{23}$ molecules. So, we may use fractions to balance the molar quantities expressed in chemical equations.

Focus 3b

To compare ΔH values, we must measure them under standard conditions. Thus, we define the **standard enthalpy change on reaction, $\Delta H^{\ominus}$**, as the enthalpy change when the mole quantities shown in the balanced chemical equation react under standard conditions of 298 K and 1 atmosphere pressure. Substances must be in their normal physical states under these conditions. Solutions must have concentrations of unit activity (1 mol dm^{-3}).

3.4 Standard enthalpy of combustion

The **standard enthalpy of combustion, $\Delta H_c^{\ominus}$**, *is the enthalpy change when one mole of a pure substance is completely burnt in oxygen under standard conditions (298 K, 1 atmosphere).*

Complete combustion is essential. For example, carbon can react with oxygen to form carbon monoxide, CO, or carbon dioxide, CO_2:

$$\text{C(graphite)} + \tfrac{1}{2}O_2(g) \longrightarrow CO(g) \qquad \Delta H^{\ominus} = -110.5 \text{ kJ}$$

$$\text{C(graphite)} + O_2(g) \longrightarrow CO_2(g) \qquad \Delta H^{\ominus} = -393.5 \text{ kJ}$$

Since complete combustion of carbon gives carbon dioxide, it is the second equation which represents the standard enthalpy of combustion of carbon, i.e.

$$\Delta H_c^{\ominus}[\text{C(graphite)}] = -393.5 \text{ kJ mol}^{-1}.$$

Often, we have to write equations which correspond to the standard thermochemical enthalpy change on combustion, e.g. **$\Delta H_c^{\ominus}$ [C_2H_6(g)] = −1560 kJ mol^{-1}** represents the equation:

$$C_2H_6(g) + \tfrac{7}{2}O_2(g) \longrightarrow 2CO_2(g) + 3H_2O(l) \qquad \Delta H^{\ominus} = -1560 \text{ kJ}$$

and **$\Delta H_c^{\ominus}$ [CH_3OH(l)] = −726 kJ mol^{-1}** represents the equation:

$$CH_3OH(l) + \tfrac{3}{2}O_2(g) \longrightarrow CO_2(g) + 2H_2O(l) \qquad \Delta H^{\ominus} = -726 \text{ kJ}$$

Activity 3.2

Write thermochemical equations from the following data (all in kJ mol^{-1}):

1 $\Delta H_c^{\ominus}$ [CO(g)] = −283
2 $\Delta H_c^{\ominus}$ [C_6H_{14}(l)] = −4195
3 $\Delta H_c^{\ominus}$ [C_2H_5OH(l)] = −1367
4 $\Delta H_c^{\ominus}$ [$C_6H_5CO_2H$(s)] = −3228
5 $\Delta H_c^{\ominus}$ [H_2(g)] = −286
6 $\Delta H_c^{\ominus}$ [Zn(s)] = −348

Standard enthalpies of combustion are accurately measured using a **bomb calorimeter** (Figure 3.2). A known mass of the sample and excess oxygen are ignited electrically. Heat is released and the temperature of the calorimeter and its contents is plotted against time (Figure 3.3). The reactants are then replaced by an electric heating coil and the current adjusted so as to reproduce the results in Figure 3.3. Then we calculate the electrical energy used, this being exactly equal to the energy change during the chemical reaction.

It should be noted that a bomb calorimeter is used to measure energy changes at constant **volume**, that is any gases formed are not able to do work, for example in expanding against atmospheric pressure. Since enthalpy changes relate to measurements at constant **pressure**, not constant volume, any value obtained from a bomb calorimeter experiment must be adjusted to give the enthalpy change under standard conditions.

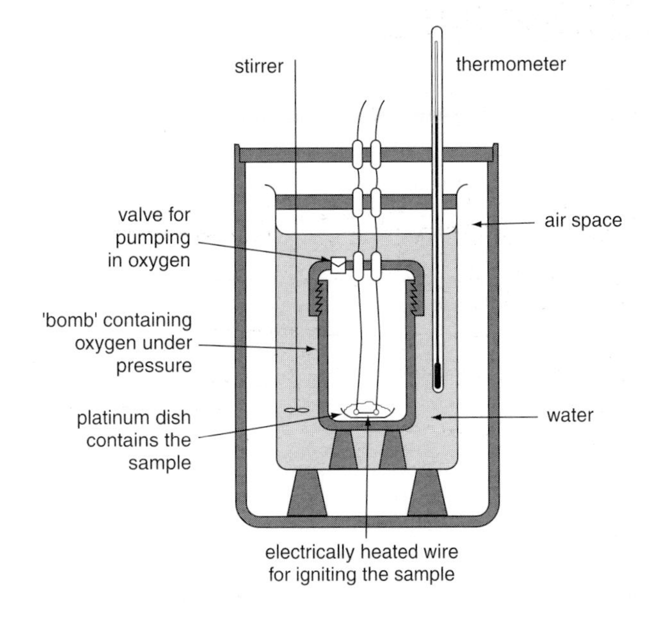

Figure 3.2
A bomb calorimeter being used to measure an enthalpy of combustion

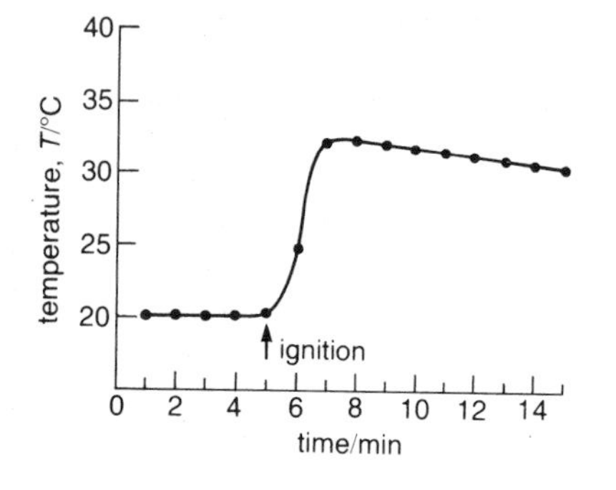

The bomb calorimeter is often used to measure the energy values of foods. Our bodies convert foods into energy using a process called **respiration**. During respiration, the food reacts with oxygen from the air, liberating energy and forming carbon dioxide and water. Since this is exactly the same exothermic reaction as that which occurs when the food is completely burnt in oxygen, the bomb calorimeter can be used to measure its energy value.

◀ **Figure 3.3**
A graph of temperature plotted against time for a typical bomb calorimeter experiment

How much energy do we need?

Most of the chemical energy in our food is converted into heat or mechanical energy. The heat keeps us warm and promotes the many bodily functions which keep us alive, whilst the mechanical energy is needed for physical activities, such as housework and sport. We obtain energy from three types of food: carbohydrates, fats and proteins.

Carbohydrates provide cells with a rapid source of energy (1 g of carbohydrate produces 16 kJ of energy). Three common carbohydrates are glucose, starch and cellulose. **Glucose** is a sweet-tasting soluble carbohydrate that is found in many plants and fruits and its structure is shown in Figure 3.4. Plants store their energy as **starch**, a tasteless, sparingly soluble carbohydrate whose molecular structure is made up of hundreds of glucose units joined together in chains. Cereal crops, root vegetables, rice and pasta have a high starch content. These foods offer us a healthy way of obtaining energy because they also contain **cellulose**, a carbohydrate whose molecule is made from several thousand glucose units. Cellulose is the substance in plant walls which holds the plant together. We can't digest cellulose and it is known as **fibre** in our diet and helps to carry away waste and toxic products, thereby preventing diseases of the intestines and bowels. Nowadays, many people have a high intake of processed sugary foods such as chocolate bars and soft drinks. Although they have a high energy content, these foods can cause tooth decay and often contain a variety of preservatives, flavourings and colourings.

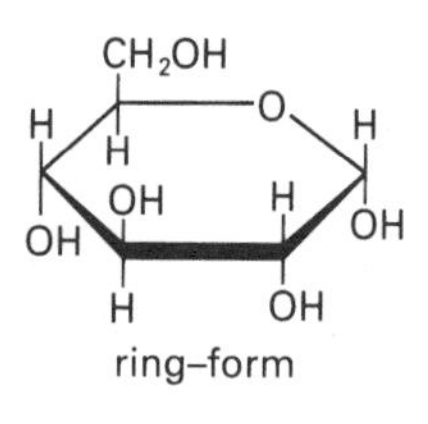

◀ **Figure 3.4**
Glucose, a simple sugar, is a carbohydrate based on a pyranose structure, that is a six-membered ring made of five carbon atoms and one oxygen atom. Carbohydrates are discussed in detail in section 12.9

Fats are found in foods like butter, milk, cheeses and meats. Fats are big sources of energy (1 g of fat yields 37 kJ of energy). A fat is described as being **saturated** or **unsaturated** according to its molecular structure. Saturated fats are less good for us as there is a clear link between their consumption and coronary heart disease. However, we do need some fat in our diet. For example, **linoleic acid** is an essential fatty acid – it must be eaten because it is needed to make prostaglandins, chemicals which play an important role in digestion and reproduction. Fats are also stored around major organs in the body, where they give some protection. They also form a protective layer under our skin and this is where excess fat is stored. On a low calorie diet, we eat less fat and carbohydrate, so excess body fat is converted into energy and we get thinner.

Proteins are the building blocks which are made into body tissues and the chemicals which control the way the body works. A teenager requires about 60 g of protein per day. If the required daily intake is exceeded, the excess protein is converted into energy (1 g protein provides 17 kJ of energy).

As you can see from Table 3.1, the composition of foods varies enormously. The overall energy value of a food can be measured using a bomb calorimeter as shown in Figure 3.2. We can also calculate the energy content from the nutritional values which are usually printed on its wrapping. For example, 100 g of peanut butter contains 25 g of protein (17 kJ g^{-1}), 10 g of carbohydrate (16 kJ g^{-1}), 50 g of fat (37 kJ g^{-1}) and 7 g of fibre. Thus, we find that

$$\text{energy in 100 g of peanut butter} = (25 \times 17) + (10 \times 16) + (50 \times 37) = 2345\ \text{kJ}.$$

Although peanut butter is high in fibre, its energy content is also very high. If the person eating it is unable to use up the energy it supplies, then the peanut butter may be said to be 'fattening'. Our daily energy requirements depend on a number of factors such as age, body weight, gender and how active a life we lead. For example, you need the same amount of energy, about 900 kJ, for half an hour's swimming or three hours of sitting down watching television – but which is the healthier activity?

Table 3.1 Composition of some foods and their typical energy values per average serving

	proteins (g)	fats (g)	carbohydrates (g)	energy value (kJ)
bacon, grilled (100 g)	25	39	0	1868
banana	2	0	34	578
beer, pint	0	0	30	480
boiled potatoes (150 g)	2	0	30	514
cabbage (80 g)	1.4	0	2	56
carrots (80 g)	0.6	0	4	74
cheese and pickle sandwich	19	26	31	1781
chips (150 g)	6	14	55	1500
choc ice	1	8	11	489
chocolate biscuits (3)	3	15	32	1118
cod, fried in batter (150 g)	30	15	11	1241
coffee, white, cup	0	0.5	0.5	26
cornflakes with semi-skimmed milk	6	3	31	709
crisps, packet	2	9	12	559
fruit squash, carton (250 cm^3)	0	0	10	160
orange	2	0	17	306
peanuts (50 g)	12	22	10	1178
peas (85 g)	4	0	8	196
pizza (cheese)	12	36	73	2704
pork chop (150 g)	44	36	0	2080
rice pudding (150 g)	6	11	24	893
sausages (2), pork	15	23	15	1346
tea, with milk, cup	0	0.5	0.5	26
toast and butter, 2 slices	7	10	28	937
yogurt (50 g)	2.5	0.5	9	205

You should note that food energy values are sometimes given in calories, where 1 calorie is equal to 4.2 kJ

Activity 3.3

1 Give two advantages of obtaining carbohydrates from fruit rather than processed sugary products.

2 Why might an excess of protein in your diet cause you to put on weight?

3 **a)** Calculate the energy content of the foods which contain the following substances per 100 g:
i) chocolate wafer biscuits: 7.2 g protein, 26.2 g fat, 59.6 g carbohydrate, trace fibre,
ii) branflakes: 10.1 g protein, 2.4 g fat, 67.7 carbohydrate, 12.7 g fibre,
iii) baked beans: 4.9 g protein, 0.4 g fat, 15.5 g carbohydrate, 5.2 g fibre.
b) Which food do you enjoy most? Which food does you the least 'good'?

4 Table 3.2 gives the daily energy requirement for women and men at various ages.
a) State the differences and similarities between the two sets of data. What explanations can you offer for your observations?
b) What do you think happens to a woman's daily energy requirement:
i) if she becomes pregnant,
ii) if she emigrates from England to Saudi Arabia?

5 Typical daily diets for Angela Grant, a 17-year-old student, and Tom Bell, a 40-year-old teacher, are shown below.

	Angela's diet	**Tom's diet**
Breakfast	Cornflakes with milk, toast and butter, coffee.	Two sausages and bacon; toast and butter, tea.
Break	Three chocolate biscuits, fruit squash.	Peanuts, coffee
Lunch	Cheese and pickle sandwich, choc ice, crisps.	Cheese pizza, banana, tea
Supper	Cod, chips, peas; orange, yogurt, coffee.	Pork chop, boiled potatoes, cabbage, carrots; rice pudding, pint of beer, coffee.

Use the data in Tables 3.1 and 3.2 to answer the following questions.

a) How much energy do Angela and Tom obtain as a result of their respective daily diets?
b) What is Angela's daily energy requirement?
c) What is Tom's daily energy requirement?
d) On the basis of these daily diets, who is the more likely to put on weight, Angela or Tom? What advice would you give her or him?

Table 3.2 Recommended daily energy requirements in kJ (DHSS)

Age	male	female
3 months	2500	2300
6 months	3700	3300
9 months	4520	4000
1 year	5000	4500
4 years	6500	6250
6 years	7250	7000
8 years	8250	8000
11 years	9500	8500
14 years	11000	9000
17 years	12000	9000
25 years	12000	9000
45 years	11500	8000
65 years	10000	8000
85 years	9000	7000

Activity 3.4

Write thermochemical equations from the following data (values in kJ mol⁻¹):

1 $\Delta H_f^\ominus[CH_3Cl(l)] = -81$
2 $\Delta H_f^\ominus[CH_3COCH_3(l)] = -248$
3 $\Delta H_f^\ominus[CH_3OH(l)] = -239$
4 $\Delta H_f^\ominus[Mg_3N_2(s)] = -461$
5 $\Delta H_f^\ominus[HI(g)] = +26$
6 $\Delta H_f^\ominus[H_2SO_4(l)] = -814$
7 $\Delta H_f^\ominus[CH_3NH_2(g)] = -23$
8 $\Delta H_f^\ominus[MgCl_2.6H_2O(s)] = -2500$

A polystyrene cup and a thermometer form an elementary calorimeter (a device for the measurement of heats of chemical reactions). Liquid reagents may be placed in the cup and any heat changes due to the reaction are monitored by reference to the thermometer. The polystyrene of the cup acts as an insulator, minimising heat changes to the ambient temperature of the laboratory. An improvement would be to add an insulating lid.

3.5 Standard enthalpy of formation

The **standard enthalpy of formation**, $\Delta H_f^\ominus$, is *the enthalpy change when one mole of a pure substance is formed from its elements in their normal physical states, under standard conditions (298 K, 1 atmosphere)*.
These enthalpy changes may be represented by thermochemical equations. For example,

$\Delta H_f^\ominus[C_2H_5OH(l)] = -277$ **kJ mol⁻¹** represents the equation:

$$2C(graphite) + 3H_2(g) + \tfrac{1}{2}O_2(g) \longrightarrow C_2H_5OH(l) \qquad \Delta H_f^\ominus = -277 \text{ kJ}$$

and $\Delta H_f^\ominus[NH_4I(s)] = -201$ **kJ mol⁻¹** represents the equation:

$$\tfrac{1}{2}N_2(g) + 2H_2(g) + \tfrac{1}{2}I_2(s) \longrightarrow NH_4I(s) \qquad \Delta H_f^\ominus = -201 \text{ kJ}$$

Another point to note is that, from the definition above, the standard enthalpy of formation of an element in its standard state is zero. For example,

$$H_2(g) \longrightarrow H_2(g) \qquad \Delta H_f^\ominus = 0 \text{ kJ mol}^{-1}$$

$$Na(s) \longrightarrow Na(s) \qquad \Delta H_f^\ominus = 0 \text{ kJ mol}^{-1}$$

Standard enthalpies of formation can sometimes be measured *directly*, for example by using a bomb calorimeter. In many cases, though, the elements will not combine under experimental conditions. For example, $\Delta H_f^\ominus[CH_4(g)]$ cannot be obtained directly because graphite and hydrogen will not form methane in a bomb calorimeter. Later on, however, we shall see that standard enthalpies of formation can be obtained *indirectly* by calculation from other thermochemical data.

3.6 Standard enthalpy of neutralisation

The **standard enthalpy of neutralisation**, $\Delta H_n^\ominus$, *is the enthalpy change when an acid reacts with a base to form one mole of water, under standard conditions (298 K, 1 atmosphere)*. Thus,

$\Delta H_n^\ominus[HCl/NaOH(aq)] = -57.9$ kJ relates to the neutralisation of hydrochloric acid by sodium hydroxide:

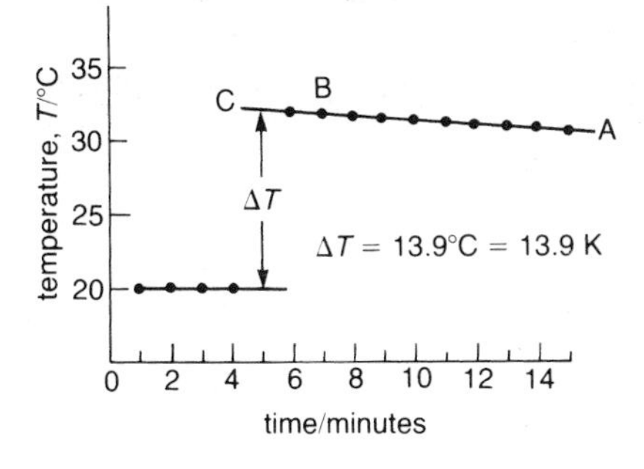

Figure 3.5
A graph of temperature plotted against time for an experiment to measure the enthalpy of neutralisation of sulphuric acid by sodium hydroxide

$$HCl(aq) + NaOH(aq) \longrightarrow NaCl(aq) + H_2O(l) \qquad \Delta H^\ominus = -57.9 \text{ kJ}$$

Standard enthalpies of neutralisation are one of the easiest enthalpy changes to measure. Known volumes of standard acid and base solution, having the same temperature, are mixed together in a simple calorimeter made from a polystyrene cup.

When aqueous sulphuric acid and sodium hydroxide react together, the equation is:

$$H_2SO_4(aq) + 2NaOH(aq) \longrightarrow Na_2SO_4(aq) + 2H_2O(l)$$

From this equation, you can see that the reactants must be mixed together in the mole ratio: **H_2SO_4(aq):NaOH(aq) = 1:2**. Consequently, we put 25 cm^3 of 1 mol dm^{-3} H_2SO_4(aq) (i.e. 0.025 mol) in the polystyrene beaker and note its temperature every minute for four minutes. On the fifth minute, 25 cm^3 of 2 mol dm^{-3} NaOH(aq) (i.e. 0.050 mol) is added and the mixture is stirred. The temperature of the reaction mixture is taken every minute for a further ten minutes.

These results are used to plot a graph of temperature against time, as shown in Figure 3.5. After making a cooling correction (by extending the line AB to C), the temperature rise during the reaction, ΔT, is found to be 13.9 K. Since the polystyrene beaker absorbs a negligible amount of heat from the solution, the enthalpy change on reaction, ΔH, is given by:

$$\begin{aligned} \Delta H &= \text{mass of aqueous solution} \times \text{its specific heat capacity} \times \text{temperature change} \\ &= 50 \text{ g} \times 4.2 \text{ J g}^{-1} \text{ K}^{-1} \times 13.9 \text{ K} \\ &= 2919 \text{ J} \\ &= 2.919 \text{ kJ} \end{aligned}$$

We have made two assumptions here. First, that 50 cm^3 of aqueous solution, like pure water, has a mass of 50 g. Second, that the aqueous solution has the same specific heat capacity as water, namely, 4.2 J g^{-1} K^{-1}.

In our experiment, 0.025 mol of H_2SO_4 reacted with 0.05 mol of NaOH to form 0.05 mol of water. Thus,

when 0.05 moles of water is formed 2.919 kJ of enthalpy is released.

when 1 mole of water is formed $2.919 \times \frac{1}{0.5} = 58.4$ kJ are released.

From this experiment, the enthalpy of neutralisation of sulphuric acid by sodium hydroxide is found to be –58.4 kJ mol^{-1} of water formed (the value is negative because the reaction is exothermic).

Specific heat capacity

Different materials require different amounts of energy to raise their temperatures by similar amounts. If, for example, the same masses of copper and water absorb equal amounts of heat energy, the temperature of each will rise but the increase for copper will be about ten times that for water. The **specific heat capacity** of a material is the heat energy required to raise the temperature of 1 g of material by 1 K. Quite often, the value is expressed in units of J g^{-1} K^{-1}. Water, for example, has a specific heat capacity of 4.2 J g^{-1} K^{-1}, whilst copper has a value of 0.38 J g^{-1} K^{-1}. So, only 0.38 J of energy are needed to heat 1 g of copper by 1 K, whereas water needs 4.2 J.

The heat energy required to raise the temperature of an object is given by the equation:

heat energy = mass of the object × its specific heat capacity × temperature rise

So, the energy required to raise the temperature of 300 cm^3 of water by 15°C (15 K) is obtained by substituting values into the above equation, as shown below:

heat required = 300 × 4.2 × 15 = 18 900 J, i.e. 18.9 kJ

Notice that the mass and volume of water have the same numerical value (300) because the density of water is 1 g per cm^3 (1 g cm^{-3}).

Focus 3c

1. The **standard enthalpy of combustion, $\Delta H_c^\ominus$**, is the enthalpy change when one mole of a substance is completely burnt in oxygen under standard conditions (298 K and 1 atm).
2. Standard enthalpies of combustion of fuels and foods are often measured using a bomb calorimeter.
3. The **standard enthalpy of formation, $\Delta H_f^\ominus$**, is the enthalpy change when one mole of a pure substance is formed from its elements in their normal physical states under standard conditions (298 K and 1 atm).
4. The **standard enthalpy of neutralisation, $\Delta H_n^\ominus$**, is the enthalpy change when an acid reacts with a base to form one mole of water under standard conditions (298 K and 1 atm).
5. Calorimetry is used to measure an enthalpy change, ΔH, where a direct reaction is feasible. This technique involves the measurement of the temperature change, ΔT, experienced by a mass, m, of specific heat capacity, C_p. Then,

$$\Delta H = m \times C_p \times \Delta T$$

Don't forget to state whether the enthalpy change is exothermic (–ve) or endothermic (+ve).

3.7 Hess's law of constant heat summation

The neutralisation of sulphuric acid by sodium hydroxide can be performed in two steps, via sodium hydrogensulphate ($NaHSO_4$):

step 1: $H_2SO_4(aq) + NaOH(aq) \longrightarrow NaHSO_4(aq) + H_2O(l)$

step 2: $NaHSO_4(aq) + NaOH(aq) \longrightarrow Na_2SO_4(aq) + H_2O(l)$

To do this, we use the method in section 3.6, but add the aqueous sodium hydroxide in two separate portions of 12.5 cm^3. A student obtained the following enthalpy changes: $\Delta H_{step\ 1} = -1.54$ kJ and $\Delta H_{step\ 2} = -1.45$ kJ. Both steps are exothermic and combining their enthalpy changes, we get $\Delta H_{total} = -1.54 + (-1.45) = -2.99$ kJ. As we saw in section 3.6, these quantities of reagents will form 0.05 moles of water. For the formation of one mole of water, then, the enthalpy released will be $-2.99 \times 1/0.05 = -59.8$ kJ. Within experimental error, this enthalpy change agrees with that obtained for the 'one-step' route,

$$H_2SO_4(aq) + 2NaOH(aq) \longrightarrow Na_2SO_4(aq) + 2H_2O(l)$$

namely, –58.4 kJ per mole of water formed.

In fact, these values illustrate the **law of constant heat summation**, proposed by Germain Hess in 1840. This law states that *the enthalpy change during a chemical reaction depends only on the nature of the reactants and the products, no matter what reaction route is followed.*

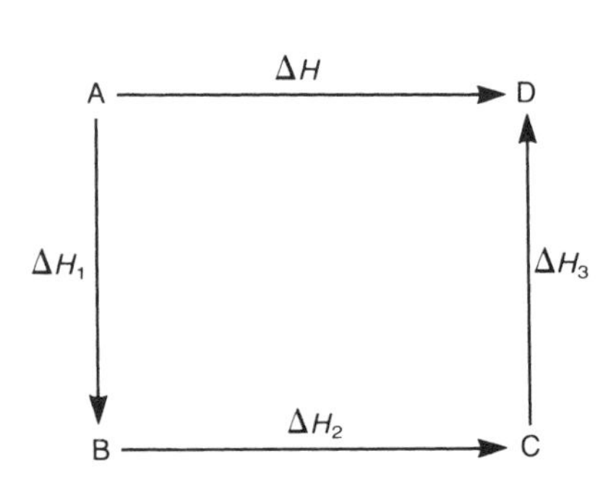

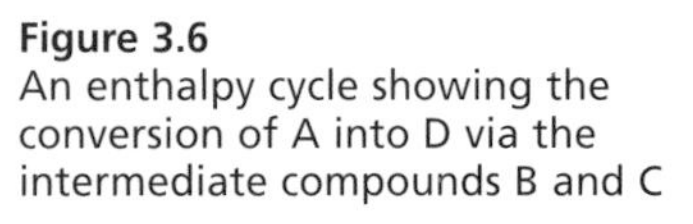
Figure 3.6
An enthalpy cycle showing the conversion of A into D via the intermediate compounds B and C

Hess's law can be represented using an **enthalpy cycle** (Figure 3.6). This enthalpy cycle links the enthalpy changes which occur when the reactants A form products D either directly or indirectly via the compounds B and C. Using Hess's law, we can equate the enthalpy changes:

$$\Delta H = \Delta H_1 + \Delta H_2 + \Delta H_3$$

A useful 'rule of thumb' in equating the terms in an enthalpy cycle is that the enthalpy changes for the clockwise arrows (AD) will equal those for the anti-clockwise arrows (AB, BC, CD).

3.8 Using Hess's law to determine standard enthalpy changes

Whilst the bomb calorimeter provides an effective way of measuring enthalpies of combustion, there are difficulties in measuring the enthalpies for reactions which do not occur under moderate conditions. For example, the standard enthalpy of formation of methane,

$$\text{C(graphite)} + 2\text{H}_2\text{(g)} \longrightarrow \text{CH}_4\text{(g)}$$

cannot be measured directly because carbon will not burn in hydrogen. In other cases, more than one reaction occurs in the bomb calorimeter, e.g.

$$\text{C(graphite)} + \tfrac{1}{2}\text{O}_2\text{(g)} \longrightarrow \text{CO(g)} \quad \text{(I)}$$

$$\text{CO(g)} + \tfrac{1}{2}\text{O}_2\text{(g)} \longrightarrow \text{CO}_2\text{(g)} \quad \text{(II)}$$

It is not possible to determine the enthalpy of formation of carbon monoxide (reaction I) by direct reaction because the carbon monoxide formed reacts with excess oxygen in the bomb calorimeter to give carbon dioxide (reaction II). So, the enthalpy change would correspond to that for the formation of carbon dioxide.

Hess's law allows us to determine standard enthalpies of formation indirectly from other thermochemical data. An enthalpy cycle for the combustion of carbon (graphite) is shown in Figure 3.7 and we can use this to find the standard enthalpy of formation of carbon monoxide, as follows. From Hess's law,

$$\Delta H_1 + \Delta H_2 = \Delta H_3$$

(clockwise arrows) (anticlockwise arrows)

So $\Delta H_1 = \Delta H_3 - \Delta H_2$ (1)

Now $\Delta H_3 = \Delta H_c^{\ominus}[\text{C(graphite)}] = -393 \text{ kJ mol}^{-1}$

and $\Delta H_2 = \Delta H_c^{\ominus}[\text{CO(g)}] = -283 \text{ kJ mol}^{-1}$

Substituting these values in equation 1,

$$\Delta H_1 = -393 - (-283) = -110 \text{ kJ mol}^{-1}$$

The standard enthalpy of formation of carbon monoxide is -110 kJ mol^{-1}. Some further examples of this type of calculation are shown below. Notice how the enthalpy changes given in the data guide us towards the structure of the enthalpy cycle.

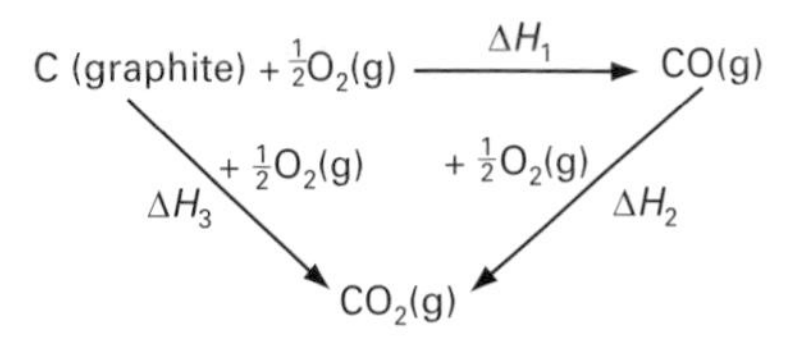

Figure 3.7
Using Hess's law and an enthalpy cycle to determine the standard enthalpy of combustion of carbon monoxide

Example 1

Calculate the standard enthalpy of formation of methanol, using the following data:

$$\Delta H_c^{\ominus}[(\text{C(graphite)})] = -393 \text{ kJ mol}^{-1}$$
$$\Delta H_c^{\ominus}[(\text{H}_2\text{(g)})] = -286 \text{ kJ mol}^{-1}$$
$$\Delta H_c^{\ominus}[(\text{CH}_3\text{OH(l)})] = -726 \text{ kJ mol}^{-1}$$

Solution

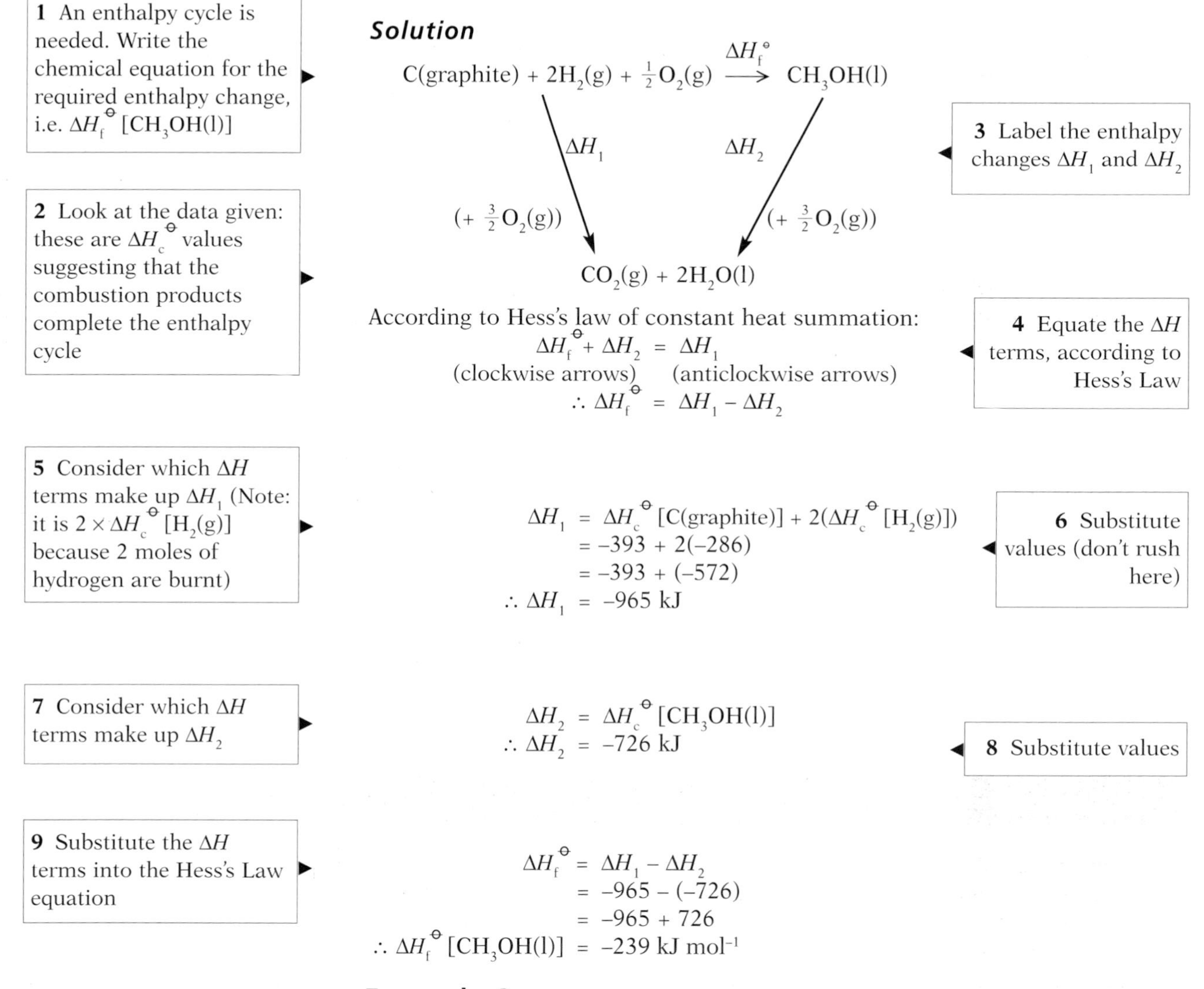

1 An enthalpy cycle is needed. Write the chemical equation for the required enthalpy change, i.e. $\Delta H_f^{\ominus}[\text{CH}_3\text{OH(l)}]$

$$\text{C(graphite)} + 2\text{H}_2\text{(g)} + \tfrac{1}{2}\text{O}_2\text{(g)} \xrightarrow{\Delta H_f^{\ominus}} \text{CH}_3\text{OH(l)}$$

ΔH_1 $(+\tfrac{3}{2}\text{O}_2\text{(g)})$ ↓ ↓ ΔH_2 $(+\tfrac{3}{2}\text{O}_2\text{(g)})$

$$\text{CO}_2\text{(g)} + 2\text{H}_2\text{O(l)}$$

2 Look at the data given: these are $\Delta H_c^{\ominus}$ values suggesting that the combustion products complete the enthalpy cycle

3 Label the enthalpy changes ΔH_1 and ΔH_2

According to Hess's law of constant heat summation:

$$\Delta H_f^{\ominus} + \Delta H_2 = \Delta H_1$$

(clockwise arrows) (anticlockwise arrows)

$$\therefore \Delta H_f^{\ominus} = \Delta H_1 - \Delta H_2$$

4 Equate the ΔH terms, according to Hess's Law

5 Consider which ΔH terms make up ΔH_1 (Note: it is $2 \times \Delta H_c^{\ominus}[\text{H}_2\text{(g)}]$ because 2 moles of hydrogen are burnt)

$$\Delta H_1 = \Delta H_c^{\ominus}[\text{C(graphite)}] + 2(\Delta H_c^{\ominus}[\text{H}_2\text{(g)}])$$
$$= -393 + 2(-286)$$
$$= -393 + (-572)$$
$$\therefore \Delta H_1 = -965 \text{ kJ}$$

6 Substitute values (don't rush here)

7 Consider which ΔH terms make up ΔH_2

$$\Delta H_2 = \Delta H_c^{\ominus}[\text{CH}_3\text{OH(l)}]$$
$$\therefore \Delta H_2 = -726 \text{ kJ}$$

8 Substitute values

9 Substitute the ΔH terms into the Hess's Law equation

$$\Delta H_f^{\ominus} = \Delta H_1 - \Delta H_2$$
$$= -965 - (-726)$$
$$= -965 + 726$$
$$\therefore \Delta H_f^{\ominus}[\text{CH}_3\text{OH(l)}] = -239 \text{ kJ mol}^{-1}$$

Example 2

Ethanol reacts with the halogenating agent phosphorus pentachloride according to the reaction:

$$\text{C}_2\text{H}_5\text{OH(l)} + \text{PCl}_5\text{(s)} \longrightarrow \text{C}_2\text{H}_5\text{Cl(l)} + \text{POCl}_3\text{(l)} + \text{HCl(g)}$$

Calculate the standard enthalpy of reaction using the following data (all in kJ mol^{-1}):

$$\Delta H_f^{\ominus}[\text{C}_2\text{H}_5\text{OH(l)}] = -277$$
$$\Delta H_f^{\ominus}[\text{PCl}_5\text{(s)}] = -443$$
$$\Delta H_f^{\ominus}[\text{C}_2\text{H}_5\text{Cl(l)}] = -136$$
$$\Delta H_f^{\ominus}[\text{POCl}_3\text{(l)}] = -597$$
$$\Delta H_f^{\ominus}[\text{HCl(g)}] = -92$$

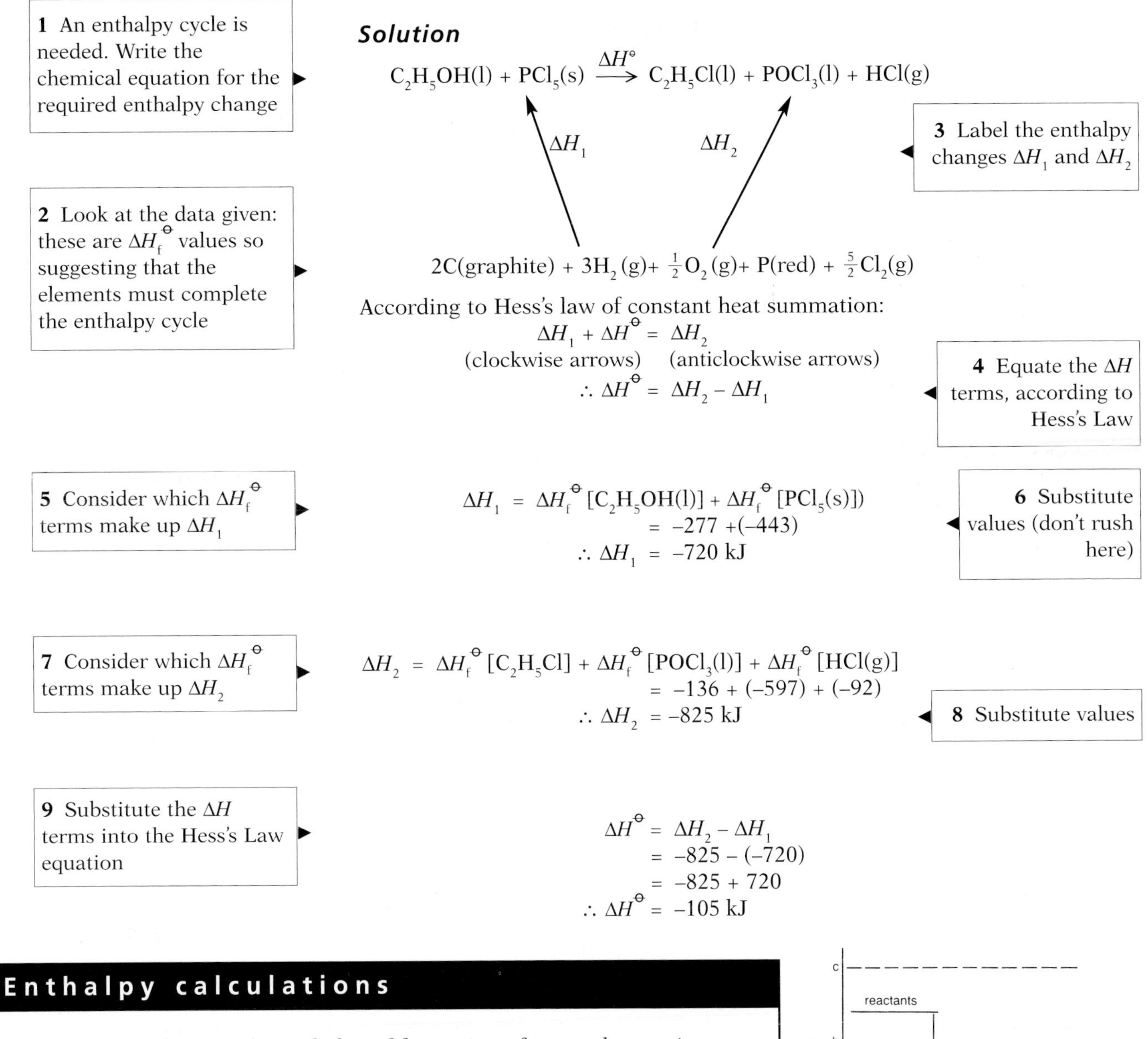

1 An enthalpy cycle is needed. Write the chemical equation for the required enthalpy change

Solution

$$C_2H_5OH(l) + PCl_5(s) \xrightarrow{\Delta H^\circ} C_2H_5Cl(l) + POCl_3(l) + HCl(g)$$

ΔH_1 $\quad$ ΔH_2

3 Label the enthalpy changes ΔH_1 and ΔH_2

2 Look at the data given: these are $\Delta H_f^\ominus$ values so suggesting that the elements must complete the enthalpy cycle

$$2C(graphite) + 3H_2(g) + \tfrac{1}{2}O_2(g) + P(red) + \tfrac{5}{2}Cl_2(g)$$

According to Hess's law of constant heat summation:

$$\Delta H_1 + \Delta H^\ominus = \Delta H_2$$

(clockwise arrows) $\quad$ (anticlockwise arrows)

$$\therefore \Delta H^\ominus = \Delta H_2 - \Delta H_1$$

4 Equate the ΔH terms, according to Hess's Law

5 Consider which $\Delta H_f^\ominus$ terms make up ΔH_1

$$\Delta H_1 = \Delta H_f^\ominus[C_2H_5OH(l)] + \Delta H_f^\ominus[PCl_5(s)])$$
$$= -277 + (-443)$$
$$\therefore \Delta H_1 = -720 \text{ kJ}$$

6 Substitute values (don't rush here)

7 Consider which $\Delta H_f^\ominus$ terms make up ΔH_2

$$\Delta H_2 = \Delta H_f^\ominus[C_2H_5Cl] + \Delta H_f^\ominus[POCl_3(l)] + \Delta H_f^\ominus[HCl(g)]$$
$$= -136 + (-597) + (-92)$$
$$\therefore \Delta H_2 = -825 \text{ kJ}$$

8 Substitute values

9 Substitute the ΔH terms into the Hess's Law equation

$$\Delta H^\ominus = \Delta H_2 - \Delta H_1$$
$$= -825 - (-720)$$
$$= -825 + 720$$
$$\therefore \Delta H^\ominus = -105 \text{ kJ}$$

Enthalpy calculations

In enthalpy calculations, the enthalpy of formation of every element is defined as zero under standard conditions (298 K and 1 atmosphere). It is reasonable to make this arbitrary definition because we are measuring enthalpy changes rather than actual enthalpies (see Figure 3.8).

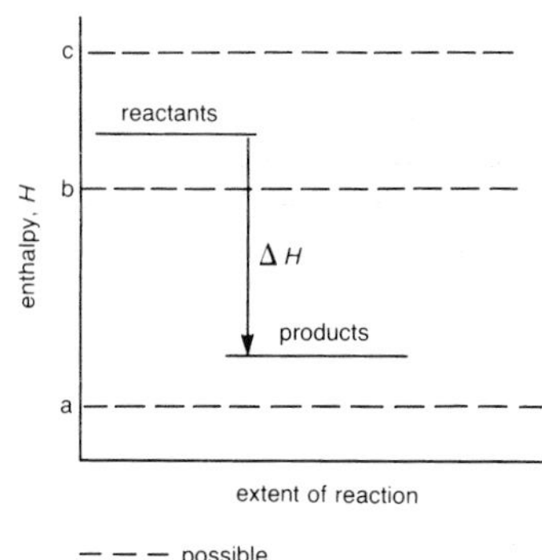

Figure 3.8
The definition of a zero enthalpy state, for example at a, b or c, does not affect the size of the enthalpy change, ΔH

Focus 3d

1. **Hess's law of constant heat summation** states that the enthalpy change during a chemical reaction depends only on the nature of the reactants and the products, no matter which reaction route is followed.
2. Enthalpy cycles based on Hess's Law can be used to estimate enthalpy changes for reactions which do not occur under standard conditions.
3. Enthalpy changes do not depend on the choice of 'zero' enthalpy level. Thus, we make the assumption that the enthalpy of formation of an element in its standard state (i.e. 298 K and 1 atm pressure) is zero.

Activity 3.5

1 Use the method from Example 1 above, and the data below to determine the following enthalpy changes:

a) the enthalpy of formation of ethane, C_2H_6,
b) the enthalpy of combustion of propan-2-ol, $CH_3CH(OH)CH_3(l)$,
c) the enthalpy of formation of benzene, $C_6H_6(l)$,
d) the enthalpy of combustion of carbon disulphide, $CS_2(l)$.

Data:

$\Delta H_c^{\ominus}$[C(graphite)] = −393 kJ mol^{-1} $\Delta H_c^{\ominus}[H_2(g)]$ = −286 kJ mol^{-1}

$\Delta H_c^{\ominus}[C_2H_6(g)]$ = −1560 kJ mol^{-1} $\Delta H_f^{\ominus}[CH_3CH(OH)CH_3(l)]$ = −318 kJ mol^{-1}

$\Delta H_c^{\ominus}[C_6H_6(l)]$ = −3268 kJ mol^{-1} $\Delta H_f^{\ominus}[CS_2(l)]$ = 88 kJ mol^{-1}

$\Delta H_f^{\ominus}[SO_3(l)]$ = −395 kJ mol^{-1}

2 The platinum-catalysed oxidation of ammonia,

$$4NH_3(g) + 5O_2(g) \longrightarrow 4NO(g) + 6H_2O(l)$$

is used in the manufacture of nitric acid. Use the method in Example 2 to calculate the enthalpy of this reaction from the following data (all in kJ mol^{-1}):

$\Delta H_f^{\ominus}[NH_3(g)]$ = −46
$\Delta H_f^{\ominus}[NO(g)]$ = +90
$\Delta H_f^{\ominus}[H_2O(l)]$ = −286
$\Delta H_f^{\ominus}[O_2(g)]$ = 0 by definition

3 Use the method from Example 2 above, and the data supplied, to calculate the enthalpy change for the following reactions.

a) Propene (CH_3CHCH_2) is obtained from the catalytic dehydration of propan-2-ol:

$$CH_3CH(OH)CH_3(l) \longrightarrow CH_3CHCH_2(g) + H_2O(l)$$

b) Sulphur dioxide, used to make sulphuric acid, is made by burning zinc sulphide in air:

$$ZnS(s) + \tfrac{3}{2}O_2(g) \longrightarrow ZnO(s) + SO_2(g)$$

c) Carbon dioxide and ammonia react together to form urea which is used to make resins:

$$CO_2(g) + 2NH_3(g) \longrightarrow CO(NH_2)_2(s) + H_2O(l)$$

Data: Standard enthalpies of formation (all in kJ mol^{-1}):

$CH_3CH(OH)CH_3(l)$ −318; $CH_3CHCH_2(g)$ + 20; $H_2O(l)$ −286,
$ZnS(s)$ −190; $O_2(g)$ 0; $ZnO(s)$ −348;
$SO_2(g)$ −297; $CO_2(g)$ −393 $NH_3(g)$ −46;
$CO(NH_2)_2(s)$ −333.

3.9 Chemical stability

The enthalpy changes during chemical reactions can be described by a **reaction pathway diagram** or **energy profile** as shown in Figure 3.9. Consider two reactant molecules which are moving towards each other. As they approach, there is an increase in repulsion between their nuclei and between their outer

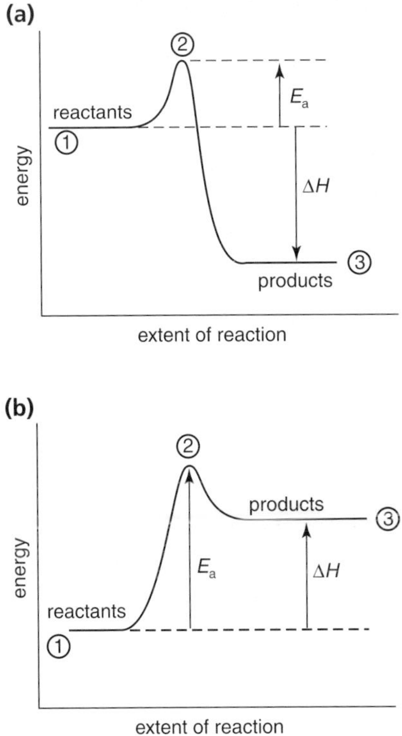

Figure 3.9
Enthalpy profiles for **(a)** an exothermic reaction (ΔH is –ve) and **(b)** an endothermic reaction (ΔH is +ve). The activation energy, E_a, is always an endothermic quantity. The activation energy acts as an energy 'barrier' which the reactant molecules must overcome if they are to form product molecules

electron clouds. If the molecules possess enough kinetic energy – i.e. they are moving fast enough – they will continue to approach each other. This causes the energy of the system to rise ①⟶② and certain bonds in the reactant molecules start to weaken. At the peak of the energy profile, ②, known as the **activated complex**, the bonds in the reactant molecules are just breaking, as the bonds in the product molecules are starting to form. The energy required to produce the activated complex from the reactant molecules corresponds to the activation energy for the reaction. Next, the energy of the system starts to fall (②⟶③) as the new bonds are formed in the products and the product molecules move away from each other.

Quite often, a chemical substance is rather loosely described as being stable or unstable. But what exactly do we mean by chemical stability? Look at the photograph of the fruit stone stuck in the jelly. The stone is said to be **energetically unstable** because it would tend to fall under gravity to the bottom of the container, a position of lower potential energy. However, the 'set' jelly acts as barrier to the stone's movement, so the stone is described as being **kinetically stable**.

This idea of energetic and kinetic stability can be applied to chemical reactions. Consider the combustion of petrol in a car's engine. When the mixture of air and petrol is ignited, there is a considerable release of energy. One of the reactions involved is the combustion of octane:

$$C_8H_{18}(l) + \frac{25}{2}O_2(g) \longrightarrow 8CO_2(g) + 9H_2O(g) \qquad \Delta H = -3498\text{ kJ}$$

Clearly, with respect to its combustion products, octane is energetically unstable. However, as long as safety procedures are followed, motorists can fill up their petrol tanks without fear of explosion. Whether a reaction is endothermic or exothermic, it will not occur unless a certain minimum amount of energy is present, and this is known as the **activation energy**, $\boldsymbol{E_a}$, for the reaction. The activation energy for the combustion of octane is quite high and octane is said to be kinetically stable, with respect to its combustion products – carbon dioxide and water at room temperature.

The size of the activation energy is one of the factors which determine whether a reaction will occur under given reaction conditions. For example, on the one hand, the decomposition of ammonia has a very high activation energy:

$$2NH_3(g) \longrightarrow N_2(g) + 3H_2(g) \qquad E_a = +335\text{ kJ}$$

so this reaction does not occur at room temperature. On the other hand, hydrogen peroxide rapidly decomposes at room temperature in the presence of an enzyme called catalase as this reaction has a very low activation energy:

$$H_2O_2(g) \longrightarrow 2H_2O(l) + O_2(g) \qquad E_a = +23\text{ kJ}$$

◄ A fruit stone stuck in a jar of jelly is energetically unstable but kinetically stable

Petrol being delivered to a service station. The petrol is energetically unstable with respect to its combustion products, carbon dioxide and water. Since the activation energy for the combustion reaction is very high, however, the petrol is kinetically stable and does not ignite spontaneously. Notice the hazard warning sign on the side of the tanker which indicates the load being carried, the dangers in handling it and the form of extinguisher that should be used in the event of a fire ►

The feasibility of a chemical reaction also depends on the **thermodynamic stability** of the reactants, which reflects:

1 The **enthalpy change** on reaction – the reaction is more likely to occur if it is exothermic, i.e. ΔH is negative.
2 The **entropy change** on reaction where entropy is a measure of the disorder in a substance. The greater the entropy, the greater the disorder. A reaction is more likely to occur if the products have a more disordered structure than the reactants. This is the case when calcium carbonate is heated:

$$CaCO_3(s) \xrightarrow{heat} CaO(s) + CO_2(g)$$

solids have ordered arrangements of particles, in this case ions

a gas has a disordered structure in which molecules move rapidly in random directions

We shall return to the concept of entropy in chapter 28.

3.10 Bond enthalpies

In the last section, we saw that reactant molecules with enough kinetic energy will get close enough together to react. As they do so, bonds weaken and eventually break. Bond breaking requires energy – it is an endothermic process. When new bonds form in the product molecules, energy is released – bond making is an exothermic process.

We define the **bond disssociation enthalpy, BDE**, as *the enthalpy change on converting one mole of a specific type of gaseous covalent bonds into the constituent atoms,* e.g.:

$$HCl(g) \longrightarrow H(g) + Cl(g) \qquad BDE(H—Cl) = +431\ kJ$$
$$H_2(g) \longrightarrow 2H(g) \qquad BDE(H—H) = +436\ kJ$$
$$Cl_2(g) \longrightarrow 2Cl(g) \qquad BDE(Cl—Cl) = +242\ kJ$$

In the case of a polyatomic molecule, the situation is a little more complex, even if the bonds present are identical. Methane, CH_4, the simplest hydrocarbon, has four C—H bonds and each has a different bond dissociation enthalpy:

$$CH_4(g) \longrightarrow CH_3(g) + H(g) \qquad BDE1 = +458\ kJ$$
$$CH_3(g) \longrightarrow CH_2(g) + H(g) \qquad BDE2 = +402\ kJ$$
$$CH_2(g) \longrightarrow CH(g) + H(g) \qquad BDE3 = +421\ kJ$$
$$CH(g) \longrightarrow C(g) + H(g) \qquad BDE4 = +371\ kJ$$

The individual values are different because each time a C—H bond is broken, the electronic environment of the remaining C—H bonds is altered. However, we may define the **bond energy term, *E***, as the average of the separate bond dissociation enthalpies for each of the four C—H bonds in methane. Thus,

$$E(C—H) \text{ in } CH_4 = \frac{BDE1 + BDE2 + BDE3 + BDE4}{4}$$

$$= \frac{1652}{4} = 413\ kJ\ mol^{-1}$$

Bond energy terms inform us about the strengths of covalent bonds – the higher the E value, the stronger the bond. To a reasonable approximation, the strength of a particular type of bond is independent of the nature of the rest of the molecule. Thus, the bond energy terms for the C—H bond are similar in a range of compounds, including:

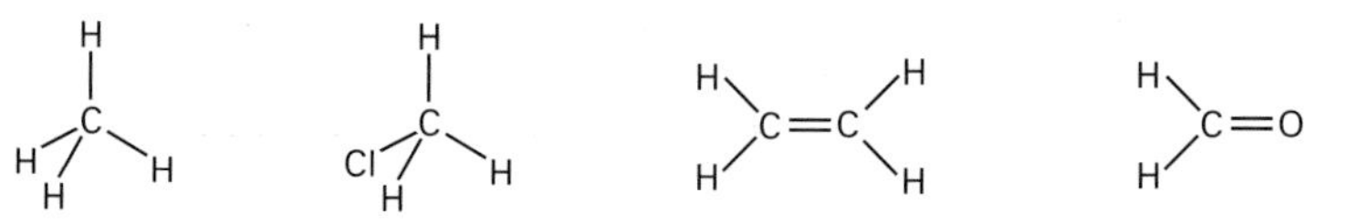

The **mean bond energy (enthalpy)** of a particular bond is taken as the average of the bond energy terms for a wide range of molecules. Since mean bond energies (enthalpies) are largely independent of molecular environment, they may be used to *estimate* enthalpy changes for reactions involving covalent substances. As an example, we shall use the mean bond energies (enthalpies) in Table 3.3 to estimate the enthalpy of hydrogenation of propene:

$$C_3H_6(g) + H_2(g) \longrightarrow C_3H_8(g)$$

The structures of the molecules are shown below:

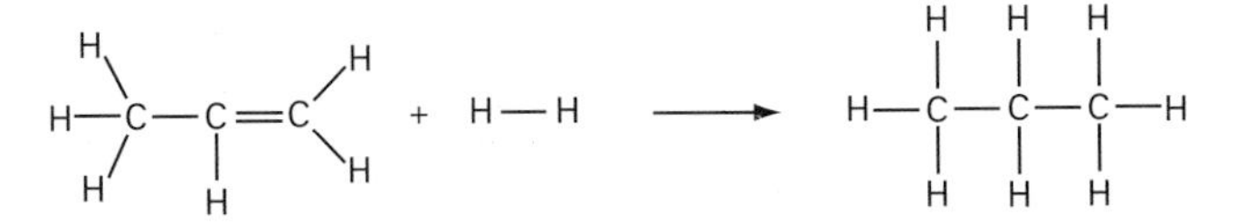

Bonds broken (moles):	1 C=C + 1 H—H	
Enthalpy required (+ve sign):	1(+612) + 1(+436)	= +1048 kJ mol^{-1}
Bonds made (moles):	1 C—C + 2 C—H	
Enthalpy released (–ve sign):	1(–348) + 2(–413)	= –1174 kJ mol^{-1}

Now,

overall enthalpy change, ΔH = enthalpy change on breaking bonds, +ve + enthalpy change on forming bonds, –ve

= +1048 + (–1174)

= –126 kJ mol^{-1}

The estimated enthalpy change is in fairly good agreement with a number of experimentally determined enthalpies of hydrogenation (in kJ mol^{-1}):

ethene	propene	but-1-ene	but-2-ene	cyclohexene
–159	–126	–127	–120	–120

The variation in these values shows the extent to which molecular environment does affect covalent bond strength. Generally, there is a small variation between values because we are adding mean bond energies and these are average values rather than being specific to any particular molecule. This point is considered further in activity 3.6.

Table 3.3 Mean bond enthalpies, *E*, for various covalent bonds

Bond	*E*/kJ mol^{-1}	bond	*E*/kJ mol^{-1}	bond	*E*/kJ mol^{-1}
H—H	436	F—F	158	Br—H	366
C—C	348	Cl—Cl	242	I—H	299
C=C	612	Br—Br	193	C—O	360
C≡C	837	I—I	151	C=O	743
C⋯C (benzene)	518			C—N	305
		C—H	413	C=N	613
Si—Si	176	Si—H	318	C≡N	890
N—N	163	N—H	388	C—F	484
N=N	409	P—H	322	C—Cl	338
N≡N	944	O—H	463	C—Br	276
P—P	172	S—H	338	C—I	238
O—O	146				
O=O	496	F—H	562	Si—O	374
S—S	264	Cl—H	431	S=O	435

Focus 3e

1. The enthalpy changes during a reaction can be represented using a reaction pathway diagram (enthalpy profile).
2. Compared to the reaction products, the reactants will be:
   a) energetically stable, if the reaction is endothermic (ΔH_r is +ve) but unstable if the reaction is exothermic (ΔH_r is –ve)
   b) kinetically stable if the **activation energy**, E_a, is high but kinetically unstable if the activation energy is low.
3. A reaction is more likely to occur if: **a)** it is exothermic, **b)** its activation energy is low and **c)** there is an increase in disorder (entropy).
4. The **bond energy term**, E, is the enthalpy change on converting one mole of a certain type of gaseous covalent bond, in a given molecule, into the constituent atoms.
5. The **mean bond enthalpy** is the average of the bond energy terms for that bond taken over a range of compounds.
6. Mean bond enthalpies are additive so they can be used to estimate the enthalpy change during a reaction.

Activity 3.6

1. Use the mean bond enthalpies in Table 3.3 to estimate the enthalpy change for the following reactions:

   a) $H_2(g) + F_2(g) \longrightarrow 2HF(g)$ (–542 kJ)

   b) $H_2(g) + \frac{1}{2}O_2(g) \longrightarrow H_2O(g)$ (–242 kJ)

   c) Do these values agree with the experimental values shown in brackets?

2. Benzene is a hydrocarbon which is used in large quantities to make certain plastics, dyes and detergents. Although benzene's molecular formula was found to be C_6H_6 in 1834, nearly thirty years passed before Friedrich August Kekulé proposed the cyclic structure shown below. The structure seemed to explain the reactions of benzene and it became widely accepted at that time.

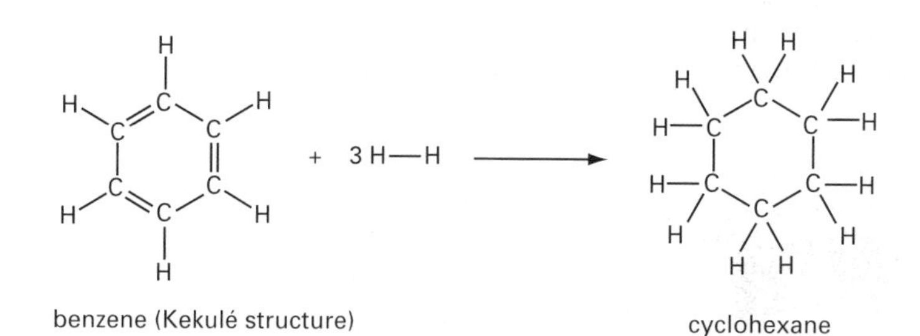

benzene (Kekulé structure) cyclohexane

In this question, you are asked to estimate the enthalpy of hydrogenation of benzene from the above structural equation. You need to use the mean bond enthalpies given in Table 3.3.

a) Find the number of moles of each type of bond which is broken.
b) Calculate the total energy required to break these bonds (endothermic, +ve).
c) Find the number of moles of each type of bond which is formed.
d) Calculate the total energy required to make these bonds (exothermic, –ve).
e) Work out the estimated enthalpy change for the reaction.
f) The experimentally determined enthalpy of hydrogenation is –208 kJ mol^{-1}. Does your value agree with this? Do you think that Kekulé's structure is correct?

Questions on Chapter 3

1 **a)** What is meant by the terms:
i) enthalpy change,
ii) exothermic reaction,
iii) endothermic reaction?

b) Draw enthalpy level diagrams which represent the enthalpy changes in:
i) an exothermic,
ii) an endothermic reaction.

2 **a)** List three factors which can affect the enthalpy change during a chemical reaction.

b) Define the term 'standard enthalpy change for a reaction, $\Delta H^{\ominus}$'.

c) Which of the following equations do not represent standard enthalpy changes? Explain your answer. (Temperature = 298 K, pressure = 1 atmosphere, unless stated)

i) $Na(g) + \frac{1}{2}Cl_2(g) \longrightarrow NaCl(s)$ at 300K

ii) $H_2(g) + \frac{1}{2}O_2(g) \longrightarrow H_2O(l)$

iii) $C(graphite) + O_2(g) \longrightarrow CO_2(s)$ at 2 atm

iv) $C(graphite) + 4H(g) \longrightarrow CH_4(g)$

3 **a)** Define the term 'standard enthalpy of combustion, $\Delta H_c^{\ominus}$'.

b) Write down the thermochemical equations represented by the following $\Delta H^{\ominus}$ terms (units in kJ mol^{-1}):

i) $\Delta H_c^{\ominus}[C_2H_6(g)] = -1560$

ii) $\Delta H_c^{\ominus}[C(graphite)] = -393$

iii) $\Delta H_c^{\ominus}[Ca(s)] = -635$

iv) $\Delta H_c^{\ominus}[C_2H_5OH(g)] = -1367$

v) $\Delta H_c^{\ominus}[C_2H_2(g)] = -1300$

vi) $\Delta H_c^{\ominus}[C_6H_6(g)] = -3268$

4 The photograph shows a simple combustion apparatus being used to estimate the enthalpy of combustion of a liquid fuel. A known mass of the fuel, m_f, is burnt on a small burner. This is surrounded by a glass water jacket containing a known mass of water, m_w. Here we see the water being stirred by the student. The rise in temperature (ΔT) of the water is measured. In a typical experiment, 0.60 g of ethanol ($C_2H_5OH(l)$) was burnt and the temperature of 320 g of water was raised from 14°C to 26°C.

a) Write an equation which gives the heat required to raise the temperature of m_w grams of water through a given temperature change, ΔT. The specific heat capacity of water is 4.2 J g^{-1} K^{-1}.

b) Use this equation to work out how much heat is needed to raise the temperature of 320 g of water from 14°C to 26°C.

c) What mass of ethanol was burnt, thereby causing this temperature rise?

d) How many moles of ethanol were burnt?

e) Use your answers from parts **b)** and **d)** to work out how much heat (in kJ) would be liberated when 1 mole of ethanol is burnt.

f) State three main sources of error in this experiment. What simple modifications would you make to the procedure and apparatus in order to minimise experimental errors?

g) The standard enthalpy of combustion of ethanol is −1367 kJ mol^{-1}. Assuming no heat is lost, what volume of ethanol would need to be burnt to boil enough water to make four cups of coffee, each 200 cm^3, from water at 18°C? (The density of ethanol is 0.79 g cm^{-3}.)

5 Table 3.4 gives the cost of some fuels.

Table 3.4 The enthalpies of combustion, and prices, of some fuels.

Fuel	$\Delta H_c^{\ominus}$/kJ mol^{-1}	cost/£ mol^{-1}
ethanol	−1367	0.9
butane	−2877	0.5
octane	−5470	3.9
decane	−6778	8.0

a) By calculating the cost of getting 10 000 kJ from each fuel, work out which offers the best value for money.

b) Apart from cost, state two desirable characteristics of a good fuel.

c) Hydrogen burns very cleanly and is relatively cheap to produce. Why do you think that it has not yet been commercially developed as a motor fuel?

6 **a)** Define the term 'standard enthalpy of formation, $\Delta H_f^{\ominus}$'.

b) Write down the thermochemical equations represented by the following $\Delta H_f^{\ominus}$ terms (units in kJ mol^{-1}):

i) $\Delta H_f^{\ominus}[C_3H_6(g)] = +20.4$

ii) $\Delta H_f^{\ominus}[CO(g)] = -110.5$

iii) $\Delta H_f^{\ominus}[CaCl_2(s)] = -795$

iv) $\Delta H_f^{\ominus}[CH_3OH(g)] = -239$

v) $\Delta H_f^{\ominus}[C_3H_3(g)] = +185$

Questions on Chapter 3 *continued*

vi) $\Delta H_f^\ominus[SO_3(g)] = -395$
vii) $\Delta H_f^\ominus[CH_3COCH_3(l)] = -218$

7 **a)** State Hess's law of constant heat summation.
b) Explain why the application of Hess's law can be very useful in determining standard enthalpies of formation.
c) An enthalpy cycle for the formation of aqueous aluminium ions from the metal is given below:

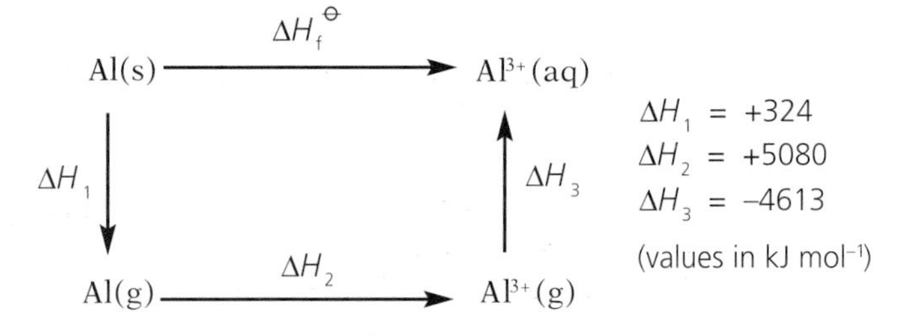

$\Delta H_1 = +324$
$\Delta H_2 = +5080$
$\Delta H_3 = -4613$
(values in kJ mol^{-1})

i) Write an equation which expresses $\Delta H_f^\ominus$ in terms of the other ΔH terms in the cycle.
ii) Calculate the value of $\Delta H_f^\ominus$.

8 **a)** Use Hess's Law and the data below to work out the standard enthalpies of formation of:
i) butan–1-ol, $CH_3CH_2CH_2CH_2OH(l)$,
ii) ethoxyethane, $C_2H_5OC_2H_5(l)$,
iii) ethanoic acid, $CH_3COOH(l)$.

Data (in kJ mol^{-1}):
$\Delta H_c^\ominus[C(graphite)] = -393$
$\Delta H_c^\ominus[H_2(g)] = -286$
$\Delta H_c^\ominus[CH_3CH_2CH_2CH_2OH(l)] = -2675$
$\Delta H_c^\ominus[C_2H_5OC_2H_5(l)] = -2761$
$\Delta H_c^\ominus[CH_3COOH(l)] = -873$

b) Which is the more energetically stable molecule with respect to the constituent elements: butan–1-ol or ethoxyethane?
c) The standard enthalpy of formation of ethanal, $CH_3CHO(g)$ is –192 kJ mol^{-1}. Use the data above to calculate its standard enthalpy of combustion.

9 Two students have been asked to determine the enthalpy of the following displacement reaction:

$$Zn(s) + CuSO_4(aq) \longrightarrow ZnSO_4(aq) + Cu(s)$$

They decide to add a known mass of powdered zinc metal to a known volume of aqueous copper sulphate contained in a polystyrene cup. The concentration of the aqueous copper sulphate solution is chosen so as to ensure that this reactant is in excess. The temperature of the copper sulphate solution is measured at one minute intervals. The zinc is added after three minutes, then the mixture is stirred vigorously and the temperature taken every minute for a further ten minutes. The students' results are shown in the tables. (Take the heat capacity of water as 4.2 J g^{-1} K^{-1}, the mass of zinc added = 2.1 g and the volume of copper sulphate solution (1 mol dm^3) = 200 cm^3.

a) Why does the polystyrene cup act as an effective calorimeter?
b) On the same axes, plot graphs of temperature (vertical axis) against time (horizontal axis) for both sets of results.

Thelma's results

Time/min	0	1	2	3	4	5	6	7
Temperature/°C	19.5	19.7	19.7	–	24.3	26.3	26.5	26.1
Time/min	8	9	10	11	12	13		
Temperature/°C	25.8	25.4	25.0	24.6	26.2	23.7		

Liam's results

Time/min	0	1	2	3	4	5	6	7
Temperature/°C	19.5	19.6	19.6	–	23.0	25.0	25.9	26.2
Time/min	8	9	10	11	12	13		
Temperature/°C	26.0	25.8	25.6	25.2	24.8	24.5		

c) By reference to Figure 3.5, estimate for each experiment the temperature rise at the point of mixing.
d) For both sets of results, calculate:
i) the enthalpy change during this experiment,
ii) how many moles of Zn(s) were added,
iii) the enthalpy change if 1 mole of powdered zinc were added to an excess of aqueous copper sulphate.

e) The standard enthalpy change for this reaction is 217 kJ mol^{-1}. Account for any discrepancy between this and the value you have calculated and suggest modifications to the procedures.

10 **a)** What do you understand by the term 'activation energy'?
b) Draw an enthalpy profile for an endothermic reaction, clearly showing the enthalpy change for the reaction and the activation energy.
c) White phosphorus, P(s) ignites spontaneously when placed in oxygen, $O_2(g)$. It is *energetically unstable* and *kinetically unstable* with respect to its combustion product, phosphorus (V) oxide, P_4O_{10}.

i) Explain the meaning of the terms in italics.

ii) Draw an enthalpy profile for this reaction

iii) Write a balanced equation for the reaction.

11 Use the data below to calculate the enthalpy changes for the following reactions:

a) $P_4O_6(l) + 2O_2(g) \longrightarrow P_4O_{10}(s)$
b) $C_2H_4(g) + HI(g) \longrightarrow C_2H_5(l)$
c) $CaCO_3(s) \longrightarrow CaO(s) + CO_2(g)$

Data: Standard enthalpies of formation, in kJ mol^{-1}:

$P_4O_6(l)$ = –1640; $P_4O_{10}(s)$ = –2984; $C_2H_4(g)$ = +52

$HI(g)$ = 26.5; $C_2H_5(l)$ = –31; $CaCO_3(s)$ = –1207

$CaO(s)$ = –635.5; $CO_2(s)$ = –393.5

12 a) Explain the meaning of the following terms: bond dissociation energy, bond energy term and mean bond energy (enthalpy).

b) Use the data below to calculate the following bond enthalpy terms:

i) a C—H bond in methane, CH_4;

ii) an N—H bond in ammonia, NH_3;

iii) an O—H bond in the H_2O molecule

c) Use the data below, and that in Table 3.3, to estimate the following bond energy terms:

i) the C—Cl bond in chloroethane, C_2H_5Cl;

iii) an N—H bond in methylamine, CH_3NH_2;

iii) a C—O bond in methoxymethane, CH_3OCH_3.

Data (all values in kJ mol^{-1}):

$C(graphite) \longrightarrow C(g)$ $\Delta H^\ominus$ = 715

$\frac{1}{2}H_2(g) \longrightarrow H(g)$ $\Delta H^\ominus$ = 218

$\frac{1}{2}N_2(g) \longrightarrow N(g)$ $\Delta H^\ominus$ = 473

$\frac{1}{2}O_2(g) \longrightarrow O(g)$ $\Delta H^\ominus$ = 248

$H_2O(l) \longrightarrow H_2O(g)$ $\Delta H^\ominus$ = 44

$\frac{1}{2}Cl_2(g) \longrightarrow Cl(g)$ $\Delta H^\ominus$ = 121

Standard enthalpies of formation: CH_4 –75; NH_3 –46; H_2O –286; C_2H_5Cl –92; CH_3NH_2 –23; CH_3OCH_3 –184.

d) Use the data in Table 3.3 to estimate enthalpy changes for the following reactions:

i) $CH_4(g) + Cl_2(g) \longrightarrow CH_3Cl(g) + HCl(g)$

ii) $C_2H_2(g) + \frac{5}{2}O_2(g) \longrightarrow H_2O(g) + 2CO_2(g)$

iii) $C_2H_4(g) + HCl(g) \longrightarrow C_2H_5Cl(g)$

Comments on the activities

Activity 3.1

1 and 4 These equations do represent standard enthalpy changes for the reactions. All the reactants and products are in their standard states and the reaction conditions are also standard (i.e. 298 K and 1 atm pressure).

The other equations do not represent standard enthalpy changes for the following reasons:

2 one of the products, water, is not in its standard state;
3 the temperature and pressure are not standard;
5 the equation is not balanced (it should be $\frac{15}{2}O_2(g)$);
6 the reactants are not in their standard states (carbon is a solid (graphite) and hydrogen exists as $H_2(g)$ molecules under standard conditions).

Activity 3.2

1 $CO(g) + \frac{1}{2}O_2 \longrightarrow CO_2(g)$ ΔH = –283 kJ

2 $C_6H_{14}(l) + \frac{19}{2}O_2(g) \longrightarrow 6CO_2(g) + 7H_2O(l)$ ΔH = –4195 kJ

3 $C_2H_5OH(l) + 3O_2(g) \longrightarrow 2CO_2(g) + 3H_2O(l)$ ΔH = –1367 kJ

4 $C_6H_5CO_2H(s) + \frac{15}{2}O_2(g) \longrightarrow 7CO_2(g) + 3H_2O(l)$ ΔH = –3228 kJ

5 $H_2(g) + \frac{1}{2}O_2(g) \longrightarrow H_2O(l)$ ΔH = –286 kJ

6 $Zn(s) + \frac{1}{2}O_2(g) \longrightarrow ZnO(s)$ ΔH = –348 kJ

Activity 3.3

1 In general, fruit is likely to contain more vitamins, fibre and less fat than processed sugary products. There is less chance of tooth decay if fruit is eaten instead of processed sugary foods.

2 Excess protein is converted into energy and, therefore, the amount of energy required from fat is lowered. Excess fat will form a layer under the skin to we put on weight.

3 **a)**

	Protein	Fat	Carbo-hydrate	Protein energy	Fat energy	Carbohydrate energy	Total energy/ 100 g food
i)	7.2	26.2	59.6	122.4	969.4	953.6	2045.4
ii)	10.1	2.4	67.7	171.7	88.8	1083.2	1343.7
iii)	4.9	0.4	59.6	15.5	14.8	953.6	983.9

Remember that fibre is not digested.

b) Although many people may enjoy eating chocolate wafers more than branflakes or baked beans, the wafers are not so good for us because of their high fat, and low fibre, content. But do you eat your branflakes with full cream milk?!

4 **a)** Similarities: initially a rapidly increasing energy requirement as the child grows and a gradual tail off later in life as the body becomes less active; the period of maximum energy requirement runs from mid-teens to mid-twenties.

Differences: on average, the body mass of a male exceeds that of a female, so the energy requirement 'peak' for females is about 25% lower than that for males.

b) i) Increased energy requirement to meet the needs of the unborn child. ii) Less energy is needed, for

Comments on the activities *continued*

example, to maintain body temperature, in the hotter climate of Saudi Arabia.

5 a) The energy values of the various meals are given below in kJ

Diet	Breakfast	Break	Lunch	Supper	Total
Angela	1672	1278	2829	3474	9253
Tom	4177	1204	3308	4123	12812

b) 9000 kJ

c) Between 11500 and 12000 kJ

d) Just on the basis of a simple comparison, Tom is the more likely to put on weight with a daily excess of between 812 and 1312 kJ. Perhaps Tom might cut down on the fried breakfasts!

Activity 3.4

1 $C(graphite) + \frac{3}{2}H_2(g) + \frac{1}{2}Cl_2(g) \longrightarrow CH_3Cl(l)$ $\quad \Delta H = -81$ kJ

2 $3C(graphite) + 3H_2(g) + \frac{1}{2}O_2(g) \longrightarrow CH_3COCH_3(l)$ $\quad \Delta H = -248$ kJ

3 $C(graphite) + 2H_2(g) + \frac{1}{2}O_2(g) \longrightarrow CH_3OH(l)$ $\quad \Delta H = -239$ kJ

4 $3Mg(s) + N_2(g) \longrightarrow Mg_3N_2(s)$ $\quad \Delta H = -461$ kJ

5 $\frac{1}{2}H_2(g) + \frac{1}{2}I_2(s) \longrightarrow HI(g)$ $\quad \Delta H = +26$ kJ

6 $H_2(g) + S(s) + 2O_2(g) \longrightarrow H_2SO_4(l)$ $\quad \Delta H = -814$ kJ

7 $C(graphite) + \frac{1}{2}N_2(g) + \frac{5}{2}H_2(g) \longrightarrow CH_3NH_2(g)$ $\quad \Delta H = -23$ kJ

8 $Mg(s) + Cl_2(g) + 6H_2(g) + 3O_2(g) \longrightarrow MgCl_2.6H_2O(s)$ $\quad \Delta H = -2500$ kJ

Activity 3.5

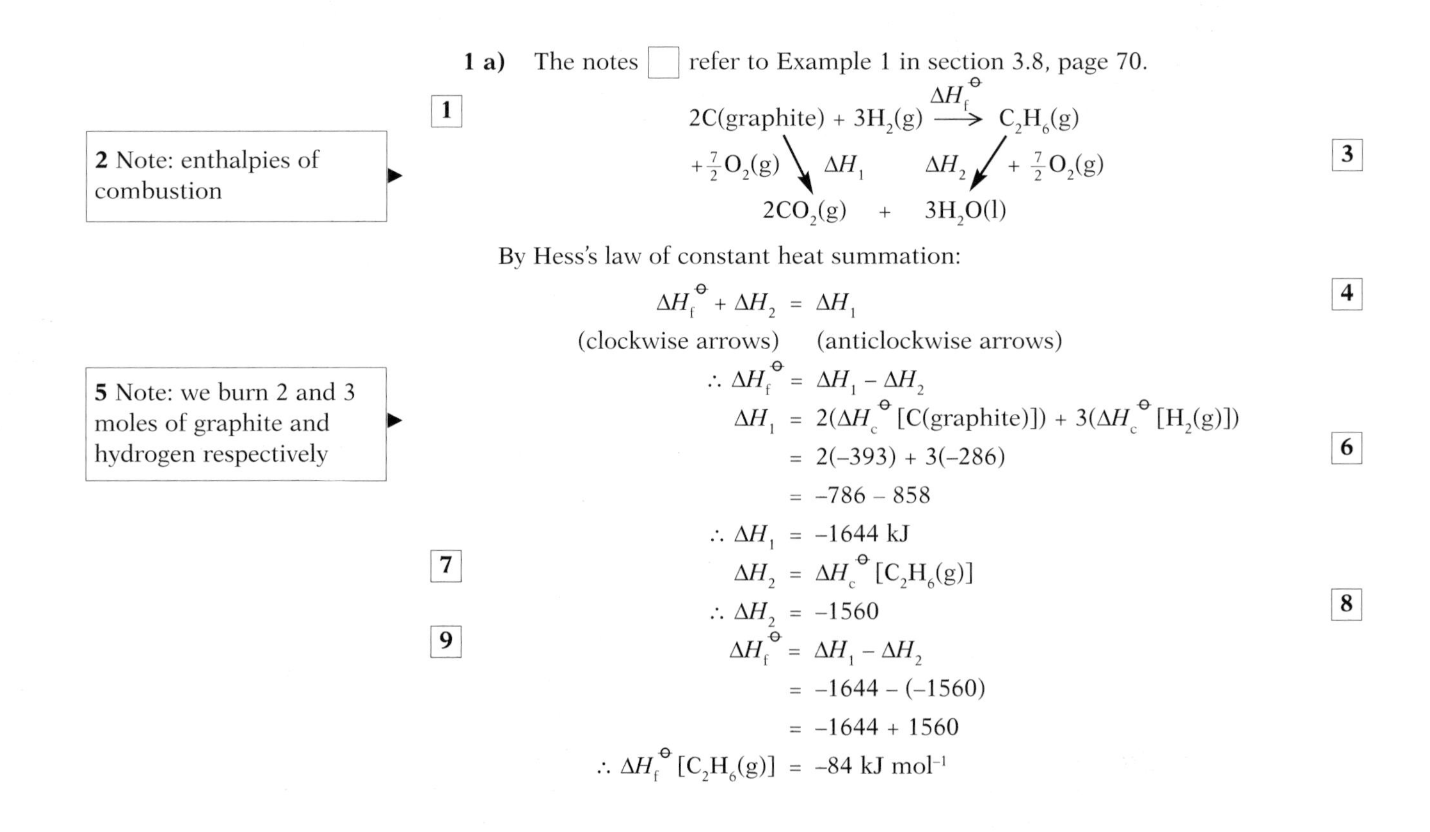

Comments on the activities *continued*

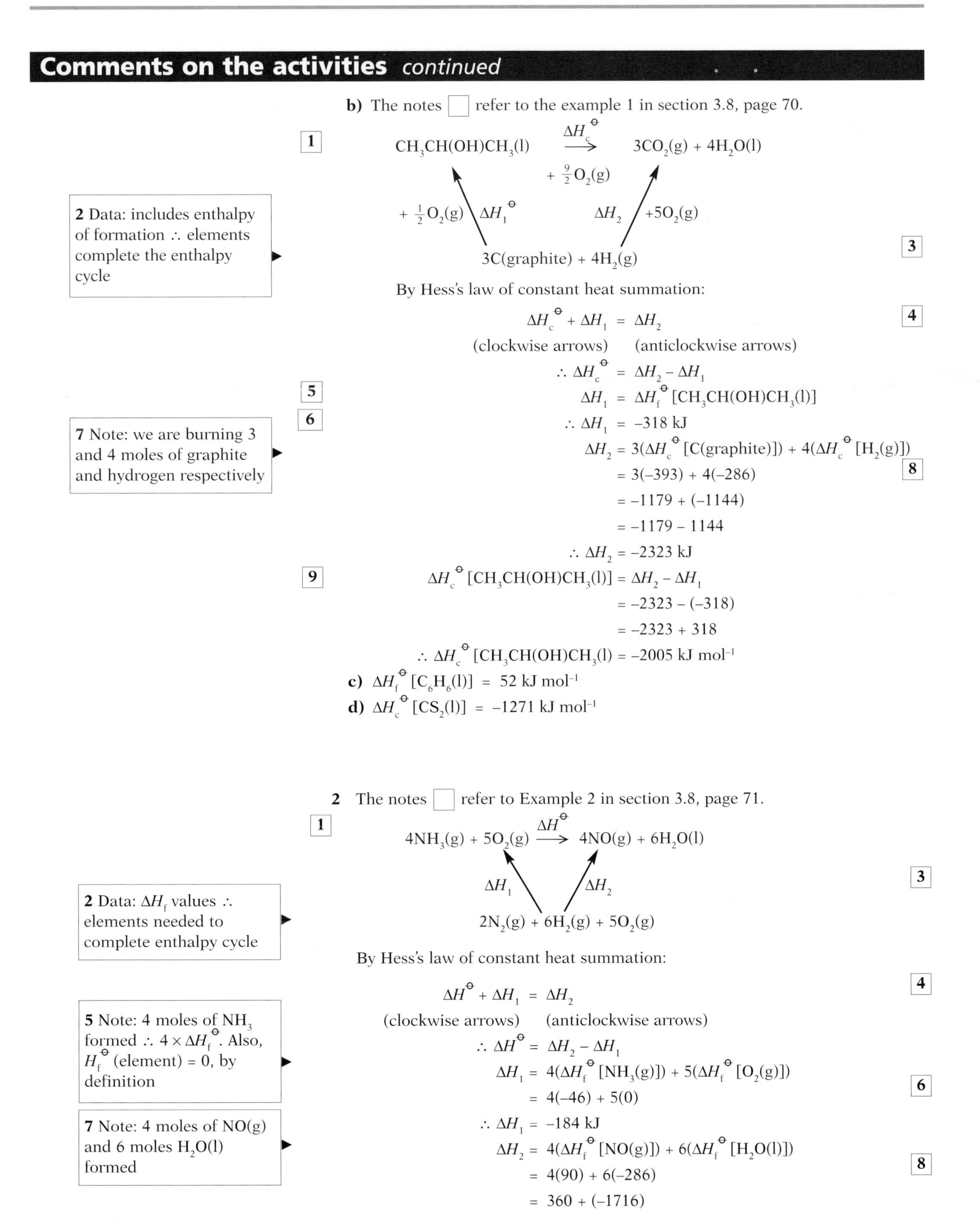

b) The notes ☐ refer to the example 1 in section 3.8, page 70.

[1]

$$CH_3CH(OH)CH_3(l) \xrightarrow{\Delta H_c^{\ominus}} 3CO_2(g) + 4H_2O(l)$$

$+ \frac{9}{2}O_2(g)$

$+ \frac{1}{2}O_2(g)$ $\Delta H_1^{\ominus}$ $\qquad$ ΔH_2 $+5O_2(g)$

$3C(graphite) + 4H_2(g)$ [3]

2 Data: includes enthalpy of formation ∴ elements complete the enthalpy cycle ▸

By Hess's law of constant heat summation:

$\Delta H_c^{\ominus} + \Delta H_1 = \Delta H_2$ [4]

(clockwise arrows) (anticlockwise arrows)

$\therefore \Delta H_c^{\ominus} = \Delta H_2 - \Delta H_1$

[5] $\Delta H_1 = \Delta H_f^{\ominus}[CH_3CH(OH)CH_3(l)]$

[6] $\therefore \Delta H_1 = -318$ kJ

7 Note: we are burning 3 and 4 moles of graphite and hydrogen respectively ▸

$\Delta H_2 = 3(\Delta H_c^{\ominus}[C(graphite)]) + 4(\Delta H_c^{\ominus}[H_2(g)])$

$= 3(-393) + 4(-286)$ [8]

$= -1179 + (-1144)$

$= -1179 - 1144$

$\therefore \Delta H_2 = -2323$ kJ

[9] $\Delta H_c^{\ominus}[CH_3CH(OH)CH_3(l)] = \Delta H_2 - \Delta H_1$

$= -2323 - (-318)$

$= -2323 + 318$

$\therefore \Delta H_c^{\ominus}[CH_3CH(OH)CH_3(l)] = -2005 \text{ kJ mol}^{-1}$

c) $\Delta H_f^{\ominus}[C_6H_6(l)] = 52 \text{ kJ mol}^{-1}$

d) $\Delta H_c^{\ominus}[CS_2(l)] = -1271 \text{ kJ mol}^{-1}$

2 The notes ☐ refer to Example 2 in section 3.8, page 71.

[1]

$$4NH_3(g) + 5O_2(g) \xrightarrow{\Delta H^{\ominus}} 4NO(g) + 6H_2O(l)$$

ΔH_1 $\qquad$ ΔH_2 [3]

$2N_2(g) + 6H_2(g) + 5O_2(g)$

2 Data: ΔH_f values ∴ elements needed to complete enthalpy cycle ▸

By Hess's law of constant heat summation:

$\Delta H^{\ominus} + \Delta H_1 = \Delta H_2$ [4]

(clockwise arrows) (anticlockwise arrows)

5 Note: 4 moles of NH_3 formed ∴ $4 \times \Delta H_f^{\ominus}$. Also, $H_f^{\ominus}$ (element) = 0, by definition ▸

$\therefore \Delta H^{\ominus} = \Delta H_2 - \Delta H_1$

$\Delta H_1 = 4(\Delta H_f^{\ominus}[NH_3(g)]) + 5(\Delta H_f^{\ominus}[O_2(g)])$

$= 4(-46) + 5(0)$ [6]

$\therefore \Delta H_1 = -184$ kJ

7 Note: 4 moles of NO(g) and 6 moles $H_2O(l)$ formed ▸

$\Delta H_2 = 4(\Delta H_f^{\ominus}[NO(g)]) + 6(\Delta H_f^{\ominus}[H_2O(l)])$

$= 4(90) + 6(-286)$ [8]

$= 360 + (-1716)$

Comments on the activities *continued*

9

$$\therefore \Delta H_2 = -1356 \text{ kJ}$$
$$\Delta H^\ominus = \Delta H_2 - \Delta H_1$$
$$= -1356 - (-184)$$
$$= -1356 + 184$$
$$\therefore \Delta H^\ominus = -1172 \text{ kJ}$$

---

3 a) +52 kJ **b)** –455 kJ **c)** +152 kJ

Activity 3.6

1 a) H—H + F—F ⟶ 2H—F

Bonds broken (moles): 1H—H + 1F—F

Enthalpy required (+ve sign): 1(+436) + 1(+158) = +594 kJ

Bonds made (moles): 2H—F

Enthalpy released (–ve sign): 2(–562) = –1124 kJ

overall enthalpy change, ΔH = enthalpy change on breaking bonds, +ve + enthalpy change on forming bonds, –ve

= (+594) + (–1124)

= –530 kJ

b) H—H + $\frac{1}{2}$O=O ⟶ H—O—H

Bonds broken (moles): 1H—H + $\frac{1}{2}$O=O

Enthalpy required (+ve sign): 1(+436) + $\frac{1}{2}$(+496) = +684 kJ

Bonds made (moles): 2O—H

Enthalpy released (–ve sign): 2(–463) = –926 kJ

overall enthalpy change, ΔH = enthalpy change on breaking bonds, +ve + enthalpy change on forming bonds, –ve

= (+684) + (–926)

= –242 kJ

c) There is very good agreement with the experimental values bearing in mind that mean bond enthalpies are average values and not specific to the molecules in the question.

2 a) 3 moles of C=C bonds and 3 moles of H—H bonds

b) Bonds broken (moles): 3C=C + 3H—H
enthalpy required (+ve sign): 3(+612) + 3(+436) = +3144 kJ

c) 3 moles of C—C bonds and 6 moles of C—H bonds

d) Bonds made (moles): 3C—C + 6C—H
Enthalpy required (+ve sign): 3(–348) + 6(–413) = –3522 kJ

e) overall enthalpy change, ΔH = enthalpy change on breaking bonds, +ve + enthalpy change on forming bonds, –ve

= (+3144) + (–3522)

= –378 kJ

f) No. The difference in values (–170 kJ) is too large to be explained by the possible error expected from using mean bond enthalpies, which are average values, rather than using values for the actual bonds in benzene. Kekulé got the structure wrong! (If you want to find out more about this, have a look at section 9.1.)

CHAPTER 4

Chemical bonding and the structure of materials

Contents

Study Checklist

After studying this chapter you should be able to:

1. Understand, in outline, how absorption spectroscopy, X-ray diffraction and electron diffraction techniques provide information about chemical bonding and structure.
2. State that ions are formed when one atom transfers one, or more, electrons to another atom and that ionic bonding results from the electrostatic attraction between these ions.
3. State that a covalent bond is formed when two atoms share a pair of electrons and that, in a co-ordinate bond (dative covalent bond), one of the bonded atoms supplies both of the bonding electrons.
4. Use dot and cross diagrams to illustrate the bonding in the following molecules and ions: $NaCl$, Li_2O, MgO, $MgCl_2$ (ionic); H_2, O_2, N_2, Cl_2, HCl, CO_2, CH_4, C_2H_4, C_2H_6 (covalent); NH_4^+, BF_4^-, BF_3NH_3 (co-ordinate).
5. Describe covalent bonding in terms of overlapping atomic orbitals. Using simple orbital diagrams, explain the difference between σ and π bonds and predict the shapes, and bond angles, in ethane and ethene.
6. Understand that bonding and non-bonding (lone) pairs of electrons are charge clouds. Explain how the shape, and bond angles, in simple molecules and ions are determined by the relative strengths of the mutual repulsion between these charge clouds, using as examples $BeCl_2$, CO_2, BF_3, CH_4, BF_4^-, NH_4^+, PF_5, SF_6, SO_2, H_2O and NH_3.
7. Understand that the distribution of the electron cloud in a covalent bond is rarely symmetrical. Define electronegativity as the power of an atom in a covalent bond to attract the bonding electron cloud towards itself. State the trends in electronegativity values around the periodic table. Relate the polarity of a covalent bond to the bond length and the difference in electronegativity of the bonded atoms.
8. Explain, using suitable diagrams, how the presence of induced or permanent dipole–dipole attractions gives rise to intermolecular forces, making particular reference to hydrogen bonding. Describe the nature of the intermolecular forces in, and their effect on the physical properties of, the following species: the noble gases and alkanes of molecular formula C_5H_{12} (van der Waals' forces); CH_3COCH_3, $CHCl_3$; the hydrogen halides, water and ammonia.
9. Describe how the particles are held together, and arranged, in simple molecular structures and giant molecular (macromolecular) structures, using as examples argon, ice, diamond, graphite and silica. Relate the physical properties (hardness, melting and boiling points, electrical conductivity and solubility in polar and non-polar solvents) and uses of these substances to the type of structure.

10 Describe how the particles are held together and arranged in giant ionic structures, using as examples sodium chloride (6:6 lattice) and caesium chloride (8:8 lattice). Define the term 'ionic radius'. Relate the physical properties of ionic substances to their structures.

11 Define the term 'lattice energy', $\Delta H_L^{\ominus}$, and explain how the lattice energy of an ionic compound depends on the charges and sizes of the ions present. Know that anions can be polarised by cations of high charge density (Fajan's rules) and that this causes some ionic compounds to have appreciable covalent character, for example in lithium iodide.

12 Describe qualitatively the changes which occur when an ionic compound dissolves in water, using suitable diagrams and making reference to the following terms: hydration, ion-dipole attractions, co-ordination number, hydration number. Define the terms 'standard enthalpy of solution', $\Delta H_s^{\ominus}$ and 'standard enthalpy of hydration', $\Delta H_h^{\ominus}$. Use Hess's law and enthalpy cycles to calculate one of the following terms, given the other two: $\Delta H_L^{\ominus}$, $\Delta H_s^{\ominus}$, $\Delta H_h^{\ominus}$.

13 Know that giant metallic structures consist of a close-packed arrangement of cations embedded in a mobile electron cloud. Use this model to account for the following physical properties of metals: hardness, melting and boiling points, electrical conductivities, solubility in polar and non-polar solvents, malleability and ductility. Interpret the uses of metals in terms of these physical properties.

14 Explain, using ionic equations, the meaning of the following terms: oxidation, reduction, redox, oxidising agent, reducing agent, oxidation number (state). Describe, and explain, redox processes in terms of electron transfer and changes in oxidation numbers.

15 Understand the structures of gases, liquids and solids in terms of the particles present, their motion and the forces acting between them. Use kinetic theory to describe melting, vaporisation and vapour pressure. Explain the energy changes associated with changes of state.

16 Explain qualitatively the concept of an ideal gas, the conditions necessary for a gas to approach ideal behaviour and the limitations of ideality at very high pressure and low temperature. State that ideal gases obey the ideal gas equation ($pV = nRT$) and use the equation in calculations, in particular to determine the molecular mass of a gas or volatile liquid.

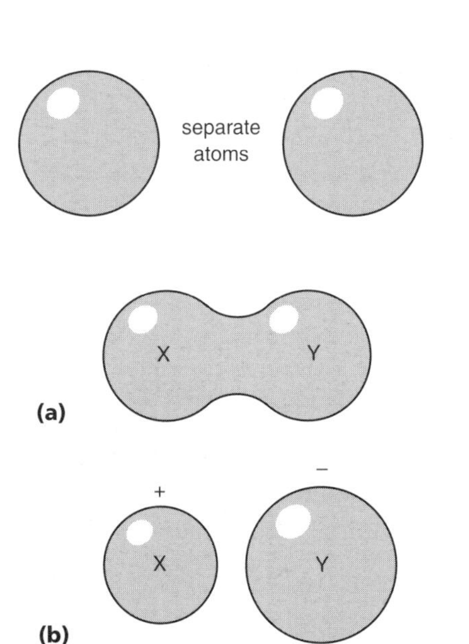

Figure 4.1
Electron rearrangement during chemical bonding. **(a)** In a covalent bond, the atoms share the bonding electrons. The nuclei of both atoms are mutually attracted to the electron cloud which is continuous between the nuclei. **(b)** An ionic bond is formed as a result of atom X transferring one, or more, electrons to atom Y. In this case, X loses one electron and forms a postively charged ion, X^+; atom Y gains one electron to form a negatively charged ion, Y^-. The electron cloud is not continuous between the nuclei of the ions

4.1 Introduction

An icebreaker is a ship with a reinforced steel hull which is capable of breaking through ice-covered seas and oceans. The icebreaker breaks the ice by ramming it and using the weight of the ship to smash it. As the icebreaker makes its passage through the icy seas, two solids, ice and steel, are brought into violent contact. Why is it that the sheet ice breaks rather than the steel hull? If equal thicknesses of steel and ice are compared, the steel is found to be much harder and this is an example of the very different physical properties of these solids. Clearly, the physical properties of a substance depend on the way its constituent particles (atoms, molecules or ions) are bonded together, and arranged throughout the structure.

In nature, separate atoms are very rare. Of the 92 natural elements, only a handful, known as the noble gases, are able to exist as separate atoms. In section 1.3, we introduced the idea that atoms form chemical bonds in order to gain more stable electron arrangements. When atoms react, they do so via two types of chemical bonding, **covalent** and **ionic** (Figure 4.1). In a covalent bond, both atoms *share* the bonding electrons. Ionic bonding involves the

▲ An icebreaker breaking through ice

transfer of one or more electrons from one atom to another atom. Quite often, the bonded atoms have electron arrangements resembling those of the noble gases, that is with 2 or 8 electrons in the outermost shell.

This chapter covers covalent and ionic bonding in detail and then focuses on how the structure and properties of a material are influenced by the type of chemical bonds that it contains.

4.2 Finding out about chemical bonding and structure

Physical properties

The physical properties of a material can provide a lot of useful information about its structure. For example, if we compare the physical properties of ice and steel, as listed in Table 4.1, the following deductions can be made:

- Steel is much harder and denser than ice, suggesting that the particles are more tightly packed together in steel's structure.
- Steel has a much higher melting point than ice, suggesting that the attractive forces between the particles in steel are much greater than those between the particles in ice.
- Steel is a much better conductor of electricity than ice, suggesting that there are a large number of charged particles moving freely throughout steel's structure.

Table 4.1 Comparing some physical properties of ice and mild steel

	ice	steel
relative hardness	hard	very hard
density/g cm^{-3}	0.92	7.86
melting point/°C	0	1427
electrical conductivity	very low	very high

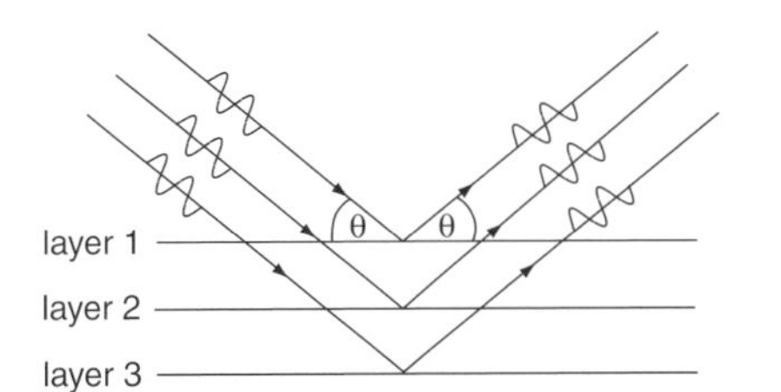

Figure 4.2 ▲
'Strong' X-ray reflections from a set of crystal planes will occur if the waves arrive and leave in phase

Absorption spectroscopy

Some molecules absorb radiation in the radio wave, infrared, visible or ultraviolet regions of the electromagnetic spectrum. In doing so, the molecule is promoted to a higher energy level. Since the frequencies absorbed are characteristic of the bonds in the molecule, information is obtained about the structure of the material. Spectroscopic analysis is covered in detail in Chapters 17 and 18.

X-ray diffraction

A crystal structure consists of a regular 3-D arrangement of atoms, molecules or ions, known as a **lattice**. The wavelength of X-rays and the gaps between the particles in crystals are of a similar size and so X-rays are strongly diffracted (reflected) if they strike the crystal lattice at certain glancing angles as shown in Figure 4.2. The X-ray diffraction pattern is photographed and used to work out the distance between the planes of particles and from these, the positions of the particles within the lattice can be established (Figure 4.3). Early workers in this field examined the 'reflection' of the X-rays from each set of crystal planes individually. Modern methods use a monochromatic X-ray beam (a single wavelength) and the crystal is rotated. This allows reflections from many sets of planes to be detected simultaneously.

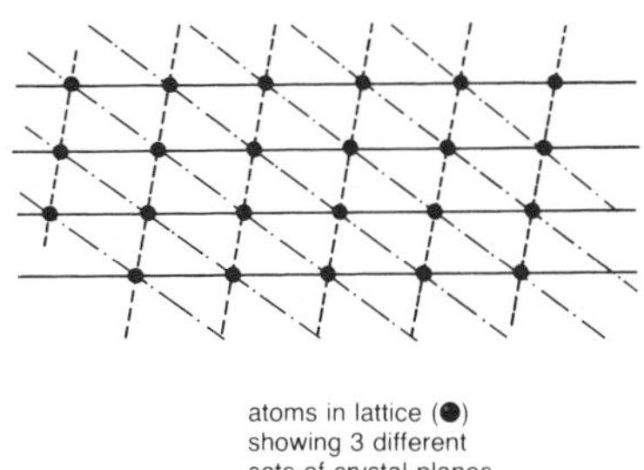

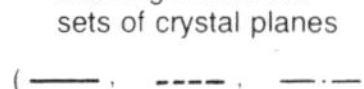

Figure 4.3 ▲
If the spacing between a sufficient number of crystal planes can be determined, then the positions of the particles within the lattice may be found

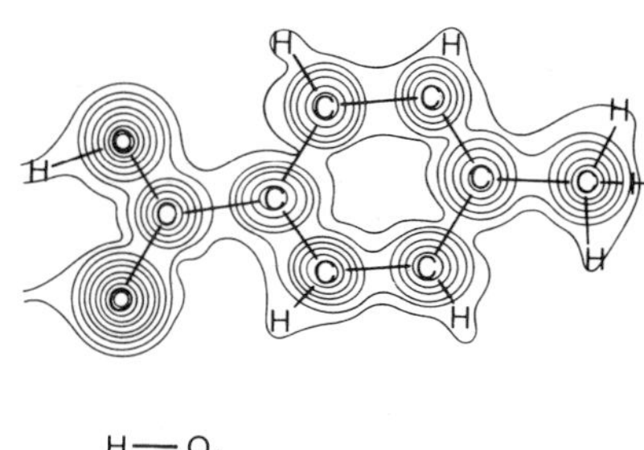

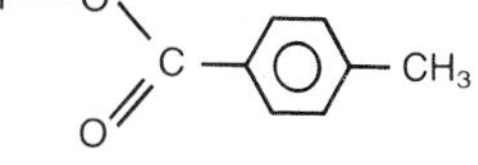

Figure 4.4
The structure of a molecule of 4-methyl benzoic acid superimposed on its electron density map. Each contour line connects points of the same electron density

Structural determination involves finding a model of the crystal lattice which will account for the observed diffraction pattern. While for simple structures this may be done by 'trial and error', advanced computer analysis of the results is required when investigating complex crystal structures. Using such methods, it is possible to produce an **electron density map** of the lattice, from which the bond lengths and bond angles may be established. A typical electron density map, that of 4-methyl benzoic acid, is shown in Figure 4.4. Since the ability of an atom to scatter X-rays depends on its atomic number, it is difficult to identify atoms of low mass, particularly hydrogen, with any accuracy. X-ray diffraction is discussed further in section 16.5.

Electron diffraction

X-ray diffraction studies are only suitable for crystalline solids. However, electrons also demonstrate wave properties and they may be used to obtain diffraction patterns from gaseous samples. In this technique, a beam of electrons strikes a thin jet of the gas at right angles and circular diffraction patterns are recorded on a photographic plate. The diameter of each circle is related to an interatomic separation within the gas molecules so a structural model may be proposed which is consistent with the diffraction pattern. Electron diffraction is capable of locating 'light' atoms, especially hydrogen, with much greater precision than X-ray diffraction.

The structure of vitamin B_{12}

In 1964, Professor Dorothy Hodgkin (1910–1994) was awarded a Nobel Prize for determining the structure of vitamin B_{12} using X-ray crystallography. An essential nutrient, vitamin B_{12} is a biochemical compound containing cobalt which is found in a wide variety of foods, such as offal, red meat, fish, dairy produce, eggs, yeast extracts, alfalfa sprouts and seaweeds. Hodgkin showed that a deficiency of vitamin B_{12} may lead to pernicious anaemia, a blood condition which involves an abnormal reduction in the number of red blood cells which are available to carry oxygen around the body.

Dorothy Crowfoot Hodgkin

Focus 4a

1. Chemical bonding involves the redistribution of electrons:
   a) in **covalent bonding**, electrons are shared by the bonded atoms
   b) in **ionic bonding**, electrons (usually up to 3) are transferred between bonding atoms.
2. By bonding, atoms achieve more stable electron arrangements often, but not always, those of the noble gas atoms.
3. Molecular structures can diffract (reflect) X-rays and electrons to produce a pattern of dots (from X-rays) or rings (from electrons). These can be recorded on a photograph and an analysis of the patterns can give a detailed, unambiguous model of the structure.

4.3 Covalent bonding: electron sharing

Figure 4.5 shows two ways of representing the electron charge clouds in the hydrogen molecule, H_2. We can see that the electron cloud is densest around the two nuclei and that there is a high concentration of electron cloud between the two nuclei. The two hydrogen atoms have formed a covalent bond by sharing a pair of electrons, as shown by the dot and cross diagram in Figure 4.6. In forming the H—H bond, each atom has achieved a more stable electron arrangement corresponding to that of a noble gas, helium.

Let us consider the energy changes involved when a single covalent bond is formed from two atoms, using Figure 4.7. As the atoms get closer together, each nucleus starts to attract the other atom's electron cloud. The covalent bond starts to form and energy is released. However, if the two atoms get too close together, there would be considerable repulsion between their nuclei and the energy of the system would rise. The covalent bond represents a position of balance in which the forces of attraction between the nuclei and the bonding electrons just match the repulsive forces between nuclei. The distance between the nuclei is known as the **bond length** (0.074 nm for H—H) and the energy required to separate the atoms in the bond is known as the **bond dissociation enthalpy** (436 kJ per mole of H—H bonds broken).

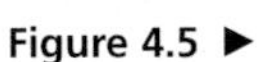

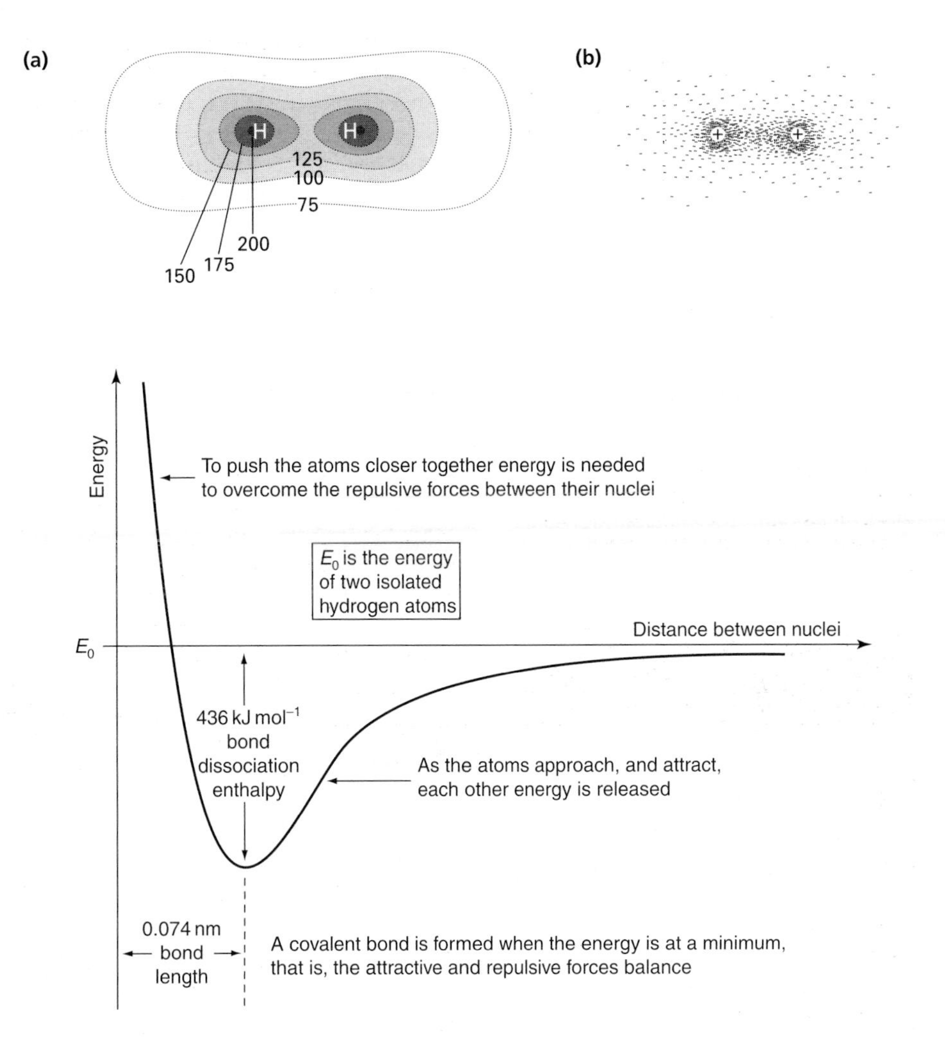

Figure 4.5 ▶
(a) An electron density map for the hydrogen molecule (the contours have units of electrons per nm^3). Each contour line joins points having the same electron density. **(b)** The charge clouds for electrons in the hydrogen molecule, H_2. The greater the density of 'dots', the greater the density of the clouds

H H

Figure 4.6 ▲
A dot and cross diagram showing the electron arrangement in a hydrogen molecule. Each atom has achieved a full outer shell of electrons

Figure 4.7 ▶
Variation in the energy of two hydrogen atoms as the distance between them is altered

Covalent bonding may be illustrated using dot and cross diagrams which show the arrangement of the outer (valence) electrons. For example, two chlorine atoms, each having seven valence electrons, may form a diatomic molecule by sharing one pair of electrons, thus achieving a more stable electron arrangement.

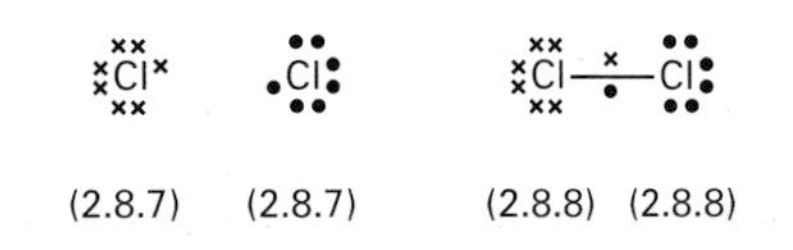

In the case of oxygen and nitrogen, two and three pairs of electrons, respectively, must be shared between the two atoms to achieve a stable electronic structure. The oxygen molecule is said to contain a **double bond** and the nitrogen molecule, **a triple bond**.

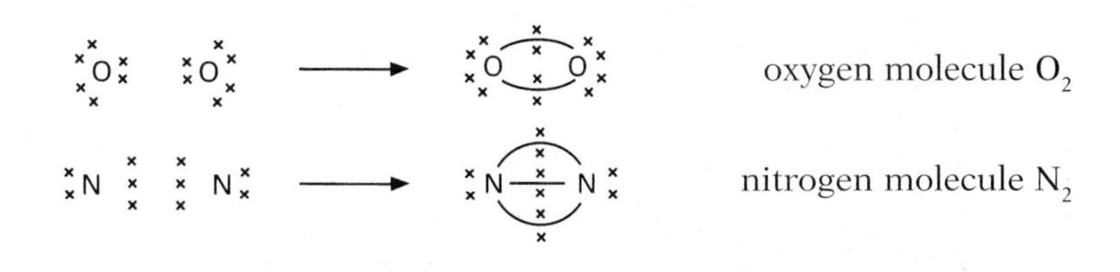

These molecules contain **lone pairs** of electrons, that is, pairs of valence electrons which are not used for bonding. We shall see later that lone pairs can have an important effect on the shape and reactivity of molecules.

Dissimilar atoms may also share electrons. Figure 4.8 gives the electron dot and cross diagrams for hydrogen chloride, methane, water, ammonia, ethane, ethene and carbon dioxide. All of the diagrams in Figure 4.8 represent molecules in which the bonded atoms have achieved the electron arrangement of a noble gas.

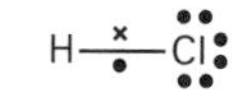
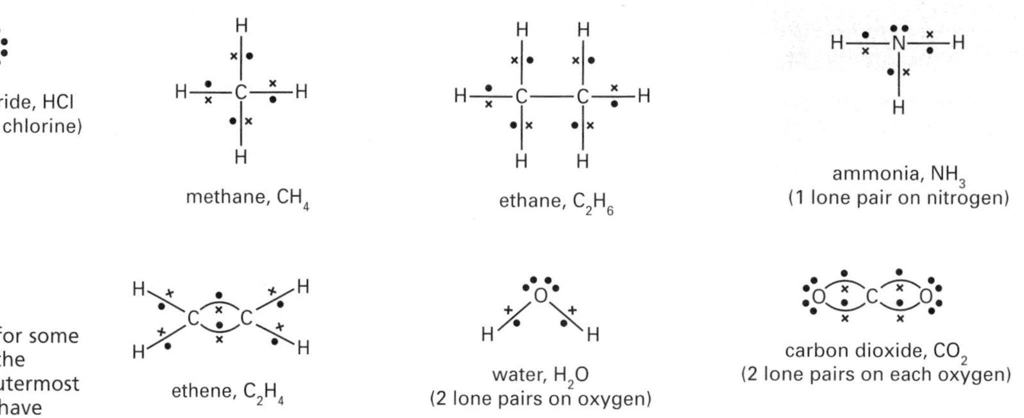

Figure 4.8 ▶ Electron dot and cross diagrams for some simple covalent molecules. Only the valence electrons (those in the outermost shells) are shown and the 'rings' have been omitted for clarity

It should be noted that the noble gas configurations are not the only stable electronic arrangements found in covalent molecules. *The third and subsequent principal energy levels may contain more than eight electrons.* Thus sulphur forms three fluorides, SF_2, SF_4 and SF_6, where the central sulphur atom has totals of 8, 10 and 12 electrons respectively in its outer (third) electron shell.

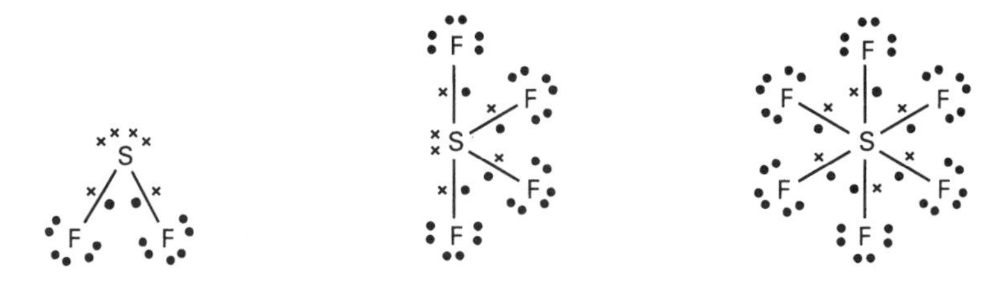

Oxygen on the other hand, although in the same group of the periodic table as sulphur, only forms a difluoride, OF_2, since *the oxygen atom can only accommodate a maximum of eight electrons in its outer (second) shell.*

Some covalent molecules contain atoms with fewer outer electrons than the corresponding noble gas. In boron trichloride, for example, the outer shell of the boron atom contains a total of just 6 electrons.

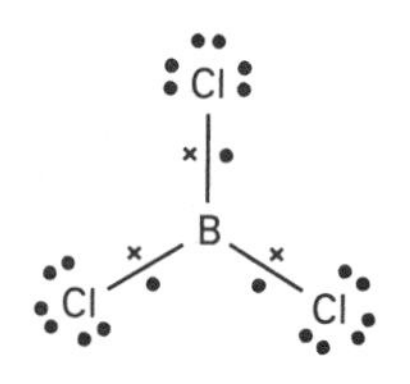

Co-ordinate (dative covalent) bonding

In some cases, both of the electrons to be shared come from the same atom, thus forming a **co-ordinate bond** or **dative covalent bond**. In dot and cross diagrams, we often denote a co-ordinate bond by an arrow pointing from the atom which donates the lone pair to the atom which receives it. For example, carbon monoxide may be drawn as shown below:

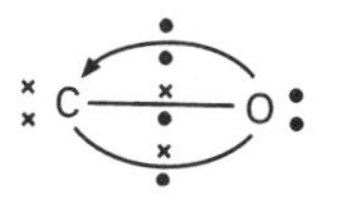

The number of electrons in principal energy levels (electron shells)

principal energy level	maximum number of electrons
n = 1	2
n = 2	8
n = 3	18
n = 4	32

This topic was introduced in section 2.7.

Co-ordinate bonding may be found in molecular addition compounds such as boron trifluoride ammonia, $BF_3.NH$.

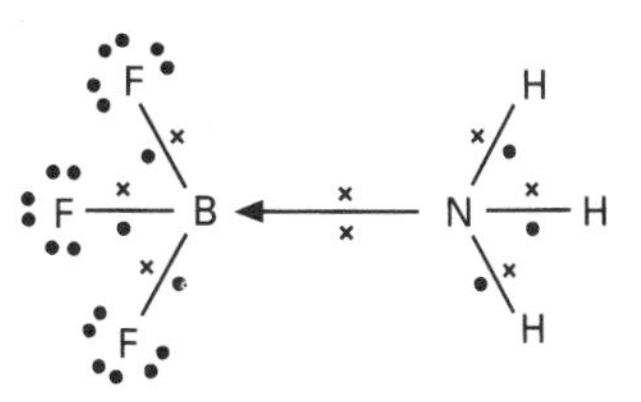

The boron atom has six electrons in its outer shell so it can hold an extra two electrons in this shell before it is full. Nitrogen donates its lone pair of electrons to the co-ordinate bond between nitrogen and boron, and both atoms achieve full outer electron shells.

Co-ordinate bonding is found in some common **polyatomic ions**. These are groups of atoms which are held together by covalent bonding and which have an overall charge. Examples are shown in Figure 4.9.

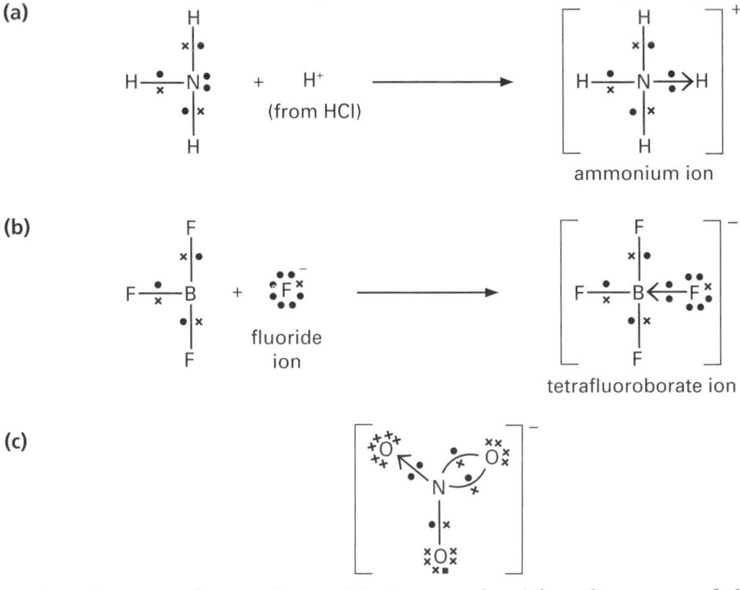

Figure 4.9
Co-ordinate (dative covalent) bonding occurs in some polyatomic ions, as shown by the following examples:
(a) Ammonia and hydrogen chloride gases react vigorously to form ammonium chloride, a white ionic solid. The ammonium ion, NH_4^+, is formed by co-ordinate bonding, with nitrogen, sharing its lone pair with a hydrogen ion.
(b) When a fluoride ion shares a lone pair with the boron atom in boron trifluoride, a tetrafluoroborate ion, BF_4^-, is formed
(c) In the nitrate ion, NO_3^-, the nitrogen atom achieves a noble gas electronic configuration by forming a co-ordinate bond with one of the oxygen atoms. The '■' electron imparts the negative charge

In all cases of co-ordinate (dative covalent) bonding, one of the atoms involved must possess a lone pair of electrons which can be shared with an electron accepting atom. Later, we shall discuss the importance of co-ordinate bonding in the mechanisms of organic reactions (section 6.9) and in the formation of transition metal complexes (section 5.6).

Activity 4.1

Draw dot and cross diagrams to show the covalent bonding in the following molecules and ions. You need only show the electrons in the valence shells.

HBr $BeCl_2$ $AlCl_3$ PF_5
SF_6 BH_4^- PCl_6^-

The physical properties of materials containing covalent bonds

Covalently bonded substances can have very different physical properties. For example, polyethene is a strong, yet flexible, material which is used to make bags and containers. It has a **simple molecular** structure (covalent bonds are only found within the molecules, not between molecules) which is characterised by its relative softness, low density (0.9 g cm^{-3}) and low melting point (210°C). Diamond, on the other hand, is a very hard solid which is used to make cutting tools, as well as jewellery. Diamond also contains covalent bonds but has a **giant covalent (macromolecular)** structure which is characterised by its hardness, higher density (2.33 g cm^{-3}) and very high melting point (3527°C). These structural types are considered in detail in sections 4.7 and 4.8.

An orbital view of covalent bonding

So far, we have seen that a covalent bond is formed when two atoms share electrons. We have used dot and cross diagrams to account for the electrons in the bonds and around the atoms. However, the atomic theory developed in Chapter 2 requires that we consider electrons as occupying **atomic orbitals** (certain regions of space around the nucleus). In the case of a hydrogen atom, the electron occupies a spherical 1s orbital. Overlap between the half-filled orbitals on two such atoms gives a **molecular orbital** containing two electrons.

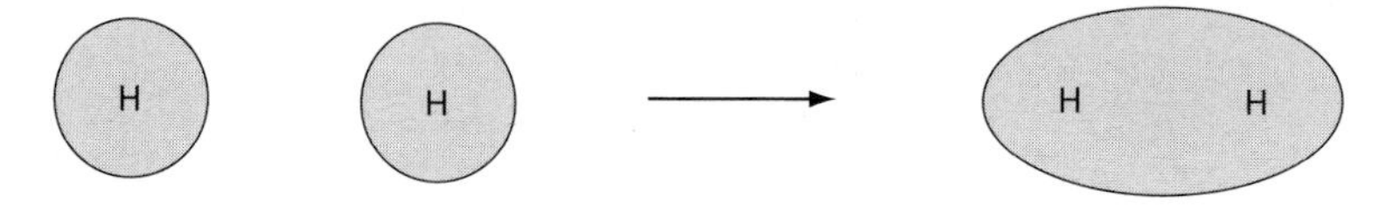

This molecular orbital treatment may also be applied to more complex systems. Simply speaking, a strong covalent bond may be formed by the overlap of two suitably placed atomic orbitals that contain between them two electrons.

For example, in the oxygen molecule, O_2, each O atom has the electron configuration [He] $2s^22p^4$.

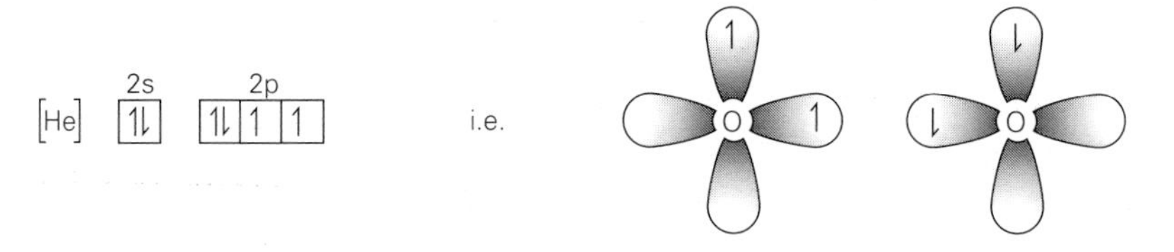

A strong bond is formed by *end-on* overlap of a half-filled p orbital on each of the atoms.

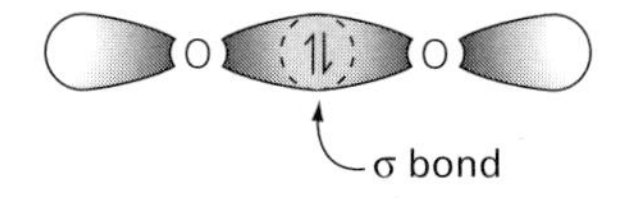

Here maximum overlap occurs directly between the two nuclei and this is known as a **σ bond**. The second bond results from *sideways* overlap of the remaining half full p orbitals above and below a line joining a nuclei. This is referred to as a **π bond**.

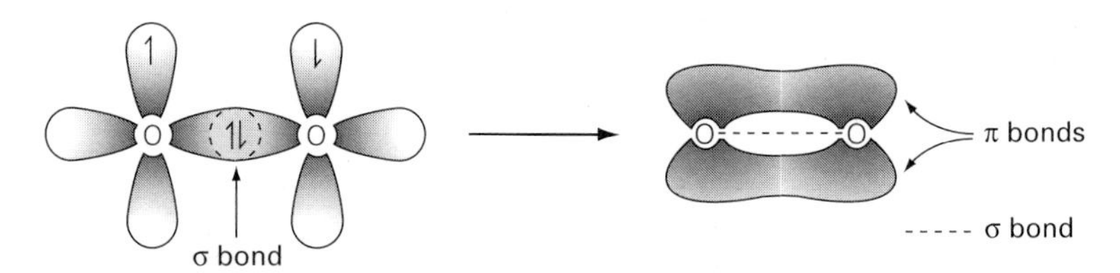

In the nitrogen molecule, N_2, each atom possesses three singly occupied p orbitals. Overlap of one pair of these orbitals gives a σ bond and overlap of the remaining two sets of p orbitals gives rise to two π bonds.

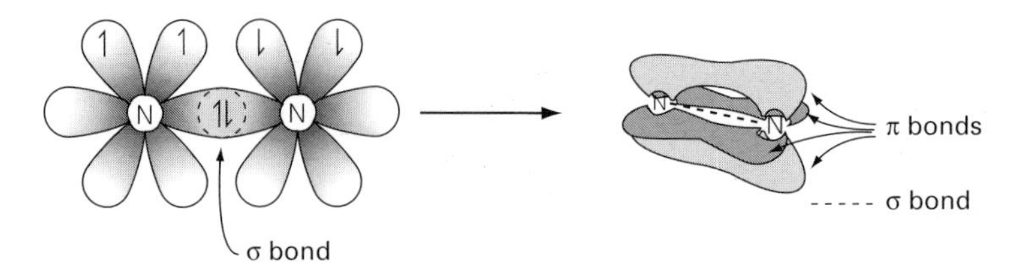

The distribution of the electron cloud in ethane and ethene

Quite often, the bonding in a covalent molecule cannot be explained by the overlap of simple atomic orbitals in their lowest energy, or **ground**, states. This is the case in ethane, C_2H_6, and ethene, C_2H_4 (refer back to Figure 4.8). In these hydrocarbons, as in almost all of its covalent compounds, carbon has a valency of *four* (section 1.4). Yet, in its ground state, $1s^22s^22p^2$, the carbon atom has only *two* unpaired electrons, these being located in the p orbitals. By absorbing a small amount of energy, an electron is promoted from a 2s orbital into the empty 2p orbital, thereby making four unpaired electrons available for covalent bonding. (Remember that the p orbitals are dumb-bell shaped and are located at 90° to each other.)

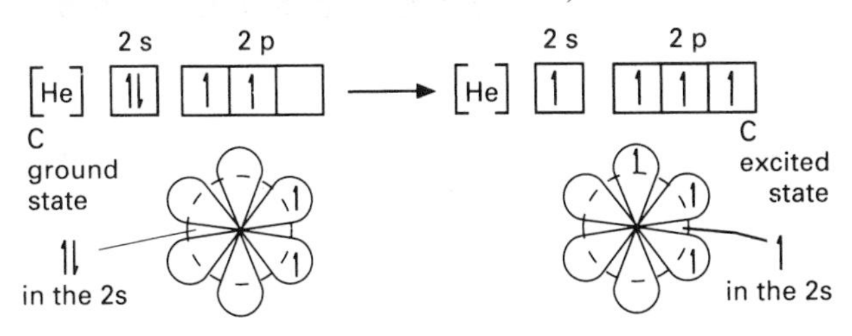

The electron cloud in some, or all, of these atomic orbitals can mix together before bonding occurs to form **hybrid** atomic orbitals. The hybrid orbitals overlap end-on with the atomic orbitals of other atoms to form a **σ bonded** molecular skeleton. Any electrons not involved in this 'mixing' remain in pure p atomic orbitals and these may overlap 'sideways' to form **π bonds**.

In ethane, each carbon atom is singly bonded to four other atoms; the other carbon and three hydrogens. The 2s and all three of the 2p orbitals are mixed together to produce four identical hybrid orbitals and these overlap end-on with the atomic orbitals in the other atoms, as shown below.

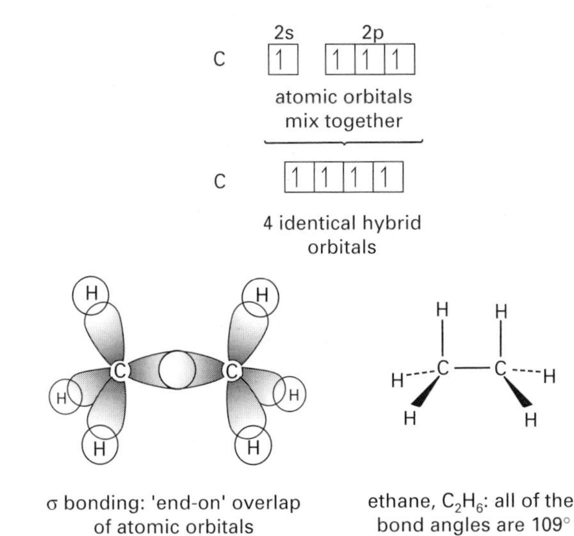

σ bonding: 'end-on' overlap of atomic orbitals

ethane, C_2H_6: all of the bond angles are 109°

In ethene, each carbon atom is σ bonded to three atoms, the other carbon and two hydrogen atoms. To do this, the 2s and two of the 2p orbitals mix together to give three identical hybrids and these overlap end-on with the atomic orbitals in the other atoms, forming a σ bond framework.

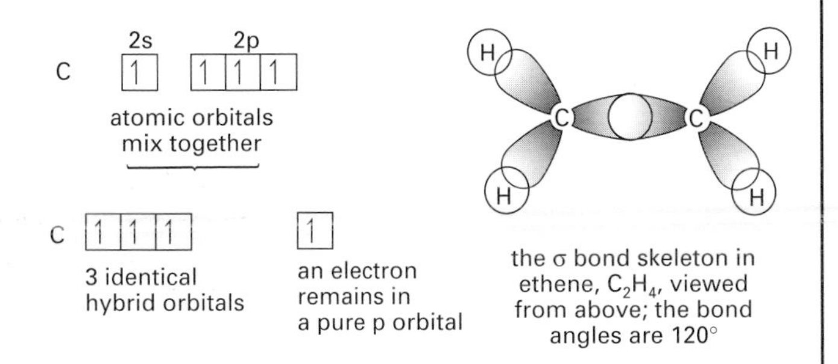

the σ bond skeleton in ethene, C_2H_4, viewed from above; the bond angles are 120°

Each carbon atom has one electron remaining in a p orbital (dumb-bell) and these orbitals π bond by overlapping sideways above and below the plane of the σ bonds.

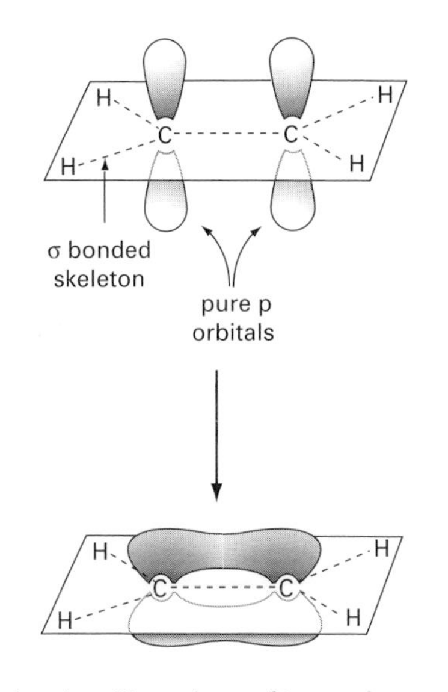

the π bond is made up of two regions of electron cloud, one above and one below the plane of the σ bonds

Focus 4b

1. In a **co-ordinate (dative covalent bond)**, both of the bonding electrons are supplied by one of the bonding atoms (e.g. in NH_4^+, $NH_3.BF_3$).
2. Covalent bonding results from the overlap of suitably placed half-full atomic orbitals producing molecular orbitals. In some cases, this may involve:
   a) promoting electrons from their ground states to higher energy levels,
   b) mixing atomic orbitals to make hybrid atomic orbitals prior to covalent bonding.
3. In **σ covalent bonds**, maximum orbital overlap occurs on a line directly between the two nuclei. They are generally stronger than **π covalent bonds** in which sideways overlap occurs above and below the σ bond framework. A double covalent bond is made up of one σ and one π bond (e.g. in ethane). A triple bond consists of one σ bond and two π bonds at right angles to each other (e.g. as in the nitrogen molecule).
4. Multiple covalent bonds are shorter and stronger than single bonds.

4.4 Shapes of covalently bonded molecules and ions

As long ago as 1894, when Fisher proposed the **'lock and key' mechanism** to account for enzyme action, it was suggested that molecular shape might be an important factor in the chemistry of some compounds. Enzymes are large protein molecules which act as catalysts in particular biochemical reactions. Without enzymes, life as we know it would be impossible. Fisher reasoned that each enzyme had **active sites** on its surface in the form of specially shaped holes (or locks). Only molecules of the correct shape to fit these holes (keys) would be acted upon by the enzyme (Figure 4.10).

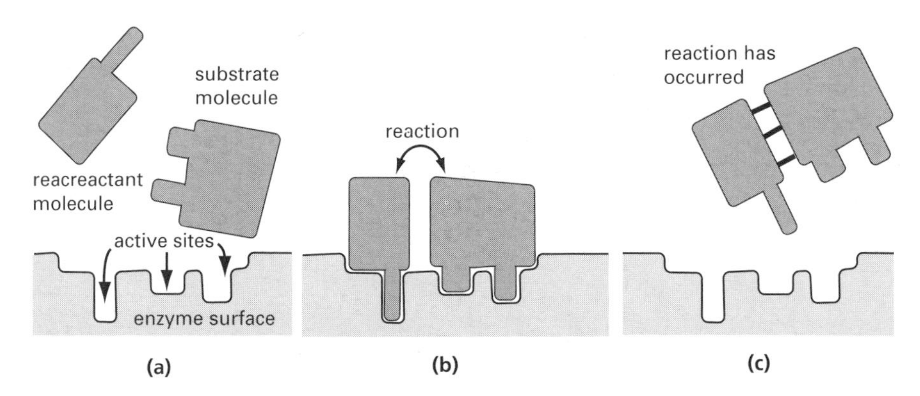

Figure 4.10 ▶
The 'lock and key' mechanism of enzyme action. Each enzyme catalyses the reaction of a specific biochemical substance, termed a substrate. Here we see molecules of the substrate and another reactant approaching the enzyme's surface **(a)**. They fit exactly into the active sites and reaction occurs **(b)**. Finally, the product molecules move away from the active sites so that further catalysis may occur **(c)**

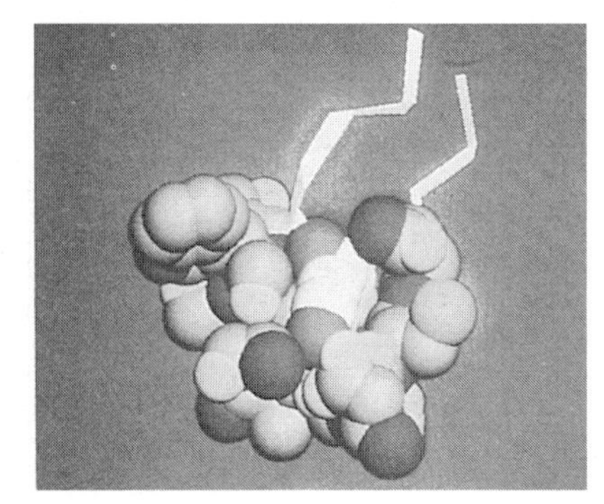

▲ The 3-D structure of trypsin, a digestive enzyme which splits complex proteins into smaller molecules

Results of X-ray diffraction experiments have helped to confirm Fisher's idea but these methods are still very time consuming and laborious. Fortunately, from such experimental results it has been possible to develop a simple theory which enables us to accurately predict the shapes of simple molecules or parts of molecules.

At first glance, there seems to be little pattern to the variety of molecular shapes. Even some molecules of the same general formula, for example boron trichloride, BCl_3 and nitrogen trichloride, NCl_3, have different shapes.

Cl B Cl Cl   N Cl Cl Cl

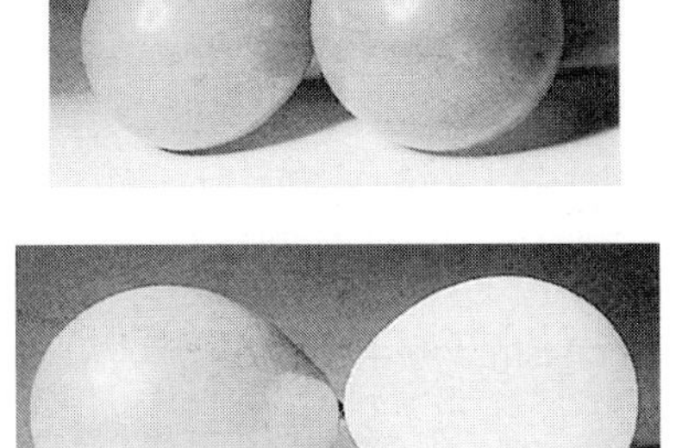

Using balloons to model molecular shapes

Note here the convention for showing the 3-D shape of molecules. Bonds in the plane of the paper are drawn —, those coming up out of the plane are drawn ◢, and those going below this plane are drawn -----.

It is only when we look at the electronic structures of the molecules in more detail that we can explain such differences.

In 1940 Sidgwick and Powell proposed that molecular shape is determined largely by the number of different sets of electrons in the valency shell of the central atom. Since electron pairs are negatively charged, they will repel each other and arrange themselves in space so as to minimise the repulsive forces. This is known as the **electron pair repulsion theory**.

Activity 4.2

The following simple experiment will help you to visualise the arrangement in space of electron sets around a central atom. You will need six balloons of the same size and shape, either round ones or the long thin type.

- First blow up two balloons, not quite fully, and join them by knotting the open ends together.
- Repeat this step with the other balloons so that you have three such pairs.
- Join the balloons by twisting their knotted centres around each other a number of times. You should now have a model with six balloons pointing roughly at right angles to each other. This represents the way in which six sets of valency electrons will arrange themselves in space so as to minimise the repulsive forces between them. If you try bending the balloons towards each other, you will find they spring back to their original position when released. This shape is referred to as octahedral, since it just fits inside a regular octahedron (an object with eight identical triangular faces).

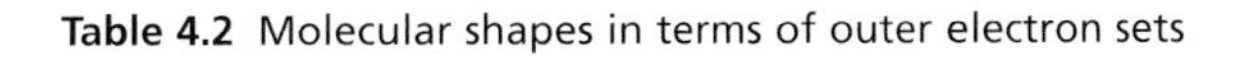

Table 4.2 Molecular shapes in terms of outer electron sets

total number of outer electron sets	spatial arrangements		
2	linear		
3	trigonal planar	V-shaped	
4	tetrahedral	trigonal pyramid	V-shaped
5	trigonal bipyramid	distorted tetrahedral	
6	octahedral	square pyramid	square planar

Activity 4.3

Use dot and cross diagrams and Table 4.3 to work out the shape of the following molecules and ions:

HCN PCl_6^- NH_4^+ PF_5
SO_2 $AlCl_3$ PH_3 IF_5

- If you now burst one of the balloons with a pin you should have the arrangement for five sets of valency electrons. If it doesn't look like the trigonal bipyramid shape shown in Table 4.2 then hold one of the balloons and gently shake the model.
- Burst the other balloons in turn, at each stage comparing the shape with that given in Table 4.2.
- The basic shapes resulting from different numbers of electron sets on the central atom can be found in Table 4.2. Some common examples of each type are given in Table 4.3.

Focus 4c

1. Molecular shape is determined by mutual repulsion of different 'sets' of electrons in the outer (valence) shell of the central atom. The basic shape of a molecule, or polyatomic ion, may be deduced from a dot and cross diagram of its outer electron structure.
2. Although the presence of non-bonding (**lone pairs**) of electrons must be taken into account, they are not included in the description of molecular shape. For example, methane, ammonia and water all have four sets of electrons around their central atoms but their shapes are respectively described as tetrahedral, trigonal pyramidal and angular.
3. Lone pairs, being closer to the nucleus, repel more strongly than bond pairs. This will cause small distortions from regular shapes and bond angles.

Table 4.3 Common molecular shapes

2 sets of electrons

Mutual repulsion gives a **linear** arrangement as in beryllium chloride, $BeCl_2$:

180°
Cl — Be — Cl

A set of electrons need not be a single pair. Atoms commonly achieve a 'noble gas' outer octet of electrons by covalent bonding. If these eight electrons are divided into two sets, two double bonds or a single and a triple bond, then again the shape of the molecules will be a straight line.

carbon dioxide, CO_2

ethyne, C_2H_2

3 sets of electrons

This gives a **trigonal planar** shape as in boron trifluoride, BF_3.

To achieve the common 'octet' electron arrangement the structure must include a double bond, as in carbonyl chloride, $COCl_2$. Note, however, that in this case, although the molecule is still planar, the bonds' angles are not identical. Since the double bond consists of four electrons, it repels more strongly than the single bonds. As a result, the chlorine atoms are squeezed together slightly, reducing the Cl—C—Cl angle to less than 120°.

Continued on p. 95

Table 4.3 Common molecular shapes *(continued from p.94)*

4 sets of electrons

This is a very common arrangement, where the outer 'octet' of electrons is divided into four pairs, giving a **tetrahedral** shape as in the following examples:

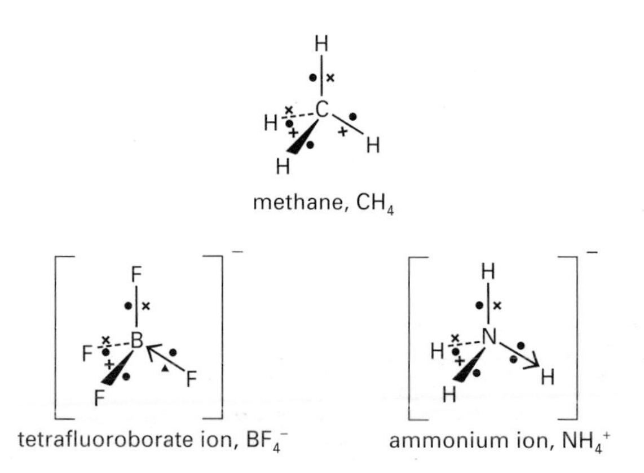

methane, CH_4

tetrafluoroborate ion, BF_4^- ammonium ion, NH_4^+

All of the bond angles in these molecules are equal to 109°. Notice that the co-ordinate bonds in BF_4^- and NH_4^+ do not affect the shapes of the ions – all of the bonding electron pairs are equivalent.

The effect of lone pairs on molecular shapes

One or more of the sets of electrons on the central atom may not be involved in bonding, as, for example, in ammonia and water:

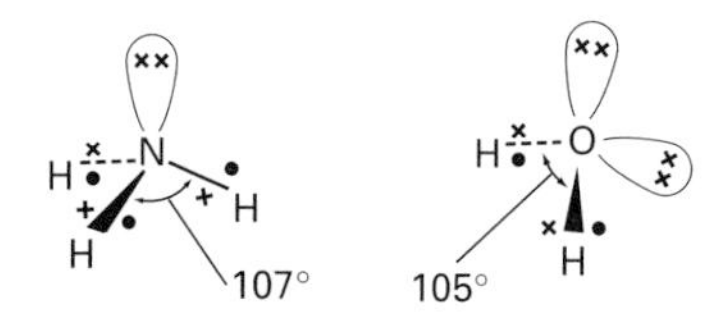

These are called **lone pairs** of electrons.

Whilst the presence of lone pairs must be taken into account when deciding upon the basic structure and 3-dimensional arrangement of the sets of electrons in space, the molecular shape describes the atomic arrangement only. Thus although the arrangement of electron sets is tetrahedral in each case, the ammonia molecule is described as **pyramidal** whilst the water molecule is **angular** or **V-shaped**.

The decrease in the bond angle on going from methane to ammonia to water is caused by the progressive replacement of bonding pairs on the central atom by non-bonding lone pairs of electrons. Whereas bond pairs are shared between two atoms, lone pairs occupy orbitals on a single atom. The lone pairs on an atom are therefore closer to the nucleus and much closer together than sets of bonding electrons. Lone pairs repel more strongly and 'squeeze' the bonds closer together.

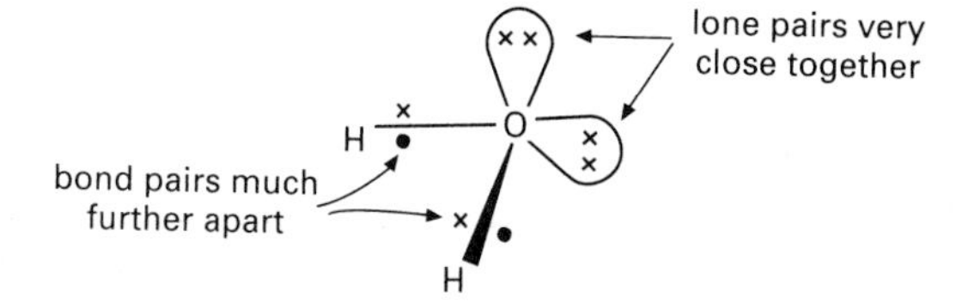

Thus, the relative strengths of the repulsive forces are as follows:

lone pair–lone pair > lone pair–bond pair > bond pair–bond pair

Atoms in the period lithium to fluorine can only accommodate a maximum of eight electrons in their valency shell. This is equivalent to four sets of electrons, at most, and their molecules are limited to the shapes covered so far.

5 sets of electrons

Five bonding electron pairs are arranged in a **trigonal pyramid** as in phosphorus pentafluoride, PF_5:

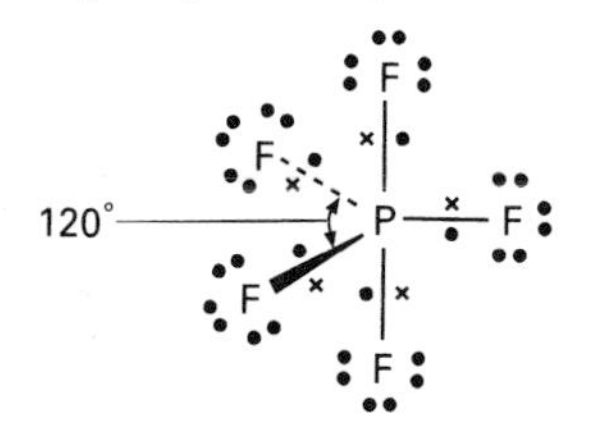

Again, one or more of the sets of electrons may be a lone pair, for example in sulphur tetrafluoride, SF_4:

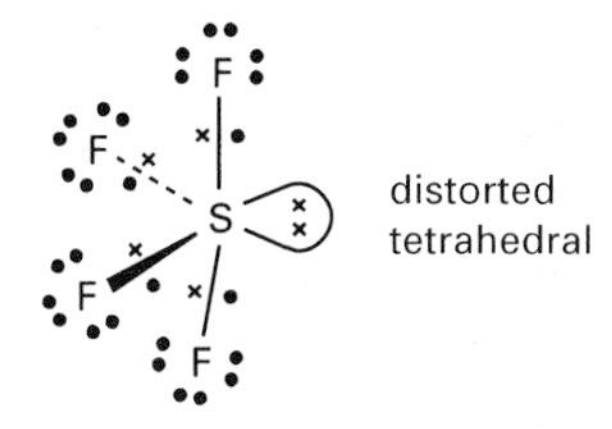

6 sets of electrons

Six bonding pairs give an octahedral shape, for example in sulphur hexafluoride:

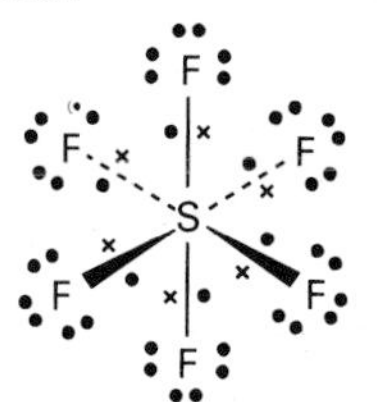

All the positions in this structure are equivalent and all the bond angles are 90°.

Inclusion of lone pairs gives the following molecular shapes.

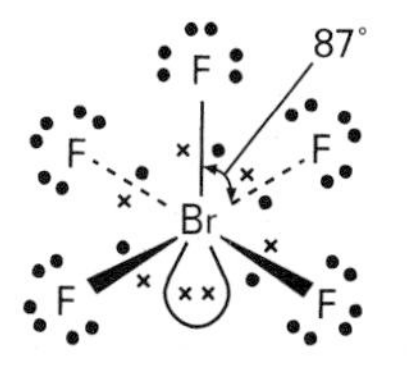

BrF_5, square pyramidal

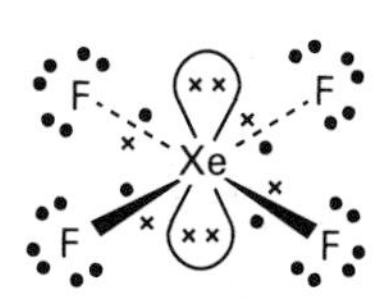

XeF_4, square planar

4.5 Polarisation of covalent bonds

The electrons in a covalent bond joining two identical atoms are equally shared. On average, they will be midway between the two nuclei and neither atom will develop a permanent charge, as is the case, for example, in the chlorine molecule:

Cl—Cl

If, however, the two atoms are dissimilar then the bonding electrons will probably not be equally shared. The power of an atom in a covalent bond to attract the bonding electron cloud to itself is given by its **electronegativity**. The greater its electronegativity, the greater the atom's ability to attract the bonding electrons. This concept was first proposed by Linus Pauling, an American scientist, in 1932. He proposed a scale of electronegativity values which are still widely accepted and shown in Table 4.4.

Table 4.4 Electronegativities of selected elements (according to Pauling).

					H	He											
					2.1												
Li	Be											B	C	N	O	F	Ne
1.0	1.5											2.0	2.5	3.0	3.5	4.0	
Na	Mg											Al	Si	P	S	Cl	Ar
0.9	1.2											1.5	1.8	2.1	2.5	3.0	
K	Ca	Sc	Ti	V	Cr	Mn	Fe	Co	Ni	Cu	Zn	Ga	Ge	As	Se	Br	Kr
0.8	1.0	1.3	1.5	1.6	1.6	1.5	1.8	1.8	1.8	1.9	1.6	1.6	1.8	2.0	2.4	2.8	
Rb	Sr	Y	Zr	Nb	Mo	Tc	Ru	Rh	Pd	Ag	Cd	In	Sn	Sb	Te	I	Xe
0.8	1.0	1.2	1.4	1.6	1.8	1.9	2.2	2.2	2.2	1.9	1.7	1.7	1.8	1.9	2.1	2.5	
Cs	Ba	La	Hf	Ta	W	Re	Os	Ir	Pt	Au	Hg	Tl	Pb	Bi	Po	At	Rn
0.7	0.9	1.1	1.3	1.5	1.7	1.9	2.2	2.2	2.2	2.4	1.9	1.8	1.8	1.9	2.0	2.2	
Fr	Ra	Ac															
0.7	0.9	1.1															

Electronegativity generally increases on passing across a period, owing to the increasing nuclear charge and decreasing atomic radius, for example:

	Li	**B**	**N**	**F**
Protons in the nucleus	3	5	7	9
Atomic radius/nm	0.123	0.080	0.074	0.072
Electronegativity	1.0	2.0	3.0	4.0

Electronegativity decreases on passing down a group since the combined effects of the increasing atomic size and the screening effect of inner electrons outweigh the effect of increasing the number of protons in the nucleus. The overall effect is to reduce the attraction of the nucleus for electrons.

In any covalent bond, the atom with the greater electronegativity will attract the bonding electrons closer to itself, so causing the bond to be **polarised**, for example:

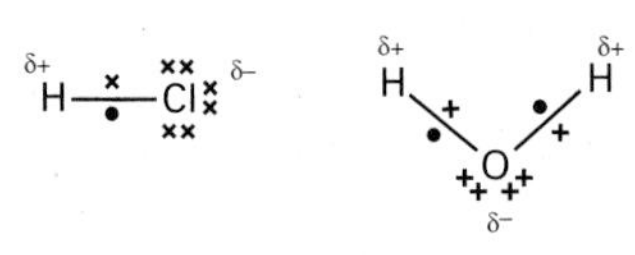

where δ indicates a partial charge and the electronegativities are:

H 2.1, Cl 3.0 and O 3.5

Table 4.5 Percentage of ionic character in bonds expressed in terms of the electronegativity difference between the elements.

Electronegativity difference	percentage ionic character
0	0
0.5	6
1.0	22
1.5	43
2.0	63
2.5	79
3.0	89

Effectively, this introduces partial ionic character into the covalent bond (see Figure 4.1). Pauling estimated the percentage of ionic character in terms of the difference in electronegativity of the atoms in the bond and this is detailed in Table 4.5. Where bond polarisation occurs, there is always some degree of **intermolecular attraction** and, as we shall see in the next section, this will have an effect on the compound's physical properties, such as melting and boiling points.

Activity 4.4

1 Use the values in Table 4.5 to sketch a graph of percentage ionic character (vertical axis) against electronegativity difference (horizontal axis).

2 Now, estimate the percentage of ionic character in the bonds formed between the following pairs of atoms:

a) O, F **b)** B, F **c)** C, Cl **d)** P, H
e) C, O **f)** O, H **g)** H, Br **h)** Al, Cl

3 The polarity of the H—F bond can be worked out from electronegativity values, as shown below:

H	F	$\overset{\delta+}{H}$—$\overset{\delta-}{F}$
2.1	4.0	

Show the polarities in the covalent bonds formed by the atoms in question 2, parts c)–g).

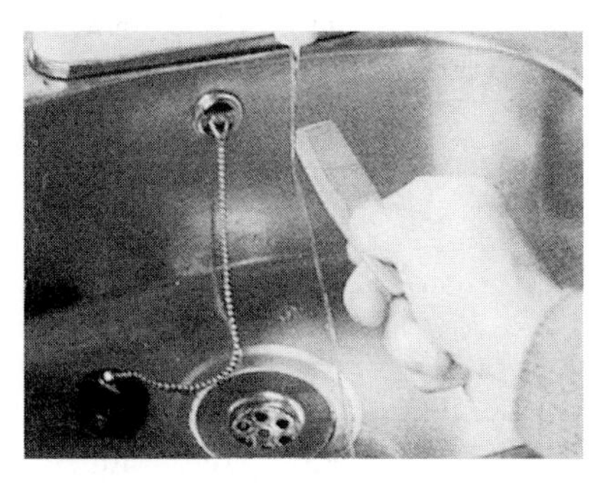

▲ Testing the polarity of a liquid. Why is the stream of water attracted by the charged comb?

4.6 The forces between covalent molecules

Polar or non-polar molecules

If you vigorously rub a dry plastic comb on some polyester fabric, the comb takes on a small electrical charge. When the comb is brought very close to a thin stream of water, from a tap or burette, the stream is deflected towards it. The results of a similar experiment using other covalent liquids, contained in a burette, are shown in Table 4.6.

Table 4.6 The effects of bringing a charged rod near to a thin stream of various covalent liquids.

liquids deflected	liquids not deflected
water	
propanone	
trichloromethane	tetrachloromethane
cyclohexene	cyclohexane

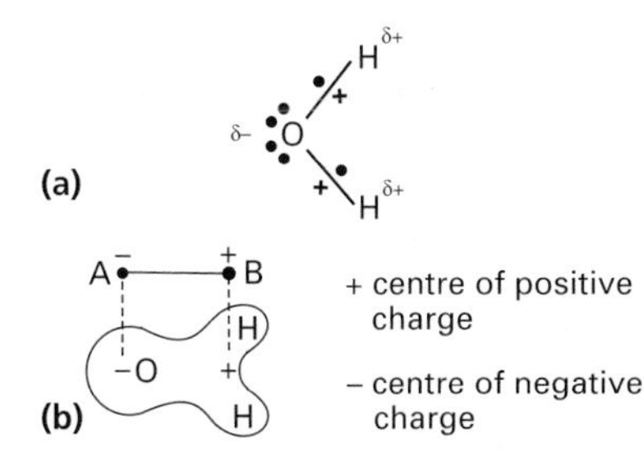

Figure 4.11 ▲
(a) An oxygen atom is much more electronegative than a hydrogen atom, thus each O—H bond in a water molecule is polarised
(b) Since the centres of positive and negative charge in a water molecule do not coincide, the molecule contains a permanent dipole, that is a charge difference separated by a small distance

In water, the O — H bonds are polarised because the electronegativities of hydrogen (2.1.) and oxygen (3.5.) atoms are very different (Figure 4.11a). Now, consider the centres of the positive and negative charges in the molecule, bearing in mind the lone pairs on oxygen. The centre of negative charge is located on the oxygen whilst the centre of positive charge is midway between the hydrogen nuclei (Figure 4.11b). So, the water molecule contains a **dipole**, which is a difference in charge separated by a small distance and this explains why it is deflected by an electric charge. Since the charges in the water molecule are permanently separated, it is said to contain a **permanent dipole**

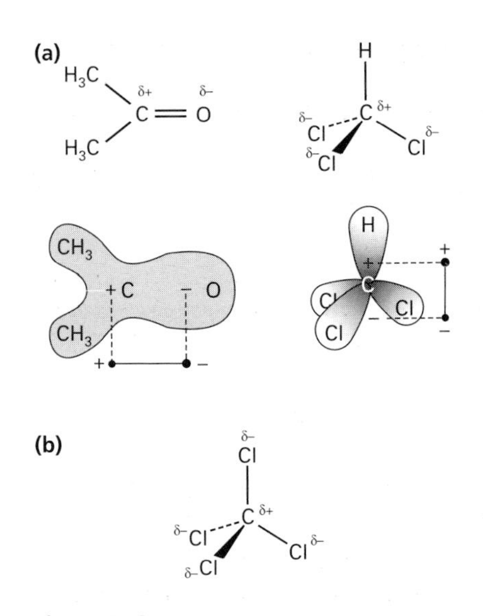

Figure 4.12
(a) Propanone and trichloromethane are polar liquids. The centres of the positive and negative charge in their molecules do not coincide.
(b) In tetrachloromethane, a highly symmetrical molecule, the centres of positive and negative charge coincide on the carbon atom. Consequently, tetrachloromethane is a non-polar liquid even though the individual C—Cl bonds are polarised

and is termed a **polar** molecule. Propanone and trichloromethane also have polar molecules as shown in Figure 4.12a.

Neither tetrachloromethane nor cyclohexane are deflected by an electric charge. They are both **non-polar** liquids. Unlike trichloromethane, tetrachloromethane is overall non-polar even though the individual C — Cl bonds are polarised. The reason for this is that CCl_4 is a highly symmetrical molecule and, as a result, the centres of the positive and negative charge coincide on the central carbon atom (Figure 4.12b), and so there is no permanent dipole in the molecule. Similarly, cyclohexane is symmetrical and the centres of positive and negative charge coincide. Some other molecules which are non-polar as a result of their symmetry are carbon dioxide and boron trichloride.

Intermolecular forces

Look at Table 4.7. This shows the boiling points of some covalent compounds which contain simple molecules. Can you say which are polar and which are non-polar? Now compare the boiling points. Is there a relationship between the boiling point and the polarity of the molecules?

Table 4.7 The boiling points of some covalent compounds which contain simple molecules. Which are polar and which non-polar? Is there a relationship between the boiling point and the polarity of the molecule?

Compound	formula	relative molecular mass, M_r	boiling point/°C
propanone	$(CH_3)_2C=O$	58	53
butane	$CH_3CH_2CH_2CH_3$	58	0
water	H_2O	18	100
methane	CH_4	16	−161

You should conclude that the non-polar compounds have lower boiling points than polar compounds of similar molecular mass. Generally speaking, when we melt, or vaporise, a simple molecular compound, the covalent bonds between atoms within the molecules are *not* broken. Instead, the energy supplied is used to overcome the attractive forces between molecules. It takes more energy to vaporise the polar compounds because their intermolecular forces are stronger than those between the non-polar molecules. Although intermolecular forces are usually much weaker than true covalent bonds, their presence can have a significant effect on the physical properties of covalent compounds.

There are two types of intermolecular forces, **van der Waals' forces** and **permanent dipole–permanent dipole attractions**.

Van der Waals' forces

Neon, one of the noble gases, is used in red fluorescent lights in shop windows and advertising displays. The noble gases have very low boiling points, for example neon boils at −246°C. Even though they exist as separate single atoms at room temperature, and are extremely unreactive, the noble gases can be liquefied at high pressure and low temperature, proving that there must be attractive forces between the atoms. Many simple molecular compounds display similar properties, for example butane is a gas at room temperature but it can be stored as a liquid under pressure for use as camping gas. So what is the nature of these forces between simple molecules?

A molecule's electron cloud is in a state of continual motion. At a given moment in time, the electron cloud may not be evenly spread over the molecule. The resulting temporary dipole induces dipoles in the neighbouring molecules and these then attract each other (Figure 4.13a). In

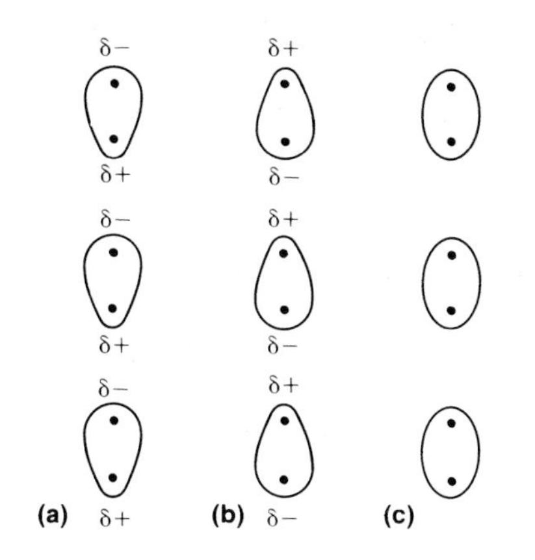

Figure 4.13 ▲
(a) and **(b)** The unequal sharing of the electron cloud in a non-polar diatomic molecule at two instants in time.
(c) Overall, the cloud is evenly distributed, that is, there is no permanent dipole

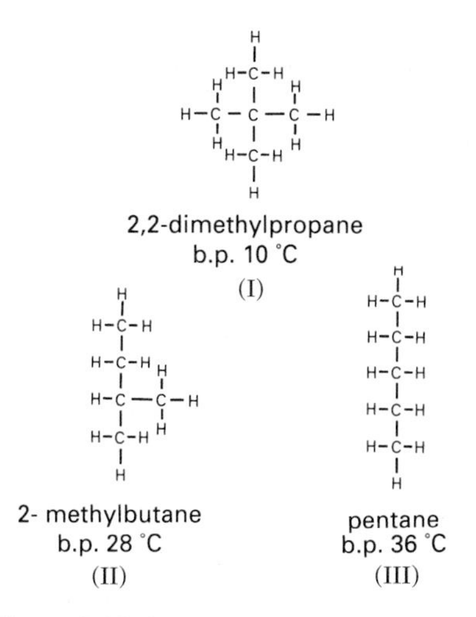

Figure 4.14 ▲
Branched alkanes have lower boiling points than straight-chain isomers

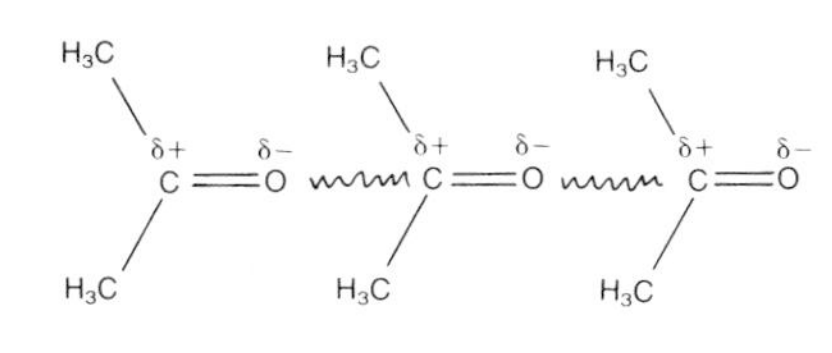

Figure 4.15 ▲
Intermolecular permanant dipole–permanent dipole attractions between propanone molecules

the next instant, the direction of polarity changes but the intermolecular attraction remains (Figure 4.13b). Van der Waals' forces result from these **induced dipole–dipole attractions**. Remember, however, that on average, the electron density remains evenly distributed within each molecule and the molecules do not have a permanent dipole (Figure 4.13c). Van der Waals' forces may be thought of as existing between closely positioned non-polar molecules.

Van der Waals' forces will be stronger if the molecules have large electron clouds as the nuclei hold on to their electrons less strongly, and these will be more easily polarised. For example, there is an increase in the boiling points of the noble gases as the atoms get larger:

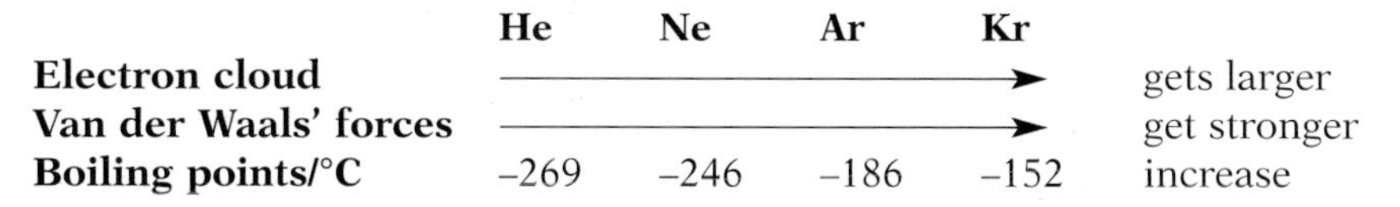

	He	Ne	Ar	Kr	
Electron cloud	⟶				gets larger
Van der Waals' forces	⟶				get stronger
Boiling points/°C	−269	−246	−186	−152	increase

The strength of the van der Waals' forces depends on the shape of the electron cloud as well as its size. The closer that the molecules can get to each other, the stronger the attraction. Consider, for example, the three alkanes of formula C_5H_{12}, as shown in Figure 4.14.

	I	II	III	
branching	⟶			decreases
electron cloud	⟶			gets less spherical
molecules	⟶			can get closer together
van der Waals' forces	⟶			increase
so boiling points/°C	10	28	36	increase

Van der Waals' forces are usually extremely weak, typically being about 1–10% of the strength of a covalent bond. However, as molecular size increases, there will be more points of contact and the van der Waals' forces can make a significant contribution to the strength of the material. This is the case in polyethene, a polymer which is made up of very long, straight-chain hydrocarbon molecules, often with 5000 carbons joined together. As a result, the van der Waals' forces are appreciable and polyethene bags and containers are strong, yet remain flexible.

Permanent dipole–permanent dipole attractions

As we have seen, polar molecules have an unsymmetrical shape and contain a permanent dipole. The dipoles in neighbouring molecules are attracted to each other forming **permanent dipole–permanent dipole** attractions, as in propanone (Figure 4.15).

Figure 4.16 shows that the boiling points of the covalent hydrides of any particular periodic group generally increase quite smoothly with relative molecular mass, as shown. This implies that van der Waals' forces are the main forces of attraction and these increase in strength as the electron cloud gets larger. You should notice, however, that the boiling points of ammonia, water and hydrogen fluoride do not fit this pattern, being surprisingly high. Clearly, there is a much stronger intermolecular attraction where the hydrogen atom is bonded to a highly electronegative atom like nitrogen, oxygen or fluorine. This particular example of permanent dipole–permanent dipole attraction is known as **hydrogen bonding**. In hydrogen fluoride, for example, zig-zag chains of molecules are held together via hydrogen bonds (Figure 4.17). Hydrogen bonds also explain the high boiling point of methanoic acid, which is not accounted for by van der Waals' forces due to small molecular size (Figure 4.18).

It is important to stress that hydrogen bonds are intermolecular forces, not chemical bonds. They are about 5–10% as strong as covalent bonds and are easily disrupted by physical methods. For example, the surface tension of water is lowered by adding a soap because of a reduction in the number of hydrogen bonds between water molecules.

There are numerous examples of hydrogen bonding in chemistry and we shall refer to some of these later in this book.

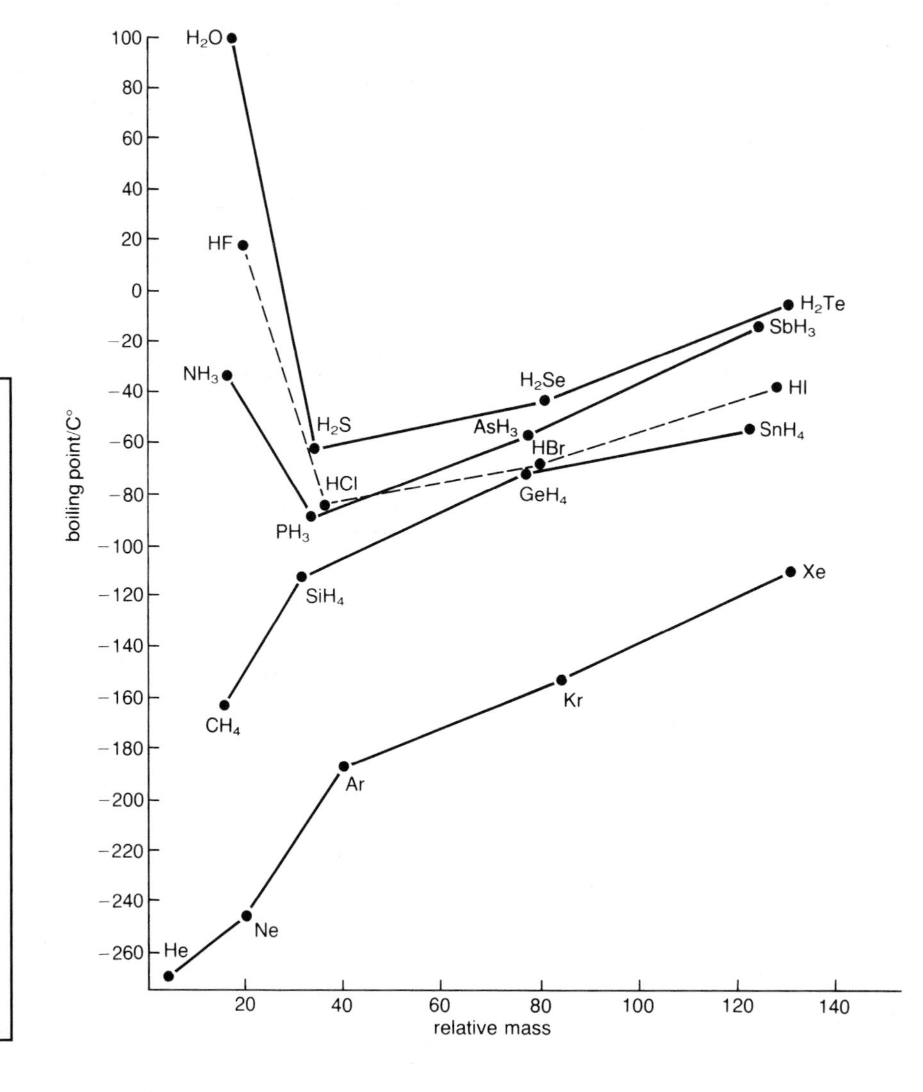

Figure 4.16 ▶ Variation in the boiling point of some p block hydrides – why are the boiling points of H_2O, NH_3 and HF so high?

Van der Waals' forces

Johanne van der Waals (1837–1923), a Dutch physicist, developed an equation which explained the behaviour of gases as they condensed. He studied the weak attractive forces between molecules and these were called van der Waals' forces in his honour. There is some inconsistency in the use of the term 'van der Waals' forces' as it is often used to describe all dipole–dipole attractions, whether they are permanent or induced.

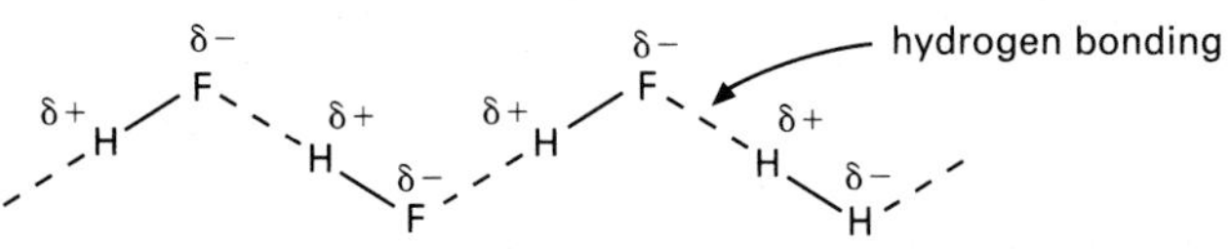

Figure 4.17 ▲ In the liquid state, hydrogen bonds hold hydrogen fluoride molecules together in a 'zig-zag' arrangement. The bond enthalpy of these hydogen bonds is about 28 kJ mol⁻¹. How many times weaker is the hydrogen bond than the covalent bond?

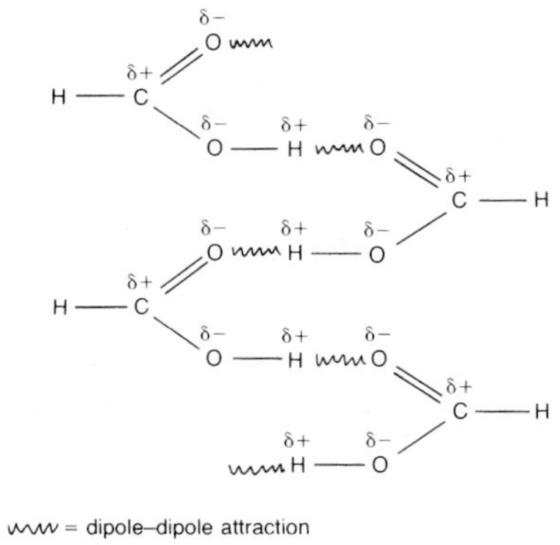

∿∿∿ = dipole–dipole attraction

Figure 4.18 ▶ Even though their relative molecular masses are similar, methanoic acid (M_r = 46, b.p. = 101°C) has a much higher boiling point than propane (M_r = 44, b.p. = −42°C). This occurs because the intermolecular forces in methanoic acid (hydrogen bonds) are much stronger than those in propane (van der Waals' forces)

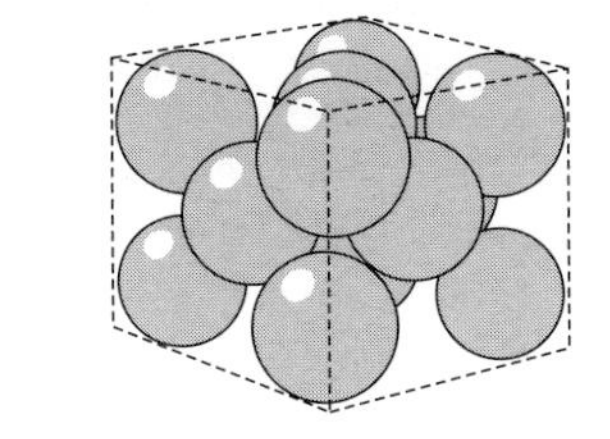

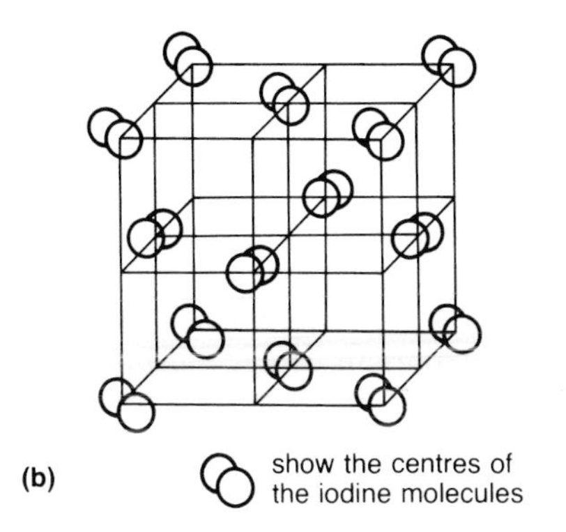

Figure 4.19 ▲
Simple molecular solid structures for **(a)** argon and **(b)** iodine. Both have a face-centred cubic unit cell

Focus 4d

1. **Electronegativity** is the power of an atom in a covalent bond to attract the bonding electron cloud to itself. The greater the electronegativity, the greater the attracting power. In the periodic table, electronegativity *increases* across a period and *decreases* down a group.
2. Most covalent bonds involving dissimilar atoms will be **polarised**, that is contain a **permanent dipole**. The more electronegative atom will have a slight negative charge, represented by δ–, the less electronegative atom will be slightly positively charged, δ+.
3. There are two types of intermolecular forces, **van der Waals' forces** and **permanent dipole–permanent dipole attractions**. Van der Waals' forces result from **induced dipole–induced dipole attractions** between closely positioned molecules; they will be strongest in compounds whose molecules have large, linear shaped, electron clouds. Permanent dipole–permanent dipole attractions occur between molecules which have an unsymmetrical shape and contain a permanent dipole. When one of the atoms in the permanent dipole is hydrogen, this type of permanent dipole–permanent dipole attraction intermolecular force is termed **hydrogen bonding**. Hydrogen bonds will be strongest when the atom is bonded to an N, O or F atom.
4. Intermolecular forces are much weaker than covalent bonds. Permanent dipole–permanent dipole interactions tend to be stronger than van der Waals' forces.

4.7 Simple molecular structures

In the solid state, most non-metal elements and their compounds consist of a lattice of covalently bonded molecules held together by weak intermolecular forces. These **simple molecular structures** may be broadly divided into two groups, depending on the type of intermolecular forces which are present.

Simple molecular solids made of non-polar molecules are held together by van der Waals' forces. These solids exist in a variety of lattice structures, for example, argon and iodine are said to have a **face-centred cubic structure** (Figure 4.19). This means that the repeating unit in the lattice, known as the **unit cell**, is a cube and that there is an atom, or molecule, at each corner and at the centre of each face.

Simple molecular solids made of polar molecules are held together by permanent dipole–permanent dipole attractions, for example as in the sulphur dioxide molecule (Figure 4.20). Hydrogen bonding is common in simple molecular solids, of which ice is an important example (Figure 4.21).

Table 4.8 lists the physical properties of some simple molecular structures. Can you see any general pattern in the values?

You should have concluded that simple molecular solids, such as argon, water, sulphur dioxide, methane and iodine:

- tend to be fairly soft materials
- have low melting points and boiling points
- are poor conductors of electricity
- are soluble in polar solvents if their structure is made up of polar molecules (e.g. sulphur dioxide is very soluble in water, ice will dissolve in ethanol, C_2H_5OH)
- dissolve in non-polar solvents if their structure is made up of non-polar molecules (e.g. iodine dissolves in hexane).

These properties are consistent with our model for the structure of a simple molecular solid. A lattice of atoms, or molecules, is held together by weak

Icebergs float because ice, having a much less compact structure than water, is less dense. However the 3-D hydrogen-bonded structure of ice is very rigid. On 14th April 1912, the luxury passenger liner 'Titanic', on her maiden voyage, struck a massive iceberg south of Newfoundland, and sank with the loss of 1500 lives. The ship was thought to be unsinkable because it had 16 watertight buoyancy compartments, yet it sank in only three hours. It transpired that the Titanic was steaming too fast in dangerous waters and that insufficient lifeboats had been provided for the 2200 crew and passengers. The disaster was made worse because a nearby ship, the Californian, had not heard the distress signals because its radio officer was off-duty and asleep.

Table 4.8 Physical properties of some substances with simple molecular solid structures

	argon	water	sulphur dioxide	methane	iodine
molecular formula	Ar	H_2O	SO_2	CH_4	I_2
physical state (25°C, 1 atm)	gas	liquid	gas	gas	solid
hardness of solids	←		soft, in comparison to other types of solid		→
melting point/°C	–189	0	–75	–182	114
boiling point/°C	–186	100	–10	–161	183
electrical conductivity	←		non-conductors		→
solubility in polar and non-polar solvents	very slightly	only in polar	high in polar	←high in non-polar	→

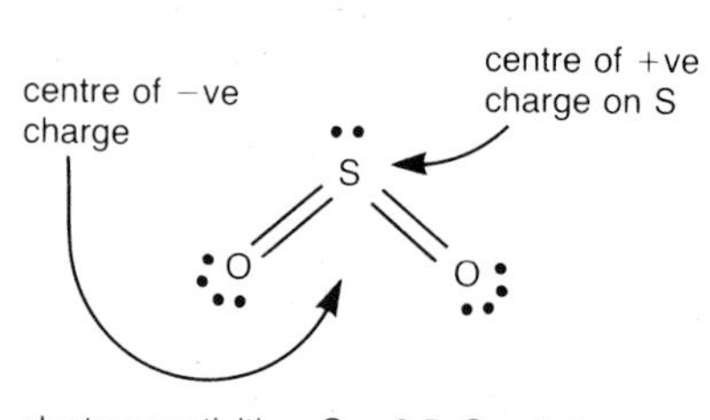

Figure 4.20 ▲
It is worth noting that sulphur dioxide and carbon dioxide (Figure 4.8) have similar formulae but different structures. As there is a lone pair of electrons on the sulphur atoms, the SO_2 molecule is V-shaped and it possesses a permanent dipole because the centres of positive and negative charge do not coincide. The CO_2 molecule is linear and symmetrical, however, making it non-polar

intermolecular forces (van der Waals' forces or permanent dipole–permanent dipole attractions). Only a small amount of energy is necessary to break up the lattice. They are usually soft and their melting points and boiling points will be low. Simple molecular structures are electrical non-conductors because they do not contain mobile ions, or electrons, which can carry the charge. In practice, many simple molecular structures are held together by a combination of van der Waals' and dipole–dipole attractions, for example in phenol as shown in Figure 4.22.

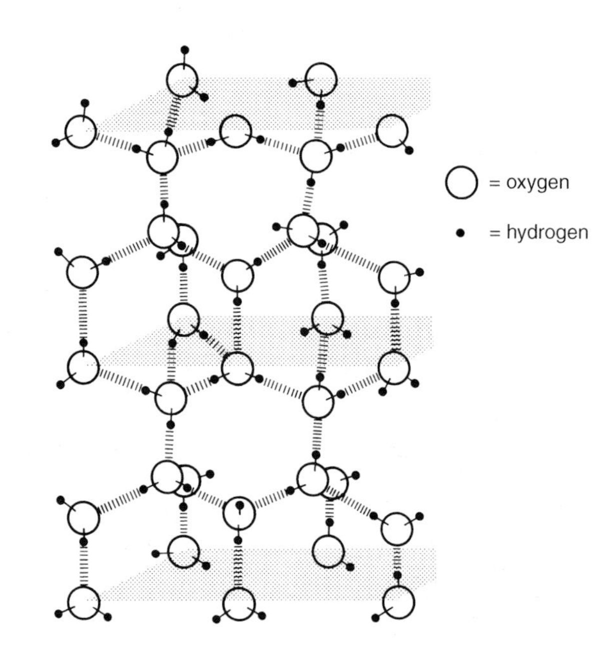

Figure 4.21 ▶
In ice, a lattice of water molecules is held together by hydrogen bonds. Each oxygen atom is surrounded tetrahedrally by four hydrogen atoms (two covalently bonded and two hydrogen bonded). Compared with the molecules in liquid water, those in ice are much less closely packed together – thus ice floats on water

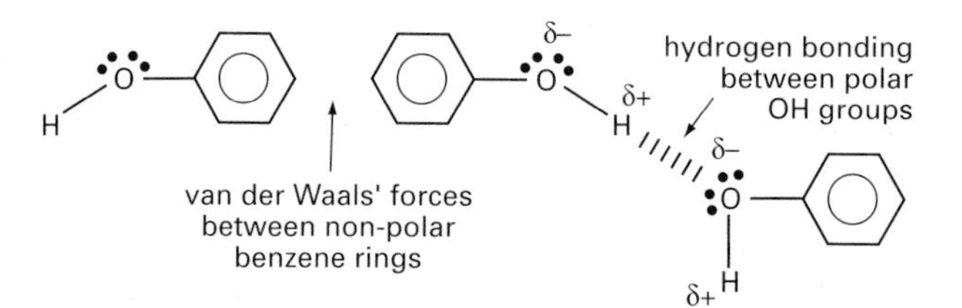

Figure 4.22 ▶
In phenol, a non-polar benzene ring is joined to a polar —OH group. This explains why phenol is soluble in polar and non-polar solvents

The solubility of solids in liquids

The solubility of solids in liquids is worth considering further. When a solution is formed, the particles from the solid mix freely with those from the liquid. The process of **dissolution** may be thought of as occurring in three stages:

1. The solid's lattice must be broken. This process will be endothermic, that is enthalpy is absorbed.
2. The intermolecular forces in the liquid, whether van der Waals' forces or dipole–dipole attractions, must be disrupted to some extent. Again, this is an endothermic process, as attractive forces are being broken and this requires energy.
3. Now if 1 and 2 above were the only processes involved, dissolution would always be an endothermic process. This is not always the case, for example when a pellet of sodium hydroxide is placed in 2 cm^3 of water, there is a considerable temperature rise. Thus, enthalpy must be released when new bonds are formed between the particles in the solid and the liquid.

Generally speaking, we find that a solid is more likely to dissolve in a liquid if the overall enthalpy change is exothermic. Putting it another way, high solubility is more likely if:

(strength of the attraction between solid and liquid particles in the solution) > (combined strengths of the attractions between particles in the pure solid and between particles in the pure liquid)

Although the thermodynamics of solubility are rather more complex than this (involving a consideration of entropy changes as described in Chapter 28), this simple 'rule of thumb' often helps us to explain patterns in solubility. For example, why is iodine soluble in hexane but not in water?

Iodine and hexane are non-polar substances. When mixed together a solution is formed because:

(strength of the iodine/hexane attractions in solution (van der Waals' forces)) > (combined strength of the attractions in iodine solid (van der Waals' forces) and hexane liquid (van der Waals' forces))

Also, water is a polar solvent, with its molecules engaged in extensive hydrogen bonding. When mixed with non-polar iodine molecules, nearly all of the water molecules continue to hydrogen bond with each other. Thus, the resulting iodine/water attractions are extremely weak in comparison to the combined strength of the hydrogen bonds in water and the van der Waals' forces in iodine. Consequently, iodine is virtually insoluble in water.

4.8 Giant molecular structures

A **giant molecular** structure may be thought of as an enormous molecule made up of covalently bonded atoms. They are also known as **macromolecular** or **giant covalent** structures. Well-known examples of giant molecular solids are diamond and graphite. These are the **allotropes** of carbon, that is, they are made up of the same element but have different structures.

Diamond is an extremely hard solid because the carbon atoms are covalently bonded in a very rigid tetrahedral arrangement (Figure 4.23a).

Graphite, on the other hand, consists of carbon atoms bonded in a hexagonal pattern (Figure 4.23b). *Within* the layers, the carbon–carbon length is

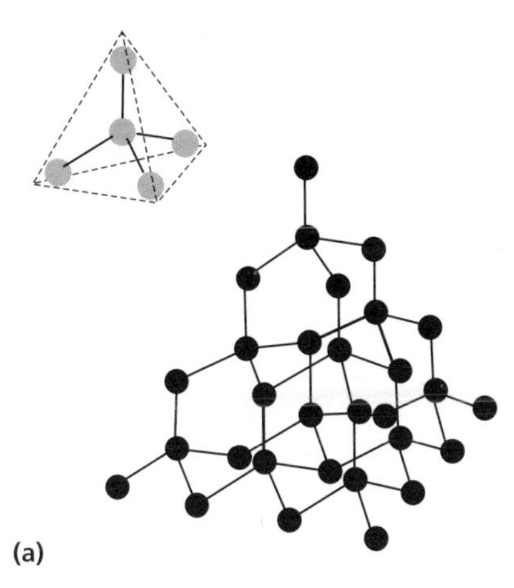

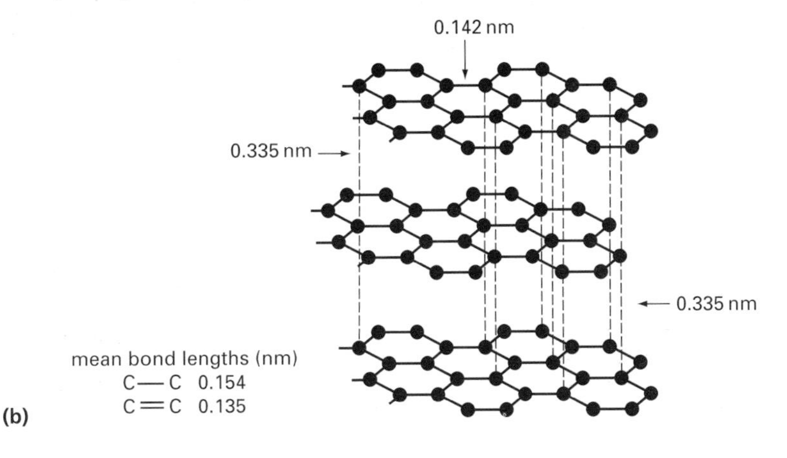

Figure 4.23
Diamond and graphite, two allotropes of carbon, are giant molecular solids.
▲ (a) In diamond, the bond angles are 109° and each carbon atom is said to have a **co-ordination number** of four because there are four neighbouring carbons near to it.
(b) In graphite, there is a trigonal planar arrangement of covalent bonds around each carbon and the bond angles are 120° ▶

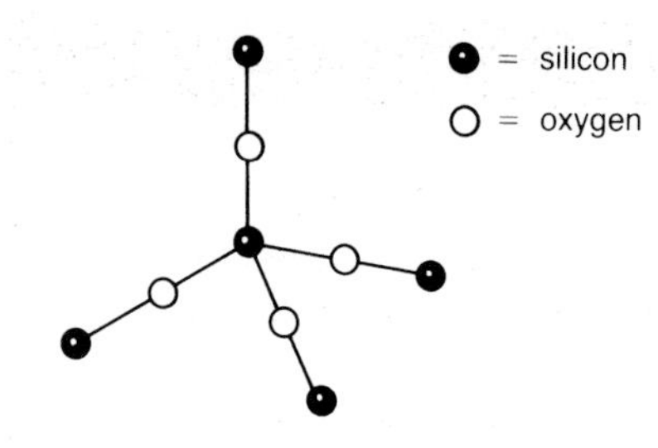

Figure 4.24
The structure of quartz

in between that of a single and a double carbon–carbon bond, suggesting that there is partial double bond character between atoms in the layers. However, the distance *between* layers is much greater than the length of a single carbon–carbon bond within the layers and this is reflected in the weakness of the attractive forces between layers. Within a layer, each carbon atom uses three of its valence electrons to form strong covalent bonds whilst the fourth is free to move over the layer, that is it is **delocalised**. The delocalised electron cloud gives rise to van der Waals' forces (induced dipole–dipole attractions) between the layers. Due to the weakness of these attractive forces, the layers are able to easily slide over each other, so graphite is soft and feels slippery to the touch, properties which make it useful as a lubricant.

Boron and silicon are other elements which exist as giant molecular solids. However, the atoms in a giant molecular solid need not be identical and **silica**, SiO_2, adopts no less than six different giant molecular crystal structures. One of these, quartz, has a structure similar to diamond in which SiO_4 tetrahedra are linked together by Si — O — Si bonds (Figure 4.24).

The properties of some giant molecular solids are shown in Table 4.9. Can you detect any patterns in the values? How does the data compare with that for simple molecular solids (Table 4.8)?

Table 4.9 Physical properties of some macromolecular solids. (Graphite **sublimes** at 3727°C, i.e. changes from solid to gas without passing through a liquid phase.)

	boron	silica	diamond	graphite
physical state (25°C, 1 atm)	solid	solid	solid	solid
hardness	hard	hard	very hard	fairly soft
melting point/°C	2027	1723	3550	sublimes
boiling point/°C	3927	2230	4827	sublimes
electrical conductivity	← non-conductors →			good
solubility	insoluble in polar and non-polar solvents			

You should have concluded that, in general, giant molecular solids, such as boron, silica and diamond:

- tend to be hard materials which are virtually insoluble
- have very high melting points and boiling points
- are poor conductors of electricity.

Our model of a giant molecular structure accounts well for these properties. Since the atoms are held together by strong covalent bonds, a considerable amount of energy is required to break apart the lattice. Hence, most giant molecular solids are hard, insoluble materials with very high melting and boiling points. As the valence electrons are firmly held in localised covalent bonds, they are poor conductors of electricity.

You will have noted that graphite is something of an exception. Its softness and electrical conductivity result from the weak forces between layers of carbon atoms and the fact that electrons between layers are delocalised.

Silicates, aluminosilicates and ceramics

We saw earlier that the structure of **quartz** – a form of silicon(IV) oxide – is a giant lattice of SiO_4 tetrahedra in which every oxygen atom is at the corner of two adjacent tetrahedra. All of the oxygen atoms are involved in Si — O — Si bonding but this is not always the case.

Chain silicates contain very long chains of SiO_4 tetrahedra (see Figure 4.25a). In each tetrahedron, two oxygen atoms are at the corners of adjacent tetrahedra whilst the other two oxygen atoms are not attached and are negatively charged and, therefore, are able to interact with positive ions. A parallel arrangement of these long chain silicate ions is held together by various types of cations (positive ions) located between the negatively charged oxygens along the chain. **Asbestos** is a chain silicate which contains magnesium, calcium or iron ions. Asbestos is fire resistant and is a poor conductor of heat, making it a valuable material in building construction and the manufacture of friction products, such as brake linings. Great care must be taken with asbestos as exposure to fibres and dust can cause asbestosis, a lung disease, and also lung and chest cancers.

In **sheet silicates**, the SiO_4 tetrahedra are arranged in sheet structures (see Figure 4.25b). Three of the four oxygens in each tetrahedra are bonded between silicon atoms and the fourth oxygen is negatively charged. The sheets are bonded together by cations which lie between them and the colour of the material depends on the type of cations which are present. **Talc**, a commonly found sheet silicate containing magnesium ions, is a soft, slippery material which is often used in soaps, tailor's chalk and talcum powder.

Many areas of Britain have **clay** soils. These soils can be recognised by their brownish colour and the fact that they are heavy and clinging when wet but become hard in dry weather. Clays are sheet silicates in which up to half of the silicon atoms have been replaced by aluminium atoms and so they are called **aluminosilicates**. When the clay is wet, water molecules are held between the sheets and the material is pliable and can be worked into various shapes. If the clay is fired (heated at a high temperature), water is expelled to produce a three dimensional macromolecular structure similar to quartz (Figure 4.24). **Ceramics**, from the Greek word *keramos*, which means potter's clay, is the general name given to materials which are fashioned from wet, pliable earthy materials and then fired in kilns. In addition to common ceramic products, such as bricks, china and pottery, the poor electrical and heat conductivity of ceramics make them suitable for use as electrical insulators and in the lining of electric furnaces. There is also a newly developing range of ceramics, containing various additives, which are designed for specific 'hi-tech' applications such as the heat shields in spacecraft, replacement for the metal parts in jet and car engines and the armour protection on military vehicles.

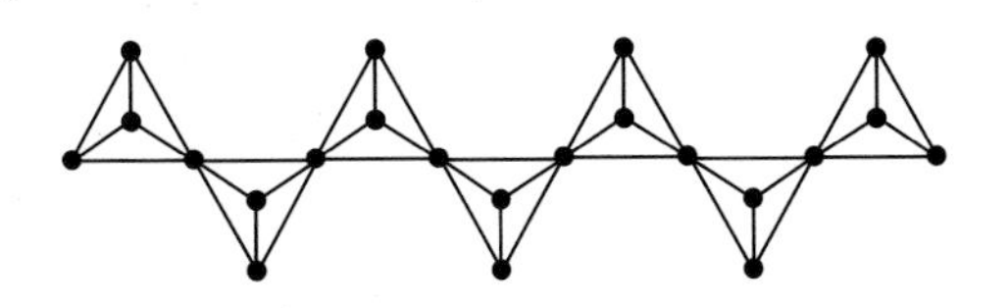

(a) a chain silicate

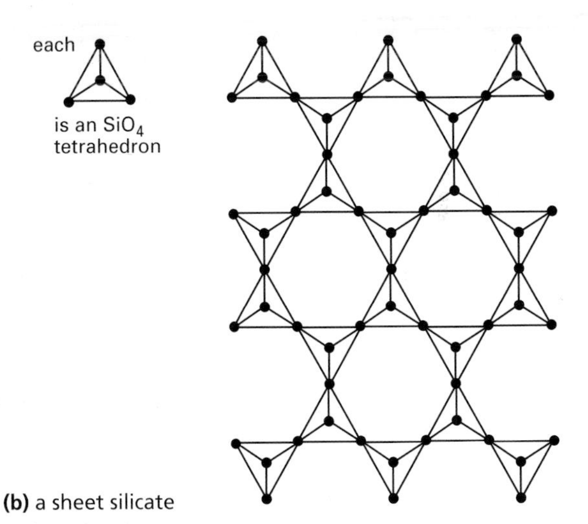

(b) a sheet silicate

Figure 4.25
Chain and sheet silicates both contain SiO_4 tetrahedra but their arrangement is different

Activity 4.5

1 **a)** Draw a dot and cross diagram to show the electron arrangement in two neighbouring SiO_4 tetrahedra joined in a chain silicate.
b) What is the charge on each tetrahedron and which atoms carry this charge?
c) Explain, in structural terms, why fibres can be pulled off a piece of asbestos.

2 **a)** Use a dot and cross diagram to work out the charge on each SiO_4 tetrahedron in sheet silicate. Which atom carries this charge?
b) Explain, in structural terms, why:
i) thin leaves of solid can be split off a large piece of talc
ii) talcum powder feels slippery to the touch.

3 Why is house buildings insurance likely to be more expensive if your house is built on a clay soil?

4.9 Giant ionic structures

Of all the elements in the periodic table, the noble gases are the most reluctant to form chemical bonds, suggesting that their electron arrangements are particularly stable. Many atoms achieve such an electron arrangement by losing, or gaining, electrons to form positive ions or negative ions, respectively.

	Na	+	Cl	$\longrightarrow$	Na^+	+	Cl^-
	$1s^22s^22p^63s^1$		$1s^22s^22p^63s^23p^5$		$1s^22s^22p^6$		$1s^22s^22p^63s^23p^6$
i.e.	$[Ne]3s^1$		$[Ne]3s^23p^5$		[Ne]		[Ar]

	Mg	+	2F	$\longrightarrow$	Mg^{2+}	+	$2F^-$
	$1s^22s^22p^63s^2$		$1s^22s^22p^5$		$1s^22s^22p^6$		$1s^22s^22p^6$
i.e.	$[Ne]3s^2$		$[He]2s^22p^5$		[Ne]		[Ne]

Whilst extremely common, the noble gas electronic structure is not the only stable arrangement found in ions, for example the following ions are stable:

Pb^{2+}	Fe^{3+}
$[Xe]5d^{10}6s^2$	$[Ar]3d^5$

In Figure 4.1b, we saw that the electron density between ions falls to zero. The mutual attraction of oppositely charged ions constitutes an **ionic** or **electrovalent bond** but it must be stressed that the crystal structure is a three-dimensional lattice and that each ion is mutually attracted to several surrounding ions of opposite charge. Thus, ionic bonding is non-directional, unlike covalent bonds where the bonding electrons are located directly between the bonded atoms.

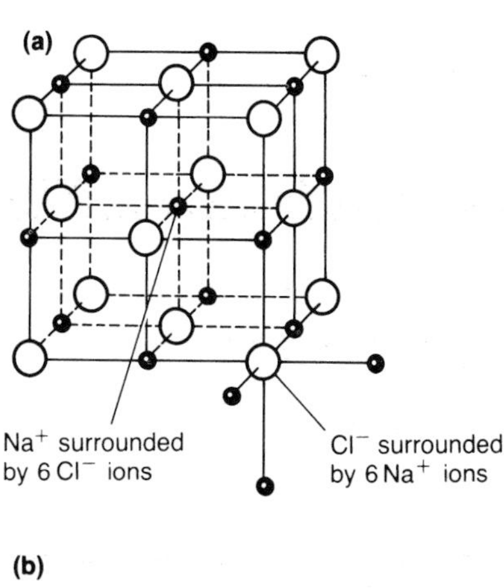

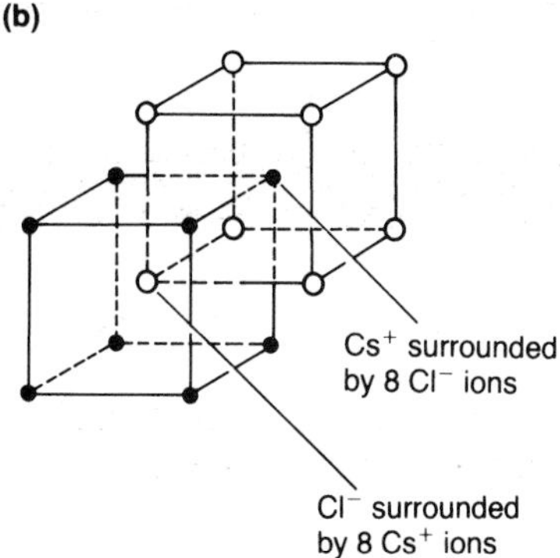

Figure 4.26
The crystal structures of **(a)** sodium chloride and **(b)** caesium chloride have cubic shapes. The centres of the ions are represented by the solid dots and circles

Activity 4.6

Scan through section 1.3, Ionic bonding, and then draw dot and cross diagrams to show the ionic bonding in the compounds below. Write the electronic configuration of each ion below its diagram (e.g. F^- will be $1s^22s^22p^6$). Note that you need show only the outer shells of electrons.

lithium fluoride, LiF	calcium oxide, CaO
sodium sulphide, Na_2S	magnesium chloride, $MgCl_2$

Cations and anions

Positive ions are often referred to as **cations** because they move towards the **cat**hode (negative electrode) when placed in an electric field. Negative ions move towards the **an**ode so they are termed **anions**.

Ionic structures

When ions come together to form an ionic lattice, the electrostatic attraction between oppositely charged ions is exactly balanced by the repulsion between the electron clouds of the ions. A single ionic crystal contains billions of ions, all arranged in regular repeating units, like those in Figure 4.26. Thus, ionic solids are said to have a 'giant' structure.

Each type of ionic structure is described by the **co-ordination numbers** of ions present. *The co-ordination number tells us how many oppositely charged ions surround a particular ion.* In a **6:6 lattice** each ion is surrounded by six oppositely charged ions, for example, sodium chloride, NaCl (Figure 4.26a). Similarly, when each ion has eight nearest neighbours, it is called an **8:8 lattice**, for example, caesium chloride, CsCl (Figure 4.26b). If the cation and anion do not combine in a 1:1 ratio, the co-ordination numbers around each ion will not be equal. In calcium fluoride, CaF_2, for example, the Ca^{2+} ion is surrounded by a cube of eight F^- ions but each F^- ion has only four near Ca^{2+} neighbours held in a tetrahedral arrangement. So, calcium fluoride is said to adopt an 8:4 ionic structure.

Although sodium and caesium are in the same periodic group, their chlorides have different structures. Clearly, the structure adopted by the ions will depend on their size or, to be more exact, their ionic radius. But how do we define **ionic radius**?

Electron density maps of ionic structures show that the ions are *sometimes* spherical (e.g. in calcium fluoride, Figure 4.27a). In many cases, though, the ions are distorted spheres (e.g. in sodium chloride, Figure 4.27b). This distortion arises from the unequal repulsion of the ions' outermost electron clouds. In defining ionic radius, therefore, we must assume that: (i) the ions are spherical and (ii) they touch each other. Then the ionic radius can be depicted using a simple diagram (Figure 4.28). X-ray diffraction can provide accurate values for the distance between the layers of ions, d. If the anion radius, r_A is known, the cation radius, r_C can be obtained from the difference:

$$r_A + r_C = d$$
$$r_C = d - r_A$$

Trends in ionic radii are discussed in section 5.1.

(a)

(b)

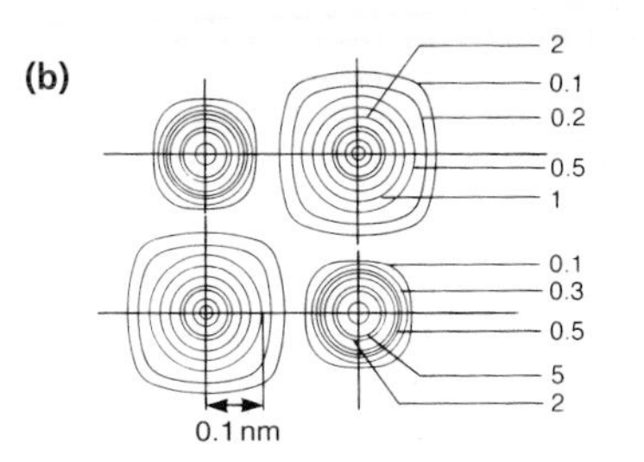

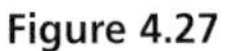

Figure 4.27
Electron density maps for **(a)** calcium fluoride and **(b)** sodium chloride (contours in electrons per cubic Ångstrom (0.1 nm))

Activity 4.7

Atomic radii and ionic radii for some elements are given in Table 4.10.

1 How do the ionic radii of the positive and negative ions compare with the radii of the atoms from which they are formed?

2 Can you explain the pattern in these values in terms of the relative numbers of protons and electrons in the atoms and ions?

Radius ratio and the type of ionic lattice

Ionic radii can be used to predict the type of lattice which will be adopted by a given pair of ions. To do this, the radius ratio of the ions is calculated, thus:

$$\text{radius ratio} = \frac{\text{radius of the cation, } r_C}{\text{radius of the anion, } r_A}$$

The relationship between the lattice type and the radius ratio is found to be:

Radius ratio	Lattice type	Example
0.225–0.414	4:4	ZnS
0.414–0.732	6:6	NaCl
0.732 and above	8:8	CsCl

This relationship is to be expected because a relatively large cation can fit a greater number of anions around itself. Thus, in sodium chloride each sodium ion (ionic radius 0.098 nm) has a co-ordination number of 6, whereas in caesium chloride each caesium ion (ionic radius 0.167 nm) has a co-ordination number of 8.

Unfortunately, radius ratios do not *always* accurately predict the type of crystal lattice. Lithium iodide, for example, has a radius ratio of 0.310, suggesting that it should exist as a 4:4 lattice. However, X-ray diffraction shows it to be a 6:6 lattice, like sodium iodide. One reason for this discrepancy is that there is a considerable degree of covalent character in lithium iodide, and this behaviour is discussed later in this section.

Table 4.10

Atom	atomic radius/nm	ion	ionic radius/nm
Na	0.186	Na^+	0.098
Mg	0.160	Mg^{2+}	0.065
Al	0.143	Al^{3+}	0.045
N	0.074	N^{3-}	0.171
O	0.074	O^{2-}	0.146
F	0.072	F^-	0.133

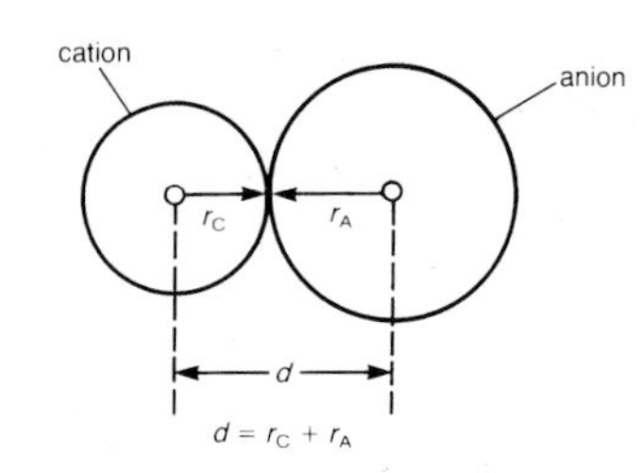

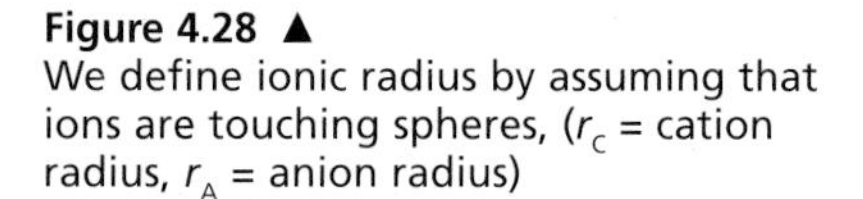
Figure 4.28 ▲
We define ionic radius by assuming that ions are touching spheres, (r_C = cation radius, r_A = anion radius)

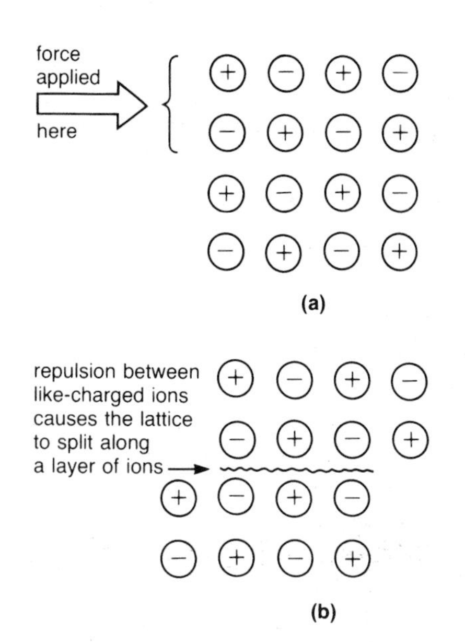

Figure 4.29 ▲
Cleavage in ionic solids: a layer of ions **(a)** before and **(b)** after cleavage

Physical properties of ionic compounds

The physical properties of ionic compounds are well defined, as follows:

- Hard, brittle solids which **cleave**, that is split cleanly when hit with a sharp chisel.
- High melting and boiling points.
- Conduct electricity when molten or dissolved in water but non-conductors in the solid state.
- Generally speaking, they are soluble in polar solvents such as water but insoluble in non-polar solvents such as hexane.

The model of a giant ionic lattice accounts very well for these physical properties. The brittle nature of an ionic crystal, and its ability to be cleaved, result from the mutual repulsion between layers when ions are displaced as seen in Figure 4.29. The large number of electrostatic attractions between oppositely charged ions cause the lattice to be strong and large amounts of energy are needed to break it up. Thus, the solids are hard and have high melting points. The strong attractive forces between ions remain in the liquid state so boiling points are also high. Since the ions are held rigidly in place, the solids cannot conduct electricity. When molten or dissolved in water the lattice is broken up and the ions are free to move, and so electricity can be conducted. Surprisingly, perhaps, there is a wide variation in the solubilities of ionic solids in water and this is discussed later in the next section.

Lattice energies and the polarisation of ions

An idea of the strength of an ionic bond is given by the **lattice energy, $\Delta H_L^\ominus$**, of the compound. This is the enthalpy change on forming one mole of an ionic solid from its isolated gas ions, at 298 K and 1 atm, for example:

$$Na^+(g) + Cl^-(g) \longrightarrow NaCl(s) \qquad \Delta H_L^\ominus = -781 \text{ kJ mol}^{-1}$$

You should note that all lattice energies are negative, that is enthalpy is released when oppositely charged ions are brought together. Some other lattice energies are given in Table 4.11. Can you detect a pattern in these values?

Table 4.11 Lattice energies of some ionic compounds. How do the values vary with the charges on the ions?

Compound	ions present	lattice energy/kJ mol^{-1}
sodium bromide	Na^+ Br^-	–1029
magnesium bromide	Mg^{2+} Br^-	–2414
calcium chloride	Ca^{2+} Cl^-	–2197
calcium sulphide	Ca^{2+} S^{2-}	–3021

Lattice energy is a measure of the strength of the interionic attractions and we find that small, highly charged ions give the strongest electrostatic attractions. Thus, lattice energies *increase* as the size of the ions *decrease* (e.g. $Rb^+ < K^+ < Na^+$) and as their charge *increases* (e.g. $Na^+ < Ca^{2+} < Al^{3+}$).

If ions are assumed to be separate, charged spheres, we can use simple electrostatic theory to calculate a value. $\Delta H^{\ominus}_{L,theor}$ for the lattice energy. By using experimental data and Hess's Law, we can obtain an experimental value for the lattice energy, $\Delta H^{\ominus}_{L,exp}$. Now, compare the values of $\Delta H^{\ominus}_{L,theor}$ and $\Delta H^{\ominus}_{L,exp}$ (in kJ mol^{-1}) shown below:

	LiI	NaI	KI	NaCl	ZnS
$\Delta H^{\ominus}_{L,theor}$	–728	–686	–632	–766	–3427
$\Delta H^{\ominus}_{L,exp}$	–753	–699	–643	–781	–3565
% difference	3.3	1.9	1.7	1.9	3.9

For many simple ionic compounds, such as NaI, KI and NaCl, the bonding may be regarded as being totally ionic, with spherical ions regularly arranged in a lattice, and the theoretical lattice energies agree to within 2% of the experimental values. In some cases, there is a larger discrepancy in these values and this indicates that appreciable covalent character is present in the bonding. Pure ionic or covalent bonding is really the exception rather than the rule and most chemical bonds may be considered to be *intermediate* between these two extremes. So, how does an ionic bond develop some covalent character?

We have so far regarded the electron distribution around the ions in the lattice as being spherically symmetrical:

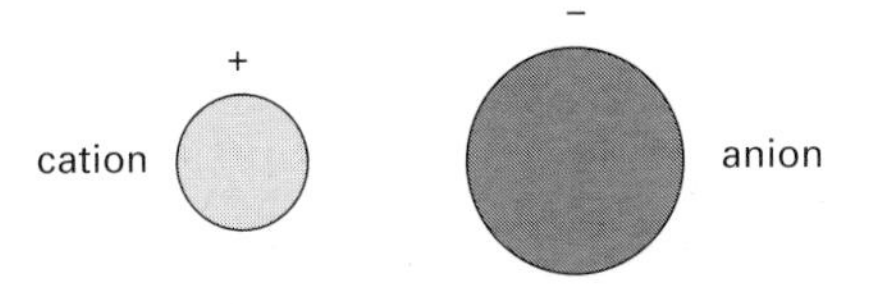

It should be clear, however, that the cation will attract the outer electrons on the anion, deforming or **polarising** it to some extent:

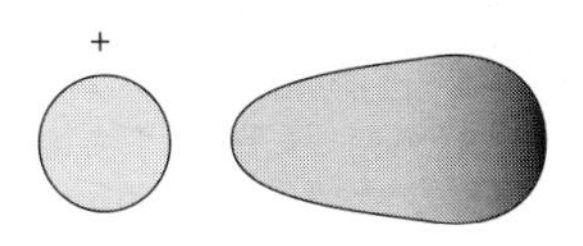

This polarisation shifts electron density from the anion towards the cation. If the deformation is sufficient, the bonding electrons may be considered to be partially shared between the ions, that is the bond develops some *covalent* character:

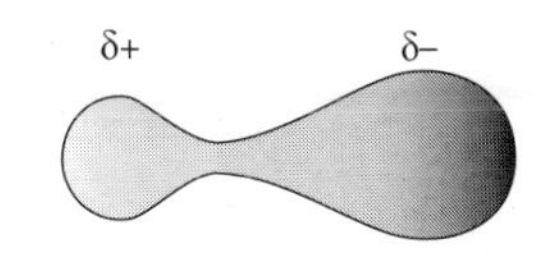

The extent to which this polarisation of ions occurs depends both upon the polarising power of the cation and the ease with which the anion may be distorted and the factors involved are summarised by **Fajan's Rules**:

CATION	ANION
high positive charge	*high* negative charge
small size	*large* size
⇒ *high* charge density	⇒ electrons more *loosely* held
∴ stronger polarising power	∴ easier distortion of the electron cloud

Introduction of appreciable covalent character into essentially ionic bonding is often demonstrated by abnormally low melting and boiling points and by solubility in covalent solvents such as ethanol.

4.10 What happens when an ionic solid dissolves in water?

When an ionic solid dissolves in water, the water molecules penetrate the lattice and attach themselves to the ions (Figure 4.30). This process is known as **hydration** and the ions are said to be **hydrated**.

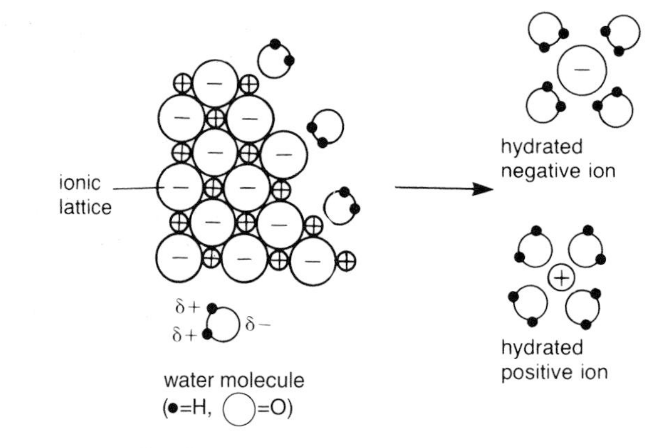

Figure 4.30 ► An ionic solid dissolving in water

Since new **ion–dipole attractions** are being formed, *hydration is an exothermic process*. Moreover, the energy released is large enough to enable the hydrated ions to move away from their oppositely charged neighbours (Figure 4.30). Eventually, the lattice is completely destroyed, and all the ions become hydrated.

A simplified example of a hydrated ion is shown in Figure 4.31. In reality, the sodium ion is surrounded by more than one layer of water molecules. *The number of water molecules in the innermost layer is termed the* **co-ordination number**. For hydrated sodium ions, then, the co-ordination number is 4. The hydration number *is the total number of water molecules which are attached, however loosely, to an ion* and some typical values are given in Table 4.12. These show that small, highly charged ions (i.e. those with the greatest charge density) are the most heavily hydrated.

Finally, let us consider further the nature of the attractive forces between an ion and the water molecules. There are two kinds of attraction:

- **Ion–dipole bonds:** These are physical attractions which exist between an ion and the oppositely charged part of a water molecule (Figure 4.31).
- **Dative covalent bonds:** If a metal ion possesses suitable empty orbitals at low enough energy it can form dative covalent bonds with the water molecules. As a result, **complex ions** are formed, for example the hexaaquaaluminium(III) ion in Figure 4.32. Some common complex 'aqua-ions' are discussed in Chapter 21.

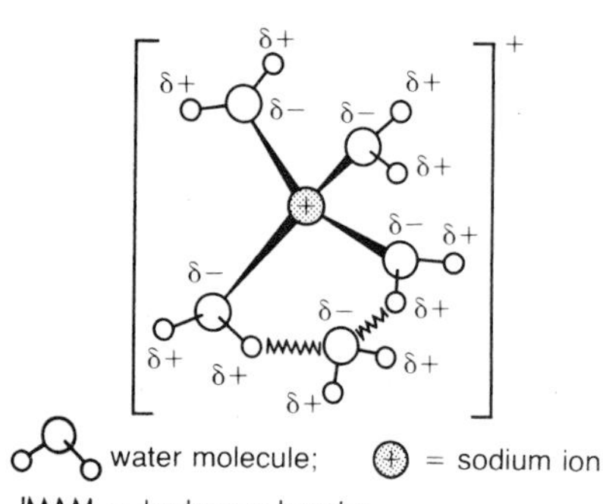

Figure 4.31 ▲ A hydrated sodium ion. Four water molecules are attached via ion–dipole bonds, whilst the fifth H_2O is co-ordinated via hydrogen bonds. The hydration number is 5

Table 4.12 Some approximate hydration numbers.

Ion	hydration number (±1 unless stated)
Li^+	5
Na^+	5
K^+	4
Cl^-	1
Br^-	1
I^-	1
Mg^{2+}	15 ± 2
Ca^{2+}	13 ± 2
Al^{3+}	26 ± 3

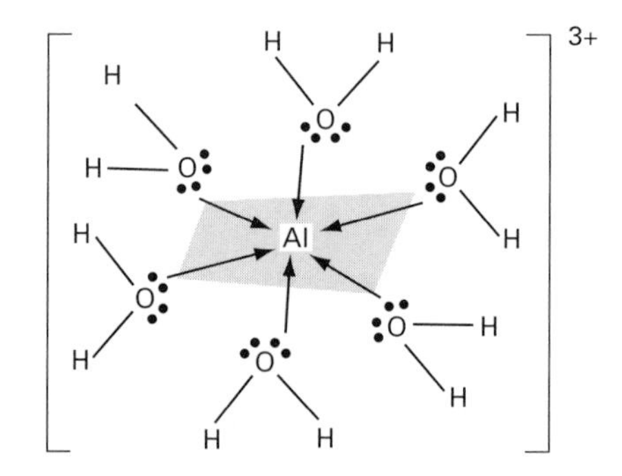

Figure 4.32 ▲ Hexaaquaaluminium(III) ion, an example of a complex aqua-ion. It has an octahedral shape with aluminium having a co-ordination number of 6. Each oxygen atom donates a lone pair of electrons to form the Al—O bond.

An enthalpy change usually occurs when an ionic solid dissolves in water, and this can be described by an **enthalpy cycle** (Figure 4.33). Each step in this enthalpy cycle has its own enthalpy term:

- $\Delta H^\ominus_s$ = **standard enthalpy change on solution**. This is the enthalpy change when one mole of solute dissolves in so much solvent that further dilution produces a negligible enthalpy change, at 298 K and 1 atm. In this case, the solute is an ionic solid and the solvent is water.
- $-\Delta H^\ominus_L$ = the **reverse of the lattice energy**. This is the enthalpy needed to convert one mole of ionic lattice into gaseous ions, e.g.

$$NaF(s) \longrightarrow Na^+(g) + F^-(g) \qquad -\Delta H^\ominus_L[NaF(s)] = 918\ kJ\ mol^{-1}$$

$$\text{and } CaCl_2(s) \longrightarrow Ca^{2+}(g) + 2Cl^-(g) \qquad -\Delta H^\ominus_L[CaCl_2(s)] = 2258\ kJ\ mol^{-1}$$

- $\Delta H^\ominus_h$ = the **standard enthalpy of hydration** of each ion, $M^+(g)$ and $X^-(g)$. It is defined as the enthalpy change when one mole of gas ions are dissolved in so much water that more dilution produces no further enthalpy change at 298 K and 1 atm pressure.

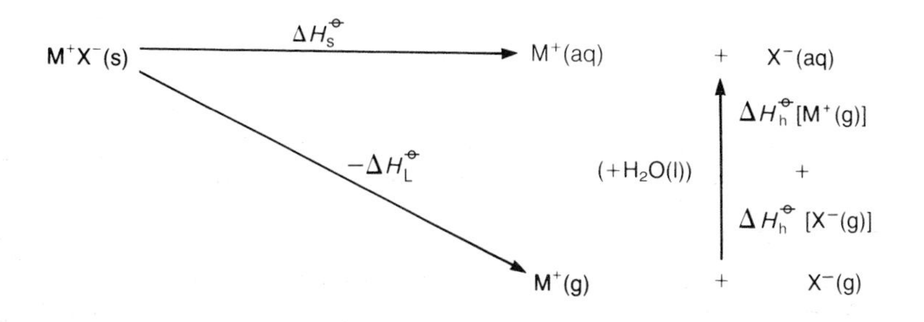

Figure 4.33
An enthalpy cycle for the dissolution of $M^+X^-(s)$ in water. You may wish to refer to sections 3.7 and 3.8 for background information

Now, Hess's law states that the enthalpy change during a reaction is independent of the route followed (section 3.7). Applying this to Figure 4.33, we get

$$\Delta H^\ominus_s[M^+X^-(s)] = \underbrace{-\Delta H^\ominus_L[M^+X^-(s)]}_{\text{endothermic (+ve)}} + \underbrace{\Delta H^\ominus_h[M^+(g)] + \Delta H^\ominus_h[X^-(g)]}_{\text{exothermic (–ve)}}$$

From this equation, we see that:

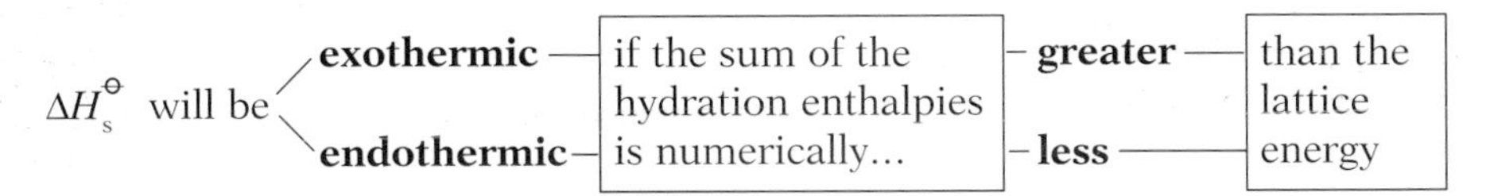

Activity 4.8

Use the following data, and an enthalpy cycle based on Figure 4.33, to calculate the enthalpy of solution for magnesium chloride.

$\Delta H^\ominus_L[MgCl_2(s)]$ = –2489 kJ mol^{-1}

$\Delta H^\ominus_h[Mg^{2+}(g)]$ = –1891 kJ mol^{-1}

$\Delta H^\ominus_h[Cl^-(g)]$ = –384 kJ mol^{-1}

Finally, we want to investigate whether there is a link between the water solubility of a salt and its enthalpy of solution. Can you draw any general conclusions from the data in Table 4.13?

Although there is a slight correlation between $\Delta H^\ominus_s$ and the solubility of the silver salts, this data suggests that *a salt's solubility cannot be predicted solely on energetic grounds*. For example, although $\Delta H^\ominus_s[LiBr(s)]$ and $\Delta H^\ominus_s[LiCl(s)]$

are both exothermic, with similar values, the chloride is a hundred times more soluble in water. In fact, the energy changes on dissolving a solid are much more complex than those in our simple enthalpy cycle of Figure 4.33 and a fuller explanation is based on the concept of 'free energy' (Chapter 28). However, we may generalise by saying that an *ionic solid is more likely to be soluble in water if (i) it has a low lattice enthalpy and (ii) the constituent ions have high hydration enthalpies*.

Table 4.13 Enthalpies of solution (ΔH_s) of some ionic solids and their solubilities in water (at 25°C). Is there a pattern here?

Solid	ΔH_s/kJ mol^{-1}	solubility/10^{-3} mol per 100 g water
LiBr	−47	20
BaI_2	−44	560
LiCl	−35	2000
NaCl	+7	650
PbI_2	+10	0.2
AgF	+21	1420
AgCl	+40	0.001
AgI	+97	0.00001

4.11 Giant metallic structures

When subjected to X-ray diffraction, all metals produce similar electron density maps, e.g. aluminium, Figure 4.34. These show that the atoms in the metallic lattice release some, or all, of their outer electrons, thereby forming positive ions (cations). This is to be expected because, in metals, the ionisation energies needed to remove the outer electrons are low. The electrons released are **delocalised**, that is they move freely as opposed to remaining in the vicinity of any one cation. Thus, a mobile electron cloud is produced. Since neighbouring cations are mutually attracted to this electron cloud, they will be attracted to each other. These attractive forces exactly balance the repulsion between the like-charged cations. The overall result, therefore, is a rigid 3-dimensional lattice of metal cations, embedded in a mobile electron cloud. This lattice of cations extends throughout the piece of metal with regular units repeating themselves; so metals are said to have 'giant' structures.

What factors affect the strength of the metallic bonds? Have a look at Figure 4.35.

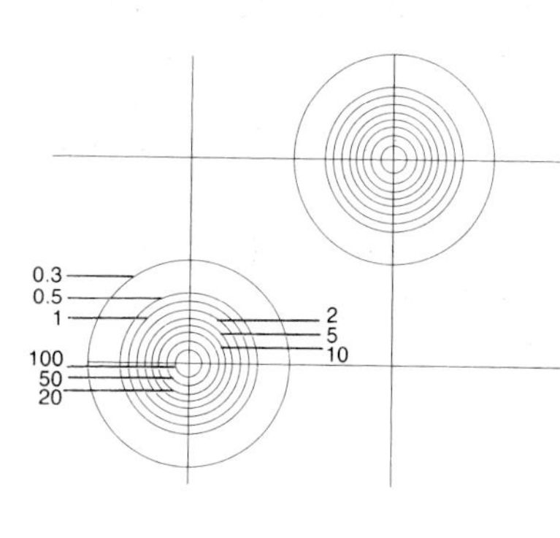

Figure 4.34 ▲
An electron density map of aluminium (contours in electrons per cubic Ångstrom, where 1 Å = 10^{-10} m). The average electron density between ions is 0.21 electrons per cubic Ångstrom

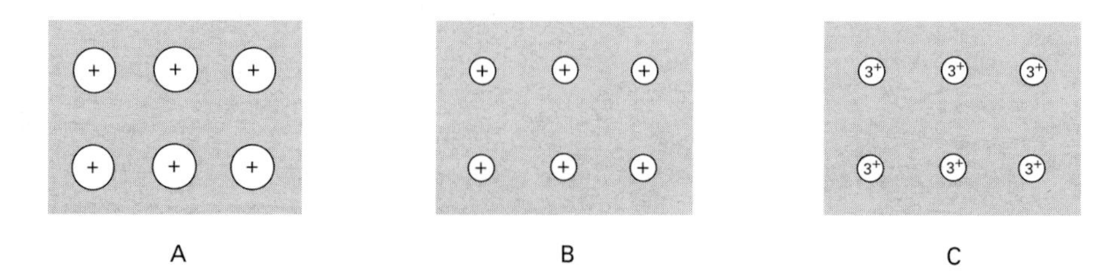

Figure 4.35 ▶
Metals with small highly-charged ions form stronger metallic bonds. Here, metallic bond strength increases
A < B < C

As we go from A ⟶ B ⟶ C, there is an increase in:

- the charge density (charge/size) of the cations
- the density of the mobile electron cloud, because the metal atoms are losing more of their outer electrons.

The strongest metallic bonds will be formed between the cations and electron cloud which have the highest charge densities. Thus, metallic bond strength increases in the order: A < B < C.

If the metallic bonds are strong, the metal will be hard and have a high melting point. For example, let us compare sodium and aluminium:

	sodium	**aluminium**
electron configuration	$[Ne]3s^1$	$[Ne]3s^23p^1$
cation charge	1+	3+
cation size/nm	0.098	0.045
cation charge density	much smaller	much larger
density of delocalised electron cloud	low	high
∴ metallic bonds are	weaker	stronger

As expected, sodium is much softer and has a much lower melting point than aluminium. The melting points are sodium 98°C and aluminium 659°C. This simple model of metallic bonding accounts well for the physical properties of metals as shown in Table 4.14.

Table 4.14 Physical properties of metals

Physical property	explanation using our model
Metals are ...	
good electrical conductors	when applied to the metal, an electromotive force (e.m.f.) can push the mobile electron cloud through the metallic lattice
good thermal conductors	when an area of the metal is heated, the kinetic energy of the electrons in that region increases; this causes them to flow quickly into the cooler parts of the metallic lattice
readily soluble in each other, when molten; on cooling alloys are formed	the mobile electron cloud can hold a variety of metal cations together, e.g. an alloy containing 1+ and 2+ cations
malleable (they can be hammered into shape when hot, e.g. as a horseshoe) and ductile (they can be drawn out into long wires)	the mobility of the electron cloud enables it to readily accommodate any distortion in the lattice as shown below:

1+ cations
2+ cations
electron cloud

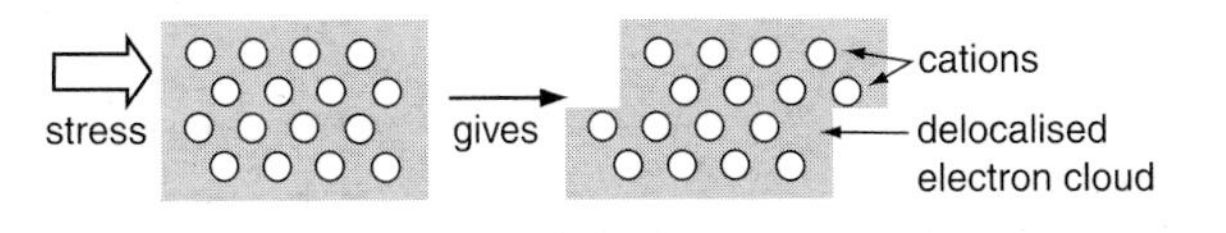

The packing of atoms in giant metallic structures

X-ray studies on metals show that the atoms are arranged in three common types of metallic lattice. You can get an idea of these arrangements using 17 identical coins, or discs, to represent the spherical atoms.

Firstly, position a layer of 11 atoms (coins) on a piece of white paper, as shown in Figure 4.36a. This structure is said to be close packed because the atoms fill as much space as possible. Now, place a second layer of 5 atoms in the 'spaces' between those in the first layer *using exactly the same packing arrangement*. (One of the atoms in the second layer is shown by dots.) If you do this correctly, you will be able to look down and see the paper through a 'hexagonal' hole in the layers. Now you will find that there are two ways of positioning a third layer of atoms, as follows:

1 Above an atom in the first layer giving a **cubic close packed (CCP)** lattice, as in calcium, aluminium and copper.

2 Above the 'hexagonal' hole in the layers to give a **hexagonal close-packed (HCP)** lattice, as in magnesium and zinc.

Now start again but make up the first layer as shown in Figure 4.36b. Notice that this is *not* a close-packed structure. Place a second layer of 4 atoms in the 'spaces' between the atoms in the first layer. Now, place the final coins in a third layer and you should see that alternate layers are directly above each other; this is termed a **body centred cubic (BCC)** lattice. Metals with a 'BCC' structure include tungsten, uranium and the group 1 metals.

(a)
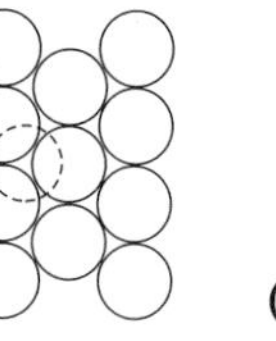
(b)
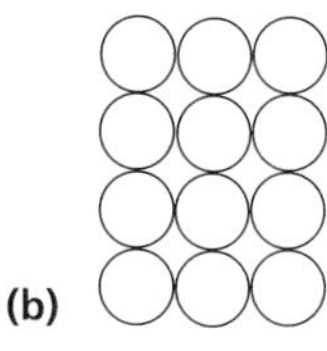

Figure 4.36 Two single layers of spheres viewed from above. Those in (a) are close-packed

Focus 4e

1. There are four main types of solid structure: **giant metallic**, **macromolecular**, **simple molecular** and **giant ionic**. Their bonding, structure and physical properties are summarised in Table 4.15.
2. The **co-ordination number** of an atom, ion or molecule tells us how many neighbouring particles are nearest to it. A **unit cell** is the smallest part of the lattice which retains the structural properties of the solid.
3. Ionic structures are often described by the co-ordination numbers of the ions present, e.g. NaCl, 6:6; CsCl, 8:8; ZnS, 4:4.
4. To define **ionic radius**, we assume that an ionic lattice consists of spherical ions which touch each other.
5. Many solids have intermediate ionic/covalent bonds. As a result, they possess intermediate ionic/macromolecular structures.
6. **Solvation** is the process in which solvent molecules penetrate a solid lattice and attach themselves to the particles present. If the solvent is water, the process is called **hydration**. The **hydration number** of an ion is the number of water molecules which are attached, however loosely, to that ion. Small, highly charged ions have high hydration numbers, and vice versa.
7. Solvent molecules can be attached to the solute by (i) **van der Waals' forces** (I_2 in hexane), (ii) **permanent dipole–permanent dipole attractions** (phenol in water), (iii) **ion–dipole attractions** (Na^+ or Cl^- ions in water) or (iv) **dative covalent bonds** (Fe^{2+} ions in water).
8. The **standard enthalpy change on solution**, $\Delta H_s^\ominus$, is the enthalpy change when one mole of solute dissolves in so much solvent that further dilution produces a negligible enthalpy change, at 298 K and 1 atm pressure. The standard enthalpy of hydration, $\Delta H_h^\ominus$, is the enthalpy change when one mole of gas ions are dissolved in so much water that more dilution produces no further enthalpy change, at 298 K and 1 atm pressure. Generally speaking, an ionic solid is more likely to be soluble in water if its lattice enthalpy is low and the hydration enthalpies of its ions are high.

Table 4.15 Comparing the natures and properties of solid structures

	giant metallic	macromolecular (giant covalent)	simple molecular	giant ionic
examples	Na, Al, Fe	graphite, diamond, SiO_2	I_2, CH_4, CO_2, HF, H_2O	Na^+I^-, $Mg^{2+}O^{2-}$, $Ca^{2+}(2F^-)$
made up of	metal **atoms**	non-metal **atoms**	**molecules**	**ions**
bonding	**cations** mutually attracted to a mobile sea of electrons	atoms held together by strong **covalent** bonds	covalently bonded molecules held together by weak intermolecular forces: **van der Waals' forces** or **permanent dipole–permanent dipole attractions**	strong **electrostatic** attraction between oppositely charged ions
Properties				
physical state at 25°C	solid, except Hg	solid	gas, liquid or solid	solid
hardness	variable: Na soft, Fe hard **malleable** and **ductile**	generally hard	soft	hard and brittle; undergoes **cleavage**
melting point	variable: Na 98°C, Fe 1500°C	very high, SiO_2 2230°C	very low/low, CH_4 −182°C, H_2O 100°C	high, NaCl 808°C
boiling point	variable: Na 890°C, Fe 2887°C	very high, diamond 5100°C	very low/low, CH_4 −181°C, H_2O 100°C	high, AgCl 1504°C
electrical conductivity	good, whether solid or liquid	non-conductors (*except graphite*)	non-conductors	solids are non-conductors but will conduct when molten or in aqueous solution
solubility	nil, but dissolves in other metals ⟶ alloys	insoluble	usually soluble in a polar or non-polar solvent	usually soluble in polar solvents; insoluble in non-polar solvents

4.12 Redox reactions and oxidation numbers (states)

When ionic bonds are formed, electrons are transferred between atoms, for example in the formation of magnesium sulphide from magnesium and sulphur:

$$\underset{2.8.2}{Mg} + \underset{2.8.6}{S} \longrightarrow \underset{2.8}{Mg^{2+}}\ \underset{2.8.8}{S^{2-}}$$

Magnesium loses two electrons and these are gained by sulphur. When chemical species lose electrons, we say that **oxidation** has occurred whereas a gain of electrons is termed **reduction**. Thus, we may split the equation above into two ionic half-equations, one involving oxidation and one reduction:

$$Mg \longrightarrow Mg^{2+} + 2e^- \quad \text{oxidation}$$
$$S + 2e^- \longrightarrow S^{2-} \quad \text{reduction}$$
$$Mg + S \longrightarrow Mg^{2+}\ S^{2-} \quad \text{redox}$$

A chemical reaction which involves **red**uction and **ox**idation is termed a **redox reaction**. A species is said to be **oxidised** if electrons have been withdrawn from it; the species which withdraws the electrons is termed an **oxidising agent**. Similarly, a species which gains electrons is said to be **reduced** and the species which supplies the electrons is termed a **reducing agent**. Get to know these terms by trying activity 4.9.

Activity 4.9

For the formation of each of the following ionic compounds from their elements:

1 calcium oxide 2 potassium sulphide 3 aluminium fluoride
4 aluminium oxide 5 magnesium nitride

a) write an overall equation showing the formation of the ions
b) write half-equations showing the oxidation and reduction processes separately
c) state which element is acting as the reducing agent and which as the oxidising agent.

At first sight the formation of a covalent compound such as hydrogen chloride, HCl, from its elements might not seem to be a redox reaction:

$$H_2 + Cl_2 \longrightarrow 2HCl$$

However, if we look more closely at the electronic rearrangement which takes place, we can see that the bonding pair of electrons is *not equally shared* in the hydrogen chloride molecule.

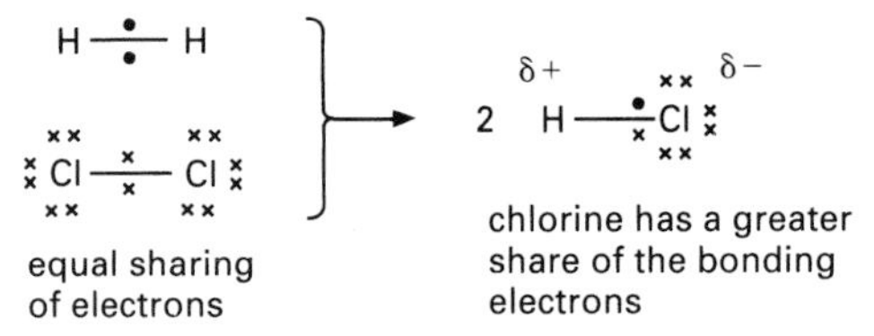

Chlorine has a higher **electronegativity** than hydrogen and attracts the bonding electrons more strongly. It may be considered to have *partially gained* an electron, whilst the hydrogen atom has *partially lost* an electron. The hydrogen has thus been partially oxidised by the chlorine. In any covalent bond the more electronegative atom will develop a slightly negative charge, δ–, whilst the other atom will become slightly positive, δ+. Knowing from section 4.5 that electronegativity generally increases on passing across a period but decreases on passing down any group in the periodic table, we may predict the partial charges present in any covalent molecule as shown by the examples in Figure 4.37. Any atom which develops a positive charge may be considered to have partially lost electrons, whilst a negative charge indicates a partial gain of electrons.

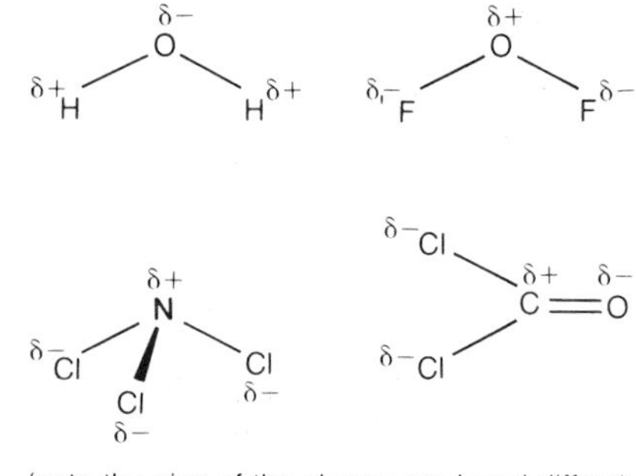

Figure 4.37
Examples of polar covalent molecules

The idea of **oxidation number** or **oxidation state** helps us to develop an understanding of redox reactions in the case of both ionic and covalent systems. *The oxidation number (state) may be described as the number of electrons which the atom loses, or tends to lose, when it forms the substance in question. If the atom gains, or tends to gain electrons, then the oxidation number is negative.*

In the case of a simple ion the oxidation number of the element is obviously equal to the ionic charge, e.g.

sodium chloride	Na^+	Cl^-
oxidation numbers	+1	–1
magnesium oxide	Mg^{2+}	O^{2-}
oxidation numbers	+2	–2
lithium nitride	$3Li^+$	N^{3-}
oxidation numbers	3(+1)	–3

For covalent compounds, the oxidation number may be regarded as the charge which the atom would develop if the electronegativity differences made the compound fully ionic, e.g.

hydrogen chloride	H	Cl	(H^+	Cl^-)
oxidation numbers	+1	−1		
phosphorus trichloride	P	Cl_3	(P^{3+}	$3Cl^-$)
oxidation numbers	+3	3(−1)		
sulphur trioxide	S	O_3	(S^{6+}	$3O^{2-}$)
oxidation numbers	+6	3(−2)		

There are a number of simple rules which may help to determine the oxidation state of an element in any compound.

1 The oxidation number of an element in its free state is always zero.

2 The sum of the oxidation numbers in a neutral molecule is always zero. In an ion, this sum equals the ionic charge, e.g.

PCl_5	oxidation numbers	P +5	= +5
		5Cl −1 each	= −5
			0 total
CO_3^{2-}	oxidation numbers	C +4	= +4
		3O −2 each	= −6
			−2 total

3 Some elements exhibit more or less fixed oxidation numbers in their compounds, e.g.
Group I metals always show an oxidation number of +1.
Group II metals always show an oxidation number of +2.
Fluorine always shows an oxidation number of −1.
Oxygen almost always shows an oxidation number of −2 (except in oxygen difluoride, OF_2, and in peroxides and superoxides, such as Na_2O_2 and KO_2).
Hydrogen usually shows an oxidation number of +1 (except in the hydrides of more electropositive elements, such as NaH).

We may often use these rules to deduce the oxidation numbers of other elements in their compounds, e.g. manganese in $KMnO_4$:

oxidation numbers	K	Mn	O_4	
	+1	+?	+4(−2)	= 0

Since the compound as a whole is neutral, the oxidation number of the manganese must be +7. This is reflected in the systematic name of the compound, potassium manganate(VII), where the Roman numerals indicate the oxidation number of the manganese.

Where there is more than one atom of a particular element present in a molecule or ion, this method will give us the average oxidation number, which may not be a whole number, e.g. sulphur in the tetrathionate ion, $S_4O_6^{2-}$

	S_4	+	O_6	
oxidation numbers	4(?)		6(−2)	= −2

The total of the oxidation numbers of the four sulphur atoms must be +10, so the average oxidation number of the sulphur is +10/4 = +2.5. This nicely illustrates the point that, whilst oxidation numbers provide a useful 'book-keeping' system for redox reactions, they must not be interpreted too literally.

If, during a reaction, the oxidation number of any atom becomes more positive then it has been oxidised. If the oxidation number becomes less positive then reduction has occurred. Consider the reaction of manganate(VII) with iron(II) in aqueous acid solution:

$$MnO_4^- + 5Fe^{2+} + 8H^+ \longrightarrow Mn^{2+} + 5Fe^{3+} + 4H_2O$$

oxidation numbers: $+7$, $5(+2)$, total $= +17$ → $+2$, $5(+3)$, total $= +17$

reduction (+7 → +2); oxidation (5(+2) → 5(+3))

The manganate(VII) has been *reduced* by the iron(II) and the iron(II) has been *oxidised* by the manganate(VII). Note that the sum of the oxidation numbers on each side of any redox equation must be the same, since the number of electrons gained by the oxidising agent equals the number lost by the reducing agent.

Activity 4.10

1 What is the oxidation number of each of the atoms in italics in the following molecules and ions?

a) $\mathit{Ca}Cl_2$ **b)** $\mathit{Si}O_2$ **c)** $K\mathit{O}_2$ **d)** $\mathit{N}H_4^+$
e) $\mathit{Cu}Cl_4^{2-}$ **f)** $Na_3\mathit{Al}F_6$

2 In each of the following equations:

$$Cl_2(aq) + 2K\mathit{Br}(aq) \longrightarrow 2KCl(aq) + Br_2(aq)$$
$$2\mathit{Cr}O_4^{2-}(aq) + 2H^+(aq) \longrightarrow Cr_2O_7^{2-}(aq) + H_2O(l)$$
$$I_2(aq) + 2\mathit{S}_2O_3^{2-}(aq) \longrightarrow 2I^-(aq) + S_4O_6^{2-}(aq)$$
$$3\mathit{Mn}O_4^{2-}(aq) + 4H^+(aq) \longrightarrow 2MnO_4^-(aq) + MnO_2(s) + 2H_2O(l)$$

a) What are the oxidation numbers of the atoms in italics?
b) Is the species concerned being oxidised? If so, what is the oxidising agent?

3 Oxidation and reduction were originally defined in terms of addition or removal of oxygen or hydrogen.

Oxidation is the addition of oxygen or the removal of hydrogen.
Reduction is the removal of oxygen or the addition of hydrogen.

On this basis, any element which combines with hydrogen to form a hydride is considered to have been reduced, e.g.

$$N_2(g) + 3H_2(g) \rightleftharpoons 2NH_3(g)$$
$$2Na(s) + H_2(g) \longrightarrow 2NaH(s)$$

a) Are both nitrogen and sodium *reduced* in terms of gain of electrons?
b) Under what circumstances may hydrogen act as an *oxidising* agent?
c) What is the only element you might expect to be able to *oxidise* oxygen?

4.13 The states of matter: solids, liquids and gases

Matter can exist in three important states: solid, liquid and gas. These states are described by the **kinetic theory of matter** which explains the extent to which the particles can move. The word kinetic is taken from the Greek word *kinesis* which means 'to move'. The main points of kinetic theory are:

1 All matter is composed of particles (atoms, molecules or ions) the size of which vary from substance to substance.
2 Solids are characterised by a **long-range order** which means that the arrangement of particles is ordered throughout the structure with the structural units repeating themselves.

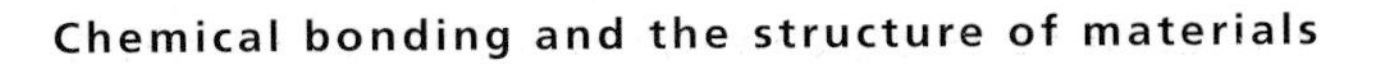

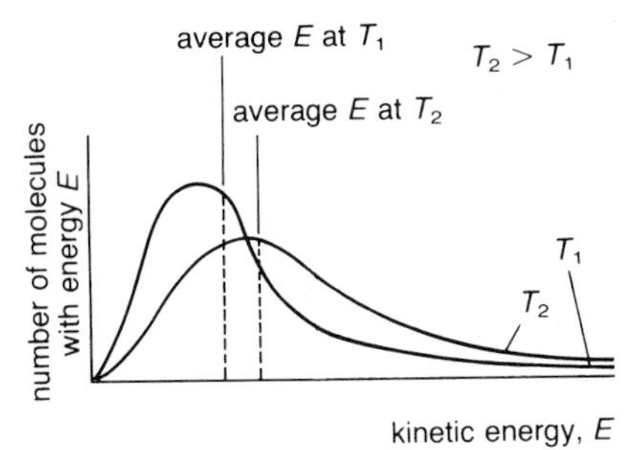

Figure 4.38 ▲
The distribution of kinetic energies of gas particles at different temperatures where $T_2 > T_1$

The particles are held in fixed positions because they do not possess enough energy to overcome the attractive forces which hold them together. However, the particles are able to vibrate and rotate about their fixed positions. If the solid is heated, the heat energy is converted into kinetic energy and the particles vibrate more vigorously. If enough energy is supplied, the lattice breaks up and a liquid is usually formed. The temperature at which this occurs is termed the **melting point**.

During melting, all of the heat energy supplied is used to break up the lattice, so the temperature remains constant. The energy required to convert one mole of a solid into a liquid at its melting point is known as the **molar enthalpy of fusion**. The stronger the attractive forces between the particles in the solid, the greater will be the molar enthalpy of fusion. In fact, patterns in the melting points and enthalpies of fusion are very similar.

Since the particles in a solid are rigidly held in place, the solid does not take the shape of a container in which it is placed nor does it expand to fill it.

3 In structural terms, gases are very much the opposite to solids. For a given pure substance, the average energy of the particles in the gas state is much greater than that in the solid state. Thus, the gas particles can overcome attractive forces and move rapidly, and randomly; there is **no order** in the gaseous state. Gases spread evenly throughout any volume in which they are placed. This process is known as **diffusion**. A good example of diffusion is an air freshener spray which can be smelt across a room some distance from where it was first sprayed.

The particles in a gas show a wide distribution of kinetic energies, Figure 4.38. If the temperature of a gas is increased, the average kinetic energy of the particles also increases and this is seen in the greater degree of vibrational and rotational motion as well as translational motion, Figure 4.39.

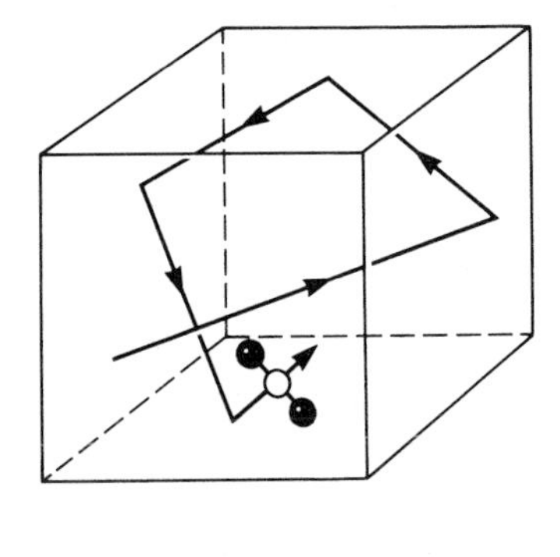

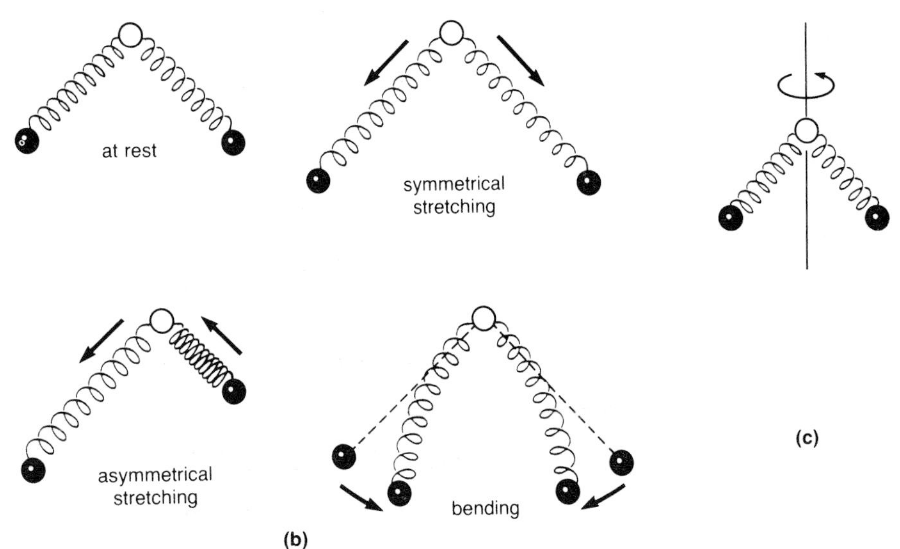

Figure 4.39 ▲
In the gas state, the average kinetic energy of the particles is great enough for the attractive forces between them to be overcome and, thus, the particles move around randomly and rapidly. In addition, gas particles display vibrational (b) and rotational (c) motion

4 The attractive forces between particles in the liquid state are weaker than they are for the same substance in the solid state. Liquids are said to have **short-range order** with ordered clusters of particles moving through areas of disorder which contain individual particles, as in gases. There is an equilibrium between areas of order and disorder, with some particles leaving a cluster as others take their place.

If a liquid is left in an open container it will **evaporate**. The reason is that, like gases, there is a wide variation in the speeds (energies) of particles in the liquid. Those moving with a very high speed possess enough kinetic energy to break through the surface of the liquid and

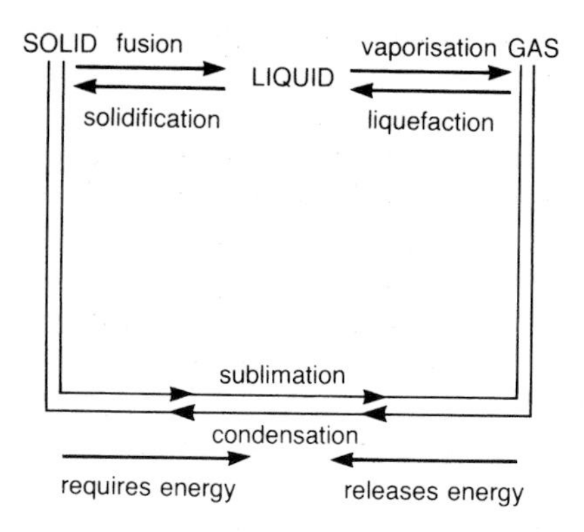

Figure 4.40 ▲
The relationship between the three main states of matter

escape against atmospheric pressure. The escaping particles exert a pressure, known as the **vapour pressure**, which increases as the temperature is raised. When the vapour pressure just above the liquid surface becomes equal to the atmospheric pressure the liquid starts to boil. At a stated pressure, the temperature at which this happens is known as the **boiling point**, for example at 1 atm water boils at 100°C, whilst at 0.5 atm it boils at about 82°C.

Whilst a liquid is boiling, the temperature remains constant because all of the heat energy supplied is being used to overcome attractive forces between the particles. So, the **molar enthalpy of vaporisation** is defined as the energy required to convert one mole of a pure liquid into a gas at its boiling point. Not surprisingly, there is a similarity in the pattern of boiling points and enthalpies of vaporisation because both quantities reflect the strength of attraction between particles in the liquid state.

Most substances can exist in the solid, liquid or gas state depending on the temperature and pressure. In some cases, all three states can exist together at a certain temperature and pressure, known as the **triple point**.

The bulk characteristics of solids, liquids and gases are summarised in Table 4.16 and the relationships between physical states are shown in Figure 4.40. It is worth noting that a few solids, when heated, change from solid to gas, without passing through a liquid state because relatively little energy is needed to overcome the remaining attractive forces once the solid has melted. Carbon dioxide provides a good example of this behaviour which is known as **sublimation**.

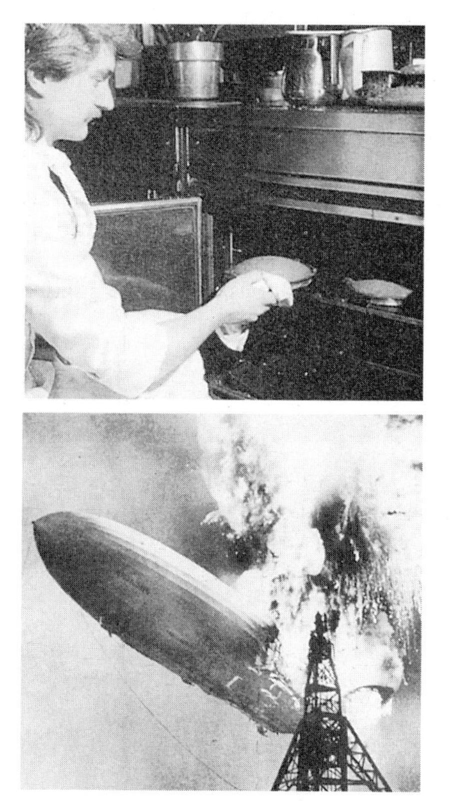

Table 4.16 Comparing some properties of solids, liquids and gases

	volume	shape	density	compressibility
solids	fixed	fixed	high	negligible
liquids	fixed	takes the shape of the container but may not fill it completely; fluid, i.e. can be poured		very slight
gases	molecules will spread evenly throughout any container; thus the volume is indefinite and shape is that of the container; fluid		low	very easily compressed

Ideal gases and the ideal gas equation

Every year massive amounts of gases are manufactured for use in our homes and in industry. A gas is often chosen for a specific task because of its chemical properties. For example, hydrogen and helium are both lighter than air and could be used to fill balloons and airships. Helium would be used,

When using gases it is important to consider both their chemical and physical properties. Although hydrogen is the lightest gas, its use in the airship 'Hindenburg' in May 1937 proved disastrous because of its ability to form explosive mixtures with air. Helium, though much more expensive, is lighter than air and chemically inert, making it suitable for children's balloons. The volume of a fixed mass of gas changes with temperature and pressure. For example, the air bubbles in a sponge cake mixture expand on heating and the sponge rises. Similarly, the inflation pressure of a car tyre will vary with the external temperature.

though, because it is chemically inert whereas hydrogen may explode when it is mixed with air. To use a gas safely, then, we need to be aware of its chemical properties, and behaviour, as well as its physical properties.

It is important to know how the volume of gas varies with temperature and pressure. Let us consider two everyday examples of this behaviour. To make a sponge cake, air bubbles are forced into the cake mixture by whisking it. On cooking, these air bubbles and carbon dioxide from the raising agent in the flour expand and the sponge 'rises'. So, when a gas is heated at a constant pressure, its volume increases. Suppose a car tyre is pumped up on a cold winter day at 0°C. If no air is lost, then on a hot summer day at 27°C, the inflation pressure in that tyre would have increased by about 10%. When a fixed volume of gas is heated, the pressure increases.

Accurate experiments with gases show that the pressure (p), volume (V) and the absolute temperature (T) of a gas are related by the **ideal gas equation**:

$$pV = nRT$$

where n is the number of gas molecules and R is known as the **gas constant** and has a value of **8.31 J K^{-1} mol^{-1}**. From its units, you can see that the gas constant enables us to work out the energy change involved when the temperature of a given amount of gas is altered. For example, 8.31 J of energy are required to raise the temperature of 1 mol of a gas by 1 K.

In fact, there are *no* ideal gases, however, under certain conditions, real gases *behave* like ideal gases. For this to happen, the attractive forces between the gas molecules must be negligible and the distance between collisions must be much greater than the size of the molecules. These conditions are most likely to apply at high temperature and low pressure, for example at 293 K (20°C) and 101 325 Pa (1 atm). Since we experience gases under these conditions in everyday life, the idea of real gases behaving like ideal gases is generally sound and we can use the ideal gas equation to perform gas calculations.

Using the ideal gas equation to work out the relative molecular mass of a gas

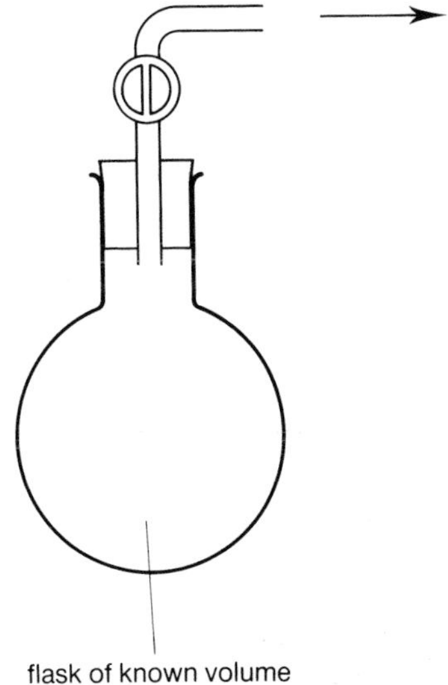

Figure 4.41
Apparatus for determining the relative molecular mass of a gas

Although relative molecular masses may be very accurately measured using a mass spectrometer (chapter 16), these instruments are expensive and not always available. Simple, yet fairly accurate, methods of measuring the relative molecular masses of gases are based on the use of the ideal gas equation, $pV = nRT$. In section 1.6, we saw that:

$$\text{number of moles of gas, } n = \frac{\text{mass of a gas, } m}{\text{relative molecular mass, } M_r}$$

Substituting in the ideal gas equation, we get

$$pV = \frac{mRT}{M_r}$$

and, rearranging,

$$\boldsymbol{M_r = \frac{mRT}{pV}}$$

So to find the relative molecular mass, M_r, we measure the mass m of a known volume of gas, V, at known absolute temperature T and pressure p. Then, the values are substituted into the above equation, taking R as 8.31 J K^{-1} mol^{-1}. An outline method, using a glass flask of known volume shown in Figure 4.41, is given below. We are working in SI units, so we measure pressure in Pascals (Pa) and volume in m^3.

Finding the relative molecular mass of gas X

1 Remove the air from a flask of known volume using a vacuum pump.
2 Close the tap, disconnect the flask from the vacuum pump and weigh it.
3 Fill the flask with the gas X and reweigh.
4 Find the mass of X.
5 Finally measure room temperature and the pressure of the gas in the flask.

Specimen results

Step 1: volume of flask	= 500 cm^3
Step 2: mass of empty flask	= 349.52 g
Step 3: mass of flask + X	= 350.44 g
Step 4: mass of X	= 0.92 g
Step 5: room temperature	= 23°C
pressure of the gas in the flask	= 99975 Pa

Calculation

1 Volume of X = 500 cm^3 = 500×10^{-6} m^3
2 Convert temperature into K, 273 + 23 = 296 K
3 Mass of X = 350.52 – 349.44 = 0.92 g

4 Use the equation $M_r = \frac{mRT}{pV}$ and substitute values

$$M_r = \frac{0.92 \times 8.31 \times 296}{99975 \times 500 \times 10^{-6}}$$

$$M_r = 45.27, \quad \text{i.e. } M_r = 45.$$

These results were obtained for carbon dioxide, M_r = 44. The method cannot be relied upon to give very accurate results but it is good enough to allow molecular formula to be calculated from empirical formula (section 6.4).

Activity 4.11

Using the above method, a student obtained the following results:

mass of gas = 2.15 g

volume of gas = 800 cm^3

temperature = 23°C

pressure = 100792 Pa

Calculate the relative molecular mass of the gas.

Focus 4f

1. When a species gains electrons it is reduced and **reduction** is said to have occurred. On **oxidation**, a species loses electrons and it is said to be oxidised. A process which involves reduction and oxidation is termed a **redox** reaction.
2. **Oxidation number (state)** may be defined as the number of electrons an atom of an element loses, or tends to lose, in a particular reaction. If the atom gains, or tends to gain, electrons the oxidation number will be negative. The rules for assigning oxidation numbers are stated in section 4.12.
3. Matter exists in three states (solid, liquid and gas) whose behaviour is described by the **kinetic theory of matter**. In solids, there is an ordered arrangement (lattice) of particles. The particles may vibrate and rotate about their fixed positions and, on heating, this movement becomes more vigorous until, at the melting point, the lattice breaks up and a liquid is usually formed. In gases, the particles move rapidly and randomly with a wide distribution of kinetic energies (Figure 4.38). At high temperature and low pressure, the attractive forces between gas particles become negligible and they move a long way, compared to their own size, between collisions. In liquids, ordered clusters of particles move through 'gas-like' areas of disorder, containing individual particles. A liquid boils when the vapour pressure just above the liquid surface becomes equal to the atmospheric pressure – this temperature is known as the **boiling point**. Table 4.16 summarises the characteristics of solids, liquids and gases.
4. The **absolute zero** of temperature, 0 Kelvin (K), is taken as –273°C. Thus,
$$X°C = (273 + X)K.$$
5. **Ideal gases** obey the equation
$$pV = nRT$$
at all temperatures and pressures (where n = number of gas moles and R = the gas constant). Real gases often behave like ideal gases at high absolute temperatures and low pressures (e.g. at 293 K and 1 atm (101 325 Pa)).
6. Methods based on the ideal gas equation can be used to determine the relative molecular mass of i) a gas and ii) a volatile liquid (section 4.13 and question 17 on page 126).

Questions on Chapter 4

1 a) Explain the difference between an 'ionic' and a 'covalent' chemical bond, in terms of the distribution of the electrons in each bond.

b) Draw dot and cross diagrams showing the electrons in the outer (valence) shell of each atom in the following compounds:
i) NaCl, MgO
ii) HCl, BF_3, CH_4, SF_6
iii) H_2O, NH_3
iv) CO_2, SO_2.

c) 'When a chemical bond is formed, the bonding atoms achieve the electronic configuration of a noble gas'. Is this statement always true, often true or always false? Use your dot and cross diagrams from part **b)** to support your answer.

d) What is a co-ordinate (dative covalent) bond?

e) Use dot and cross diagrams to illustrate the co-ordinate bonding in the following species:
i) an ammonium ion, NH_4^+
ii) boron trifluoride ammonia, BF_3NH_3
iii) carbon monoxide, CO.

2 Explain how covalent bonds are formed as a result of the overlap of atomic orbitals. You should refer to σ and π bonds, taking as examples the bonding in molecules of oxygen, ethane and ethene.

3 a) Explain how the shape of a simple covalent molecule can be predicted by considering the repulsion between electron pairs. Illustrate your answer by referring to the structure, and shapes, of methane, ammonia and water molecules.

b) Give the shapes of, and bond angles in, the following molecules:
$BeCl_2$ BF_3 $SiCl_4$ PCl_5 SF_6

4 Simple covalent molecules, and ions, display a variety of molecular shapes including:
i) linear ii) trigonal planar iii) tetrahedral
iv) octahedral v) planar vi) V-shaped
By drawing dot and cross diagrams, select the shape which most accurately describes the shape of the following species:
a) PCl_3 **b)** HCN **c)** BF_4^- **d)** C_2H_4
e) PCl_6^- **f)** C_2H_2 **g)** HCHO **h)** $AlCl_3$

Questions on Chapter 4 *continued*

i) AlH_4^- **j)** AlF_6^{3-} **k)** C_2H_6 **l)** H_2O

5 How do you account for the following observations?

a) The sulphur dioxide molecule, SO_2, is V-shaped but the carbon dioxide molecule, CO_2, is linear.

b) The bond angles in ammonia are smaller than those in an ammonium ion.

c) The H—C—H bond angle in methanal, HCHO, is slightly less than 120°.

d) The bond angles in phosphine, PH_3, are 94° whilst those in ammonia, NH_3, are 107°.

6 a) What is meant by the term 'electronegativity'?

b) How and why do electronegativity values vary:
i) across a period and
ii) down a group of the periodic table?

c) Use δ^+ and δ^- signs to show the polarities, if any, in the following bonds:
C—O H—N H—Cl O—S C—C Cl—C
Si—F Cl—B

d) 'A molecule will always be polar if it contains polarised covalent bonds'. Is this statement true or false? Support your answer by:
i) explaining what a permanent dipole is and how it arises in a covalent molecule,
ii) referring to the structure and polarity of molecules of trichloromethane, tetrachloromethane, propanone, cyclohexane and water.

e) Which of the following molecules would you expect to have a permanent dipole?
i) NH_3 ii) BCl_3 iii) CH_2Cl_2 iv) CO_2
v) SO_2 vi) HBr vii) SiF_4

7 There are two main types of intermolecular forces: permanent dipole–permanent dipole attractions and van der Waals' forces.

a) Using suitable diagrams and examples, explain how these intermolecular forces arise.

b) In which of the following substances will permanent dipole–permanent dipole attractions be found?
i) Methanol, CH_3OH
ii) Ethane, C_2H_6
iii) Propanone, CH_3COCH_3
iv) Carbon disulphide, CS_2
v) Bromine, Br_2
vi) Methylamine, CH_3NH_2

c) Two of the compounds in part **b)** contain hydrogen bonds.
i) Identify the two compounds.
ii) Using the structure of these two compounds as examples, explain the meaning of the term 'hydrogen bond'.

d) How does the presence of hydrogen bonds explain the following observations:
i) The trend in the boiling points of the hydrides of the group 6 elements (O, S, Se and Te).
ii) The relative molecular mass of gaseous methanoic acid, HCOOH, appears to be 92.
iii) Ice floats on water.
iv) The energy required to vaporise one mole of water is over twenty times greater than that of tetrachloromethane, CCl_4, a molecule of much higher relative molecular mass.

8 a) Describe the nature of the attractive forces which hold the particles together in a simple molecular solid structure, illustrating your answer by referring to propanone, ice and iodine.

b) The physical properties of some simple molecular substances are shown in the table below.
i) Classify the substances as being polar or non-polar.
ii) Which two substances are most likely to contain hydrogen bonds? Explain your answer.
iii) Substances **A** and **B** are used as industrial solvents. Explain, in energetic terms, why substance **E** is almost insoluble in **A** but is readily soluble in **B**.
iv) Substance **F** is one of the following three compounds. Identify which one it is.

$CH_3CH_2CH_2CHO$ butanal
$CH_3CH_2CH_2CH_2CH_3$ pentane
CH_3CH_2COOH propanoic acid

Substance	M_r	m.p./°C	b.p./°C	solubility in	
				polar solvents	non-polar solvents
A	74		118	high	low
B	72		36	negligible	high
C	332	92		low	high
D	90	157		high	very low
E	283	37		negligible	high
F	72		74	good	good

9 a) Graphite and diamond are said to have 'giant molecular' structures. Explain what this means.

b) Sketch the structures of diamond and graphite and explain why:
i) Graphite conducts electricity, whereas diamond is an insulator.
ii) Graphite is used as a lubricant but diamond is used to make drill bits.
iii) Graphite is much less dense than diamond.

c) List the characteristic physical properties of giant molecular structures.

10 The chlorides of the group 1 metals, lithium to caesium, are ionic solids. As such, they are soluble in water and the aqueous solution conducts electricity.

a) Use dot and cross diagrams to show how an ionic bond in lithium chloride is formed from lithium and chlorine atoms.

b) Draw diagrams to show how the ions are arranged in a crystal of:
i) sodium chloride
ii) caesium chloride

c) Explain the meaning of the following terms: co-ordination number, 6:6 lattice, 8:8 lattice.

Questions on Chapter 4 *continued*

d) A giant ionic lattice is held together by 'non-directional electrostatic forces'. Explain what this means.

e) Ionic solids can be cleaved when struck with a sharp chisel. What does this mean? How do you account for this property in structural terms?

f) Why do ionic substances not conduct electricity in the solid state, but do so when molten or dissolved in aqueous solutions?

g) Unlike the other chlorides, lithium chloride reacts with water (hydrolyses) to produce an acidic solution.

i) Give the name of the acidic substance which is formed and write an equation for the reaction.

ii) Many covalent chlorides hydrolyse in water. This suggests that there is a degree of covalent character in lithium chloride. Explain how this may arise.

11 a) Explain, using suitable diagrams, what structural changes occur when an ionic solid dissolves in water.

b) Define the following terms: lattice energy, hydration enthalpy and standard enthalpy of solution.

c) Write an enthalpy cycle which links these energetic quantities.

d) The hydration enthalpies for selected ions are (in kJ mol^{-1}):

Li^+ −499 Na^+ −390 K^+ −305 Rb^+ −281
Cs^+ −248 Cl^- −384 Pb^{2+} −1449 Ag^+ −464

i) Why does each value have a negative sign?

ii) Are bonds being made or broken when hydration occurs? Which bonds are involved?

iii) Explain the pattern in the values of the hydration enthalpies of the group 1 metal ions, in terms of the charge density (charge/size) on the ions.

iv) Calculate the standard enthalpy of solution for each group 1 metal chloride. (Lattice energies: LiCl −849, NaCl −781, KCl −710, RbCl −685, CsCl −641 kJ mol^{-1}.)

e) Lead chloride and silver chloride are sparingly soluble in cold water. The lattice enthalpies are −2234 and −890 kJ mol^{-1}, respectively. Calculate standard enthalpies of solution for these solids and compare them with those for the group 1 chlorides. Are there any general conclusions that you can draw?

12 a) Outline a model which explains how the atoms are held together in a giant metallic structure.

b) Use this model to explain why metals:

i) are good conductors of heat and electricity,

ii) readily form alloys,

iii) are malleable and ductile.

c) Sodium is a soft metal which melts at 98°C. Iron is very hard and melts at 1539°C. Account for these properties by considering:

i) the relative strengths of the metallic bonding based on their electronic configurations and ionisation energies,

ii) the closeness of the packing of the metal atoms in the giant metallic lattice.

d) Use your model of metallic bonding to predict the trend in:

i) the melting points of group 1 metals,

ii) the melting points of potassium, calcium and scandium,

iii) the electrical conductivities of the group 1 metals,

iv) the electrical conductivities of sodium, magnesium and aluminium.

13 The properties of various substances, A–K, are shown below. Which of the following structures seem to fit the atomic properties?

i) Simple molecular ii) Giant molecular
iii) Giant ionic iv) Giant metallic

Substance	m.p. (°C)	b.p. (°C)	hardness of the solid	conductivity	relative solubility
A	635	1300	hard, brittle	when molten or in aqueous solution	good, in water
B	2700	(decomposes)	very hard	none	insoluble
C	1539	2887	hard, malleable	good	insoluble
D	−112	−108	soft	none	very low
E	1350	(decomposes)	hard, brittle	when molten	almost insoluble
F	3727	(sublimes)	soft	good	insoluble
G	−39	357	soft	good	insoluble
H	7	81	soft	none	good, in non-polar solvents
I	−98	64	soft	none	good, in water
J	1610	2230	very hard	none	insoluble
K	772	1600	hard, brittle	when molten or in aqueous solution	good, in water

Questions on Chapter 4 *continued*

14 a) What do you understand by the terms: oxidation, reduction, redox reaction and oxidation number?

b) What are the oxidation numbers of:

i) Sulphur in Na_2S, SO_4^{2-}, SO_3^{2-}, $S_4O_6^{2-}$, $S_2O_3^{2-}$, SO_2, SF_6.

ii) Nitrogen in NO_2, N_2O, NO_3^-, NH_4^+, NH_3, N_2O_5, NO, N_2H_4.

iii) Chlorine in Cl_2, HCl, ClO_3^-, ClO^-, ClO_4^-, ClO_2^-, ClF, $AlCl_4^-$,

iv) Iron in $FeSO_4$, $FeCl_3$, $Fe(NO_3)_2$, $FeCl_6^{3-}$, Fe_2O_3, FeO.

c) In each of the following equations, state which atom has been oxidised, which has been reduced and the total number of electrons which have been transferred.

i) $Cl_2 + 2KBr \longrightarrow 2KCl + Br_2$

ii) $Mg + 2HCl \longrightarrow MgCl_2 + H_2$

iii) $4NH_3 + 5O_2 \longrightarrow 4NO + 6H_2O$

iv) $2S_2O_3^{2-} + I_2 \longrightarrow S_4O_6^{2-} + 2I^-$

v) $5C_2H_5OH + 2MnO_4^- + 6H^+ \longrightarrow 5CH_3CHO + 2Mn^{2+} + 8H_2O$

15 a) By referring to the kinetic theory of matter, describe how the particles are arranged in solids, liquids and gases and the extent to which the particles can move in each state. You should use simple diagrams to represent the arrangement of particles.

b) Use the structural models from part **a)** to explain why:

i) Gases can be easily compressed, but it is often impossible to compress liquids or solids.

ii) The temperature of a liquid at its boiling point remains constant, even if you continue to heat it.

iii) You can smell someone's perfume, or aftershave, some distance away from them.

iv) The water in a reservoir will evaporate, to some extent, well below its boiling point, with the rate of evaporation depending on the temperature, the pressure and the surface area of the water.

16 a) What is meant by the term 'vapour pressure'?

b) Explain why the boiling point of a liquid varies with the total pressure above the liquid.

c) Ethoxyethane, $C_2H_5OC_2H_5$, is a valuable solvent used in the preparation of organic chemicals. It is a very volatile and flammable substance. The vapour pressure of ethoxyethane varies with its temperature, as shown below:

temperature (K)	225	246	262	291	308
vapour pressure (atm)	0.013	0.053	0.132	0.530	1.000

i) Plot a graph of vapour pressure (vertical axis) against temperature (horizontal axis).

ii) Use your graph to estimate the vapour pressure of ethoxyethane when the temperature is 265 K and 300 K respectively.

iii) At what temperatures would ethoxyethane boil if the external pressure were 0.010 and 0.600 atm respectively?

iv) A chemist has prepared a valuable new drug which is a non-volatile solid. It is obtained as a solution in ethoxyethane and the chemist now wishes to remove the ethoxyethane. This poses a problem because ethoxyethane does not boil until 35°C at normal atmospheric pressure and the drug decomposes at 30°C. Give suggestions as to how to remove the ethoxyethane without decomposing the drug.

17 The apparatus shown in Figure 4.42 can be used to determine the relative molecular mass of a liquid which boils below about 80°C. A known mass of the volatile liquid is injected into the gas syringe, whereupon it vaporises and the syringe is pushed out. The volume of the vapour produced, the temperature of the steam jacket and the atmospheric pressure are noted. Determine the relative molecular mass of liquids A, B and C from the following data.

	A	B	C
Volume of vapour produced/cm³	83	46	63
Mass of liquid used/g	0.24	0.18	0.27
Temperature/°C	100	100	98
Pressure/Pa	100680	101325	100900

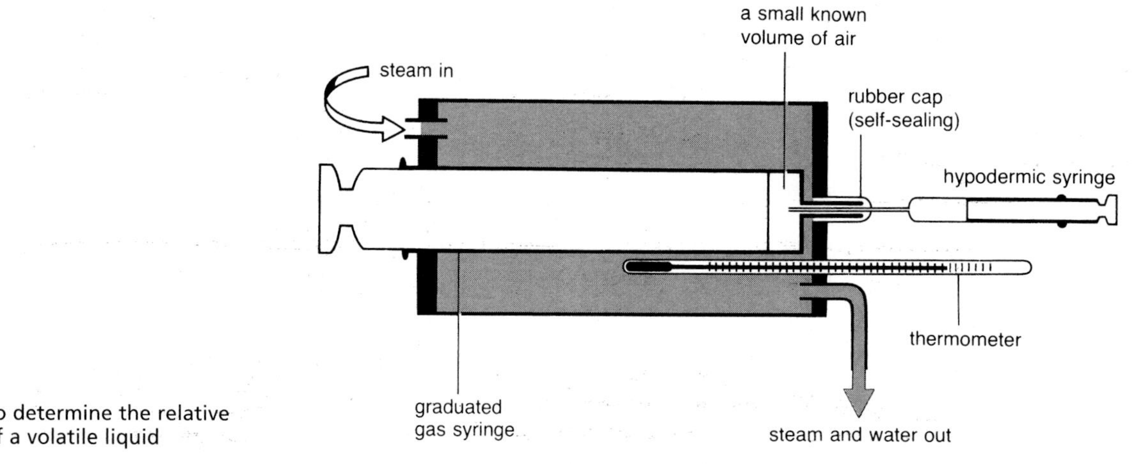

Figure 4.42
Apparatus used to determine the relative molecular mass of a volatile liquid

Comments on the activities

Activity 4.1

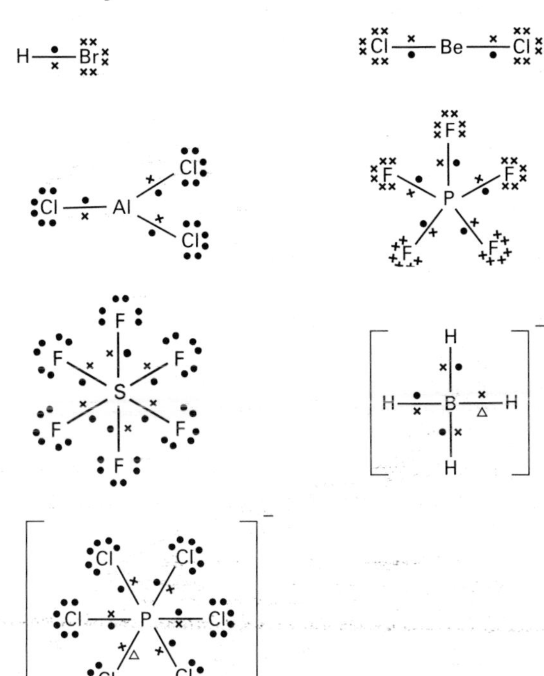

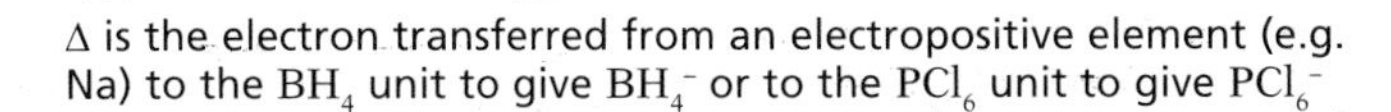

Δ is the electron transferred from an electropositive element (e.g. Na) to the BH_4 unit to give BH_4^- or to the PCl_6 unit to give PCl_6^-

Activity 4.3

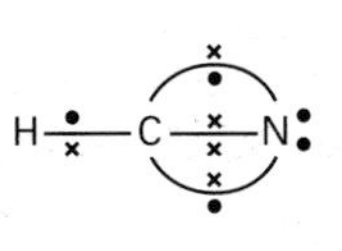

2 sets of electrons around C
∴ **linear**

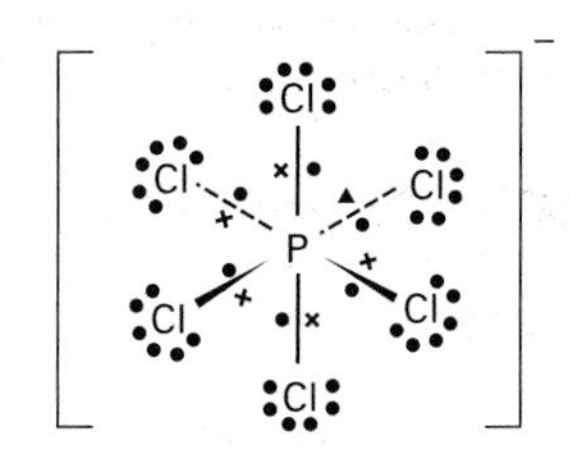

6 sets of electrons around P
∴ **octahedral**

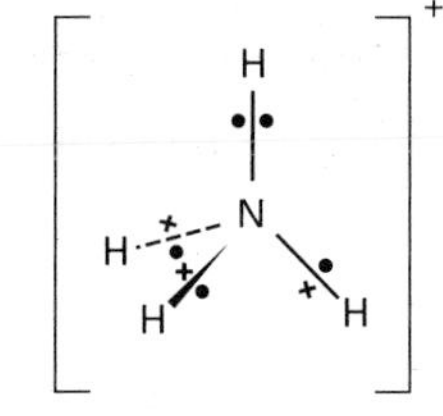

4 sets of electrons around N
∴ **tetrahedral**

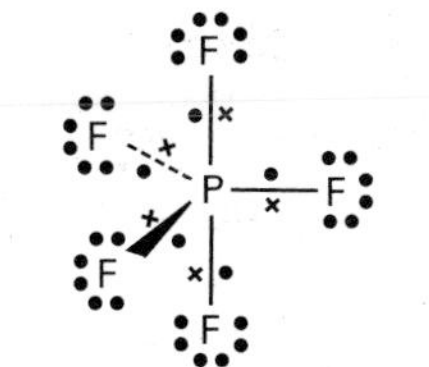

5 sets of electrons around P
∴ **trigonal bipyramid**

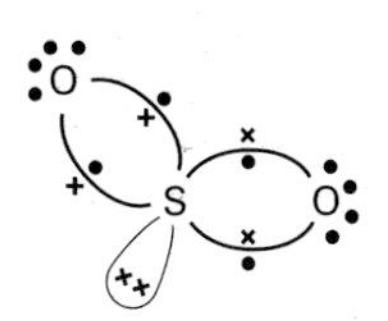

3 sets of electrons around S, 1 lone pair and 2 bonding pairs; the molecule is V-shaped

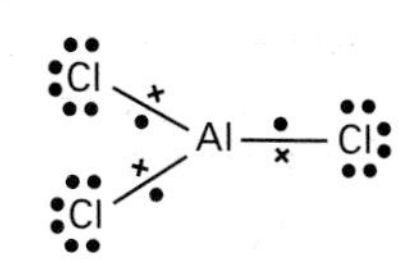

3 sets of electrons around Al
∴ **trigonal planar**

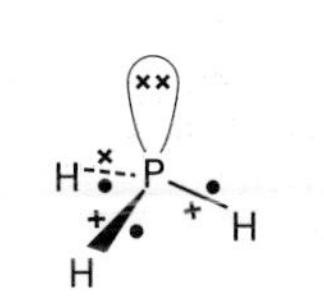

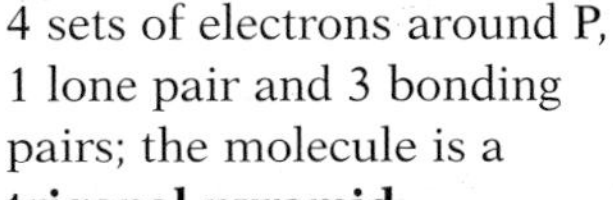

4 sets of electrons around P, 1 lone pair and 3 bonding pairs; the molecule is a **trigonal pyramid**

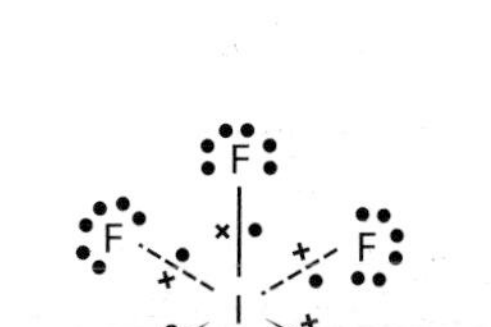

6 sets of electrons around I; 1 lone pair and 5 bonding pairs; the molecule is a **square pyramid**

Note: Lone pairs affect the shape of the molecule but they are not part of the molecular shape. Thus, PH_3 is trigonal pyramidal, *not* tetrahedral.

Activity 4.4

1

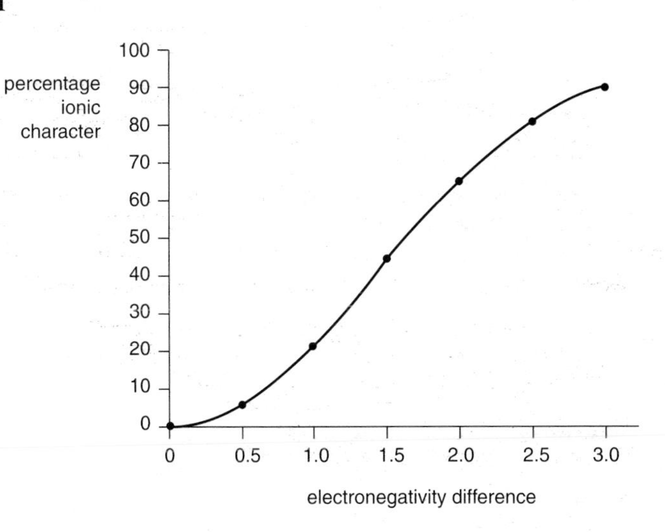

2 **a)** O, F: electronegativity difference 0.5, ionic character 6%;
b) B, F: 2.3, 74;
c) C, Cl: 0.5, 6;
d) P, H: 0, 0;
e) C, O: 1.0, 22;
f) O, H: 1.4, 39;
g) H, Br: 0.7, 12;
h) Al, Cl: 1.5, 43.

3 $\overset{\delta+}{C}—\overset{\delta-}{Cl}$ P—H $\overset{\delta+}{C}—\overset{\delta-}{O}$ $\overset{\delta-}{O}—\overset{\delta+}{H}$ $\overset{\delta+}{H}—\overset{\delta-}{Br}$

Comments on the activities *continued*

Activity 4.5

1 a)

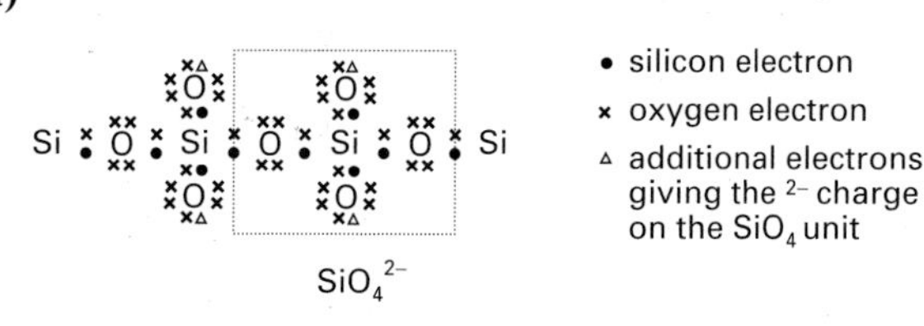

b) SiO_4^{2-}; the 2− charge is spread over the two oxygen atoms which are not shared with the silicon atoms in the 'chain'.

c) The covalent bonds between the silicon and oxygen atoms in a particular chain are much stronger than the electrostatic forces holding neighbouring chains together (via cations such as Mg^{2+} or Ca^{2+}).

2 a)

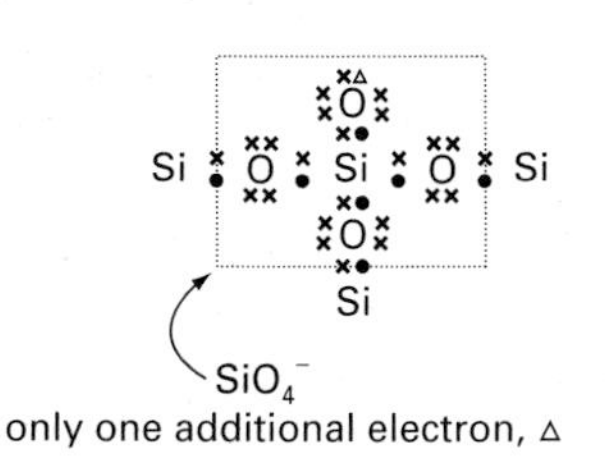

b i) and ii) are both explained by the strength of the covalent bonds within the individual sheet structure being much stronger than the attractive forces between neighbouring sheet (electrostatic forces, via cations such as Mg^{2+} or Al^{3+}).

3 When wet, the clay soil absorbs a large amount of water and expands. Then, as the soil dries, the clay contracts, which may cause the foundations to move. Houses which have been affected by this behaviour, known as subsidence, are characterised by cracked walls and badly fitting doors. As subsidence is covered by household buildings insurance, homes built on clay foundations are more expensive to insure.

Activity 4.6

Only the outer shells of electrons are shown.

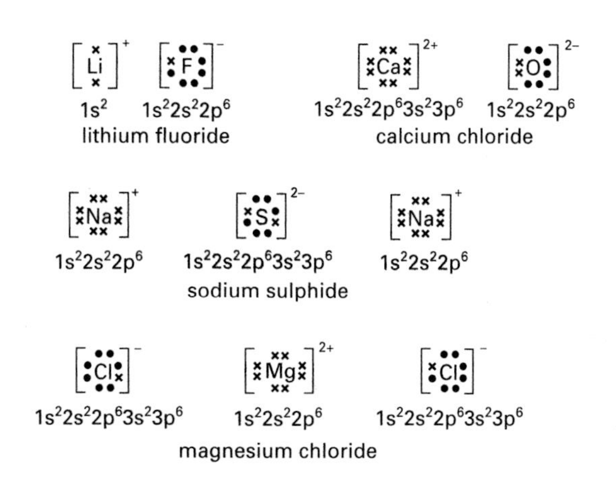

Activity 4.7

1 The ionic radii of the positive ions (cations) are smaller than those of the atoms from which they are formed. Negative ions (anions) are larger than the parent atoms. The relative difference in size between the atom and its ion increases with the charge on the ion.

2 A cation has fewer electrons than the parent atom, so the nucleus attracts them more strongly causing a reduction in size. An anion has more electrons than the parent atom, so the attraction of the nucleus for each electron is reduced, resulting in an expansion. These effects will become greater as more electrons are lost, or added, so the relative difference in size increases as the ions become more highly charged.

Activity 4.8

From Figure 4.33 and Hess's Law,

$$\Delta H^{\ominus}_{s}[MgCl_2(s)] = -\Delta H^{\ominus}_{L}[MgCl_2(s)] + \Delta H^{\ominus}_{h}[Mg^{2+}(g)] + 2\Delta H^{\ominus}_{h}[Cl^{-}(g)]$$

$$\Delta H^{\ominus}_{s}[MgCl_2(s)] = -(-2489) + (-1891) + 2(-384)$$
$$= +2489 - 1891 - 768$$
$$= -170 \text{ kJ mol}^{-1}$$

Note that, since 2 moles of $Cl^{-}(g)$ ions are hydrated, we must take $2 \times \Delta H^{\ominus}_{h}[Cl^{-}(g)]$.

In these calculations, beware the common error of failing to use multiples of the hydration enthalpies (where necessary). For example, when 1 mole of iron(III) sulphate, $Fe_2(SO_4)_3$, is dissolved in water, the sum of the enthalpies is: $2 \times \Delta H^{\ominus}_{h}[Fe^{3+}(g)]$ and $3 \times \Delta H^{\ominus}_{h}[SO_4^{2-}(g)]$.

Activity 4.9

1 $2Ca + O_2 \longrightarrow 2Ca^{2+} + 2O^{2-}$

$2Ca \longrightarrow 2Ca^{2+} + 4e^{-}$ oxidation

$O_2 + 4e^{-} \longrightarrow 2O^{2-}$ reduction

Oxygen is the oxidising agent and calcium is the reducing agent.

2 $2K + S \longrightarrow 2K^{+} + S^{2-}$

$2K \longrightarrow 2K^{+} + 2e^{-}$ oxidation

$S + 2e^{-} \longrightarrow S^{2-}$ reduction

Potassium is the reducing agent and sulphur is the oxidising agent.

3 $2Al + 3F_2 \longrightarrow 2Al^{3+} + 6F^{-}$

$2Al \longrightarrow 2Al^{3+} + 6e^{-}$ oxidation

$3F_2 + 6e^{-} \longrightarrow 6F^{-}$ reduction

Aluminium is the reducing agent and fluorine is the oxidising agent.

4 $4Al + 3O_2 \longrightarrow 4Al^{3+} + 6O^{2-}$

$4Al \longrightarrow 4Al^{3+} + 12e^{-}$ oxidation

$3O_2 + 12e^{-} \longrightarrow 6O^{2-}$ reduction

Aluminium is the reducing agent and oxygen is the oxidising agent.

5 $3Mg + N_2 \longrightarrow 3Mg^{2+} + 2N^{3-}$

$3Mg \longrightarrow 3Mg^{2+} + 6e^{-}$ oxidation

$N_2 + 6e^{-} \longrightarrow 2N^{3-}$ reduction

Comments on the activities *continued*

Magnesium is the reducing agent and nitrogen is the oxidising agent.

Activity 4.10

1 Using the 'rules' given in section 4.12,

a) Ca Cl_2: +2, 2(−1) **b)** Si O_2: +4, 2(−2) **c)** K O_2: +1, 2(−0.5)

d) N H_4^+: −3, 4(+1) **e)** Cu Cl_4^{2-}: +2, 4(−1) **f)** Na_3 Al F_6: 3(+1) +3 6(−1)

2 $Cl_2(aq) + 2KBr(aq) \longrightarrow 2KCl(aq) + Br_2(aq)$

2(−1) → 2(0)

Bromide has been oxidised by chlorine.

$2CrO_4^{2-}(aq) + 2H^+(aq) \longrightarrow Cr_2O_7^{2-}(aq) + H_2O(l)$

2(+6) → 2(+6)

No oxidation has occurred.

$I_2(aq) + 2S_2O_3^{2-}(aq) \longrightarrow 2I^-(aq) + S_4O_6^{2-}(aq)$

4(+2) → 4(+2.5)

The thiosulphate, $S_2O_3^{2-}$, has been oxidised by the iodine.

$3MnO_4^{2-}(aq) + 4H^+(aq) \longrightarrow$

3(+6)

$2MnO_4^-(aq) + MnO_2(s) + 2H_2O(l)$

2(+7) +4

The manganate(VI), MnO_4^{2-}, has been oxidised by itself to manganate(VII).

3 **a)** Nitrogen is more electronegative than hydrogen and so tends to gain electrons during the formation of ammonia. In terms of change in oxidation numbers, we can see that the nitrogen is reduced.

$N_2(g) + 3H_2(g) \rightleftharpoons 2NH_3(g)$

2(0) 6(0) 2(−3) 6(+1)

Sodium, however, is less electronegative than hydrogen. Group 1 metals only show an oxidation number of +1 in their compounds.

$2Na(s) + H_2(g) \rightleftharpoons 2NaH(s)$

2(0) 2(0) 2(+1) 2(−1)

In this case the addition of hydrogen has oxidised the sodium.

b) Hydrogen will only act as an oxidising agent if it gains or tends to gain an electron. It will only do this when combining with an element of lower electronegativity.

c) In order to act as a reducing agent oxygen must lose or tend to lose electrons. Since it has a very high electronegativity this will be very rare. In fact only fluorine has a greater electronegativity than oxygen. Oxygen does in fact show an oxidation state of +2 in oxygen difluoride, OF_2, but this compound cannot be made by direct combination of the elements.

Activity 4.11

1 Volume of gas = $800\ cm^3$ = $800 \times 10^{-6}\ m^3$

2 Temperature = 23 + 273 K = 296 K

3 Mass of gas = 2.15 g

4 Using the equation

$$M_r = \frac{mRT}{pV}$$

$$M_r = \frac{2.15 \times 8.31 \times 296}{100792 \times 800 \times 10^{-6}}$$

$$M_r = 65.6$$

CHAPTER

5

Periodicity

Contents

Study Checklist

After studying this chapter you should be able to:

1. Classify an element as being in the s, p or d block according to its position in the periodic table.
2. Describe and explain the trends in atomic radius, first ionisation energy and electronegativity on passing across a period and down a group in the periodic table.
3. State that 'bulk' properties, such as melting point, depend upon the way in which the particles are arranged and the attractive forces within the structure of a material.
4. Describe and explain the variation in melting point and electrical conductivity on passing across period 3 (sodium to argon).
5. Describe any reactions of the elements in period 3 (sodium to argon) with oxygen, chlorine and water, giving the formulae of any products.
6. Describe and explain the trends in melting point and any reaction with water of the oxides and chlorides of the elements in period 3 (sodium to argon).

Periodicity deals with the variation in properties of the elements (and their compounds) according to position in the periodic table. This chapter can only deal with some of the main trends but more detailed information is given in the topics dealing with the s, p and d blocks. The requirements of different examination boards vary, and you should check your syllabus in order to identify which extra material is needed.

At first glance, the chemical elements seem to show a bewildering variety of properties. For example, if we arrange the elements in order of increasing atomic number, then we find the totally 'inert' gas neon sandwiched between two of the most reactive elements: fluorine and sodium.

The first suggestion of a link between the properties of elements and their atomic masses came from **Johann Wolfgang Dobereiner**, a German chemist, in 1829. He found that elements with similar properties occurred in groups of three, called **triads**, in which the atomic mass of the middle element was just about the average of the masses of the other two elements, e.g. lithium (A_r 6.9), sodium (A_r 23.0) and potassium (A_r 39.1). Also, the properties of the middle element were found to be intermediate between those of the other triad members. The next important contribution came in 1866 from **John Alexander Newlands**, an English chemist. He arranged the elements in order of atomic mass and found that every eighth element had similar properties, like the eighth note of a musical octave. This idea, which Newlands termed a 'law of octaves', was ridiculed by his peers because of the musical analogy and the way it grouped together some dissimilar elements, e.g. iron with oxygen and sulphur, Figure 5.1.

H	Li	Be	B	C	N	O
F	Na	Mg	Al	Si	P	S
Cl	K	Ca	Cr	Ti	Mn	Fe

Figure 5.1
The first three rows of Newlands' octaves

On 17 February 1869, **Dmitri Mendeléev**, a professor of chemistry at the University of St Petersburg in Russia, was preparing a lecture. He had a series of cards with the symbols, and properties, of the elements written on them. Whilst ordering the cards for his lecture, Mendeléev noted that if the elements were arranged in order of their atomic masses, a pattern in their properties repeated itself on several occasions. Within a month, Mendeléev had written a paper on the subject and presented it, to considerable acclaim, to the Russian Chemical Society. He went on to produce a periodic table which was the forerunner of the modern version which is shown in Figure 5.2.

Only 63 elements had been discovered at the time of Mendeléev's paper and one of the most persuasive features of his table was that he left spaces for elements which were yet to be discovered, such as gallium, scandium and germanium. In many cases, he accurately predicted their properties. Mendeléev's work focused scientists' attention on the discovery of the missing elements and, by 1885, another seven elements had been discovered and had been allotted their places in the periodic table. Just after Mendeléev's initial discovery, **Julius Lothar Meyer**, a German professor of chemistry, plotted various physical properties against atomic mass and obtained a series of curves with similar elements occurring at equivalent points on the curves. He produced a periodic table similar to Mendeléev's and also left spaces for undiscovered elements, an amazing coincidence bearing in mind that there had been no communication between the two men.

One difficulty with Mendeléev's periodic table was that tellurium (Te), with an atomic mass of 128, was placed before iodine (I), atomic mass 127. Mendeléev assumed that the mass of tellurium was in error and that, with more sophisticated measuring equipment, his order would be shown to be correct. Indeed, in a later version of his table, Mendeléev listed the mass of Te as 125 without explaining why the value had changed! It was not until work by Henry Moseley in 1812 showed that the elements should be ordered by their atomic numbers and not masses, that the modern form of the **periodic table** took shape and tellurium, atomic number 52, justifiably took its place before iodine, atomic number 53.

When the elements are arranged in order of atomic number, many of their physical and chemical properties show a regular pattern which repeats itself periodically, this behaviour is known as **periodicity**. This chapter explains the form and construction of the modern periodic table and the variation in the properties of the elements in terms of atomic structure, which was unknown to the pioneers in this field.

Without doubt, the periodic classification of the elements is one of the greatest achievements in chemistry. A working knowledge of the ideas presented in this chapter gives a theoretical framework within which to build up the detailed chemistry of the elements. Knowing and understanding the well-defined trends makes it possible to predict, with a fair degree of confidence, the properties of any unfamiliar element simply by considering its position in the periodic table.

Time spent in studying periodicity certainly helps to make inorganic chemistry seem clearer, more logical and therefore easier to understand and recall. You must, however, be familiar with the ideas on atomic structure covered in Chapter 2.

Figure 5.2
Sub-divisions of the periodic table

Period \ Group	1	2											3	4	5	6	7	0
1	1 H Hydrogen 1.0																	2 He Helium 4.0
2	3 Li Lithium 6.9	4 Be Beryllium 9.0											5 B Boron 10.8	6 C Carbon 12.0	7 N Nitrogen 14.0	8 O Oxygen 16.0	9 F Fluorine 19.0	10 Ne Neon 20.2
3	11 Na Sodium 23.0	12 Mg Magnesium 24.3											13 Al Aluminium 27.0	14 Si Silicon 28.1	15 P Phosphorus 31.0	16 S Sulphur 32.1	17 Cl Chlorine 38.5	18 Ar Argon 39.9
4	19 K Potassium 39.1	20 Ca Calcium 40.1	21 Sc Scandium 45.0	22 Ti Titanium 47.9	23 V Vanadium 50.9	24 Cr Chromium 52.0	25 Mn Manganese 54.9	26 Fe Iron 55.9	27 Co Cobalt 58.9	28 Ni Nickel 58.7	29 Cu Copper 63.5	30 Zn Zinc 65.4	31 Ga Gallium 69.7	32 Ge Germanium 72.6	33 As Arsenic 74.9	34 Se Selenium 79.0	35 Br Bromine 79.9	36 Kr Krypton 83.8
5	37 Rb Rubidium 85.5	38 Sr Strontium 87.6	39 Y Yttrium 88.9	40 Zr Zirconium 91.2	41 Nb Niobium 92.9	42 Mo Molybdenum 95.9	43 Tc Technetium 99.0	44 Ru Ruthenium 101.1	45 Rh Rhodium 102.9	46 Pd Palladium 106.4	47 Ag Silver 107.9	48 Cd Cadmium 112.4	49 In Indium 114.8	50 Sn Tin 118.7	51 Sb Antimony 121.8	52 Te Tellurium 127.6	53 I Iodine 126.9	54 Xe Xenon 131.3
6	55 Cs Caesium 132.9	56 Ba Barium 137.3	57 La Lanthanum 138.9	72 Hf Hafnium 178.5	73 Ta Tantalum 180.9	74 W Tungsten 183.9	75 Re Rhenium 186.2	76 Os Osmium 190.2	77 Ir Iridium 192.2	78 Pt Platinum 195.1	79 Au Gold 197.0	80 Hg Mercury 200.6	81 Tl Thallium 204.4	82 Pb Lead 207.2	83 Bi Bismuth 209.0	84 Po Polonium 210.0	85 At Astatine 210.0	86 Rn Radon 222.0
7	87 Fr Francium 233.0	88 Ra Radium 226.0	89 Ac Actinium 227.0	104 Rt Rutherfordium 261	105 Db Dubnium 262	106 Sg Seaborgium 263	107 Bh Bohrium 262	108 Hs Hassium 265	109 Mt Meitnerium 266	110 Uun Ununnilium 269	111 Uuu Unununium 272	112 Uub Ununbium 277						

s block

d block

p block

58–71 lanthanum series	58 Ce Cerium 140.1	59 Pr Praseodymium 140.9	60 Nd Neodymium 144.2	61 Pm Promethium 145.0	62 Sm Samarium 150.4	63 Eu Europium 152.0	64 Gd Gadolinium 157.3	65 Tb Terbium 158.9	66 Dy Dysprosium 162.5	67 Ho Holmium 164.9	68 Er Erbium 167.3	69 Tm Thulium 168.9	70 Yb Ytterbium 173.0	71 Lu Lutetium 175.0
90–103 actinium series	90 Th Thorium 232.0	91 Pa Protoactinium 231.0	92 U Uranium 238.1	93 Np Neptunium 237	94 Pu Plutonium 242	95 Am Americium 243	96 Cm Curium 247	97 Bk Berkelium 249	98 Cf Californium 251	99 Es Einsteinium 254	100 Fm Fermium 253	101 Md Mendelevium 256	102 No Nobelium 254	103 Lr Lawrencium 257

f block

Key

Atomic number
Symbol
Name
Relative atomic mass

5.1 The periodic table and atomic structure

The elements are laid out in the commonly used form of the periodic table (Figure 5.2) in the following way:

- In order of increasing atomic number.
- Starting a new row when electrons begin to fill a new principal energy level or shell.
- Placing elements whose atoms have similar outer electronic structures in vertical columns.

The resulting table may be sub-divided in a number of ways:

- Into s, p, d and f **blocks**. Within any block, the final electron to be added to the atom enters a sub-level of the type shown by the block letter.
- Into vertical columns, referred to as **groups**. Elements within any group have similar outer electronic structures. Sub-division into groups is particularly useful in the s and p blocks.
- Into horizontal rows, known as **periods**. In any period all the atoms have their 'outer' electrons in the same principal energy level (electron shell). Periods 1–3 are often referred to as the 'short' periods of the table.

Activity 5.1 Blocks, groups and periods

State the block, group (if any) and period to which each of the following elements belong in the periodic table shown in Figure 5.2:

lithium (Li), silicon (Si), bromine (Br), iron (Fe), helium (He), oxygen (O), calcium (Ca), uranium (U), lead (Pb), gold (Au).

Alternative group numbering systems

An alternative system of numbering vertical groups is now beginning to gain favour. Under the new system proposed by the IUPAC (International Union of Pure and Applied Chemistry), the d block is included in the group numbering system. Groups 1 and 2 retain the same number, the d block accounts for the 'new' groups 3–12, and in the p block, groups 3–0 are renumbered 13–18. (At A/AS level however, the original group numbering system is preferred.)

Atomic size

This is important in predicting and explaining the variation in many other properties. In this section, we will use the idea of atomic size without stating precisely what we mean. In fact, atomic size may be defined and measured in more than one way, as explained in the accompanying extension box.

The 'size' of an atom will depend upon the 'position' of the electrons in the outer principal energy level. This, in turn, will depend upon the strength of the attraction from the positively charged nucleus.

- *On crossing from left to right across a period, atomic size generally decreases*. At first glance this might seem strange. After all, since more particles are successively being added to the atoms, shouldn't they be getting bigger? The reason is that the increasing positive charge on the nucleus exerts a stronger pull on the outer electrons, causing the atom to shrink.
- *On passing down a group, atomic size generally increases*. Although the nuclear charge is increasing, the outer electrons enter new shells which are effectively screened from nuclear attraction by a greater number of inner electrons.

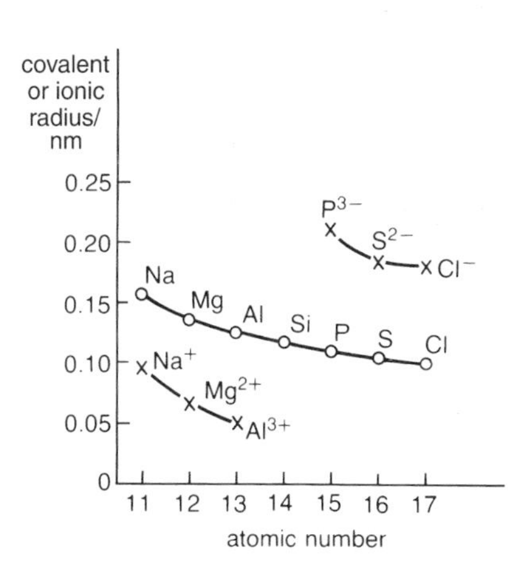

Figure 5.3
Comparison of covalent and ionic radii in period 3. Ar, Cl^-, S^{2-}, K^+ and Ca^{2+} have the same arrangement of electrons. Place these particles in order of increasing size.

Before leaving this subject, it is worth considering what happens when an atom forms an ion. Positive ions are always smaller than their parent atoms, since there are fewer electrons to 'share' the attraction of the nucleus. The greater the number of electrons lost, the greater the shrinkage will be. Conversely, negative ions are always bigger than their parent atoms since the same nuclear attraction must be shared with a greater number of outer electrons. Figure 5.3 shows the relative effect on particle size on forming common ions of some elements in period 3.

Measurement of atomic 'size'

We cannot measure the size of an isolated atom because we can never be exactly sure where its electrons will be. In practical terms, we must describe the size of an atom in terms of the space it *appears* to occupy in any structure. Since atomic nuclei are relatively heavy, we may establish their positions quite accurately within any structure by methods which involve deflection of moving particles or waves. From such experiments, we can construct 'space-filling' models, where the atoms appear as spheres just touching their nearest neighbours. A typical result for a metallic element is shown in Figure 5.4.

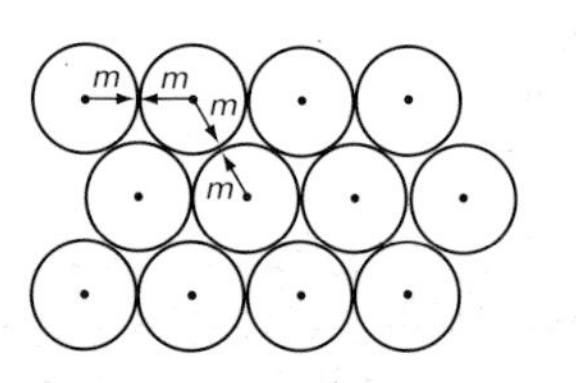

◀ **Figure 5.4** The position of nuclei in a metallic lattice. The metallic radius, *m*, is half the shortest internuclear distance in such a structure

Each atom appears to occupy a sphere of radius *m* which is known as its **metallic radius**. In the case of a non-metal however, the result may not be quite so straightforward. The pattern of nuclei in solid iodine, for example, is shown in Figure 5.5. The nuclei appear in pairs where the radius of each atom seems to be *c*. However, in order to fill the gaps between each of these pairs, we would need to increase the radius of each atom to *v*. The reason for this difference is that within each pair, the atoms are linked by a single covalent bond, whilst the attraction between the atoms in different pairs is limited to van der Waals' forces. In effect the iodine atom seems to have two different sizes: a **covalent radius** (*c*) when it is bonded to other atoms, and a **van der Waals' radius** (*v*) when non-bonded.

We thus have three possible estimates of atomic size: metallic radius, covalent radius and van der Waals' radius. Each of these depends upon the way in which a particular atom is attached to its neighbours in any particular structure and will consequently have a different value. It is important, therefore, when comparing atomic sizes to compare 'like with like'. Figure 5.6 shows how the different measures of atomic radius vary with an element's atomic number and position in the periodic table.

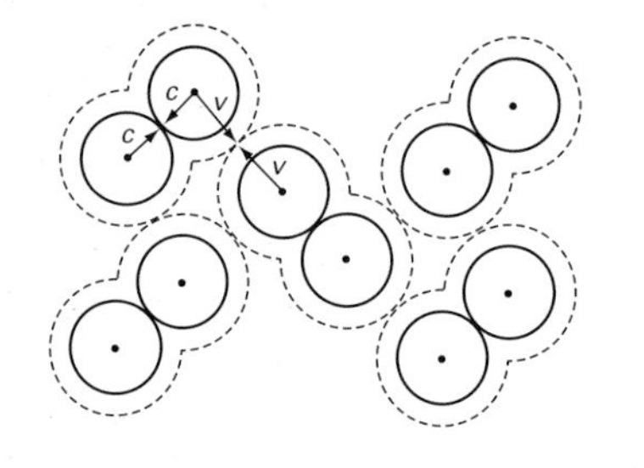

◀ **Figure 5.5** The structure of solid iodine showing the covalent radius, *c*, between bonded atoms within I_2 molecules and the van der Waals' radius, *v*, between non-bonded atoms

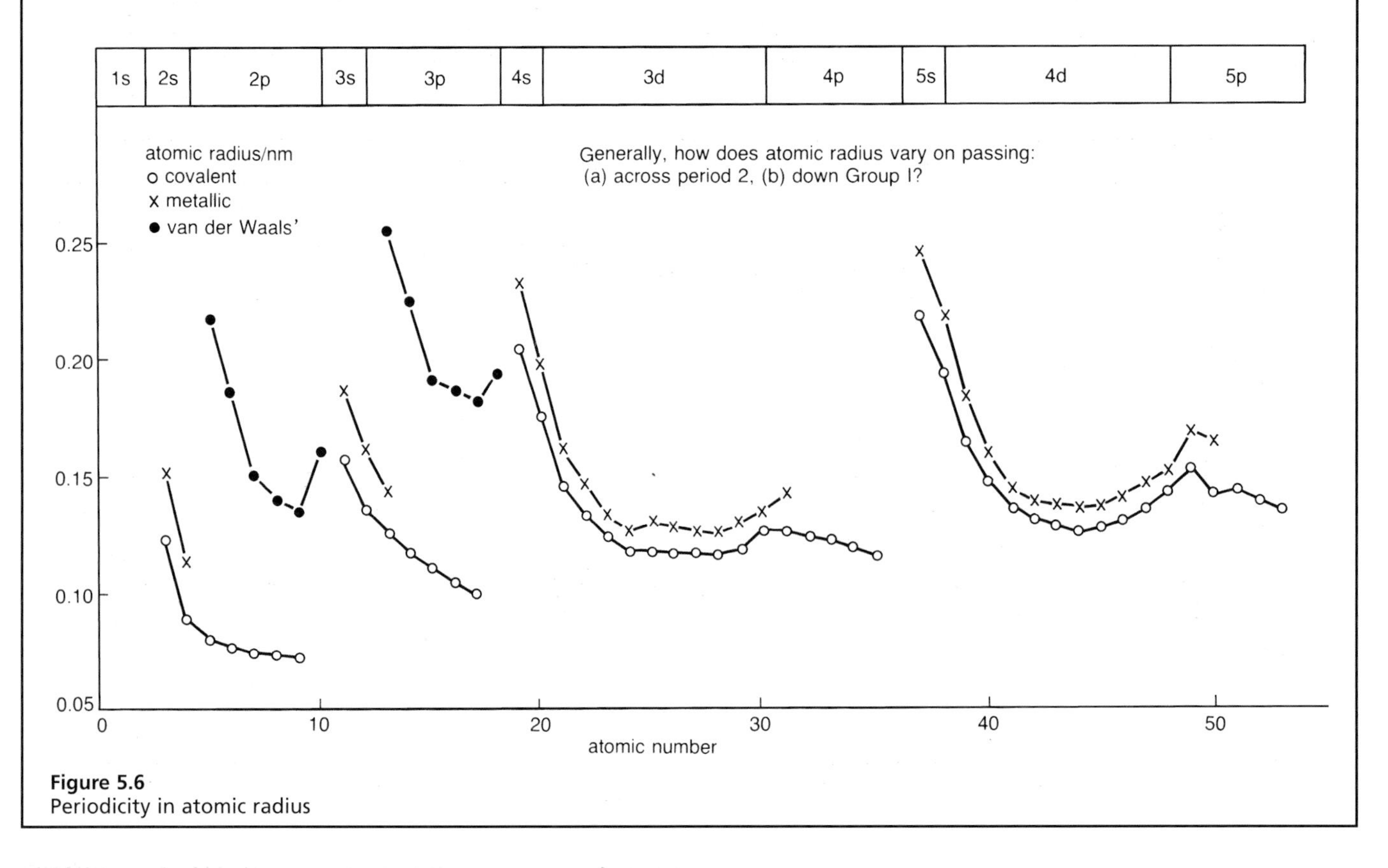

Figure 5.6
Periodicity in atomic radius

5.2 Atomic properties of elements

The properties described in this section depend only upon the nature of the individual atoms and not on how they are arranged in the structure of the element.

First ionisation energies

The **first ionisation energy** of an element may be defined as the enthalpy change on converting one mole of isolated gaseous atoms into gaseous ions, each of which has a single positive charge, i.e.:

$$M(g) \longrightarrow M^{+}(g) + e^{-}$$

In effect, it is a measure of the energy required to remove the most loosely bound electron from the atom. It therefore depends upon the attraction of the nucleus for that electron which, in turn, will depend upon the following factors:

- the size of the positive charge on the nucleus,
- the distance of the electron away from the nucleus,
- the number of inner screening electrons present.

On passing from left to right across a short period, the number of inner screening electrons does not alter, but nuclear charge increases and atomic size falls. Both of these changes increase the attraction of the nucleus for the most loosely bound electron so that first ionisation energy rises.

On passing down a group, the increase in nuclear charge would tend to increase first ionisation energy but, since the atoms get bigger, the outer electron is further away from the nucleus and also better screened by more inner electrons. First ionisation energy therefore falls on passing down a group.

Both of these general trends can be seen in the graph of first ionisation energy against atomic number shown in Figure 5.7. However, the increase in first ionisation energy on passing across a period is not smooth. These irregularities provide evidence for the existence of different sub-levels within the principal electron energy levels of atoms, as explained below.

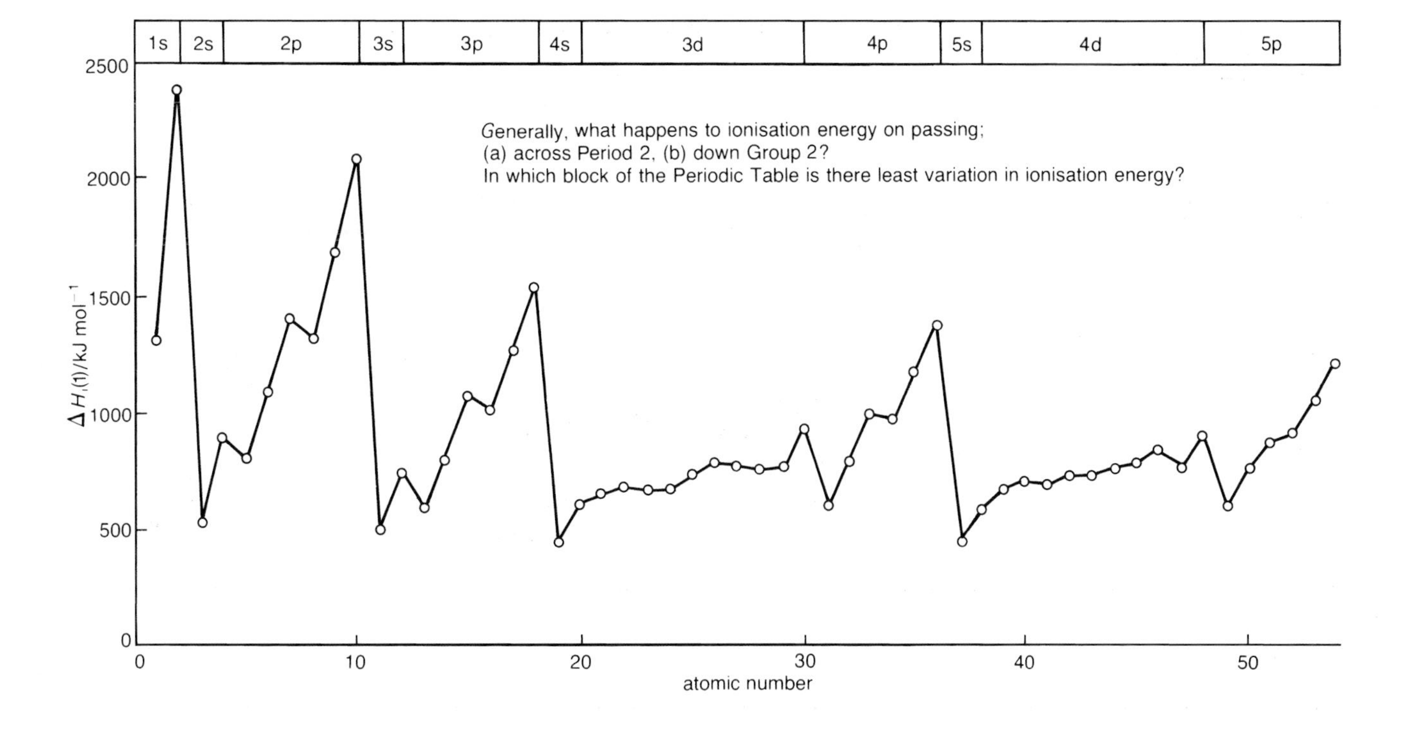

Figure 5.7
Variation in first ionisation energy $\Delta H_i(1)$ for the first 54 elements

Figure 5.8 shows the variation in first ionisation energy on crossing period 2 (lithium to neon) in detail. There are two 'breaks' in a generally smooth increase in ionisation energy. At the start of the period, electrons fill into the 2s sub-level. After this is filled at beryllium, the next electron enters the 2p sub-level, which on average is a little further from the nucleus and less strongly attracted by the nucleus. For this reason, there is a drop in ionisation energy between beryllium and boron. The second drop in ionisation energy, which occurs between nitrogen and oxygen, is not caused by starting a new sub-level. In the nitrogen atom, each of the three orbitals in the 2p sub-level contains one electron. At oxygen, the extra electron must pair up in a p orbital. As this is repelled by the first electron in that orbital, it is easier to remove, causing a drop in ionisation energy.

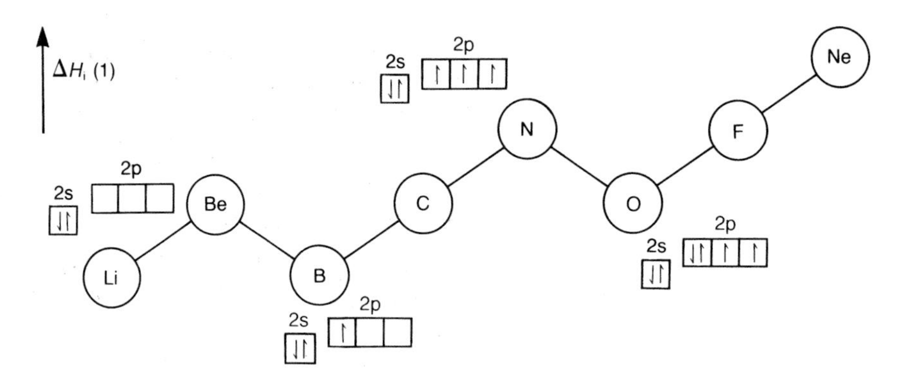

Figure 5.8
First ionisation energy $\Delta H_i(1)$ does not increase smoothly across a period

Whilst a similar pattern in ionisation energies occurs in period 3, for the same reasons, the above effects are much less marked in the d block elements. Their fairly constant first ionisation energies across a row result from the following factors:

- Atomic size decreases only slightly on passing across a row of d block elements.
- Electrons are filling into an inner d sub-level, which helps to screen the most loosely bound outer s electron from the increased attraction of the nucleus. (Remember that the 4s sub-level fills before the 3d sub-level but the 4s electrons are removed first, see section 2.9).

In section 4.11, we saw that metallic structures consist of a lattice of positive ions held together by a delocalised cloud of valence (outer) electrons. Clearly, the more loosely bound the outer electrons are, i.e. the lower the ionisation energy, the easier it will be for a metal to adopt a metallic lattice structure. As expected, metallic character decreases on passing across a period, but increases on passing down a group, see Table 5.1.

Electronegativity

Another atomic property, closely related to first ionisation energy, is **electronegativity**. This is a measure of the relative attraction of an atom for the electrons shared in a covalent bond. Again, this will depend directly upon the nuclear 'pull' felt by the electrons.

Earlier in this section, we saw that the attraction of the nucleus for outer electrons will increase on passing across a period, but decrease on passing down a group. Figure 5.9 shows that electronegativity values follow these same trends.

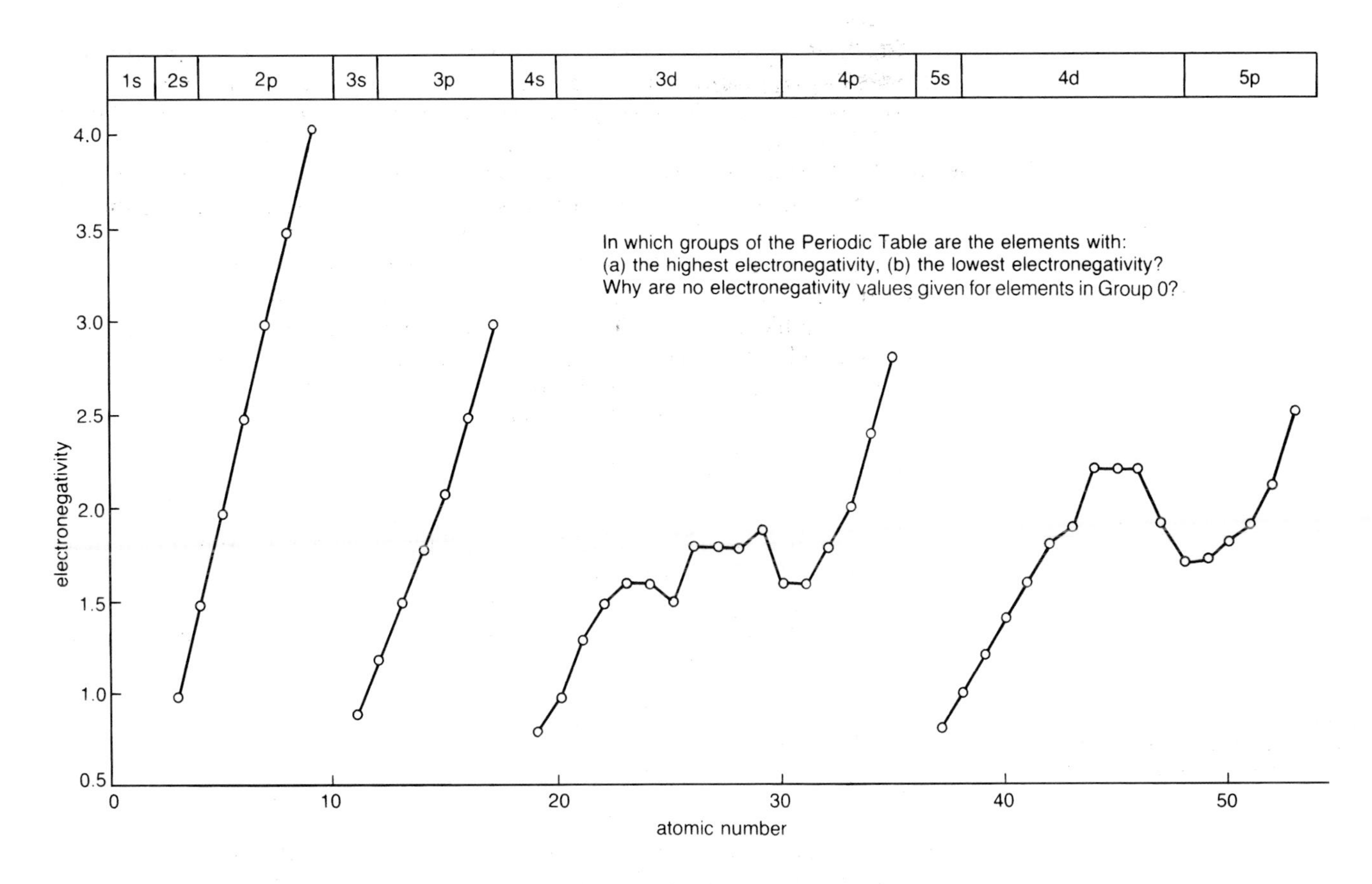

Figure 5.9 ▲
Electronegativities of the first 54 elements

Figure 5.10 ▼
Common oxidation states (numbers) shown by the first 54 elements

Oxidation state

We defined **oxidation state** or **number** in section 4.12 as the number of electrons which an atom loses, or tends to lose, when it forms a particular compound. Negative values indicate a tendency to gain electrons. Some of the common oxidation states of the first 54 elements are shown in Figure 5.10. Of course, all elements show an oxidation state of zero when uncombined.

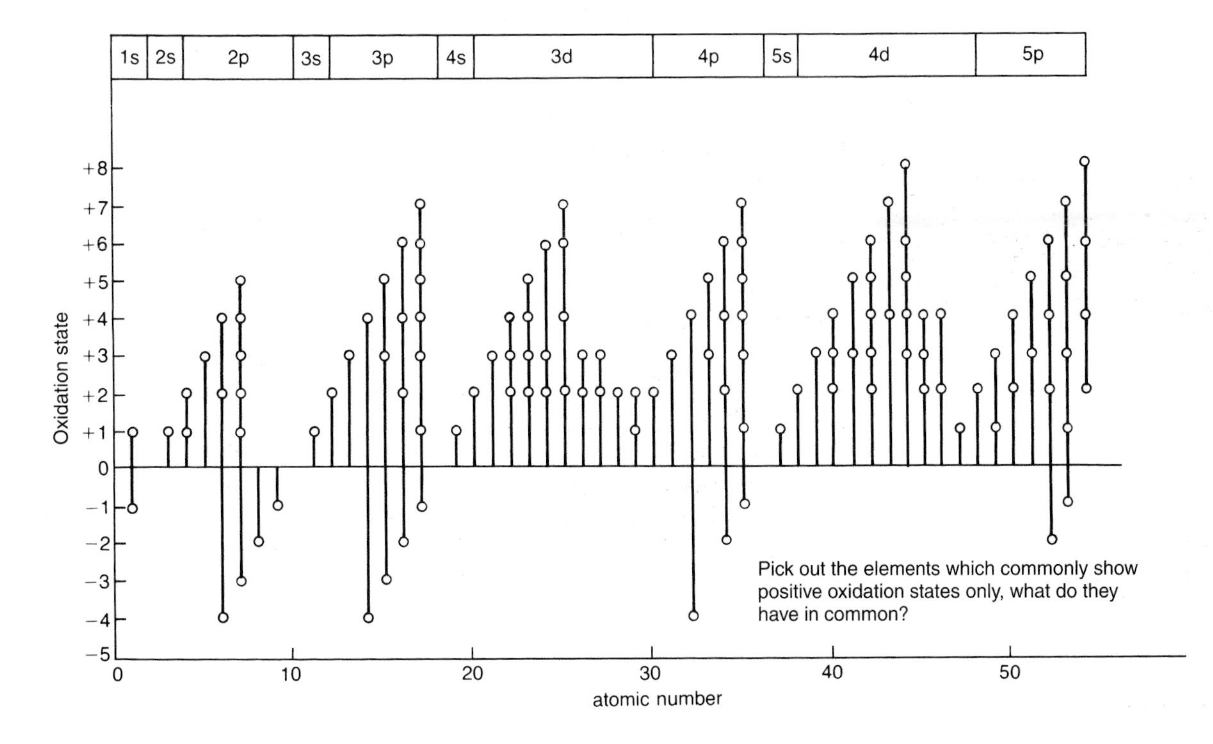

Oxidation state is related to the atom's electron arragement and thus the oxidation states shown by atoms in the same group tend to be similar. On passing across a period, the atoms tend to show a trend for greater variation in oxidation state, and negative values become more common.

Metals commonly show positive states only, i.e. their atoms lose, or tend to lose, electrons on forming compounds. This is to be expected because the ionisation energies and electronegativities of metals are low. Whilst the metals in the s block only show an oxidation state equal to the group number, a wide variation in oxidation states is shown by the d block metals (and to a lesser extent the p block metals). The d block elements may use electrons from the inner d sub-level as well as the outer s orbital in forming compounds. Manganese, for example, commonly shows oxidation states between +2 (only the 4s electrons used) and +7 (all 3d electrons are used as well as 4s electrons):

	$MnCl_2$	MnO_2	K_2MnO_4	$KMnO_4$
oxidation state	+2	+4	+6	+7

Due to their variable oxidation states, d block elements are involved in a wide range of redox reactions.

Non-metals may exhibit positive or negative oxidation states depending on the electronegativity of the element with which they combine. Thus, hydrogen shows an oxidation state of +1 and –1:

	NaH	HBr
oxidation state	–1	+1
	(Na is less electronegative)	(Br is more electronegative)

Most non-metals show a variety of oxidation states and so take part in many redox reactions. For example, sulphite ions, SO_3^{2-}, are oxidised by iodine, I_2, in aqueous solution:

oxidation states

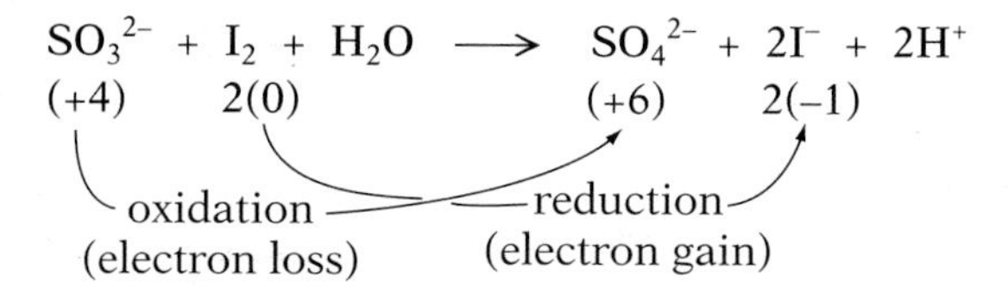

Oxidation states and Roman numerals

When naming compounds, oxidation states are often noted using Roman numerals. Thus, K_2MnO_4 is potassium manganate(VI) because the oxidation state of the manganese is +6 whereas $KMnO_4$ is potassium manganate(VII) because the oxidation state of the manganese is +7.

Focus 5a

1. In general, atomic size:
   - decreases on moving from left to right across a period,
   - increases on passing down a group.
2. Loss of electrons gives a positive ion which is smaller than the parent atom. Gain of electrons produces a negative ion which is larger than the parent atom.
3. **First ionisation energy** and **electronegativity** are atomic properties which depend upon the power with which the positively charged nucleus attracts electrons in the outer shell. This depends upon the charge on the nucleus, atomic size and the number of inner screening electrons. In general, first ionisation energy and electronegativity:
   - increase on passing from left to right across a period,
   - decrease on passing down a group.
4. **Metallic 'character'** decreases on passing across a period, but increases on passing down a group. Metals exhibit positive oxidation states. Most non-metals may have positive or negative oxidation states depending on the electronegativity of the atom to which they are bonded. A variable oxidation state, and thus redox chemistry, is a property of most d block elements and many non-metals.

5.3 Bulk properties of elements

These properties, which include melting point, boiling point, density and electrical conductivity, depend largely upon the way that the atoms of an element are arranged into a large-scale structure.

Melting point

The graph of the melting point against atomic number for the first 54 elements (Figure 5.11) shows a periodic variation.

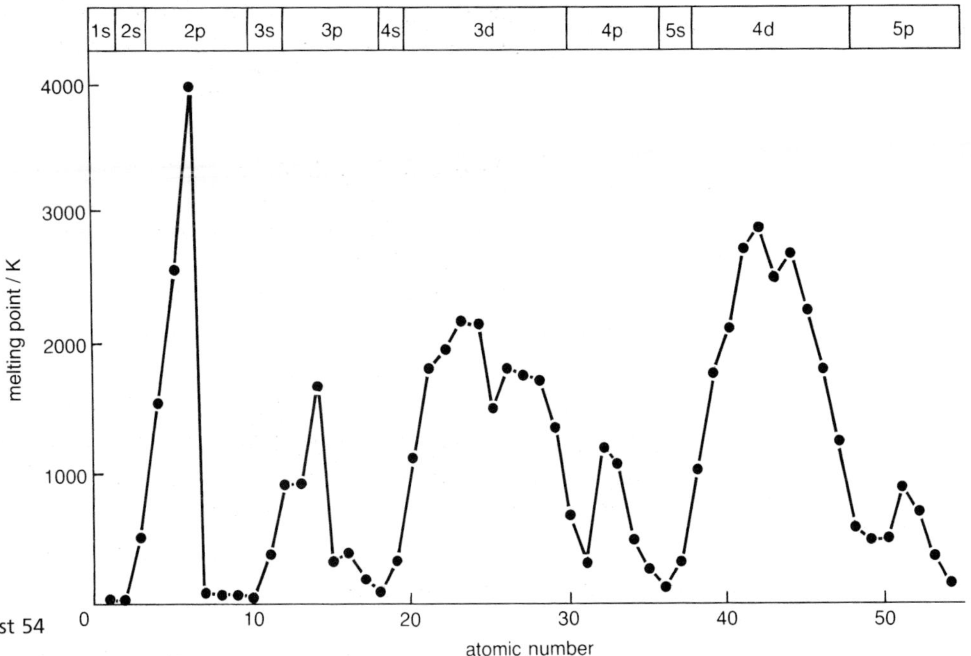

Figure 5.11
Variation in melting point for the first 54 elements

In periods 2 and 3, melting point generally rises to a maximum near the middle of the period and then drops away again sharply. We can explain this trend by considering the bonding and structure of the elements. For example, period 2 starts with lithium, an element with a **giant metallic lattice** structure containing Li^+ ions held together by their mutual attraction for the 'sea' of mobile electrons.

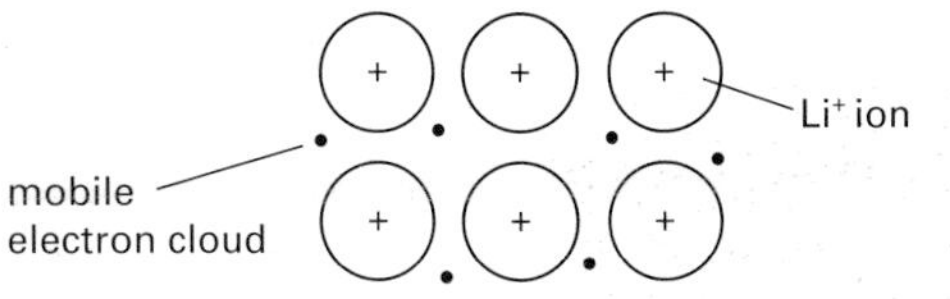

In order to melt lithium, energy must be supplied to overcome the attraction between the metal ions and the 'sea' of electrons. Lithium actually melts at 180°C.

The next element, beryllium, is also a metal. The only difference is that its ions are smaller and carry a 2+ charge.

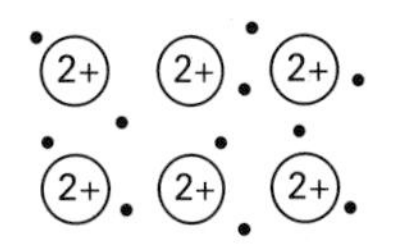

The attractive forces, i.e. the strength of the metallic bonding, in beryllium is therefore greater than that in lithium. More energy is needed to break the lattice and so the melting point of beryllium (1280°C) is much greater than that of lithium.

Boron, the next element in the period, is not a metal. It has a **giant molecular (macromolecular)** structure (section 4.8) in which each atom is connected to three of its neighbours by covalent bonds. These bonds are very strong and must be broken before any of the particles in the structure can move freely. Boron therefore has a very high melting point, which is 2030°C.

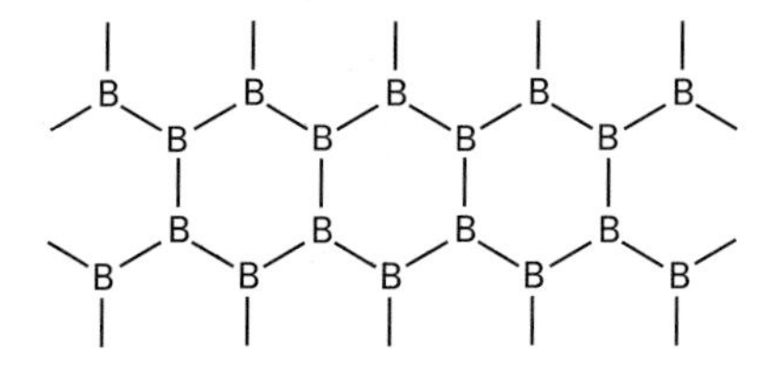

Carbon (like boron) forms covalent macromolecules. In diamond, for example, each carbon atom is connected to four others by single covalent bonds, again leading to a very high melting point of 3730°C.

The next element is nitrogen, a gas at room temperature. It has a melting point of –210°C. Like boron and carbon, nitrogen is covalent but, unlike these elements, it does not have a macromolecular structure. The atoms are connected together in pairs by strong triple covalent bonds, but the only attraction between the separate molecules is provided by very weak van der Waals' forces (section 4.6). Since in this case melting only involves overcoming weak intermolecular attractions, such **simple molecular structures** have low melting points (section 4.7). Oxygen and fluorine, which also consist of diatomic molecules, melt at –218°C and –220°C respectively. Neon, the last element in the period, is a noble gas which consists of separate atoms and has the lowest melting point in the period at –249°C. Table 5.1 summarises the trends in structures and melting points of the elements in periods 1–3.

Table 5.1 Patterns in the structures and melting points (°C) of the elements in periods 1–3.

Structure	Melting points
Giant metallic	mostly high melting points
Giant molecular (macromolecular)	very high melting points
Simple molecular	low melting points

H –259							He –269
Li 180	Be 1280	B 2030	C 3600	N –210	O –218	F –220	Ne –249
Na 98	Mg 650	Al 659	Si 1410	P 44	S 119	Cl –101	Ar –189

There is no general trend in melting points on passing down the periodic table, but we can explain the changes which occur in particular groups.

The alkali metals (group 1: lithium to caesium) have the same type of metallic structure, i.e. M^+ ions held together by a 'sea' of mobile electrons. The only difference is the size of the metal ion. In a bigger ion, the nucleus is further away from the electron 'sea' and better screened from it by more inner electrons, resulting in weaker attraction between the nucleus and the electron cloud. The strength of the metallic bonding, and therefore melting point, decreases on passing down group 1.

In the halogens (group 7: fluorine to astatine) however, melting point increases with atomic number. These elements consist of simple diatomic covalent molecules, X_2, and melting involves overcoming the weak van der Waals' intermolecular forces. These forces increase with the size of the electron cloud and therefore its molecular mass, causing the increase in melting point on passing down the group.

A prediction of melting point trends becomes much more difficult if the type of structure changes on passing down a group, e.g. in group 4 where the structure changes from giant molecular to giant metallic.

In comparison to the s block metals, those of the d block have relatively low successive ionisation energies, Table 5.2. Therefore more than one electron is often involved in the metallic bonding and so the d block elements consist of a lattice of small, highly charged cations within a dense, mobile electron cloud. As a result, the metallic bonds will be strong and the d block metals tend to be hard and have high melting points. Tungsten, for example, has the highest melting point (3377°C) of all metals. However, there are exceptions, notably mercury, which is the only metallic element to exist as a liquid at room temperature.

Table 5.2 Successive ionisation energies for the removal of the four outermost electrons of selected metals and their melting points.

Metal	block	ionisation energies/kJ mol^{-1}				melting point/°C
		1st	2nd	3rd	4th	
potassium	s	418	3070	4600	5900	64
calcium	s	590	1150	4940	6500	850
scandium	d	632	1240	2390	7110	1400
iron	d	762	1560	2960	5400	1549
molybdenum	d	680	1600	2600	4500	2617

Activity 5.2 Some other 'bulk' properties

1 Use the element datasheets from Chapters 20 and 21 to plot a graph of the boiling points of the first 20 elements (from hydrogen to calcium) against atomic number. How closely does the shape of your graph resemble the variation in the melting points of these elements shown in Table 5.3? Is this what you would expect? Explain your answer.

2 Account for the following changes in density of the elements in their solid or liquid states, which occur on passing across period 2:

	Li	Be	B	C (diamond)	N	O	F	Ne
density/g cm^{-3}	0.53	1.85	2.34	3.51	0.81	1.15	1.11	1.20

3 Structures which contain mobile charged particles conduct electricity, whilst those which do not are good electrical insulators. How and why would you expect the electrical conductivity of the elements to vary on passing across period 3 from sodium to argon?

Focus 5b

Bulk properties depend largely upon the strength of the attractive forces between the particles and how these are arranged within an element's large-scale structure. These trends are summarised in Figure 5.11 and the answers to activity 5.2.

5.4 Patterns in the reactivity of the period 3 elements

The way in which elements react and the nature of the products they form also show periodic variation. Within a group, a well-known example of this is the steadily increasing reactivity of the group 1 (alkali) metals as the group is descended. For example, when added to water, lithium floats on the surface and reacts steadily, producing a stream of hydrogen gas bubbles:

$$Li(s) + H_2O(l) \longrightarrow LiOH(aq) + \tfrac{1}{2}H_2(g)$$

If a piece of potassium is added to water however, it moves rapidly across the surface, reacting violently and producing hydrogen gas which ignites as it is formed:

$$K(s) + H_2O(l) \longrightarrow KOH(aq) + \tfrac{1}{2}H_2(g)$$

On passing down group 1, the first ionisation energy decreases. As a result, the metals lose their outermost electron more easily and thus reactivity increases down the group.

On passing across period 3, we can spot a well-defined pattern in the formulae of the common oxides and chlorides, their structures and methods of preparation from the elements (Table 5.3).

Table 5.3 Periodicity in the formulae, structure and methods of preparation of **(a)** common chlorides and **(b)** common oxides of the period 3 elements. (The standard enthalpy of formation, $\Delta H_f^\ominus$, is the enthalpy change when one mole of the compound is formed from its elements at 298 K and 1 atmosphere pressure.)

(a) Common chlorides

Element	sodium Na	magnesium Mg	aluminium Al	silicon Si	phosphorus P	sulphur S	chlorine Cl
group	1	2	3	4	5	6	7
formula	$NaCl$	$MgCl_2$	Al_2Cl_6	$SiCl_4$	PCl_3 PCl_5	S_2Cl_2	Cl_2
state (at 20°C)	s	s	s	l	l s	l	g
structure	ionic	ionic	← simple molecular → (except PCl_5 which, surprisingly, is ionic)				
$\Delta H_f^\ominus$ per mol of Cl atoms / kJ	−411	−321	−235	−160	−107 −89	−30	0

(b) Common oxides

Element	sodium Na	magnesium Mg	aluminium Al	silicon Si	phosphorus P	sulphur S	chlorine Cl
group	1	2	3	4	5	6	7
formula	Na_2O	MgO	Al_2O_3	SiO_2	P_4O_6 P_4O_{10}	SO_2 SO_3	Cl_2O_7
state (at 20°C)	s	s	s	s	s	g l	l
structure	ionic	ionic	ionic	giant molecular	simple molecular	simple molecular	simple molecular
ΔH_f per mol of O atoms / kJ	−416	−602	−559	−456	−298 −273	−149 −132	+80

With the exception of sulphur trioxide and chlorine(VII) oxide, the oxides and chlorides can be made by the direct combination of the elements. For example, aluminium chloride may be prepared by passing dry chlorine gas over hot aluminium, and sulphur dioxide is made by heating sulphur in oxygen:

$$2Al(s) + 3Cl_2(g) \longrightarrow Al_2Cl_6(s)$$
$$S(s) + O_2(g) \longrightarrow SO_2(g)$$

Sulphur trioxide is made in the **contact process** by heating the dioxide with oxygen in the presence of a vanadium pentoxide catalyst, section 23.3:

$$2SO_2(g) + O_2(g) \rightleftharpoons 2SO_3(g)$$

Also, where an element has more than one oxidation state, the chloride or oxide which forms will depend on the relative proportions of the reactants used, e.g.:

$$2P(s) + \underset{\text{limited supply}}{3Cl_2(g)} \longrightarrow 2PCl_3(l) \text{ but } 2P(s) + \underset{\text{in excess}}{5Cl_2(g)} \longrightarrow 2PCl_5(s)$$

Since chlorine and oxygen are electronegative elements, they react most vigorously with the more electropositive, metallic elements at the left of the period. This behaviour is reflected by the standard enthalpies of formation ($\Delta H_f^{\ominus}$ values), which become less exothermic as we move across the period. Thus for example, whilst sodium and oxygen react violently together, chlorine and oxygen (two very electronegative elements) will not combine directly.

Looking again at Table 5.3, you should see that there is a periodic variation in the structures of the chlorides and oxides. As expected, the s block metals, being very electropositive form ionic bonds with the electronegative chlorine and oxygen. A magnesium atom, for example, loses both of its loosely bound outer electrons to form a Mg^{2+} ion and therefore it must combine with two chlorine atoms:

$$\underset{[Ne]3s^2}{Mg} \longrightarrow \underset{[Ne]}{Mg^{2+}} + 2e^- \quad \text{and} \quad \underset{[Ne]3s^23p^5}{2Cl} + 2e^- \longrightarrow \underset{[Ar]}{2Cl^-}$$

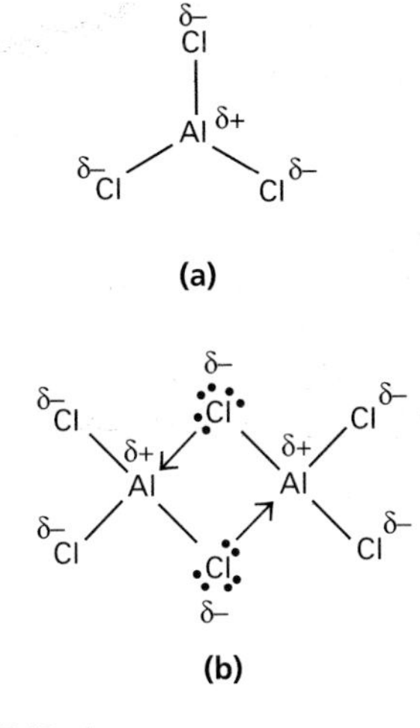

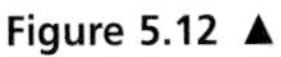
Figure 5.12 ▲
Aluminium chloride exists as trigonal planar molecules in the gas phase **(a)** but as a polar covalent dimer in the solid states owing to co-ordinate covalent bonding **(b)**

Due to their mutual attraction, the magnesium and chloride ions are held together in a giant ionic lattice structure (section 4.9).

Whilst aluminium oxide has an ionic structure, aluminium chloride is, surprisingly, not ionic but polar covalent. The reason for this is that the aluminium ion is so small and highly charged that it is able to polarise anions of all but the most electronegative elements, oxygen and fluorine (you may wish to refer to Fajan's rules in section 4.9). Aluminium chloride has a simple molecular structure. In the vapour state, it contains trigonal planar molecules in which each of the three electrons in the outer shell of the aluminium atom forms a single covalent bond with a chlorine atom, Figure 5.12(a). Since the chlorine is more electronegative than the aluminium, the chlorine atoms develop a partial negative charge, whilst the aluminium is somewhat positive. On cooling to room temperature, because of the bond polarities, aluminium chloride forms polar covalent dimers, Figure 5.12(b).

With the exception of silicon dioxide, which has a giant molecular (macromolecular) structure (section 4.8), and phosphorous(V) chloride which is ionic, the other elements form polar covalent chlorides and oxides, which have similar molecular structures, Figure 5.13.

The formulae of the chlorides and oxides of the period 3 elements also show a periodic pattern based on the element's highest oxidation state (which is equal to the number of electrons in the outer shell of the atom), e.g.:

chlorides	NaCl	$MgCl_2$	Al_2Cl_6	$SiCl_4$	PCl_5		
oxidation state	+1	+2	+3	+4	+5		
oxides	Na_2O	MgO	Al_2O_3	SiO_2	P_4O_{10}	SO_3	Cl_2O_7
oxidation state	+1	+2	+3	+4	+5	+6	+7

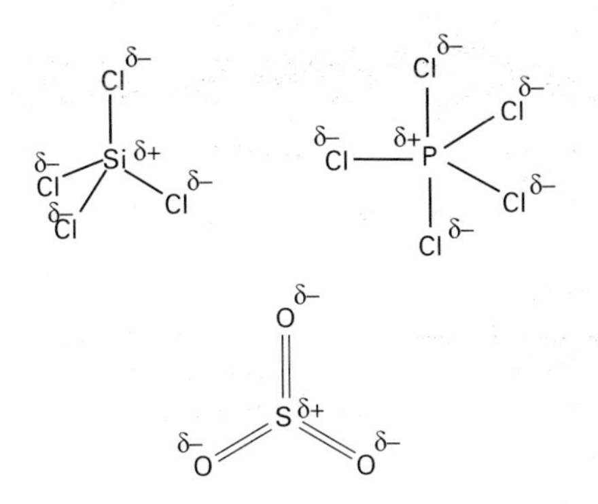

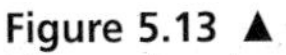
Figure 5.13 ▲
The molecular structures of silicon(IV) chloride, phosphorus(V) chloride (vapour state) and sulphur(VI) oxide. In the solid state, phosphorus(V) chloride is ionic, being made of PCl_4^+ and PCl_6^- ions

A more detailed look at periodicity in oxidation states

Naturally, we might expect sulphur to form the chloride SCl_6, but its highest chloride is only SCl_4. Since sulphur does form a hexafluoride (SF_6) it may be that the sulphur atom is simply not big enough to accommodate more than four chlorine atoms around it. In fact, SCl_4 itself is extremely unstable and decomposes above −31°C. The most stable chloride of sulphur is S_2Cl_2, an orange liquid of revolting smell. What is the oxidation state of sulphur in S_2Cl_2?

Phosphorus provides another example of a variable oxidation state in forming two chlorides: PCl_5 and PCl_3; and two oxides: P_4O_{10} and P_4O_6. Such behaviour is quite common in the p block, especially towards the bottom of each group where it is known as the **inert pair effect** (section 20.1).

Metals in the d block frequently show a variety of positive oxidation states by using some or all of their inner d electrons in bonding, as well as the outer s electrons (Figure 5.10).

In the main text of this section, we have only considered positive oxidation states, in which the element in question must be combined with a more electronegative element. Clearly fluorine, the most electronegative element of all, can only have a negative oxidation state. Since in all its compounds a fluorine atom gains, or tends to gain, a single electron, completing its noble gas electronic structure, fluorine can only show an oxidation state of −1 in its compounds. Other electronegative elements also show negative oxidation states in which the atoms gain (or tend to gain) enough electrons to obtain the electron arrangement of a noble gas atom, e.g. in the following polar covalent hydrides:

	H_3P	H_2S	HCl
oxidation state	−3	−2	−1

Clearly, negative oxidation states will become more common on passing across any period, as the electronegativity of the elements increases.

Activity 5.3

Write balanced equations for the formation of the following compounds by direct combination of the elements:

NaCl $MgCl_2$ $SiCl_4$ Na_2O

MgO Al_2O_3 SiO_2 P_4O_{10}

Focus 5c

There are periodic trends in the formulae, structure and methods of preparation of the common chlorides and oxides of the period 3 elements:

1. With the exception of sulphur trioxide and chlorine(VII) oxide, the oxides and chlorides can be made by direct combination.
2. In general, the chlorides and oxides have giant ionic structures (if metals) and simple molecular structures (if non-metals). The exceptions are aluminium chloride (simple molecular), silicon dioxide (macromolecular) and phosphorous(V) chloride (ionic).
3. In their highest oxides and chlorides, the elements use all of their outer electrons to bond with more electronegative atoms. By definition, therefore, the maximum positive oxidation state of these elements is equal to the number of electrons in the outer shell of the atom. (Some exceptions to this general rule are given in the oxidation state extension box above.)

5.5 Patterns in the properties of the chlorides and oxides of the period 3 elements

As well as showing well-defined trends in their formulae, the oxides and chlorides of period 3 also show variations in physical and chemical properties which in turn depend upon changes in bonding and structural type. Tables 5.4 and 5.5 give information on the main chlorides and oxides, respectively, of period 3.

At the start of the period, both the chlorides and oxides show properties typical of ionic compounds, e.g. quite high melting points and good conduction of electricity when molten. The change in electrical conductivity marks the point at which the bonding becomes essentially covalent. This is

Table 5.4 Properties of some chlorides of period 3. NOTE: pH values in this table are guide values – the actual pH will depend on the relative quantities of chloride and water used.

	NaCl	$MgCl_2$	Al_2Cl_6	$SiCl_4$	PCl_5
melting point/°C	808	714	178(sub)	–70	160(sub)
electrical conductivity in the liquid state	good	good	no liquid state	none	no liquid state
soluble in water?	yes	yes	vigorous reaction; fumes of HCl(g)	vigorous reaction; fumes of HCl(g)	vigorous reaction; fumes of HCl(g
electrical conductivity of aqueous solution	good	good	good	good	good
acid/base nature of aqueous solution	neutral	very slightly acidic	acidic	strongly acidic	strongly acidic
pH	7	6.5	3	2	2

Table 5.5 Properties of some oxides of period 3. NOTE: pH values in this table are guide values – the actual pH will depend on the relative quantities of oxide and water used.

	Na_2O	MgO	Al_2O_3	SiO_2	P_4O_{10}	SO_3	Cl_2O_7
melting point/°C	920	2900	2040	1610	300	17	–59
electrical conductivity in the liquid state	good	good	good	poor	none	none	none
soluble in water?	yes	slightly	no	no	vigorous reaction	vigorous reaction	vigorous reaction
electrical conductivity of aqueous solution	good	good	–	–	good	good	good
acid/base nature of aqueous solution	strongly alkaline	alkaline	–	–	acidic	strongly acidic	strongly acidic
pH	13	11				3	2 2

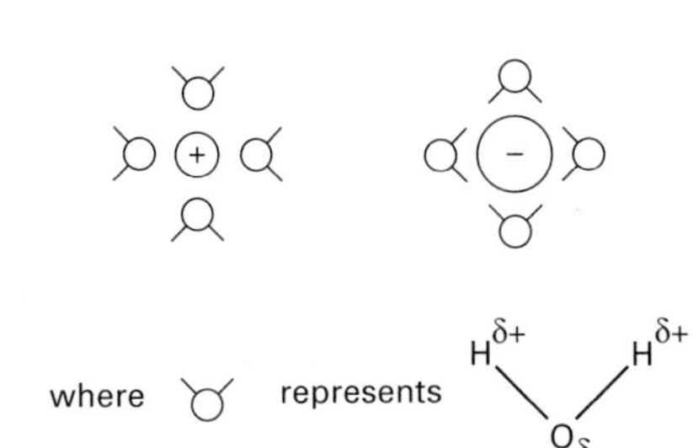

Figure 5.14 ▲
In hydrated positive ions (cations) and negative ions (anions) a layer of water molecules is held around the central ion by ion–dipole attractions

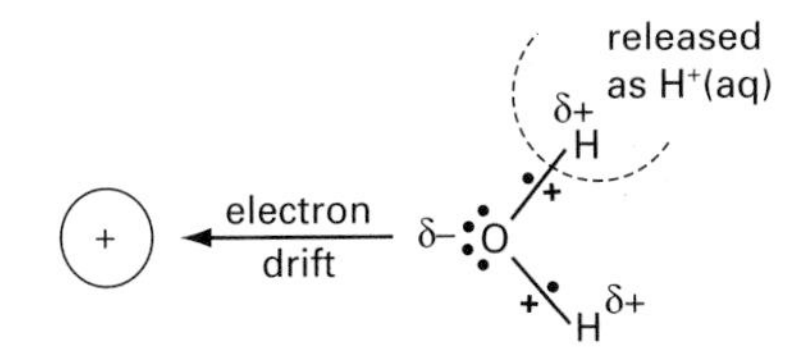

Figure 5.15 ▲
A positive ion, or positively charged atom in a covalent molecule, may increase the polarisation in a water molecule by attracting one of the lone pairs on the oxygen atoms. If this effect becomes strong enough, the O—H bond will break and a hydrated hydrogen ion is formed, thereby increasing the acidity of the solution

generally marked by a drop in the melting point, as strong ionic attractions are replaced by weaker intermolecular forces. Note however, that giant molecular (macromolecular) covalent structures (such as silicon dioxide) still have high melting points.

Ionic compounds generally dissolve in water to give aqueous solutions which are good conductors of electricity. Since water molecules are polar covalent in nature, they become attached to the ions by **ion–dipole attractions**. This process provides the energy needed to break down the lattice and eventually a solution of **hydrated ions** is produced, Figure 5.14. (Note: you may also wish to refer to section 4.10). Unless the ions chemically interact with the water molecules, the resulting solution will be neutral, as for example in aqueous sodium chloride. However, if the cation is small and highly charged, it may attract electrons from the water molecules, thereby releasing H^+(aq) ions and producing an acidic solution, Figure 5.15. Clearly this effect will become more marked as the charge on the cation increases and

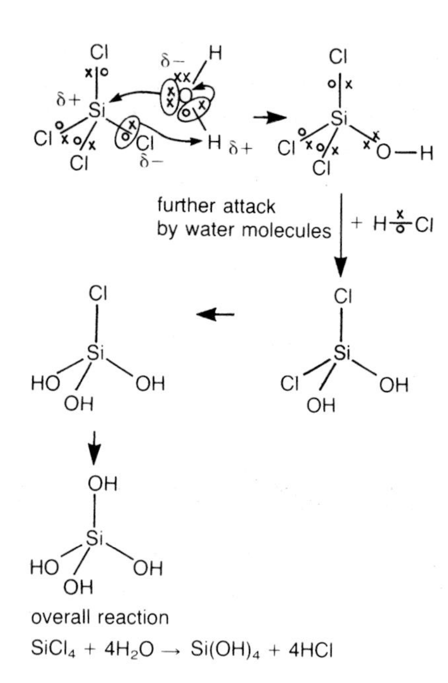

Figure 5.16
Mechanism of the hydrolysis of silicon tetrachloride, $SiCl_4$. Silicon tetrachloride dissolves exothermicaly in water giving a *strongly acidic* solution. The silicon atom carries a slight positive charge and can accept a lone pair of electrons from the oxygen atom on a water molecule into an empty d orbital in its valency shell. This brings a positive hydrogen atom close to a negative chlorine atom and hydrogen chloride is eliminated. This mechanism is repeated until all the chlorine atoms have been replaced by hydroxide groups

as its size falls. This explains, for example, why Al^{3+}(aq) solutions are quite strongly acidic:

$$[Al(H_2O)_6]^{3+} (aq) \rightleftharpoons [Al(H_2O)_5OH]^{2+} (aq) + H^+(aq)$$

The ability of a cation (or a positively charged atom in a covalent molecule) to polarise water molecules explains the increasing acidity of the solutions on passing water across the period. For example, the δ+ sulphur atom in sulphur(VI) oxide polarises a water molecule, thereby causing the release of a hydrogen ion and so an acidic solution is produced:

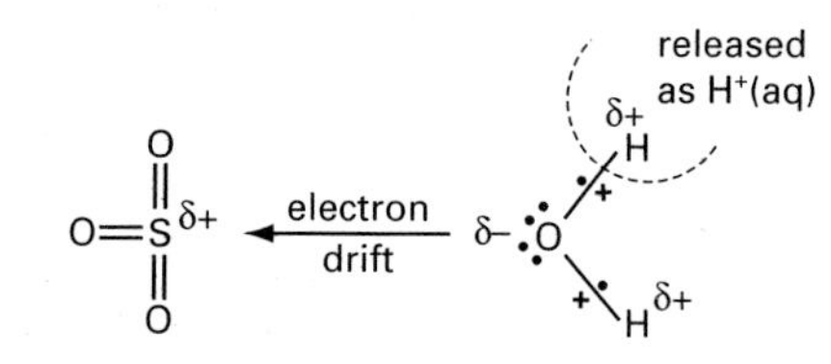

The mechanism of the **hydrolysis** (reaction with water) of a covalent chloride such as silicon (VI) chloride, is similar, Figure 5.16.

Acids, bases and the pH scale

In 1923, Johannes Nicolaus Brönsted, a Danish chemist, and (independently) Thomas Martin Lowry, a British chemist defined an acid as a proton (hydrogen ion, H^+) donor and a base as a proton acceptor. Thus, when hydrogen chloride reacts with ammonia, an acid–base reaction occurs with a proton being donated by the acid (HCl) to the base (NH_3):

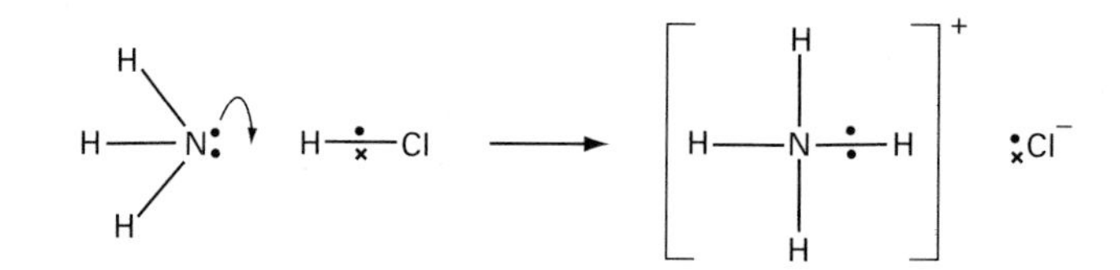

The acidity of an aqueous solution depends on the concentration of H^+(aq) ions present, the higher the concentration, the greater the acidity. In section 26.2, we shall explain how the pH scale provides a simple way of expressing the acidity of an aqueous solution. For now, you should note that, at 25°C, the pH scale runs from pH = 0 (extremely acidic), through pH = 7 (neutral) to pH = 14 (extremely alkaline). Put another way, an acidic solution will have a pH less than 7 and an alkaline solution will have a pH greater than 7.

Whilst the non-metal oxides are acidic, those at the start of the period are alkaline, since they contain O^{2-} ions which react with water to form aqueous hydroxide ions, OH^-(aq):

$$O^{2-}(s) + H_2O(l) \longrightarrow 2OH^-(aq)$$

The oxides therefore change from basic to acidic on passing across the period. In fact, aluminium oxide marks the changeover point and is **amphoteric**, i.e. it shows both basic and acidic properties, e.g.:

$$\underset{\text{base}}{Al_2O_3(s)} + \underset{\text{acid}}{6HCl(aq)} \longrightarrow 2AlCl_3(aq) + 3H_2O(l)$$

$$\underset{\text{acid}}{Al_2O_3(s)} + \underset{\text{base}}{2NaOH(aq)} + 3H_2O(l) \longrightarrow \underset{\text{sodium aluminate}}{2Na[Al(OH)_4](aq)}$$

It is worth noting that although they do not react with water, aluminium and silicon oxides do take part in acid/base reactions. As we have just seen, aluminium oxide is amphoteric, whereas silicon dioxide is acidic. For example, in the manufacture of soda glass, silicon dioxide (from sand) acts as an acid and displaces carbon dioxide from sodium and calcium carbonates (both basic substances):

$$Na_2CO_3(s) + SiO_2(g) \longrightarrow Na_2SiO_3(s) + CO_2(g)$$
$$CaCO_3(s) + SiO_2(g) \longrightarrow CaSiO_3(s) + CO_2(g)$$

More detail on the chemistry of the elements and compounds mentioned in this chapter may be found in the sections dealing with the s and p blocks, Chapters 20 and 21 respectively.

Activity 5.4

Write balanced equations for the reaction of the following chlorides and oxides with water, given the formulae of the products in brackets:

a) Al_2Cl_6 ($Al(OH)_3$, HCl)
b) PCl_5 ($POCl_3$, HCl)
c) MgO ($Mg(OH)_2$)
d) P_4O_{10} (H_3PO_4)
e) SO_3 (H_2SO_4)

Activity 5.5 Some hydrides of period 3

Sodium, phosphorus and chlorine all form stable compounds with hydrogen, but their properties vary considerably. Identify each of these hydrides from the following brief descriptions and discuss the structure and bonding types which give rise to their properties.

- **Hydride A** is a crystalline solid which melts at about 800°C to give a liquid which readily conducts electricity. It reacts with water forming an alkaline solution and hydrogen gas.
- **Hydride B** is a gas at room temperature and has a melting point of −133°C. It is almost insoluble in water and, in its liquid state, does not conduct electricity.
- **Hydride C** is also a gas at room temperature and has a melting point of −114°C. It does not conduct electricity in its liquid state but readily dissolves in water to give a strongly acidic solution which conducts readily.

Focus 5d

On moving from left to right across period 3, the chlorides and oxides generally:

- change in bonding type from ionic to covalent
- become more acidic in nature.

5.6 Periodic patterns and the d block elements

Atomic properties

Although the d sub-level starts in the third principal quantum level (shell), the first row of the d block is in period 4. This is because electrons enter the 4s sub-level before filling the 3d. On passing across a row in the d block, therefore, an extra screening electron is added every time an extra proton is added to the nucleus. As a result, those properties that depend mainly upon the strength of nuclear attraction, for example atomic radius, first ionisation energy and electronegativity, remain roughly constant as shown in Figure 5.17. Table 5.6 gives the electron arrangements and some physical properties of iron and copper. On going from iron to copper, three protons are added to the nucleus. This would significantly increase the nuclear 'pull' experienced by the 4s electrons, were it not for the increased shielding caused by adding electrons to the 3d sub-level. As a result, the outermost electron in each atom experiences a similar nuclear pull and the atomic radii, first ionisation energies and electronegativity values are comparable. Notice also that the outer electron structure in copper is $3d^{10}4s^1$ which, having a full 3d sub-level, is more stable than the $3d^94s^2$ arrangement.

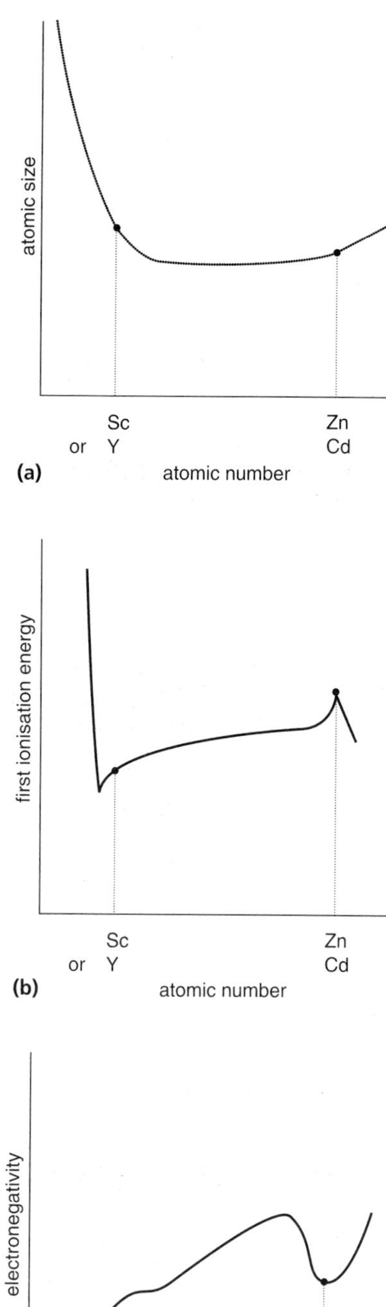

Figure 5.17
Schematic diagrams to show the general variation in the following properties of the d block elements, scandium to zinc *or* yttrium to cadmium: **(a)** atomic size; **(b)** first ionisation energy; **(c)** electronegativity. The detailed patterns can be seen in Figures 5.6, 5.7 and 5.9, respectively

Table 5.6 The electronic arrangement in atoms of iron and copper and some atomic properties.

	electron arrangement	atomic radius /nm	first ionisation energy/kJ mol^{-1}
iron	$[Ar]3d^64s^2$	0.126	762
copper	$[Ar]3d^{10}4s^1$	0.128	745

Variable oxidation states (states)

In the s and p blocks there is a steady increase in successive ionisation energies until a noble gas electron configuration is reached. There is then a large jump as an inner shell is broken. For a d block element, the outer s electrons are lost relatively easily but there is no large jump in ionisation energy on removing electrons from the inner d sub-level. As a result, elements of the d block generally show more than one oxidation state. The lowest oxidation state corresponds to loss of the outer s electrons only, whereas the higher states also involve loss of some, or all, of the inner d electrons. Vanadium for example, which has the atomic electron configuration $[Ar]\ 3d^3\ 4s^2$, shows all oxidation states between +2 and +5 in its compounds, each having different colours.

V(+2)	$[Ar]3d^3\,4s^0$	for example in VCl_2	violet
V(+3)	$[Ar]3d^2\,4s^0$	for example in VCl_3	green
V(+4)	$[Ar]3d^1\,4s^0$	for example in $VOCl_2$	bright blue
V(+5)	$[Ar]3d^0\,4s^0$	for example in V_2O_5	pale yellow

Since different oxidation states usually have different colours, so a colour change often accompanies a redox reaction. When a piece of iron is placed into aqueous copper sulphate solution for example, it displaces copper from the solution; the colour of which changes from blue to green.

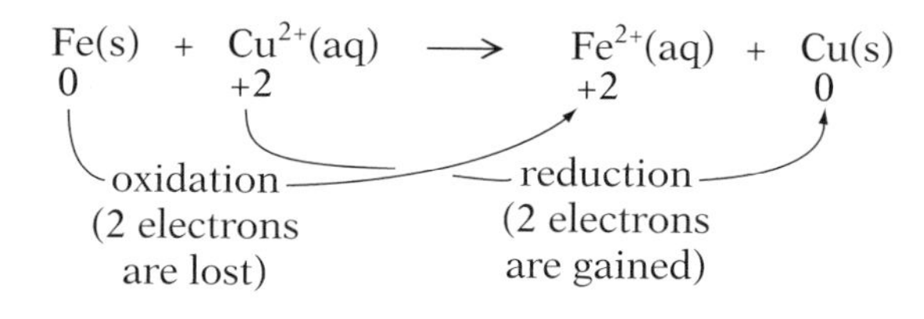

Coloured compounds and ions

Many, but not all, of the compounds formed by d block elements are coloured. At least for lower oxidation states, colour seems to require a partially filled inner d sub-level. Thus, for example, scandium(III) compounds with no electrons in the 3d sub-level are colourless as are the compounds of zinc(II) with a completely full 3d sub-level. On the other hand titanium(III) compounds, $3d^1$, and copper(II) compounds, $3d^9$, are generally coloured. **Transition elements** are defined as those containing a partially filled d sub-level in either the atom or a common oxidation state. It is important to realise that not ali members of the d block are transition metals.

All isolated gaseous d block ions are colourless, even those with a partially filled d sub-level. It is only when the ions are present in compounds that colour develops. A full treatment of the origins of colour in d block compounds may be found in Chapter 18. Table 5.7 gives the colours of some common d block ions and compounds.

As mentioned earlier a redox reaction is often accompanied by a colour change. For example, in acidified aqueous solution, manganate(VII) ions and dichromate(VI) ions are powerful oxidising agents and can oxidise iron(II) ions:

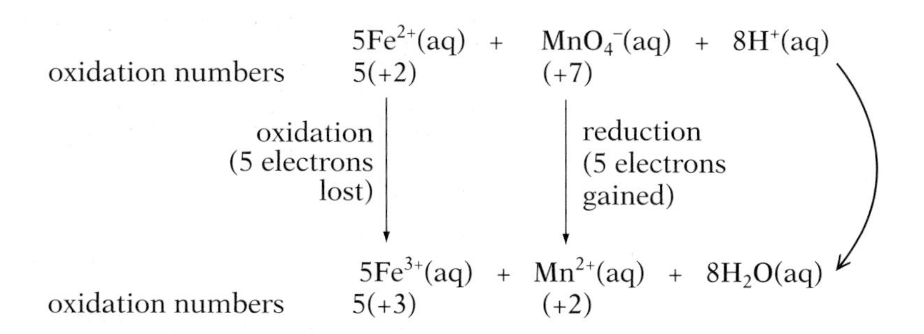

Once again, the oxidation states balance on each side (+17), with each manganese(VII) atom gaining five electrons and each iron(II) atom losing one electron. Also the ionic charges (superscripts) equate, with $^{17+}$ on each side. Thus, an *exact* number of electrons are lost or gained in a redox reaction, *none are created or destroyed*. The reaction produces a colour change: light green (Fe^{2+}) and purple (MnO_4^-) solutions are mixed, yielding a brown solution (Fe^{3+}).

Table 5.7 The oxidation states and colours of some common species containing iron, manganese and chromium.

Name of the compound or ion	formula	oxidation state	colour
iron(II)	$Fe^{2+}(aq)$	+2	light green
iron(III)	$Fe^{3+}(aq)$	+3	brown
manganese(II)	$Mn^{2+}(s)$; $Mn^{2+}(aq)$	+2	pink colourless
manganese(IV) oxide	$MnO_2(s)$	+4	brown
manganate(VII)	$MnO_4^-(aq)$	+7	purple
chromium(III)	$Cr^{3+}(aq)$	+3	dark green
dichromate(VI)	$Cr_2O_7^{2-}(aq)$	+6	orange

Activity 5.6

In an acidified aqueous solution, iron(II) ions [$Fe^{2+}(aq)$] are oxidised by dichromate(VI) ions [$Cr_2O_7^{2-}(aq)$]. By reference to Table 5.7 and the above example, write a balanced ionic equation for the reaction using the method suggested below:

1 How many electrons will be lost by each Fe^{2+} ion?

2 What is the oxidation state of each chromium atom in **a)** a $Cr_2O_7^{2-}$ ion and **b)** a Cr^{3+} ion?

3 How many electrons does each $Cr_2O_7^{2-}$ ion gain when it forms two Cr^{3+} ions?

4 From **1** and **3**, how many Fe^{2+} ions will react with one $Cr_2O_7^{2-}$?

5 Write the reacting number of ions into the equation:

$$? \ Fe^{2+}(aq) + ? \ Cr_2O_7^{2-}(aq) \longrightarrow$$

6 Now, write the corresponding number of product ions into the equation:

$$\longrightarrow \ ? \ Fe^{3+}(aq) + ? \ Cr^{3+}(aq)$$

7 This reaction takes place in acid solution, so $H^+(aq)$ ions must be added on the reactant side. What product will they form?

8 Balance the equation ensuring that **a)** the oxidation states and **b)** the ionic charges are equal on each side of the equation.

Formation of complex ions

We saw in section 5.3 that sodium ions are hydrated in aqueous solution, i.e. water molecules form a layer around the positively charged ions as a result of ion–dipole attractions (Figure 5.14). In hydrated d block **metal** ions however, water molecules are attached to the central ion via **co-ordinate covalent bonds** (dative covalent bonds), formed when one of oxygen's lone pair of electrons is donated to an empty orbital in the metal atom. The water molecules are said to act as **ligands**. Thus, in aqueous iron(II) ions, six water molecules are co-ordinated around the iron atom in an octahedral structure, Figure 5.18, and the formula is written as $[Fe(H_2O)_6]^{2+}$. (Notice that the oxidation state of iron is +2 because water is a neutral molecule.) Similarly aqueous copper(II) ions exist as an octahedral complex ion $[Cu(H_2O)_6]^{2+}$, whereas only four water molecules are able to co-ordinate around the zinc atom in an aqueous zinc(II) ion, $[Zn(H_2O)_4]^{2+}$. Water ligands are fairly easily replaced by ligands which have a greater affinity for the central metal atom, e.g. when an excess of aqueous ammonia is added to $[Cu(H_2O)_6]^{2+}(aq)$ ions:

$$\underset{\text{light blue}}{[Cu(H_2O)_6]^{2+}(aq)} + 4NH_3(aq) \longrightarrow \underset{\text{deep royal blue}}{[Cu(NH_3)_4(H_2O)_2]^{2+}(aq)} + 4H_2O(l)$$

Similarly, aqueous iron(III) ions react with thiocyanate ions, SCN^-, to form a blood-red thiocyanate complex $[FeSCN(H_2O)_5]^{2+}(aq)$:

$$\underset{\text{brown}}{[Fe(H_2O)_6]^{3+}(aq)} + SCN^- \longrightarrow \underset{\text{blood-red}}{[Fe(SCN)(H_2O)_5]^{2+}(aq)} + H_2O(l)$$

The blood-red colour is so intense that this reaction may be used to identify very small concentrations of $[Fe(H_2O)_6]^{3+}(aq)$ ions.

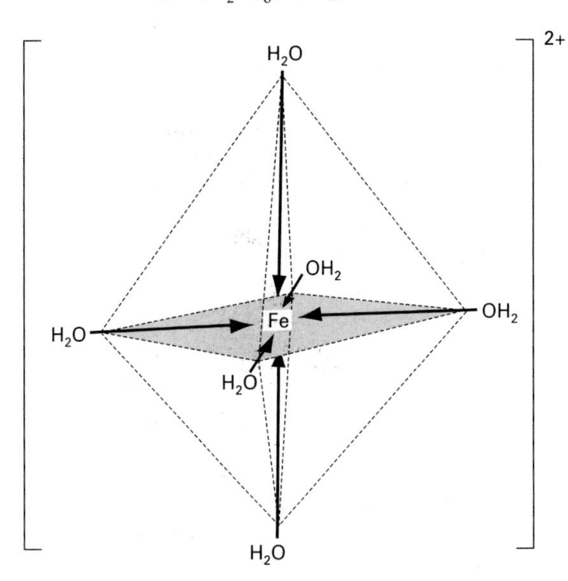

Figure 5.18
The structure of the $[Fe(H_2O)_6]^{2+}$ ion. Six water ligands are held by co-ordinate covalent bonds in an octahedral arrangement around the central iron(II) ion.

Catalysis

The d block elements and their compounds are widely used in industry as **catalysts**, that is substances which speed up the rate of a chemical reaction whilst remaining chemically unchanged at the end of the reaction. Thus iron filings are used as a catalyst in the Haber process for the manufacture of ammonia and vanadium(V) oxide V_2O_5 is used in the contact process in the manufacture of sulphuric acid (section 23.3). A more complete treatment of catalysis must wait until Chapter 24 but this behaviour is often assoociated with the ability of d block elements to show a variety of oxidation states.

Questions on Chapter 5

1 Give the block, period and group (if any) in which each of the following elements are found in the periodic table. Write the detailed electron configuration of each of the atoms: krypton (Kr), chromium (Cr), fluorine (F), caesium (Cs), plutonium (Pu), arsenic (As)

2 Place the particles within each of the following sets in order of increasing size. In each case explain why size varies in the way you have indicated.
- **a)** Li, Na, K
- **b)** Li, Be, B
- **c)** F, Ne, Na
- **d)** Cl^-, Ar, K^+
- **e)** Fe, Fe^{2+}, Fe^{3+}

3 Define first ionisation energy and write an equation which represents the process involved for chlorine. How and why does first ionisation energy vary on passing:
- **a)** across period 2, from lithium to neon,
- **b)** down group 2, from beryllium to barium.

4
- **a)** What is meant by the oxidation state (number)?
- **b)** What is the oxidation state of fluorine in all of its compounds? Explain your answer.
- **c)** Give the oxidation state of the other element in each of the following fluorine compounds: NaF, AlF_3, SF_6, XeF_4, OF_2.
- **d)** How and why does the maximum positive oxidation state of the elements vary on passing across period 3 from sodium to argon?

5 What determines the melting point of an element? Explain how and why the melting points of the elements vary on passing:
- **a)** across period 2 from lithium to neon,
- **b)** down group 0 from helium to xenon,
- **c)** down group 1 from lithium to caesium.

6 Two elements, A and B, each in period 2 of the periodic table, form oxides on burning in excess oxygen. Element A forms an oxide of formula A_2O which is a crystalline solid, soluble in water and with a high melting point. When melted or dissolved in water, A_2O readily conducts electricity. The oxide of element B has the formula BO_2, which is a gas at room temperature. BO_2 is moderately soluble in water, giving a weakly acidic solution.
- **a)** Discuss the bonding type and structure of each of the oxides and suggest the identity of the elements A and B.
- **b)** What kind of a reaction might you expect if gaseous BO_2 is bubbled through an aqueous solution of A_2O? Explain your answer.

7 Here are the electron configurations of three atoms:

X [noble gas core]d^0s^1
Y [noble gas core]$d^{10}s^2p^6$
X [noble gas core]$d^{10}s^2p^2$

- **a)** To which block, and which group (if any), of the periodic table do each of the elements belong?
- **b)** Predict the formula of the highest oxide of each of the elements.
- **c)** Which of these oxides will conduct electricity when molten? Explain why.
- **d)** Explain why the oxide of X dissolves in water to give an alkaline solution.
- **e)** What is meant by the term 'amphoteric' oxide? Which of the oxides given in your answer to **b)** above is most likely to be amphoteric? Explain your answer.

8 Choose from the elements boron (atomic number 5) to zinc (atomic number 30) and their compounds the substances which meet the following criteria and answer the extension question:

	Criteria	Extension question
a)	The element with the greatest atomic radius.	Explain why.
b)	The two elements with the lowest first ionisation energies.	Explain why.
c)	The two elements with the highest first ionisation energies.	Explain why.
d)	Four elements, next to each other, having very similar electronegativities.	Explain why.
e)	Two metals having variable oxidation states.	Explain why.
f)	The two elements with the highest melting points.	Describe their structures.
g)	Two elements having very low densities.	Describe their structures.
h)	A metal which forms a covalent chloride.	Draw its structures in the gaseous and solid states.
i)	Two ionic chlorides.	Give their formulae and an equation for their preparation from the elements.
j)	Three covalent oxides.	Give their formulae and an equation for their preparation from the elements.
k)	An amphoteric oxide.	Give equations for its reaction with HCl(aq) and NaOH(aq).

Questions on Chapter 5 *continued*

l)	Two oxides which react with water to produce basic solutions.	Give balanced equations for the reactions.
m)	Two oxides which react with water to produce solutions having a pH less than 7.	Give balanced equations for the reactions.
n)	A crystalline hydride which conducts electricity when molten.	Give the equation for the reaction of the hydride with water.
o)	A coloured ion in which ligands are attached to a central metal atom via co-ordinate bonds.	Draw the ion's structure and explain the origin of the colour.

9 When S_2Cl_2 reacts with water, one of the sulphur atoms is reduced and the other is oxidised. Simultaneous oxidation and reduction of the same element is known as **disproportionation**. If two of the products are sulphur and sulphur dioxide:

a) Write a balanced equation for the reaction.
b) Show which sulphur atom is oxidised and which is reduced.
c) State how many electrons are transferred between sulphur atoms.

Extension Question

10 Using the method in activity 5.6, balance the following ionic equations:

a) $Fe^{3+}(aq) + I^-(aq) \longrightarrow Fe^{2+}(aq) + I_2(aq)$
b) $Cl_2(aq) + Br^-(aq) \longrightarrow Cl^-(aq) + Br_2(aq)$
c) $IO_3^-(aq) + Fe^{2+}(aq) + H^+ \longrightarrow I_2(aq) + Fe^{3+}(aq) + H_2O(l)$
d) $S_2O_3^{2-}(aq) + I_2(aq) \longrightarrow S_4O_6^{2-}(aq) + I^-(aq)$
e) $MnO_4^-(aq) + SO_3^{2-}(aq) + H^+(aq) \longrightarrow Mn^{2+}(aq) + SO_4^{2-}(aq) + H_2O(l)$

(**Note:** You may need to refer back to section 4.12).

Comments on the activities

Activity 5.1

	Li	Si	Br	Fe	He	O	Ca	U	Pb	Au
block	s	p	p	d	p	p	s	f	p	d
group	1	4	7	–	0	6	2	–	4	–
period	2	3	4	4	1	2	4	7	6	6

Activity 5.2

1 Figure 5.19 shows the variation in boiling point for the first 20 elements. The shape of this graph is very similar to that for melting points shown in Figure 5.11. This is to be expected, as both melting point and boiling point depend upon the strength of the attractive forces holding particles together within a structure.

2 From lithium to carbon the elements have structures in which all the particles are packed closely together. Since the size of the atoms falls whilst their mass increases, density rises. The sudden drop at nitrogen is caused by a change in structural type from macromolecular carbon to simple N_2 molecules which are quite widely spaced, even in the liquid or solid state. Thus although the individual atoms are smaller and heavier, they occupy a greater volume and the density falls. Oxygen and fluorine also consist of diatomic molecules, whilst neon exists as well-separated single atoms.

3 Only the first three elements in period 3 are good electrical conductors. Sodium, magnesium and aluminium are all metals with a structure that contains

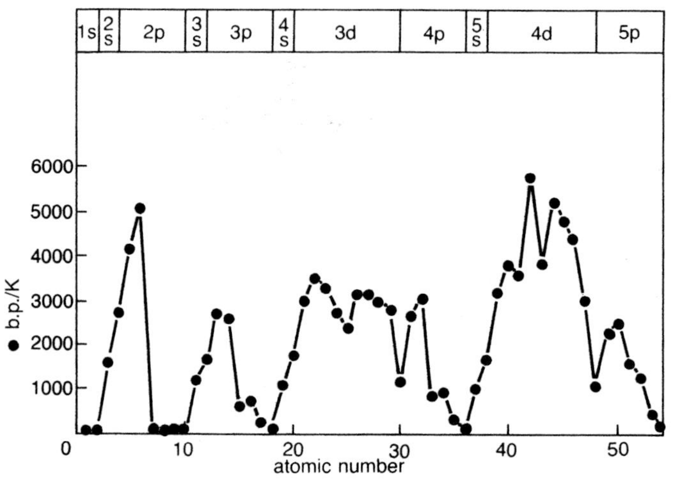

Figure 5.19
The pattern in boiling points of the first elements in the periodic table

Comments on the activities *continued*

positive ions embedded in a 'sea' of electrons. Since these electrons are free to move between the positive ions, metals are good conductors of electricity. The remaining elements all possess covalent structures in which all the outer electrons are tightly bound. Since there are no charged particles able to move easily through the structure, these elements are good electrical insulators. Note however, that in the case of elements such as silicon, the introduction of impurities can lead to the production of a semi-conductor.

Activity 5.3

$Na(s) + \frac{1}{2}Cl_2(g) \longrightarrow NaCl(s)$

$Mg(s) + Cl_2(g) \longrightarrow MgCl_2(s)$

$Si(s) + 2Cl_2(g) \longrightarrow SiCl_4(l)$

$2Na(s) + \frac{1}{2}O_2(g) \longrightarrow Na_2O(s)$

$Mg(s) + \frac{1}{2}O_2(g) \longrightarrow MgO(s)$

$2Al(s) + 3O_2(g) \longrightarrow Al_2O_3(s)$

$Si(s) + O_2(g) \longrightarrow SiO_2(s)$

$4P(s) + 5O_2(g) \longrightarrow P_4O_{10}(s)$

Activity 5.4

$Al_2Cl_6 + 6H_2O \longrightarrow 2Al(OH)_3 + 6HCl$

$PCl_5 + H_2O \longrightarrow POCl_3 + 2HCl$

$MgO + H_2O \longrightarrow Mg(OH)_2$

$P_4O_{10} + 6H_2O \longrightarrow 4H_3PO_4$

$SO_3 + H_2O \longrightarrow H_2SO_4$

Activity 5.5

The relatively high melting point and electrical conductivity whilst molten indicate ionic bonding in **A**. This compound must be sodium hydride, which contains Na^+ and H^- ions. When this dissolves in water, the hydride ions react with water molecules to give hydrogen gas and aqueous hydroxide ions which make the solution alkaline:

$$H^- + H_2O(l) \longrightarrow H_2(g) + OH^-(aq)$$

The low melting points and lack of conductivity in the molten state of the other hydrides indicates covalent molecular structures. Since on passing across a period acidic nature becomes more pronounced, **C** is hydrogen chloride (HCl) whereas **B** is phosphorus hydride (PH_3). When hydrogen chloride dissolves in water, the polar covalent bond breaks, giving an ionic solution:

$$HCl(g) \longrightarrow H^+(aq) + Cl^-(aq)$$

It is the $H^+(aq)$ ions which makes the solution acidic.

Activity 5.6

1 Fe^{2+} forms Fe^{3+} by losing 1 electron.

2 **a)** Each O atom is –2. The sum of the oxidation states equals the charge on the ion, so 2(Cr) + 7(–2) = –2, from which each Cr must have an oxidation state of +6.
b) The oxidation state of Cr^{3+} is +3.

3 The oxidation state of each chromium atom changes from +6 to +3, so when one $Cr_2O_7^{2-}$ ion forms two Cr^{3+} ions, 6 electrons are gained.

4 Six Fe^{2+} ions will react with each $Cr_2O_7^{2-}$ ion.

5 $6Fe^{2+}(aq) + Cr_2O_7^{2-}(aq) \longrightarrow$

6 $\longrightarrow 6Fe^{3+}(aq) + 2Cr^{3+}(aq)$

7 The seven oxygen atoms in the $Cr_2O_7^{2-}$ ion form seven H_2O molecules, requiring 14 H^+ ions.
$6Fe^{2+}(aq) + Cr_2O_7^{2-}(aq) + 14H^+(aq) \longrightarrow 6Fe^{3+}(aq) + 2Cr^{3+}(aq) + 7H_2O(l)$

8 **a)** Oxidation states balance at +24; **b)** ionic charges balance at $^{24+}$.

C H A P T E R

Carbon compounds – an introduction to organic chemistry

Contents

Study Checklist

After studying this chapter you should be able to:

1. State that carbon atoms are able to catenate, that is form strong covalent bonds with each other and that, as a result, a range of organic compounds are formed which contain carbon–carbon single, double and triple bonds arranged in chains and rings.
2. Explain the meaning of the following terms, giving examples in each case: aliphatic, alicyclic and aromatic compounds; functional group; empirical formula, molecular formula, short structural formula, displayed structural formula, skeletal formula; homologous series.
3. Describe the main stages and techniques involved in organic synthesis under the following headings: methods of mixing, stirring, heating and cooling the reactants; safety procedures; solvent extraction, distillation, recrystallisation, filtration and chromatography.
4. Calculate the percentage yield of a product given the quantities of reactants used and the balanced equation for the reaction.
5. Describe, in outline, how the purity of a product may be determined from its melting or boiling point, its infrared absorption or its mass spectrum.
6. Calculate the empirical formula of a compound from (i) combustion data and/or (ii) the percentage elemental composition. Derive the molecular formula of the compound from its empirical formula and its relative molecular mass. Recall that the structure of a compound is usually determined using a combination of instrumental techniques.
7. Define the term 'isomer'. Explain the difference between structural and stereo isomerism in terms of the structural features of molecules which exhibit this behaviour. Describe, and explain, the techniques which may be used to distinguish between pairs of each type of isomer.
8. Explain, using appropriate diagrams, the meaning of the following terms: geometrical *(cis–trans)* isomerism, plane polarised light, polarimeter, optical isomer, dextrorotatory, laevorotatory, enantiomers, chirality, asymmetric carbon atoms, racemic mixture. Describe, in outline, how optical isomers may be separated.
9. Make predictions about the existence of isomers from the structures of molecules.
10. Use electronegativities to deduce the polarity of covalent bonds and hence identify the reactive centres in an organic molecule.

11 State that when homolytic bond fission occurs each atom in the covalent bond takes one electron from the bond and free radicals (very reactive atoms, or groups of atoms, which possess an unpaired electron) are formed. Know that in heterolytic bond fission, the more electronegative atom takes both bonding electrons and ions are formed. Write equations for the homolytic or heterolytic fission of a given bond.

12 Describe a nucleophile as an electron-rich species (δ–) which has at least one lone pair (non-bonding pair) of electrons which will attack an electron-poor (δ+) reactive centre. Describe an electrophile as an electron-poor (δ+) species that will accept a pair of electrons from a nucleophile to form a dative covalent bond. State examples of simple nucleophiles and electrophiles and write down their dot and cross diagram.

13 Regard organic molecules as having two main types of reactive centre, the functional group(s) and a hydrocarbon skeleton made up of a combination of C—H, C—C, C=C, C≡C bonds and the partial π bonds between the carbon atoms in the benzene ring. State that the nature of the hydrocarbon skeleton can modify the reactivity of a functional group, and vice versa.

14 Understand that a reaction mechanism explains how the electron clouds move around during a chemical reaction and that a 'curly arrow' shows the movement of a pair of electrons.

6.1 Organic chemistry – an overview

In 1807, the famous Swedish chemist Jöns Jakob Berzelius proposed that chemicals should be divided into two distinct categories, organic and inorganic. Substances which existed in, or were created by, plants or animals were termed **organic**. It was thought that these materials could only be formed in living organisms via some kind of mystical '**vital force**'. On the other hand, **inorganic** compounds were far less mysterious since they could easily be obtained from mineral sources. This division largely resulted from chemists' lack of knowledge of naturally occurring compounds. At that time, it was very difficult to extract organic chemicals in a pure enough form for analysis.

Up to 1828, only a handful of organic chemicals had been identified. It was in that year, though, that the German chemist Friedrich Wöhler made a remarkable discovery. He prepared one of these organic compounds, urea, from an inorganic starting material, ammonium cyanate:

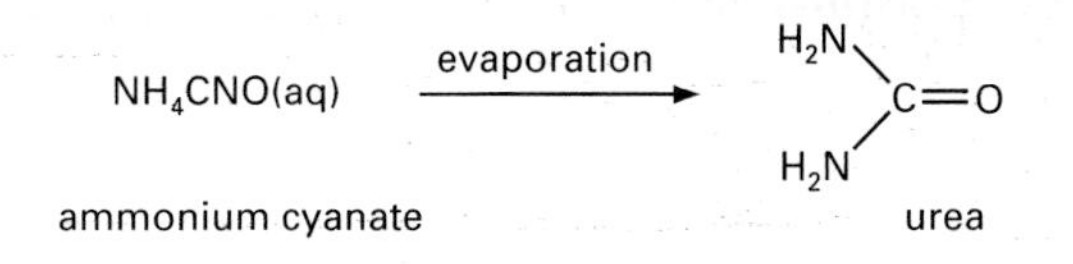

Wöhler proved that this synthetic urea was identical to the urea which he had extracted from urine. At first, Wöhler's research made little impact on the supporters of the 'vital force' theory. Indeed, it was not until the 1850s that the idea of a vital force was finally discredited. By that time, many more organic compounds had been prepared and successfully analysed. Consequently, with a few exceptions which are not actually regarded as organic, e.g. CO, CO_2 and carbonates, organic chemistry was re-defined as *the chemistry of carbon compounds*.

During the last 140 years, our knowledge of organic compounds, and their uses, has increased many times over. Indeed, a recent estimate suggested that there are upwards of 2.5 million known organic compounds.

In the late nineteenth century, the main source of organic chemicals was coal tar, a by-product of the coal industry. This was used as starting material in the manufacture of a variety of household and industrial products, such as dyes, paints, disinfectants and solvents. Nowadays, about 80% of organic substances are made, directly or indirectly, from one of the components of crude oil or **petroleum** (section 7.4). Petroleum, then, is the world's major organic chemical feedstock.

A large number of organic products are now a common, and often irreplaceable, feature of everyday life. Some are prepared in vast amounts, for example plastics, detergents and solvents, others are manufactured in small quantities for more specialist uses, such as drugs and pesticides.

Although many organic compounds occur naturally, most commercial products are synthetic. Some synthetic pathways are short and straightforward, e.g. the manufacture of ethanol by the hydration of ethene:

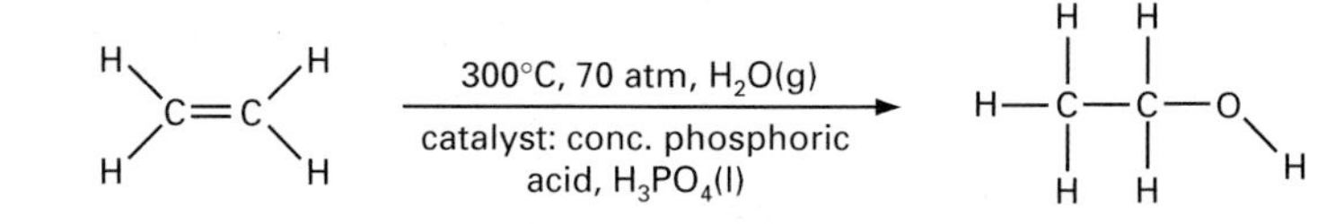

More complex organic substances often need longer synthetic routes, for example the manufacture of aspirin:

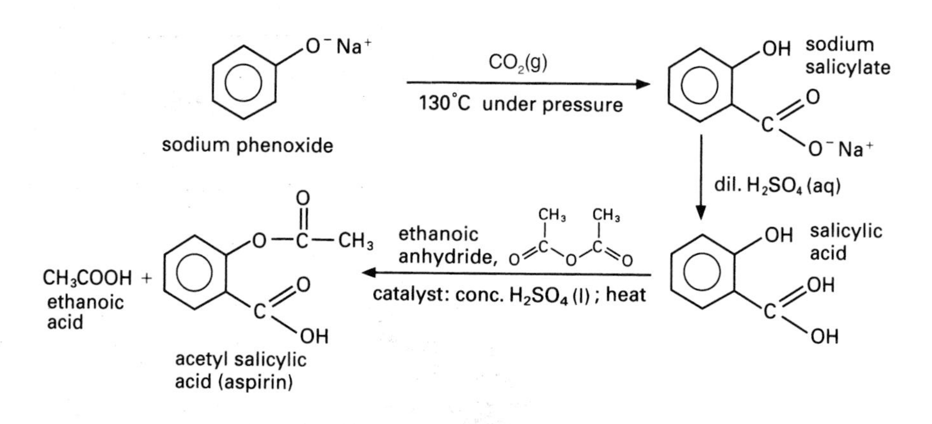

Aspirin – versatility is its second name!

Large quantities of **aspirin** are made daily by pharmaceutical companies all round the world. Aspirin is an analgesic drug, i.e. a painkiller, which also has the ability to reduce inflammation and lower the patient's temperature, so it is very effective as a treatment for colds and flu. Aspirin works by blocking the formation of **prostaglandins**, chemicals which cause pain and inflammation to develop and also cause the blood to clot.

Recent studies have shown that taking a small dose of aspirin, say 0.1 g per day, may significantly reduce the risk of a heart attack by preventing the coronary artery from becoming blocked by a blood clot, or **thrombus**. Unfortunately, aspirin also affects the blood flow in the stomach lining and this may lead to ulcers or internal stomach bleeding. So, although aspirin may not be universally safe as a long-term treatment for heart disease, it can help to stabilise patients whilst they await a coronary by-pass operation.

Aspirin can also make plants feel better! Over the last twenty years, scientists have found that the level of **salicylic acid** (used in making aspirin) in a diseased plant can build up until it triggers the formation of disease-fighting proteins. It is not particularly surprising that plants generate their own aspirin; indeed, the name 'salicylic' comes from the willow tree, Salix, which North American Indians used to make potions for relieving headaches. So, we find that cut flowers tend to keep a little longer if a small amount of aspirin is added to the water in the vase and that certain plant viruses may be controlled by injecting the plant with aspirin.

The benzene ring

Aspirin is a derivative of benzene, a cyclic hydrocarbon of molecular formula C_6H_6, whose structure is represented by the symbol:

The 'circle' represents a system of delocalised π electrons which extends over all of the carbon atoms (see section 9.1 for a full explanation).

Aspirin, one of the first synthetic drugs, was developed in 1899. Since then, tens of thousands of drugs have been prepared and, in many cases, one particular drug may be sold under a variety of trade names. In the USA alone, 50 million aspirin are swallowed every day.

As you can imagine, the business of selling pharmaceuticals (drugs) and other organic chemicals is highly profitable. These profits are needed, though, to finance the enormous capital investment associated with the research, development, testing and production of a new chemical product. For example, in 1990 the Association of the British Pharmaceutical Industry (ABPI) observed that a major new medicine takes about 10 years to develop at a cost of around £125 million!

The purpose of this section is to give you an insight into the world of the organic chemist. Whether in the school or industrial laboratory, organic chemistry is based on common principles and techniques. Although you may not have access to 'hi-tech' industrial equipment, the way you will tackle problems requires an approach almost identical to that used by industrial organic chemists.

When working through Chapters 6–14, you will find it useful to focus on a few general points, as follows.

Organic synthesis

Organic synthesis changes a starting material into the required product via one, or more, stages. Usually, each stage will involve a chemical reaction between the **starting material** (or an intermediate product) and a **chemical reagent**.

For example, in the manufacture of ethanol, given earlier, the reagent is steam in the presence of concentrated phosphoric acid as a catalyst. This reagent does a certain 'job', namely it adds water to a C=C bond; consequently, $H_2O(g)$/conc. $H_3PO_4(l)$ is called a **hydrating agent**.

The aim of any synthesis is to cause a desired alteration to a molecule's structure. To do this, we must use a suitable reagent, often under *carefully controlled conditions*. You would not use, for example, a toothbrush to paint a door in the pouring rain!

One key aspect of an organic chemist's work, then, is the planning of synthetic pathways. When you have to do this:

- Remember that reagents do specific 'jobs'. Thus, get to know these reagents and, perhaps, compile a list as you work through the text.
- Always state the temperature, pressure and, where relevant, physical states of the reactants, reaction times and any special reaction techniques.

The systematic nature of organic chemistry

There are over 2.5 million organic chemicals. It would be a very large book or computer file which could store the chemical properties of all these compounds. Yet, after studying Chapters 6–14, you will be able to comment on the probable chemical reactions of almost any organic substance just by looking at its molecular structure.

The mean bond enthalpy of the carbon–carbon bond is much higher than that of most other single bonds, as shown in Table 6.1. Due to the strength of these bonds, carbon atoms have the ability to **catenate**, that is form strong covalent bonds to each other, thereby forming an almost infinite number of compounds containing **chains** or **rings** of carbon atoms. These may be linked

Table 6.1 Mean bond enthalpies of some X—X type bonds.

Bond	mean bond enthalphy/ kJ mol^{-1}
C—C	348
N—N	163
O—O	146
F—F	158
Si—Si	176

together via **single (C—C)**, **double (C=C)** or **triple bonds (C≡C)**. Each catenated carbon displays a valency of four by also bonding to hydrogen atoms or functional groups (see below). There are three main types of **hydrocarbon skeleton** in organic compounds: aliphatic, alicyclic, and aromatic.

Aliphatic compounds contain straight-chain or branched-chain hydrocarbon skeletons, for example:

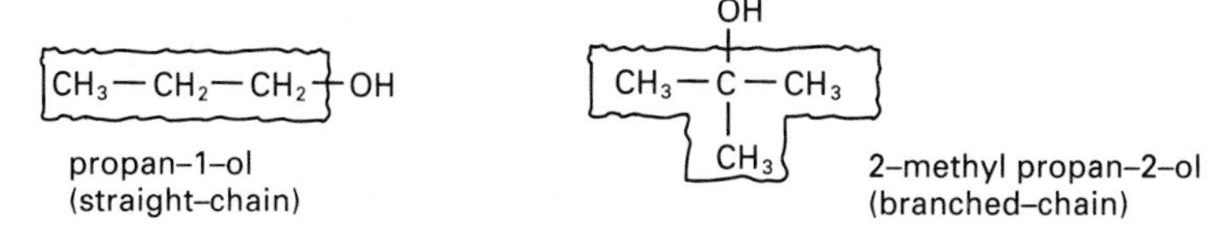

Alicyclic hydrocarbon skeletons contain rings of carbon atoms yet still possess chemical properties similar to those of the corresponding aliphatic skeletons, for example:

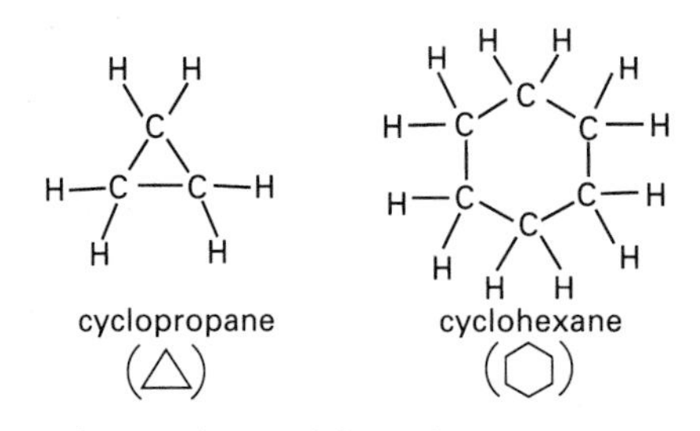

Aromatic compounds are derived from benzene, a hydrocarbon of formula C_6H_6 which contains a ring of delocalised π electrons. Some simple aromatic compounds are shown below:

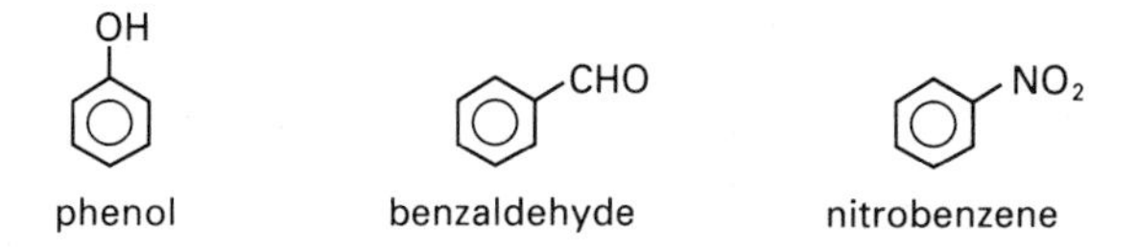

In Chapters 7 to 9, you will see that the chemical properties of aromatic compounds are very different from those of aliphatic and alicyclic compounds.

Chapters 10 to 14 are arranged under functional group headings. A **functional group** is an atom, or group of atoms, which imparts characteristic chemical properties to any organic compound in which it appears. We group organic compounds into families, or **homologous series**, in which each hydrocarbon skeleton increases in size by a —$\mathbf{CH_2}$— structural unit. For example, the first three members of the homologous alcohol series are methanol CH_3OH, ethanol CH_3CH_2OH and propanol $CH_3CH_2CH_2OH$; each molecule contains the hydroxy, —OH, functional group. The members of a homologous series have a general formula, in this case, $C_nH_{2n+1}OH$. They have similar chemical properties and their physical properties show a gradual change as the size of the molecule increases.

It is important that you do not view each organic chapter as an isolated mass of information. Look out for the links between functional groups which are stressed in the text and the activities. In this way, you will develop a systematic approach which leads to a better understanding of organic chemistry.

Identification of organic compounds

To determine the molecular structure of an organic compound, we need to use the modern instrumental techniques described in Chapter 17. However, where these are not available, we can still identify the presence of a particular

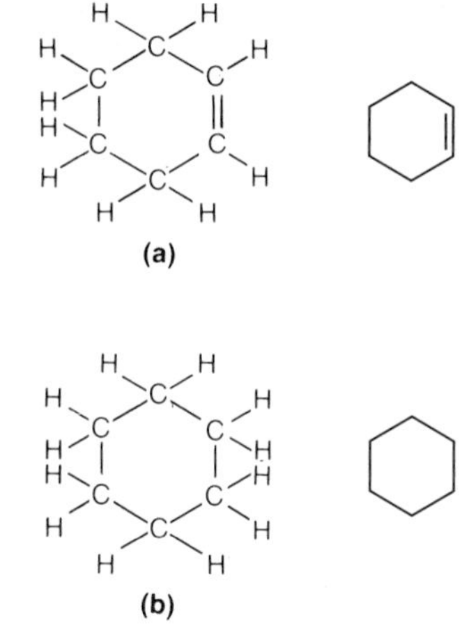

Figure 6.1 ▲
Structural and skeletal (on the right) formulae of **(a)** cyclohexene and **(b)** cyclohexane

functional group by studying simple chemical reactions of the compound. Suppose, for example, that you were asked to distinguish between the liquids cyclohexene (a) and cyclohexane (b), Figure 6.1. From their structural formulae, we can see that both molecules possess C—H and C—C bonds but only the cyclohexene has a C=C bond. Hence, we test each liquid with a reagent which reacts with a C=C bond but not with a C—H or C—C bond. A suitable reagent would be bromine water:

cyclohexane + bromine water ⟶ no reaction
BUT cyclohexene + bromine water ⟶ reaction (orange bromine colour disappears)

Thus, the colour change, or lack of it, distinguishes the compounds. The identification of functional groups using simple laboratory tests is another important aspect of the study of organic chemistry.

Writing the structures of organic compounds

It is important that the 3-D shape, or stereochemistry, of an organic molecule can be represented on paper in a simple and accurate way. Methoxyethane has the **molecular formula**, C_3H_8O – this means that each molecule contains 3 carbon atoms, 8 hydrogen atoms and 1 oxygen atom. The molecular formula does not indicate the way in which the atoms are joined together. A model of the methoxyethane molecule is shown in the photograph and there are various ways of drawing this structure.

▲ A molecular model of methoxyethane, molecular formula C_3H_8O

- You can see from these models that the molecule contains a C—O—C structure, so we can use a **short structural formula** to show the grouping of atoms:

$$CH_3OCH_2CH_3$$

- To show the relative placing of the atoms and the number of bonds between them, we use a **displayed structural formula**, Figure 6.2a
- The complete **stereochemical formula** provides a 3-D view of the molecule by showing the orientation of the bonds relative to the plane of the paper, Figure 6.2b.
- Some other conventions are used to represent formulae. Thus, a **skeletal structure** only shows the bonding between carbon atoms and with the functional groups, Figure 6.2c. Skeletal structures are frequently used to represent cyclic molecules, such as cyclopropane and cyclohexane, Figure 6.1b. When working with skeletal structures, remember that each carbon atom forms four covalent bonds and that the C—H bonds have been omitted for clarity.

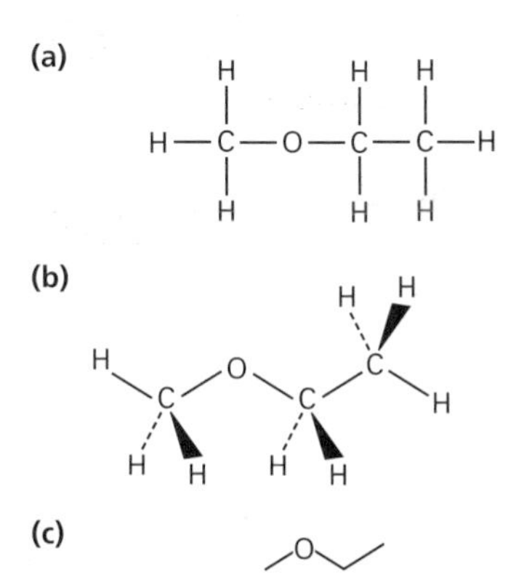

Figure 6.2 ▲
Various ways of representing the structure of methoxyethane, a molecule of short structural formula $CH_3OCH_2CH_3$: **(a)** displayed structural formula, **(b)** stereochemical formula, **(c)** skeletal structure.

Activity 6.1

The molecular formulae of some organic molecules are shown below. Build molecular models for each molecule and then write down:

a) the short structural formula,
b) the displayed structural formula,
c) the stereochemical formula,
d) the skeletal structure.

You should note that there are TWO possible structures for each molecule.

1 C_4H_{10} **2** C_3H_7Cl **3** C_2H_6O

Focus 6a

1 In organic synthesis, chemical **reagents** are used to convert the starting material into the final product. The choice of conditions, such as temperature and pressure, may also determine which products are obtained.

2 Organic chemistry has a systematic nature. The compounds are classified according to the hydrocarbon skeleton and the functional groups which are present.

3 The most common features of the **hydrocarbons skeletons** are C—H, C—C, C=C and C≡C bonds, and the delocalised π bonds in the benzene ring. The hydrocarbon skeleton may be classified as being **aliphatic**, **alicyclic** or **aromatic**.

4 A **functional group** is an atom, or group of atoms, in a molecule which imparts characteristic chemical properties to that molecule. A **homologous series** is a family of compounds which contain the same functional group and in which the molecular formula of each successive member increases by a $-CH_2-$ unit. The members of a homologous series have similar chemical properties and gradually changing physical properties.

5 Although the structure of an organic compound can be obtained with certainty using expensive modern instrumental techniques, we can obtain a lot of information about the structure using simple laboratory tests.

6 It is important to accurately represent the structure of an organic compound. This can be done in various ways, as summarised below for propanone.

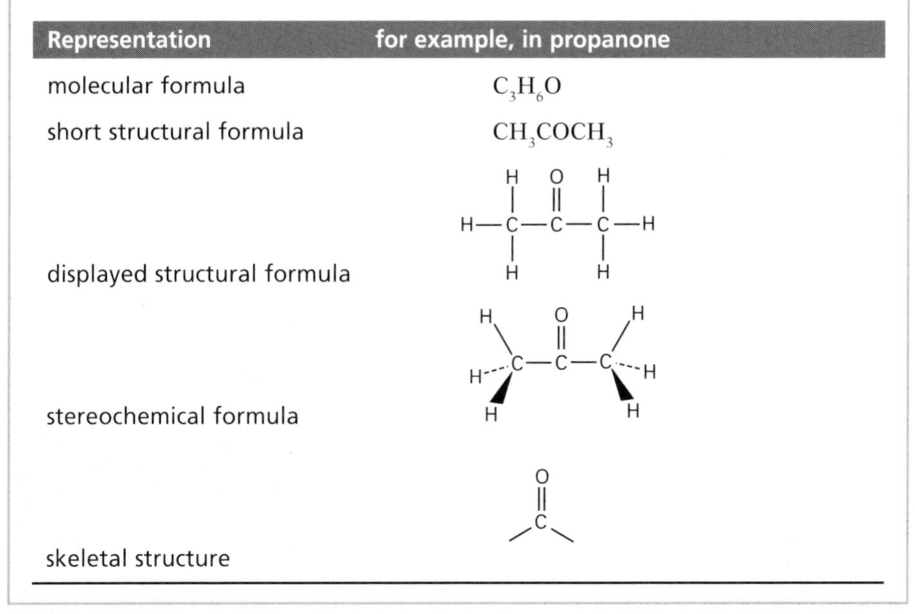

Representation	for example, in propanone
molecular formula	C_3H_6O
short structural formula	CH_3COCH_3
displayed structural formula	
stereochemical formula	
skeletal structure	

Organic synthesis

Converting an organic starting material into the final pure product may require a number of synthetic steps. Each step involves:

- planning and performing the chemical reaction,
- separating the crude product and purifying it,
- analysing the product to ensure that it is the desired compound.

Sections 6.2–6.4 look at some of the ways you might carry out these tasks.

6.2 Planning and performing an organic synthesis

If the reaction has been performed before, you will be able to find the procedure in a reference text. Some modifications may be needed; for example, if double the quantity of product is required, then the amounts of reactants will also need to be doubled. However, if you are the first person to try the synthesis, it may be necessary to repeat the reaction under varied experimental conditions, for example using different reactant concentrations, temperature and pressure, ways of mixing the reactants and so on. Eventually, you will be able to work out which set of conditions gives the most economical yield of product.

Organic chemicals are often volatile and flammable; many are also toxic. These hazards must be borne in mind when planning the experimental procedure and designing the apparatus. Nowadays, small-scale organic syntheses are performed in hard glassware which fits closely together via ground glass joints. Before use, these joints are lightly smeared with high melting point grease. This improves the seal and allows the glassware to be easily dismantled after use. In industry, large-scale preparations are usually carried out in stainless steel vessels.

Whether it is performed in school or industry, there are four important aspects of any organic synthesis, and these are described below.

Heating and cooling the reaction mixture

When heating is necessary, a **reflux apparatus** is used (Figure 6.3). As the reaction mixture boils, vapour passes into the condenser where it cools, changes into a liquid and falls back into the flask. No vapour is able to escape into the atmosphere.

Another consideration is the source of heat. Obviously, using a Bunsen burner could present a danger of fire or explosion. Thus, we use (i) a bath containing water (up to 100°C) or liquid paraffin (up to 220°C) or (ii) an electric heater (up to 300°C).

For highly exothermic reactions, cooling may be necessary. To do this, the reaction flask is placed in one of the following: (i) cold water (room temperature), (ii) an ice/salt/water mixture (–10°C) or (iii) a mixture of solid carbon dioxide and ethanol (–40°C).

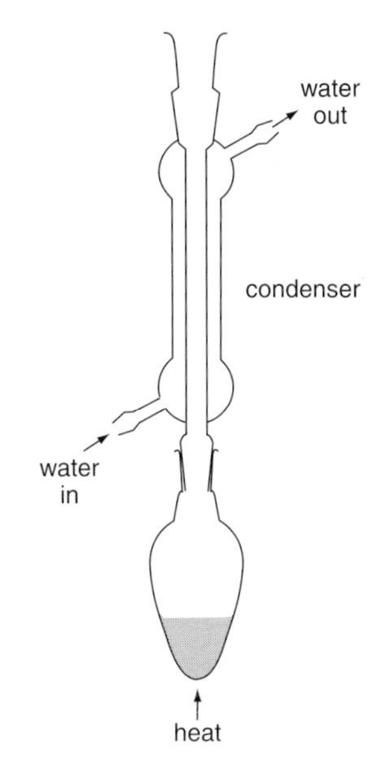

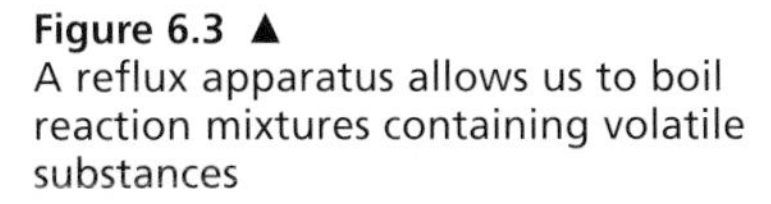

Figure 6.3 ▲
A reflux apparatus allows us to boil reaction mixtures containing volatile substances

Mixing the reactants together

Often, it is necessary to mix the reactants together in small portions whilst the reaction is proceeding. This is achieved by using a double- (or triple-) necked flask fitted with a dropping funnel (see the photograph below).

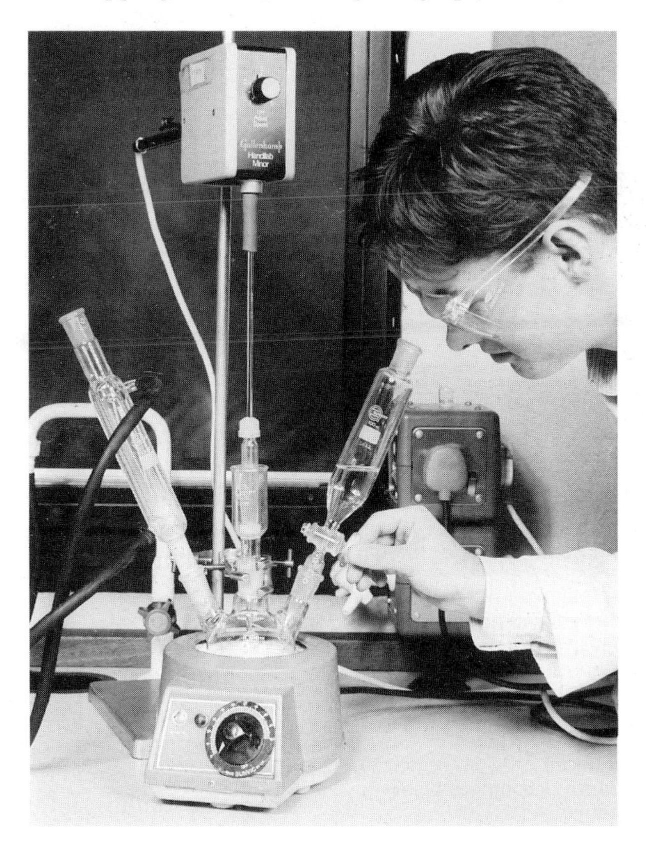

A triple-necked flask fitted with a dropping funnel and an overhead mechanical stirrer being refluxed on an electric heating mantle. The student is adding one of the reactants in portions from the dropping funnel. ►

One interesting example of the value of gradually mixing reactants is the preparation of ethanal by using acidified aqueous potassium dichromate(VI) solution to oxidise ethanol. If the reactants are mixed before heating, any ethanal which forms is further oxidised to ethanoic acid:

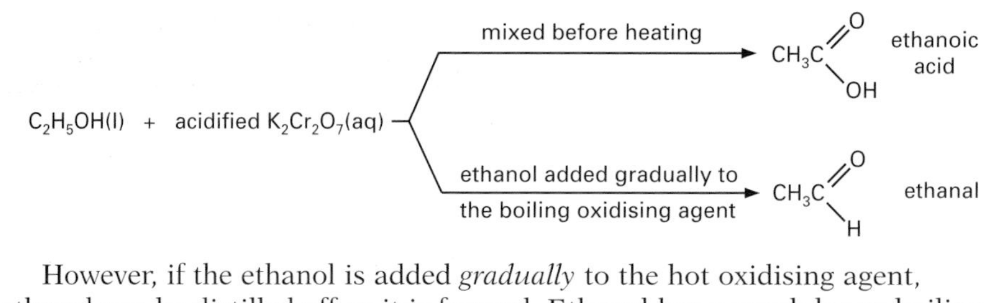

However, if the ethanol is added *gradually* to the hot oxidising agent, ethanal can be distilled off as it is formed. Ethanal has a much lower boiling point (20°C) than ethanol (78°C), since the latter contains fairly strong hydrogen bonds. So, by adding the ethanol dropwise to the oxidising agent (at 70°C), the ethanal distils off and further oxidation is minimised. Using this method, only a small amount of ethanoic acid (b.p. = 118°C) is formed and this remains in the reaction flask.

Stirring the reaction mixture

Stirring is used to help the reactants mix together and to distribute heat evenly throughout the reactant mixture. Two types of stirrer are available: (i) a mechanical overhead stirrer or (ii) a magnetic stirring bead which spins round in time with a larger magnet rotating below the reaction flask.

We can also make reaction mixtures boil more 'evenly' by adding a few small bits of porcelain called **'anti-bumping' granules**.

Safety

Finally, we must stress the need for stringent **safety precautions** when preparing organic compounds. Where possible, reactions should be performed in a fume cupboard, away from any naked flames. Of course, safety clothing and goggles should be worn. It is also important to know the location of fire-fighting equipment and how to use it.

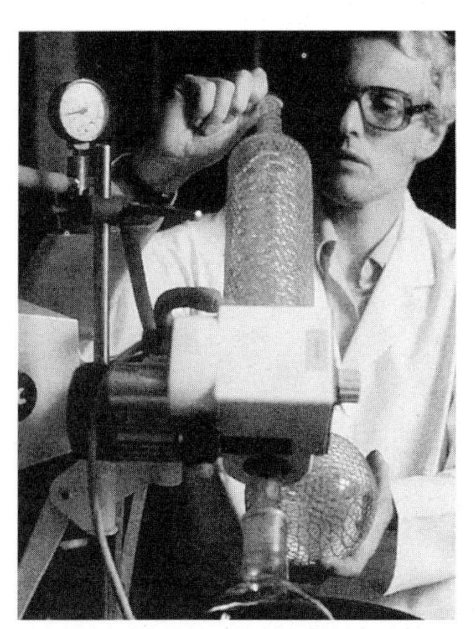

Distillation and fractional distillation can be performed at reduced pressure, by using a **rotary evaporator**. A liquid boils when its vapour pressure equals the external pressure. Therefore, by lowering the external pressure, the liquid will boil at a lower temperature. Hence, distillation at reduced pressure is used to purify liquids which decompose below their boiling points.

6.3 Separating the crude product, purifying it and calculating the percentage yield

When the reaction has finished, the desired product is described as being in a **crude state**. Often, it is heavily contaminated, perhaps by unused reactants, a solvent or a reaction by-product. Thus, the synthetic chemist's next job is to separate the product from these impurities.

Some of the most commonly used separation techniques are listed in Table 6.2. As you can see, the choice of technique is governed by the physical states of the product and its impurities.

We shall now look briefly at each method.

Distillation and fractional distillation

Distillation is used to separate a liquid from either a dissolved non-volatile solid or another liquid. In the latter case, the two liquids must have very different boiling points, e.g. the separation of phenylamine (b.p. = 184°C) from its solution in ethoxyethane (b.p. = 25°C). Mixtures of miscible liquids with similar boiling points can often be separated by using **fractional**

distillation, Figure 6.4. This technique is widely used in the chemical industry, for example to separate the hydrocarbon components in petroleum (section 7.4).

Table 6.2 Some common physical separation techniques

Technique	used to separate
distillation	a liquid from a dissolved non-volatile solid or liquids with very different boiling points
fractional distillation	liquids which have similar boiling points
steam distillation	a high boiling point liquid from a non-volatile solid
solvent extraction	a solid or liquid from its solution
recrystallation	a solid from other solid impurities
filtration	an insoluble solid from a liquid
chromatography	various types of mixtures

Which technique(s) could be used to separate the following mixtures:
a) propan-1-ol, b.p. 97°C and 1-iodopropane, b.p. 103°C;
b) phenylamine, b.p. 184°C and tin(II) chloride (s);
c) pentane, b.p. 36°C, hexane, b.p. 69°C and heptane, b.p. 99°C;
d) benzoic acid (s) contaminated by charcoal dust?

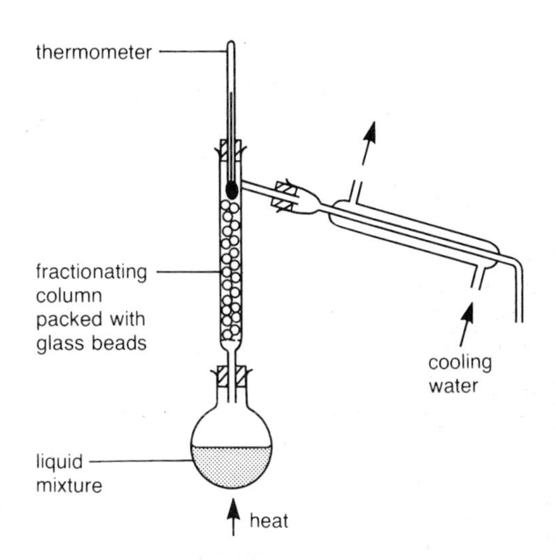

Figure 6.4
Fractional distillation is used to separate liquids which have similar boiling points. The fractionating column is a long tube packed with small glass beads. As such, it provides a large surface area for condensation. Vapour rising from the flask warms up the bottom of the column and condenses back into liquid. As it falls, it meets fresh hot vapour moving upwards. This process is repeated continually up the column and each time the vapour becomes richer in the more volatile component, that is, the one with the lower boiling point. If the column is long enough, the vapour which escapes at the top is the pure, more volatile component, and this may be confirmed by checking its boiling point on the thermometer. Whilst simple distillation is based on identical principles, a column is not required to separate the components because they should have very different boiling points.

Steam distillation

An organic liquid which has a high boiling point and very low solubility in water may be removed from a mixture by **steam distillation**. As steam is passed through the hot mixture, a mixture of the two components distils out of the flask and the condensate is collected as shown in Figure 6.5. On standing, the condensate forms two layers; one is the organic product, the other is aqueous. These are separated and any product in the aqueous layer is removed by **solvent extraction**.

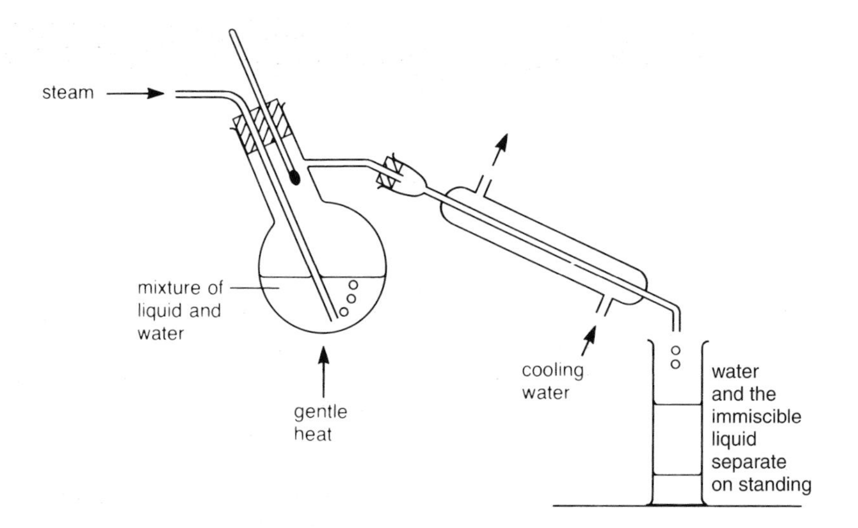

Figure 6.5
Steam distillation apparatus. To separate an organic liquid using steam distillation, the liquid must be immiscible with water, yet be volatile when boiled with water.

Solvent extraction

In this technique, the 'crude' product is shaken up with a small quantity of a volatile solvent in which only the product is soluble. After filtering off any solid impurities, if necessary, the mixture is transferred to a separating funnel. On standing, the mixture divides into two distinct layers, which can be run off separately. We keep the layer containing the product; this is called an **extract**. Since small amounts of product still remain in the other layer, the shaking and separation procedure is repeated twice, each time using fresh solvent. All the product extracts are combined.

Solvent extraction is often used to separate an organic product which is contaminated by an aqueous solution. Such extracts will always contain a small amount of water and this is removed by 'drying' it with an anhydrous ionic salt, such as magnesium sulphate. Usually, drying takes a few hours, after which time the ionic salt, which is now partially hydrated, is filtered off.

Finally, we must remove the volatile solvent from the dried extracts. This is easily achieved by distillation. The resulting product is then further purified by fractional distillation (for liquids), or recrystallisation (for solids).

Recrystallisation

Recrystallisation is used to separate a solid product from other solid impurities. In this technique, we choose a solvent in which:

- the product readily dissolves when hot but hardly does so when cold,
- the impurities must either not dissolve or remain dissolved at low temperatures.

Often, the selection of this solvent is simply a case of 'trial and error'. A solid product can sometimes be purified by recrystallisation from water. The impure solid is dissolved in the minimum quantity of hot water. Impurities which are insoluble in water can be removed by quickly filtering through a fluted filter paper held in a hot funnel. What would happen if the funnel was cold?

As the filtrate cools, any soluble impurities remain in solution while the pure product crystallises out. The crystals are filtered off using a **vacuum filtration** apparatus. Apart from giving a rapid filtration, vacuum filtration has the added advantage that the flow of air past the crystals helps them to dry out.

The final drying of an organic solid is carried out in either a desiccator at low pressure or an oven. The method chosen will depend on the properties of the solvent and the solid, for example the latter might decompose in an oven.

Fruit juice fraud

There's a lot of money to be made from selling fruit juice! Over £800 million worth of juice is sold in Britain each year whilst in the USA the market is a staggering $12 billion. In recent years there have been several cases of fruit juice fraud. In 1991, for example, the Ministry of Agriculture, Fisheries and Food tested 21 popular brands of orange juice sold in Britain – they found that 16 contained substances which should not have been there.

To call a product 'orange juice', it has to be made from the fleshy part of the orange not from 'pulpwash', a bitter tasting product which is produced from the remains of the squeezed oranges. When 'pulpwash' is added to orange juice, the juice develops a slight bitter taste, so beet sugar is added to improve the taste. If you buy a carton of pure orange juice, then that is what it should be, with no additives, including sugars. Fortunately, analytical chemists can identify this fraud using **chromatography**. By comparing a chromatogram of the pure juice and the doctored sample, it is possible to identify substances which have been added.

Apple juice contains three natural sugars: fructose, glucose and sucrose. A common fraud is to dilute the juice slightly and add cane or beet sugar to it. Since cane and beet sugar contain only sucrose, the proportion of sugars in the diluted juice differs from that in pure juice and, once again, the fraud can be detected by chromatography.

Many juices are marketed as containing fruit from a particular area, such as English apple juice or Florida orange juice. The raw materials for these 'speciality' products are expensive, so it is tempting for a producer to defraud the customer by using cheaper fruit from other regions. Chromatography is used to identify chemicals, called **flavonoids** and **polyphenols**, which give the fruit its characteristic flavour and colour, and this makes it possible to determine its place of origin.

Another common fraud surrounds the treatment of the juice. Sometimes the producer concentrates the orange juice by evaporation. It is then transported to a wholesaler who dilutes the juice with water. Orange juice produced in this way must be labelled 'made from concentrated orange juice'. Due to the heating during concentration, fruit juice made from concentrated and freshly squeezed fruit juice contain different chemicals and this allows them to be distinguished.

Chromatography

Chromatography is a particularly valuable technique since it enables us to separate very complex mixtures, even where the components are present in small amounts.

In chromatography, the components can be separated because of their different strengths of attraction for a **stationary phase** and a **moving phase**, Figure 6.6. As a result, the moving phase carries the components through the stationary phase at different rates and a separation is achieved.

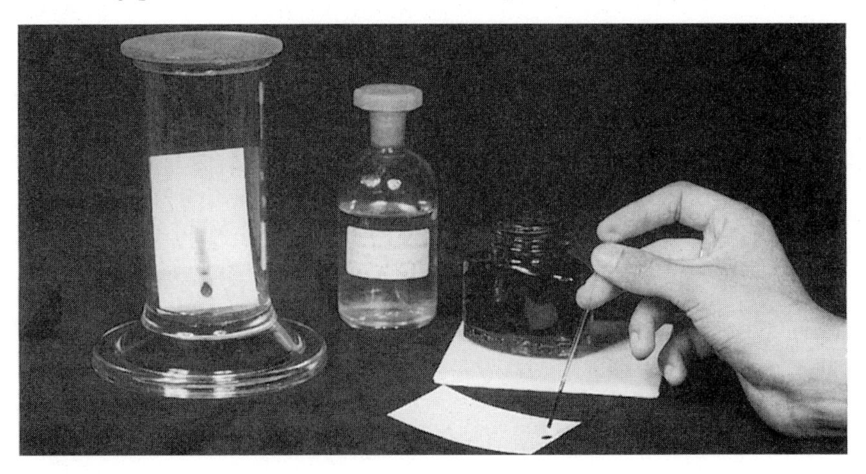

Thin-layer chromatography is used to separate a mixture of solids. The stationary phase is a thin-layer of alumina, Al_2O_3, and the moving phase is a liquid. A drop of black ink is applied to a thin layer plate using a fine capillary tube. The thin layer plate is allowed to dip into the solvent (in this case a mixture of butan-1-ol, ethanol and aqueous ammonia) contained in a gas jar. After a while the different components in the ink form separate 'spots' as the solvent front rises up the plate. Why do we choose a mixture of butan-1-ol, ethanol and aqueous ammonia? Simply because it gives a good separation! Very often the choice of stationary and moving phases is based on 'trial and error' experiments

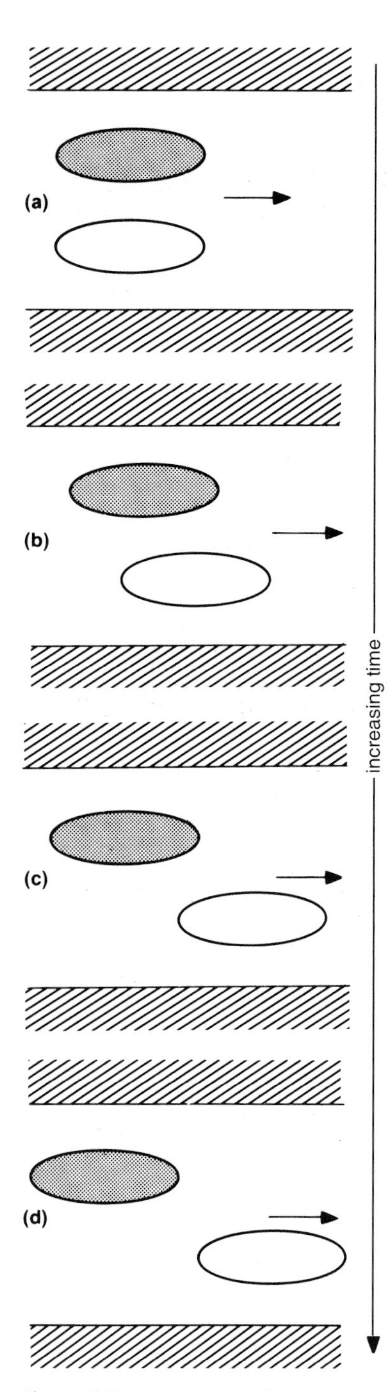

Figure 6.6
Chromatographic separation of two substances ◯ and ●, as time passes from (a)⟶(d). Since ◯ is less strongly attracted to the stationary phase than ●, it moves along more readily with the moving phase. The arrows show the direction of the moving phase

There are various chromatographic methods, the most common being thin-layer chromatography and gas–liquid chromatography. These are described in section 15.2. Full experimental details can be found in a practical chemistry textbook.

Calculating the percentage yield of product

Whenever we write a chemical equation, it is understood that the reactants will form the theoretical quantities of products, that is there is a 100% yield.

In real life this is never the case, and in organic chemistry in particular the actual yield of purified product is often well below the theoretical yield. Thus, we say that:

$$\textbf{\% yield} = \frac{\textbf{actual yield of purified product}}{\textbf{theoretical yield}} \times \textbf{100}$$

A worked example is shown below.

Example

When 15.0 g of butan-1-ol (A) and 10.0 g of ethanoic acid (B) were refluxed together in the presence of concentrated sulphuric acid, 17.8 g of 1-butylethanoate (C) were formed.

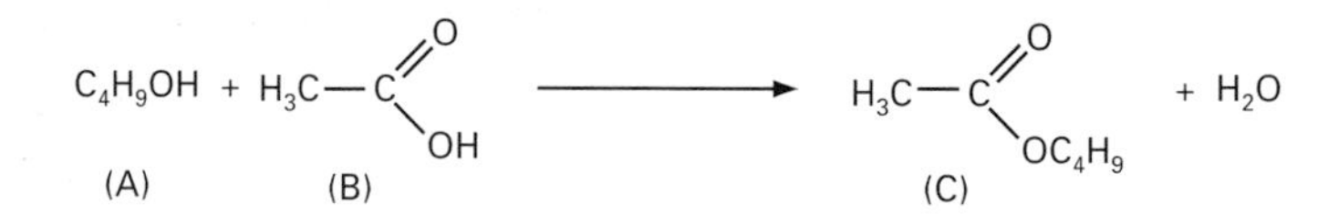

Calculate the percentage yield of C.

Solution

First of all, we must work out which reactant is present in excess because this will not affect the yield of product. Thus, we use the expression,

$$\text{number of moles used} = \frac{\text{mass used}}{\text{relative molecular mass } (M_r)}$$

Now, M_r(A) = 74 and M_r(B) = 60 and the equation is

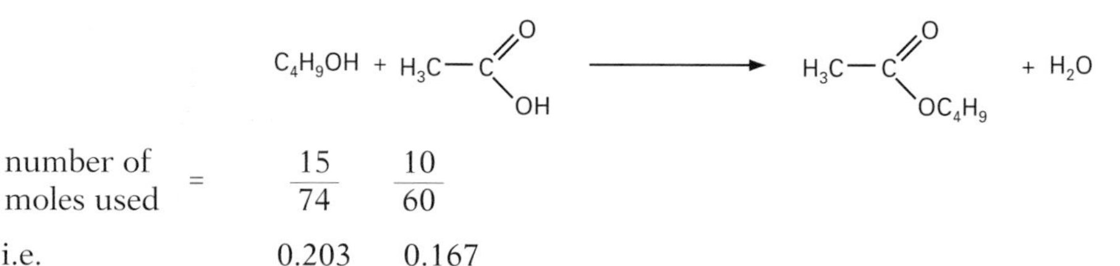

number of moles used $= \quad \frac{15}{74} \quad \frac{10}{60}$

i.e. $\quad 0.203 \quad 0.167$

The equation above shows that 1 mole of butan-1-ol reacts with 1 mole of ethanoic acid. Hence, the amount of ethanoic acid will determine the product yield: butan-1-ol is in excess.

Now, 1 mole of ethanoic acid forms 1 mole of 1-butylethanoate (C)

$\therefore$ 0.167 moles of ethanoic acid forms 0.167 moles of 1-butylethanoate

Thus,

theoretical mass of 1-butylethanoate $= 0.167 \times M_r = 0.167 \times 116 = 19.4$ g

Finally,

$$\% \text{ yield} = \frac{\text{actual yield of purified product}}{\text{theoretical yield}} \times 100$$

$$= \frac{17.8}{19.4} \times 100$$

giving a yield of C = 91.8%, say 92%.

In reality, yields rarely closely approach 100%. Indeed, a 90% yield would often be thought of as being most satisfactory. There are a number of reasons why the percentage yield may be lower, e.g.:

- the reactants may not be completely pure, e.g. some moisture may be present;
- loss of product during purification, e.g. small quantities of solid left on the filter paper or funnel during recrystallisation;
- side reactions which give rise to by-products.

Often, product yields can be improved by carrying out the synthesis using an excess of one, or more, reactant.

Sometimes, we have to accept a low yield. For example, there may be no other method of making the compound. Also, a low yield does not necessarily mean that a reaction is uneconomical, since a popular commercial product may still be sold at a profit.

Activity 6.2

1 A chemist prepared 5.90 g of aspirin by warming 5.20 g of 2-hydroxybenzoic acid with 7.50 g of ethanoic anhydride at 60°C. The equation is:

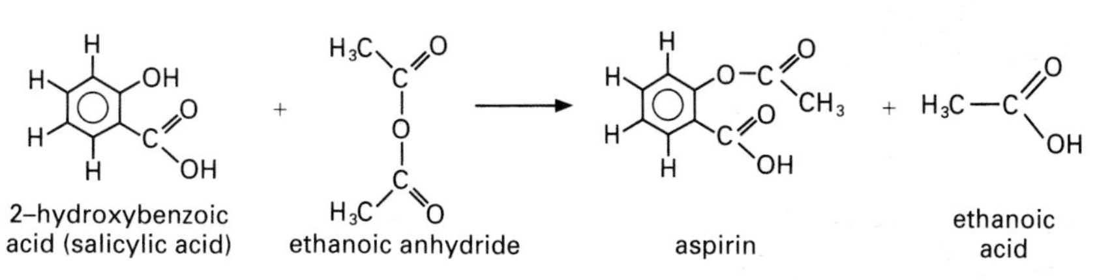

Calculate the percentage yield of aspirin.

2 When 6.20 g of anhydrous 2-methylpropan-2-ol (A) is treated with excess concentrated hydrochloric acid at room temperature, 6.20 g of 2-chloro-2-methylpropane (B) is formed:

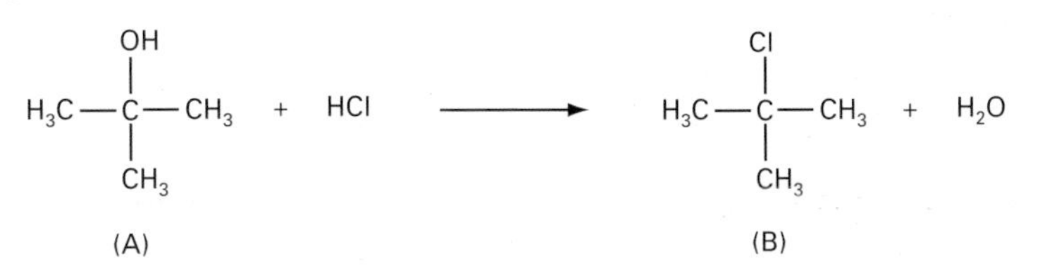

Calculate the percentage yield of B.

Focus 6b

1 Organic synthesis involves three main tasks: **a)** planning and performing the chemical reaction; **b)** separating the crude product and purifying it; **c)** checking the purity of the final product and determining its percentage yield.

2 Any organic preparation that you devise should give information about the amounts and concentrations of the reactants and the way they are mixed, and the temperature and pressure of the reaction. In industry, these factors would be varied in order to obtain the most economical yield of product.

3 Organic compounds are often volatile, inflammable and toxic. Hence, organic reaction mixtures are heated under **reflux**, Figure 6.3, and stringent safety procedures are followed.

4 Crude organic products may be purified by: distillation techniques, solvent extraction, recrystallisation and chromatography, Table 6.2.

5 $$\text{\% yield of pure product} = \frac{\text{actual yield of pure product}}{\text{theoretical yield}} \times 100$$

Any reactant which is present in excess amounts does not usually govern the yield of the product (section 6.3). Percentage yields are always less than 100%. The reasons for this might be impure reactants, loss of product during purification or the formation of reactant by-products.

6.4 Testing the purity of a synthetic product

A commercial apparatus for determining melting point

Once the synthetic product has been made and purified, it is important to check just how pure it is. Sometimes small amounts of impurities may be tolerated, for example if the product is to be used as the starting material in another synthesis. However, many products, such as drugs and food additives, have to be as pure as possible.

If the product was prepared using a well-known synthetic route, then its purity can be quickly tested by measuring its **melting point** or **boiling point** and checking the value against those in a data book. The purity of the compound can then be confirmed using a variety of spectroscopic techniques (discussed more fully in Chapters 15–18).

One of the most common techniques is **infrared absorption spectroscopy**. Most organic molecules absorb infrared radiation, with each bond in the molecule absorbing radiation of a characteristic frequency, Figure 6.7. Thus, the infrared spectrum of a pure compound is like a '**fingerprint**' and we can compare it with the spectrum of our synthetic product. Organic compounds may also give characteristic absorption spectra in the visible and ultraviolet regions.

In section 1.10, we saw how the **mass spectrometer** is used to measure the mass of atoms. The main principle of the mass spectrometer is that a high energy electron strikes a gaseous atom, or molecule, forming a positive ion:

M(g)	+	e^-	⟶	M^+(g)	+	$2e^-$
atom or molecule		high energy electron		positive ion		lower energy electrons

The positive ions are detected and their masses, and relative abundances, are measured. Since the mass of an electron is negligible compared to the mass of atomic nuclei, the relative mass of the molecule will be virtually equal to that of the heaviest positive ion, known as the **parent ion** or **molecular ion**. When organic molecules are placed in a mass spectrometer, they break down into smaller ions, called **fragments**. The abundance of each fragment is plotted against is relative mass, as illustrated in Figure 6.8. Pure organic compounds give characteristic fragmentation patterns, so these can be used to determine the purity of a synthetic product.

Determining the structure of a synthetic product

In chemical research, things don't always go according to plan and chemists who work on novel synthetic pathways sometimes end up with an unknown product, or by-product. Usually, the substance can be identified from spectra; occasionally it can't, and then real mystery and excitement surround its existence! In such cases, the structure of the unknown product needs to be determined. A typical sequence for structure determination is shown in

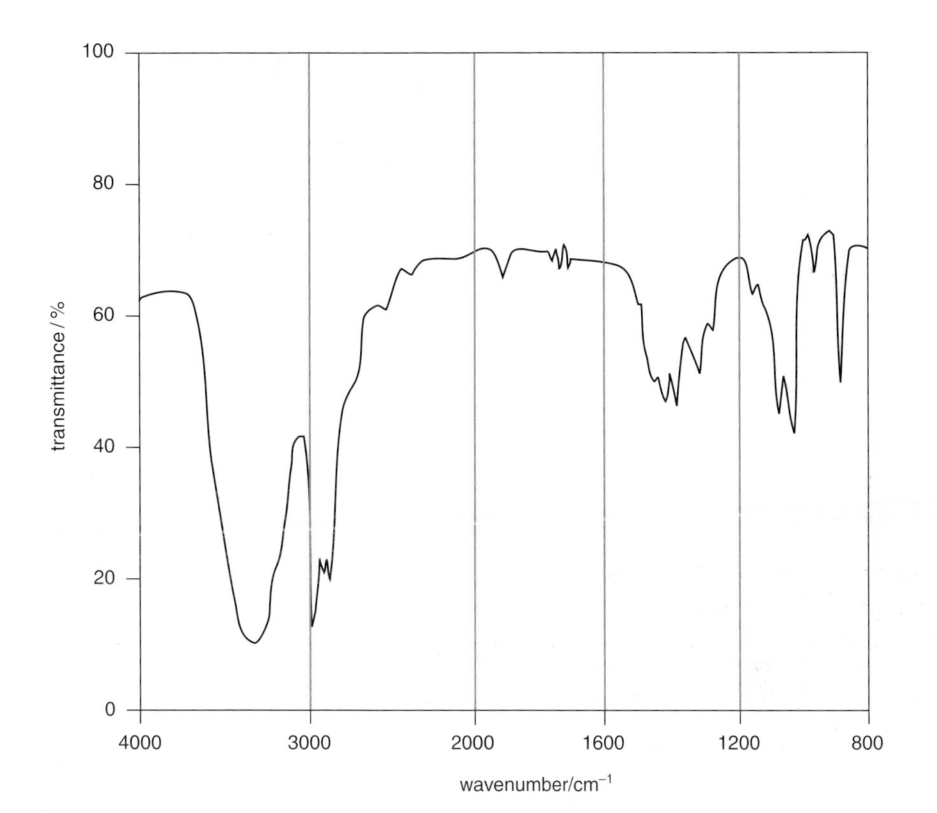

Figure 6.7 ▲
A fairly straightforward infrared absorption spectrum, that of ethanol, CH_3CH_2OH. The lower the transmittance, the stronger the absorption of a given frequency, traditionally referred to by its wavenumber (where wavenumber = 1/wavelength). Each absorption peak corresponds to a certain mode of vibration of the bonds in the molecule. For example, in this spectrum:

absorption/cm^{-1}	870	1030/80	1400	3000	3300
bond present	C — H	C — O	O — H	C — H	O — H

This structural information is consistent with the displayed structural formula for ethanol:

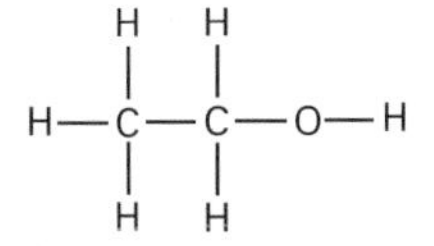

The region of an infrared spectrum from 600 to 1500 cm^{-1} is often referred to as the '**fingerprint**' region as it is unique to each molecule. A pure compound will exactly reproduce this 'fingerprint'; thus, impurities in a sample may be detected.

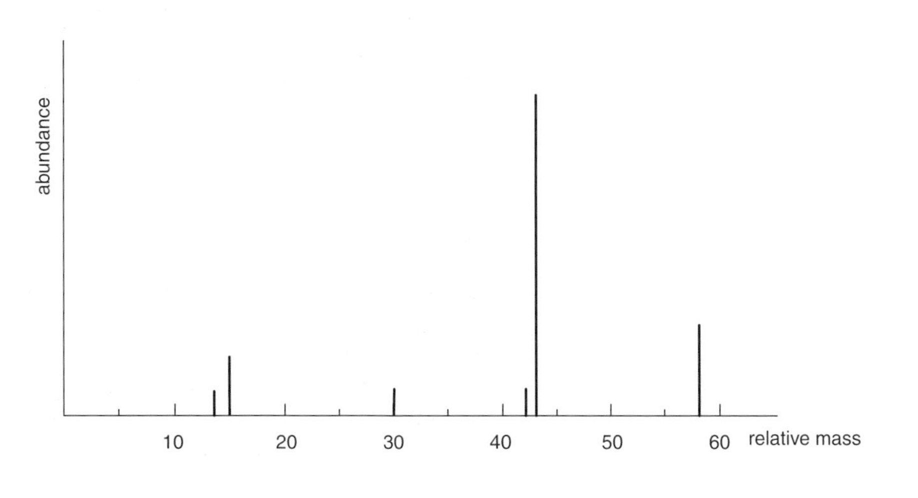

Figure 6.8 ▶
A simplified mass spectrum of propanone, of molecular formula C_3H_6O. The parent ion, or molecular ion, observed at a relative mass of 58, is formed when the molecule is struck by a high energy electron:

$$C_3H_6O + e^-_{fast} \longrightarrow C_3H_6O^+ + 2e^-_{slow}$$

This molecular ion confirms that the relative molecular mass of propanone is 58. The most abundant fragments appear at relative masses of 15 and 43, corresponding to the ions CH_3^+ and CH_3CO^+, respectively. The information provided by the mass spectrum of propanone is consistent with the displayed structural formula:

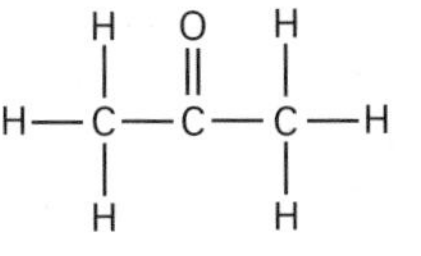

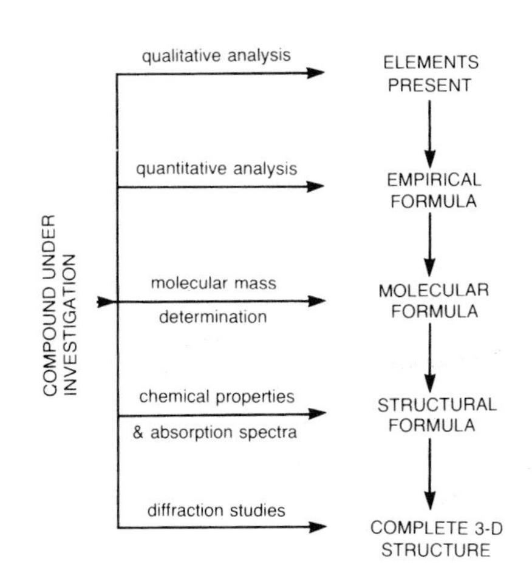

Figure 6.9
A typical sequence for structure determination

Figure 6.9 and examples of its use are given in Chapters 15–17. Here we shall focus on the determination of a compound's empirical and molecular formulae.

Firstly, the molecule is broken down into its constituent elements and these are detected using unique tests. For example, any sulphur atoms present in the compound may be converted into aqueous sulphate ions and these can be identified by adding acidified barium chloride solution:

$$Ba^{2+}(aq) + SO_4^{2-}(aq) \xrightarrow{\text{dilute nitric acid}} BaSO_4(s)$$

in colourless solutions — dense white precipitate

Working out which elements are present in the compound is known as **qualitative analysis**. The next stage is to perform a **quantitative analysis**, to determine the proportions by mass of the elements present. From this data, we can calculate the **empirical formula**, that is the simplest whole number atomic ratio of the elements present.

Calculating the empirical formula from combustion data

The masses of carbon and hydrogen in a given mass of organic compound are often found using combustion analysis. Complete combustion of the compound will convert carbon into carbon dioxide and hydrogen into water and these gaseous products are passed through tubes containing potassium hydroxide, to absorb the carbon dioxide, and concentrated sulphuric acid, to absorb the water. The mass of each product equals the increase in mass of the relevant absorption tube. Work through the following example and then try activity 6.3, question 1.

Example

On complete combustion, 1.000 g of compound X gave 2.382 g of carbon dioxide and 1.215 g of water. Qualitative analysis shows that X only contains carbon, hydrogen and oxygen atoms. Find the empirical formula of X.

Calculate:

1 The number of moles of carbon dioxide and water which are formed.

$$\text{mol } CO_2 = \frac{\text{mass of } CO_2}{M_r(CO_2)} = \frac{2.382}{44} = 0.0541$$

$$\text{mol } H_2O = \frac{\text{mass of } H_2O}{M_r(H_2O)} = \frac{1.215}{18} = 0.0675$$

2 The number of moles of carbon atoms in the 1.000 g sample of X

1 mol of CO_2 gas is obtained from 1 mol of C atoms in X, thus, 0.0541 of CO_2 is obtained from 0.0541 mol of C atoms in X.

3 The mass of carbon atoms in the 1.000 g sample of X.

$$\begin{aligned}\text{mass of carbons atoms} &= \text{moles of carbon atoms} \times A_r(C)\\ &= 0.0541 \times 12\\ &= 0.6492 \text{ g}\end{aligned}$$

4 The number of moles of hydrogen atoms in the 1.000 g sample of X.

1 mol of H_2O gas is obtained from 2 mol of H atoms in X, thus, 0.0675 mol of H_2O is obtained from 0.1350 mol of H atoms in X.

5 The mass of hydrogen atoms in the 1.000 g sample of X.

$$\begin{aligned}\text{mass of H atom} &= \text{mol of H atoms} \times A_r(\text{H}) \\ &= 0.1350 \times 1 \\ &= 0.1350\ \text{g}\end{aligned}$$

6 The mass of oxygen atoms in the 1.000 g sample of X.

$$\begin{aligned}\text{mass of O atoms} &= \text{mass of sample} - \text{mass of carbon} - \text{mass of hydrogen} \\ &= 1.000 - 0.6492 - 0.1350 \\ &= 0.2158\ \text{g}\end{aligned}$$

7 The number of moles of O atoms in the sample of X.

$$\text{mol O atoms} = \frac{\text{mass of O}}{A_r(\text{O})} = \frac{0.2158}{16} = 0.0135$$

Using the mole ratios in 2, 4 and 7 above, we can write a formula for X:

C 0.0541 H 0.1350 O 0.0135

Now divide through by the smallest number to get the empirical formula:

$$\text{C}\ \frac{0.0541}{0.0135} = 4.007$$

$$\text{H}\ \frac{0.1350}{0.0135} = 10$$

$$\text{O}\ \frac{0.0135}{0.0135} = 1$$

So, the empirical formula of X is $\mathbf{C_4H_{10}O}$.

Calculating the empirical formula from percentage composition by mass

Data from quantitative analysis may be supplied as percentages and the empirical formula can be calculated as shown in the example below. Work through it and then try activity 6.3, question 2.

Example

Compound Y is found to contain 55.81% carbon, 6.980% hydrogen and 37.21% oxygen by mass. Find the empirical formula of compound Y.

In 100 g of Y, there are	55.81 g C	6.980 g H	37.21 g O
Divide by relative atomic masses	$\frac{55.81}{12}$	$\frac{6.980}{1}$	$\frac{37.21}{16}$
This gives the moles of each element	4.651	6.980	2.326

Using these mole ratios, we can write a formula for X by dividing through by the smallest number.

$$\text{C}\ \frac{4.651}{2.326} = 2.000$$

$$\text{H}\ \frac{6.980}{2.326} = 3.001$$

$$\text{O}\ \frac{2.326}{2.326} = 1$$

This suggests that the empirical formula of Y is $\mathbf{C_2H_3O}$.

Working out the molecular formula and the structure of the compound

The relative molecular mass of an organic compound can be obtained from the mass of the molecular ion in its mass spectrum, Figure 6.8, or, if it can be vaporised, by measuring the volume of a given mass of gas at a known temperature and pressure, section 4.13. Once the empirical formula and relative molecular mass are known, the molecular formula can be determined.

Since the empirical formula is the simplest whole number atomic ratio of the elements present, the molecular formula will be always a simple multiple of the empirical formula. For example, if compound Z has an empirical formula of CH_2 and a molecular mass 84, then,

$$\frac{\text{molecular mass}}{\text{empirical mass}} = \frac{84}{14} = 6$$

and the molecular formula of the compound is $6(CH_2)$, i.e. C_6H_{12}. Now try activity 6.3 question 3.

Having obtained the molecular formula, the structure of the molecule is determined using various instrumental techniques: mass spectrometry, absorption spectroscopy, nuclear magnetic resonance spectroscopy and X-ray crystallography. Structural analysis involving these techniques is covered in Chapters 15–17.

Activity 6.3

1 An organic compound, A, contains carbon, hydrogen and oxygen only. On complete combustion, 1.202 g of A gave 2.640 g of carbon dioxide and 1.442 g of water. Find the empirical formula of A.

2 Calculate the empirical formula of the following compounds from the analytical data supplied:
- **a)** B 15.40% carbon; 3.20% hydrogen; 81.40% iodine
- **b)** C 19.99% carbon; 6.71% hydrogen; 26.64% oxygen; 46.66% nitrogen
- **c)** D 48.65% carbon; 8.10% hydrogen; 43.25% oxygen

3 Use the information supplied to work out the molecular formula of each of the following compounds:

	empirical formula	relative molecular mass
a) E	CH	78
b) F	CH_2O	60
c) G	$CHCl$	97
d) H	C_7H_{16}	100
e) I	CHO_2	90

4 The relative molecular masses of compounds A, B and C in the above examples are 60, 156 and 60, respectively. Compound C contains a $C{=}O$ bond. Work out the molecular formulae of these compounds and draw their displayed structural formulae.

Focus 6c

1. We can test the purity of a compound simply by taking its melting or boiling point and checking the value in a data book.
2. Spectra of pure organic compounds are like '**fingerprints**'. Infrared, ultraviolet, nuclear magnetic resonance and mass spectra may be used to identify an organic compound and determine its purity. This is covered in detail in Chapters 15–18.
3. The **empirical formula** of a compound is the simplest whole number ratio of the atoms present. The empirical formula can be obtained from combustion data. You need to be able to carry out calculations using: a) the masses of the combustion products and b) the percentage composition by mass of the elements present.
4. If the relative molecular mass is known, the **molecular formula** can be determined because the molecular mass is a multiple of the empirical formula mass.

6.5 Isomerism

Propanoic acid and ethyl methanoate have the same molecular formula, $C_3H_6O_2$. However, their structure and properties are different:

These molecules are **isomers**, that is they have the same molecular formula but their atoms are arranged in a different way.

Molecular models of ethyl methanoate and propanoic acid, a pair of structural isomers both having the molecular formula $C_3H_6O_2$

There are very many examples of isomerism in chemistry and these may be divided into two main classes:

STRUCTURAL ISOMERISM		STEREOISOMERISM
Isomers whose molecules have their atoms linked together in different bonding arrangements	and	Isomers whose molecules have their atoms linked together in the same bonding arrangement but are arranged differently in space.

Structural isomerism

There are three main types: chain isomerism, positional isomerism and functional group isomerism.

Chain isomers have different hydrocarbon chains. For example, C_5H_{12} is the molecular formula of three alkanes, pentane, 2-methylbutane and 2,2-dimethylpropane.

Chain isomers have similar chemical properties but different physical properties, as shown by the melting points, boiling points and densities of the C_5H_{12} alkanes in Table 6.3. Thus, we can distinguish between chain isomers by measuring their melting or boiling points.

Table 6.3 Physical properties of the isomers of molecular formula C_5H_{12}. Can you explain why the boiling points decrease in the order shown?

Property	pentane	2-methylbutane	2,2-dimethylpropane
structural formula			
melting point/°C	−130	−160	−16
boiling point/°C	36	28	10
density/g cm^{-3}	0.63	0.62	0.59$_{(liquid)}$

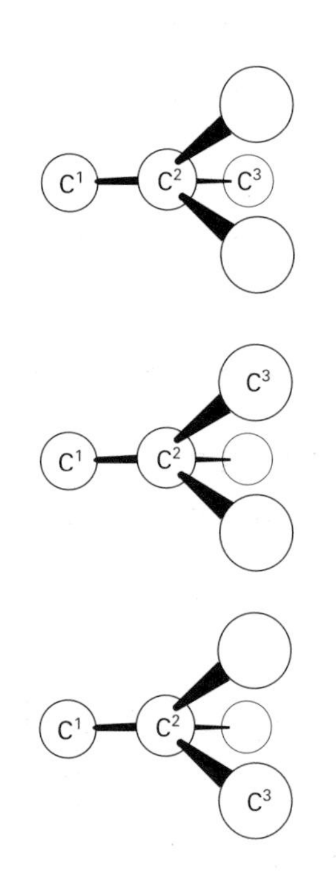

Figure 6.10 ▲
Due to the rotation about a single bond, these displayed formulae are all equivalent – they are not isomers

Before carrying on, a brief word about drawing structural formulae. In all molecules, groups of atoms are able to rotate around a single bond. For example, consider the displayed formulae of propane, Figure 6.10. Due to the continual rotation around the $C^1—C^2$ bond, the hydrogen atoms and the $—CH_3$ group on C^2 are in equivalent positions. Hence, the three structural formulae in Figure 6.10 are also equivalent.

Positional isomers have a particular atom, or bond, or group of atoms, in different molecular positions. For example, there are two positional isomers of molecular formula C_3H_7Br.

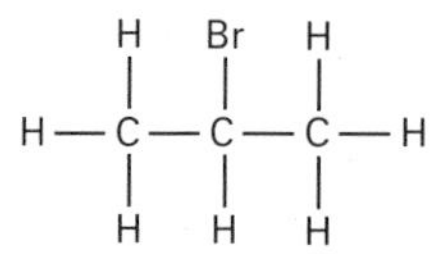

2-bromopropane
(boiling point 60°C)

1-bromopropane
(boiling point 71°C)

Molecular models of the positional isomers, 2-bromopropane and 1-bromopropane ▶

Positional isomers have different physical properties. They take part in similar chemical reactions, but often at different rates. For example, 2-bromopropane reacts faster with water than does 1-bromopropane, section 10.5. Thus, we can identify positional isomers by physical, and sometimes chemical, methods.

Functional group isomers contain different organic functional groups, section 6.1. Alcohols and ethers exist as functional group isomers, for example.

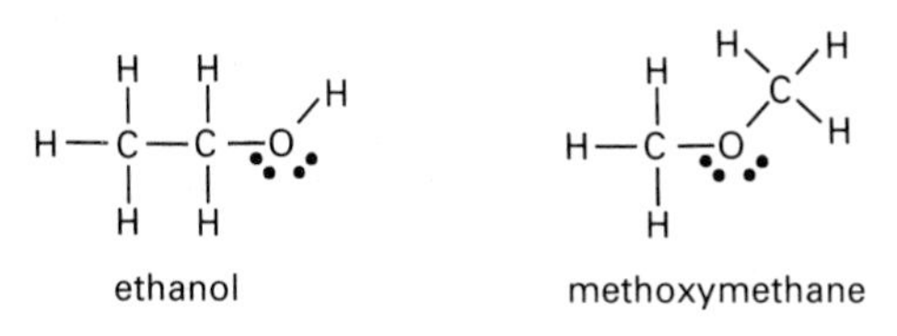

ethanol methoxymethane

Functional group isomers often display very different physical and chemical properties. Consequently, they are easily distinguished by measuring melting or boiling points or by performing simple chemical tests.

Activity 6.4

1 Draw displayed structural formulae for:
 a) four structural isomers of formula $C_3H_6Cl_2$,
 b) five structural isomers of formula C_4H_8,
 c) four structural isomers of C_4H_9Br.

2 Several compounds have the molecular formula $C_2H_4O_2$.
 a) Three of these are functional group isomers containing a carbonyl group, C=O. Give their structure.
 b) Three other isomers contain a C=C bond. Draw their structural formulae. How are these isomers related to those in part (a)?

3 Give structural formulae for the isomers of molecular formula:
 a) $C_4H_{10}O$ (ethers) **b)** $C_3H_6O_2$ (esters) **c)** $C_5H_{10}O$ (ketones)

 Refer to Table 6.4 on page 184 for help, if necessary.

Stereoisomerism I: geometrical isomerism

Stereoisomers are molecules which have different properties even though the same atoms are bonded together in the same way.

Consider the molecular models of 1,2-dichloroethene shown in the photograph.

Molecular models of 1,2-dichloroethene

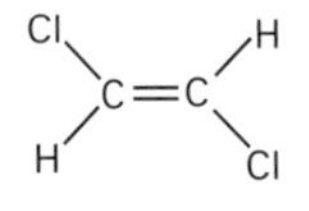

cis-1,2-dichloroethene *trans*-1,2-dichloroethene

Activity 6.5

The molecular formulae of four compounds are:

1 $C_2H_4Br_2$
2 C_2H_3Cl
3 C_4H_7Br
4 $C_4H_6O_2$ (carboxylic acids)

Each contains a C=C bond. Draw out the possible structural formulae and classify the isomers (if any) as structural or geometrical.

Each molecule has the same atoms linked together in the same way. However, their 3-D structures differ because the π electron cloud prevents rotation about the double bond. Thus, **geometrical isomers** have identical atoms, or groups, arranged either 'on the same side of' (**cis-**) or 'across' (**trans-**) a rigid section of the molecule's structure, such as C=C bond. Geometrical isomers are also known as ***cis–trans*** isomers; they can be identified by their different physical properties.

Stereoisomerism II: optical isomerism

Like geometrical isomers, **optical isomers** have the same atoms linked together. They have identical physical and chemical properties except in their behaviour towards plane polarised light. Have a look at Figure 6.11.

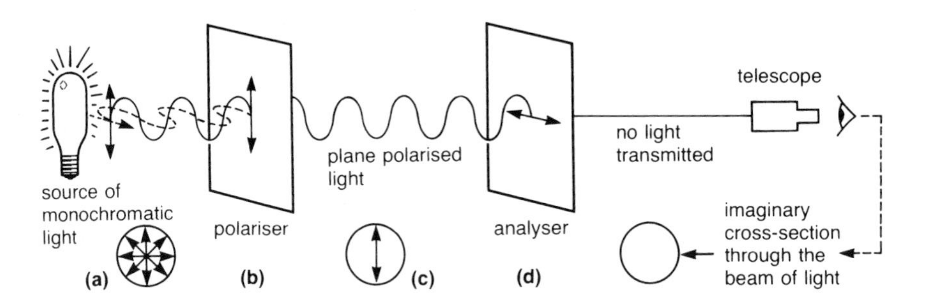

Figure 6.11 ▶
The production of plane polarised light. Light will not pass through two polaroid sheets whose polarising axes are at right angles to each other

a) A beam of light consists of an infinite number of waves vibrating in all planes at right angles to the direction of the beam's movement.
b) Certain materials will let through light which vibrates in only one plane, called **plane polarised light**. One well-known **polariser** is the **polaroid sheet** used in sun-glasses.
c) When it emerges from the polariser, the beam of light has vibrations in only one plane. It is plane polarised.
d) To check for plane polarisation, the light is directed onto another polaroid sheet, known as the **analyser**. When the polarising axes of the polariser and analyser are at right angles, no light will be transmitted.

The effect of passing plane polarised light through a chemical substance can be studied using a polarimeter. Its main components resemble Figure 6.11, except that the plane polarised light passes through a long glass tube, containing the sample, before it reaches the analyser. First, the apparatus is adjusted so that no light is transmitted. Then the polarimeter tube is filled with a solution of the sample and replaced in the polarimeter. If light can now be seen through the telescope, the sample must have rotated the plane of polarisation of the light. Thus, the compound is said to be **optically active**. To find the **angle of rotation**, **α**, and its direction, the analyser is rotated once again until no light is transmitted.

An isomer is said to be **dextrorotatory (+)** or **laevorotatory (–)**, depending on whether it rotates the plane of polarisation to the right (clockwise) or left (anticlockwise), respectively. Under the same conditions, a compound's dextrorotatory and laevorotatory isomers produce the same angle of rotation.

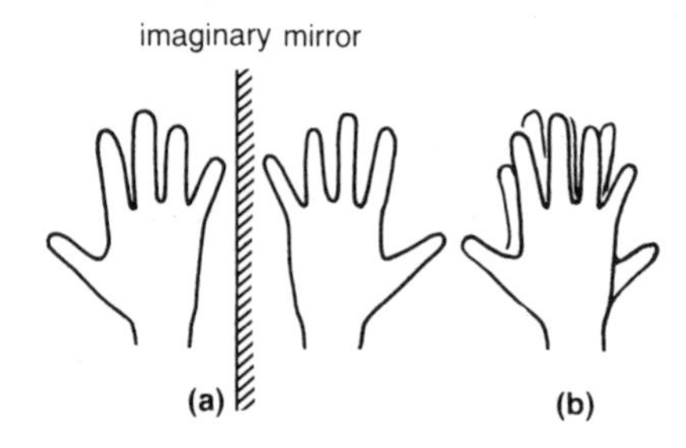

Figure 6.12 ▲
Your hands form non-superimposable mirror images

Do optical isomers have a common structural feature?

If you hold up your hands, palms facing, they will form mirror images of each other (Figure 6.12). However, if you place one hand on top of the other, palms down, the fingers and thumbs do not line up. They are non-superimposable. Our hands are examples of structures which form **non-superimposable mirror images**. Research has shown that the molecules of optical isomers exist as non-superimposable mirror images, and we call these **enantiomers**. Such molecular structures display the property of **chirality**, that is, they have no centre, axis or plane of symmetry.

Consider, for example, the mirror-image molecular models having four different groups arranged around a central carbon atom.

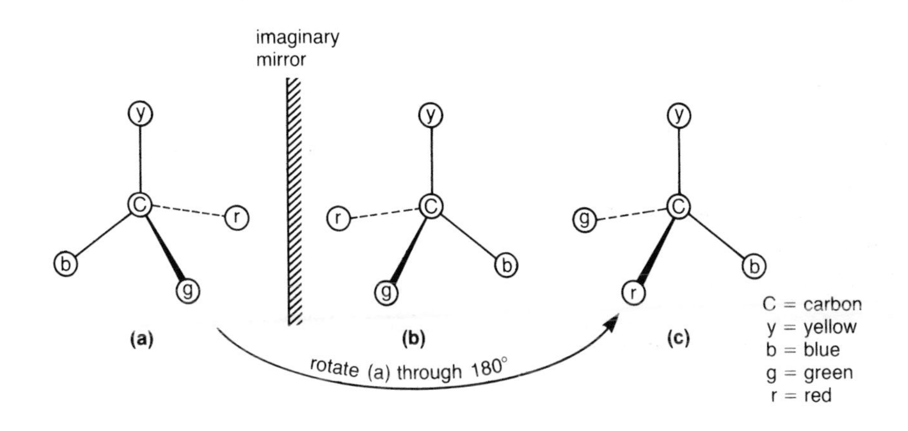

Figure 6.13
When **(a)** is rotated through 180° it gives **(c)**. Structure **(c)** can not be superimposed on **(b)** because, whilst atoms y and b do occupy the same positions, atoms r and g do not coincide. Thus, **(a)** and **(b)** are non-superimposable mirror-images. They also display the property of **chirality**, that is they do not contain a centre, axis or plane of symmetry.

Figure 6.13 shows these mirror-image structures, (a) and (b). Now, if (a) is rotated through 180°, giving (c), it cannot be superimposed on (b). Thus, structures (a) and (b) represent a pair of enantiomers. Two examples of this type of optical isomer are:

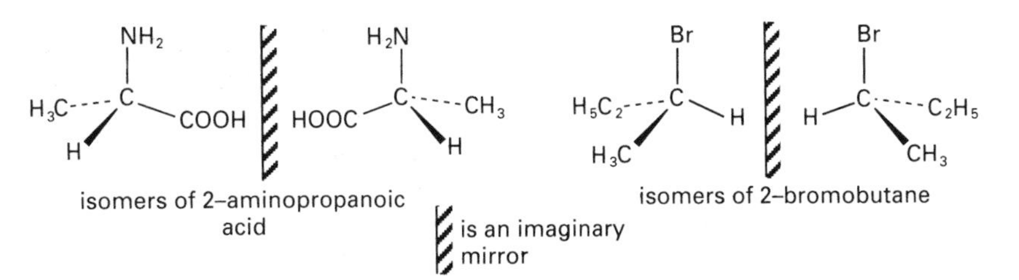

In fact, any organic compound possessing at least one **asymmetric carbon atom**, that is, one bonded to four different atoms, or groups, must exist as optical isomers.

2-aminopropanoic acid, or alanine, contains an asymmetric carbon. The molecule exists an enantiomers with the (+) isomer occurring naturally in muscle and proteins.When 2-aminopropanoic acid is being prepared in the laboratory, however, the products are found to be optically inactive. In fact, we find that the products are an equimolar mixture of the (+) and (–) isomer. Both isomers rotate the plane of polarisation to the same extent but in opposite directions. Consequently, their individual effects on the plane of polarisation cancel out. Any equimolar mixture of optical isomers is termed a **racemic mixture**. It will be optically inactive. Although optical isomers are common in plant and animal life, they are seldom found as a racemic mixture.

Separating the isomers in a racemic mixture

Optical properties apart, enantiomers show exactly the same physical and chemical behaviour. Thus, we can not separate them using physical methods, such as distillation, crystallisation and chromatography.

Some racemic mixtures can be separated, or resolved, by hand-picking the differently shaped crystals (these are mirror-images of each other). This technique was first used in 1848 by Louis Pasteur, the French chemist, who laboriously obtained the (+) and (–) forms of tartaric acid. Other techniques involve the removal of one isomer, for example by using a bacterium to specifically feed on it.

However, by far the most common procedure for resolving a racemic mixture involves the preparation of **diastereoisomers**. These are obtained when a racemic mixture reacts with a pure enantiomer, for example:

(+) acid / (–) acid racemic mixture	+	(+) base pure enantiomer
	↓	
(+) / (+) salt	two diastereoisomers	(–) / (+) salt

Now, unlike optical isomers, diastereoisomers have different physical properties. Thus they can be separated by the physical methods mentioned above. Then, a chemical reaction is carried out to recover each isomer, (+) acid and (–) acid, from the diastereoisomers.

Activity 6.6

Draw mirror-image structures for the optical isomers of formula

1 $C_3H_6O_3$

2 $C_4H_{10}O$

3 C_3H_5NO

4 C_4H_9Cl

These structures have a common feature. What is it?

Focus 6d

1 **Isomers** are different compounds having the same molecular formula but different physical or chemical properties. It is worth noting that isomerism is a property of organic *and* inorganic compounds. Isomers may be divided into two main categories: structural isomers and stereoisomers. Stereoisomerism falls into two categories: geometrical (*cis–trans*) isomerism and optical isomerism.

2 Molecules of **structural isomers** have the same number and types of atoms joined together in different bonding arrangements. Molecules of **stereoisomers** have the same atoms joined together in identical bonding arrangements but they have different 3-D structures.

3 **Geometrical (*cis–trans*) isomers** have similar atoms, or groups of atoms, arranged either 'on the same side' of (*cis*-) or 'across' (*trans*-) a rigid section of the molecule's structure, usually a C=C bond. Geometrical isomers have different physical properties.

4 **Optical isomers** have identical physical properties except in their behaviour towards plane polarised light, that is light which vibrates in only one plane.

5 An optical isomer is said to be **dextrorotatory** or **laevorotatory**, depending on whether it rotates the plane of polarisation to the right (clockwise) or left (anti-clockwise), respectively. Dextrorotatory isomers are denoted by a (+) sign and laevorotatory isomers by a (–) sign. Under constant conditions, (+) and (–) isomers produce the same angle of rotation.

6 All optical isomers have a common structural feature, **chirality**, which means that there is no centre, axis or plane of symmetry in the molecule. The isomers exist as enantiomers, that is non-superimposable mirror images. Any molecules which contain a carbon atom bonded to four different groups, termed an **asymmetric** or **chiral carbon atom**, will be optically active.

7 A **racemic (±) mixture** contains (+) and (–) isomers in equimolar quantities and is optically inactive. The laboratory synthesis of an optical isomer often leads to a racemic mixture, so the required isomer must then be separated from the reaction mixture.

6.6 The reactivity of organic compounds

When ethene is bubbled through bromine water, at room temperature, the orange colour of the bromine quickly disappears. However, if the experiment is repeated, this time using ethane instead of ethene, there appears to be no reaction.

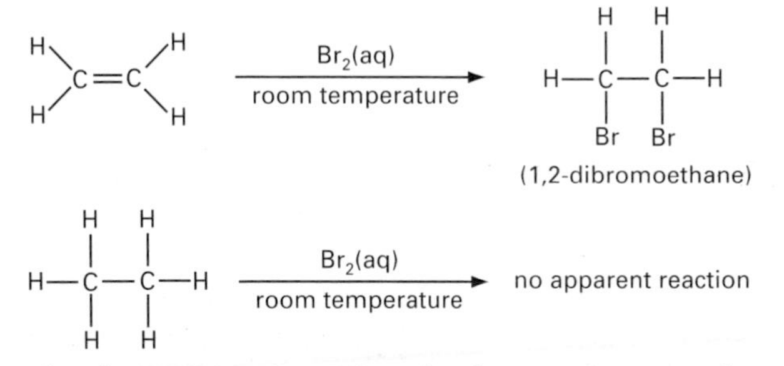

Since each molecule contains C—H bonds, the reaction must be characteristic of the C=C bond present in ethene, but not in ethane. This is an example of an **addition** reaction. In fact, we find that bromine can be added to all molecules containing a C=C bond, and we say that the C=C bond is a **reactive centre** in the molecule.

Although ethane will not react with bromine water, a gas phase reaction will occur in the presence of ultra violet light:

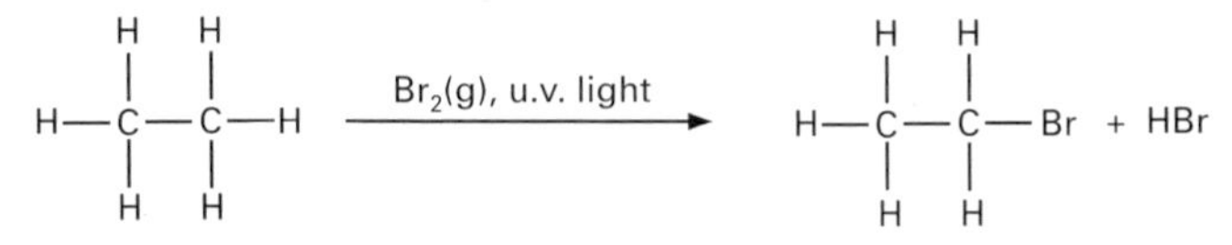

In this case, a bromine atom has **substituted** for a hydrogen atom and the reactive centre in the alkane is the C—H bond. As expected, other compounds containing C—H bonds will react with bromine in the same way, under these conditions.

In the examples above, bromine's behaviour depends on the experimental conditions. The reason for this is that the bromine molecule can split up in two ways and these give rise to the different mechanisms for the addition and substitution reactions. **Reaction mechanisms** explain how the electron clouds move around during a chemical reaction. An understanding of why and how these clouds shift will help you to predict the outcome of mixing organic chemicals together.

The purpose of this section is to lay the foundations upon which our study of organic reactions is based. To do this, we shall:

- review the concept of **electronegativity** and use this to identify the reactive centres in a variety of molecules,
- discuss **bond fission**, that is the ways in which covalent bonds can be broken,
- survey some common types of organic reaction mechanism.

After working through the following sections, you will be better able to detect and analyse the many patterns of behaviour that run through Chapters 7–14.

Electronegativity and bond polarity

In section 4.5 we saw that **electronegativity** may be defined as the power of an atom in a covalent bond to attract the bonding electron cloud to itself. The values suggested by Pauling are listed in Table 4.4, to which you should now refer. If the atom has a high electronegativity, it will attract the bonding electron cloud very strongly, and vice versa. When two atoms of unequal electronegativity are bonded together, a **permanent dipole** will result, for example,

	$\delta+$ $\delta-$	$\delta+$ $\delta-$	$\delta+$ $\delta-$
	C—F	C=O	N—H
electronegativities	2.5 4.0	2.5 3.5	3.0 2.1

Electronegativity values, therefore, allow us to work out the polarity of the bonds in an organic molecule.

Activity 6.7

Look again at Table 4.4, and then answer these questions.

1 Use $\delta+$ and $\delta-$ signs to show the polarity of the following bonds:

C—Br O—H N—H
S—O C=O C=N
C—Cl C—F

2 The greater the difference in the electronegativities of the bonded atoms, the more polarised the bond will be. Hence, place the bonds above in order of increasing polarity.

Breaking covalent bonds

When a molecule absorbs energy, the bonded atoms vibrate more vigorously around their mean positions. If enough energy is supplied **bond fission** will occur. This can happen in two ways, known as homolytic and heterolytic bond fission.

In **homolytic bond fission** each atom takes an equal share of the bonding electron cloud:

$$A{:}B \longrightarrow A^{\times} + {\cdot}B$$

At the instant of their formation, A$^{\times}$ and •B are very excited atoms, or groups of atoms, possessing an unpaired electron; these are known as **free radicals**. Free radicals are especially reactive because the previously bonded electrons have not yet returned to their lower energy ground states.

Two examples of free radical formation are given below:

$$Cl{:}Cl \longrightarrow Cl^{\times} + {\cdot}Cl \qquad \Delta H = +242\ kJ$$
$$H_3C{:}H \longrightarrow H_3C^{\times} + {\cdot}H \qquad \Delta H = +435\ kJ$$

Generally speaking, homolytic fission is more likely to occur in non-polar, or only slightly polar, bonds.

Heterolytic bond fission is a very common process in which one atom takes both bonding electrons and ions are formed:

$$A{:}B \longrightarrow [A]^{+} + [{:}B]^{-}$$

Before fission, atoms A and B may each be thought of as effectively 'owning one electron in the shared pair. When the bond breaks:

- Atom A loses its 'owned' electron ($^{\times}$). By losing a negative charge, atom A becomes a **positive ion**.
- Atom B takes back its own electron (•) but also gains an electron ($^{\times}$). Thus, atom B becomes a **negative ion**.

Clearly, this approach is simplistic since we know that chemical bonds are formed by the redistribution of electron clouds, not particles. Also, by using this idea of 'owned' electrons, we have ignored polarity in the chemical bonds. However, as long as you realise its limitations, this idea of an atom 'owning' a shared electron will prove very useful when we discuss reaction mechanisms.

Heterolytic bond fission tends to occur when the bonded atoms have very different electronegativities. After fission, the more electronegative atom will form the negative ion, for example.

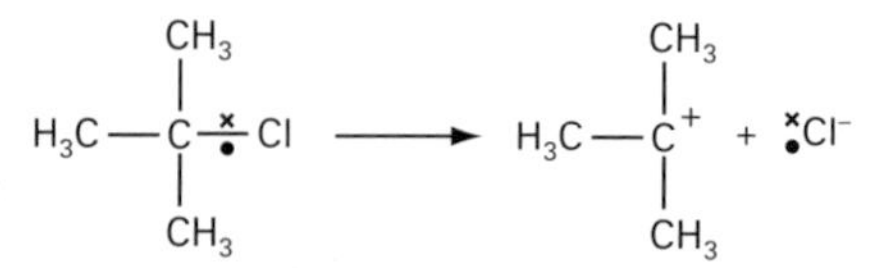

The chlorine atom takes both of the bonding electrons because it is more electronegative than carbon.

The $(CH_3)_3C_3^{+}$ ion is an example of a **carbocation** or **carbonium ion** that is an ion which contains a *positively charged carbon atom*.

Some organic reactions involve the formation of a **carbanion**, an ion containing a negatively charged carbon atom, for example:

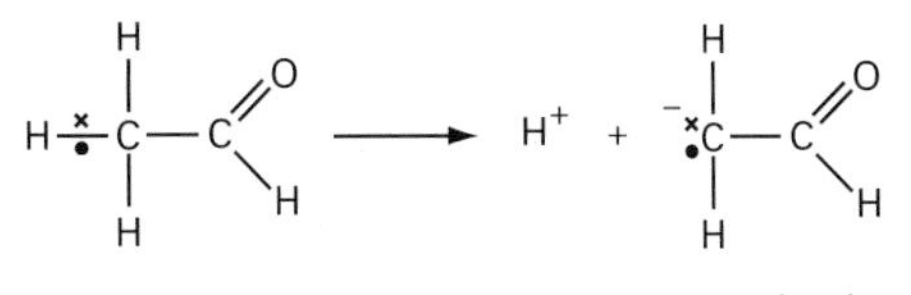

Since carbocations and carbanions are usually unstable and highly reactive particles, they only exist briefly as reaction intermediates.

6.7 Reagents in organic chemistry

From a knowledge of bond polarities, we can predict which types of reagent might attack an organic molecule. **Chemical reagents** are substances which bring about a desired structural change in another molecule. For example, phosphorus pentachloride, $PCl_5(s)$, is a reagent which is used to substitute a —Cl atom for the —OH group in an alcohol or carboxylic acid, e.g.

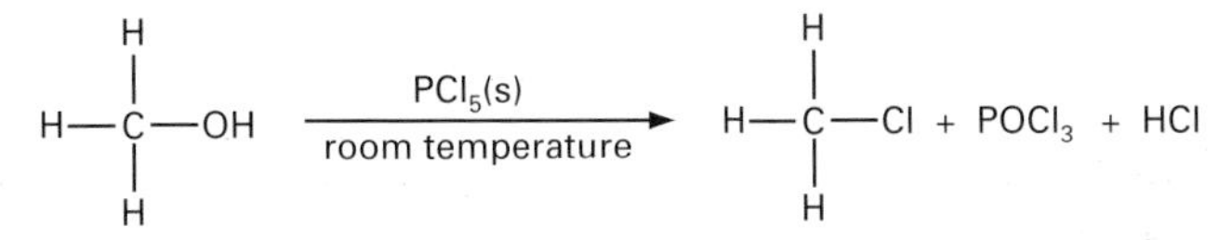

Reagents contain, or generate, chemical species which attack the organic molecule. There are three main types of reactive species: free radicals, electrophiles and nucleophiles.

Free radicals

As mentioned earlier, free radicals are highly reactive atoms or groups of atoms. They are usually formed by the fission of non-polar bonds.

Electrophiles

These are 'electron-loving' species in which one atom can accept a pair of electrons from an electron-rich atom in the organic molecule, resulting in the formation of a co-ordinate (dative covalent) bond. Electrophiles may be positive ions, such as H^+, NO_2^+, Cl^+, Br^+ and CH_3^+, or neutral molecules, such as SO_3. Dot and cross diagrams show why these molecules, and ions, can act as electron acceptors, Figure 6.14.

H^+ (NO_2^+) Cl^+ Br^+ (CH_3^+) (SO_3)

Figure 6.14
Dot and cross diagrams of some common electrophiles. These are electron acceptors because they contain an atom, in bold print, which has a + or δ+ charge

Nucleophiles

These are 'nucleus-loving' species which are able to donate a lone pair of electrons to an 'electron-poor' atom in an organic molecule, thereby forming a co-ordinate bond with it. Nucleophiles may be negative ions, such as OH^-, CN^-, Cl^- and CH_3O^-, or neutral molecules, such as H_2O and NH_3, Figure 6.15.

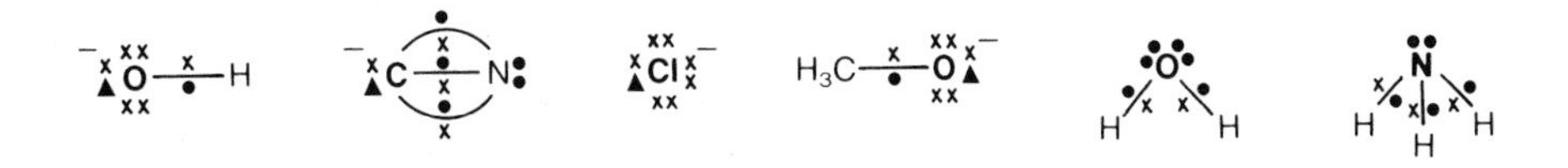

Figure 6.15
Dot and cross diagrams of some common nucleophiles. These are electron donors because they contain an atom, in bold print, which has at least one lone pair of electrons (▲ is the electron left behind following the loss of a positive ion)

Activity 6.8

1 Using dots and crosses, show what happens to the bonding electrons on:
  a) homolytic fission of the following bonds:
  i) O—O ii) Br—Br iii) H—H iv) C—C
  b) heterolytic fission of the following bonds:
  i) C—O ii) Cl—Cl iii) H—Br iv) N—C v) C—Cl
  vi) N—H vii) N—O

2 Classify the following reactive species as:
  A free radical **B** electrophile **C** nucleophile **D** carbocation **E** carbanion
  i) $CH_3\bullet$ ii) H^+ iii) $CH_3CH_2^+$ iv) CH_3^- v) Br^- vi) SH^- vii) I^+
  viii) $CH_3O\bullet$

Why do we grow older?

Over the years, film makers have produced countless science fiction movies based on the search for the 'elixir of eternal life'. Quite often, an eccentric scientist is seen drinking a steaming potion, the after-effects of which are usually rather nightmarish! These films owe their success to our own, natural, wish to extend our lives. Some people go to extremes, for example in the use of cryogenics to deep freeze their corpses, in the hope that they can be revived in the future when medical knowledge would allow their illnesses to be treated.

For most people, the human body goes into a slow, but steady, decline after our teenage years. Our skin starts to wrinkle, our hair may turn grey and even fall out. Physical and mental tasks take longer. Scientists have been searching for the reasons why we get older and, over the last few years, some ideas are starting to take shape.

One theory is that aging results from cell damage caused by **free radicals** which are highly reactive atoms, or groups of atoms, that are formed as a by-product of normal cell chemistry. Certainly, free radicals are suspected of causing some heart diseases and cancers. The problem with free radicals is that they have the ability to oxidise other biochemical molecules, causing these to be damaged and perform incorrectly. A natural defence to this cell damage is provided by enzymes, such as catalase and superoxide dismutase, and vitamins C and E. These substances are **antioxidants** and render free radicals harmless by removing their oxidising power.

Can the aging process be reversed or slowed down? At the start of this century, the average life expectancy for men and women was 46 and 49 years, respectively. By the year 2001, on average, men will live for 75 years and women for 80 years. Not only do these figures reflect the advances in medical science, they also provide some evidence that aging can be slowed, mainly through improved diet and exercise. Experiments on rats show that those fed with meagre, low calorie diets tend to live longer because these foods contain more catalase and superoxide dismutase. Worms bred on foods containing free radical 'trappers' have lived up to 65% longer than those bred on foods lacking these 'trappers'.

Scientists have now discovered a chemical substance called PBN which traps free radicals. When elderly gerbils are injected with PBN, they soon perform tasks as well as their young.

Will we reach a time when aging can be controlled? Would it be morally correct to prevent the aging process? What would be the effects worldwide on population and the provision of health services?

6.8 Types of reactive centre in organic molecules

Generally speaking, an organic molecule may be pictured as having two types of reactive centre:

- a **hydrocarbon framework** or **skeleton**, that is one made of carbon and hydrogen atoms;
- one, or more, **functional groups**, which are attached to the hydrocarbon skeleton, Figure 6.16. Functional groups are single atoms, or groups of

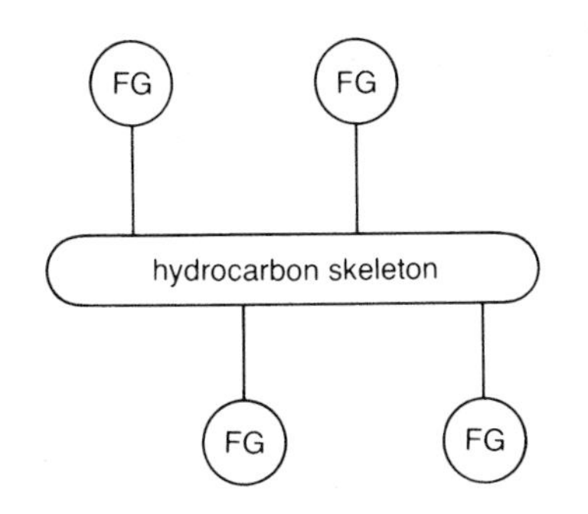

Figure 6.16 ▲
Organic compounds may be thought of as a hydrocarbon skeleton which may be attached to one, or more, functional groups (FG)

atoms, which give characteristic chemical properties to any organic molecule in which they are found.

Let us survey the main types of hydrocarbon skeleton and functional groups.

The hydrocarbon skeleton

In section 6.1, we saw that the ability of carbon atoms to form strong single, and multiple, bonds with each other results in organic compounds having a wide variety of hydrocarbon skeletons. These may be broadly classified as being aliphatic, alicyclic or aromatic. (You may wish to review the meaning of these terms.)

A compound may contain more than one type of hydrocarbon skeleton. If so, each will impart its own unique chemical properties to the compound. For example, consider the hydrocarbon skeleton in phenylethene, Figure 6.17. This gives reactions typical of:

- a C=C bond, for example by adding hydrogen bromide at room temperature:

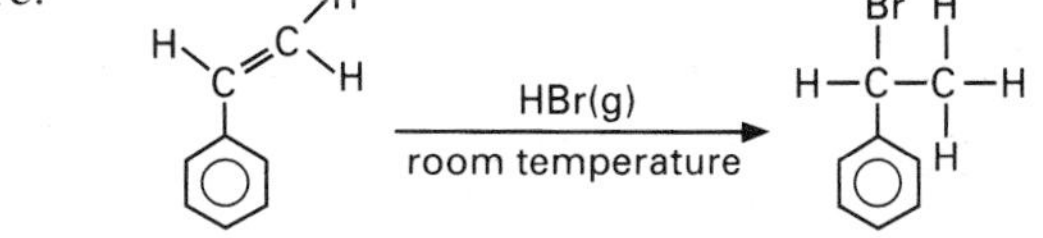

- an aromatic C—H bond, for example in its substitution by a methyl group:

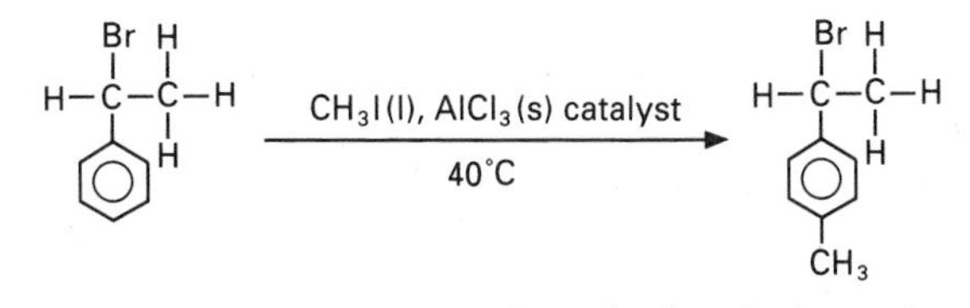

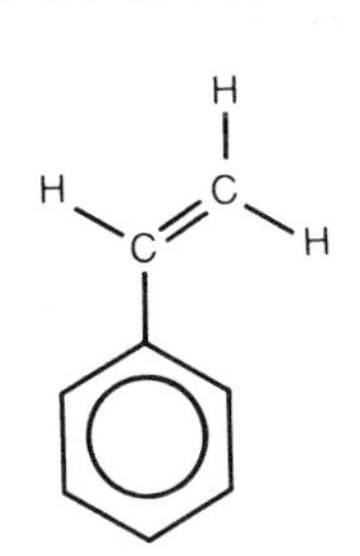

Figure 6.17 ▲
The phenylethene molecule

Organic compounds are sometimes described as being primary, secondary or tertiary compounds according to the hydrocarbon skeletons which are present, as shown below:

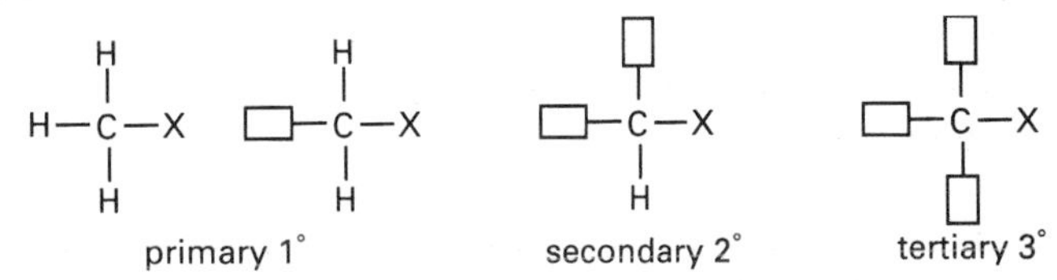

Here □— represents a hydrocarbon skeleton, whether aliphatic, alicyclic or aromatic, and X is a functional group, often a halogen atom or an OH group. They are known as **primary**, **secondary** or **tertiary** compounds because they have one (or zero), two or three hydrocarbon skeletons, respectively, bonded to the carbon which is attached to the functional group. Isomers which can exist as primary, secondary and tertiary compounds have similar chemical reactions but they usually occur at different rates.

The reactivity of a hydrocarbon skeleton is often modified by the presence of functional groups. For example, consider the different reactivities of benzene and phenol towards aqueous bromine:

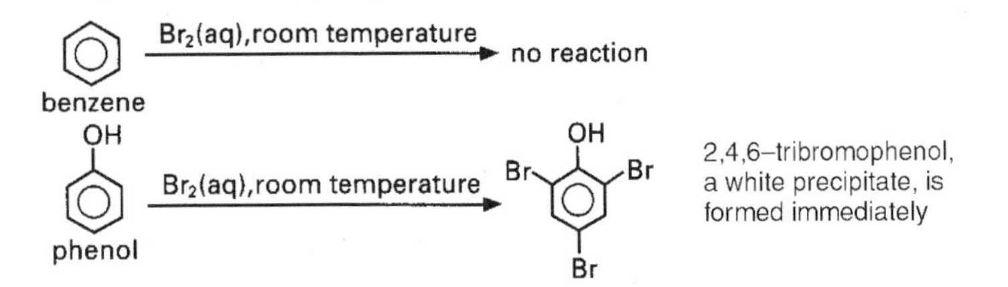

Here the OH functional group makes the benzene ring more reactive towards electrophiles.

Functional groups

Functional groups are atoms, or groups of atoms, which have been substituted for a hydrogen in a hydrocarbon skeleton. Table 6.4 lists the main types of functional group and some compounds in which they are found.

Table 6.4 The main types of functional group. (The way we name these compounds is discussed in Chapters 7–14; ☐— is a hydrocarbon skeleton)

Functional group	...is found in...	aliphatic example		aromatic example	
halogen	organic halogen compounds	$H-CH_2-CH_2-Br$	bromoethane	C_6H_5-I	iodobenzene
OH	alcohols	$H-CH_2-CH_2-CH_2-OH$	propan-1-ol	$C_6H_5-CH_2-OH$	phenylmethanol
	phenols (if the OH is joined directly to a benzene ring)			C_6H_5-OH	phenol
$-O-$☐	ethers	$H_3C-O-CH_3$	methoxymethane	$C_6H_5-O-CH_3$	methoxybenzene
$-C(=O)-H$	aldehydes	$H_3C-C(=O)-H$	ethanal	$C_6H_5-C(=O)-H$	benzaldehyde
$-C(=O)-$☐	ketones	$H_3C-C(=O)-CH_3$	propanone	$C_6H_5-C(=O)-CH_3$	phenylethanone
$-C(=O)-OH$	carboxylic acids	$H_3C-C(=O)-OH$	ethanoic acid	$C_6H_5-C(=O)-OH$	benzoic acid
$-C(=O)-Cl$	acid chlorides	$H_3C-C(=O)-Cl$	ethanoyl chloride	$C_6H_5-C(=O)-Cl$	benzoyl chloride
$-C(=O)-NH_2$	acid amides	$H_3C-C(=O)-NH_2$	ethanamide	$C_6H_5-C(=O)-NH_2$	benzamide
$-C(=O)-O-$☐	acid esters	$H_3C-C(=O)-O-CH_3$	methyl ethanoate	$C_6H_5-C(=O)-O-CH_3$	methyl benzoate
$-NO_2$	nitro compounds	H_3C-NO_2	nitromethane	$C_6H_5-NO_2$	nitrobenzene
$-NH_2$	amines	H_3C-NH_2	methylamine	$C_6H_5-NH_2$	phenylamine
$-CN$	nitriles	H_3C-CN	ethanonitrile	C_6H_5-CN	benzonitrile

In section 6.6, we saw that electronegativities can be used to work out bond polarities. From these, it is possible to predict the type of chemical reagent (free radical, electrophilic or nucleophilic) which might attack a given bond. We can do this for a functional group and its neighbouring carbon atom, as shown below:

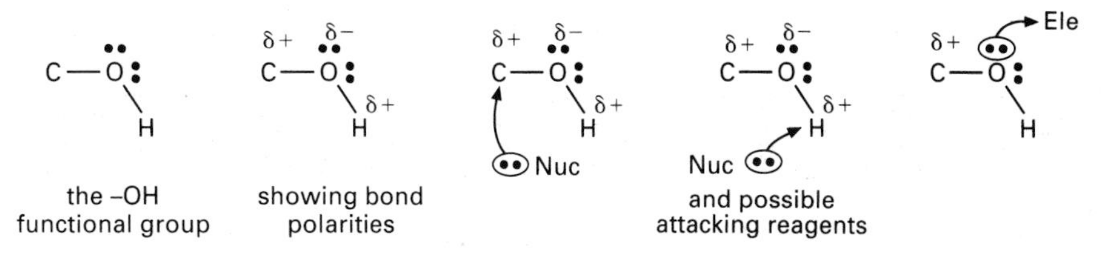

There are three possible reactive centres $C^{\delta+}$, $H^{\delta+}$ and $O^{\delta-}$. The δ+ and δ– centres will be targets for nucleophiles (Nuc) and electrophiles (Ele), respectively. Note the use of '**curly arrows**' to show the movement of a pair of electrons.

When predicting the chemical properties of a molecule, remember to consider each functional group in the context of the molecule as a whole. The reactivity of a functional group is often modified by the type of hydrocarbon skeleton to which it is attached, as is the case, for example, with the relative rates of hydrolysis of chloromethane and chlorobenzene:

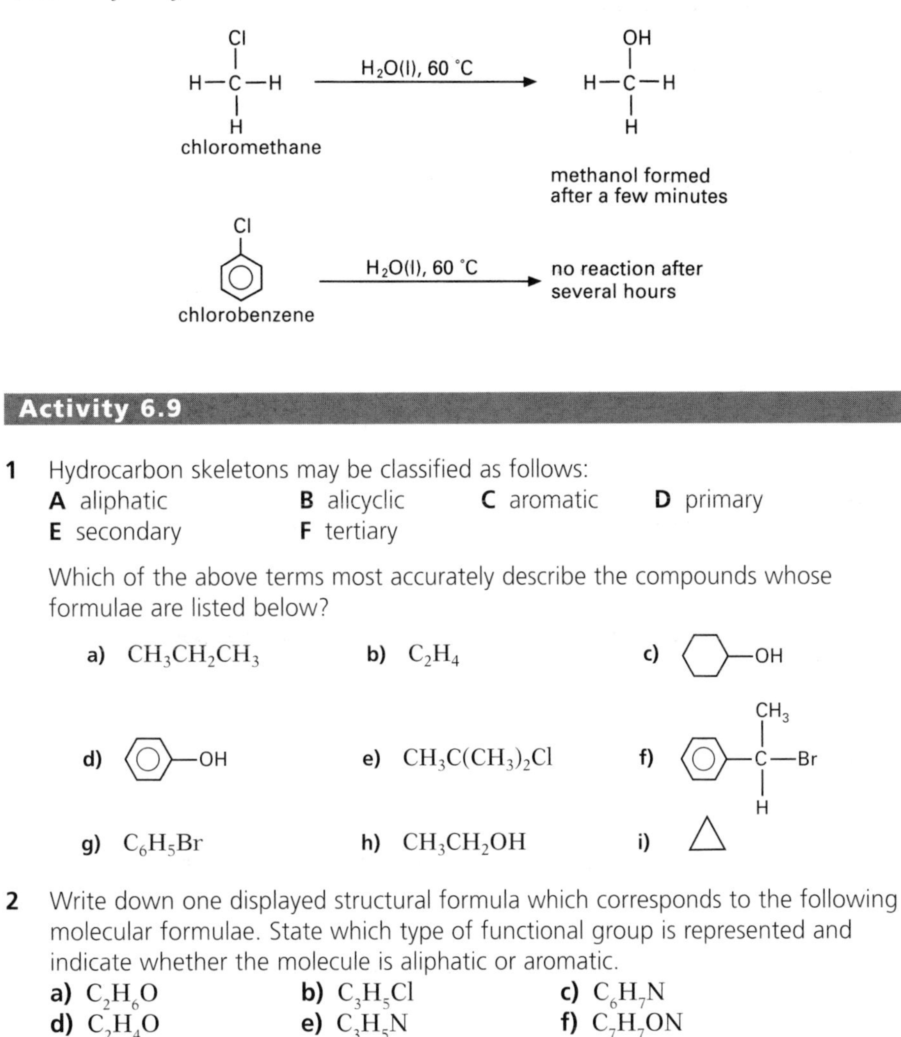

Activity 6.9

1 Hydrocarbon skeletons may be classified as follows:

A aliphatic **B** alicyclic **C** aromatic **D** primary
E secondary **F** tertiary

Which of the above terms most accurately describe the compounds whose formulae are listed below?

a) $CH_3CH_2CH_3$ **b)** C_2H_4 **c)** (cyclohexane ring)–OH

d) (benzene ring)–OH **e)** $CH_3C(CH_3)_2Cl$ **f)** (benzene ring)–C(CH$_3$)(H)–Br

g) C_6H_5Br **h)** CH_3CH_2OH **i)** △

2 Write down one displayed structural formula which corresponds to the following molecular formulae. State which type of functional group is represented and indicate whether the molecule is aliphatic or aromatic.

a) C_2H_6O **b)** C_3H_5Cl **c)** C_6H_7N
d) C_2H_4O **e)** C_3H_5N **f)** C_7H_7ON

Activity 6.9 continued

3 Use δ+ and δ– signs to show the bond polarities in the following structural units:

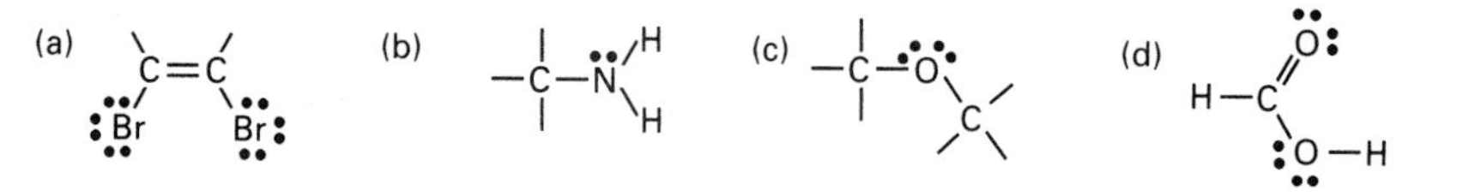

In theory at what positions might these structures be attacked by electrophiles or nucleophiles?

6.9 Mechanisms of organic reactions

Reaction mechanisms describe the way in which electron clouds shift during the reaction. Although there are millions of organic reactions, many have one of a small group of well-known mechanisms and these are outlined below.

Organic reaction mechanisms are usually classified according to:

- the nature of the reagent species, which *first* attacks the organic compound. Reagents may be **free radical**, **electrophilic** or **nucleophilic** (section 6.7)
- the type of structural change that the organic compound undergoes. Frequently, this involves

  i) **addition** to a C=C, C≡C or C=O bond, for example,

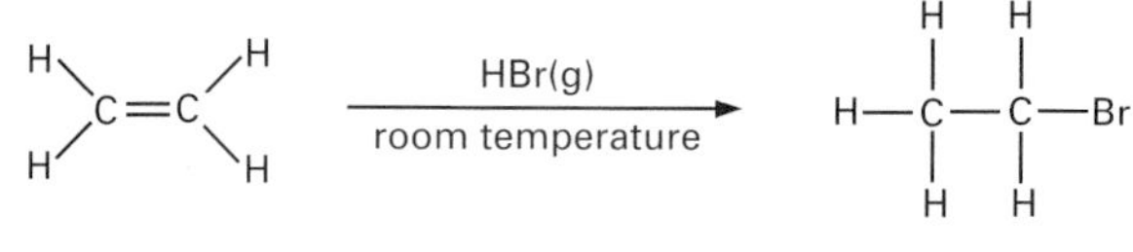

ethene

1–bromoethane

  ii) **substitution** of one functional group for another, for example,

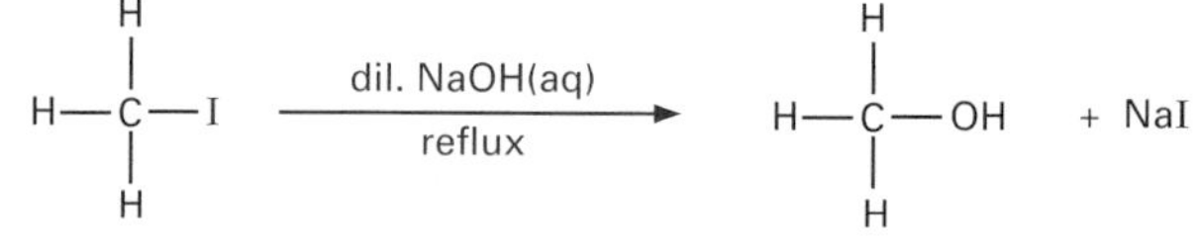

iodomethane methanol

  iii) **elimination** of a small molecule, for example,

conc. NaOH(aq), reflux → + NaBr + H_2O

2-bromo-2-methylpropane 2-methylpropene

Finally, you should note that organic compounds take part in many important redox reactions, for example,

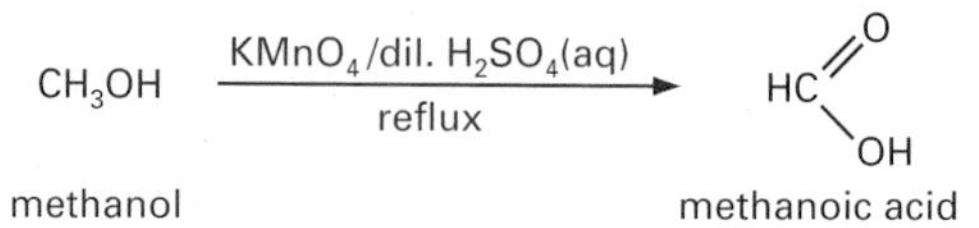

methanol methanoic acid

The mechanisms of redox reactions are rather complex, and go beyond this study.

Redox reactions and ionic equations

When a chemical species loses electrons, it is said to be **oxidised**. If electrons are gained, the species is said to be **reduced**.

When methanol reacts with manganate(VII) ions, the methanol is oxidised, losing two electrons and two hydrogen ions in the process:

$$CH_3OH \longrightarrow HCHO + 2e^- + 2H^+$$

The manganate(VII) ions are an **oxidising agent**, that is, they cause another species to be oxidised. Thus, the manganate(VII) ions are reduced to manganese(II) ions in acidic solution:

$$MnO_4^- + 8H^+ + 5e^- \longrightarrow Mn^{2+} + 4H_2O$$

Since there must be an exact transfer of electrons between the reactants, we find that 5 moles of methanol are oxidised by 2 moles of manganate(VII) ions, according to the ionic equation:

$$5CH_3OH + 2MnO_4^- + 6H^+ \longrightarrow 5HCHO + 2Mn^{2+} + 8H_2O$$

Since reduction and oxidation is occurring, this is termed a **redox** reaction.

Focus 6e

1 **Electronegativity** is the power of an atom in a covalent bond to attract the bonding electron cloud to itself. Electronegativities can be used to deduce the polarity of covalent bonds and, hence, to identify the reactive centres in an organic molecule.

2 In **homolytic bond fission**, each atom in the covalent bond takes one electron and free radicals are formed. Free radicals are very excited atoms, or groups of atoms, which possess an unpaired electron. In **heterolytic bond fission**, the more electronegative atom takes both bonding electrons and ions are formed.

3 Chemical reagents generate three main types of reactive species, free radicals via homolytic bond fission and nucleophiles and electrophiles often via heterolytic bond fission. A **free radical** is a highly reactive atom, or group of atoms. A **nucleophile** is an electron-rich species which has at least one lone pair (non-bonding pair) of electrons; it will attack an electron-poor (δ+) reactive centre. An **electrophile** is an electron-poor species which will accept a pair of electrons from a nucleophile to form a dative covalent bond.

4 An organic molecule can be considered as having two main types of reactive centre, the **functional group(s)** and the **hydrocarbon skeleton**, and these display characteristic reactions. The nature of the hydrocarbon skeleton can modify the reactivity of a functional group and vice versa.

5 The most common features of the hydrocarbon skeletons are C—H, C—C, C=C, C≡C bonds and the partial π bonds between the carbon atoms in the benzene ring.

6 A **reaction mechanism** explains how the electron clouds move around during a chemical reaction. A '**curly arrow**' shows the movement of a pair of electrons.

Questions on Chapter 6

1 a) Explain what is meant by the term 'functional group'.

b) For each of the compounds below:

i) draw the displayed structural formula

ii) use Table 6.4 to identify the functional group present and name the homologous series in which the compound is found

iii) use δ+ and δ– to show the presence of polar bonds

iv) indicate where a nucleophile or an electrophile might attack the functional group.

A CH_3CH_2Br B $CH_3CH(OH)CH_3$
C CH_3COCH_3 D CH_3CONH_2
E $CH_3CH_2COOCH_3$ F $HCHO$
G CH_3COOH H $CH_3CH_2CH_2NH_2$

2 a) What are the typical properties of a 'homologous series'?

b) The boiling points and relative molecular masses of eight organic compounds are shown below. By drawing a sketch graph, suggest which of the compounds appear to be part of the same homologous series.

Compound	A	B	C	D	E	F	G	H
relative molecular mass	130	60	86	46	102	114	74	62
boiling points/K	468	370	375	248	430	399	391	470

c) The molecules in this homologous series contain only the elements carbon, hydrogen and oxygen. State a *possible* general formula of the homologous series.

d) Identify the *two* functional groups which would form a homologous series having this general formula.

3 Read through the following method for the laboratory preparation of bromoethane, C_2H_5Br, and then answer the questions which follow.

I) Weigh out 25 g of powdered potassium bromide, KBr, into a 100 cm^3 beaker. Add 20 cm^3 of water to the beaker and stir the contents until nearly all the solid has dissolved; then add 25 cm^3 of ethanol.

II) Transfer the mixture to a 250 cm^3 round-bottomed flask and set up the apparatus for reflux. Ensure that water is running briskly through the condenser.

III) Carefully measure out 20 cm^3 of concentrated sulphuris acid, H_2SO_4. Add the acid to the aqueous potassium bromide, a few drops at a time, by pouring it down the condenser and then swirling the flask. The mixture should get slightly hot after each addition. Avoid adding the acid too quickly as the contents of the flask will start to boil. If this happens, cool the flask down. The reactions involved here are:

- generating the reagent, concentrated hydrobromic acid, HBr
  $KBr + H_2SO_4 \longrightarrow KHSO_4 + HBr$
  $CH_3CH_2OH + HBr \longrightarrow CH_3CH_2Br + H_2O$
- bromination of the ethanol to give bromoethane

IV) When all of the acid has been added, change the apparatus set-up from reflux to distillation. Bromoethane is a volatile liquid, boiling point 39°C, so its distillate must be collected below the surface of about 30 cm^3 of ice-cold water contained in a conical flask. Heat the round bottomed flask gently and distil off the bromoethane at about 39°C; this will appear as an oily layer at the bottom of the conical flask.

V) When all of the bromoethane has been collected, transfer the contents of the conical flask to a separating funnel. Allow the layers to separate, then run off the lower 'bromoethane' layer into another conical flask and stopper this flask. Discard the aqueous layer into a 'residues' bottle.

VI) Take the conical flask and add to it a similar volume of aqueous sodium carbonate solution. (This removes any unreacted sulphuric acid which might have distilled over with the bromoethane.) Swirl the mixture and then transfer it to the separating funnel.

VII) Stopper the separating funnel, invert it carefully and shake the mixture, opening the tap every now and again to release any pressure which has built up.

VIII) Allow the layers to separate. Save the 'bromoethane' layer and discard the aqueous layer. Return the bromoethane to the separating funnel and repeat the procedure in VI) and VII).

IX) Run the bromoethane into a clean conical flask, add a spatula full of anhydrous calcium chloride. Stopper the flask firmly and swirl it for a few minutes. Leave the flask for at least an hour in a cool place.

X) Filter the bromoethane into a small 'pear-shaped' distillation flask. Add a few anti-bumping granules and distil off the bromoethane into a very cold conical flask. Collect the fraction which boils in the range 37–40°C. Store the bromoethane in a well-stoppered flask in a cold place.

a) Draw a diagram of the reflux apparatus used in stage II.

b) How would you cool the flask in stage III?

c) Draw a diagram of the distillation apparatus used in stage IV. How would you heat the flask and its contents?

d) In stage V, how will you know that all the bromoethane has been collected?

e) Why is powdered anhydrous calcium chloride added at stage IX?

f) Write an equation for the reaction of the excess sulphuric acid with sodium carbonate and, hence, explain why pressure may build up in the separating funnel.

g) Why should bromoethane be stored in a cool place?

h) What are the 'anti-bumping' granules used for?

i) How could you test the purity of the bromoethane?

j) What safety precautions would you take in carrying out this preparation?

k) Is the bromination of ethanol an addition or a substitution reaction?

l) i) The density of ethanol is 0.79 g cm^{-3}. Calculate the mass of ethanol that is used in this experiment.

Questions on Chapter 6 *continued*

ii) Calculate the relative molecular mass of ethanol, C_2H_5OH, and then work out the number of moles of ethanol that are used.

iii) In theory, how many moles of bromoethane should be produced?

iv) What is the theoretical mass of bromoethane formed?

v) In fact, only 36 g of pure bromoethane was obtained. Assuming that the hydrobromic acid is present in excess, calculate the percentage yield of bromoethane.

4 Phenylamine is an aromatic amine which is used to make dyes and pharmaceutical compounds. It is prepared in the laboratory by reducing nitrobenzene with tin and concentrated hydrochloric acid:

C_6H_5–NO_2 —Sn(s)/conc.HCl(aq),reflux→ C_6H_5–NH_2

nitrobenzene phenylamine

a) Steam distillation is used to extract the phenylamine from the reaction mixture. Explain the principles upon which this technique is based and draw a diagram of the apparatus which is used.

b) The steam distillate separates into two layers.

i) Will the phenylamine (density 1.022 g cm^{-3}) be in the top or bottom layer?

ii) Explain how solvent extraction may be used to obtain the phenylamine from this mixture.

c) In one experiment, 4.2 g of phenylamine was prepared from 6.1 g of nitrobenzene. Calculate the percentage yield of product.

d) What reason(s) might you give for not achieving 100% yield?

5 Benzoic acid and its alkali metal salts are used as preservatives in soft drinks, canned fruit and fruit juices. The acid may be prepared by oxidising methylbenzene with an excess of acidified potassium manganate(VII):

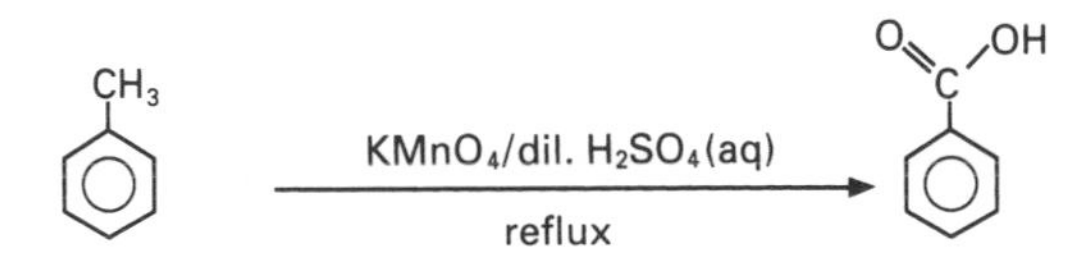

The benzoic acid is soluble in the hot aqueous solution but it crystallises out as the solution cools down.

a) Describe how benzoic acid may be purified by recrystallisation from water.

b) Describe a technique which can be used to determine the purity of your product.

c) When 10 cm^3 of methyl benzene is oxidised, 9.5 g of benzoic acid crystals is produced. Calculate the percentage yield of benzoic acid. (The density of methylbenzene is 0.867 cm^{-3}.)

6 On complete combustion, 0.981 g of compound K gave 0.792 g of carbon dioxide and 0.405 g of water. Further analysis indicates that K has a molecular mass of 74 and contains only the elements, carbon, hydrogen and oxygen.

a) Find the molecular formula of K.

b) Draw the displayed structural formulae of the *seven* possible structures for K.

7 Combustion analysis of compound P shows that it contains 8.4% hydrogen, 78.5% carbon and 13.1% nitrogen. The molecular mass of P is found to be 107.

a) Find the molecular formula of P.

b) Draw *five* possible structures for P.

8 Sorbic acid, a member of the 'carboxylic' acid family of compounds, and its salts are used as food preservatives since they kill fungi and inhibit moulds. The composition of sorbic acid is 64.3% carbon, 7.1% hydrogen and 28.6% oxygen and its relative molecular mass is 112.

a) Calculate the molecular formula of sorbic acid.

b) The sorbic acid molecule has a straight-chain hydrocarbon skeleton and contains one carboxyl functional group and two carbon-carbon double bonds. Write down the skeletal structure of sorbic acid.

9 a) What is meant by the term 'isomer'?

b) Explain the difference between:

i) a structural isomer and a stereoisomer

ii) a geometrical isomer and an optical isomer.

c) Draw suitable structural diagrams of the isomers of the compounds whose molecular formulae are given below. State which type of isomerism is present and classify the isomers as being aliphatic, alicyclic or aromatic.

i) C_5H_{12}

ii) C_3H_6

iii) $C_6H_4Br_2$ (this molecule does *not* contain C=C or C≡C bonds)

iv) $C_2H_2I_2$

d) How would you distinguish between the isomers in part c)?

10 Methadone is a synthetic drug used in the treatment of heroin addiction. Heroin is an illegal drug because, for many people, its use will lead to addiction. In some cases, the addiction may occur within a week of its use. When treating heroin addiction, doctors often prescribe methadone, Figure 6.18, to control the withdrawal symptoms. Methadone and heroin produce similar sensations but it takes about a month for methadone to become addictive. By that time, it would be hoped that the patient is well advanced in their withdrawal from the heroin addiction. Methadone is an optically active compound whose molecule can exist in two non-superimposable mirror-image forms. Only the laevorotatory (–) isomer is physiologically active.

a) What is meant by the terms 'optically active' and 'laevorotatory (–) isomer'?

b) What conditions must be fulfilled in order for a compound to be optically active?

c) Draw the non-superimposable mirror-image forms of methadone. Simplify the structure by taking X to be the

$_2$CCOCH$_2$CH$_3$ group

d) Identify the asymmetric (chiral) carbon atom in the molecule.

e) Depending on the method of preparation, a racemic mixture of the isomers may be obtained. What is a racemic mixture and outline the principle by which the (–) isomer be obtained from it?

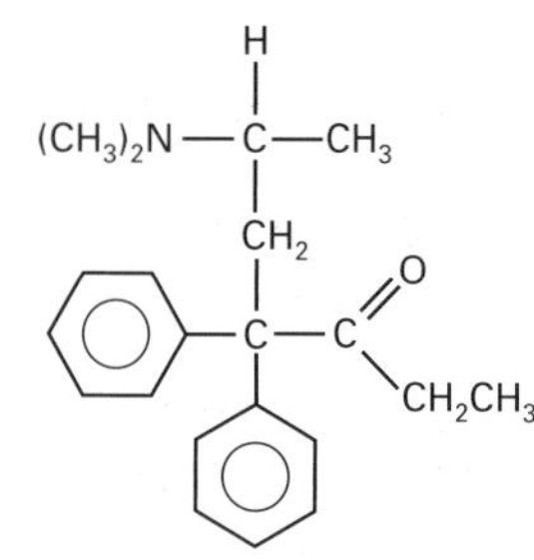

Figure 6.18
Methadone, a synthetic drug used in the treatment of heroin addiction

11 a) 'Whereas homolytic fission of a covalent bond produces free radicals, heterolytic fission yields ions.' Explain this statement, giving suitable examples.

b) Use dot and cross diagrams to show the difference between a nucleophile and an electrophile.

c) Consider the three reactions below:

A $CH_3CH{=}CH_2 + Cl_2 \longrightarrow CH_3CHClCH_2Cl$

B $CH_3CH_2Br + OH^- \longrightarrow CH_3CH_2OH + Br^-$

C $CH_3CH_2I + OH^- \longrightarrow CH_2{=}CH_2 + I^- + H_2O$

From the species which are involved in these reactions, give one which:

i) behaves as a nucleophile,
ii) is an addition product,
iii) undergoes an elimination reaction,
iv) behaves as an electrophile,
v) undergoes a substitution reaction.

Comments on the activities

Activity 6.1

1 a) $CH_3CH_2CH_2CH_3$ butane — $CH_3CH(CH_3)CH_3$ 2-methylpropane

b)

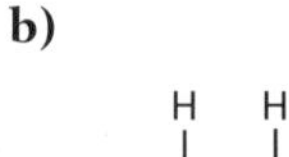

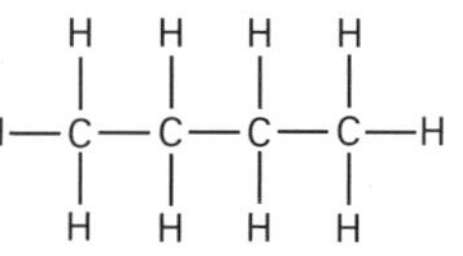

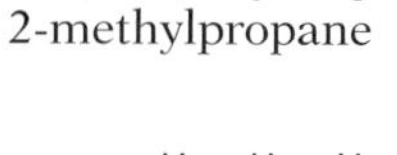

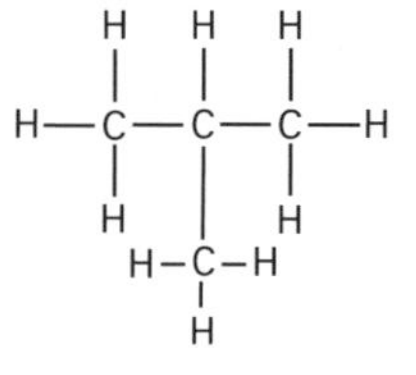

c)

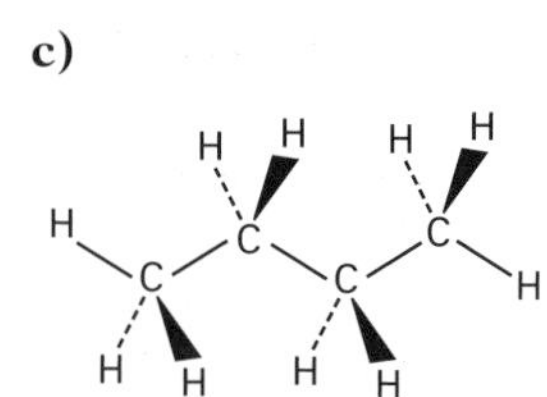

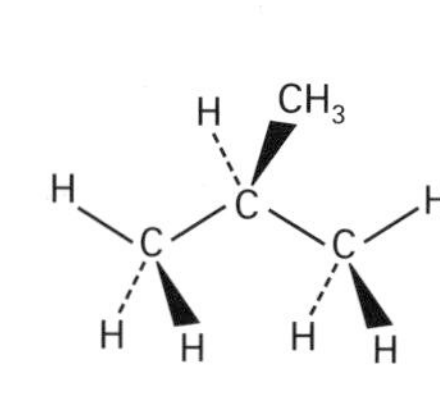

d)

2 a) $CH_3CH_2CH_2Cl$ 1-chloropropane — $CH_3CHClCH_3$ 2-chloropropane

b)

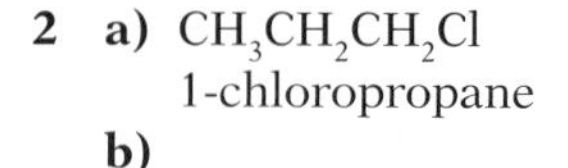

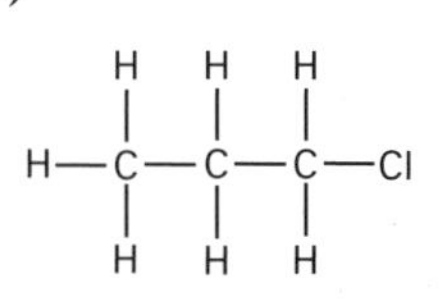

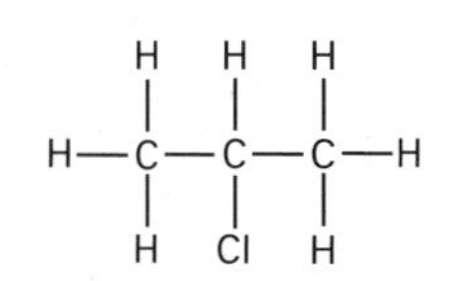

c)

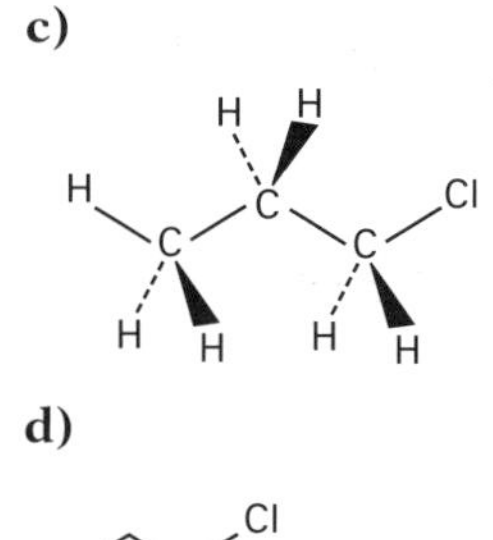

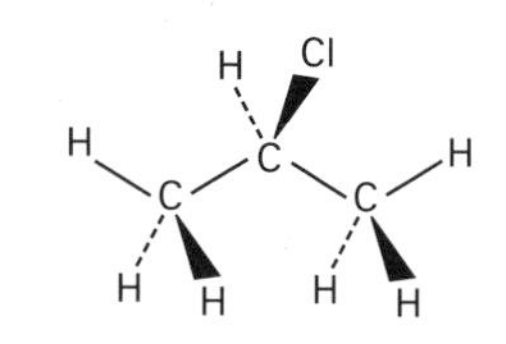

d)

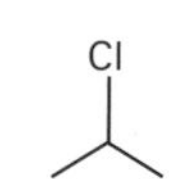

3 a) CH_3CH_2OH ethanol — CH_3OCH_3 methoxymethane

b)

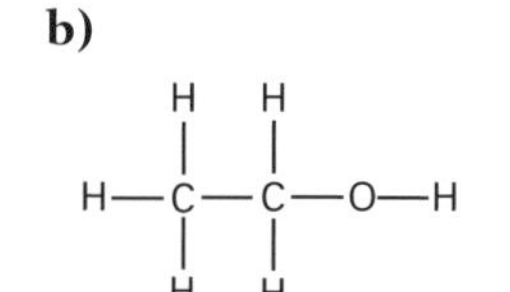

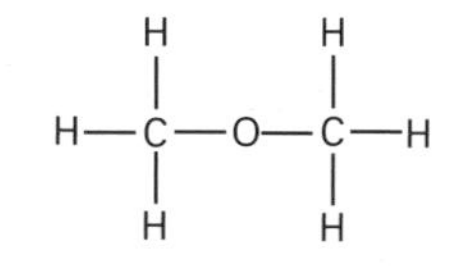

c)

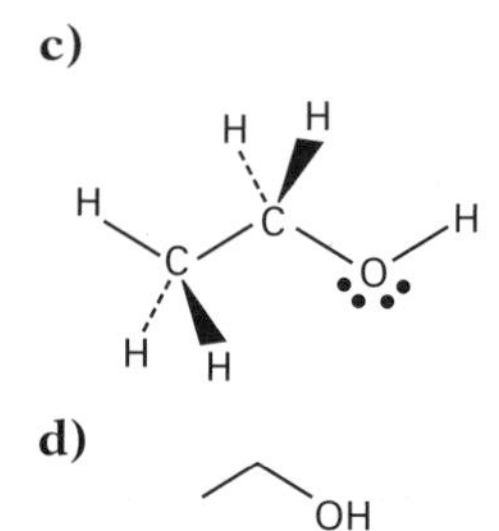

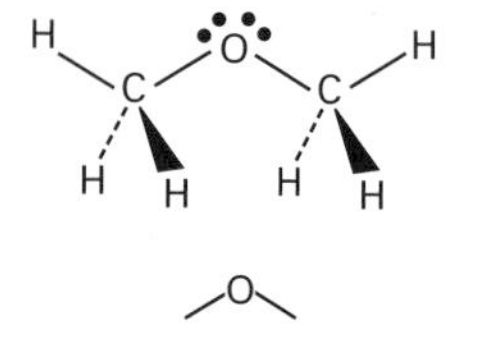

d)

Comments on the activities *continued*

Activity 6.2

1 2-hydroxybenzoic acid (salicylic acid) + ethanoic anhydride ⟶ aspirin + ethanoic acid

moles used: $\frac{5.20}{138} = 0.0377$ $\quad \frac{7.50}{102} = 0.0735$

Hence, the amount of salicylic acid will determine the product yield; ethanoic anhydride is in excess.
Now, 1 mole of salicylic acid forms 1 mole of aspirin.
∴ 0.0377 moles of salicylic acid forms 0.0377 moles of aspirin.
So the theoretical yield of aspirin= $0.0377 \times M_r$
$= 0.0377 \times 180 = 6.79$

Finally,

$$\% \text{ yield} = \frac{\text{actual yield of purified product}}{\text{theoretical yield}} \times 100$$

$$= \frac{5.90}{6.79} \times 100 = 86.9$$

% yield = 87%

2 You can use the same method as in question 1. Here is another way of tackling this problem.
In theory, 74.0 g will form 92.5 g

∴ 1 g will form $\frac{92.5}{74.0}$ g

and 6.20 g will form $\frac{92.5}{74.0} \times 6.20 = 7.75$ g

So, the theoretical yield of product = 7.75 g and the actual yield = 6.20 g. Thus,

$$\% \text{ yield} = \frac{\text{actual yield of purified product}}{\text{theoretical yield}} \times 100$$

$$= \frac{6.20}{7.75} \times 100 = 80.0$$

∴ % yield of B = 80%

Activity 6.3

1 $\text{mol } CO_2 = \frac{\text{mass of } CO_2}{M_r(CO_2)} = \frac{2.640}{44} = 0.0600$

$\text{mol } H_2O = \frac{\text{mass of } H_2O}{M_r(H_2O)} = \frac{1.442}{18} = 0.0801$

1 mol of CO_2 gas is obtained from 1 mol of C atoms in A, thus, 0.0600 mol of CO_2 is obtained from 0.0600 mol of C atoms in A.
Mass of carbons atoms in 1.202 g of A
= moles of carbon atoms × A_r(C)
= 0.0600 × 12
= 0.7200 g

1 mol of H_2O gas is obtained from 2 mol of H atoms in A, thus, 0.0801 mol of H_2O is obtained from 0.1602 mol of H atoms in X.

Mass of H atoms in 1.202.g of A = mol of H atoms × A_r(H)
= 0.1602 × 1
= 0.1602 g

Mass of O atoms in 1.202 g of A = mass of sample – mass of C – mass of H
= 1.202 – 0.7200 – 0.1602
= 0.3218 g

So mol O atoms present = $\frac{\text{mass of O}}{A_r(O)} = \frac{0.3218}{16}$
= 0.0201

Using the mole ratios above, we can write a formula for A:

C 0.0600 H 0.1602 O 0.0201

Dividing through by the smallest number to get the empirical formula:

$$C\,\frac{0.0600}{0.0201} \quad H\,\frac{0.1602}{0.0201} \quad O\,\frac{0.0201}{0.0201}$$

= C 2.985 H 7.970 O 1.000

The empirical formula of compound A is C_3H_8O.

2 **a)** In 100 g of B, there are

15.4 g C 3.20 g H 81.4 g I

Divide by the relative atomic masses

$$\frac{15.4}{12} \quad \frac{3.20}{1} \quad \frac{81.4}{127}$$

So obtaining the moles of each element

1.28 3.20 0.641

Using these mole ratios, we can write a formula for B:

C 1.28 H 3.20 I 0.641

Now divide through the smallest number to get the empirical formula:

$$C\,\frac{1.28}{0.641} \quad H\,\frac{3.20}{0.641} \quad I\,\frac{0.641}{0.641}$$

= C 1.99 H 4.99 I 1.00

The empirical formula of compound B is C_2H_5I.

b) In 100 g of C, there are

19.99 g C 6.710 g H 26.64 g O 46.66 g N

Divide by relative atomic masses

$$\frac{19.99}{12} \quad \frac{6.710}{1} \quad \frac{26.64}{16} \quad \frac{46.66}{14}$$

So obtaining the moles of each element

1.666 6.710 1.665 3.333

Using the mole ratios, we can write a formula for C:

C 1.666 H 6.710 O 1.665 N 3.333

Now divide through by the smallest number to get the empirical formula:

$$C\,\frac{1.666}{1.665} \quad H\,\frac{6.710}{1.665} \quad O\,\frac{1.665}{1.665} \quad N\,\frac{3.333}{1.665}$$

The empirical formula of compound C is CH_4ON_2.

c) In 100 g of D, there are

48.65 g C 8.100 g H 43.25 g O

Divide by relative atomic masses

$\frac{48.65}{12}$ $\frac{8.100}{1}$ $\frac{43.25}{16}$

So obtaining the moles of each element

4.054 8.100 2.703

Using these mole ratios, we can write a formula for D:

C 4.054 H 8.100 O 2.703

Now divide through by the smallest number to get the empirical formula:

C $\frac{4.054}{2.703}$ H $\frac{8.100}{2.703}$ O $\frac{2.703}{2.703}$

= C 1.500 H 2.997 O 1.000

This suggests a mole ratio of $C_{1.5}H_3O$. Since an empirical formula is the simplest *whole* number ratio of the atoms present, the empirical formula of D is $C_3H_6O_2$.

3 a) Empirical formula mass = 13.

Using relative molecular mass/empirical formula mass = 78/13 = 6, the molecular formula of E = 6(CH) = C_6H_6.

Using the same method:

b) F is $C_2H_4O_2$

c) G is $C_2H_2Cl_2$

d) H is C_7H_{16}

e) I is $C_2H_2O_4$.

4 A: molecular formula = C_3H_8O giving the displayed formulae

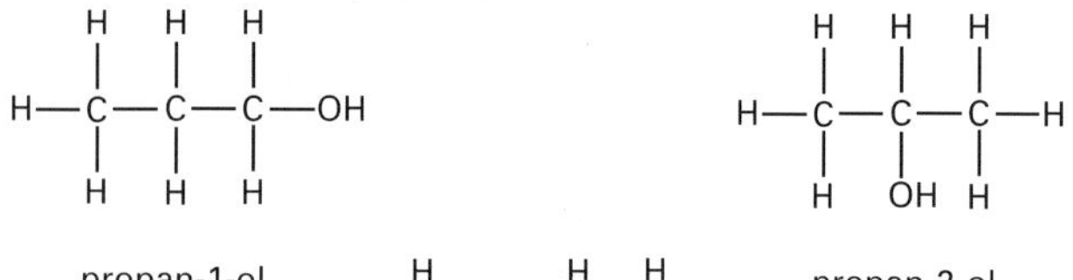

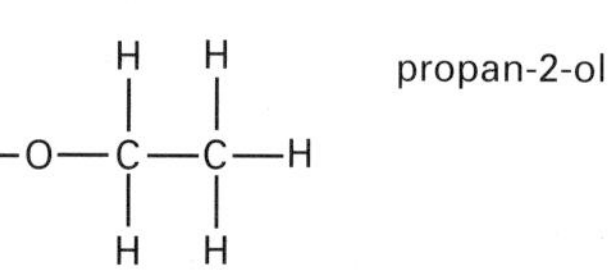

B: molecular formula = C_2H_5I giving the displayed formula

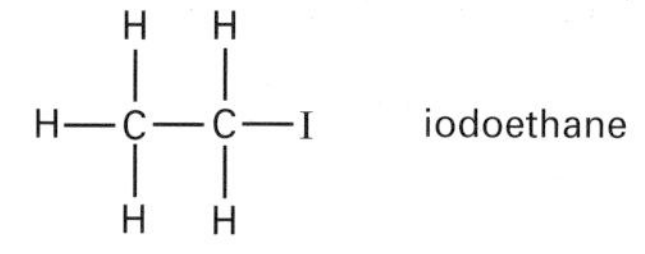

C: molecular formula = CH_4ON_2 giving the displayed formula

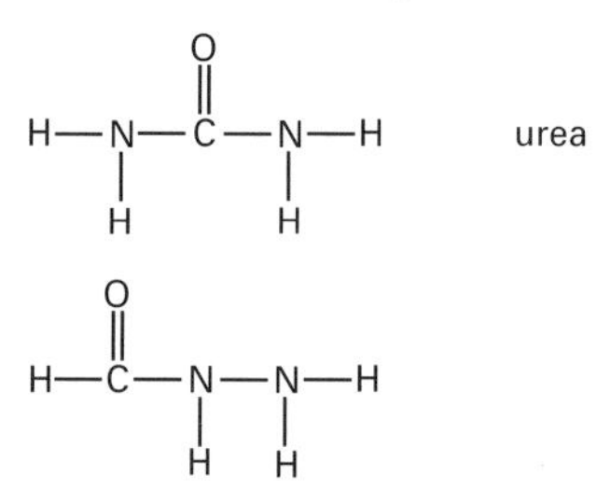

You may have different, but equivalent, displayed formulae – have a quick look at Figure 6.10.

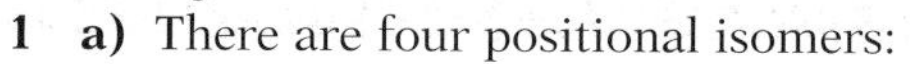

Activity 6.4

1 a) There are four positional isomers:

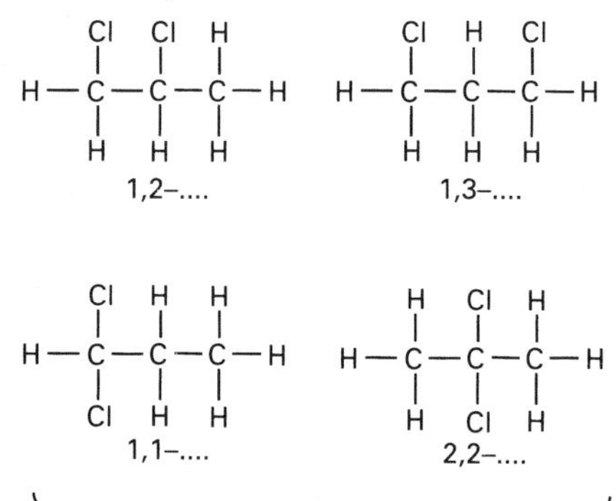

Once again, remember that rotation around the single bonds means that structural formulae such as

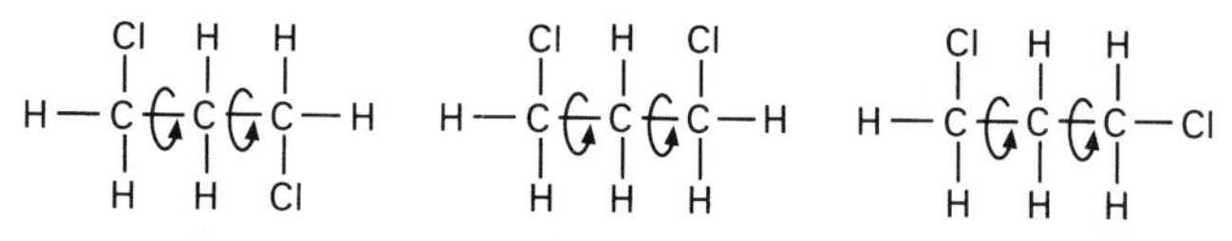

are all equivalent. They are not positional isomers.

b) There are five chain isomers, two of which are also positional isomers:

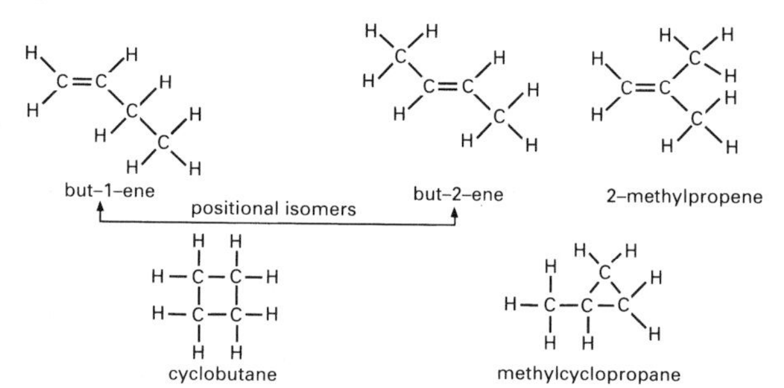

c) There are *two pairs* of positional isomers:

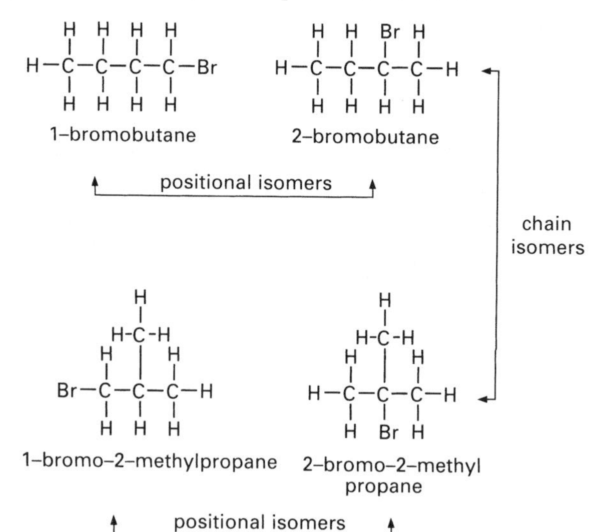

As you can see, each pair are chain isomers of the other pair.

Comments on the activities *continued*

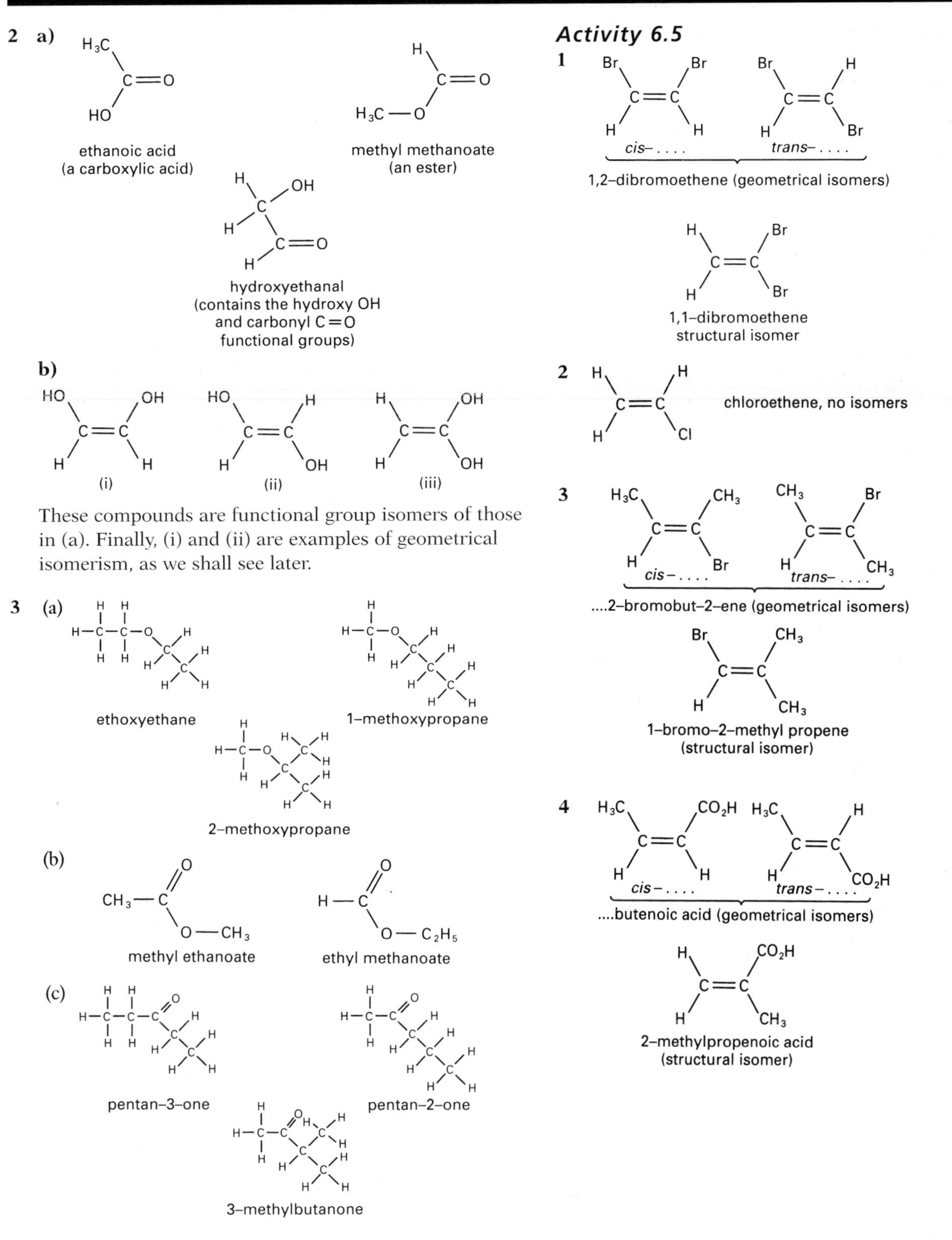

These compounds are functional group isomers of those in (a). Finally, (i) and (ii) are examples of geometrical isomerism, as we shall see later.

Comments on the activities *continued*

Activity 6.6

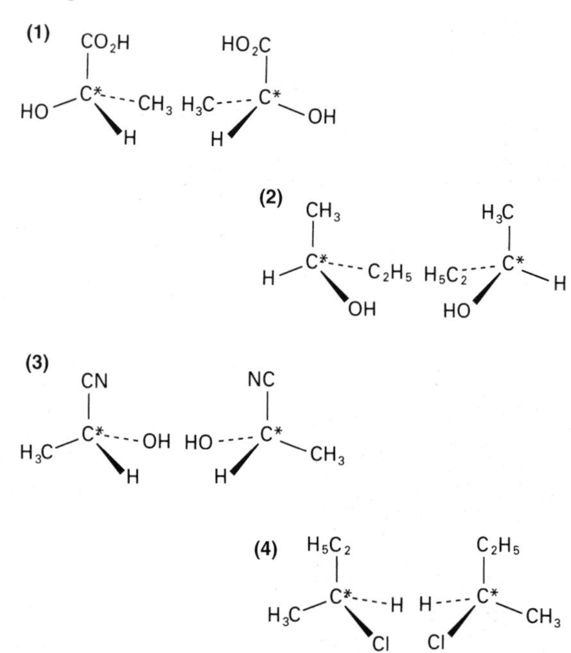

They are all enantiomers because they possess an **asymmetric** or **chiral carbon atom**, marked *, and possess the property of **chirality** (no axis of symmetry).

Answers to Activity 6.7

1

δ+ δ– C—Br	δ– δ+ O—H	δ– δ+ N—H	δ+ δ– S—O
δ+ δ– C=O	δ+ δ– C=N	δ+ δ– C—Cl	δ+ δ– C—F

2 Increasing polarity:

	δ+ δ–	δ+ δ–	δ+ δ–	δ+ δ–	δ+ δ–
electronegativity	C—Br	C—Cl	C≡N	N—H	C=O
difference	0.3	0.5	0.5	0.9	1.0

	δ+ δ–	δ– δ+	δ+ δ–
electronegativity	S—O	O—H	C—F
difference	1.0	1.4	1.5

Answers to Activity 6.8

1 a)	i)	O:O	⟶	O•	+	•O
	ii)	Br:Br	⟶	Br•	+	•Br
	iii)	H:H	⟶	H•	+	•H
	iv)	C:C	⟶	C•	+	•C
b)	i)	C:O	⟶	C^+	+	$:O^-$
	ii)	Cl:Cl	⟶	Cl^+	+	$:Cl^-$
	iii)	H:Br	⟶	H^+	+	$:Br^-$
	iv)	N:C	⟶	$N:^-$	+	C^+
	v)	C:Cl	⟶	C^+	+	$:Cl^-$
	vi)	N:H	⟶	$N:^-$	+	H^+
	vii)	N:O	⟶	N^+	+	$:O^-$

2 i) A ii) B iii) B, D iv) C, E v) C vi) C vii) B viii) A

Answers to Activity 6.9

1 **a)** A **b)** A **c)** B, E **d)** C **e)** A, F **f)** B, C, E **g)** C **h)** A, D **i)** B

2 There are many possible answers. Some are shown below:

a)

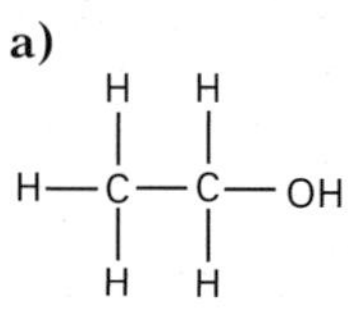

alcohol, aliphatic

b)

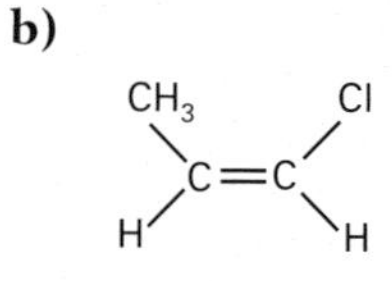

chloro compound, aliphatic

c)

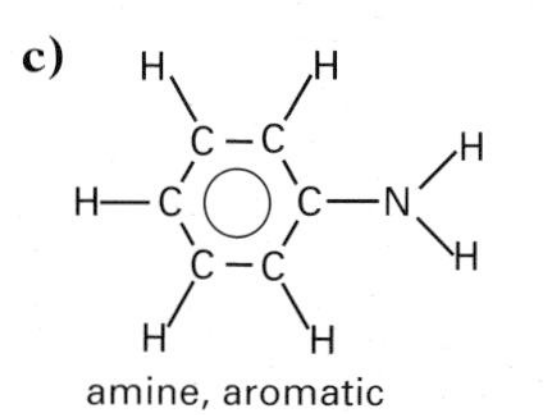

amine, aromatic

d)

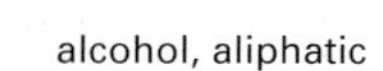

alcohol, aliphatic

e)

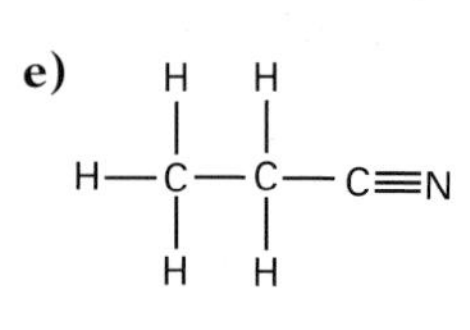

nitrile, aliphatic

f)

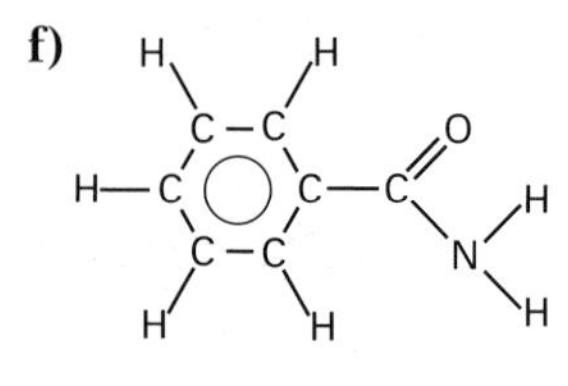

amide, aromatic

3 The reactive centres and possible types of attacking reagents are shown below:

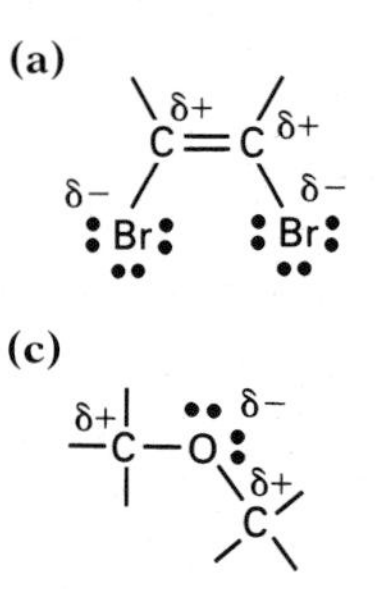

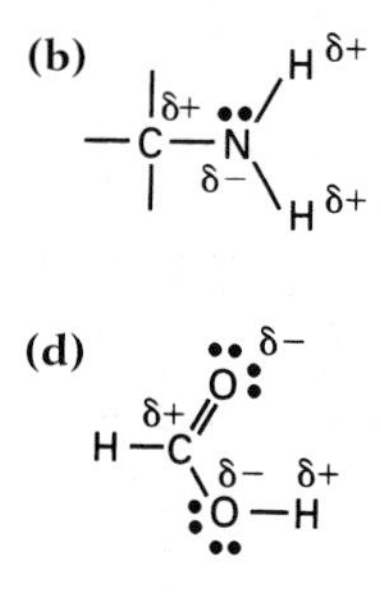

THEME B

The chemistry of organic chains and rings

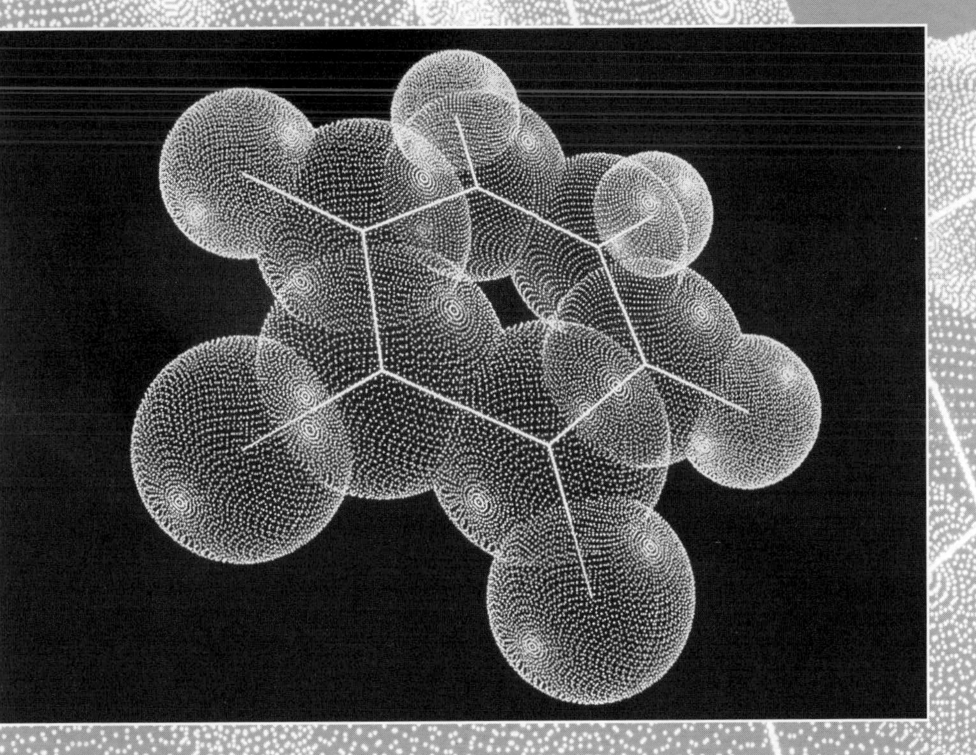

C H A P T E R

7

Alkanes

Contents

Study Checklist

By the end of this chapter you should be able to:

1. Name alkanes, including cycloalkanes, given their short structural, displayed structural or skeletal formula.
2. Understand the meaning of the following terms: saturated hydrocarbon, unbranched hydrocarbon, branched hydrocarbon.
3. Describe, in outline, how petroleum (crude oil) is converted into more useful alkanes and alkenes, with particular reference to fractional distillation, cracking and reforming.
4. Describe how the combustion reactions of alkanes make them suitable for use as fuels, for example in the home and in transport. Explain why various additives are found in petrol, with particular reference to knocking, octane number, tetraethyl lead, unleaded fuel.
5. Explain how the internal combustion engine produces a number of pollutants, for example NO_x, CO and unburnt hydrocarbons, and describe the environmental implications of these exhaust emissions. Describe how these pollutants may be removed from motor vehicle exhaust gases using a catalytic converter.
6. Explain, in terms of van der Waals' forces, why the boiling points of alkanes increase as their molecules get larger and those of isomeric alkanes decrease as the molecule becomes more branched.
7. Write chemical equations for the combustion of alkanes (containing up to 8 carbon atoms per molecule) and explain why $\Delta H_c^\ominus$ for unbranched alkanes shows a steady increase with the number of carbon atoms in the molecule.
8. State that alkanes are attacked by free radicals but not by nucleophiles of electrophiles.
9. Describe the reaction mechanism of methane with chlorine as a free radical substitution reaction in terms of initiation, propagation and termination steps and use 'curly arrows' to show how the bonding electron clouds shift during the reaction.
10. Summarise the chemical reactions of alkanes as shown in Figure 7.12.

The miner's safety lamp was invented by Humphry Davy in 1815. It dramatically reduced the number of deaths caused by methane explosions in coal mines. Nowadays the methane is collected and burnt to provide heat for the surface buildings

In the early days of coal mining, many lives were lost because of explosions caused by combustible gases in the coal seam. One of these gases is **methane**, $\mathbf{CH_4}$, the simplest alkane. Alkanes are hydrocarbons in which each carbon is bonded to four other atoms via single covalent bonds. They are said to be **saturated** hydrocarbons because they contain the highest possible ratio of hydrogen atoms to carbon atoms.

Methane is also found in stagnant ponds, swamps and marshes where it is formed by the bacterial decay of plants and animals. In some countries, the controlled decay of sewage material is used to produce small quantities of methane, and this can be used by the local community as cheap fuel (e.g. for heating, cooking or gas-powered cars).

A more plentiful source of methane is **natural gas**. Typically, this contains about 75% methane, 21% other gaseous alkanes and 4% of other gases, such as hydrogen sulphide. Often, natural gas is found together with **crude oil**. This is a complex mixture of many thousands of different hydrocarbons, most of which are alkanes. The actual composition of crude oil varies with its place of origin.

Environmentally safe alternatives to fossil fuels: a wind-powered electrical generator and a tidal power scheme. Although 'safe', they are not without environmental implications. It might be argued that an array of wind 'blades' is noisy and lessens the beauty of the landscape. Tidal barriers often influence the variety of sealife and birds found in the locality.

Oil and gas from the North Sea

Commercial amounts of natural gas were first found in the North Sea in 1965 by British Petroleum (BP) drilling about 40 miles off the Humber Estuary. The find, named, West Sole, produced 10 million cubic feet of gas per day. Although the search for gas was the main target of offshore drilling, steeply rising oil prices in the early 1970s focussed attention on Amoco's discovery of the Arbroath oil field about 135 miles off the Scottish coast.

Because of North Sea production, the United Kingdom has been self-sufficient in gas and oil for over 15 years. Some oil has been exported and through taxation on its sales, this has brought in over £100 000 million in revenue for the nation.

In order to maintain this position over the next twenty years, the UK Offshore Operators Association (UKOOA) predicts that between 100 and 300 new fields will be needed.

Even using 'state-of-the-art' seismic equipment to find possible new fields, only 1 in 6 of the wells drilled finds gas or oil so the risk of a 'dry hole' is high. It is not surprising then that nearly 40% of the money raised from each barrel of oil is reserved for exploration and development. Oil companies often get together when applying for new drilling sites and then share the cost of exploration and development.

Natural gas and crude oil are formed from the decay of animals and plants which inhabited shallow seas many millions of years ago. As these died, they sank to the sea-bed where they were gradually covered with layers of sand, mud and weathered rocks. This 'burial' caused the plant and animal remains to decompose under pressure and at high temperature. Natural gas and crude oil, the decay products, were then absorbed by porous rocks, such as limestone. Crude oil is known more accurately as **petroleum**, which comes from the Greek words for rock (*petra*) and oil (*oleum*). The refining of petroleum is described in section 7.4.

Because they take such a long time to form, natural gas and petroleum are **finite resources**, i.e. they cannot be replaced. At present, about 65% of the world's energy comes from oil and natural gas, yet it is generally agreed that these supplies may only last about another 45 years. What do you think could, or should, replace them?

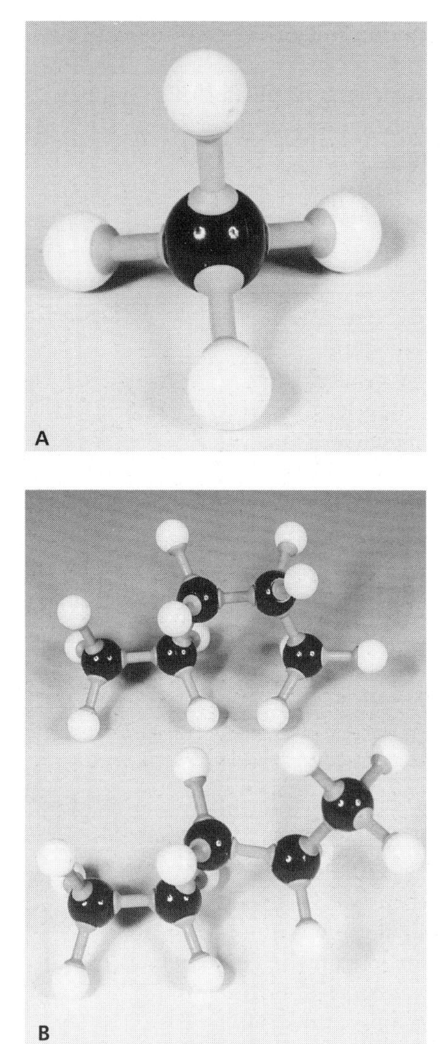

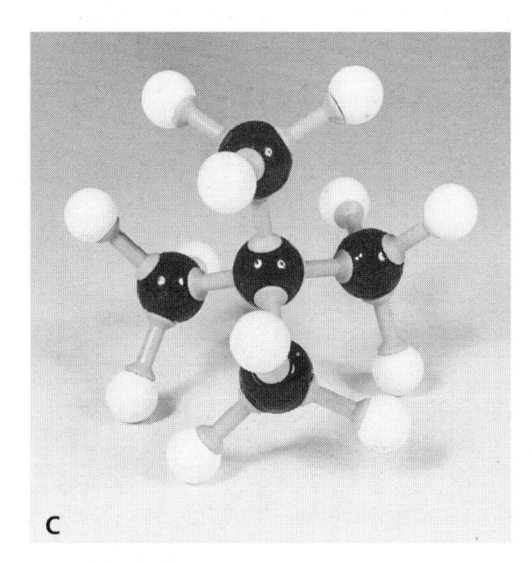

A methane, **B** pentane,
C 2,2-dimethylpropane

7.1 The structure of alkanes

A molecular model of methane is shown in photograph A. It consists of four hydrogen atoms arranged tetrahedrally around one carbon atom. The electron clouds in the C—H bonds are mutually repulsive and this results in bond angles of 109.5°, *not* 90° as might be suggested by its displayed formula (A).

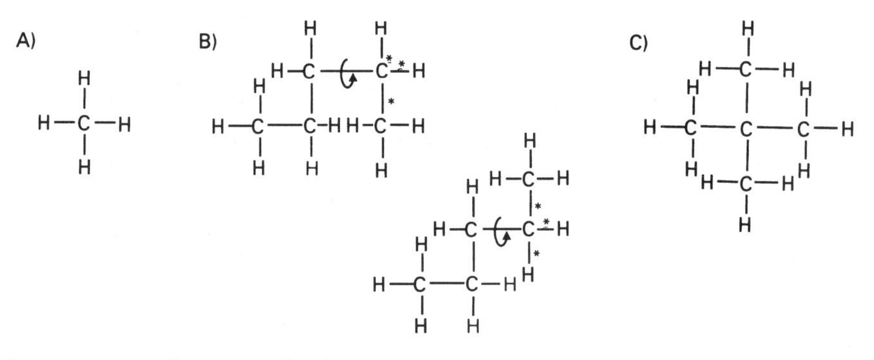

When naming alkanes, and other organic compounds, remember that a displayed structural formula does not show the molecule's 3-D shape and this must be taken into account when identifying structural isomers. For example, consider the molecular models of pentane, C_5H_{12}, a straight-chain alkane, shown in photograph B. Both models, and their displayed formulae (B), are equivalent and correct. They are not structural isomers (section 6.5) because the bonds marked * are equivalent due to the continuous rotation around single bonds. However, photograph C shows the molecular model of 2,2-dimethylpropane. This is one of two **branched-chain** alkanes which are structural isomers of pentane. Can you write down the structural formula of the other isomer?

7.2 Naming alkanes

To name an alkane, we follow a set of five rules.

1 Identify and name the longest straight chain. The name consists of two parts:

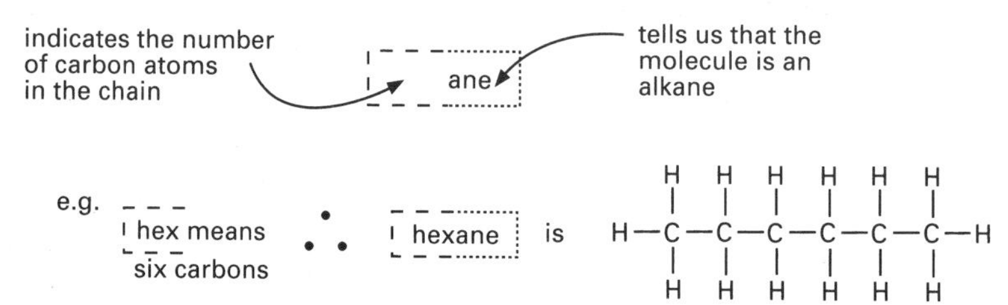

Table 7.1 lists the names of the first eight straight-chain alkanes.

2 For branched-chain alkanes, name each branch and state its point of attachment to the longest straight chain and then add this to the name of the longest straight-chain alkane. These 'branches' are known as **alkyl groups** (Table 7.2). Some examples:

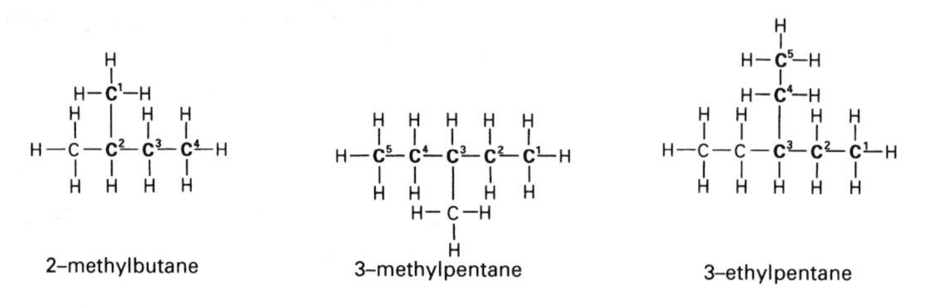

Where the point of attachment could be described by two numbers, we choose the smaller, for example:

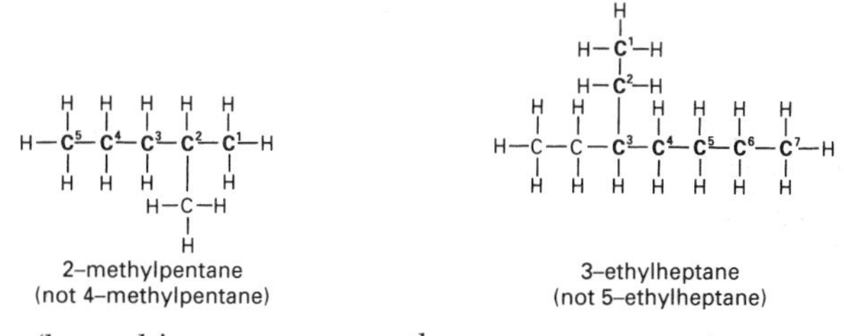

Table 7.1 Names and molecular formulae of the first eight straight-chain alkanes

Name	molecular formula
methane	CH_4
ethane	C_2H_6
propane	C_3H_8
butane	C_4H_{10}
pentane	C_5H_{12}
hexane	C_6H_{14}
heptane	C_7H_{16}
octane	C_8H_{18}

Can you work out a link between the number of carbon and hydrogen atoms in each molecule?

Table 7.2 Names of some alkyl groups used, for example, to name branched alkanes

Name of group	molecular formula
methyl	$—CH_3$
ethyl	$—C_2H_5$
propyl	$—C_3H_7$
butyl	$—C_4H_9$

3 If the same 'branch' appears more than once, we must
i) prefix the alkyl group name with di-, tri-, tetra-, penta-, etc. and
ii) give the positions of each alkyl group on the longest straight chain, keeping the numbers as small as possible.

Some examples are:

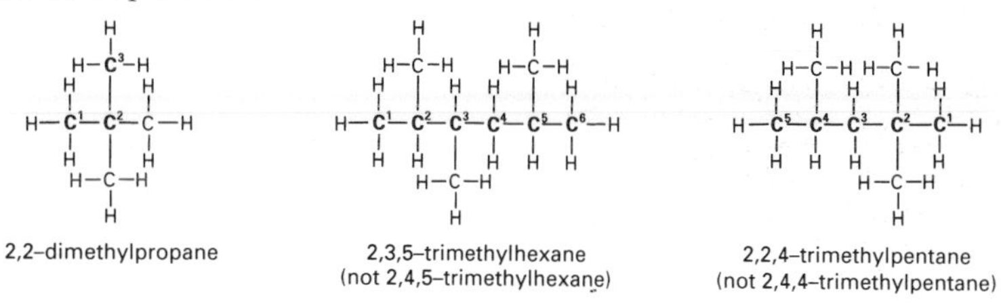

4 If a branched-chain alkane contains more than one type of alkyl group, we give them in alphabetical order, for example:

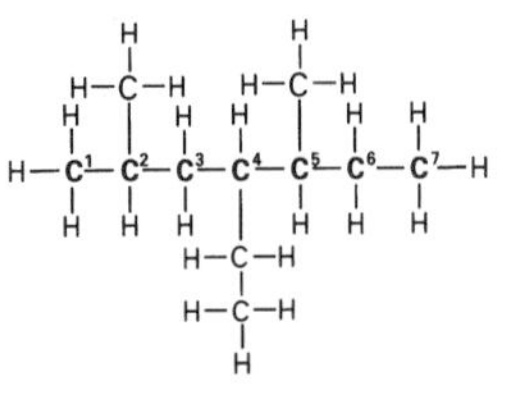

4-ethyl-2,5-dimethylheptane

5 Finally, when naming cycloalkanes we use the prefix cyclo- and the rules above, for example:

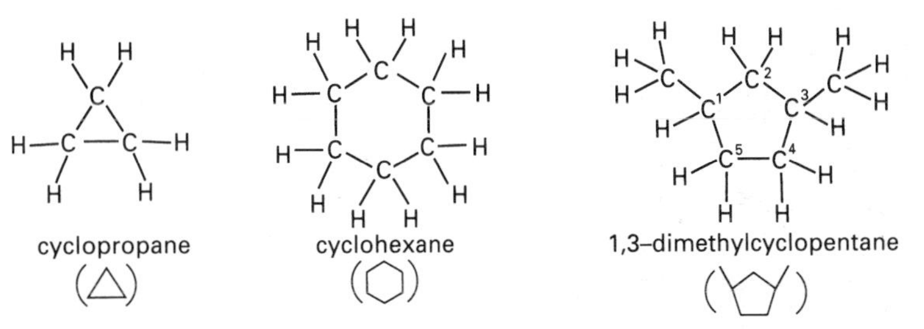

In the skeletal formulae of these molecules (shown in brackets), the presence of the 'ring' carbon and hydrogen atoms is assumed.

Focus 7a

1 Alkanes are **saturated** hydrocarbons, that is they contain only single bonds and have the highest ratio of hydrogen to carbon atoms. Each carbon is tetrahedrally surrounded by four other atoms, thereby giving bond angles of 109.5°.

2 Alkanes are named by using the set of five rules given in section 7.2. Structural isomers are a common feature.

Activity 7.1

1 From the displayed structural formulae shown below, select those which are a) equivalent and b) structural isomers (see section 6.5).

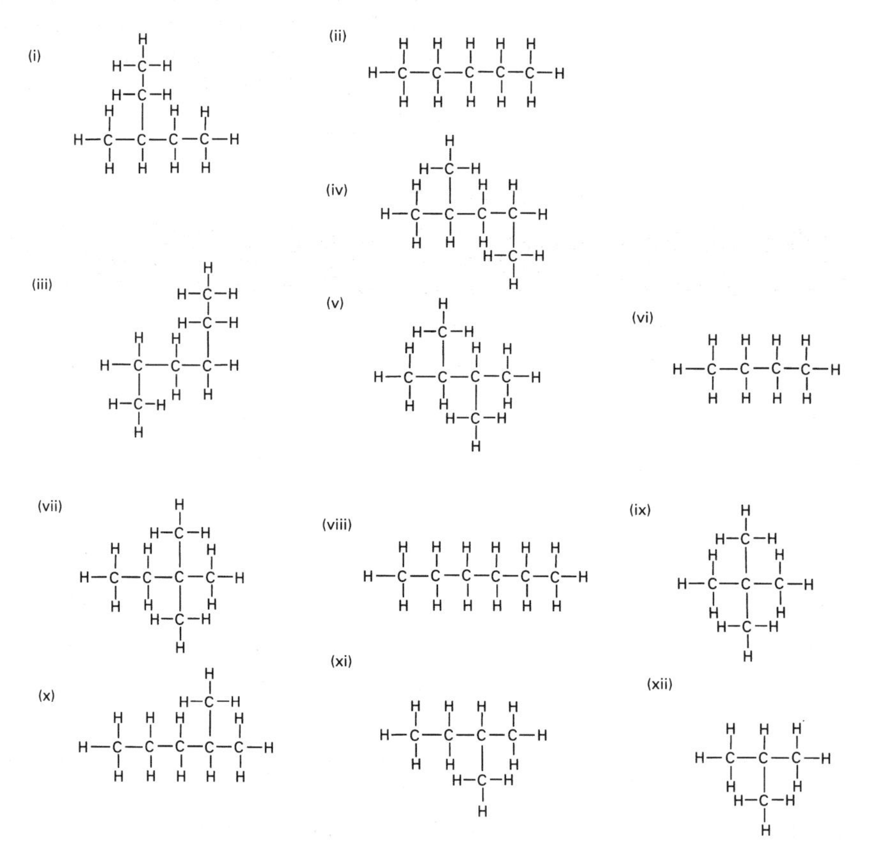

2 Now name all the alkanes drawn in question 1.

3 Name the following branched-chain alkanes. (Hint: first draw the displayed structural formula; bracketed groups are attached to the previous carbon atom.)

a) $CH_3C(CH_3)_2CH_2CH_3$
b) $CH_3CH(CH_3)CH_2CH_3$
c) $CH_3CH_2C(C_2H_5)_2CH_2CH_2CH_3$
d) $CH_3CH(CH_3)CH_2CH(CH_3)CH_3$
e) $CH_3CH_2CH(CH_3)CH(C_2H_5)CH_2C(CH_3)_2CH_3$

4 Draw displayed structural formulae for the following alkanes:

a) 2-methylpropane
b) 3-methylheptane
c) 2,3-dimethylbutane
d) 2,2,6-trimethyloctane
e) 4-ethyl-2,3-dimethylheptane
f) 1,3-diethylcyclobutane

5 There are nine structural isomers which have the molecular formula C_7H_{16}. Draw their displayed structural formulae and name them.

7.3 Physical properties of the alkanes

Some properties of straight-chain alkanes are given in Table 7.3. Notice that the molecular formulae of successive members of the alkane family *differ by a —CH_2— unit*. Such compounds are said to form a **homologous series**. Moreover, they have a general formula, namely C_nH_{2n+2} (and C_nH_{2n} for cycloalkanes). So, when $n = 4$, the alkane will have the formula $C_4H_{(2\times 4)+2} = C_4H_{10}$, i.e. butane.

Table 7.3 Physical properties of some straight-chain alkanes

Name	formula	state (at 298 K)	melting point/°C	boiling point/°C	density (as liquids)/g cm^{-3}	$\Delta H_c^\ominus$ /kJ mol^{-1}
methane	CH_4	gas	−183	−162	0.424	−890
ethane	C_2H_6	gas	−172	−89	0.546	−1560
propane	C_3H_8	gas	−188	−42	0.582	−2220
butane	C_4H_{10}	gas	−135	−0	0.579	−2877
pentane	C_5H_{12}	liquid	−130	−36	0.626	−3509
hexane	C_6H_{14}	liquid	−95	69	0.659	−4195
heptane	C_7H_{16}	liquid	−91	98	0.684	−4853
octane	C_8H_{18}	liquid	−57	126	0.703	−5512
nonane	C_9H_{20}	liquid	−54	151	0.718	−6124
decane	$C_{10}H_{22}$	liquid	−30	174	0.730	−6778

$\Delta H_c^\ominus$ is the standard enthalpy of combustion (section 3.4).

Is there any pattern in the boiling points, densities and standard enthalpies of combustion of these alkanes?

As you work through the text, you will see that the members of a homologous series have:

- similar chemical properties,
- gradually changing physical properties.

The alkanes provides an excellent example of this behaviour.

Alkanes are non-polar substances. As such, their molecules are attracted to each other by **van der Waals' forces** (section 4.6). These intermolecular forces result from the mutual attraction between molecules which contain induced dipoles. Molecules with larger electron clouds make a greater number of contacts with their neighbours, thereby giving rise to stronger van der Waals' forces. As the alkane gets larger, therefore, more energy is needed to separate neighbouring molecules and the boiling points of the straight-chain alkanes show a steady increase. Indeed plotting their boiling points against molecular mass gives a smooth curve (Figure 7.1).

Similar reasoning explains why branched-chain alkanes have lower boiling points and densities than their straight-chain isomers (Table 7.4). Branching prevents the close contact of neighbouring molecules, thereby producing a more open structure. Consequently, the strength of the van der Waals' forces is reduced, and so is the density.

Another feature of this homologous series is that the enthalpies of combustion of successive members increase fairly steadily by about 600–650 kJ mol^{-1} (see Figure 7.2). This is to be expected since we are burning an extra —CH_2— unit each time.

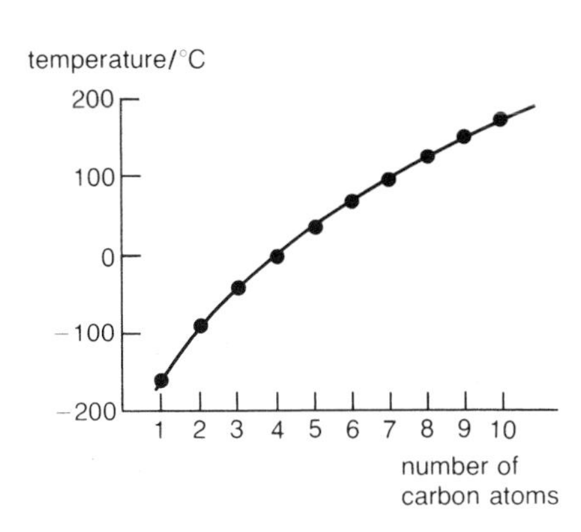

Figure 7.1
Graph of the boiling points of straight-chain alkanes plotted against the number of carbon atoms in the molecule

Table 7.4 Physical properties of some isomeric alkanes

Name	structure (H atoms omitted)		state (at 298 K)	boiling point/°C	density (as liquid)/gcm^{-3}
butane	C—C—C—C	isomers of C_4H_{10}	gas	0	0.579
2-methylpropane	C—C(—C)—C		gas	–12	0.557
pentane	C—C—C—C—C	isomers of C_5H_{12}	liquid	36	0.626
2-methylbutane	C—C—C(—C)—C		liquid	28	0.620
2,2-dimethylpropane	C—C(—C)(—C)—C		gas	10	0.591

What happens to i) the boiling point and ii) the density of the isomers as their molecules become more branched?

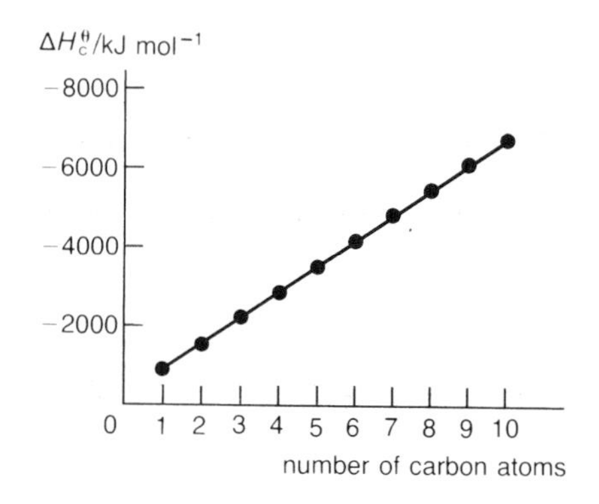

Figure 7.2 ▲
Graph of the standard enthalpies of combustion, $\Delta H_c^\ominus$, of the straight-chain alkanes plotted against the number of carbon atoms in the molecule. Can you predict the $\Delta H_c^\ominus$ value for the straight-chain alkane $C_{12}H_{26}$?

Focus 7b

Alkanes form a **homologous series** of general formula C_nH_{2n+2}. They have very similar chemical properties and their physical properties change gradually with increasing molecular mass. Alkanes have non-polar molecules, which are weakly attracted to each other via van der Waals' forces.

7.4 Refining petroleum

In its crude state, petroleum is a virtually useless material. However, when refined, the hydrocarbons it contains:

- supply almost half of the world's current energy needs (Figure 7.3),
- are the starting materials from which about 90% of the world's organic chemicals are made.

Refined petroleum, then, is highly valued as a source of fuel and a chemical feedstock.

In the UK, underwater pipelines or tankers carry the crude oil to the refinery. After the removal of water and any insoluble impurities, the petroleum is heated to about 400°C and then fractionally distilled. The fractionating column is about 60 metres high and contains layers of trays fitted with **bubble caps** (Figure 7.4). Five main fractions are taken off the column during the primary distillation.

At the bottom of the column, the temperature is high enough to vaporise all but the highest boiling point hydrocarbons, that is those with boiling points greater than about 350°C. These hydrocarbons, known as the **residue**, remain as liquids at the bottom of the column and can be run off. Petroleum also contains very volatile alkanes, having boiling points less than 20°C. These are removed as **refinery gases** at the top of the column. The three other fractions, termed **naphtha**, **kerosene** and **gas-oil**, are removed at intervals along the column.

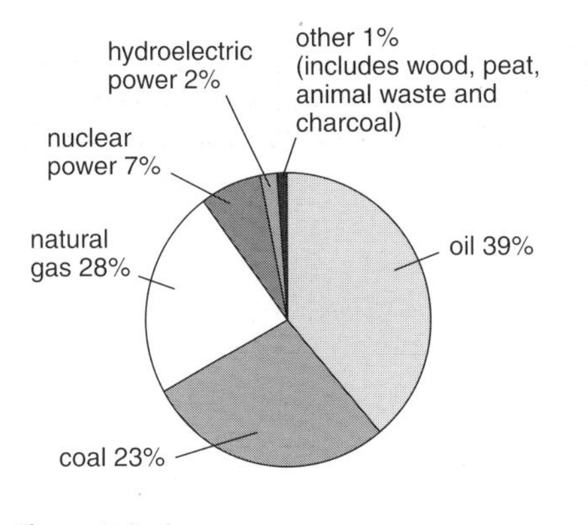

Figure 7.3 ▲
A 'pie chart' showing the use of the world's energy resources (1994). How might this pie chart differ from that for our use of energy in the year 2100?

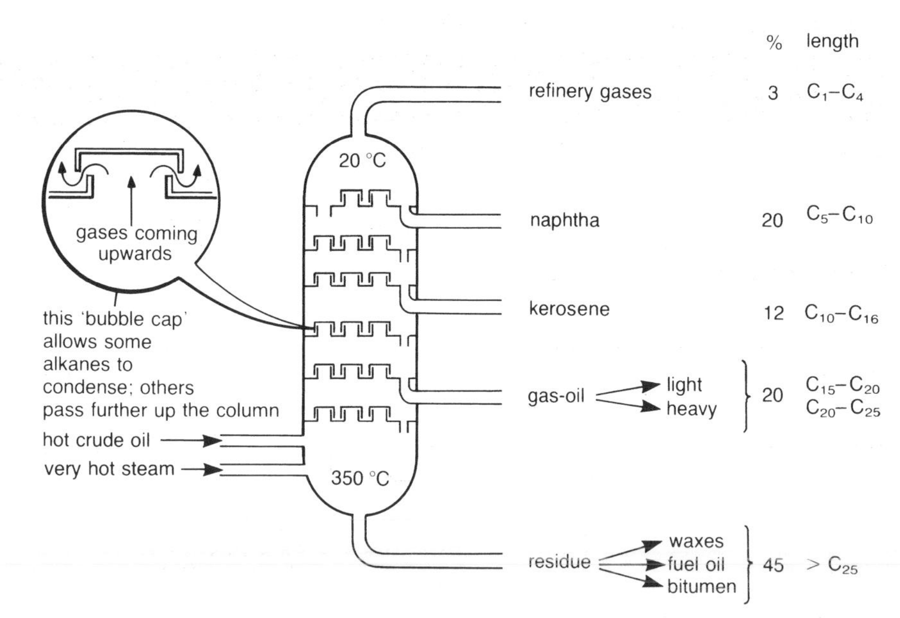

Figure 7.4 ▲
The primary fractionation of petroleum. A fractionating column contains many layers of 'bubble caps'. (% = approximate % by weight of that fraction in petroleum; length = number of carbon atoms in the molecules of the alkanes found in that fraction)

Before the fractions can be used, sulphur compounds (e.g. hydrogen sulphide) must be removed, and this is done by washing them with aqueous sodium hydroxide solution. Since massive amounts of petroleum and natural gas are purified in this way, we can extract appreciable amounts of **sulphur** and it is a valuable by-product.

The further treatment and uses of the petroleum fractions are described in the next two sections.

▲ A 1200 litre tank provides butane gas for a whole house – central heating, cooking and a living-flame fire – and is particularly useful where there is no mains gas supply

7.5 Combustion of alkanes and their use as fuels

Alkanes burn readily in excess air to form carbon dioxide and water, e.g.

$$C_4H_{10}(g) + \tfrac{13}{2}O_2(g) \longrightarrow 4CO_2(g) + 5H_2O(l) \; \Delta H = -2877 \text{ kJ mol}^{-1}$$

Although the combustion of an alkane is a highly exothermic process, no reaction will occur unless a certain minimum amount of energy, termed the **activation energy**, is supplied (Figure 7.5). Thus, with respect to the reaction products, an alkane/air mixture is energetically unstable, but kinetically stable (section 3.9). As a result, given sensible precautions, alkanes can safely be used as fuels.

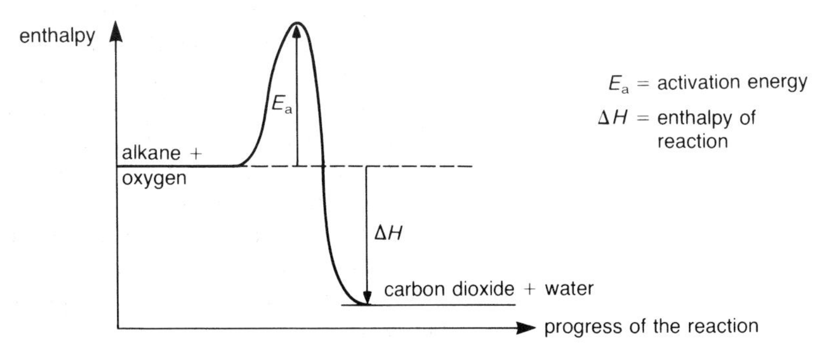

Figure 7.5 ►
Enthalpy profile for the combustion of an alkane. Compared to the combustion products, the alkane and oxygen are energetically unstable (i.e. ΔH is large and exothermic), but kinetically stable (i.e. E_a is large)

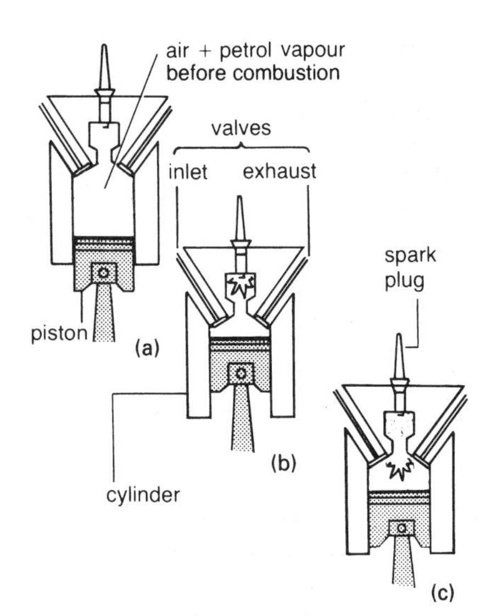

Figure 7.6 ▲
A piston and cylinder from an internal combustion engine. **(a)** The piston compresses the mixture of petrol vapour and air. **(b)** Maximum compression of the petrol/air mixture coincides with ignition by an electrical spark. **(c)** Spontaneous combustion occurs before maximum compression is obtained; this is termed 'knocking'

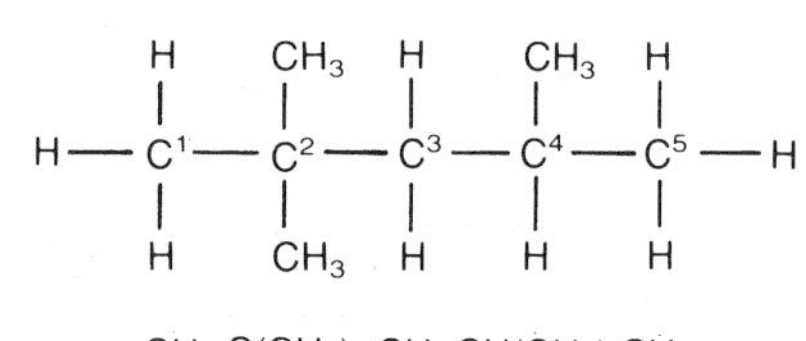

Figure 7.7 ▲
2,2,4-Trimethylpentane, an alkane with a very high resistance to knocking

The **refinery gases** are valuable fuels. In some cases, the mixture is used directly at the refinery, for example to heat the incoming crude petroleum. Alternatively, the refinery gases can be liquefied and then fractionally distilled to produce the individual gases (methane, ethane, propane and butane). Since propane and butane are easily condensed under pressure, they are stored, and sold, as a liquid fuel, such as 'Calor' gas.

Further fractionation of **naphtha** yields **petrol** (C_5–C_8 alkanes). In the internal combustion engine, a mixture of air and petrol vapour is compressed by a piston (Figure 7.6). At the point of maximum compression, the petrol/air mixture is ignited by an electrical spark and rapid combustion occurs (Figure 7.6b). A typical reaction would be the combustion of octane:

$$C_8H_{18}(g) + \frac{25}{2}O_2(g) \longrightarrow 8CO_2(g) + 9H_2O(g) \quad \Delta H = -5470 \text{ kJ mol}^{-1}$$

The hot gaseous products expand against the piston and force it downwards. This mechanical energy is transmitted to the drive wheels of the car, thus enabling it to move.

Petrol also contains various additives such as lubricants, rust inhibitors and **anti-knock agents**. Some hydrocarbons have a tendency to ignite spontaneously before maximum compression is achieved (Figure 7.6c). This premature explosion, known as **knocking**, still forces the piston downwards and powers the vehicle. However, the chemical energy in the petrol is less efficiently converted into mechanical energy. As a result, the vehicle will do fewer miles per gallon. Moreover, knocking causes a rough ride and excessive engine wear because the explosion is occurring as the piston is still moving up in its cylinder.

Nowadays, the anti-knock properties of petrol must conform to an international standard known as the **octane scale**. Branched-chain alkanes have been found to be much more resistant to knocking than their straight-chain isomers. In particular, **2,2,4-trimethyl pentane** (Figure 7.7) has been identified as having an exceptionally high resistance to knocking. Thus we say that pure 2,2,4-trimethylpentane has an octane number of 100. On the other hand, pure heptane is given an octane number of 0 because it is the straight-chain alkane with the lowest resistance to knocking. By comparing the anti-knock performance of 2,2,4-trimethylpentane/heptane mixtures with that of petrol, we can give the latter an octane number (Table 7.5).

Table 7.5 Octane numbers of some petrol samples. Which sample would you use in your new sports car?

Sample ...	behaves like mixture of .. % 2,2,4-tmp	+ .. % heptane	its octane number is
A	55	45	55
B	95	5	95

(2,2,4-tmp = 2,2,4-trimethylpentane)

There are two important ways of increasing the petrol's octane number:

1 By adding about 2% by mass of **tetraethyl lead**, $(C_2H_5)_4Pb$, an anti-knock agent. This breaks up during the combustion to form tiny lead(IV) oxide particles which ensure even combustion of the fuel at a moderate rate. To prevent a build-up of lead (IV) oxide in the cylinder, a small amount of 1,2-dibromoethane ($BrCH_2CH_2Br$) is added to the petrol. This reacts with the residual lead oxide to produce lead bromide, a fairly volatile compound. Because it is volatile, lead bromide passes with the other exhaust gases into the atmosphere.

   Unfortunately, lead compounds are known to be very toxic and the dangers of the atmospheric lead pollution are well known. Inhalation of lead-polluted air causes a build-up of lead in the blood, and this is thought

Kerosene, a mixture of C_{10}–C_{16} alkanes, is used as aviation fuel.

All petrol-driven cars being manufactured in the UK today must be able to run on unleaded petrol

to impair the brain's activity. In acute cases, lead poisoning can cause permanent brain damage.

Such is the concern over the effects of lead pollution that many countries are phasing out the use of tetraethyl lead. In Britain, for example, all new petrol-driven vehicles have to run on unleaded petrol. Indeed from January 2000, leaded petrol will be banned from general sale. Alternatives to lead will be available, for example, lead replacement fuel (LRF) union contains a high proportion of branched alkanes, branched chain ethers and cyclic hydrocarbons – these substances have a high resistance to knocking. Also, Anti-Wear Additives (AWAs) will be available for mixing with petrol when filling up the car. Whilst these measures protect the engine under normal driving conditions, if the car is put to hard use (e.g. regular motorway use or towing a trailer) the engine may need to be modified to run on unleaded petrol.

2 By using an unleaded petrol which contains an increased proportion of branched alkanes, cycloalkanes and aromatic hydrocarbons in the petrol. These compounds have excellent anti-knock properties but they only occur in relatively small amounts in crude oil. Thus, the refinery makes them from the other crude oil fractions using **cracking** and **reforming** processes (section 7.6).

The other petroleum fractions are also used as fuels. **Kerosene** is used as fuel in jet engines and, domestically, as paraffin for lamps and heaters. **Light gas-oil** is used in oil-fired central heating systems. **Heavy gas-oil** finds widespread use as **diesel** fuel. In the diesel engine, compression of the air/fuel vapour mixture causes it to ignite spontaneously. Thus, no sparking plugs are needed and the common problems associated with an electrical ignition are avoided. Unfortunately, unlike petrol, diesel may freeze in very cold weather! Most of the thick oily **residue** from the primary fractionation can be used as fuel oil to power ships or electrical generators.

Focus 7c

1. Alkanes occur naturally in **natural gas** and **petroleum**. These were formed by the bacterial decay of marine plants and animals which lived millions of years ago. Alkanes are used as fuels and as a feedstock for the chemical industry.
2. Alkanes are good fuels because they readily burn in excess air to form mainly carbon dioxide and water. A small amount of incomplete combustion, giving C(s) and CO(g), may occur, especially where the oxygen supply is restricted.

Motor vehicle fuels and the environment

Whenever alkanes are burnt in a limited supply of oxygen, some **incomplete combustion** occurs. In a petrol engine, for example, incomplete combustion causes a deposit of carbon to build up on the valves and pistons, and this eventually decreases the engine's efficiency. Another product of incomplete combustion is carbon monoxide, a very poisonous gas which bonds irreversibly with the haemoglobin in blood, thereby blocking the uptake of oxygen and preventing its transport around the body (page 390). In an enclosed space, exposure to carbon monoxide can be fatal. Car exhaust fumes may contain up to 2% by volume of carbon monoxide, together with other pollutants, most notably the oxides of nitrogen and sulphur, unburnt hydrocarbons and lead compounds (from anti-knock agents).

Even after purification, petrol contains small amounts of sulphur compounds and these burn to the sulphur oxides (SO_2 and SO_3, collectively termed SO_x). During combustion, the temperature is high enough for the nitrogen and oxygen in air to react, forming nitrogen oxides, NO_x:

$$N_2(g) + O_2(g) \longrightarrow 2NO(g)$$
$$N_2(g) + 2O_2(g) \longrightarrow 2NO_2(g)$$

Motor vehicle fuels and the environment *continued*

Sulphur and nitrogen oxides are formed whenever large quantities of fossil fuels are burnt, for example in power stations, and they are involved in two major environmental problems: acid rain and smog.

In 1872, Robert Angus Smith used the term **acid rain** to describe the acidic precipitation that fell on Manchester at the start of the Industrial Revolution. At that time, a greater reliance on machinery increased the demand for energy, so more coal and oil were burnt. This caused a corresponding increase in atmospheric pollution from SO_x, NO_x and hydrocarbons. In the atmosphere, these substances take part in a complex series of reactions, forming NO_2 and SO_3 which then react with moisture to form acids:

$$2NO_2(g) + H_2O(l) \longrightarrow \underset{\text{nitrous acid}}{HNO_2(aq)} + \underset{\text{nitric acid}}{HNO_3(aq)}$$

$$SO_3(g) + H_2O(l) \longrightarrow \underset{\text{sulphuric acid}}{H_2SO_4(aq)}$$

The effect of 'acid rain' on the stonework figures at York Minster and on trees in a Czechoslovakian forest

Often, acidic clouds are carried by air currents hundreds of miles away from the source of the pollution before coming to ground in mist, rain or snow. In certain areas, the acidity of this precipitation has steadily increased with rain having a pH as low as 4 falling over Scandinavia, Germany and Canada. Acid rain has taken its toll of natural areas destroying forests and making lakes and rivers unable to support aquatic life. In cities, acid rain has damaged stone and metal structures, in many cases disfiguring statues and carvings. Fortunately, over the last 15 years many industrialised nations have introduced legislation to reduce SO_x and NO_x emissions to acceptable levels.

Vehicle exhaust emissions are also responsible for **photochemical smog**. This is a mixture of dust particles, nitrogen oxides, ozone (O_3), hydrocarbons and a variety of unpleasant organic compounds which results from a series of reactions that occur in sunlight. Certain cities, such as Los Angeles in California, are prone to photochemical smog because of their geographical location (Figure 7.8). In many other large cities, the inhabitants experience some degree of poor air quality due to the atmospheric pollution from vehicle exhausts, fuel evaporation and industrial plants. This pollution is thought to have increased the incidence of respiratory diseases, such as asthma, and allergies, such as hay fever. In January 1993, the EEC placed strict limits on levels of pollutants from car exhausts and testing this is now part of the Ministry of Transport (MOT) Test.

One way of controlling exhaust emission is to fit a **three-way catalytic converter** into the exhaust system, as shown in Figure 7.9. Inside it is a metal or ceramic honeycomb which is coated with the finely divided catalyst mesh of the precious metals platinum, palladium and rhodium. In this structure, the typical surface area of the catalyst that is available to the

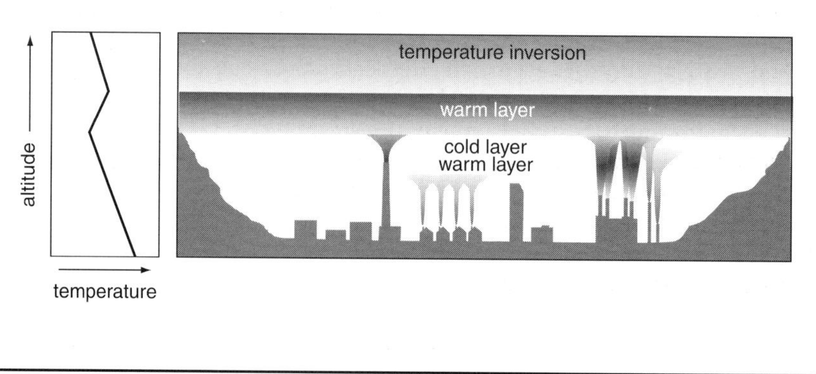

Figure 7.8 ▶
Whenever a photochemical smog occurs, certain geographical and weather conditions are present. Since photochemical reactions are involved, the sun needs to be shining. The day needs to be windless so that a mass of air is trapped above the source of pollution, that is the vehicle fumes from a city. Also, the air is trapped more effectively where the city is surrounded by mountains. The inability of the ground air to move results in a **thermal inversion** in which warm air acts like a 'lid' keeping the cooler, more dense air nearer ground level. As the photochemical reactions proceed, there is an increase in the concentration of pollutants at ground level and thus the city dwellers start to experience respiratory difficulties, sore throats and watery eyes

Motor vehicle fuels and the environment *continued*

exhaust gases is about the size of two football pitches and it manages to convert 95% of the pollutants into carbon dioxide, nitrogen, and water vapour. The catalytic converter, which fits into a long-life stainless steel exhaust, costs several hundred pounds to produce because the precious metals it contains are very expensive (fortunately, they can be recycled from old converters). Since the catalyst is poisoned by lead, cars fitted with converters must be run on lead-free petrol.

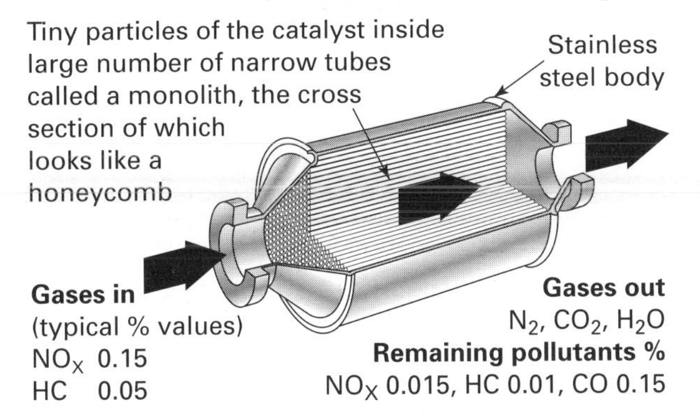

Figure 7.9 ▲
A three-way catalytic converter converts CO into CO_2, NO_x into N_2 and unburnt hydrocarbons (HC) into CO_2 and H_2O. Although the 'honeycomb' is only the size of a soft drink can, the total surface area of the catalyst is equivalent to that of two football pitches

Controls on exhaust emission in the USA are extremely strict. In California, for example, Ultra-Low-Emission Vehicle (ULEV) requirements were introduced in 1998 (Table 7.6) as a prelude to a Zero Emissions standard early in the next century. The European Union usually follows US legislation and tighter standards for petrol and diesel engines were introduced in 1998 with further legislation in 2000. To meet such exacting standards, motor vehicle manufacturers are focusing on combustion technology and the introduction of cleaner fuels. Most of the emissions from catalytic converters occur during the first two minutes of the journey so it is important that the catalyst quickly reaches its optimum operating temperature. This could be aided by placing the converter closer to the engine's exhaust outlet or by using a heater to prewarm the catalyst. Another way forward may be to supply hot air to the exhaust valves so that excess hydrocarbons are burnt as they leave the cylinder and the converter reaches operating temperature more rapidly. Also, many modern engines are based on 'lean-burn' technology, that is the fuel is burnt in a large excess of air, thereby minimising the emission of unburnt hydrocarbons and carbon monoxide.

Table 7.6 Emissions from Volvo's 850 Bi-Fuel car running on compressed natural gas (CNG) compared to statutory limits, in g/km travelled

	CO	NO_x	HC
850 Bi-Fuel	0.30	0.07	0.020
ULEV	1.06	0.125	0.025
European 1996	2.2	total 0.50	

HC represents the amount of unburnt hydrocarbons. ULEV is the Ultra-Low-Emission requirement which came into force in California in 1998.

Volvo, the Swedish vehicle manufacturer, has a far-reaching environmental policy and has been fitting catalytic converters since 1974! In 1992, Volvo produced its Environmental Concept Car (ECC), a hybrid car having a battery-powered engine for city use and a diesel-driven gas turbine for motorways. Gas turbines are adaptable to most types of fuel, liquid and gas, and produce very low emissions of pollutants. The ECC is a environmentally friendly car, produced without the use of chlorofluorocarbons (CFCs) and made with recyclable materials wherever possible. Although the ECC was not produced commercially, Volvo have since introduced the 850 Bi-Fuel, a car which runs on petrol *or* methane (in **natural gas**). The driver can switch from one fuel to the other at a flick of a switch, using petrol on open roads and natural gas in the city. When the 850 Bi-Fuel is run on natural gas, it exceeds California's 1998 ULEV requirements. Whilst natural gas is more difficult to distribute, store and handle than petrol, it is a much cleaner fuel with a very low emission of hazardous substances (Figure 7.10). The reserves of methane, as natural gas, equal those of crude oil and significant amounts of methane are also found in the biogas emitted from landfills and sewage works. There are about 1 million natural gas powered vehicles in the world today and such vehicles are expected to play an increasingly important role in the future.

Of course, the cleanest fuel we could burn would be hydrogen. Although a tiny amount of nitrogen oxides are produced, the main combustion product, water, is a harmless substance. However, hydrogen is difficult to store and its use would require engine modifications, so it is not yet an economically viable option.

Motor vehicle fuels and the environment *continued*

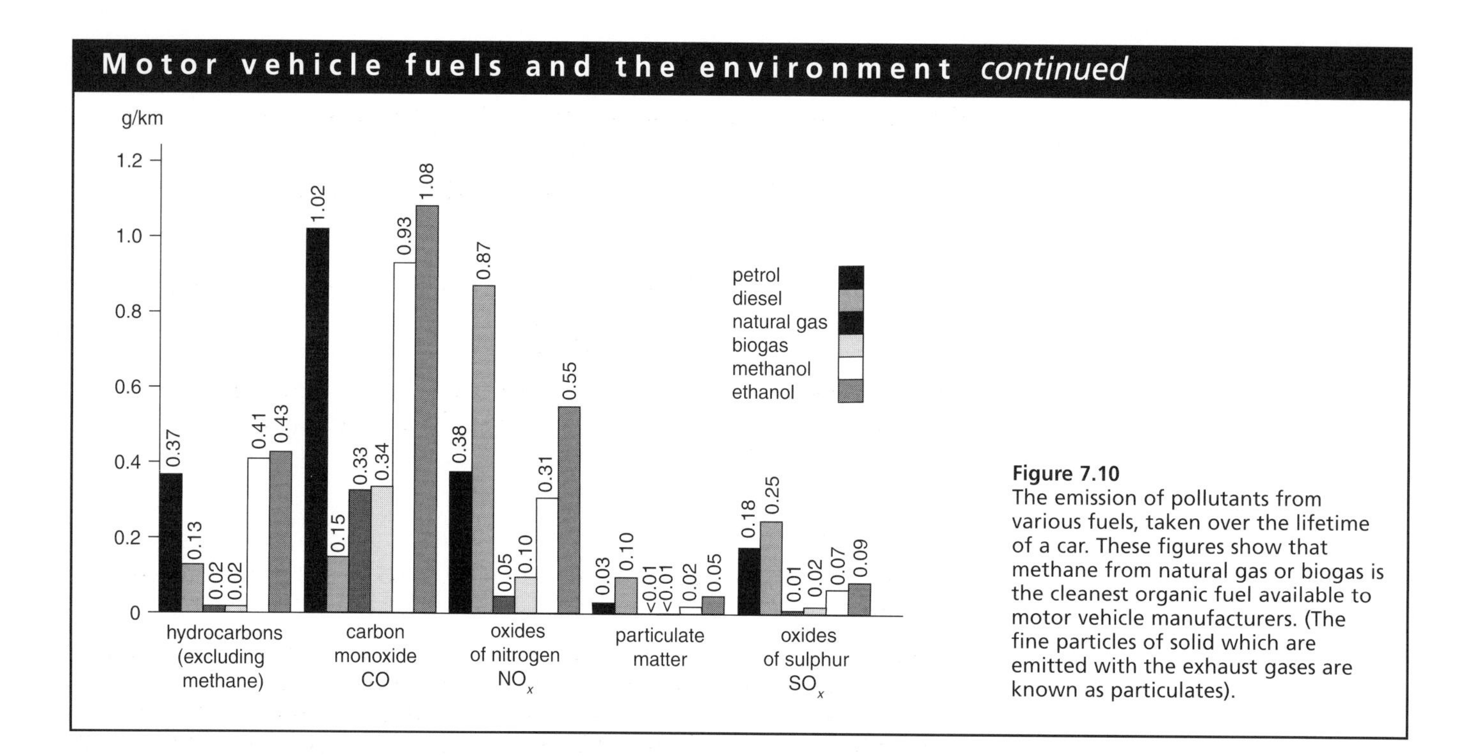

Figure 7.10
The emission of pollutants from various fuels, taken over the lifetime of a car. These figures show that methane from natural gas or biogas is the cleanest organic fuel available to motor vehicle manufacturers. (The fine particles of solid which are emitted with the exhaust gases are known as particulates).

An oil refinery in the Netherlands, where a great number of heavy chemical plants have been sited to take advantage of North Sea oil and gas and the excellent port facilities

7.6 Cracking and reforming the petroleum fractions

Although 90% of petroleum is used as fuel, fractionating petroleum is only part of the refinery's work. Many other process plants will be based on site (see photograph, left). Two important types of process are **cracking** and **reforming**.

Cracking

Only about a fifth of crude oil is gasoline and this does not produce enough petrol to satisfy world demands. However, there is a surplus of kerosene and light gas oil. Thus, chemists have developed a process called cracking, which converts large alkane molecules into smaller molecules (alkanes, alkenes or hydrogen).

Since strong C—C and C—H bonds must be broken, cracking is carried out at high temperature, often using a catalyst. For example, kerosene can be cracked at 500°C in the presence of powdered silicon and aluminium oxides, e.g.

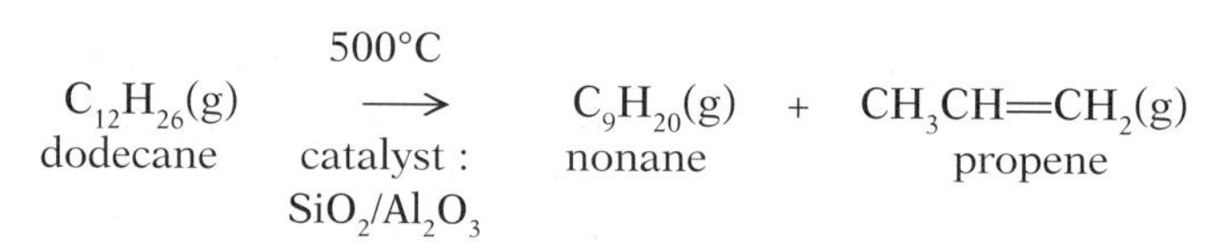

$$\underset{\text{dodecane}}{C_{12}H_{26}(g)} \xrightarrow[\text{catalyst : } SiO_2/Al_2O_3]{500°C} \underset{\text{nonane}}{C_9H_{20}(g)} + \underset{\text{propene}}{CH_3CH{=}CH_2(g)}$$

An added benefit of cracking is that the products are often rich in branched alkanes. As we have seen, these are valued for their high resistance to 'knocking'. Cracking nearly always gives a mixture of organic products. These are then passed to a fractionating column for separation.

Apart from supplementing petrol supplies, cracking also produces alkenes. Because they contain C=C bonds, alkenes are more reactive than alkenes. Hence, they are a convenient starting point from which to make other organic chemicals. In particular, the manufacture of plastics and organic solvents requires vast amounts of ethene and propene. These can be made by cracking the ethane and propane obtained from refinery and natural gas:

$$H_3C{-}CH_3(g) \xrightarrow{700°C} H_2C{=}CH_2(g) + H_2(g)$$

$$H_3C{-}CH_2{-}CH_3(g) \xrightarrow[\text{catalyst: } Cr_2O_3/Al_2O_3]{500°C} H_3C{-}CH{=}CH_2(g) + H_2(g)$$

Note here that (i) no alkanes are produced, and (ii) the hydrogen gas is a valuable by-product used, for example, to make ammonia in the Haber process (section 23.3).

Reforming

In these processes, alkane molecules are either restructured or combined to form larger molecules.

Isomerisation and alkylation When heated in the presence of an aluminium chloride catalyst, straight-chain alkanes can be converted into their branched-chain isomers. This process is called **isomerisation**. For example, butane from refinery or natural gas yields the rather more useful, branched isomer, 2-methylpropane:

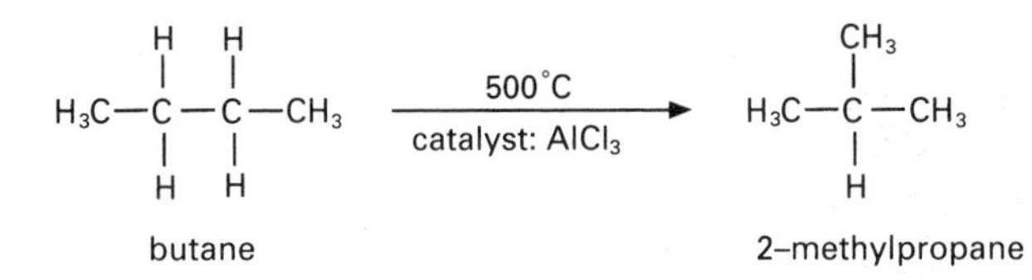

Branched-chain alkanes are valuable because they can be added to petrol, thereby increasing its octane number. Alternatively, they can be made to react with alkenes (from cracking) to form longer chain branched alkanes. This process, known as **alkylation**, occurs at low temperatures and is catalysed by concentrated sulphuric acid, e.g.

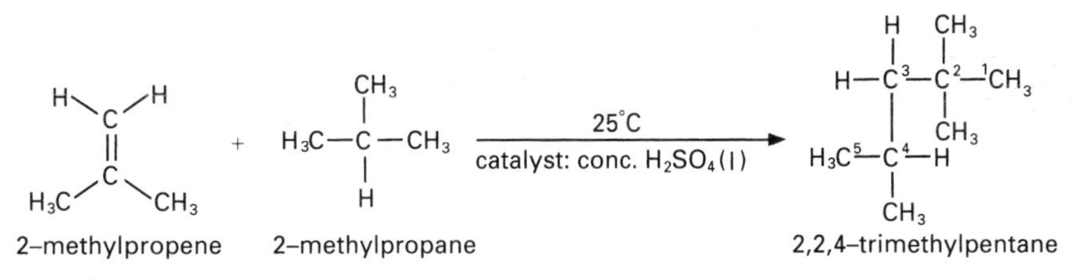

As we have seen, 2,2,4-trimethylpentane has excellent 'anti-knock' properties.

Cyclisation and aromatisation In these processes, cycloalkanes and aromatic hydrocarbons are made from the straight-chain C_6–C_{10} alkanes in naptha. The naphtha vapour is heated to 500°C and then passed, at high pressure, over a catalyst of platinum metal absorbed on aluminium oxide. A typical reaction would be:

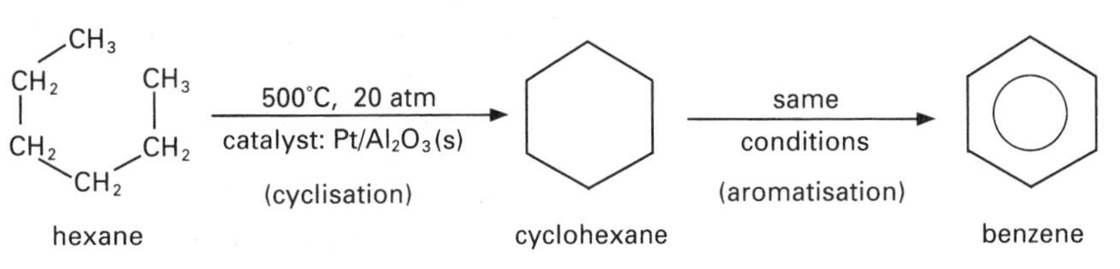

Cyclic hydrocarbons can be added to petrol to increase its octane number. Benzene is used as a solvent and as starting material in the manufacture of many important aromatic compounds.

7.7 Some other uses of the alkanes in petroleum

Large quantities of the **methane** in refinery and natural gases are treated with steam at 450°C, using a nickel catalyst:

$$CH_4(g) + H_2O(g) \longrightarrow CO(g) + 3H_2(g)$$

After separation, the hydrogen may be used in the manufacture of ammonia (refer to the Haber process, section 23.3).

Some of the **residue** is redistilled under reduced pressure. This yields a variety of **lubricating oils**, each containing dissolved **paraffin wax** which is removed by solvent extraction. After purification, paraffin wax is used to waterproof paper cartons and in the manufacture of candles and petroleum jelly (Vaseline).

Like naphtha, paraffin wax can be cracked by mixing it with steam at 500°C. The products are valuable alkenes containing 5 to 18 carbon atoms and a terminal double bond, $CH_3(CH_2)_{14}CH{=}CH_2$. After fractional distillation, the smaller alkenes are passed to the alkylation plant. The longer chain alkenes are used in the manufacture of **detergents** (page 349).

After vacuum distillation of the original residue, we are left with a thick tarry solid. Even this substance, called **bitumen**, is not discarded but is used to surface roads and in the manufacture of roofing felt and damp-proof coursing for buildings.

The key aspects of petroleum refining are summarised in Figure 7.11, in Focus 7d.

7.8 Free radical substitution reactions

Because of the overall non-polarity of an alkane molecule, the C—H bonds will undergo **homolytic fission**, with each bonded atom taking one of the bonding electrons, for example:

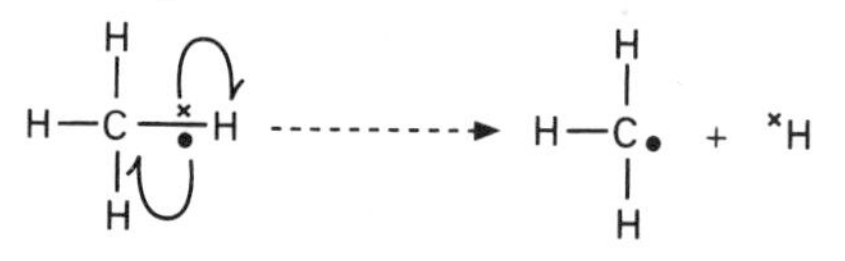

The products are **free radicals**, that is, highly reactive atoms, or groups of atoms, which have an unpaired electron. Each ↷ arrow represents the movement of the charge cloud due to one electron.

Alkanes take part in **free radical substitution** reactions with halogens, in which one or more halogen atoms substitute for hydrogen atoms, for example:

$$\underset{\text{ethane}}{C_2H_6(g)} + Cl_2(g) \xrightarrow{\text{dull sunlight}} \underset{\text{chloroethane}}{C_2H_5Cl(g)} + HCl(g)$$

Focus 7d

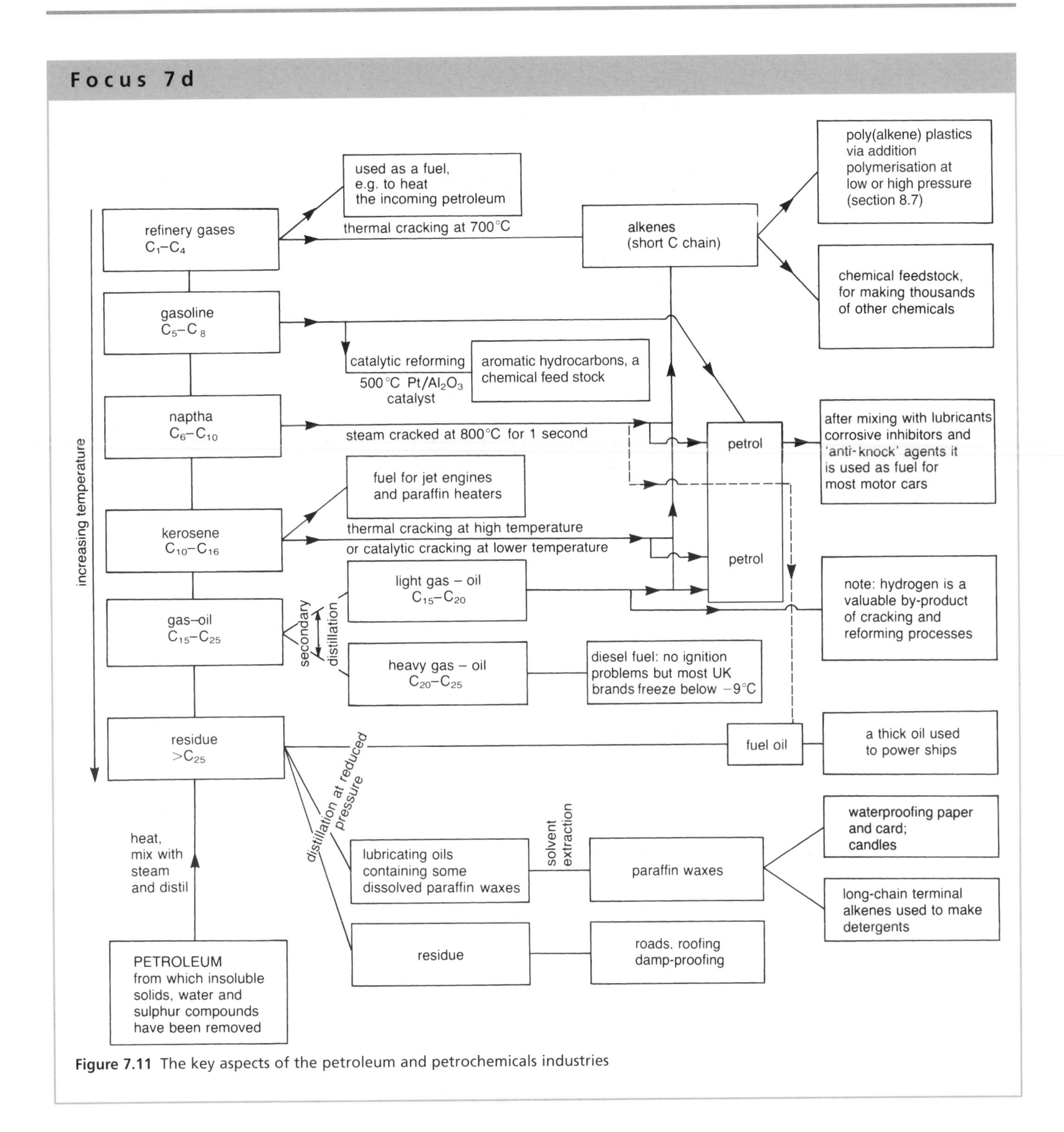

Figure 7.11 The key aspects of the petroleum and petrochemicals industries

Table 7.7 summarises the effects of mixing each halogen with an excess of methane.

Table 7.7 Observations on mixing a halogen with an excess of methane

Halogen used	conditions	
	30°C in the dark	30°C in ultraviolet (sun) light
fluorine	explosion	explosion
chlorine	no reaction	green colour of chlorine quickly disappears
bromine	no reaction	orange colour of bromine slowly disappears
iodine	no reaction	no reaction

The results show that:

- the reactivity of the halogens decreases down the group, $F_2 > Cl_2 > Br_2 > I_2$,
- where a reaction occurs, the reactant molecules obtain their activation energies by absorbing ultraviolet light. Thus, they are termed **photochemical** reactions.

Free radical substitution in alkanes take place via three distinct stages, known as **initiation**, **propagation** and **termination**. The chlorination of methane has been well researched and its mechanism is shown below.

Initiation

Chlorine molecules absorb the ultraviolet light. Some homolytic bond fission takes place, and a small amount of Cl free radicals are produced:

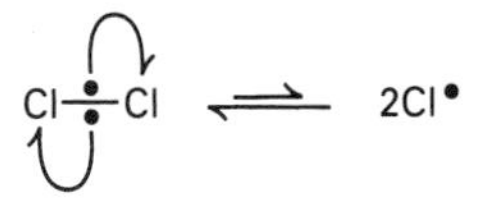

Propagation

The Cl• free radical is a highly reactive atom and this 'pulls' an H atom off the methane molecule. Hydrogen chloride and a methyl radical are formed.

The methyl radical may then react with an undissociated chlorine molecule, thereby forming chloromethane and providing a Cl• to take part in another reaction (a). This leads to a rapid repetition of the propagating reactions ((a)... (b)... (a)... (b)... and so on); this is termed a **free radical chain reaction**.

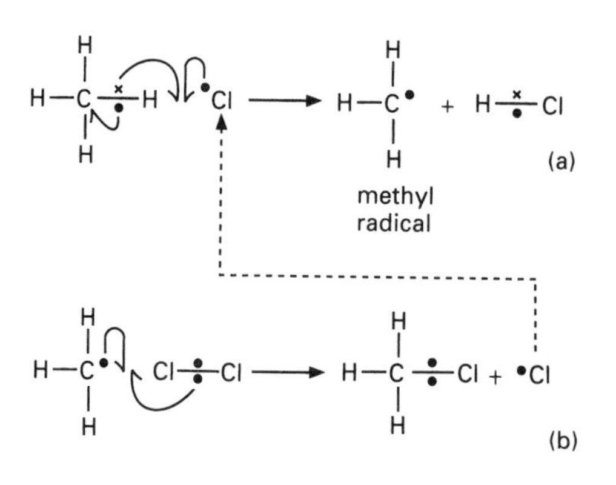

Termination

When one of the reactants has been used up, the chain reaction is terminated by the combination of free radicals:

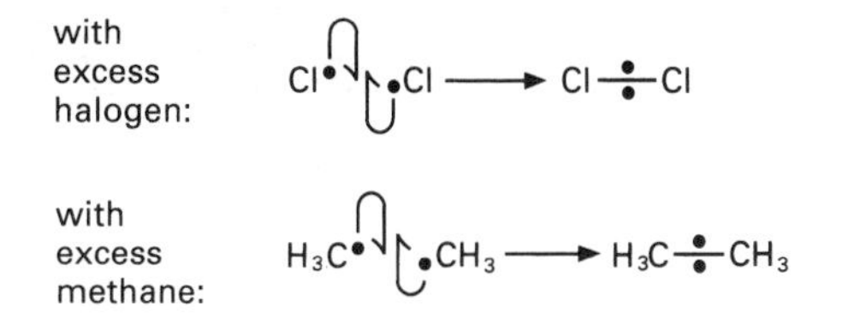

Further free radical substitution may occur, particularly if an excess of chlorine is used, leading to a mixture of products:

$$CH_3Cl(g) + Cl_2(g) \longrightarrow CH_2Cl_2(g) + HCl(g)$$
$$CH_2Cl_2(g) + Cl_2(g) \longrightarrow CHCl_3(g) + HCl(g)$$
$$CHCl_3(g) + Cl_2(g) \longrightarrow CCl_4(g) + HCl(g)$$

The halogenation of larger alkanes proceeds through an identical mechanism, for example:

$$\underset{\text{propane (in excess)}}{CH_3CH_2CH_3(g)} + Br_2(g) \xrightarrow[\text{u.v. light}]{} \underset{\text{1-bromopropane}}{CH_3CH_2CH_2Br(g)} + \underset{\text{2-bromopropane}}{CH_3CHBrCH_3(g)}$$

Focus 7e

1. Alkanes take part in three main types of reaction: (i) **free radical substitution** of the hydrogen atoms, (ii) **combustion** and (iii) **cracking**. These are summarised in Figure 7.12.
2. Substitution in alkanes occurs via a free radical chain reaction. The mechanism involves **initiation**, **propagation** and **termination**. Remember there are at least two propagating steps.
3. The conversion of a longer chain alkane into a shorter chain alkane and/or an alkene is known as **cracking**. This is carried out at high temperature, often in the presence of a catalyst. It is a **free radical** reaction.

Activity 7.2

1 Give the name and structural formula of the main organic product formed in each of the following reactions.

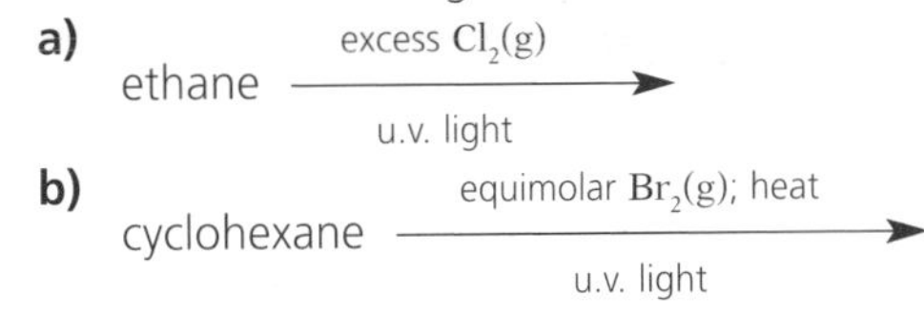

a) ethane $\xrightarrow[\text{u.v. light}]{\text{excess } Cl_2(g)}$

b) cyclohexane $\xrightarrow[\text{u.v. light}]{\text{equimolar } Br_2(g)\text{; heat}}$

2 **a)** Write balanced equations for the complete combustion of the following alkanes: (i) ethane, (ii) 3-methylpentane and (iii) cyclobutane.

b) What other products would be formed if these alkanes were burnt in a poor supply of oxygen?

3 Write down the mechanism for the following reaction:

$$CHBr_3(g) + Br_2(g) \longrightarrow CBr_4(g) + HBr(g)$$

Focus 7f

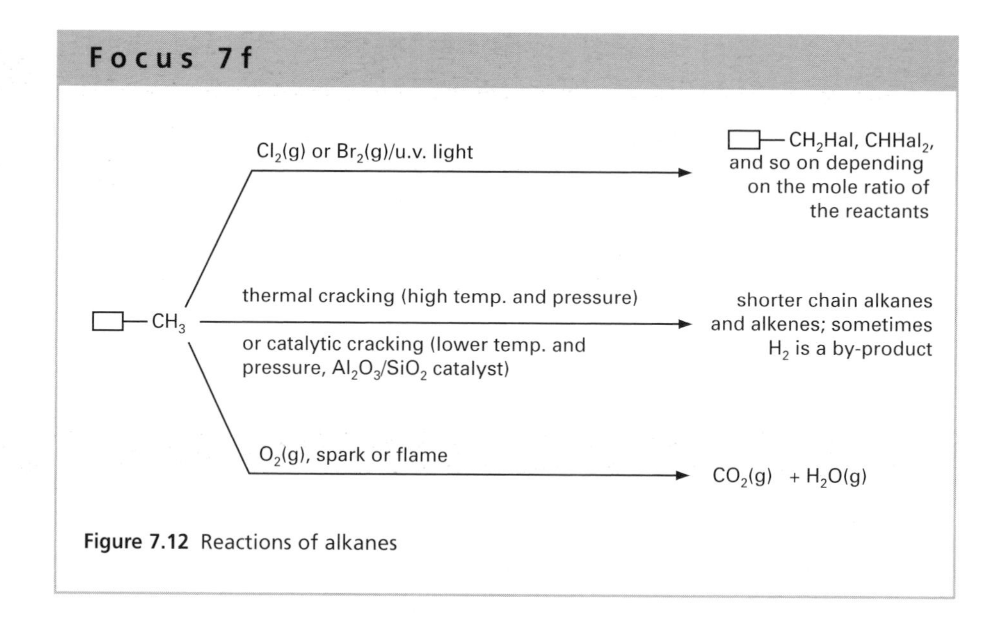

Figure 7.12 Reactions of alkanes

Questions on Chapter 7

1 This question concerns the following alkanes:

A $CH_3CH_2CH(CH_3)CH_3$ **B** $(CH_3)_3CH$
C $CH_3CH_2CH_2CH_2CH_2CH_3$
D $CH_3CH_2CH(CH_3)CH_2CH_3$
E $CH_3CH(CH_3)CH(CH_3)CH_3$
F $CH_3CH(C_2H_5)CH_2CH_3$ **G** $CH_3C(CH_3)_2CH_3$

H (cyclohexane ring) **I** (methylcyclopentane ring)

a) Give the names and molecular formulae of the alkanes.
b) Divide the alkanes into groups of structural isomers.
c) Write down the displayed formulae for the structural isomers of molecular formula C_6H_{14} which are *not* shown above.
d) The cycloalkanes may be represented by skeletal formulae (section 6.1). Write down similar formulae for the other structural isomers of compounds H and I.
e) Arrange compounds A–E in order of increasing boiling point and explain your reasoning.
f) Which compounds do not have the general formula for alkanes C_nH_{2n+2}?
g) Which compounds might undergo cyclisation reactions to form compound I?

2 Give the displayed structural formulae and empirical formulae of the following alkanes:
a) propane **b)** 2-methylbutane **c)** cyclobutane
d) 2,2-dimethylbutane **e)** butane

3 a) On complete combustion, 1.10 g of a hydrocarbon X gave 3.30 g of carbon dioxide and 1.80 g of water. Calculate the molecular formula, and state the name of compound X.
b) Write an equation for the complete combustion of X.
c) Use the data in Table 7.3 to calculate the amount of energy released when 50 g of X is completely burnt.
d) X may be used as camping gas. At a scout camp, 20 mugs of tea, each holding 250 cm³ of liquid, need to be made by boiling water over a camping gas burner. The cylinder of gas has only got 50 g of X in it and the starting temperature of the water is 17°C.
i) By referring to section 3.6, and assuming *no* heat loss, state whether the cylinder needs to be replaced before the water boils and explain your reasoning.
(ii) How might heat energy be lost during the boiling of the water and what precautions could be taken to minimise these losses?

4 a) One of the main uses of alkanes as fuels is in car engines. Summarise this topic using the following prompts and no more than 400 words: the alkanes in petrol, a typical combustion equation, knocking, octane scale, anti-knock agents, incomplete combustion, environmental issues, three-way catalytic converters.
b) Write balanced equations for the reactions which occur in a three-way catalytic converter, namely: (i) the reduction of NO_2 by CO, (ii) the reduction of NO by CO, (iii) the oxidation of CO by O_2 and (iv) the oxidation by O_2 of octane, a typical hydrocarbon in petrol.
c) The number of cars in Britain continues to increase and, by the year 2000, over 75% of households will be running one, or more, cars. What are the implications of this trend? If you were the Minister for Transport, what policies would you suggest to meet any potential problems?

5 a) Under what conditions do alkanes react with chlorine and bromine?
b) Write a chemical equation for the main reaction which occurs when ethane is mixed with bromine in a 1:1 molar ratio.

Questions on Chapter 7 *continued*

c) When methane is treated with *excess* bromine, under suitable conditions, four main products may be isolated. Name each of the products and write a balanced equation to show how it is formed.

d) Explain the mechanism of the reaction in part (b) with particular reference to the terms homolytic fission, initiation, propagation and termination. Why is a trace of butane produced when bromine is mixed with *excess* ethane?

e) Until recently, large quantities of chlorinated alkanes were used as solvents. Explain why their use is gradually being phased out.

Comments on the activities

Activity 7.1

1 **a)** Equivalent structures: (iii) and (viii), both hexane
(iv) and (x),
both 2-methylpentane.

b) Isomeric structures: (i), (iii) or (viii), (iv) or (x)
(v) and (vii) are all C_6H_{14}
(ii), (ix) and (xi) are all C_5H_{12}
(vi) and (xii) are both C_4H_{10}

2 (i) 3-methylpentane; (ii) pentane; (iii) and (viii) hexane; (iv) and (x) 2-methylpentane; (v) 2,3-dimethylbutane; (vi) butane; (vii) 2,2-dimethylbutane; (ix) 2,2-dimethylpropane; (xi) 2-methylbutane; (xii) 2-methylpropane.

3

(a) 2,2-dimethylbutane

(b) 2-methylbutane

(c) 3,3-diethylhexane

(d) 2,4-dimethylpentane

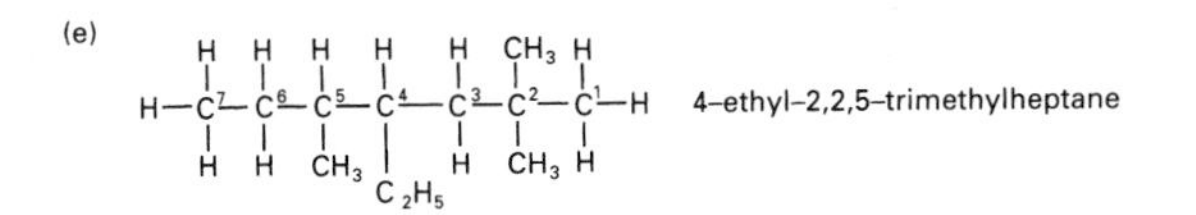

4

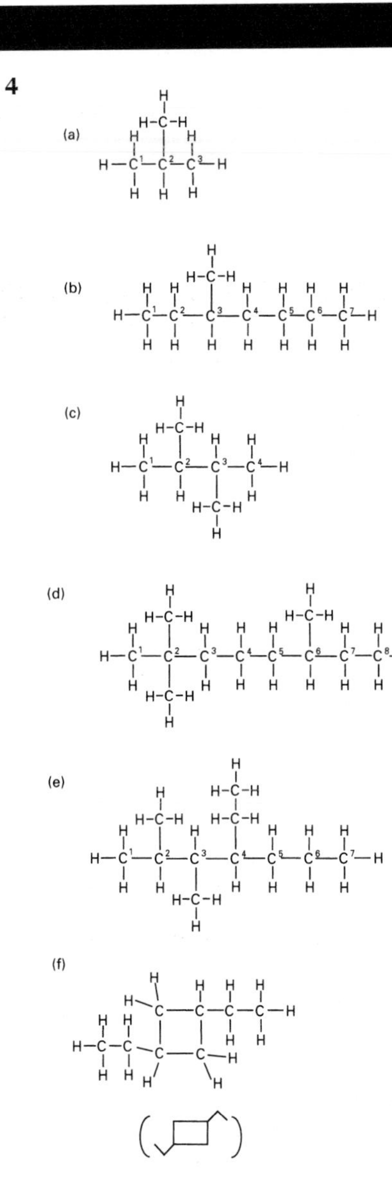

Comments on the activities *continued*

5

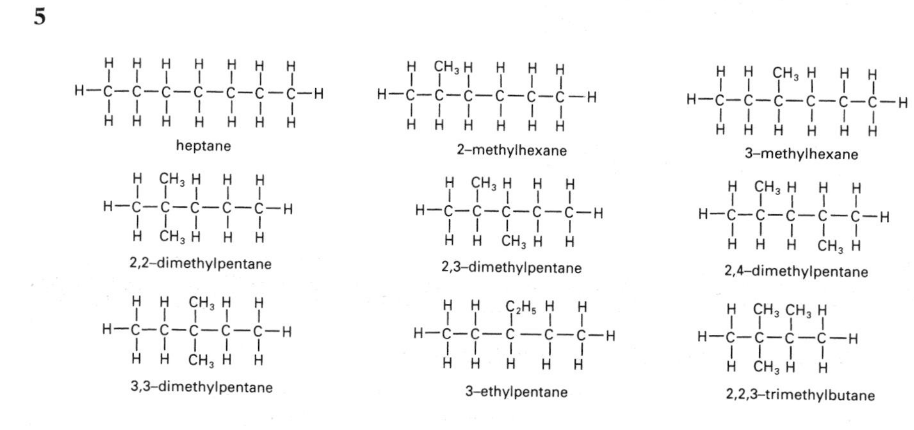

Activity 7.2

1 **a)** **b)**

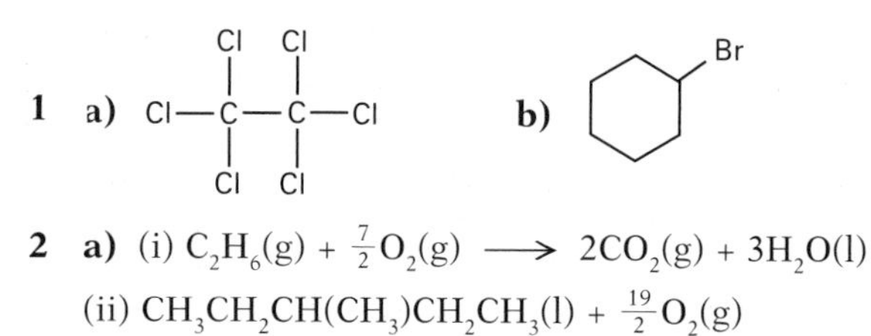

2 **a)** (i) $C_2H_6(g) + \frac{7}{2}O_2(g) \longrightarrow 2CO_2(g) + 3H_2O(l)$

(ii) $CH_3CH_2CH(CH_3)CH_2CH_3(l) + \frac{19}{2}O_2(g)$

$\longrightarrow 6CO_2(g) + 7H_2O(l)$

(iii) $C_4H_8(g) + 6O_2(g) \longrightarrow 4CO_2(g) + 4H_2O(l)$

3

Initiation: $Br\!:\!Br \rightleftharpoons 2Br^\bullet$

Propagation: $Br_3C\!:\!H + {}^\bullet Br \longrightarrow Br_3C^\bullet + H\!:\!Br$

$Br_3C^\bullet + Br\!:\!Br \longrightarrow Br_3C\!:\!Br + Br^\bullet$

Termination: $Br^\bullet + {}^\bullet Br \longrightarrow Br\!:\!Br$

$Br_3C^\bullet + {}^\bullet Br \longrightarrow Br_3C\!:\!Br$

C H A P T E R

Alkenes

Contents

Study Checklist

By the end of this chapter you should be able to:

1 State that alkenes are unsaturated hydrocarbons which are manufactured by cracking the fractions produced from petroleum.

2 Name alkenes given their short structural, displayed structural or skeletal formulae.

3 Explain, in terms of van der Waals' forces, why the boiling points of unbranched-chain alkenes gradually increase as the molecules get larger.

4 State that alkenes may be prepared in the laboratory (a) by refluxing an alcohol with concentrated sulphuric acid at 180°C or (b) by refluxing a halogenoalkane with ethanolic potassium hydroxide.

5 Describe the electronic structure of the double bond in alkenes in terms of orbital overlap, σ and π bonding. Explain, in terms of lack of freedom of rotation about a multiple bond, why the $C{=}C$ structure is planar and how this arrangement gives rise to geometrical (*cis–trans*) isomerism, e.g. in but-2-ene.

6 State that the $C{=}C$ bond in an alkene is a centre of high electron density and, as such, it will be attacked by electrophiles and oxidising agents.

7 Describe the chemistry of alkenes, as exemplified by the following reactions of ethene and propene:

a) electrophilic addition of halogens, hydrogen halides, hydrogen, concentrated sulphuric acid and, indirectly, water;

b) oxidation by cold, dilute acidified potassium manganate(VII) to form a diol;

c) free radical addition polymerisation to form plastics.

8 Describe the mechanism of free radical polymerisation and draw the polymer structures produced by a given monomer.

9 Describe the mechanism of the electrophilic addition to alkenes using, as examples, the reactions of (a) ethene with bromine and (b) propene with hydrogen bromide. State Markovnikov's rule and explain it in terms of the stability of the possible carbocations formed during the addition reaction. Use 'curly arrows' to show how the bonding electron clouds shift during the reaction.

10 Outline the process, and significance, of the catalytic hydrogenation of vegetable oils in the manufacture of margarine.

11 Describe the importance of alkenes in the production of organic chemicals, for example alcohols, synthetic rubber, poly(alkene) plastics and antifreeze. Outline the manufacture of epoxyethane and its conversion into antifreeze and polyesters.

12 Show an awareness of the issues surrounding the disposal of poly(alkene) plastics due to non-biodegradability, and making reference to recycling, incineration and landfill.

13 Summarise the chemical reactions of alkenes as shown in Figure 8.4.

14 State, and understand, the use of the reagents shown in Table 8.6

Carrier bags, antifreeze, 'throwaway' plastic cups and washing powder all have something in common – they contain materials made from **alkenes**. Alkenes are hydrocarbons which contain at least one C=C bond, e.g. ethene, $H_2C{=}CH_2$. In these compounds, the hydrogen to carbon ratio is less than that in alkanes; thus, alkenes are said to be **unsaturated** hydrocarbons.

8.1 Naming alkenes

Like alkanes, the names of alkenes consist of two parts:

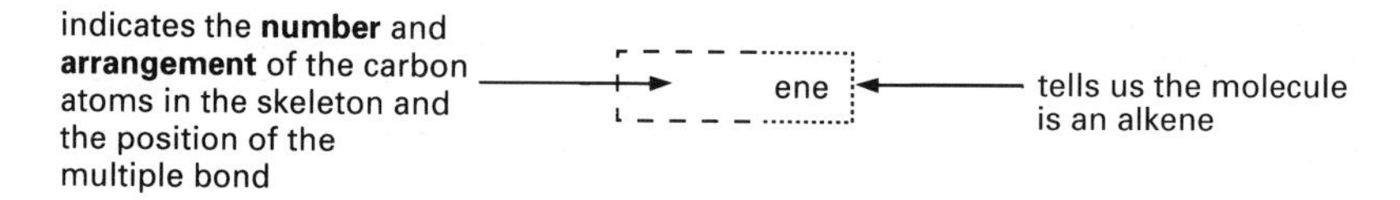

For example:

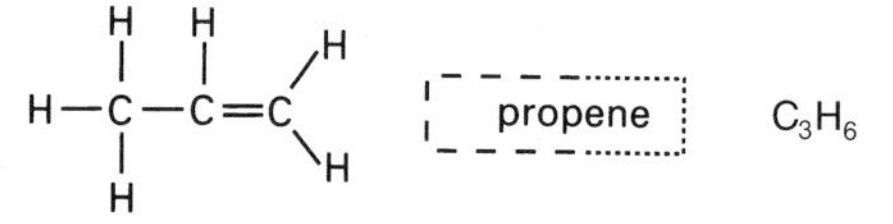

For alkenes with four, or more, carbon atoms, there is the possibility of **structural isomerism** (section 6.5):

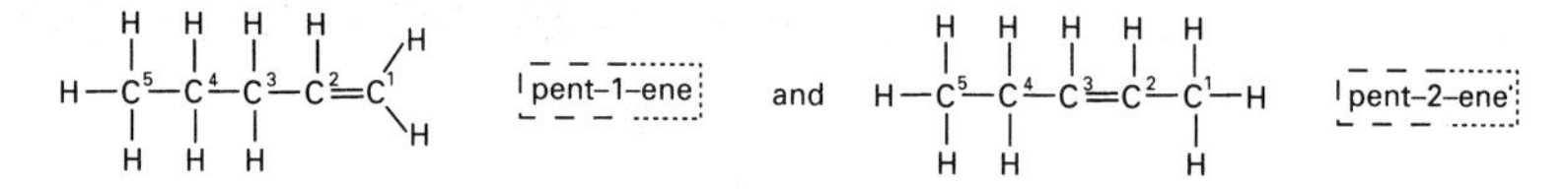

The numbers refer to the position of the first unsaturated carbon. Where two numbers are possible, the smaller must be used, e.g.

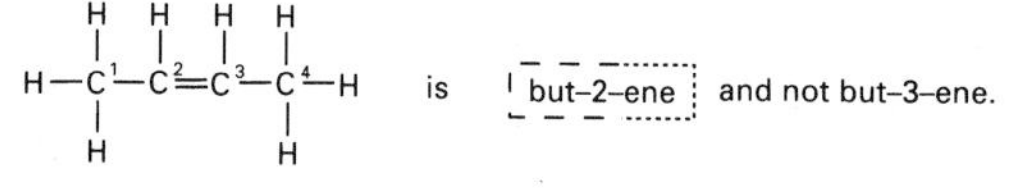

If a molecule contains more than one C=C, then the position of each must be stated, for example:

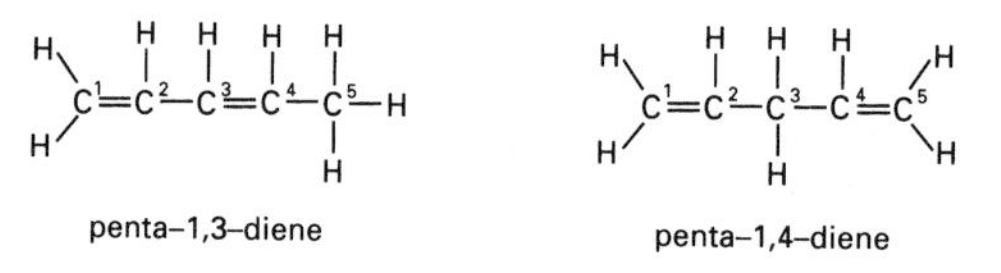

Penta-1,3-diene contains alternating single and double bonds and is termed a **conjugated alkene**.

Focus 8a

1. Alkenes are **unsaturated** hydrocarbons, that is they contain at least one C=C bond.
2. When naming alkenes, you must state the position of the multiple carbon–carbon bond(s).
3. The presence of the C=C bond enables some alkenes to exist as **geometrical** or ***cis–trans*** **isomers**.

Plastics

Every year, around 100 million tonnes of plastics are made worldwide and now life without plastics would be almost unthinkable. We have become dependent on them because they are:

- **Safe**
  Most plastics are inert and discourage contamination (bottles, disposable syringes). They are tough and shatterproof (contact lenses, children's toys) and tend to be insulators (electrical products, saucepan handles)
- **Lightweight and durable**
  Plastics are replacing heavier construction materials such as wood, metal and glass. This has caused a big decrease in the weight of domestic waste over the last twenty years. Using plastic materials in a car can decrease its mass by 10% and produce a fuel saving of up to 7%.
- **A designer's dream!**
  Plastics can be moulded and stretched, given various textures and easily coloured. They are often mixed with additives which make them flame resistant, stronger or fluorescent. Thus, designers can create products which are not only suitable for use but also look and feel good!

Plastics are safe to use, light, durable and a designer's dream

Activity 8.1

1. Name the alkenes whose formulae are shown below:
   a) $CH_3CH_2CH_2CH{=}CHCH_3$ **b)** $CH_3CH(CH_3)CH{=}CHCH_3$
   c) $CH_3CH_2CH{=}C(CH_3)_2$ **d)** H and H on one carbon side, H_7C_3 and C_2H_5 below: $(H)(H_7C_3)C{=}C(H)(C_2H_5)$, with both H atoms on the same side
   e) $(H_3C)(H)C{=}C(C_2H_5)(CH_3)$, with H_3C and C_2H_5 on the same side
   f) $(CH_3)_2C{=}CHCH(CH_3)_2$
   (You may find it helpful to draw out the structural formulae, where necessary.)
2. Draw structural formulae for the following alkenes:
   a) 2,3-dimethylbut-2-ene; **b)** *trans*-pent-2-ene
   c) 2-methyl-3,4-diethylhex-2-ene **d)** *cis*-hex-3-ene; **e)** phenylethene
   f) 4-ethylcyclohex-1-ene **g)** 2-methylbuta-1,3-diene
3. Which of the following compounds exhibit geometrical isomerism?
   a) hex-2-ene **b)** but-1-ene **c)** 1-bromopropene ($CH_3CH{=}CHBr$)
   d) 2-methylpropene **e)** penta-1,3-diene
   (Hint: in each case, first draw out the structural formulae.)

Rubber

For nearly two hundred years, people have extracted natural rubber from a tropical tree called *Hevea brasiliensis*. Natural rubber is a polymer of isoprene (methylbuta-1,3-diene):

$$H_2C{=}C(CH_3){-}C(H){=}CH_2 \quad \text{isoprene}$$

At first, rubber had limited usefulness because it was rather soft and became sticky in hot weather. However, in 1839, Goodyear solved this problem by heating the natural rubber with a small amount of sulphur. During this process, known as **vulcanisation**, the sulphur atoms form cross-linkages between neighbouring rubber molecules. This produces a stronger and more elastic product.

Sap being drained from a rubber tree in Malaysia. It is a slow laborious process and it is easy to see why synthetic rubber had to be developed

By the start of this century, demand for rubber began to exceed the natural supply and chemists turned their attention to the synthesis of rubber substitutes. One of the first synthetic rubbers, called **Buna rubber**, was obtained from the sodium catalysed polymerisation of buta-1,3-diene:

$$n\,H_2C{=}C(H){-}C(H){=}CH_2\,(g) \xrightarrow{\text{Na(g) catalyst}} \left[CH_2{-}C(H){=}C(H){-}CH_2 \right]_n$$

Buna rubber

Here n may be up to 2000. Nowadays, the most popular synthetic rubber is a co-polymer of 80% buta-1,3-diene and 20% phenylethene.

8.2 Physical properties of alkenes

Some properties of alkenes are listed in Table 8.1.

Table 8.1 Some properties of alkenes

Name	formula	state at 298 K	boiling point/°C	$\Delta H_c^\ominus$ /kJ mol^{-1}
ethene	C_2H_4	gas	–102	–1411
propene	C_3H_6	gas	–48	–2058
but-1-ene	C_4H_8	gas	–6	–2717
but-2-ene	C_4H_8	gas	3	–2710
hex-1-ene	C_6H_{12}	liquid	64	–4004
cyclohexene	C_6H_{10}	liquid	83	–4128

$\Delta H_c^\ominus$ = standard enthalpy of combustion (section 3.4)

What effect does increasing chain length have on the boiling points and standard enthalpies of combustion of the straight-chain alkenes?

What value would you predict for the standard enthalpy of combustion of pent-1-ene?

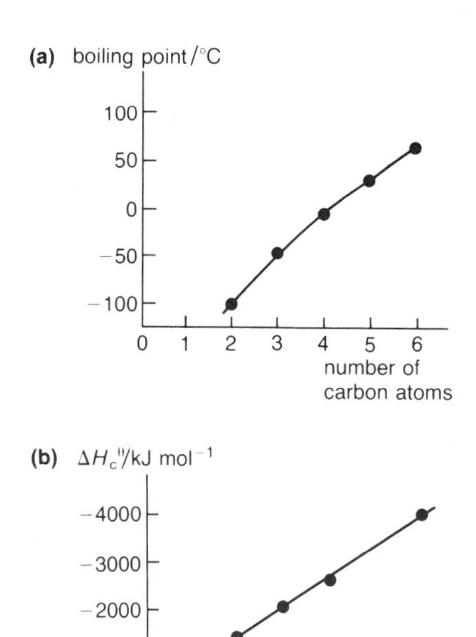

Figure 8.1
Graphs of **(a)** the boiling points of straight-chain alkenes and **(b)** the standard enthalpies of combustion of straight-chain alkenes, $\Delta H_c^\ominus$, each plotted against the number of carbon atoms in the molecule

Alkenes form a homologous series and as a result have:

- a general formula, namely C_nH_{2n};
- gradually increasing boiling points (Figure 8.1a) and enthalpies of combustion (Figure 8.1b);
- similar chemical properties.

As expected, the lower members of the series are gases. With increasing molecular size, alkenes are found as liquids and, eventually, as solids. Can you explain why?

Since alkenes are non-polar substances, they are insoluble in water. However, they are readily soluble in many organic solvents.

8.3 Making alkenes

Although alkenes are the most abundant hydrocarbons in petroleum, they are involved in only a narrow range of chemical reactions and this limits their use as a chemical feedstock. Since alkenes are much more reactive than alkanes, large quantities are produced and then converted into other organic compounds, Table 8.2.

Table 8.2 Some examples of the use of alkenes as a chemical feedstock (the numbers in brackets refer to the sections where they are discussed)

Starting material	commercial product	use
ethene	poly(ethene) (8.7)	construction, packaging
	ethanol (11.4)	solvent, cosmetics
	epoxyethane (8.6)	manufacture of antifreeze (8.6) polyester and detergents (13.7)
propene	propanone, as a by-product of the manufacture of phenol (9.5)	solvent
buta-1,3-diene	synthetic rubber (8.1)	tyres, electrical insulation

Ethene and propene are manufactured by cracking (i) the naphtha or gas-oil fractions of petroleum or (ii) the ethane and propane from natural gas (section 7.6). In the laboratory, an alkene may be prepared from:

- an alcohol, by refluxing it with excess concentrated sulphuric acid:

$$CH_3CH_2OH \xrightarrow[180^\circ C]{\text{excess conc. } H_2SO_4(l)} CH_2{=}CH_2 + H_2O$$

  Since water is eliminated from the alcohol, this is termed a **dehydration** reaction. Concentrated sulphuric acid is acting as a **dehydrating agent**.
- a halogenoalkane, by refluxing it with potassium hydroxide dissolved in ethanol to eliminate the hydrogen halide:

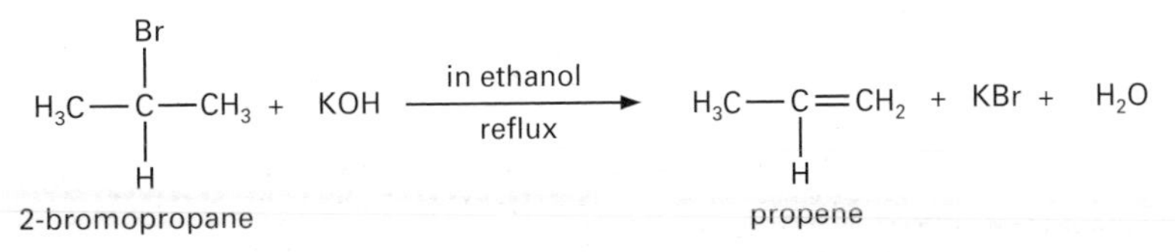

8.4 Electronic structure of alkenes

The mean bond enthalpy of a covalent bond (section 3.10) gives us an idea of the strength of that bond in a molecule; the *greater* the mean bond enthalphy, the *stronger* the bond. Taking the mean bond enthalpy of the C—C bond to be 348 kJ mol^{-1}, what value might you expect for the C=C bond? If the C=C bond is made of identical single bonds, we would expect the mean bond enthalpy to be about 2 × 348, that is 696 kJ mol^{-1}. In fact, the mean bond enthalpy for the C=C is much lower, at 611 kJ mol^{-1}, suggesting that different types of covalent bond contribute to the double bond.

C [He] 2s 2p
mix together
three hybrid orbitals
remaining p orbital

A carbon atom in its ground state has two unpaired p electrons, suggesting that its probable **valency** is 2. In fact, carbon normally forms **four** covalent bonds. To do this, an electron is promoted from the 2s orbital to the vacant orbital. Then, the electron clouds in the s and p orbitals mix together to form **hybrid orbitals** which then overlap with the orbitals of other atoms (section 4.3).

In each of the carbon atoms in ethene, the electron clouds in the s orbital and two of the p orbitals mix to form three hybrid orbitals. These overlap end-on with the orbitals of hydrogen and the other carbon atom, thereby forming a planar framework of **σ bonds**.

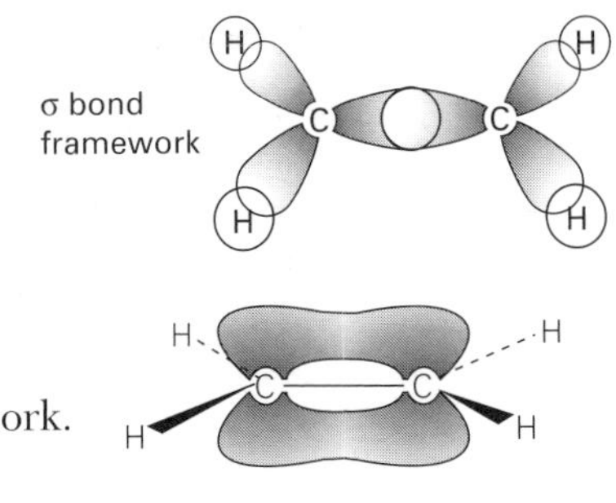

The remaining 2p orbital on each carbon atom is at right angles to the σ bond framework. These overlap sideways to give a **π bond**.

Since the electron cloud of a σ bond is concentrated between the bonding nuclei, it holds them firmly together. In a π bond, the electron cloud is spread out further from the nuclei, making it weaker than the corresponding σ bond. Hence, π bonds make a somewhat smaller contribution than σ bonds to the

Focus 8b

1. Alkenes form a **homologous series**, with the general formula C_nH_{2n}.
2. Alkenes are manufactured by cracking the naphtha or gas-oil fractions from petroleum. Ethene and propene are extremely valuable starting materials in industrial organic synthesis.
3. A C=C bond is made up of a **σ bond (end-on overlap of orbitals)** and a **π bond (sideways overlap of orbitals)**.
4. Since C=C bonds are electron-rich reactive centres, they will add on **electrophiles** and react with **oxidising agents**. They can also 'link' together to form **addition polymers**.

strength of a multiple bond, and this explains why the mean bond enthalpy of a C=C bond is 611 kJ mol^{-1} and not about 696 kJ mol^{-1}.

The π bond prevents rotation around the carbon–carbon bond. (Try making a molecular model). Thus, ethene is a planar molecule as is the $>C=C<$ structure. The bonding electron clouds repel each other giving an H—C—H angle of slightly less than 120°.

Since the π bonds in alkenes are centres of high electron density, they will be attacked by **electrophiles**, such as H^+ and Br^+ (section 6.7), and **oxidising agents**, such as alkaline potassium manganate(VII) (section 6.9). Being weaker than σ bonds, π bonds readily break, and this enables alkenes to take part in a wide range of **addition** and **polymerisation** reactions.

8.5 Addition reactions of alkenes

Reactions with halogens

When alkenes react with halogens, the order of increasing reactivity is $I_2 < Br_2 < Cl_2 < F_2$.

Thus, at room temperature, ethene and fluorine explode on contact whilst iodine gives a very slow addition reaction:

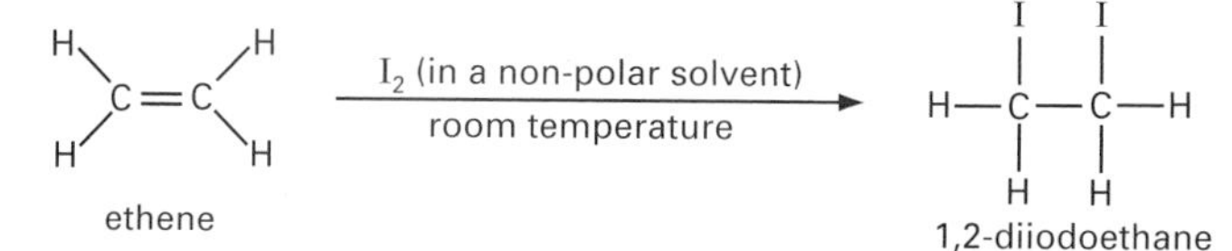

We can use this addition reaction as a **test for unsaturation**. Usually the unknown hydrocarbon is shaken vigorously with a solution of bromine dissolved *either* in a non-polar solvent, such as hexane, *or* in water. If the sample is unsaturated, the orange colour of the bromine will disappear on shaking. Why wouldn't an alkane react rapidly under these conditions? (See section 7.8.)

Mechanism

Useful information about the reaction mechanism comes from the reaction of ethene with bromine water. *Two* products are formed:

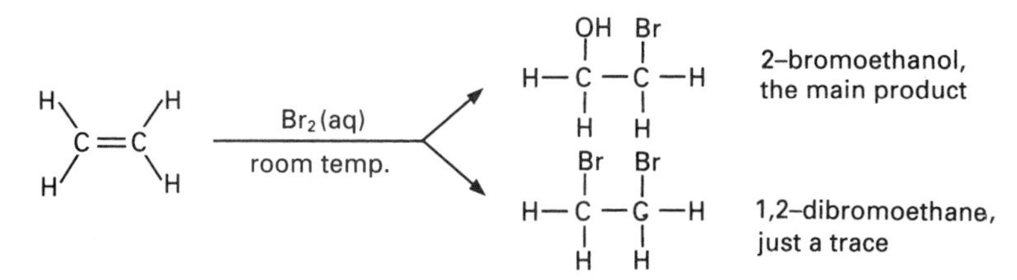

Since ethene does not react directly with $OH^-(aq)$ ions, the reaction's first step must involve only ethene and bromine. However, if this were a simple sideways attack

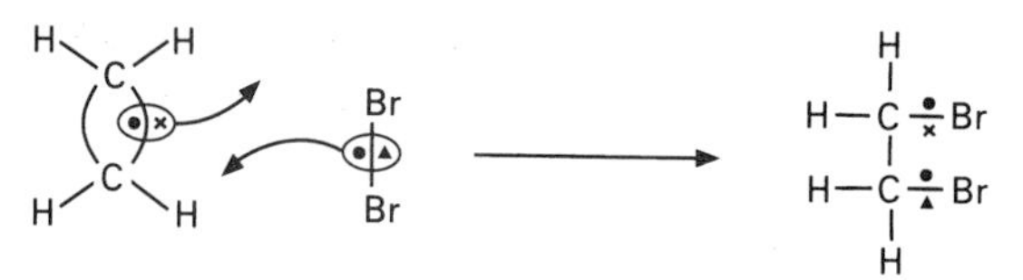

only *one* product, 1,2-dibromoethane, would be formed. In fact, our results support the following step-wise mechanism:

i) Reactant molecules approach and polarise each other. The induced dipoles hold the molecules loosely together.

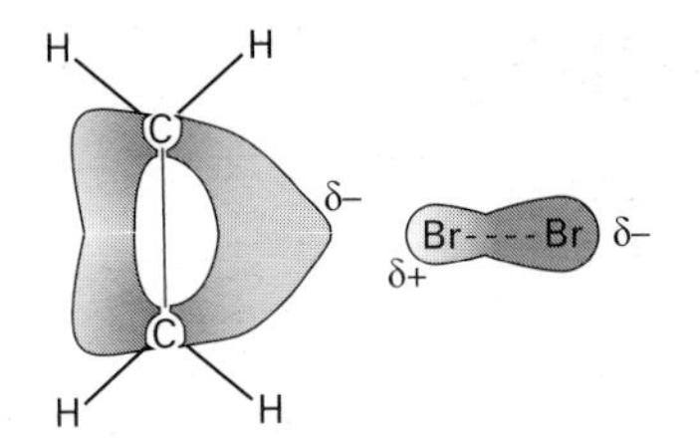

ii) Electron cloud from the π bond moves into the space between a carbon atom and the positively charged bromine atom. A **co-ordinate (dative covalent) bond** is formed; at the same time, **heterolytic fission** of the Br—Br bond occurs.

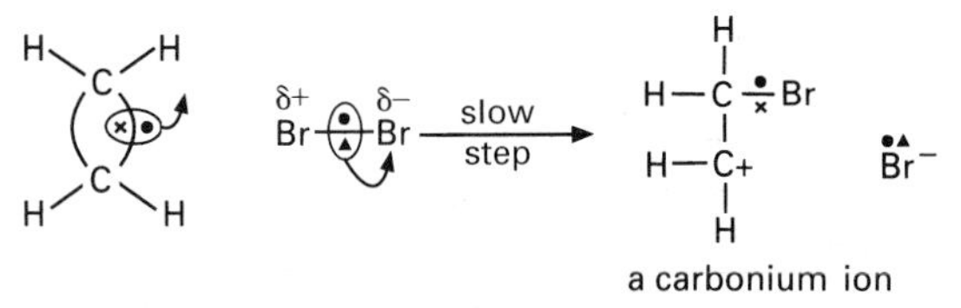

Now, counting up the electrons, we can see that we have formed a Br^- ion and a **carbonium ion** or **carbocation** (that is, a species with a positively charged carbon). The carbon has a positive charge because it has lost an electron, ×.

iii) Finally, the carbonium ion is rapidly attacked by a nucleophile, Br^- or H_2O, and since there are many more H_2O molecules than Br^- ions, the main product is 2-bromoethanol.

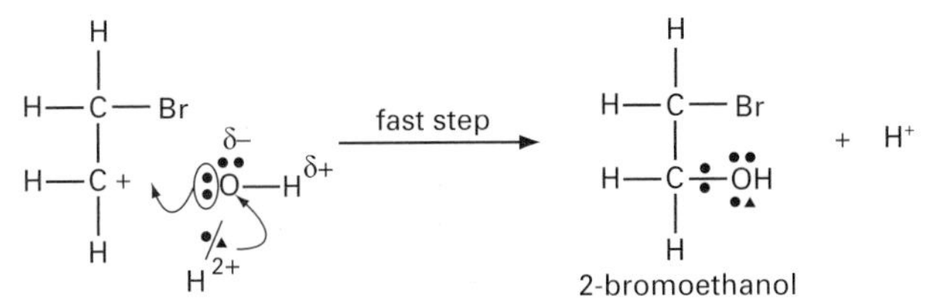

Since an *addition* reaction has occurred, via the initial attack of the *electrophile*, the entire mechanism is known as **electrophilic addition**.

Reaction with hydrogen halides

Alkenes are attacked by hydrogen halides, either as gases or dissolved in a non-polar solvent such as hexane:

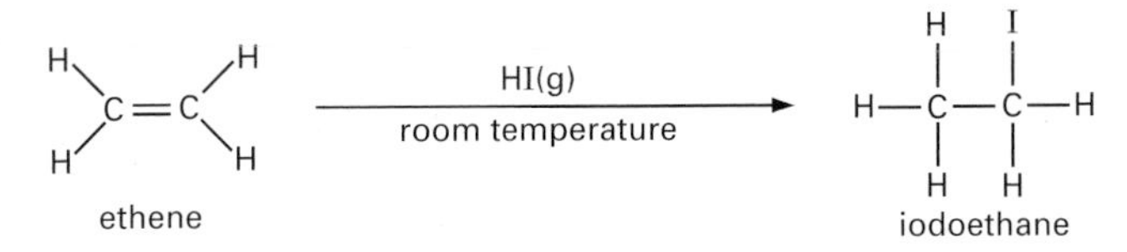

The reactivity increases in the order HCl < HBr < HI. This is explained by the corresponding ease with which the H—Hal bond is broken:

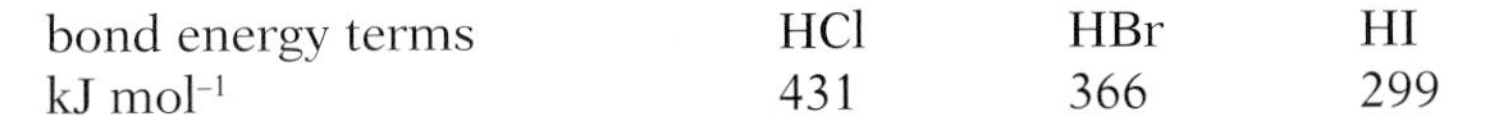

bond energy terms	HCl	HBr	HI
$kJ\ mol^{-1}$	431	366	299

The addition of hydrogen halides to propene produces interesting results. For example, with hydrogen bromide we obtain two products:

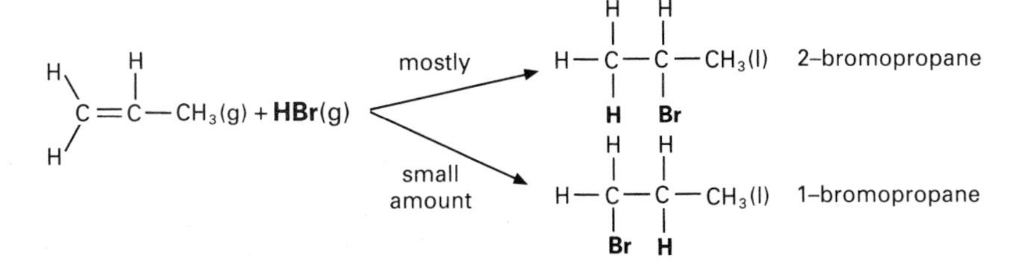

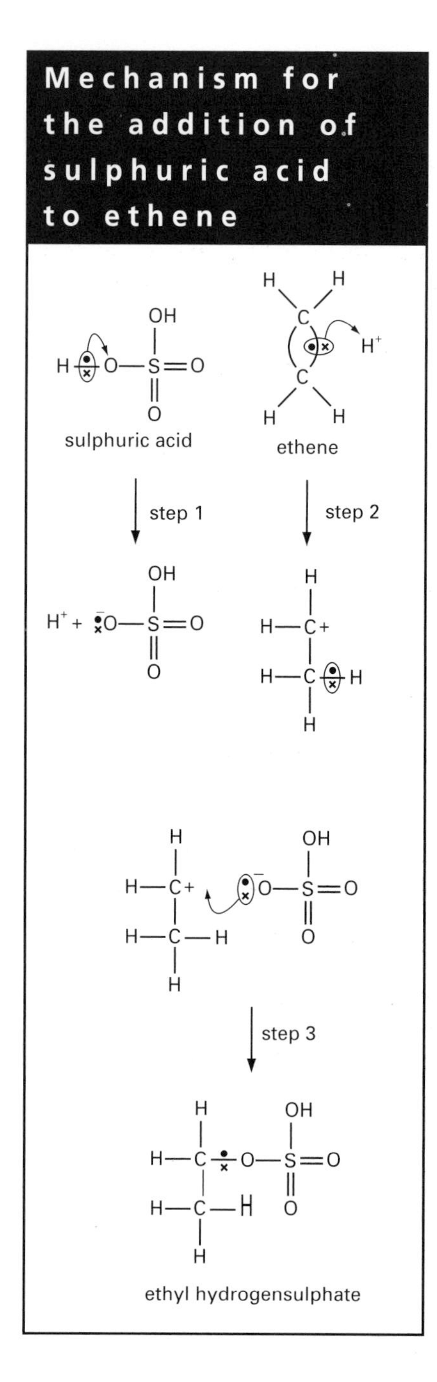

Once again, the mechanism is electrophilic addition (see above). However, in this case, step (ii) of the mechanism can give two different carbonium ions via shift of electron cloud from the π bond to a or b:

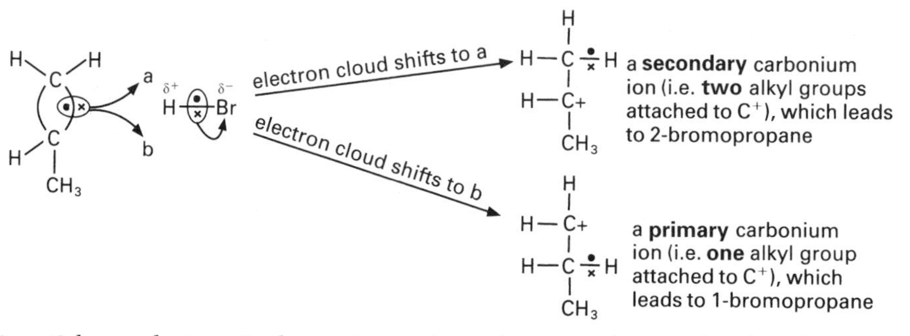

Since 2-bromobutane is the main product, the secondary carbonium ion must be formed in preference. It is the more stable. But why?

Well, alkyl groups are said to have a **positive inductive effect**. This means that they tend to push electron cloud away from themselves. In so doing, the carbon's positive charge can be **delocalised**, that is, spread out over the whole molecule:

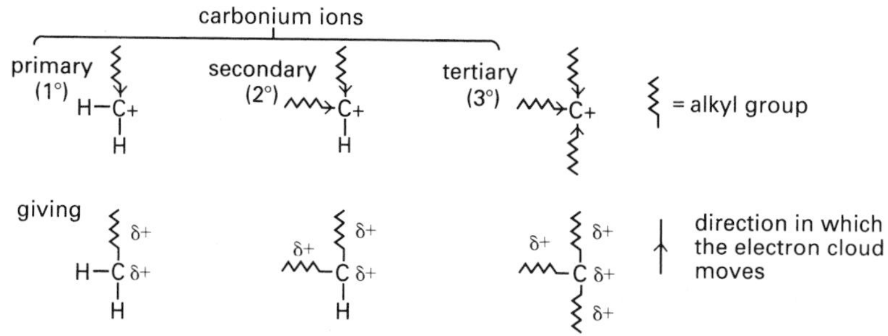

This delocalisation of charge stabilises the carbonium ion, with the stability increasing as more alkyl groups are attached to the C^+, i.e. $1° < 2° < 3°$.

Markovnikov found that other asymmetric alkenes behaved in this way, and he produced a useful rule:

When an HX molecule adds to an asymmetric alkene the hydrogen atom is more likely to add to the unsaturated carbon which already has the most hydrogen atoms.

Typically H—X can be HCl, H—Br, H—I, H—OH(H_2O) and H—OSO_3H(H_2SO_4). (It is important to note that it is the relative stabilities of the carbonium ions, *not* Markovnikov's rule which explains the experimental results.)

Reaction with sulphuric acid

Alkenes slowly add concentrated sulphuric acid at room temperature, for example:

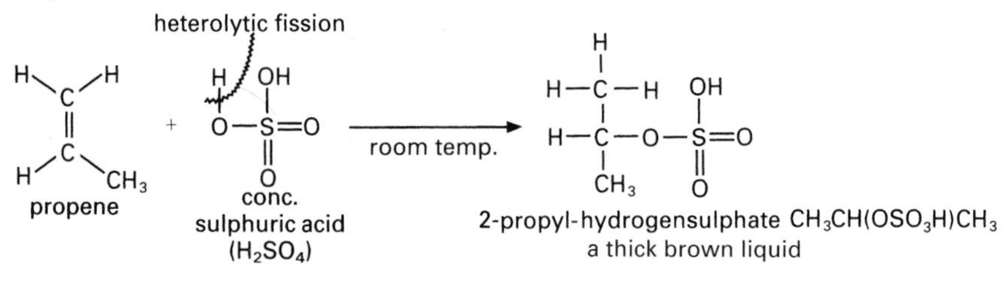

Once again, the mechanism is electrophilic addition and Markovnikov's rule is followed. Alkyl hydrogensulphates are useful products because they can be readily hydrolysed to alcohols:

Making margarine

Nowadays, a major use of catalytic hydrogenation is in the margarine industry. In the middle of the last century, there was a world shortage of butter and animal fats. Thus, scientists investigated ways of converting vegetable oils into solid fats. Commercially, the first real success came in 1910 when it was found that vegetable oils could be hydrogenated at 200°C in the presence of finely divided nickel. Vegetable oils, such as those from sunflower seeds, soya beans and ground nuts, have molecules which contain a number of C = C bonds. Hydrogenation of these unsaturated oils produces oils which are less unsaturated and these form soft solid fats on cooling. These fats are separated, purified and usually shaken up with milk. Vitamins A and D, food colouring and various other additives are mixed in before the margarine finds its way to the customer.

Plant for the manufacture of margarine. Nowadays more sophisticated catalysts are used instead of finely divided nickel

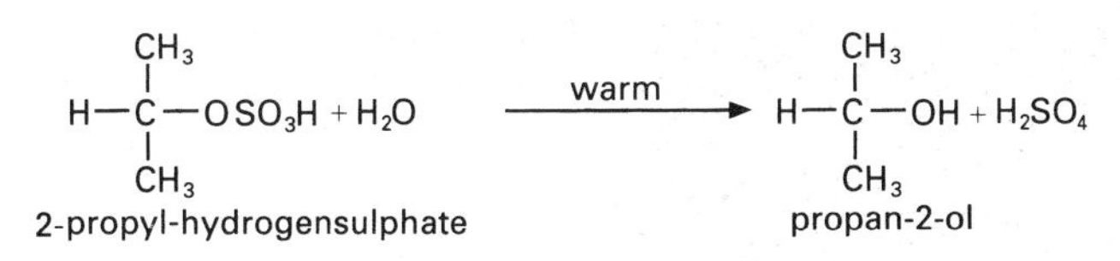

The net result here is **hydration**, that is the addition of water across the C=C bond and the product is an **alcohol**. Alcohols are manufactured by the catalytic hydration of alkenes obtained by 'cracking' petroleum fractions, for example:

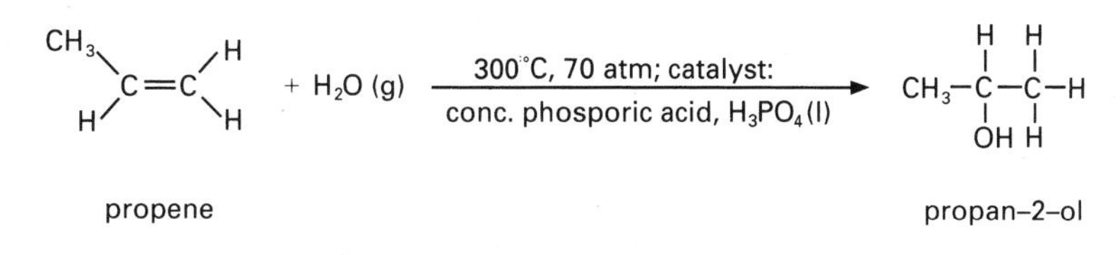

With hydrogen

Alkenes are reduced by hydrogen at 200°C in the presence of a finely divided nickel catalyst, e.g.

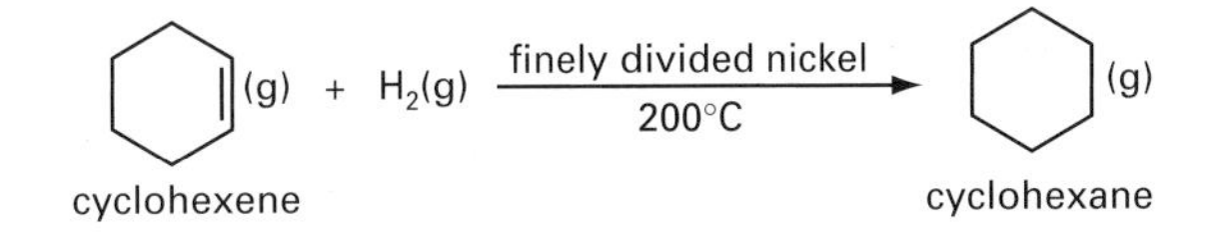

Focus 8c

1. Alkenes readily add halogens, hydrogen halides and sulphuric acid at room temperature.
2. **Markovnikov's rule** says that 'when an H—X molecule adds to an alkene, the hydrogen atom is more likely to add to the unsaturated carbon which already has the most hydrogen atoms'. The relative stabilities of carbonium ions (i.e. 3° > 2° > 1°) explain this behaviour.
3. The rapid decolorisation of bromine is a test for unsaturation.
4. The catalytic reduction of alkenes to alkanes by hydrogen gas is important in the margarine industry.

8.6 Oxidation of alkenes

Reaction with aqueous potassium manganate(VII)

Alkenes are oxidised by aqueous acidified potassium manganate(VII) (potassium permanganate) at room temperature to form alkanediols, for example:

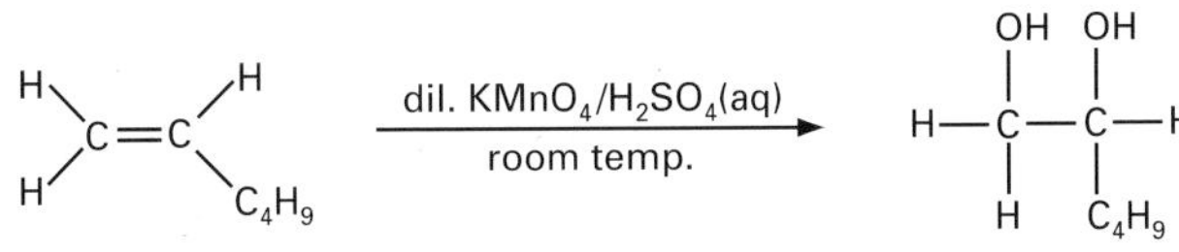

During these reactions, a colour change is observed:

$$\underset{\text{purple}}{MnO_4^-(aq)} \longrightarrow \underset{\text{colourless}}{Mn^{2+}(aq)}$$

Since alkanes do not react in this way, acidified potassium manganate (VII) can be used to distinguish alkenes from alkanes.

Combustion

Alkenes burn readily in air or oxygen. Ethene, for example, can react explosively, e.g.

$$C_2H_4(g) + 3O_2(s) \longrightarrow 2CO_2(g) + 2H_2O(g) \qquad \Delta H_c^{\ominus} = -1411 \text{ kJ mol}^{-1}$$

Although the combustion reactions are highly exothermic alkenes are *not* useful fuels because:

- they are more expensive than alkanes (from which they are made);
- they are more valuable as a **chemical feedstock**;
- **incomplete combustion** occurs as a result of their relatively high carbon content (e.g. ethene contains 86% carbon by mass). Compared to alkanes, therefore, alkenes burn in air with a more luminous, sooty flame.

Manufacture of epoxyethane

E

In industry, the addition of oxygen to ethene is catalysed by finely divided silver at 200°C:

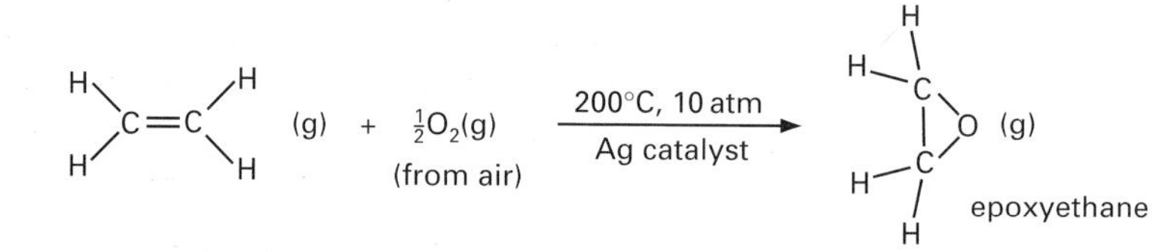

Epoxyethane is an unstable molecule which is involved in a variety of important organic syntheses (section 11.11). For example, it can easily be hydrolysed to form ethane-1,2-diol:

$$H_2C(O)CH_2(g) + H_2O(l) \xrightarrow[60°C]{\text{dilute } H_2SO_4(aq)} HOH_2C{-}CH_2OH(aq)$$

ethane-1,2-diol

Ethane-1,2-diol is used as **antifreeze** and in the manufacture of polyesters (section 13.7).

Activity 8.2

1 You have been given unlabelled bottles containing the liquids hexane and hex-1-ene. What chemical tests could you use to distinguish them?

2 Describe the reagents and reaction conditions that you would use to convert pent-1-ene into:
a) 1,2-dibromopentane **b)** 2-bromopentane **c)** pentan-2-ol
d) pentane **e)** pentane-1,2-diol

3 Give the name(s) and structural formula(e) of the organic product(s) formed by the following reactions:

a) but-1-ene $\xrightarrow[\text{room temp.}]{Cl_2(g)}$?

b) propene in hexane (l) $\xrightarrow[\text{room temp.}]{I_2}$?

c) *cis*-pent-2-ene $\xrightarrow[\text{room temp.}]{HI(g)}$?

Activity 8.2 continued

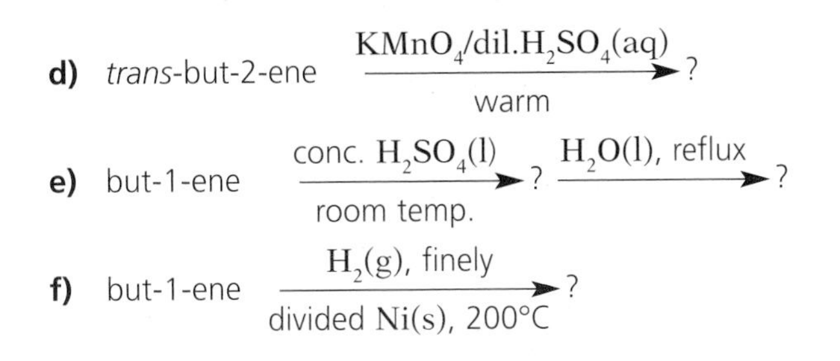

8.7 Polymerisation of alkenes

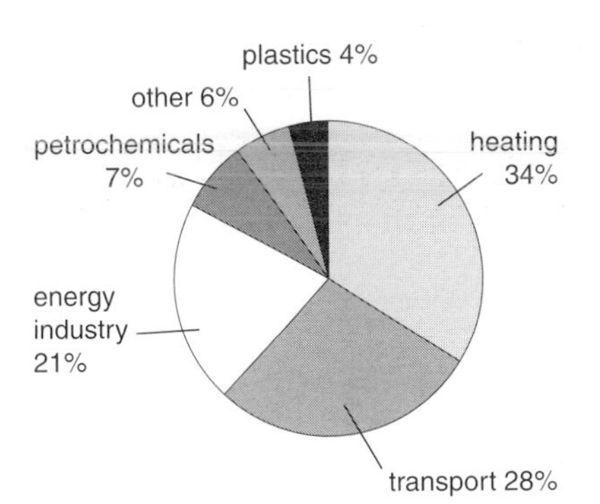

Figure 8.2 ▲
Uses of petroleum products in Western Europe

Over the last fifty years, one particular range of synthetic materials has made a dramatic contribution to our lifestyle. What are these materials? Plastics. No matter where you look, plastic objects are everywhere. Even so, the manufacture of plastics accounts for only about 4% of the petroleum products consumed in Western Europe (Figure 8.2).

Plastics are made up of massive molecules, called **polymers**, which have been formed by joining together many small molecules, called **monomers**. This process, known as **polymerisation**, can occur in two ways.

In **condensation polymerisation**, a small molecule, usually water, is liberated each time a monomer—monomer linkage is formed. Thus,

many monomer molecules ⟶ *one* polymer molecule + small molecules

The manufacture of **polyester** and **nylon** (section 13.7) are examples of condensation polymerisation.

However, we are interested here in **addition polymerisation**. In this case, the polymer is the only product, that is:

many monomer molecules ⟶ *one* polymer molecule

Alkenes, and substituted alkenes, undergo addition polymerisation to form a wide variety of plastics. The general reaction is:

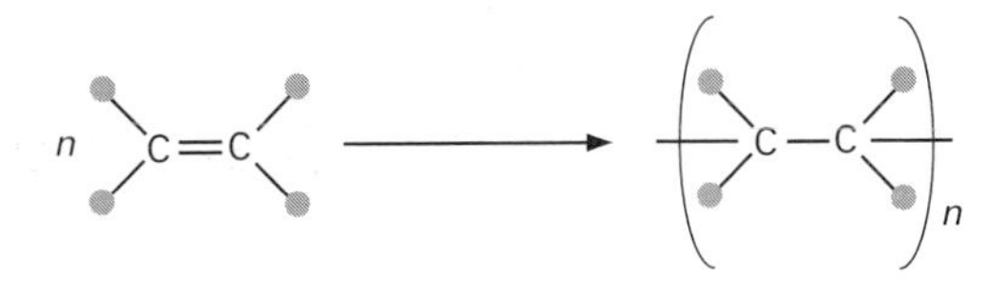

where n is typically between 500 and 5000. The properties and uses of the plastic will depend on (i) the value of n, (ii) the nature of the —● group and (iii) the reaction conditions. ● may be identical or different.

Some common addition polymers are described in Table 8.3. Although they are made from fairly similar monomers, the polymers show a wide variety in physical properties. These properties can be related to the polymer's 3-D structure as shown in Table 8.4.

Nowadays, plastics have taken the place of many natural materials, e.g. in the construction and clothing industries. Generally speaking, plastics are more versatile, last longer and are cheaper than alternative natural materials (Figure 8.3).

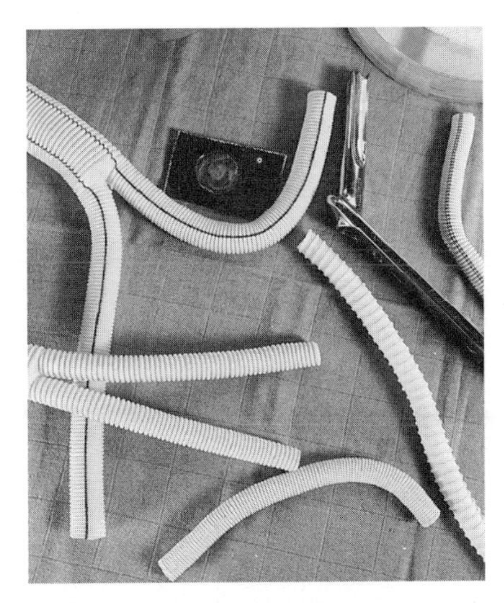

▲ An assortment of artificial arteries. The branched tubes on the left are aorta arteries. These implants are mainly made of 'tergal' a type of strong, synthetic fibre composed of polyester. These arteries are lined with 'Gore-tex', another type of synthetic fibre made of nylon and Teflon. The Gore-tex acts as an impermeable barrier which prevents blood clots from forming on the artery. Arteries made in this way can be as small as 5 mm across.

Table 8.3 Addition polymers of some alkenes and their derivatives (*n* is usually between 500 and 5000)

Monomer	**ethene**	**phenylethene**	**chloroethene** (vinyl chloride)	**propenonitrile**
	$CH_2{=}CH_2$	$C_6H_5CH{=}CH_2$	$ClCH{=}CH_2$	$NCCH{=}CH_2$
Polymer	poly(ethene)	poly(phenylethene)	poly(chloroethene)	poly(propenonitrile)
Common name	**polythene**	**polystyrene**	**PVC** (polyvinyl chloride)	**Acrilan**
Formula	$-(CH_2-CH_2)_n-$	$-(CH(C_6H_5)-CH_2)_n-$	$-(CHCl-CH_2)_n-$	$-(CH(CN)-CH_2)_n-$
Manufacture	(I) 200°C, 1000 atm, free radical initiator; or use (II)	(II) 50–100°C, 5–20 atm, Ziegler catalyst (a mixture of titanium(IV) chloride and an organo-aluminium compound)		
Properties	(I) produces low density polyethene which easily softens at 105°C. It is a thermoplastic, i.e. one which melts on gentle heating and can then be reshaped (II) gives high density polyethene. It is fairly rigid with a higher melting point of 135°C	Thermoplastic which is rigid but fractures easily. Polystyrene can be made into a foam by dissolving it and then distilling off the solvent – the product is expanded polystyrene	Thermoplastic whose properties can be adjusted by mixing in various additives. More rigid than polyethene, yet still a flexible product	Thermoplastic of great strength; thus it is used to make fibres
Combustion		Nearly all plastics will burn, often giving off a poisonous vapour		
Uses	I – plastic bags II – bottles, cups, buckets	cups, wall insulation packaging foam, egg boxes	shoes and clothing (imitation leather), electrical insulation, records	clothes, carpets

poly(propene)
poly(ethene), low density
poly(ethene), high density
PVC, poly(chloroethene)
poly(phenylethene)
steel
copper
aluminium
relative energy consumption

Figure 8.3 ▲
The energy needed to produce an equal mass of various materials

▲ This chemical engineer is standing in front of BP's Plastic Feedstock Recycling Pilot Plant at Grangemouth, Scotland. He is holding a sample of an oil product derived from the plant. This product can be used as a primary feedstock for refineries or the petrochemical industry.

▲ These discarded plastic drinks bottles are being loaded into a compactor at a recycling facility. The bottles are compacted, then shredded. The resulting plastic chips can be reused in new bottles.

Table 8.4 Comparing polymer structures

Polymer made up of ...	branched chains irregularly packed	unbranched chains regularly packed	unbranched chains with cross-linking bonds
∴ the chains will be ...	further apart ◄	————	► closer together
Consequently, density will be	LOW	HIGHER	HIGH
Also, the attractive forces between chains will be	WEAK (van der Waals')	STRONGER (van der Waals')	STRONG (covalent)
∴ melting points will be ...	LOW	HIGHER	HIGH
and the polymers will be ...	SOFT and FLEXIBLE	HARDER	VERY HARD and RIGID
Example	low density polythene	high density polythene	phenol methanal 'Bakelite' (section 12.8)
Uses	'plastic' bags	bottles, cups, bowls	fuse boxes, switches, handles

Activity 8.3

Write down the structures of the polymers formed from:

a) tetrafluoroethene, $F_2C{=}CF_2$
b) but-1-ene
c) 2-methylpropene
d) 2,3-dimethylbut-2-ene.

Free radical polymerisation

Addition polymerisation often takes place in the presence of a **free radical initiator**. These chemicals are unstable, especially when exposed to ultraviolet light, and break apart giving free radicals:

The free radicals are very reactive and attack the alkene's π bond, for example:

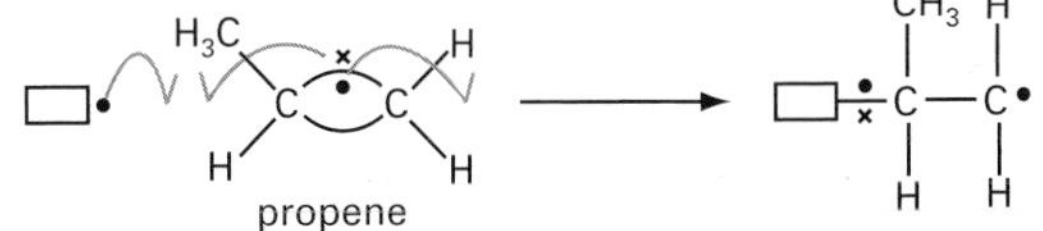

Since the product is still a free radical, the addition polymerisation propagates, and chains containing several thousand monomer molecules are formed:

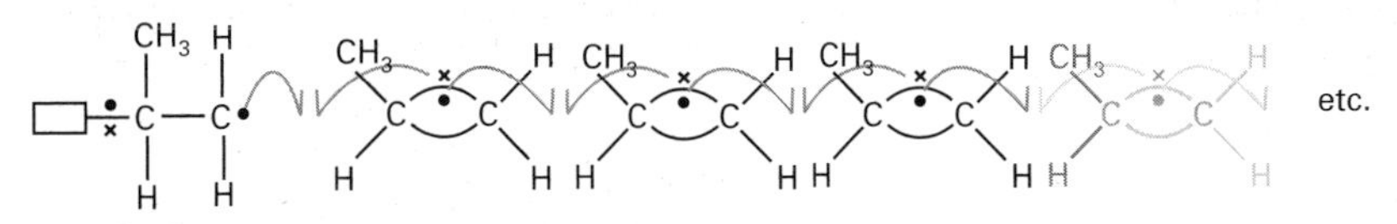

The reaction terminates when 'long-chain' free radicals combine with a similar radical or with a radical from the initiator forming the polymer molecule, in this case poly(propene):

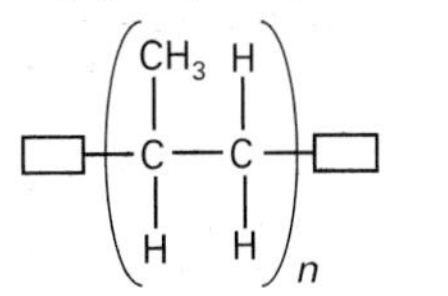

(*n* is between 500 and 5000)

Notice that a portion of free radical initiator's molecule is attached to each end of the polymer chain.

How should we dispose of our waste plastic?

There are three main ways of dealing with our waste plastics: they can be recycled, burnt or buried.

If plastics are recycled, the environment is conserved because we use up less raw materials and save energy. All plastics can be recycled using two types of process: material recycling and feedstock recycling. Unfortunately, **material recycling** is not always an economic option due mainly to the cost of collecting and separating the plastics. Larger items can be separated by hand as they are coded (Figure 8.5). Some plastics are separated by **flotation** – they are cut into flakes and mixed with a solvent in which some float and others sink. Passing the plastics through an electric field allows those which take up a charge, such as PVC, to be removed. After cleaning, and drying, the waste plastic is melted and forced through small holes – this process is called **extrusion**. The plastics strips are allowed to cool and then chopped into small pellets which are made into new products.

The aim of **feedstock recycling** is to convert the plastic's polymeric molecules into **feedstock chemicals** for other processes. In one example, called **gasification**, the plastics are heated in air giving carbon monoxide and hydrogen gases which are then used to make methanol, CH_3OH. In **pyrolysis**, the plastics are heated in a vacuum to give a mixture of gaseous and liquid hydrocarbons.

In Western Europe, plastics account for about 7% of the domestic waste produced. Domestic plastic waste is difficult to recycle because it is usually heavily contaminated and difficult to collect in bulk. Another option is **incineration**, or **energy recovery**, as plastics contain, weight for weight, 1.3 times as much energy as coal. Plastics burn with such an intense heat that they cause the ignition of less combustible waste materials, such as aluminium foil. One criticism of waste incineration is that it may produce toxic fumes. Incineration is a high-tech process, however, with computer-controlled furnaces operating at about 850°C. This ensures the complete combustion of carbon to carbon dioxide and minimises the emission of poisonous chemicals such as dioxins. A further concern is that the combustion of PVC (polychloroethene) produces hydrogen chloride gas, HCl, which contributes to acid rain. In fact, only 0.5% of the acidity in the atmosphere is from waste incineration because 'scrubbers' remove the acidic gases before emission so as to comply with strict European Union standards.

In the UK, about 3% of our waste is converted into energy. A successful waste incinerator in Edmonton, North London, produces steam which powers an electrical generator and the electricity is sold. Other plants process the waste to remove non-combustible items, such as metal and glass, which are recycled. The remainder, known as **refuse derived fuel (RDF)**, is shredded and sold to generating stations. Several countries with the highest environmental standards already recover considerable amounts of energy by burning RDF. For example, Switzerland and Denmark burn 72% and 62% of their waste, respectively. If we could burn 40% of our waste, we would save 4 million tonnes of coal per year. Waste incineration appears to be safe and provides a way of recovering energy. It also reduces the bulk of the waste products that must eventually be buried.

In some countries, it is impractical to recycle or burn domestic waste, so it is buried in huge pits, which quite often are disused quarries. This process, termed **landfill**, is a waste of resources. It also often produces toxic liquids, which leach into the surrounding water system, and a potentially explosive mixture of methane and carbon dioxide gases. In many landfill sites this gas is collected and burnt to generate heat or electricity. 90% of municipal waste in the UK is buried. A major disadvantage of most plastics is that they are **non-biodegradable**, that is their giant molecules are not broken down by microorganisms, for example, in soil. Most of our plastic waste is buried, but are we creating problems for future generations? As you can imagine, biodegradable plastics are the subject of considerable interest. One of these, 'Biopol' from Zeneca Group is made by the fermentation of a sugar feedstock by naturally occurring microorganisms – it is described in section 13.7.

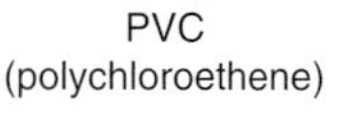

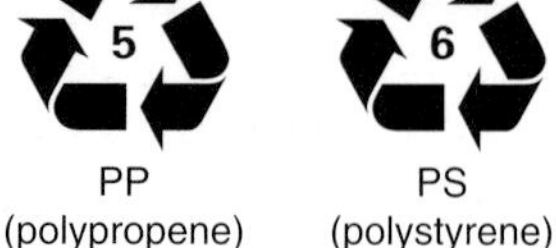

Figure 8.4
Codes which indicate the type of plastic which is present

Activity 8.4

1 **a)** What is the difference between material recycling and feedstock recycling?
   b) Give one advantage and one disadvantage of each of the following ways of waste plastics disposal: materials recycling, energy recovery and landfill.

2 The total plastic waste in the European Union in 1992 was 15.23 million tonnes. Of this, 1.036 million tonnes were recycled and 2.422 million tonnes were burnt for energy. The remainder was used for landfill.
   a) How many tonnes of waste plastic were recovered (recycled or burnt)?
   b) What percentage of the waste plastics were buried as landfill?

3 In 1992, the total domestic waste in the European Union was 132 million tonnes.
   a) What is the mass of plastic material in the domestic waste?
   b) The average calorific values of domestic waste, coal and plastic waste are 9, 30 and 40 GJ per tonne, respectively. Use this data to answer the following questions.
   i) Calculate the amount of energy that could be obtained if all the domestic waste was incinerated.
   ii) How much coal would have been saved if all the domestic waste was burnt?
   iii) How many tonnes of coal would be saved if the waste plastics were separated from the domestic waste and then burnt?
   c) What are the arguments for and against separating the plastics from the waste before incineration?

4 From a conservation point of view, recycling of plastic waste makes a lot of sense. Write a brief account of the economic factors that might affect the demand for recycled plastics.

Focus 8d

1 Some important reactions of alkenes are summarised in Figure 8.4

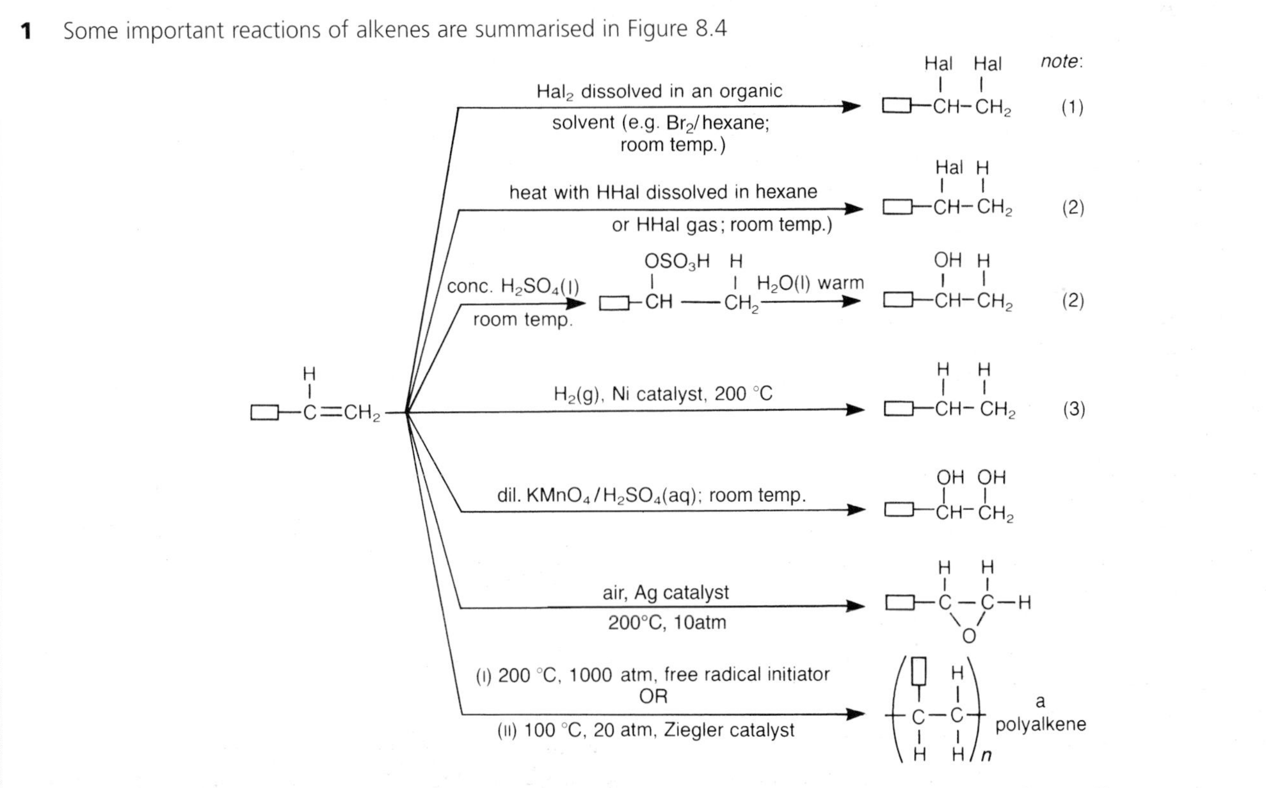

Figure 8.4 Some important reactions of alkenes. Notes: (1) rapid decolorisation of Br_2 is used as a test for an alkene; (2) Markovnikov's rule is followed; (3) used in the margarine industry

2 **Polymers** are formed by the joining together of many small molecules, called **monomers**. This process is called **polymerisation**. The general formula of an addition polymer is written:

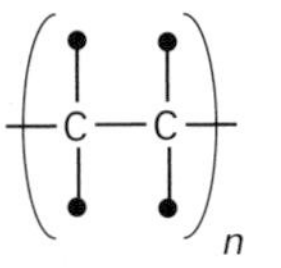

3 The important new reagents used in this chapter are shown in Table 8.5.

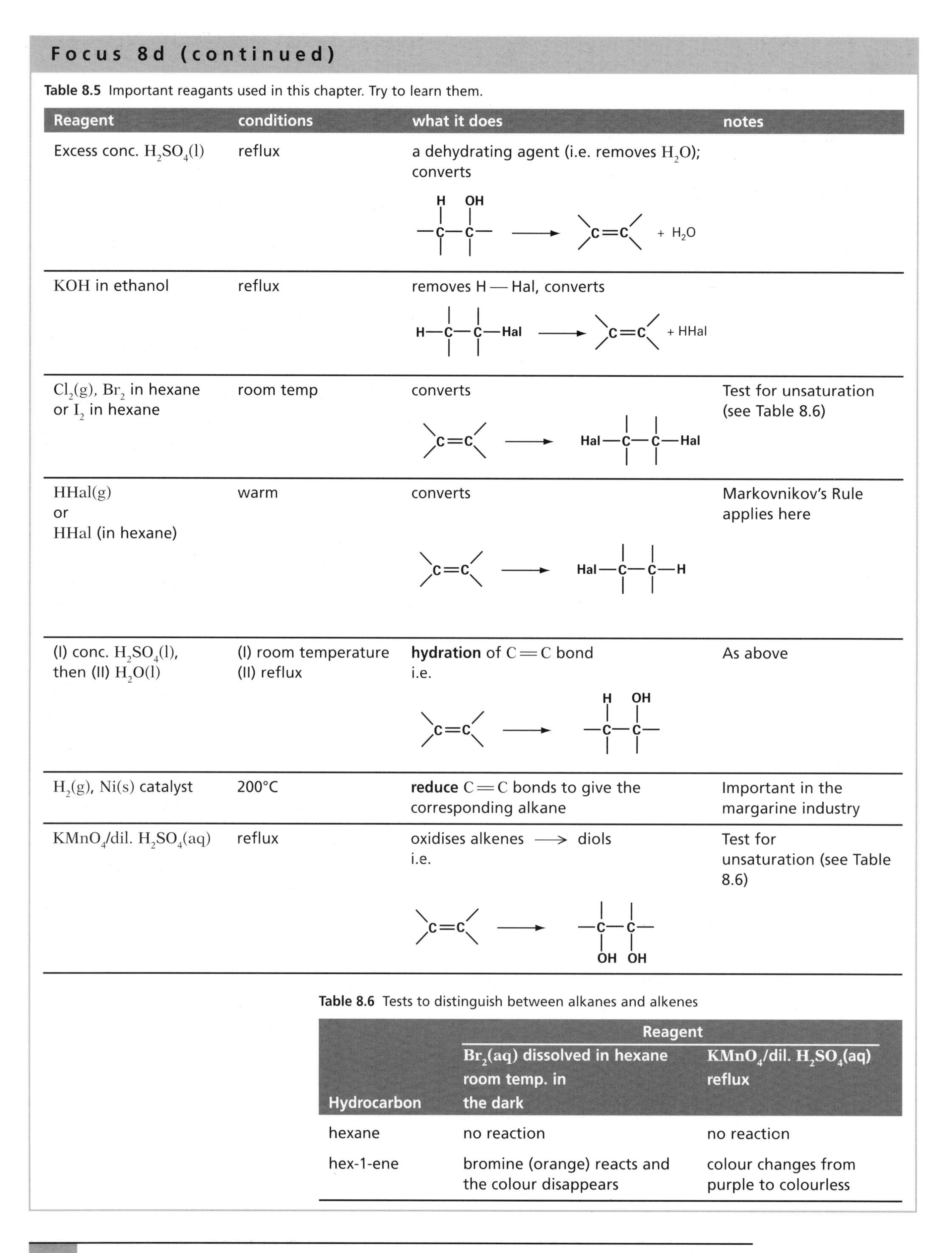

Focus 8d (continued)

Table 8.5 Important reagants used in this chapter. Try to learn them.

Reagent	conditions	what it does	notes
Excess conc. $H_2SO_4(l)$	reflux	a dehydrating agent (i.e. removes H_2O); converts $-\overset{H}{C}-\overset{OH}{C}- \longrightarrow >C{=}C< + H_2O$	
KOH in ethanol	reflux	removes H — Hal, converts $H-C-C-Hal \longrightarrow >C{=}C< + HHal$	
$Cl_2(g)$, Br_2 in hexane or I_2 in hexane	room temp	converts $>C{=}C< \longrightarrow Hal-C-C-Hal$	Test for unsaturation (see Table 8.6)
HHal(g) or HHal (in hexane)	warm	converts $>C{=}C< \longrightarrow Hal-C-C-H$	Markovnikov's Rule applies here
(I) conc. $H_2SO_4(l)$, then (II) $H_2O(l)$	(I) room temperature (II) reflux	**hydration** of C=C bond i.e. $>C{=}C< \longrightarrow -\overset{H}{C}-\overset{OH}{C}-$	As above
$H_2(g)$, Ni(s) catalyst	200°C	**reduce** C=C bonds to give the corresponding alkane	Important in the margarine industry
$KMnO_4$/dil. $H_2SO_4(aq)$	reflux	oxidises alkenes ⟶ diols i.e. $>C{=}C< \longrightarrow -\underset{OH}{C}-\underset{OH}{C}-$	Test for unsaturation (see Table 8.6)

Table 8.6 Tests to distinguish between alkanes and alkenes

	Reagent	
Hydrocarbon	$Br_2(aq)$ dissolved in hexane room temp. in the dark	$KMnO_4$/dil. $H_2SO_4(aq)$ reflux
hexane	no reaction	no reaction
hex-1-ene	bromine (orange) reacts and the colour disappears	colour changes from purple to colourless

Questions on Chapter 8

1 Name the following alkenes:

a) $CH_3CH{=}C(CH_3)_2$ **b)** **c)** **d)** **e)**

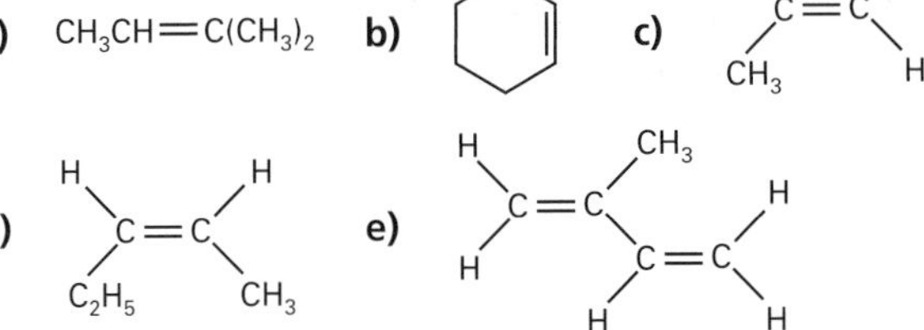

2 Write displayed structural formulae for these alkenes
a) propene **b)** *cis*-but-2-ene
c) cyclopentene **d)** *trans*-pent-2-ene

3 This question concerns the following alkenes:
A $CH_3CH{=}CH_2$ **B** $CH_3CH_2CH{=}CHCH_3$
C $CH_3CH{=}CHCH_3$ **D** C_2H_4

E CH_3 **F** $CH{=}CH_2$

G $CH_3CH{=}CHCH{=}CH_2$

a) Give the names of the hydrocarbons.
b) Which have the general formulae C_nH_{2n}?
c) Write down the displayed formula for a structural isomer of C.
d) Which can exist as geometrical isomers?
e) Which can be prepared by heating propan-2-ol with concentrated sulphuric acid at 180°C?
f) Which react with bromine to form products of empirical formula CH_2Br?
g) Which may react with excess bromine to form two additional products which are *cis–trans* isomers? Explain why the isomers are formed.
h) On reaction with HCl gas, which forms a product having the following % composition by mass:
51.9% C 9.7% H 38.4% Cl?
i) Write an equation for the complete combustion of G.
j) Give the structure of the addition polymer formed by F.

4 a) Use suitable diagrams to describe the bonding and shape of the ethene molecule.
b) Explain the difference between structural isomerism and geometrical (*cis–trans*) isomerism using the alkenes of formula C_4H_8 as examples. How might you distinguish between these isomers?
c) How does the presence of the C=C bond in a molecule give rise to *cis–trans* isomerism?

5 a) What is an 'electrophilic addition' reaction?
b) Why is the C=C bond susceptible to this type of reaction?
c) Write balanced equations for the electrophilic addition to ethene of
i) bromine
ii) hydrogen iodide.
d) What conditions are needed for these reactions to occur?
e) Describe a simple laboratory test which distinguishes hexane from hex-1-ene.
f) Write down the mechanism of the gaseous reaction of ethene with chlorine, using 'curly arrows' to explain how the electron cloud moves during the reaction.

6 a) State Markovnikov's rule.
b) Use suitable diagrams to explain the chemical theory behind this rule, taking the reaction of pent-2-ene with hydrogen iodide as an example.

7 Describe the reaction of alkenes with hydrogen and its importance in the manufacture of margarine.

8 a) Write chemical equations for the complete combustion of hexane and hex-1-ene.
b) Look back at the data tables to find out the standard enthalpy changes for these reactions.
c) Work out the relative molecular masses of hexane and hex-1-ene.
d) Now calculate the amount of energy liberated by 1 kg of each hydrocarbon if it is completely burnt in oxygen.
e) Which is the better fuel, hexane or hex-1-ene? Give your reasons.

9 Plastics are now an extremely important part of our lives. Many of them are formed by the addition polymerisation of alkenes or their substituted compounds. Nearly five million tonnes of polypropene (PP) were manufactured in Western Europe in 1993 for numerous applications including margarine tubs, garden furniture, crisp packets and children's toys.
a) What is meant by the term 'addition polymerisation'?
b) Give an equation for the formation of polypropene from propene.
c) State a set of conditions for the polymerisation.
d) Polypropene, and most addition polymers, are thermoplastics. What does this mean?
e) How would the polypropene be formed into a garden chair?
f) How might you tell if a child's plastic hammer is made from polypropene?
g) Describe *three* ways of disposing of waste polypropene focusing on the conservation of resources, environmental factors and economic viability.

Questions on Chapter 8 *continued*

10 Epoxyethane is the starting point for a number of industrial syntheses.

a) Write down the displayed formula of epoxyethane and draw in the approximate bond angles. Why is epoxyethane a reactive molecule?

b) Describe how epoxyethane is made from ethene, noting any essential safety precautions.

c) Explain how epoxyethane is made into ethane-1,2-diol. State two uses of the product.

d) i) Describe the formation of 1-methoxyethanol, $HOCH_2CH_2OCH_3$, from epoxyethane, write a chemical equation for the process and state one use of this alkoxyalcohol.

ii) How would you modify the process in part (b) to produce an alkoxyalcohol for use as a non-ionic detergent?

11 Natural rubber is found as latex, an emulsion of rubber particles in water. This sticky liquid oozes from the *Hevea brasiliensis* tree when it is cut. Natural rubber is a polymer of a conjugated alkene called **isoprene** which, on analysis, is found to contain by mass 88.2% carbon and 11.8% hydrogen. The mass spectrum of isoprene has a molecular ion with a relative mass ratio of 68.

a) Work out the molecular formula of isoprene and draw its possible structures.

b) Isoprene does not display geometrical isomerism. Hence, write down its displayed formula.

c) When isoprene is treated with an excess of hydrogen bromide, four products result from the addition of HBr across the localised C=C bonds.

i) Give the structures and names of these four addition products.

ii) Which compound would you expect to be the main product? Explain your answer.

d) In isoprene polymers, the repeating structural unit has the general formula:

$$-CH_2-C(CH_3)=CH-CH_2-$$

i) By drawing displayed formulae of this structural unit, show that it can exist in *cis* and *trans* forms.

ii) Natural rubber is poly-*cis*-isoprene. Draw the displayed formula of three monomer molecules linked together in part of a natural rubber molecule.

Comments on the activities

Activity 8.1

1 **a)** hex-2-ene*; **b)** 4-methylpent-2-ene*; **c)** 2-methylpent-2-ene **d)** *cis*-hept-3-ene; **e)** *trans*-3-methylpent-2-ene; **f)** 2,4-dimethylpent-2-ene (* may be *cis* or *trans*-isomers).

2 Remember to draw the longest unbranched chain first.

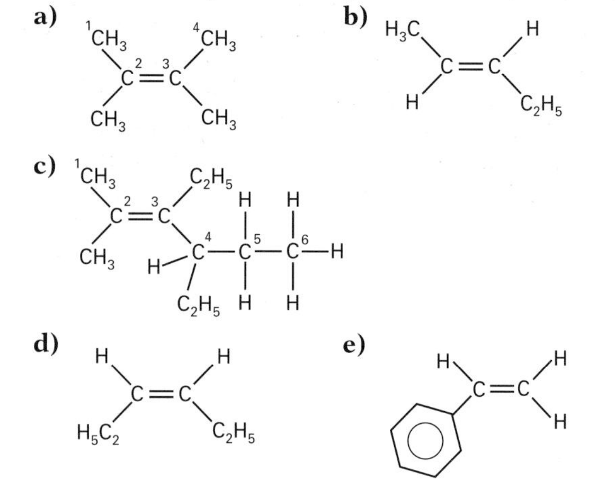

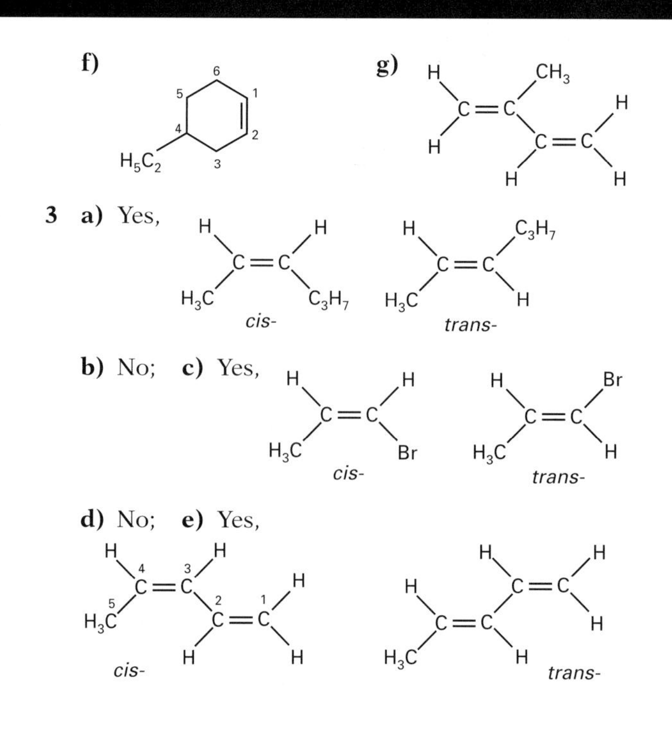

Comments on the activities *continued*

Activity 8.2

1 You could use any reaction which gives a different observation with saturated hydrocarbons (Table 8.6).

2 **a)** Br_2 dissolved in hexane(l), at room temperature;
b) HBr(g), warm;
c) Two steps: firstly, conc. H_2SO_4(l) at room temperature; then H_2O(l) warm
d) H_2(g), finely divided nickel, 200°C;
e) $KMnO_4$/dil. H_2SO_4(aq), reflux.

3

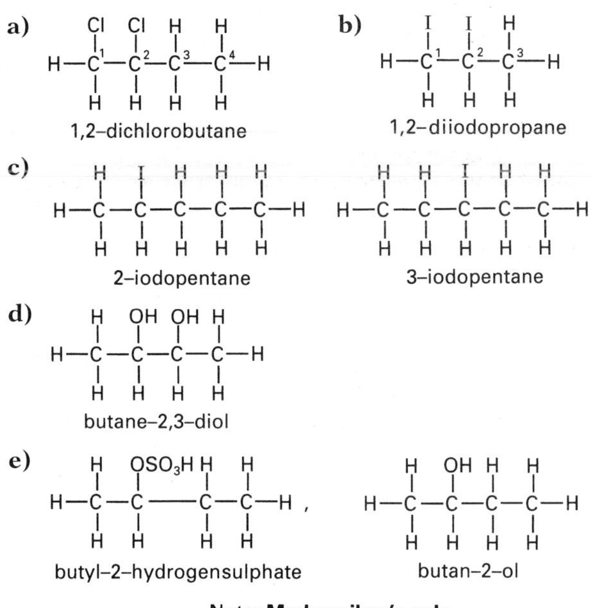

f)

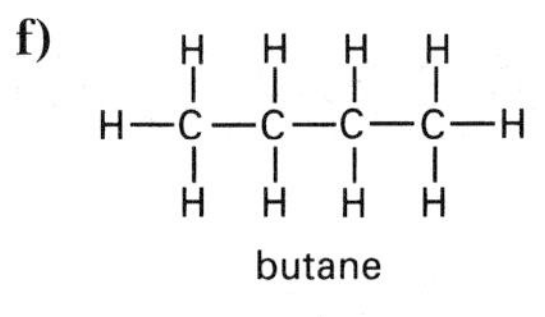

Activity 8.3

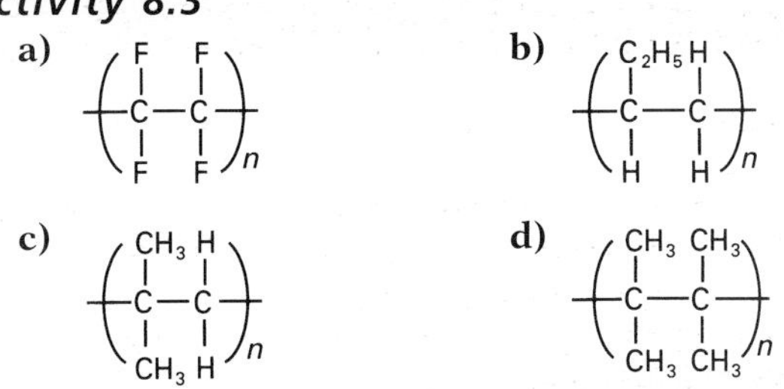

Note: When drawing structures derived from the monomer, remember that each repeating unit contributes only two carbon atoms to the chain.

Activity 8.4

2 **a)** Mass of the recovered plastic

$= 1.036 + 2.422 = 3.458$ tonnes

b) Mass of the landfill plastics

$= 15.23 - 3.458 = 11.77$ tonnes

$$\% \text{ of landfill plastics in waste} = \frac{11.77}{15.23} = 77.28\%$$

3 **a)** From the text plastics make up about 7% of domestic waste.

$$\text{Mass of plastic waste} = \frac{7}{100} \times 132$$

$= 9.24$ million tonnes

b) i) Energy obtained from domestic waste

$=$ mass (tonnes) $\times$ calorific value

$= 132 \times 10^6 \times 9 = 1.19 \times 10^9$ GJ

ii) $$\text{Mass of coal saved} = \frac{\text{energy saved}}{\text{calorific value for coal}}$$

$$= \frac{1.19 \times 10^9}{30}$$

$= 3.97 \times 10^8$ (397 million) tonnes

iii) Energy obtained from plastics

$= 9.24 \times 10^6 \times 40 = 3.70 \times 10^8$ GJ

$$\text{Mass of coal saved} = \frac{3.70 \times 10^8}{30}$$

$= 1.23 \times 10^7$ (12.3 million) tonnes

c) Mass for mass, separated plastic waste produces over four times as much energy as domestic waste and reduces the bulk of the material to be incinerated. However, the cost of separating out the plastic materials is very high. Thus it is more economical to burn the domestic waste without separating out the plastic material.

C H A P T E R

9

Aromatic hydrocarbons

Contents

Study Checklist

By the end of this chapter you should be able to:

1. Describe, and explain, the structure of benzene in terms of delocalisation of electrons. Use thermochemical data from (a) the enthalpies of hydrogenation of cyclohexene and benzene and (b) the formation of benzene from its gaseous atoms as evidence for this structure.
2. Understand why benzene has a planar structure and a carbon-carbon bond length intermediate between that of a single and double bond.
3. Draw structural formulae for the following compounds: benzene, methylbenzene, nitrobenzene, chlorobenzene, bromobenzene, phenol, phenylethanone and benzoic acid.
4. Describe the chemistry of arenes, focussing on the following reactions of benzene and methylbenzene: (a) nitration, (b) substitution reactions with chlorine and bromine,(c) sulphonation, (d) Friedel-Crafts alkylation and acylation reactions and (e) oxidation of the side chain by acidified potassium manganate(VII).
5. Summarise the reactions of benzene and methylbenzene as shown in Focus 9c.
6. Compare the reactions of cyclohexane, cyclohexene and benzene with (a) chlorine and bromine, (b) nitrating mixture and (c) concentrated sulphuric acid, as shown in Table 9.2.
7. Describe the mechanism of electrophilic substitution in arenes, using the mononitration of benzene as an example. Use 'curly arrows' to show the movement of electron clouds during the reaction and explain how the intermediate carbocation is stabilised by electron delocalisation.
8. Predict whether halogenation will occur in the side-chain or aromatic nucleus in arenes depending on the reaction conditions.
9. State the positions of electrophilic substitution in methylbenzene. Understand that a substituent may alter the density of the π electron ring in benzene, influencing its reactivity and the preferred positions(s) for further electrophilic substitution.

During the early years of organic chemistry, chemists were puzzled by the structures of some hydrocarbons which seemed to be unsaturated, yet resisted addition reactions. Originally, these compounds were called **aromatic hydrocarbons** because of their pleasant smell.

In 1825, the English scientist Michael Faraday isolated **benzene**, the simplest aromatic hydrocarbon, during the distillation of whale oil. Its value as a solvent for waxes, fats and resins was soon realised and within twenty years large amounts were being obtained from the distillation of coal tar. Once its molecular formula, C_6H_6, was worked out in 1834, various 'alkene-like' structures were proposed, for example:

$$CH_2{=}C{=}CH{-}CH{=}C{=}CH_2 \qquad HC{\equiv}C{-}CH{=}C{=}CH{-}CH_3$$

▲ Although it is accepted that Friedrich August Kekulé was the first to publish the structure of benzene, in 1865, an Austrian schoolteacher named Josef Loschmidt had proposed a ring of six carbon atoms in 1861. Whilst Loschmidt's work was not widely available, Kekulé had read Loschmidt's paper and perhaps this influenced the dream in which he claimed to have visualised benzene's structure. Apparently, this dream occurred whilst Kekulé dozed in front of the fire and he described it as follows: 'I turned the chair to the fireplace and sank into a half sleep. The atoms flitted before my eyes. Long rows, variously, more closely, united; all in movement, wriggling and turning like snakes. And see, what was that? One of the snakes seized its own tail and the image whirled scornfully before my eyes. As though from a flash of lightning I awoke; I occupied the rest of the night in working out the consequences of the hypothesis.' (*The Art of Scientific Investigation*, W.I.B. Beveridge, Heinemann, 3rd Edn 1957)

What puzzled chemists was that unsaturated hydrocarbons like those above would be extremely reactive, giving addition reactions at room temperature – yet benzene seemed to be almost inert! Benzene's structure remained unclear for many years until, in 1865, Friedrich August Kekulé, a German professor of chemistry at the University of Ghent, proposed that its molecule had a hexagonal ring of six carbon atoms joined by alternating double and single bonds. To explain the lack of chemical reactivity, Kekulé suggested that the double and single bonds in this cyclic triene were rapidly oscillating (Figure 9.1), *an idea which we now know to be incorrect*. However, at the time Kekulé's model did explain the chemistry of benzene very well and it was accepted for many years.

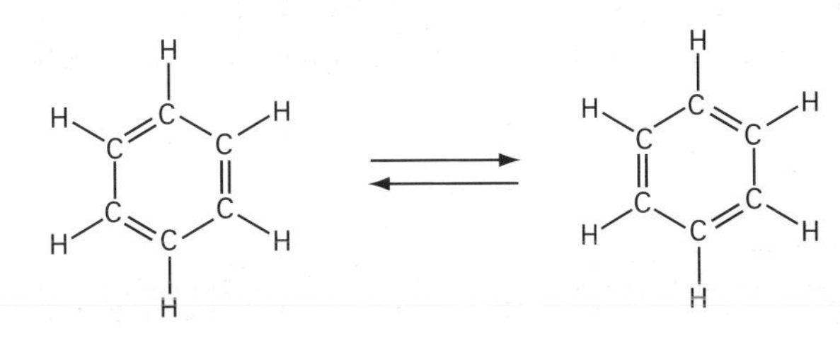

Figure 9.1 ▲
In order to explain benzene's lack of reactivity, Kekulé proposed that its structure consisted of two cyclic trienes which were rapidly oscillating, an idea which was later shown to be incorrect

9.1 A modern view of benzene's structure

The main problem with Kekulé's model is that the cyclic triene, like other alkenes, should give addition reactions because of its high level of unsaturation. In fact, benzene resists addition and most of its reactions involve substitution of one or more hydrogen atoms. Our modern view of benzene's structure is based on three main pieces of evidence:

1 Double bonds are shorter than single bonds between the same atoms, for example, the average C—C bond is 0.154 nm long whereas the bond length of the C=C bond is 0.134 nm. An electron density map obtained from **X-ray diffraction studies** (section 4.2) shows that all the carbon–carbon bond lengths in benzene are equal and, at 0.139 nm, are intermediate between a single and a double bond length.
2 Like cyclohexene, benzene adds hydrogen at 200°C in the presence of a nickel catalyst and the product is cyclohexane. If benzene is a cyclic triene, we would expect the enthalpy change for this reaction to be about *three times* that of the hydrogenation of cyclohexene. In fact, the enthalpy change is much *less* exothermic than predicted, Figure 9.2. As less energy is released it appears that 152 kJ more energy is needed to break the π bonds in one mole of benzene than to break those in one mole of the hypothetical cyclic triene. Thus, the carbon–carbon bonds in benzene are, on average, much stronger than those in the cyclic triene.

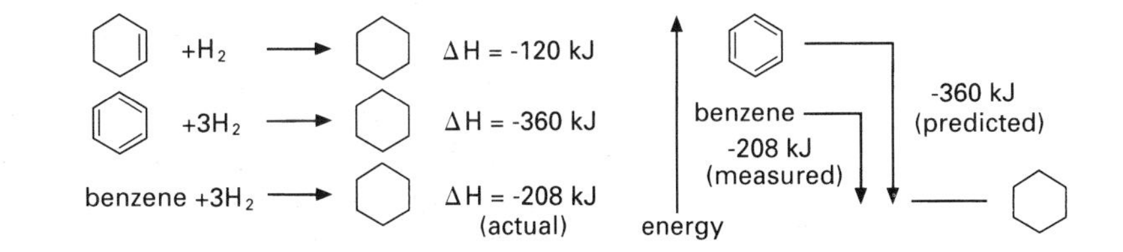

Figure 9.2 ►
Enthalpies of hydrogenation of benzene and cyclohexene suggest that the structure of benzene is about 150 kJ mol^{-1} more stable than Kekulé's cyclic triene

3 More thermochemical evidence comes from the formation of benzene from its gaseous atoms:

$$6C(g) + 6H(g) \longrightarrow C_6H_6(l)$$

The reaction is highly exothermic with a measured enthalpy change, $\Delta H_{\text{measured}}$, of –5516 kJ mol^{-1}. If we assume that benzene is a cyclic triene, we can use mean bond enthalpies (section 3.10) to calculate an enthalpy change, $\Delta H_{\text{calculated}}$, for the reaction:

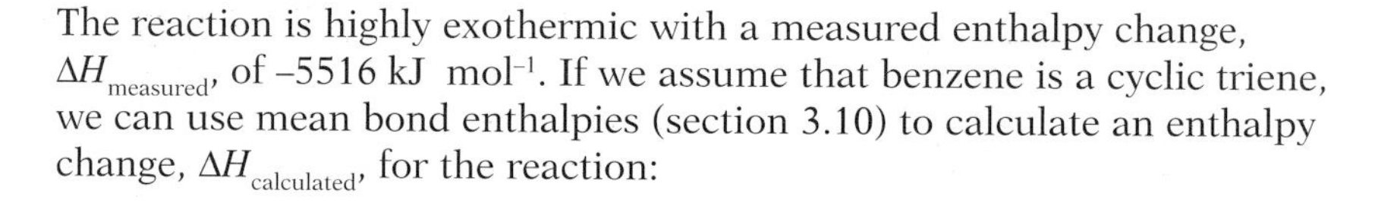

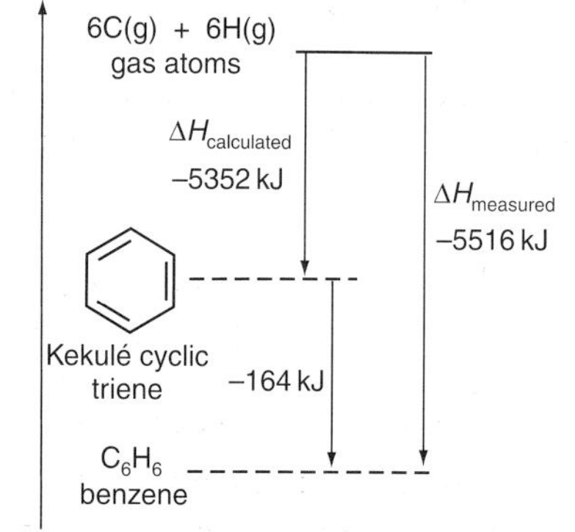

Figure 9.3
Comparing the measured enthalpy of formation of benzene from its gaseous atoms with that for the Kekulé structure calculated using bond energy terms. This shows that benzene is 164 kJ mol^{-1} *more* stable than the cyclic triene

The Kekulé model consists of 6 C—H, 3 C—C and 3 C=C bonds, so:

bonds made (moles): 6C—H + 3 C—C + 3 C=C
$\Delta H_{\text{calculated}}$ (–ve sign): 6 (–412) + 3(–348) + 3(–612) = – 5352 kJ mol^{-1}.

The 'minus' sign arises because bond formation is an exothermic process. Now, if Kekulé's structure for benzene is correct, $\Delta H_{\text{measured}}$ should equal $\Delta H_{\text{calculated}}$ within experimental error. In fact, when benzene is formed from gas atoms, 164 kJ mol^{-1} more energy is released than would be the case if the cyclic triene was formed. Once again, we see that benzene's structure is more stable than the cyclic triene (Figure 9.3).

If we consider the structure of benzene in terms of orbital overlap, then the true situation becomes clearer. As in ethene, each carbon forms three hybrid orbitals by mixing atomic orbitals:

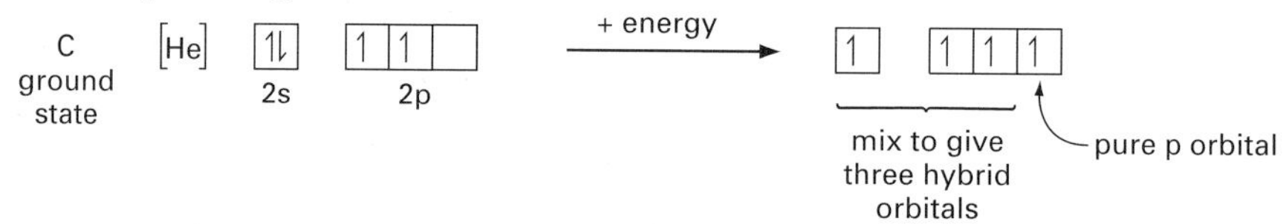

The hybrid orbitals overlap with orbitals of the neighbouring carbon and hydrogen atoms to form a planar hexagonal ring of σ bonds:

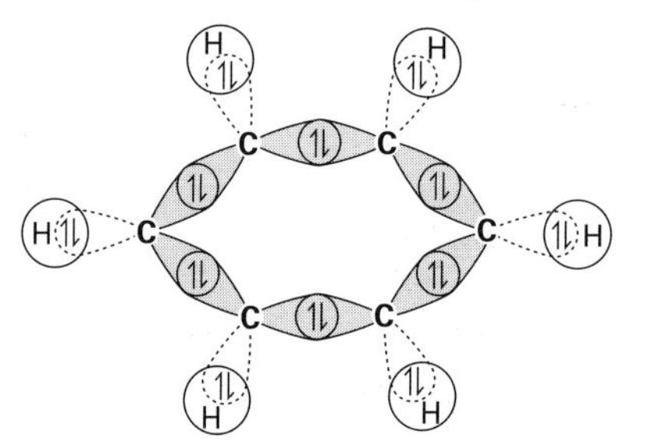

The remaining unpaired electron on each carbon atom occupies a pure p orbital at right angles to the plane of the ring. These p orbitals overlap sideways with their neighbours, above and below the σ bonded ring, to form a circular π molecular orbital which contains six electrons and extends over all the carbon atoms in the ring:

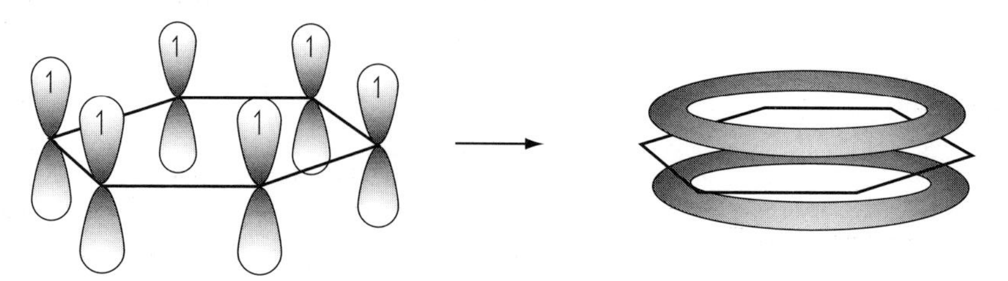

The electrons in the π bonds are not associated with any particular carbon atom but are spread around the ring – they are said to be **delocalised**. The effect of delocalisation is to even out the electron charge and this explains why benzene is so stable (evidence 2 and 3). Each carbon–carbon bond is identical, consisting of a single (σ) bond and a partial double (π) bond – thus, the bonds are equal in length with a value intermediate between that of a single and double bond (evidence 1).

9.2 Naming aromatic compounds

Nowadays, thousands of organic compounds are known to contain a benzene ring. In the structural formula, a benzene ring is indicated by a circle within a hexagon, for example:

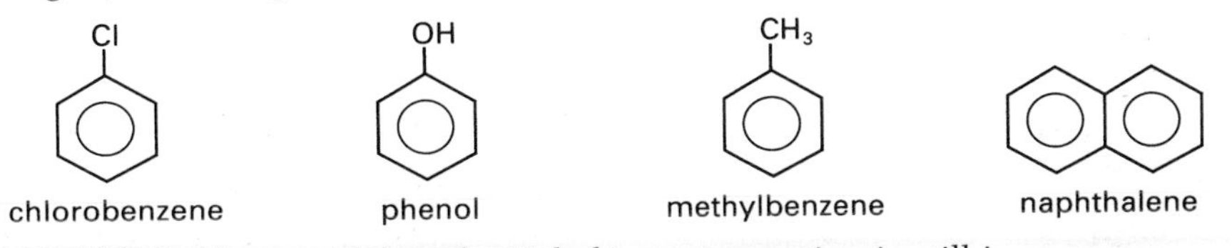

chlorobenzene phenol methylbenzene naphthalene

(It should also be noted that the Kekulé representation is still in common use). Even though many of these compounds are foul-smelling and toxic, we still use the term aromatic to describe them. Aromatic hydrocarbons, such as methylbenzene and naphthalene, are often called **arenes**.

Aromatic compounds can be named in three ways:

- by identifying the functional groups or hydrocarbon chains which are attached to the benzene ring, for example:

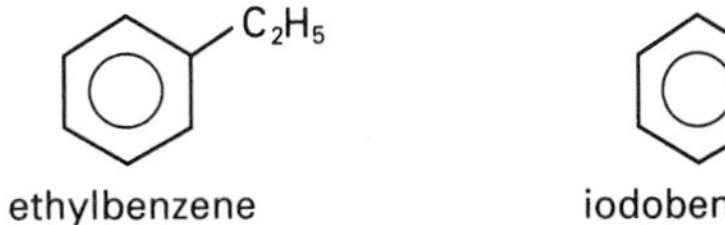

ethylbenzene

iodobenzene

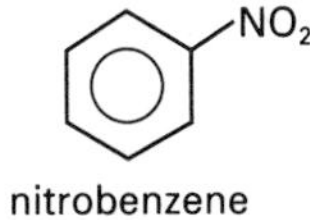

nitrobenzene

- by using the name **phenyl** to identify each ⌬– group in the molecule, for example:

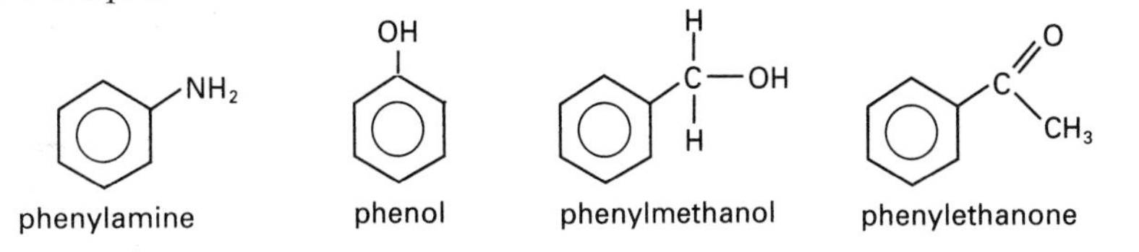

phenylamine phenol phenylmethanol phenylethanone

- sometimes a molecule will also be known by a historical (trivial) name, for example:

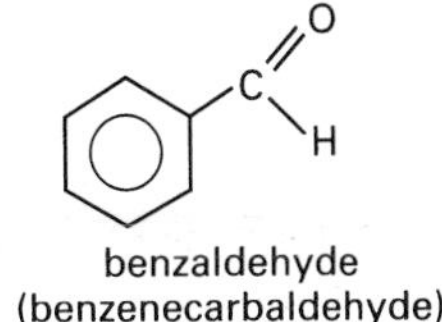

benzaldehyde
(benzenecarbaldehyde)

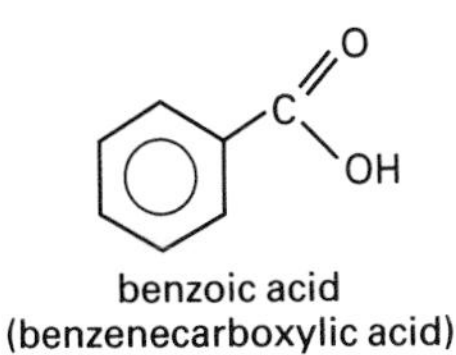

benzoic acid
(benzenecarboxylic acid)

Structural isomerism can occur in substituted benzenes. When naming these compounds, we must number the relative positions of the substituent groups, for example:

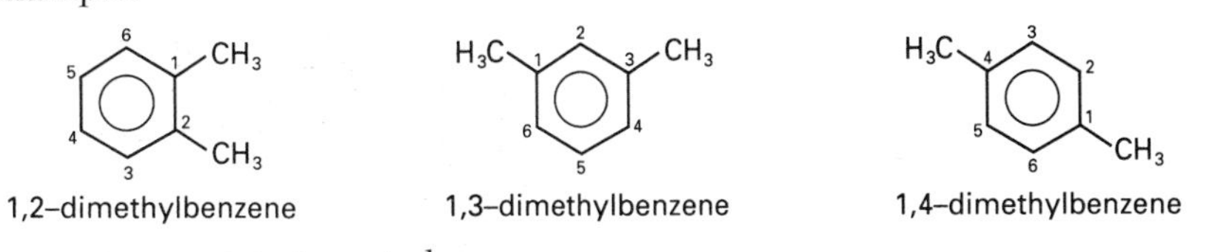

1,2-dimethylbenzene 1,3-dimethylbenzene 1,4-dimethylbenzene

There are two points to note here:

- 1,5- and 1,6-dimethylbenzene are incorrect names because they are the 1,3- and 1,2-isomers, respectively.

- Avoid thinking of the benzene ring solely in terms of clockwise numbering with position 1 at the top.

Often a substituted benzene has a name based on a common 'parent' molecule for example:

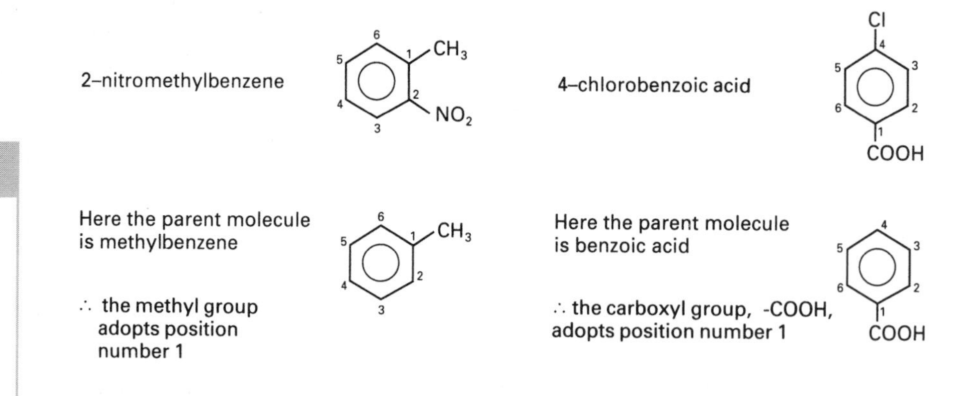

Focus 9a

1. Aromatic compounds have a common structural feature, the benzene ring. Aromatic hydrocarbons are known as **arenes**.
2. Benzene's molecule contains a **delocalised π bond system** above and below the σ bond framework. Evidence for this structure comes from bond lengths (measured using X-ray diffraction) and a comparison of the enthalpies of hydrogenation of benzene and cyclohexene.
3. **Positional isomerism** can occur in 'multi-substituted' benzene derivatives. Thus, you must always number the relative positions of the substituent groups in these compounds.

Activity 9.1

1 Name the aromatic compounds whose structural formulae are shown below:

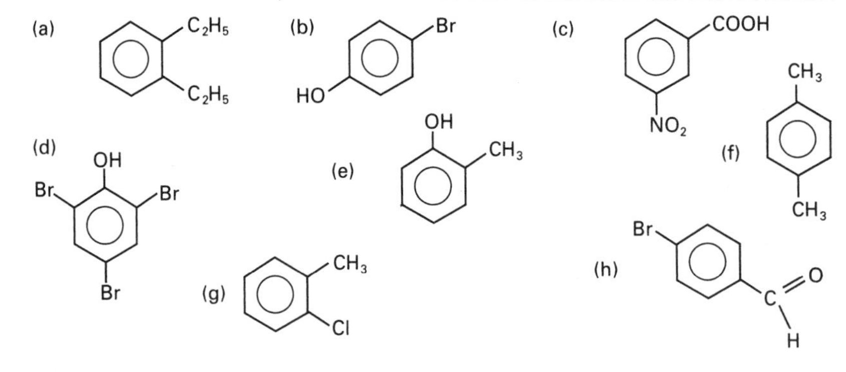

2 Draw structural formulae for the following compounds.

a) methylbenzene **b)** 1,2-dihydroxybenzene
c) 4-nitrophenol **d)** 4-bromophenylamine
e) 3-chlorobenzaldehyde **f)** 3-chlorobenzoic acid

9.3 Uses and physical properties of benzene and methylbenzene

Benzene and methylbenzene are widely used as non-polar solvents and starting materials in the manufacture of many commercially important aromatic compounds, such as polystyrene and trinitrotoluene (TNT). Arenes are not found naturally in large amounts. There are two main methods of producing benzene and methylbenzene: (i) by cracking and reforming the naphtha fraction from petroleum (section 7.4) and (ii) by heating coal in the absence of air.

Benzene and methylbenzene are colourless liquids with boiling points of 80 and 111°C respectively. Both liquids are almost insoluble in water and float on the surface. They are good solvents for non-polar compounds, the one drawback being that benzene is very carcinogenic.

9.4 The reactivity of benzene

Like alkenes, benzene has an **electron-rich reactive centre** namely, its delocalised ring of π electrons. Thus, benzene will also be a target for **electrophilic** attack, for example by Cl^+ and Br^+. Also, since benzene is an unsaturated molecule, we would expect it to undergo addition reactions. In fact, benzene resists most of the addition reactions shown by alkenes, for example:

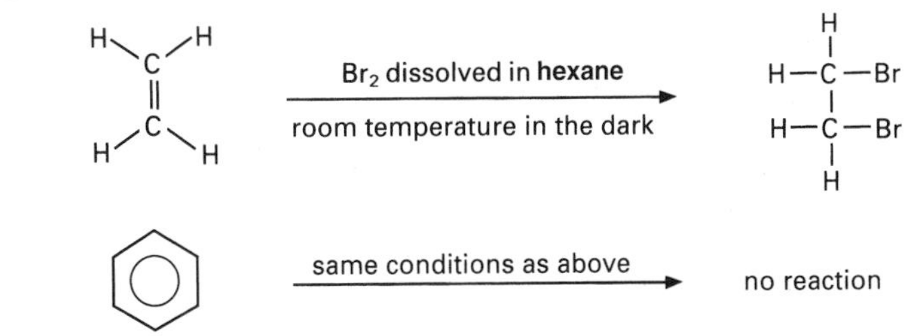

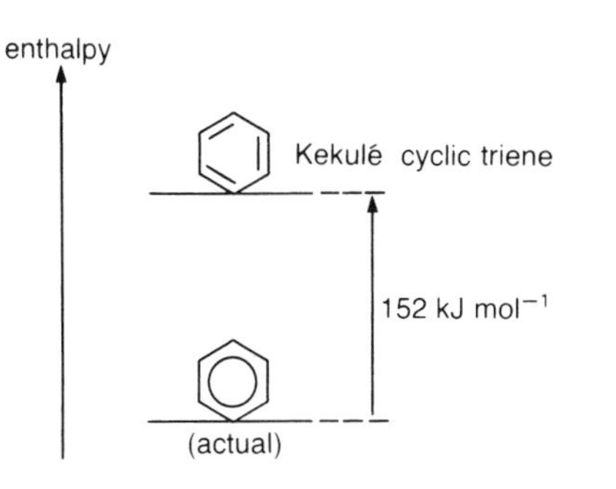

Figure 9.4 ▲
The relative energetic stabilities of benzene and Kekulé's cyclic triene

As we have seen, benzene has a delocalised π bond system, rather than three separate double bonds. Delocalisation of the π electron cloud stabilises the molecule, by about 152 kJ mol^{-1}, in comparison to Kekulé's cyclic triene, Figure 9.4. On reaction, a benzene molecule needs to absorb this amount of energy in order to localise the π bonds. If benzene were to undergo an electrophilic addition, at one of these localised π bonds, the aromatic molecule formed would have a less extensive delocalised cloud and, hence, less stability than benzene itself. Thus, electrophilic addition reactions are energetically unfavourable.

A few addition reactions take place but these occur via a free radical mechanism, for example:

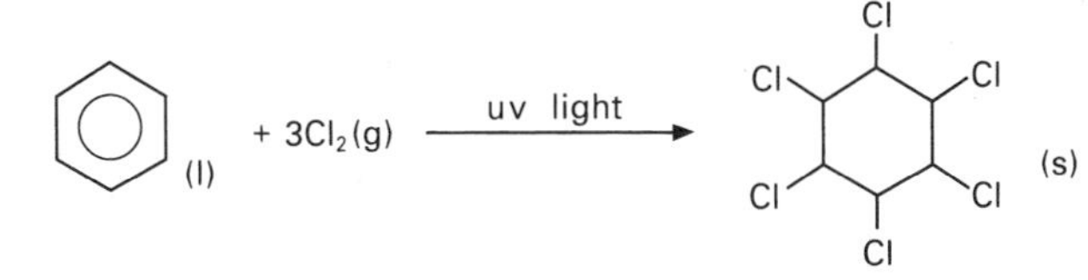

The product 1,2,3,4,5,6-hexachlorocyclohexane can exist as eight geometrical isomers. Can you explain why? One of these isomers, known as Gammexane, is used as an insecticide.

Focus 9b

1. Benzene and methylbenzene are widely used in industry as (i) solvents and (ii) starting materials in the manufacture of other aromatic compounds. They are mainly obtained from the catalytic reforming of naphtha.
2. Benzene, though unsaturated, *resists* electrophilic addition reactions (e.g. like those of alkenes). Such reactions remove part, or all, of delocalisation of the π electron cloud which stabilises the benzene molecule. Some free radical addition reactions are known, and these demonstrate the unsaturated nature of benzene.

9.5 Electrophilic substitution in benzene

In these reactions, a C—H bond is converted into a C—X bond, where X may be $-NO_2$ (nitration), –Cl or –Br (halogenation), $-SO_3H$ (sulphonation), $-CH_3$ (alkylation) or $-COCH_3$ (acylation).

Nitration

Nitrobenzene is an intermediate product in the manufacture of explosives and dyes (section 14.1). It is made by warming benzene with a **nitrating mixture** of concentrated nitric acid and concentrated sulphuric acid at 50°C:

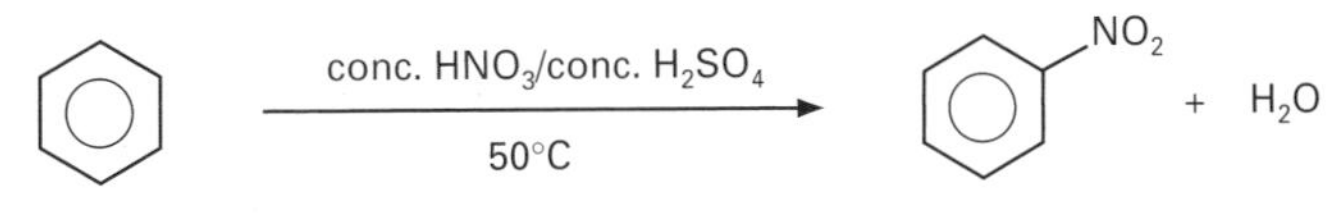

On diluting the reaction mixture with water, a dense oily layer of nitrobenzene appears and this can be separated by solvent extraction. After drying with

▲ A TNT explosion

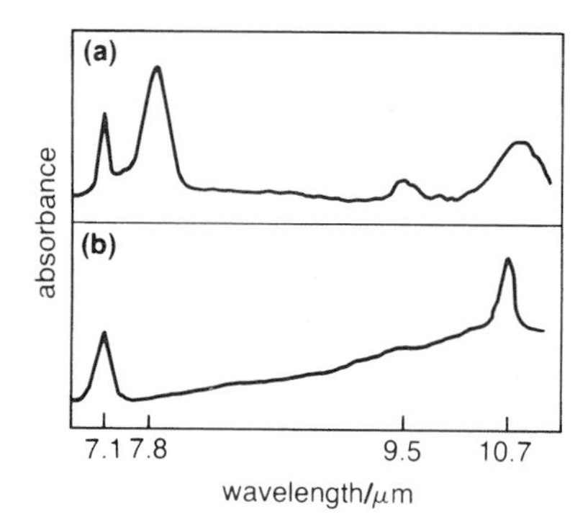

Figure 9.5 ▲
Infrared spectra of **(a)** 'nitrating' mixture and **(b)** aqueous nitronium chlorate(VII), $NO_2^+\ ClO_4^-$ (aq). In both, the peak at 7.1 μm is due to the nitryl cation, NO_2^+

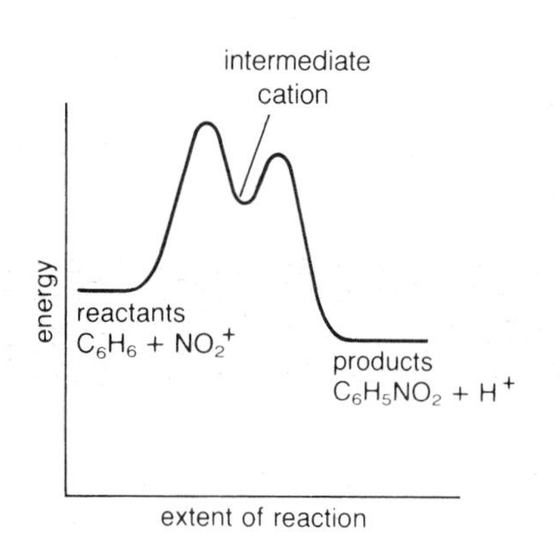

Figure 9.6 ▲
Energy profile for the reaction between benzene and the nitryl cation

anhydrous calcium chloride, the pure nitrobenzene (b.p. = 210°C) is obtained by distillation.

Mechanism of the nitration of benzene

Individually, concentrated nitric acid or sulphuric acids do *not* react with benzene at 50°C. On mixing, though, they form the **nitryl cation** or **nitronium ion**, NO_2^+. Evidence for this comes from a similar infrared absorption spectra of the 'nitrating mixture' and nitronium chlorate(VII), $NO_2^+ClO_4^-$(s) (Figure 9.5). Furthermore, like 'nitrating mixture', nitronium chlorate(VII) will nitrate benzene. Thus, we conclude that the nitryl cation is responsible for the nitration of benzene, according to the following mechanism.

Nitryl cation formation Sulphuric acid is so strong an acid that it can **protonate** a nitric acid molecule, forming an $H_2NO_3^+$ ion. This breaks up to give a nitryl cation, NO_2^+:

$$HNO_3 + 2H_2SO_4 \longrightarrow NO_2^+ + H_3O^+ + 2HSO_4^-$$

Attack of the NO_2^+ electrophile on the benzene ring The nitryl cation is attracted to the benzene's delocalised π electron cloud, and it bonds with one of the carbon atoms. Notice that this removes the stability of the π electron cloud. However, the reaction is feasible because the intermediate cation is also stabilised, to some extent, by the delocalisation of the remaining four π electrons over the other five carbon atoms (Figure 9.6).

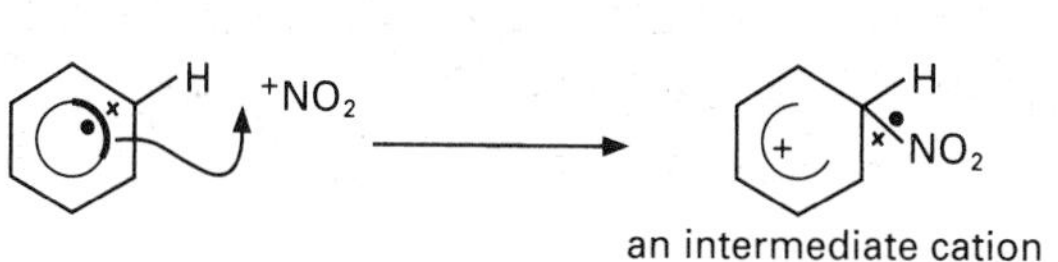

(here the '+' means that the positive charge from the $^+NO_2$ is now spread over carbon atoms 2–6)

Proton loss The intermediate cation rapidly ejects a proton, H^+, reforming the ring of delocalised π electrons and stabilising the product:

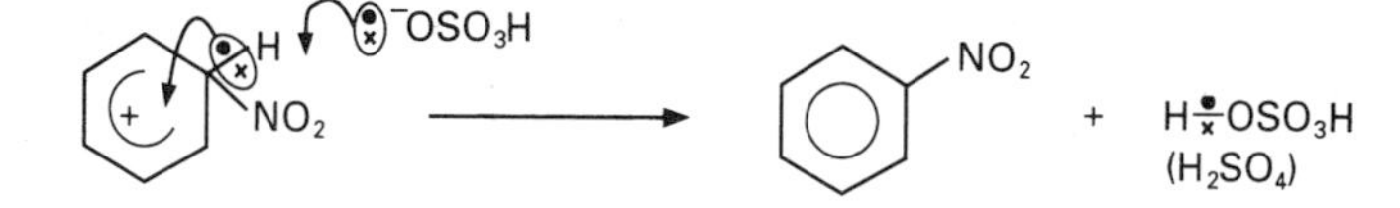

Notice here that sulphuric acid is acting as a **homogenous catalyst**.

Halogenation, alkylation and acylation of arenes

Although benzene is an unsaturated hydrocarbon, it does not react with halogens unless a catalyst is present, for example:

$$C_6H_6(l) + Br_2(l) \xrightarrow[\text{in the dark}]{AlCl_3\text{, room temp.}} C_6H_5Br(l) + HBr(g)$$

The catalyst, known as a **halogen carrier**, promotes the formation of the Br^+ electrophile:

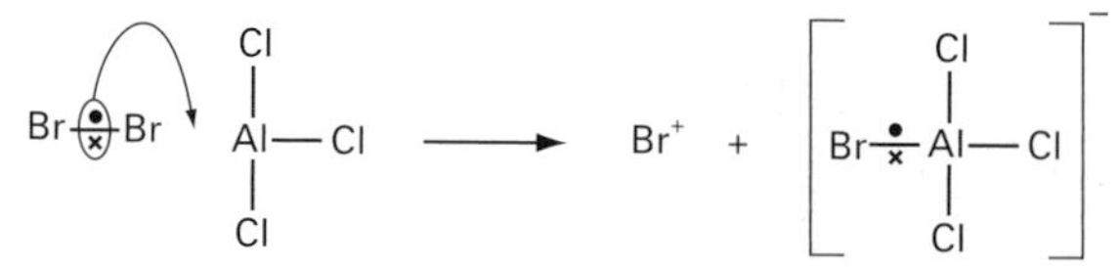

The reaction of methylbenzene with halogens is interesting because the reaction conditions determine which products are formed. In ultraviolet light or strong sunlight, **free radical substitution occurs** in the methyl side chain (as in alkanes, section 7.8). For example:

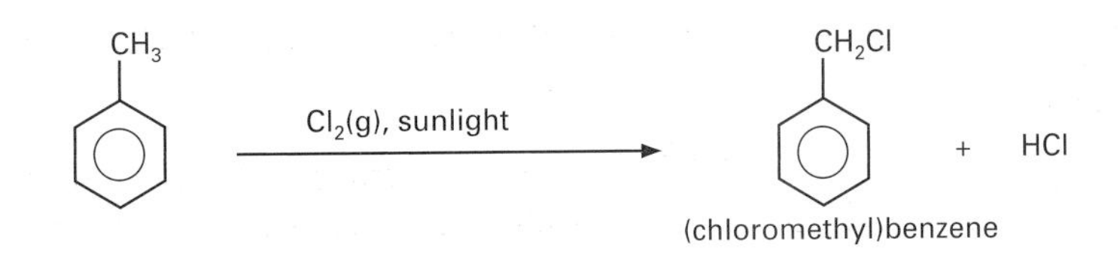

(Notice how we use a bracket to clarify the name of a compound with a substituent in its side chain.) In the dark, however, in the presence of a halogen carrier, **electrophilic substitution** occurs in the aromatic ring, for example:

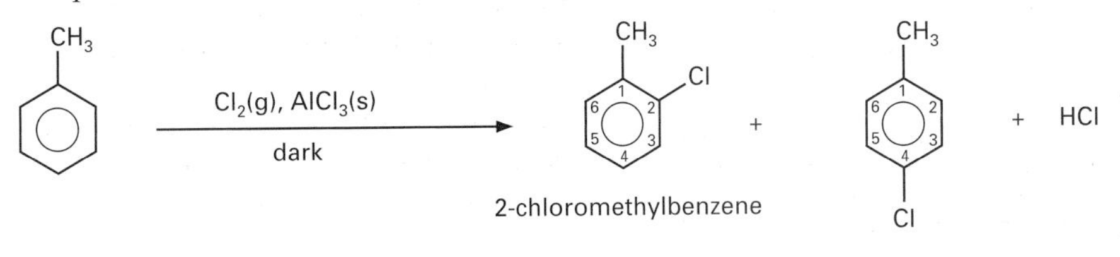

These products are structural (positional) isomers. Due to the methyl group's ability to release electrons, (page 224), the methylbenzene molecule becomes slightly polarised and a partial negative charge builds up at the 2 and 4 ring positions respectively. Thus, an incoming electrophile will be directed towards these positions and the 1,2- and 1,4- products will be formed.

The cumene process: an industrial application of a Friedel–Crafts reaction

In 1877, the French chemist Charles Friedel and his American colleague James Mason Crafts discovered that aluminium chloride catalyses the attachment of an alkyl group to a benzene ring. Since then, Friedel-Crafts reactions have played an important role in synthetic chemistry, for example in the cumene process for the manufacture of phenol and propanone, as shown below.

The electrophile for the Friedel-Crafts alkylation is a carbocation, $(CH_3)_2CH^+$, which may be generated by using $AlCl_3$ or H^+ ions (from concentrated phosphoric acid) as a catalyst.

Phenol is used to make antiseptics and plastics, such as Bakelite and nylon (section 11.10) whilst propanone is a valuable solvent and is used to make perspex (section 12.8).

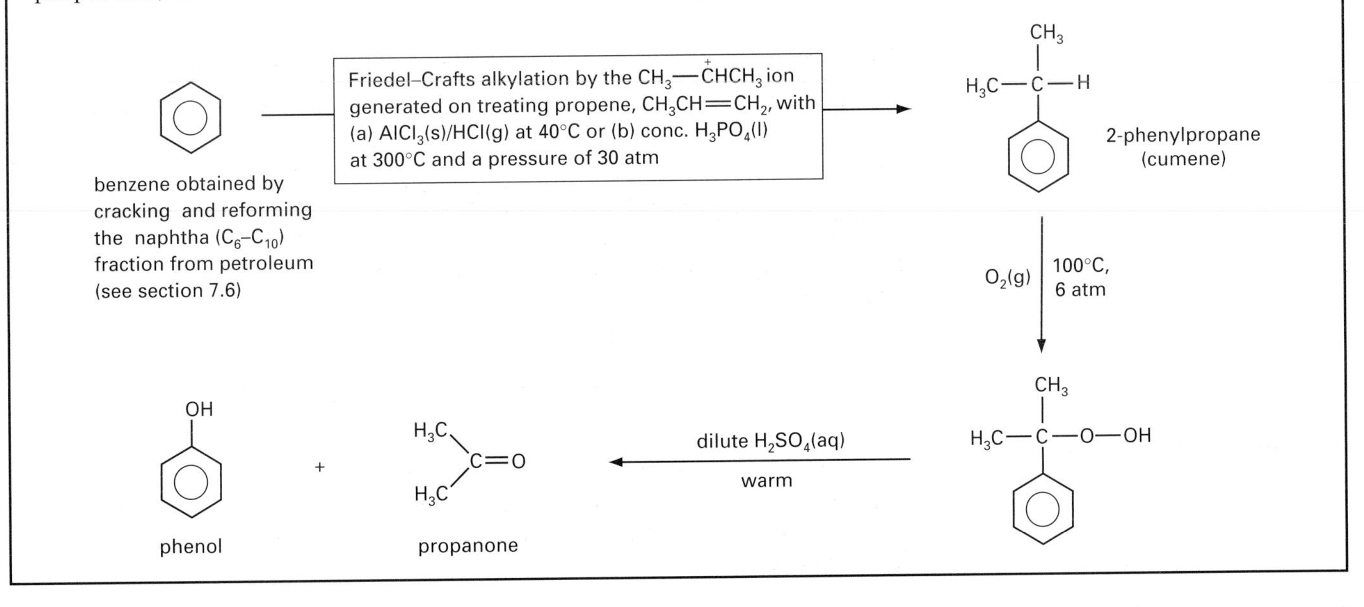

Alkylation and acylation involve the following structural changes:

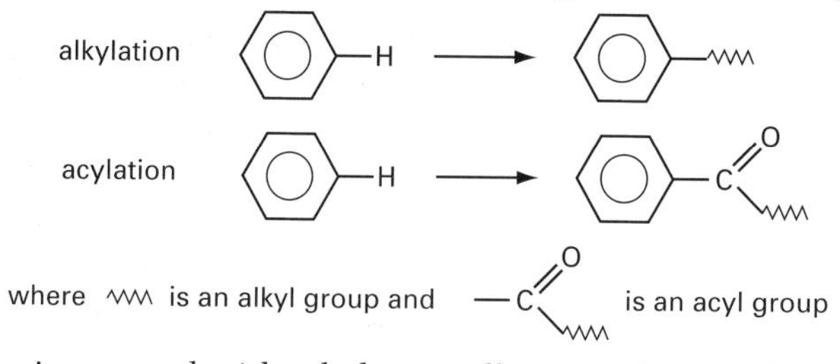

When benzene is warmed with a halogenoalkane, and a Friedel–Crafts catalyst, such as $AlCl_3(s)$, **alkylation** occurs, and an alkylbenzene is formed, for example:

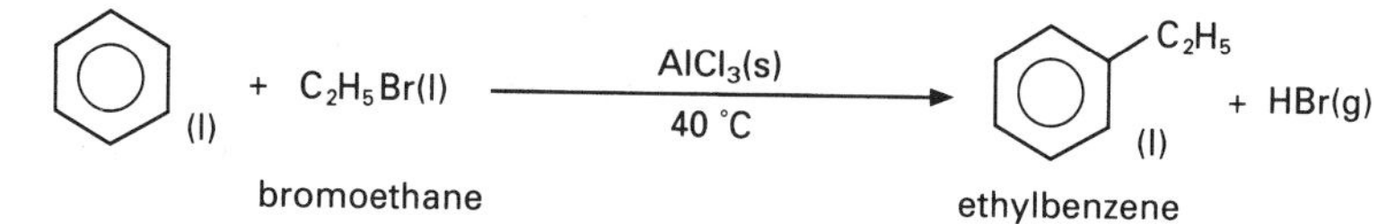

Alkylation of benzene is an important synthetic step in the manufacture of polystyrene and detergents (section 7.7). In industry, though, it is cheaper to use an alkene direct from the cracking of a petroleum fraction instead of a halogenoalkane. For example, ethylbenzene is produced by reacting benzene with ethene in the presence of hydrogen chloride and aluminium chloride. The ethylbenzene can then be catalytically dehydrogenated to phenylethene (styrene), the monomer from which polystyrene is made (section 8.7).

On warming benzene with an acid chloride and a Friedel–Crafts catalyst, **acylation** occurs, and the product is an aromatic ketone, for example:

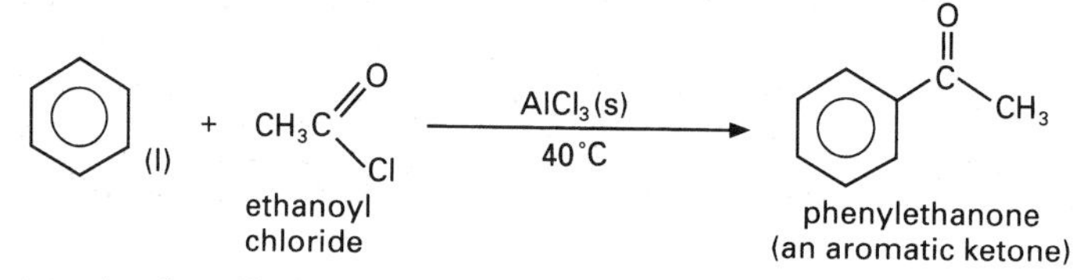

Not surprisingly, the alkylation, acylation and halogenation reactions of benzene have almost identical mechanisms. Of most importance is the ability of the Friedel–Crafts catalyst (halogen carrier) to behave as an electrophile, that is, an electron-pair acceptor. Thus the δ^+ aluminium atom in the catalyst attracts the electron cloud in the C—Hal or Hal—Hal bond:

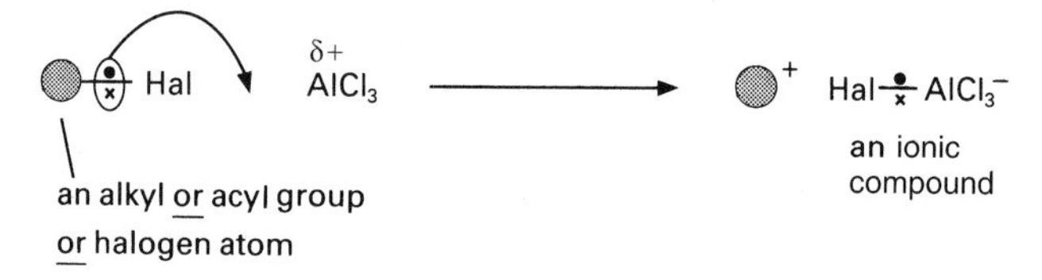

A ionic compound is formed. Since this possesses an electron-poor centre (the positive ion), it can act as an electrophile. As in the nitration of benzene, the electrophile attacks benzene's π electron ring (1) and, finally, a proton is lost (2):

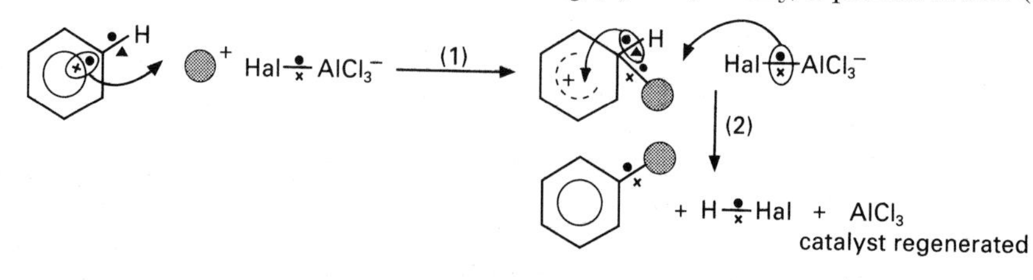

Sulphonation

The structural change C—H ⟶ C—SO_3H is called **sulphonation.**

Benzene is sulphonated by refluxing it with concentrated sulphuric acid for several hours:

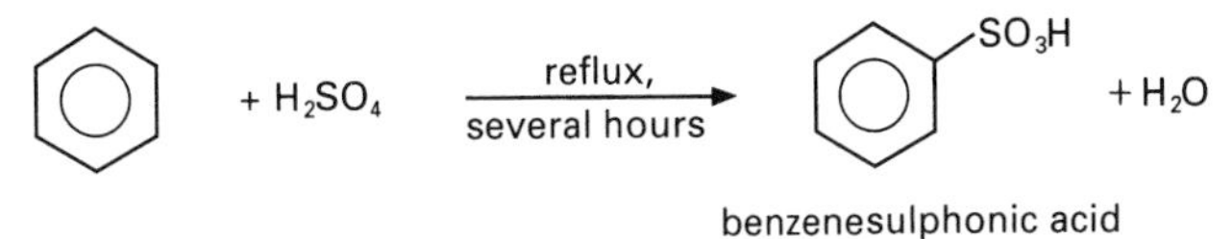

Benzenesulphonic acid, a strong acid, is a crystalline solid which is very soluble in water. Thus, by sulphonating an aromatic compound we can make it water soluble. For this reason, a sulphonation reaction is often used in the manufacture of water-soluble dyes and detergents. (Sections 18.4 and 13.7 respectively.)

Benzenesulphonic acid is also an intermediate compound in the synthesis of phenol:

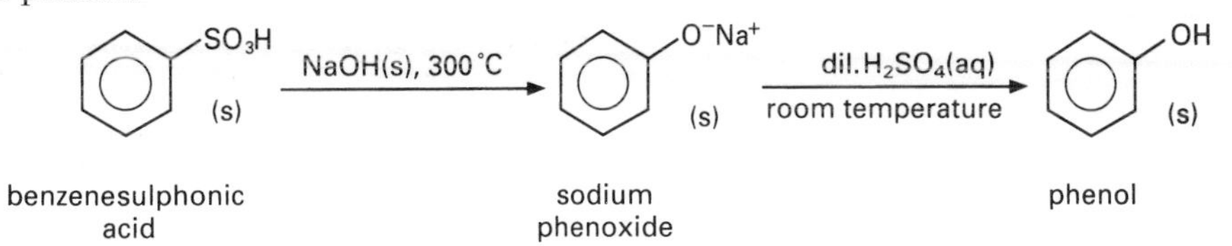

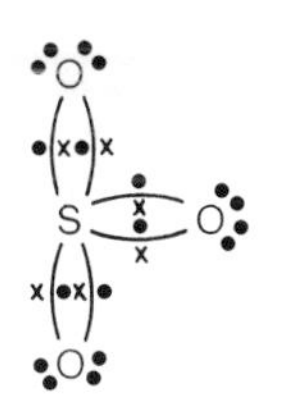

Figure 9.7
Sulphur trioxide can act as an electrophile because the sulphur atom is able to accept a lone pair of electrons

The mechanism of sulphonation resembles that for nitration except that, this time, the electrophile is **sulphur trioxide, SO_3**, a neutral molecule (Figure 9.7). Indeed, the sulphonation is very much faster if we use **fuming sulphuric acid**, that is, a solution of sulphur trioxide in concentrated sulphuric acid.

How do substituent groups affect the reactivity of the benzine ring?

If one of the hydrogens in benzene is replaced by another atom, or group of atoms, the density of the electron cloud around the ring is altered. Some groups such as —OH,—NH_2 and —CH_3 activate the benzene ring towards further electrophilic substitution, that is speed up the reaction. Thus, phenol is readily brominated at room temperature:

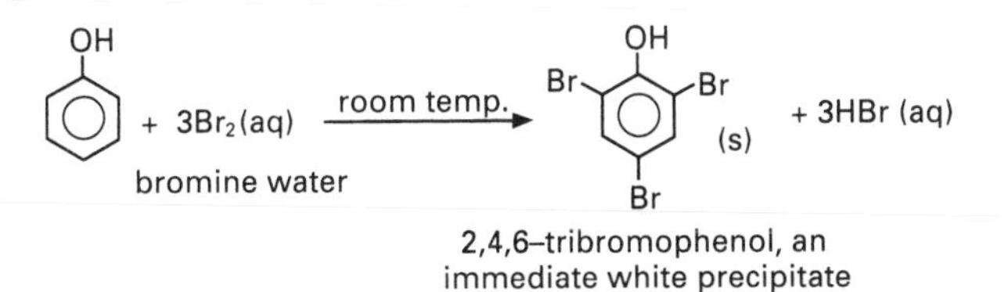

2,4,6–tribromophenol, an immediate white precipitate

Activating groups often have a lone pair of electrons on the atom adjoining the benzene ring and these become delocalised into the π electron cloud. This causes a slight excess of negative charge to develop at ring positions 2 and 4, so the electrophile is more likely to be attached at these carbons, rather than at the 3 position. Hence the product of the above reaction is 2,4,6-tribromophenol.

Some substituent groups can withdraw electrons from benzene's π cloud thereby deactivating the ring towards further electrophilic substitution. Thus, the nitration of nitrobenzene needs a temperature of 100°C whereas benzene can be nitrated at 50°C:

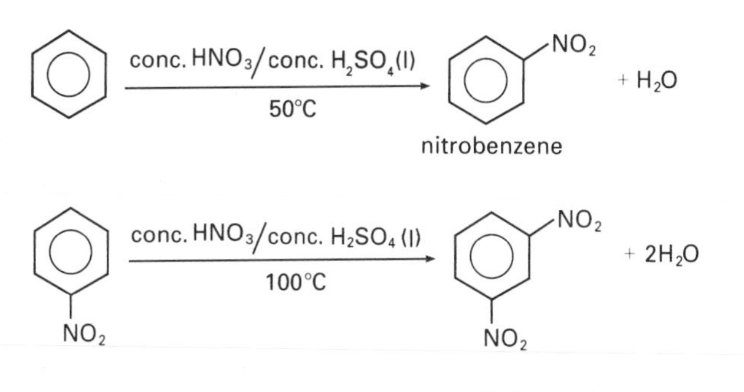

Deactivating groups, such as —NO_2 and —COOH, cause ring positions 2 and 4 to develop a less negative charge making these positions less susceptible to attack by an electrophile (itself electron-poor). Hence, the product of the above reaction is 1,3-dinitrobenzene.

Obtaining substitution at specific positions in the benzene ring is important in industrial synthetic chemistry. An interesting example is the methylation of phenol in which the choice of catalyst determines the percentage of each product formed:

How do substituent groups affect the reactivity of the benzine ring? *continued*

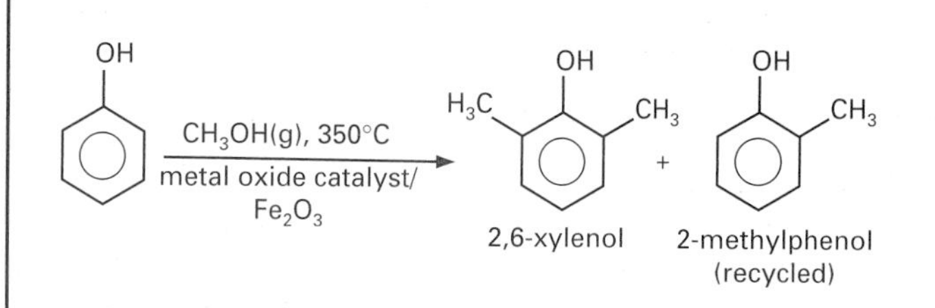

Synthetic Chemicals Ltd, a division of the Shell Group, use this process to make 2,6-xylenol (2,6-dimethylphenol) which is the starting material for the manufacture of local anaesthetics and antidiarrhoea drugs. There are a number of by-products, the most significant being 2-methylphenol which is recycled back to the reaction vessel. The vaporised reactants are passed through the pellets of the catalyst containing about 87% of iron(III) oxide and ideally about 1% of another metal oxide. Early on in the development, various metal oxides were tested to see which gave the best yield of 2,6-xylenol. In the chemical industry, when a major discovery is made, the company usually seeks a **patent** for its 'invention' – this means that a competitor may not use the process without gaining a licence from the 'inventor'. The European Patent Office granted a patent on this process in 1981 and some details are given in Table 9.1.

Table 9.1 Data from the catalysed methylation of phenol

Metal oxide (1% by weight unless otherwise stated) mixed with Fe_2O_3 to make the catalyst	Zr (1%) zirconium	Zr (2%)	Zr (3%)	Zr (10%)	Ti titanium	Pb lead	Cd cadmium	Sn tin
percentage of phenol which reacts when it passes through the catalyst	94	99	60	54	99	68	100	94
percentage of 2,6-xylenol in the products	69.1	82.1	23.7	15.0	81.5	20.9	89.4	76.5
percentage of 2-methylphenol in the products	28.6	16.0	71.2	77.4	12.7	76.8	6.8	20.6
percentage of other products	2.3	1.9	5.1	7.6	5.8	2.3	3.8	2.9

Activity 9.2

1 You have been asked to prepare a sample of 3-nitromethylbenzene from benzene. Bearing in mind the effect of a substituent groups on ring substitution, outline a synthetic route.

2 Use the data in the Table 9.1 to answer this question.
   a) Which catalyst is most effective in causing the phenol to be methylated? Which catalyst is the least effective?
   b) Which catalyst is most specific in methylating phenol to 2,6-xylenol? Which catalyst is the least specific?
   c) What is the effect of increasing the percentage of zirconium oxide in the catalyst on:
      i) the percentage of phenol which is methylated
      ii) the percentage of 2,6-xylenol in the product?
   d) Which zirconium oxide mixture gives the optimum yield of 2,6-xylenol?
   e) Two tonnes of phenol are passed through the 'zirconium 1%' catalyst at 350°C. What mass of 2,6-xylenol will be formed?
   f) Name one of the other products that you would expect from these reactions.
   g) Zirconium 1%, titanium and cadmium give a similar performance. What economic and environmental factors might you take into consideration when deciding which to use?

9.6 Oxidation reactions of benzene and methylbenzene

Benzene, and other aromatic hydrocarbons, burn readily when ignited. In theory, the complete combustion products are carbon dioxide and water, e.g.

$$C_6H_6(l) + \frac{15}{2}O_2(g) \longrightarrow 6CO_2(g) + 3H_2O(l) \quad \Delta H = -3268 \text{ kJ mol}^{-1}$$

However, the very high carbon content in aromatic hydrocarbons (benzene, for example, is 92% C by mass) causes a lot of **incomplete combustion**. Thus, aromatic hydrocarbons burn with a bright orange, sooty flame and they are not used as fuels though small quantities are added to unleaded petrol to prevent pre-ignition or knocking (section 7.5).

Unlike alkenes, the benzene ring is not attacked by common oxidising reagents, such as acidified $KMnO_4(aq)$. However, one important synthetic reaction is the oxidation of the methyl side chain in methylbenzene:

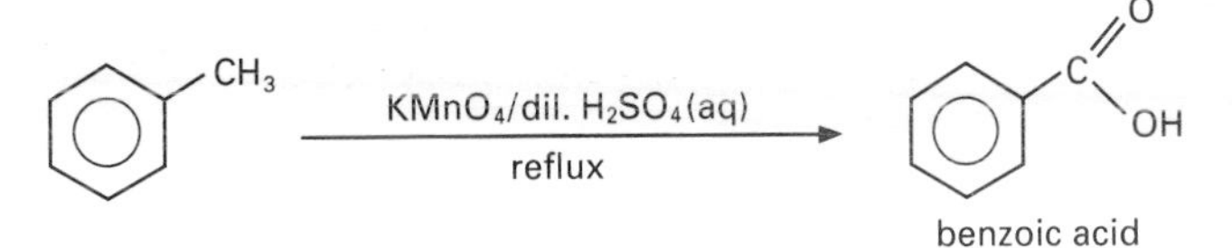

Under similar conditions, methane is unreactive.

Since this reaction also produces a colour change, purple $MnO_4^-(aq)$ to colourless $Mn^{2+}(aq)$, we can use it to tell the difference between methylbenzene and benzene.

Activity 9.3

1 Describe the reagents and reaction conditions that you would use to convert benzene into:
a) nitrobenzene; **b)** chlorobenzene
c) bromobenzene; **d)** benzenesulphonic acid

2 Give the names and formulae of the products formed when benzene reacts with:
a) conc. HNO_3/H_2SO_4 at 50°C **b)** Cl_2, $AlCl_{3(s)}$ catalyst, room temp
c) H_2, Ni catalyst, 200°C **d)** CH_3I, $AlCl_3$ catalyst, 40°C.
e) conc. H_2SO_4, reflux for several hours

3 How would you convert methylbenzene into:
a) (bromomethyl)benzene **b)** 2-bromomethylbenzene **c)** benzoic acid

9.7 Comparing the reactions of hydrocarbons

In an examination, you may well be asked to compare chemical reactions involving alkanes, alkenes and benzene. Thus, it is a valuable revision exercise to make a table summarising these properties. Activity 9.4 suggests how you might do this.

Activity 9.4

Draw up a table which compares, and contrasts, the reactions of hexane, hex-1-ene and benzene with the following reagents:

1 $O_2(g)$ 2 Cl_2, Br_2 3 $HCl(g)$, $HBr(g)$ and $HI(g)$
4 conc. $H_2SO_4(l)$ 5 $KMnO_4$/dil. $H_2SO_4(aq)$ 6 $H_2(g)/Ni(s)$

Focus 9c

1 The reactions of benzene, and other arenes, often involve **electrophilic substitution** of the ring hydrogen atoms.

2 Some important reactions involving the benzene ring are shown below:

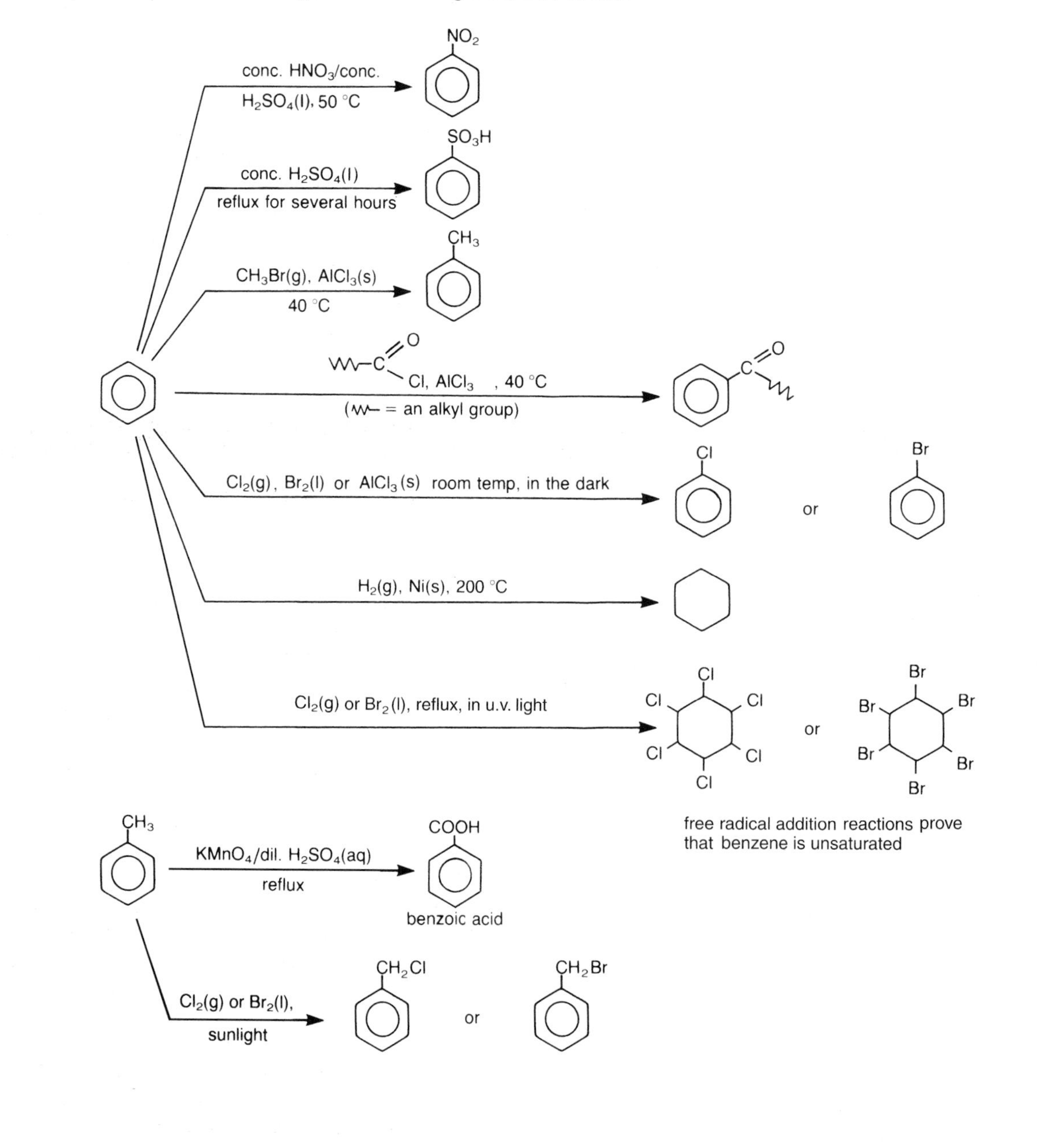

Questions on Chapter 9

1 Write down structural formulae for the following compounds:

a) bromobenzene **b)** 3-nitrophenol **c)** phenylethanol

2 Study the following structural formulae:

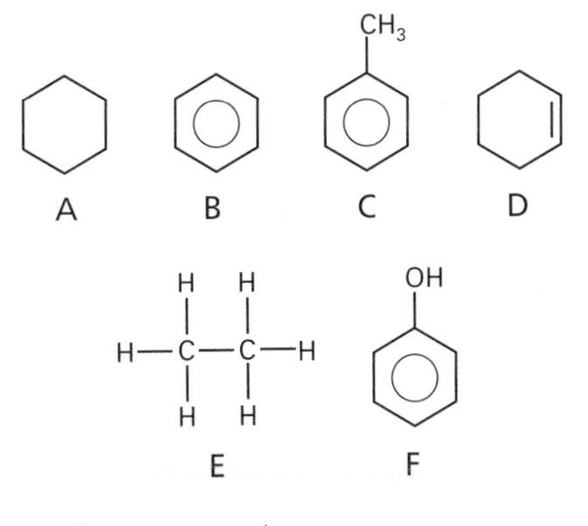

a) Name the compounds.
b) State which of these compounds:
i) are unsaturated
ii) react with bromine in the dark
iii) will decolourise acidified potassium manganate(VII) solution
iv) can be hydrogenated in the ratio, *three* moles H_2 to *one* mole of compound
v) is soluble in water
vi) are cyclic
vii) react with concentrated sulphuric acid
viii) produce fumes of HBr when treated with bromine and aluminum chloride.

c) How would you distinguish between A, B, C and D?

3 Butadiene has the molecular formula C_4H_6.
a) Write down its structural formula.
b) Draw a diagram to show how the π bonds are formed from carbon's p orbitals. Use the diagram to explain how delocalisation of electrons might occur in butadiene.
c) What evidence would you gather to prove whether delocalisation occurs?

4 Give reagents and reaction conditions for the following syntheses:
a) benzene ⟶ cyclohexane
b) benzene ⟶ nitrobenzene
c) benzene ⟶ ethylbenzene
d) benzene ⟶ benzoic acid (two steps)
e) methylbenzene ⟶ 2-methylbenzene sulphonic acid
f) methylbenzene ⟶ (bromomethyl)benzene
g) methylbenzene ⟶ 2-bromomethylbenzene
h) benzene ⟶ 2-nitromethylbenzene (two steps)
i) methylbenzene ⟶ 3-chlorobenzoic acid (two steps)

5 Write down the mechanism for nitration of benzoic acid to 3-nitrobenzoic acid.

Comments on the activities

Activity 9.1

1 **a)** 1,2-diethylbenzene **b)** 4-bromophenol
c) 3-nitrobenzoic acid **d)** 2,4,6-tribromophenol
e) 2-methylphenol **f)** 1,4-dimethylbenzene
g) 2-chloromethylbenzene **h)** 4-bromobenzaldehyde

2

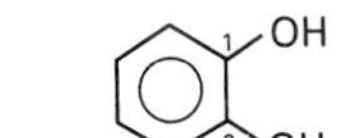

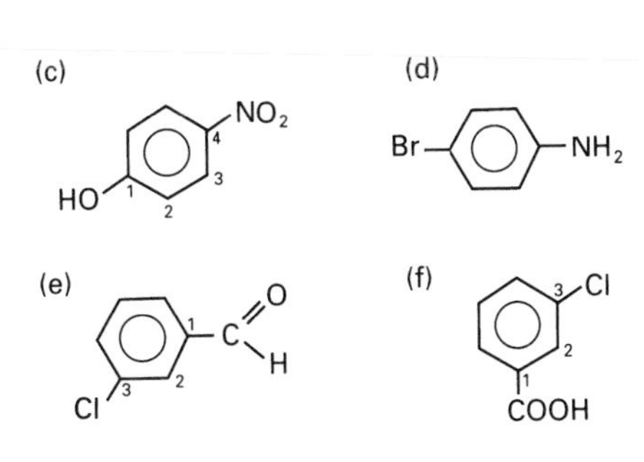

Activity 9.2

1 Nitration using a mixture of concentrated nitric and sulphuric acids at 50°C. Then, alkylation with a halogenomethane and a Friedel–Crafts catalyst at 40°C.

If the steps are reversed, the nitro and methyl groups will be found in the 1,2- and 1,4- positions.

2 **a)** Most specific: 2% Zr oxide mixture. Least effective: 10% Zr oxide mixture.
b) Most specific: Cd oxide mixture. Least specific: 10% Zr oxide mixture.
c) (i) 1 ⟶ 2% Zr oxide increases the percentage of phenol which is methylated;
2 ⟶ 3 ⟶ 10% Zr oxide decreases the percentage.
ii) 1 ⟶ 2% Zr oxide increases the percentage of 2,6-xylenol;
2 ⟶ 3 ⟶ 10% Zr oxide dramatically decreases the percentage.
d) 2% Zr oxide mixture.
e) Mass of 2,6-xylenol = 2 × 69.1/100 = 1.38 tonnes.
f) 2,4-dimethylphenol or 2,4,6-trimethylphenol.
g) Some of the questions you might ask: How often does the catalyst need to be replaced? How much does the catalyst cost? How easy is it to recycle the catalyst? What effect does temperature have on the catalyst's performance (energy considerations)? What safety issues are raised by the use of these metal oxides (toxicity, disposal)?

Comments on the activities *continued*

Activity 9.3

1 **a)** conc. HNO_3/conc. H_2SO_4(l), 50°C
b) Cl_2(g), $AlCl_3$(s) catalyst, 40°C
c) Br_2(l), $AlCl_3$(s) catalyst, 40°C
d) conc. H_2SO_4(l), reflux.

2 **a)** nitrobenzene —NO_2
b) chlorobenzene —Cl
c) cyclohexane
d) methylbenzene —CH_3
e) benzenesulphonic acid —SO_3H

3 **a)** Br_2(l), uv light
b) Br_2(l), $AlCl_3$(s), 40°C
c) $KMnO_4$/dil. H_2SO_4; reflux.

Activity 9.4

Compare your table with Table 9.2

Table 9.2 Comparing the reactions of hydrocarbons

Reagent	alkanes, e.g. C_6H_{14} hexane	alkenes e.g. $C_4H_9-CH=CH_2$ hex-1-ene	arenes e.g. benzene
O_2(g), ignited	all burn; main products CO_2(g) and H_2O(l); more orange flame and soot as % of carbon increases		
Cl_2, Br_2	**free radical substitution** in u.v. light ⟶ a mixture of **halogenoalkanes**	**electrophilic addition** in the dark at room temp. ⟶ $C_4H_9-CH(Hal)-CH_2(Hal)$	**electrophilic substitution** in the dark at room temp. ⟶ (ring)—Hal
$AlCl_3$ catalyst	e.g. $C_6H_{13}Cl$	Test for C=C bonds	
HCl, HBr, HI gases	no reaction	**addition** on warming ⟶ $C_4H_9-CH(Hal)-CH_3$ Markovnikov's rule applies	no reaction
conc. H_2SO_4(l)	no reaction	**addition** at room temp. Markovnikov's rule will apply ⟶ $C_4H_9-CH(OSO_3H)-CH_3$	**substitution** on refluxing ⟶ (ring)—SO_3H (section 9.5)
$KMnO_4$/dil. H_2SO_4(aq) no reaction reflux	no reaction	**oxidised** ⟶ diols	no reaction (but see section 9.6)
H_2(g), Ni(s) at 200°C	no reaction	**reduction** ⟶ C_6H_{14}, hexane	**reduction** ⟶ cyclohexane

CHAPTER 10

Organic halogen compounds

Contents

Study Checklist

By the end of this chapter you should be able to:

1. State that there are three main types of organic halogen compounds: halogenoalkanes, halogenoarenes and carboxylic acid halides.
2. Interpret, and use, the names, short structural formulae and displayed formulae of simple halogenoalkanes and classify them as being primary, secondary or tertiary compounds. State that monohalogenoalkanes form a homologous series of general formula $C_nH_{2n+1}Hal$ and describe the gradual change in physical properties that occurs as molecular mass is increased.
3. Name simple halogenoarenes and explain how polysubstituted compounds may exist as structural (positional) isomers.
4. State that the chemical properties of a halogen compound depend on: (a) which halogen atom is present, (b) its molecular environment.
5. Explain the polar nature of the C—Hal bond in terms of the relative electronegativities of the atoms present.
6. Describe the following nucleophilic substitution reactions of halogenoalkanes: (a) the formation of alcohols by reaction with water or $OH^-(aq)$ ions (hydrolysis), (b) the formation of nitriles by reaction with alcoholic potassium cyanide, (c) the formation of primary amines by reaction with excess ammonia.
7. Explain the relative rates of hydrolysis by primary halogenoalkanes in terms of the relative strengths and polarities of the C—Hal bonds (C—F, C—Cl, C—Br, C—I).
8. Interpret the different reactivities of primary, secondary and tertiary halogenoalkanes towards hydrolysis in terms of the different types of reaction mechanism (S_N1 and S_N2). Write down examples of these mechanisms using curly arrows to show the movement of pairs of electrons.
9. Explain the resistance of chlorobenzene towards hydrolysis in terms of delocalisation of electrons. State that the ethanoyl chloride is rapidly hydrolysed due to the presence of the highly polar—COCl group.
10. Explain the meaning of the term 'elimination reaction' and know that when halogenoalkanes react with nucleophiles which are also strong bases, substitution and elimination reactions will be in competition. State the names of the reaction products formed on elimination of hydrogen bromide from a halogenoalkane e.g. 2-bromobutane. Predict the reaction products which predominate when a primary, secondary or tertiary halogenoalkane reacts with sodium hydroxide in an aqueous or an ethanolic solution.
11. Outline the preparation and uses of Grignard reagents as intermediates in organic synthesis.

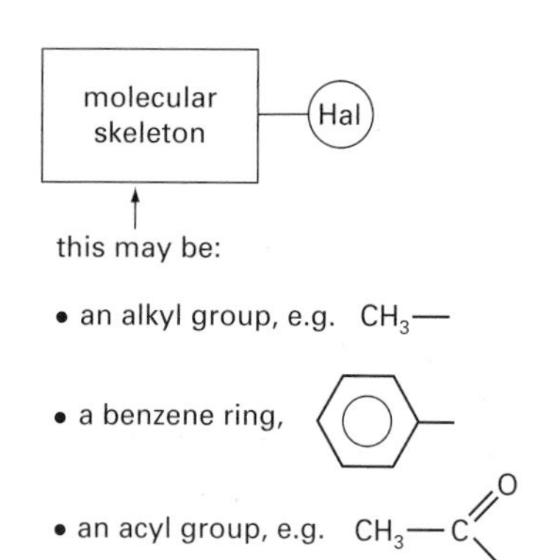

Figure 10.1 ▲
The properties of an organic halogen compound depend on (i) the type of halogen atom and (ii) its molecular location

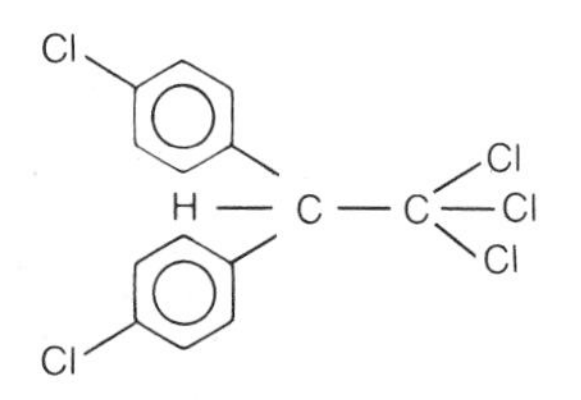

Figure 10.2 ▲
The molecular structure of dichlorodiphenyltrichloroethane (DDT). DDT is a colourless crystalline solid which is insoluble in water. Would benzene be a suitable solvent for DDT?

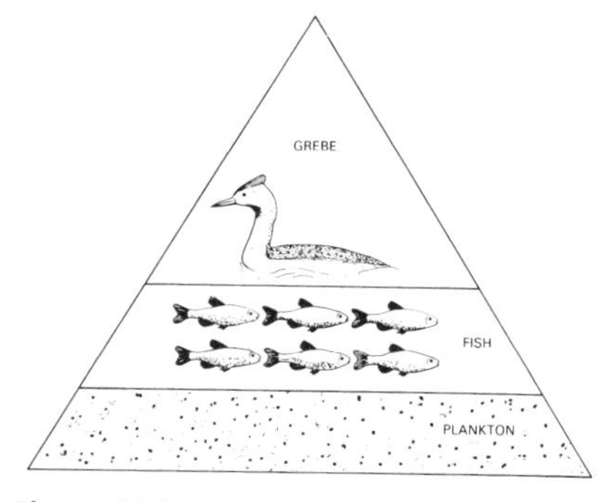

Figure 10.3 ▲
DDT affected not only the insects that it was intended to kill, but also those organisms that fed on the insects. In this case the concentration of insecticide in the animal's body gradually increases as it passes from the plankton to the fish and then to the grebes

12 Summarise the main reactions of organic halogen compounds, as shown in Figure 10.13. State the reagents in Table 10.5 and use them in simple synthetic routes.

13 Explain the uses of fluoroalkanes and fluorohalogenoalkanes, for example as anaesthetics, aerosol propellants and plastics, in terms of their relative chemical inertness.

14 Outline the reasons for concern about the use of:
(a) chlorofluorocarbons (CFCs) in terms of the effects of the products of their breakdown on the ozone layer and (b) hydrofluorocarbons (HFCs) in terms of their contribution to global warming.

An **organic halogen compound** has a halogen functional group attached to a hydrocarbon skeleton (Figure 10.1). A most notorious example is the insecticide **DDT** (Figure 10.2).

Although DDT was first prepared by Othmar Zeidler in Germany in 1874, it was not widely used until the Second World War when Paul Muller of Switzerland discovered its potential as an insecticide. Due to the poor sanitation in wartime, soldiers and civilians were perfect hosts for parasites such as body lice. Apart from causing discomfort, these insects were a threat to life because they carried typhus, a deadly micro-organism. Fortunately, the lice could be killed by dusting with DDT powder.

Similarly, when the Allies liberated the Pacific islands, preinvasion spraying of DDT was used to destroy the mosquitoes which spread malaria.

After the war, DDT was used by farmers to kill insects while leaving plants and animals unharmed, and almost doubling their yield. The incidence of malaria in many countries in Asia, South America and Africa was greatly decreased. In India, for example, over a decade, DDT reduced malaria from 75 million to 5 million cases per year, saving nearly 10 million lives. At first, the treatment appeared to be extremely successful and other organochlorine insecticides were developed. However, before long, certain insects were showing a resistance to DDT, and there was growing concern about the increased dosage that was being applied. In 1962, the American marine biologist Rachel Carson's publication 'Silent Spring' showed that the use of DDT caused birds to lay infertile or deformed eggs. Even more worrying was the discovery of appreciable amounts of DDT in dead birds found in regions where the insecticide had been used. This happened because the DDT molecule is extremely stable, and remains unchanged for a long time after spraying. It is also soluble in body fat and so builds up in the bodies of organisms that eat it. When it was used to kill midges over American lakes, the DDT was taken in by plankton in the water, which were then eaten by fish, which were then eaten by birds. As each fish ate millions of plankton and each bird ate many fish, the concentration of DDT in the bird's body became high enough to kill it (Figure 10.3). By 1970, the concern over the use of DDT caused many countries, including the UK, to restrict its use to major health emergencies.

DDT is just one of the many organic halogen compounds that have made an impact on everyday life. Nowadays, they have numerous uses, ranging from solvents to anaesthetics, from aerosol propellants to plastics. Organic halogen compounds also display a wide variety of chemical reactions, and this makes them useful in organic synthesis.

DDT was used during the Second World War to kill lice at home and mosquitoes at the battlefront ▶

10.1 The different types of organic halogen compounds

There are three main types of organic halogen compound: halogenoalkanes, halogenoarenes and carboxylic acid halides.

Halogenoalkanes

Here one, or more, hydrogens in the alkane have been replaced by halogen atoms:

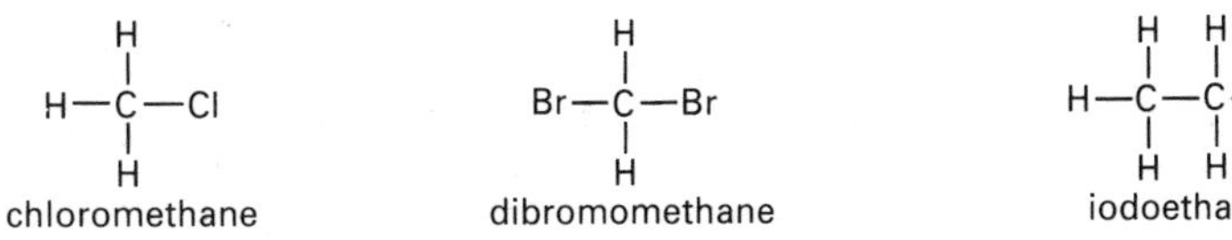

If the halogenoalkane can exist as **structural isomers**, you must indicate the position(s) of the halogen atom(s) along the hydrocarbon chain:

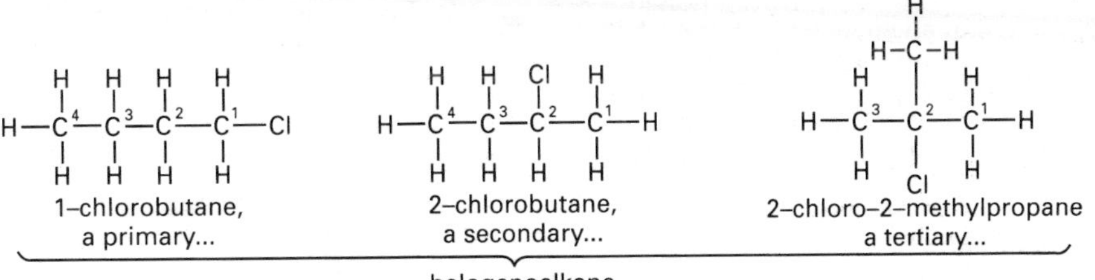

Notice here the classification of isomeric halogenoalkanes as **primary**, **secondary** and **tertiary** compounds. Thus, we say that

- PRIMARY (1°) halogenoalkanes have the general formula:
- SECONDARY (2°) halogenoalkanes have the general formula:

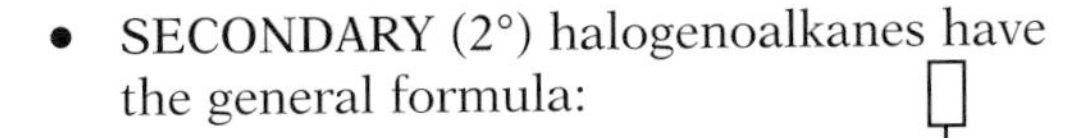

- TERTIARY (3°) halogenoalkanes have the general formula:

where ☐— is a hydrocarbon skeleton, ● is ☐— or an H atom and Hal = F, Cl, Br or I.

Halogenoalkanes with a **chiral centre** will exist as **optical isomers** (Figure 10.4).

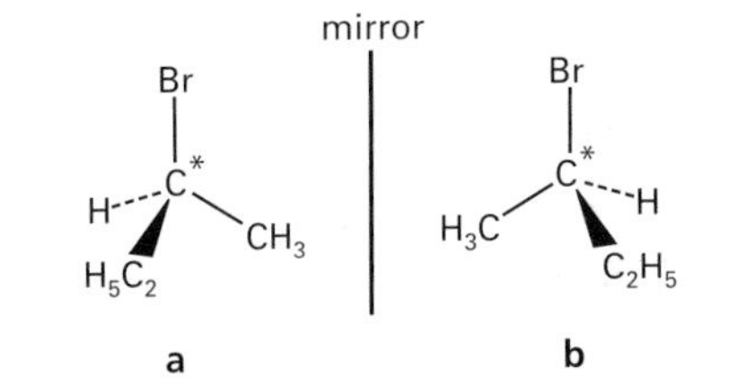

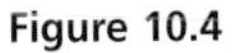

Figure 10.4
These two forms of 2-bromobutane are **enantiomers**, that is non-superimposable mirror images. Rotating **a** through 180° does not give **b**. There is no plane of symmetry through the carbon atom, C*, and it is said to be a **chiral centre**. Enantiomers have the same physical and chemical properties except that they are able to rotate the plane of plane polarised light in opposite directions (section 6.5)

Halogenoarenes

Here, halogen atoms have been substituted for one, or more, of the hydrogens in benzene. For example,

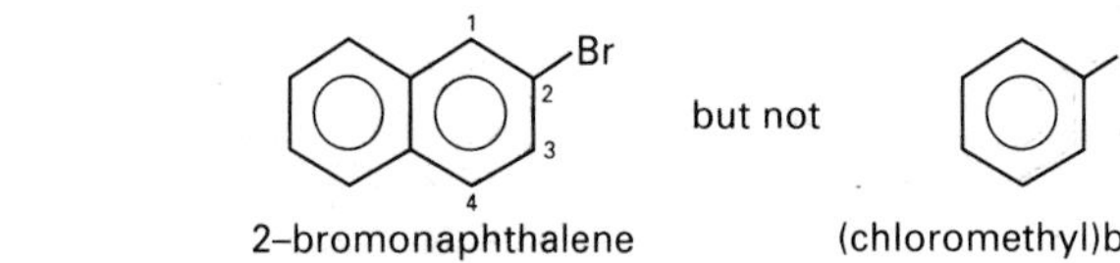

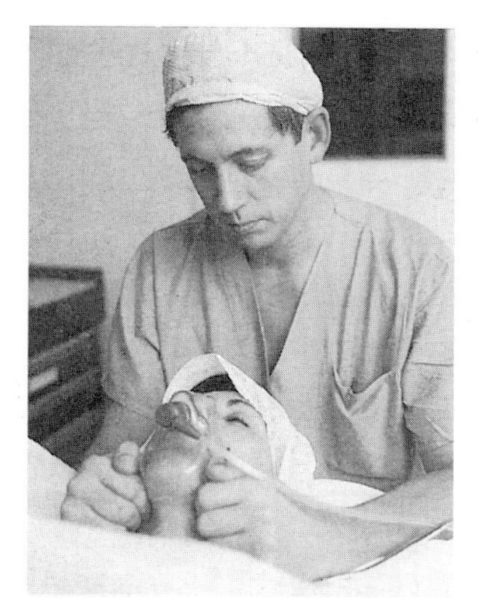

▲ Halothane, an organic halogen compound, formula $CF_3CHBrCl$, is used in hospitals as an anaesthetic. It is mixed with oxygen in precise amounts and inhaled by the patient. Halothane, which was first used in 1956, provides deep, lasting anaesthesia with an acceptable level of side effects. In halothane, chemical name 2-bromo-2-chloro-1,1,1-trifluoroethane, each halogen atom is present for a reason. Substituting a chlorine atom for a hydrogen atom gives the molecule its anaesthetic properties – more than one chlorine would make the compound too toxic. The presence of very stable C—F bonds makes the compound more inert, non-toxic and non-flammable. Introducing the bromine atom increases the boiling point of the compound, thereby making it easier to store and use

(Chloromethyl)benzene is not a halogenoarene because the halogen atom is not directly attached to the ring but is located in an alkyl side chain. In naming this compound, note the use of the brackets to indicate that the chlorine atom is in an alkyl(methyl) side-chain. Do you think that the chemical properties of (chloromethyl)benzene would resemble those of chlorobenzene or chloromethane?

Structural isomers may occur in halogenoarenes, for example:

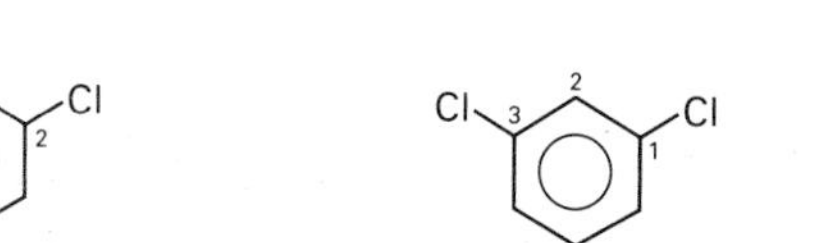

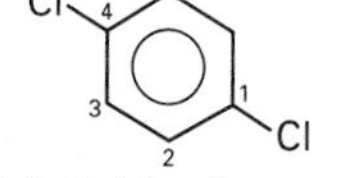

1,2-dichlorobenzene 1,3-dichlorobenzene 1,4-dichlorobenzene

Carboxylic acid halides

Here the —OH group in the carboxylic acid has been replaced by a halogen atom, for example:

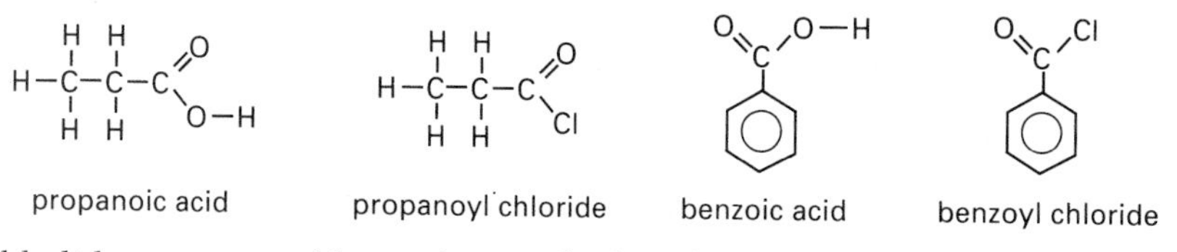

propanoic acid propanoyl chloride benzoic acid benzoyl chloride

Acid halides are named by replacing the '**-oic**' of the acid name by '**oyl halide**'. The chemistry of acid halides is also covered in section 13.6.

Activity 10.1

1 Name the following organic halogen compounds;

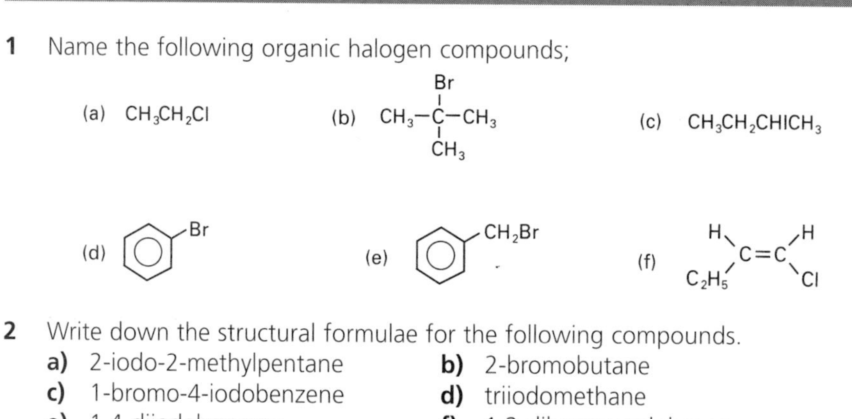

(a) CH_3CH_2Cl (b) $CH_3-C(Br)(CH_3)-CH_3$ (c) $CH_3CH_2CHICH_3$

(d) (e) (f)

2 Write down the structural formulae for the following compounds.
a) 2-iodo-2-methylpentane **b)** 2-bromobutane
c) 1-bromo-4-iodobenzene **d)** triiodomethane
e) 1,4-diiodobenzene **f)** 1,2-dibromocyclohexane

3 Of the saturated compounds given in questions 1 and 2, which are:
i) primary, ii) secondary and iii) tertiary?

4 Seven *non-cyclic* compounds are isomers of molecular formula $C_3H_4Cl_2$. Draw their structural formulae and name them.

5 **a)** Identify the type of isomerism in both (i) 1,2-dibromoethene and (ii) 1,2-dichlorocyclohexane. (For help, see section 6.5.)
b) Draw the isomers.
c) What structural feature is responsible for this isomerism?

6 Does 'halothane' exhibit optical isomerism? Explain your answer.

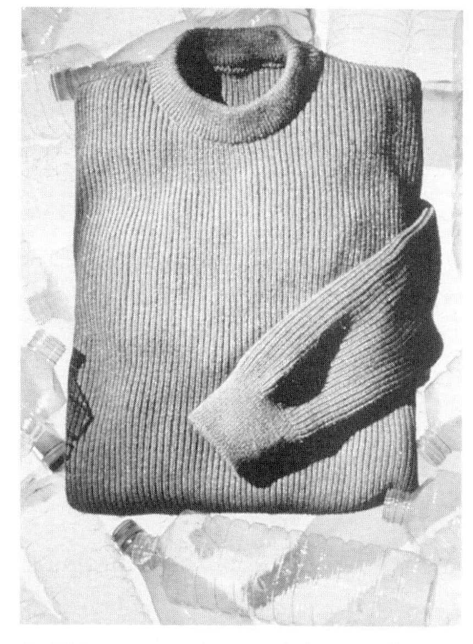

▲ This sweater was made by recycling mineral-water bottles made of poly(chloroethene), commonly known as PVC, and mixing the recycled product with wool fibres

10.2 Some physical properties of organic halogen compounds

Some properties of organic halogen compounds are shown in Table 10.1. Activity 10.2 will help you to analyse this data and explain any trends which are found.

Table 10.1 Some physical properties of organic halogen compounds

Name	formula	state (at 25°C)	melting point/°C	boiling point/°C	density as liquids/g cm^{-3}	$\Delta H_c^{\ominus}$ /kJ mol^{-1}
chloromethane	CH_3Cl	gas	−98	−24	0.92	−687
chloroethane	C_2H_5Cl	gas	−136	12	0.90	−1325
1-chloropropane	C_3H_7Cl	liquid	−123	47	0.89	−2001
1-chlorobutane	$CH_3(CH_2)_3Cl$	liquid	−123	79	0.89	−2704
2-chlorobutane	$C_2H_5CHClCH_3$	liquid	−131	68	0.87	
2-chloro-2-methylpropane	$(CH_3)_2CClCH_3$	liquid	−25	51	0.84	−2693
chlorobenzene	C_6H_5Cl	liquid	−45	132	1.11	−3112
ethanoyl chloride	CH_3COCl	liquid	−112	51	1.10	
bromomethane	CH_3Br	gas	−93	4	1.68	−770
bromoethane	C_2H_5Br	liquid	−118	38	1.46	−1425
bromobenzene	C_6H_5Br	liquid	−32	156	1.49	
iodomethane	CH_3I	liquid	−66	33	2.28	−815
iodoethane	C_2H_5I	liquid	−111	72	1.94	−1490
iodobenzene	C_6H_5I	liquid	−30	189	0.90	−3193

Can you predict the standard enthalpies of combustion of the following compounds: (a) 1-chloropentane and (b) bromobenzene?

Activity 10.2

Briefly study Table 10.1 and then work through the following questions.

1. For the primary chloroalkanes only, plot graphs of (a) boiling point and (b) standard enthalpy of combustion, $\Delta H_c^{\ominus}$, against the number of carbon atoms in the chain. Account for the shape of the graphs. (For help here, and in questions 3 and 4, see section 7.3.)
2. Do these halogenoalkanes form a homologous series? If so, what is their general formula?
3. Compare (a) the boiling points and (b) the densities of isomeric primary, secondary and tertiary chloroalkanes. Do you detect any trends in the values? If so, what explanation can you give for this behaviour?
4. Compare the boiling points of the halogenoalkanes which have different halogen atoms attached to the same hydrocarbon skeleton. Once again, can you detect, and explain, any trends in the values?

Focus 10a

1. There are three main types of halogen compound: **halogenoalkanes**, **halogenoarenes** and **acid halides**.
2. To name a halogenoalkane, put a prefix (i.e. fluoro-, chloro-, bromo- or iodo-) before the name of the parent hydrocarbon. If structural isomers are possible, you must state the numbered position(s) of the halogen atom(s) along the hydrocarbon skeleton.
3. Halogenoalkanes are described as being **primary (1°)**, **secondary (2°)** or **tertiary (3°)**, depending on whether there are one, two or three alkyl groups attached to the carbon of the C—Hal bond.
4. Monohalogenoalkanes form a **homologous series** of formula $C_nH_{2n+1}Hal$. Thus, their physical properties change gradually as the molecular mass is increased.
5. To name a halogenoarene, put the halogen prefix (e.g. iodo-) before the name of the arene. If **positional isomers** are possible, you must state the ring position of each halogen atom.
6. Acid halides are named by replacing the '**–oic**' of the carboxylic acid name by '**–oyl halide**'.

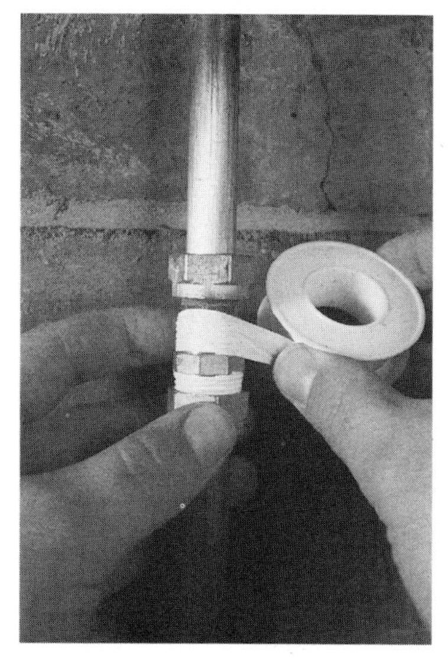

PTFE tape being used on a pipe joint providing a waterproof seal

10.3 Making halogenoalkanes and halogenoarenes

Halogenoalkanes can be made in two main ways:

1. from **alkenes** by reaction with a halogen, or hydrogen halide (section 8.5),
2. from **alcohols** by the action of a halogenating agent (section 11.4).

Halogenoarenes: Chloro and bromoarenes are obtained by treating the arene with the halogen in the presence of a Friedel–Crafts catalyst (section 9.5).

Activity 10.3

By researching the section references above, explain how you would perform the following conversions.

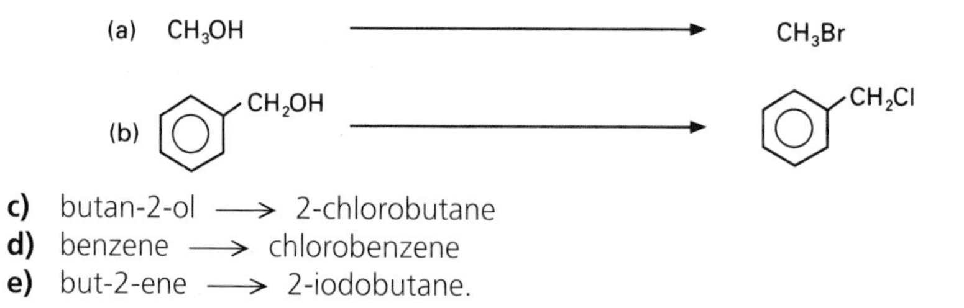

c) butan-2-ol ⟶ 2-chlorobutane
d) benzene ⟶ chlorobenzene
e) but-2-ene ⟶ 2-iodobutane.

10.4 Predicting the reactivity of organic halogen compounds

The chemical properties of an organic halogen compound will depend on:

- Which halogen atom (F, Cl, Br or I) is present. Looking at the mean bond enthalpies in Table 10.2, we can see that the C—F bond is very strong. It is not surprising, therefore, that organic fluorine compounds are found to be

Table 10.2 Some mean bond enthalpies

Bond	mean bond enthalpies /kJ mol^{-1}
C—F	484
C—Cl	338
C—Br	276
C—I	238
C—H	412

Can you explain the trend in C—Hal bond strength?

extremely unreactive. We shall study, then, the reactions of chloro, bromo and iodo compounds.

- The location of the C—Hal bond within the molecule. It will be interesting to see how the chemical reactivity of a given C—Hal bond, say C—Cl, depends on its molecular environment.

Due to the C—Hal bond polarities, **nucleophiles** should attack the $C^{\delta+}$, thereby displacing a halogen ion, e.g.

$$^{\delta-}Hal—C^{\delta+} + \underset{\text{a nucleophile}}{Nuc^-} \longrightarrow Hal^- + C—Nuc$$

Apart from **nucleophilic substitution**, organic halogen compounds undergo three other important types of reaction:

1 **Elimination** of hydrogen halide from neighbouring carbon atoms in a halogenoalkane to form an alkene:

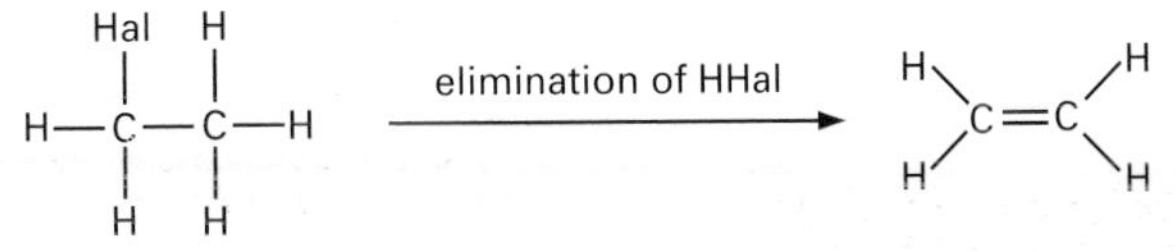

2 **Electrophilic substitution** of the 'ring' hydrogens in halogenoarenes (section 9.5).

3 The reaction with magnesium turnings in dry ethoxyethane solvent ($C_2H_5OC_2H_5$ – an ether) to form **Grignard reagents**. These **organometallic** compounds are very useful in organic synthesis.

10.5 Nucleophilic substitution reactions

Substitution of —Hal by —OH (hydrolysis)

The reactivity of an organic halogen compound depends on (i) the type of halogen atom and (ii) its molecular environment. These effects can be easily studied by measuring the relative rates of hydrolysis of various organic halogen compounds, i.e.

$$\rangle C—Hal + H_2O \longrightarrow C—OH + HHal$$

Experiment 1

To study the effect of different halogen atoms on the rate of hydrolysis

Method Three test-tubes (labelled Cl, Br and I), each containing ethanol (2 cm^3), are loosely stoppered and placed in a water bath at 60°C. A fourth test-tube containing aqueous silver nitrate solution (5 cm^3, 0.1 mol dm^{-3}) is also placed in the water bath. The tubes, and their contents, are allowed to warm up to 60°C. Then, the labelled tubes are removed from the water bath and 2 drops of 1-chlorobutane and 1 cm^3 of the aqueous silver nitrate are added to tube 'Cl' and the tube is recorked and shaken. This procedure is repeated for tubes 'Br' and 'I' but using 1-bromobutane and 1-iodobutane, respectively. A clock is started and the tubes are observed for ten minutes. (Notes: (i) the nucleophile is H_2O from the aqueous silver nitrate not the NO_3^- ion and (ii) the ethanol acts as a solvent, not a reactant.)

Observations A precipitate of silver halide appears in each tube. We find that:

AgCl(s) white — AgBr(s) cream — AgI(s) yellow
→ precipitate forms more rapidly

Explanation As soon as the halide ion is displaced from the halogenoalkane, it reacts with the $Ag^+(aq)$ ion to give the precipitate of silver halide:

$$Ag^+(aq) + Hal^-(aq) \longrightarrow AgHal(s)$$

Thus, the rates of hydrolysis are proportional to the amount of AgHal that is formed in a given time. The results show that

1-chlorobutane → 1-bromobutane → 1-iodobutane

increasing rate of hydrolysis

Similar trends are found when these halogen compounds react with other nucleophiles, such as NH_3 and CN^-.

Two opposing factors affect the rate of nucleophilic substitution in halogenoalkanes:

- The polarity of the C—Hal bond. Electronegativity decreases on going from chlorine to iodine. This suggests that 1-chlorobutane should react fastest with nucleophiles because it contains the most electron-poor ($\delta+$) carbon atom.
- The C—Hal bond strength. For substitution to occur, the C—Hal bond must be broken, and this will happen more readily if the bond is weak. The C—Hal bond energy terms in Table 10.2 suggest that 1-iodobutane should react fastest because it has the weakest C—Hal bond.

Since 1-iodobutane did hydrolyse most rapidly, *the effect of the bond strength seems to outweigh that of the bond polarity*. Hence, for the same hydrocarbon skeleton, we find that:

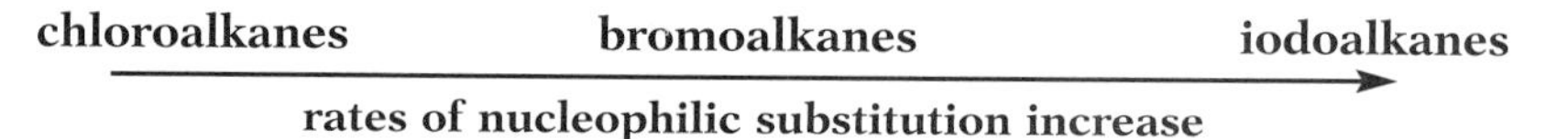

Experiment 2

To study the effect of changing the molecular location of the halogen atom on the rate of hydrolysis

Method Here we compare the rates of hydrolysis of five types of chlorocompound:

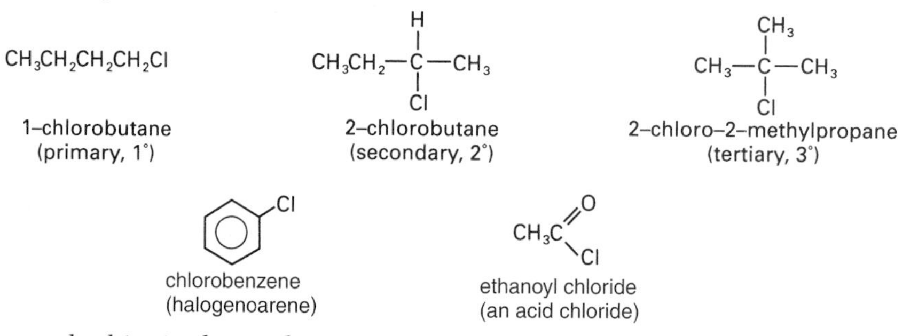

The method is similar to that in experiment 1 except that ethanoyl chloride is mixed with 1 cm^3 of aqueous silver nitrate at room temperature – no ethanol is added because it reacts with enthanoyl chloride.

Observation Figure 10.5 describes the appearance of the reaction mixtures.

These results show that the rate of hydrolysis does depend on the molecular location of the halogen atom. For a given halogen we find that:

halogenoarenes → halogenoalkanes ($1° < 2° < 3°$) → acid halides

increasing rate of hydrolysis

Explanation The increase in rate of nucleophilic substitution as we go from 1° to 2° to 3° halogenoalkanes occurs because these reactions have different mechanisms, and these are described in section 10.6.

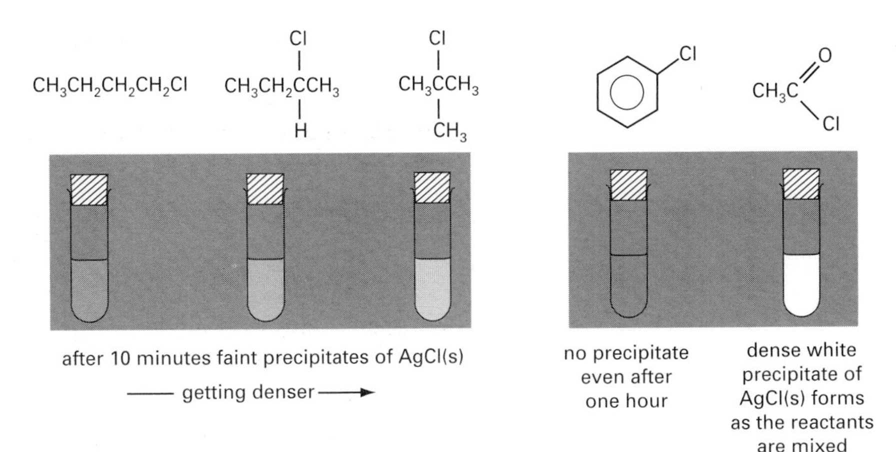

Figure 10.5 ▶
Observations from experiment 2: the appearance of the reaction mixtures after 10 minutes

Halogenoarenes do not give nucleophilic substitution reactions under normal laboratory conditions. The reason for this is that the p orbitals on the carbon and halogen atoms overlap sideways to form a **delocalised π bond system** (Figure 10.6).

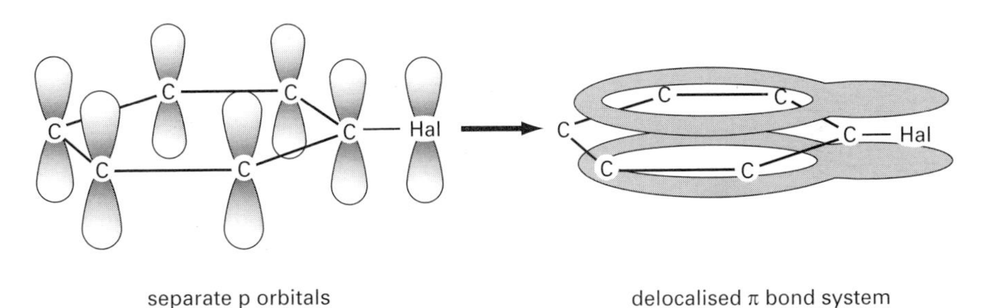

Figure 10.6 ▶
The delocalised π bond system in halogenoarenes both strengthens the C—Hal bond and decreases its polarity

This has two effects:

- The C—Hal bond gains some π bond character, making it stronger, and less reactive, than the same C—Hal bond in a halogenoalkane;
- The C—Hal bond polarity is decreased, making the $C^{\delta+}$ atom much less susceptible to nucleophilic attack.

Compared to halogenoalkanes of similar mass, **acid halides** are vigorously hydrolysed at room temperature, e.g.

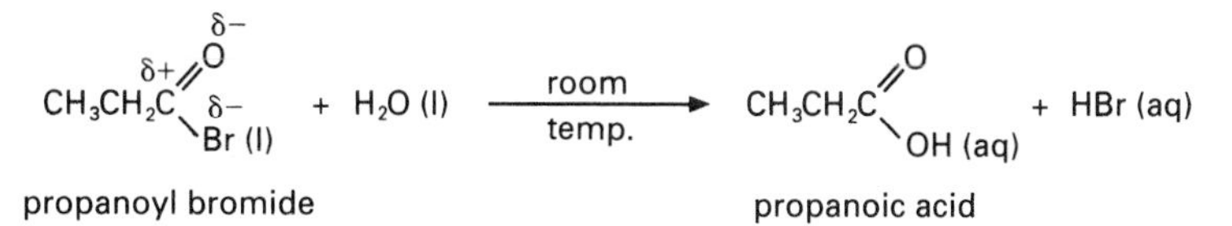

In acid halides, the polarisation of the $^{\delta+}C{=}^{\delta-}O$ and the $^{\delta+}C{-}^{\delta-}Hal$ bonds produces a highly electron-poor (δ+) carbon atom, towards which nucleophiles are strongly attracted (Figure 10.7). Consequently, we find that acid halides undergo nucleophilic substitution much faster than the halogenoalkanes of similar mass.

To summarise, then, the tendency of organic halogen compounds to undergo nucleophilic substitution is described by the following trends:

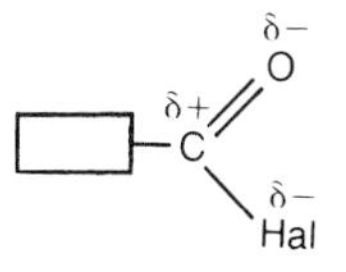

Figure 10.7 ▲
In acid halides, both the O and Hal atoms can withdraw electron cloud from the neighbouring C atom, making it highly susceptible to nucleophilic attack

chloro- → **bromo-** → **iodoalkanes or arenes**

GET FASTER

primary → **secondary** → **tertiary halogenoalkanes**

GET FASTER

halogenoarenes → **halogenalkanes** → **acid halides**

GET MUCH FASTER

Primary alcohols may be prepared, in good yields, by the alkaline hydrolysis of primary halogenoalkanes, for example:

$$CH_3CH_2CH_2I \xrightarrow[\text{reflux}]{\text{dil. NaOH(aq)}} CH_3CH_2CH_2OH + NaI$$

Under the same conditions, secondary and tertiary halogenoalkanes do form the alcohols, but the yields are poorer, for example:

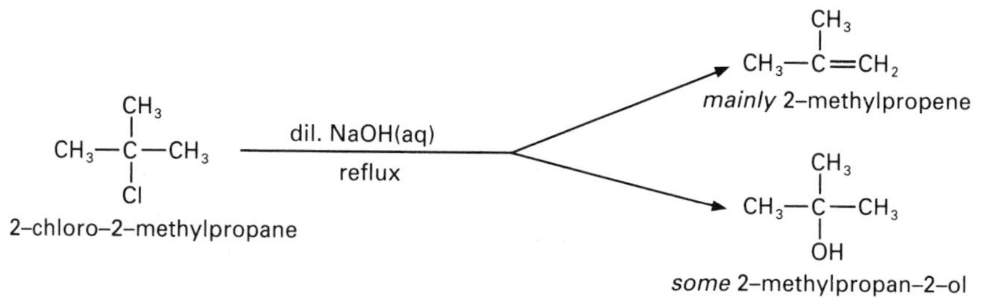

Here the alkene is produced by an **elimination** reaction. *Whenever halogenoalkanes react with hydroxide ions, substitution and elimination reactions are in competition*. The factors that determine which reaction will predominate are discussed in section 10.7.

Substitution of —Hal by —CN

When an ethanolic solution of a halogenoalkane and potassium cyanide is refluxed, a **nitrile** is formed, e.g.

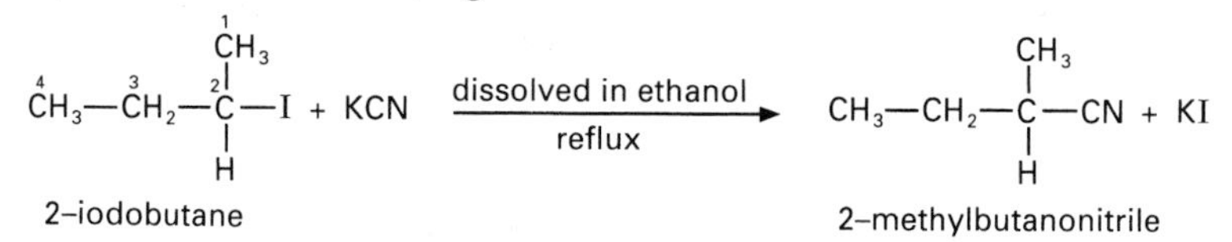

Here the nucleophile is the cyanide ion, CN^-, and the mechanism is described in section 10.6. Cyanide ions do not react with halogenoarenes.

A very important aspect of this reaction is that it allows us to lengthen the molecule's hydrocarbon skeleton, that is to add another —CH_2 unit. This is done in two stages:

1 The nitrile is converted into a carboxylic acid by refluxing with dilute sulphuric acid:

$$CH_3-CH_2-CH(CH_3)-CN + 2H_2O \xrightarrow[\text{reflux}]{\text{catalyst: dil. } H_2SO_4(aq)} CH_3-CH_2-CH(CH_3)-COOH + NH_3$$

2 The carboxylic acid is reduced by refluxing with lithium tetrahydrido-aluminate $Li^+[AlH_4]^-$ dissolved in ethoxyethane, $C_2H_5OC_2H_5$, as solvent:

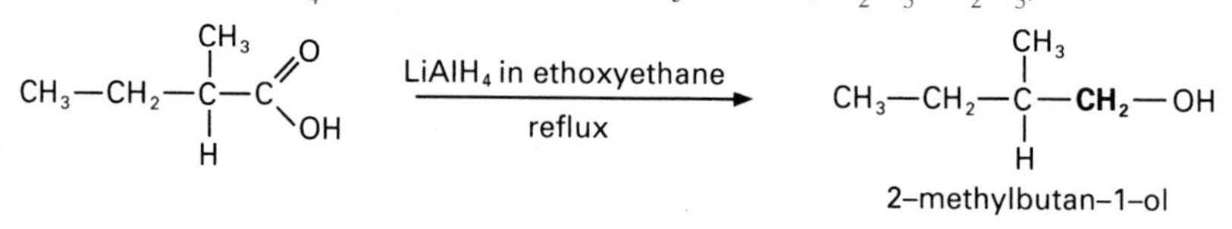

Substitution of —Hal by —NH_2

When a halogenoalkane and an excess of ammonia dissolved in ethanol are heated in a sealed metal tube (this prevents the loss of ammonia and creates a high pressure), an **amine** is formed, for example:

$$\underset{\text{iodoethane}}{C_2H_5I} \xrightarrow[\text{high pressure}]{NH_3 \text{ in ethanol solvent; heat;}} \underset{\text{ethylamine}}{C_2H_5NH_2}$$

Since it has one hydrocarbon skeleton attached to the nitrogen atom, ethylamine is termed a **primary (1°) amine**.

Like ammonia, amines have a lone pair of electrons on the nitrogen atom, and they can behave as nucleophiles. Thus, if the reaction above is carried out with the halogenoalkane in excess, we obtain a mixture of **secondary (2°)** and **tertiary (3°) amines**, and a **quaternary ammonium salt**, for example:

$$\underset{\substack{\text{ethylamine}\\(1°)}}{C_2H_5\ddot{N}H_2} \xrightarrow{C_2H_5I} \underset{\substack{\text{diethylamine}\\(2°)}}{(C_2H_5)_2\ddot{N}H} \xrightarrow{C_2H_5I} \underset{\substack{\text{triethylamine}\\(3°)}}{(C_2H_5)_3\ddot{N}} \xrightarrow{C_2H_5I} \underset{\substack{\text{tetraethylammonium}\\\text{iodide}}}{(C_2H_5)_4N^+I^-}$$

Since such mixtures are hard to separate, secondary and tertiary amines are not made this way (see section 13.6). Secondary and tertiary amines have two and three hydrocarbon skeletons, respectively, attached to the nitrogen atom. Quaternary ammonium salts are used as detergents.

Halogenoarenes do not react with ammonia, or amines, under similar conditions. However, acid halides react violently with ammonia at room temperature and pressure to form **acid amides**, for example:

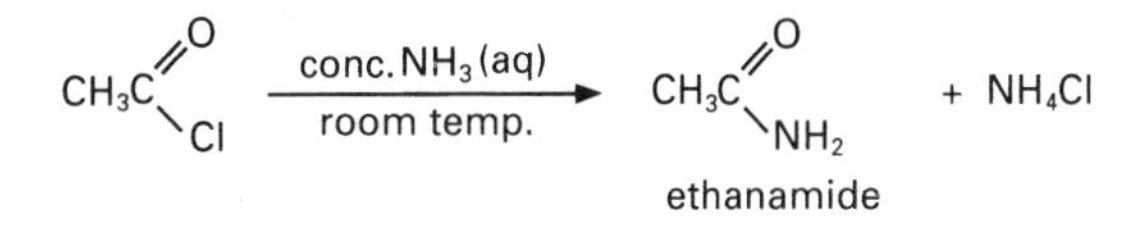

Primary, secondary and tertiary compounds

Amines are classified as being primary, secondary or tertiary compounds in a different way to that used to describe halogenoalkanes. In amines, the classification relates to the number of hydrocarbon skeletons which are attached to the *nitrogen* atom. Halogenoalkanes are classified according to the number of hydrocarbon skeletons which are attached to the *carbon* which is bonded to the halogen atom.

Activity 10.4

1 Draw the structural formulae of: 1-chloropropane, bromobenzene, 1-iodopropane, propanoyl chloride, 1-bromopropane.

Number these structures in order of their *increasing* reactivity with water.

2 Give the structural formula of the organic product(s) formed in the following reactions:

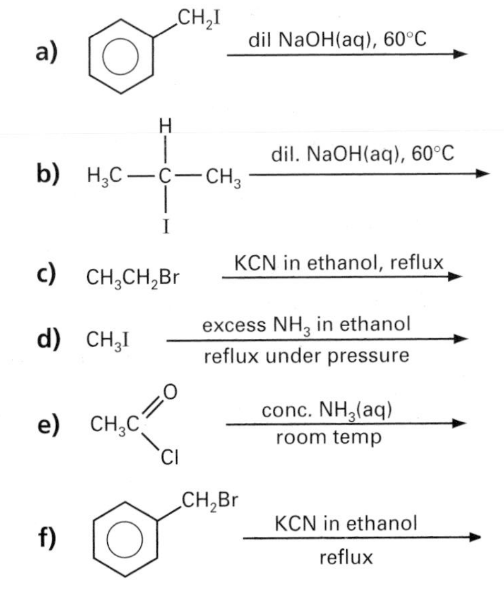

Activity 10.4 continued

3 Give the reagents and reaction conditions needed for the following conversions:

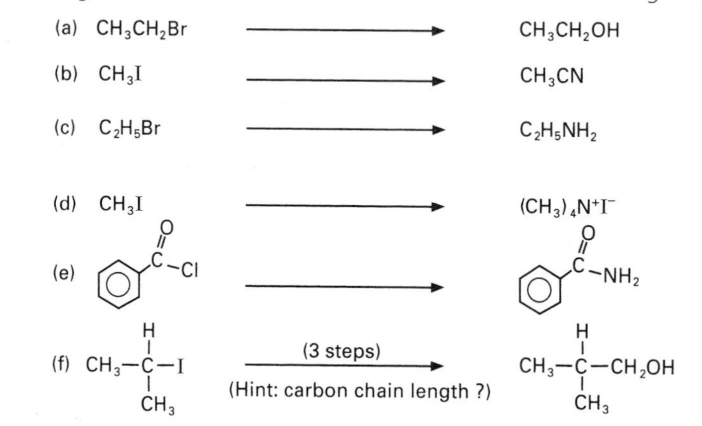

10.6 The mechanism of nucleophilic substitution in halogenoalkanes

One way of investigating the mechanism of a chemical reaction is to study its **reaction kinetics** (Chapter 24). By finding out how the rate of reaction varies with different reactant concentrations, we can get an idea of which reactant molecules are involved in each step of the mechanism. The kinetics of the hydrolysis of halogenoalkanes indicate that nucleophilic substitution in primary and tertiary halogenoalkanes proceeds via *different* mechanisms.

Nucleophilic substitution in tertiary halogenoalkanes

First, the C—Hal bond breaks heterolytically, giving a **carbocation (carbonium ion)** and a Hal^- ion. This step is slow:

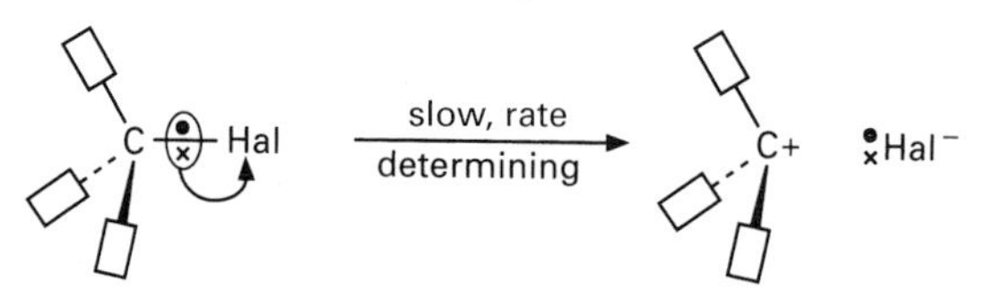

The second step is very much faster since it involves the combination of a nucleophile (an electron-rich particle) with the positively charged (electron-poor) carbocation:

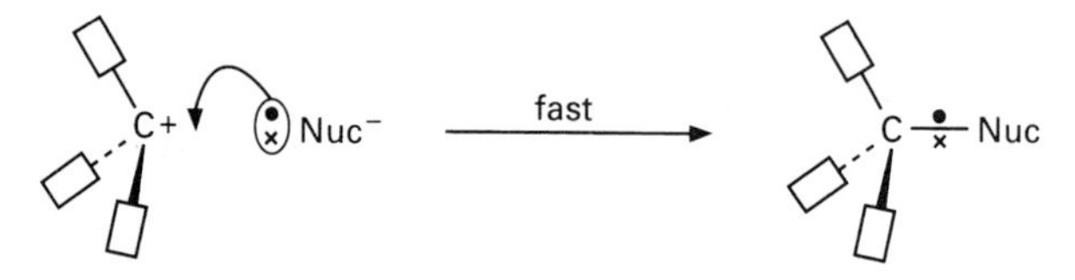

Since the slowest step involves only one reactant molecule, we describe this mechanism as **unimolecular nucleophilic substitution**. It is given the symbol $\mathbf{S_N1}$. Figure 10.8 shows the enthalpy profile for an exothermic S_N1 reaction.

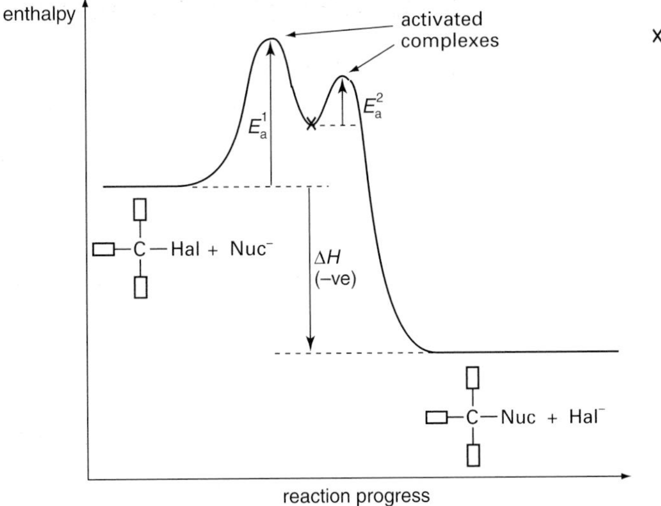

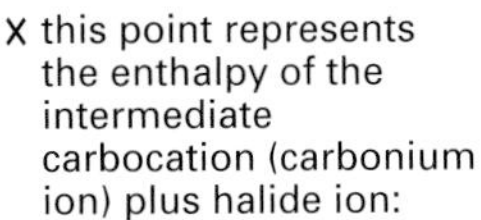

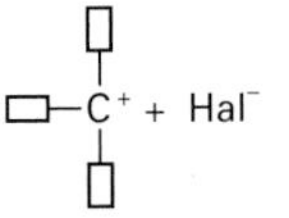

E_a^1 and E_a^2 are the activation energies for the first (slow) step and the second (fast) step respectively

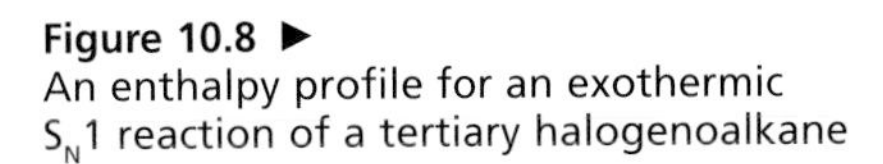

Figure 10.8 ▶
An enthalpy profile for an exothermic S_N1 reaction of a tertiary halogenoalkane

(E_a = activation energy; ΔH = enthalpy of reaction; □ = hydrocarbon skeleton)

Nucleophilic substitution in primary halogenoalkanes

Electron cloud from the nucleophile shifts towards $C^{\delta+}$, and a **co-ordinate (dative covalent) bond** starts to form. As this happens, the C—Hal bond is weakened; eventually, this breaks heterolytically (●— is an alkyl group or an H atom):

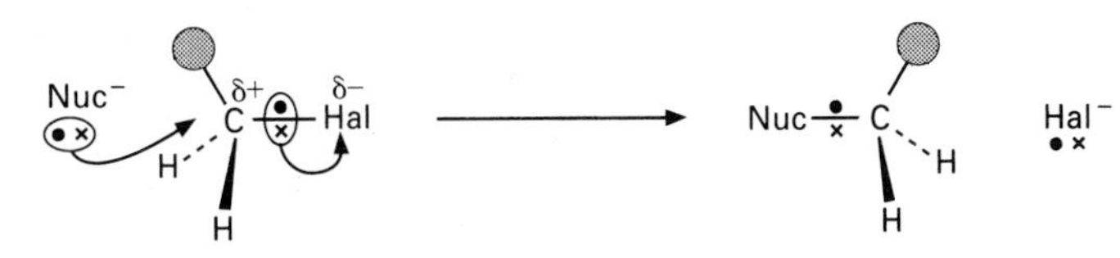

There is only one step. Since it involves two molecules, the reaction mechanism is termed **bimolecular nucleophilic substitution** and given the symbol $\mathbf{S_N2}$. Figure 10.9 shows the **enthalpy profile** for a typical S_N2 reaction. The **activated complex** is the structure at the instant in time when the new C—Nuc bond forms just as the old C—Hal bond breaks.

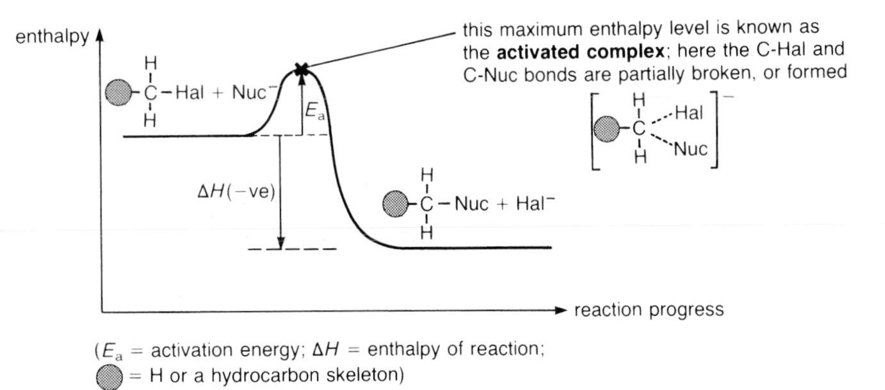

Figure 10.9 ▶
An enthalpy profile for an exothermic S_N2 reaction of a primary halogenoalkane.

(E_a = activation energy; ΔH = enthalpy of reaction; ● = H or a hydrocarbon skeleton)

Why does nucleophilic substitution in primary and tertiary halogenoalkanes occur via different mechanisms? We saw on page 221 that the **inductive effect** of alkyl groups causes stability of carbocations to increase in the order

primary **secondary** **tertiary** →

carbocations increase in stability

Thus, primary halogenoalkanes do not react via the S_N1 route because the primary carbocation is highly unstable and does not form. Tertiary halogenoalkanes do not give an S_N2 reaction because the three bulky hydrocarbons skeletons hinder the approach of the nucleophile. Experiments with isomeric halogenoalkanes show that S_N1 reactions are faster than S_N2 reactions. Nucleophilic substitution in secondary halogenoalkanes proceeds via a combination of S_N1 and S_N2 mechanisms.

Focus 10b

1. The chemical properties of an organic halogen compound depend on (i) which halogen atom is present and (ii) its molecular environment.
2. Halogenoalkanes readily undergo nucleophilic substitution reactions; halogenoarenes do not. The relative rates of nucleophilic substitution show a clear pattern.

   (a) chloro- bromo- iodo- alkanes
   → GET FASTER

   (b) halogenoarenes halogenoalkanes (1°<2°<3°) acid halides
   → GET MUCH FASTER

3. Brief explanations of the trends in 2:

   a) is caused by the decrease in bond strength as we go from C—Cl to C—Br to C—I.

   b) occurs because the reactions have different mechanisms: primary is $\mathbf{S_N2}$, tertiary is $\mathbf{S_N1}$ and secondary is a combination of S_N1 and S_N2. For isomeric halogenoalkanes, the S_N1 route is faster than the S_N2 route. Halogenoarenes resist nucleophilic substitution because **partial π bonding** (i) strengthens the C—Hal bond and (ii) decreases the C—Hal bond polarity (Figure 10.6). In acid halides, polarisation of the C=O and the C—Hal bonds produces a highly electron-poor (δ+) carbon atom, towards which nucleophiles are strongly attracted (Figure 10.7). Thus, acid halides react readily with nucleophiles.

4. Lengthening the hydrocarbon chain in bromo- or iodoalkanes by reaction with KCN/ethanol is an important synthetic step.

10.7 Elimination reactions

When halogenoalkanes react with nucleophiles which are also strong bases, such as OH^- and $C_2H_5O^-$ ions, there is *competition* between nucleophilic substitution and elimination reactions. A mixture of products is obtained:

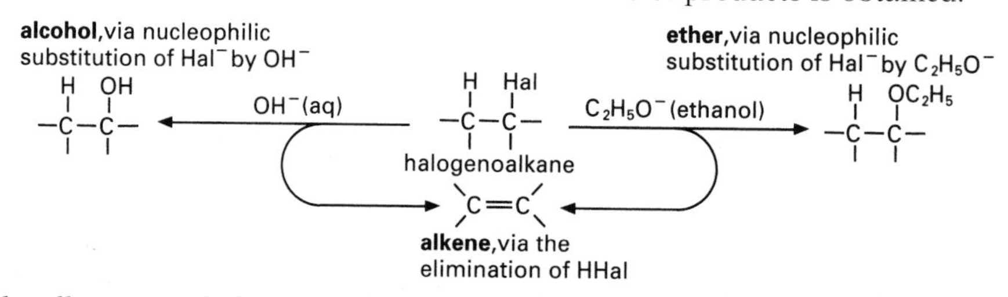

The alkenes result from the elimination of a molecule of hydrogen halide, e.g.

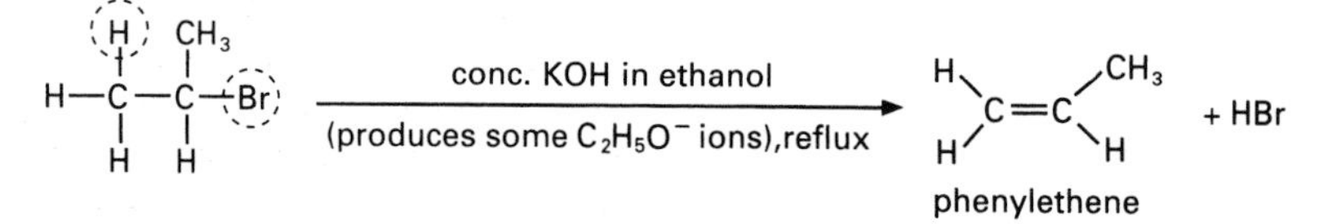

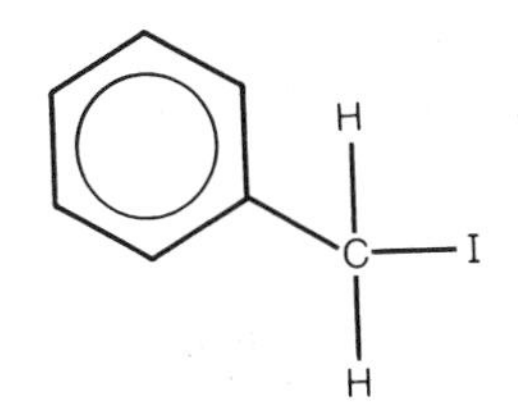

Figure 10.10
(Iodomethyl)benzene – can it give an elimination reaction?

Notice how the H and Hal atoms come off adjacent carbon atoms. So, could (iodomethyl)benzene eliminate an HI molecule (Figure 10.10)?.

When using these reactions in organic synthesis, we need to know which set of reactants and conditions favour elimination, and which favour substitution. This information is summarised in Table 10.3.

Table 10.3 The effect of the reagent on the competing nucleophilic substitution and elimination reactions of halogenalkanes (the main product is stated). Note that potassium hydroxide, KOH, can also be used in these reactions

	Reagents (reflux in each case)		
Type of halogenoalkane	**$H_2O(l)$**	**dil. NaOH(aq)**	**conc. NaOH dissolved in ethanol**
primary	alcohol	alcohol	ether
secondary	alcohol	alcohol	alkene
tertiary	alcohol	alkene	alkene

☐ substitution ☐ elimination

Which type of halogenoalkane is most resistant to elimination reactions? And the least resistant?

In the mechanism of elimination reactions, the OH^- or $C_2H_5O^-$ ions act as strong bases by removing an H^+ ion from the hydrocarbon skeleton, e.g.

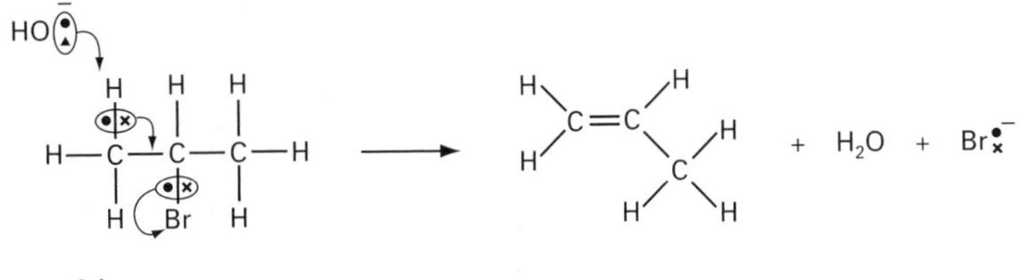

This is a one-step process involving two reacting molecules – thus, it is termed **bimolecular elimination** and given the symbol **E2**.

Elimination reactions can also proceed through a two-step mechanism involving only the halogenoalkane in the slowest step, for example:

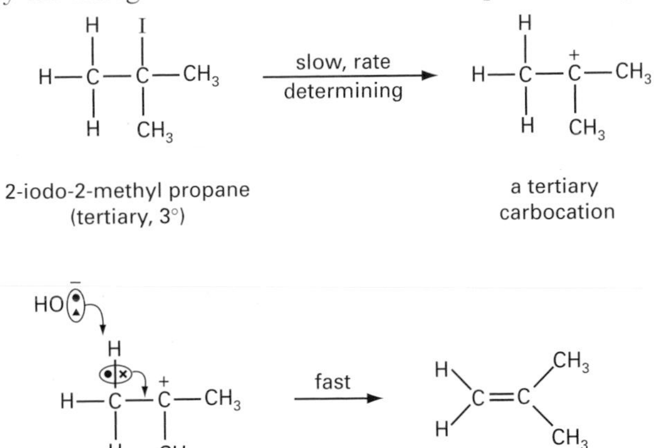

This is an example of a **unimolecular** (one molecule in the slowest step) **elimination** reaction and given the symbol **E1**.

Tertiary halogenoalkanes are more likely to undergo E1 reactions because tertiary carbocations are stabilised by the inductive effect, that is the delocalisation of electrons from the alkyl groups towards the C^+ (page 126). Primary halogenoalkanes tend to give E2 reactions due to the instability of the primary carbocations. The solvent also has an effect on which mechanism predominates. In aqueous sodium hydroxide, the E1 mechanism is favoured because the carbocation is further stabilised by the formation of ion–dipole attractions with the water molecules. When ethanol is used as solvent, however, the E2 mechanism is favoured.

Activity 10.5

1 By reference to Table 10.3, arrange the following compounds in order of increasing tendency towards elimination reactions:

2-bromo-2-methylbutane, 1-bromopentane, 2-bromopentane.

2 **a)** 2-iodo-2-methylbutane gives *two* elimination products. Write down their formulae and name them.
b) Halogenoarenes and acid chlorides do not give simple elimination reactions. Can you explain why?

3 Give the reagents and reaction conditions needed for the following conversions.
a) 2-iodobutane ⟶ butan-2-ol
b) 2-iodopropane ⟶ propene
c) bromoethane ⟶ ethoxyethane ($H_5C_2OC_2H_5$, an ether)
d) 2-chloro-2-methylpropane ⟶ 2-methylpropene
e) 2-chloro-2-methylpropane ⟶ 2-methylpropan-2-ol

Grignard reagents

In 1900, the French chemist Victor Grignard introduced synthetic reagents with the general formula ☐—Mg—Hal, where ☐— is a hydrocarbon skeleton and Hal is a halogen atom, often bromine. The structure of these **organometallic** compounds is very interesting because a hydrocarbon skeleton and a halogen are covalently bonded to magnesium, a metal which usually forms ionic bonds. As you might expect, then, **Grignard reagents** are unstable compounds and cannot be stored. They must be prepared in dry conditions in a non-polar solvent, such as ethoxyethane, and used immediately. For example:

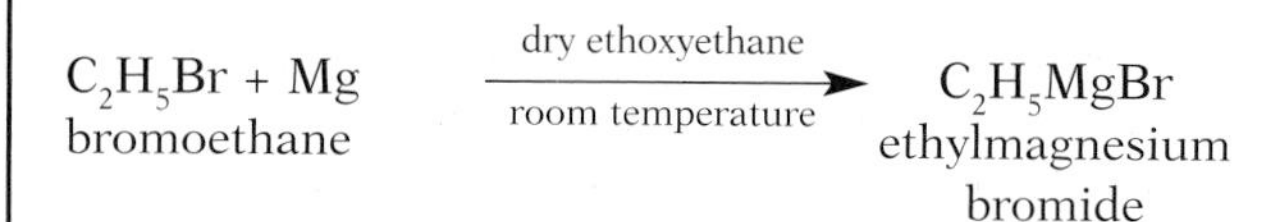

A wide range of organic compounds can be prepared from Grignard reagents. To make a primary alcohol, the Grignard reagent is treated with methanal, HCHO, and the resulting magnesium complex is decomposed by carefully adding dilute hydrochloric acid, for example:

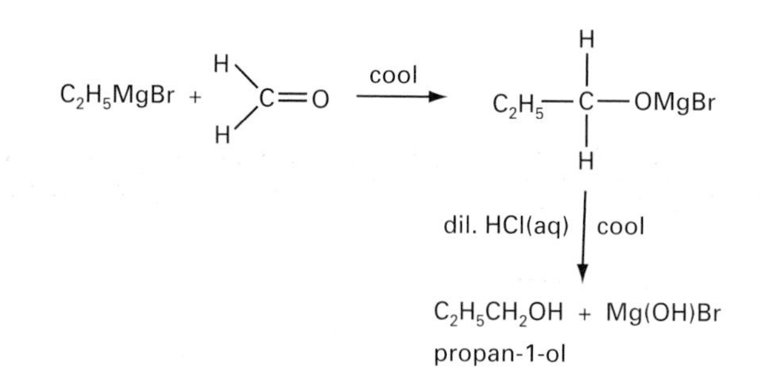

Notice that the hydrocarbon chain length has increased. Treating a Grignard reagent with solid carbon dioxide ('dry ice') gives a carboxylic acid, for example:

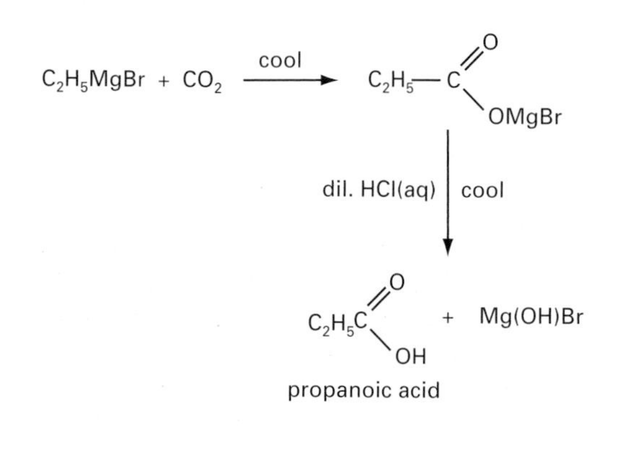

Grignard reagents *continued*

Once again, the hydrocarbon chain is lengthened, and this is one reason for using a Grignard synthesis. Provided that all the reactants, and the glassware, are completely dry, these exothermic reactions proceed rapidly and are easily controlled by cooling (compare this method with the 'nitrile' route in section 10.5). During the reaction, the Grignard reaction splits into $MgBr^+$ ions and a **carbanion**, that is a hydrocarbon skeleton with a negatively charged carbon atom. The carbanion, a nucleophile, attacks the $C^{\delta+}$ in the C=O group resulting in the addition of the Grignard reagent across the C=O bond.

Activity 10.6

1 Ethylmagnesium bromide is prepared by *slowly* adding bromoethane to *excess* magnesium turnings in ethoxyethane solvent. The reactants, the solvent and the glassware must be completely *dry*. The reaction is exothermic and the reaction mixture is *cooled* so that the ethoxyethane gently boils at about 35°C. Bromoethane and ethoxyethane are *flammable*. By using the information above and researching sections 6.2–6.3, write an outline method for the preparation of ethylmagnesium bromide, sketching the apparatus that you need and noting any safety precautions.

2 **a)** Give reagents and conditions for the preparation of the following compounds from ethylmagnesium bromide: i) butan-2-ol, ii) pentan-3-ol, iii) 2-methylbutan-2-ol.

b) If 14.8 g of butan-2-ol is to be made using a Grignard synthesis, what mass of ethylmagnesium bromide should you aim to make?

Organic halogen compounds and the environment

Organic halogen compounds are found in numerous industrial and domestic applications, many of which do not pose any threat to the environment. For example, we have already mentioned **halothane**, an anaesthetic, and **polytetrafluoroethane (PTFE)**, a very strong, 'slippery' plastic which is used to make bearings and as a coating for clothes and non-stick cookware. In this section, we briefly look at two types of organic halogen compound which have hit the headlines in the environmental press.

Polychloroethene (PVC) is an organohalogen plastic made by the addition polymerisation of chloroethene (section 8.7). Although it was discovered in 1872, PVC was not produced in large quantities until the 1940s when the supply of traditional construction materials was restricted by the effects of the Second World War. With a worldwide annual production of over 18 million tonnes, PVC is now the second most popular plastic after polyethene. About 750 companies in the UK, employing at least 50 000 people, make products from PVC-based materials. Various additives allow the PVC to be 'tailor-made' to fit the application. It can be flexible, for example in clothing and wiring insulation, or extremely rigid and impact resistant when used as frames for double-glazed windows and doors. It can be opaque and coloured, in insulation for electrical wiring, or clear and colourless, in bottles for mineral water or blood.

In recent years, some environmental groups have called for PVC to be banned mainly because of concern that its combustion might produce dangerous chemicals called dioxins, shown in Figure 10.11. Evidence suggests that burning pure PVC produces no more dioxins that would be released by burning wood. Research continues, though, on the environmental effects of some of the additives in PVC materials, such as phthalates (plasticisers) and antimony and molybdenum salts (heat stabilisers). Since PVC is a thermoplastic (it melts when heated) it is an ideal material for recycling. Due to its high chlorine content, the polymer remains inert when

Organic halogen compounds and the environment *continued*

buried and actually increases the stability of landfill sites. Overall, PVC compares favourably with other plastics – it is cheap, versatile, fairly inert and may be recycled – and the slight environmental benefit gained from its withdrawal would not appear to compensate for its loss as a construction material.

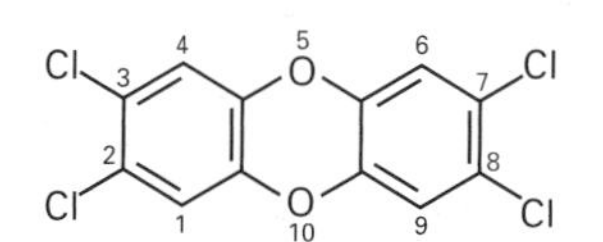

2,3,7,8-tetrachlorodibenzo-dioxin
(2,3,7,8-TCDD)

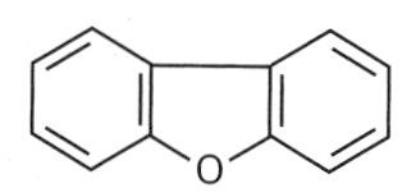

dibenzo-furan

Figure 10.11 ▲
Dioxin is the general name given to over two hundred compounds in which two benzene rings are connected via one, or two, oxygen atoms. Of the dibenzo-dioxins, seven are toxic with the most dangerous being 2,3,7,8-tetrachlorodibenzo-dioxin (2,3,7,8-TCDD). The toxicity of a dioxin depends on its ability to exactly fit the receptor site on an enzyme and then block its action. Thus, even slight changes to the structure of 2,3,7,8-TCDD significantly alter its toxicity (e.g. moving a chlorine atom from atom 7 to atom 1 reduces the toxicity to 1000th of its original value). Most dioxins are produced by municipal waste incinerators, with a milligram of dioxins escaping for every tonne of waste which is burnt. The World Health Organisation has fixed the tolerable daily intake of dioxins at 10 picograms (10^{-11} g) per kilogram of body weight. The average person probably takes in between 1 and 3 picograms per kilogram of body weight and, in their body, has around 70 picograms (per kilogram of body weight). Whilst there have been no reported human deaths due to 2,3,7,8-TCDD, it is certainly very toxic to small animals, for example an oral dose of 1000th of a gram would be enough to kill a rat

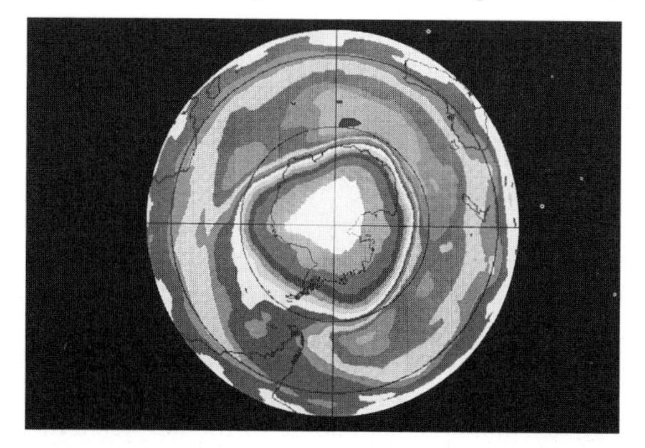

▲ A map of total atmospheric ozone concentration in the Southern Hemisphere on 6 October 1993 recorded by a spectrometer on the Russian Meteor-3 satellite. The Antarctic 'ozone-hole' is at the centre and represented the worst recorded depletion of ozone to that date, down to about 40% of pre-1975 levels.

As an organochlorine compound, PVC is sometimes grouped (unjustifiably) with some other organic halogen compounds which are known to deplete the **ozone layer**. Ozone is produced in the upper atmosphere by the effect of sunlight on oxygen molecules:

$$O_2 \xrightarrow{\text{sunlight}} 2O^{\bullet}$$
$$O^{\bullet} + O_2 \longrightarrow O_3$$

Atmospheric ozone, O_3, absorbs up to 99% of the ultraviolet radiation emitted by the Sun. If this radiation were not blocked by the ozone layer, we would suffer more sunburns, skin cancers and cataracts, and the air we breathe would be even more polluted by the products of photochemical reactions.

From 1978 to 1988, the worldwide concentration of ozone decreased by 10%. Due to meteorological factors, the reduction became more acute in certain geographical areas, such as Antarctica where a 'hole' in the ozone layer was observed in 1986. Scientists identified a link between ozone depletion and the use of certain organic halogen compounds, such as **chlorofluorocarbons (CFCs)**, **halons** and **chlorinated alkanes** (Figure 10.12), and they recommended a 95% reduction in the use of these 'ozone depleters'.

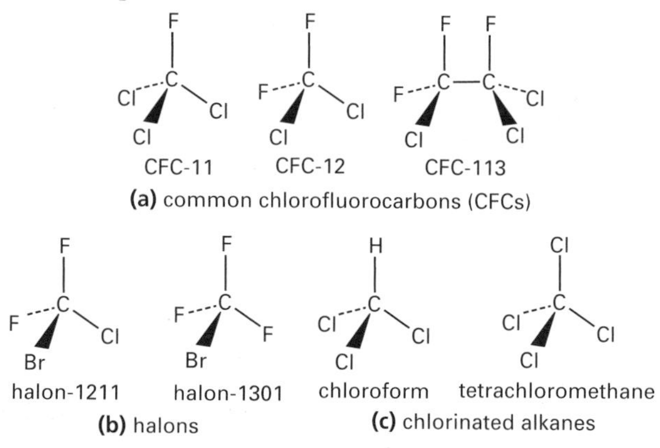

Figure 10.12
There appears to be a link between atmospheric ozone depletion and the presence of the following organic halogen compounds: (a) chlorofluorocarbons, CFCs, are compounds with one or two carbon atoms to which are bonded only fluorine or chlorine atoms, for example, CFC-11, CFC-12 and CFC-113. Until recently, large amounts of these chemicals were produced for use as aerosol propellants, refrigeration fluids, degreasing solvents and in fire extinguishers; (b) halons are halogenated hydrocarbons containing carbon–bromine bonds, for example, halon-1211 and halon-1301, which are used in fire extinguishers; (c) at one time, chloroalkanes, such as trichloromethane (chloroform) and tetrachloromethane (carbon tetrachloride), were widely used as solvents in the laboratory and in industrial cleaning. Due to their toxicity, as much as their effect on the ozone layer, these substances are no longer used in large quantities

Organic halogen compounds and the environment *continued*

Some 65 countries are now party to the Montreal Protocol, first signed in 1987. This seeks a reduction in the production, and consumption, of several organic halogen compounds which are emitted into the atmosphere, many of which are cheap and widely used. The main culprits are CFCs such as CFC-11 and CFC-12 which have been used in aerosols and, especially, in refrigeration. The Montreal agreement calls for a worldwide phasing out of CFCs and this poses a particular problem for developing countries which will struggle to meet the cost of the more expensive, but less damaging, alternatives, such as alkanes.

The problem with CFCs is that C—F bonds tend to stabilise the molecules so that these remain in the atmosphere for some time. In sunlight, chlorine radicals are formed due to photochemical fission of the C—Cl bonds and these react with ozone, forming chlorine oxide radicals and oxygen molecules, for example:

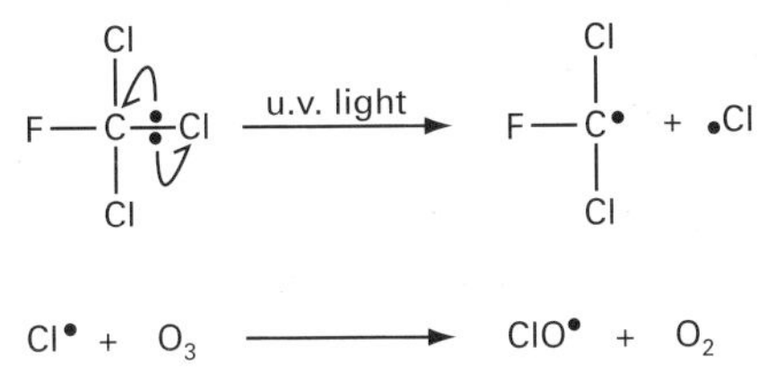

$$Cl^{\bullet} + O_3 \longrightarrow ClO^{\bullet} + O_2$$

If this were the only reaction involved, it would pose little danger to the ozone layer. However, the $ClO^{\bullet}$ radicals produced react with oxygen radicals to reform chlorine radicals and a **chain reaction** is established. The problem is exacerbated by the use of certain fire-fighting compounds, called halons. These can dissociate in sunlight, releasing $Br^{\bullet}$ radicals which also react with ozone molecules.

In addition to causing a depletion of atmospheric ozone, many volatile organic halogen compounds contribute to the greenhouse effect, page 206. In 1990, the Montreal Protocol brought forward its targets for the banning of CFCs and halons, and added further chemicals to the list shown in Table 10.4. It also allowed the use of hydrofluorocarbons (HFCs) as an interim measure. Whilst hydrofluorocarbons are not ozone depleters, they are powerful greenhouse gases and some decompose to produce trifluoroethanoic acid, CF_3COOH, a potentially harmful strong acid. Thus, HFCs will only be allowed as a replacement for CFCs until 2040. Well before then it is hoped that technological developments will allow the widespread use of environmentally safer alternatives.

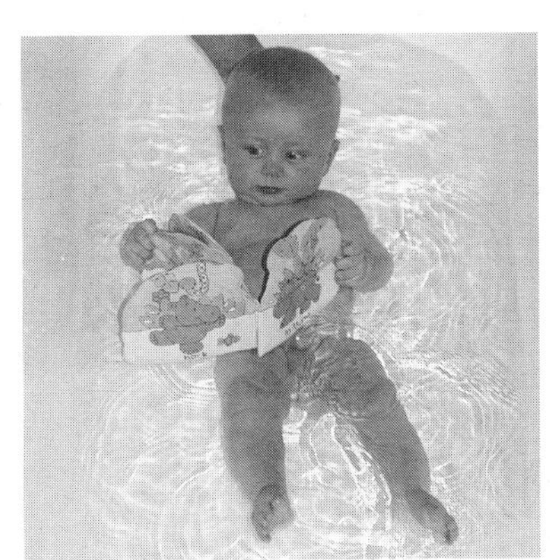

"I'm just doing a little light reading from my PVC book!"

Table 10.4 Montreal Protocol proposals for reducing the use of ozone-depleting chemicals (1987)

Compound	uses	restriction
chlorofluorocarbons (CFCs)	making insulating foam, refrigerant liquids	banned by 1996
halons	fire extinguishers	banned by 1994
tetrachloromethane, CCl_4 (carbon tetrachloride)	making pesticides and rubberised paints	banned by 1996
1,1,1-trichloroethane, CH_3CCl_3 (methylchloroform)	solvent for cleaning and degreasing	banned by 1996
bromomethane, CH_3Br (methyl bromide)	fumigant used to kill pests	frozen at 1992 level of consumption
hydrofluorocarbons (HFCs)	temporary replacement for CFCs	banned by 2040

Activity 10.7

Give the reagents and reaction conditions which are needed to perform the following conversions.

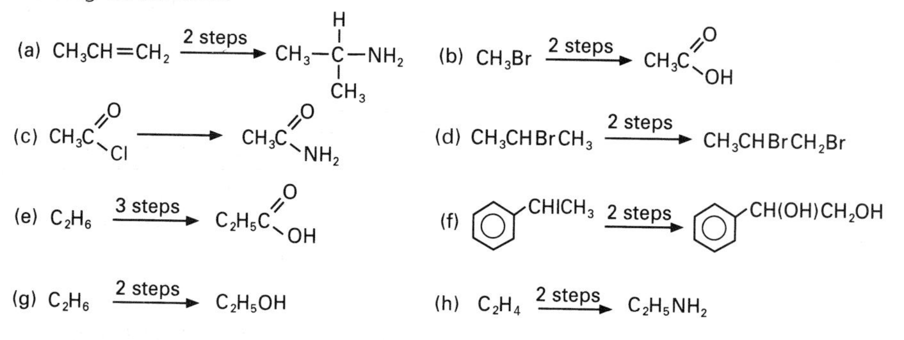

(Note: conversions (e) to (h) need synthetic steps from earlier chapters.)

Focus 10c

1 When halogenoalkanes react with nucleophiles which are also strong bases (such as OH^- and $C_2H_5O^-$ ions), **nucleophilic substitution** and **elimination** reactions will be in competition. The tendency of a halogenoalkane to undergo elimination increases in the order: primary<<secondary<tertiary. Elimination is also more likely when ethanol is used as a solvent (rather than water).

2 **Grignard reagents** are versatile synthetic reagents with the formula ☐– Mg – Hal where ☐– is a hydrocarbon skeleton and Hal is a halogen atom. They are used to prepare compounds containing new carbon–carbon bonds.

3 Some important synthetic steps are shown in Figure 10.13.

4 Important new reagents used in this chapter are shown in Table 10.5.

Table 10.5 Important reagents used in this chapter. Try to learn them (☐– is a hydrocarbon skeleton)

Reagent	conditions	what it does to a halogenoalkane
dil. NaOH(aq) or $H_2O(l)$	reflux	converts C – Hal ⟶ C – OH, an alcohol (acid halides give similar reactions but much more vigorously)
KCN in ethanol	reflux	converts C – Hal ⟶ C – CN, a nitrile (this is one step in the lengthening of the hydrocarbon skeleton)
excess NH_3 in ethanol	heated in a sealed tube	converts C – Hal ⟶ C – NH_2, a primary amine
conc. NaOH (or KOH) in ethanol	reflux	converts: 2° or 3° C – Hal into one, or two, alkenes; 1° C–Hal give mainly alcohols

5 Organic halogens compounds have many uses, ranging from anaesthetics (halothane) to plastics (PTFE and PVC). Until the early 1990s, large quantities of **chlorofluorocarbons** (CFCs) were used as aerosol propellants and refrigerant liquids. CFCs have now been banned (Montreal Protocol) because they react with atmospheric ozone.

Focus 10c *continued*

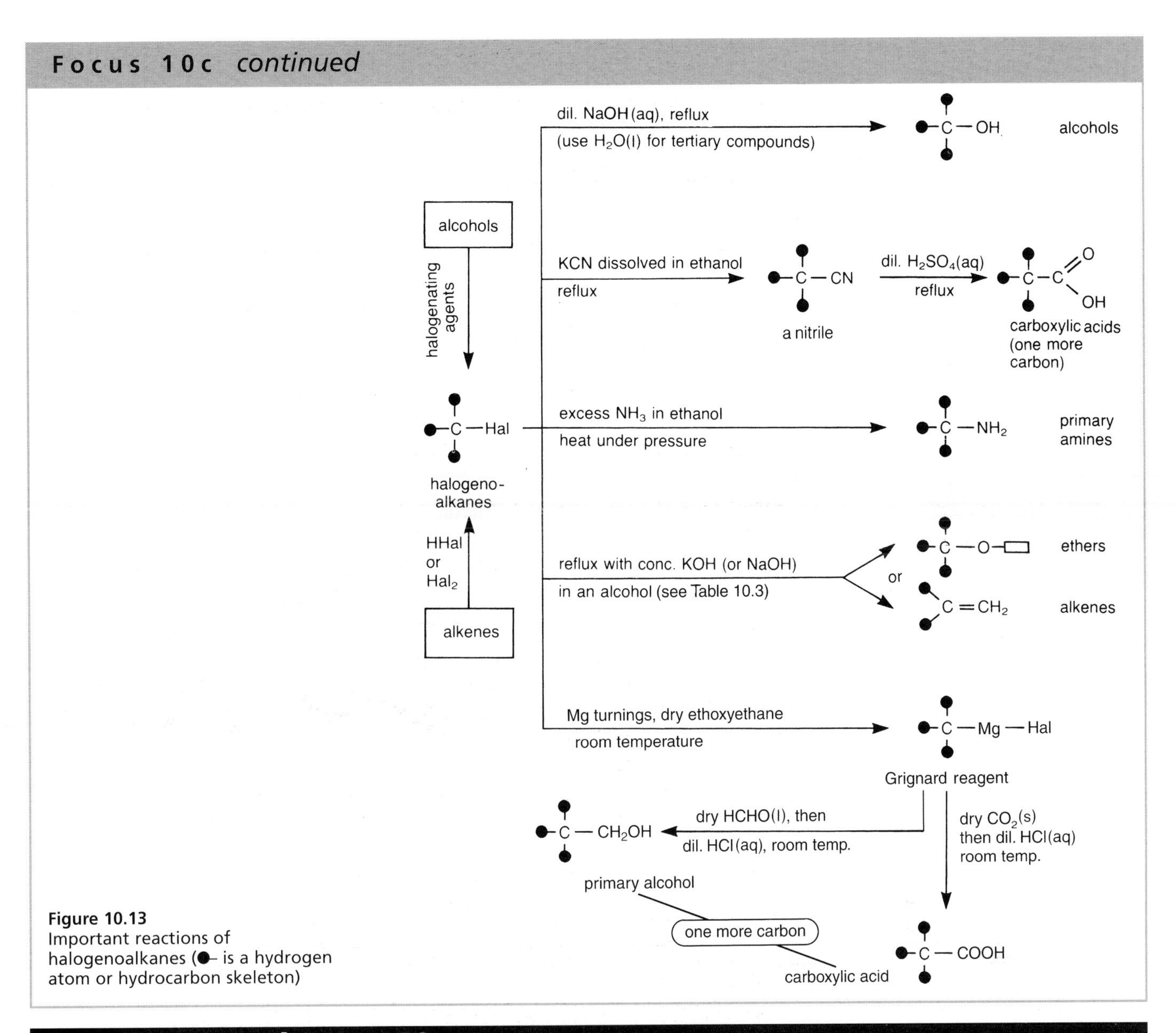

Figure 10.13
Important reactions of halogenoalkanes (●– is a hydrogen atom or hydrocarbon skeleton)

Questions on Chapter 10

1 There are five isomers of molecular formula C_4H_9I.

a) Draw structural formulae for the isomers and state which types of isomerism are present.

b) Describe how to distinguish between the isomers.

2 This question concerns the organic halogen compounds whose structural formulae are shown below:

$CH_3CH_2C(CH_3)_2I$ **A** $CH_3CH_2CH_2CH_2CH_2Br$ **B**

$CH_3CH_2CH_2CH_2CH_2Cl$ **C** C_6H_5–I **D**

$CH_3CH_2CH_2CH_2COCl$ **E** $CH_3CH_2CH_2CH_2CH_2I$ **F**

a) Name the compounds A–F.

b) Classify each as a halogenoalkane (primary or tertiary), halogenoarene or acid halide.

c) How would you measure the relative rates of hydrolysis of these compounds?

d) Place the compounds in order of increasing reactivity with water at 60°C.

e) Explain this trend in reactivity in terms of bond energy and electron delocalisation.

f) Compare the mechanisms of the hydrolysis reactions of A and B.

g) Draw up a table which compares the reactions of B and D with:

i) dilute NaOH(aq),
ii) concentrated NaOH in ethanol,
iii) potassium cyanide in ethanol,
iv) excess ammonia dissolved in ethanol,
v) a nitrating mixture of concentrated nitric and sulphuric acids.

Questions on Chapter 10 *continued*

3 Give the names and structural formulae of the main organic product(s) from the following reactions and identify the mechanism in each case:

a) 1-chloropropane and dilute NaOH(aq)
b) 2-iodobutane and concentrated NaOH in ethanol
c) bromoethane and $CH_3O^-Na^+$ in ethanol
d) (cyclohexyl)–Cl and KCN in ethanol
e) (cyclohexyl)–Cl and excess concentrated ammonia in ethanol
f) CH_3CH_2OH and $PCl_5(s)$
g) $CH_3CH{=}CH_2$ and HBr(g)
h) benzene and $Cl_2(g)$ (with a Friedel–Crafts catalyst)

4 Give the reagents and reaction conditions needed for the following conversions:

a) $CH_3CH_2CH_2Br \longrightarrow CH_3CH_2CH_2OH$
b) $CH_3CH(CH_3)CH_2Cl \longrightarrow CH_3CH(CH_3)CH_2CN$
c) $CH_3CH_2CH_2I \longrightarrow CH_3CH_2CH_2COOH$ (2 steps)
d) $CH_3Br \longrightarrow CH_3OCH_3$
e) C_6H_5–$CH_3 \longrightarrow C_6H_5$–$CH_2COOH$
f) $CH_3CH_2Br \longrightarrow CH_3CH_2CH_2Br$ (4 steps)
g) $CH_3CH{=}CH_2 \longrightarrow CH_3CH(OH)CH_3$ (2 steps)
h) $CH_3OH \longrightarrow CH_3NH_2$ (2 steps)
i) C_6H_5–$CHBrCH_3 \longrightarrow C_6H_5$–$CHBrCH_2Br$ (2 steps)

5 Explain how aqueous silver nitrate solution may be used to distinguish between:

a) 1-chlorobutane and chlorobenzene
b) 2-chloropropane and 2-iodopropane
c) ethanoyl chloride and 2-bromobutane
d) 1-iodopentane and 2-iodo-2-methylbutane

In each case, give the reaction conditions and describe your expected observations.

6 Compound T is a colourless gas with the elemental composition: 38.4% carbon, 4.8% hydrogen and 56.8% chlorine. When 0.275 g of T is heated to 150°C at a pressure of 101 000 Pa, the gas occupies a volume of 150 cm^3.

a) Calculate the empirical formula of T.
b) By reference to section 4.13, if necessary, use the ideal gas equation to calculate the relative molecular mass of T.
c) Write down the molecular and displayed formulae of T and state its name.
d) When T is heated, under pressure, with a catalyst, a polymer V is formed.
i) State the name of V and give the structure of its repeat unit.
ii) State two uses of V, relating these to the physical properties of the polymer.
iii) How might you dispose of waste objects made of V?
e) When T and hydrogen gas are passed over a nickel catalyst at 200°C, compound W is formed. State the name of W and draw its displayed structural formula.
f) Write down the names, and formulae, of the products formed when W is treated with:
i) dilute aqueous potassium hydroxide
ii) an excess of concentrated alcoholic ammonia
iii) sodium cyanide dissolved in ethanol
iv) sodium hydroxide dissolved in ethanol

Comments on the activities

Activity 10.1

1 **a)** chloroethane **b)** 2-bromo-2-methylpropane
c) 2-iodobutane
d) bromobenzene **e)** (bromomethyl)benzene
f) *cis* -1-chlorobut-1-ene (both H's on the same side of the C=C bond)

2

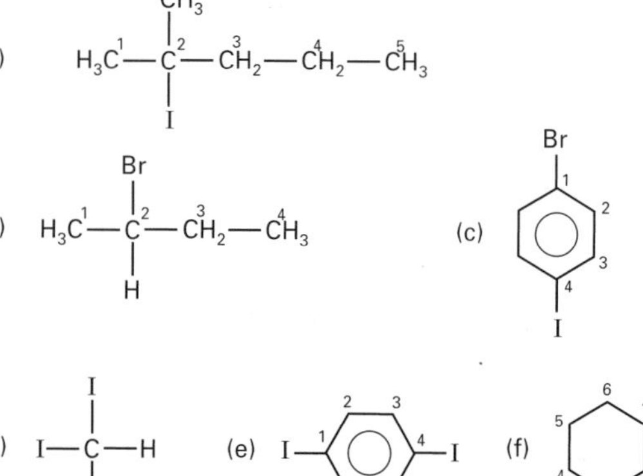

3 i) 1°: 1a, 2d; ii) 2°: 2c, 2b, 2f iii) 3°: 1b, 2a

4

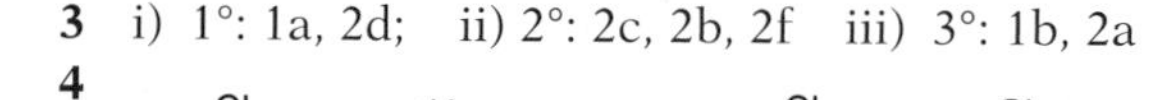

1,1–dichloropropene *cis*–1,2–dichloropropene

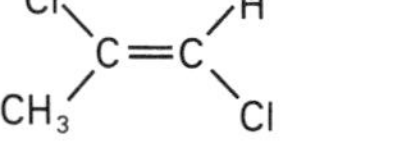

trans–1,2–dichloropropene *cis*–1,3–dichloropropene

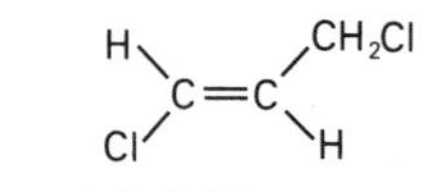

trans-1,3–dichloropropene 2,3–dichloropropene

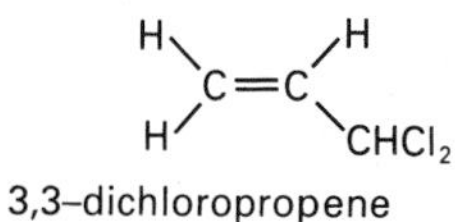

3,3–dichloropropene

Comments on the activities *continued*

5 **a)** Geometrical (*cis–trans*) isomerism

b)

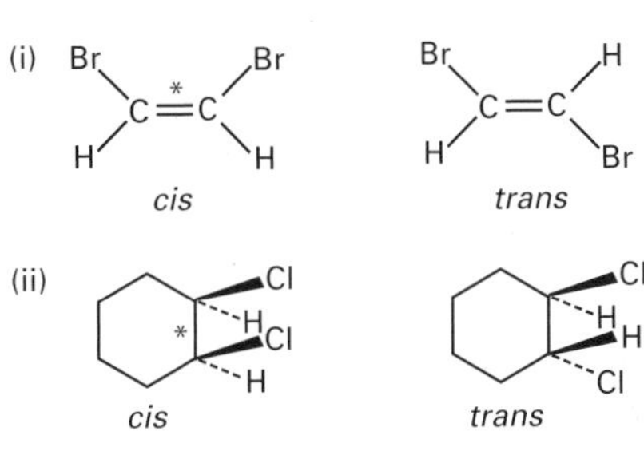

c) A rigid part of the molecule which does not allow rotation of the bonds marked *.

6 Halothane will exist as optical isomers as it contains a chiral centre (the asymmetric carbon atom*).

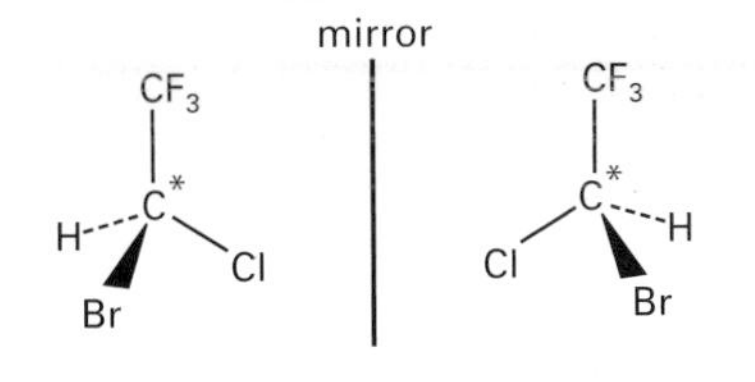

Activity 10.2

1 **a)** For the primary chloroalkanes, graphs of boiling point against the number of carbon atoms in the chain are smooth curves (Figure 10.14). As the hydrocarbon chain length is increased, the molecule's electron cloud gets larger and makes more contact with neighbouring molecules. Thus, the **van der Waals' forces** (section 4.6) will be strongest, and boiling points will be highest, in halogenoalkanes with long straight chains.

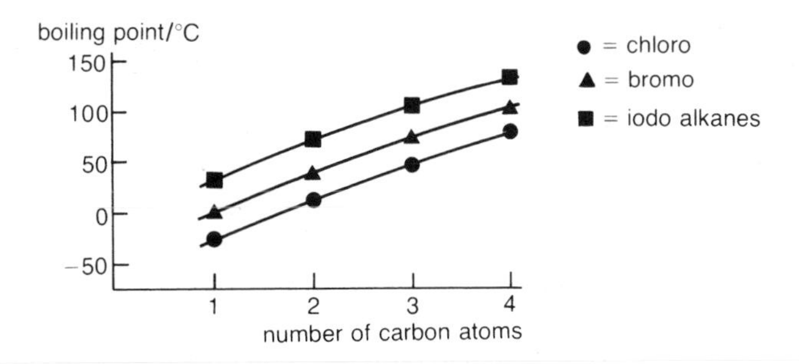

Figure 10.14 ▲
The relationship between the boiling points of some primary chloroalkanes

b) The regular increase in $\Delta H_c^\ominus$ values is typical of compounds which form a homologous series (Figure 10.15). Roughly speaking, the increase corresponds to the enthalpy of combustion of a CH_2 group.

2 The halogenoalkanes in Table 10.1 form a homologous series with the general formula C_nH_{2n+1} Hal.

3 For a given halogen atom, we find that boiling points and densities decrease in the order:
1° halogenoalkane > 2° halogenoalkane > 3° halogenoalkane.

On going from 1° ⟶ 2° ⟶ 3° halogenoalkanes, the molecules' electron clouds become more spherical and, therefore, they make less contact with their neighbours. A more open structure is produced and, hence, the densities decrease. Thus, the van der Waals' forces become weaker, thereby causing the boiling points to decrease.

4 For organic halogen compounds with the same hydrocarbon skeleton, the boiling points increase in the order: chloro- < bromo- < iodo- compound, e.g.:

As molecular mass increases, so does the size of the molecular electron cloud, making it more easily polarised by neighbouring molecules. Thus, the van der Waals' forces will be strongest in the iodo compounds and these will have the highest boiling points.

Activity 10.3

a) P(red) + Br_2(l)/reflux or $NaBr$(s) + conc. H_3PO_4(l)/reflux.

b) and **c)** PCl_5(s)/room temp., PCl_3(l)/room temp., $SOCl_2$(l)/reflux or conc. HCl(aq)/reflux.

d) Cl_2(g) + $AlCl_3$(s)/40°C.

e) HI(g), warm.

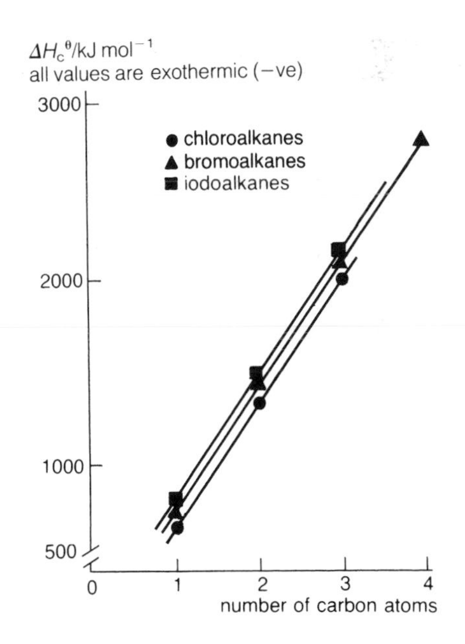

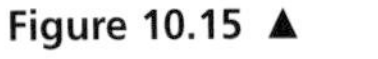

Figure 10.15 ▲
The relationship between the standard enthalpies of combustion, $\Delta H_c^\ominus$, of some primary chloroalkanes

Activity 10.4

1 The structural formulae are:

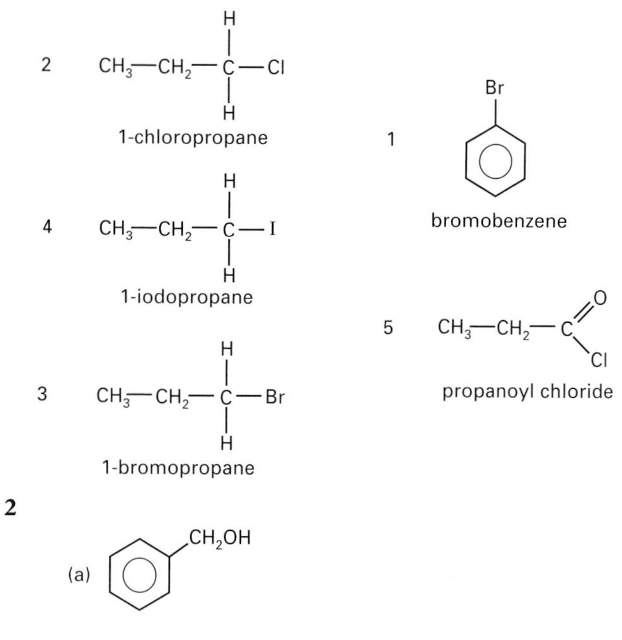

(b) mainly $CH_3CH{=}CH_2$ (propene) by elimination;

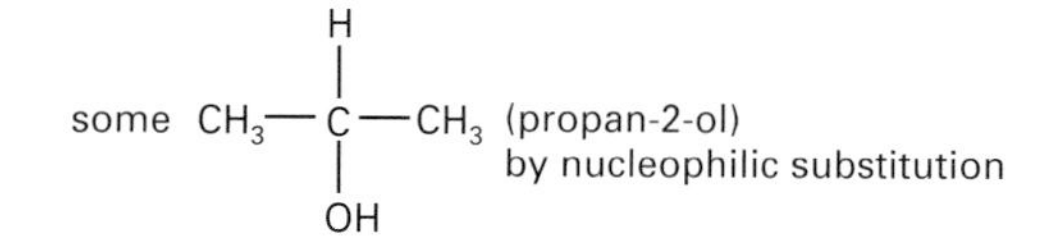

(c) CH_3CH_2CN

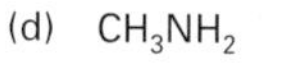

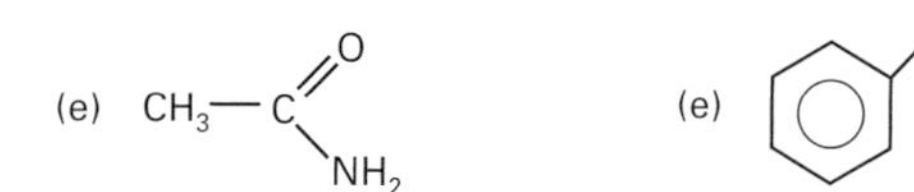

(e) $C_6H_5CH_2CN$

3 a) dil. NaOH(aq); reflux.
b) KCN in ethanol; reflux.
c) excess conc. NH_3 in ethanol; heat in a sealed tube.
d) as **c)** but iodoethane in excess.
e) conc. NH_3(aq), room temperature.
f)

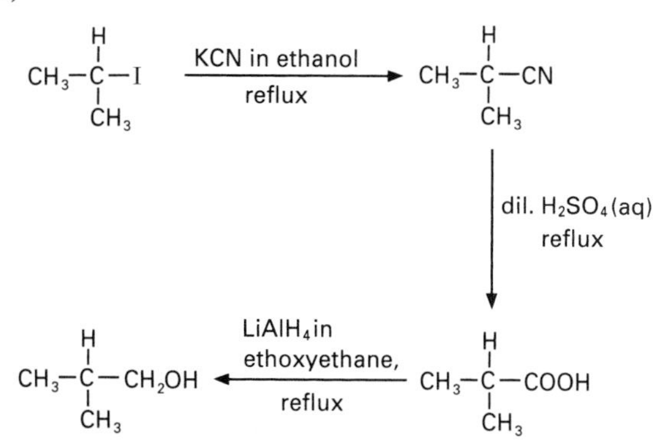

Activity 10.5

1

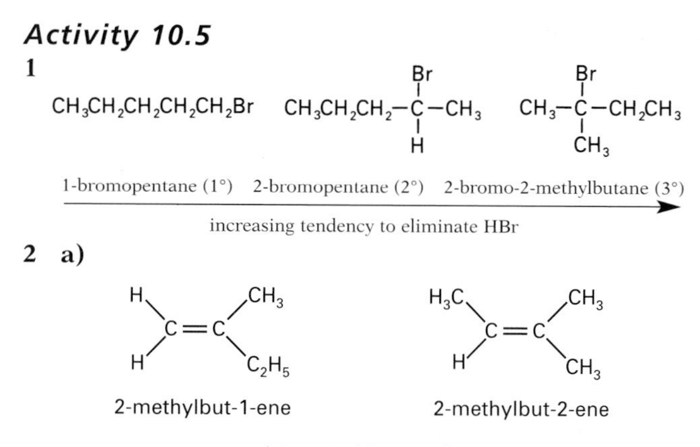

b) These molecules do not have H and Hal atoms on adjacent carbon atoms.

3 a) dilute NaOH(aq) or H_2O(l); reflux.
b) concentrated NaOH (ethanol); reflux.
c) concentrated NaOH (ethanol); reflux (1° halogenoalkanes resist elimination reactions).
d) dilute NaOH(aq); reflux. (3° halogenoalkanes are very susceptible to elimination reactions, even under these conditions.)
e) H_2O(l); reflux.

Activity 10.6

2 a) i)

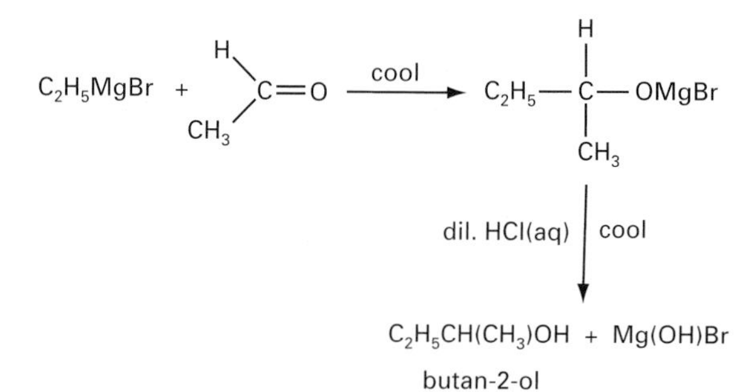

ii) and iii) have exactly the same method except that the starting materials are propanal (CH_3CH_2CHO) and propanone (CH_3COCH_3), respectively.

b) 14.8 g of butan-2-ol (M_r = 74) is 0.2 mol. Since 1 mole of butan-2-ol is formed from 1 mole of C_2H_5MgBr, we shall require *at least* 0.2 mol of the C_2H_5MgBr. Thus,

mass of $C_2H_5MgBr = 0.2 \times M_r(C_2H_5MgBr)$
$= 0.2 \times 133 = 26.6$ g

In fact, a slight excess of the Grignard reagent should be used as this will maximise the yield of butan-2-ol.

Activity 10.7

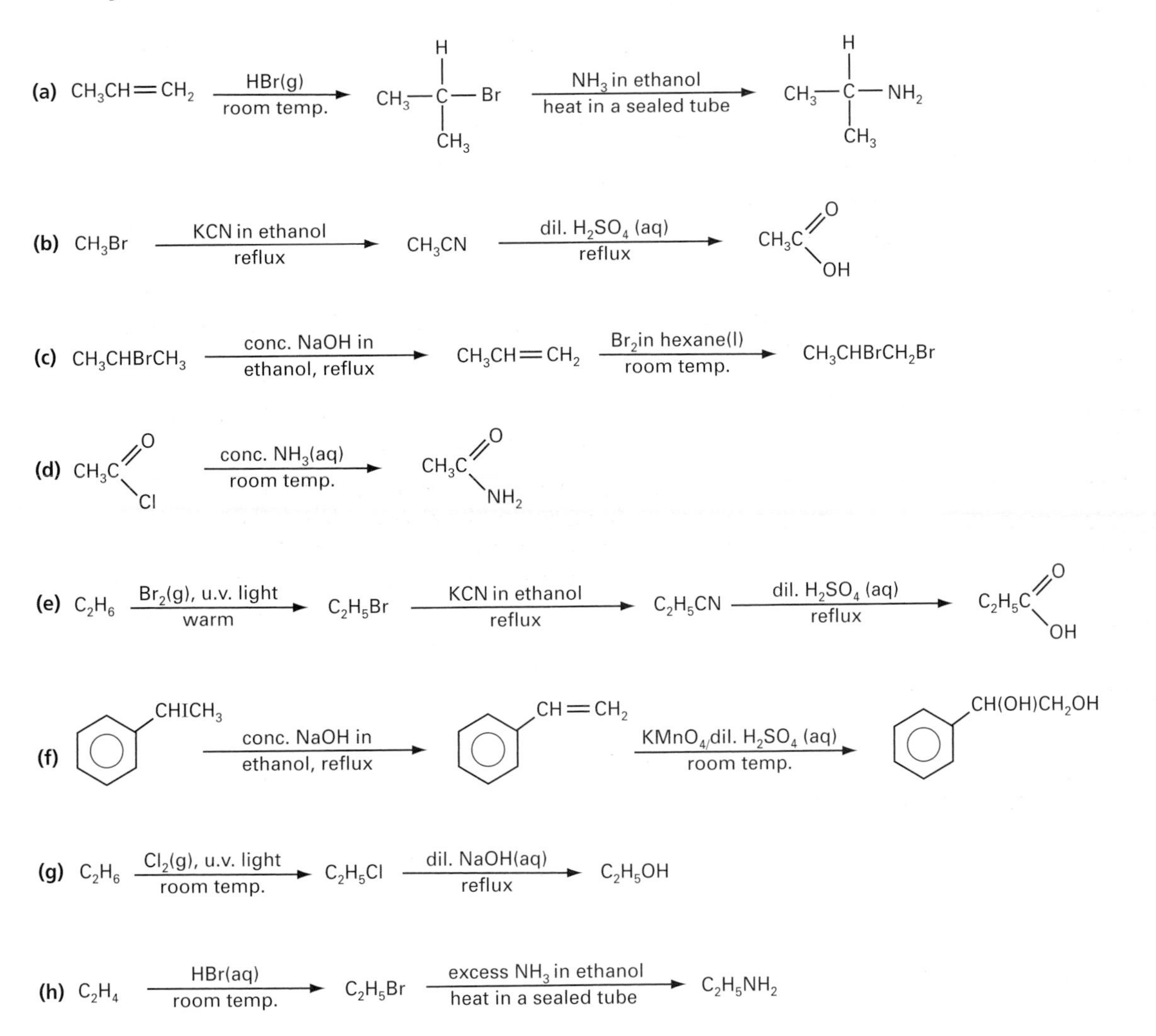

CHAPTER 11

Alcohols, phenols and ethers

Contents

Study Checklist

After studying this chapter you should be able to:

1 Interpret, and use, the names, short structural formulae and displayed formulae of simple aliphatic alcohols and classify them as being primary, secondary or tertiary compounds; state that aliphatic alcohols form a homologous series of general formula $C_nH_{2n+1}OH$ and describe the gradual change in physical properties that occurs as molecular mass is increased.

2 State that phenols have at least one —OH group directly bonded to a benzene ring. Name simple phenols and state that polysubstituted compounds may exist as structural (positional) isomers. Describe the structural difference between an aromatic alcohol (e.g. phenylmethanol) and its isomeric phenol (e.g. 4-methylphenol).

3 State that there is strong intermolecular hydrogen bonding in alcohols and explain its origin in terms of the relative electronegativities of the atoms present. Describe the effects of hydrogen bonding on the physical properties of alcohols, as follows:

a) the boiling points of alcohols are much higher than those of alkanes of similar molecular mass,

b) the lower alcohols are extremely soluble in water but solubility decreases as the non-polar hydrocarbon skeleton gets longer,

c) the lower aliphatic alcohols are valuable solvents which are often used to mix polar and non-polar reactants.

4 Outline the manufacture of alcohols by the acid-catalysed hydration of the alkene fractions obtained from cracking petroleum. State that some ethanol is produced by the fermentation of glucose, catalysed by yeast, and that this method is economically viable where glucose is available in large quantities, for example in sugar cane in Brazil. State the reagents and conditions for the laboratory preparation of alcohols from alkenes or halogenoalkanes.

5 State that there are four main types of reactions of alcohols and phenol: **a)** as acids, **b)** as nucleophiles, **c)** oxidation reactions and **d)** reactions of the hydrocarbon skeleton.

6 Explain that the alcohols and phenol (XO—H) are weak acids which dissociate in aqueous solution, thus:

$$XO—H(aq) + H_2O(l) \rightleftharpoons H_3O^+(aq) + XO^-(aq)$$

Explain, in terms of electron delocalisation in the anion $XO^-(aq)$, the following trend in acidity:

alcohols — water — phenols

→ increasing acid strength

7 Give the reagents and reaction conditions for the conversion of alcohols to: **a)** halogenoalkanes and thence to amines, **b)** esters and **c)** ethers, via nucleophilic substitution reactions.

8 State that when alcohols react with nucleophiles which are also strong bases, substitution and elimination reactions will be in competition.

9 Explain how the reaction of ethanol with concentrated sulphuric acid may produce two different reaction products depending on the conditions.

10 Describe the oxidation reactions of alcohols and distinguish between 1°, 2° and 3° alcohols by analysing their oxidation products. State that tertiary alcohols and phenols tend to resist oxidation reactions because a C—C bond must be broken in the process.

11 State that all alcohols having the —$CH(OH)CH_3$ group will give a positive iodoform (triiodomethane) test.

12 Recall the chemistry of phenols, as exemplified by the following reactions: **a)** with sodium, **b)** with aqueous sodium hydroxide, **c)** nitration of, and bromination of, the benzene ring and **d)** the reaction with neutral iron(III) chloride solution.

13 State the general formula of a simple ether and give the reagents and conditions for its preparation from an alcohol. State that, although they are structural isomers, alcohols have much higher boiling points than their isomeric ethers due to the presence of hydrogen bonding. State that ethers are fairly unreactive compounds which, if used with care, make useful solvents for non-polar substances.

14 Summarise the main reactions of alcohols and phenols as shown in Table 11.10. Recall the reagents in Table 11.9 and use them in simple synthetic routes.

11.1 Fermentation

Over the last twenty years, making wine and beer at home has become a popular hobby. Nowadays, you can buy kits which contain all the necessary ingredients and equipment and, providing the instructions are followed, these can give good results. Brewing and wine-making are chemical processes in which **ethanol** is formed by the **fermentation** of the **glucose** obtained from fruit or grain:

$$\underset{\text{glucose}}{C_6H_{12}O_6(aq)} \longrightarrow \underset{\text{ethanol}}{2C_2H_5OH(aq)} + 2CO_2(g) \qquad \Delta H = \text{negative}$$

The reaction is catalysed by zymase, an enzyme present in **yeast**. Yeast is a micro-organism which obtains its energy from what is available during the reaction. All alcoholic drinks are products of this reaction. They are, in effect, flavoured aqueous solutions containing different concentrations of ethanol, Table 11.1. When the ethanol content of the brew reaches a particular concentration, the yeast is poisoned and the fermentation stops. Different strains of yeast can survive at different alcohol concentrations. Some people enjoy stronger drinks such as whisky, gin and rum. These contain 40% ethanol and are known as 'spirits' because they are distilled off the fermented liquids.

Table 11.1 Typical percentages by volume of ethanol in various alcoholic drinks

Drink	beer	wine	sherry	brandy
% ethanol	4	11	15	40

◄ Making wine at home. Yeast is mixed with sugar and water in a starter bottle and kept warm until fermentation starts. The starter yeast mixture is then added to the fruit extract in a large jar fitted with a bubble trap. The trap allows carbon dioxide to escape but no air to get in – if air enters the jar the ethanol would be oxidised to ethanal and ethanoic acid. After about three weeks the fermentation is complete and the wine is filtered. After standing for a few more weeks, the filtered wine becomes clear and can be bottled

Alcohol and health

Alcohol has been brewed for centuries! Beer was first made around 5000 BC in ancient Egypt and it reached Britain some 2000 years later. There has always been concern, it seems, about the consumption of alcohol and various laws have been passed down the centuries to limit its use. In 1643, the sale of alcohol became subject to a tax called excise duty. Since then, successive governments have received a significant revenue from taxing alcoholic drinks. To fall in line with European Union legislation, public houses are now allowed to stay open all day and young children may be allowed in.

In Britain, more than 90% of adults drink alcohol to varying degrees, most with considerable enjoyment, and this, together with the taxes raised, has to be balanced against the negative effects of alcohol on health and society. For example, Eurocare, an alliance of 23 European Union alcohol agencies, has determined that, in 1995, one in five male hospital admissions and one in three road traffic deaths followed alcohol consumption. Also, two out of three violent offenders were drinking at the time of the crime.

Alcoholism is addiction to drinking alcohol. An alcoholic person has an overwhelming desire to drink, but often does not admit to having a problem. The need to drink can take precedence over everything else, including work, family and friends. Alcoholism is thought to result from a combination of hereditary and sociological factors. Where it runs in a family, there is some evidence that the presence of a rare gene, called the dopamine-receptor gene, may impose the alcohol dependency.

Apart from the social problems associated with alcohol, over-consumption has serious effects on health, Table 11.2. Long-term heavy drinking often leads to cirrhosis of the liver, cancers of the throat, hypertension and strokes. Pregnant women who drink are risking their child with foetal alcohol syndrome, a condition which is characterised by retarded growth before and after birth, delayed mental development and physical defects in facial structure. Finally, we should not overlook the calorific value of alcohol; there are over 200 calories in a pint of beer and heavy beer drinkers can often be identified by their 'pot bellies'. So what is the sensible limit for drinking alcohol?

When consumed, the ethanol, C_2H_5OH, in the alcoholic drink is oxidised by air, via the action of an enzyme, alcohol dihydrogenase, present in our liver. The product is **ethanal, CH_3CHO**, an aldehyde. The ethanal then decomposes, via a series of reactions, to form carbon dioxide and energy is released. Whilst the liver *can* deal with large amounts of alcohol

Table 11.2 A rough guide to the effects of alcohol on physical and mental behaviour. The effects on an individual will vary with sex, age, body weight and that person's tolerance to alcohol

Number of drinks		units of alcohol	alcohol level in blood	effects
beer (pints)	spirits (singles)	(1 unit is about 10 g)	(mg per 100 cm³)	
1.5	3	3	50	a relaxed, warm feeling; impairment of inhibitions and judgement
2.5	5	5	80	loss of licence if found driving a vehicle
4	8	8	130	loss of balance and self-control; may be argumentative; slurring of speech
6	12 (½ bottle)	12	200	blurred vision; falling over; drowsy and confused
9	18 (¾ bottle)	18	300	drunken stupor; oblivion; possible coma
12	24 (1 bottle)	24	400	coma, maybe dying
18	36 (1½ bottles)	36	600	certain death

Alcohol and health *continued*

through this series of reactions, it cannot do so quickly. For example, since the liver can only process about 12 cm^3 or 10 g of alcohol per hour, a heavy night's drinking may still lead to a positive test for alcohol the following morning.

A standard '**unit**' of alcohol is taken to be 10 g and this is typically the amount in one glass of wine, a half-pint of beer or a single measure of a spirit. The actual rate at which the alcohol is broken down depends on sex, size and weight and the condition of the liver. Of particular concern is the consumption of alcohol by younger people and 'binge' drinkers, that is those who get 'plastered' on a Saturday night rather than spreading their drinking throughout the week.

The situation has become more complicated for health educators recently. Research has confirmed that a moderate intake of alcohol can have a beneficial effect, a view which has been around since the 1920s. American studies have shown that moderate drinkers reduced their chance of having a heart attack because the alcohol prevents platelets from clumping together to form a blood clot. Danish scientists have shown that people who consume six units a week are the least likely to die of any cause and the risk only rose when 42 units are consumed.

There is a difficulty in trying to set sensible levels for alcohol consumption because different groups in society run different risks. Whilst a man in his fifties can reduce the chance of heart disease by moderate drinking, a young woman drinker will increase her risk of breast or bowel cancer. In December 1995, the government published a report called *Sensible Drinking*. It focused on daily limits and emphasised the view that moderate amounts of alcohol may be beneficial for some people. A regular intake of over 4 units, for men, and over 3 units, for women is not advisable. For men over 40, or for women who have been through the menopause, one or two units a day will provide some protection from heart disease. Not more than one or two units a day, once or twice a week, is the healthy limit for pregnant women.

A driver's view of a policeman reaching into a car with a breathalyser in his hand. Electronic breathalysers use alcohol to produce an electric current. Breath from the tube is drawn into a battery-like cell where it touches a platinum electrode. The platinum causes any alcohol present to oxidise to ethanoic acid, releasing electrons as it does so. These electrons flow to the positive electrode producing a current. The strength of the current is used to estimate how much alcohol the subject has drunk.

Activity 11.1

1. **a)** Write an equation for the oxidation by air of ethanol to ethanal.
   b) How many grams of ethanal will be formed when 10.0 g of ethanol are oxidised?
2. After an evening's drinking, a man has the equivalent of 10 units of alcohol in his blood when he walks home at 11 pm. At about what time would he be within the sensible limit, say 3 units, to drive?
3. Why is it inappropriate to drink alcohol:
   a) in the workplace
   b) when taking medication.

11.2 Naming alcohols and phenols

Ethanol is a **hydroxy** compound that is, an organic molecule containing an **—O—H** functional group. Two important types of hydroxy compounds are alcohols and phenols.

Alcohols

Alcohols have at least one hydroxy group attached to the aliphatic, or alicyclic, part of a hydrocarbon skeleton. They are named by changing the final '**e**' of the hydrocarbon name to '**ol**'.

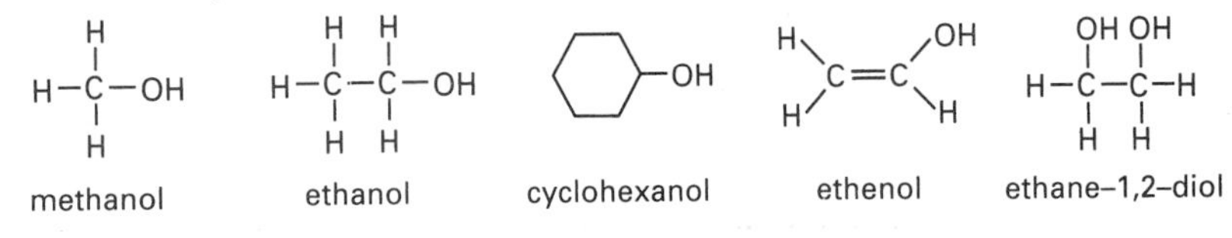

Ethane-1,2-diol is a dihydric alcohol, or **diol**, meaning that it contains two —OH groups.

Where structural isomers are possible, the position of the —OH group must be stated:

$CH_3CH_2CH_2-C(OH)(H)-H$ butan-1-ol (primary,1°)

$CH_3CH_2-C(OH)(H)-CH_3$ butan-2-ol (secondary,2°)

$CH_3-C(OH)(CH_3)-CH_3$ 2-methylpropan-2-ol (tertiary,3°)

Like halogenoalkanes, alcohols are classified as being **primary** (1°), **secondary** (2°) or **tertiary** (3°), depending on whether they have no or one, two or three hydrocarbon skeletons, respectively, attached to the carbon bonded to the —OH group. There is one more structural isomer of the three alcohols above – can you draw its structure?

Phenols

Phenols are compounds in which an —OH group is *directly* bound to a benzene ring:

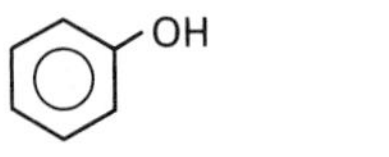

phenol

naphthalene-2-ol

Substituted phenols can exist as structural (positional) isomers, for example:

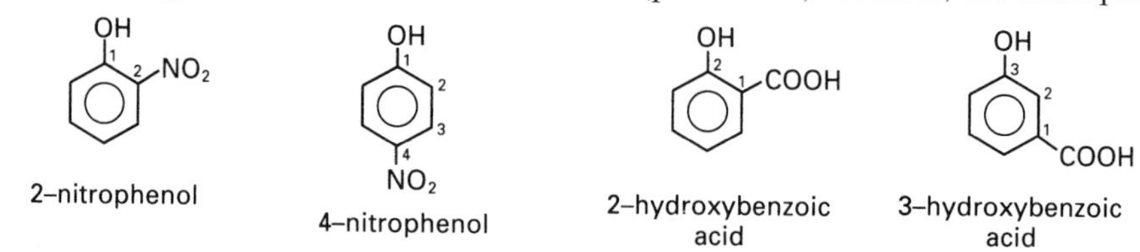

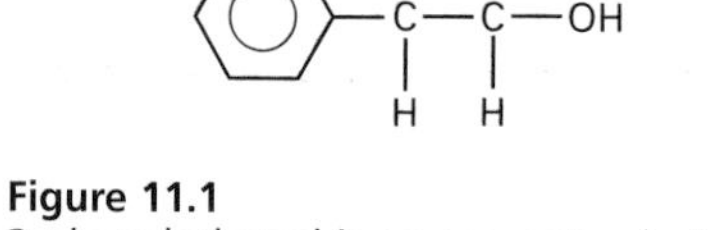

Figure 11.1
2-phenylethanol is an aromatic alcohol because the —OH group is attached to the aliphatic part of the molecule

It is important to remember that an alcohol can be aromatic without being a phenol, for example 2-phenylethanol, Figure 11.1.

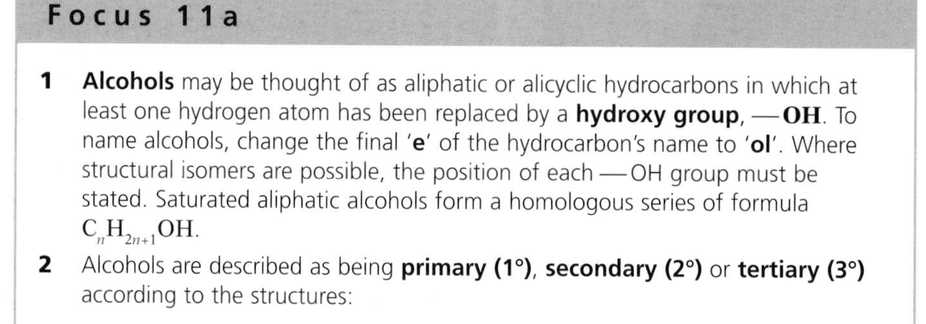

Focus 11a

1. **Alcohols** may be thought of as aliphatic or alicyclic hydrocarbons in which at least one hydrogen atom has been replaced by a **hydroxy group, —OH**. To name alcohols, change the final '**e**' of the hydrocarbon's name to '**ol**'. Where structural isomers are possible, the position of each —OH group must be stated. Saturated aliphatic alcohols form a homologous series of formula $C_nH_{2n+1}OH$.
2. Alcohols are described as being **primary (1°)**, **secondary (2°)** or **tertiary (3°)** according to the structures:

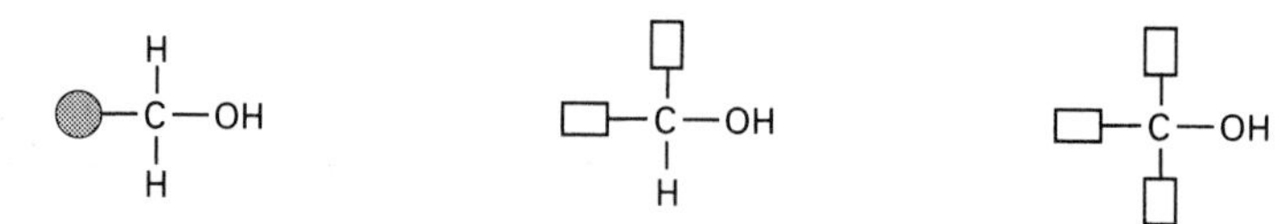

where □— is an alkyl or phenyl group, ● is □— or an H atom.

3. Phenols are compounds which have at least one —OH group directly bonded to a benzene ring. Remember that phenylmethanol, $C_6H_5-CH_2OH$, is an aromatic alcohol not a phenol.

Activity 11.2

1 Name the following hydroxy compounds:

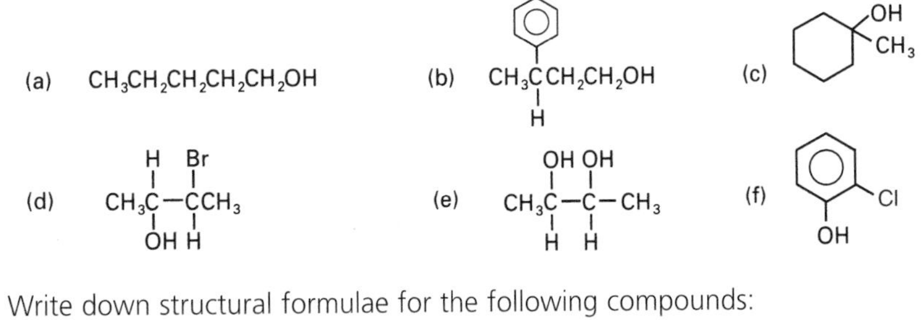

2 Write down structural formulae for the following compounds:
a) pentan-2-ol **b)** 3-methylhexan-1-ol **c)** 2-phenylpropan-2-ol
d) cyclohexane-1,3-diol **e)** 4-chlorophenol **f)** 1,2-dihydroxybenzene

3 Of the alcohols in questions 1 and 2 which are
a) aliphatic/alicyclic **b)** aromatic **c)** primary
d) secondary **e)** tertiary

4 Three diols have the same molecular formula, $C_2H_4O_2$. Draw their displayed formulae and name them.

5 The two stereoisomers of butan-2-ol exist as non-superimposable mirror images.
a) Draw diagrams to illustrate this stereochemistry.
b) What effect would the stereoisomers have on plane polarised light (i) individually and (ii) when present in an equimolar mixture?

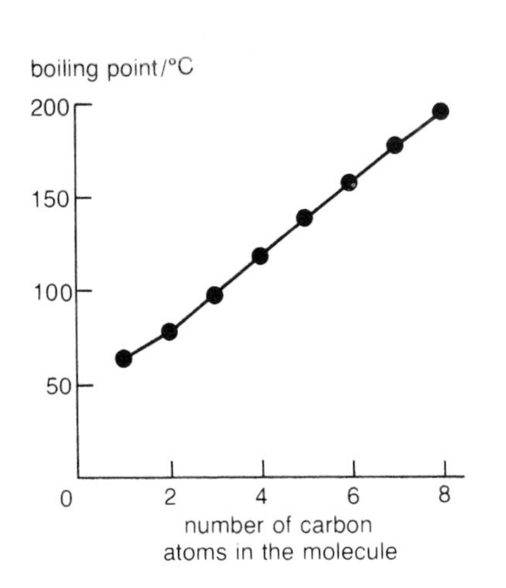

Figure 11.2 ▲
A graph of the boiling points of some straight-chain primary alcohols plotted against the number of carbon atoms in the molecule

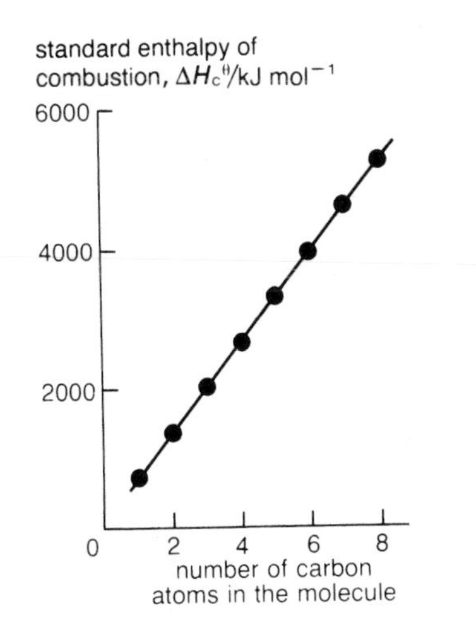

Figure 11.3 ▲
A graph of the standard enthalpies of combustion, $\Delta H_c^\ominus$, of some straight-chain primary alcohols plotted against the number of carbon atoms in the molecule

11.3 Physical properties of alcohols and phenol

The short formulae and boiling points of some alcohols and phenol are listed in Table 11.3.

For the straight-chain primary alcohols, a graph of boiling point plotted against the number of carbon atoms in the molecule is a smooth curve, Figure 11.2. As the hydrocarbon skeleton gets longer, the size of the molecular electron cloud also increases. Larger electron clouds make a greater amount of contact with those of neighbouring molecules. Thus, the intermolecular forces will be the strongest, and boiling points will be highest, for alcohols with long hydrocarbon skeletons. The steady increase in boiling points and standard enthalpies of combustion, Figure 11.3, is typical of compounds which form a homologous series. Their general formula is $C_nH_{2n+1}OH$. As in other homologous series, the successive increase in $\Delta H_c^\ominus$ values roughly corresponds to the enthalpy of combustion of a $—CH_2—$ unit.

The boiling points of the isomeric alcohols generally decrease in the order: primary (1°) > secondary (2°) > tertiary (3°). The molecular electron clouds become more spherical as the branching increases, and make less close contact with their neighbours. Consequently, the intermolecular forces become weaker, and there is a decrease in the boiling points. This trend is similar to those already seen for alkanes, alkenes and halogenoalkanes, namely greater branching in the hydrocarbon skeleton produces lower boiling points.

Alcohols have much higher boiling points than the alkanes of similar molecular mass, for example: propane, $M_r = 44$, b.p. = –42°C; ethanol, $M_r = 46$, b.p. = 78°C. This behaviour results from the differing strengths of the

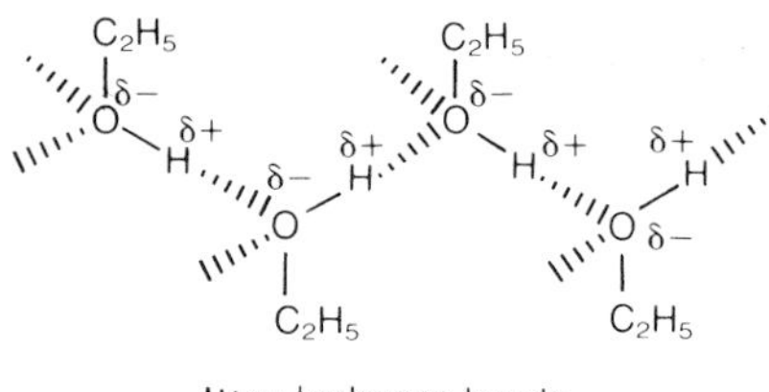

IIIII hydrogen bonds

Figure 11.4 ▲
Hydrogen bonding between molecules of ethanol

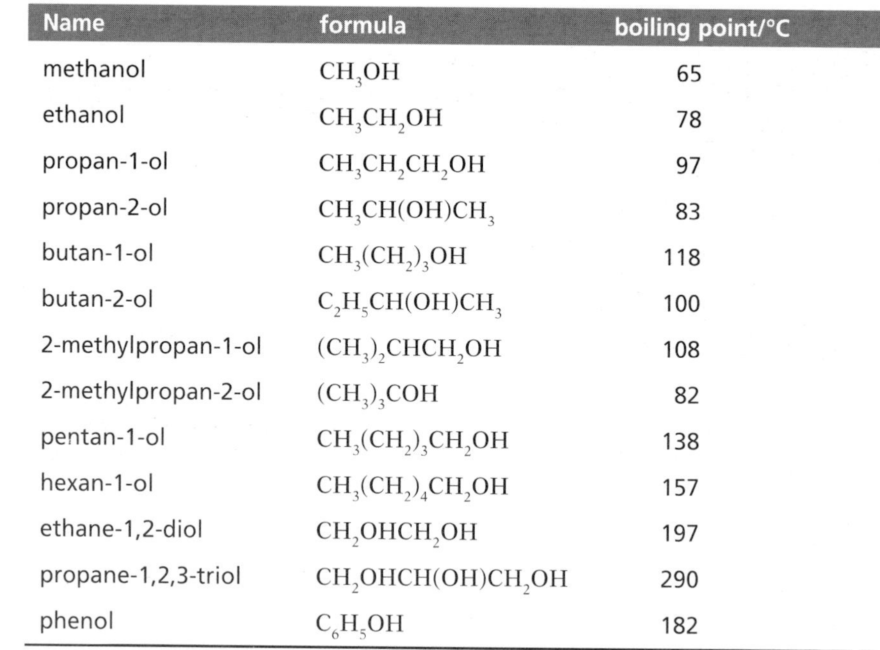

Table 11.3 Short formulae and boiling points of some alcohols and phenol. At 25°C, the alcohols are liquids and phenol is a solid.

Name	formula	boiling point/°C
methanol	CH_3OH	65
ethanol	CH_3CH_2OH	78
propan-1-ol	$CH_3CH_2CH_2OH$	97
propan-2-ol	$CH_3CH(OH)CH_3$	83
butan-1-ol	$CH_3(CH_2)_3OH$	118
butan-2-ol	$C_2H_5CH(OH)CH_3$	100
2-methylpropan-1-ol	$(CH_3)_2CHCH_2OH$	108
2-methylpropan-2-ol	$(CH_3)_3COH$	82
pentan-1-ol	$CH_3(CH_2)_3CH_2OH$	138
hexan-1-ol	$CH_3(CH_2)_4CH_2OH$	157
ethane-1,2-diol	CH_2OHCH_2OH	197
propane-1,2,3-triol	$CH_2OHCH(OH)CH_2OH$	290
phenol	C_6H_5OH	182

Is there a pattern in the boiling points of the primary aliphatic alcohols? What values would you predict for the boiling points of heptan-1-ol and octan-1-ol?

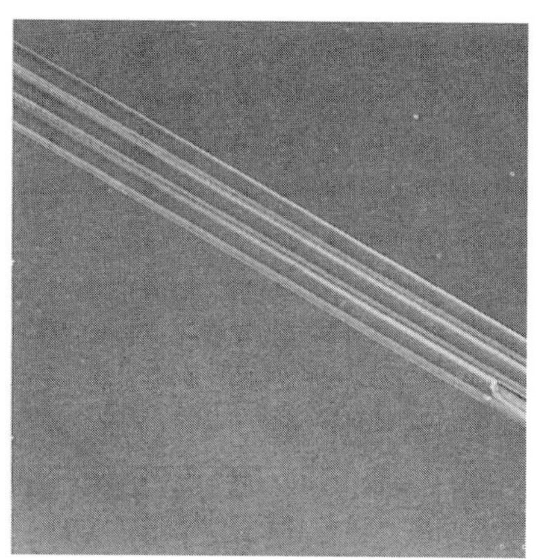

▲ These three sealed tubes are filled with (from left to right) propan-1-ol (I), propane-1, 2-diol (II) and propane-1, 2, 3-triol (III), and a bubble of air is trapped at the top of each tube. When the apparatus is inverted, the air bubbles move upwards through the liquids at different speeds: I > II > III. This shows that the viscosity increases as the degree of hydrogen bonding increases: propan-1-ol (one —OH group) < propane-1, 2-diol (two —OH groups) < propane-1, 2, 3-triol (three —OH groups)

intermolecular forces between neighbouring molecules. Whilst fairly weak van der Waals' forces operate between the molecules in both propane and ethanol, the ethanol molecules are also strongly attracted to each other by **hydrogen bonds**, Figure 11.4. As the number of —OH groups is increased, there is a greater degree of hydrogen bonding (section 4.6). As a result, in going from propan-1-ol ⟶ propane-1,2-diol ⟶ propane-1,2,3-triol:

- the boiling points rise dramatically (Table 11.4)
- the alcohols become much more viscous. You can prove this by timing how quickly a bubble of air moves through each liquid.

Table 11.4 Boiling points of propane and some related alcohols

Compound	boiling point/°C
propane	–42
propan-1-ol	97
propane-1,2-diol	189
propane-1,2,3-triol	290

What happens to the boiling points as the number of —OH groups increases?

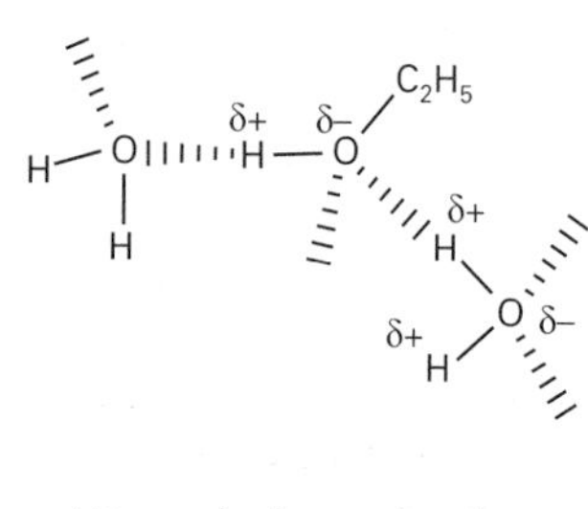

IIIII = hydrogen bonds

Figure 11.5 ▲
Hydrogen bonding between molecules of ethanol and water

Since alcohols form hydrogen bonds, we might expect them to dissolve in water and other polar solvents (section 4.6). Certainly, this is true for the lower members of the series (see Table 11.5). Methanol, ethanol and propan-1-ol are completely miscible with water, and 'new' hydrogen bonds are formed between alcohol and water molecules (Figure 11.5). As the non-polar hydrocarbon chain gets longer, the water solubility of alcohols drops markedly. At the same time, though, the alcohols become better solvents for

▲ A polyvinylalcohol, PVA, based adhesive being used to glue a wooden joint. PVA is a useful glue because its molecule is water soluble, making it safe and convenient to use.

PVA is a water-soluble polymer because its molecule contains numerous hydroxy groups and these form strong hydrogen bonds with the water molecules. PVA gums are also used to coat the backs of postage stamps because they do not cause the gummed sheets to curl up on the printing press

non-polar substances. Polyhydric alcohols form more hydrogen bonds. Thus, they are extremely soluble in water.

Table 11.5 Solubilities of some alcohols in water

Compound	formula	solubility (g in 100 cm³ water at 20°C)
methanol	CH_3OH	miscible in all proportions
ethanol	C_2H_5OH	miscible in all proportions
butan-1-ol	C_4H_9OH	8.0
hexan-1-ol	$C_6H_{13}OH$	0.6
heptan-1-ol	$C_7H_{15}OH$	0.1
hexane-2,3-diol		miscible in all proportions

What happens to the solubility as (i) the hydrocarbon chain gets longer and (ii) the number of —OH groups increases?

The lower alcohols are valuable solvents. Ethanol, for example, is often used to enable non-polar and polar chemical reactants to mix together:

$$C_2H_5I + KCN \xrightarrow{\text{in ethanol, reflux}} C_2H_5CN + KI$$

Generally speaking, phenols are colourless, low-melting point, solids. Like alcohols, phenols have higher melting and boiling points than hydrocarbons of comparable molecular mass because the intermolecular forces in phenols are due to hydrogen bonds as well as van der Waals' forces.

Although they can form hydrogen bonds with water, phenols also contain the non-polar benzene nucleus. As a result, many phenols tend to be slightly soluble in cold water but very soluble in hot water.

▲ This computer-controlled reactor is being used to make ethanol by fermenting glucose using the yeast *Saccharomyces cerevisiae*. The aqueous ethanol produced is dried by refluxing with quicklime (CaO) and the pure (absolute) ethanol is then extracted by fractional distillation. Making ethanol by fermentation is only economically viable where petroleum resources are limited and there is an abundant supply of cheap 'fermentable' materials, such as grain, sugar cane and molasses. In Africa, the fermentation of fuel alcohol from sugar cane has been pioneered by the Triangle Distillery in south-east Zimbabwe which has been operating for seventeen years. The alcohol is used as a 12–15% blend with petrol

11.4 Making alcohols and phenol

Many alcohols are manufactured from the alkenes produced by cracking petroleum fractions (see section 7.4). The alkenes are hydrated by steam at about 300°C and 70 atm pressure using a catalyst of concentrated phosphoric acid, for example:

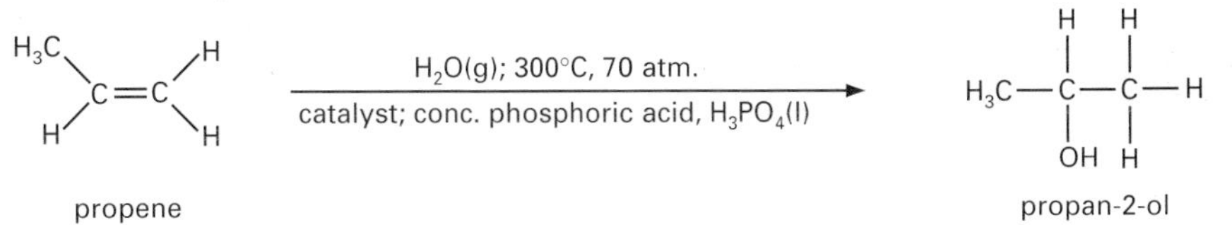

Some industrial ethanol is also produced by fermentation but the process is only economical where there is a plentiful supply of sugar cane or starch. Enzymes catalyse the conversion of starch into glucose, which is then fermented.

Both of these processes yield ethanol in aqueous solution. Fractional distillation of this solution gives a distillate which contains about 96% ethanol and boils at 78.1°C. Pure ethanol, commonly known as **'absolute' ethanol**, is obtained by refluxing the distillate for several hours with quicklime (calcium oxide). This drying agent absorbs the water so that further distillation give the 'absolute' ethanol, boiling point 78.5°C.

Alcohols are best prepared in the laboratory by hydrating alkenes, section 8.5, or by hydrolysing halogenoalkanes, section 10.5.

Activity 11.3

Give reagents and conditions for the laboratory preparation of the following:

1. butan-2-ol from but-2-ene,
2. propan-1-ol from 1-chloropropane
3. propan-1-ol from ethanol (4 steps)

(Hint: for part 3, see section 10.5.)

Phenol used to be obtained from the lower boiling point fractions of coal tar distillation. Nowadays, most phenol is obtained from petroleum using the 'cumene' process. Phenol may be prepared in the laboratory from benzenesulphonic acid (section 9.5) or diazonium salts (section 14.6).

Focus 11b

1. There is strong intermolecular **hydrogen bonding** in alcohols. Thus, we find that:
   i) the boiling points of alcohols are much higher than those of alkanes of similar molecular mass
   ii) the lower alcohols are extremely soluble in water.

   The solubility in water decreases as the non-polar hydrocarbon skeleton gets longer. The lower alcohols are valuable solvents which are often used to mix polar and non-polar reactants.
2. Most phenols are colourless, low melting point, solids. They are only slightly soluble in water.
3. Most alcohols are manufactured from the alkene fractions obtained from cracking petroleum. These are hydrated with steam at high temperature and pressure, in the presence of a phosphoric acid catalyst. Some ethanol is produced by the **fermentation** of glucose, catalysed by yeast; this method is economically viable where glucose is available in large quantities, for example from sugar cane in Brazil. In the laboratory, alcohols can be made from alkenes (section 8.5) or halogenoalkenes (section 10.5). Most phenol is made from petroleum.

11.5 Reactions of alcohols and phenol

All alcohols and phenol contain the C—O—H structure. Electronegativities indicate that these will be polar bonds (Figure 11.6). We might predict, then, the following possible types of reaction:

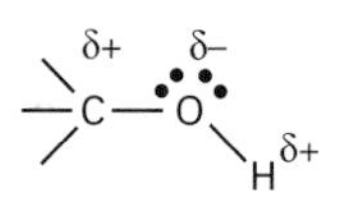

Figure 11.6
The polar nature of the C—O—H structure

- **Fission of the O—H bond**, caused by the attack of a base (an electron pair donor) at the $H^{\delta+}$ atom:

  Here the alcohol or phenol would be acting as an acid, a proton donor.

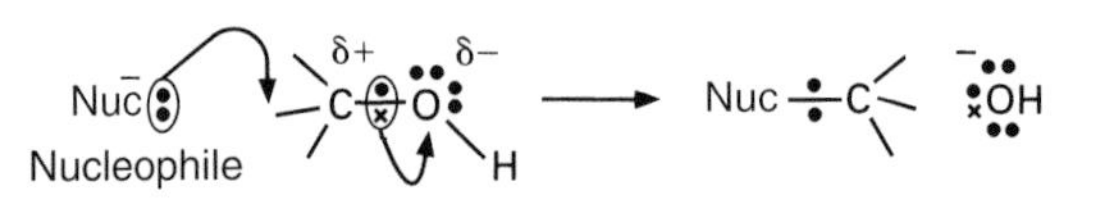

- A nucleophilic substitution reaction at $C^{\delta+}$ atom, resulting in **fission of the C—O bond**:

- In strongly acidic solutions, **protonation of the $O^{\delta-}$ atom** may occur:

  $$\geq C—O(\delta-)—H + H^+ \longrightarrow \geq C—O^+(H)—H$$

  This process makes the $C^{\delta+}$ atom more susceptible to nucleophilic substitution, as shown above, and also promotes the **elimination of a water molecule** across a C—O bond.

We shall look at these reactions together with the oxidation reactions of alcohols and substitution reactions in phenol's aromatic ring.

As you work through sections 11.6–11.9, notice how the reactivity of the hydroxy group, —OH, is affected by its molecular location in an alcohol or a phenol.

11.6 Reactions involving fission of the O—H bond

Acidity

If a reactive metal, such as sodium, is treated with an acid, hydrogen gas is liberated, for example:

$$Na(s) + HCl(aq) \longrightarrow NaCl(aq) + \tfrac{1}{2}H_2(g)$$

In practice, this reaction would be inadvisable – it would be extremely violent! Ethanol, water and phenol also react with sodium, indicating that they, too, can behave as acids, Table 11.6. Unlike ethanol and water, however, phenol also forms a salt with aqueous sodium hydroxide:

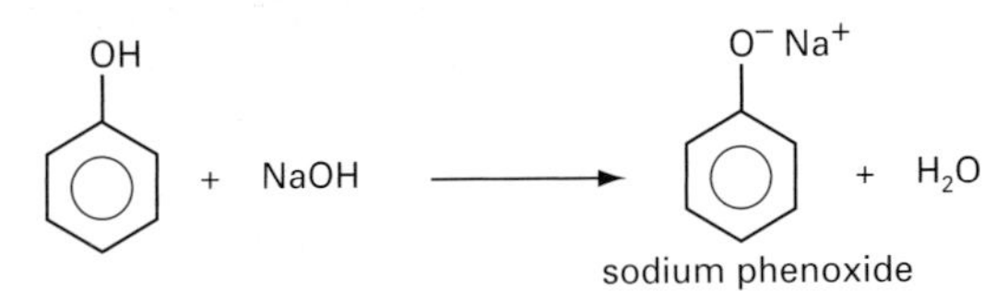

Table 11.6 Observations when ethanol, water and phenol react with a small pellet of sodium at room temperature

Reactant	observations	products: $H_2(g)$ +
ethanol	sodium sinks; steady evolution of hydrogen gas (gives a squeaky pop when tested with a lighted splint)	sodium ethoxide $C_2H_5O^- Na^+$
water	sodium moves quickly around the surface of the water; vigorous evolution of hydrogen	sodium hydroxide Na^+OH^-
phenol (solution in ethanol)	sodium sinks; vigorous evolution of hydrogen	sodium phenoxide $C_6H_5O^- Na^+$

These experiments suggest that:

ethanol **water** **phenols** →

increasing acid strength

Although it is the strongest of the three acids, phenol is still over a hundred thousand times weaker as an acid than a carboxylic acid, such as benzoic acid. Thus, benzoic acid *but not phenol* will liberate carbon dioxide from a hydrogencarbonate:

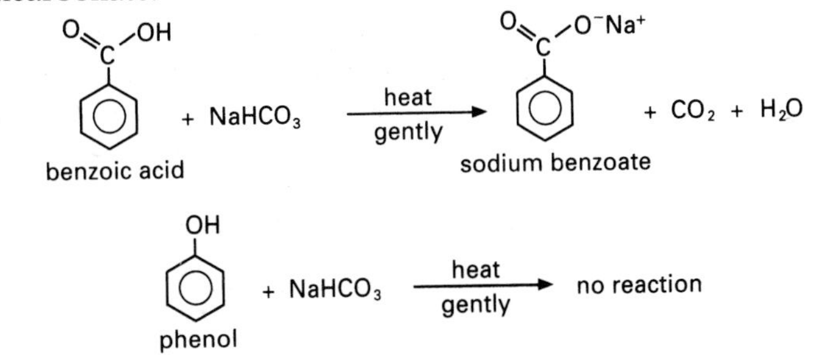

Note how the acidity of the hydroxy group varies with its molecular location, i.e. in an alcohol, phenol or carboxylic acid.

Alcohols and phenols are **Brønsted–Lowry acids**, that is a substance which can donate protons (H^+ ions) to a base. On dissolving a weak acid, XO—H, in water an equilibrium mixture is formed (section 26.2):

$$\underset{\text{undissociated molecules}}{XO\text{—}H(aq)} \rightleftharpoons \underset{\text{ions}}{XO^-(aq) + H^+(aq)}$$

In an equilibrium, the reaction proceeds forwards and backwards, at the same time, and the concentrations reach constant values. However, in the case of a weak acid, we find that:

concentration of undissociated molecules >> concentration of ions

The greater the proportion of acid molecules which dissociate, the stronger the acid.

The acid will be stronger, that is more likely to dissociate:

- if the O—H bond is weak and breaks easily,
- if the negative charge on the oxygen in XO^- is delocalised over the entire anion. This will stabilise the anion by lowering its ability to attract a proton and reform an undissociated molecule.

Both these factors are influenced by the **inductive effect** of the X group, that is its ability to attract or repel the electron cloud.

In the ethanol molecule and the ethoxide ion, the ethyl group pushes electrons away from itself:

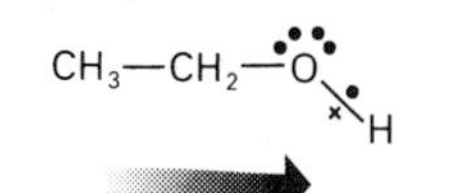

the electron cloud shift **strengthens** the O—H bond, making the loss of the H^+ **more difficult**

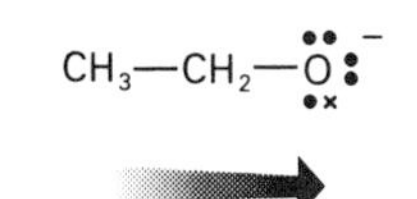

the electron cloud shift localises the negative charge on the O^-, making it **more easily protonated**, that is **less stable**

Overall, these effects decrease the degree of dissociation of ethanol. Consequently, ethanol, and other alcohols, are very weak acids.

However, in the phenol molecule and the phenoxide ion, the lone pair in oxygen's p orbital overlaps sideways with benzene's π electron ring (Figure 11.7).

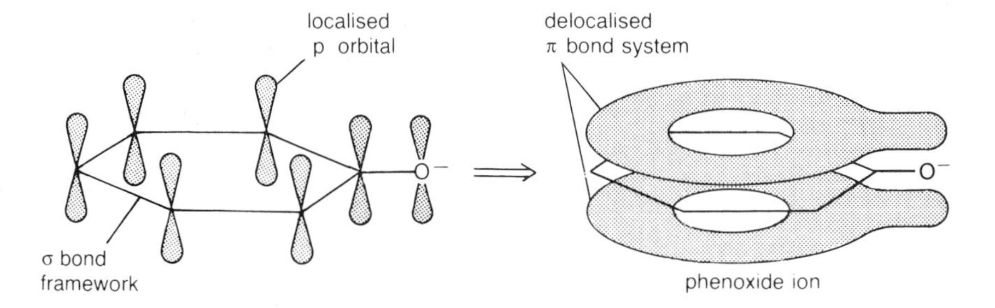

Figure 11.7
The delocalised π bond system in the phenoxide ion is formed by the sideways overlap of the p orbitals on the carbon and oxygen atoms. A similar π bond system is found in phenol itself

As a result, electron cloud from the oxygen and the O—H bond is drawn towards the benzene ring:

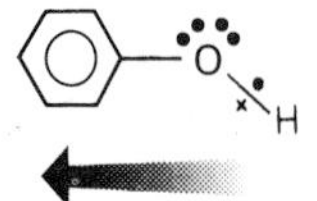

the electron cloud shift **weakens** the O—H bond making it **easier** to lose the H^+

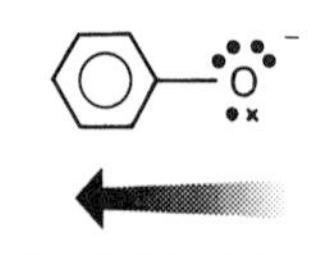

the electron cloud shift **delocalises** the negative charge on the O^-, making it **less** easily protonated that is **more stable**

The effect of substituent groups on the acidity of phenol

The presence of an electron withdrawing substituent group, such as the nitro group, —NO_2, increases the acidity of phenol. Such a group enhances the drift of the electron cloud away from the oxygen atom and the O—H bond. This has the effect of slightly weakening the O—H bond and stabilising the oxyanion. Thus, we find that 2,4,6-trinitrophenol is a fairly strong acid and, like the mineral acids, it is able to displace carbon dioxide from carbonates. As expected, an electron donating ring substituent, such as the amino group, —NH_2, causes the electron cloud to move towards the benzene ring and into the O—H bond, thereby decreasing the acid strength. Thus 2,4,6-triaminophenol is a weaker acid than phenol.

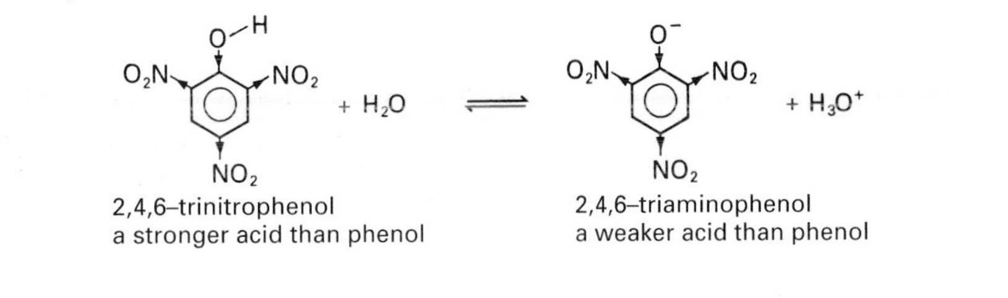

Although phenol is a stronger acid than ethanol, remember that they are both weak acids. Thus, they are both displaced from their salts by stronger acids. For example, even water is a strong enough acid to displace ethanol, a weaker acid, from its salt:

$$CH_3CH_2O^-Na^+(s) + H_2O(l) \longrightarrow CH_3CH_2OH(l) + NaOH(aq)$$
sodium ethoxide

In section 13.5, we shall compare the acid strengths of alcohols and phenols with those of carboxylic acids.

Esterification

Due to bond polarisation and the presence of lone pairs, a dense electron cloud surrounds the oxygen atom in an alcohol or phenol. Hence, these molecules act as **nucleophiles**, via the $O^{\delta-}$ atom, and attack electron-poor (δ^+) centres. An important example of this type of reaction is **esterification**.

When an alcohol is refluxed with a carboxylic acid for several days, an **acid ester** is formed in an equilibrium mixture, for example:

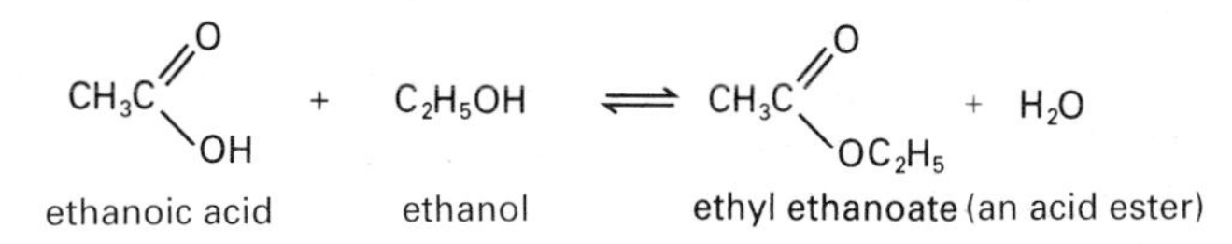

Fortunately, small amounts of concentrated sulphuric acid, or dry hydrogen chloride, catalyse the reaction, giving an equilibrium mixture after a few hours. Also, if an excess of alcohol is used, the equilibrium shifts towards the ester, thereby increasing its yield. Even so, good yields are only obtained with primary alcohols.

The mechanism of esterification is interesting. At first glance, it may appear to be a simple **acid/base** reaction:

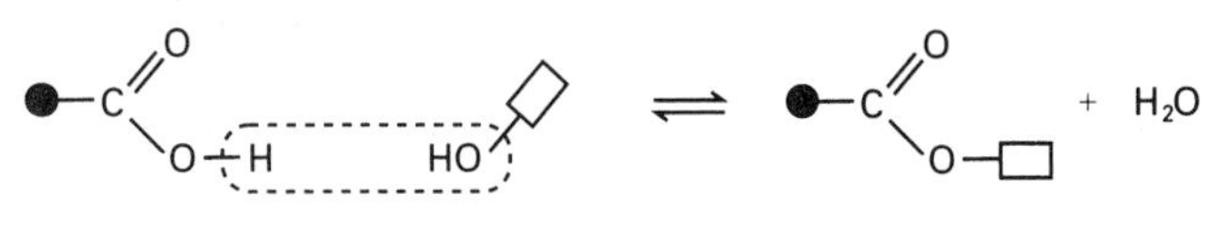

In theory though, bond fission may occur in two ways:

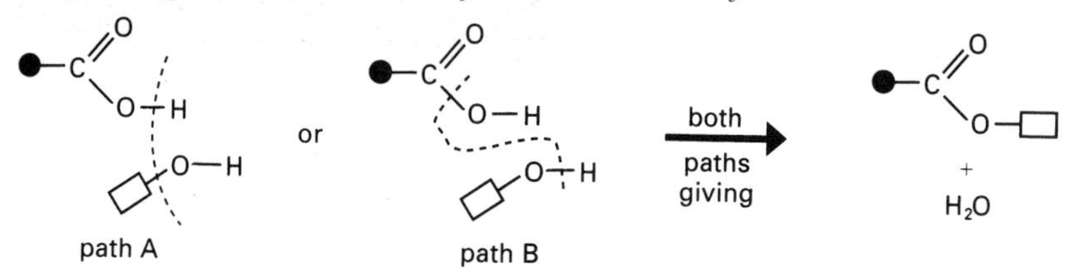

For the simple acid/base reaction, path A, the oxygen from the alcohol would end up in the water molecule. Alternatively, **nucleophilic attack** by the alcohol on the carboxyl carbon, path B, would leave the alcohol's oxygen in the ester. So which mechanism is correct?

This problem provides a good example of how **atomic isotopes** are used to help work out a reaction mechanism. To follow the fate of a particular atom during a reaction, we '**label**' it by using a particular isotope. After the reaction, we can analyse the products to find out which contains the labelled atom. In this case, we want to know what has happened to the alcohol's oxygen. Thus, we prepare an alcohol labelled with ^{18}**O** and use it in the esterification.

Consider, for example, the reaction of ethanoic acid with labelled ethanol, $C_2H_5{}^{18}OH$. Two esters are possible:

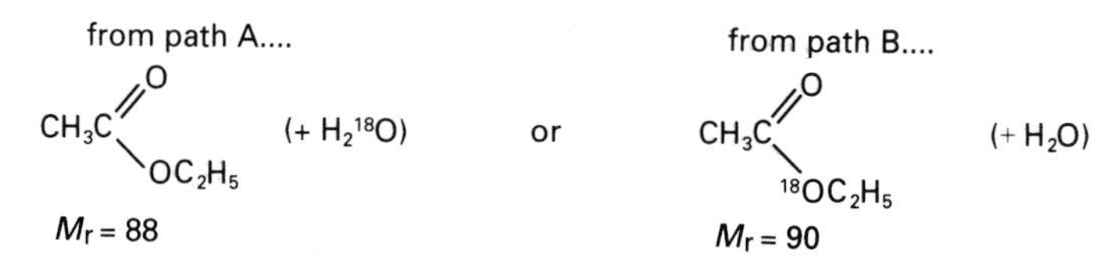

A **mass spectrum** of the purified ester gives a molecular ion of $M_r = 90$ (section 6.4). Hence, the alcohol's oxygen ends up in the ester and the reaction must proceed via path B.

A carboxylic acid ester is best prepared by reacting an alcohol with an acid chloride, for example:

$$CH_3COCl + HOC_2H_5 \xrightarrow[\text{temp.}]{\text{room}} CH_3COOC_2H_5 + HCl$$

ethanoyl chloride ethyl ethanoate

Due to the delocalisation of the oxygen's lone pair of electrons (Figure 11.7), phenols are weaker nucleophiles than alcohols. Thus, phenols do not form esters by direct reaction with carboxylic acids, though they will react with acid chlorides, for example:

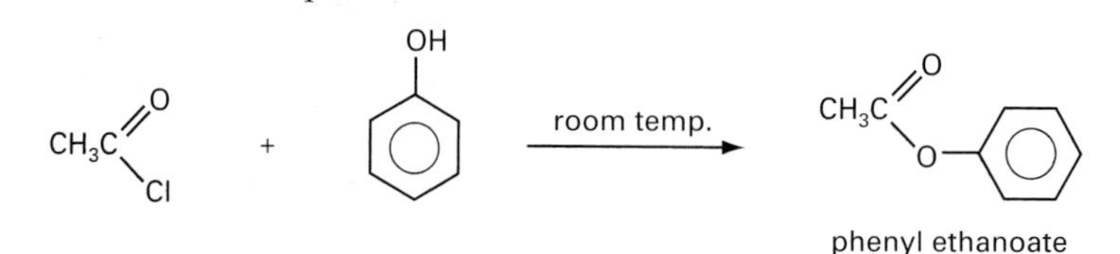

Mechanism of the esterification

When methanol reacts with methanoic acid, the structural change involves the substitution of the acid's hydroxy group, —OH, by a methoxy group, $—OCH_3$, from the alcohol. This is not a simple nucleophilic substitution like the S_N1 and S_N2 mechanisms in halogenoalkanes, section 10.6. Instead, the mechanism is a nucleophilic addition–elimination mechanism, as follows.
In the first step, an H^+ ion from the concentrated sulphuric acid (catalyst) is attracted to the $O^{\delta-}$ of the acid's carbonyl group and a carbocation (or carbonium ion) is formed:

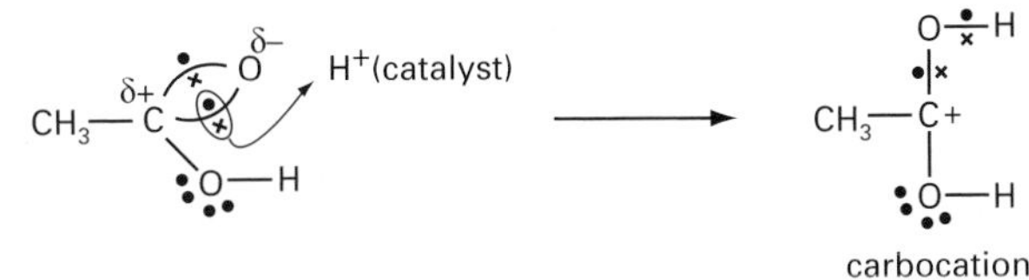

The carbocation is then attacked by the alcohol molecule acting as a nucleophile due to the lone pair of electrons on oxygen atom:

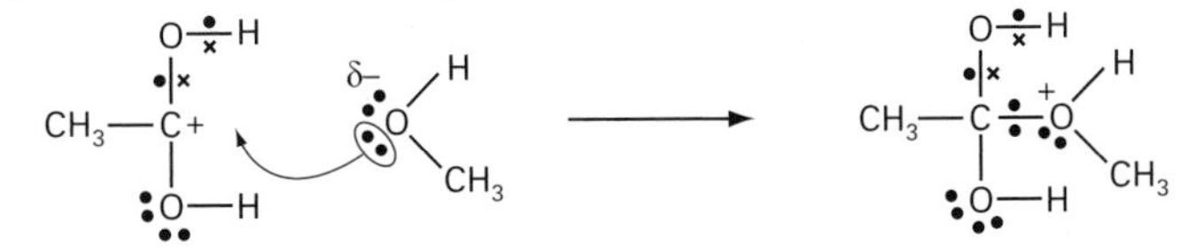

After a proton shift, a molecule of water is eliminated across the C—O bond, reforming the double bond, and an ester is produced:

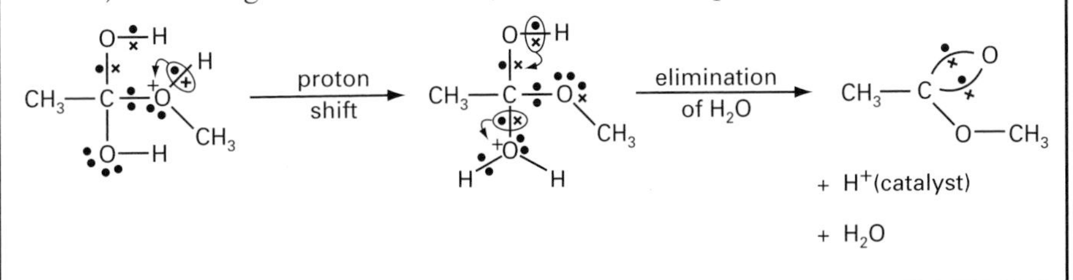

11.7 Reactions involving fission of the C—O bond

With concentrated sulphuric acid

When alcohols react with concentrated sulphuric acid, different products are formed according to the reaction conditions.

At 140°C, in the presence of excess alcohol, the main product is an **ether**, for example:

$$\underset{\text{in excess}}{2C_2H_5OH} \xrightarrow[\text{with conc. } H_2SO_4(l)]{\text{reflux at 140°C}} \underset{\text{ethoxyethane (a useful non-polar solvent)}}{C_2H_5—O—C_2H_5} + H_2O$$

Here an —OH group has been replaced by an $—OC_2H_5$ group. However, at 170°C, in the presence of excess concentrated sulphuric acid, a molecule of water is **eliminated** and the main product is an **alkene**, for example:

$$CH_3CH_2OH \xrightarrow[\text{excess conc. } H_2SO_4(l)]{\text{reflux at 170°C}} \underset{\text{ethene}}{CH_2{=}CH_2} + H_2O$$

There are two points to note here:

- concentrated sulphuric acid acts as a **dehydrating agent** in both reactions;
- like the halogenoalkanes, alcohols can become involved in competing substitution and elimination reactions. Figure 11.8 shows the mechanisms of these reactions.

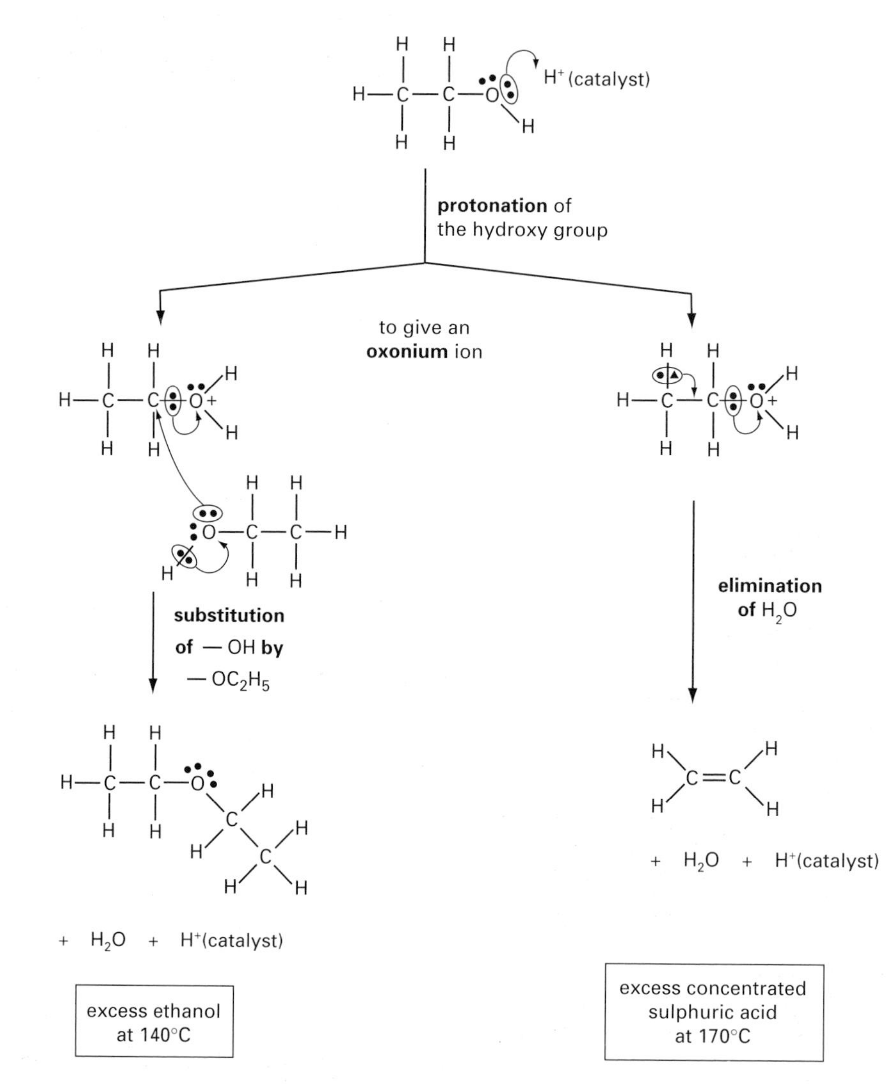

Figure 11.8 ▶
When ethanol is treated with concentrated sulphuric acid, the nature of the reaction products and the mechanism depend on the conditions and the relative concentrations of the reactants

Phenol does not give these dehydration reactions. Unlike benzene, phenol *readily* reacts with concentrated sulphuric acid, resulting in electrophilic substitution in the benzene ring, Figure 11.9.

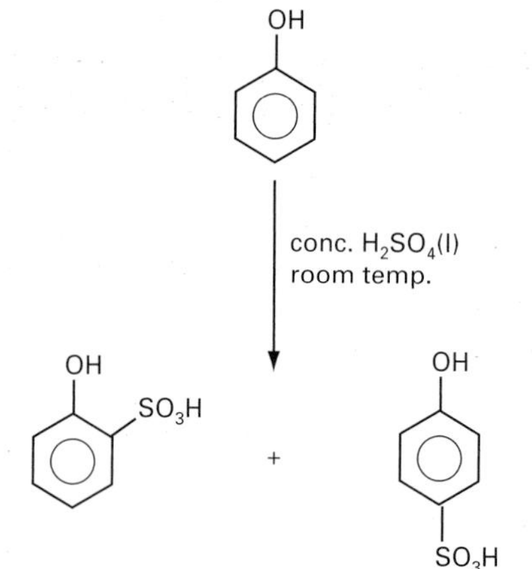

Figure 11.9 ▲
Whereas phenol is readily sulphonated at room temperature, benzene will not react under these conditions; thus in phenol the — OH group must make the ring more reactive towards electrophiles

With halogenating agents

Halogenoalkanes are prepared by treating an alcohol with a halogenating agent (Table 11.7), for example:

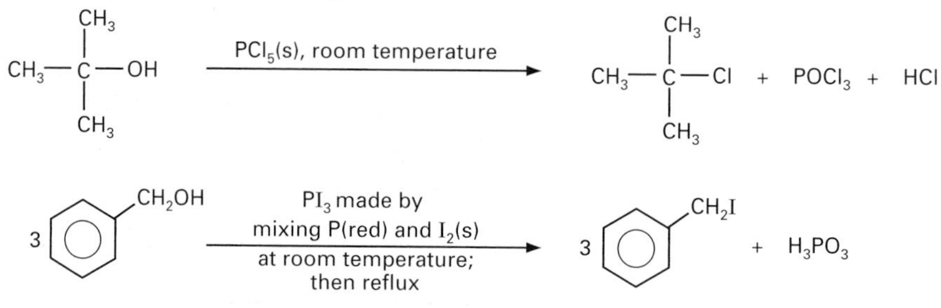

These are nucleophilic substitution reactions in which a halogen atom replaces the hydroxy group.

Phenol does not give nucleophilic substitution reactions with these halogenating agents because partial π bonding strengthens the C—O bond and decreases its polarity (Figure 11.7).

Table 11.7 Some common halogenating agents

Reagent	reaction conditions
PCl_5(s)	room temperature
PCl_3(l)	
P(red) + Br_2(l)	reflux
P(red) + I_2(s) (giving $PHal_3$ *in situ*)	
$SOCl_2$(l)	
conc. HCl(aq)	
KBr(s)/conc. H_3PO_4(l)	
KI(s)/conc. H_3PO_4(l) (giving HHal *in situ*)	

Notes:

1) one mole of PCl_5 can only halogenate one mole of —OH bonds, whereas $PHal_3$ can halogenate three moles of —OH bonds
2) $SOCl_2$ is a useful reagent because the unwanted products, being gases (SO_2 and HCl), are easily removed
3) *in situ* means making one reactant in the presence of the other reactants or just before they are added

11.8 Oxidation reactions

The ethanol in brandy burns with a clean blue flame

Although lower alcohols burn in air with a clean flame, the cost of production limits their use as a 'bulk' fuel. However, in some countries where it is economically produced by the fermentation of sugar, or starch, ethanol *is* used as a fuel. Due to their high carbon/hydrogen ratio, phenols burn in air with a sooty flame.

Primary and secondary alcohols are easily oxidised by refluxing them with a solution of potassium dichromate, $K_2Cr_2O_7$, in dilute sulphuric acid. An acidified dichromate ion readily accepts electrons:

$$\underset{\text{orange}}{Cr_2O_7^{2-}(aq)} + 14H^+(aq) + 6e^- \longrightarrow \underset{\text{green}}{2Cr^{3+}(aq)} + 7H_2O(l)$$

and this makes it a powerful **oxidising agent**. Also, the colour change

Oxidation and reduction

In section 4.12, we defined **oxidation** as a process which involves the loss of electrons, whereas a gain of electrons is termed **reduction**. A chemical reaction which involves **red**uction and **ox**idation is termed a **redox reaction**. A species is said to be **oxidised** if electrons have been withdrawn from it; the species which withdraws the electrons is termed an **oxidising agent**. Similarly, a species which gains electrons is said to be **reduced** and the species which supplies the electrons is termed a **reducing agent**.

You will be able to recognise many organic redox reactions from a change in molecular structure, as follows:

oxidation	gain of oxygen atoms	*or*	loss of hydrogen atoms
reduction	loss of oxygen atoms	*or*	gain of hydrogen atoms

Sometimes the symbol [O] is used to represent oxidation and to balance redox reactions.

indicates that a reaction is occurring.
On oxidation, a **primary alcohol** first forms an **aldehyde**:

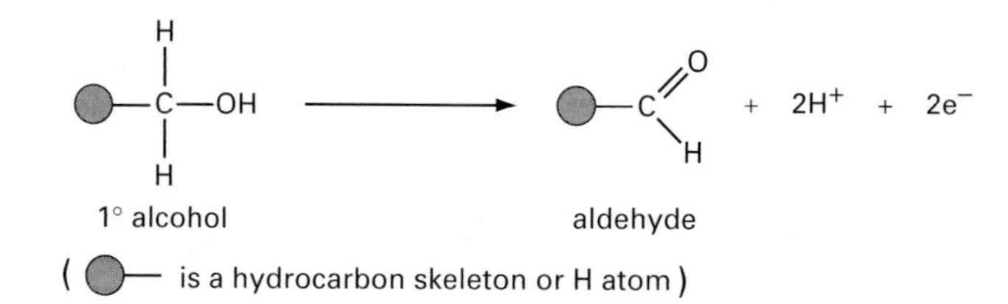

Then the aldehyde is further oxidised to a **carboxylic acid**:

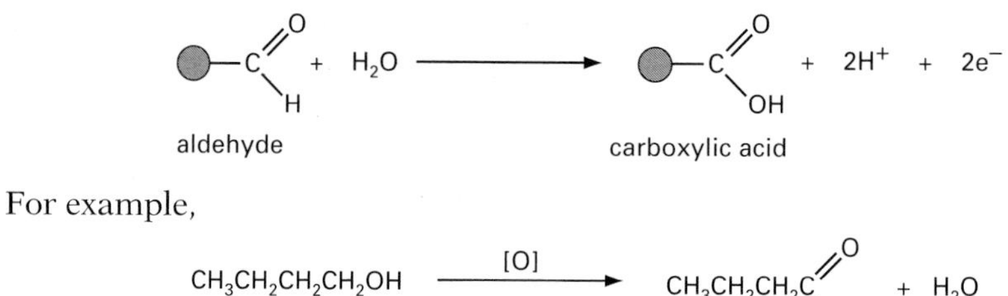

For example,

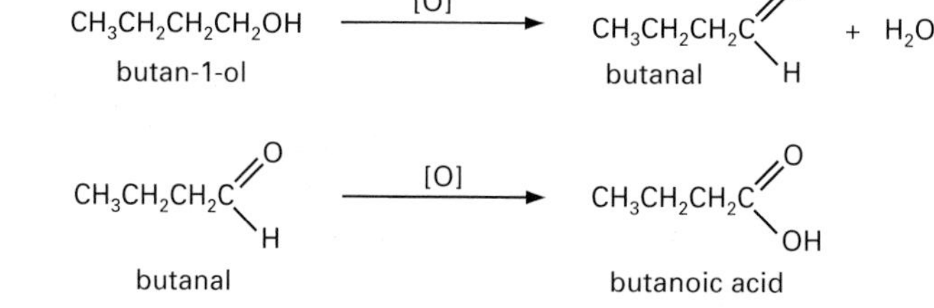

If the aldehyde is the desired product, it should be distilled off as it is formed, using the apparatus shown in Figure 11.10. Unlike alcohols, aldehydes do not form hydrogen bonds, so they boil at much lower temperatures than the alcohols of similar molecular mass (for example, butanal, b.p. = 72°C; butan-1-ol, b.p. = 118°C). As a result, the aldehyde molecules may be distilled from the reaction mixture as soon as they are formed.

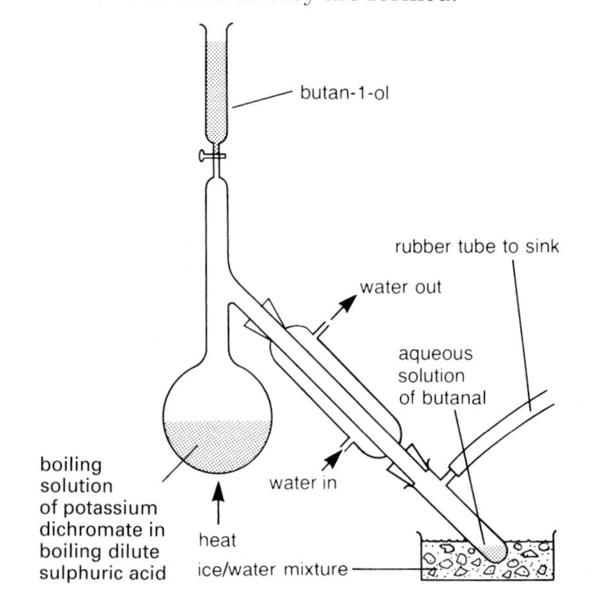

Figure 11.10
Apparatus used to prepare an aldehyde such as butanal. Butanal, and some water, distil off. How would you obtain the butanal (b.p. 72°C) from its aqueous solution?

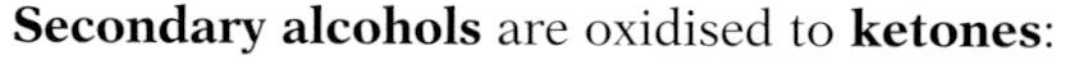

Secondary alcohols are oxidised to **ketones**:

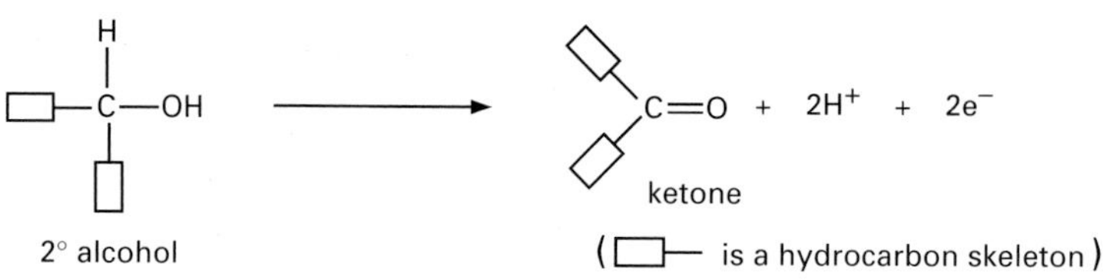

For example,

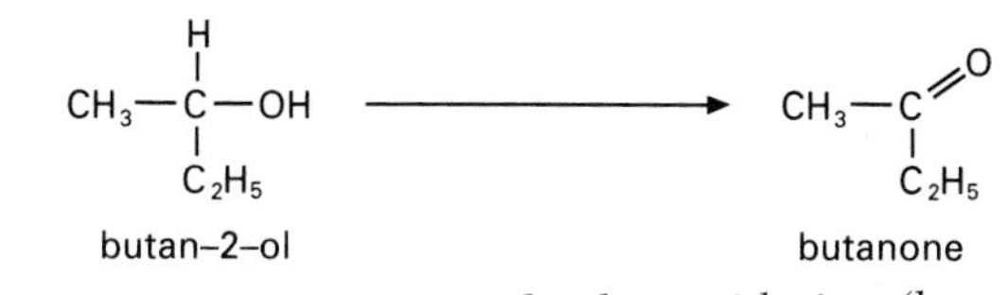

Under these conditions, ketones resist further oxidation (but see section 12.6).

During the oxidation of primary and secondary alcohols, hydrogen atoms are removed from the carbon atom bearing the —OH group. Since they have no such hydrogen atoms, **tertiary alcohols** are not readily oxidised by hot acidified potassium dichromate. However, prolonged treatment with a very strong oxidising agent, such as hot concentrated nitric acid, converts them into a mixture of ketones, carboxylic acids and carbon dioxide.

A variety of oxidising agents are used, both in the laboratory and industrially, to convert alcohols into aldehydes, ketones and carboxylic acids. Some examples are given in Table 11.8.

Table 11.8 Some reagents commonly used to oxidise alcohols

Reagent	reaction conditions	notes
potassium dichromate(VI), $K_2Cr_2O_7$, in dilute $H_2SO_4(aq)$	reflux	colour change: **orange Cr(VI)** to **blue-green Cr(III)** confirms the reaction
potassium manganate(VII), $KMnO_4$, in dilute $H_2SO_4(aq)$	reflux	colour change: **purple Mn(VII)** to **colourless Mn(II)** confirms the reaction
air + silver oxide catalyst	600°C	manufacture of methanal from methanol
air + copper catalyst	500°C	manufacture of propanone from propan-2-ol

We can distinguish between primary, secondary and tertiary alcohols by analysing their oxidation products. Simple chemical tests are used to identify any aldehydes, ketones or carboxylic acids which are formed (see sections 12.6, 12.7 and 14.6). Alternatively, the oxidation products might be separated, purified and then analysed from infrared, NMR and mass spectra (Chapters 15–17).

Finally, you should note that the —OH group in phenol is not readily oxidised because the ring carbon to which it is attached cannot form a C=O group.

The iodoform test

The iodoform test provides a simple way of distinguishing between certain alcohols. When treated with iodine dissolved in potassium iodide and aqueous sodium hydroxide, alcohols which contain the structure

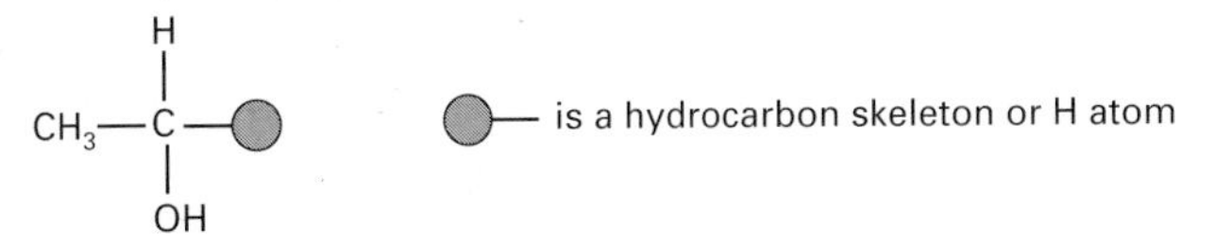

form a yellow precipitate of triiodomethane, CHI_3. This compound is commonly known as **iodoform**. Ethanol and propan-2-ol, for example, will form iodoform:

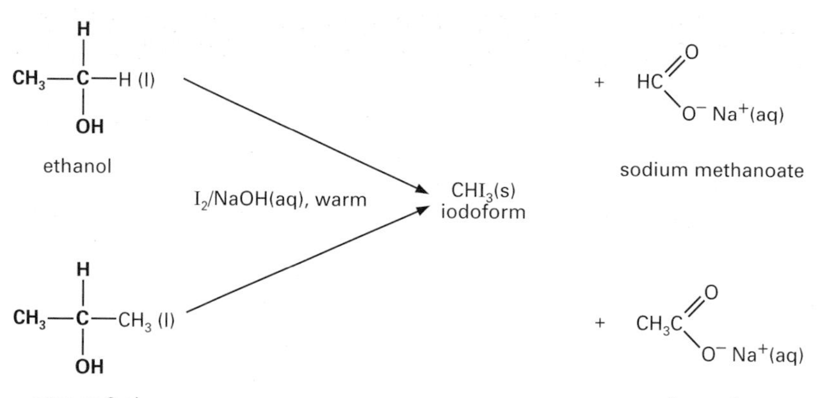

However, the propan-1-ol molecule, $CH_3CH_2CH_2OH$, does not contain the $CH_3CH(OH)$ group, so it gives a negative iodoform test.

In the iodoform reaction, the iodine reacts with the aqueous sodium hydroxide to produce iodate(I) ions, $IO^-(aq)$. These act as an oxidising agent and, then, as an iodinating agent as shown below:

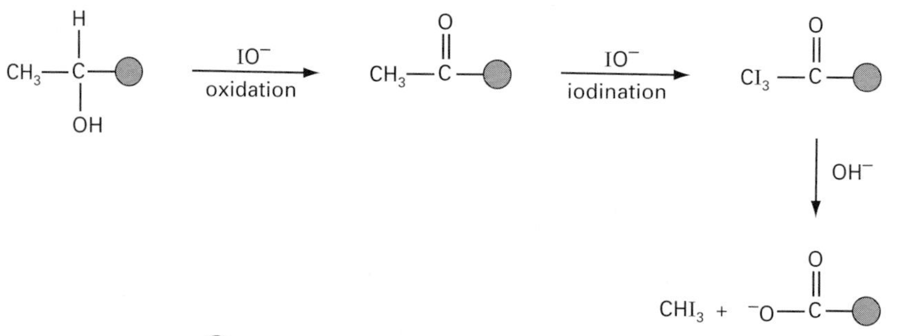

Since the CH_3CO— structure occurs in certain aldehydes and ketones, they will also undergo the iodoform reaction (section 12.7).

Focus 11c

1. There are four main types of reactions of alcohols and phenols: i) as acids, ii) as nucleophiles, iii) oxidation reactions and iv) reactions of the hydrocarbon skeleton.
2. All organic hydroxy compounds can act as **acids** which dissociate in aqueous solution, thus:

$$XO—H(aq) + H_2O(l) \rightleftharpoons H_3O^+(aq) + XO^-(aq)$$

The acids become stronger, that is, dissociate into ions to a greater extent, as the X group becomes more able to draw the electron cloud towards itself (section 11.6), thus:

alcohols   water   phenols   carboxylic acids

→ **increasing acid strength**

3. Primary alcohols are readily oxidised to aldehydes and then to carboxylic acids (Figure 11.10). Secondary alcohols are oxidised to ketones. Tertiary alcohols resist oxidation because a C—C bond must be broken in the process. We can distinguish between 1°, 2° and 3° alcohols by analysing their oxidation products.
4. All alcohols having the —$CH(OH)CH_3$ group will give a positive iodoform (triiodomethane) test.
5. The main reactions of alcohols and phenols are summarised in Table 11.10 in the comments on activity 11.6 at the end of this chapter.

Ethanol – fuel for thought?

Although alcohols were first tried as fuels for internal combustion engines in the 1870s, their use has only been commercially viable when petrol was in short supply. Alcoline and agrol, two blends of petrol with ethanol, were introduced in the UK in the 1930s but, at that time, oil was abundant and cheap, so the blends were uncompetitively priced. When the oil producers increased prices in the early 1970s, interest in ethanol fuels was rekindled.

Gasohol is a blend of 90% unleaded petrol and 10% ethanol or methanol. In fact, up to 17% ethanol can be blended without modification to the carburettor and ignition systems, though the engine seals need to be modified to resist attack by ethanol as it is a polar solvent. Gasohol fuel consumption compares well with petrol and there is no increase in engine wear. Since it has a higher octane value than petrol, anti-knocking additives like tetraethyl lead and benzene are not required (section 7.5). Gasohol burns more cleanly than petrol and pollution from unburnt hydrocarbons and CO emissions are significantly reduced. However, emission of nitrogen oxides, NO_x, is increased and some ethanol is oxidised to ethanal, CH_3CHO, an aldehyde. Nitrogen oxides and ethanal are a worry because they contribute to acid rain and photochemical smog – fortunately, catalytic converters can convert them to less harmful products (section 7.5). Whilst the US limits the ethanol component in gasohol to 10%, blends containing up to 85% ethanol have been used in Los Angeles, causing a significant reduction in atmospheric pollutants and photochemical smog above the city.

Of course, where oil is plentiful and prices are low, ethanol is not a viable economic option, even though it may be more environmentally friendly. About half of the industrial ethanol produced worldwide comes from the fermentation of the sugars in potatoes, corn, sugar cane and other raw materials. In countries where these grow readily, such as Brazil, ethanol is a real alternative to petrol. Brazil has no significant oil reserves so, in 1975, the government established a national programme aimed at replacing petroleum with alcohol-based fuels, such as **hydrated ethanol**, a mixture of 95% ethanol and 5% water. Although an engine modification is needed, fuel consumption is respectable and the decreased level of atmospheric pollutants is a definite bonus. Strong hydrogen bonding in hydrated ethanol means the fuel is less volatile than petrol and, early on, this caused some problems with starting. The oil companies overcame this difficulty by adding small amounts of very volatile ethers, such as 2-methoxy-2-methylpentane, which also increased the octane value. As a result of these efforts, over two million cars in Brazil are now running on ethanol fuels.

In some ways, hydrated ethanol is an environmentalist's dream! The combustion of fossil fuels, like petrol, contribute to **global warming** by creating a layer of dust and gases which reflects back the Earth's heat. Scientists predict a 1–3°C rise in temperature over the next 50 years with the concern that this may melt polar ice, causing the water levels to rise and so flood low-lying lands. The amount of carbon dioxide produced by burning fossil fuels is so great that plants are not able to 'fix' it all during **photosynthesis**:

$$6CO_2(g) + 6H_2O(l) \longrightarrow \underset{\text{plant sugar}}{C_6H_{12}O_6(aq)} + 6O_2(g)$$

This therefore leads to ever increasing levels of carbon dioxide. However, if the plant material, or **biomass**, is converted into ethanol and burnt, it gives off the same amount of carbon dioxide as it would if it had decomposed naturally. Thus, the overall CO_2 balance from these processes would remain steady. There are three main concerns with a biomass fuel industry based on hydrated ethanol.

First, the waste aqueous liquid from the fermentation has to be treated to remove aerobic (oxygen-using) microbes before discharging into rivers and lakes, which is an important cost factor. A second worry is that the intensive farming of soil to produce massive yields of fermentable plants removes essential micronutrients from the soil and also causes its texture to disintegrate. The fine soil is washed into streams, and rivers, settling on the bottom as silt and making them difficult to navigate. There is also international concern about the destruction of the Amazon forests as land is cleared for sugar cane production because less forestation is available to 'fix' CO_2 by photosynthesis.

The third consideration is the morality of converting food into fuel. Over the last 60 years, there has been a reversal in the direction of grain trade. Latin America, Asia and Africa used to sell grain to Western Europe – now they import it. This is because the population increases in these developing countries have not been matched by advances in crop production techniques and machinery. There is a dilemma here – should grain and sugar crops be grown for ethanol production if this contributes to a food shortage, higher prices and the possibility of malnutrition? One solution is to increase the yield of non-edible fermentable material such as eucalyptus and poplar trees, which grow quickly and need little attention.

Ethanol – fuel for thought? *continued*

Activity 11.4

1 Gasohol is more volatile than petrol, causing greater evaporation from the fuel tank and carburettor. Explain why this happens by considering the intermolecular forces in gasohol and its pure components.

2 How can hydrated ethanol be converted into pure ethanol? Give one advantage and one disadvantage of using pure ethanol as a fuel.

3 Use the following data in this question.
- A small family car is driven 7000 miles per year on hydrated ethanol and averages 28 miles per gallon.
- Corn and poplar trees can be fermented to produce 200 and 625 gallons of hydrated ethanol per acre, respectively.
- An annual subsistence diet requires the corn grain from 0.2 acres of cropland.

a) How many gallons of hydrated ethanol does the car use each year?
b) How many acres of corn must be grown to produce this amount of fuel?
c) How many people could be supported on a subsistence diet using this acreage of corn?
d) Assuming the same acreage calculated in part (b) is used to plant both poplar trees for fuel and corn for food, how many people could be fed as well as providing fuel for the car?

11.9 Reactions of the hydrocarbon skeleton

When predicting the reactions of an alcohol or phenol, remember that the hydrocarbon skeleton also has some characteristic reactions. Thus, a C—H bond will undergo photocatalysed bromination like the C—H bond in methane:

$$CH_3OH \xrightarrow{Br_2(l) \text{ in u.v. light}} CBr_3OH$$

This is a **free radical substitution** reaction (section 7.8).

Often, the reactivity of the hydrocarbon skeleton is modified by the presence of the —OH group. The hydroxyl group activates the benzene ring in phenol making it very susceptible to electrophilic attack. Even under mild conditions, substitution will readily occur in the 2-, 4- and 6- positions:

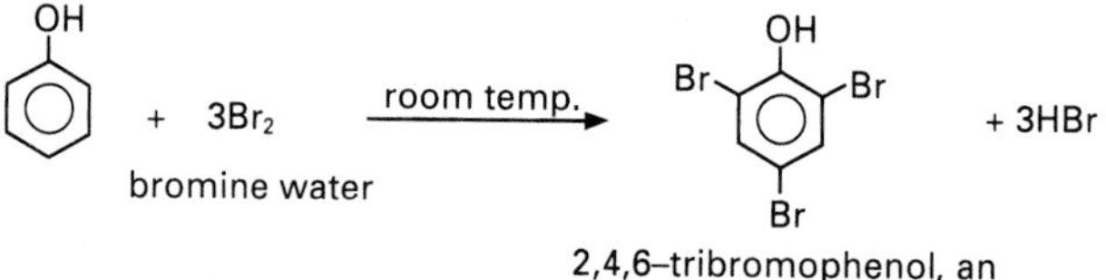

Similarly, phenol can be nitrated with *dilute* nitric acid at room temperature, to give 2- and 4-nitrophenols:

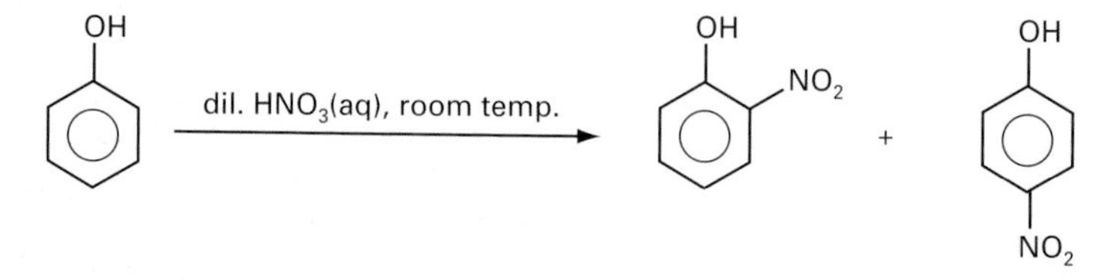

What conditions would you use to brominate or nitrate benzene (section 9.5)? Finally, we must mention the industrial hydrogenation of phenol vapour:

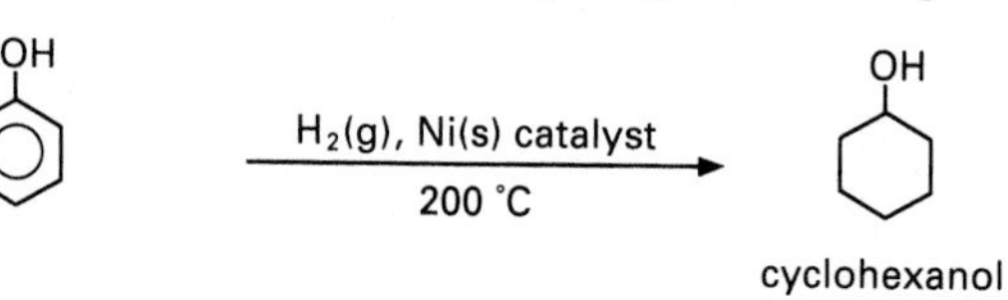

Simple chemical tests for phenol

Phenol reacts with a neutral solution of iron(III) chloride to give a violet-coloured complex ion. Phenol may also be identified by testing with (i) dilute sodium hydroxide; phenol dissolves forming sodium phenoxide *and* then (ii) sodium hydrogencarbonate; there is no reaction.

This is an important reaction because large quantities of cyclohexanol are used to make **nylon** (section 13.6).

Activity 11.5

1 Give the name(s) and formula(e) of the organic product(s) formed by the following reactions:

a) sodium methoxide $\xrightarrow[\text{room temperature}]{\text{dil. } H_2SO_4(aq);}$

b) $CH_3-C(=O)Cl$ (ethanoyl chloride) $\xrightarrow{CH_3OH(l)}$

c) butan-2-ol $\xrightarrow[\text{reflux at 170°C}]{\text{conc. } H_2SO_4(l) \text{ in excess;}}$

d) methanol (excess) $\xrightarrow[\text{reflux at 140°C}]{\text{conc. } H_2SO_4(l);}$

e) $CH_3CH_2OH \xrightarrow[\text{then reflux}]{P_{red}, Br_2(l) \text{ mixed at room temp.;}}$

f) ethanol $\xrightarrow[\text{reflux}]{KMnO_4/\text{dil. } H_2SO_4(aq);}$

g) $CH_3-CH(OH)-CH_3$ (propan-2-ol) $\xrightarrow[\text{distil}]{K_2Cr_2O_7/\text{dil. } H_2SO_4(aq);}$

h) hexan-2-ol $\xrightarrow{I_2/NaOH(aq); \text{ warm}}$

2 Give the reagents and reaction conditions needed for the following conversions:

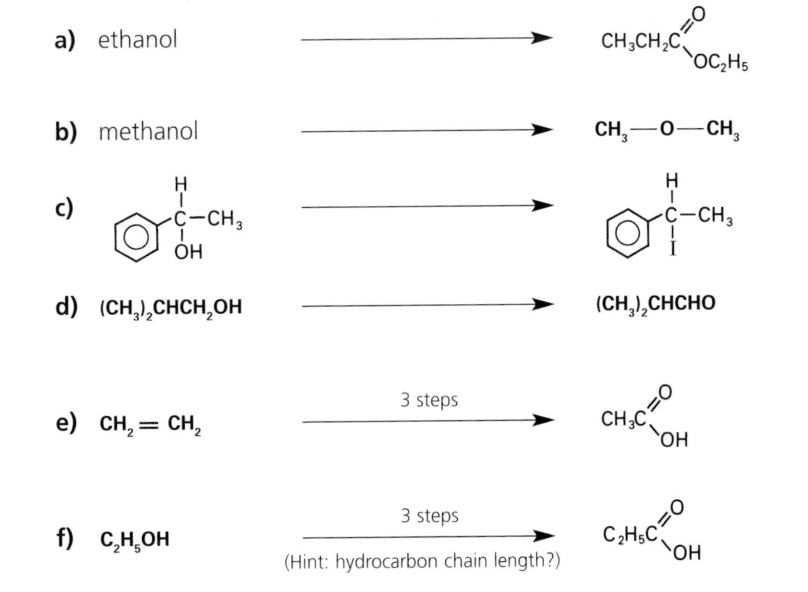

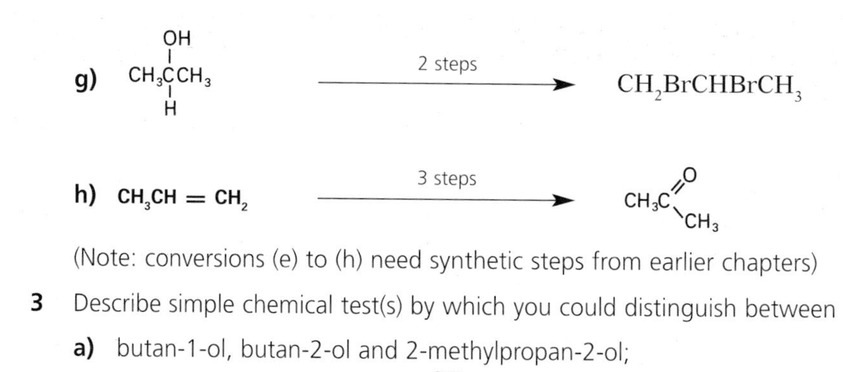

(Note: conversions (e) to (h) need synthetic steps from earlier chapters)

3 Describe simple chemical test(s) by which you could distinguish between

a) butan-1-ol, butan-2-ol and 2-methylpropan-2-ol;

b) phenol and phenylmethanol, C_6H_5—CH_2OH .

11.10 Use of alcohols and phenol

Large quantities of the lower alcohols are used as solvents, e.g. in the manufacture of drugs, cosmetics and varnishes. Alcohols are also used as a **chemical feedstock** in the synthesis of other organic compounds. Two of these are mentioned below.

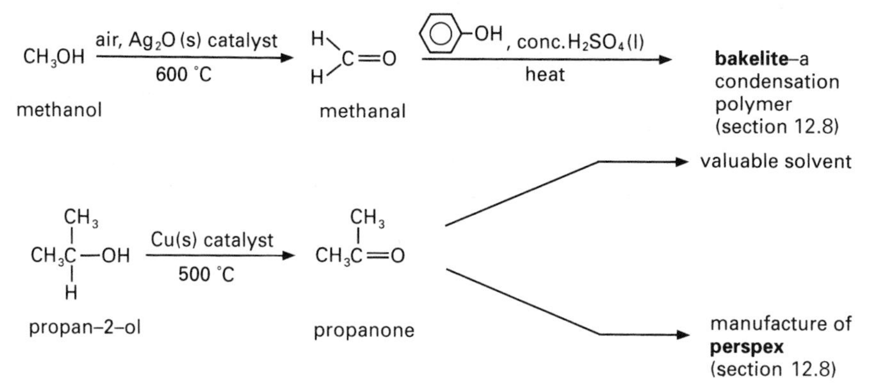

Dihydric and polyhydric alcohols also have their uses. Ethane-1,2-diol is used as antifreeze and in the manufacture of polyester (section 13.6).

Propane-1,2,3-triol is used to make nitroglycerine the explosive material in dynamite.

A dilute aqueous solution of phenol has powerful antiseptic properties. Known as 'carbolic acid', this solution was first used in 1867 by Joseph Lister, a famous English surgeon. By spraying a fine mist of aqueous phenol near a wound, Lister was able to hinder the development of bacterial diseases during the operation or the patient's recovery period. Although the use of 'carbolic acid' increased the chances of the patient surviving surgery, it tended to burn the skin around the wound. In fact, we now know that phenol is poisonous by skin absorption. Nowadays, derivatives of phenol are used as antiseptics, for example:

A surgical operation in the early days of antiseptic surgery. Carbolic spray was first used by Joseph Lister in 1865 as an antiseptic

Not only are these much more active anti-bacterial agents than phenol, but they are also much less toxic and do not irritate the skin. Phenol is also used to make nylon (section 13.6).

Airports and antifreeze

When the weather is cold and damp, planes 'ice-up', that is their wings and fuselage become covered with a layer of ice. This increases the weight of the craft and lessens the lift on take-off and, needless to say, could have disastrous consequences. In addition, if ice forms on a runway, there is the danger that a plane may skid on take-off or landing. Thus, over 50 million litres of de-icing fluids are sprayed over aircraft and runways every year to keep airports operational and planes flying.

The de-icing liquids are very effective but they do create an environmental problem. One of the most commonly used liquids is ethane-1,2-diol, commonly known as **ethylene glycol**. This substance, which can be used to depress the freezing point of water as far as –13°C, is also used as an antifreeze in car coolant systems. Ethane-1,2-diol is miscible with water so, when it runs off the plane and runway, it is carried into the waterways and ground water. Once there, it is oxidised by bacteria found in streams and rivers to carbon dioxide. This process lowers the oxygen content in the water, thereby reducing the amount of oxygen available to aquatic life.

Sweden was one of the first countries to detect the problem. The Arlanda Airport in Stockholm now uses vehicles with giant suction devices to suck up the de-icing fluid and transport it to storage tanks where it is broken down by bacteria to carbon dioxide. Other European airports collect the glycol in underground gullies which run along the edge of the runway. At Copenhagen airport, they use the 'RoMat', a massive rubber mat, reinforced with steel, upon which the plane parks to be de-iced just before take-off. The excess de-icing fluid is collected through grooves in the mat and passed to storage tanks, where it undergoes biological degradation.

Whilst the de-icing liquids are very effective, a problem can arise if the plane does not take off immediately after spraying, as ice may still form. Thus, it is desirable to make the de-icer adhere to the plane's fuselage by introducing chemicals which make it stick on.

BP in Britain has developed an alternative de-icer called 'Clearway 1'. This is an aqueous solution of potassium ethanoate solution which is now used at fifty five airports worldwide, including Heathrow and Gatwick. The advantage of Clearway 1 is that it breaks down to carbon dioxide and water without using so much oxygen, and it is much more effective at lowering the freezing point of water, working down to –60°C. The disadvantage of Clearway 1 is that it is more expensive to produce than ethylene glycol de-icers.

The wings of an airliner being de-iced with antifreeze

Activity 11.6 Comparing the reactions of alcohols and phenol

We are interested in how the chemical reactivity of the hydroxy group depends on its molecular location. This activity will help you to compare the behaviour of ethanol (1°), propan-2-ol (2°), 2-methylpropan-2-ol (3°) and phenol. It is a very useful revision exercise.

Prepare a summary table which compares the reactivity of the compounds listed above. You should consider their reactions with the following reagents:

1. sodium metal
2. conc. $H_2SO_4(l)$
3. halogenating agents, e.g. $PCl_5(s)$
4. $K_2Cr_2O_7$/dil. $H_2SO_4(aq)$
5. I_2/NaOH(aq)
6. $Br_2(aq)$
7. electrophiles such as NO_2^+ and Br^+
8. neutral $FeCl_3(aq)$

11.11 Ethers

Ethers have the general structure:

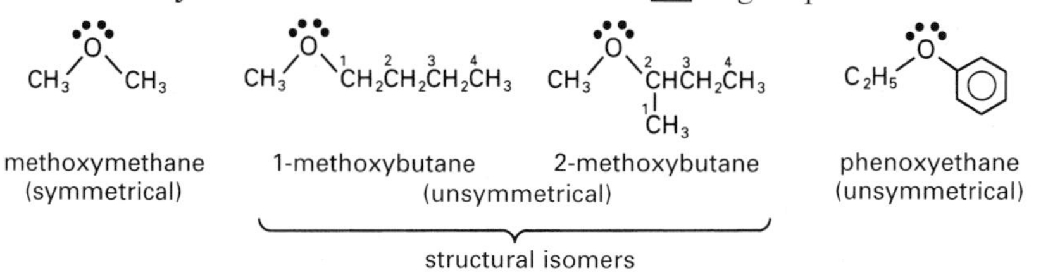

where ☐— is a hydrocarbon skeleton. Ethers are named as alkoxy- or phenoxy- derivatives of the longest hydrocarbon chain or benzene ring. Since structural isomerism can occur, we must number the carbon atom which bears the oxy group. A **symmetrical ether** has identical ☐— groups whereas an **unsymmetrical ether** has different ☐— groups.

methoxymethane (symmetrical) | 1-methoxybutane | 2-methoxybutane (unsymmetrical) | phenoxyethane (unsymmetrical)

structural isomers

We can prepare symmetrical ethers by refluxing concentrated sulphuric acid with an excess of an alcohol, section 11.7. An unsymmetrical ether is made by heating a sodium alkoxide, or sodium phenoxide, with an iodoalkane, for example:

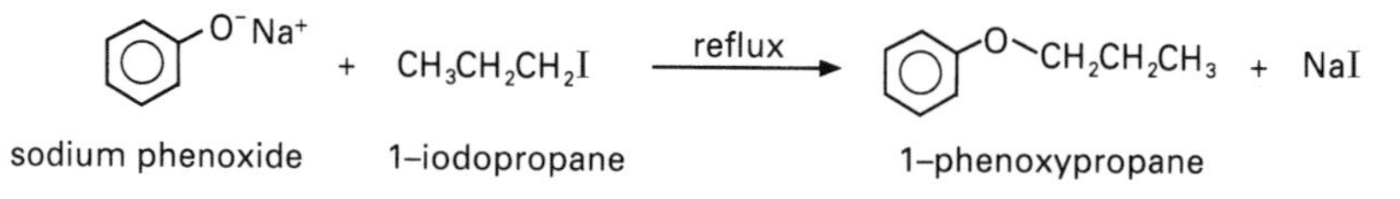

Corresponding ethers and alcohols are structural isomers, having different functional groups:

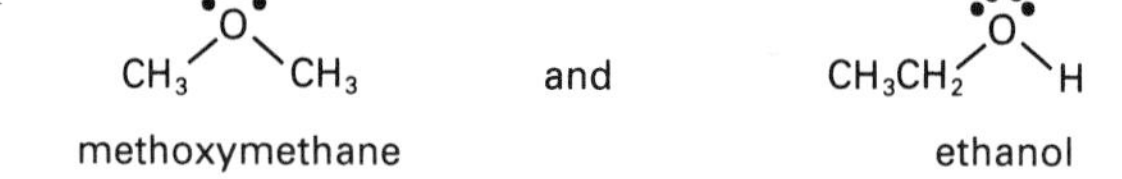

Since they are unable to form hydrogen bonds, ethers have much lower boiling points than the isomeric alcohols. Indeed, ethers are as volatile as alkanes of similar relative molecular mass. This would suggest that the dipole–dipole attractions in ethers, via their polar C—O bonds, must be very weak. Hence, ethers are only slightly water-soluble and are good solvents for non-polar substances. The most commonly found ether, **ethoxyethane, $C_2H_5OC_2H_5$**, is used in large quantities as a solvent in organic chemistry. As well as helping reactants to mix together, ethoxyethane is frequently used to solvent extract crude organic products from aqueous reaction mixtures, section 6.3.

A concern when using ethoxyethane as a solvent is that it is a highly flammable and volatile liquid, b.p. 35°C, which forms explosive mixtures with air. Since it is denser than air, ethoxyethane vapour can diffuse through the laboratory at bench or floor level, creating a danger of fire well away from the actual source of the 'ether'.

Combustion apart, ethers are rather inert compounds. Although the oxygen atom bears two lone pairs of electrons, ethers are poor nucleophiles and only very weak bases, that is proton acceptors. Protonation of the oxygen atom does make the molecule more reactive towards nucleophilic attack, for example:

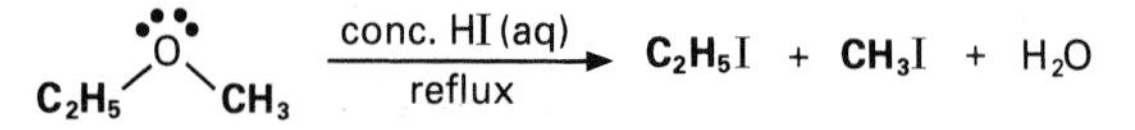

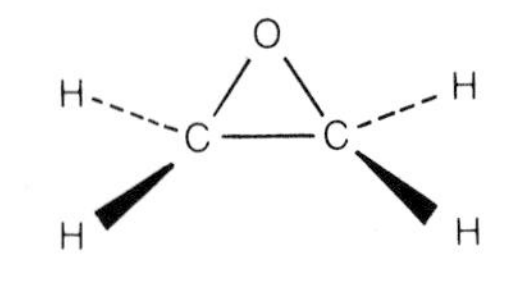

Figure 11.11
Epoxyethane, an industrially important ether. Would you expect it to be very reactive?

Apart from ethoxyethane, another ether of industrial importance is epoxyethane, an alicyclic ether, Figure 11.11. Owing to the strain in its three-membered ring, epoxyethane is far more reactive than aliphatic ethers and this makes it an important intermediate in industrial syntheses, section 8.6.

Focus 11d

1 The hydroxy group in a phenol makes the benzene ring more reactive towards attacking electophiles, such as NO_2^+ and Br^{+}.

2 A simple chemical test for phenol: reaction with neutral iron(III) chloride solution gives a violet-coloured complex ion.

3 **Ethers** may be prepared from alcohols or iodoalkanes. Although ethers and alcohols are structural isomers, only the alcohols display hydrogen bonding. Thus, an alcohol has a much higher boiling point than its isomeric ether. Ethers are fairly unreactive compounds which, if used with care, make useful solvents for non-polar substances.

4 Important new reagents used in this chapter are listed in Table 11.9.

Table 11.9 Important reagents used in this chapter

Reagent	conditions	what it does to a hydroxy compound
ethanoyl chloride, $CH_3C(=O)Cl$	room temp.	converts alcohol or phenol into an **ester**
conc. $H_2SO_4(l)$ (a dehydrating agent)	(i) excess alcohol, 140°C (ii) excess acid, 170°C	(i) converts 1°, 2° alcohol $\longrightarrow$ **ether** (ii) converts 1°, 2°, 3° alcohol $\longrightarrow$ **alkene**
PCl_5 P(red) + $Br_2(l)$, P(red) + $I_2(s)$	room temp. reflux	converts alcohol (–OH) $\longrightarrow$ **halogen compound** (–Hal)
$K_2Cr_2O_7$/dil. $H_2SO_4(aq)$ *or* $KMnO_4$/dil. $H_2SO_4(aq)$	reflux	oxidises 1°alcohol $\longrightarrow$ **aldehyde** $\longrightarrow$ **carboxylic acid** oxidises 2° alcohol $\longrightarrow$ **ketone**
$I_2(aq)$/dil. $NaOH(aq)$	warm	**iodoform, CHI_3,** is formed if the alcohol has this group: H—C(CH₃)(OH)—
neutral $FeCl_3(aq)$	room temp.	a test for **phenol**: if present, the mixture turns a violet colour

How can we ensure that a primary alcohol is oxidised to an aldehyde rather than to a carboxylic acid?

Questions on Chapter 11

1 This question is about the following compounds:

A 2-methylbutan-2-ol **B** pentan-2-ol
C 4-methylphenol **D** pentan-1-ol
E cyclopentanol **F** 1-methoxypropane
G phenylmethanol

a) Draw structural formulae for these compounds.
b) Classify the compounds as being i) aliphatic, ii) alicyclic or iii) aromatic.
c) Identify the alcohols and classify them as being i) primary, ii) secondary or iii) tertiary.
d) Which is a phenol?
e) Which is an aromatic alcohol?
f) Which is an ether?

2 Again, refer to the compounds in question 1.
a) Which are structural isomers?
b) Which can exist as enantiomers?
c) Which has the lowest boiling point?
d) Which react with sodium metal?
e) Which is the strongest acid?
f) Which could react with concentrated sulphuric acid to form pent-1-ene?
g) Which react with phosphorus pentachloride?
h) Which could be oxidised to an aldehyde?
i) Which could be oxidised to a ketone?
j) Which are resistant to oxidation by acidified potassium dichromate(VI)?
k) Which react with iodine and sodium hydroxide to form iodoform?
l) Which gives a brightly coloured compound when treated with neutral iron(III) chloride solution?

3 Using appropriate diagrams, explain the following observations.
a) Butan-1-ol is much more soluble than ethoxyethane in water.
b) Ethane-1,2-diol has a much higher boiling point than ethanol.
c) Although they are both weak acids, phenol is stronger than ethanol.

Questions on Chapter 11 *continued*

d) When butan-2-ol is refluxed with concentrated sulphuric acid at 170°C, *three* isomeric alkenes are formed.
e) The iodoform test can be used to distinguish between propan-1-ol and propan-2-ol.
f) 4-nitrophenol has a higher melting point than 2-nitrophenol.

4 Give the reagents and reaction conditions that you would use to make propan-2-ol from:
a) 2-bromopropane
b) propene (2 steps)

5 A technician has found four unlabelled bottles of liquid. She knows that the bottles contain:

A $CH_3CH_2CH_2OH$ **B** $CH_3CH_2CH(OH)CH_3$
C $(CH_3)_3COH$ **D** $CH_3CH_2CH_2OCH_2CH_3$

What *chemical* tests should she use to find out which liquid is in each bottle?

6 Give the names and structural formulae of the organic products formed in the following reactions:
a) Methanol is treated with sodium metal at room temperature.
b) Ethanoyl chloride is added to methanol at room temperature.
c) Propan-1-ol is heated with excess concentrated sulphuric acid at 170°C.
d) Cyclohexanol is refluxed with acidified potassium manganate(VII).
e) Ethanol is added slowly to boiling acidified potassium dichromate(VI) and distilled.
f) Methanol is treated with phosporus pentachloride.
g) Phenol is treated with ethanoyl chloride.
h) Dilute nitric acid is added to phenol at room temperature.

7 Give the reagents and reaction conditions that you would use to carry out the following conversions, giving the structural formulae of any intermediate products. Parts i) to o) require a knowledge of the work from earlier chapters.
a) methanol to methoxymethane
b) propan-2-ol to propene
c) ethanol to sodium ethoxide
d) methanol to methylethanoate
e) propan-2-ol to 2-iodopropane
f) propan-1-ol to propanoic acid
g) propan-2-ol to propanone
h) ethanol to ethanal
i) C_2H_4 to CH_3CH_2CN (2 steps)
j) CH_3CH_2I to CH_3CHO (2 steps)
k) CH_3CH_2OH to CH_3CH_2CN (2 steps)
l) CH_3CH_2OH to CH_2BrCH_2Br (2 steps)
m) CH_3CHCH_2 to CH_3COCH_3 (3 steps)
n) C_2H_4 to CH_3COOH (3 steps)
o) C_2H_4 to $CH_3CH_2COOC_2H_5$ (3 steps)

8 Compound A is produced in large quantities as the starting material for making resins and plasticisers. It contains, by mass, 62.1% carbon, 10.3% hydrogen and 27.6% oxygen. When 0.36 g of A was vaporised at 423 K and 99 300 Pa, the vapour occupied a volume of 220 cm^3.
a) Calculate the empirical formula of A.
b) Use the ideal gas equation (section 4.13), to calculate the relative molecular mass of A. (The gas constant should be taken as 8.31 J K^{-1} mol^{-1}.)
c) What is the molecular formula of A?
d) When A is treated with phosporous pentachloride at room temperature, hydrogen chloride vapour is evolved. What functional group is present in A?
e) Draw the possible non-cyclic structural formulae of A.
f) Compound A is found to react with bromine dissolved in hexane to produce 2,3-dibromopropan-1-ol. Write down the displayed formula of A.
g) Give the structure of the addition polymer formed by A (section 8.7).

9 Aromatic compounds C, D and E are structural isomers containing, by mass, 77.8% carbon, 7.4% hydrogen and 14.8% oxygen. Compounds C and D are liquids, whereas E is a solid. The relative molecular mass of the isomers is found to be 108.
a) Calculate the empirical formulae of each isomer.
b) What is the molecular formula of each isomer?
c) Write down the five possible structural formulae of the compounds.
d) When compound C is refluxed with acidified potassium dichromate(VI), the colour of the reaction mixture changes from orange to green and compound F, molecular formula $C_7H_6O_2$, is formed. Give the names of C and F.
e) D is insoluble in water, whereas E is sparingly soluble. The aqueous solution of E is slightly acidic. Give the name of D.
f) What simple procedure could be used to distinguish between the possible structures of E?

10 a) Write down the names and displayed formulae of the ethers of molecular formula $C_4H_{10}O$.
b) Give the reactants and reaction conditions you would use to prepare:
i) ethoxyethane
ii) 1-methoxypropane.
c) Describe one physical method and one chemical method of distinguishing between ethanol and methoxymethane.

Comments on the activities

Activity 11.1

1 **a)** $C_2H_5OH + \frac{1}{2}O_2 \rightarrow CH_3CHO + H_2O$

b) $M_r(C_2H_5OH) = 46$;
mol of ethanol oxidised = 10.0/46 = 0.217;
From the equation, 1 mole of ethanol gives 1 mole of ethanal, thus 0.217 mol of ethanal is formed;
$M_r(CH_3CHO) = 44$;
mass of ethanal formed = 0.217 × 44 = 9.55 g.

2 Before driving, the alcohol level needs to drop from 10 to 3 units, that is, 7 units must be metabolised. 7 units of alcohol correspond to 70 g of ethanol. If his liver breaks down 10 g of ethanol per hour, it will take 7 hours for the ethanol to be metabolised. Thus, the man should not drive until 6 am at the earliest the following morning.

3 Some possible reasons:
a) Danger to self and others, efficiency impaired, smell of alcohol on breath.
b) Ethanol may modify/intensify the action of the drug. For example, alcohol intensifies the drowsiness caused by some 'hayfever' drugs.

Activity 11.2

1 **a)** pentan-1-ol **b)** 3-phenylbutan-1-ol
c) 1-methylcyclohexanol **d)** 3-bromobutan-2-ol
e) butane-2,3-diol **f)** 2-chlorophenol

2

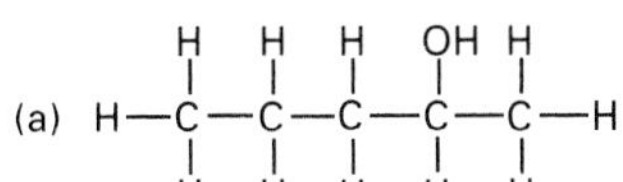

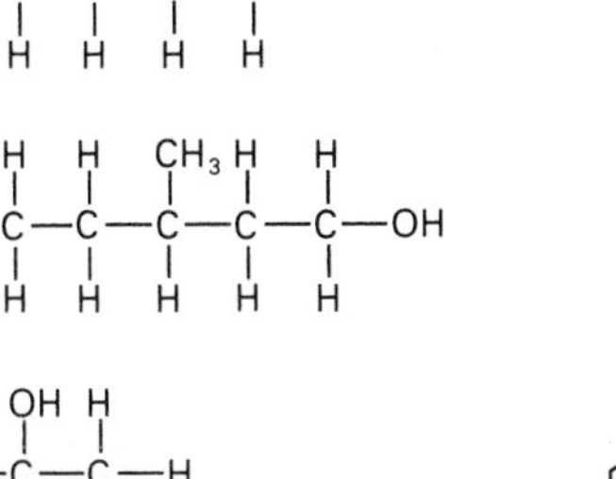

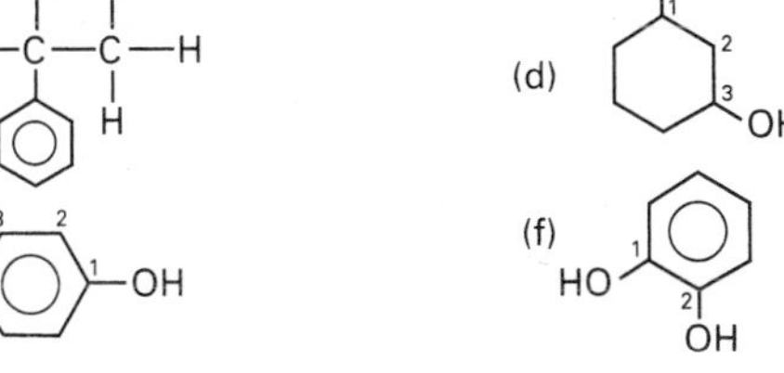

3 **a)** aliphatic/alicyclic: 1a, 1c, 1d, 1e, 2a, 2b, 2d
b) aromatic: 1b, 2c **c)** 1°: 1a, 1b, 2b
d) 2°: 1d, 1e, 2a, 2d **e)** 3°: 1c, 2c

4

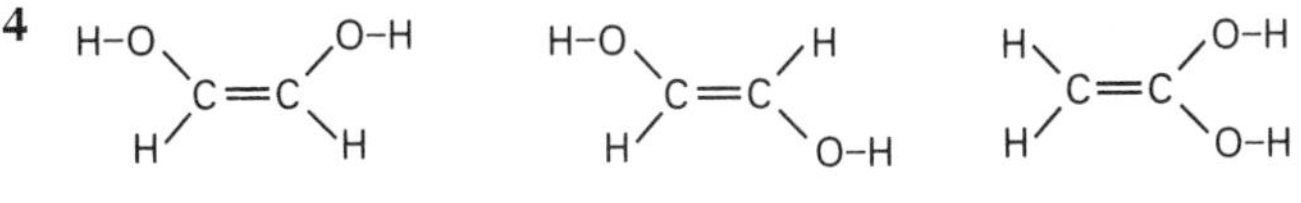

cis–ethene–1,2–diol *trans*–ethene–1,2–diol ethene–1,1–diol

5 **a)**

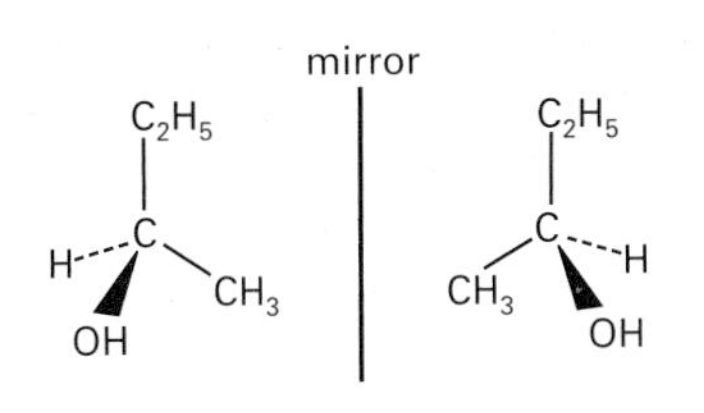

b) (i) an individual stereoisomer will rotate the plane of plane polarised light *either* in a clockwise direction (dextrorotatory, +) *or* an anti-clockwise direction (laevorotatory, –).
(ii) This is termed a racemic mixture. Each isomer produces the same degree of optical rotation but in opposite directions; thus, the mixture is optically inactive.

Activity 11.3

1 Two ways:

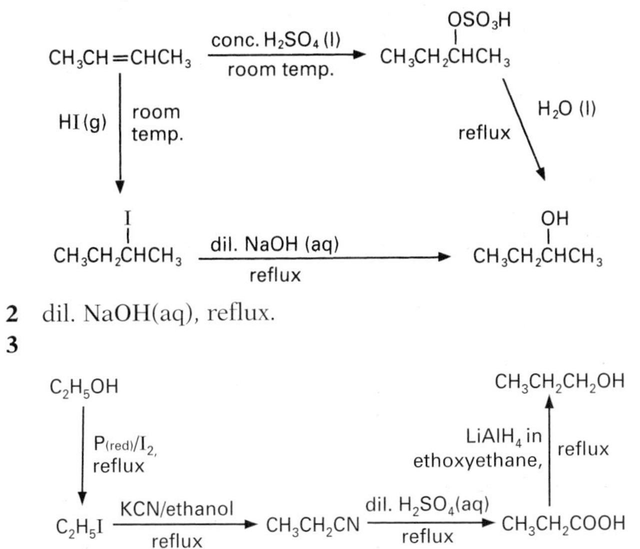

2 dil. NaOH(aq), reflux.

3

Activity 11.4

1 The intermolecular forces between the compounds in petrol (largely alkanes) are van der Waals' forces whilst the attractions between ethanol molecules result from hydrogen bonding (section 4.6). On mixing the liquids, the forces between ethanol and alkane molecules are weaker than those which existed in the pure liquids. Thus, the gasohol is more volatile than either of its individual components, petrol or ethanol.

2 Dry by refluxing with quicklime, CaO. Then by fractional distillation and collecting the liquid which boils at 78.5°C. Advantage: significant reduction in atmospheric pollution. Disadvantages: expensive, depending on the supply of fermentable materials; the effects of intensive farming on soil quality; fermentation of edible crops may decrease the food supply.

3 **a)** 7000 miles/28 miles per gallon = 250 gallons
b) 250 gallons/200 gallons per acre = 1.25 acres
c) 1.25 acres/0.2 acres per person = 6.25 (6) people

Comments on the activities *continued*

d) Fuel required = 250 gallons. Obtain this from poplar trees:

250 gallons/625 gallons per acre of trees	=	0.400 acres.
Acreage left for corn production	=	1.25 – 0.4
	=	0.85 acres.
0.85 acres/0.2 acres per person can also be fed.	=	4.25 (4) people

Activity 11.5

1 a) methanol, CH_3OH; this very weak acid is displaced from its salt by the strong acid, $H_2SO_4(aq)$.

b) methyl ethanoate $CH_3{-}C({=}O)OCH_3$

c) a mixture of but-1-ene, $CH_3CH_2CH = CH_2$ (via 1,2 elimination) and *cis*- and *trans*-but-2-ene, $CH_3CH{=}CHCH_3$ (via 2,3 elimination).

d) methoxymethane, $CH_3{-}O{-}CH_3$

e) bromoethane, CH_3CH_2Br

f) ethanal $CH_3{-}C({=}O)H$

and then ethanoic acid $CH_3{-}C({=}O)OH$

g) propanone, $CH_3{-}C({=}O){-}CH_3$

h) triiodomethane (iodoform), CHI_3

and sodium pentanoate $CH_3CH_2CH_2CH_2C({=}O)O^-Na^+$

2 a) propanoyl chloride $C_2H_5C({=}O)Cl$; room temperature

b) reflux at 140°C with conc. $H_2SO_4(l)$; alcohol in excess.

c) reflux with either i) P(red) + $I_2(s)$ or ii) NaI, conc. $H_3PO_4(l)$.

d) boiling $K_2Cr_2O_7$/dil. $H_2SO_4(aq)$ or $KMnO_4$/dil. $H_2SO_4(aq)$; distil off the aldehyde as it forms (Figure 11.10).

e)

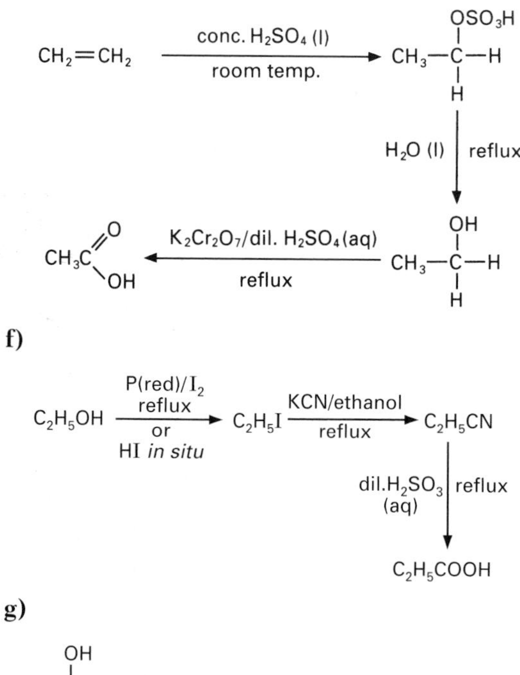

g)

$CH_3C(OH)(H)CH_3$ —conc.H_2SO_4(l) in excess, reflux at 170°C→ $CH_2{=}CHCH_3$ —Br_2 in hexane(l), room temp.→ $CH_2BrCHBrCH_3$

h)

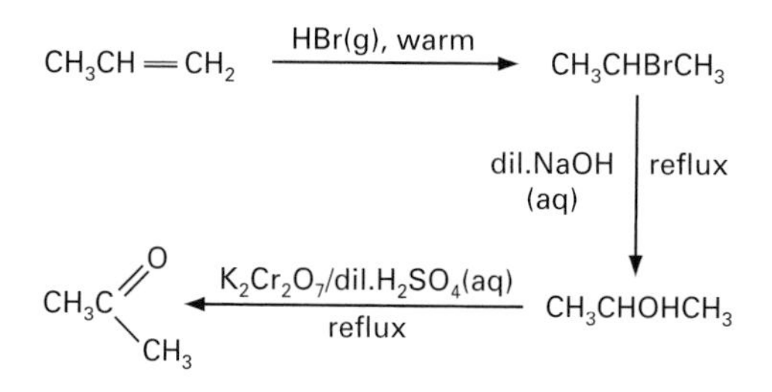

Comments on the activities *continued*

3 a)

$$CH_3CH_2CH_2CH_2OH\ (1°) \xrightarrow[\text{test}]{\text{iodoform}} \text{negative}$$

$$\xrightarrow[\text{reflux for five minutes}]{K_2Cr_2O_7/\text{dil.}H_2SO_4(aq)} \text{butanoic acid formed; colour change orange to green}$$

$$CH_3CH_2CH(OH)CH_3\ (2°) \xrightarrow[\text{test}]{\text{iodoform}} \text{positive = identified}$$

$$(CH_3)_3C{-}OH\ (3°) \xrightarrow[\text{test}]{\text{iodoform}} \text{negative}$$

b) We need to differentiate between phenol and a *primary* aromatic alcohol (phenylmethanol). A variety of tests are possible, as summarised in Table 11.9. In particular, you might try one of the following reagents: dil. NaOH(aq), PCl_5(s), $K_2Cr_2O_7$/dil. H_2SO_4(aq), Br_2(aq) or neutral $FeCl_3$(aq).

Comments on the activities *continued*

Activity 11.6

Table 11.10 Comparing the reactions of alcohol and phenols

Reagent	1°alcohol, e.g. C_2H_5OH	2° alcohol, e.g. $CH_3CH(OH)CH_3$	3° alcohol, e.g. $CH_3C(OH)(CH_3)_2$	phenol C_6H_5OH
Na(s)	H_2(g) evolved at room temperature (phenols are more acidic than alcohols ∴ react faster) $C_2H_5O^-Na^+$	$CH_3CH(O^-Na^+)CH_3$	$CH_3C(O^-Na^+)(CH_3)_2$	$C_6H_5O^-Na^+$
conc. H_2SO_4(l)	← excess alcohol at 140°C gives ethers → $C_2H_5OC_2H_5$ ← excess acid at 180°C gives alkenes → C_2H_4	$(CH_3)_2CH{-}O{-}CH(CH_3)_2$ $CH_3CH=CH_2$	very susceptible to dehydration, giving an alkene $CH_3C(CH_3)=CH_2$	electrophilic substitution in the benzene ring when refluxed 2-hydroxybenzenesulfonic acid (SO_3H ortho to OH) and 4-hydroxybenzenesulfonic acid (SO_3H para to OH)
halogenating agents, e.g. PCl_5(s)	← –OH group; replaced by –Hal group → C_2H_5Cl	$CH_3CHClCH_3$	$CH_3CCl(CH_3)_2$	–OH not replaced by –Hal
$K_2Cr_2O_7$/dil. H_2SO_4(aq)	on refluxing, oxidation occurs readily 1° alcohol ⟶ aldehyde (CH_3CHO), then acid (CH_3COOH) 2° alcohol ⟶ ketone (CH_3COCH_3)		no reaction; tertiary alcohols and phenols resist oxidation	
I_2(aq)/dil. NaOH(aq)	on warming, molecules having the structure: $CH_3{-}CH(OH){-}$ will form triiodomethane (iodoform), a yellow precipitate		no $CH_3{-}CH(OH){-}$ unit ∴ no iodoform reaction	
Br_2(aq)	← no reaction →			immediate reaction at room temp. 2,4,6-tribromophenol (Br, Br, Br on ring with OH) a white precipitate
electrophiles, e.g. NO_2^+, Br^+	← no substitution reaction →			ring substitution at room temperature
neutral $FeCl_3$(aq)	← no reaction →			a violet-coloured mixture; used as a test for phenol

CHAPTER 12

Aldehydes and ketones

Contents

Study Checklist

After studying this chapter you should be able to:

1. State that aldehydes and ketones contain the carbonyl bond, C=O, and, hence, that they are also known as carbonyl compounds.
2. Interpret, and use, the names, short structural formulae and displayed formulae of aldehydes and ketones, including benzaldehyde and phenylethanone. State that the aliphatic compounds form a homologous series of general formula $C_nH_{2n}O$ and describe the gradual change in physical properties that occurs as molecular mass is increased.
3. Recall that aldehydes and ketones have higher boiling points and greater solubility in water than the alkanes of similar molecular mass and explain these properties in terms of permanent dipole–dipole attractions between neighbouring molecules.
4. Give the reagents, reaction conditions and a sketch of the apparatus used in the preparation of aliphatic aldehydes and ketones from alcohols.
5. Contrast the reactions of the C=C bond in alkenes (electrophilic addition) with those of the C=O bond in aldehydes and ketones (nucleophilic addition).
6. State that there are three main types of reactions of aldehydes and ketones: (a) nucleophilic addition to the carbonyl bond, (b) reactions involving an α-hydrogen and (c) redox reactions.
7. State the reagents, and reaction conditions, for the conversion of aldehydes and ketones to: (a) hydroxynitriles and thence to hydroxycarboxylic acids, (b) 2, 4-dinitrophenylhydrazones and (c) alcohols. Explain the significance of: reaction 6(a) in terms of the lengthening of the hydrocarbon skeleton and reaction 6(b) in terms of the identification of the aldehyde or ketone.
8. Explain why the α-hydrogen atoms in ethanal and propanone are readily halogenated. Describe the iodoform (triiodomethane) reaction and the structural features that are required for it to occur.
9. Recall that aldehydes are easily oxidised to carboxylic acids, even by mild oxidising agents such as Tollen's reagent and Fehling's solution whereas, under the same conditions, ketones resist oxidation. Distinguish between aldehydes and ketones using the results of these reactions.
10. Outline the uses of aldehydes and ketones in the manufacture of Bakelite and Perspex, explaining the meaning of the terms; 'condensation polymer', 'thermosetting plastic' and 'thermoplastic'. Know that aldehydes and ketones are often used as fragrances in perfumes.
11. Summarise the main reactions of aldehydes and ketones and know the uses of the reagents in Table 12.5.
12. Explain the following terms; 'monosaccharide', 'disaccharide', 'polysaccharide', 'glycosidic bond' and 'sugar'. Describe the structures of: (a) the chain and ring forms of glucose, (b) sucrose, (c) the amylose and amylpectin forms of starch and (d) cellulose. State one example each of a triose, pentose and hexose sugar.

13 Explain why some sugars (e.g. glucose) are reducing agents whereas others (e.g. sucrose) are not and outline the use of Benedict's solution for the qualitative and quantitative determination of a reducing sugar.

14 Explain the existence of the D- and L- series of sugars. Describe, and compare, the structure of α-D- and β-D-glucose.

15 Explain the following properties in terms of intramolecular, or intermolecular, hydrogen bonding: (a) the solubilities of monosaccharides and polysaccharides and (b) the ability of cellulose and glycogen to act as structural, or storage, polymers in plants and animals.

At first glance, slug-bait and sweets don't seem to have much in common! However, both can be made from compounds which contain the carbonyl functional group:

(electronegativities, C = 2.5, O = 3.5)

Since oxygen is much more electronegative than carbon, the carbonyl bond is strongly polarised and, as we shall see, this has a big effect on the properties of these compounds.

Two important types of carbonyl compound are aldehydes and ketones, whose structural formulae are:

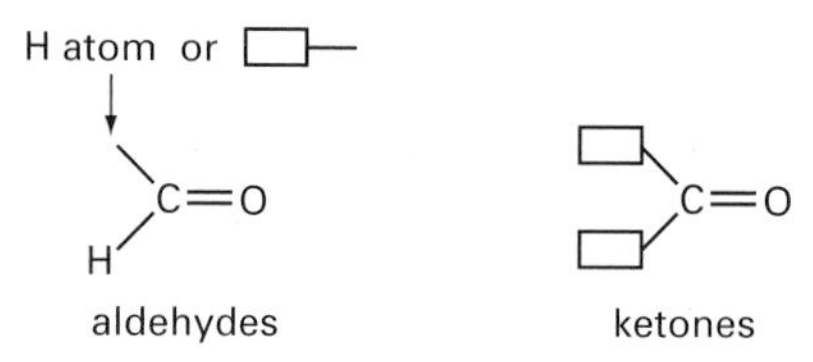

Here □— is an aliphatic or aromatic hydrocarbon skeleton. In this chapter, we shall see how the chemical properties of the carbonyl group are affected by its different molecular locations.

Carbonyl group

Slug-bait and sweets are both made from compounds containing the carbonyl group, C=O. **Metaldehyde**, a commonly used slug-bait, is formed when ethanal is added to concentrated sulphuric acid at 0°C:

$$4\ CH_3CHO \xrightarrow[0°C]{\text{conc. } H_2SO_4(aq)} \text{metaldehyde}$$

metaldehyde

So-called 'energy tablets' contain **glucose**, a carbohydrate which may exist as a mixture of open-chain and ring forms in aqueous solution:

open-chain form (an aldehyde) ⇌ ring-form

glucose

The open-chain molecule contains an aldehyde functional group (Figure 12.6).

12.1 Naming aldehydes and ketones

Aliphatic aldehydes and ketones are named as derivatives of the hydrocarbon with the same carbon 'skeleton'. Thus, the final '**e**' of an alkane name is replaced by '**al**' for an aldehyde and '**one**' for a ketone. For example:

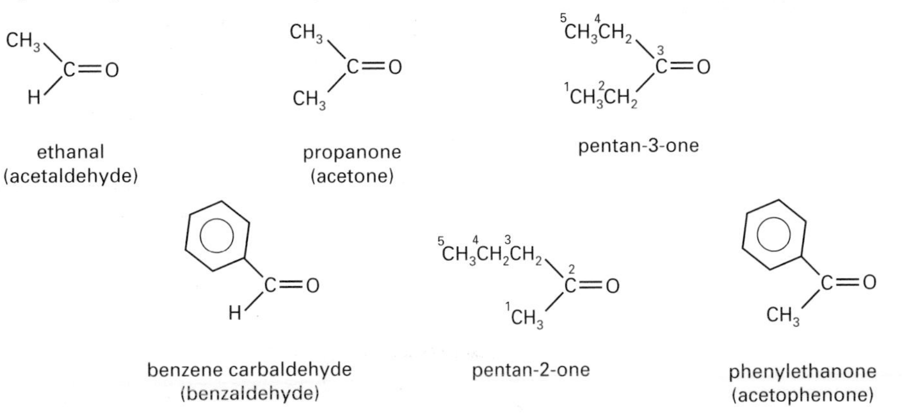

Some carbonyl compounds are often referred to by their historical names, as shown in brackets.

Whilst the carbonyl group in an aldehyde is always at the end of the chain, ketones can exist as **structural isomers**. So, when naming a ketone, you may need to number the 'carbonyl' carbon. In aromatic aldehydes and ketones, the carbonyl group is bonded to a benzene ring.

12.2 Physical properties

Table 12.1 lists short structural formulae and boiling points of some aldehydes and ketones. Once again, there is a steady rise in boiling points and standard enthalpies of combustion, $\Delta H_c^{\ominus}$, values with increasing carbon chain length (Figure 12.1). These graphs are typical of those given by compounds which form a **homologous series**. Aliphatic aldehydes and ketones have the same general formula, $C_nH_{2n}O$.

When comparing compounds of similar molecular mass, we find that

alkanes → aldehydes/ketones → alcohols

boiling points increase

Table 12.1 Formulae and boiling points of some aldehydes and ketones

Name	formula	boiling point/°C	$\Delta H_c^{\ominus}$/kJ mol^{-1}
methanal	HCHO	−19	−550
ethanal	CH_3CHO	21	−1167
propanal	CH_3CH_2CHO	48	−1817
butanal	$CH_3CH_2CH_2CHO$	72	−2497
benzaldehyde	C_6H_5CHO	178	−3520
propanone	CH_3COCH_3	56	−1821
butanone	$CH_3CH_2COCH_3$	80	−2438
pentan-2-one	$CH_3CH_2CH_2COCH_3$	102	−3078
phenylethanone	$C_6H_5COCH_3$	198	−4137

Activity 12.1

1 Name the following carbonyl compounds:

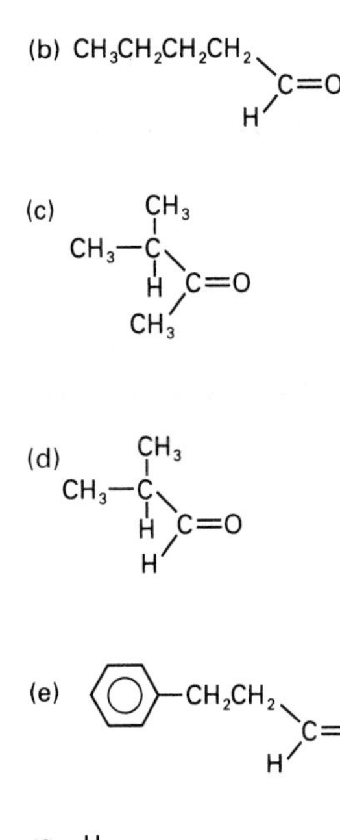

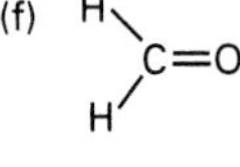

2 Write down the structural formulae for the following compounds:

a) ethanal
b) cyclohexanone
c) 3-methylbutanal
d) dimethylpropanal
e) butanone
f) 4-phenylpentanal

3 Of the compounds in 1 and 2, which are structural isomers?

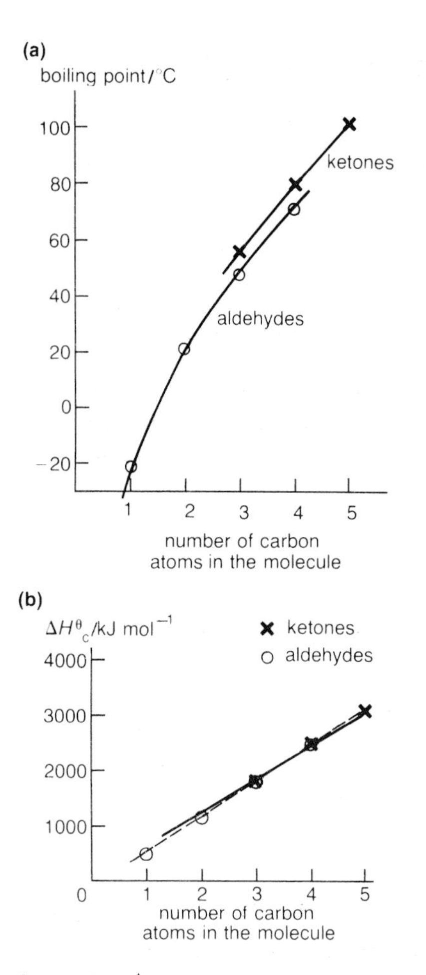

Figure 12.1 ▲
Steady increases in **(a)** the boiling points and **(b)** the $\Delta H^{\ominus}_{c}$ values of some straight-chain aldehydes (o) and ketones (✖) as the hydrocarbon chain gets longer. What does this suggest?

For example,

propane (M_r = 44)	ethanal (M_r = 44)	ethanol (M_r = 46)
b.p. –42°C	b.p. 21°C	b.p. 78.5°C

This pattern reflects the increasing strength of the intermolecular attractions (section 4.6):

strength of the intermolecular attractions increases →

alkanes	**aldehydes or ketones**	**alcohols**
molecules experience only van der Waals' forces, that is *weak* intermolecular forces resulting from the attraction between temporary induced dipoles	molecules experience van der Waals' forces plus *fairly weak* permanent dipole–dipole attractions, that is	molecules experience van der Waals' forces plus hydrogen bonds (*strong* permanent dipole–dipole attractions)

$\delta+$ $\delta-$ $\delta+$ $\delta-$ C=O ∿∿ C=O

∿∿ = a dipole–dipole attraction

$\delta-$ $\delta+$ $\delta-$ O—H⋯O $\delta+$ H

⋯ = hydrogen bonds

Due to their polarity, aldehydes and ketones of low molecular mass are very soluble in water. They are also able to dissolve a wide range of polar and non-polar solutes. Indeed, propanone is an important industrial solvent. Of course, as the hydrocarbon skeleton gets larger, the polar C=O group has less effect on the physical properties of the compound. Thus, benzaldehyde, C_6H_5CHO, is almost insoluble in water.

Aldehydes and ketones are strongly smelling compounds. The pungent smell of formalin, a 40% aqueous solution of methanal, for example, will be well known to students who work with preserved biological specimens. On the other hand, in small amounts, benzaldehyde has a pleasant smell of almonds and these nuts actually contain a benzaldehyde derivative.

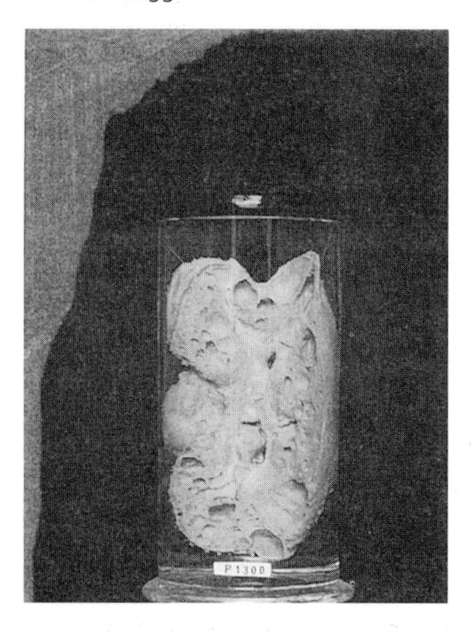

▲ Formalin, a 40% aqueous solution of methanal, is used to preserve biological specimens and as embalming fluid. The specimen in this jar is a lung

Focus 12a

1. Aldehydes and ketones contain the **carbonyl group**, C=O.
2. To name an aldehyde or ketone, replace the final **e** of the parent hydrocarbon's name by **al** for an aldehyde or **one** for a ketone. If the ketone exists as isomers, number the 'carbonyl' carbon.
3. Aldehydes and ketones form **homologous series** which have the same general formula $C_nH_{2n}O$.
4. The carbonyl bond is strongly polarised, $^{\delta+}C=O^{\delta-}$, and this gives rise to **permanent dipole–dipole attractions** between neighbouring molecules. As a result, aldehydes and ketones have higher boiling points than the alkanes of similar molecular mass and those of low molecular mass are very soluble in water.

12.3 Making aldehydes and ketones

Aldehydes and ketones are usually made in the laboratory by oxidising alcohols with acidified dichromate(VI) ions. In section 11.8, we saw that primary alcohols are oxidised to aldehydes and secondary alcohols to ketones:

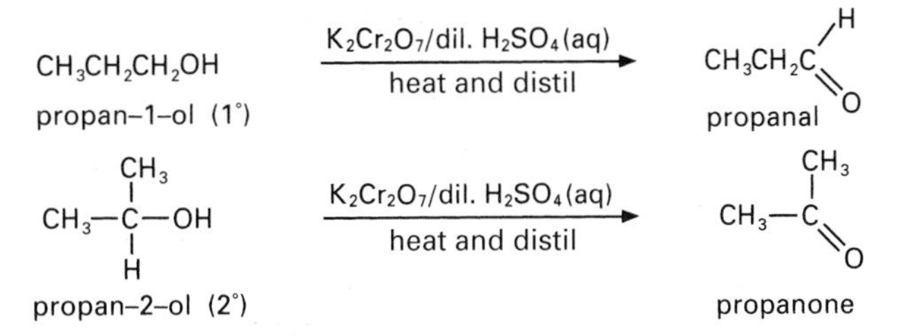

To avoid the possible problems of further oxidation to a carboxylic acid, the aldehyde or ketone should be distilled off as it forms using the apparatus shown in Figure 11.10.

Aldehydes and ketones are manufactured industrially by the catalytic oxidation of alcohols. A mixture of oxygen and the gaseous alcohol are passed over a catalyst of silver or copper at 500°C. Ethanal is also produced by the oxidation of ethene in the Wacker process (section 12.6). Most propanone is obtained as a by-product in the manufacture of phenol (section 9.5).

Activity 12.2

Give the names and formulae of the alcohols which can be oxidised to give the following carbonyl compounds:

- **a)** butanal
- **b)** 3-methylbutanone
- **c)** 2-methylpropanal
- **d)** cyclohexanone
- **e)** 3-phenylbutanal
- **f)** benzaldehyde
- **g)** pentan-2-one
- **h)** phenylethanone

12.4 The reactivity of aldehydes and ketones

Firstly, let us look more closely at the electronic structure of the carbonyl group, taking methanal as an example. In its ground state, a carbon atom has the electron arrangement:

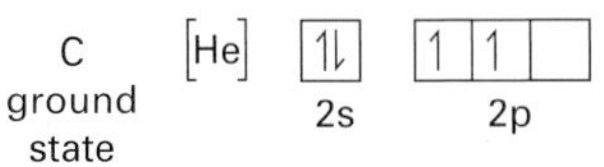

To form the C = O bond, the electron clouds in the 2s orbital and one of the 2p orbitals mix together to form three **hybrid orbitals** (section 8.4):

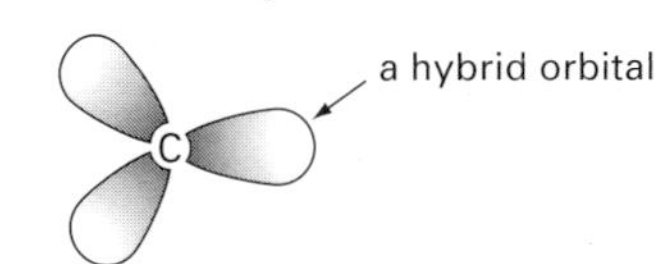

Two of these orbitals form **σ bonds** with the 1s orbitals of hydrogen. The third hybrid orbital overlaps end-on with a 2p orbital on oxygen, giving another σ bond:

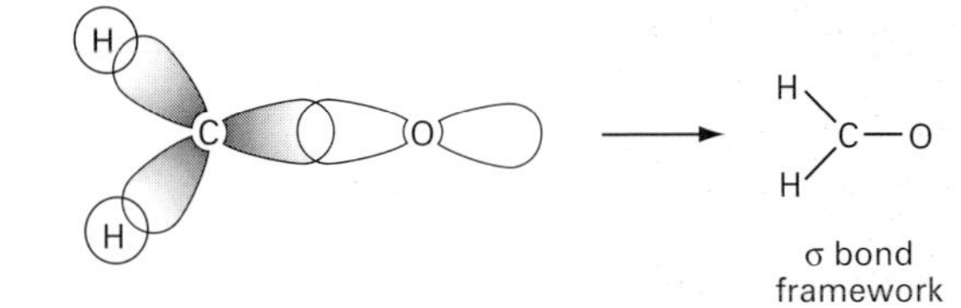

The remaining 2p orbitals on carbon and oxygen overlap sideways-on to form a **π bond**:

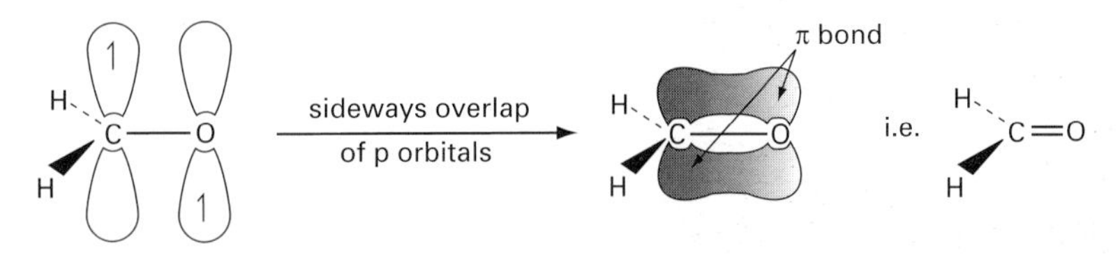

Now, compare the electronic structure of the C = O bond with that of the C = C bond (Figure 12.2). At first glance, these bonds appear to be closely related: both are planar, each consisting of a σ and a π bond. But does this similarity extend to their chemical properties?

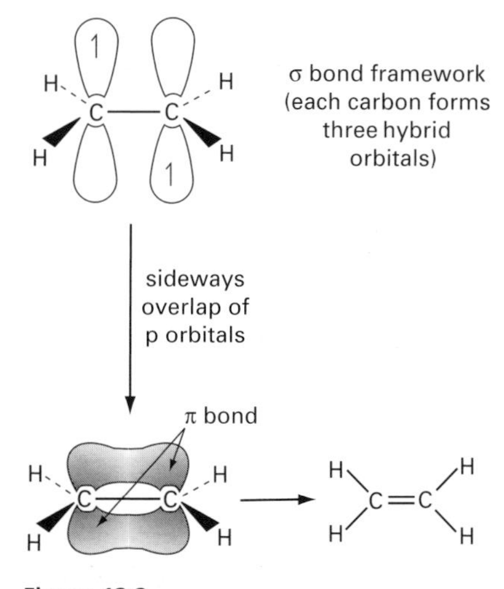

Figure 12.2
Electronic structure of the C$=$C bond in ethene

Due to their unsaturated nature, both C$=$C and C$=$O bonds undergo **addition reactions.** Generally speaking, though, they do not react with the same reagents (Table 12.2). This interesting behaviour is explained by the different distributions of the bonding electron clouds:

Table 12.2 Comparing the reactions of the C$=$O bond in aldehydes and ketones with the C$=$C bond in alkenes

Reaction with HBr(g) at room temperature	
aldehydes or ketones	no reaction
alkenes	addition (electrophilic) to form bromoalkanes e.g. $CH_3CH{=}CH_2 \longrightarrow CH_3CHBrCH_3$
Reaction with HCN(aq) at room temperature	
aldehydes or ketones	addition (nucleophilic) to form hydroxynitriles e.g. $(CH_3)_2CO \longrightarrow (CH_3)_2C(OH)CN$
alkenes	no reaction

In aldehydes and ketones, the carbonyl group reacts in two ways: **nucleophilic addition reactions** (section 12.5) and **redox reactions** (section 12.6).

Focus 12b

1 Aldehydes and ketones are made by oxidising primary or secondary alcohols, respectively. In the laboratory, hot acidified sodium (or potassium) dichromate(VI) is used as oxidising agent. To prevent further oxidation (to a carboxylic acid), an aldehyde must be distilled off as it forms (see Figure 11.10). Aromatic ketones can also be made via a Friedel–Crafts reaction (section 9.5).

2 Aldehydes and ketones take part in three main types of reaction:
- **nucleophilic addition** to the carbonyl bond (section 12.5)
- **redox** reactions (section 12.6)
- reactions involving the α-**hydrogen(s)** (section 12.7).

Also, remember that the hydrocarbon skeleton may display its own typical reactions.

3 Although both the C$=$O bond and C$=$C bond are planar and consist of one σ and one π bond, they differ in their polarities:

C$=$C is non-polar and undergoes **electrophilic** addition (e.g. of Hal_2, HHal, H_2SO_4) but $^{\delta+}C{=}O^{\delta-}$ is permanently polarised and undergoes nucleophilic addition (e.g. HCN).

12.5 Nucleophilic addition reactions

With hydrogen cyanide

Aliphatic aldehydes and all ketones will add hydrogen cyanide at room temperature, using an excess of potassium cyanide dissolved in dilute hydrochloric acid, for example:

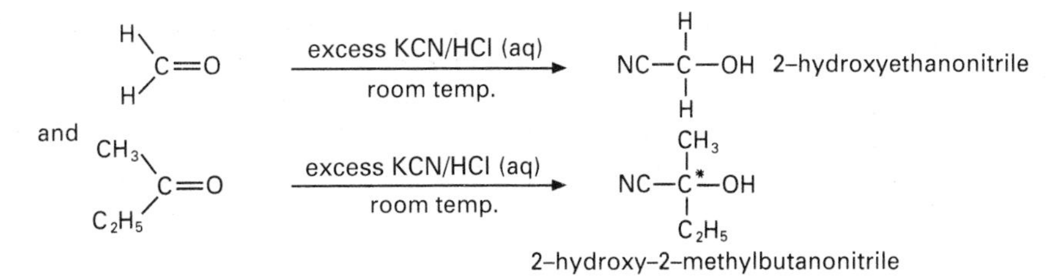

The reaction is a useful synthetic step because it increases the length of the carbon skeleton. Also, note that some hydroxynitriles will possess a **chiral centre** resulting from the **asymmetric carbon atom, C***. Where this occurs, the products of the above reaction will form a **racemic mixture** which may be separated to yield two optical isomers (section 6.5).

The stereochemistry of the nucleophilic addition to a carbonyl bond

When hydrogen cyanide is added to ethanal, the nucleophilic cyanide ion may attack the carbonyl carbon from above, *or* below, the plane of the ethanal molecule.

Consequently, the product, 2-hydroxypropanonitrile, will be an equimolar mixture of molecules which are non-superimposable mirror images, or **enantiomers**. The enantiomers have the ability to rotate the plane of plane-polarised light, to the same degree, but in opposite directions. As a result, each isomer cancels out the other's effect on the plane polarised light and the mixture is found to be optically inactive, that is it causes no overall rotation of the plane of polarisation. An equimolar mixture of optical isomers is termed a **racemic mixture**.

If HCN adds to a symmetrical ketone, such as propanone, $(CH_3)_2CO$, a racemic mixture will not be formed because the product, 2-hydroxy-2-methylpropanonitrile $(CH_3)_2C(OH)CN$, has no asymmetric carbon atom.

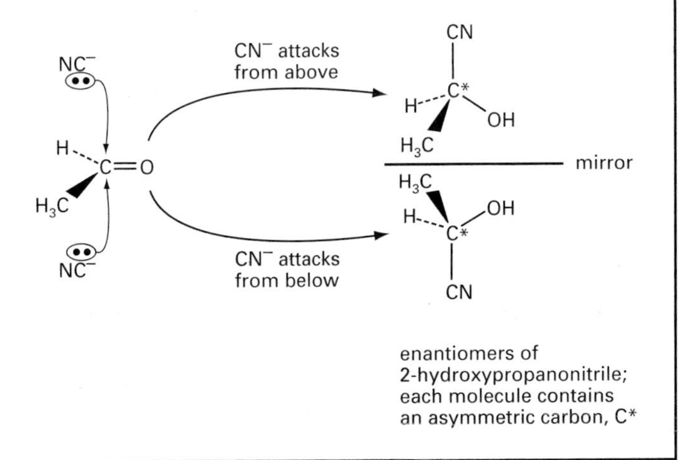

Figure 12.3
When HCN adds to ethanal, a racemic mixture is formed because it is equally likely that the CN⁻ nucleophile will attack from above or below the plane of the molecule

enantiomers of 2-hydroxypropanonitrile; each molecule contains an asymmetric carbon, C*

The mechanism of the reaction is nucleophilic addition and involves two steps. In the slower step, the electron cloud from the cyanide ion (the nucleophile) is used to form a dative covalent bond with the 'carbonyl' carbon. At the same time, the π bond weakens and then breaks heterolytically with the oxygen atom taking both bonding electrons:

$$NC^- \; + \; >C{=}O \xrightarrow{\text{slow}} NC{-}\overset{|}{\underset{|}{C}}{-}O^-$$

an **oxoanion**

This oxoanion rapidly accepts a proton from an H_3O^+ ion, thereby forming the hydroxynitrile:

$$NC{-}\overset{|}{\underset{|}{C}}{-}O^- \; + \; H{-}O^+H_2 \xrightarrow{\text{fast}} NC{-}\overset{|}{\underset{|}{C}}{-}O{-}H \; + \; H_2O$$

With 2,4-dinitrophenylhydrazine – an addition–elimination reaction

In these reactions, the aldehyde or ketone reacts with a solution of 2,4-dinitrophenylhydrazine to give a yellow, or orange, crystalline solid, for example:

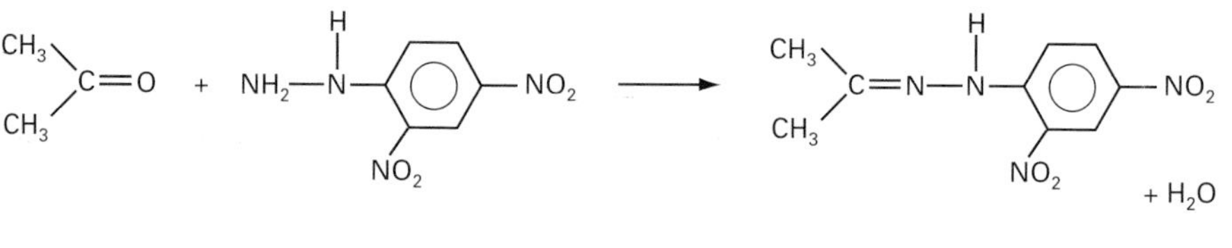

propanone   2,4-dinitrophenylhydrazine   propanone-2,4-dinitrophenylhydrazone (an orange solid)

When purified by recrystallisation, the 2,4-dinitrophenylhydrazones have sharp melting points which can be used to identify the parent carbonyl compound. Suppose, for example, you were asked to distinguish between samples of pentanal (b.p. 104°C) and pentan-2-one (b.p. 102°C). Clearly, a boiling point determination would be inconclusive. Thus, a 2,4-dinitrophenylhydrazone is prepared from each sample, purified and its melting point taken (section 6.4). A firm identification is possible because pentanal 2,4-dinitrophenylhydrazone melts at 98°C but pentan-2-one 2,4-dinitrophenylhydrazone melts at 141°C.

In this type of reaction, a molecule of water is **eliminated** from the initial **addition** product:

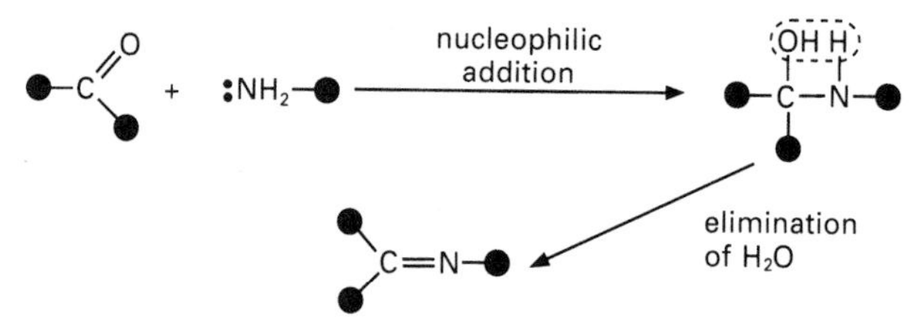

Thus, this is termed an **addition–elimination** or **condensation** reaction.

Activity 12.3

1 Give the name and structural formula of the organic product formed in each of the following reactions:
   a) ethanal + excess KCN/dil. HCl(aq)
   b) pentan-2-one + excess KCN/dil. HCl(aq)
   c) pentan-3-one + excess KCN/dil. HCl(aq)
   d) ethanal + 2,4-dinitrophenylhydrazine
   e) butanone + 2,4-dinitrophenylhydrazine
   Which reactions will yield a racemic mixture?

2 Give the reagents and reaction conditions needed for the following conversions:
   a) propanal → 2-hydroxybutanonitrile

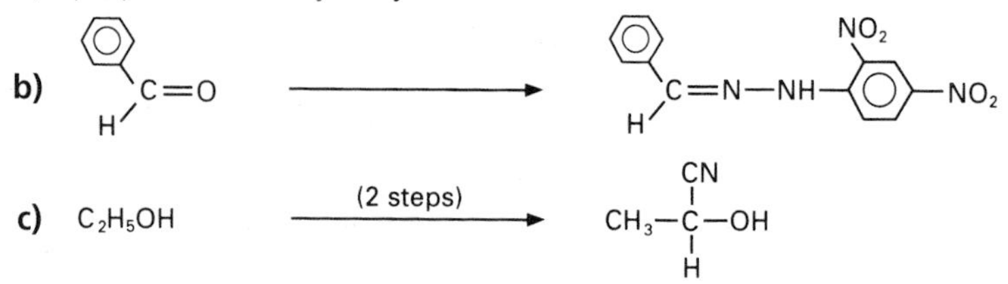

3 Describe how you would distinguish between butanal and 2-methylpropanal without using spectroscopy.

12.6 Redox chemistry of aldehydes and ketones

Oxidation

Aldehydes are readily oxidised to carboxylic acids, even by mild oxidising agents, for example:

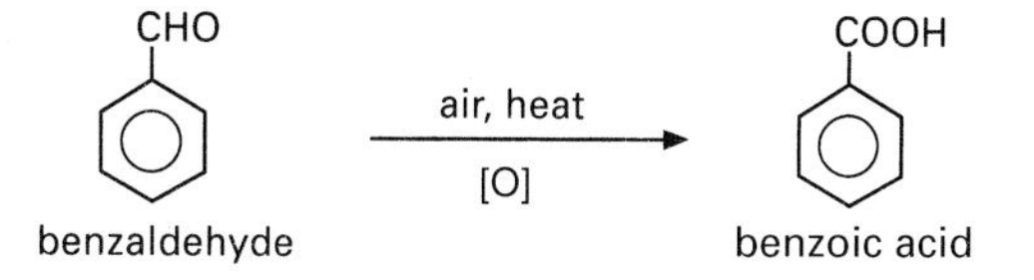

On the other hand, ketones resist oxidation, and will only react after prolonged treatment with a powerful oxidising agent, for example:

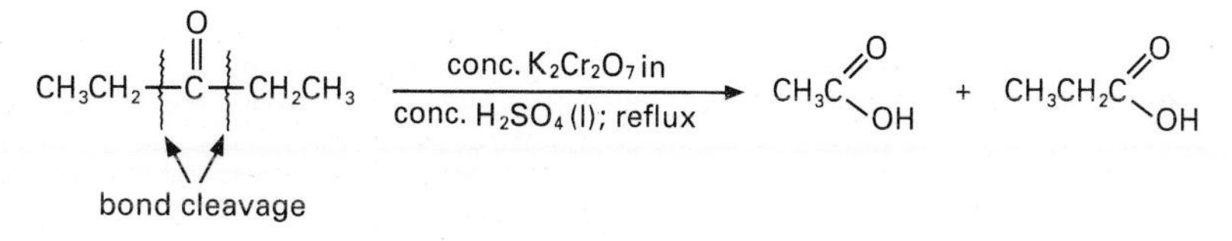

The greater reactivity of the aldehydes results from the carbonyl group's ability to activate the adjacent hydrogen towards oxidising agents.

We can distinguish between an aldehyde and a ketone by testing each sample with a mild oxidising agent – the aldehyde will react, the ketone will not. Two mild oxidising agents are in common use, Tollen's reagent and Fehling's solution.

Tollen's reagent is an aqueous solution of the complex **diamminesilver(I) ion, $Ag(NH_3)^{2+}(aq)$**. On reaction with an aldehyde, the diamminesilver(I) ions are reduced to silver metal and this often appears as a silver 'mirror' on the sides of the test-tube, for example:

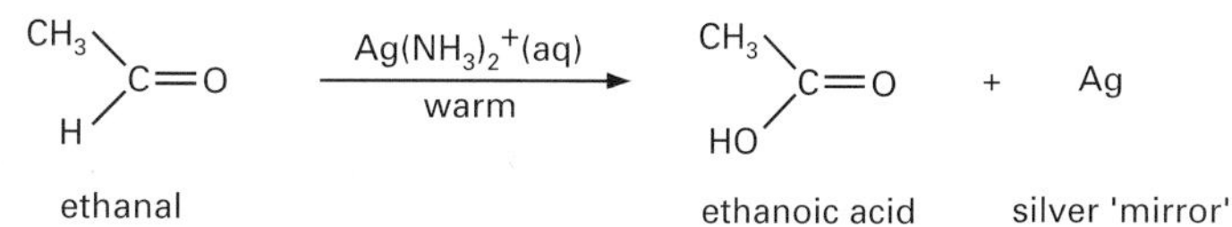

With ketones, there will be no reaction.

Fehling's solution has a blue colour and contains a complex ion of copper(II). Most aldehydes reduce it to give a red-brown precipitate of copper(I) oxide, Cu_2O, for example:

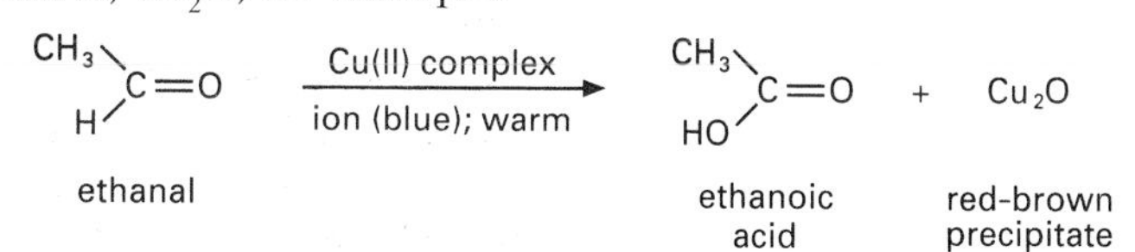

Fehling's solution is too mild an oxidising agent to react with ketones. Similar behaviour is observed with Benedict's reagent which contains a copper(II) complex.

Reduction

Aldehydes and ketones are easily reduced to primary and secondary alcohols, respectively, by refluxing with sodium tetrahydridoborate(III) in ethanol, for example:

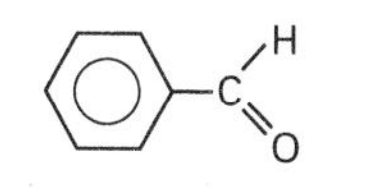

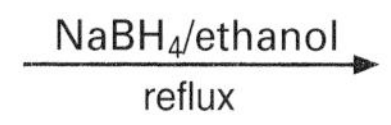

$-CH_2OH$

benzaldehyde

phenylmethanol (a primary alcohol)

or with lithium tetrahydridoaluminate(III) in ethoxyethane, for example:

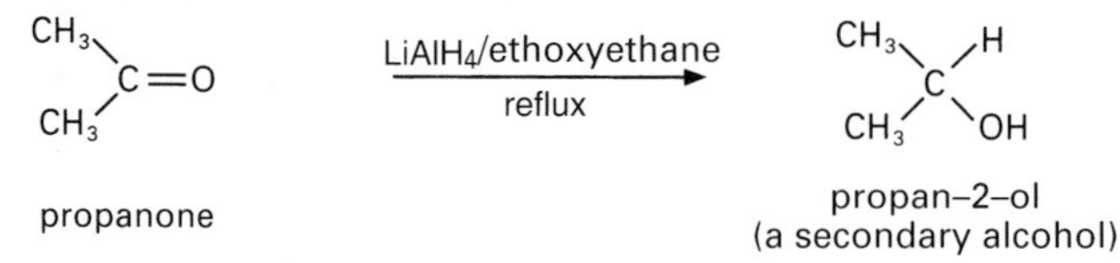

It is worth noting that $LiAlH_4$ and $NaBH_4$ will reduce a C=O bond *but not* a C=C bond. The tetrahydrides act as a source of hydride ions, H^-, and these attack the carbonyl carbon due to its partial positive charge on the carbonyl carbon bond, $^{\delta+}C=O^{\delta-}$. Hydrides will not be attracted to the non-polar C=C bond. Of the two reducing agents, $LiAlH_4$ is the more powerful and, unlike $NaBH_4$, it can also reduce carboxylic acids and their derivatives (section 13.5).

Organic electrochemistry

By the mid-1970s, most ethanal was made by the direct oxidation of ethene using the **Wacker process**. In this process, a mixture of ethene and oxygen is bubbled through an aqueous solution of copper(II) and palladium(II) chlorides at temperatures from 20 to 60°C.

The process is fairly complex. Firstly, Pd^{2+} ions oxidise the ethene to ethanal:

$$CH_2=CH_2 + Pd^{2+} + H_2O \longrightarrow CH_3CHO + 2H^+ + Pd$$

Then, Cu^{2+} ions oxidise Pd, thereby regenerating the Pd^{2+} ions:

$$Pd + 2Cu^{2+} \longrightarrow Pd^{2+} + 2Cu^+$$

The Cu^+ ions are themselves regenerated by reaction with the oxygen:

$$2Cu^+ + \tfrac{1}{2}O_2 + 2H^+ \longrightarrow 2Cu^{2+} + H_2O$$

Overall, the reaction is

$$CH_2=CH_2(g) + \tfrac{1}{2}O_2(g) \xrightarrow[20\text{–}60°C]{PdCl_2(aq)/CuCl_2(aq)} CH_3CHO(aq)$$

The palladium(II) and copper(II) chlorides, being in the same chemical states at the start and the end of the reaction, act as catalysts.

With time, two main problems with the process were identified. Firstly, aqueous copper(II) chloride is hydrolysed by water to produce an acidic solution and this slowly reacts with some metal surfaces. The second concern was the potential danger posed by forcing oxygen into the reaction mixture. As a result of these difficulties, the Wacker process lost ground to other ways of manufacturing ethanal, such as the catalytic oxidation of ethanol by silver metal.

In recent years, however, there has been renewed interest in the Wacker process. An organic electrochemical called **benzoquinone** can be used to regenerate the palladium(II) catalyst:

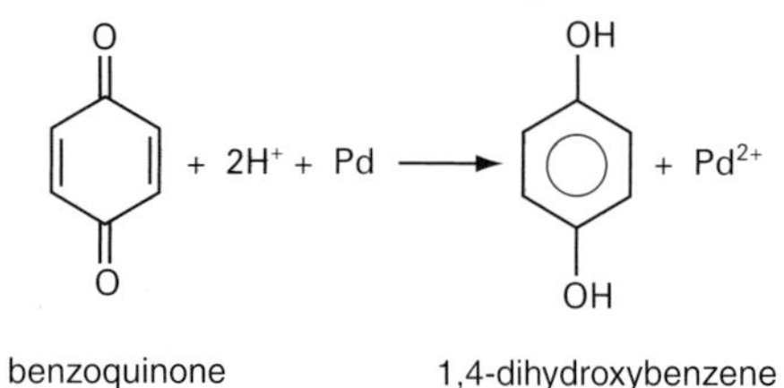

As you can see, benzoquinone is a cyclic molecule having two carbonyl bonds and, as expected, these are reduced to the hydroxy groups in 1,4-dihydroxybenzene. Benzoquinone is itself regenerated when each 1,4-dihydroxybenzene molecule loses two electrons at an inert anode (positive electrode) which dips into the reaction mixture:

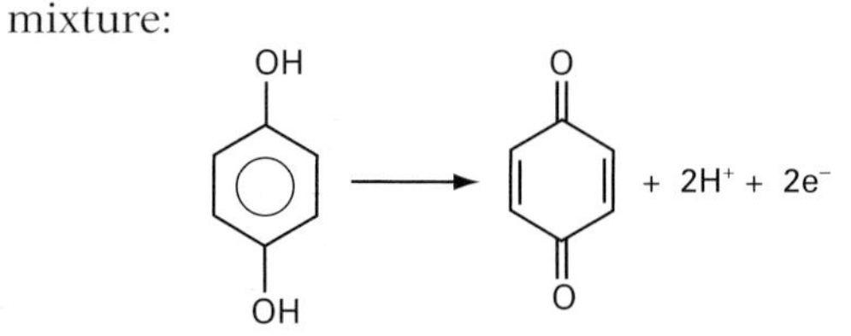

Organic electrochemicals, like benzoquinone, have several advantages over traditional inorganic oxidising and reducing agents. For example, they often work well at low temperatures, thereby minimising energy costs, and they are relatively cheap to make.

Activity 12.4

1 The oxidation of ethene to ethanal involves the nucleophilic attack of a water molecule on one of the carbon atoms in the C=C bond.
   a) What structural feature allows the water molecule to act as a nucleophile?
   b) Would you expect a water molecule to readily attack the C=C bond? Explain your answer.
   c) The Pd^{2+} ions activate the ethene molecule towards nucleophilic attack. How do you think that this 'activation' occurs?

2 The Wacker process can also be used to make propanone. What alkene would be used as the starting material instead of ethene?

3 Describe a simple chemical test to prove that benzoquinone contains carbon–carbon double bonds.

12.7 Halogenation reactions – the iodoform test

A carbon atom which is attached to the carbonyl group is termed an α-carbon atom, Figure 12.4a. Each hydrogen atom on the **α-carbon, $^{\alpha}$C**, is termed an **α-hydrogen, H$^{\alpha}$**. Since the carbonyl group is strongly polarised, it can withdraw electron density from the α-carbon and its C—H bonds, making them more polar than usual, Figure 12.4b. As a result, the α-hydrogen atoms are more easily replaced than those in alkanes. For example, ethanal and chlorine readily react in darkness to form trichloroethanal, CCl_3CHO, a chemical which is used to make insecticides, such as DDT (see page 252).

$$CH_3CHO \xrightarrow[\text{room temperature, in darkness}]{Cl_2(g)} CCl_3CHO$$

(a)

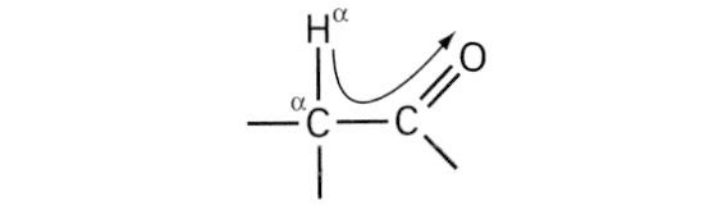

(b) Electron cloud shifting towards the electronegative oxygen atom ...

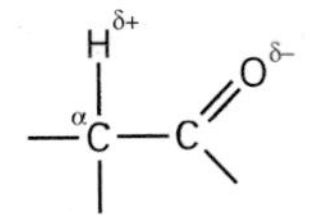

... causes the α-hydrogen atoms to become more electropositive. This makes them more susceptible to substitution, for example by halogen atoms

Figure 12.4
Reactivity of an α-hydrogen atom, H$^{\alpha}$, in an aldehyde or ketone

Since ethane and chlorine will not react under these conditions, section 7.8, the α-hydrogens in ethanal must be activated by the presence of the carbonyl group.

Of particular interest to analytical chemists are the reactions of ethanal and methylketones with iodine in alkaline solution, for example:

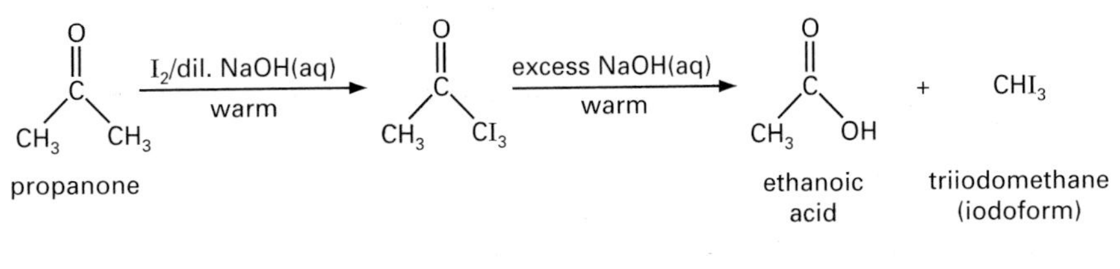

This reaction only occurs if the carbonyl compound contains the $CH_3C(=O)$ group.

Focus 12c

1. Addition of HCN to an aldehyde or ketone: **a)** gives a **hydroxynitrile** which may be optically active and **b)** is a step in lengthening the hydrocarbon chain (section 12.5.)
2. An unknown aldehyde or ketone can be identified by preparing a pure sample of its **2,4-dinitrophenylhydrazone** and taking its melting point. This value is then compared with those in a data book.
3. Aldehydes are easily oxidised to carboxylic acids, even by mild oxidising agents (such as **Tollen's reagent** and **Fehling's solution**); under the same conditions, ketones resist oxidation. These reagents can be used, therefore, to distinguish between an aldehyde and a ketone.
4. A positive **iodoform** reaction proves that a molecule contains a $CH_3—C{=}O$ or $CH_3—CH(OH)—$ structural unit. The reagent for the iodoform reaction is I_2/dil. NaOH(aq).

Thus,

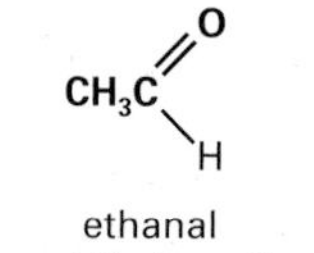

ethanal forms triiodomethane **but** $CH_3CH_2C(=O)CH_2CH_3$ pentan-2-one does not form triiodomethane

Since triiodomethane (iodoform), CHI_3 is a yellow solid which is insoluble in water, it is easily identified as a reaction product. Thus, we often use the I_2/dilute NaOH(aq) reagent to test for the presence of a $CH_3C(=O)—$ group and this is known as the **iodoform** reaction. As alcohols containing the CH_3CHOH group are oxidised to the $CH_3C(=O)—$ group by I_2/dilute NaOH(aq), they will also give a positive iodoform reaction. For example, butan-2-ol, $CH_3CH_2CH(OH)CH_3$, gives a positive iodoform reaction.

Activity 12.5

1. From the following compounds select those that will react with **a)** Tollen's reagent and **b)** Fehling's solution: ethanal, propanone, phenylethanone, propanal, methanal.
2. Give the name(s) and formula(e) of the organic product(s) formed in each of the following reactions.

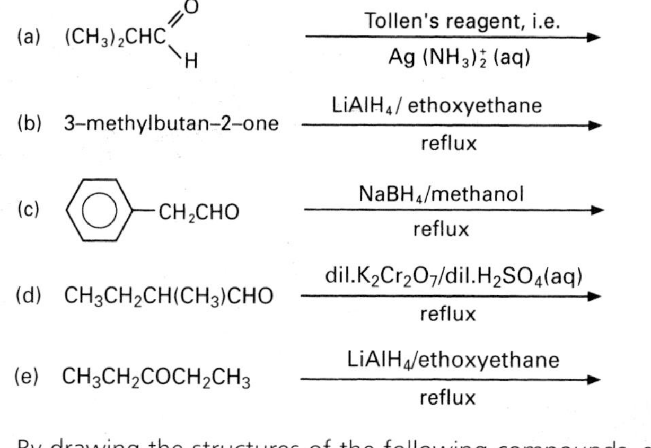

3. By drawing the structures of the following compounds, state which will give a positive iodoform reaction:
   a) benzaldehyde **b)** propanal **c)** ethanol
   d) butanone **e)** 2-methylpropan-2-ol **f)** phenylethanone
   g) propan-2-ol

12.8 Using aldehydes and ketones: plastics and perfumes

Aldehydes and ketones are widely used as bulk chemicals, for example in the manufacture of plastics and resins, and in specialist applications, such as perfumes and flavourings.

Large quantities of methanal (formaldehyde) are used to make plastics and resins. One of these, phenol-formaldehyde, was the first completely synthetic plastic. It was invented by Leo Baekeland, a Belgian chemist, in 1909, and is produced under its trade name '**Bakelite**'. When gently heated in the presence of an acid, phenol reacts with methanal in a two-step process. Firstly, the phenol is substituted in the 2 and 4 positions:

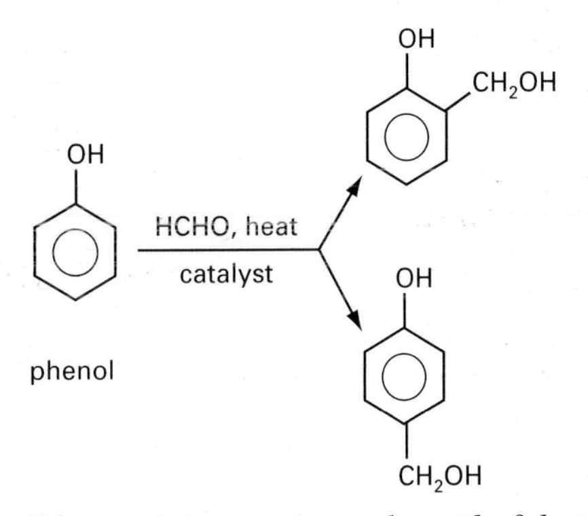

At this stage, various fillers, such as mica and textile fabrics, may be added to the resin to increase the strength and resistance properties of the final product. The hot resin is then poured into moulds and, as it hardens, the substituted phenol molecules link together to form a three-dimensional network in which each aromatic ring is substituted in the 2, 4 and 6 positions:

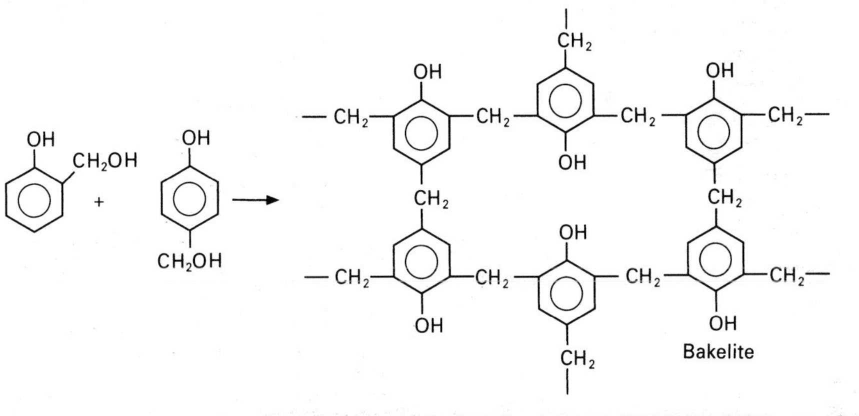

A squash court made of Perspex – an ideal application for this polymer as it can be formed into transparent, rigid yet lightweight sheets

Each time a cross-link is formed a molecule of water is eliminated; hence, this is described as a **condensation polymerisation**.

Bakelite is a **thermosetting plastic**, that is once set, it cannot be heated and remoulded. Like other methanal plastics, Bakelite is a hard solid, which burns with difficulty and has a low electrical conductivity. These properties make the methanal plastics very suitable for use in electrical fittings and in the handles of domestic irons and saucepans. They are also relatively cheap to make. The main problem in using methanal is that it presents health hazards because it is toxic and carcinogenic.

Propanone is used in bulk as an industrial solvent and in the manufacture of '**Perspex**' acrylic sheets, a process which involves a series of simple organic reactions:

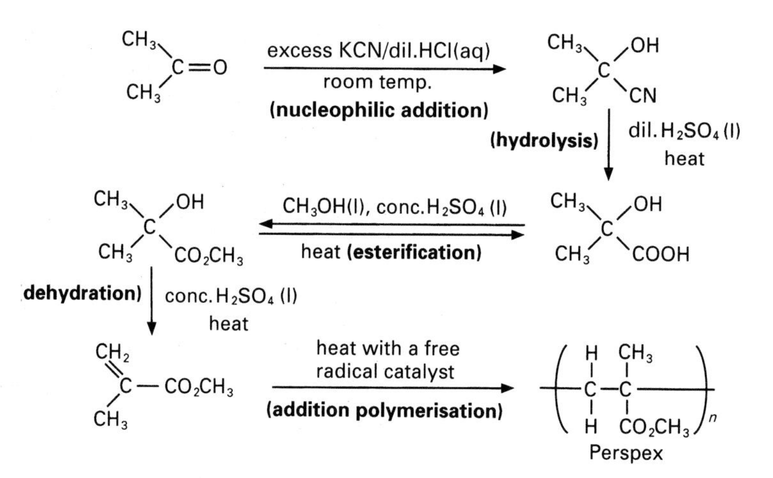

Perspex is a **thermoplastic**, that is it softens when it is heated and can be remoulded into different shapes. It is a lightweight, yet strong, transparent material and this makes it a valuable alternative to glass, for example in signs, windshields and aircraft canopies.

Since they tend to have pleasant smells, a wide variety of aldehydes and ketones are used in perfumes. A typical perfume contains at least three components of decreasing volatility, known as the **top**, **middle** and **base** notes, Table 12.3. The top note is the odour which predominates when the perfume is first worn, whilst the middle one provides a full, solid character. To achieve the 'total effect' of the perfume, the top and middle notes are brought together by the base note, the least volatile component with the most persistent odour. Sometimes a 'sex attractant', such as androsterone, is added to the base note but there is considerable doubt about its effectiveness!

At the start of the twentieth century, perfumes were only available to the rich because the fragrances had to be extracted from natural products, using techniques such as **steam distillation** and **solvent extraction**. The 'natural' oils were expensive because only small amounts are obtained from large quantities of the raw material. For example, the jasmine blossoms gathered by one worker, picking all day, only produce about 1.5 g of jasmine oil, with a market value of about £10. With the advances in analytical chemistry, however, it has become possible to identify the components in a natural fragrance. This type of analysis is no easy task, for example jasmine was found to contain over 200 components! In 1921, Chanel No 5 became the first perfume to contain synthetic fragrances. Nowadays, most perfumes contain synthetic fragrances because they are cheaper to produce than natural extracts.

After-shave lotions and other perfumes are complex mixtures of aroma substances – nowadays many of them are synthesized rather than extracted from natural sources

Table 12.3
The basic fragrances of perfumery. A perfume is characterised by its combination of top, middle and base notes, just like the arrangement of notes in a musical composition. When the perfume is worn the top notes are smelt first because they are the most volatile components, evaporating within 15 minutes. The base notes are the least volatile components and they may be smelt for up to 5 hours

Group name	examples	use		
		top	middle	base
floral	rose, jasmine, lilac			
green	lavender, basil, pine		range of use	
spicy-wood	myrrh, cinnamon, cedar			
animal	musk, ambergris, civet			

Fine perfumes contain a minimum of 15% of the fragrance diluted with alcohol and a fixative to prevent the essential oils evaporating too quickly in use or storage. Spray-on colognes and after-shave lotions are perfumes diluted with water and they typically contain up to 3% of the fragrance.

Activity 12.6 Comparing the reactions of aldehydes and ketones

Looking back over the reactions of aldehydes and ketones, would you describe these compounds as having mainly similar or different properties? This activity should help you to decide. It is also a useful revision exercise.

Prepare a summary table which compares the reactions of aldehydes and ketones. You should consider their reactions with the following reagents:

1 excess KCN/dil. HCl(aq)
2 2,4-dinitrophenylhydrazine
3 $LiAlH_4$/ethoxyethane
4 Tollen's reagent
5 Fehling's solution
6 $K_2Cr_2O_7$/dil. H_2SO_4(aq)
7 I_2/dil. NaOH(aq)

In your opinion, do aldehydes and ketones have similar, or different, chemical properties?

12.9 Carbohydrates

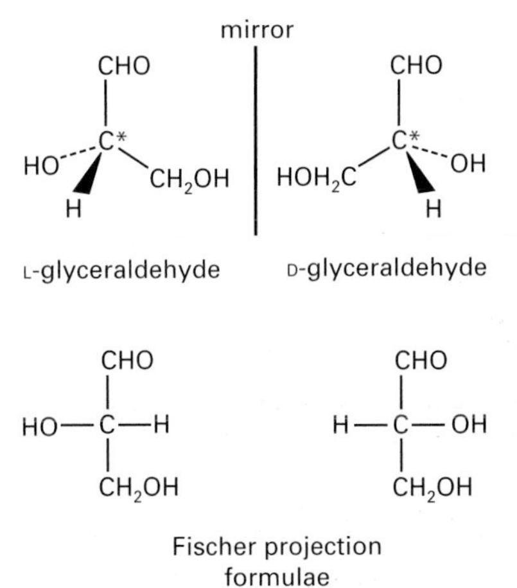

Figure 12.5
Fischer projection formulae of glyceraldehyde, showing the D and L forms. In these formulae, the vertical bonds go behind the plane of the paper whilst horizontal bonds come outwards. Since these isomers have an asymmetric carbon atom, C*, they will be optically active, that is, each can rotate the plane of plane polarised light through the same angle but in opposite directions

Carbohydrates are compounds which provide an organism with the energy it needs to live. Many carbohydrates have the general formula $C_x(H_2O)_y$. Whilst this formula may suggest that carbohydrates are hydrates of carbon (hence the name), nothing could be further from the truth! There are three main types of carbohydrate: **monosaccharides**, **disaccharides** and **polysaccharides**.

Monosaccharides

Monosaccharides are colourless, crystalline solids of low molecular mass which taste sweet and are very soluble in water. The most important monosaccharide is undoubtedly **glucose**, $C_6H_{12}O_6$, which our bodies convert into energy during respiration:

$$C_6H_{12}O_6 + 6O_2 \longrightarrow 6CO_2 + 6H_2O + \text{energy}$$

The physical and chemical properties of biological molecules, like glucose, are heavily dependent on their 3-D molecular structure. Thus, we need an unambiguous way of showing these structures on paper and this is provided by **Fischer projection formulae**. Consider for example, such projections for glyceraldehyde, Figure 12.5. It is accepted that the 'horizontal' bonds come out of the plane of the paper whilst the 'vertical' bonds go backwards, behind the plane. By convention, we write the most reactive group at the top, in this case, the aldehyde group, and we number the 'end' carbon nearest that group, number 1. As you can see from the photo, the two isomeric structures of

▲ The D and L forms of glyceraldehyde are enantiomers, that is, non-superimposable mirror images. Thus they will rotate the plane of plane polarised light to the same degree but in opposite directions. (Note : D-isomers are not always dextrorotatory; L-isomers are not always levorotatory.)

glyceraldehyde are mirror images and are designated as the D and L forms. Notice, also, that the mirror images are non-superimposable, having an asymmetric carbon atom, C*, section 6.5. Thus, each isomer will rotate the plane of plane polarised light to the same degree but in opposite directions, clockwise (+) for D-glyceraldehyde and anti-clockwise (–) for L-glyceraldehyde. Glyceraldehyde is a **triose**, that is a monosaccharide which contains three carbon atoms.

Figure 12.6 shows the possible structures of glucose; as you can see, they are also non-superimposable mirror images. Each isomer has four asymmetric carbon atoms, C^2 to C^5, in the open-chain form, but five asymmetric carbon atoms, C^1 to C^5, in the ring form. For carbohydrates, the structure in which the asymmetric carbon atom furthest from the most reactive group (C^5) has the same configuration as the asymmetric carbon atom in D-glyceraldehyde is taken to be the D-isomer; the mirror image will be the L-isomer.

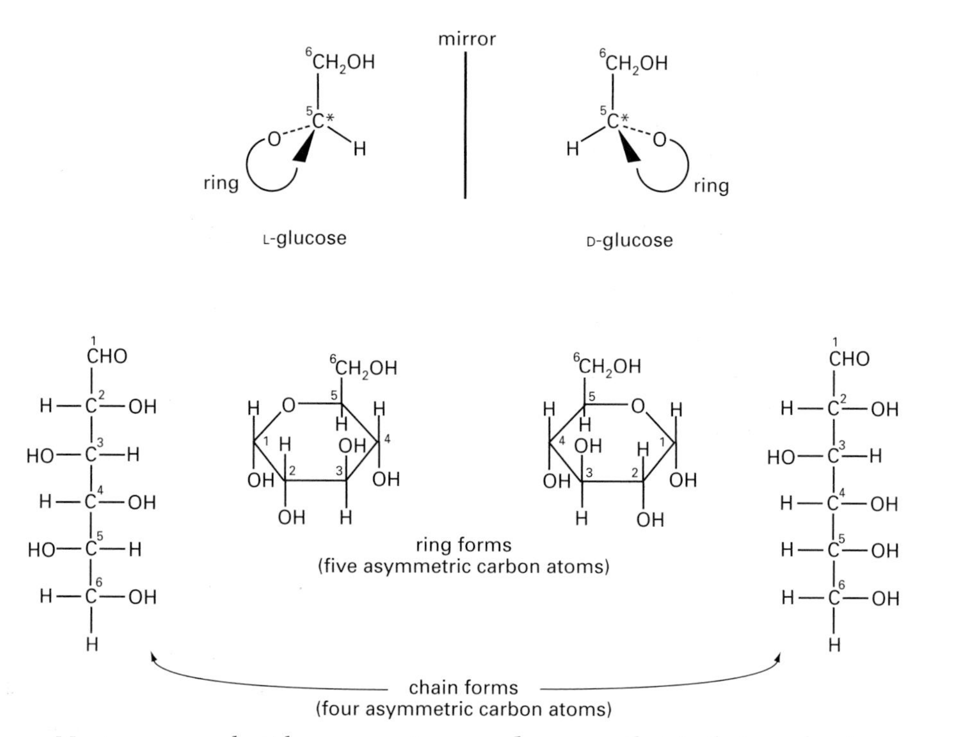

Figure 12.6 ▶
The D and L forms of glucose are identified by comparing the arrangement of groups around C^5 with the asymmetric carbon atom in glyceraldehyde. Glucose is termed a hexose, that is it is a monosaccharide having six carbon atoms. The six atom heterocyclic ring present in glucose is known as a **pyranose** ring

Most monosaccharides are **pentoses** or **hexoses**, that is their molecules contain five or six carbon atoms, respectively. Thus, ribose is a pentose (Figure 12.7) and glucose is a hexose. An important property of pentoses and hexoses is their ability to exist as chain and ring forms. To get an idea of how this happens, try making a molecular model of glucose's structure. Let's look

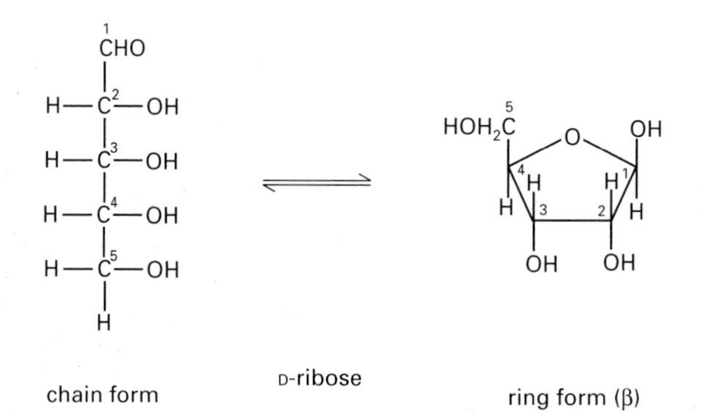

Figure 12.7 ▶
D-ribose is a pentose, that is a monosaccharide containing five carbon atoms. How many asymmetric carbon atoms are present?

more closely at carbon atoms C^1 and C^5 in the D-glucose molecule, Figure 12.6. The carbonyl group on C^1 is polarised and is susceptible to nucleophilic addition (section 12.4). The oxygen of the hydroxy group on C^5 has two lone pairs, making it a potential nucleophile, and this adds across the carbonyl bond. However, the hydroxy group may approach the carbonyl group from two different directions, above or below its plane, respectively (Figure 12.8). Thus, D-glucose has two different ring structures known as α-D-glucose and β-D-glucose which differ according to the position of the hydroxy group on carbon, C^1. This group may be below, or above, the plane of the ring; if below, it is termed the α isomer, if above, it is the β isomer. In the solid state, D-glucose may be obtained in the two crystalline forms, **α-D-glucose** and **β-D-glucose**. When either is dissolved in water, though, an equilibrium is set up between the ring and chain forms, with the equilibrium heavily favouring the ring isomers. We shall see later that the existence of the α and β structural isomers leads to different properties of polysaccharides, such as starch and cellulose.

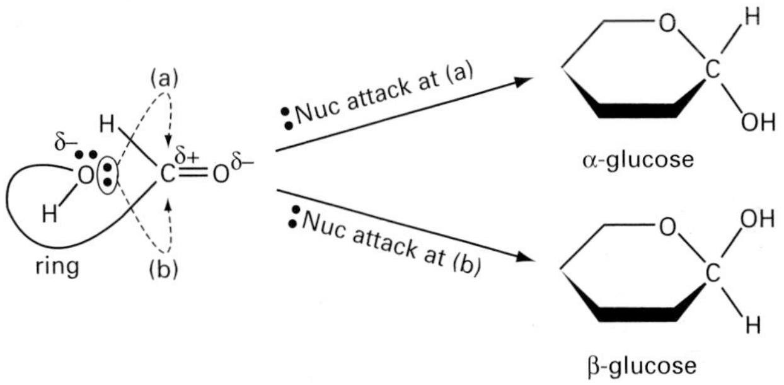

Figure 12.8 ▶
Due to its lone pairs of electrons and high electronegativity, the oxygen atom of a hydroxy group can act as a **nucleophile**. It may attack the carbon atom of a carbonyl group, this being slightly positively charged (δ^+) and the O—H group adds across the C=O bond, thereby yielding a ring structure. Since the hydroxy group may attack the $C^{\delta+}$ atom from above (a) or below (b) the plane of the >C=O structure, two isomeric products may be formed. These are known as α-D-glucose and β-D-glucose.

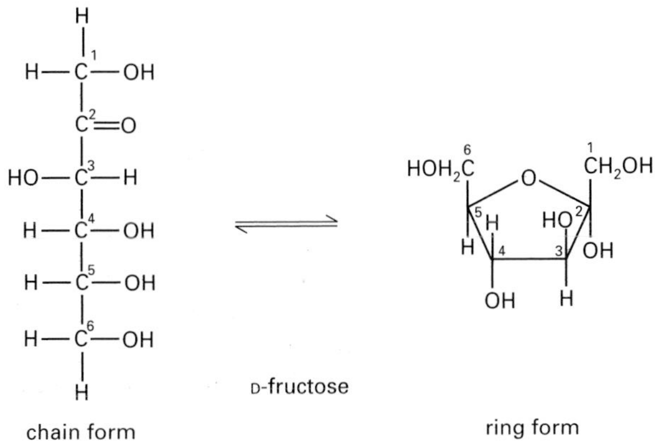

Figure 12.9 ▶
D-fructose is a ketose because its open-chain form contains a 'ketone' structure on carbons 1-3. The five atom heterocyclic ring present in fructose and ribose (Figure 12.7) is known as a furanose ring. Is this the α or β form of D-fructose?

As we have seen, monosaccharides have similar molecular structures. The presence of many hydroxy groups enables their molecules to form extensive hydrogen bonds: i) with each other, causing them to have high melting and boiling points, ii) with water, hence their high solubility. The chemical properties of the monosaccharides are consistent with the presence of hydroxy groups and the carbonyl bond in the open-chain form, Table 12.4.

Monosaccharides which have an open-chain aldehyde structure, for example compounds like glucose, are called **aldoses** whilst those with a ketone structure, like fructose, are called **ketoses**. Unlike ketones, however, ketoses are oxidised by Tollen's reagent and Fehling's solution. Thus, monosaccharides are termed **reducing sugars**.

Table 12.4 Some reactions of glucose (an aldose) and fructose (a ketose)

Reagent/conditions	product formed by glucose or fructose
$CH_3COOH(l)$ + conc. $H_2SO_4(l)$; reflux	a pentaethanoate (i.e. five ester groups are present)
$LiAlH_4$ in ethoxyethane, reflux	reduction of the CHO group occurs, giving another –OH group
KCN/dil. HCl(aq); room temp.	hydroxynitrile formed
2,4-dinitrophenylhydrazine solution; room temp.	addition–elimination to give a 2,4-dinitrophenylhydrazone
Tollen's reagent	silver 'mirror' formed
Fehling's solution	red-brown precipitate of $Cu_2O(s)$

Would you expect these reactions from the 'ring' structures of glucose and fructose?

A reducing sugar may also be detected using Benedict's reagent which, like Fehling's solution, contains copper(II) sulphate. On gently heating an aqueous sugar solution with Benedict's reagent, the initial blue colour of the mixture turns green, then yellowish, and a red-brown precipitate of copper(I) oxide may be obtained. **Benedict's reagent** is more sensitive than Fehling's solution towards reducing sugars. Moreover, the test is semi-quantitative – the colour of the mixture gives us an idea of how much reducing sugar is present:

blue **green** **yellow** **brown precipitate**

→ **increasing amount of reducing sugar**

Monosaccharides are the building blocks from which are formed larger carbohydrates called **disaccharides** and **polysaccharides**.

Activity 12.7

You will need a flexible, molecular model kit for this activity.

1. Look at the photo on page 323 and Figure 12.7 and, hence, make a model of α-D-ribose.
2. Draw the ring structure of α-D-ribose.
3. Open up the ring of your model by breaking the C^1—O bond and draw the Fischer projection formula for the chain form of α-D-ribose.
4. Alongside, draw the chain form of the α-L-ribose.
5. What type of isomerism exists between α-D-ribose and α-L-ribose?
6. What would you expect to be the effect of passing plane polarised light through:
   a) each isomer in turn
   b) an equimolar mixture of the two isomers?

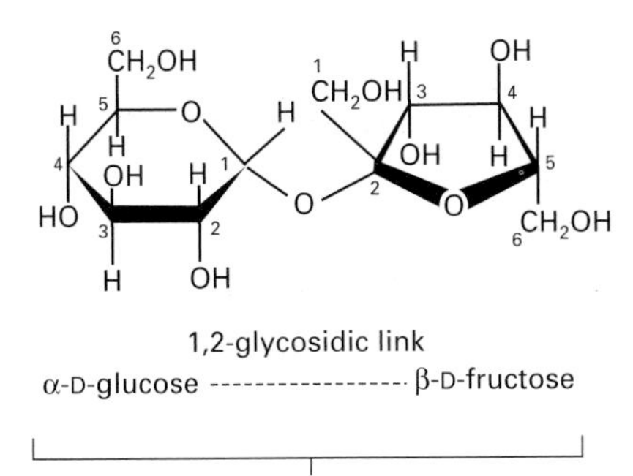

Figure 12.10
In sucrose, an α-D-glucose molecule is bonded to a β-D-fructose molecule via a 1,2-glycosidic bond. This bond is formed via the nucleophilic attack on the fructose C^2 atom by of the oxygen in the hydroxy group on the glucose C^1 atom, followed by the loss of a water molecule. This is an example of a **condensation reaction**

Disaccharides

Disaccharides are formed when two monosaccharides join together. Many disaccharides are possible, the most abundant of which is **sucrose**, $C_{12}H_{22}O_{11}$, or cane sugar, Figure 12.10. It is formed when molecules of α-D–glucose and β–D–fructose are linked together via a condensation reaction, that is a reaction in which a bond is formed by eliminating a small molecule, in this case, water. The oxygen 'bridge' between the monosaccharides is known as a **glycosidic bond** and as you can see it is actually an ether group, C—O—C. We often state which carbon atoms are joined by a glycosidic bond, for example, sucrose has a 1,2-glycosidic bond. The monosaccharides which make up a disaccharide, or polysaccharide, are often termed 'residues'.

When sucrose is hydrolysed by hot dilute hydrochloric acid, it yields equimolar amounts of glucose and fructose, as expected from the structure shown in Figure 12.10. The rate of this reaction, known as the **inversion of sucrose**, may be determined by following the change in the optical rotation (section 6.5) of the reaction mixture with time:

$$\underset{\substack{(+)\text{ sucrose}\\ \text{dextrorotatory}}}{C_{12}H_{22}O_{11}(aq)} + H_2O(l) \longrightarrow \underset{(+)\text{ glucose}}{C_6H_{12}O_6(aq)} + \underset{(-)\text{ fructose}}{C_6H_{12}O_6(aq)}$$

overall laevorotatory

Glucose and fructose rotate the plane of plane polarised light in opposite directions but fructose has the greater angle of rotation, so the final reaction product is laevorotatory.

Unlike individual monosaccharides, sucrose will not react with Benedict's reagent; it is not a reducing sugar. This confirms that the monosaccharide residues in sucrose are joined via the aldehyde group of the glucose and the ketone group of fructose, thereby preventing the formation of the open-chain structures. Thus, there is no discrete carbonyl group available to reduce the copper(II) ions in the Benedict's reagent.

Like monosaccharides, disaccharides are colourless solids, which taste sweet. With so many hydroxy groups, their molecules possess an enormous capacity for hydrogen bonding and this makes them extremely soluble in water. Monosaccharides and disaccharides are commonly referred to as **sugars**.

Polysaccharides

Polysaccharides are tasteless powders whose molecules have very high molecular mass. Surprisingly, they are only slightly soluble in water but do dissolve in some organic solvents, the reason being that their many OH groups tend to be involved in strong hydrogen bonding, *either* within the polysaccharide molecule (see Figure 12.11) *or* with OH groups in neighbouring polysaccharide molecules. Thus, the OH groups are not able to attract water molecules. Polysaccharides are condensation polymers in which thousands of monosaccharide residues are joined together via glycosidic bonds. Our study will take in three important polysaccharides: **starch**, **cellulose** and **glycogen**. You will see that the properties of these polysaccharides depend both on the nature of the monosaccharide from which they are formed and the relative positions of the atoms linked by the glycosidic bonds.

Starch

In plants, most of the energy is stored as starch. Plants convert the starch into energy where photosynthesis is not possible, for example during the underground sprouting of a potato tuber or daffodil bulb. When starch is hydrolysed in acid solution, the only product is α-D-glucose, suggesting that starch is a polymer of α-D-glucose. Spectroscopic studies show this to be the case with starch existing in two forms: amylose and amylpectin.

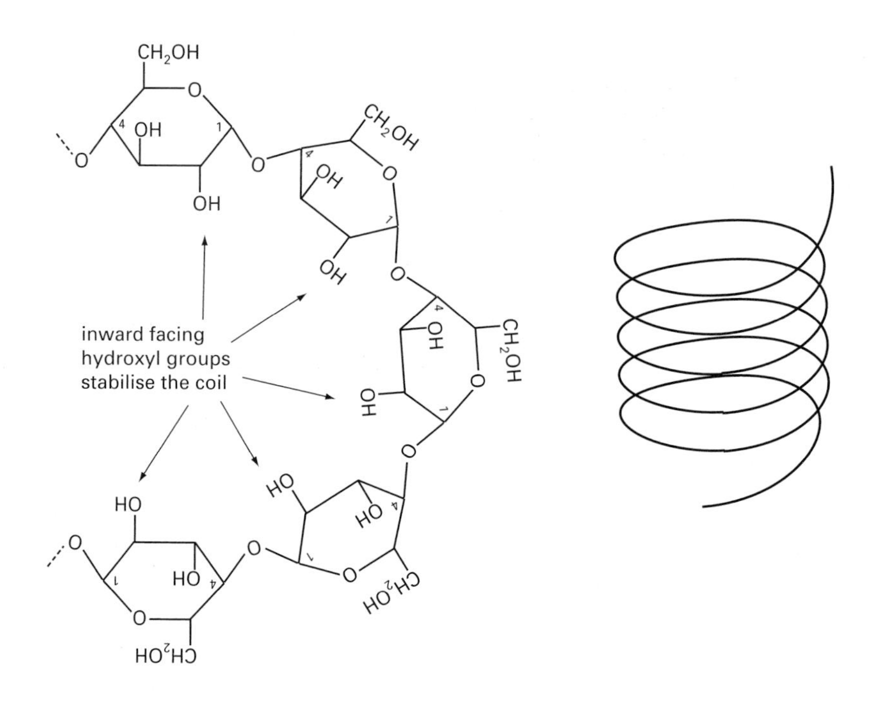

Figure 12.11
In amylose, α-D-glucose residues are held together in long chains via 1,4-glycosidic bonds. Due to intramolecular hydrogen bonding between the inward facing hydroxyl groups along the polymer chain, the chains curl up into a compact helical structure

In **amylose**, long unbranched chains of glucose residues, held together by 1,4-glycosidic links, coil up to form a compact helical structure, Figure 12.11. The coil is stabilised by the hydrogen bonding between the hydroxy groups situated within the helix. The structure of **amylpectin** also has chains of α-D-glucose molecules held together by 1,4-glycosidic bonds but about every 24–30 residues a 1,6-glycosidic bond is formed and this causes the chain to branch, Figure 12.12. Thus, amylpectin has closely packed, branched chains which are held in a 'brush' shape by hydrogen bonding between the hydroxy groups located along the chains. Since amylose and amylpectin have compact molecular shapes, plants are able to store many glucose residues in a small space within their cells.

As expected, amylose and amylpectin have different properties as a result of their different 3-D structures. For example, amylpectin molecules are larger than those of amylose and more of their —OH groups are involved in *intramolecular* hydrogen bonding (unlike amylose which has a greater tendency to form *intermolecular* bonds with water molecules). Consequently, amylose is much more soluble in water than amylpectin. Starch may be detected using a solution of iodine in aqueous potassium iodide. Amylose gives a blue-black colour whereas amylpectin gives a red-violet colour.

Starch is the main carbohydrate in our diet, present in foods such as potatoes, cereals and vegetables. In its natural state, starch molecules accumulate into grains which, when stained with iodine, can be easily seen with a microscope. On digestion, starch is converted in stages into glucose by enzyme-catalysed hydrolyses. The way in which amylose and amylpectin react, though, is quite different.

Amylose is hydrolysed to glucose via the action of two enzymes, **amylase** and **maltase**. In the first reaction, amylase causes random cleavage of the 1,4-glycosidic links, with the formation of glucose and maltose, a disaccharide unit, Figure 12.13. Then, maltase catalyses the hydrolysis of maltose to glucose. However, when amylpectin is hydrolysed by the action of amylase, successive maltose units are set free until a branch point is reached, Figure 12.14. The enzyme is unable to hydrolyse the 1,6-glycosidic bond found at the branching point and so the reaction stops. The polysaccharide fragment of the amylpectin which remains is known as a **limit dextrin**.

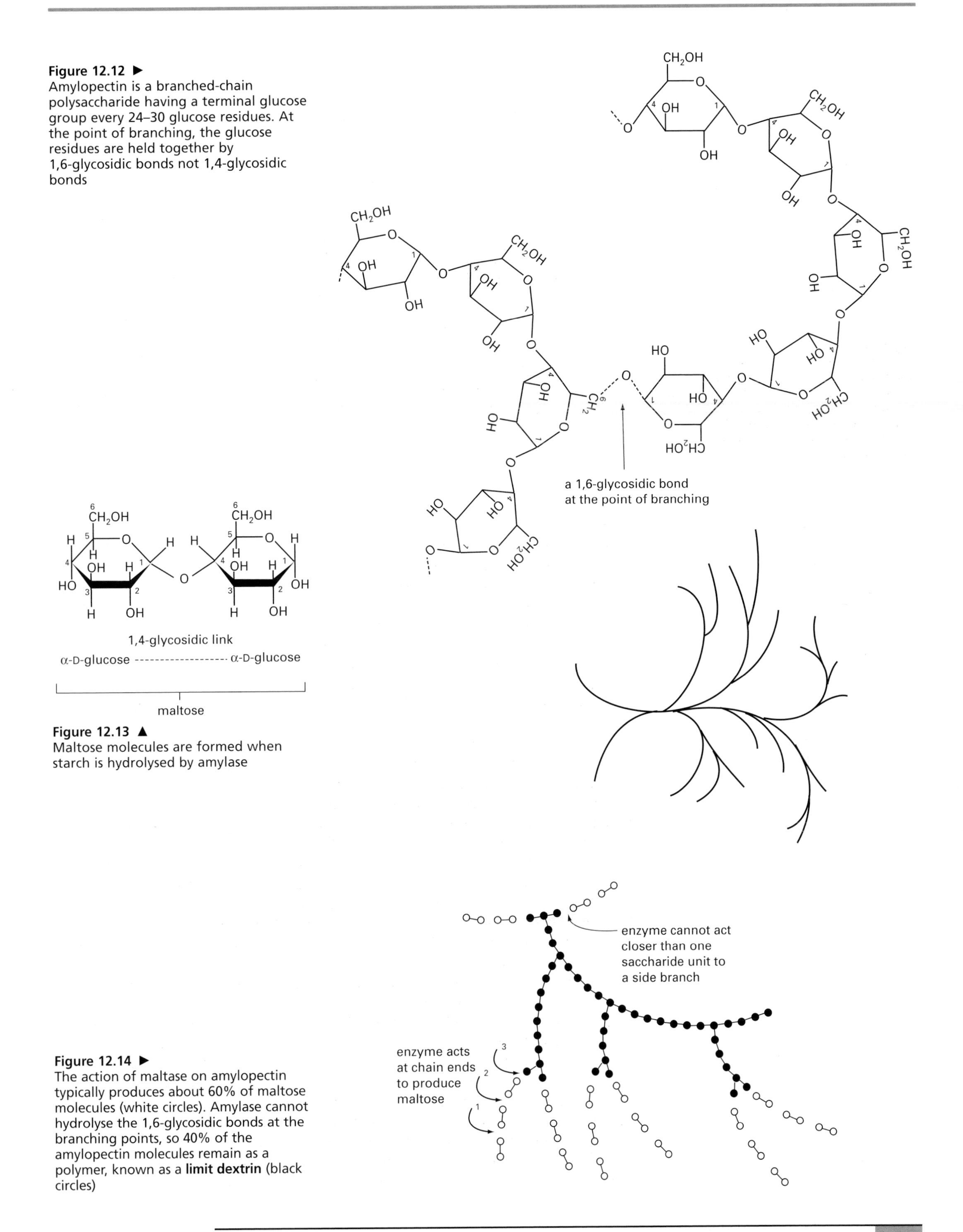

Figure 12.12 ▶
Amylopectin is a branched-chain polysaccharide having a terminal glucose group every 24–30 glucose residues. At the point of branching, the glucose residues are held together by 1,6-glycosidic bonds not 1,4-glycosidic bonds

Figure 12.13 ▲
Maltose molecules are formed when starch is hydrolysed by amylase

Figure 12.14 ▶
The action of maltase on amylopectin typically produces about 60% of maltose molecules (white circles). Amylase cannot hydrolyse the 1,6-glycosidic bonds at the branching points, so 40% of the amylopectin molecules remain as a polymer, known as a **limit dextrin** (black circles)

Polysaccharides may also be hydrolysed in dilute acid conditions, in which case all of the glycosidic bonds are hydrolysed. Chromatography of the hydrolysis products enables us to determine which monosaccharides were present in the polysaccharide.

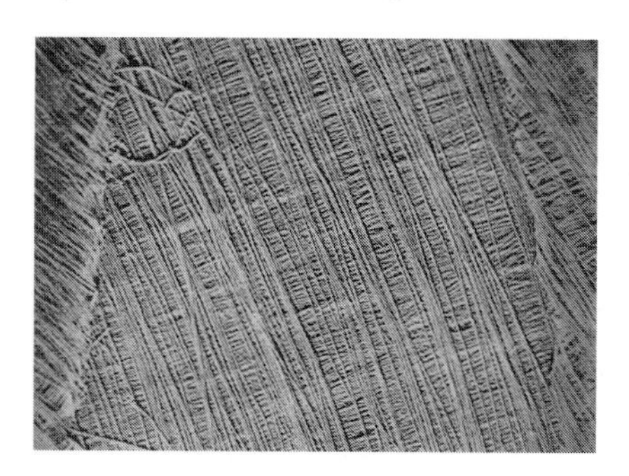

Electron micrograph showing the cross-linking of microfibrilsin cellulose

Cellulose

Cellulose is the main material in the cell walls of plants and only certain herbivorous animals can digest it. When cellulose is hydrolysed in dilute acid solution, there is only one product, β-D-glucose, suggesting that cellulose is a polymer of β-D-glucose, Figure 12.15. In cellulose, the β-D-glucose residues are held together in straight, unbranched chains by 1,4-glycosidic bonds with successive residues arranged at 180° to each other. This strengthens the attraction between adjacent chains, as it extends the hydrogen bonding between them. The chains pack together in 'bundles' of 60–70 known as **microfibrils** and these are, in turn, arranged in parallel fashion to form fibres of cellulose.

The cell wall of a plant is like a 'matting' of cellulose fibres running in different directions, producing a material of considerable strength. Even so, cellulose is permeable and allows water and dissolved substances to reach the cells.

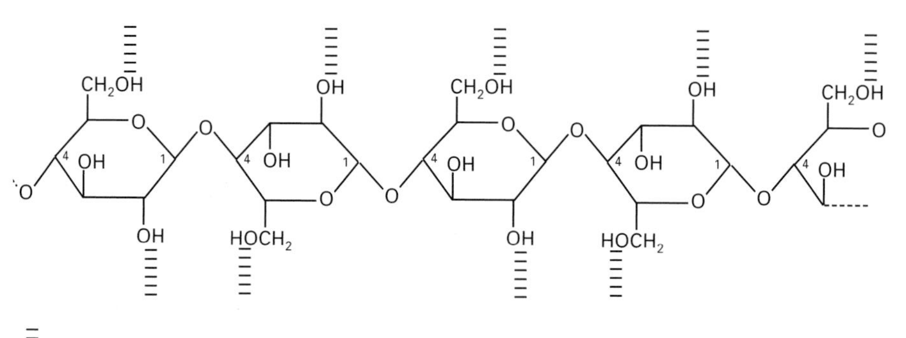

Figure 12.15 ▶ Cellulose is a polymer of β–D-glucose molecules held together by 1,4-glycosidic bonds in which every other residue is turned 'upside down'. The polymer forms straight chains rather than the helical structures found in starch (amylose and amylopectin). The individual polymer chains are strongly attracted 'sideways' to each other by intermolecular hydrogen bonds, forming strands known as **microfibrils**. The cellulose microfibrils are very strong and this makes them a suitable building material for plant cell walls

Cellulose is an important food source for some animals, bacteria and fungi. Bearing in mind the abundance of cellulose throughout the world, surprisingly few animals possess the enzyme cellulase which catalyses the digestion of cellulose. Cows are able to digest the cellulose in grass because they have bacteria living in their guts. Once again, we can see how the chemistry of a biomolecule depends on its 3-D shape. Although cellulose and starch have very similar structures, amylase will not catalyse the hydrolysis of cellulose because its 3-D shape does not match that of the enzyme, section 24.2.

Glycogen

Our bodies convert excess glucose into a polysaccharide called **glycogen** which is stored in the liver and the muscles. Glycogen is an energy store, the animal equivalent of starch. Indeed, it has a similar structure to the amylpectin form of starch but has more 1,6-glycosidic bonds, every 8–12 residues. Thus, glycogen is even more branched and compact than amylpectin. When required, the glycogen in the liver is converted into glucose via **aerobic** (oxygen needed) hydrolysis catalysed by an enzyme called **phosphorylase**. The balance between glucose in the bloodstream and glycogen in the liver is maintained by hormones, such as insulin. Glycogen is also stored in the muscles. When these are exercised very vigorously, there may not be enough time to get oxygen to the muscles so the muscle glycogen breaks down to glucose via an **anaerobic** (oxygen not needed) hydrolysis.

Activity 12.8

Lactose is a sugar which is found exclusively in mammalian milk. It is a disaccharide formed when β-D-galactose and β-D-glucose link together via a 1,4-glycosidic bond. The Fischer projection formula of D-galactose is given below.

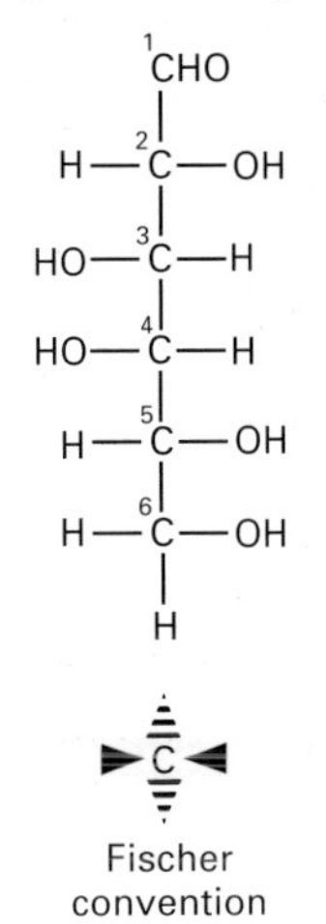

Fischer convention

1. Use a flexible bond model kit to construct the open-chain form of D-galactose.
2. Now convert this to the β-ring form of D-galactose (remember that the —OH group on C^1, like the $^6CH_2OH$ group, will be above the plane of the ring).
3. Construct a model of β-D-glucose from the photograph on page 323.
4. Now, form lactose by making the glycosidic link between C^1 in glucose and C^4 in galactose.
5. Draw the structural formula of lactose.
6. When lactose is dissolved in water, is it possible for the galactose residue to form its open-chain isomer? Try it on your model!
7. Can the glucose residue form its open-chain isomer?
8. Is lactose a reducing sugar? Explain your answer.

Questions on Chapter 12

1 This question is about the following compounds:

A methanal B butanone C cyclohexanone
D benzenecarbaldehyde E ethanal
F phenylethanone

a) Draw *displayed* formulae for these compounds.
b) Classify the compounds as being: i) aliphatic, ii) alicyclic, iii) aromatic, iv) an aldehyde, v) a ketone.
c) Which may be prepared by oxidising secondary alcohols?
d) Which react with hydrogen cyanide in cold, acidified solution, to produce a mixture of optical isomers?
e) Which would give a crystalline product when treated with a solution of 2,4-dinitrophenylhydrazine?
f) Which would react with Fehling's solution to produce a red precipitate of copper(I) oxide?
g) Which would give a 'silver mirror' on treatment with Tollen's reagent?
h) Which are reduced by sodium tetrahydridoborate(III) to primary alcohols?
i) Which would be readily oxidised by acidified potassium dichromate(VI)?
j) Which would produce a yellow crystalline precipitate when treated with iodine in alkaline solution?

2 Explain the following observations:

a) Although their molecules have identical relative molecular masses, butanone boils at 80°C and pentane boils at 36°C.
b) Methanal and ethene both contain double bonds, yet only methanal will react with an excess of potassium cyanide in cold, acidified solution.
c) Propanone and pentan-3-one are both ketones, yet only one of them reacts with iodine in alkaline solution to produce a yellow crystalline solid.
d) If ethane and ethanal are separately treated with chlorine, in darkness, only the ethanal will react.

Questions on Chapter 12 *continued*

3 Give the names, and structural formulae, of the organic products formed in the following reactions:

a) propanone + excess KCN/dilute HCl(aq)
b) ethanal + 2,4-dinitrophenylhydrazine solution
c) butanal + $K_2Cr_2O_7$/dilute H_2SO_4(aq)
d) propanal + Tollen's reagent
e) benzenecarbaldehyde + Fehling's solution
f) pentan-2-one + $NaBH_4$/methanol
g) ethanal + I_2/dilute NaOH(aq)

4 Four unlabelled bottles contain propan-1-ol, propanal, butanal and propanone, respectively. How might the compounds be identified using simple chemical tests?

5 Give the reagents and reaction conditions needed for the following conversions. Note that parts **f)** to **i)** may require a reference to earlier chapters.

a) $CH_3CH_2CHO \longrightarrow CH_3CH_2CH_2OH$
b) $CH_3CHO \longrightarrow CH_3COOH$
c) $CH_3CH_2COCH_3 \longrightarrow CH_3CH_2CH(OH)CH_3$
d) $CH_3CHO \longrightarrow CH_3CH(OH)CN$
e) $CH_3CH_2CHO \longrightarrow CH_3Cl_2CHO$
f) $(CH_3CH_2)_2CO \longrightarrow (CH_3CH_2)_2C(OH)COOH$ (2 steps)
g) $CH_3CH_2CHO \longrightarrow CH_3CH_2CH_2Br$ (2 steps)
h) $CH_3COCH_3 \longrightarrow CH_3CHCH_2$ (2 steps)
i) $C_2H_4 \longrightarrow CH_3CHO$ (3 steps)
j) $CH_3CH_2Br \longrightarrow CH_3CH_2CHO$ (3 steps)

6 Write down the mechanism of the reaction between propanone and hydrogen cyanide.

7 Compound X contains 55.8% carbon, 7.0% hydrogen and 37.2% oxygen. When 0.4 g of X is heated to 127°C at a pressure of 101 325 Pa, the vapour occupies a volume of 150 cm^3.

a) Work out the molecular formula of X. (You will need to use the ideal gas equation, section 4.13; the gas constant, R, is 8.31 J K^{-1} mol^{-1}.)
b) X reacts with Tollen's reagent to give a 'silver mirror'. What structural feature is present in X?
c) On treatment with iodine in alkaline solution, X yields a yellow crystalline precipitate. Deduce the structural formula of X.
d) Write down the structural formulae of the products formed when X reacts with:
i) lithium tetrahydridoaluminate(III) in ethoxyethane
ii) potassium dichromate(VI) and dilute sulphuric acid
iii) excess potassium cyanide in acidic solution.
e) The product of reaction d)iii) contains a mixture of optical isomers. Explain what this means and identify the structural feature of the product which is responsible for this behaviour.

8 This question is about the following carbohydrates:

A

B

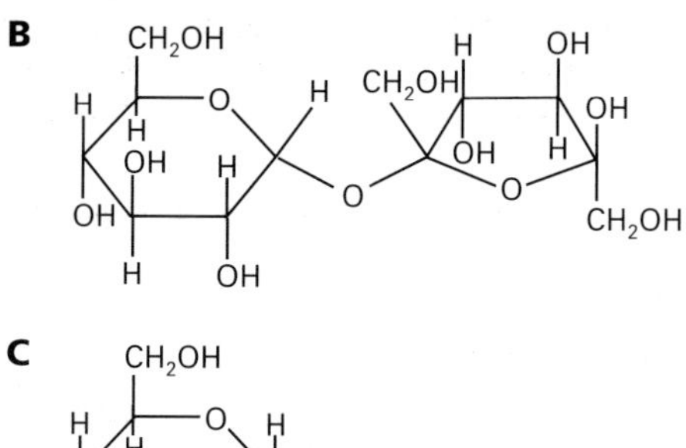

C

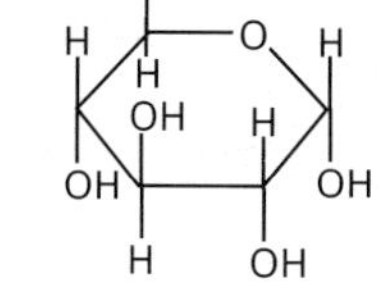

D $CH_2(OH)CH(OH)CHO$

E starch

F cellulose

G

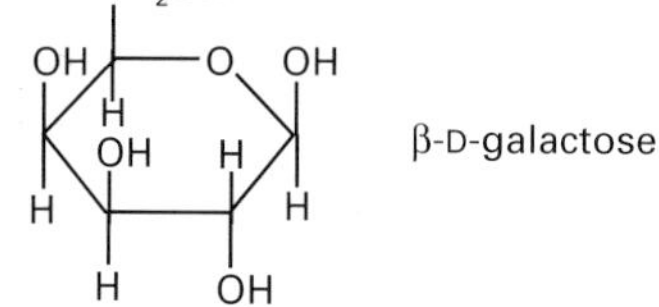

β-D-galactose

a) Which of these are: (i) monosaccharides, (ii) disaccharides, (iii) polysaccharides, (iv) sugars.
b) Using **D** as an example, explain the meaning of the terms D- and L- as applied to carbohydrates.
c) **A** is α-D-ribose, a pentose. **C** is β-D-glucose, a hexose. Explain the meanings of the terms: α, β, D-, pentose, hexose.
d) Draw the open-chain forms of **A** and **C** and identify the asymmetric carbon atoms which are present in each molecule.
e) Two bottles containing aqueous solutions of **B** and **C**, respectively, have lost their labels. How would you distinguish between them?
f) Sucrose is *disaccharide* in which a glucose molecule is joined to a fructose molecule by a *1,2-glycosidic bond* which is formed in a *condensation* reaction.
Explain the terms in italics, referring as necessary to the appropriate structures **A** to **G**.
g) Draw the structure of lactose, a disaccharide of glucose and galactose containing 1,4-glycosidic bonds. How might the results of enzymatic hydrolysis confirm that this is the structure of lactose?
h) Draw diagrams to illustrate the structures of the two forms of **E**, showing the different types of glycosidic bonds that are present.
i) Which can act as (a) a structural polymer and (b) a storage polymer in plants? Explain your answer in terms of molecular structure.

Extension questions on Chapter 12

9 In section 10.7, we saw that **Grignard reagents** are valuable in organic synthesis because they allow the formation of carbon–carbon bonds under mild laboratory conditions. The reagents have the general formula RMgHal where R is an alkyl group and Hal is Br or I. On reaction they generate the R^- nucleophile and this is able to attack the carbonyl bond in aldehydes and ketones, thus:

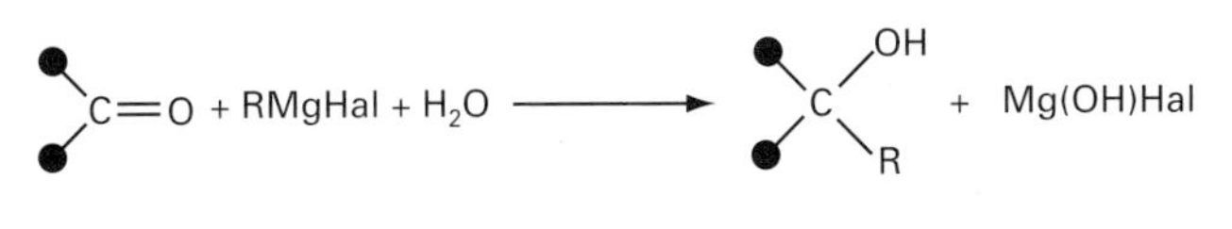

● = H or a hydrocarbon skeleton

Use this general scheme to work out which alcohols are produced by the following Grignard syntheses:

a) ethyl magnesium bromide with ethanal
b) methyl magnesium iodide with butanone
c) ethyl magnesium iodide with methanal

10 Nowadays, most perfumes consist of a mixture of synthetic aroma substances which smell just like the natural products. Cinnamon was the first natural aroma substance to be synthesised, by Luigi Chiozza in 1856. The main component in cinnamon is cinnamic aldehyde, or **cinnamaldehyde** for short.

a) Use the following information to deduce the structure of cinnamaldehyde and explain your reasoning.
i) Cinnamaldehyde contains 81.8% carbon, 6.1% hydrogen and 12.1% oxygen, by mass. It has a relative molecular mass of 132.
ii) A silver precipitate is obtained when cinnamaldehyde is treated with Tollen's reagent.
iii) A solution of bromine in hexane is decolourised when cinnamaldehyde is added.
iv) The mass spectrum shows the presence of fragments of relative masses 29, 55, 77 and 103.

b) There are two stereoisomers of cinnamaldehyde. Write down their structures and explain how they may be distinguished.

c) Cinnamaldehyde may be made by treating benzenecarbaldehyde with ethanal in alkaline solution. The mechanism involves three stages: i) the formation of a carbanion (a molecule with a negatively charged carbon atom) by the attack of an OH^- ion on an α-hydrogen; ii) nucleophilic addition and iii) elimination of a water molecule. Write down the mechanism for the reaction.

Comments on the activities

Activity 12.1

1 **a)** propanal **b)** pentanal
c) 3-methylbutanone **d)** 2-methylpropanal
e) 3-phenylpropanal **f)** methanal

2

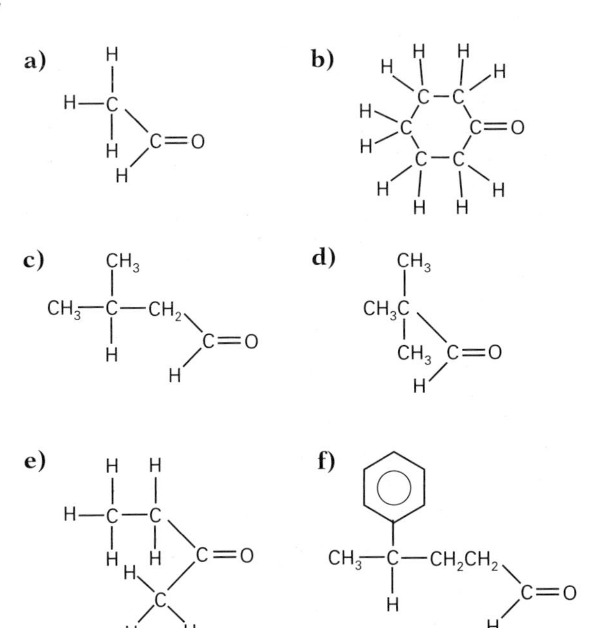

3 pentanal, 3-methylbutanone, 3-methylbutanal, dimethylpropanal are all $C_5H_{10}O$; 2-methylpropanal and butanone are both C_4H_8O.

Activity 12.2

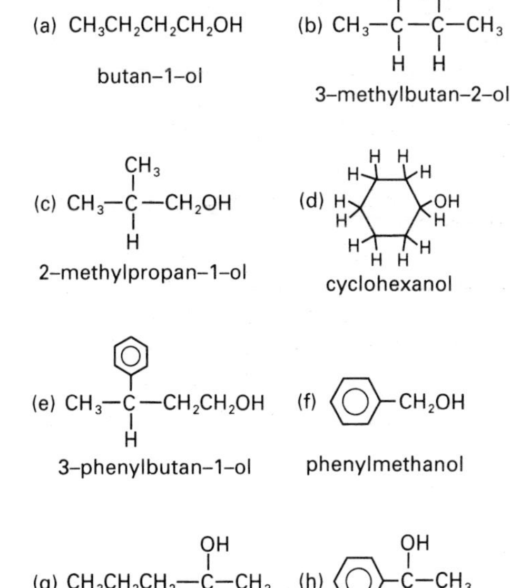

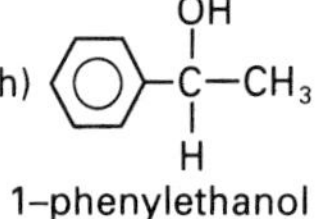

1-phenylethanol

Comments on the activities *continued*

Activity 12.3

1 **a)** 2-hydroxypropanonitrile $CH_3—C^*(OH)(H)—CN$

b) 2-hydroxy-2-methylpentanonitrile

$CH_3CH_2CH_2—C^*(OH)(CH_3)—CN$

c) 2-hydroxy-2-ethylbutanonitrile

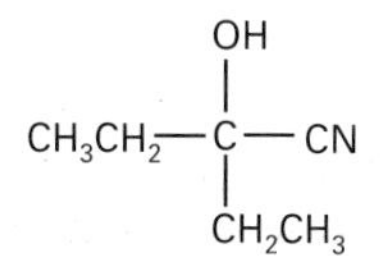

d) ethanal-2,4-dinitrophenylhydrazone

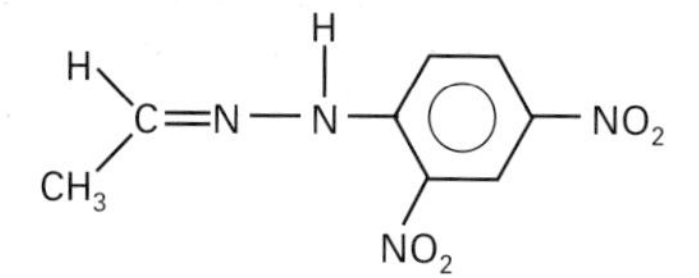

e) butanone-2,4-dinitrophenylhydrazone

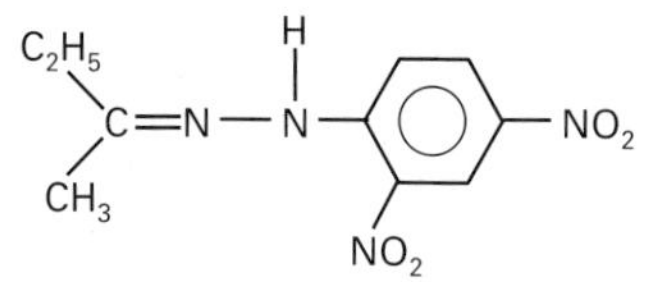

Optically active: **a)** and **b)**; each molecule of product contains a **chiral centre** due to the **asymmetric carbon atom, C***. Equal amounts of the D- and L-isomers will be formed giving a racemic mixture

2 **a)** excess KCN/dil.HCl(aq) at room temperature

b)

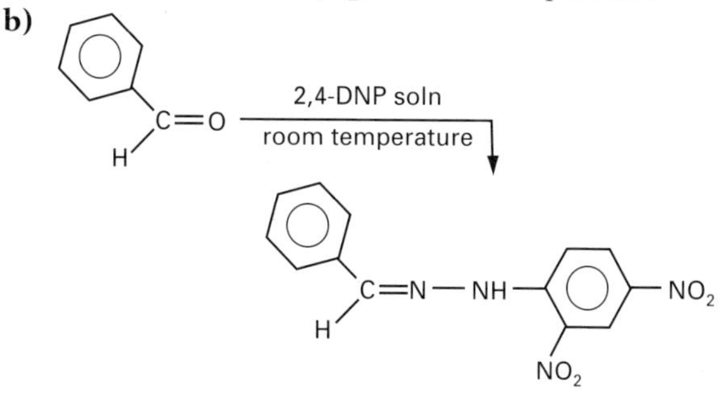

c)

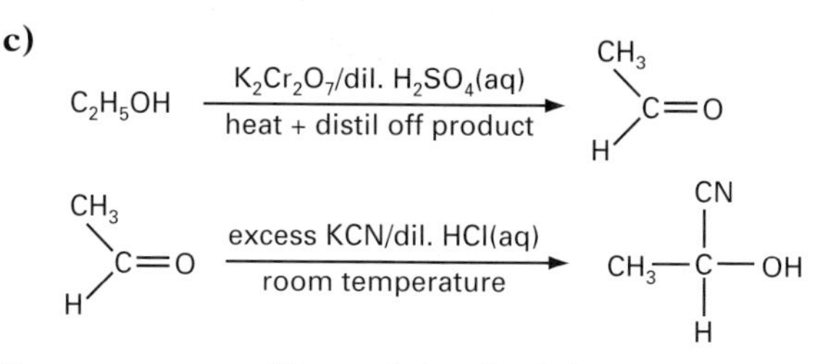

3 Prepare, recrystallise and dry the 2,4-dinitrophenylhydrazones of each compound. Take the melting points and check these in a book of data, e.g. butanal-2,4-dinitrophenylhydrazone m.p. 123°C; 2-methylpropanal-2,4-dinitrophenylhydrazone m.p. 187°C. Thus, the compounds can be *characterised* (identified).

Activity 12.4

1 **a)** The presence of *at least* one lone pair of electrons (the oxygen atom has two) which is used to form a co-ordinate (dative covalent) bond with an 'electron-poor' atom, for example the C^+ atom in a carbocation.

b) The C=C bond is an 'electron-rich' centre in a molecule and this would not attract a nucleophile which is itself an 'electron-rich' species. The C=C bond will be attacked by electrophiles, such as H^+ or Br^+ (section 8.5).

c) The Pd^{2+} ion bonds loosely to the C=C bond causing it to become polarised, thereby encouraging the attack of a nucleophile at the $C^{\delta+}$:

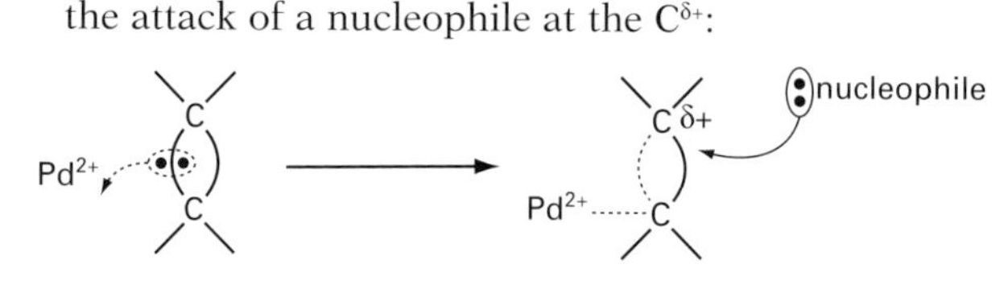

2 Propene

3 On adding benzoquinone to bromine water at room temperature, the orange colour of the bromine rapidly disappears (section 8.5).

Activity 12.5

1 **a)** ethanal, propanal, methanal; **b)** as (a).

2 **a)** 2-methylpropanoic acid $(CH_3)_2CHCOOH$

b) 3-methylbutan-2-ol $CH_3—CH(CH_3)—CH(CH_3)—OH$

Comments on the activities *continued*

c) 2-phenylethanol $C_6H_5CH_2CH_2OH$

d) 2-methylbutanoic acid, $CH_3CH_2CH(CH_3)COOH$

e) pentan-3-ol, $CH_3CH_2CH(OH)CH_2CH_3$

3 Positive iodoform test:

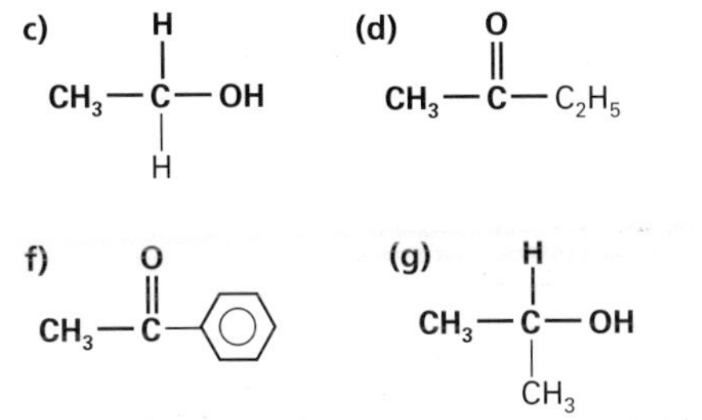

because they contain the required structural units (shown in bold print).

Activity 12.6

See Table 12.5 on the next page. In general aldehydes and ketones have similar chemical properties largely governed by the presence of the carbonyl bond. However, aldehydes are less resistant to oxidation.

Activity 12.7

2

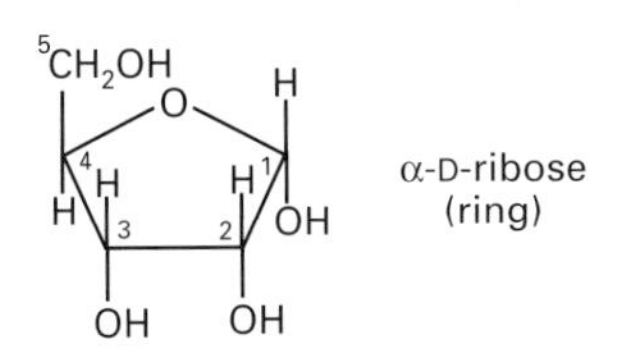

α-D-ribose (ring)

3, 4

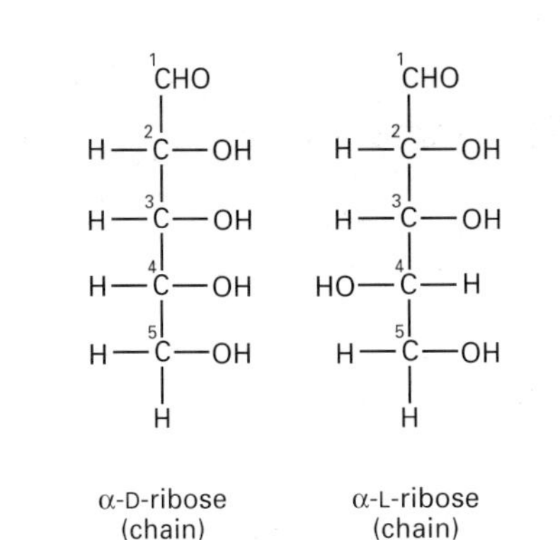

α-D-ribose (chain) α-L-ribose (chain)

5 Optical isomerism (see section 12.5).

6 a) The plane of polarisation would be rotated in different directions – one isomer would rotate it clockwise, the other anti-clockwise – though the angle of rotation would be the same.

b) A racemic mixture is formed in which the rotation caused by one isomer cancels out the rotation caused by the other – overall, the mixture would be optically inactive.

Activity 12.8

5

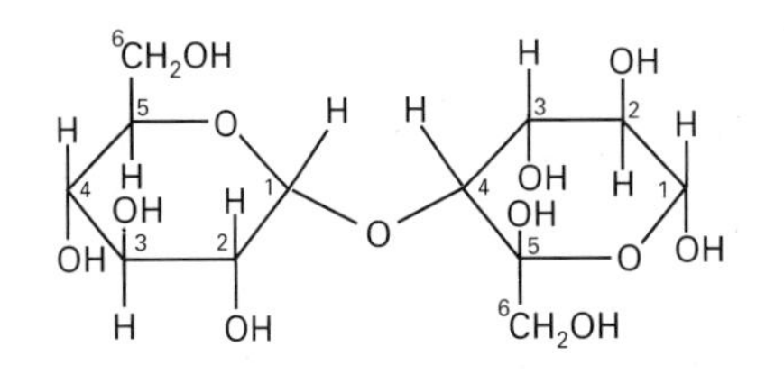

6 Yes.

7 No.

8 Since the galactose residue in lactose can exist in its chain form, a carbonyl group is available in the aldehyde group on C^1. The carbonyl group is easily oxidised (section 12.6), for example by Benedict's solution. Thus, lactose is said to be a reducing sugar.

Comments on the activities *continued*

Table 12.5 Comparing some important reactions of aldehydes and ketones

Reagent/conditions	aldehydes e.g. $CH_3(H)C{=}O$ ethanal	ketones e.g. $(CH_3)_2C{=}O$ propanone	notes
excess KCN/dil. HCl(aq) room temp.	**hydroxynitriles** are formed and the hydrocarbon chain is lengthened $\longrightarrow CH_3{-}C(H)(CN){-}OH$	or $CH_3{-}C(CH_3)(CN){-}OH$	(1)
$:NH_2{-}NH{-}C_6H_3(NO_2)_2$ solution, room temp.	addition–elimination reactions to give **2,4-dinitrophenylhydrazones** $\longrightarrow CH_3(H)C{=}N{-}NH{-}C_6H_3(NO_2)_2$	or $(CH_3)_2C{=}N{-}NH{-}C_6H_3(NO_2)_2$	(2)
$LiAlH_4$/ethoxyethane, reflux or $NaBH_4$/ethanol, reflux	reduction to an **alcohol** (aldehyde $\longrightarrow$ 1°, ketone $\longrightarrow$ 2°) $\longrightarrow CH_3CH_2OH$	$\longrightarrow CH_3CH(OH)CH_3$	
Tollen's reagent, warm	**silver 'mirror'** and a carboxylic acid are formed $\longrightarrow CH_3COOH + Ag(s)$	no reaction	(3)
Fehling's solution, warm	**red-brown $Cu_2O(s)$** and a carboxylic acid are formed $\longrightarrow CH_3COOH + Cu_2O(s)$	no reaction	(3)
$K_2Cr_2O_7$/dil. H_2SO_4(aq)	rapidly oxidised on warming to a **carboxylic acid** $\longrightarrow CH_3COOH$	refluxing for several hours with *concentrated* oxidising agent gives a **mixture of carboxylic acids** $\longrightarrow CH_3COOH + HCOOH$	
I_2(aq)/dil. NaOH(aq)	only ethanal and methyl ketones will react to form **iodoform**, $CHI_3(s)$		(4)

Notes:
(1) Which of these products will be formed as a racemic mixture?
(2) Why do we prepare these compounds?
(3) Why do ketones resist oxidation?
(4) What structural feature of the molecules is responsible for the reaction?

CHAPTER 13

Carboxylic acids and their main derivatives

Contents

Study Checklist

After studying this chapter you should be able to:

1. Recall that carboxylic acids contain the carboxyl group, —COOH.
2. Explain that, in carboxylic acid derivatives, the —OH group is replaced by —Hal (acid halides), —NH_2 (amides), —O—□ (esters) or —O—CO—□ (acid anhydrides), where □ is a hydrocarbon skeleton.
3. Interpret, and use, the names, short structural formulae and displayed formulae of monocarboxylic acids, including benzoic acid, and dicarboxylic acids. Know that the aliphatic compounds form homologous series of general formula $C_nH_{2n}O_2$ and describe the gradual changes in physical properties that occur as molecular mass is increased.
4. Recall that carboxylic acids have much higher boiling points and greater solubility in water than the alkanes of similar molecular mass. Using molecular diagrams, explain these properties in terms of hydrogen bonding between neighbouring molecules.
5. Give the reagents and reaction conditions needed to prepare: (a) aliphatic carboxylic acids from (i) primary alcohols, (ii) aldehydes and (iii) nitriles; (b) benzoic acid from methylbenzene. Know that carboxylic acids are manufactured by the catalytic oxidation of petroleum fractions and that ethanoic acid, as vinegar, is produced in large quantities by fermentation of sucrose.
6. Recall that there are three main types of reactions of carboxylic acids: (a) reduction of the C=O bond, (b) fission of the O—H bond which gives rise to the acidic character and (c) fission of the C—O bond resulting in the substitution of the —OH group.
7. State the reagents, and reaction conditions, for the conversion of carboxylic acids to: (a) primary alcohols, (b) salts of the s-block metals, (c) acid chlorides, (d) esters and (e) amides. Know that carboxylate ions may be identified by the red-brown solution they form on reaction with iron(III) chloride.
8. Describe, and explain, the acid- and base-catalysed hydrolysis of esters, outlining the mechanism for acid-catalysed route. State that esters are only sparingly soluble in water, often have 'fruity' smells and that they are used in perfumes and as flavourings.
9. State that carboxylic acids are weak acids. Explain the relative acidity of alcohols, phenols, carboxylic acids and chlorine-substituted ethanoic acids in terms of their structures.
10. Explain the relative ease of hydrolysis of acid chlorides, halogenoalkanes and halogenoarenes in terms of their structures (also see section 10.5).

11 State the reagents, and reaction conditions, for the interconversion of carboxylic acid derivatives (see 2 above), including the formation of N-substituted amides.

12 Give the meaning of the terms 'acylation' and 'acylating agent'. Explain why a mixture of ethanoic and its anhydride is normally used as an acylating agent in preference to ethanoyl chloride (e.g. in the manufacture of cellulose ethanoate).

13 State the reagents, and reaction conditions, for the conversion of acid amides into (a) nitriles, (b) amines having the *same* number of carbon atoms and (c) amines having *one less* carbon atom.

14 Write an outline account of the production of 'Terylene'. In your account you should be able to: (a) state the names and give the molecular structures of the raw materials, (b) explain the terms 'condensation polymerisation' and 'thermoplastic', (c) draw the structure of the repeating unit of the polymer, (d) explain how the strength of the fibres is increased by the processes of extrusion and cold-drawing, (e) give two uses of the polymer.

15 Write an account of the production of soaps and soapless detergents. In your account, you should be able to: (a) explain the meaning of the terms 'fatty acid' and 'triglyceride', (b) give the short molecular structure of a typical triglyceride which contains one unsaturated and two saturated fatty acid residues, (c) state the reagents, and reaction conditions, needed to convert this triglyceride into soap, and give the formulae of products, (d) explain how soaps remove dirt from fabrics, (e) know how soapless detergents are made from petroleum 'fractions' and describe the structure of a soapless detergent, (f) compare the advantages and disadvantages of soaps and soapless detergents.

16 Recall the existence of biodegradable plastics, such as poly(hydroxybutyrate/valerate), PHB-V.

17 Summarise the main reactions of carboxylic acids and know the uses of the reagents in tables 13.4 and 13.5.

Every time you drink orange juice, eat a pickled onion or annoy an ant, you will come into contact with a carboxylic acid! These organic acids contain the **carboxyl group**:

$$-C(=O)-O-H$$

As you can see, this consists of a carbonyl group, C$=$O, joined to a hydroxyl group, —OH. Later on, we shall see whether these groups behave in isolation or influence each other during physical and chemical changes. Firstly, let us look at the way we name these compounds.

Aboriginal honey ants

13.1 Naming carboxylic acids

Monocarboxylic acids contain one carboxyl group. To name them, work out the name of the hydrocarbon which has the same carbon chain, and then replace the final '**e**' by '**oic acid**'. For example,

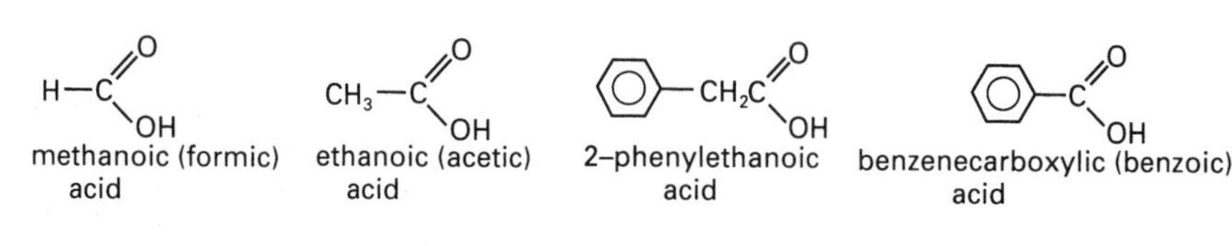

methanoic (formic) acid   ethanoic (acetic) acid   2–phenylethanoic acid   benzenecarboxylic (benzoic) acid

Of the common names shown in brackets, only benzoic acid will be used in this text. Dicarboxylic acids contain two carboxyl groups, for example:

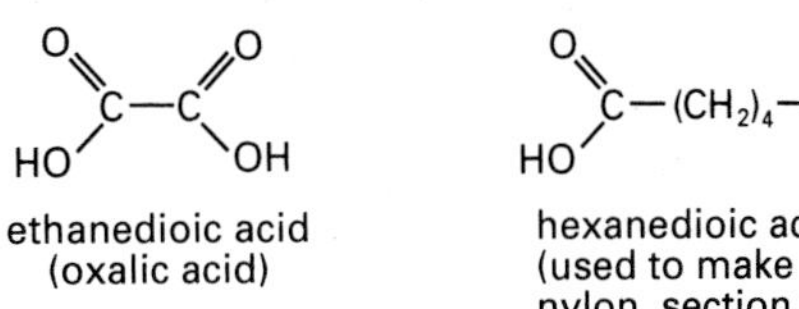
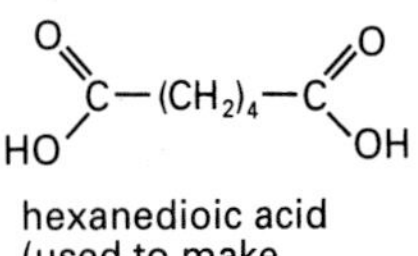
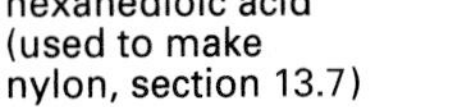
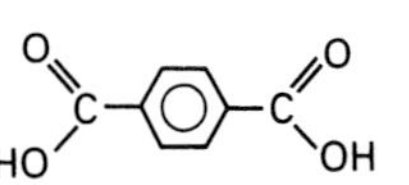

ethanedioic acid (oxalic acid)

hexanedioic acid (used to make nylon, section 13.7)

benzene–1, 4–dicarboxylic acid (used to make 'Terylene', a polyester, section 13.7)

Activity 13.1

1 Name the following carboxylic acids:

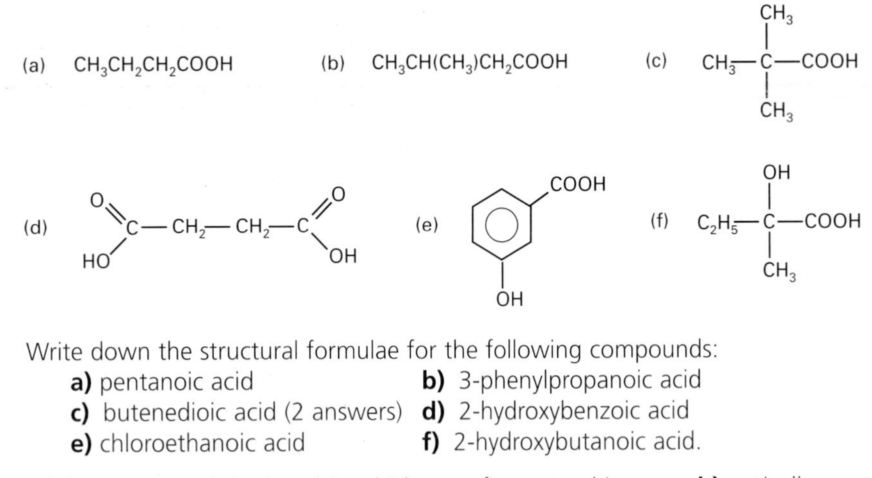

2 Write down the structural formulae for the following compounds:
a) pentanoic acid **b)** 3-phenylpropanoic acid
c) butenedioic acid (2 answers) **d)** 2-hydroxybenzoic acid
e) chloroethanoic acid **f)** 2-hydroxybutanoic acid.

3 Of the compounds in 1 and 2, which are: **a)** structural isomers; **b)** optically active (see section 6.5)?

13.2 Physical properties

Monocarboxylic acids of low molecular mass are pungent smelling, colourless liquids. As expected, the higher members are solids. Some properties of carboxylic acids are given in Table 13.1.

Table 13.1 Physical properties of some carboxylic acids

Name	formula	melting point/°C	boiling point/°C
methanoic acid	$HCOOH$	9	101
ethanoic acid	CH_3COOH	17	118
propanoic acid	CH_3CH_2COOH	–21	141
butanoic acid	$CH_3CH_2CH_2COOH$	–4	164
2-methylpropanoic acid	$(CH_3)_2CHCOOH$	–47	154
ethanedioic acid	$(COOH)_2$	157	189
benzoic acid	C_6H_5COOH	122	249

Although they have very similar relative molecular masses, butanoic acid is a liquid but ethanedioic acid is a solid (at 25°C). Can you explain why?

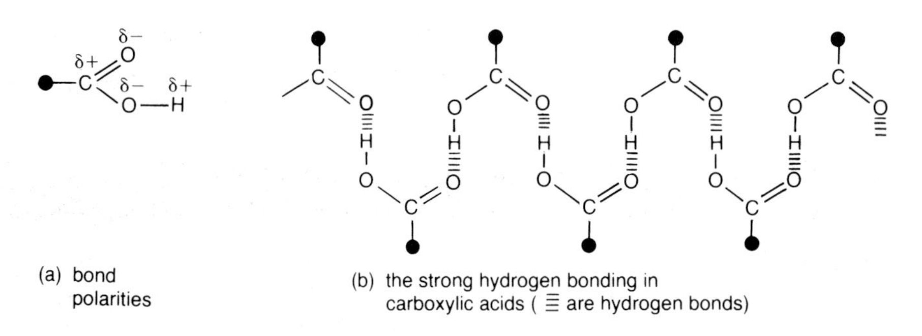

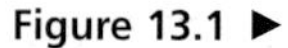
Figure 13.1 ▶
The polarity of the bonds in carboxylic acids and their ability to form hydrogen bonds with neighbouring molecules. (● = H atom or a hydrocarbon skeleton)

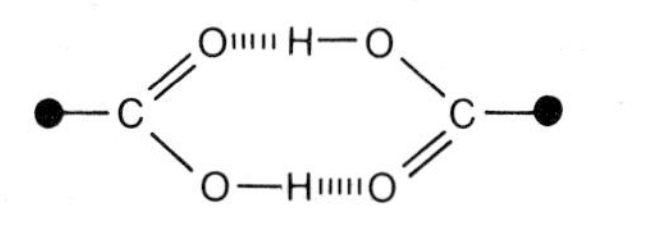

Figure 13.2 ▲
Carboxylic acids form dimers in the vapour state and when dissolved in organic solvents. As a result of this dimerisation, gas density measurements of the relative molecular mass of a carboxylic acid often produce a value which is about twice the actual mass

Aliphatic monocarboxylic acids form a homologous series of general formula $\mathbf{C_nH_{2n+1}COOH}$, and this is reflected by the steady increases in their boiling points. Electronegativity values suggest that the bonds in the carboxyl group are polar (Figure 13.1a). As a result, the acids form strong intermolecular hydrogen bonds (Figure 13.1b). Hence, their melting and boiling points are much higher than those of the alkanes of comparable molecular mass, for example ethanoic acid, $M_r = 60$, b.p. 118°C but butane $M_r = 58$, b.p. 0°C. Indeed, their great ability to form hydrogen bonds enables the lower acids to:

- dimerise in the vapour state or when dissolved in an organic solvent (Figure 13.2);
- dissolve readily in water and other polar solvents. As expected, their water solubility decreases as the hydrocarbon skeleton gets larger. Benzoic acid, C_6H_5COOH, for example, is only sparingly soluble in cold water, but it is very soluble in hot water.

Since dicarboxylic acids can form hydrogen bonds via both carboxyl groups, they exist as high melting points solids, for example ethanedioic acid melts at 157°C.

13.3 Making carboxylic acids

Manufacture

In general, carboxylic acids are manufactured by the catalytic oxidation of fractions from the distillation of petroleum, for example:

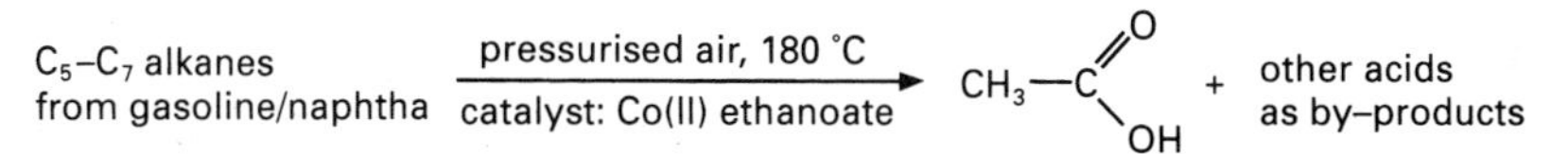

Here the main product is ethanoic acid, with methanoic and propanoic acids also obtained as useful by-products. Most ethanoic acid is converted into ethanoic anhydride, $(CH_3CO)_2O$, which as the name suggests is ethanoic acid (2 moles) less water (1 mole). Ethanoic anhydride is used to make aspirin and rayon, a synthetic fibre obtained from cellulose.

Ethanoic acid is also produced in large quantities as **vinegar** for use in sauces, pickles and as a condiment. Vinegar – the name comes from *'vin aigre'*, the French for sour wine – is made using two microbial processes. Firstly, sucrose, derived from fruit or grain, is fermented to ethanol in the absence of air, using a yeast called *Saccharomyces cerevisiae*. Then the ethanol is exposed to air in the presence of *Acetobacter*, an aerobic bacteria, and vinegar is obtained as an aqueous solution containing from 4 to 10% by volume of ethanoic acid. The vinegar's flavour depends on the source of the ethanol, for example from wine or malt, and it is sometimes flavoured with herbs or fruit extracts.

A number of carboxylic acids which have long hydrocarbon chains are known as **fatty acids**. Many natural fats and oils are triesters of these fatty acids and propane-1,2,3-triol (known as glycerol), section 13.7.

Benzoic acid is manufactured by oxidising methylbenzene obtained from the catalytic reforming of naphtha, section 7.4. Benzoic acid is mainly used as a preservative in canned fruit and drinks.

Linoleic acid – an essential fatty acid

Of all the fatty acids which are present as esters in fats and oils, only one, **linoleic acid**, is essential for human life. Linoleic acid is an unsaturated fatty acid containing 18 carbon atoms with double bonds at carbons 9 and 12; thus, it is also known as 9,12-octadecadienoic acid:

COOH

linoleic acid – note the *cis,cis* arrangement of the double bonds

A high content of linoleic acid is found in margarine and many vegetable oils. Our bodies convert linoleic acid into a series of other compounds, notably arachidonic acid which is a precursor for the production of **prostaglandins**. Found in glands and the liver, prostaglandins control the actions of certain hormones and nerve transmitters affecting, for example, blood pressure, ulcer formation, the smoothness of muscle contraction and the clotting ability of blood.

Laboratory preparation

Various methods are available, including:

- oxidation of a primary alcohol, or aldehyde, by refluxing with acidified aqueous dichromate(VI) or manganate(VII) ions, for example:

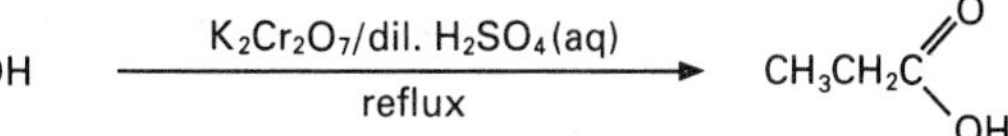

- hydrolysis of a nitrile by refluxing with dilute sulphuric acid, for example:

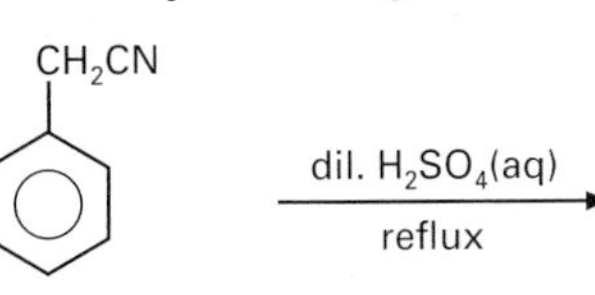

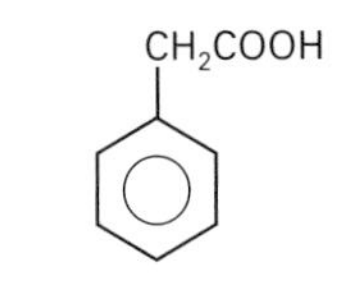

phenylethanonitrile phenylethanoic acid

We saw earlier that nitriles are prepared by refluxing a halogenoalkane with potassium cyanide in ethanol, thereby lengthening the molecule's hydrocarbon skeleton, section 10.5.

- oxidation of methylbenzene to give benzoic acid:

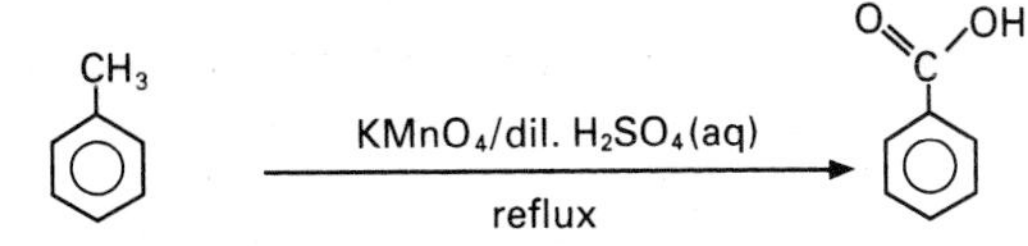

Activity 13.2

State the reagents and reaction conditions which are used in the laboratory synthesis of propanoic acid from:

1. propan-1-ol
2. propanonitrile
3. propanal
4. bromoethane (2 steps)
5. bromoethane (2 steps, using a Grignard reagent, see study question 12.8).

Focus 13a

1. Carboxylic acids contain the **carboxyl group, —COOH**.
2. To name the aliphatic carboxylic acid, name the hydrocarbon with the same carbon chain, and then replace the final '**e**' by '**oic acid**'. The simplest aromatic carboxylic acid is benzoic acid, $C_6H_5COOH(s)$.
3. Aliphatic carboxylic acids form a homologous series of general formula $\mathbf{C_nH_{2n+1}COOH}$. There is a steady rise in boiling points as the hydrocarbon chain length increases.
4. Because of the polarisation in the COOH group the acids form strong intermolecular **hydrogen bonds**. Hence, (i) they dimerise when dissolved in an organic solvent and (ii) the lower acids are very soluble in water.
5. Aliphatic carboxylic acids are prepared by (i) oxidising a primary alcohol with hot $K_2Cr_2O_7$/dil.$H_2SO_4(aq)$ or (ii) hydrolysing a nitrile with hot dil. $H_2SO_4(aq)$. Benzoic acid is prepared by oxidising methylbenzene with hot acidified $KMnO_4(aq)$.

13.4 Reactivity of carboxylic acids

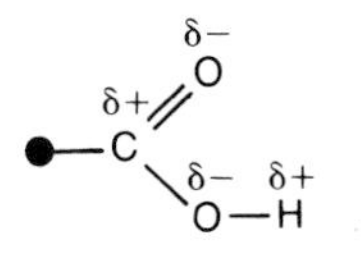

Figure 13.3
General structure of a carboxylic acid
●— is a hydrogen atom or hydrocarbon skeleton

Looking at the general structure of a carboxylic acid in Figure 13.3 we can identify the three main reactive centres:

- the **carbonyl bond, C=O**, which, like that in aldehydes and ketones, may give **nucleophilic addition** reactions:

- the **O—H bond**, fission of which gives rise to the **acidic character** of the carboxyl group and the formation of ionic salts:

●—C(=O)—O(δ−)—H(δ+) + :Base ⟶ ●—C(=O)—O⁻ + [H :Base]⁺

A similar process occurs in alcohols and phenols.

- the **C—O bond**, which may break heterolytically during the **substitution of the —OH group**:

Once again, alcohols take part in similar substitution reactions.

It is interesting to compare the chemistry of the carboxyl group with that of the C=O group in aldehydes/ketones and the O—H bond in alcohols. Try to keep this in mind as you work through the following sections.

13.5 Reactions of carboxylic acids

Nucleophilic addition to the C=O bond

In section 12.5, we saw that the carbonyl bond in aldehydes and ketones readily adds nucleophiles, such as HCN and 2,4-dinitrophenyl hydrazine. However, when carboxylic acids are treated with these reagents there are no addition reactions. So, why is the C=O bond in carboxylic acids resistant to nucleophilic addition?

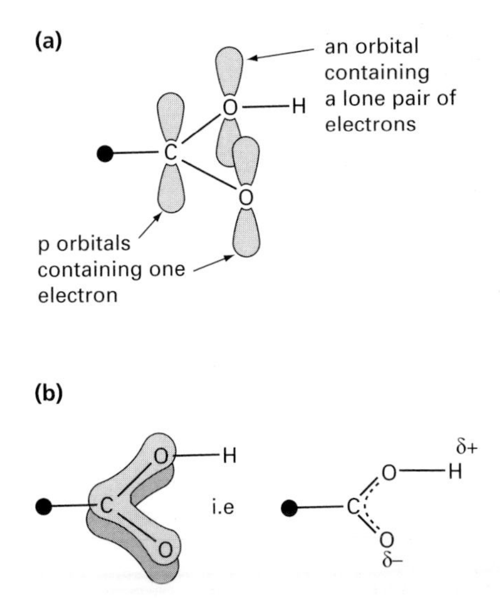

Figure 13.4
Some electron delocalisation occurs in carboxylic acids. Compare this electronic structure with that of the localised C=O bond in aldehydes and ketones, section 12.4

Figure 13.4a shows the σ bond framework of the carboxyl group, together with the p orbitals which are available for π bonding. Sideways overlap of these p orbitals produces a **delocalised π bond system** (Figure 13.4b). A molecule, or ion, becomes more energetically stable if it contains delocalised electrons, for example as in benzene, section 9.1. An addition reaction would remove the electron delocalisation in the carboxyl group, thereby destabilising the addition product. Hence, carboxylic acids resist addition to their C=O bond. Aldehydes and ketones, on the other hand, contain a localised double bond and this readily adds other molecules.

In fact, carboxylic acids will give one addition reaction: **reduction.** Hydrogen is added by refluxing the acid with lithium tetrahydridoaluminate(III) in ethoxyethane, for example:

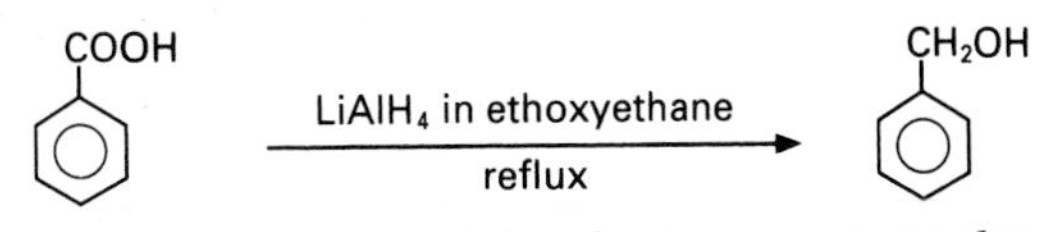

This reaction requires a very powerful reducing agent, $LiAlH_4$; it will not work, for example, with $NaBH_4$.

Fission of the O—H bond: behaviour as acids

Three types of organic hydroxy compounds have an acidic nature and some examples are shown below:

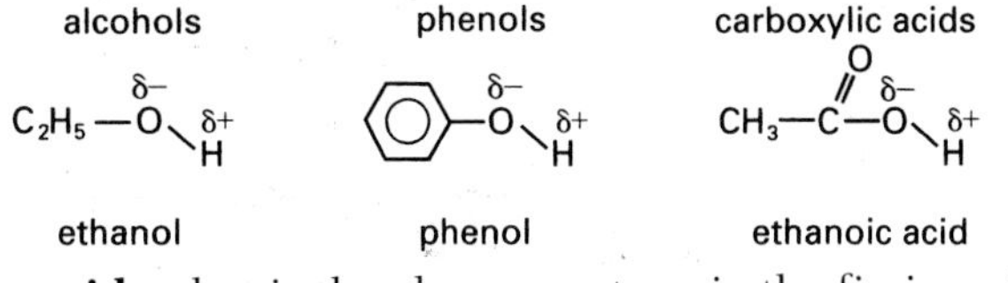

They are all **oxoacids**, that is they lose a proton via the fission of an O—H bond. By comparing their acid strength, therefore, we should be able to see how the acidic nature of the O—H group is affected by its molecular location. A similar comparison of the acidic strengths of alcohols and phenols was carried out in section 11.6. You will find it helpful to scan through that section before carrying on.

Carboxylic acids are weak acids, dissociating only slightly in aqueous solution (● = H atom or a hydrocarbon skeleton):

A mixture of reactants and products, at constant concentrations, is obtained. In this particular **equilibrium** mixture, the concentrations of the ions (right hand side) will be much lower than the concentrations of the undissociated molecules (left hand side). For example, in 0.1 mol dm^{-3} ethanoic acid, only about 1.3% of acid molecules dissociate into ions.

Carboxylic acids, like phenols, to some extent react with metals and alkalis to give **ionic** salts, for example:

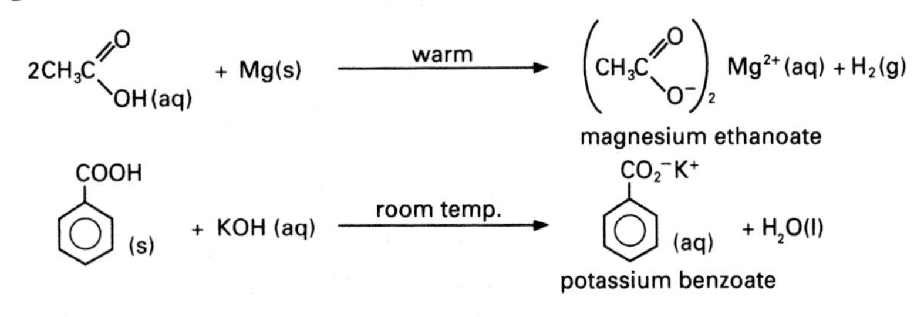

However, carboxylic acids are many thousands of times stronger acids than phenols, and this is shown by their ability to liberate carbon dioxide from a hydrogencarbonate (a weak base):

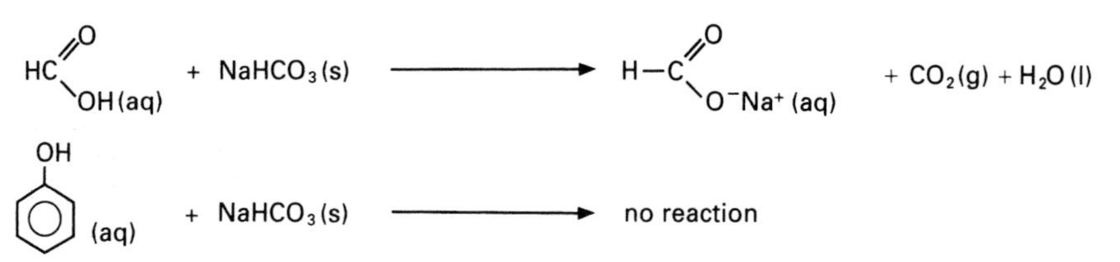

Indeed, this reaction may be used to distinguish carboxylic acids from phenols.

This behaviour suggests that the acidic strength increases in the order:

alcohols **phenols** **carboxylic acids**

→

increasing acid strength

An oxoacid, X—O—H, dissociates in aqueous solution:

$$X—O—H(aq) + H_2O(l) \rightleftharpoons X—O^-(aq) + H_3O^+(aq)$$

The *stronger* the acid, the *greater* the degree of dissociation, that is the greater the concentration of the oxoanions, the X—O⁻, and hydroxonium ions, H_3O^+. This is more likely to happen if the negative charge on the oxoanion, X—O⁻, is delocalised, thereby increasing its stability and encouraging the loss of an H^+ ion. In the phenol and ethanoic acid molecules, the X group has the ability to pull electron cloud towards itself:

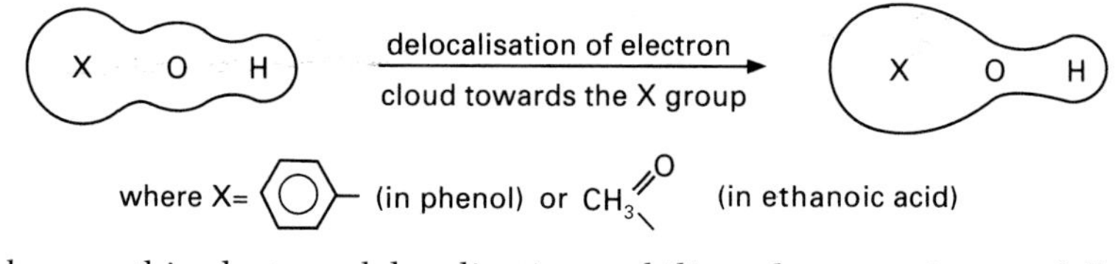

In each case, this electron delocalisation stabilises the oxoanion, as follows:

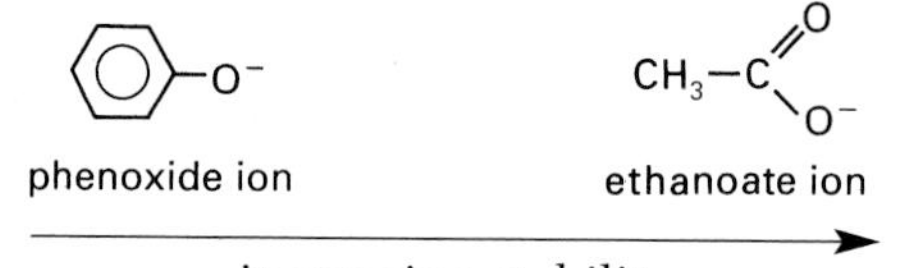

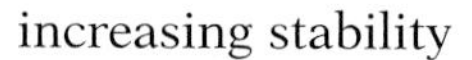

increasing stability

The electron delocalisation in each oxoanion spreads out its negative charge, Figure 13.5. However, we find that the greatest oxoanion stability (i.e. resistance to protonation) is achieved when there are more oxygen atoms to share the negative charge, due to the high electronegativity of oxygen. Thus, the trend in acid strength is directly related to stability of the oxoanion – the ethanoate ion is more stable and ethanoic acid is the stronger acid.

Ethyl groups, like other alkyl groups, have a positive **inductive effect**, that is they push electron density away from themselves. This localises negative charge on the oxygen of the ethoxide ion, making it easier for it to be protonated. Thus, the equilibrium

$$C_2H_5OH(aq) + H_2O(l) \rightleftharpoons C_2H_5O^-(aq) + H_3O^+(aq)$$

consists almost entirely of undissociated molecules and ethanol is an extremely weak acid.

A similar approach can be used to explain how the strength of a carboxylic acid, Y—COOH, depends on the nature of the Y group.

As the electron-attracting power of Y increases the acids get progressively stronger. For example, the strengths of the monohalogenated ethanoic acids increase as the halogen atom becomes more electronegative.

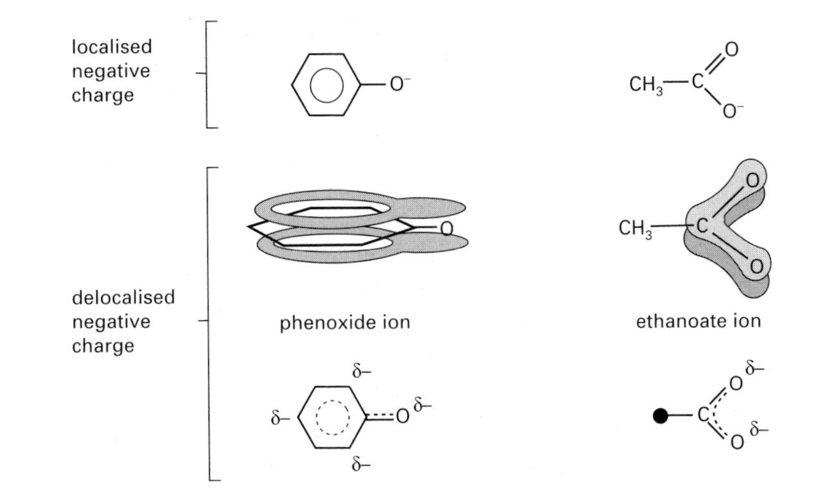

Figure 13.5
Delocalisation of the negative charge on the oxygen stabilises the phenoxide and ethanoate anions

$$\text{H–CH}_2\text{COOH} \quad \text{I–CH}_2\text{COOH} \quad \text{Br–CH}_2\text{COOH} \quad \text{Cl–CH}_2\text{COOH}$$

increasing acid strength

Similarly, the electron-attracting power of the Y group increases as more halogen atoms are attached:

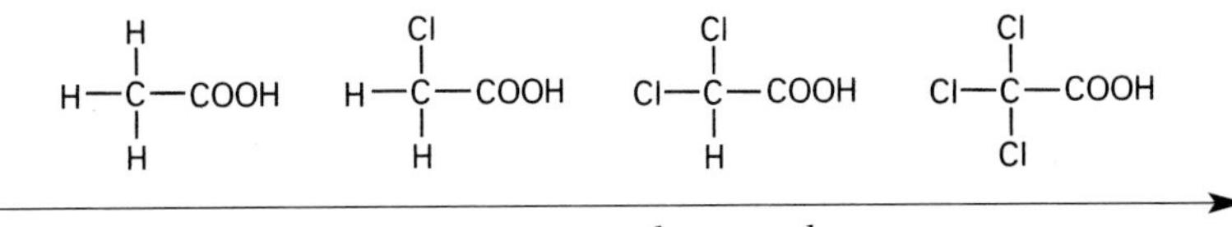

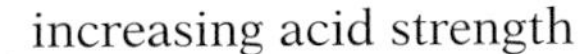

increasing acid strength

Increasing the electron-attracting power of the Y group stabilises the resulting oxoanion, thereby assisting the loss of a proton, thus there is a corresponding increase in acid strength.

However, if the Y group has an electron-releasing effect, the O—H bond will be strengthened and the acid oxoanion will be encouraged to accept a proton. Thus, the acid will dissociate into ions to a lesser extent, that is it will be weaker. For example, the acid strength decreases as the alkyl chain is lengthened:

$$\text{H–COOH} \quad \text{CH}_3\text{–COOH} \quad \text{CH}_3\text{CH}_2\text{–COOH}$$

increasing acid strength

Fission of the C—O bond

Halogenation

The hydroxyl groups in a carboxylic acid and an alcohol have similar reactions with halogenating agents. Thus, carboxylic acids may be halogenated to give **acid halides**:

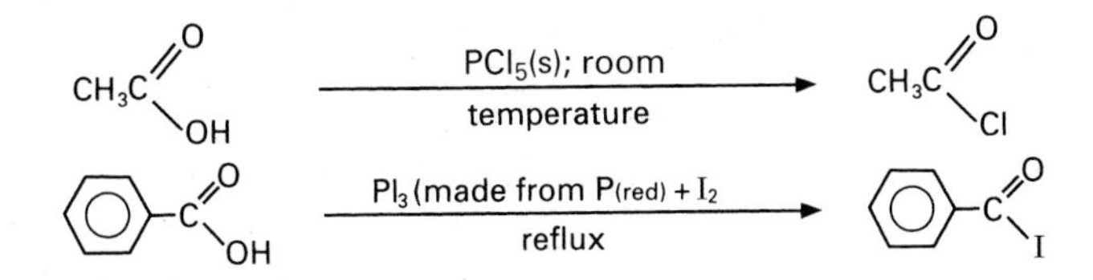

Table 11.7 lists some other halogenating agents used to convert carboxylic acids into their halides.

Esterification

Carboxylic acids react with alcohols in the presence of H^+ ions, from concentrated sulphuric acid or dry hydrogen chloride gas, to produce an ester and water, for example:

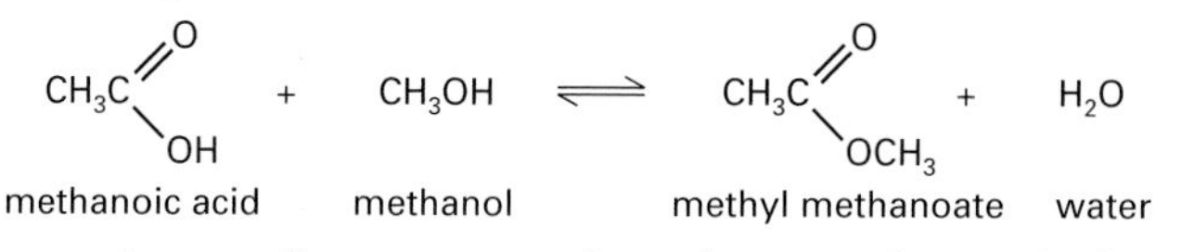

An equilibrium mixture of reactants and products is obtained after refluxing for a few hours, whereas without the H^+ ions to catalyse it, the reaction would take several days. Also since concentrated sulphuric acid has a high affinity for water, it reacts with the water formed, thereby pushing the reaction to the right and increasing the yield of the ester. Isotopic studies show that the reaction mechanism involves the replacement of the methanoic acid's hydroxy group, —OH, by a methoxy group,—OCH_3, from methanol (section 11.6).

Esters are volatile compounds which are slightly soluble in water. They often have characteristic sweet, 'fruity' smells and are used to make perfumes and to flavour foods. Ethyl ethanoate is widely used as a solvent, particularly in the manufacture of glues.

Activity 13.3

1 Arrange the following compounds in order of increasing acid strength:
 a) CH_3COOH, C_2H_5COOH, $HCOOH$
 b) $ClCH_2CH_2COOH$, $ClCH_2COOH$, $ClCH_2CH_2CH_2COOH$
 c) 2-iodopropanoic acid, 2-methylpropanoic acid, propanoic acid, 2-bromopropanoic acid, 2-chloropropanoic acid
 d) ethanoic acid, trichloroethanoic acid, phenol, benzoic acid, ethanol, bromoethanoic acid

2 Write balanced chemical equations, with state symbols, for the following aqueous reactions and name the salt which is formed:
 a) ethanoic acid + zinc metal
 b) methanoic acid + potassium hydroxide
 c) benzoic acid + calcium hydroxide
 d) ethanoic acid + potassium hydrogencarbonate

3 Give the names and structural formula of the organic product formed in each of the following reactions.

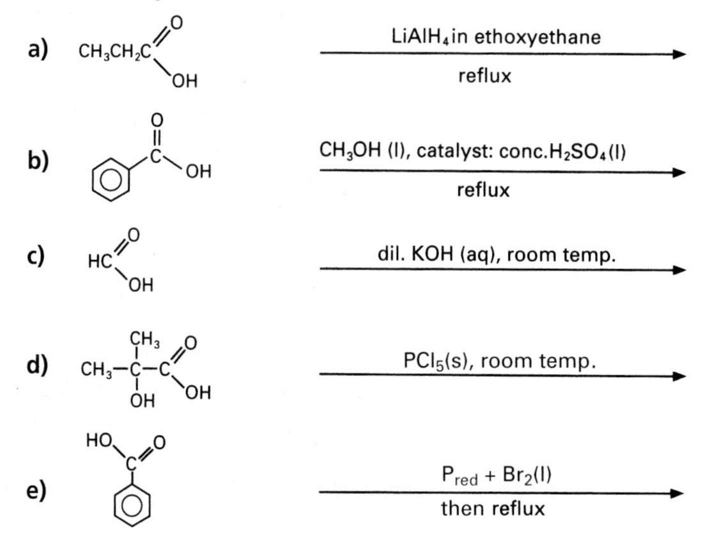

Activity 13.3 continued

4 Give the reagents and reaction conditions needed for the following conversions:

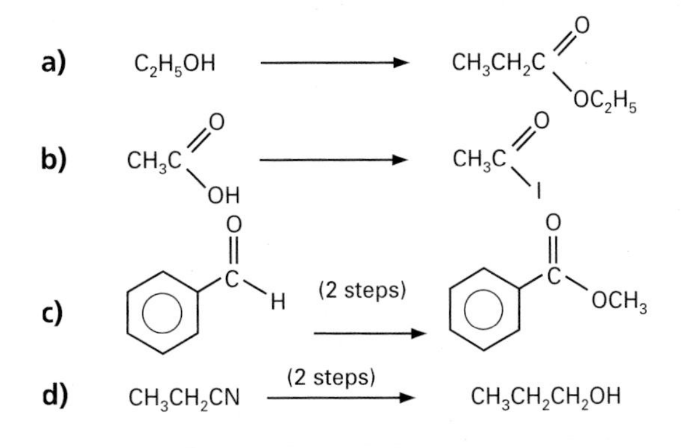

Parts (e) to (g) cover the material in sections 13.3–13.5 and earlier chapters.

e) $CH_3CH_2COOH \xrightarrow{\text{(3 steps)}} CH_3CHBrCH_2Br$

f) $CH_3CH_2Cl \xrightarrow{\text{(3 steps)}} ClCH_2COOH$

g) $CH_3CH{=}CH_2 \xrightarrow{\text{(3 steps)}} CH_3CH(COOH)CH_3$

Focus 13b

1 Carboxylic acids may take part in three main types of reaction:

- **nucleophilic addition** to the C$=$O bonds; this is uncommon, one example being reduction (attack by H^- ions).
- as an **acid**, via heterolytic fission of the O—H bond
- **substitution of the —OH group** by a nucleophile.

Some important examples of these reactions are given in Table 13.4, page 356. Also remember that the hydrocarbon skeleton will display its own typical reactions.

2 The C$=$O bond in a carboxylic acid resists addition reactions because these would remove the stability resulting from the partial delocalisation of the π electron cloud over the COO^- anion (Figure 13.4b). However, the localised C$=$O bond in aldehyde and ketones readily undergoes nucleophilic addition, for example with HCN and H_2.

3 Oxoacids dissociate in aqueous solution:

$$XO{-}H(aq) + H_2O(l) \rightleftharpoons H_3O^+(aq) + XO^-(aq)$$

The oxoacids become stronger, that is they dissociate to a greater extent, as the X group becomes more able to draw electron density towards itself. Thus, we observe the pattern:

increasing acid strength →

alcohols	**water**	**phenols**	**carboxylic acids**
(X = alkyl group)	**(X = H)**	**(X = –C₆H₅ ring)**	**(X = –C(=O)–)**

The strength of the carboxylic acid Y— COOH also increases as the electron withdrawing power of Y is increased, for example as more halogen atoms are added.

4 Carboxylic acids react with metals and alkalis to form ionic salts.

Table 13.2 Naming carboxylic acid derivatives

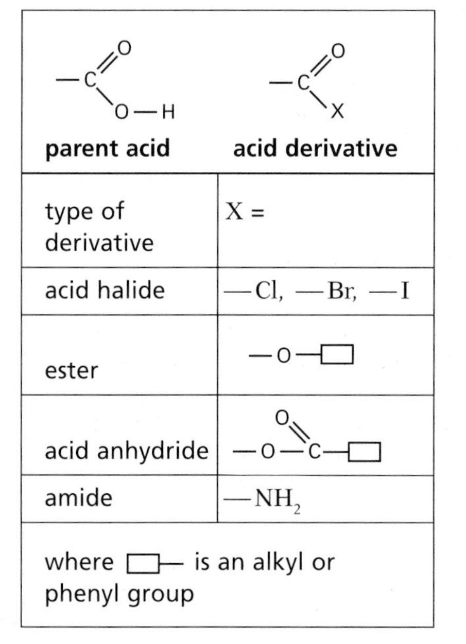

parent acid	acid derivative
type of derivative	X =
acid halide	—Cl, —Br, —I
ester	—O—☐
acid anhydride	—O—C(=O)—☐
amide	$—NH_2$
where ☐— is an alkyl or phenyl group	

13.6 Carboxylic acid derivatives

Carboxylic acid derivatives may be thought of as a parent acid in which the hydroxy group has been replaced by another functional group, as shown in Table 13.2.

Naming acid derivatives

Acid halides

Name the parent acid and replace '**oic acid**' by '**oyl halide**', for example:

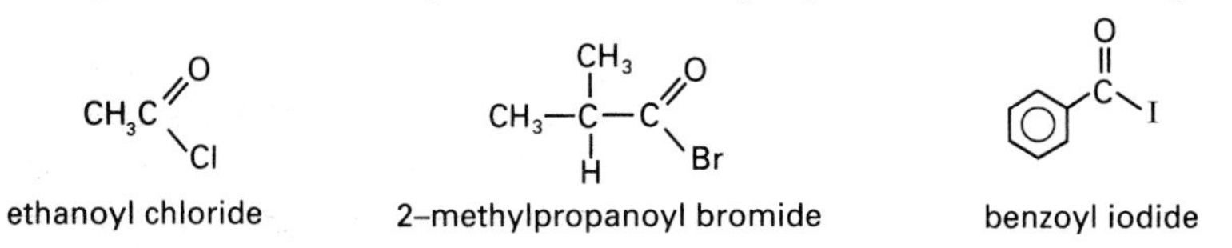

Esters

Esters are named as if they are **alkyl** or **phenyl** '**salts**' of the parent acid, for example:

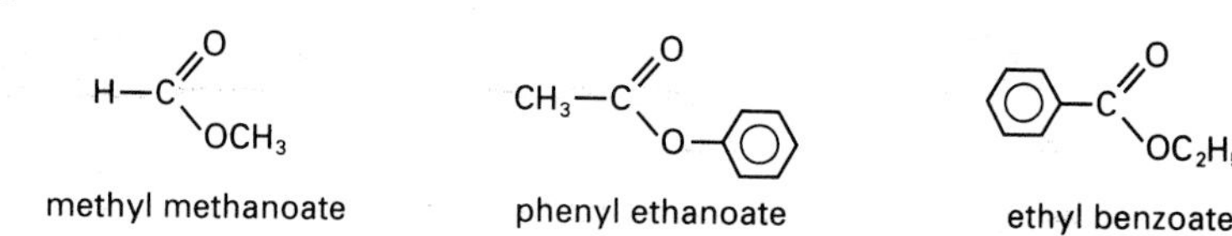

Acid anhydrides

Anhydrides consist of two carboxylic acid molecules which have been linked together with the loss of a water molecule. Thus, we just place the name(s) of the parent acid(s) in front of the word **anhydride**, for example:

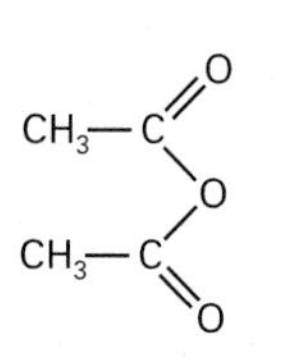

ethanoic anhydride

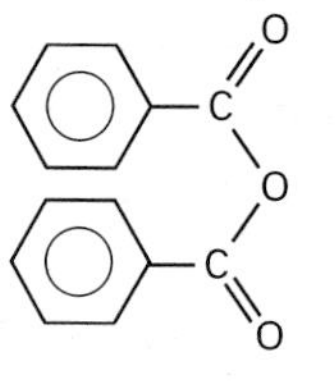

benzoic anhydride

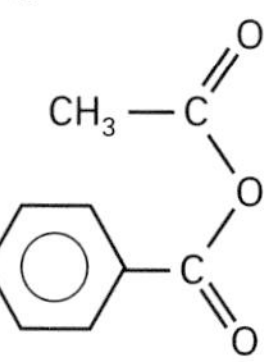

ethanoic benzoic anhydride

Acid amides

Name the parent acid and replace '**oic acid**' by '**amide**', for example:

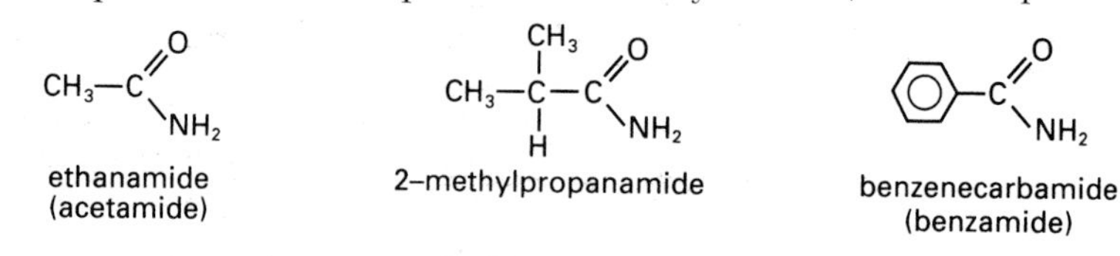

Physical properties

Whilst we need not discuss these in detail, certain trends are worth noting. The lower mass esters, halides and anhydrides are all fairly volatile liquids. They have strong odours, many esters having well-known 'fruity' smells.

For compounds of similar molecular mass, we find that:

esters/halides — anhydrides — acids — amides →
boiling points increase

This is explained by a corresponding increase in the strengths of the intermolecular attractions. Acid amides, for example, form strong intermolecular hydrogen bonds, Figure 13.6.

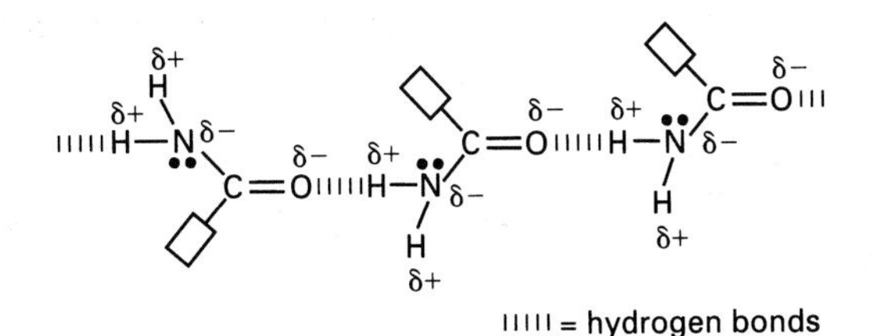

Figure 13.6
Intermolecular hydrogen bonding in amides

Consequently, apart from methanamide, which is a liquid, the amides are white crystalline solids of fairly high melting point. When purified, an amide can often be identified from its sharp melting point.

The water solubility of the aliphatic acid derivatives, of low molecular mass, increases with the strength of their intermolecular attraction with water molecules, i.e.

esters (insoluble)	anhydrides	acids	amides (very soluble)

→ increasing solubility in water

Thus, esters are insoluble because they are unable to hydrogen bond with water whereas amides do so readily and this makes them very soluble in water.

Reaction of acid derivatives

Acid halides are easily prepared by treating the carboxylic acid with a halogenating agent (Table 11.7), such as phosphorus pentachloride, for example:

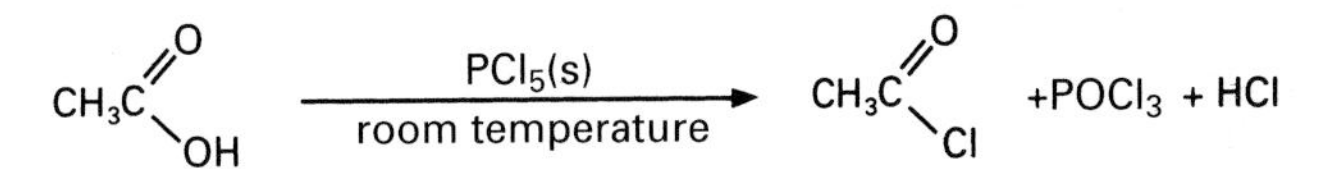

The other acid derivatives are conveniently prepared from acid chlorides, as shown in Figure 13.7. These conversions occur at very different rates. For example, ethanoyl chloride is vigorously hydrolysed by cold water at room temperature,

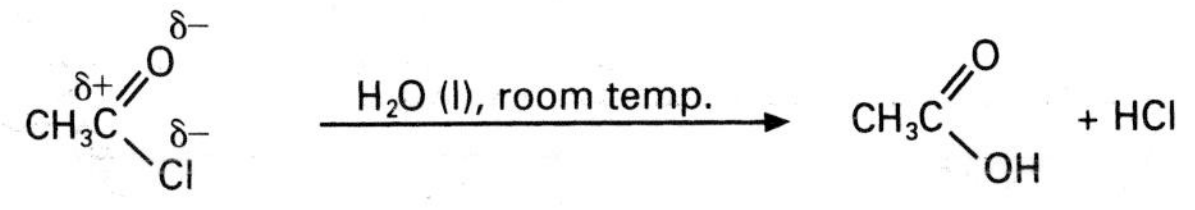

because of the polarity of the —COCl group. Under similar conditions, chloroalkanes hydrolyse very slowly and chlorobenzene is unreactive, section 10.5. Acid chlorides also react with:

- ammonia to form **amides**, for example:

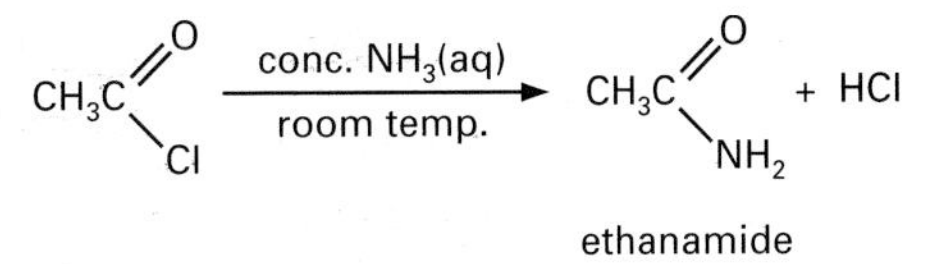

- primary amines to form '**nitrogen-substituted' amides**, section 14.7, for example:

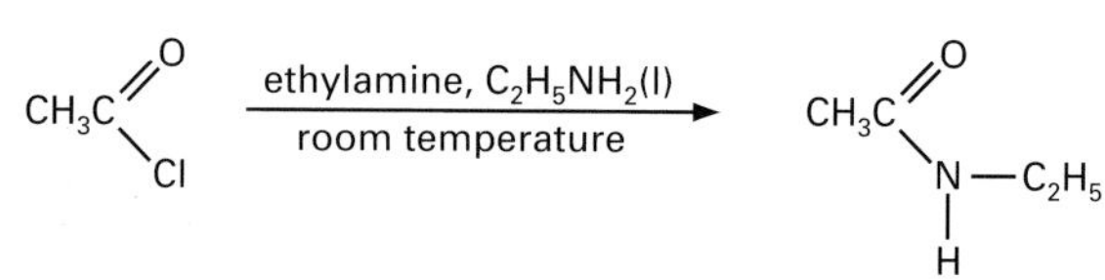

- alcohols and phenols, to form **esters**:

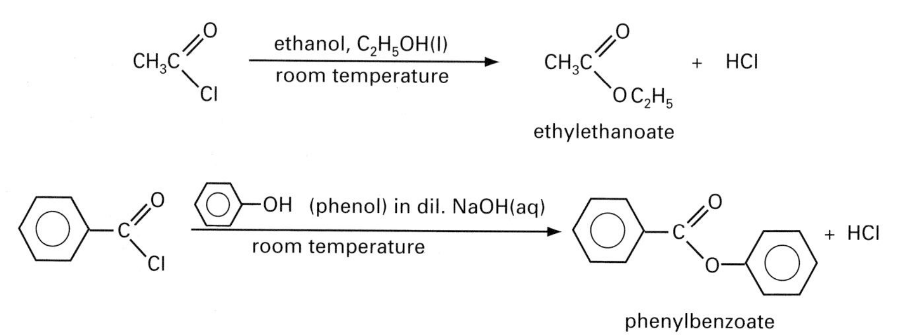

Note that phenol is a much weaker nucleophile than ethanol, so the formation of phenyl esters from acid chlorides is much slower and a base catalyst is required.

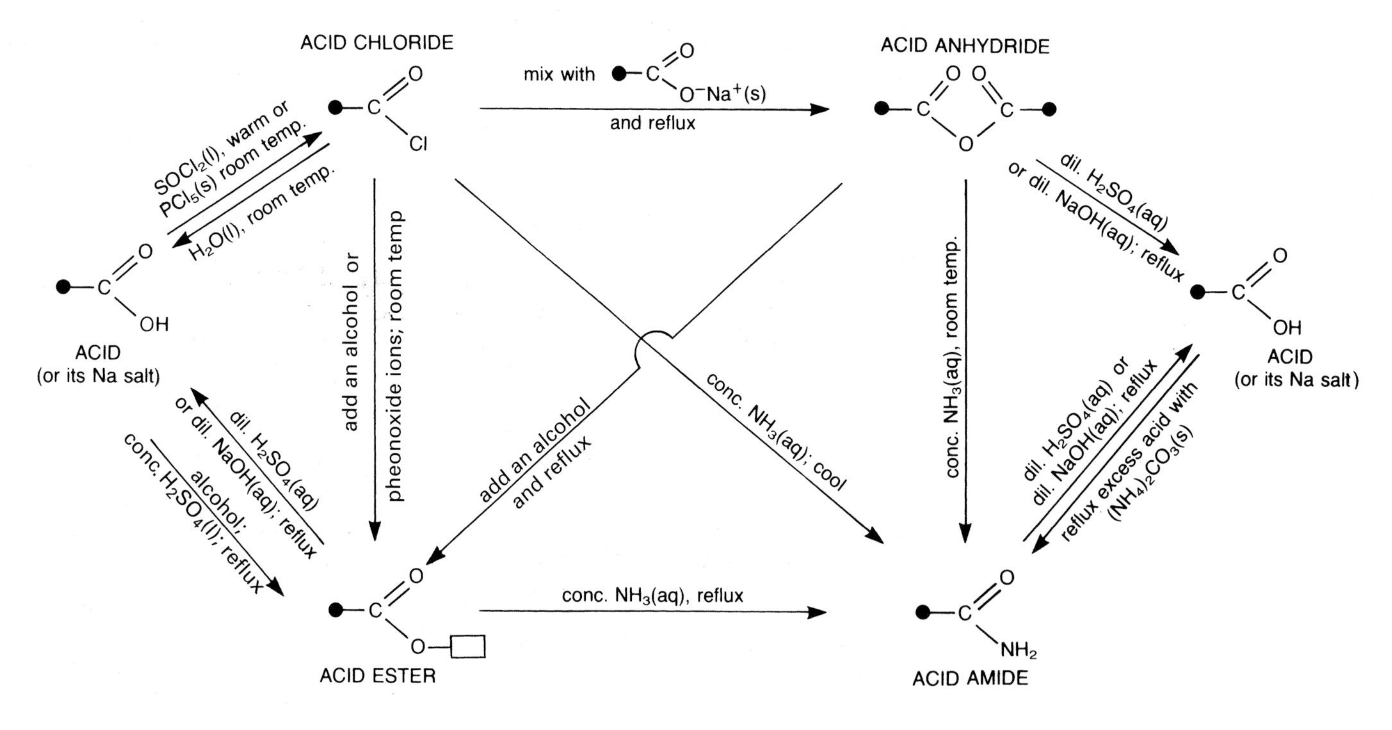

Figure 13.7
Acid derivatives take part in many nucleophilic substitution reactions.
Here the ☐— symbol is a hydrocarbon skeleton; ● is a hydrogen atom or ☐—

Compared to acid chlorides, acid anhydrides and esters hydrolyse slowly – a catalyst of $H^+(aq)$ or $OH^-(aq)$ ions is needed and the reaction mixture must be refluxed, for example:

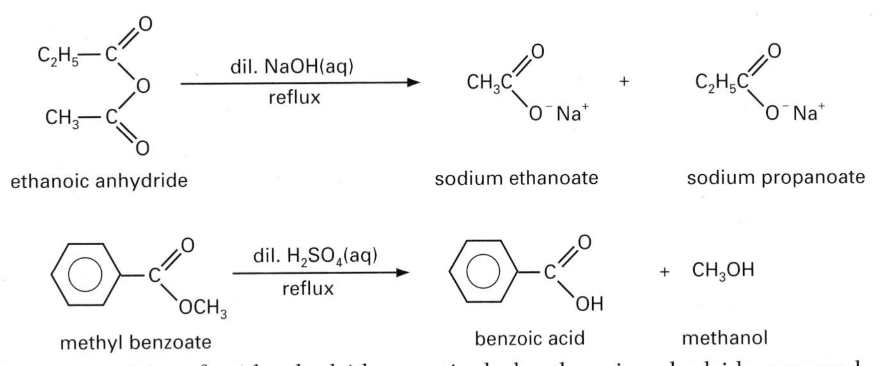

Large quantities of acid anhydrides, particularly ethanoic anhydride, are used industrially as **acylating agents**, that is to introduce the acyl group, □—C(=O)—, into another molecule (□— is an alkyl group). Acid anhydrides are preferred to acid chlorides for industrial acylation processes because the chlorides tend to be too reactive for safe operation and they are also more expensive. Of major importance is the manufacture of **cellulose ethanoate**, produced by acylating cellulose with a mixture of ethanoic acid and ethanoic anhydride. Cellulose ethanoate is used to make films, lacquers, varnishes and a synthetic material called rayon.

The conversions in Figure 13.7 are substitution reactions which occur in two stages. Consider, for example, the alkaline hydrolysis of a methyl ester. Firstly, the nucleophile, OH^-, attacks the carbonyl carbon (this bears a slight positive charge):

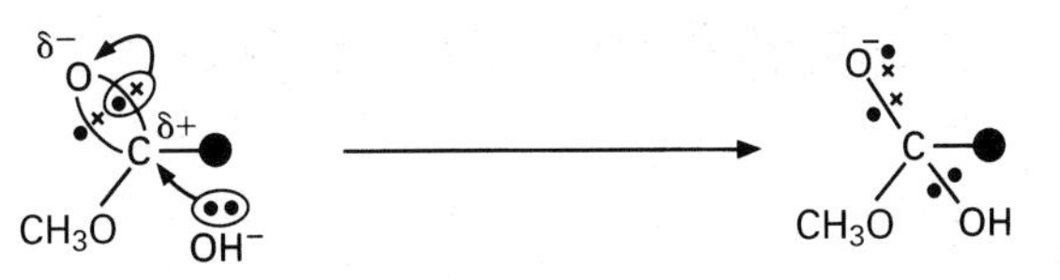

Then the carbonyl bond reforms and the leaving group CH_3O^- departs:

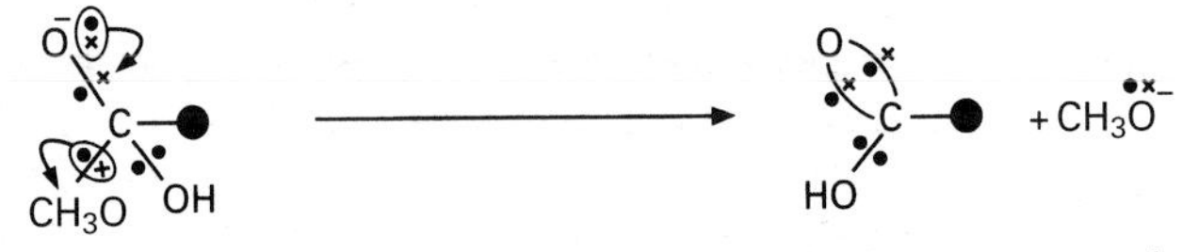

The carboxylic acid then reacts with the excess alkali to give a salt (e.g. ●—COO^-K^+).

Synthetic routes using acid amides

Amides are resistant to nucleophilic attack because of electron delocalisation, as shown in Figure 13.8. This decreases the δ^+ charge on the carbonyl carbon, making it less attractive to the attacking nucleophile, whilst also imparting some π bond character to the carbon–nitrogen bond, making it more difficult to break. Another effect of the delocalisation of the lone pair on nitrogen is that amides are much weaker bases than amines, section 14.5.

Although the —NH_2 group in amides resists substitution by other nucleophiles, amides undergo three useful preparative reactions:

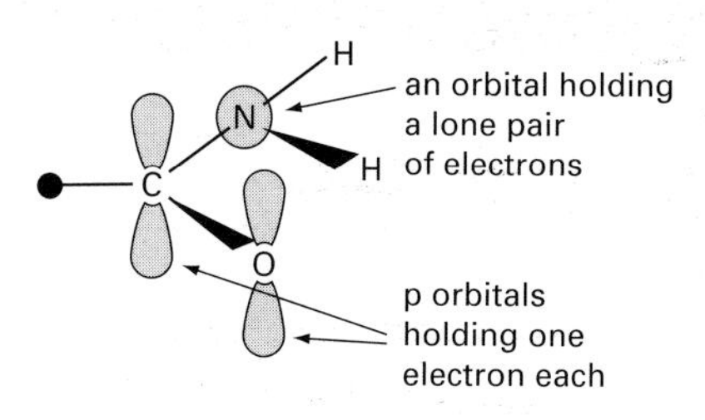

(a) σ bond framework and the p orbitals which are available for π bonding

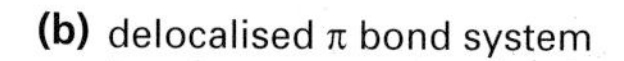

(b) delocalised π bond system

Figure 13.8
Electron delocalisation in acid amides makes them (i) resistant to nucleophilic attack and (ii) extremely weak bases

- **Dehydration**
  A **nitrile** may be prepared by heating a mixture of an amide and phosphorus(V) oxide, a powerful dehydrating agent, for example:

$$CH_3CH_2CONH_2 \xrightarrow[\text{distil off product}]{P_4O_{10}\text{(s); heat and}} CH_3CH_2CN + H_2O$$

propanamide → propanonitrile

- **Hofmann's degradation**
  When warmed with bromine and concentrated potassium hydroxide, an amide yields an **amine** containing *one less* carbon atom, for example:

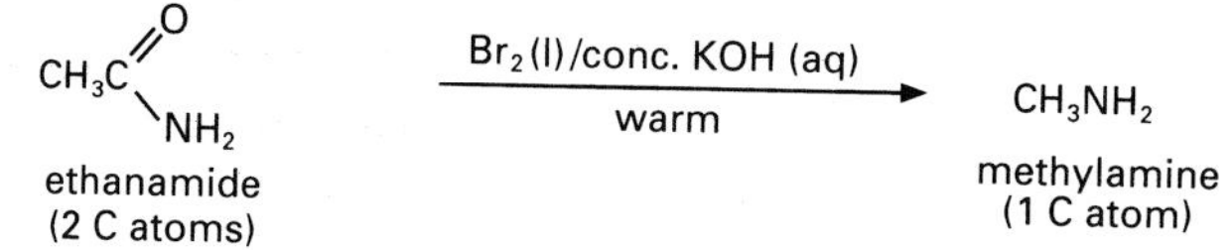

- **Reduction**
  An amide may be reduced to an **amine** having the *same number* of carbon atoms:

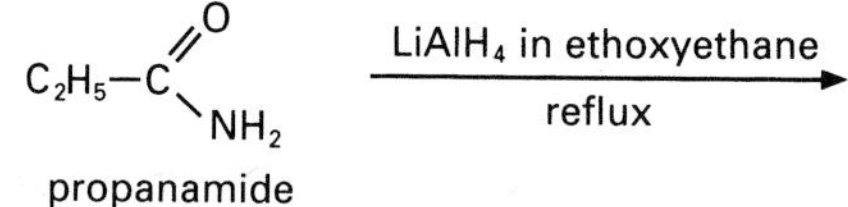

$$\xrightarrow[\text{reflux}]{LiAlH_4 \text{ in ethoxyethane}} C_2H_5CH_2NH_2$$

propylamine

Activity 13.4

1 Name the following carboxylic acid derivatives:

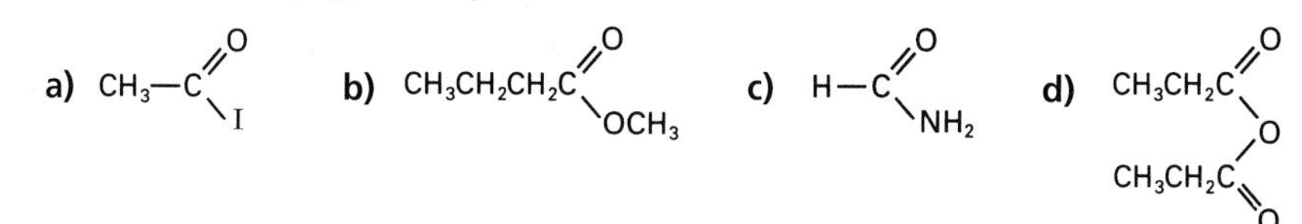

2 Give the name(s) and structural formula(e) of organic product(s) formed in the following reactions:

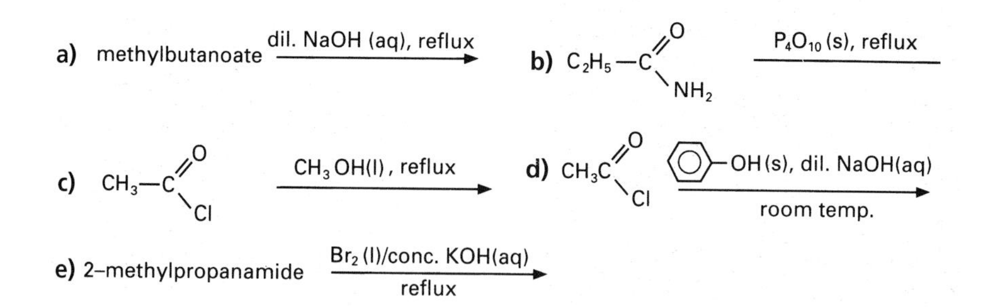

3 This question covers the material in sections 13.4 to 13.6. Give the reagents and reaction conditions needed for the following conversions:
a) ethyl ethanoate ⟶ ethanol + ethanoic acid

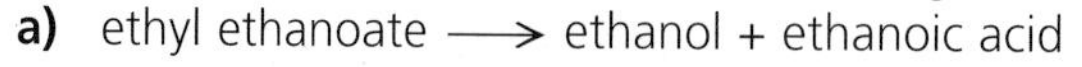

b)

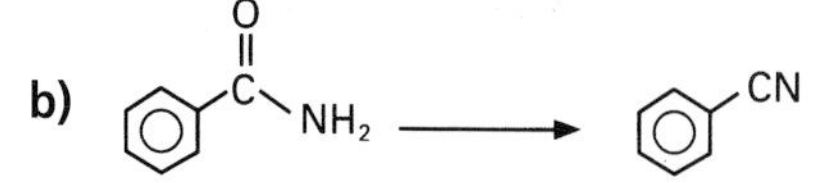

c) propanoyl chloride ⟶ propanamide
d) methanol ⟶ methyl benzoate
e) $CH_3COOH \xrightarrow{\text{2 steps}} CH_3NH_2$

Activity 13.4 continued

f) $CH_3CONH_2 \longrightarrow CH_3CH_2NH_2$

Parts (g) to (j) cover material from this chapter and earlier chapters:

g) ethanoic acid $\xrightarrow{\text{3 steps}}$ ethane-1,2-diol

h) C_6H_5–CN $\xrightarrow{\text{3 steps}}$ C_6H_5–C(=O)–O–C_6H_5

i) $CH_3COOH \xrightarrow{\text{3 steps}} CH_3CN$

j) $CH_3CH_2CH_2OH \xrightarrow{\text{3 steps}} CH_3CH_2NH_2$ (Hint: hydrocarbon chain lengths?)

4 Explain how conc. $NH_3(aq)$ can be used to identify an unknown acid chloride.

13.7 Some important applications of carboxylic acid derivatives and salts

Unless you are reading this book in the bath, there is a good chance that you are wearing a carboxylic acid derivative! Indeed, when you take a bath or shower, you will probably use a substance which was made from an acid derivative. Puzzled? Take a look at the photographs.

In this section, we shall discuss the industrial applications of acid derivatives in polyesters and soaps. Nylon is discussed later in section 14.7.

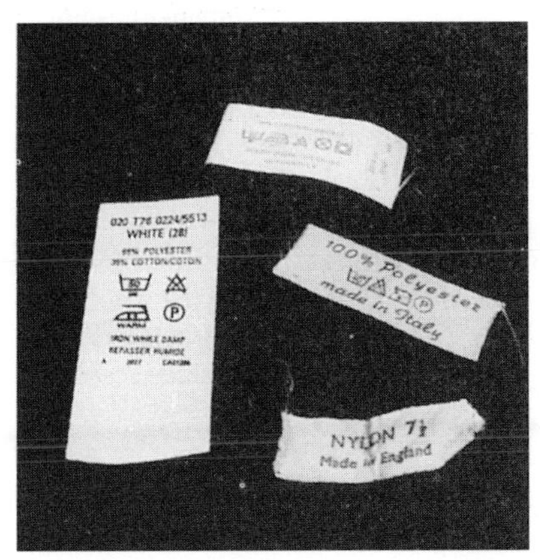

Nowadays, a lot of clothing is made from fabric containing a mix of natural fibres, such as cotton and wool, with synthetic fibres, such as **nylon** and **polyester**. Synthetic fabrics, though not as soft as the natural materials, are harder wearing, wash well and are often cheaper to produce. A common example, poly(ethylene terephthalate) or PET, is discussed in this section. Nylon is covered on page 377.

A recently developed polyester with enormous potential is poly(hydroxybutyrate), PHB, which is a natural and totally biodegradable thermoplastic that can be made by fermenting an agricultural feedstock

Soaps are the salts of carboxylic acids which have long hydrocarbon skeletons, such as sodium stearate, $CH_3(CH_2)_{16}CO_2^-Na^+$. They are made by hydrolysing various oils and fats which are naturally occurring carboxylic acid esters

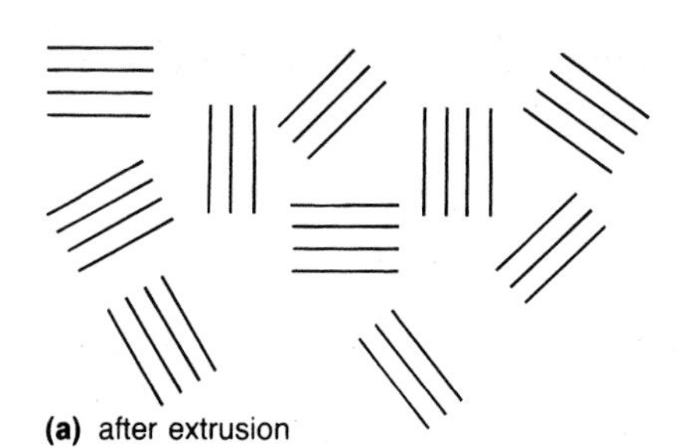

(a) after extrusion

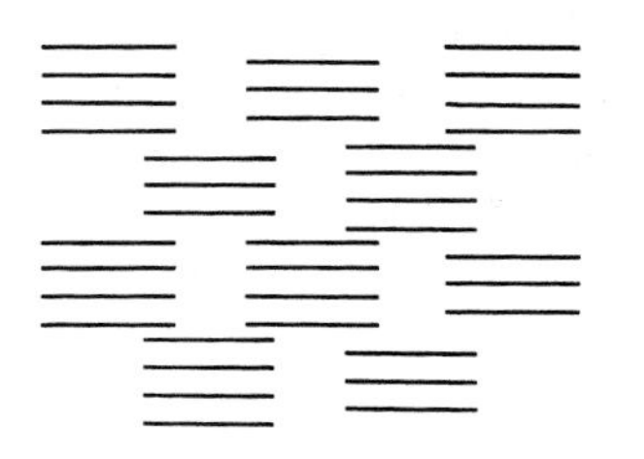

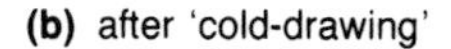

(b) after 'cold-drawing'

Figure 13.9
The polymer chains of PET are **(a)** disordered after extrusion (i.e. forcing through small holes). On 'cold drawing' (i.e. gentle stretching) **(b)**, the chains adopt the same orientation

Poly(ethylene terephthalate) – PET

PET is a **polyester** which first made its commercial impact in the 1940s. It is formed by the co-polymerisation of ethane-1,2-diol (ethylene glycol) and benzene-1,4-carboxylic acid (terephthalic acid) – hence, the name poly(ethylene terephthalate):

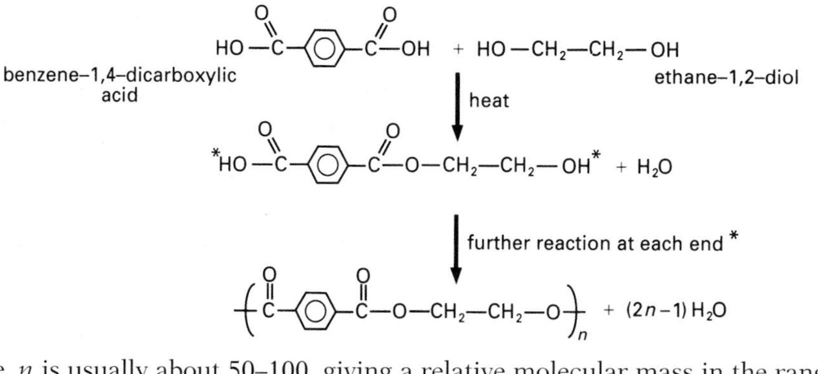

Here, n is usually about 50–100, giving a relative molecular mass in the range 10 000 to 20 000. Since each ester linkage produces a water molecule, this is termed a **condensation polymerisation**. Both monomers are obtained from petroleum, as shown below:

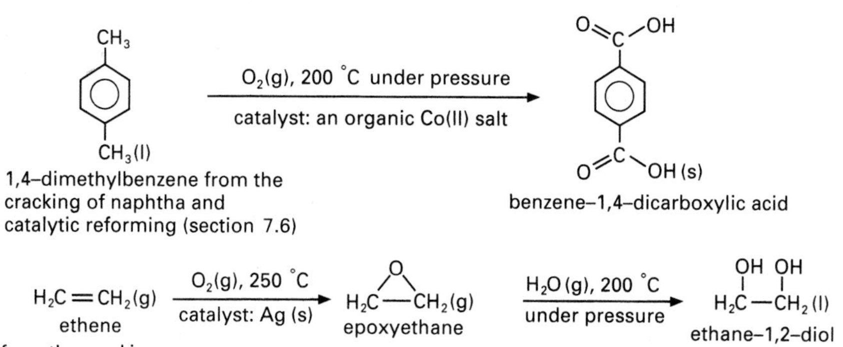

PET is a **thermoplastic**, that is it melts without decomposition. Thus, the hot polymer can be extruded, that is forced through small holes called spinarets. As they emerge, the streams of liquid polymers are cooled and they solidify. In this state, the fibres have little strength because the long chain molecules of the polymer are disordered. However, when the fibres are gently stretched, the molecular chains adopt the same orientation, Figure 13.9. This enables the fibres to 'cross-link' via dipole–dipole interactions:

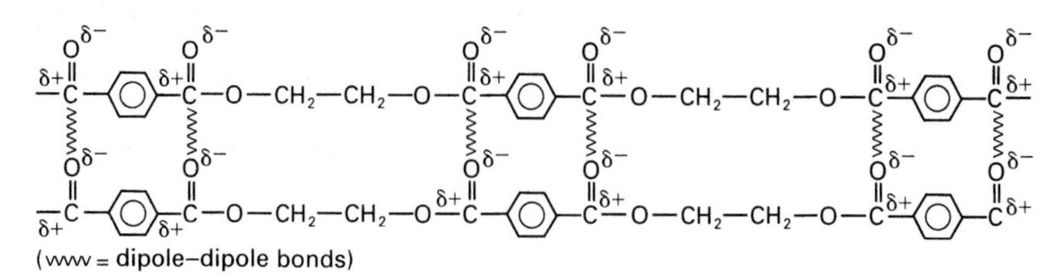

Due to the extrusion and stretching processes, the PET fibres become much stronger and more elastic. These properties, and their ability to wash well, make PET fibres very popular in the clothing industry and they appear under tradenames such as '**Dacron**' and '**Terylene**'.

As well as being used in clothing fabrics, PET can be rolled into sheets one-thirtieth the thickness of a human hair for use in audio and video tapes. Its inertness and non-toxic nature makes PET very suitable for food packaging and soft drink bottles, identified by the symbol 'PET' or '1' on the bottle. 'Dacron' is used as a surgical material, for example to make temporary arteries during heart by-pass operations.

Poly(hydroxybutyrate-valerate) PHB–V plastics

A market leader in the field of biodegradable PHB plastics is '**Biopol**' which was developped in the UK by Imperial Chemical Industries (ICI) Ltd in the mid-1970s. 'Biopol' is a polyester composed of hydroxybutyrate (HB) units with hydroxyvalerate (HV) units randomly distributed along the polymer chain, Figure 13.10. 'Biopol' is produced by fermenting sucrose from sugar beet or cereal crops with a specific micro-organism *Alcaligenes eutrophus*, which is often found in water and soil. Once purified, the polymer is granulated prior to forming it into plastic articles such as bottles, films and fibres.

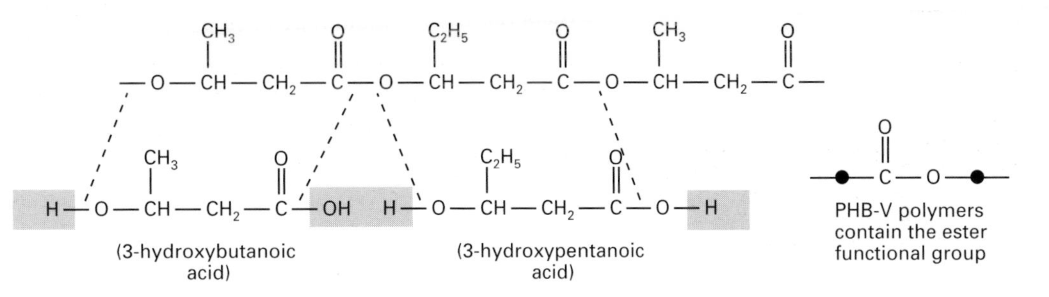

Figure 13.10 PHB-V polymers have a random arrangement of 'hydroxybutyric' and 'hydroxyvaleric' acid residues along the polymer chain. A water molecule is released every time two molecules of the monomer join together

The fascination of PHB–V plastics, such as 'Biopol', is that they are stable and durable under normal conditions of storage and use yet, in a suitable microbiological environment, they biodegrade to carbon dioxide and water. PHB plastics also degrade under anaerobic conditions, making them safe to bury, or compost, and their incineration produces only carbon dioxide and water.

At present, PHB–V plastics are more expensive than most other common plastics. However, a current research project seeks to insert genes from *Alcaligenes eutrophus* into plants which produce large quantities of starch or sugar, so that the production of poly(hydroxybutyrate) PHB occurs within the plant itself. One day, we may see huge plantations of genetically engineered plants which yield PHB at a cost comparable with that of petroleum-based plastics!

Soaps and soapless detergents

The first written references to soaps are found on clay tablets which were made about 4500 years ago by the Sumerians who lived in the areas now known as Iraq and Iran. The use of soaps as cleansing agents and medicinal products was spread by the ancient Romans and in the Middle Ages centres of soap-making developed throughout Western Europe. By the end of 18th century, the results of soap-making became more consistent and with the industrial revolution soap-making became a competitive industry.

Soaps are made by **saponification**, which is the alkaline hydrolysis of naturally occurring fats and oils. These fats and oils are known as **triglycerides** because they are the triesters of propane-1,2,3-triol (glycerol) and aliphatic carboxylic acids with long hydrocarbon skeletons, called **fatty acids**. The general reaction is:

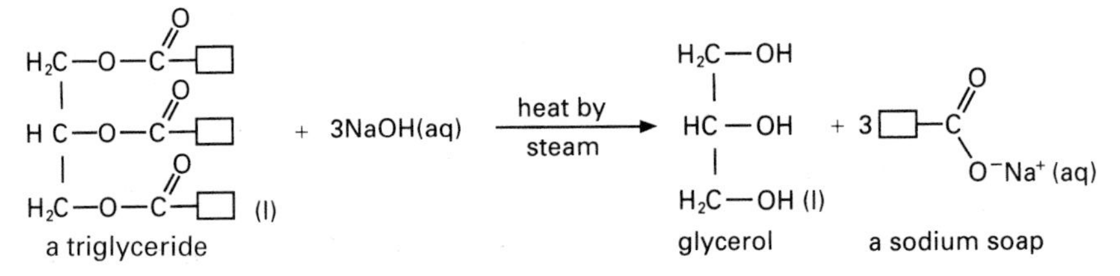

where the hydrocarbon chain, □—, usually contains 14–16 carbon atoms. The sodium and potassium salts of these long-chain acids are soaps, three examples of which are given in Table 13.3.

Table 13.3 Some common soaps and the fats/oils from which they are made

Source	soap	formula
animal fats	sodium stearate	$CH_3(CH_2)_{16}COO^-Na^+$
palm oil	sodium palmitate	$CH_3(CH_2)_{14}COO^-Na^+$
olive oil	potassium oleate	$CH_3(CH_2)_7CH{=}CH(CH_2)_7COO^-Na^+$

Activity 13.5

Soaps and detergents work by helping water to remove the thin layer of grease which holds particles of dirt on our bodies, clothes, plates, etc. To observe this effect, try the following simple experiment.

1 Place 10 drops of cooking oil in a test-tube and gently swirl it round. Add 2 cm^3 of distilled water. Cork the test-tube, shake vigorously and leave the test-tube to stand for a couple of minutes. What do you see?

2 Empty the test-tube and look at its walls. Are they clean?

3 Repeat 1 and 2. Then, add 3 drops of washing liquid and shake again. What do you see this time?

In the manufacture of sodium soaps, superheated steam is passed for several hours through a mixture of the molten fats, oils and aqueous sodium hydroxide. During saponification, some soap floats to the top of the reaction mixture where it is skimmed off. However, much of the soap, □—$CO_2^-Na^+$, remains dissolved in the aqueous solution. This may be **salted out**, that is precipitated from the solution by adding sodium chloride, because soap is less soluble in salt solution than in water.

After removing the soap, the glycerol is obtained from the aqueous solution by distillation. Glycerol is a marketable by-product, having many uses, for example in medicines and the manufacture of nitroglycerine, an explosive.

Identical methods may be used to make potassium soaps. These are softer and more water soluble than the sodium soaps. Most commercial products are a mixture of various soaps, disinfectants, perfumes and colouring agents.

Soaps work by enabling water to form an emulsion with the grease which holds dirt on to our bodies, clothes, and cooking equipment. Looking at the structure of a soap, we can see why this might happen:

Whilst the long hydrocarbon chain readily dissolves non-polar molecules, such as grease, the ionic COO^- group enables the soap to dissolve in water, Figure 13.10.

Soapless detergents have similar structures to soaps, having a long hydrocarbon skeleton, which is oil soluble, and a water-soluble ionic group, usually a sulphonate, $—SO_3^-$ group:

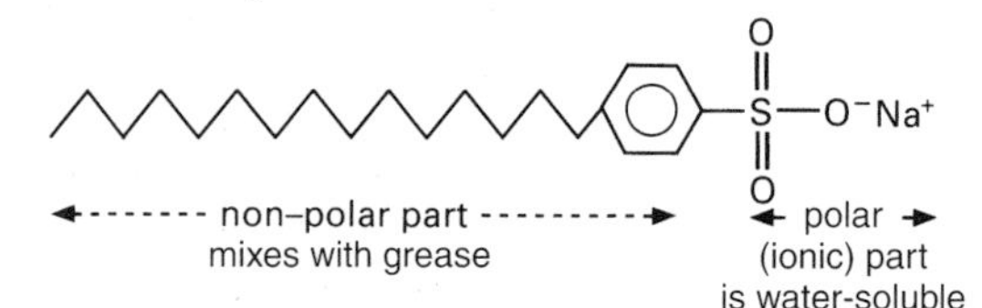

Figure 13.10
Water-soluble substances, such as a sticky sugar solution and some inks, can be washed off your hands with an excess of water. However, since water is a polar substance and oil is non-polar, they will not mix together and oily dirt cannot be removed by water alone. Soaps and soapless detergents are **surfactants**, that is surface-active cleansing agents which work at the water/oil interface. The surfactant molecule consists of a polar, or ionic, group which is **hydrophilic** ('water-loving'). This is attached to a long hydrocarbon chain which is non-polar and **hydrophobic** ('water-hating'), as shown below:

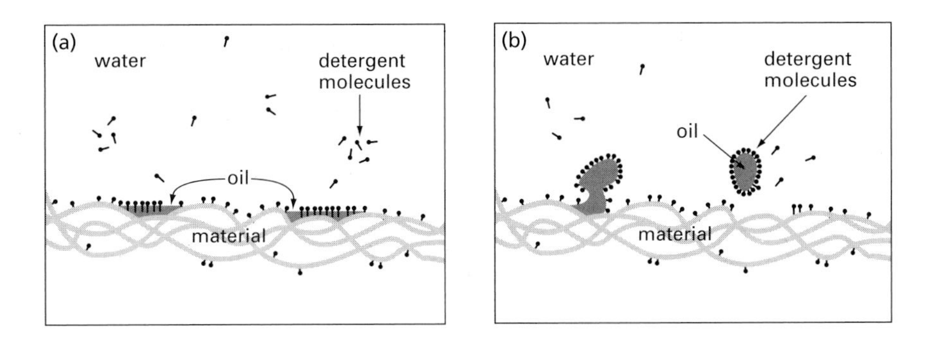

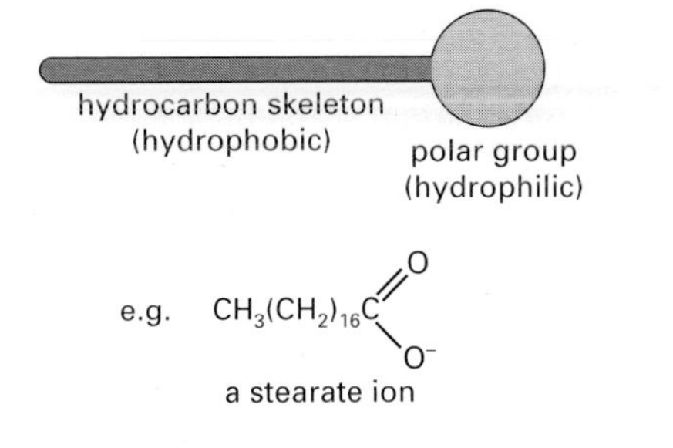

The surfactant works by lying across the water/oil interface, the polar group entering the water layer whilst the hydrocarbon skeleton adheres to oil on the material, (a). As the material is agitated, small droplets of oil break off the material and these are immediately surrounded by surfactant molecules, (b). These aggregates, of oil surrounded by surfactant molecules, are known as **micelles**. Since each micelle has a 'polar' shell, they repel each other so the 'imprisoned' oil droplets are unable to recombine. Providing that enough surfactant is available, the micelles will form an emulsion with water and the oil will be carried away.

Nowadays, various types of surfactant are available with hydrophilic groups which may be anionic (negatively charged), cationic (positively charged) or non-ionic (neutral). Non-ionic surfactants have several advantages over ionic surfactants. They foam less and, as they are uncharged, they do not form salts with the calcium and magnesium ions in hard water

Soapless detergents are synthesised from the alkylbenzenes produced in the petroleum industry, for example:

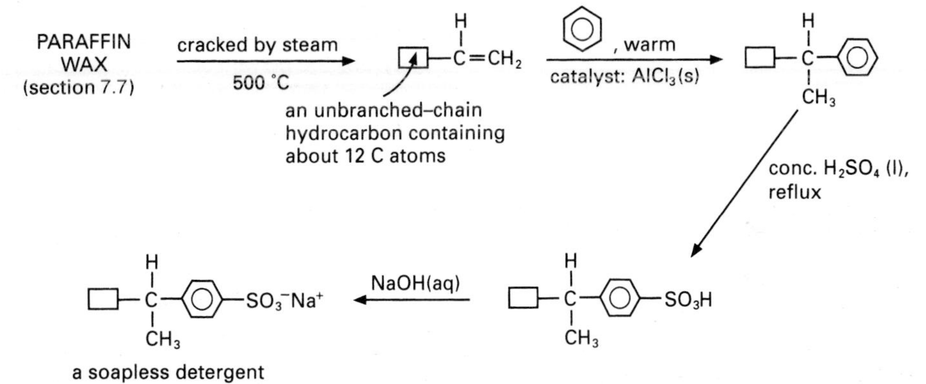

Soaps have a disadvantage in that they form a precipitate, or 'scum', with the calcium and magnesium ions that are formed in hard water, for example:

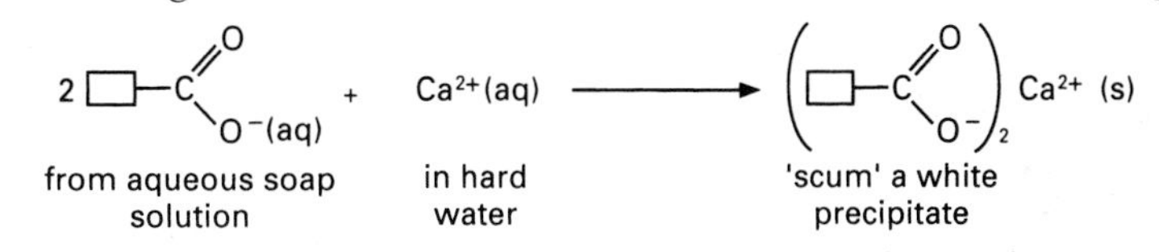

Apart from its unpleasant appearance, the formation of 'scum' wastes a lot of soap and leaves deposits on laundered clothing. Soapless detergents, on the other hand, do not form 'scums' because the calcium and magnesium sulphonates are soluble in water. Synthetic detergents became popular just before World War II but some early products, having branched-chain hydrocarbon skeletons, were slow to biodegrade and this led to foaming in some streams and sewage treatment works. However, in the early 1960s, Karl Ziegler and Giulio Natta developed catalysts which enabled non-branched alkenes to be polymerised into long-chain alkylbenzenes. These alkylbenzenes were converted into detergents which were fully biodegradable.

There is still some environmental concern surrounding synthetic detergents as complex polyphosphate 'builders' are added to enhance the cleaning power. Since phosphorus is one of the nutrients needed by aquatic plant life and algae, the discharge of phosphates into streams and rivers causes aquatic plant life to grow too rapidly. Eventually, the plants consume more oxygen than they produce and a 'swamp' is formed in which fish cannot survive. Due to this process, known as **eutrophication**, several national governments have placed limits on the use of phosphates. In turn, this has encouraged manufacturers to develop phosphate-free detergents whose cleaning action is improved by the addition of digestive **enzymes** such as protease and amylase. As well as the cleaning agent, builder and enzymes, a typical 'washing powder' will contain bleaches and fluorescers which make the clothes look whiter after washing (section 18.5).

Focus 13c

1. Carboxylic acid derivatives have the general formula (□— is a hydrocarbon skeleton):

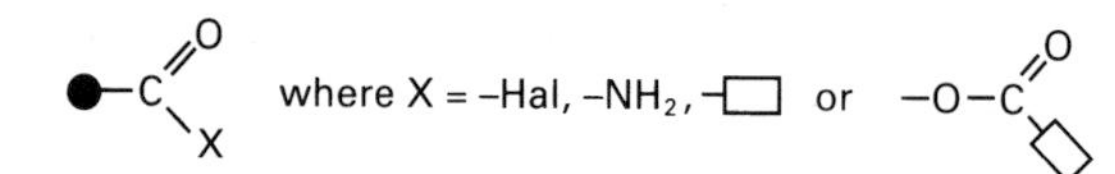

2. Acid halides are formed when a carboxylic acid reacts with a halogenating agent (Table 11.7). Other acid derivatives are made from acid chlorides (Figure 13.7).
3. Some important reactions of the acid derivatives are also summarised in Figure 13.7. Reactivity increases in the order:

   amides ~ esters < anhydrides ~ carboxylic acids << acid halides.

4. Acid derivatives play an important role in everyday life. Three examples: (i) **nylons** are polyamides, used to make clothing, carpet pile and ropes (see section 14.7), (ii) **polyesters**, such as Terylene are also used as synthetic fibres in clothing and the manufacture of bottles for soft drinks; (iii) the triesters of glycerol and various long-chain carboxylic acids are used to make **soaps**. Soaps are the sodium or potassium salts of these acids, for example, **sodium stearate**, $CH_3(CH_2)_{16}CO_2^-Na^+(s)$.
5. Important reagents used in this chapter are listed in Tables 13.4 and 13.5.

Table 13.4 Important reactions of carboxylic acids

Reagent/conditions	products formed by a carboxylic acid	notes
$LiAlH_4$/ethoxyethane; reflux	Reduction to a **primary alcohol** (addition of H_2 across the C=O bond); eg. $CH_3COOH \longrightarrow C_2H_5OH$	(1)
various bases, e.g. $NaHCO_3(s)$	Form **salts**, e.g. $CH_3COOH + NaHCO_3 \longrightarrow CH_3COONa + CO_2 + H_2O$	
$PCl_3(l)$, $P_{red} + Br_2(l)$, $P_{red} + I_2(s)$, $SO_2Cl_2(l)$, reflux; *or* PCl_5 room temp.	**Halogenation** , e.g. $CH_3C(=O)OH \xrightarrow[\text{agent}]{\text{halogenating}} CH_3C(=O)Hal + HHal$	(2)
alcohol, catalyst of conc. $H_2SO_4(l)$ *or* dry $HCl(g)$; reflux	**Esterification**, e.g. $CH_3C(=O)OH + C_2H_5OH \rightleftharpoons CH_3C(=O)OC_2H_5 + H_2O$	(3)
$(NH_4)_2CO_3(s)$; reflux with excess acid	Forms an **acid amide**, e.g. $CH_3C(=O)OH \longrightarrow CH_3C(=O)NH_2$	

Notes: (1) Addition to the carboxyl C=O bond is uncommon.
(2) Methanoic acid does not form acid halides.
(3) Isotopic labelling is used to work out the mechanism (section 11.6).

Table 13.5 Important reagents used in this chapter

Reagent/conditions	what it does
$K_2Cr_2O_7$/dil. $H_2SO_4(aq)$; reflux	oxidises **primary alcohols** ⟶ **carboxylic acids** e.g. $C_2H_5OH \longrightarrow CH_3COOH$
dil. $H_2SO_4(aq)$, reflux	hydrolysis of **nitrile** (CH_3CN) to **carboxylic acid** (CH_3COOH)
$KMnO_4$/dil. $H_2SO_4(aq)$ reflux	oxidises **methylbenzene** $C_6H_5CH_3$ to **benzoic acid** C_6H_5COOH valuable in organic synthesis
alcohol, room temp., or phenol, reflux with $OH^-(aq)$	converts **acid chloride** into **acid ester** e.g. $CH_3C(=O)Cl \xrightarrow{\square\text{-}OH} CH_3C(=O)O\text{-}\square + HCl$
conc. $NH_3(aq)$, room temp.	converts (i) **acid** ⟶ **ammonium salt** (ii) **acid chloride** ⟶ **acid amide** e.g. $CH_3C(=O)Cl \longrightarrow CH_3C(=O)NH_2$
dil. $NaOH(aq)$, reflux	hydrolyses **acid derivative** ⟶ **sodium salt of the carboxylic acid** e.g. $CH_3C(=O)X \longrightarrow CH_3C(=O)O^-Na^+$
$P_4O_{10}(s)$, heat	a dehydrating reagent; converts **amide** ⟶ **nitrile** e.g. $CH_3C(=O)NH_2 \longrightarrow CH_3CN + H_2O$
$Br_2(l)$/conc. $KOH(aq)$; reflux	Hofmann's degradation; converts **amide** ⟶ **amine with one less C atom** e.g $CH_3C(=O)NH_2 \longrightarrow CH_3NH_2$ *(2 carbon atoms)* *(1 carbon atom)*

Questions on Chapter 13

1 This question is about the following compounds:

A propanoyl chloride **B** ethanoic anhydride
C benzoic acid **D** ethanamide
E methyl ethanoate **F** propanoic acid

a) Draw structural formulae for these compounds and identify which functional group is present.
b) Which are able to form intermolecular hydrogen bonds?
c) Which is reduced to phenylmethanol by lithium tetrahydridoaluminate(III) in ethoxyethane?
d) Which is rapidly hydrolysed by water to produce an acidic gas?
e) Which may be made by oxidising propan-1-ol?
f) Which will form a 'fruity' smelling liquid when warmed with ethanol and concentrated sulphuric acid?
g) Which is formed by the reaction of ethanoyl chloride with concentrated aqueous ammonia?

2 Give the names, and structural formulae, of the organic products formed in the following reactions:

a) methanoic acid + dil. $NaOH(aq)$
b) propanoic acid + $P_{red}/Br_2(l)$
c) 2-methylbutanoic acid + $LiAlH_4$/ethoxyethane
d) excess ethanoic acid + $(NH_4)_2CO_3(s)$
e) benzoic anhydride + dil. $H_2SO_4(aq)$
f) benzoyl chloride + conc. $NH_3(aq)$
g) methyl methanoate + dil. $NaOH(aq)$
h) ethanamide + $LiAlH_4$/ethoxyethane
i) ethanamide + Br_2/conc. $NaOH(aq)$
j) propanamide + P_4O_{10}

3 Explain the following observations:

a) A student who measured the relative molecular mass of propanoic acid vapour obtained a value of 149.

Questions on Chapter 13 *continued*

b) Although ethanoic acid and phenol both react with dilute sodium hydroxide solution, only the ethanoic acid will react with solid sodium hydrogencarbonate.

c) Hydrogen cyanide readily adds across the C=O bond in propanone but it has no effect on the C=O bond in ethanoic acid.

d) The boiling points of hexane, C_6H_{12}, and butanoic acid, $CH_3CH_2CH_2COOH$, are 69°C and 164°C, respectively.

e) Solutions of ethanoic acid and dichloroethanoic acid, both of concentration 0.1 mol dm^{-3}, have pH values of 2.9 and 1.1, respectively.

f) Ethanoyl chloride reacts vigorously with water at room temperature whereas ethanamide will only undergo hydrolysis if it is refluxed with dilute acid or alkali.

5 Give the reagents and reaction conditions needed for the following conversions. Note that parts **g)** to **j)** may require reference to earlier chapters.

a) $CH_3CH_2COOH \longrightarrow CH_3CH_2COCl$

b) $CH_3COOH \longrightarrow CH_3CH_2OH$

c) $HCOOH \longrightarrow HCOO^-Na^+$

d) $CH_3COOCH_3 \longrightarrow CH_3OH + C_2H_5OH$

e) C_6H_5–COOH $\longrightarrow$ C_6H_5–$CONH_2$

f) $C_2H_5OH \longrightarrow CH_3COBr$ (2 steps)

g) $CH_3CHO \longrightarrow CH_3CH(OH)COOH$ (2 steps)

h) $CH_3CH_2OH \longrightarrow CH_3NH_2$ (3 steps)

i) $C_2H_4 \longrightarrow CH_3COOH$ (3 steps)

j) $CH_3CH_2Br \longrightarrow CH_3CH_2COCl$ (3 steps)

h) $CH_3OH \longrightarrow CH_3NH_2$ (2 steps)

i) C_6H_6 (benzene) $\longrightarrow$ C_6H_5–$CONH_2$ (3 steps)

j) $C_2H_6 \longrightarrow CH_3CH_2COCl$ (4 steps)

6 Describe chemical tests which would enable you to distinguish between the following pairs of substances, clearly stating i) the reagents and reactions conditions that you would use, and ii) the observations that you would expect to make.

a) $(CH_3CH_2)_2O$ and CH_3CH_2COOH

b) CH_3COCl and CH_3CONH_2

c) benzoic acid and phenol

7 Compound M is a carboxylic acid which is found in apples. It contains 35.8% carbon, 4.5% hydrogen and 59.7% oxygen by mass and its molecular mass is 134.

a) Work out the molecular formula of M.

b) In a titration, 10 cm^3, of an aqueous solution of M, of concentration 0.1 mol dm^{-3}, was found to be equivalent to 20 cm^3 of 0.1 mol dm^{-3} aqueous sodium hydroxide. How many carboxyl groups are present in one molecule of M?

c) When M is treated with phosphorus pentachloride, *three* moles of PCl_5 react with *one* mole of M. How many —OH groups are present in one molecule of M?

d) Write down **three** possible structural formulae for M.

e) A freshly prepared sample of M may be separated into two optical isomers. Write down the 3-D structure of M and indicate which structural feature is responsible for its optical properties.

f) On warming M with methanol in concentrated sulphuric acid, a 'fruity' smelling compound N, of molecular formula $C_6H_{10}O_5$, is formed. What is the structural formula of N and what types of functional groups does it contain?

g) What types of functional group make M suitable for use in the manufacture of a polyester? Write down the structure of a theoretical polyester which may be made by copolymerisation of M with ethane-1,2-diol.

h) Compound M may be made from ethene in a synthetic pathway which has *five* steps. Give the reagents and conditions that you would need to carry out this synthesis and draw the structures of the intermediate products.

8 Compound B contains 41.4% carbon, 3.4% hydrogen and 55.2% oxygen by mass and its mass spectrum indicates a molecular ion of mass 116.

a) Work out the molecular formula of B.

b) Write down *three* possible structural formulae for compound B.

c) Of the molecules you have drawn, two display *cis-trans* isomerism.
- i) Identify these molecules and name them.
- ii) How might the isomers be distinguished using physical methods?

d) When heated, each isomer forms the same products; water and a molecule of formula $C_4H_2O_3$. The isomers undergo dehydration at different temperatures, the '*cis*-' at 150°C and the '*trans*-' at 250°C.
- i) Name the organic product of this reaction and draw its structural formula.
- ii) Explain why the isomers undergo dehydration at dif ferent temperatures.

e) Write down the structures of the organic products formed when the isomers are treated with:
- i) bromine dissolved in hexane
- ii) dilute aqueous sodium hydroxide
- iii) lithium tetrahydridoaluminate(III) in ethoxyethane
- iv) phosphorus pentachloride.

Comments on the activities

Activity 13.1

1 **a)** butanoic acid
b) 3-methylbutanoic acid
c) 2,2-dimethylpropanoic acid
d) butanedioic acid
e) 3-hydroxybenzoic acid
f) 2-hydroxy-2-methylbutanoic acid

2

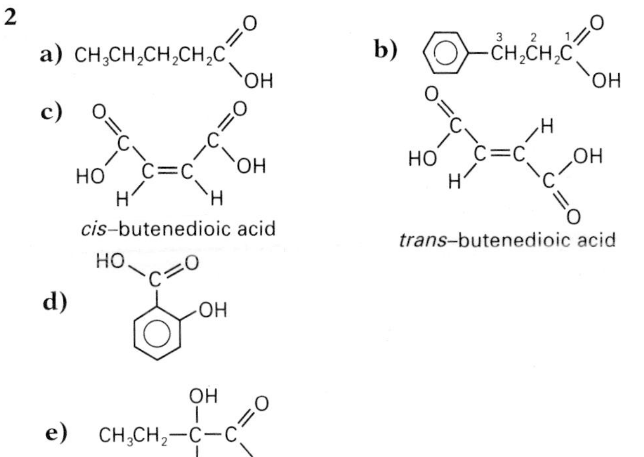

3 **a)** pentanoic acid, 3-methylbutanoic acid and 2,2-dimethylpropanoic acid are all $C_5H_{10}O_2$;
3-hydroxybenzoic acid and 2-hydroxybenzoic acid, both $C_7H_6O_3$;
2-phenylpropanoic acid and 3-phenylpropanoic acid, both $C_9H_{10}O_2$.
b) 2-hydroxy-2-methylbutanoic acid and 2-hydroxybutanoic acid; both have an asymmetric carbon atom.

Activity 13.2

1 $K_2Cr_2O_7$/dil. H_2SO_4(aq), reflux; *or* $KMnO_4$/dil. H_2SO_4(aq), reflux.
2 dil. H_2SO_4(aq), reflux.
3 as **1**
4

$$C_2H_5Br \xrightarrow[\text{reflux}]{\text{KCN in ethanol}} C_2H_5CN \xrightarrow[\text{reflux}]{\text{dil } H_2SO_4(aq)} C_2H_5COOH$$

Notice that the number of carbon atoms in the molecule has increased.

5 A Grignard synthesis:

$$C_2H_5Br \xrightarrow[\text{room temp.}]{\text{Mg(s); dry ethoxyethane}} C_2H_5MgBr$$

$$\downarrow \ CO_2(s), \text{ then dil. HCl(aq)}$$

$$C_2H_5COOH$$

Activity 13.3

1 **a)** C_2H_5COOH, CH_3COOH, HCOOH
Reason: Electron releasing effect of the alkyl group destabilises the oxoanion. The smaller the alkyl group, the stronger the carboxylic acid.
b) $ClCH_2CH_2CH_2COOH$, $ClCH_2CH_2COOH$, $ClCH_2COOH$.
Reason: increasing acidity as the halogen atom gets closer to the carboxyl group and thus become better able to attract electron cloud to itself.
c) 2-methylpropanoic acid, propanoic acid, 2-iodopropanoic acid, 2-bromopropanoic acid, 2-chloropropanoic acid.
Reason: increasing acidity as the Y group in Y—COOH becomes better able to attract electron cloud to itself.
d) ethanol, phenol, ethanoic acid, benzoic acid bromoethanoic acid, trichloroethanoic acid.

2 **a)** $2CH_3CO_2H(aq) + Zn(s) \longrightarrow (CH_3CO_2)_2Zn(aq) + H_2(g)$; zinc ethanoate
b) $HCO_2H(aq) + KOH(aq) \longrightarrow HCO_2K(aq) + H_2O(l)$; potassium methanoate
c) $2C_6H_5CO_2H(s) + Ca(OH)_2(aq) \longrightarrow (C_6H_5CO_2)_2Ca(aq) + 2H_2O(l)$; calcium benzoate
d) $CH_3CO_2H(aq) + KHCO_3(aq) \longrightarrow CH_3CO_2K(aq) + CO_2(g) + H_2O(l)$; potassium ethanoate

3 **a)** $CH_3CH_2CH_2OH$
propan-1-ol

b)

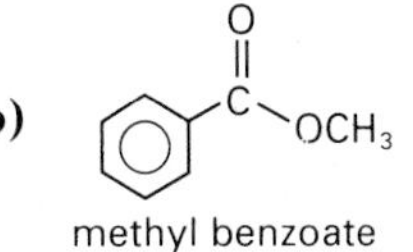

methyl benzoate

c)

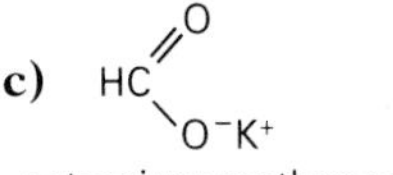

potassium methanoate

d)

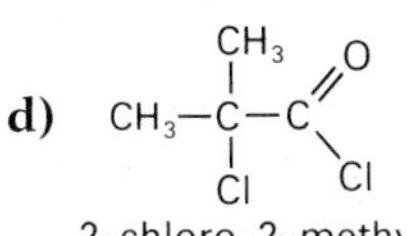

2-chloro-2-methyl propanoyl chloride

e)

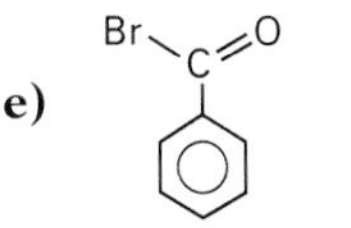

benzoyl bromide

Comments on the activities *continued*

4

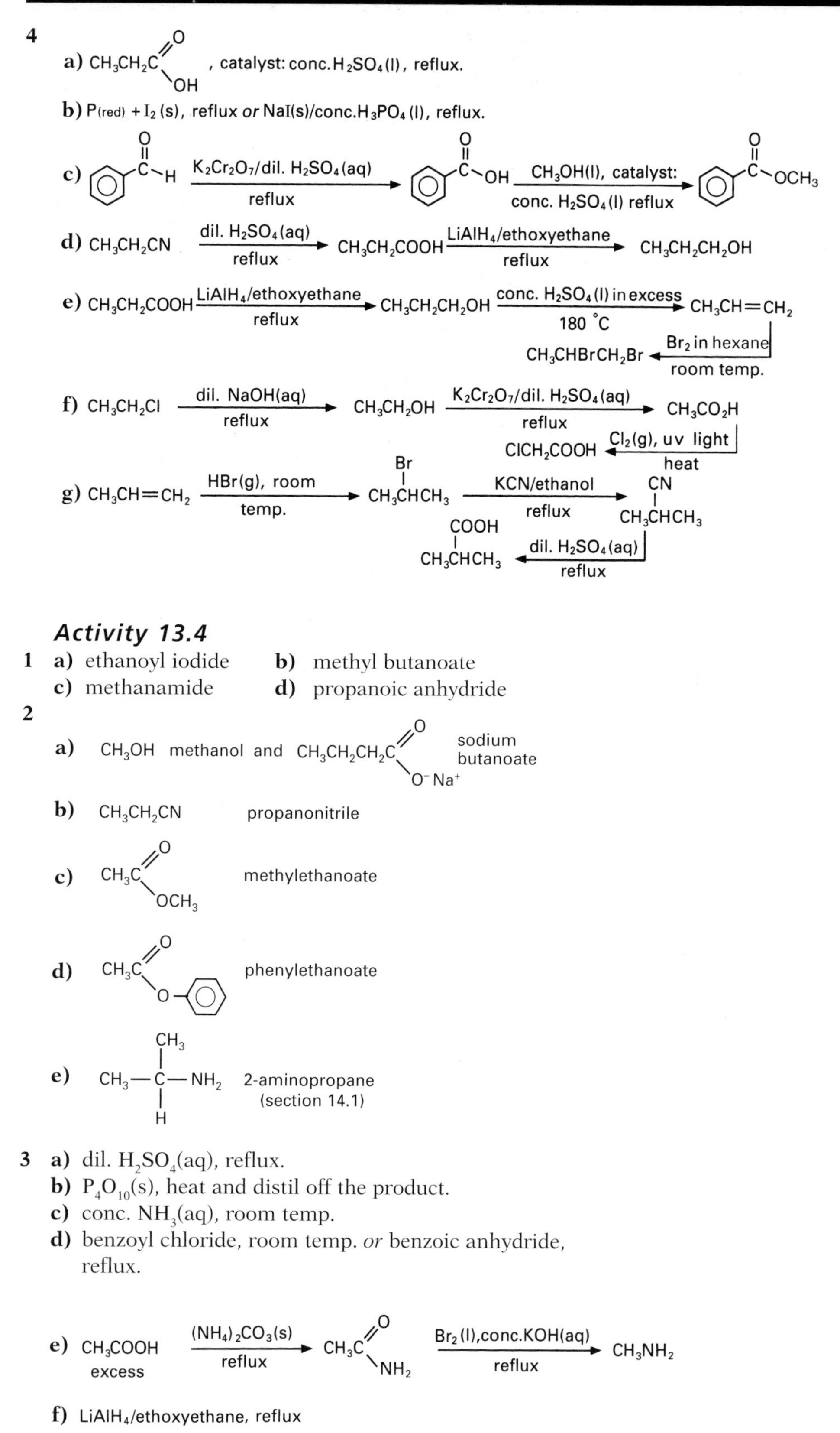

Activity 13.4

1 **a)** ethanoyl iodide **b)** methyl butanoate
c) methanamide **d)** propanoic anhydride

2

3 **a)** dil. H_2SO_4(aq), reflux.
b) P_4O_{10}(s), heat and distil off the product.
c) conc. NH_3(aq), room temp.
d) benzoyl chloride, room temp. *or* benzoic anhydride, reflux.

Comments on the activities *continued*

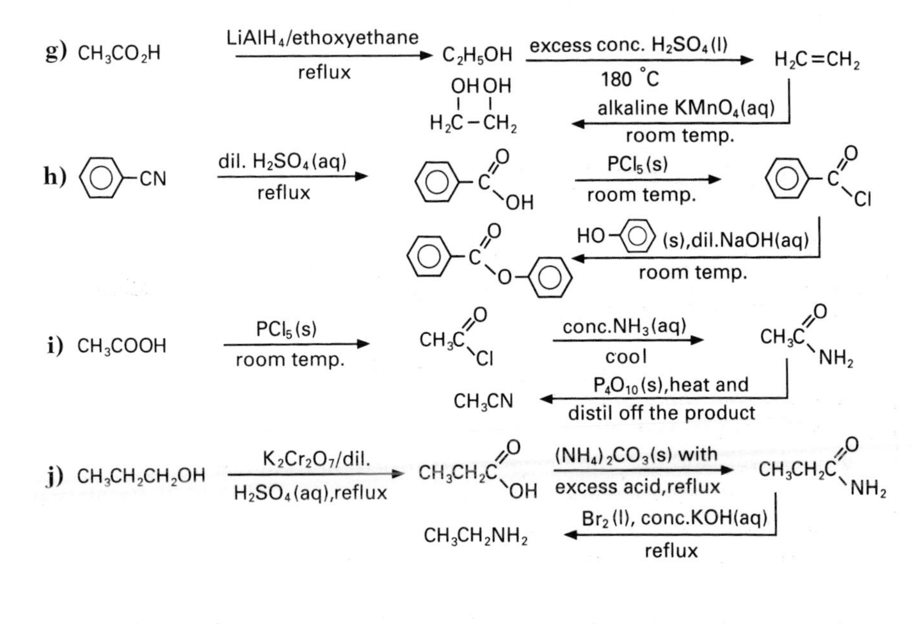

4 Prepare the amide,

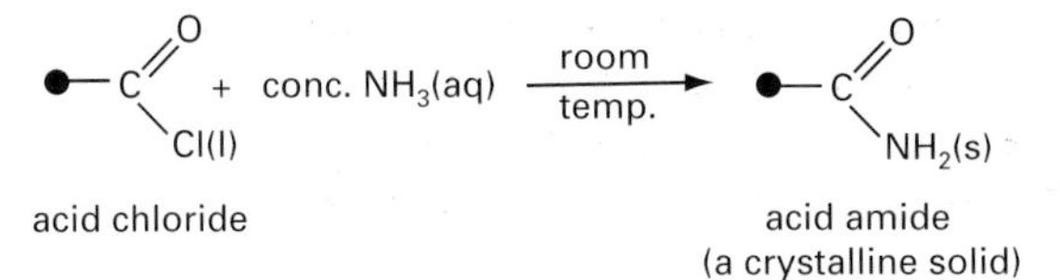

purify it by recrystallisation and take its melting point. Look up the value in a book of data and identify the acid amide. Hence work out the acid chloride from which it was made.

Activity 13.5

You have probably made the following observations:

1 When the oil is shaken with distilled water, a cloudy emulsion is formed. On standing, this separates to give two layers: oil on top, water below.

2 Although we can now remove some of the oil by emptying the test-tube, oily drops still remain on its walls. It is still 'greasy'.

3 In the presence of soap, though, the oil and water give a cloudy mixture which does not form layers on standing. Moreover, on emptying the tube its walls are found to be clean, not 'greasy'.

CHAPTER

14

Organic nitrogen compounds

Contents

Study Checklist

By the end of this chapter you should be able to:

1. Recall that there are five main types of organic nitrogen compound: **nitro compounds, nitriles, amines, amides** and **amino acids**.
2. Interpret, and use, the names, short structural formulae and displayed formulae of simple organic nitrogen compounds (Table 14.2). Recall that the aliphatic compounds form homologous series and describe the gradual change in physical properties that occurs as molecular mass is increased.
3. Recall that organic nitrogen compounds contain polar functional groups and form strong intermolecular dipole–dipole attractions. State that, as a result, they tend to have much higher boiling points and greater solubility in water than the alkanes of similar molecular mass.
4. Recall that the introduction of a nitrile group into an organic molecule is an important step in adding a $—CH_2—$ unit to the hydrocarbon skeleton, section 14.2.
5. Classify an amine as being a primary (1°), secondary (2°) or tertiary (3°) compound, depending on whether its nitrogen atom is bonded to 1, 2 or 3 hydrocarbon skeletons, respectively.
6. Explain that, due to the lone pair of electrons on the nitrogen atom, amines react as nucleophiles and bases and summarise the important reactions of primary amines as shown in Table 14.4.
7. State that amines are weak bases, that is they ionise slightly in aqueous solution to give an equilibrium mixture in which the concentrations of molecules and ions are constant, yet very different:

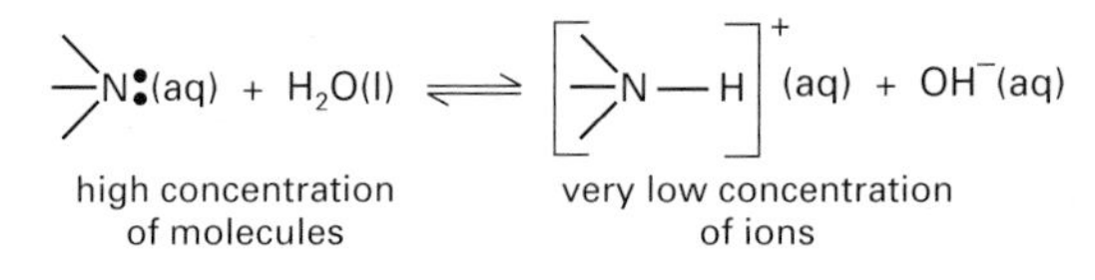

8. Recall that experiments reveal a pattern in amine basicity:

increasing base strength →

primary aromatic amines << ammonia < primary aliphatic amines
e.g. $C_6H_5NH_2$ e.g. $C_2H_5NH_2$

Explain this pattern in terms of the greater localisation of the lone pair on the amine's nitrogen atom (moving left to right) and, thus, its increasing tendency to accept a proton.

9. Recall that the reactions of amines with nitrous acid are important because they allow us to distinguish between a primary aliphatic and a primary aromatic amine. State that the latter form diazonium salts which are very valuable in organic synthesis.
10. Recall that aromatic diazonium salts give two types of reaction: i) those in which nitrogen gas is evolved and a nucleophilic group, such as OH^-

or I^-, is attached to the aromatic ring; ii) coupling reactions in which the azo group, —N=N—, is retained, the products often being brightly coloured compounds which may be used as dyes.

11 Recall that amides are white crystalline solids which, when pure, have very sharp melting points. Explain how an unknown acid derivative may be identified by reacting it with a particular amine and taking the melting point of the amide which is formed.

12 Recall that amides are extremely weak bases and very poor nucleophiles because, although they contain the NH_2 group, the nitrogen lone pair is delocalised over the N—C—O σ bond framework.

13 Recall important synthetic routes involving organic nitrogen compounds and the reagents/reaction conditions used in these (Table 14.7).

14 State the generalised formula of an α-amino acid as $RCHNH_2COOH$ and describe the nature of the functional group (e.g. —COOH, —NH_2, —OH, —SH and non-polar groups) which may be present in R. Draw suitable diagrams to show that, except for glycine, the α-amino acids can exist as optical isomers. Appreciate that all such naturally-occuring amino acids have the L-configuration (Fischer convention).

15 Explain the term 'zwitterion' and interpret the physical, and chemical, properties of α-amino acids in terms of their structure.

16 Explain the effect of pH on amino acid ionisation, and solubility, and account for the existence of the isoelectric point for an amino acid.

17 Explain the formation of the peptide bond between α-amino acids leading to the idea that polypeptides and proteins are condensation polymers.

18 Explain the term 'primary structure of a protein' and describe, in outline, how this may be determined by hydrolysing the protein, separating the products by chromatography and electrophoresis and using end-group analyses.

19 Describe the secondary structures of proteins (α-helix in α-keratin, β-pleated sheet in silk fibroin) and the stabilisation of these structures by hydrogen bonds.

20 State the importance of the tertiary structure of a protein and explain the stabilisation of the tertiary structure with regard to the interactions between the R groups in the amino acid residues (hydrophobic bonds, ionic bonds, disulphide bonds, hydrogen bonds) and the role of proline.

21 Understand the term 'quaternary structure of a protein' and describe, in outline, the structure of haemoglobin.

22 Explain denaturation of proteins by: extremes of temperature, pH changes, the addition of heavy metal ions and detergents.

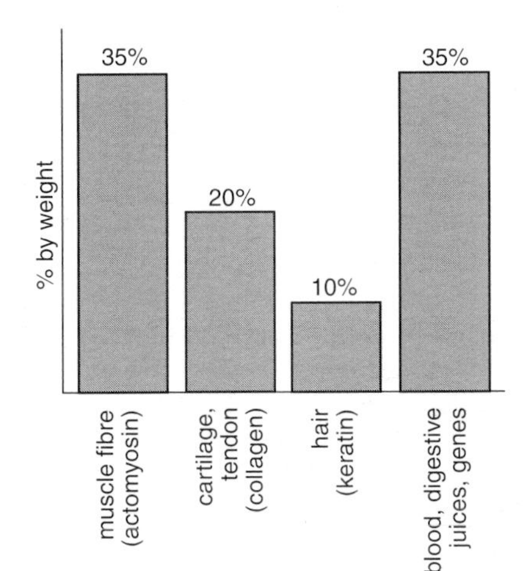

Figure 14.1 ▲
The importance of proteins in the human body

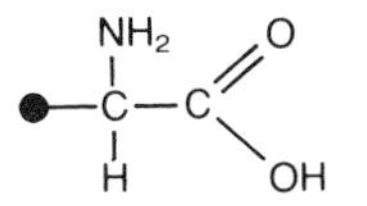

Figure 14.2 ▲
General formula of a 2-amino acid. ● is a hydrogen atom or hydrocarbon skeleton which might have a wide variety of functional groups attached to it, for example, –COOH, –NH_2, –OH and –SH. Note that the amino group, –NH_2, and carboxyl group, –COOH, are attached to the same carbon and that 2-amino acids are also known as **α-amino acids** (section 14.9)

Nitrogen atoms are found in a variety of organic molecules. Of most importance are **proteins**. These make up about a sixth of our bodyweight and are found, for example, in muscle, hair, skin and blood (Figure 14.1). We need proteins for the growth and repair of body tissue, and the maintenance of our normal bodily functions.

All naturally occurring proteins are polymeric molecules containing a unique combination of **2-amino acids** (Figure 14.2). A living organism can control the exact sequence in which these 2-amino acids join together to form the protein. This behaviour is almost incredible, especially when you bear in mind that over 3 million different molecules can be made by arranging 10 different amino acid residues in different orders.

We obtain our supply of 2-amino acids from the proteins in meat, fish and vegetables. Digestive enzymes in the stomach and intestines catalyse the hydrolysis of these proteins. After hydrolysis, the amino acids are carried by the bloodstream to the various organs and body tissues where they are converted into the necessary proteins.

Proteins are discussed further in section 14.10–14.12. This chapter covers the chemistry of amino acids and four other types of organic nitrogen compound: **nitro compounds**, **nitriles**, **amines** and **amides**.

14.1 Nitro compounds

Nitromethane is used as a fuel in 'Top Fuel' dragster racing, a popular sport in the USA. Due to their powerful engines and aerodynamic design, it only takes drag cars about 6 seconds to accelerate from 0 to 200 miles per hour over the quarter of a mile run

Nitro compounds contain the **nitro group**, **—NO_2**, and are named by placing '**nitro**' before the hydrocarbon's name:

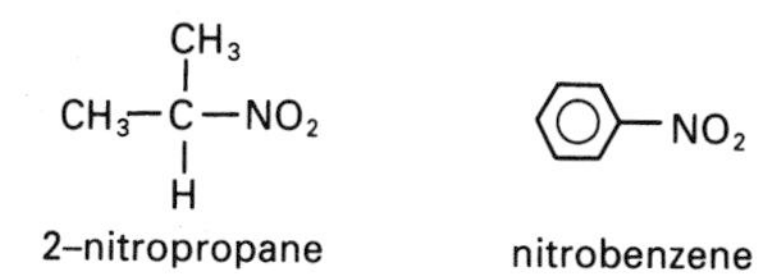

2–nitropropane nitrobenzene

Aliphatic nitro compounds are used as to make specialist fuels and industrial solvents. Aromatic nitro compounds are of more importance, especially **nitrobenzene**, which is used in large amounts to make dyes (section 14.7). Due to the polarity of the nitro group, intermolecular dipole–dipole attractions are strong and nitrobenzene is a colourless liquid of high boiling point (Figure 14.3). It is almost insoluble in water, however, due to the presence of the benzene ring.

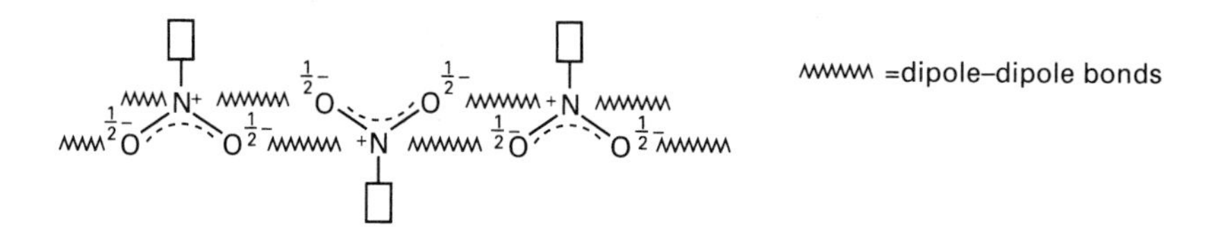

Figure 14.3 ▶
The strong dipole–dipole attractions in nitro compounds cause them to have much higher boiling points than the alkanes of similar molecular mass (nitromethane b.p. 101°C; butane b.p. 0°C)

Nitrobenzene is made by treating benzene with a nitrating mixture of concentrated nitric and sulphuric acids at 50°C (section 14.5). Most nitrobenzene is reduced to phenylamine, an important chemical in the manufacture of dyes and sulphonamide drugs. In the laboratory, the reducing agent is tin in concentrated hydrochloric acid:

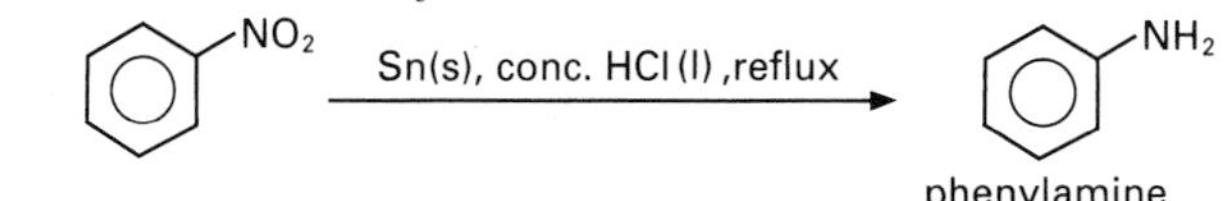

phenylamine

In industry, iron is used instead of tin because it is cheaper and almost as effective.

14.2 Nitriles

Nitriles have a **cyano group**, **—C≡N**, bonded to a hydrocarbon skeleton. To name them, use the name of the hydrocarbon with the same number of carbon atoms but change the final '**-e**' to '**-onitrile**' *or* if another functional group modifies the name ending, then the prefix **'cyano-'** is used, for example:

CH_3CN — ethanonitrile

benzonitrile (not benzenonitrile)

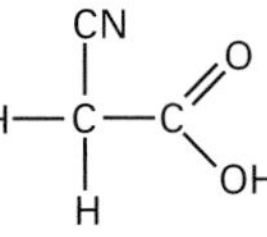

cyanoethanoic acid

The lower mass aliphatic nitriles are colourless, pleasantly smelling liquids which are fairly soluble in water due to the polarity of the C≡N bond. Benzonitrile is an oily liquid, insoluble in water. Although hydrogen cyanide and cyanide salts are extremely poisonous, nitriles are only mildly toxic because they do not form hydrogen cyanide on contact with water.

Nitriles are valuable intermediaries in organic synthesis since their preparation from halogenoalkanes and carbonyl compounds *lengthens the hydrocarbon chain* (section 14.5). The nitrile is then hydrolysed to an acid and reduced to an alcohol,

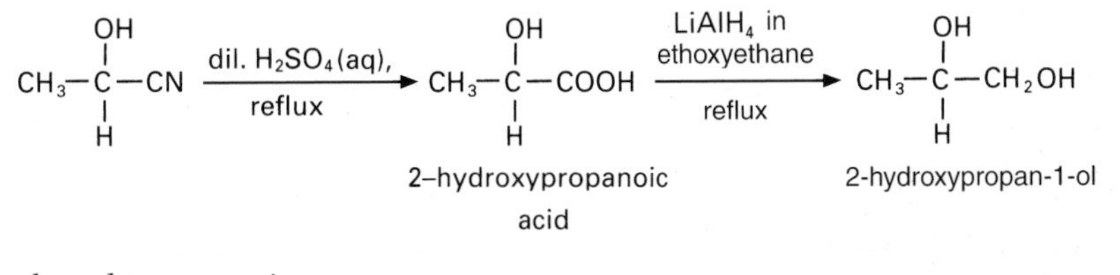

or reduced to an amine,

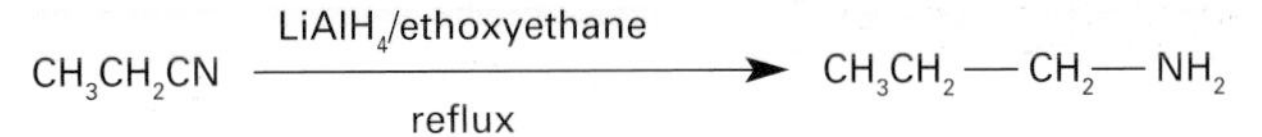

Both pathways result in the introduction of a CH_2 unit into the hydrocarbon skeleton.

Activity 14.1

1 Name the following organic nitrogen compounds

a) $CH_3CH_2CH_2CH_2NO_2$

b) CH_3CH_2CN

c) $CH_3CH(CH_3)CN$

d) CH_3–C_6H_4–NO_2

2 Write down the structural formulae for the following compounds:

a) 2-nitro-3-methylbutane

b) 2-methylbutanenitrile

c) phenylethanenitrile

d) 1,3,5-trinitrobenzene

3 State the reagents and the reaction conditions which are used to synthesise:

a) 2-methylbutanenitrile from 2-bromobutane

b) 2-hydroxyethanoic acid from methanal

4 Give the names and structural formulae of the organic product(s) formed in the following reactions:

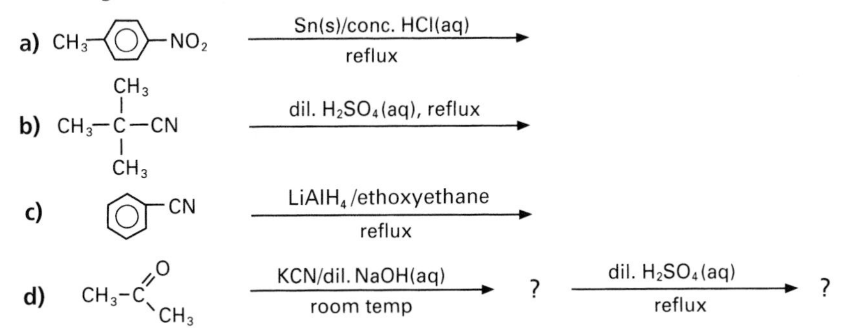

Activity 14.1 continued

5 Give the reagents and reaction conditions needed for the following conversions, each of which involves two steps and may require reference to earlier chapters:

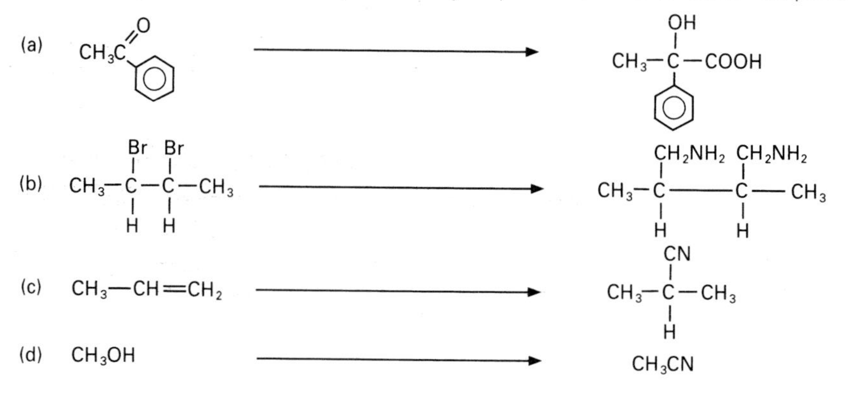

14.3 Amines: names and physical properties

Most of this chapter will be about the chemistry of amines. These are classified as **primary**, **secondary** or **tertiary** compounds, where □— is a hydrocarbon skeleton:

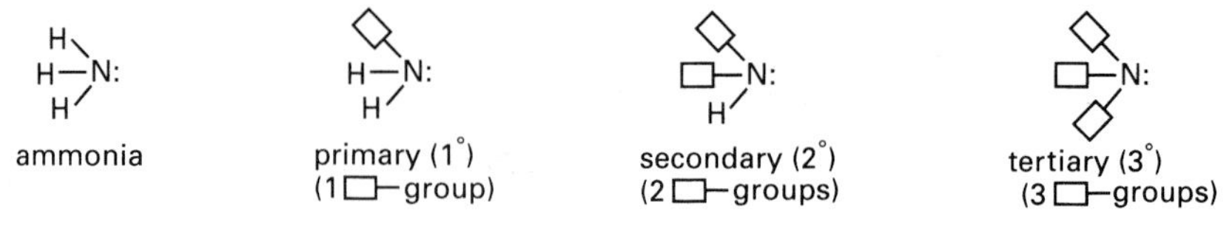

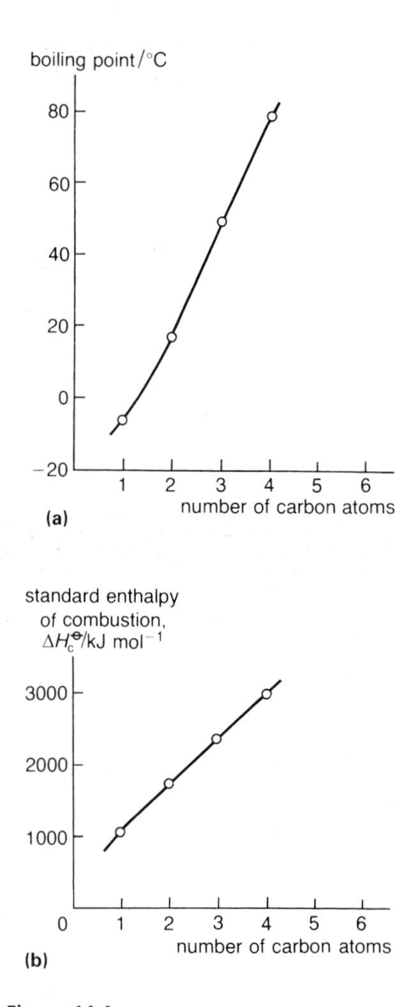

Figure 14.4
Steady increases in **(a)** the boiling points and **(b)** the $\Delta H_c^\ominus$ values of some straight-chain primary aliphatic amines

Notice how these structures resemble that of ammonia. Perhaps this similarity will extend to their chemical properties?

Naming an amine is not always straightforward. Usually, we place the names of its alkyl, or phenyl, groups in front of the word '**-amine**', naming the shortest first, for example,

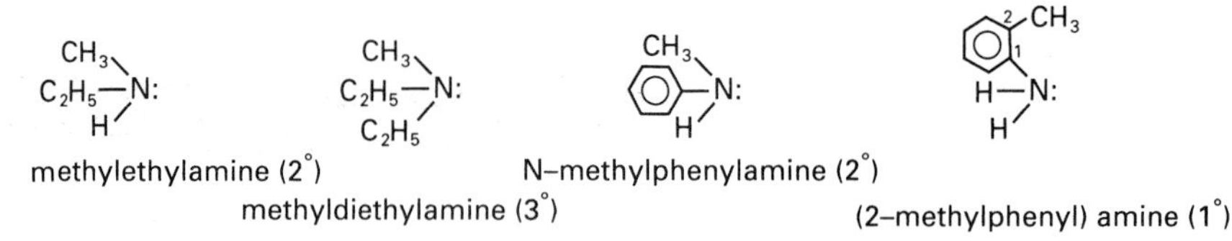

The last two amines are **structural isomers**. To distinguish them, place '**N–**' in front of each group which is attached to the nitrogen atom. Also, note the use of a bracket to clarify the name of a substituted □— group.

Sometimes '**amino-**' is used to denote an —NH_2 group, for example:

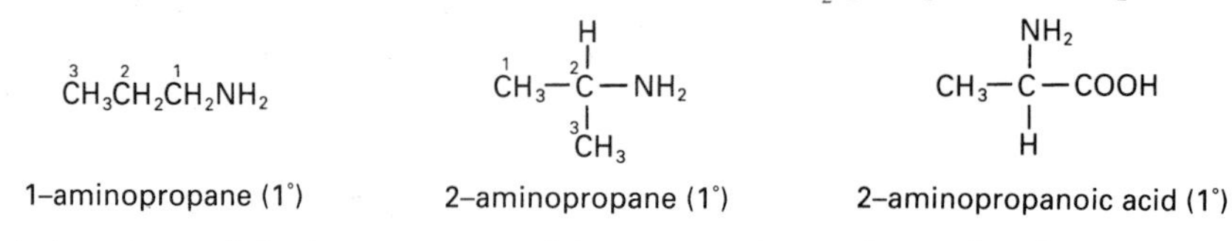

This is very useful where structural isomers exist or the molecule contains more than one functional group.

The boiling points of common amines are listed in Table 14.1. The lower aliphatic amines are gases or volatile liquids which often have a fishy smell. Indeed, various amines are found in rotting fish. Each type of amine (primary, secondary or tertiary) forms a homologous series; thus, we observe a typical

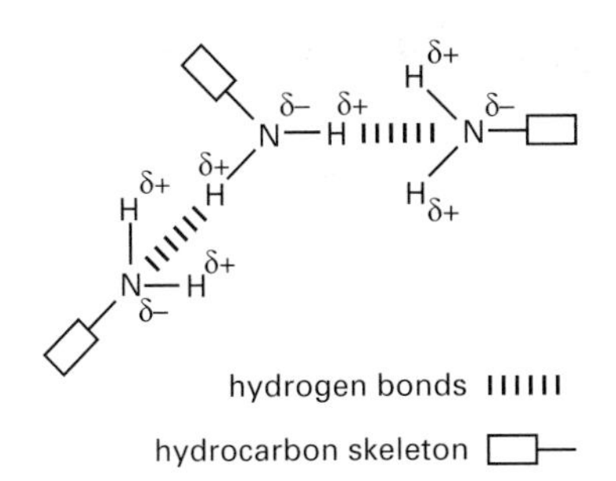

Figure 14.5 ▲
Strong intermolecular hydrogen bonding causes the boiling points of primary amines to be much higher than those of alkanes of similar molecular mass (CH_3NH_2, b.p. = −6°C; C_2H_6, b.p. = −89°C)

steady increase in boiling points and standard enthalpies of combustion, $\Delta H_c^{\ominus}$, of primary amines (Figure 14.4).

Due to the existence of intermolecular hydrogen bonds, the boiling points of primary and secondary amines are higher than those of hydrocarbons of similar molecular mass, Figure 14.5.

However, since tertiary amines have no N—H bonds, they are unable to form hydrogen bonds. Consequently, their molecules experience only weak intermolecular forces, and their boiling points do not differ greatly from branched alkanes of similar molecular mass; ($(CH_3)_3N$, b.p. 3°C; $(CH_3)_3CH$, b.p. −15°C).

Like ammonia, the lower aliphatic amines are readily soluble in water because they interact forming strong hydrogen bonds, Figure 14.6.

Since their molecules contain large hydrocarbon skeletons (non-polar), aromatic amines are virtually insoluble in water. They do dissolve, though, in organic solvents.

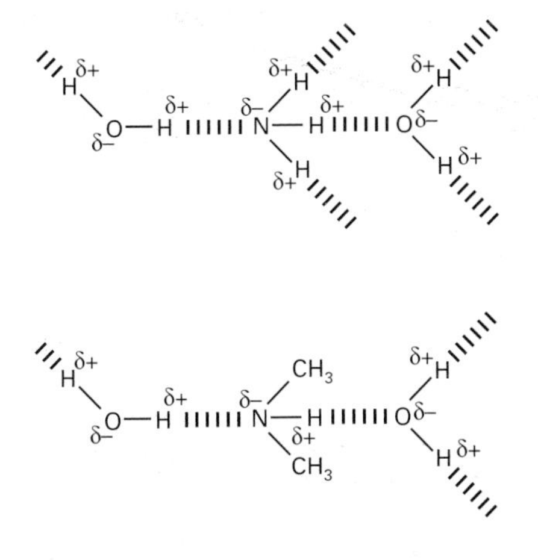

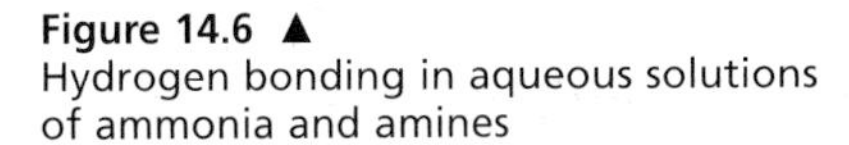

Figure 14.6 ▲
Hydrogen bonding in aqueous solutions of ammonia and amines

Table 14.1 Short structural formulae and boiling points of some amines

Name	formula	boiling point/°C
methylamine	CH_3NH_2	−6
dimethylamine	$(CH_3)_2NH$	7
trimethylamine	$(CH_3)_3N$	3
ethylamine	$C_2H_5NH_2$	17
1-aminopropane	$CH_3CH_2CH_2NH_2$	49
2-aminopropane	$CH_3CHNH_2CH_3$	33
1-aminobutane	$CH_3(CH_2)_3NH_2$	78
2-aminobutane	$CH_3CHNH_2CH_2CH_3$	68
phenylamine	$C_6H_5NH_2$	184

Primary, secondary and tertiary amines

You should note that amines are classified as being primary, secondary or tertiary compounds in a different way to that used to describe halogenoalkanes (section 10.1) and alcohols (11.2). In amines, the classification relates to the number of hydrocarbon skeletons which are attached to the *nitrogen* atom, as shown in the main text. In halogenoalkanes and alcohols, however, the classification is based on the number of hydrocarbon skeletons attached to the *carbon* which is bonded to the halogen atom or hydroxy group, for example:

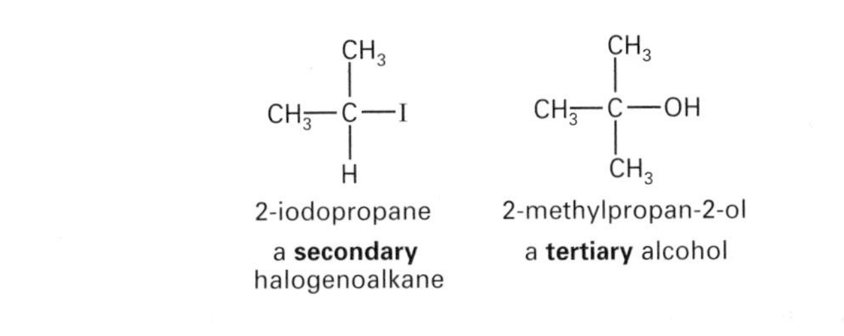

14.4 Preparation of amines and their uses

Primary aliphatic amines are prepared in the laboratory by the reduction of nitriles (section 14.2) or amides:

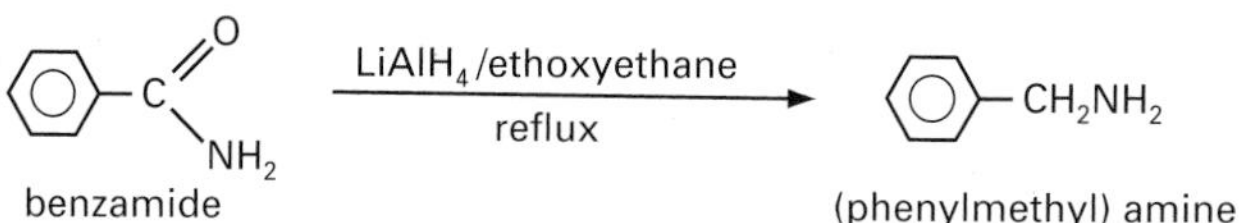

The Hofmann degradation of an amide produces a primary amine and also shortens the hydrocarbon skeleton (section 13.6).

Primary aromatic amines are formed by the reduction of nitro compounds with tin and concentrated hydrochloric acid (section 14.1).

Secondary and tertiary amines are best prepared by reducing N-substituted amides, for example:

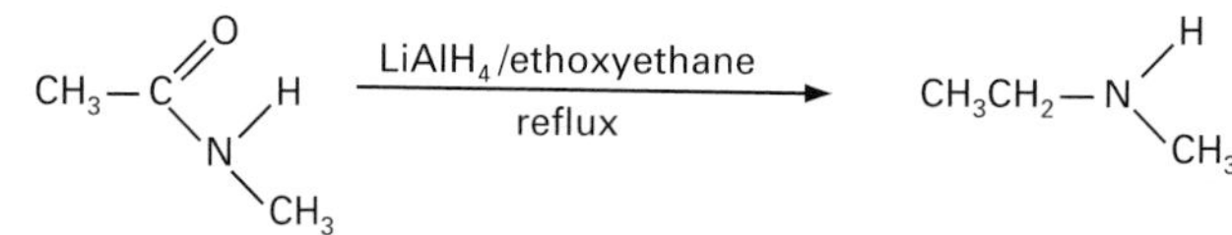

Industrially, amines are made by reducing a nitro compound with iron in hydrochloric acid, reacting a halogenoalkane with ammonia (section 10.5) or heating an alcohol with ammonia under pressure. Their most important uses, by far, are the manufacture of dyes from primary aromatic amines (section 14.6), and the production of nylon from 1,6-diaminohexane (section 14.9).

Hay fever – how does one amine cause so much misery?

Well over 10% of people in Britain have suffered from hay fever at some stage in their lives. Hay fever is an allergy, that is an abnormal response of the body towards a foreign substance which, for most people, is regarded as being harmless. The foreign substance is termed an **allergen**. In this case, the allergens are the pollen from wind-pollinated plants such as grasses, ragweed and birch trees. Most of these allergens are proteins, some are polysaccharides (section 12.9). For example, the main allergen in ragweed, named **antigen E**, is a protein with a relative molecular mass of about 38 000. An injection of just 10^{-12} g of antigen E, will produce a response in an allergic person!

The surface of a pollen grain has tiny hollows in which a variety of proteins are stored. When the pollen lands on a damp surface, such as the stigma of a plant or a mucous membrane, the grain swells and the proteins are rapidly released. Some of these proteins are powerful allergens and cause our immune system to release an antibody called **immunoglobin E**, or **IgE**, to defend against the allergen. Each IgE antibody molecule, itself a protein of relative mass about 196 000, is attached to specific cells, called **mast cells**, in the nose and bronchial passages. The IgE antibody and antigen combine at the surface of the mast cell, triggering the release of allergy mediators from granules within the mast cell. One of these mediators, an amine called **histamine**, Figure 14.8, is responsible for nearly all of the symptoms of hay fever – swelling of body tissue, nasal congestion, a runny nose and itchy, watery eyes.

For most people, the immune system gets to know a variety of harmless substances and infectious diseases in our early years and we are said to be desensitised towards them. Many hay fever sufferers inherit the allergy from their parents, though for some the symptoms may not appear until adulthood. There is considerable evidence that hay fever is more pronounced in areas of heavy traffic use. Exhaust fumes, particularly nitrogen dioxide, are thought to destroy **cilia**, the microscopic hairs that line the nostrils and windpipe and which carry foreign substances out of the body through the nose and mouth. This allows pollen to remain on the mucous membranes for longer and the allergens to more easily pass into the body.

Hay fever – how does one amine cause so much misery? *continued*

Allergies may be treated by avoiding the allergen but this is not feasible with substances like pollen, house dust or mites. Some sufferers gain relief through desensitisation, a process in which repeated doses of the allergen are injected just below the skin until the body 'gets used' to their presence. Antihistamine drugs, first introduced in 1945, alleviate the symptoms by blocking the receptor sites which are normally occupied by histamine, Figure 14.8. Sprays containing steroids can help reduce inflammation of the nasal linings and nasal decongestants provide short-term relief, though they are counter-productive over a long period because they deactivate the cilia.

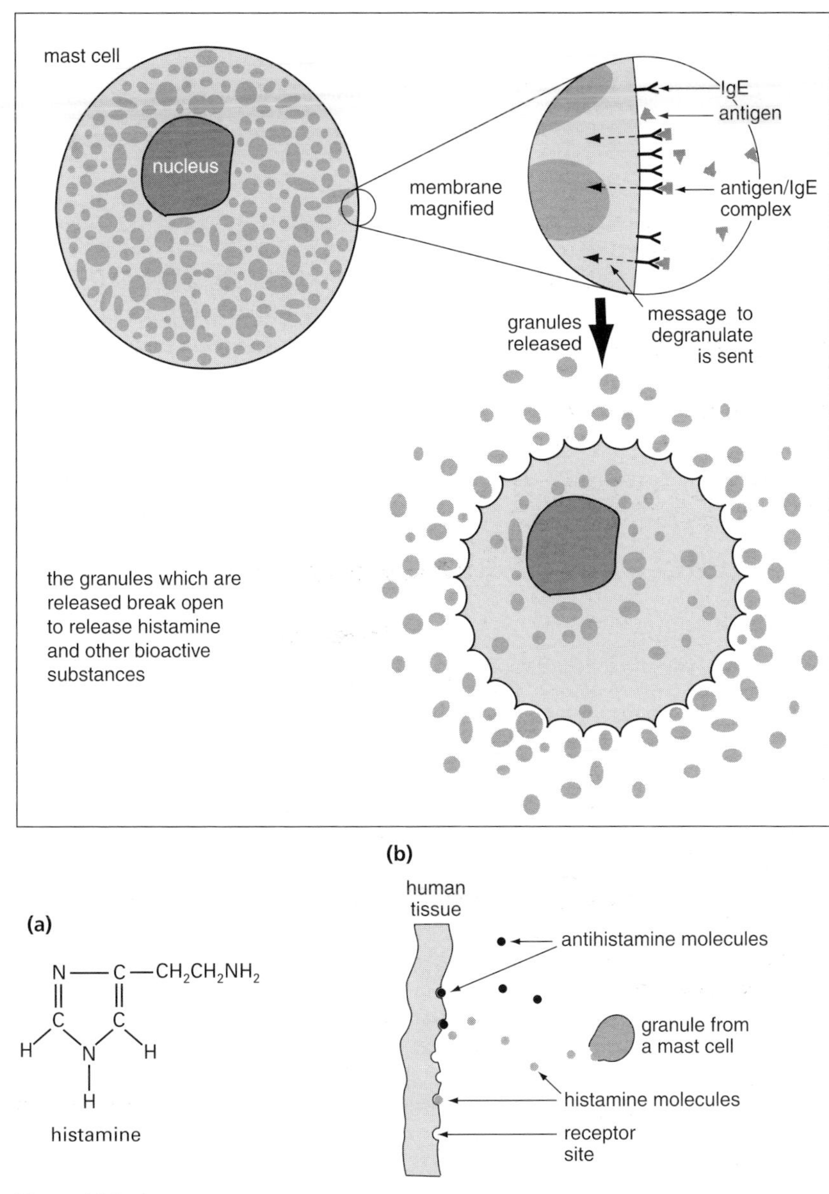

◄ **Figure 14.7**
Mast cells release mediators which attack invading parasites and, sometimes, harmless allergens such as pollen

Figure 14.8 ▲
(a) Histamine is a heterocyclic amine. **(b)** Antihistamine drugs are widely used to treat allergies and each contains, as does histamine, the ethylamine group ($-CH_2CH_2NH_2$). The drugs act by blocking the receptor sites on cells that are normally occupied by histamine

Activity 14.2

1. Explain why histamine is said to be a 'heterocyclic' amine.
2. Is histamine a primary or a secondary amine, or both?
3. In Britain, why are the effects of hay fever most pronounced in late spring and early summer?

Focus 14a

1. There are five main types of organic nitrogen compounds: **nitro compounds, nitriles, amines, amides** and **amino acids**, Table 14.2.

Table 14.2 The functional groups in some organic nitrogen compounds

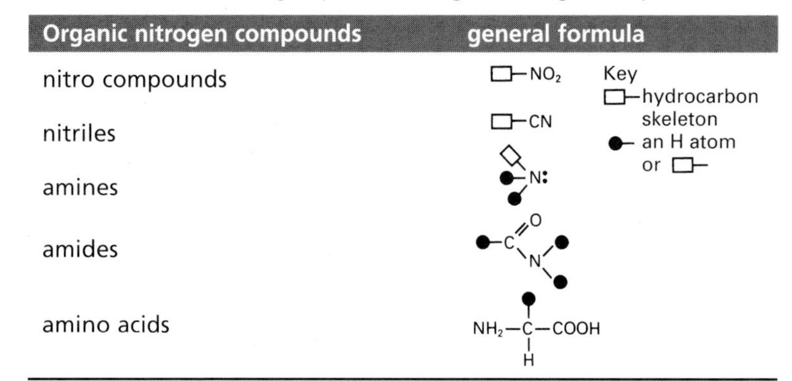

Organic nitrogen compounds	general formula
nitro compounds	□–NO_2
nitriles	□–CN
amines	(●)₃N:
amides	●–C(=O)–N(●)●
amino acids	NH_2–C(●)(H)–COOH

2. The aliphatic members of each type of organic nitrogen compound form **homologous series**. Since they contain polar functional groups, most organic nitrogen compounds form strong intermolecular dipole–dipole attractions. Thus, we find that i) their boiling points are usually higher than the alkanes of similar mass; ii) the lower aliphatic members are fairly soluble in water.
3. An amine may be **primary**, **secondary**, or **tertiary**, depending on whether its nitrogen atom is bonded to 1, 2 or 3 hydrocarbon skeletons, respectively.
4. Introducing a nitrile group into an organic molecule is an important step in adding a —CH_2— unit, section 14.2.

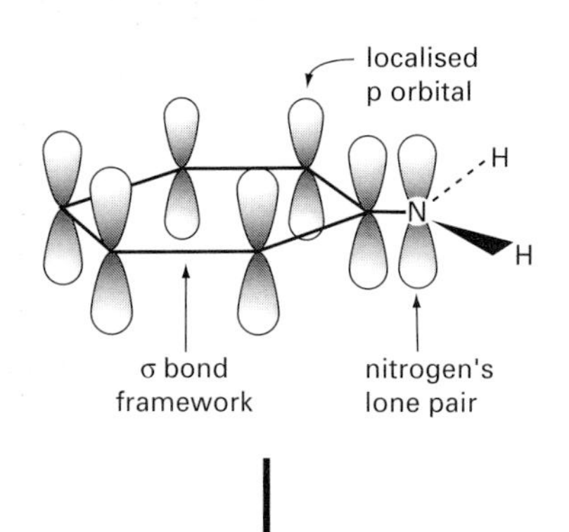

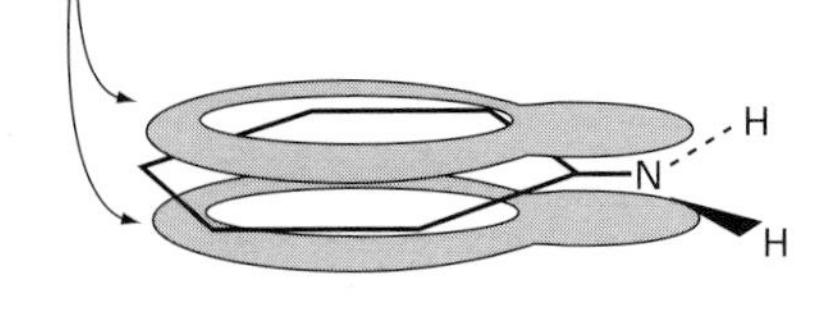

Figure 14.9
In phenylamine, sideways overlap of orbitals produces a **delocalised electron cloud** over the carbon and nitrogen atoms

14.5 Reactions of amines as bases

When aqueous solutions of aliphatic amines, and ammonia, are tested with universal indicator, a blue or purple colour is obtained, indicating that the solutions are **alkaline**. Amines, like ammonia, are said to be **Brønsted–Lowry bases** which means that they can accept protons from other molecules, such as water:

$$\gt N\!:\!(aq) + H_2O(l) \rightleftharpoons [\gt N{-}H]^+(aq) + OH^-(aq)$$

Notice that the nitrogen atom uses its lone pair of electrons to form a dative covalent bond with a proton obtained from the water molecule. An equilibrium mixture is formed in which there are very many more undissociated molecules than ions. Since they are only slightly dissociated in aqueous solution, amines are described as being **weak bases**. Quantitative experiments show that there is a general pattern in the basic strengths of amines:

increasing base strength →

primary aromatic amines	**ammonia**	**primary aliphatic amines**
(e.g. phenylamine, $C_6H_5NH_2$)	NH_3	(e.g. ethylamine, $C_2H_5NH_2$)

In phenylamine, the direct attachment of the —NH_2 group to the benzene ring considerably reduces its basic strength, that is its ability to accept protons, H^+ ions. Nitrogen's lone pair of electrons forms an extended delocalised cloud with benzene's π electron ring, Figure 14.9. Due to this delocalisation, the lone pair is much less available for dative covalent bonding

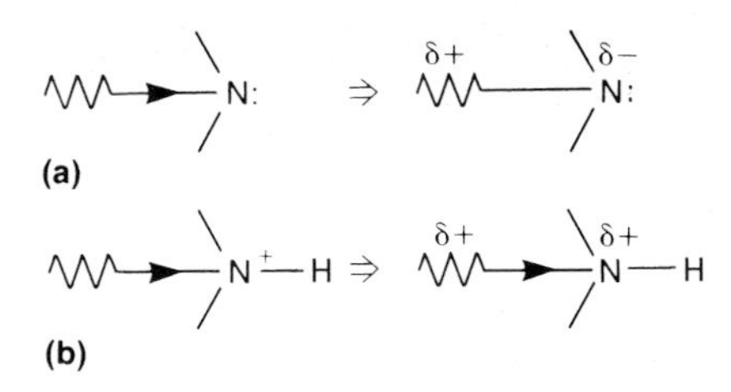

Figure 14.10 ▲
The **positive inductive effect** of an alkyl group, that is its electron releasing ability both **(a)** localises electron cloud on the nitrogen atom of an amine and **(b)** delocalises the positive charge on the alkylammonium cation

with a proton. Thus, phenylamine is a poor proton acceptor, making it a very weak base (it is about 40 000 times weaken than ammonia).

Exactly the opposite case can be made for aliphatic amines. As we saw on page 224, alkyl groups are able to push electrons away from themselves and this has two effects. First, the alkyl group is able to assist protonation by localising electron density on nitrogen, Figure 14.10a. Second, an alkyl group will help stabilise the resulting cation by delocalising its positive charge, Figure 14.10b. Both effects assist proton acceptance, making the base stronger, and this explains why primary aliphatic amines are stronger bases than ammonia. Methylamine, CH_3NH_2, for example, is about 25 times stronger a base than ammonia.

Amines react with acids at room temperature to form ionic salts, for example:

$$2C_2H_5NH_2(g) + H_2SO_4(aq) \longrightarrow [CH_3NH_3^+]_2SO_4^{2-}(aq)$$
methylamine → methylammonium sulphate

$$C_6H_5NH_2(g) + HCl(aq) \longrightarrow [C_6H_5NH_3^+]Cl^-(aq)$$
phenylamine → phenylammonium chloride

On evaporating their aqueous solutions, these salts are obtained as white crystalline solids, typically ionic, with fairly high melting points and good solubility in water. Stronger bases will displace amines from their ionic salts, for example:

$$[C_2H_5NH_3^+]Cl^-(s) + KOH(aq) \longrightarrow C_2H_5NH_2(aq) + KCl(aq) + H_2O(l)$$
strong base (KOH) — weak base ($C_2H_5NH_2$)

The base strength of secondary and tertiary aliphatic amines

Since alkyl groups have a positive inductive effect, we would expect the basic strength of an amine to increase with the number of alkyl groups attached to the nitrogen atom. This is true for secondary amines, for example diethylamine, $(C_2H_5)_2NH$, is nearly three times stronger a base than ethylamine, $C_2H_5NH_2$. However, tertiary aliphatic amines are *weaker* bases than primary and secondary aliphatic amines! Clearly, another factor must be involved here.

When an alkylammonium ion is hydrated, it is stabilised by **hydrogen bonding** with water molecules, Figure 14.11. As the number of N — H bonds decreases, the alkylammonium ion becomes less able to form hydrogen bonds and the stability of the ions decreases in the order primary > secondary > tertiary. Tertiary amines are surprisingly weak bases, therefore, because they have the least stable aqueous alkylammonium ions.

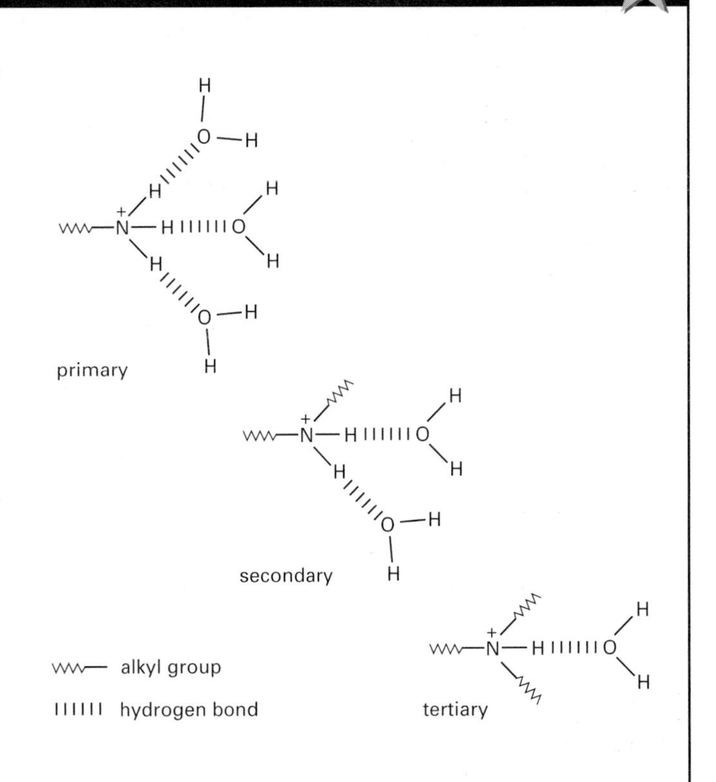

Figure 14.11 ►
Primary, secondary and tertiary alkylammonium ions forming hydrogen bonds with water molecules

Activity 14.3

1 Name the following amines
   a) CH_3NH_2
   b) $CH_3CH_2CH(NH_2)CH_3$
   c) $(C_2H_5)_3N$

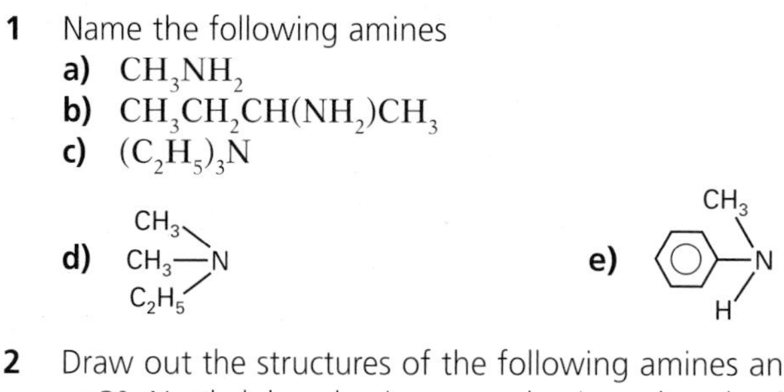

2 Draw out the structures of the following amines and then classify them as 1°, 2° or 3°: N-ethylphenylamine, propylamine, phenylamine, diethylamine, 4-methylphenylamine, 1,4-diaminobenzene, triethylamine.

3 Complete the following equations:
   a) $2C_2H_5NH_2(l) + H_2SO_4(aq) \longrightarrow$
   b) $(C_2H_5)_3N{:}(l) + HCl(aq) \longrightarrow$
   c) $[CH_3NH_3^+]Cl^-(s) + NaOH(aq) \longrightarrow$
   d) $[C_6H_5NH_3^+]_2SO_4^{2-}(s) + 2KOH(aq) \longrightarrow$

14.6 Amines as nucleophiles

Reaction with acid chlorides and acid anhydrides

Amines react vigorously with acid chlorides at room temperature to give N-substituted amides:

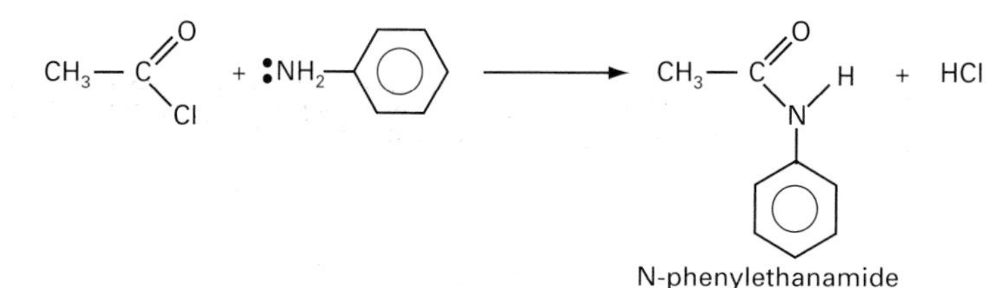

N-phenylethanamide

This reaction is an example of **acylation**, that is the introduction of an acyl group, $\sim\sim C(=O)-$, into a molecule ($\sim\sim$ is an alkyl group).

In industry, acid anhydrides are used as **acylating agents** in preference to acid chlorides. The anhydrides are reactive enough for reaction to occur but there is much less danger of the reaction running out of control. Also, the anhydrides cost less to produce than acid chlorides, thereby making the process more economical. For example, ethanoic anhydride is used as an ethanoylating agent in the manufacture of aspirin (section 6.1).

When a $C_6H_5-C(=O)-$ group is introduced into a molecule, the process is termed **benzoylation**, for example:

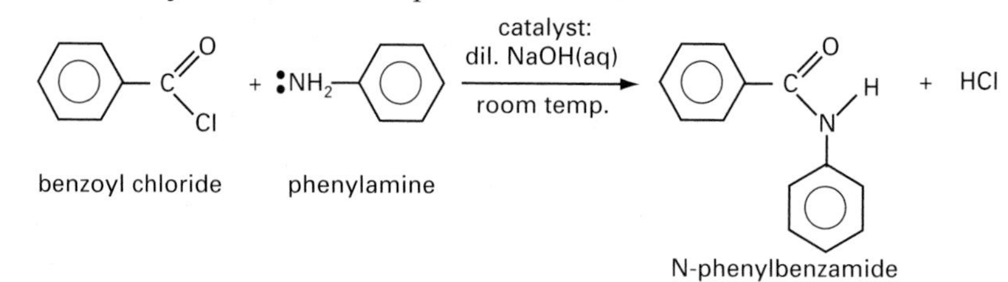

benzoyl chloride   phenylamine

N-phenylbenzamide

Amides are crystalline solids which, when purified, have sharp melting points. As a result, chemists can often identify an amine by preparing an amide. To do this, the unknown amine is acylated; then the resulting amide is purified, usually by recrystallisation, and dried. Its melting point is taken and the value is compared with the accepted melting points of amides found in a data book.

In practice, when faced with an unknown compound, an organic chemist will suggest a molecular structure based on mass, infrared and NMR spectra. By preparing a derivative and taking its melting point, we can confirm the structure. Some typical characterisation reactions are:

- aldehyde or ketone $\longrightarrow$ 2,4-dinitrophenylhydrazone (section 12.5);
- acid, acid derivative or amine $\longrightarrow$ amide.

Reaction with halogenoalkanes

When an ethanolic solution of an aliphatic amine and a halogenoalkane is heated under pressure, a mixture of secondary, tertiary and quaternary alkylammonium salts is formed (section 10.5). The composition of this mixture depends on the mole quantities of the reactants used, for example:

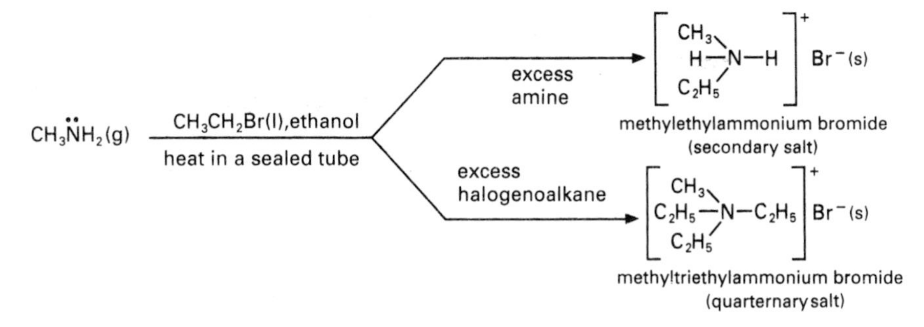

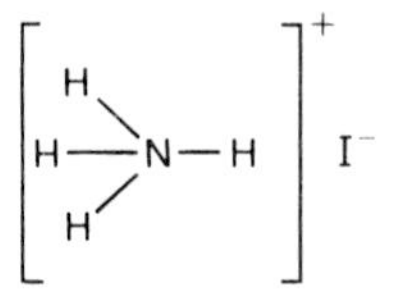

Figure 14.12
Ammonium iodide, $NH_4^+I^-$

Notice the structural similarity between alkylammonium salts and simple ammonium salts, such as ammonium iodide (Figure 14.12). Since the free amines are weak bases, they can be liberated from these salts by a strong base (section 14.6).

This synthesis is of limited value because the mixture of amines produced is difficult to separate, so secondary and tertiary amines are usually prepared by reducing acid amides (section 14.4).

Reaction with nitrous acid, HNO_2

Nitrous acid is unstable and exists only in solution so to test its reactions, we prepare small quantities from ice-cold solutions of sodium nitrite and dilute hydrochloric acid:

$$NaNO_2(aq) + HCl(aq) \xrightarrow{\text{at about 5°C}} HNO_2(aq) + NaCl(aq)$$

When formed, the nitrous acid solution is kept below 5°C by cooling in an ice/water mixture. Table 14.3 describes the reactions of primary amines with nitrous acid at 5°C.

When primary amines react with nitrous acid at about 5°C, **diazonium salts** are formed:

$$\square\text{—}\ddot{N}H_2 \xrightarrow[\text{5°C}]{NaNO_2/HCl(aq)} \square\text{—}\overset{+}{N}\equiv N \;\; Cl^-$$

a diazonium chloride

These ionic compounds are unstable and even at this low temperature, *aliphatic* diazonium ions decompose quickly, giving nitrogen gas and an alcohol as the main products, for example:

$$[C_2H_5\overset{+}{N}\equiv N]\,Cl^-(aq) \xrightarrow{\text{rapid decomposition}} \underset{\text{ethanol}}{C_2H_5OH(aq)} + N_2(g)$$

However, aromatic diazonium salts are more stable and can exist in aqueous solution, providing that the temperature does not rise above 5°C. Although solid diazonium salts have been isolated, they are very unstable and most are explosive.

Aromatic diazonium salts take part in two types of reaction: (i) those in which nitrogen gas is evolved and (ii) those in which the nitrogen atoms are retained.

Table 14.3 Results obtained when primary amines are added dropwise to nitrous acid at 5°C.

With ethylamine, $C_2H_5NH_2$, a primary aliphatic amine	
observation	**product**
colourless solution, gas bubbles released	A diazonium salt, $[C_2H_5 - N^+ \equiv N]\,Cl^-$, which rapidly decomposes giving $N_2(g)$, and ethanol, C_2H_5OH
With phenylamine, C_6H_5–NH_2, a primary aromatic amine	
observation	**product**
colourless solution, no gas bubbles	$[C_6H_5-\overset{+}{N}\equiv N]\,Cl^-$ A diazonium salt, benzenediazonium chloride

Reactions in which nitrogen gas is evolved

Loss of N_2 from the benzenediazonium ion gives a **phenyl cation**,

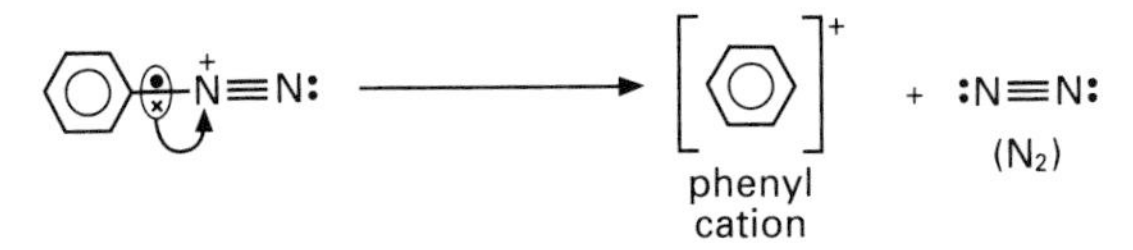

which readily reacts with a nucleophile. The overall effect is the attachment of a nucleophilic group to the benzene ring. Since the benzene ring normally reacts with electrophiles (section 9.5), this is an extremely useful synthetic step. Some examples are given below:

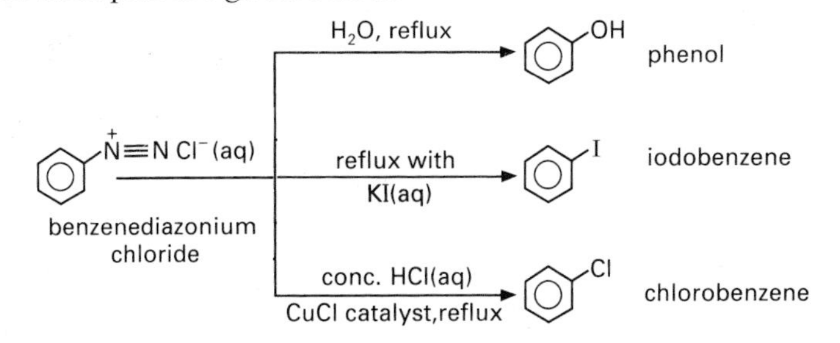

Reactions in which the nitrogen atoms are retained

Aromatic diazonium salts react with cold alkaline solutions of aromatic amines and phenols to give coloured **azo-compounds**, for example:

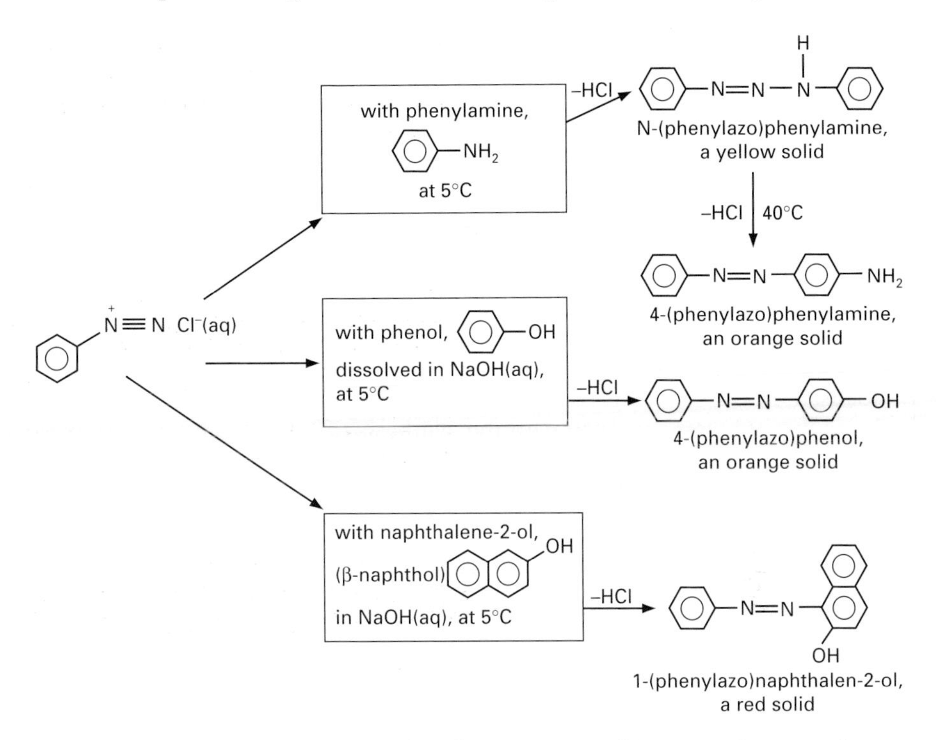

Figure 14.13
Delocalisation of the electron cloud from the N≡N bond towards benzene's π electron ring produces a δ+ charge on the end nitrogen atom. This enables it to act as an electrophile in coupling reactions

Because they give such clear results, these **azo-coupling reactions** can be used to detect primary aromatic amines, as their diazonium salts. Azo-coupling reactions have three common features:

- the diazonium ion acts as an **electrophile**, see Figure 14.13
- the — NH_2 and — OH groups **activate** the benzene ring towards nucleophilic attack by giving it a slight negative charge at the 2- and 4-positions enabling the coupling reaction to occur rapidly even at low temperature

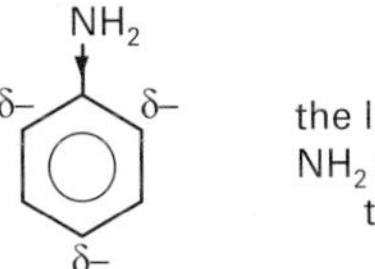

the lone pairs of electrons on the NH_2 and OH groups are attracted towards the benzene ring

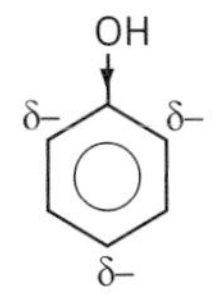

- each product contains two benzene rings linked by an **azo group**, — **N = N** —. This group is a **chromophore**, that is it is a structural arrangement which has the potential to create colour in a molecule, section 18.1.

Azo-coupling reactions are used industrially to make **azo-dyes**. For example, an azo-dye known commercially as **Acid Orange 7**, is manufactured by a coupling reaction with naphthalene-2-ol (β-naphthol):

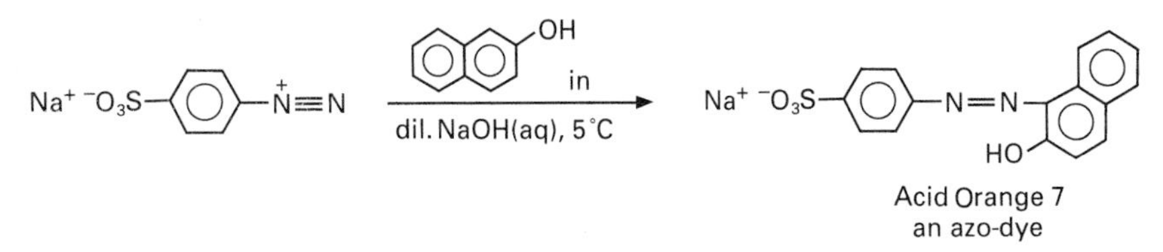

Activity 14.4

Describe two chemical tests which may be used to tell the difference between

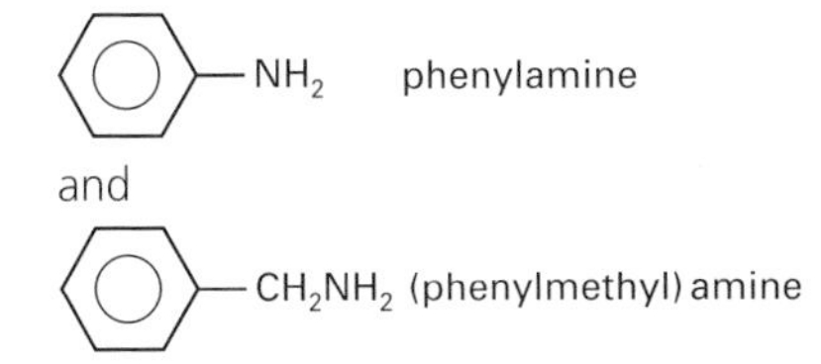

Dyes

In 1856, a fifteen-year-old British chemist named William Henry Perkin was trying to synthesise **quinine**, a drug used to fight malaria, by oxidising phenylamine with potassium dichromate. To his surprise, he obtained a violet substance that could be used as a dye. Perkin soon realised the commercial potential of the dye, which he called **mauve**, and he set up factories around the country to make it. Fortunately for Perkin, the mauve colour was very fashionable at that time and he made so much money that he was able to retire at the age of thirty six!

Following Perkin's discovery, attention became focused on the synthesis of natural dyes. A major breakthrough came in 1880 with the synthesis of **indigo**, a blue dye made by fermenting the leaves of a tropical plant called Indigofera. **Indigo** is well-known as the dye used to colour denim jeans. As you know, the colour usually fades every time the jeans are washed! Natural dyes tend to be fairly dull colours which fade rapidly and which often need a **mordant** to make the dye bind to the fabric's structure.

Nowadays, nearly all of the 3000 dyes available are synthetic, with different dyes chosen for particular applications. The dyes are classified according to the chemical nature of the **chromophore**, that is the unsaturated group of atoms which imparts the colour. In azo dyes, the chromophore is an **azo group,** **—N=N—**, which links two benzene rings, as in Acid Orange 7. Quite often, synthetic dyes are modified forms of natural dyes, for example, the blue and green phthalocyanine dyes, whose chemical structure is based on that of chlorophyll.

Dyes may be attached to the fabric via physical interactions (non-reactive dyes) or chemical bonds (reactive dyes). The earliest dyes were non-reactive, for example the **vat dyes** are water-insoluble substances which may be reduced, in a vat, to water-soluble dyes. After the material has been dyed, it is exposed to the air and the dye is oxidised to its water-insoluble form which then adheres to the fibres of the material. Indigo is a vat-dye. Many azo dyes are **direct dyes** because they can be applied directly to the fabric. Unfortunately, the use of non-reactive (vat and direct) dyes to colour natural fibres, such as cotton and wool, sometimes produces disappointing results. A major breakthrough came in 1956 when two chemists working at ICI produce the first **reactive dyes**, that is dyes in which the chromophore is attached to a functional group which can form a *chemical* bond with the fabric. An example would be the reaction of acid chloride group, COCl, in the dye with a OH or NH_2 group in the fabric, forming an ester or amide group.

Perkin's accidental discovery of mauve marked the birth of the synthetic organic chemicals industry. When he sold his factories in 1876, he turned his attention to research and his contribution to the development of organic chemistry was recognised by a knighthood in 1906.

The chemistry of dyes is covered in detail in section 18.4.

14.7 Amides

Amides are derivatives of carboxylic acids, named by replacing the '**-oic acid**' of the parent's name with '**-amide**'.

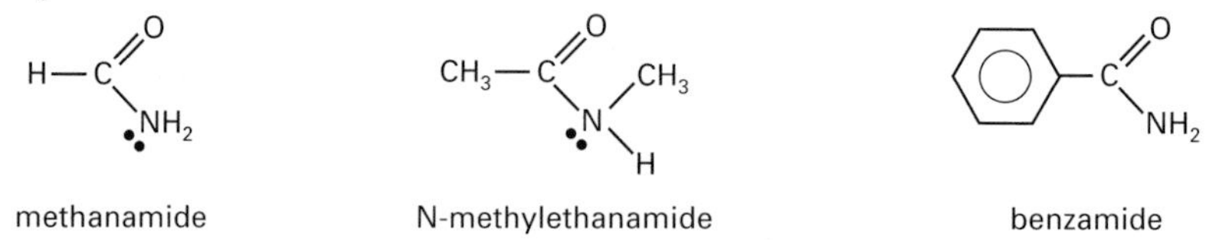

N-methylethanamide is an example of an **N-substituted amide** where the N-prefixes the group(s) attached to the nitrogen. Methanamide is a liquid, but the other amides are white crystalline solids with fairly high melting points. The lower amides are extremely soluble in water due to their ability to form hydrogen bonds with the water molecules (Figure 13.6).

Amides are most conveniently prepared from acid chlorides:

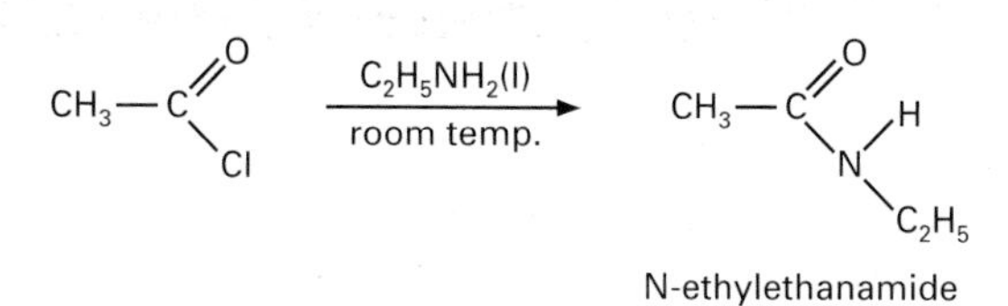

Although the nitrogen of the amide group has a lone pair of electrons, it is a very weak base and a poor nucleophile because the lone pair is delocalised over the N — C — O structure and less available, therefore, for donation to electron-poor species, such as the H^+ ion (Figure 13.8). The reactions of amides were discussed in section 13.6.

Nylon – a synthetic polyamide

Nylon, the first completely synthetic fibre, was developed by Wallace Hume Carothers in America during the 1930s and was first marketed in 1939 for stockings. There are various types of nylon, the most common being **Nylon 66**, which is made by the copolymerisation of **hexanedioic acid** and 1,6-diaminohexane:

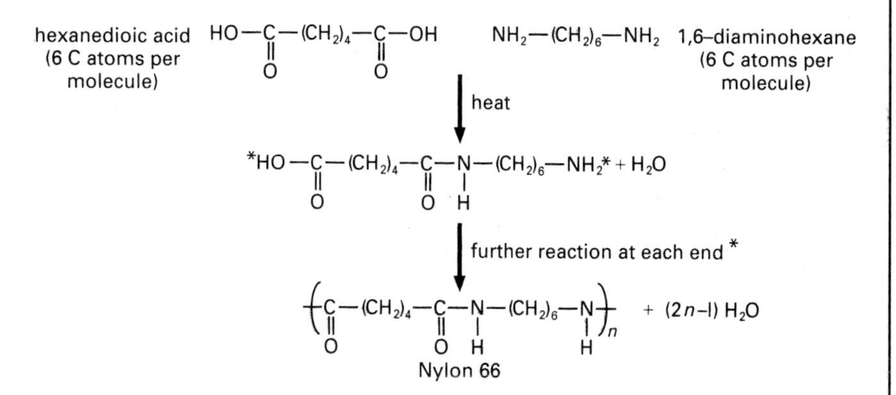

The product is known as Nylon **66** because both of the monomers have molecules which contain **6** carbon atoms. The monomers are made from benzene, as shown below:

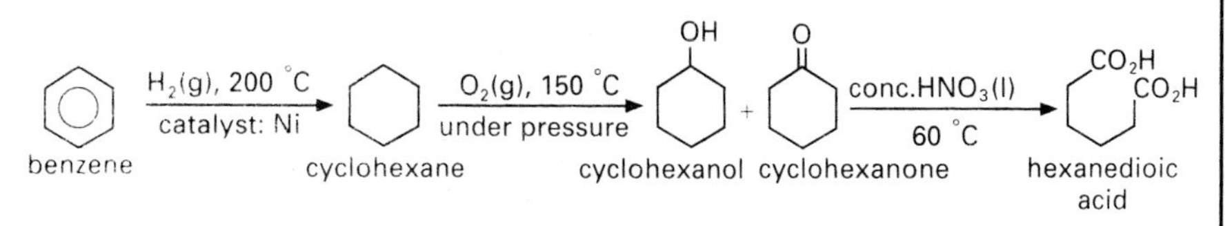

Some of the hexanedioic acid is converted into 1,6-diaminohexane:

$$HO_2C—(CH_2)_4—CO_2H \xrightarrow{\text{conc. } NH_3(aq)} NH_4^{+\,-}O_2C—(CH_2)_4—CO_2^{-}NH_4^{+}$$

$$\downarrow \text{heat catalyst}$$

$$\underset{\text{1,6-diaminohexane}}{H_2N—(CH_2)_6—NH_2} \xleftarrow[\text{catalyst: Ni}]{H_2(g),\ 200°C} NC—(CH_2)_4—CN$$

Nylon – a synthetic polyamide *(continued)*

Since nylons are **thermoplastics**, that is, they melt without decomposition, the hot polymer can be extruded, that is, forced through small holes called spinnerets. As they cool, the streams of polymer solidify, forming fibres. However, in this state, the fibres have little strength because the long-chain polymer molecules are disordered. As with polyester fibres, by gently stretching the nylon fibres, the molecular chains adopt the same orientation (Figure 13.9). This enables the fibres to cross-link via an extensive network of intermolecular **hydrogen bonds**:

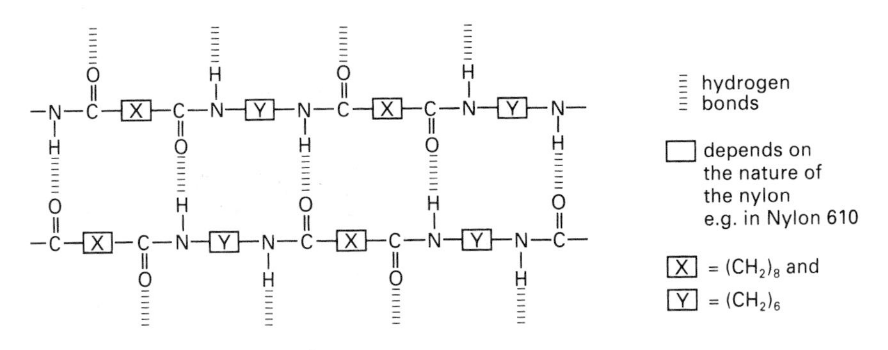

As a result, the nylon fibres become much stronger and more elastic, which is hardly surprising when one considers the similarity between this structure and that of silk (Figure 14.25).

A closer look at nylon's structure also shows that each monomer molecule is attached to its neighbour by a **peptide bond**, –C(=O)–N(H)–, and every time a peptide bond is formed, a water molecule is released. Thus, this is a **condensation polymerisation** (i.e. one in which a small molecule, such as H_2O, is released every time two molecules of the monomer(s) join together). The peptide bond is a characteristic structural feature in N-substituted amides (section 14.7) and proteins (section 14.10). Nylons and proteins are both described, therefore, as being **polyamides** and consequently the physical properties of nylon fibres bear some similarity to those of natural 'protein' fibres, such as wool. Although they are not as comfortable to wear as wool, nylons are harder wearing and dry more rapidly – thus nylons are popular materials in the clothing industry. Nylons are also used in a wide variety of other applications where strength and flexibility are required, for example dinghies, fishing line, climbing ropes and brushes.

What properties of nylon make it suitable for use in inflatable boats?

Focus 14b

1 Due to the lone pair on nitrogen, amines react as nucleophiles and bases. Some important reactions of primary amines are summarised in Table 14.4.

Table 14.4 Important reactions of amines

Reagents/conditions	1° amines, e.g. $C_2H_5NH_2$ and $C_6H_5NH_2$
acid (e.g. HCl(aq)), room temp.	**ionic salts** are formed $\longrightarrow C_2H_5NH_3^+Cl^-$
acid chloride, e.g. 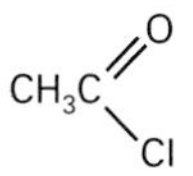 CH_3COCl room temp. or warming (base catalyst needed if aromatic amine used)	**N-substituted amides** formed 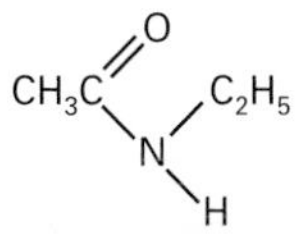 $\longrightarrow CH_3CONHC_2H_5$
halogenoalkane heat under pressure (e.g. CH_3I with excess amine)	$\longrightarrow$ **amines** 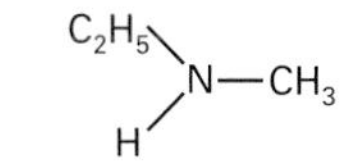 $C_2H_5NHCH_3$
$NaNO_2$/dil. HCl(aq), which forms nitrous acid, HNO_2, at 5°C	aliphatic: $\longrightarrow$ **alcohol** + $N_2(g)$ (C_2H_5OH); aromatic: $\longrightarrow$ **diazonium salt** 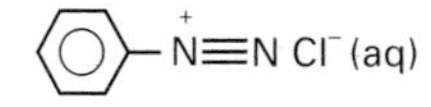 $C_6H_5\overset{+}{N}{\equiv}N\ Cl^-(aq)$

2 **Amines** are weak bases. Thus, they ionise slightly in aqueous solution to give an equilibrium mixture in which the concentrations of molecules and ions are constant, yet very different:

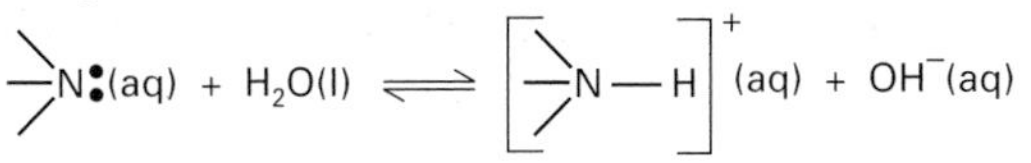

$$\geq N\!:\!(aq) + H_2O(l) \rightleftharpoons [\geq N-H]^+(aq) + OH^-(aq)$$

high concentration of molecules — very low concentration of ions

3 Experiments reveal a pattern in amine basicity:

increasing base strength →

primary aromatic amines << ammonia < primary aliphatic amines

e.g. $C_6H_5NH_2$ — e.g. $C_2H_5NH_2$

This pattern is explained by the greater localisation of the lone pair on the amine's nitrogen atom (moving left to right) and, thus, its increasing tendency to accept a proton.

4 The reactions with nitrous acid are important because they allow us to distinguish between a primary aliphatic and primary aromatic amine. Primary aromatic amines form more stable **diazonium salts** which are very valuable in organic synthesis.

5 Aromatic diazonium salts give two types of reactions: i) those in which nitrogen gas is evolved and a nucleophilic group, such as OH^- or I^-, is attached to the aromatic ring; ii) coupling reactions in which the **azo group**, $-N=N-$, is retained.

6 **Amides** are white crystalline solids which, when pure, have very sharp melting points. Thus, an unknown acid derivative may be identified by reacting it with ammonia or, a particular amine, and taking the melting point of the amide which is formed.

7 Although amides contain the NH_2 group, the nitrogen lone pair is delocalised over the $N-C-O$ σ bond framework. Thus, amides are extremely weak bases and very poor nucleophiles.

14.8 Why do we need proteins and what are they made of?

The name protein, taken from the Greek word *proteios*, or primary, was first used by the Dutch chemist G.J. Mulder in 1835 to describe a wide range of organic nitrogen compounds which were found in the cells of animals and plants. They are, indeed, the primary ingredient for life, making up about 50% of the dry mass of animals. Some of the many important roles of proteins within organisms are given below.

- **Structural proteins** are the main components of hair, skin, nails, teeth and connective tissue.
- All **enzymes** are proteins. Enzymes are substances that catalyse specific reactions in and around cells (section 24.4).
- **Transport proteins** enable substances to move around the body, for example haemoglobin is the protein in red blood cells which transports oxygen molecules.
- Proteins are a major component of **cell membranes**. Some proteins strengthen the membrane's structure whilst others control the transfer of materials across the membrane.
- **Immunoproteins** are the antibodies which form a defence mechanism against invading microorganisms and foreign substances.
- Some proteins act as **hormones**, substances which control growth rate and regulate metabolic functions.

It is estimated that there are around 30 000 different proteins in the human body, with molecular masses ranging from several thousand to more than a million. Although there are so many proteins, each with different functions, all proteins are made from the same building blocks – these are the **2-amino acids**.

Proteins in our diet

When eaten, proteins are hydrolysed under acidic or enzymic conditions into amino acids. Cells then bring these amino acid residues together in a very specific fashion, thereby forming the tens of thousands of proteins we require for the growth, repair and maintenance of our bodies. Various reports show that the daily dietary requirement for protein varies with age, sex and physical activity. Thus, a very active teenage girl, weighing 55 kg, needs around 60 g of protein per day, about the same amount as a sedentary, middle aged man who weighs 70 kg. Weight for weight, children require more protein than adults because of their rapid growth.

The proteins in our bodies are made from 20 naturally occurring amino acids, Table 14.5. In 1955, W. Rose, an American biochemist, placed volunteer students on a diet in which one of these amino acids was omitted, in turn. He found that the omission of 8 amino acids had an immediate detrimental effect and these were termed **essential amino acids** (they are Iso, Leu, Lys, Met, Phe, Thr, Val). Adults can synthesise the other 12 naturally occurring amino acids from glucose and almost any source of nitrogen.

However, the essential amino acids *must* be eaten. In addition, young children cannot synthesise histidine, an amino acid which is needed to bind the haem unit to the globin in haemoglobin, the red oxygen carrier in blood. Histidine deficiency in children may lead to impaired haemoglobin production and eczema.

Table 14.5 The twenty naturally occurring 2-amino acids. Glycine apart, the acids occur in the L-configuration. Adults can only synthesise twelve of these amino acids, the other eight must be eaten and they are termed essential amino acids.

Of the 20 commonly occurring amino acids 19 are based on the structure shown below; the particular side group for each amino acid is listed opposite.

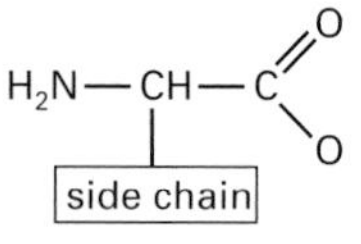

Proline, which can be considered a heterocyclic amino acid, has a structure closely related to the other 19. It is shown below in full :

proline (Pro)
pI = 6.3

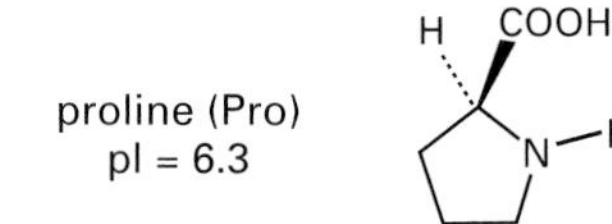

Proteins in our diet *continued*

Table 14.5 continued

	symbol	pI	side chain
side chain: hydrogen or alkyl group			
glycine	(Gly)	6.0	—H
alanine	(Ala)	6.0	$—CH_3$
valine*	(Val)	6.0	$—CH(CH_3)_2$
leucine*	(Leu)	6.0	$—CH_2CH(CH_3)_2$
isoleucine*	(Ile)	6.0	$—CH_2CH_2CH_3$ (with CH_3 on first carbon)
side chain: contains hydroxy group			
serine	(Ser)	5.7	$—CH_2OH$
threonine*	(Thr)	5.7	$—CHOH$ (with CH_3)
side chain: contains hydroxy group			
cysteine	(Cys)	5.1	$—CH_2SH$
methionine*	(Met)	5.7	$—CH_2CH_2SCH_3$
side chain: dicarboxylic			
aspartic acid	(Asp)	2.8	$—CH_2COOH$
glutamic acid	(Glu)	3.2	$—CH_2CH_2COOH$
side chain: cyclic (aromatic and heterocyclic)			
phenylalanine*	(Phe)	5.5	$—CH_2—C_6H_5$
tyrosine	(Tyr)	5.7	$—CH_2—C_6H_4—OH$
tryptophan*	(Trp)	5.9	$—CH_2$–indole (N–H)
histidine (* for children)	(His)	7.6	$—CH_2$–imidazole (N, N–H)
proline (see opposite)			
side chain: diamino			
lysine*	(Lys)	9.7	$—CH_2CH_2CH_2CH_2NH_2$
arginine	(Arg)	10.8	$—CH_2CH_2CH_2NHCNH_2$ (C=NH)
side chain: acid amide			
asparagine	(Asn)	5.4	$—CH_2CONH_2$
glutamine	(Gln)	5.7	$—CH_2CH_2CONH_2$

(* are essential amino acids;
pI is the isoelectric point, see page 383.)

The ability of a given diet to satisfy the body's nitrogen requirement depends on its digestibility, the total protein content and the composition of the amino acids in the proteins. The fraction of the dietary nitrogen which is retained after the protein has been eaten is termed the **net protein utilisation (n.p.u.)** or **biological value**, Table 14.6. Proteins from animal sources, such as milk and eggs, have a high n.p.u. value and are rich sources of protein, whereas seed proteins, such as maize and wheat, must be consumed in much larger quantities if they are to satisfy a person's protein requirement. In addition, seed proteins provide a narrower range of amino acids, for example showing a deficiency in methionine, lysine or tryptophan. Thus, a daily diet, whether vegetarian or otherwise, should contain protein from a variety of sources so that amino acids which are lacking in one food are supplied by another.

Excess protein is broken down by our bodies, providing about the same amount of energy as we get from carbohydrates. Some athletes believe that a high protein diet improves performance and, in fact, an athlete does use about 10% more protein than a moderately active person. However, a high protein diet, coupled with insufficient carbohydrate intake can lead to **uremia**, a form of blood-poisoning in which an over-production of urea causes kidney failure.

The **Biuret test** is often used to detect the presence of proteins, for example in food. The sample in solution is added to an equal volume of aqueous sodium hydroxide (2 mol dm^{-3}) and 1 drop of copper sulphate is added (0.1 mol dm^{-3}). A violet complex is formed between the metal ions and the peptide bonds as a result of dative (co-ordinate) covalent bonding (section 4.3).

Table 14.6 The value of some foods as sources of dietary nitrogen, as published by the World Health Organisation (1973). The percentage of the nitrogen which is retained after feeding is termed the net protein utilisation

Type of food	net protein utilisation
human milk	95
hen's egg	87
cow's milk	81
rice	63
peanuts	57
wheat	49
maize	36

All of these foods are rich in protein. Digestive enzymes catalyse the hydrolysis of the proteins forming polypeptides and individual amino acids

14.9 Amino acids

Most of the proteins found in living organisms are made of 20 naturally occurring amino acids, Table 14.5. Proline apart, these amino acids have the general formula:

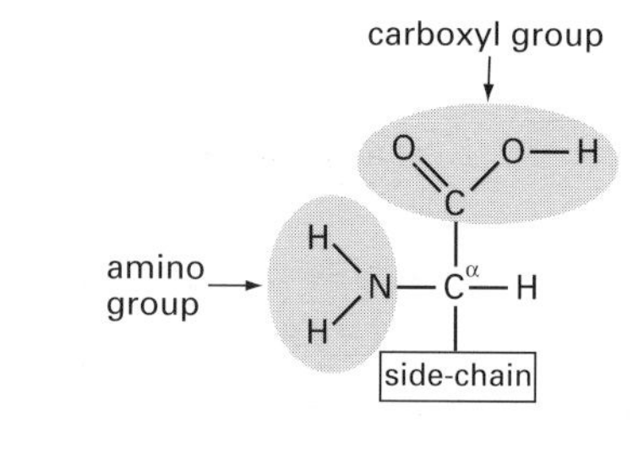

Since the amino group is attached to the carbon atom next to the carboxyl group, they are termed **α-amino acids** or **2-amino acids**. The nature of the side-chain has a substantial effect on the 3-D structure of a protein and its resulting properties (sections 14.11–14.14).

With the exception of glycine, the 2-amino acids possess a **chiral centre**, that is an atom through which there is no plane of symmetry. Thus, they are chiral molecules and exist as **enantiomers**, that is non-superimposable mirror images. Figure 14.14, for example, shows the Fischer projection formulae of the enantiomers of alanine (2-aminopropanoic acid). These two structures, termed the **L- and D- configurations**, are optical isomers. The dextrorotatory (+)-isomer rotates the plane of plane polarised light clockwise, the laevorotatory (–)-isomer rotates it anti-clockwise. Note that the symbols D and L refer to the 3-D arrangement of atoms, *not* to the direction of optical rotation. Thus, L-alanine is actually dextrorotatory and is fully described by the name, (+)-L-alanine! (You might like to review the topics of chirality and optical activity – these were covered in section 6.5).

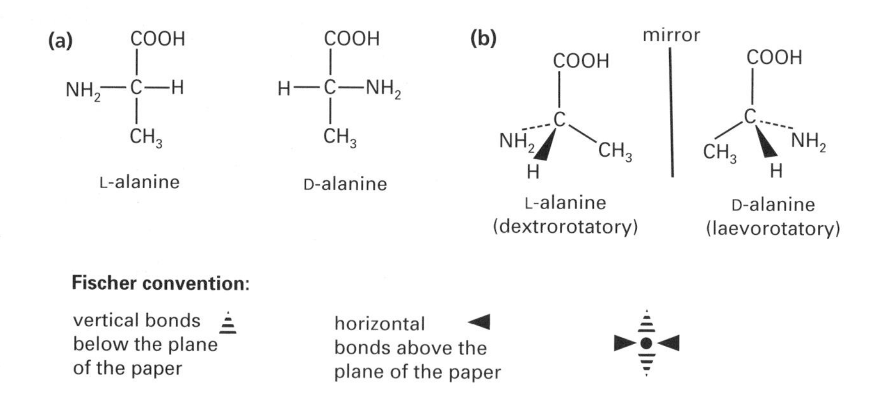

Figure 14.14
(a) Fischer projection formulae of the enantiomers of alanine. In Fischer projections, the vertical bonds go down below the plane of the paper whereas those placed horizontally come out of the plane of the paper. By convention, the carboxyl group is placed at the top and the $-NH_2$ group and the H atom are placed horizontally. In such an arrangement, the L-configuration will have the $-NH_2$ to the left of the α-carbon atom whereas in the D-configuration, the $-NH_2$ group is located to the right of the α-carbon. **(b)** The molecules of L- and D-alanine are enantiomers, that is, non-superimposable mirror images. Enantiomers are said to be 'optically active' because they have the ability to rotate the plane of plane polarised light. The dextrorotatory (+)-isomer rotates the plane of plane polarised light clockwise, the laevorotatory (–)-isomer rotates it anti-clockwise.

All naturally occurring 2-amino acids adopt the L-configuration. Only twelve of these amino acids can be synthesised by the adult human body, the other eight must be eaten and these are known as the essential amino acids (Table 14.5). Histidine is also an essential amino acid for children.

Amino acids have high melting points, for example glycine melts at 234°C. Generally speaking, they are very soluble in water but only dissolve slightly in organic solvents. These physical properties are typical of those in an ionic salt, such as sodium chloride. In fact, in the solid state, amino acids exist as internal ionic salts, called **zwitterions** (Figure 14.15).

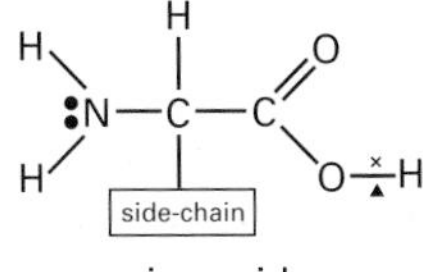

side-chain

zwitterion

Figure 14.15 ▲
A **zwitterion** is a species which contains a positive group and a negative group. Amino acids form zwitterions due to the transfer of a proton from the carboxyl group (acidic, proton-donor) to amino group (basic, proton-acceptor). The nitrogen atom uses its lone pair to form a dative covalent bond with the proton.

Activity 14.5 Stereochemistry of some amino acids

You will need a 'flexible bond' molecular model kit for this activity. By reference to Table 14.5, make molecular models for the L-form of the following amino acids:

Valine	Serine	Aspartic acid
Cysteine	Asparagine	Proline

Acid–base behaviour of amino acids

In aqueous solutions, amino acids can behave as acids (i.e. proton donors) via the carboxyl group:

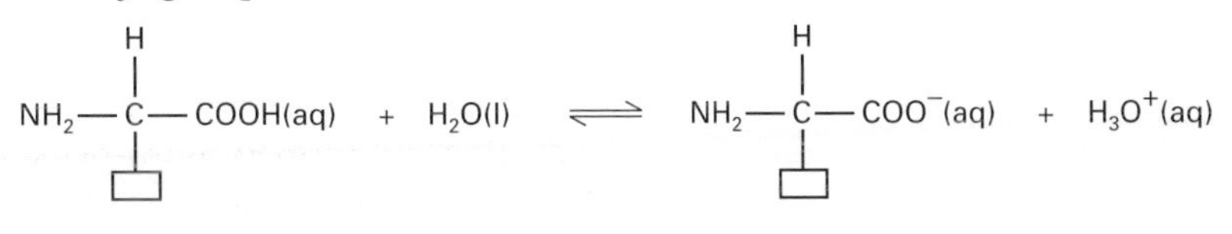

The higher the pH, the greater the concentration of OH^- ions, and the more the reaction will shift to the right-hand side to restore the balance (Le Chatelier's principle, section 25.5). Thus, at high pH, amino acids tend to exist as their carboxylate ions. If an electric field is applied to this solution, the carboxylate ions (–ve) will move to the anode (+ve electrode).

Aqueous amino acids can also act as weak bases (i.e. proton acceptors), via the amino group:

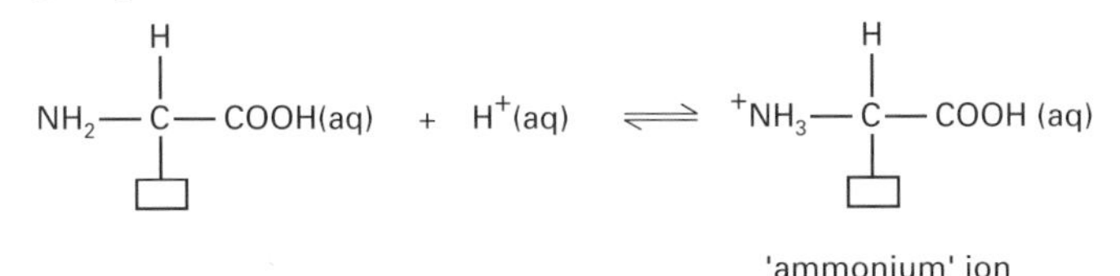

The lower the pH, the greater the concentration of H^+ ions, and the more the reaction will shift to the right-hand side. Thus, at low pH, amino acids tend to exist as their 'ammonium' ions. In an electric field, the 'ammonium' ions (+ve) will move to the cathode (–ve electrode).

At a certain pH, the number of protons leaving the — COOH group will exactly equal the number of protons becoming attached to the — NH_2 group. At this pH, known as the **isoelectric point**, almost all of the amino acid molecules will exist as **zwitterions**, that is species which contain a positive group (— NH_3^+) and a negative group (— COO^-):

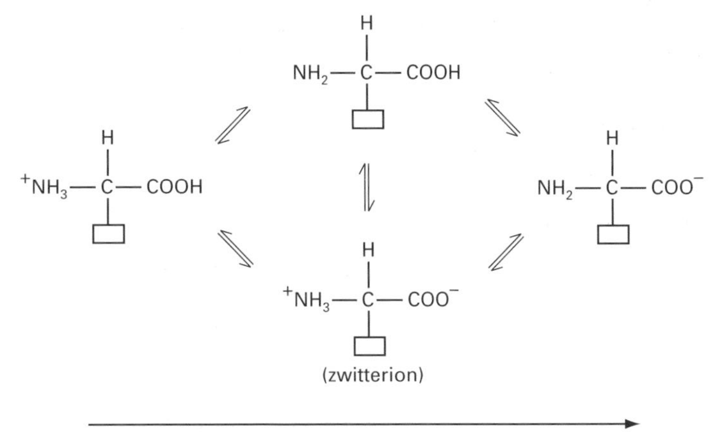

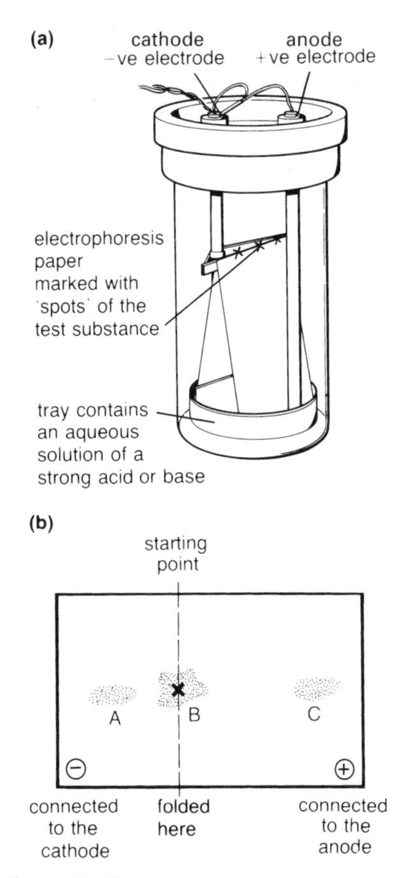

Figure 14.16 ▲
(a) Electrophoresis is a variation on paper chromatography. **(b)** This electrophoretogram was obtained at pH 5.6. Amino acid 'spots' were placed on the paper, as follows: A, glycine; B, threonine; C, cysteine. The 'spot' at B has not moved indicating that B is threonine and its isoelectric point is 5.6. The 'spot' at A, glycine,has moved towards the cathode (–ve) indicating the presence of some $^+NH_3CH_2COOH$ cations and suggesting that the isoelectric point of glycine is greater than 5.6. By similar reasoning, at pH 5.6, cysteine forms some $NH_2CH(CH_2SH)COO^-$ anions. These results suggest that the isoelectric point of cysteine is less than 5.6. In practice, pI values are measured by electropheresis over a pH gradient in a gel and we measure pH at which the 'spot' stops moving.

Zwitterions will *not* move in an electric field. Due to the formation of zwitterions, amino acids have high melting points and most are soluble in water.

Isoelectric points may be determined using **electrophoresis**, a technique in which an electric field is passed across a paper chromatogram, Figure 14.16. The pH of the solution in the tray is varied until the amino acid 'spot' remains stationary at the point of application. Under these conditions, the amino acid molecules must be present as zwitterions and the pH is the isoelectric point. The isoelectric points, pI, of some amino acids are given in Table 14.5. Electrophoresis may also be used to separate amino acids which have different isoelectric points.

Proteins are polymers made up of 2-amino acid residues and, like 2-amino acids, proteins exist as zwitterions at their isoelectric points. Indeed, proteins are least soluble at their isoelectric points because the attractive forces between zwitterions will be at a maximum, thereby causing the protein to precipitate:

$$\text{IIII } {}^+NH_3\text{—∿∿—}COO^- \text{ IIIIII } {}^+NH_3\text{—∿∿—}COO^- \text{ IIIIII } {}^+NH_3\text{—∿∿—}COO^- \text{ IIII}$$

IIIIII represents attraction

protein zwitterons

At a pH greater, or less, than the isoelectric point the protein molecules will exist as ions. Since these will be mutually repulsive, they tend to resist precipitation:

$$HOOC\text{—∿∿—}\overset{+}{N}H_3 \text{ <≡> } {}^+NH_3\text{—∿∿—}COOH \qquad H_2N\text{—∿∿—}COO^- \text{ <≡> } {}^-OOC\text{—∿∿—}NH_2$$

protein cations — pH < isoelectric point

protein anions — pH > isoelectric point

<≡> represents repulsion

Isoelectric precipitation occurs, for example, when milk 'goes off'. Milk has a white colour because it contains emulsified fats and the calcium salts of **casein**, the main protein in milk. Fresh milk has a pH of about 6.7. On ageing, the fats in milk become rancid and milk turns sour, due to the formation of carboxylic acids. As a result, the acidity increases and the pH falls. When the pH reaches 4.7, the casein precipitates leaving a thin watery liquid above it. In the production of some types of cheese and yogurt, the lactic acid produced during fermentation causes the proteins to precipitate, thereby forming the dairy product.

Focus 14c

1. Amino acids contain both the $-NH_2$ and $-COOH$ groups. They are crystalline solids whose molecules exist as **zwitterions** (internal salts):

$$ {}^+NH_3\text{—}\underset{\displaystyle H}{\overset{\displaystyle \bullet}{C}}\text{—}COO^- $$

● = H or a hydrocarbon skeleton

Apart from aminoethanoic acid, 2-amino acids are **optically active.** The naturally occurring compounds exist as the L-configuration. Amino acids show the chemical properties of amines *and* carboxylic acids.

2. Some of the reagents mentioned in this chapter so far, are listed in Table 14.7.

Activity 14.6

A student has produced a mixture of four amino acids by hydrolysing a peptide. The amino acids are known to be in this list: aspartic acid, cysteine, methionine, leucine, histidine, lysine and arginine. Electrophoresis of the mixture was carried out at a variety of pH values.

1 Use the results shown in Figure 14.17 and the data in Table 14.5 to work out the identity of the amino acids and explain your reasoning.

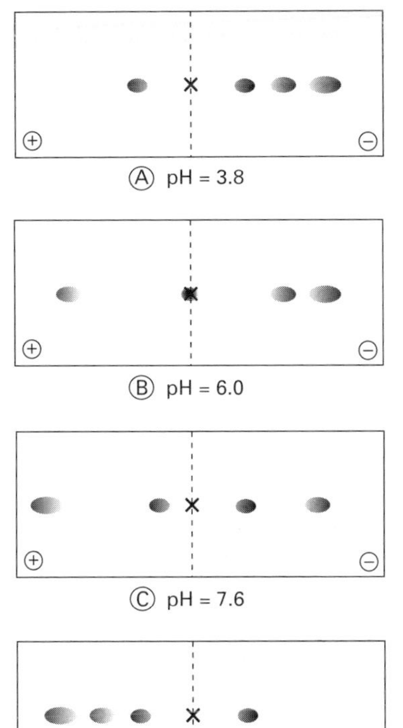

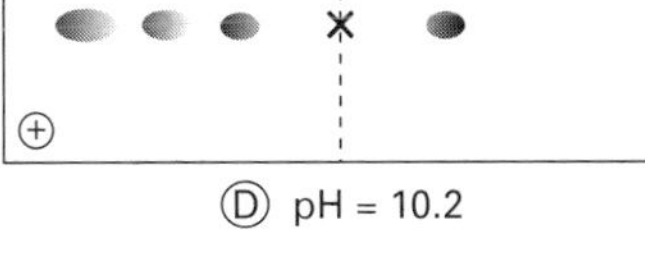

Figure 14.17 ▲ Electrophoretograms of a mixture of four amino acids taken at various pH values. The mixture was applied as a 'spot' at point X.

2 Draw the structural formulae of the amino acid molecules which will predominate in the following aqueous solutions:
   i) Asp at pH 2.8
   ii) Leu at pH 1
   iii) Met at pH 13.

Table 14.7 Important reagents used in this chapter, plus some (*) used earlier with carboxylic acids

Reagent/conditions	what it does
conc. $HNO_3(l)$, conc. $H_2SO_4(l)$, 50°C	**nitrates** a benzene ring $C_6H_6 \longrightarrow C_6H_5NO_2$
$Sn(s)$, conc. $HCl(l)$ reflux	**reduces** an aromatic nitro compound to a **primary amine** $C_6H_5NO_2 \longrightarrow C_6H_5NH_2$
KCN/ethanol, reflux	converts $\rangle C{-}Hal \longrightarrow \rangle C{-}CN$ thereby adding another C atom to the hydrocarbon skeleton
$LiAlH_4$/ethoxyethane, reflux	reducing agent; **nitrile** ⟶ **amine** $\rangle C{-}CN \longrightarrow \rangle C{-}CH_2NH_2$
KI(aq), reflux	converts **diazonium salt** ⟶ **iodobenzene** $[C_6H_5{-}\overset{+}{N}{\equiv}N]Cl^- \longrightarrow C_6H_5{-}I$
conc. HCl(aq) CuCl(s), reflux	converts diazonium salt ⟶ **chlorobenzene** $C_6H_5{-}Cl$
$H_2O(l)$, boil	converts diazonium salt ⟶ **phenol** $C_6H_5{-}OH$
aromatic amine or phenol, 5°C	converts **diazonium salts** ⟶ **azo-dyes** (section 14.6)
*$PCl_5(s)$, room temp.	**halogenates** an amino acid, e.g. $NH_2CH_2COOH \longrightarrow NH_2CH_2COCl$
*alcohol (e.g. $(CH_3OH(l))$, conc. $H_2SO_4(l)$, reflux	**esterification** of an amino acid, e.g. $NH_2CH_2COOH \longrightarrow NH_2CH_2COOCH_3$

14.10 The formation of polypeptides and proteins

Proteins and polypeptides are formed from amino acids by **condensation polymerisation**, that is a reaction in which a small molecule is lost every time a monomer molecule is added to the polymer's chain. In this case, a water molecule is released every time two amino acid molecules are joined together. If the polymer has less than about 50 amino acid residues, it is termed a **polypeptide**, if more than about 50 it is a **protein**. Each amino acid in the chain, known as a **residue**, is joined to its neighbour by a **peptide bond**, formed by the nucleophilic attack of the amino group on the carboxyl group of the other, Figure 14.18a. In the peptide bond, the carbon–nitrogen bond length is between that of a C — N bond and a C = N bond, suggesting that there is a partial π bond between the carbon and nitrogen atoms, Figure 14.18b. Due to the partial π bond character, rotation around the carbon–nitrogen bond is restricted. Thus, the peptide bond is almost planar and is able to exist in *cis-* and *trans-* forms. In most protein structures, the *trans-* form is preferred because it maximises the room available for the bulky side-chains, Figure 14.18c.

When two different amino acid molecules are joined, there are only two possible dipeptides, depending on the orientation of the molecules as the peptide bond is formed. For example, glycine and alanine may form glycylalanine, Gly-Ala or alanylglycine, Ala-Gly:

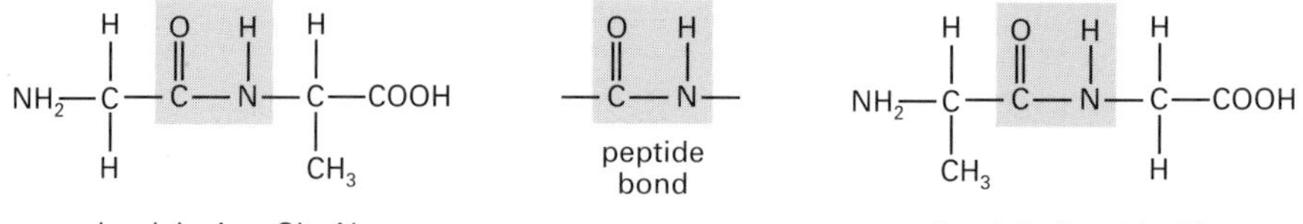

As the number of amino acids increases, however, the number of possible combinations increases massively. Thus, 10 different amino acid molecules used once each will form 3 628 800 possible polypeptides! Bearing this in mind, it is a quite astounding fact that human cells selectively create only about 30 000 proteins from the 20 naturally occurring amino acids which are available. Although many proteins are common to all human beings, some are unique to an individual because their production is genetically controlled by the deoxyribonucleic acid (DNA) of the cell in which the protein is made. The behaviour of each protein may be related to its 3-D structure which is described in terms of four levels of organisation, known as the **primary**, **secondary**, **tertiary** and **quaternary** structures.

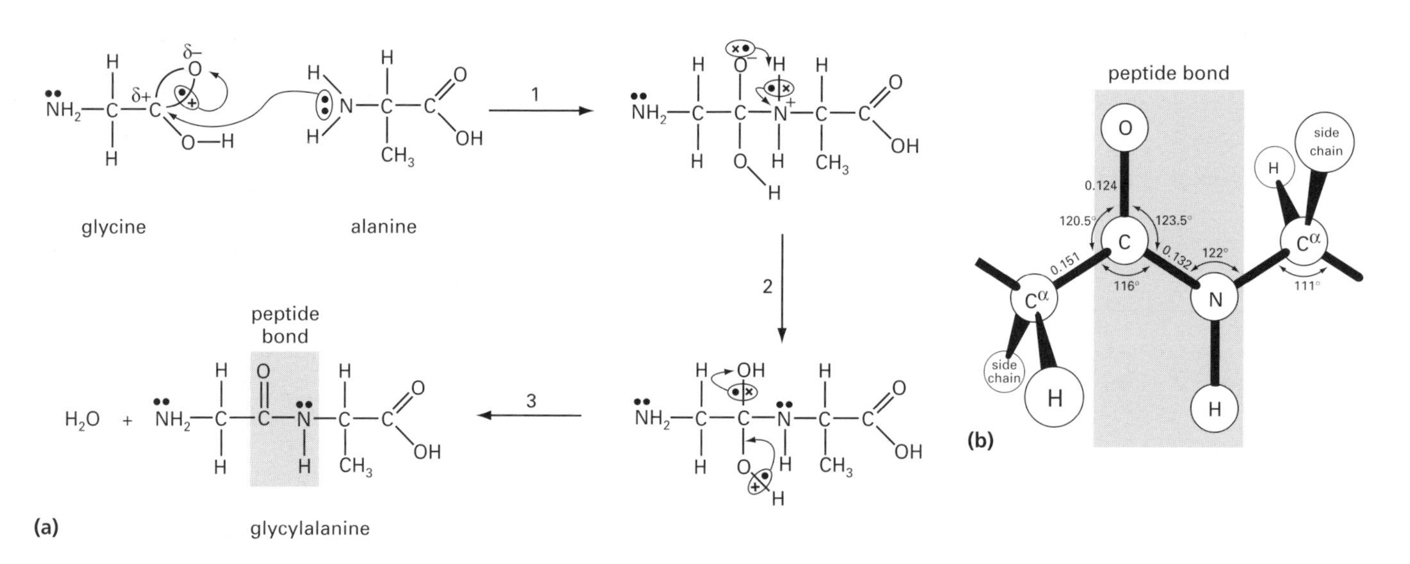

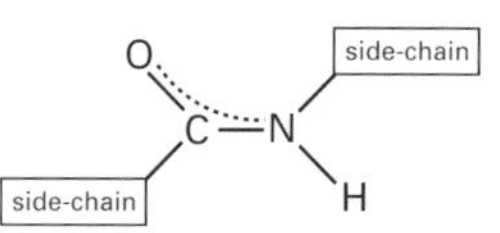

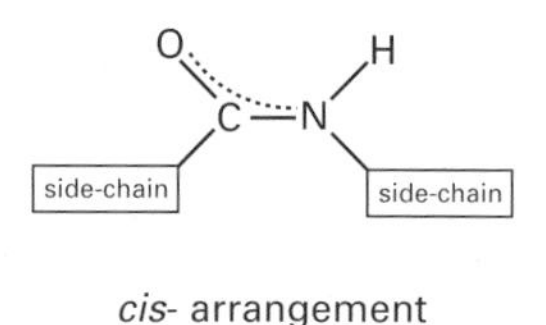

Figure 14.18
(a) Formation of the peptide bond, via a condensation reaction between glycine and alanine. In this reaction, the amino group in alanine acts as a nucleophile (section 6.7), attacking the slightly positively charged carbon of glycine's carbonyl group, step 1. Then, a proton shifts from the nitrogen to oxygen, thereby removing the positive charge on the nitrogen, step 2. Finally, a water molecule is eliminated across a C—O bond, thereby reforming the carbonyl group, step 3. **(b)** At 0.132 nm, the carbon–nitrogen bond length is much less than the mean bond length of a C—N single bond, 0.147 nm, indicating the appreciable double bond (π bond) character between the carbon and nitrogen atoms in the peptide bond. The partial π bond results from the sideways overlap of p orbitals on the oxygen, carbon and nitrogen atoms (section 4.3). **(c)** Due to its partial π bond character, there is no rotation about the carbon–nitrogen bond, so the peptide bond may adopt a *cis-* or *trans-* arrangement. In the *trans-* arrangement, the side-chains are placed *across* the C—N bond; in the *cis-* arrangement, they are located *on the same side* of the C—N bond. In which arrangement are the side-chains the furthest away from each other?

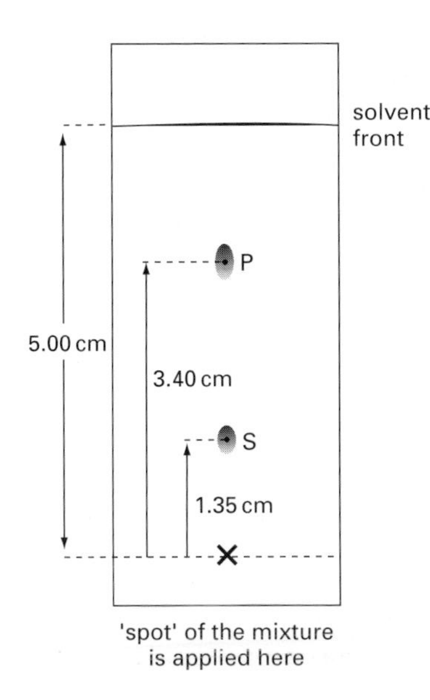

Figure 14.19 ▲
Thin-layer chromatogram of a mixture of phenylalanine, P, and serine, S, using an aqueous mixture of butan-1-ol and ethanoic acid as solvent. The solvent moves up the thin-layer plate carrying the amino acid molecules with it *but* at different rates of movement. The **retardation factor** (R_F value) is defined by the equation:

$$R_F = \frac{\text{distance moved by the 'spot'}}{\text{distance moved by the solvent}}$$

Thus,

$$R_F \text{ (phenylalanine)} = \frac{3.40}{5.00} = 0.68$$

Similarly, the R_F value for serine is found to be 0.27

14.11 Primary structure of proteins

The primary structure of a protein describes the sequence in which the amino acid residues are joined together. Determining this sequence is a complex process, which we can only briefly describe here.

The first step involves an analysis of the protein's **amino acid composition**. The protein is usually refluxed with a proteolytic enzyme or aqueous hydrochloric acid (6 mol dm^{-3}) for a least 24 hours in an oxygen-free atmosphere. Under these conditions, all of the peptide bonds are hydrolysed and the individual amino acids are produced.

The free amino acids in the mixture are then separated using chromatographic methods (section 6.3). If paper, or thin-layer, chromatography is used, the positions of the amino acid 'spots' are determined by spraying the chromatogram with ninhydrin, whereupon the 'spots' turn a purple colour. The amino acids may then be identified from their **retardation factors** or **R_F values**, Figure 14.19. If electrophoresis is used, the amino acids are identified from their isoelectric points.

Ion-exchange chromatography (section 15.2) is widely used for the separation and quantitative estimation of amino acids. The mixture of amino acids is eluted down a column packed with a sulphonated polystyrene resin at constant pH < 7. Under these conditions, the sulphonic acid groups will be ionised to sulphonate ions,

$$\text{RESIN} - SO_3H + H_2O(l) \rightleftharpoons \text{RESIN} - SO_3^- + H_3O^+(aq)$$

and most of the amino acids exist as cations (positive ions):

$$NH_2CHRCOOH + H_3O^+ \rightleftharpoons {}^+NH_3CHRCOOH + H_2O$$

The strength of attraction of each amino acid cation for the sulphonate ions will depend on the nature of the side-chain, R. Thus, a separation can be achieved. As the eluant leaves the column, it is automatically mixed with ninhydrin and a purple colour develops. The eluant is then passed through a light beam which is focused on to a photocell, itself connected to a chart recorder or a computerised display. A typical chromatogram is shown in Figure 14.20. Each amino acid gives an absorption peak, the area under which is proportional to the relative amount of the amino acid in the mixture.

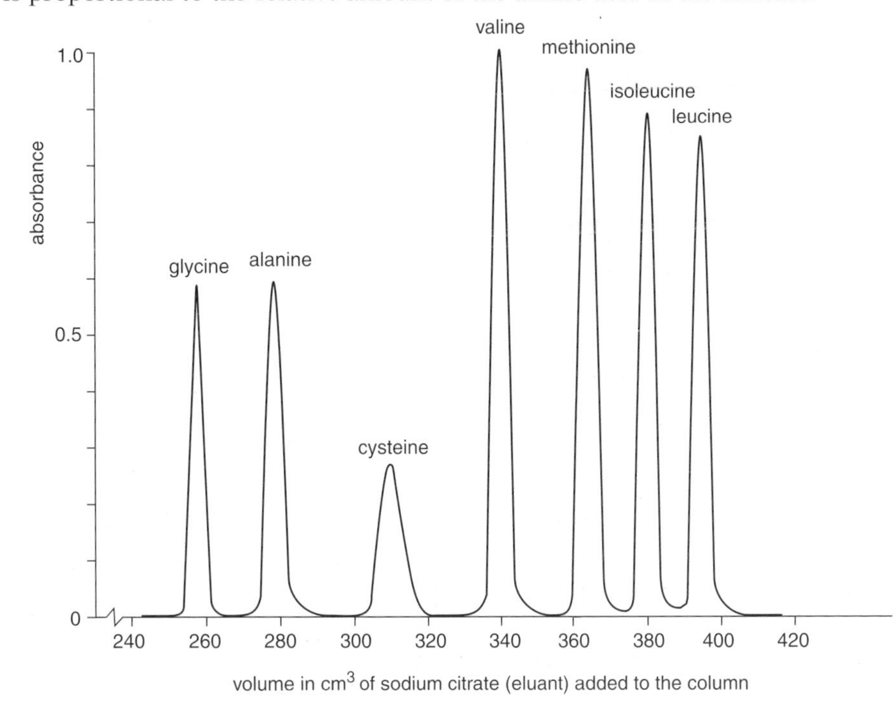

Figure 14.20
Part of an automatically recorded chromatographic analysis of a mixture of amino acids on a sulphonated polystyrene support. The column is 150 cm long and a buffer solution (section 26.2) is used to keep the pH constant at 4.25

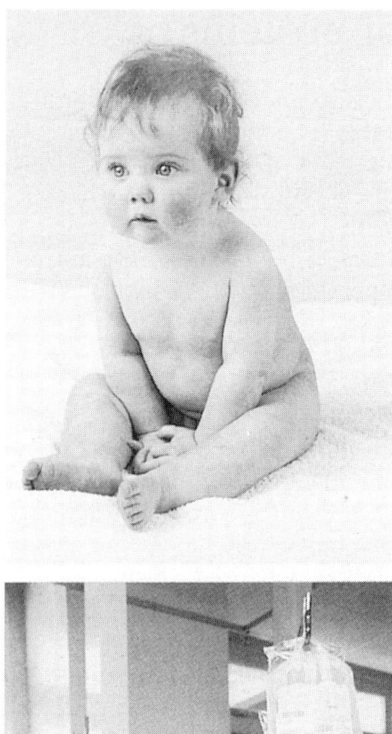

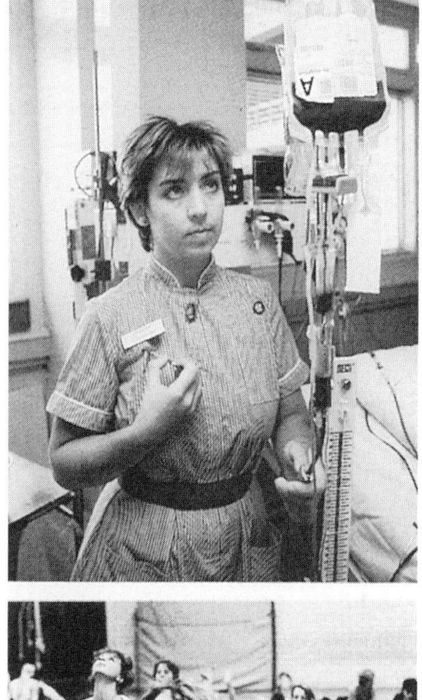

Growth, repair and maintenance – all depend on proteins

Once the relative amounts of each amino acid are known, it is possible to calculate the number of those residues in the protein molecule, much in the same way as we can calculate an empirical formula from elemental composition, section 6.4.

Example

On partial hydrolysis, cattle insulin yields a polypeptide which has a relative molecular mass of 1028 and has the amino acid composition shown in the table below. Calculate the number of each type of amino acid residue in the polypeptide.

Amino acid	symbol	% by mass of the amino acids forming the polypeptide
serine	Ser	9.1
leucine	Leu	22.7
glutamic acid	Glu	12.7
tyrosine	Tyr	31.4
asparagine	Asn	11.4
glutamine	Gln	12.7

Solution

	Ser	Leu	Glu	Tyr	Asn	Gln
In 100 g of polypeptide, there will be:	9.10	22.7	12.7	31.4	11.4	12.7 g
that is, in moles	$\frac{9.10}{105}$	$\frac{22.7}{131}$	$\frac{12.7}{147}$	$\frac{31.4}{181}$	$\frac{11.4}{132}$	$\frac{12.7}{146}$ mol
	0.0867	0.173	0.0864	0.173	0.0864	0.0870 mol
giving the simplest whole number ratio:	1	2	1	2	1	1

Thus, the polypeptide contains the following residues: Ser-1, Leu-2, Glu-1, Tyr-2, Asn-1, Gln-1. Note that the sum of the relative molecular masses of these acids (1154) does not equal the relative molecular mass of the polypeptide (1028). The difference (126) corresponds to the relative mass of the 7 molecules of water which are released on forming seven peptide bonds.

Having worked out which amino acids are present, we now need to determine the order in which they are joined together, that is, the protein's **primary structure**. Firstly, we want to know how many chains of amino acid residues are present in the protein and which amino acids are at the ends of the chains. To find out, an **end-group analysis** is performed, as outlined below:

- **Identification of the amino end-group**
  The protein is treated with **1-fluoro-2,4-dinitrobenzene (FDNB)**, a rather caustic substance which attaches itself to the amino end-group and which cannot be removed by acid hydrolysis. On hydrolysing the protein, then, each 'DNP-amino acid' may be identified by chromatography, as they have characteristic R_F values.

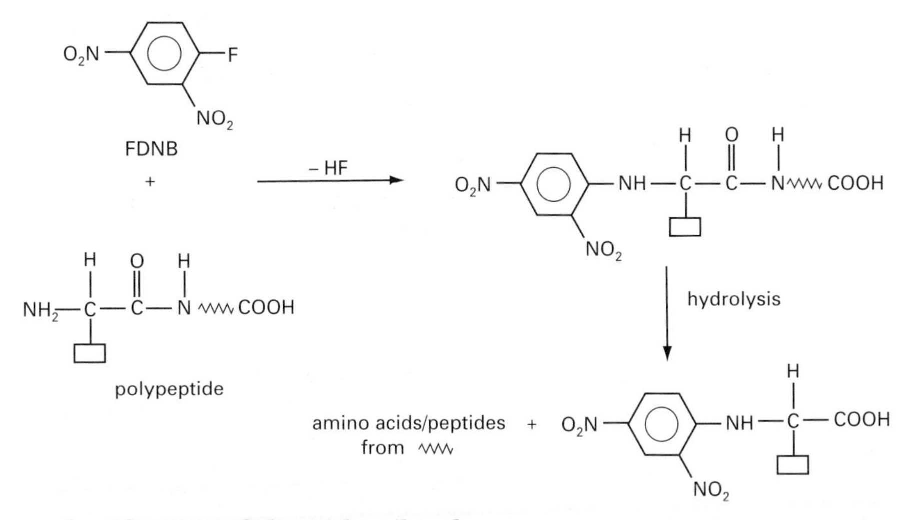

- **Identification of the carboxyl end-group**
  The protein is treated with *either* hydrazine, NH_2NH_2, in dry conditions at 100°C *or* an enzyme called carboxypeptidase. Both reagents liberate the amino acid residue which contains the carboxyl end-group. This amino acid is then isolated, and identified, using chromatography or electrophoresis.

An end-group analysis of human insulin, for example, indicates that there are two amino end-groups, glycine and phenylalanine, and two carboxyl end-groups, asparagine and alanine. This suggests that human insulin contains *two* polypeptide chains, Figure 14.21.

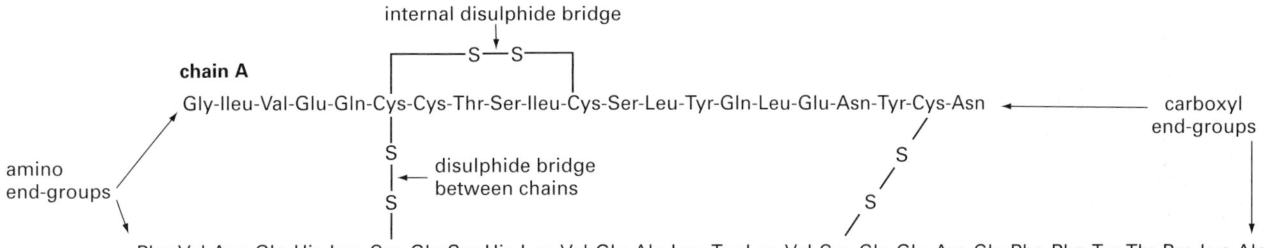

Figure 14.21
The primary structure of human insulin contains two polypeptide chains (A and B) which are joined together by **disulphide bridges** at the cysteine residues. There is also an internal disulphide bridge in chain A. Cattle insulin only differs slightly from human insulin in that the Thr-Ser-Ileu section in chain A is replaced by Ala-Ser-Val.

We now know i) which amino acids residues are present in the protein, ii) how many polypeptide chains it contains and iii) which amino acid residues are at the end of these chains. Next, the protein is partially hydrolysed to small peptides, using **proteolytic enzymes** to catalyse the hydrolysis of specific peptide bonds (e.g. the enzyme trypsin only catalyses the hydrolysis of peptides bonds formed via the carboxyl groups of lysine and arginine). Nowadays, automated chromatographic methods and electrophoresis are used to separate the small peptides and then the amino acid sequence in each is determined from its amino acid composition and end-group analyses. By strategically choosing the proteolytic enzyme to be used, it is possible to build up overlapping sequences of amino acid residues which can then be pieced together like a jigsaw to give the entire primary structure. For example, partial hydrolysis of cattle insulin indicates the presence of the following peptides:

Ser – Leu   Gln – Leu   Cys – Asn
Leu – Tyr   Leu – Glu
Ser – Leu – Tyr   Leu – Glu – Asn – Tyr – Cys
Tyr – Gln – Leu
Gln – Leu – Glu   Tyr – Cys – Asn

This is consistent with the sequence:

Ser – Leu – Tyr – Gln – Leu – Glu – Asn – Tyr – Cys – Asn

which is found in chain A of cattle insulin, Figure 14.21.

Activity 14.7 Working out the sequence of amino acids in a polypeptide

Lysozyme is a protein which is found in nasal mucus and tears. It destroys bacteria by breaking down one of the polysaccharides which are found in the bacterial cell wall. A biochemist carried out a partial hydrolysis of lysozyme and obtained a polypeptide, A, which had ten amino acid residues and a relative molecular mass of 1302. The composition of the polypeptide is shown below:

Amino acid	symbol	Mr	% by mass in A
arginine	Arg	174	23.8
asparagine	Asn	132	18.0
tryptophan	Trp	204	27.9
alanine	Ala	89	12.2
valine	Val	117	7.90
methionine	Met	149	10.2

1 Work out the composition of the amino acids in polypeptide A.

2 End-group analysis of the polypeptide shows that the amino acid end-group is methionine and the carboxyl end-group is arginine. Enzymatic hydrolysis of A produces the following peptides, with the amino acid residue on the left having the free amino group:

Met – Asn – Ala   Arg – Asn – Arg   Trp – Val – Ala
Ala – Trp – Val   Ala – Trp – Arg

Work out the sequence of the amino acids in A.

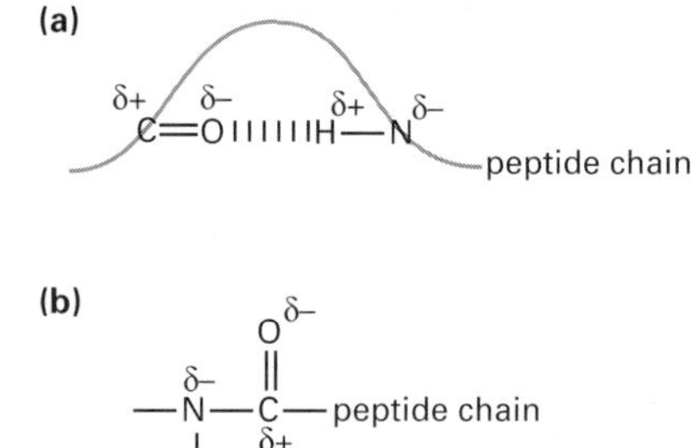

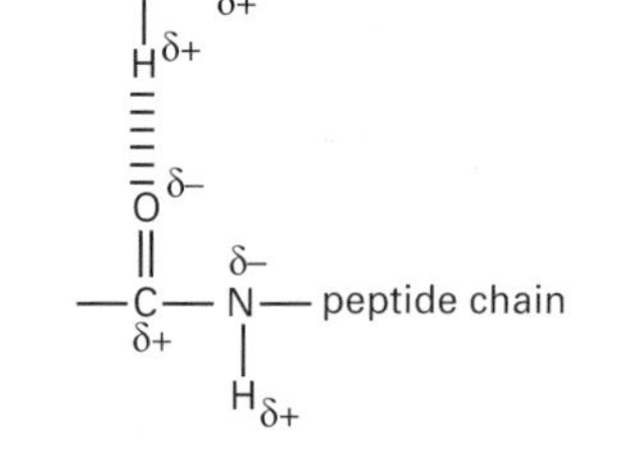

||||| hydrogen bonds

δ+ δ– C=O   δ+ δ– H—N

electronegativities: C < O and H < N

Figure 14.22
Hydrogen bonding in peptides. A dipole will exist in a covalent bond formed between two atoms of unequal electronegativity. Electronegativity is the power of an atom in a covalent bond to attract the bonding electron cloud towards itself (section 4.5). Thus, C = O and N — H bonds contain dipoles.

A hydrogen bond is a dipole–dipole interaction in which at least one dipole contains a hydrogen atom. The hydrogen bonds in peptides may occur **(a)** within the chain (intramolecular) or **(b)** between chains (intermolecular).

14.12 The secondary, tertiary and quaternary structures of proteins

Secondary structure

If peptide bonds were the only structural linkage between the amino acid residues in proteins, these molecules would have randomly coiled 3-D structures. In fact, the polypeptide chains are usually folded in a regular arrangement, mainly as a result of the **hydrogen bonding** between the oxygen of C = O bond and the hydrogen of the N — H bond, Figure 14.22. X-ray diffraction studies have shown that there are two common arrangements, the **α-helix** and the **β-pleated sheet**, and these are described below.

α-helix

In the α-helix, the polypeptide chain is coiled. The helix is maintained by the intramolecular hydrogen bonding between the C = O group in one residue and the N — H group in the fourth residue along the chain, this being positioned in an adjacent turn, Figure 14.23. There are 3.6 amino acid residues per turn of the helix. The α-helix has a diameter about 1.05 nm and the distance between corresponding points on the α-helix is about 0.54 nm. If viewed end-on, the coil turns in an anti-clockwise direction as it moves towards the viewer, Figure 14.24. Most structural proteins, such as the keratins, found in hair, nails and bones, have an entirely α-helical structure and this accounts for their strength.

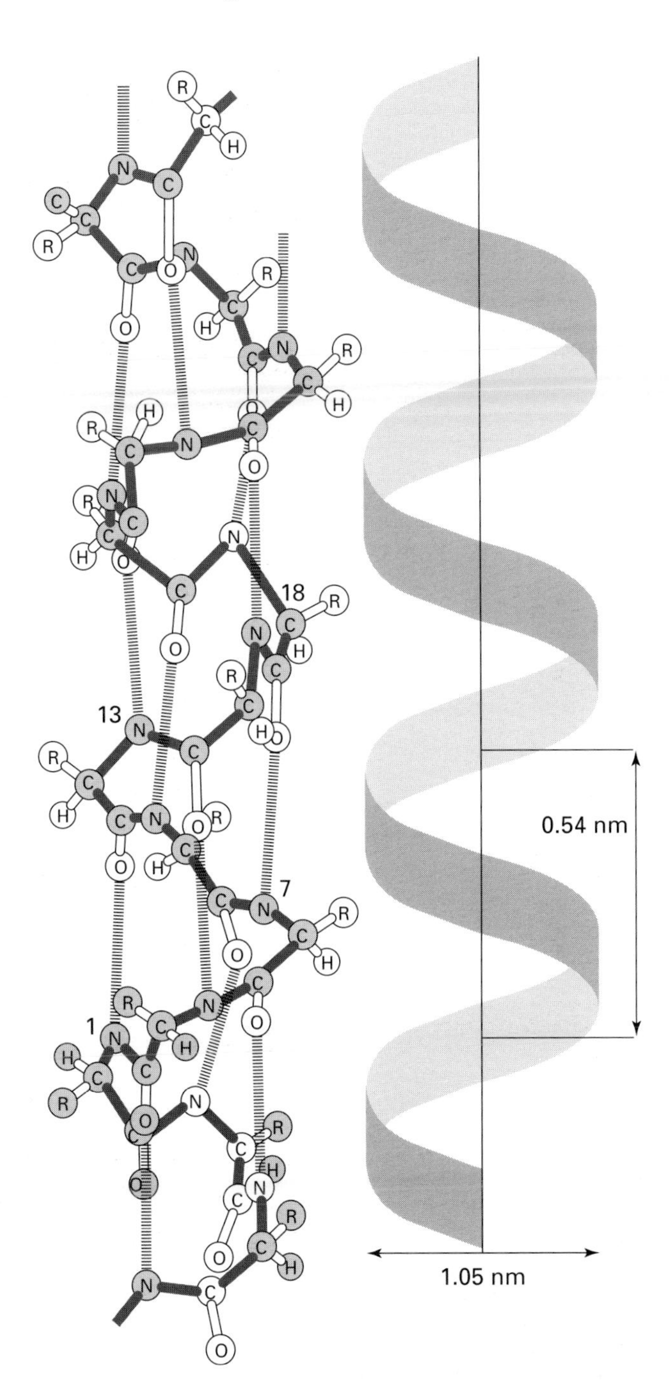

Figure 14.23
Sideways view of an α-helix in a peptide showing the 3.6 amino acid residues per turn. (Note that certain N atoms along the chain have been numbered from 1,7,13 and 18 to assist you with Activity 14.8)

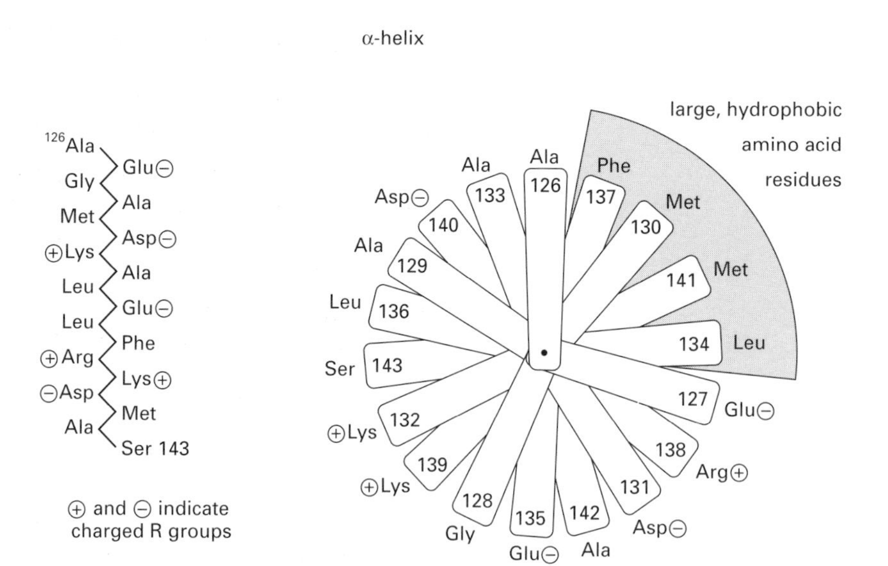

Figure 14.24 ▶
An amino acid sequence from human myoglobin viewed from above. Serine (residue) is furthest from the viewer at the 'bottom' of the α-helix, alanine is at the 'top'. Notice how the α-helix turns anti-clockwise as it approaches the viewer. Since there are 3.6 residues per turn of the α-helix the residues are spaced at 100° intervals when viewed from the end of the α-helix. There is an intramolecular hydrogen bond between the C = O and N — H groups in every fourth residue, for example, between the C = O bond in leucine (134) and the N—H group in arginine (138). The α-helix also brings four hydrophobic (water-hating) side-chains closely together (Phe, Met, Met and Leu). This compact hydrophobic section makes contact with other hydrophobic regions along the chain, thereby causing the α-helix to fold up in a characteristic fashion.

β-pleated sheet

In this structure, the polypeptide chain does not coil as in the α-helix. Instead, chains are arranged alongside each other and held strongly together via extensive intermolecular hydrogen bonding. Figure 14.25 shows three adjacent chains forming part of a β-pleated sheet structure. You can see that the side-chains are on opposite 'sides' of the sheet, forming a sort of zig-zag pattern. If extended, this structure gives us the pleat effect. Silks have the β-pleated sheet structure.

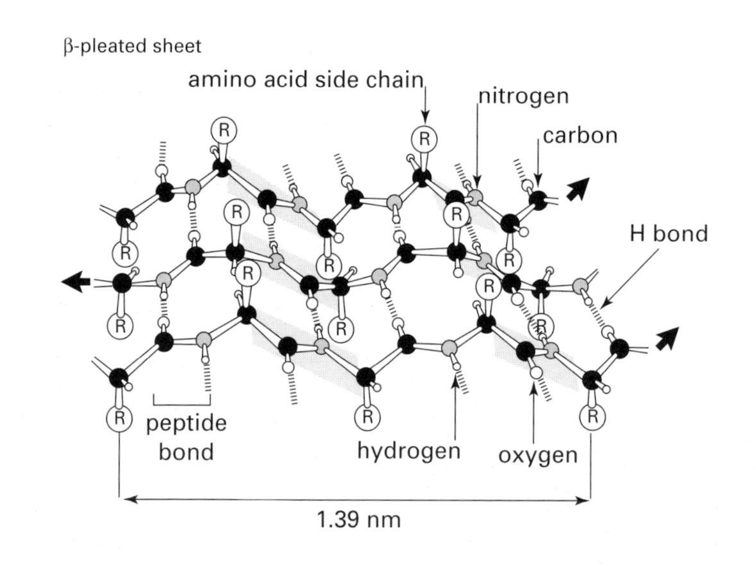

Figure 14.25 ▶
View of a β-pleated sheet. The peptide chains are held together by intermolecular hydrogen bonds

Activity 14.8 Modelling the α-helix

You will need a flexible bond modelling kit for this activity.

1. Write down the Fischer projection formulae for L-alanine, $NH_2CH(CH_3)CO_2H$, and L-serine, $NH_2CH(CH_2OH)CO_2H$. Explain the convention used in these formulae.
2. Make models of L-alanine and L-serine which are consistent with the projection formulae. Use single 'balls', of different colours, to represent the — CH_3 and — CH_2OH side-chains.

Activity 14.8 Modelling the α-helix continued

3 Write down the structural formula of the dipeptide, Ala–Ser. Remember that the first amino acid in a written sequence is the amino end-group. Here it is Ala.

4 Now carry out a 'peptization' by making a model of Ala–Ser and satisfy yourself that there are two possible arrangements around the C — H bond (see Figure 14.18). Draw a 3-D view of each arrangement and explain the difference between them.

5 The stereochemistry of the peptide group in proteins is given in Figure 14.18c. Make *three* identical models of the Ala–Ser dipeptide, each having this arrangement.

6 Next, join up the models to form the hexapeptide, Ala–Ser–Ala–Ser–Ala–Ser, ensuring that the correct stereochemistry of the peptide link is maintained, as in Figure 14.18c.

7 Look again at the diagram of the α-helix in Figure 14.23. By using a retort stand and clamps, make a model of the portion of the helix from chain atom 1 to atom 18. Note that there should be about 3.6 amino acid residues per turn.

8 The α-helix is held together by intramolecular hydrogen bonds. Use elastic bands to show the hydrogen bonds, as follows:

$^{4}N — H^{\delta+}$ to $^{\delta-}O = C^{14}$ and $^{7}N — H^{\delta+}$ to $^{\delta-}O = C^{16}$

9 Looking down on your model, satisfy yourself that the α-helix does turn anti-clockwise as it approaches you.

Tertiary structure

Some proteins have structures which consist solely of α-helical chains *or* β-pleated sheets; these are termed **fibrous proteins**. Fibrous proteins have structural roles in the organism, being very durable and insoluble in water, Table 14.8. Collagen has a unique structure in which three very long polypeptide molecules are intertwined. To form this structure, every third residue in the polypeptides must be glycine, the smallest amino acid, because hydrogen bonds are formed between the C = O and N — H bonds in every *third* residue – not every fourth residue as in the α-helix.

Table 14.8 Some examples of fibrous proteins

Name	structural type	found in
keratin	α-helix	hair, feathers and nails
myosin	two entwined α-helical chains	muscle
collagen	three entwined α-helical chains	cartilage, tendons, ligaments
fibroin (silk)	β-pleated sheets	thread produced by silk worms used to make cocoon

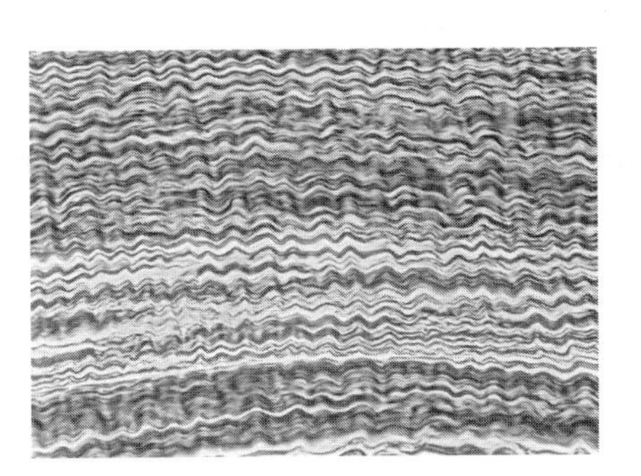

A light micrograph of a longitudinal section through connective tissue from a tendon. Tendon connects muscle to bone or muscle to muscle. It is a dense fibrous connective tissue made up of parallel collagen (protein) fibres. Here the collagen appears as wavy bands. Tendons are extremely strong, flexible, yet do not stretch.

X-ray diffraction studies show that many proteins have a globular structure in which sections of the α-helix, and sometimes β-pleated sheets, are folded around each other in an ordered structure. An example of a **globular protein** is myoglobin which is present as an oxygen store in muscles, Figure 14.26. About 75% of the 153 amino acid residues in myoglobin are arranged in eight different α-helical sections. The other residues form non-helical sections which tend to be at the 'corners' of the molecule, where the chain

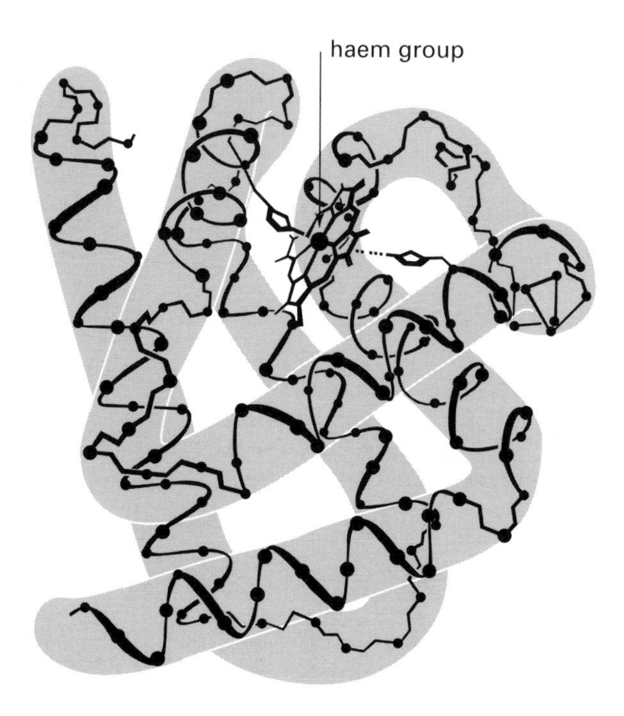

Figure 14.26
The tertiary structure of myoglobin, a globular protein, consists of eight sections of α-helix held together by non-helical sections at the 'corners' of the molecule. The iron-containing haem group can bond reversibly with an oxygen molecule

turns back on itself. These **reverse turns** occur when the amino hydrogen of one residue forms a hydrogen bond with the carbonyl group of the third following residue (rather than the fourth following residue, as in an α-helix, Figure 14.23). This tight folding is more likely to occur where the residue's side-chain is small, for example in glycine or alanine. Reverse turns also result from the presence of **proline** residues because carbon-5 of the heterocyclic ring occupies the position of a peptide hydrogen, preventing the hydrogen bonding between the N — H and O = C groups which is needed to maintain an α-helical structure:

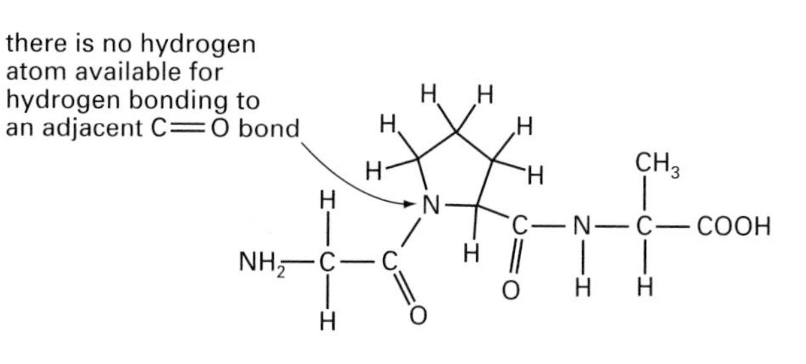

Some globular proteins contain sections of α-helix and β-pleated sheets held in a specific 3-D arrangement. For example, a segment of alcohol dehydrogenase, an enzyme which catalyses the oxidation of ethanol to ethanal in the liver, contains four α-helical and six β-pleated sections.

The properties of a globular protein are heavily dependent on its tertiary structure, that is, the unique way in which the chain folds up. Apart from hydrogen bonding, there are four other types of attraction between groups along a polypeptide chain (Figure 14.27):

- **hydrophobic bonds** The side-chains in a number of amino acid residues are essentially hydrophobic, that is they have little or no attraction for water molecules. However, they will be attracted to each other by van der Waals' forces (section 4.6). These forces will be stronger where aromatic rings of phenylalanine, tyrosine or tryptophan are stacked side by side, thereby allowing their π electron clouds to interact with each other.

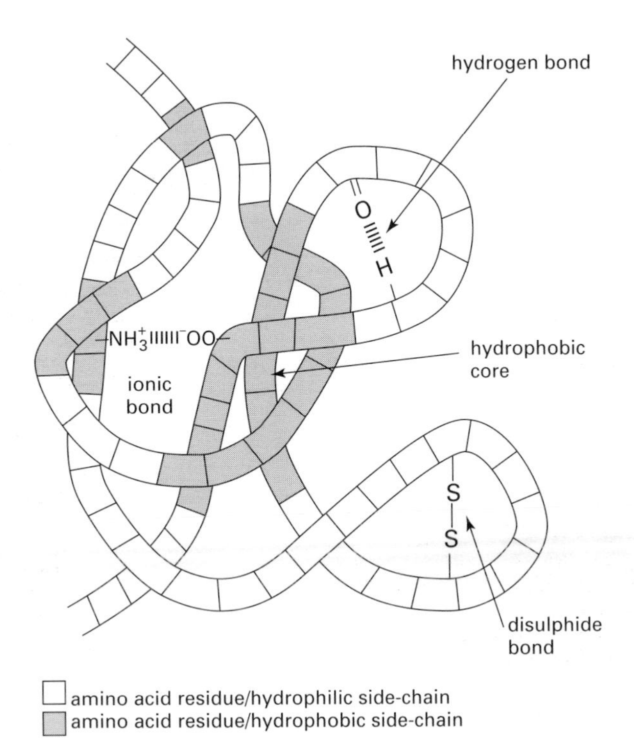

Figure 14.27
A schematic representation of part of a globular protein molecule showing the various types of bonds in a tertiary structure: hydrogen bonds, hydrophobic bonds, ionic bonds and disulphide bonds. Hydrophobic groups are attracted to each other by van der Waals' forces and try to expose a minimum amount of their surface to a polar solvent; thus, they form the central core of a globular protein molecule. Since ionic bonds would break in a polar solvent, they are found in the hydrophobic core

- **ionic bonds** Two ions can attract, or repel, each other depending on the charge. For example, if the positively charged ammonium group in a arginine residue is located close to the negative carboxyl ion in a glutamic acid residue, strong attractive forces will occur. Such forces will only be strong in the absence of water which would destroy the bond. However, hydrophobic portions of the chain can 'squeeze' out a water molecule, thereby protecting an ionic bond.
- **disulphide bonds** These strong covalent bonds are formed between neighbouring cysteine residues. Disulphide bonds are broken by reducing agents.

In globular proteins, the chains fold up so that the hydrophobic groups point towards the centre of the molecule whilst the hydrophilic, 'water-loving' groups, are on the outside. Thus, globular proteins usually dissolve in water, giving a colloidal suspension.

Quaternary structure

Many proteins are made up of two or more **sub-units**, that is polypeptide chains, each of which has its own tertiary structure. The sub-units are held together by a combination of hydrogen, disulphide, hydrophobic and ionic bonds.

Haemoglobin, the oxygen carrier in blood, is a globular protein which has four sub-units meshed together in a very compact molecule. Two of the chains are identical and are called the **α-chains**; the other two are identical and are called the **β-chains**, Figure 14.28a. The hydrophobic portion of each chain is directed towards the centre of the molecule; the polar and ionic groups are outside, enabling the molecule to dissolve in water. Each sub-unit contains a **haem group**, Figure 14.28b. This forms oxyhaemoglobin, a stable complex in which an oxygen molecule is held in the crevice between the haem group and the surrounding polypeptide chain, known as the **globin**. Also, the bonding site is specific for oxygen; thus, although O_2 and N_2 are both present in air, N_2 will not complex with haemoglobin.

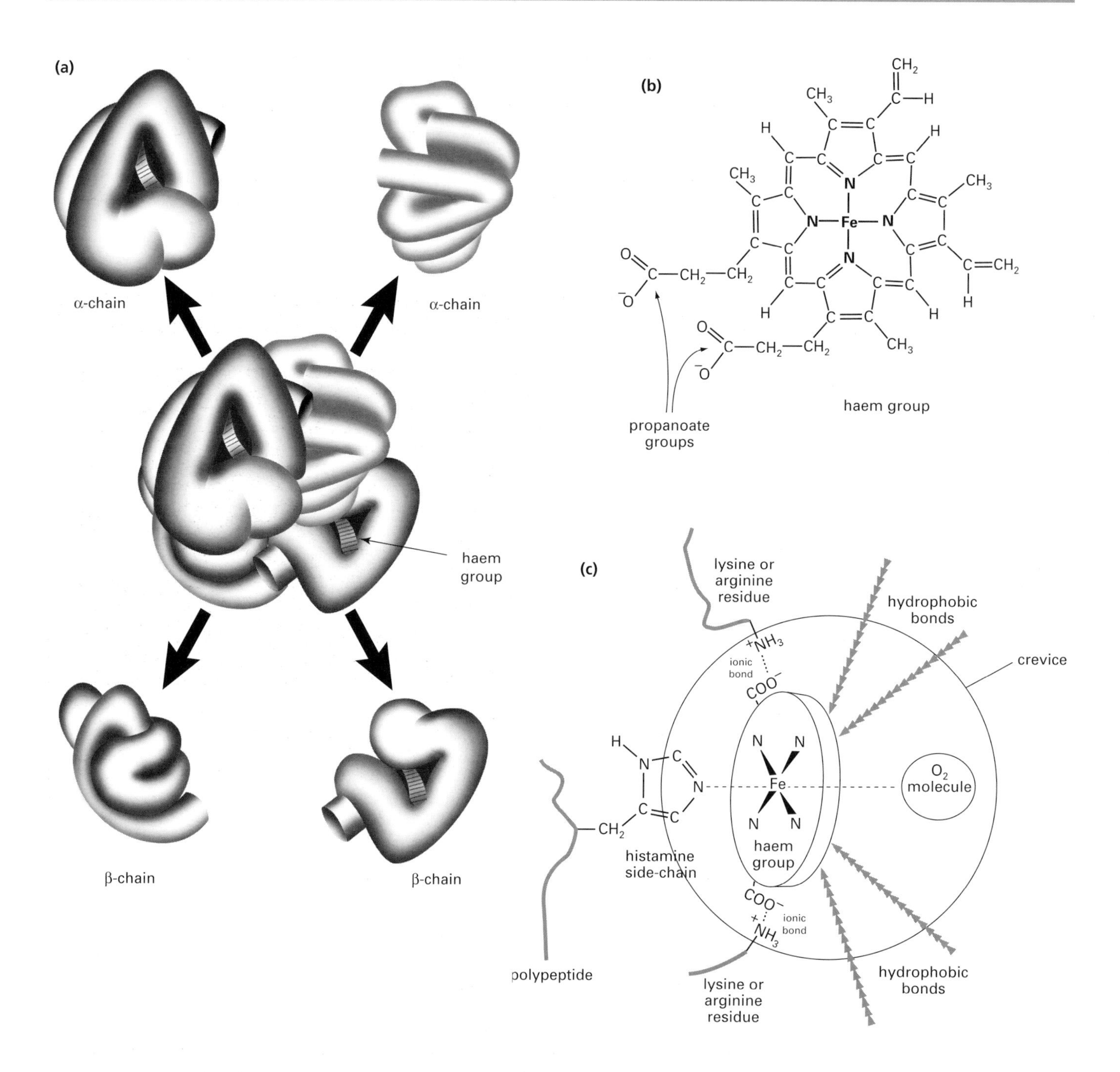

Figure 14.28
(a) Haemoglobin is a globular protein whose quaternary structure is made up of two pairs of polypeptide chains, termed the **α-chain** and **β-chain** sub-units. These sub-units are held 'snugly' together by the intermolecular forces between adjacent amino acid side-chains (Figure 14.27), particularly those between hydrophobic groups. Overall, the haemoglobin molecule is almost spherical in shape, having a hydrophobic core and hydrophilic exterior.
(b) There are 141 amino acid residues in an α-chain and 146 residues in a β-chain. Each sub-unit contains a haem group. A haem group is essentially a 6-coordinate complex of an iron(II) ion (section 21.2). Four of the coordination sites are taken up by the nitrogen atoms of the **protoporphoryn ring**, a heterocyclic ring which is planar due to the extensive delocalisation of π-bonds.
(c) Each haem group is held in a crevice within its sub-unit due to: (i) hydrophobic bonds between the hydrocarbon skeletons of the protoporphoryn ring and adjacent amino acid residues; (ii) ionic bonds between the protoporphoryn's negatively charged propanoate groups and the positively charged nitrogens in nearby lysine or arginine residues; (iii) a coordinate bond formed by the imidazole side-chain of histidine and the iron(II) ion. The sixth co-ordination site of the iron(II) ion is used to bind the oxygen molecule, thus forming **oxyhaemoglobin**.

As oxygen is attached to haemoglobin, the colour of the complex changes, from the purple, deoxygenated, form to the bright red of oxyhaemoglobin. The oxyhaemoglobin moves through the bloodstream carrying oxygen to the cell tissues. The iron is not oxidised by the oxygen but remains as Fe(II) during oxygen transport. Of most importance is the variable strength of attraction between haemoglobin and the oxygen molecule. The oxygen molecule is held tightly when it is picked up at the lungs but loosely when it reaches the cell tissue, thus allowing it to be released.

Sickle cell anaemia

Sickle cell anaemia is a disease in which a mutation in the DNA causes an alteration in the amino acid sequence on the β-chain sub-unit, namely the replacement of the glutamic acid residue at position 6 by a valine residue. A glutamic acid residue is hydrophilic and, in haemoglobin, this residue is located on the surface of the β-chain sub-unit. Valine, however, is hydrophobic and, when introduced at the β6 position, its side-chain is attracted towards the hydrophobic central core of the sub-unit. This brings about a change in the overall shape of the haemoglobin molecule which decreases its solubility. As a result, the haemoglobin molecules aggregate into long fibres and the affected red blood cells adopt a 'sickle' or crescent shape. Sickled cells have an average life of 17 days compared to 120 days for normal red blood cells. More importantly, the sickle cells can clog up small blood vessels and this may lead to life-threatening symptoms, such as cardiac or kidney failure.

A scanning electron micrograph of red blood cells in a sickle cell anaemia patient. Here normal blood cells (rounded) contrast with a single elongated sickle shaped cell. As yet no cure is known for this disease

Denaturation of proteins

We have seen each that protein has a unique, and highly ordered, 3-D structure and this largely determines its biological function. Under certain conditions, however, the protein's 3-D structure may be disrupted into randomly arranged peptide chains, with a corresponding loss of its biological activity. This process, known as **denaturation**, may occur in various ways as described below.

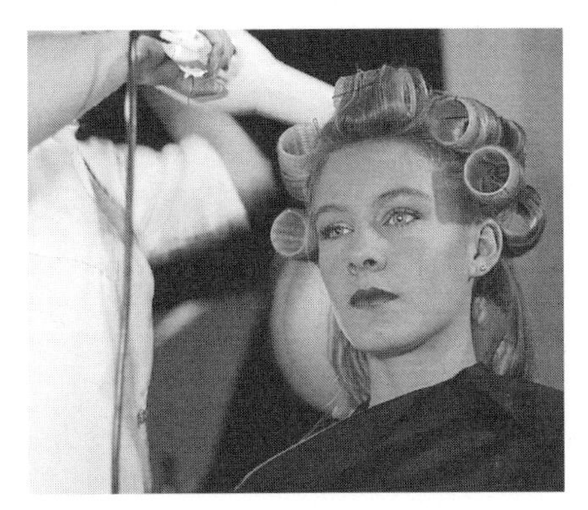

If you have your haired permed, your hairdresser will be carrying out a chemical reaction on the top of your head! On wetting hair, it can be stretched to one and a half times is length because the water weakens some of the ionic bonds in keratin causing it to expand. When a permanent wave is put in hair, tension is developed in the disulphide bridges between the protein chains by winding the hair on rollers. These bridges are broken, using a reducing agent such as aqueous thioglycolic acid, $HSCH_2COOH$, thereby relaxing the tension in the hair. On treating the hair with an aqueous oxidising agent, typically hydrogen peroxide, H_2O_2, or sodium perborate, $NaBO_2$, the disulphide bridges are regenerated and the hair adopts the shape of the rollers. As the hair dries, hydrogen bonds between adjacent polymer are reformed and the styling is complete. As you can imagine, frequent 'perming' will cause the hair to become lifeless and break easily

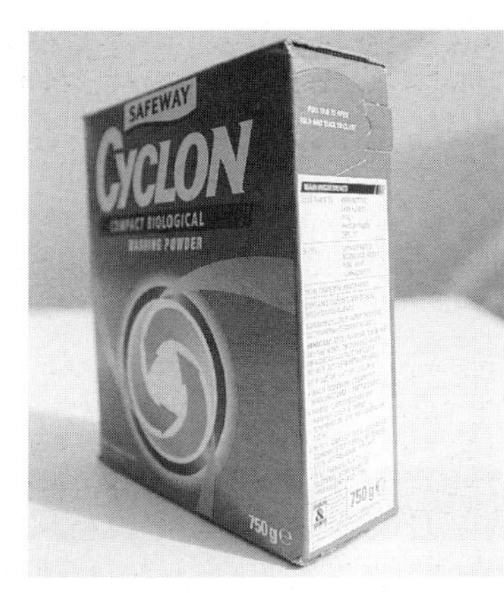

Detergent molecules have long hydrocarbon skeletons which will bond with the hydrophobic region of a protein thereby altering its 3-D shape

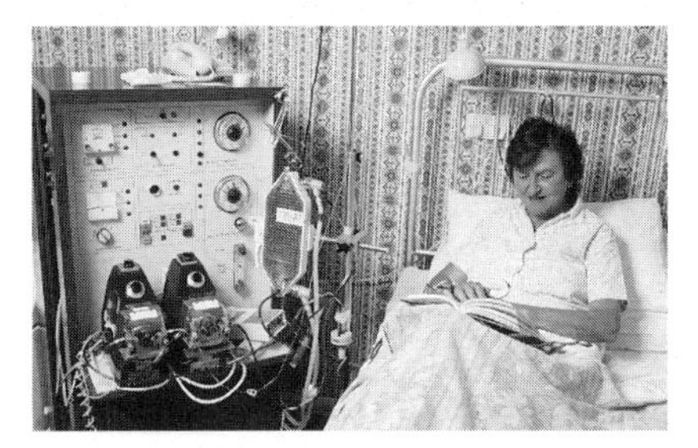

A woman undergoing renal dialysis on a kidney machine installed in her home. Haemodialysis removes waste material from the bloodstream of a person whose kidneys function insufficiently. A stream of blood from an artery is circulated trough the machine on one side of a semi-permeable membrane, while a solution of comparable electrolytic concentration circulates on the other side. Water and waste products pass across the membrane but blood cells and proteins are too large to do so. Purified blood is returned to the body through a vein

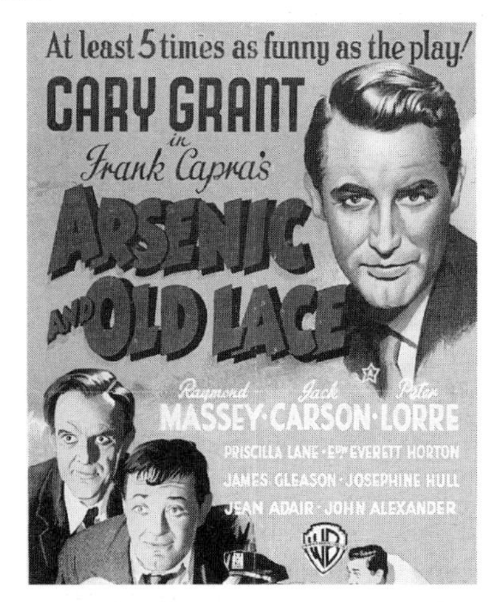

In the classic comedy 'Arsenic and Old Lace' two elderly women use arsenic to poison their lodger who said they wished they were dead!

Heating

A well-known example would be the cooking of egg-white. This is a solution of egg albumen which is a globular protein. On strong heating, the protein's globular structure is disrupted and polypeptide chains are formed. Due to the intermolecular forces between these chains, a white solid is formed. For most chemical reactions, activation energies range from about 60 to 100 kJ mol^{-1}, whilst activation energies for the thermal denaturation of proteins are much higher, up to about 400 kJ mol^{-1}. These values reflect the stability of the tertiary structures of proteins and the very large number of weak interactions (hydrogen bonds, disulphide bonds, hydrophobic interactions and ionic bonds) which are present.

Changing the pH

A protein's tertiary structure, and its properties, are often critically affected by a change in pH because the addition of an excess of H^+ or OH^- ions affects side-chains forming ionic bonds. Enzymes, for example, display optimum activity at a given pH. Thus, pepsin, a digestive enzyme, causes the hydrolysis of proteins at pH 1–2 (strongly acidic), whereas 'biological' washing powders contain proteolytic enzymes, such as trypsin, which work best at a pH of around 8 (weakly alkaline). These 'biological' detergents can function at lower temperatures than conventional detergents because of their enzyme action. In enzymic studies, it is often necessary to use a buffer solution to maintain a constant pH.

Vigorously shaking the protein

Attractions within the globular protein are broken so that it forms a foam, for example, when an egg white is whisked to form a meringue.

Adding certain detergents

Detergents derived from petrochemicals (section 13.7) have long hydrophobic regions which will become attracted to similar regions in the protein, thereby altering its structure.

Adding high concentrations of urea, $(NH_2)_2CO$

Urea disrupts the bonding between polypeptide chains by forming hydrogen bonds with them. This is an example of reversible denaturation since the protein will return to its original configuration once the urea is removed, for example by dialysis.

Heavy metal ions, such as lead, mercury and arsenic

Arsenic and heavy metals act as poisons by reacting with — SH groups present in cysteine residues in enzymes, such as those involved in the generation of cellular energy. For example, arsenic ions react with glutathione, a tripeptide of glutameic acid, cysteine and glycine residues which is present in most tissues:

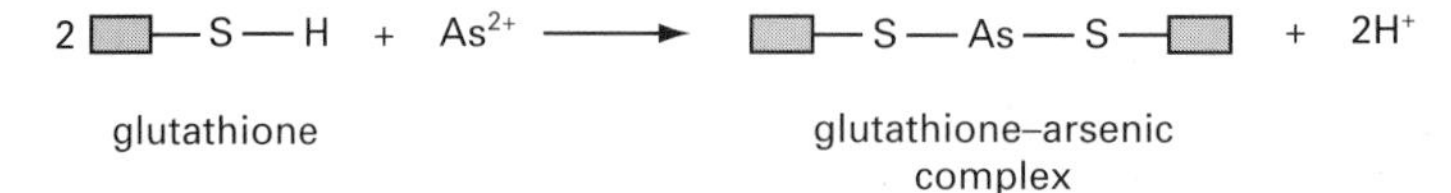

Due to the great strength of the As — S bond, the two glutathione molecules are denatured and prevented from taking part in any further biochemical reaction. In the Biuret test for proteins, the protein molecule is denatured when it forms a violet complex with a copper(II) ion. Apart from losing its biological activity, a denatured protein will often exhibit very different physical properties such as increasing viscosity, decreased solubility and a change in absorption spectra. However, if the conditions causing denaturation are not extreme, for example a small change in pH, the protein molecule will often recover its structure when such conditions are removed.

Questions on Chapter 14

1 This question is about the compounds whose structural formulae are shown below:

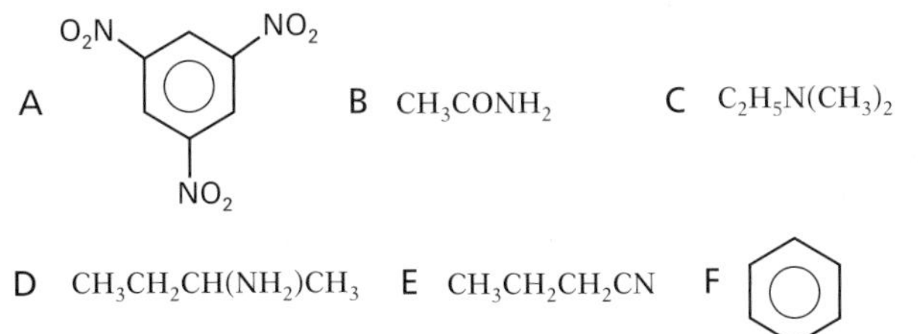

D $CH_3CH_2CH(NH_2)CH_3$ E $CH_3CH_2CH_2CN$ F C_6H_5–NHC_2H_5

G $CH_3CH_2CH(NH_2)CO_2H$ H C_6H_5–NH_2

a) Name each compound and identify the functional groups present.
b) Which is a tertiary amine?
c) Which are able to form intermolecular hydrogen bonds?
d) Which may have an effect on plane polarised light?
e) Which is reduced to 1-aminobutane by lithium tetrahydridoaluminate(III) in ethoxyethane?
f) Which is made by treating benzene with a mixture of concentrated sulphuric and nitric acids?
g) Which may be dehydrated to ethanonitrile by treatment with phosphorus(V) oxide?
h) Which forms a fairly stable diazonium salt when treated with nitrous acid at 5°C?

2 Give the names, and structural formulae, of the organic products formed in the following reactions:
a) 1,3-dinitrobenzene + tin/concentrated hydrochloric acid
b) 2-methylpropanonitrile + dilute $H_2SO_4(aq)$
c) ethanonitrile + $LiAlH_4$/ethoxyethane
d) benzoyl chloride + CH_3NH_2
e) methanamide + dilute $H_2SO_4(aq)$
f) N-ethylethanamide + $LiAlH_4$/ethoxyethane
g) ethylamine (excess) + iodomethane
h) ethylamine + bromoethane (excess)
i) phenylamine + $NaNO_2$/dilute $HCl(aq)$ at 5°C
j) benzenediazonium chloride + $KI(aq)$
k) 2-aminoethanoic acid + phosphorus pentachloride

3 a) Place the following compounds in order of increasing base strength and explain your reasoning: methylamine, phenylamine, ammonia, dimethylamine.
b) Write a balanced equation for each of the following reactions and name the organic products which are formed:
i) dimethylamine with hydrochloric acid,
ii) ethylamine with sulphuric acid,
iii) ethylammonium bromide with potassium hydroxide.

4 Explain the following observations:
a) When phenylamine is added to distilled water, two immiscible layers form. When concentrated hydrochloric acid is added dropwise, and the mixture shaken, a solution, D, is obtained
b) On adding solution D dropwise to an aqueous solution of sodium nitrite at 5°C, a slightly yellow solution, E, is obtained. If solution E is slowly warmed up, there is an antiseptic smell and bubbles of gas are evolved.
c) If a solution of phenol in sodium hydroxide is added dropwise to solution D, at 5°C, a yellow precipitate, F, is formed.

5 a) Write down the general structure of a 2-amino acid.
b) Explain the following observations:
i) Aminoethanoic acid has a molecular mass of 75 and melts at 234°C; propanoic acid has a molecular mass of 74 and melts at –21°C.
ii) An aqueous solution of aminoethanoic acid conducts electricity.
iii) When an electric current is passed through an aqueous solution of aminoethanoic acid at pH1, the aminoethanoic acid molecules move to the cathode (–ve electrode) whereas at pH = 13, the aminoethanoic acid molecules move to the anode (+ve electrode).
c) Describe how you would use simple test-tube reactions to prove that aminoethanoic acid contains two functional groups.
d) What is meant by the terms: peptide link, protein, polypeptide and tripeptide?
e) Explain, using suitable diagrams, how three molecules of aminoethanoic acid may join together to form a tripeptide.
f) In what way is the structure of the tripeptide in e) similar to that of nylon 66?

6 Give the reagents and reaction conditions needed for the following conversions. Note that parts j) to o) may require reference to earlier chapters.
a) $CH_3CH_2CN \longrightarrow CH_3CH_2CH_2NH_2$
b) $CH_3CH_2CONH_2 \longrightarrow CH_3CH_2CN$
c) C_6H_5–$NO_2 \longrightarrow C_6H_5$–$NH_2$
d) $CH_3CONH_2 \longrightarrow CH_3COOH$
e) $CH_3CH(CH_3)CN \longrightarrow CH_3CH(CH_3)COOH$
f) $CH_3COCl \longrightarrow CH_3CON(CH_3)_2$
g) $NH_2CH_2COOH \longrightarrow NH_2CH_2COOCH_3$
h) $CH_3CH_2CH_2NH_2 \longrightarrow CH_3CH_2CH_2NHCH_3$
i) $C_2H_5Br \longrightarrow [(C_2H_5)_4N^+]\,Br^-$
j) C_6H_5–$NH_2 \longrightarrow C_6H_5$–$OH$ (2 steps)
k) C_6H_5–$NH_2 \longrightarrow C_6H_5$–$Cl$ (2 steps)
l) $C_2H_4 \longrightarrow CH_3CH_2NH_2$ (2 steps)
m) $CH_3CH_2Br \longrightarrow CH_3CH_2CH_2OH$ (3 steps)
n) $CH_3CHO \longrightarrow CH_3CH(OH)CONH_2$ (3 steps)
o) $CH_4 \longrightarrow CH_3CH_2NH_2$ (3 steps)
p) $C_6H_6 \longrightarrow C_6H_5$–$OH$ (4 steps)
q) $C_2H_6 \longrightarrow CH_3CH_2CONHC_6H_5$ (5 steps)

Questions on Chapter 14 *continued*

7 Describe chemical tests which would enable you to distinguish between the following pairs of substances, clearly stating i) the reagents and reactions conditions that you would use and ii) the observations that you would expect to make:

a) ethanamide and ethylamine,

b) phenylamine and phenylmethylamine, $C_6H_5CH_2NH_2$,

c) benzoic acid and benzamide,

d) ethylamine and dimethylamine (note, this also involves a physical technique).

8 Compound A is a nitrile which is used to make plastic fibres for rugs and fabrics. It was accidentally discovered in the late 1950s by American scientists who were trying to find a use for the waste propane gas from petroleum refining. Nowadays, A is manufactured in large quantities by heating propene with ammonia and air at 400°C in the presence of a molybdenum catalyst.

a) A contains 67.9% carbon, 5.70% hydrogen and 26.4% nitrogen by mass. When 0.300 g of A are heated to 102°C at a pressure of 101 000 Pa, the vapour occupies a volume of 176 cm^3.

i) Find the empirical formula of A.

ii) Work out the molecular formula of A. (You will need to use the ideal gas equation, section 4.13; the gas constant, R, is 8.31 J K^{-1} mol^{-1}).

b) Name A and draw its displayed formula.

c) Write an equation for the synthesis of A from propene, ammonia and oxygen.

d) The polymerisation of A produces fibres which are sold under the name '**Orlon**'. Write down the structure of Orlon.

d) State whether the manufacture of Orlon involves addition, or condensation, polymerisation of A and explain your reasoning.

e) Give the structures, and the names, of the organic products you would expect to be formed when A reacts with:

i) hydrogen gas in the presence of a nickel catalyst,

ii) dilute sulphuric acid,

iii) bromine dissolved in hexane.

9 a) Complete the following reaction sequence by giving the names and structural formulae of compounds B to G and stating the reagents and reaction conditions which need to be used.

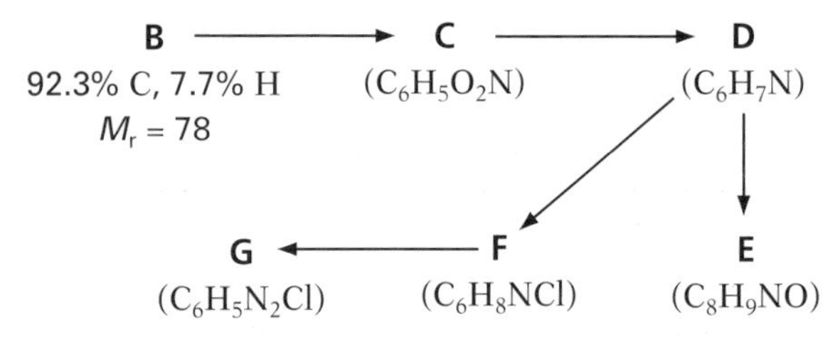

b) Adding G slowly to a solution of phenol in sodium hydroxide yields a bright orange precipitate H, $C_{12}H_{10}N_2O$.

i) Give a possible structure for H and explain why it is brightly coloured.

ii) Explain the significance of this type of reaction and the products which are formed.

iii) Why can't the reaction be carried out at room temperature?

c) When D is treated with ethanoyl chloride at room temperature, a new compound, E, is formed. Give the name and structure of E.

d) Compound E may be converted to **paracetamol**, $C_8H_9NO_2$, a very well-known painkiller. Paracetamol dissolves sparingly in water, forming a slightly acidic solution. When paracetamol is tested with neutral iron(III) chloride solution an intense blue/purple colour is produced.

i) Write down three possible structures for paracetamol.

ii) Although the monobromination of paracetamol might, in theory, produce two isomeric products, only *one* forms in practice. Deduce the structure of paracetamol.

iii) How might paracetamol by synthesised from E in three steps?

10 This question concerns the following compounds:

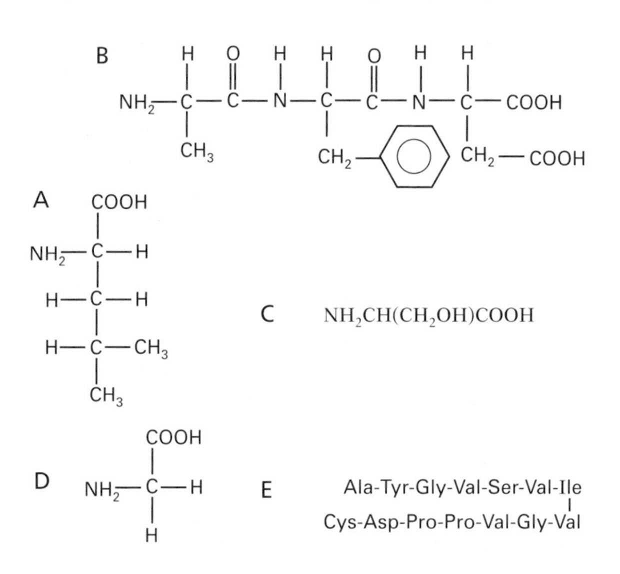

a) Which compounds are amino acids and which are polypeptides?

b) Which of the amino acids has the most hydrophobic side-chain? Which has the most hydrophilic side-chain?

c) Using Fischer projection formulae, explain the difference between the L- and D- forms of C. Which form is naturally occurring?

d) Which compounds will not be optically active?

e) Explain the meaning of the term 'essential amino acid'. Which of the amino acid residues in B and E residues are derived from essential amino acids?

f) Write down the structures of the zwitterions of the amino acids.

g) Explain the meaning of the term 'isoelectric point'. The isoelectric point of C is 5.7. Write down the structure of the C when it is dissolved in an aqueous solution at i) pH 9.8, ii) pH 5.7 and iii) pH 2.5.

Questions on Chapter 14 *continued*

h) Using symbols to denote residues (Table 14.5), write down the possible tetrapeptides which may be formed from two molecules of A and one molecule each of C and F.

i) Draw the structural formula of the tetrapeptide having the amino acid sequence: D–A–C–A. Draw circles around the peptide bonds. Explain why these bonds are said to be formed via 'condensation' reactions.

j) B is completely hydrolysed to give a mixture of free amino acids. Sketch the electrophoretogram of this mixture which would be produced at pH 5.5 (Table 14.5).

k) Partial enzymic hydrolysis of E yields *five* polypeptides, the primary structures of which confirm the amino acid sequence in E. Write down *possible* structures of these five polypeptides. (Note that there are many possible answers).

l) Which of the amino acid residues in E are likely to form: i) hydrogen bonds, ii) hydrophobic bonds, iii) disulphide bridges?

m) Which of the amino acid residues in E often cause 'reverse turns' in the tertiary structure?

n) Which compounds will give a violet complex in the Biuret test?

11 F is a 2-amino acid which has a relative molecular mass of 89 and the following elemental composition: 40.4% carbon, 7.90% hydrogen, 36.0% oxygen and 15.7% nitrogen. F exists in two optically active forms, of which only one is found naturally. In aqueous solution, at pH 6, F exists as a zwitterion. Thin-layer chromatography, using an aqueous mixture of butan-1-ol and ethanoic acid, gives an R_F value for F of 0.380.

a) Calculate the empirical and molecular formulae of F.

b) Draw the displayed structural formulae of F and its structural isomer, G.

c) Draw 3-D representations of the D- and L- forms of F. Which form is found naturally? Use the diagrams to explain why isomers are optically active.

d) What is a zwitterion? Draw the structure of the zwitterion of F. Write ionic equations for the reactions of this zwitterion with i) $H^+(aq)$ ions and ii) $OH^-(aq)$ ions.

e) Using a suitable diagram, *briefly* describe the technique of thin-layer chromatography and explain the meaning of the term 'R_F value'. On a particular chromatogram, developed in aqueous butan-1-ol/ethanoic acid, the solvent front moved 4.50 cm. How far did the spot of F move?

12 Figure 14.29 shows a schematic representation of a two-dimensional paper chromatogram of an amino acid mixture.

a) Calculate the R_F values of

i) histidine, glutamic acid and isoleucine in aqueous butan-1-ol/ethanoic acid;

ii) asparagine, threonine and methionine in aqueous phenol/ammonia.

b) Which amino acids were completely separated by running the chromatogram in i) dimension 1 and ii) dimension 2?

c) Describe, briefly, how the percentages of isoleucine, phenylalanaine and leucine in the mixture may be obtained using ion-exchange chromatography.

13 Give an explanation of the following statements, giving your answers in note form and using suitable diagrams.

a) The solubility curve of β-lactoglobin, a protein, as a function of pH is U-shaped with a minimum at pH 5.3.

b) The secondary structure of keratin is an α-helix whilst that of Tussore silk is a β-pleated sheet.

c) Proteins are classified as being 'fibrous' or 'globular', as a result of different types of weak attractions within their structures.

d) Proline and glycine residues are often found in non-ordered sections of a polypeptide chain and they are often responsible for 'reverse turns'.

e) The biological properties of a protein may be severely affected by changing just one amino acid residue in its primary structure.

f) Proteins may be denatured by rapid agitation, heat and changes in the pH.

14 A sample of cattle insulin was subjected to acid hydrolysis and then the constituent amino acids were separated and estimated using ion-exchange chromatography. The results are shown in Table 14.9.

a) By considering each amino acid as an 'element', use the method in section 14.10 to calculate the amino acid composition of cattle insulin.

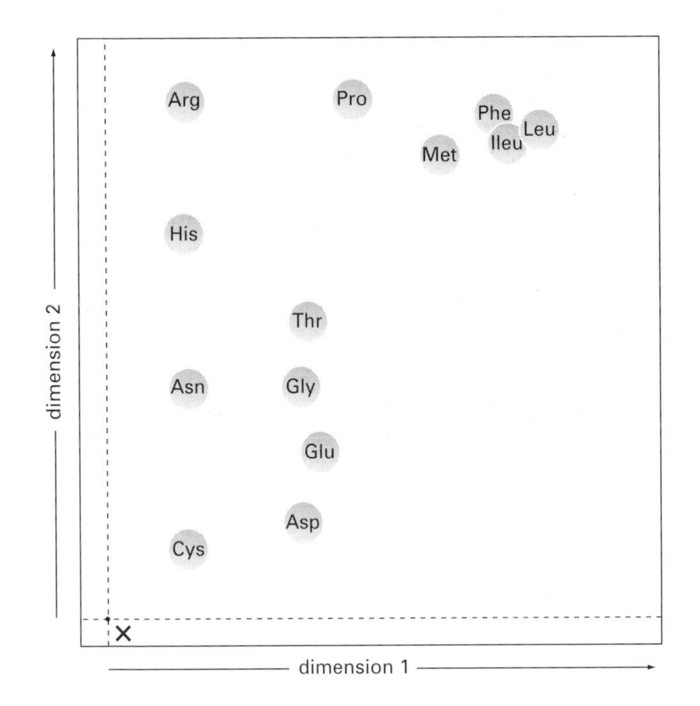

Figure 14.29
A schematic representation of a 2-dimensional paper chromatogram. A 'spot' of the amino acid mixture was originally placed at point X. This was developed with aqueous mixture of butan-1-ol and ethanoic acid (in dimension 1). The chromatogram was allowed to dry and developed with an aqueous mixture of phenol and ammonia (in dimension 2). In both cases, the solvent front moved to the edge of the paper

Questions on Chapter 14 *continued*

Table 14.9 Amino acid composition of cattle insulin

Amino acid	symbol	M_r	% by mass in cattle insulin
Alanine	Ala	89	4.17
Arginine	Arg	174	2.67
Asparagine	Asn	132	6.20
Cysteine	Cys	107	10.8
Glutamine	Gln	146	6.63
Glutamic acid	Glu	147	8.78
Glycine	Gly	75	4.53
Histidine	His	155	4.75
Isoleucine	Ileu	131	1.98
Leucine	Leu	131	11.7
Lysine	Lys	146	2.24
Phenylalanine	Phe	165	7.44
Proline	Pro	115	1.79
Serine	Ser	105	4.75
Threonine	Thr	119	1.78
Tyrosine	Tyr	181	11.1
Valine	Val	117	8.69

(Note: This question provides an ideal application for using an IT spreadsheet, using the 'cell' layout shown below. Here M_r is the relative molecular mass and the simplified mole ratio is the value in a given D cell divided by the lowest value in the D cells (this should come out as 0.0150 for threonine).

	A	B	C	D	E
1	Symbol	% present	M_r	mole ratio	simplified mole ratio
2	Ala	4.17	89	= B2/C2	= D2/0.0150
3	Arg	2.67	174	0.0153	1.02

b) The partial hydrolysis of cattle insulin produces a hexapeptide made of the following residues: Asn, Phe, Leu, Val, Gln, His. Explain how the primary structure of the hexapeptide may be determined.

15 The variation in the ionic composition of an amino acid with pH may be investigated by titrating the amino acid against a standard acid or alkali. A student carried out the experiment using glycine. Her results are shown in Table 14.10.

Table 14.10 Variation in pH during the titration of 10 cm^3 of glycine (0.01 mol dm^{-3}) with (a) hydrochloric acid (0.01 mol dm^{-3}) and (b) aqueous potassium hydroxide (0.01 mol dm^{-3}).

volume of HCl(aq) added/cm^{-3}	0	0.5	1.0	2.0	3.0	4.0	6.0	8.0
pH	6.0	3.7	3.4	3.0	2.8	2.6	2.2	1.8
volume of KOH(aq) added/cm^{-3}	0	0.5	1.0	2.0	3.0	5.0	7.0	9.0
pH	6.0	8.6	8.9	9.2	9.4	9.6	10.1	10.8

a) Plot these results on one set of axes, with 'Volume of solution added' on the vertical axis and the 'pH' on the horizontal axis.

b) i) How many titratable groups are there in glycine?
ii) Write an equation for the reaction of each titratable group with HCl(aq), or KOH(aq), as appropriate.

c) Over what pH range do the glycine molecules exist almost completely as zwitterions?

d) At what pH is the number of moles of glycine equal to the number of moles of added HCl(aq)? In what form(s) are the glycine molecules at this pH?

e) At what pH is the number of moles of glycine equal to the number of moles of added KOH(aq)? In what form(s) are the glycine molecules at this pH?

f) What form of glycine predominates at the following pH values: i) 2.1, ii) 10.2, iii) 4.6?

g) Does there appear to be a relationship between the pH values in your answers to questions c), d) and e)?

Comments on the activities

Activity 14.1

1 **a)** 1-nitrobutane
b) propanonitrile
c) 2-methylpropanonitrile
d) 4-methylnitrobenzene

2 **a)** **b)** **c)** **d)**

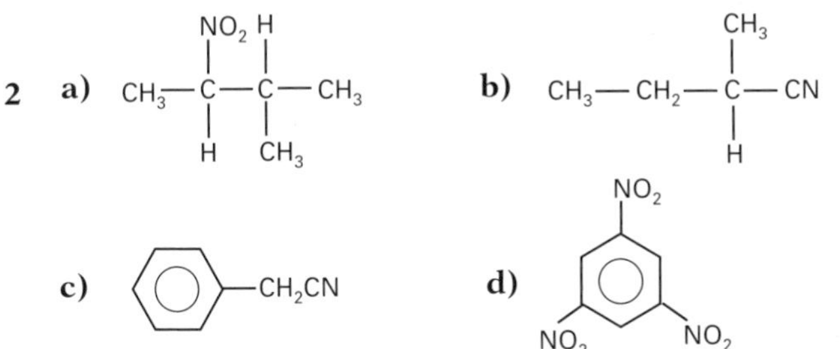

3 **a)** KCN (ethanol), reflux; **b)** excess KCN/dil. HCl(aq) room temp ⟶ 2-hydroxyethanonitrile; then, dil. H_2SO_4(aq) reflux ⟶ 2-hydroxyethanoic acid.

4 **a)** 4-methylphenylamine

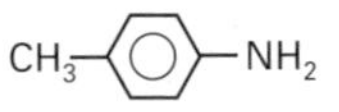

b) 2,2-dimethylpropanoic acid CH_3—C(CH_3)(CH_3)—COOH

c) (phenylmethyl) amine C_6H_5—CH_2NH_2

d) first step: CH_3—C(OH)(CH_3)—CN, 2-hydroxy-2-methylpropanonitrile

second step: CH_3—C(OH)(CH_3)—COOH, 2-hydroxy-2-methylpropanoic acid

5 **a)**

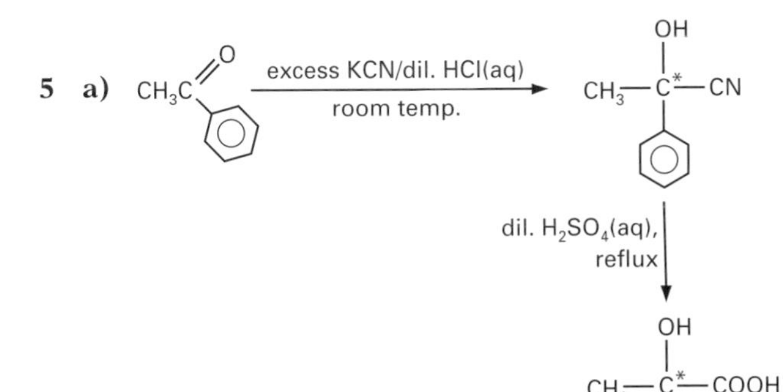

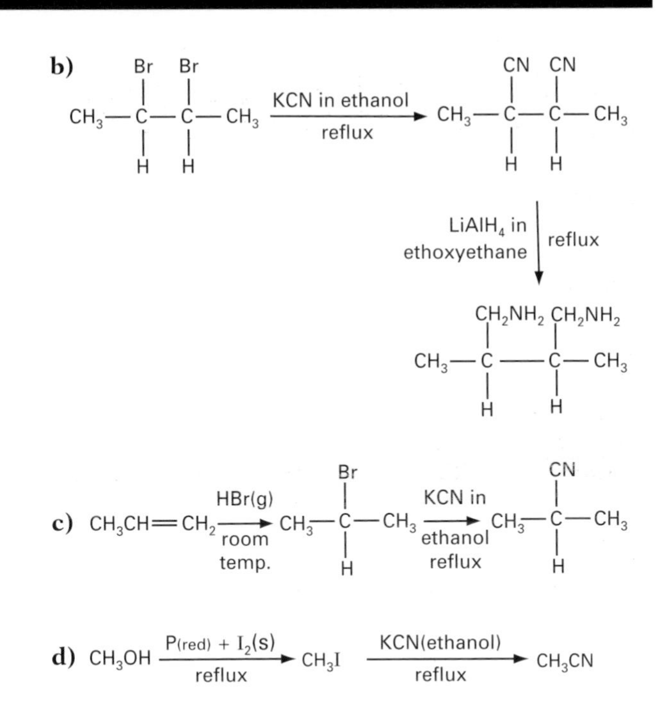

Note: various halogenating agents can be used in c) and d) step 1.

Activity 14.2

1 All the atoms in the ring are not the same.
2 A primary amine at the $\ddot{N}H_2$ group and a secondary amine at the ring $\ddot{N}H$ group.
3 More pollen in the air as flowers and blossom are formed.

Activity 14.3

1 **a)** methylamine
b) 2-aminobutane
c) triethylamine
d) dimethylethylamine
e) N-methylphenylamine

2 1°: $CH_3CH_2CH_2NH_2$ propylamine; C_6H_5—NH_2 phenylamine; CH_3—C_6H_4—NH_2 4-methylphenylamine; H_2N—C_6H_4—NH_2 1,4-diaminobenzene

2°: C_2H_5—N(H)—C_6H_5 N-ethylphenylamine; $(C_2H_5)_2NH$ diethylamine

3°: $(C_2H_5)_3N$ triethylamine

Comments on the activities *continued*

3 a) $2C_2H_5NH_2(l) + H_2SO_4(aq) \longrightarrow [C_2H_5NH_3]_2SO_4(aq)$
b) $(C_2H_5)_3N{:}(l) + HCl(aq) \longrightarrow [(C_2H_5)_3NH]Cl(aq)$
c) $[CH_3NH_3]Cl + NaOH(aq) \longrightarrow CH_3NH_2(aq) + NaCl(aq)$
d) $[C_6H_5NH_3]_2SO_4 + 2KOH(aq) \longrightarrow 2C_6H_5NH_2(l) + K_2SO_4(aq)$

Activity 14.4

Method I: With an acid chloride, for example, ethanoyl chloride:

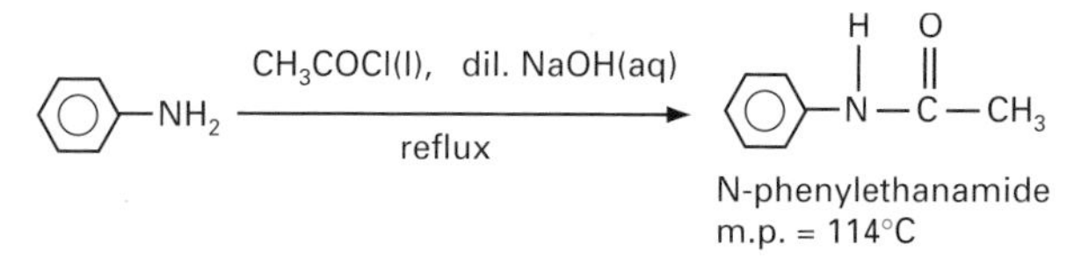

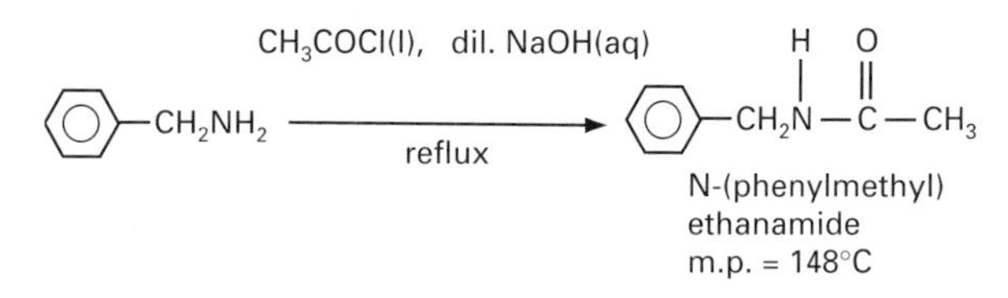

As we saw in section 14.7 pure amides are crystalline solids which can be identified from their sharp melting points. Thus, we can distinguish between the amides and, hence, their parent amines.

Method II: With nitrous acid at 5°C:

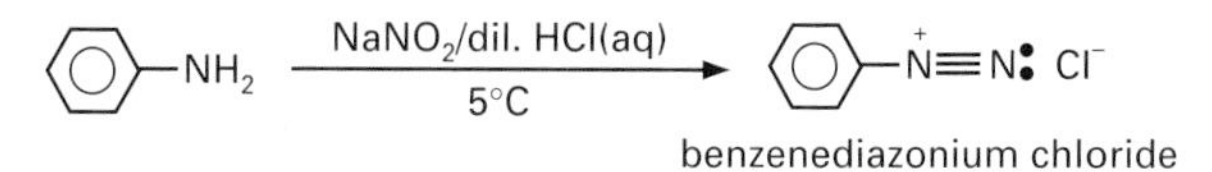

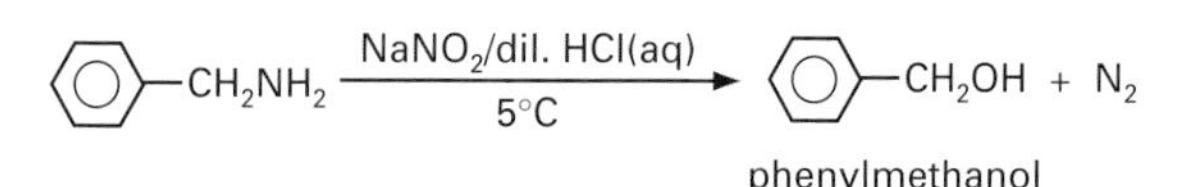

Vigorous evolution of nitrogen gas will only be seen in the reaction with phenylmethylamine. A coupling reaction of benzenediazonium chloride with phenol, in alkaline solution would give a yellow azo-dye. Phenylmethanol would not react in this way.

Activity 14.5

All of the models should have the same 3-D arrangements of bonds around the α-carbon as shown in Figure 14.14(b). The amino acid side-chain occupies the same position as the —CH_3 group in these diagrams.

Activity 14.6

1 The isoelectric points of the amino acids are:

Asp	Cys	Met	Leu	His	Lys	Arg
2.8	5.1	5.7	6.0	7.6	9.7	10.8

The electrophoretograms provide the following information:

A At pH 3.8, one amino acid moves to the anode (+ve), that is it is negatively charged. Its isoelectric point must occur, then, at pH < 3.8. Thus, aspartic acid is present in the mixture. The other three acids exists as 'ammonium' ions at pH 3.8, suggesting that their isoelectric points are greater than 3.8.

B One spot has not moved at pH 6.0, indicating that leucine is present. Once again, the Asp spot moves to the anode. The two spots which move to the cathode (–ve) have isoelectric points greater than 6.0. Thus, Cys and Met are not present.

C Two spots have moved towards the cathode. If His had been present, its spot would not have moved at pH 7.6. Thus, the two other amino acids are Lys and Arg.

D At pH 10.2, three amino acids (Asp, Leu and Lys) exist as anions and move to the anode. Arginine will exist as the 'ammonium' ion at this pH, so its 'spot' moves to the cathode.

Thus, the amino acids in the mixture are Asp, Leu, Lys and Arg.

2

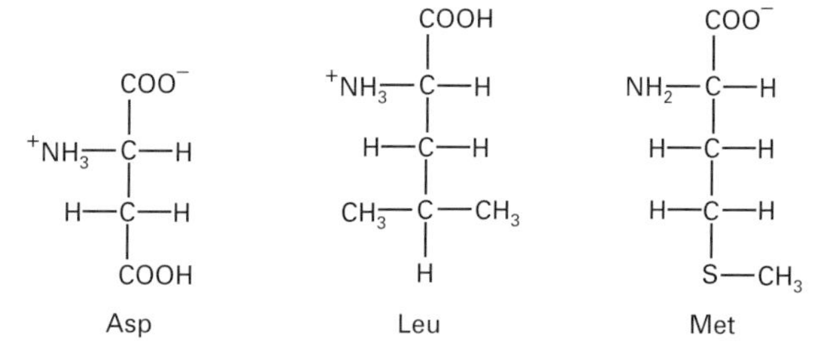

Activity 14.7

1

	Arg	Asn	Trp	Ala	Val	Met
In 100 g of polypeptide, there will be:	23.8	18.0	27.9	12.2	7.90	10.2 g
that is, in moles:	$\frac{23.8}{174}$	$\frac{18.0}{132}$	$\frac{27.9}{204}$	$\frac{12.2}{89}$	$\frac{7.90}{117}$	$\frac{10.2}{149}$
	0.137	0.136	0.137	0.137	0.067	0.068
giving the simplest whole number ratio:	2	2	2	2	1	1

The amino acid composition is:
Arg 2, Asn 2, Trp 2, Ala 2, Val 1, Met 1.

2 The amino acid sequence is:
Met – Asn – Ala – Trp – Val – Ala – Trp – Arg – Asn – Arg.

Comments on the activities *continued*

Activity 14.8

1

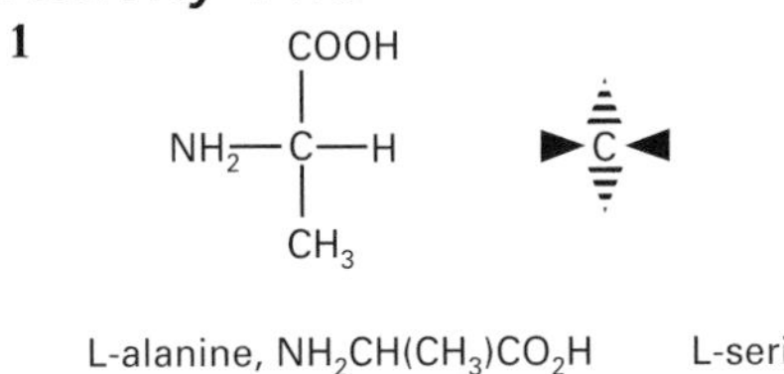

L-alanine, $NH_2CH(CH_3)CO_2H$ L-serine, $NH_2CH(CH_2OH)CO_2H$

3

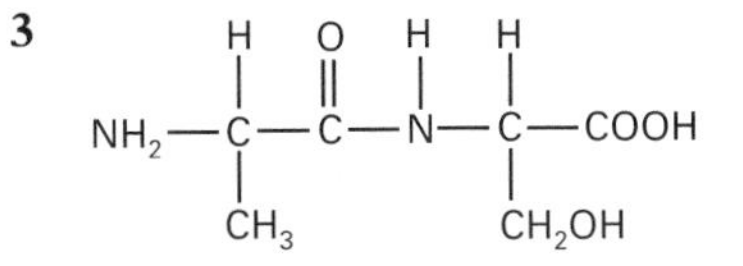

Ala–Ser dipeptide

4 Due to the considerable π bond character in the peptide bond, *cis*- and *trans*- isomers may be formed on peptization:

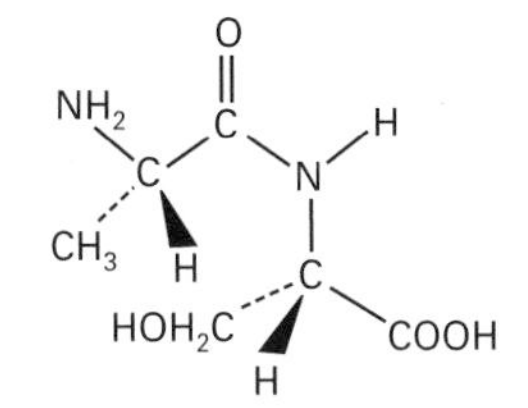

peptide bond: *cis*- conformation

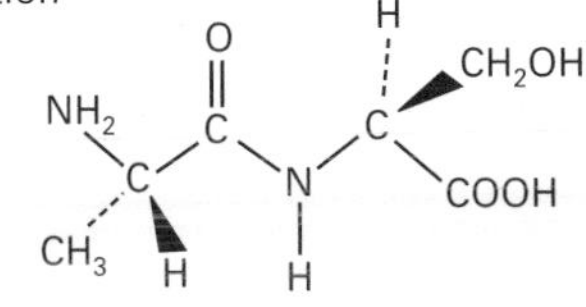

peptide bond: *trans*- conformation

THEME C

Methods of analysis and detection of the chemistry of colour

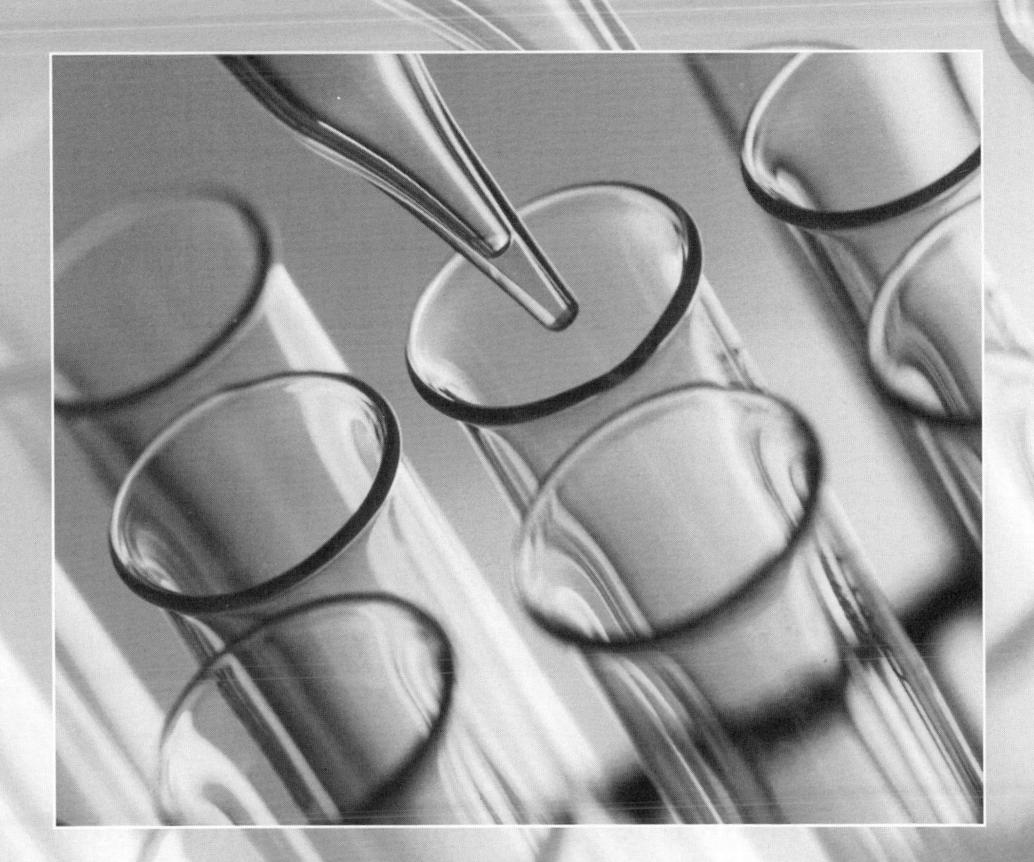

CHAPTER 15

Separation and purification

Contents

Study Checklist

After studying this chapter you should be able to:

1. Appreciate the importance of a 'pure' product in chemical analysis and in areas such as food and drugs.
2. Explain how melting and/or boiling behaviour may be used as a rough guide to the purity of an organic compound.
3. Describe and explain the principles involved in paper, thin-layer and gas–liquid chromatography.
4. Define the terms 'R_F value' and 'retention time' and explain their use.
5. Given suitable data, identify the components of a mixture from a paper or thin-layer chromatogram.
6. Show how a gas–liquid chromatogram may be used in the quantitative analysis of a mixture.
7. Outline the principles involved in electrophoresis and explain the importance of pH control in this technique.
8. Describe important analytical applications of chromatography and/or electrophoresis including the identification of proteins and genetic (DNA) fingerprinting.
9. Appreciate that the amount of material that can be separated by chromatography/electrophoresis is usually limited.
10. Describe the following methods for the larger scale separation of mixtures (see also Chapter 27): recrystallation; steam distillation; solvent extraction; distillation and fractional distallation.
11. Outline how carboxylic acids, phenols and amines may be separated from neutral organic compounds by the formation of water-soluble salts.

15.1 Chemical 'purity' and its importance

In many people's minds a '**pure**' material is regarded as desirable, whereas an '**impure**' material must be inferior. Even in the case of food and drink, this is not necessarily true. Pure water, for example, contains no beneficial trace elements such as calcium, which is needed in bone formation. Indeed water companies may deliberately add impurities such as chlorine to kill harmful bacteria or extra fluoride to reduce the incidence of tooth decay. Water that is **potable** or safe to drink may not, therefore, in the strictest sense be entirely pure. In the same way many natural foods contain trace impurities which may be beneficial.

When contaminated food can be good for you

Essential trace elements are being lost from the diet in Western societies because of an obsession with purity in food and food preparation. Consumer demand has just gone too far, said Conor Reilly, head of the school of public health at the Queensland University of Technology and an expert on trace elements.

'The quality that we demand and have a right to in our food does not have to be one of absolute purity,' Reilly told the ANZAAS conference. 'There is no need to have an excessive fear of all non-food components of our diet.'

Reilly said that food laws in Australia could be traced back to a British parliamentary act passed in 1875, in reaction to uncontrolled selling of food that had been adulterated to increase its weight or change its appearance. These 'toxic cocktails' included bread made from flour that had been mixed with chalk and alum, wine and vinegar topped up with sulphuric acid and beer laced with copper sulphate.

However, there was also wisdom in the old saying that 'every man must eat a peck of dirt before he dies', The body needs what Reilly called 'adventitious material' – nutrients ingested accidentally.

In Tasmania, the unrinsed residues of detergents in milk containers had added iodophores – chemicals containing iodine – to the milk consumed by children. This had doubled the intake of the element. As a result, there had been a decrease in endemic goitre, a severe swelling of the thyroid gland cause by iodine deficiency.

'Unfortunately for Tasmanians, the use of iodophores is now declining and intake of iodine by Tasmanian children is once again falling,' Reilly said. Tasmania has low iodine levels in its soils and consequently in locally grown foods.

Cast-iron cooking pots reduce the risk of iron deficiency

A study of children in Brisbane has shown that food provides less than the daily requirement of chromium, which is necessary for insulin function and carbohydrate metabolism. The children take in between 16 and 51 micrograms of chromium a day from food but require between 50 and 200 micrograms. However, said Reilly, the children's needs are supplemented from an adventitious source – tinplate used in cans and the steel of cooking utensils and processing equipment. Cooking in a stainless steel saucepan can cause a sevenfold increase the level of chromium in vegetables.

By the same token, Reilly said that a return to cast-iron cooking pots might be one way of reducing the 'intractable problem' of iron deficiency that afflicts most advanced countries. Evidence from rural southern Africa shows the value of iron cooking pots. The people there consume little red meat, which is rich in iron, but anaemia, caused by iron deficiency, is rare. The staple food, maize porridge, is boiled in cast-iron cooking pots. Reilly warned, however, that there are dangers here. Some Africans, mostly men, suffer from iron overload, which can lead to liver failure. These men drink large amounts of beer made in crude equipment. The beer picks up iron and other impurities from recycled containers used for fermentation.

This Week, *New Scientist*, 26 September 1992, Vol. 135 No. 1840 Page 10 by Ian Anderson

Why then are chemists so concerned about purity? Well, of course, not all impurities are beneficial or even harmless. In the case of food the law permits only certain chemicals at certain levels to be added to improve its appeal or shelf-life. Even some approved food additives may cause problems for some consumers. The orange food colouring tartrazine (E102), for example, is suspected of causing hyperactivity in some children.

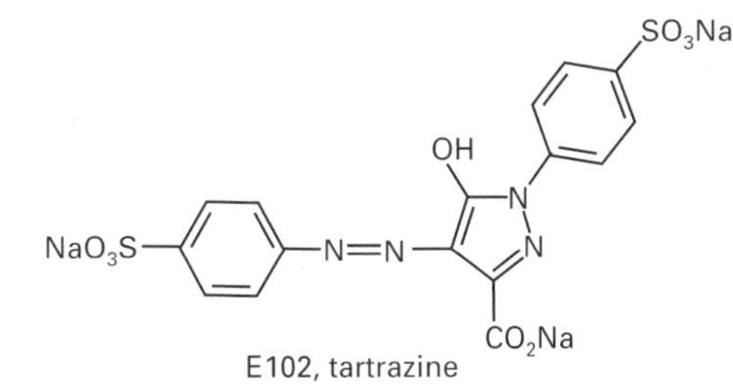

E102, tartrazine

Purity is of particular importance in the drug industry. Most chemical reactions give a mixture of products, possibly together with unchanged starting materials. If a particular compound is to be screened as a potential drug it must be separated from such a mixture so that any effects are known to be due to this material rather than any impurities. Thalidomide provides a tragic example of the use of a drug that was a mixture of two components. The thalidomide molecule can exist in two different mirror-image forms or optical isomers (see section 6.5). Whilst chemically identical, the two forms of thalidomide had dramatically different biochemical effects. One relieved the symptoms of morning sickness during early pregnancy, whilst the other was eventually shown to have caused severe birth defects.

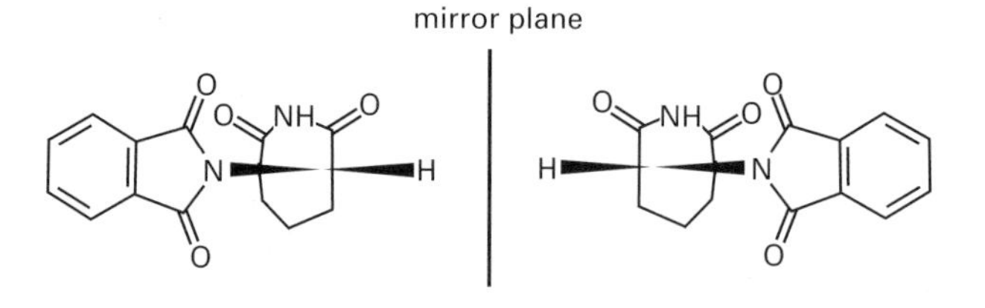

Additionally, as we shall see later in this chapter, we need a pure sample in order to establish the formula and structure of an unknown compound.

15.2 Determination of purity

How then do we set about establishing the purity, or otherwise, of a sample? One quick, but not infallible, indication is given by the behaviour of the sample on melting or boiling. Pure substances generally have sharp melting and boiling points, i.e. they change state over a very narrow temperature range, whilst mixtures melt or boil gradually over a much wider temperature range. If a sharp melting or boiling point is obtained then this may be used to help establish the identity of the sample. For example, aspirin may be made by treating 2-hydroxybenzoic acid with ethanoic anhydride (see section 6.1).

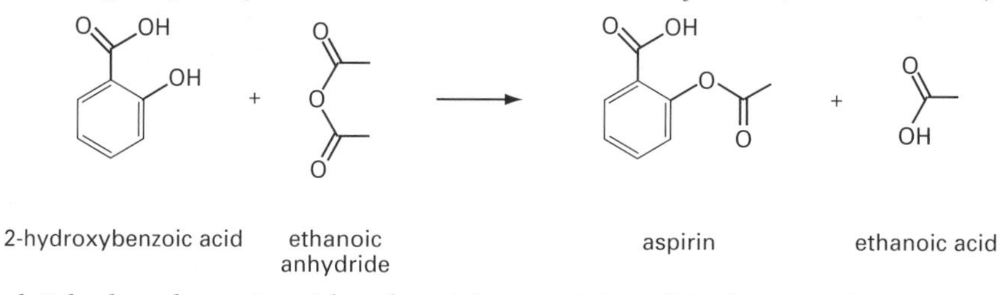

Both 2-hydroxybenzoic acid and aspirin are white solids that are almost insoluble in water but we can check that the reaction has been successful by separating the solid product and measuring its melting point. 2-hydroxybenzoic acid melts at 158°C, whereas the melting point of aspirin is 135°C.

Activity 15.1

1 What conclusion would you draw in each case if the product from the above reaction melted as follows:
   a) between 134 and 135°C
   b) between 156 and 158°C
   c) gradually over the range 120–126°C?

2 How could you use pure aspirin to test whether an unknown solid with a similar melting point was in fact also pure aspirin?

Focus 15a

1. Pure substances generally have fixed, sharp melting points.
2. The identity of a substance may be checked by comparing its melting point to that of a mixture with an authentic sample.

The most reliable way of ensuring that a sample is pure is to show that it is not a mixture, i.e. that it cannot be separated into two or more distinct components. This is often conveniently checked by **chromatography**, a technique originally developed to separate mixtures of dyes or similar coloured components. If you spot a little black ink from a fountain pen or felt-tipped pen onto a piece of filter paper or blotting paper you may notice that it starts to look coloured around the edges as it soaks into the paper. If water is slowly dripped onto the centre of the blot the different coloured inks that together make up black spread out further on the paper, forming a pattern of concentric rings of different colours known as a **chromatogram**. Each ring corresponds to a different dye in the mixture. If you try other ink colours you may find one that forms only a single ring of colour, suggesting it contains only one dye.

In this form of chromatography the dyes in the sample adhere to the paper but as the water soaks past the ink spot, it tends to dissolve each component. Those dyes that stick only weakly to the paper but dissolve readily in water will travel the fastest and therefore spread out furthest.

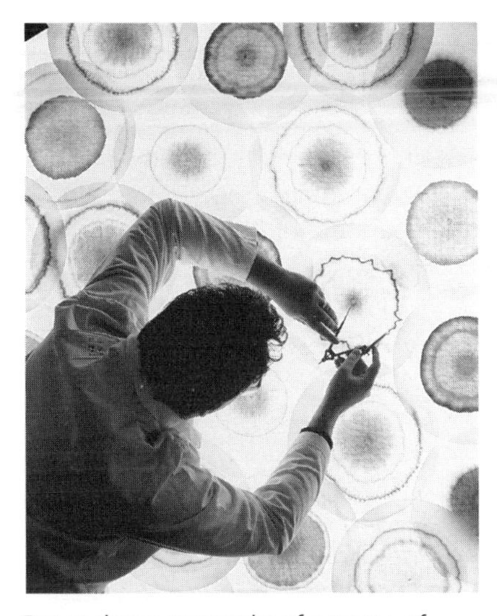

Paper chromatographs of a range of industrial dyes. The distance moved by each 'ring' is being measured.

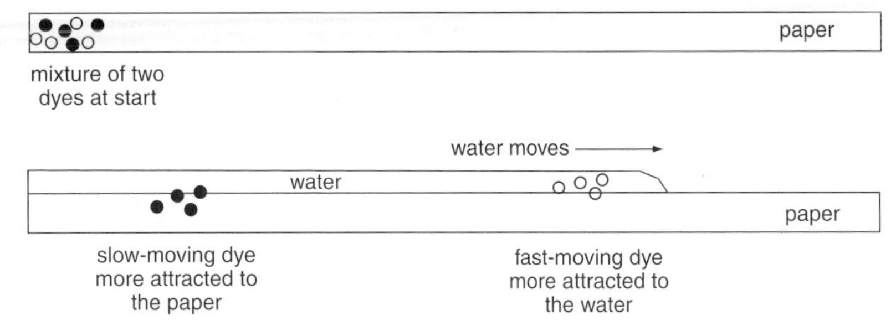

Each dye is therefore distributed between the paper, known as the **stationary phase**, and the water, known as the **mobile phase**. Of course, if you used waterproof inks in the above experiment, none of the dyes would move at all. To get a good separation in such a case you would need to find another mobile phase in which the ink dissolves.

Although this technique was originally devised to separate dyes, it has been extended to cover many other types of mixture. For instance, it is sometimes possible to convert colourless materials into coloured spots by developing the dried chromatogram with a particular reagent. Alternatively, the position of certain substances may be detected by fluorescence under ultraviolet light.

The separation of a mixture by chromatography relies upon the fact that each component is attracted to different extents by the stationary phase and the mobile phase. In **partition chromatography**, the stationary phase is actually also a liquid and is adsorbed onto the surface of a solid support. The components of the mixture have differing solubilities in the two solvents forming the stationary phase and the mobile phase and so will distribute, or partition, themselves accordingly. In **adsorption chromatography** the solid, stationary phase does not rely upon the action of an adsorbed layer of solvent but directly attracts each component by surface adsorption.

Paper chromatography

This is an example of partition chromatography since the stationary phase is actually water that is adsorbed onto the paper. If paper chromatography is carried out under controlled conditions, as, for example, using the apparatus shown in Figure 15.1, then any particular component will always travel the same distance relative to the distance moved by the solvent front after it passes the sample spot. The $\boldsymbol{R_F}$ **value** obtained from such a chromatogram may be used to help identify the components of a mixture. The R_F value can be defined as:

$$R_F \text{ value} = \frac{\text{distance travelled by component}}{\text{distance travelled by solvent front}}$$

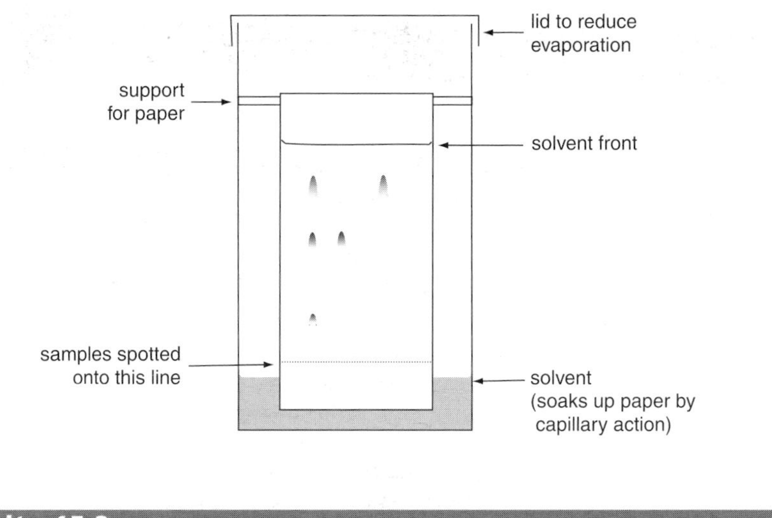

Figure 15.1 ▶
Ascending paper chromatography
N.B. the distance moved by any component is generally taken from the starting point to the middle of its spot

Activity 15.2

Paper chromatography is carried out on three green food colourings, A, B and C, using the apparatus in Figure 15.1. The resultant chromatogram is drawn to scale in Figure 15.2. The experiment was performed at a temperature of 25°C using a solvent consisting of 60% by volume of aqueous propanone.

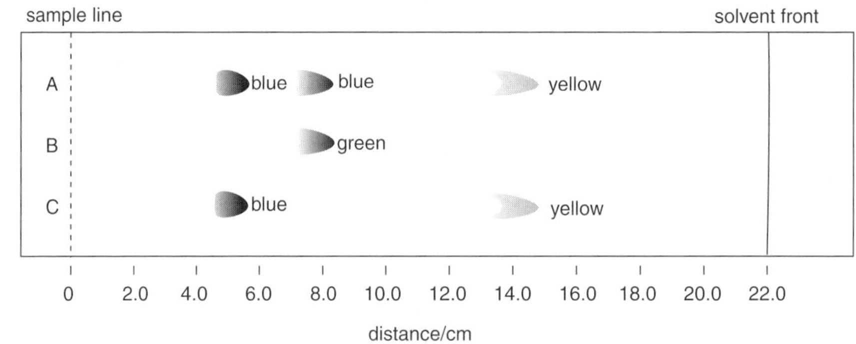

Figure 15.2 ▶
Chromatogram of the three green food colourings for activity 15.2

1. Which of the food colourings could be a pure compound?
2. What is the R_F value of this material under the conditions of this experiment?
3. Why might a different R_F value be obtained for this substance if the experiment were repeated without the lid covering the tank?
4. The R_F value of the food colouring 'Sunset Yellow' under these conditions is known to be 0.70. Explain whether any of the samples could contain this compound.
5. It has been suggested that A could be a mixture of B and C. Explain whether you think this is likely.

Thin-layer chromatography

Paper chromatography is a quick and convenient way of checking whether a particular sample is a pure compound or a mixture. However, results can be variable as the properties of the paper depend upon its moisture content. Rather more reproducible results are usually obtained using thin-layer chromatography or TLC. This is an example of adsorption chromatography since the stationary phase is a thin layer of an adsorbent powder such as alumina. The powder is mixed to a smooth paste with water and a binding agent such as starch and spread uniformly over a glass plate. After drying, the plate may be used in place of the paper in the apparatus shown in Figure 15.1.

Two-dimensional chromatography

Use of a single mobile phase moving in one direction may only give a partial separation of some complex mixtures. However, if the chromatogram is dried and then treated with a different mobile phase moving at right angles to the first, a complete separation may be obtained. This technique, illustrated in Figure 15.3, has been used in the analysis of proteins. The metabolism of sugars in our bodies is controlled by insulin, a protein made from 17 different amino acids (see section 14.8). A mixture of these amino acids, obtained by hydrolysing insulin, is partially separated by thin-layer chromatography using a mixture of butan-1-ol and ethanoic acid as the first mobile phase. Treatment with aqueous phenol moving at right angles to this completes this separation and the position of each amino acid spot is revealed by a purple coloration on spraying with ninhydrin solution.

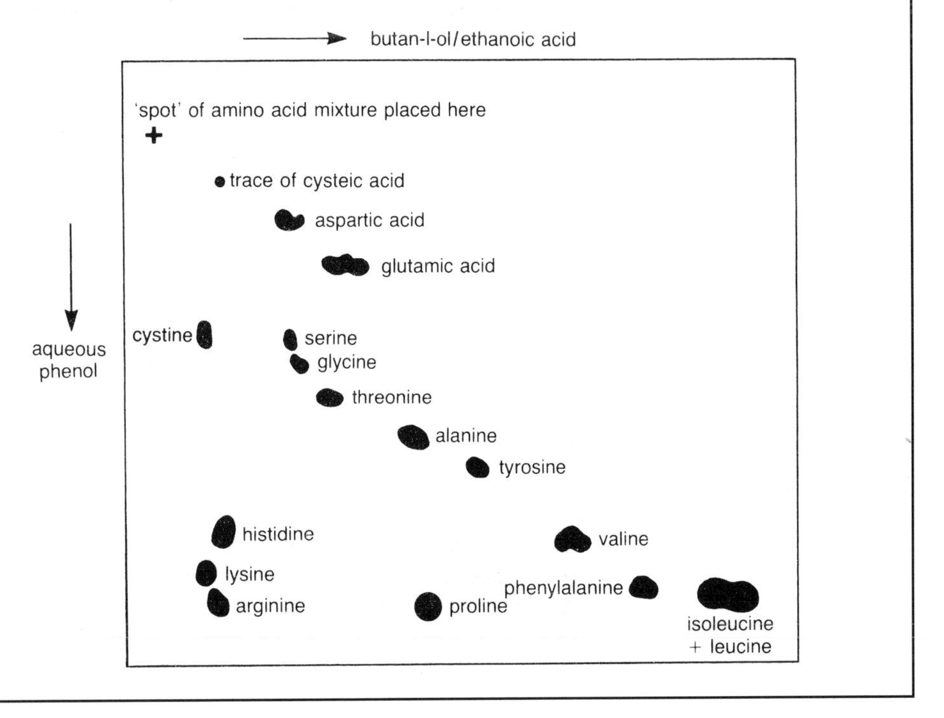

Figure 15.3
A mixture of amino acids, obtained by hydrolysing insulin, is partially separated by thin-layer chromatography using a mixture of butan-1-ol and ethanoic acid as the first mobile phase. Treatment with aqueous phenol moving at right angles to this completes the separation and the position of each amino acid spot is revealed by a purple colouration on spraying with ninhydrin solution

Thin-layer chromatography is designed for the separation of quite small amounts of a mixture. Larger quantities of each component can be obtained by packing the stationary phase into a glass column. The mixture is placed on the top of the stationary phase and the mobile phase is allowed to flow down the column at a steady rate. Each component may be identified as it leaves the column by passing the liquid through an infrared spectrometer. A modern extension of **column chromatography** is **high pressure liquid chromatography**, **HPLC**. Here the column is made of stainless steel, rather than glass, and the mobile phase is pumped through the stationary phase in the column under pressures of up to 100 atmospheres. Use of such high pressure allows the separation to be carried out more quickly than with simple gravity-fed column chromatography.

Gas–liquid chromatography

This type of partition chromatography is a very powerful and sensitive method of analysis, particularly in organic chemistry. The components of a typical **gas-liquid chromatography (GLC)** set up are shown schematically in Figure 15.4. Here the stationary phase-up is a non-volatile liquid, often adsorbed onto the surface of a solid packing material, contained within a long narrow coiled metal tube, or **column**, which is kept at a controlled temperature in an oven. The mobile phase is an inert **carrier** gas, such as nitrogen or a noble gas, which is passed through the column at a constant measured rate. The sample is introduced into the system via an injection port and is swept into the column by the carrier gas. Here the components distribute themselves between the liquid and gas and pass through the column at different rates. Generally, the more volatile components tend to remain largely in the gas phase and, therefore, pass more quickly through the system. The gases leaving the column may be passed through a hydrogen flame that will burn any sample material present, producing ions. These ions carry an electric current between two charged plates and this, in turn, may be used to produce a response on a chart recorder. The size of the signal depends upon the number of ions produced and, therefore, on the amount of the material entering the detector. In more sophisticated systems, the **flame ionisation detector** may be replaced by an infrared or mass spectrometer giving a 'fingerprint' for the component as it leaves the column (see Chapters 16 and 17 for more details on the use of infrared and mass spectra in analysis.)

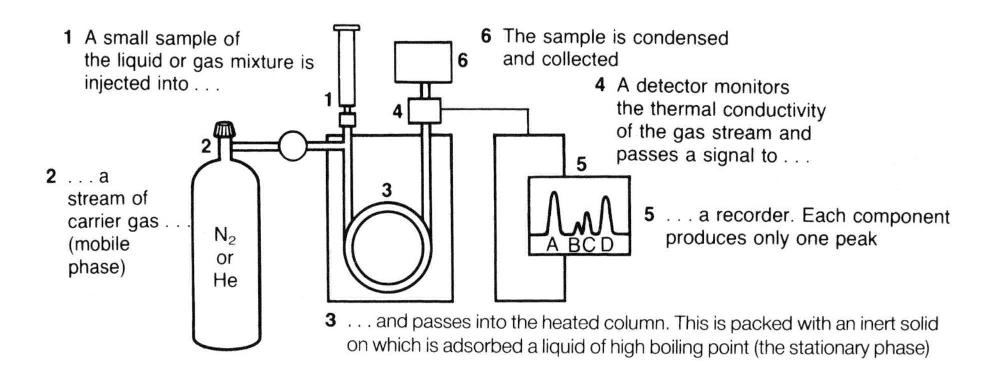

Figure 15.4
The main features of a gas–liquid chromatograph and a typical recorder trace for a mixture of four volatile liquids, A, B, C and D

Just as with paper and thin-layer chromatography the rate at which any material passes through the system depends upon the conditions used. For any particular column at a given temperature with a specified flow rate of a certain carrier gas, the time taken for a particular component to travel through the system, known as its **retention time**, is fixed.

This technique is very sensitive and may be used to determine the amount of any particular component present in a mixture. An example of this is in the determination of alcohol in urine or blood samples taken from drivers who have failed a roadside breathalyser test. The retention time of ethanol is determined by injecting a pure sample into the column. This is repeated with different known amounts of ethanol so that the relationship between peak size and mass of ethanol can be established.

Activity 15.3

Figure 15.5 shows the GLC traces obtained using fixed volumes of solutions containing known concentrations of ethanol in water. A sample of urine with the same volume as the calibration solutions gave the trace shown in Figure 15.6.

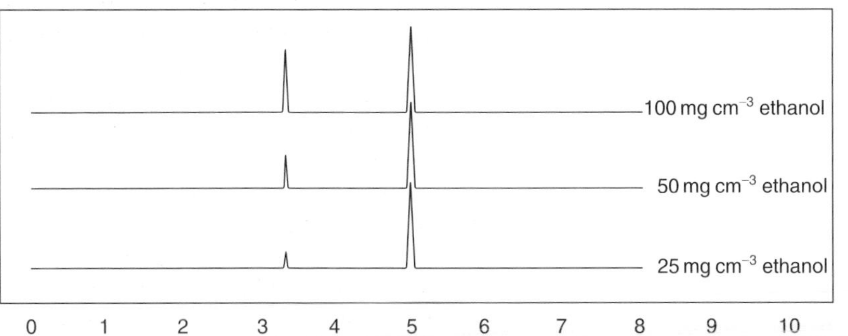

Figure 15.5 ▶
GLC calibration traces for aqueous ethanol

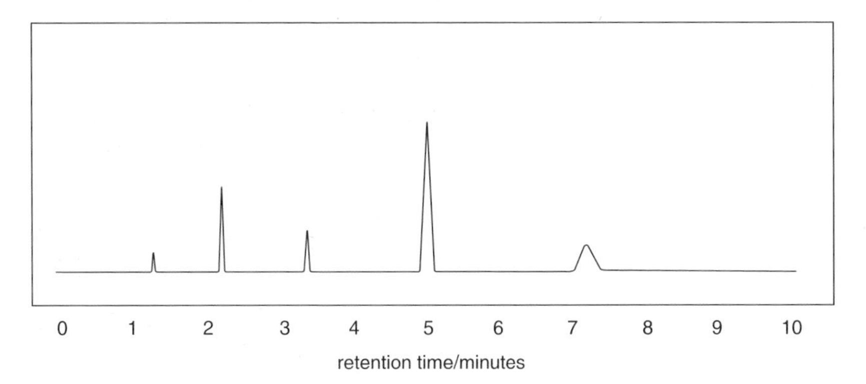

Figure 15.6 ▶
GLC trace for urine sample

1. What is the retention time for ethanol under these conditions?
2. Estimate the concentration of ethanol in the sample of urine.
3. Why does the GLC trace for urine show multiple peaks?
4. What substance do you think is responsible for the peak corresponding to a retention time of 5 minutes in all these chromatograms?
5. Another testing centre has asked to use the ethanol calibration trace shown in Figure 15.5. Explain why this request should be refused.

Focus 15b

1. Chromatography involves separation of a mixture by passing a mobile phase over a stationary phase.
2. In **adsorption chromatography**, e.g. TLC and HPLC, the mixture is adsorbed directly onto a solid stationary phase.
3. In **partition chromatography**, e.g. paper and GLC, the stationary phase is a liquid, usually supported on an inert solid.
4. Under set conditions the rate at which any substance moves through the system is fixed. R_F **values** in paper chromatography/TLC and **retention times** in GLC make use of this fact as an aid to component identification.

Some specialised chromatographic techniques

In an **ion-exchange column** the stationary phase consists of small beads of resin which have a giant lattice structure containing replaceable ions. When a solution containing ions of a similar charge are passed through the resin, exchange of ions can take place. The diagram below illustrates how a cation exchange resin can be used to soften hard water which contains dissolved calcium and/or magnesium ions. As the water moves down the column, the metal ions are progressively removed from solution and replaced by sodium ions. Of course, after prolonged use, all the original sodium ions will eventually be replaced by calcium and the resin will be 'exhausted'. However, the above process is quite easily reversed. Can you suggest a cheap, readily available material that could be used to regenerate spent resin? If different positive ions are present in the water they will exchange with sodium ions in the resin at different rates and so be separated into different bands on passing down the column.

Anion exchange resins, which contain replaceable negatively charged ions embedded in a positively charged fixed lattice, can be used to remove negatively charged ions, such as nitrates, from contaminated water supplies.

Gel permeation chromatography separates components according to molecular size. The stationary phase contains tiny pores, typically 20–200 nm in size, which can trap small molecules more easily than big ones. Larger molecules therefore pass through the column more quickly than smaller ones. In effect, the stationary phase acts as a molecular sieve but unlike most sieves this one allows larger particles to pass through it more easily than smaller ones.

Na^+	Na^+	Na^+
Na^+	Na^+	Na^+
Na^+	Na^+	Na^+

lattice containing removable sodium ions attached to fixed negatively charged sites

$Ca^{2+}(aq)$ $\rightleftharpoons$

calcium ions cause water hardness

Ca^{2+}		Na^+
Na^+	Na^+	Na^+
Na^+	Na^+	Na^+

\+ $2Na^+(aq)$

calcium ions trapped in lattice and so removed from solution

Electrophoresis

In chromatography the components of a mixture are separated because they travel along the stationary phase at different rates. This movement is achieved by the 'sweeping' effect of the mobile phase. In **electrophoresis** charged particles are made to move through a stationary phase under the influence of an electric field. This can be demonstrated using the apparatus shown in Figure 15.7.

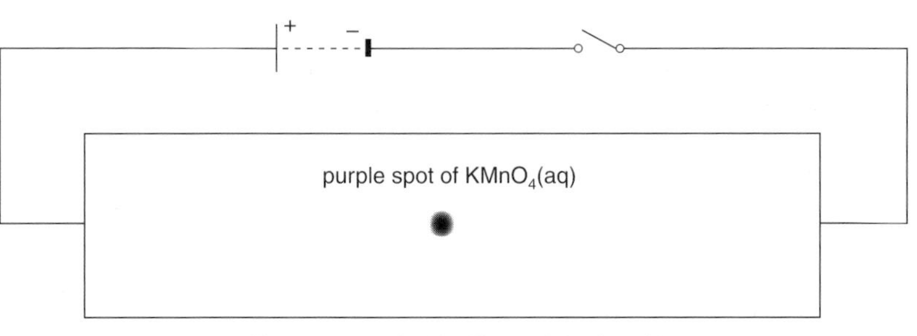

Figure 15.7
A simple demonstration of electrophoresis

Potassium manganate(VII) solution is deep purple due to the MnO_4^- ion. When the electricity supply is switched on the ions present migrate towards the oppositely charged electrode and so the purple spot in Figure 15.7 is seen to move to the left towards the positive electrode. Although you cannot see it, the K^+ ions also move but towards the negative electrode on the right.

The electric charge on some species depends upon the conditions used. For example, an amino acid produces a negatively charged ion under alkaline conditions but a positively charge ion when acidified.

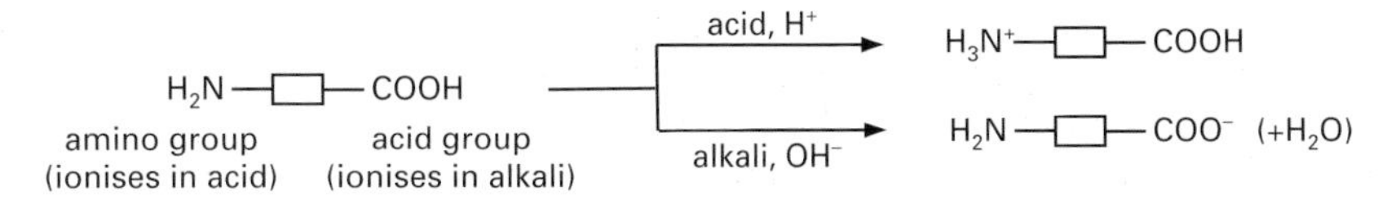

If the amino acid is used in place of the $KMnO_4$ in Figure 15.7, the acidic conditions mean it will exist as a positive ion which will therefore migrate to the right. If, however, the paper is soaked in an alkaline solution, an anion is formed which moves to the left. For each amino acid, there is a certain pH (which is dependent upon its molecular structure) at which it will not move at all under the influence of an electric field. At this **isoelectric point** the amino acid exists as an internal salt with the following structure (see also section 14.9):

$$H_3N^+—□—COO^-$$

One way of separating and identifying a mixture of amino acids is to carry out electrophoresis on a layer of gel which has a uniform pH gradient from one side to the other. Each component will move until it reaches that part of the plate where the pH is the same as its isoelectric point where it will come to rest. Spraying the plate with a reagent such as ninhydrin will reveal the position of each amino acid as a coloured spot as illustrated in Figure 15.8.

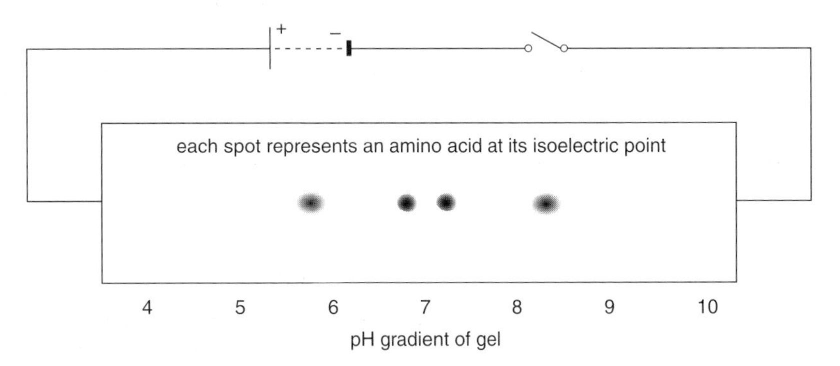

Figure 15.8
Separation of amino acids according to isoelectric point by electrophoresis

Electrophoresis is widely used in the analysis of proteins that are natural polymers formed from amino acids and in DNA fingerprinting.

How does DNA fingerprinting work?

Each cell in our body carries the genetic information passed on from our parents in the form of very long double-stranded molecules known as deoxyribonucleic acid or DNA. This consists of a backbone polymer chain made from a sugar, deoxyribose, and phosphoric acid. Attached to this chain are four types of amine base which are represented by the letters G, C, A and T. All DNA molecules have an identical backbone but it is the sequence of amine bases which carries the genetic code which is unique to every individual (except identical twins!).

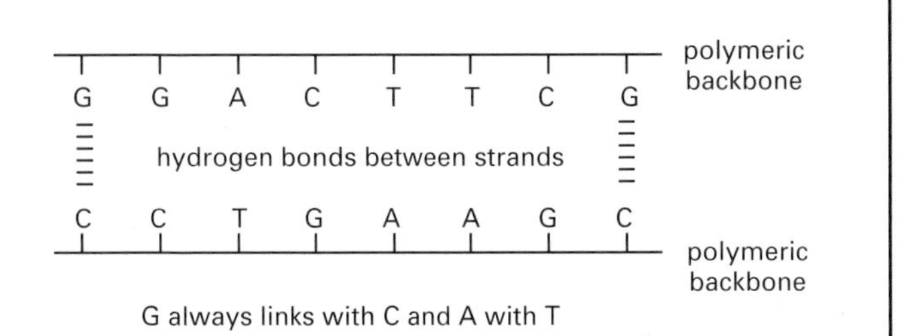

Some parts of the DNA, known as **minisatellites**, carry no specific genetic information but consist of a certain base sequence repeated several times.

How does DNA fingerprinting work? *continued*

Everyone's DNA contains many of these regions but their positions in the DNA chain vary. We inherit half our minisatellite positions from each parent. The chance of two unrelated people having one minisatellite in the same position is 1 in 4 or 0.25 and for ten such regions on DNA the chance of a perfect match for all positions is 0.25^{10} or about 1 in 1 000 000. So to identify an individual from their DNA we must find some way of checking the minisatellite positions.

In practice the DNA sample is sliced up into shorter lengths by **restriction enzymes** which cut the molecules only at points which have a certain base sequence. Since the base sequence varies with the individual, the number of the fragments and their sizes will also vary. **Gel electrophoresis** is then used to separate the DNA fragments. The problem now is to identify those fragments which contain a minisatellite.

The **electrophoretogram** is sprayed with a solution containing prepared DNA fragments labelled with the radioactive isotope ^{32}P that have exactly the right sequence of bases to attach to a minisatellite region by hydrogen bonding. The excess of this DNA probe is then washed off and the plate covered with photographic film. Only those DNA fragments containing minisatellites will have absorbed the radioactive probe and this darkens the film. When it is processed these regions show up as a series of dark bands known as a DNA fingerprint. Comparison of such fingerprints from genetic material, such as blood, hair or semen, obtained from the scene of a crime may be compared with samples taken from a suspect in order to provide evidence in court.

Focus 15c

1. **Electrophoresis** separates the components of a mixture according to their movement in an electric field.
2. This technique is especially useful for amino acids which form anions *and* cations depending upon the pH. The pH at which an amino acid does not move in an electric field is known as its **isoelectric point** and may be used for identification purposes.
3. DNA fingerprinting involves electrophoresis.

Activity 15.4

DNA fingerprinting can also be used to identify the parents of a child since the offspring will inherit half of its DNA from the father and half from the mother. Figure 15.9 shows the DNA fingerprints of a mother and daughter together with those of two men. Do you think either of these fathered the child?

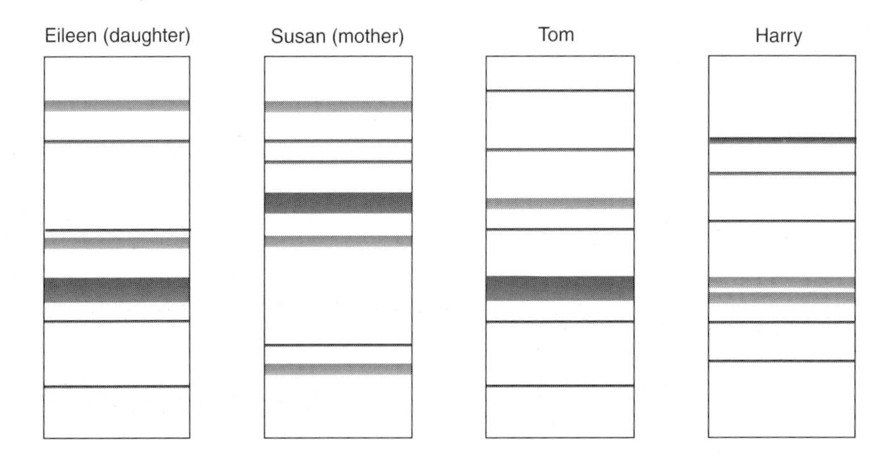

Figure 15.9
Who is Eileen's father?

15.3 Purification on a larger scale

Chromatography and electrophoresis are mainly used on relatively small samples to check for purity or to identify the components of a mixture. Although they can be used to separate larger quantities of material, the amounts which can be obtained in this way are limited and alternative methods are often available which are simpler, more convenient and cheaper.

Chapter 27 gives details of separation methods, such as fractional distillation, steam distillation and solvent extraction, which depend upon differences in the distribution of the components of a mixture between two phases.

Differences in chemical properties may also be exploited in the separation of some types of mixture. Most organic compounds are non-polar and, as a result, are relatively insoluble in water. However, if a functional group is present which can be converted into an ionic salt this will readily dissolve. Examples of functional groups that form salts are shown in Figure 15.10.

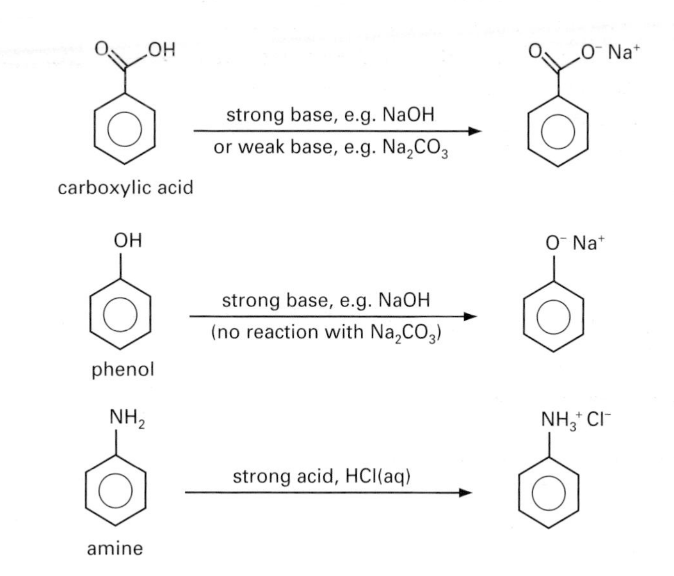

Figure 15.10
Organic functional groups which readily form water-soluble ionic salts

Thus if we have a neutral organic liquid contaminated by a little carboxylic acid, the latter may be removed by shaking with sodium hydroxide or sodium carbonate solution. The acid will be extracted into the aqueous layer that is immiscible with the organic liquid. After separation, the acid may be recovered from the aqueous solution of its salt by acidification.

Similarly an amine impurity may be removed from an organic product by extracting with dilute hydrochloric acid.

Focus 15d

1. Acidic or alkaline components can often be separated from neutral organic compounds by converting them into a water-soluble salt.
2. Such methods are often more convenient for larger scale separation than chromatographic methods.

Activity 15.5

A neutral organic liquid is thought to be contaminated with both a phenol and an aromatic carboxylic acid. Outline how you might separate these three components using sodium hydroxide solution, sodium carbonate solution and dilute hydrochloric acid.

Questions on Chapter 15

1 Do you think the labelling of orange juice in supermarkets as 'pure' is justified? Are so-called 'pure' foods necessarily better for you than less pure foods? In each case explain your answer.

2 When a new drug is being developed, why is it particularly important to use pure samples during testing?

3 Propanedioic acid, $HOOCCH_2COOH$, and decanedioic acid, $HOOC(CH_2)_8COOH$, both melt at 133°C. How, using melting point apparatus and known pure samples of each of these compounds, could you find out whether an unknown solid was either, or neither, of these acids?

4 Paper chromatograms formed by two orange food colourings, A and B, are shown drawn to scale in Figure 15.11.
a) Which of these must be a mixture of dyes?
b) Is the other food colouring necessarily a pure substance? Explain your answer.
c) Is its possible that the same dye is present in each of the food colourings? Explain your answer.
d) Use the information in Figure 15.11 to calculate the R_f value of the substance responsible for the red spot in the chromatogram of A.

5 Outline how gas–liquid chromatography might be used in analysing urine samples from athletes in order to detect the use of illegal drugs. You should include: a summary of the principles involved; a description of the apparatus used; and a list of the preliminary work you would have to carry out in order to use the technique to measure the concentration of any particular drug in the urine sample.

6 **a)** Draw the structure of 2-aminoethanoic acid.
b) What species will be present in a strongly acidic solution of this compound?
c) What species will be present in a strongly alkaline solution of this compound?
d) Explain what is meant by the isoelectronic point of an amino acid and outline how could you determine the isoelectric point of 2-aminoethanoic acid using electrophoresis.

7 The DNA fingerprint of blood obtained from a broken window at the scene of a robbery is shown in Figure 15.12 together with DNA fingerprints of the occupants of the house and two possible suspects.
a) Explain the principles underlying this analytical technique.
b) What are the chances of an 'accidental' match of three bands in two such 'fingerprints'?
c) Explain any conclusions you draw from a comparison of the DNA fingerprints in Figure 15.12.

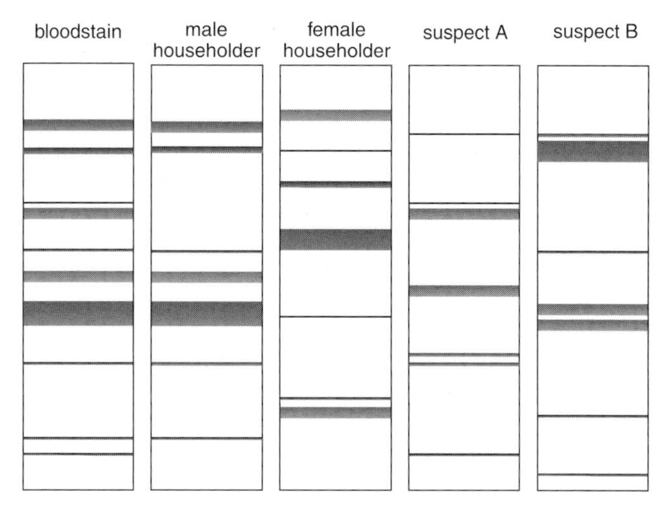

Figure 15.12 ▲
DNA fingerprints used in a robbery investigation

8 Phenylalanine is an amino acid that may be obtained from benzaldehyde by following this two-step process

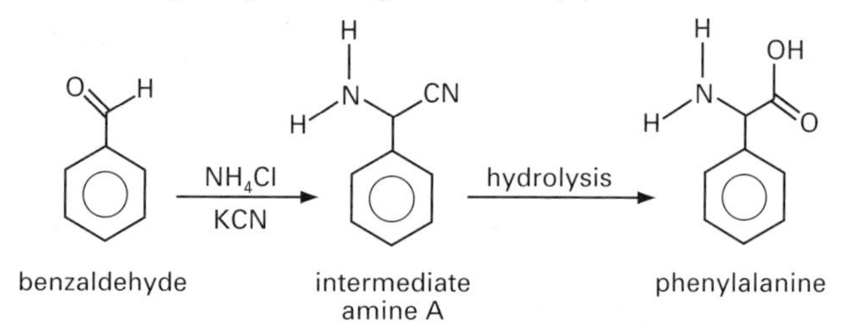

The product may be contaminated with both unreacted benzaldehyde and the intermediate amine, A. Explain how, using dilute hydrochloric acid and sodium hydroxide solution, you could obtain:
a) phenylalanine free from benzaldehyde and intermediate A,
b) any unreacted benzaldehyde free from intermediate A and phenylalanine.

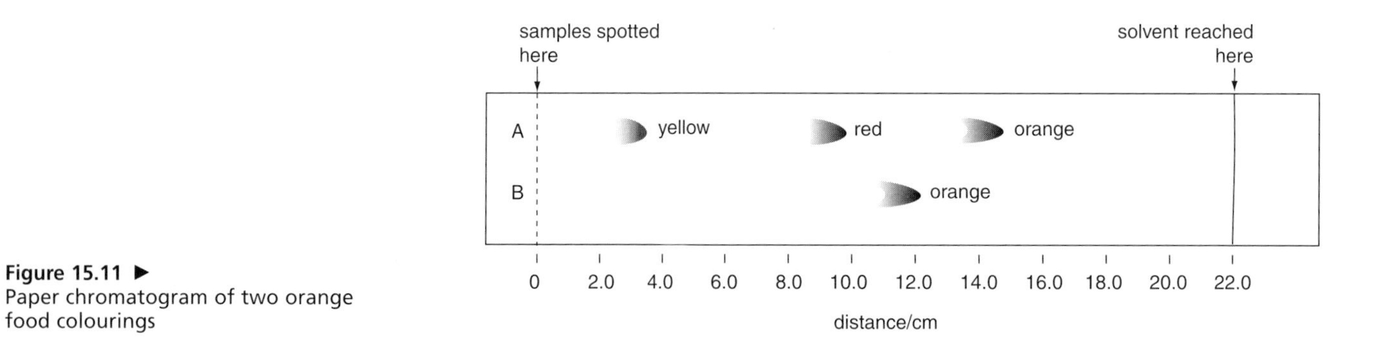

Figure 15.11 ▶
Paper chromatogram of two orange food colourings

Comments on the activities

Activity 15.1

1 **a)** A quite sharp melting point between 134 and 135°C indicates fairly pure aspirin.
b) A quite sharp melting point between 156 and 158°C indicates largely unchanged 2-hydroxybenzoic acid.
c) Gradual melting over the range 120–126°C, which is lower than the melting point of either of the pure solids, indicates a mixture containing significant amounts of both 2-hydroxybenzoic acid and aspirin.

2 If you mix the two solids together, then if both are pure aspirin the melting point will not change. If the solids are different the melting point will fall considerably and become less sharp. This **mixed melting point** method is a common way of checking the identity of a solid.

Activity 15.2

1 Only sample B might be a pure compound as it gives a single spot in the chromatogram.

2 R_f value of B $= \dfrac{\text{distance travelled by B}}{\text{distance travelled by solvent front}}$

$$= \frac{7}{20} = 0.35$$

3 If the lid is left off the tank then the more volatile propanone could evaporate, changing the composition of the mobile phase.

4 R_f value $= 0.70 = \dfrac{\text{distance travelled by component}}{\text{distance travelled by solvent front}}$

$$= \frac{\text{distance travelled by component}}{20\text{ cm}}$$

So, in this experiment, Sunset Yellow should travel $0.70 \times 20 = 14$ cm. As you can see from Figure 15.2, food colourings A and C could both contain this compound.

5 A could contain C as it has spots with identical colour and position. However, it cannot contain B since, although the spot is in the correct position, it has a different colour and so cannot be the same dye.

Activity 15.3

1 The two peaks in the calibration traces must be due to water and ethanol. Since the peak at retention time 3.3 minutes increases in size with ethanol concentration, it is this peak that must be due to ethanol.

2 By comparison of the relative size of the ethanol peak in the urine with those in the calibration traces you can see that the ethanol concentration in the urine is about 50 mg cm^{-3}. If a more exact figure is needed you can compare the areas under the peaks. Assume that the concentration of the component is directly proportional to the area under its peak which is given by

$$\frac{\text{base} \times \text{height}}{2}$$

3 The multiple peaks in the GLC trace for urine show it contained several other substances.

4 Apart from ethanol, the only substance *known* to be present in all the samples is water so this must be the substance with a retention time of 5 minutes.

5 It would not be appropriate to use this calibration graph on any other apparatus as the retention times and detector responses will vary with the column characteristics and conditions used.

Activity 15.4

If you ignore the three bands in Eileen's DNA fingerprints which occur in the same position as her mother's then you will see that all four of the remaining bands correspond with those obtained from Tom, whilst only one matches those from Harry. It is highly likely, therefore, that Tom is Eileen's father.

Activity 15.5

1 Shake the mixture with sodium carbonate solution. Only the carboxylic acid will form a salt and so dissolve in the aqueous layer. Separate this layer and acidify with dilute hydrochloric acid to reform the carboxylic acid.

2 Shake the remaining organic liquid with sodium hydroxide solution. The phenol will form a salt and dissolve. Again separate the aqueous layer and acidify to reform the phenol.

3 The remaining organic liquid should contain no acid or phenol.

CHAPTER 16

Determination of formulae

Contents

Study Checklist

After studying this chapter you should be able to:

1. Explain, with examples, what is meant by empirical formula, molecular formula and structural formula.
2. State that atomic spectroscopy can be used to detect the presence of certain elements in a sample.
3. Describe how the amount of carbon and hydrogen in an organic compound may be estimated by analysing for the products of complete combustion (carbon dioxide and water).
4. Describe an experiment to determine the molecular mass of a volatile liquid by measuring the volume of its vapour under known conditions.
5. Recall the kinds of information that may be obtained from both low and high resolution mass spectrometry and be able to interpret supplied examples.
6. Give examples of the use of 'spot tests' for the detection of certain organic functional groups (by reference to other chapters if necessary).
7. Appreciate that unambiguous determination of structure, including molecular size and geometry, may often be obtained by X-ray crystallography but that this is relatively time-consuming compared to spectroscopic methods which are covered in Chapter 17.

16.1 What's in a name? Different kinds of formulae

In 1928 when Fleming accidentally discovered the first antibiotic drug, penicillin, in a bread 'mould', he had little, if any, idea why it was so effective at killing bacteria. Before large quantities of this material could be made synthetically, its detailed chemical structure had to be established. It was not until the 1940s that this task was completed and it has led to the development of the wide range of similar drugs in use today.

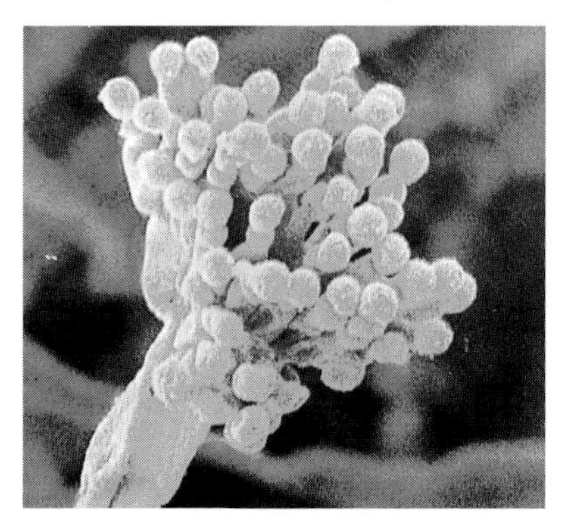

Scanning electron micrograph of the printing body of a *Penicillium* mould

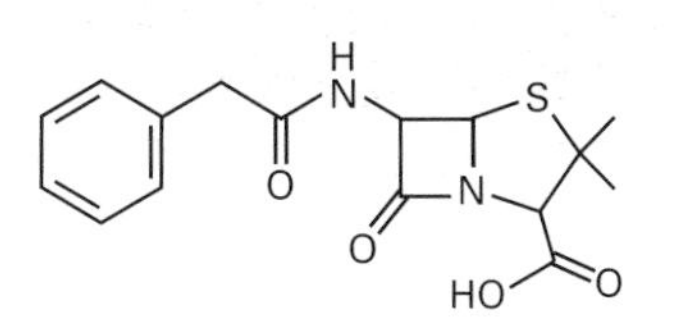

penicillin G

This is the common shorthand way of drawing the structure of an organic molecule. Carbon atoms are not shown but are assumed to be at the end of each bond (straight line) unless another atomic symbol is indicated. By convention, hydrogen atoms attached to carbon atoms are not shown but are assumed to be present in order to complete the normal valency of carbon, i.e. 4.

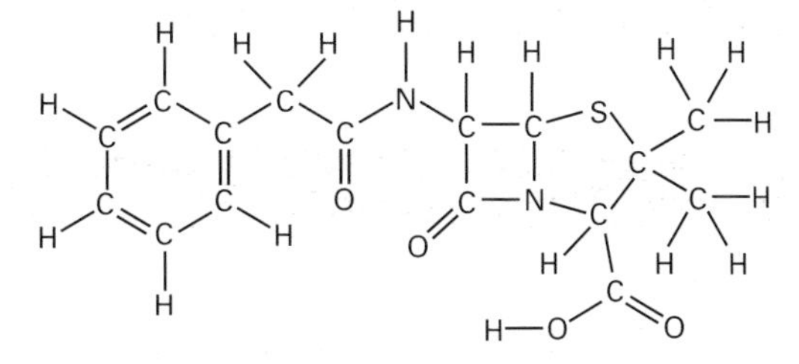

This shows all the atoms present in the above structure

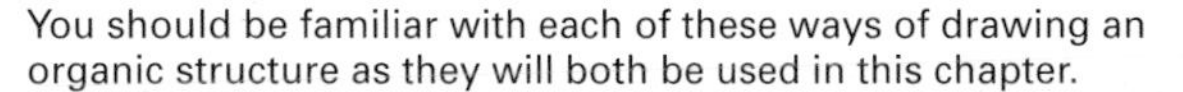

You should be familiar with each of these ways of drawing an organic structure as they will both be used in this chapter.

The most complete information about any particular compound is given by its **structural formula**. This shows exactly how the different atoms present in a molecule of the compound are bonded together and may even provide information on molecular geometry, i.e. bond lengths and angles. For example ethanedioic acid has the structure

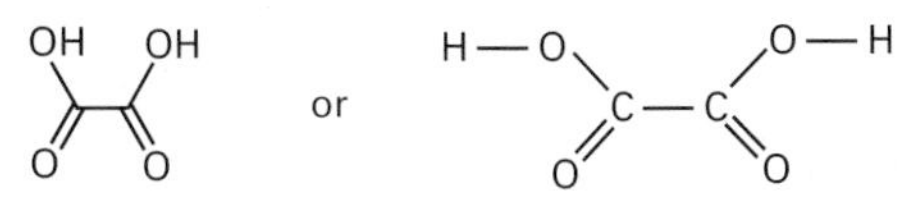

The **molecular formula** simply gives the number and type of each kind of atom present in the molecule. The molecular formula of ethanedioic acid, for example, is $C_2H_2O_4$. Often there are two or more structures that are compatible with a single molecular formula. Thus, given the normal valencies of the elements present (C = 4, O = 2 and H = 1), we may draw at least two other possible structures with the same molecular formula as ethanedioic acid.

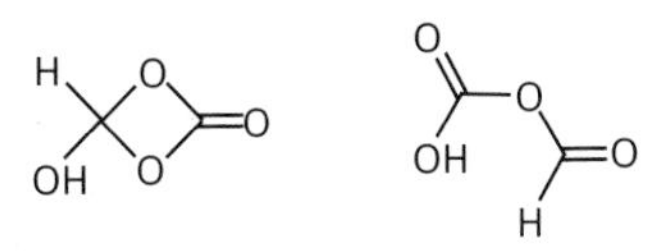

Yet another step down, in terms of information, comes the **empirical formula**, which gives only the simplest whole number ratio of the different types of atom present. Ethanedioic acid, therefore, has the empirical formula CHO_2. The molecular formula must be either the same as the empirical formula or a whole number multiple of it. In this case we cannot write a feasible structure with molecular formula CHO_2 but we can write several structures with molecular formula $(CHO_2)_4$ i.e. $C_4H_4O_8$, including those shown below.

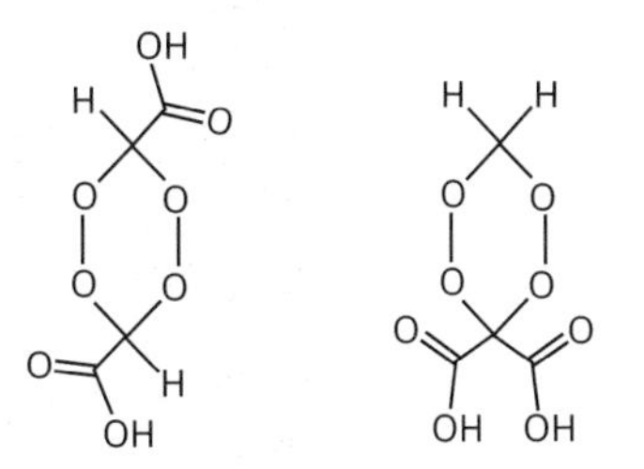

Thus, whereas the structural formula of ethanedioic acid completely and unambiguously describes the molecule, molecular and empirical formulae are progressively less precise. Why then should we bother with a less than complete description of the structure? Well the answer lies in the simple fact that the empirical formula is generally the quickest and easiest to determine, whilst a complete unambiguous structure determination is often far more difficult and time-consuming. In the rest of this chapter we shall look at a selection of practical techniques which can be used to determine the various types of formula.

Focus 16a

1. The **empirical formula** shows only the simplest whole number atomic ratio of the elements present in a compound.
2. The **molecular formula** gives the actual number of each type of atom in a molecule. It may be the same as the empirical formula or a whole number multiple of it.
3. The **structural formula** shows how the atoms are bonded together in the molecule. A full 3-D structure also indicates values for bond lengths and angles.

Activity 16.1

The structural formulae of ethanal and dioxan are shown below.

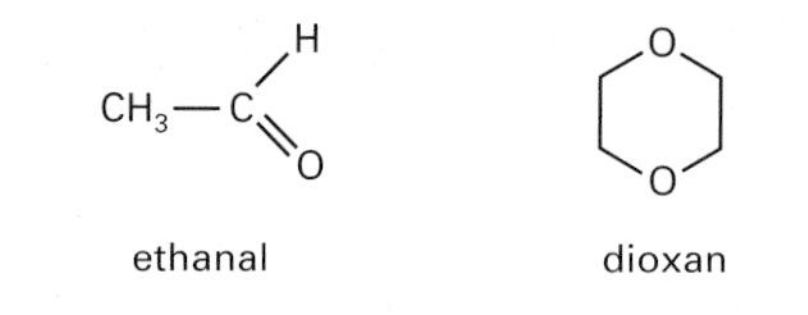

1. What are the molecular formulae of these two compounds?
2. They share a common empirical formula. What is this?
3. For each of the above compounds, draw at least **two** different feasible structures that have the same molecular formula. What name is given to compounds that have the same molecular formula but different structures? (see Chapter 6.5)
4. Draw **one** possible structure which has the same empirical formula as ethanal and dioxan but has the molecular formula of neither.

16.2 What have we got? Analysing for elements

Modern systematic methods for the detection of elements include atomic spectroscopy. When salt (sodium chloride) is heated in a Bunsen burner it colours the flame bright yellow-orange. The same flame colour is observed with all sodium compounds and so can be used as a test for the presence of this element. Sodium lamps are also commonly used for street lighting.

Some other metals that also give characteristic **flame tests** are shown in Table 16.1.

Table 16.1 Characteristic flame colours of some metals

Metal	flame colour
lithium	crimson
sodium	yellow-orange
potassium	lilac
rubidium	red
caesium	blue
calcium	brick red
strontium	crimson
barium	apple green
lead	pale blue
copper	green

When light is passed through a prism, it is separated into different colours. This occurs naturally when sunlight is split into a rainbow on passing through rain-drops. Unlike sunlight, however, when light from a flame test is separated in this way it is seen to consist largely of just a few sharp lines of characteristic colour or wavelength. The **atomic emission spectrum** of sodium is shown in Figure 16.1.

The flame provides the energy to promote electrons within the atom to higher energy levels. As an electron drops back to a lower energy state, radiation of characteristic colour and wavelength is emitted. Since the pattern of electronic energy levels is unique to each atom, each element has its own unique pattern of lines in its emission spectrum and this may be used to identify it. A more detailed examination of the atomic hydrogen spectrum can be found in Chapter 17.

◄ Sodium lamps used for road lighting

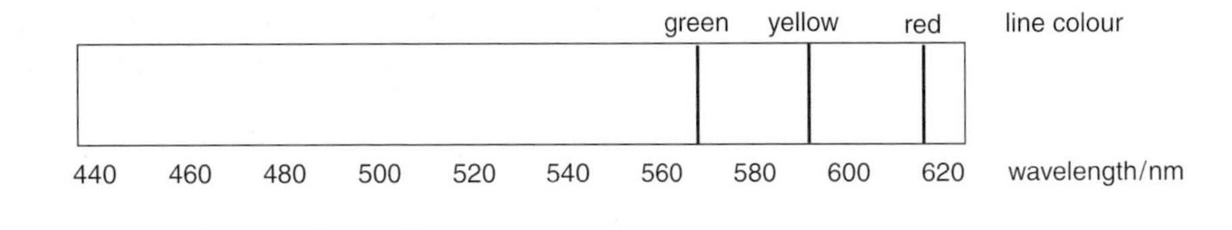

Figure 16.1 ▲
Main lines in the visible atomic emission spectrum of sodium

Atomic absorption spectrometry

In atomic emission spectroscopy, energy is supplied by a flame or electric discharge to promote electrons to higher energy levels. Emission of radiation of characteristic wavelengths then occurs when the electrons drop back to lower levels. This mixture of radiation must be separated and examined before the elements present may be identified.

In atomic absorption spectroscopy, the sample is vaporised in a low temperature flame so that fewer electrons are promoted than in atomic emission spectroscopy. Light is then shone through the atomised sample. Only those wavelengths that have exactly the right energy to promote an electron to a higher energy level will be absorbed. Since the pattern of energy levels determines the energy jumps an electron can make, atomic absorption for any element occurs at exactly the same wavelengths as atomic emission. This is shown in Figure 16.2.

Modern atomic absorption spectrometers are designed to detect particular elements. The source produces light of wavelength corresponding to a prominent line in the atomic spectrum of the target element. The intensity of the beam before and after passing through the sample is compared as shown in Figure 16.3 and any reduction must be caused by the presence of the target element.

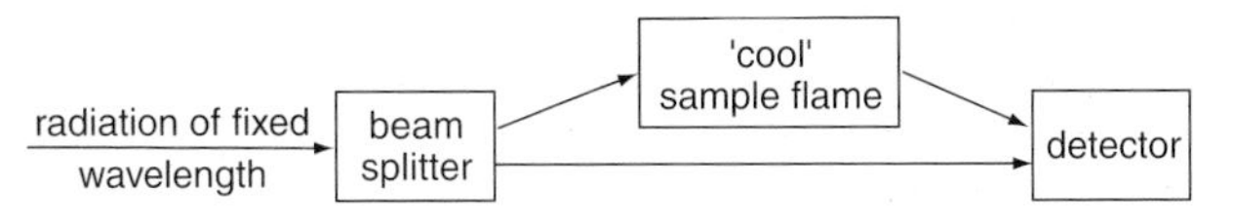

Figure 16.3 ▲
Principle of a modern atomic absorption spectrometer

This technique is extremely sensitive and may be used to measure trace metallic impurities in water supplies.

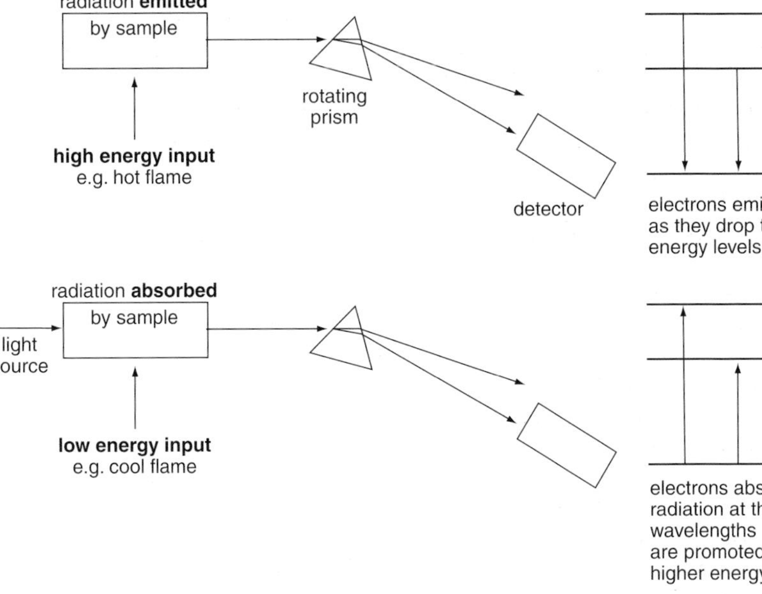

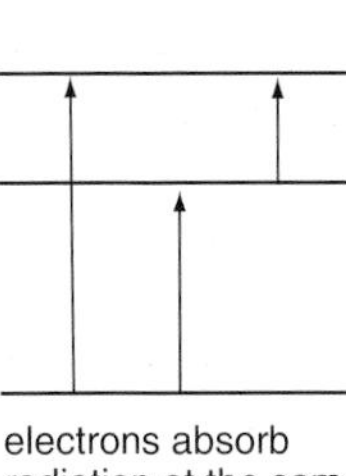

Figure 16.2 ◄
A comparison of (a) atomic emission spectroscopy and (b) atomic absorption spectroscopy

Atomic spectroscopy and 'new' elements

Since each element has its own unique pattern of lines in its atomic spectrum, any 'new' lines may indicate the presence of an unknown material. Although there have been many false alarms, several elements have been discovered using atomic spectroscopy.

As early as 1860 two previously unobserved blue lines were spotted in the emission spectrum of a mineral water sample. The new element was named caesium, from the Latin meaning 'sky-blue'.

blue line colour

440 460 480 500 520 540 560 580 600 620
wavelength/nm

When sunlight is passed through a spectrometer, it produces an almost continuous 'rainbow' spectrum. However, Fraunhofer observed several 'dark' lines in the solar spectrum which are caused by absorption by atoms in the outer layer of the Sun and in the Earth's atmosphere. Most of these 'Fraunhofer' lines could be accounted for by known elements but, during a solar eclipse in 1868, a prominent new line was observed in the yellow region at about 588 nm. This was ascribed to a new element helium named after helios, the Greek meaning 'Sun'.

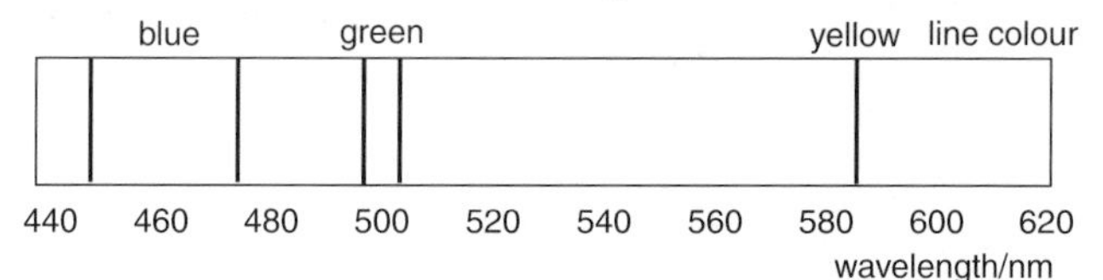

Of course, at the time nothing else was known about the new element but it was given the name ending '-ium', suggesting a metal. However, when helium was first identified on Earth by spectroscopic analysis of gases emitted by uranium, it was found to be a non-metal. Though the name helion would perhaps fit in rather better with the rest of the noble gases, the element has not been renamed.

Atomic spectroscopy remains one of the most powerful methods for cosmological research. It not only provides information on the composition of stars but also allows an estimate to be made of their distance from the earth. If you are interested in this then try to find out more about the Doppler effect and 'red shift'.

The presence of certain atoms or groups within a compound may often be identified without sophisticated apparatus by carrying out simple tests. If, for example, an aqueous solution of a substance gives a white precipitate with silver nitrate solution that dissolves in dilute ammonia but not in dilute nitric acid, chloride ions must be present:

$$Ag^+(aq) + Cl^-(aq) \longrightarrow AgCl(s)$$

$$AgCl(s) + 2NH_3(aq) \longrightarrow [Ag(NH_3)_2]^+(aq) + Cl^-(aq)$$

Similarly, the formation of a white precipitate on adding barium chloride solution acidified with dilute hydrochloric acid indicates the presence of sulphate ions:

$$Ba^{2+}(aq) + SO_4^{2-}(aq) \longrightarrow BaSO_4(s)$$

In the case of organic compounds, carbon and hydrogen may be detected by oxidation with hot dry copper(II) oxide. Any hydrogen is converted into water which, after condensing, will turn anhydrous cobalt(II) chloride from blue to pink, whilst carbon is converted into carbon dioxide which will form a milky white precipitate with calcium hydroxide solution (lime-water). Fusion with sodium metal will convert any nitrogen into cyanide (providing enough carbon is present), sulphur into sulphide and halogens into halide ions, all of which are easily identified by test-tube reactions. As with inorganic substances, specific 'spot-tests' are often useful in order to identify certain functional groups. For instance the carboxyl group –COOH shows typical acidic properties and will liberate carbon dioxide from a carbonate:

$$2X\text{–}COOH(aq) + CO_3^{2-}(aq) \longrightarrow 2X\text{–}COO^-(aq) + H_2O(l) + CO_2(g)$$

(Details of tests for other organic functional groups may be found in Theme B).

Once the elements contained in a compound have been identified, the next step is to determine the percentage by mass of each one present. When this has been done, the empirical formula may be deduced.

Activity 16.2

1 This question concerns a gaseous compound A.

a) When A is passed over heated copper(II) oxide, and the products are cooled, a colourless liquid condenses out which turns blue cobalt chloride paper pink. The remaining gaseous product, when passed through calcium hydroxide solution, produces a milky white precipitate. What information does this give concerning the composition of A?

b) When 0.700 g of A is fully burned in oxygen, 0.900 g of water and 2.200 g of carbon dioxide are produced. Calculate the percentage composition by mass of compound A and write a general equation to illustrate the combustion reaction.

c) Calculate the empirical formula of A.

d) When A is bubbled through bromine solution, the orange colour of the latter disappears. What information does this give regarding the structure of A?

e) Write **two** possible molecular formulae for A. (You may find examples given in section 6.4 helpful here.)

2 This question concerns a liquid compound B that has been shown to contain carbon, hydrogen and oxygen only.

a) Elemental analysis shows that B contains 38.7% by mass of carbon and 9.8% by mass of hydrogen. What is the empirical formula of B?

b) Suggest a feasible structure for B and describe a simple chemical test that might provide indirect evidence to support this (see Chapter 11).

3 Compound C has the following composition by mass: 30.6% carbon, 3.8% hydrogen, 20.4% oxygen and 45.2% chlorine.

a) How could you show that C contains chlorine?

b) Calculate the empirical formula of C.

c) An aqueous solution of C liberates carbon dioxide from sodium carbonate. Suggest **one** possible functional group that C might contain (see section 13.6).

d) Draw a feasible structure for C containing the functional group you have identified in **c)**.

Focus 16b

1 The **empirical formula** can be deduced from the percentage composition by mass of elements present in a compound.

2 The percentage of carbon and hydrogen in an organic compound can be determined by finding the mass of carbon dioxide and water produced on complete combustion of a weighed sample.

16.3 How heavy? Molecular masses

Empirical formula allows us to make educated guesses about possible alternative structures but, if we can establish the molecular formula, this will reduce the number of feasible alternatives. In order to do this we need to know the relative molecular mass, M_r. For gases or easily vaporised liquids, we may use the **ideal gas equation**, $\boldsymbol{pV = nRT}$. If we measure the volume, V, of a known mass of gas at a given absolute temperature, T, and pressure, p, then given the value of the gas constant, R, we may calculate the number of moles, n, and so calculate the mass of one mole.

Finding the relative molecular mass of a gas by the evacuated flask method

The apparatus for this experiment is shown in Figure 16.4.

1 The air in the flask is first removed using a vacuum pump.
2 The tap is closed and the flask is disconnected from the pump and weighed accurately.
3 The flask is then connected to a supply of the gas and allowed to fill to atmospheric pressure by opening the tap.
4 Again the tap is closed and the flask reweighed after disconnecting from the gas supply.
5 Finally the temperature and atmospheric pressure under which the experiment was carried out are measured.

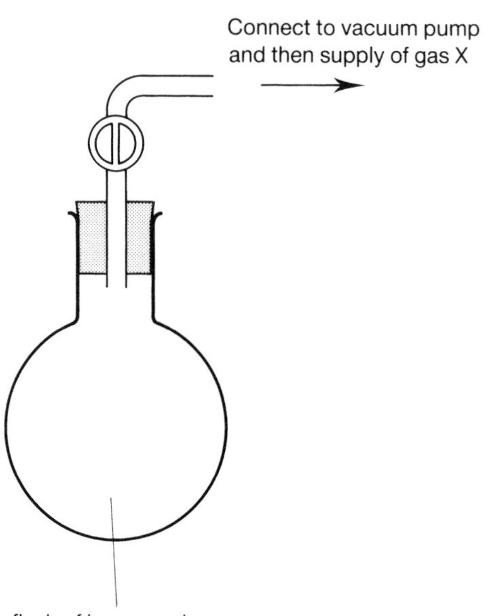

Figure 16.4
Apparatus for determining the relative molecular mass of a gas

Activity 16.3

A set of results for a gas X of empirical formula CH_2 is given below.

volume of flask	= 500 cm^3
mass of evacuated flask	= 349.52 g
mass of flask containing X	= 350.13 g
room temperature	= 23°C
atmospheric pressure	= 100.0 kPa

1 Care must be taken when using the ideal gas equation, $pV = nRT$, to use units of p, V and T which match those of the gas constant, R. Using SI units, R is 8.31 J mol^{-1} K^{-1}, p must be expressed in Pa, volume in m^3 and T in K. Convert the pressure, volume and temperature measurements above into these units.

2 Use the ideal gas equation to calculate n, the number of moles of gas X contained in the flask.

3 Calculate the mass of gas X contained in the flask.

4 Use your answers above to calculate the mass of 1 mole of X and hence derive its molecular formula.

Finding the relative molecular mass of a volatile liquid using a gas syringe

The apparatus for this experiment is shown in Figure 16.5.

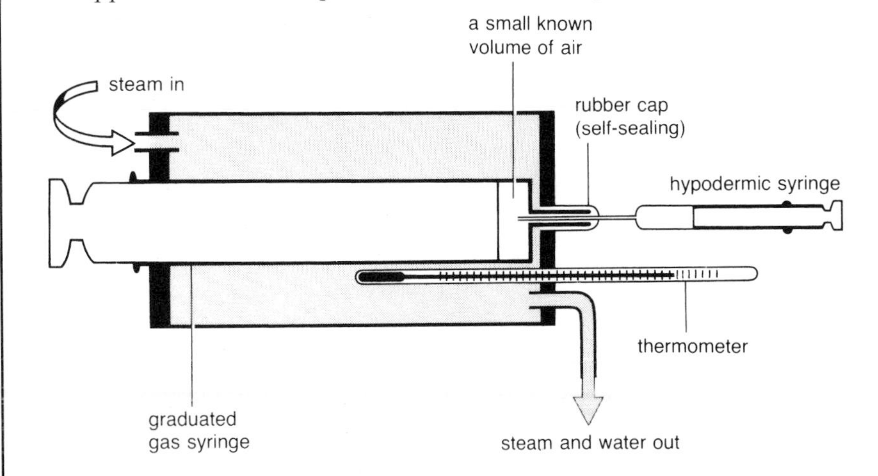

Figure 16.5
Apparatus used to determine the relative molecular mass of a volatile liquid

1. A gas syringe containing a little air is placed into the steam jacket. If a steam jacket is not available a tall beaker of boiling water may be used as a bath to heat the gas syringe. As it heats up, the air expands and the volume is noted when a steady reading is obtained.
2. An injection syringe containing the liquid under test (which should boil below 80°C) is then weighed accurately.
3. The liquid is carefully injected into the gas syringe through the rubber sealing cap and the empty injection syringe weighed. As the liquid vaporises the gas syringe barrel is pushed out and the total volume of air plus vapour is measured when the reading is steady.
4. Finally the temperature in the steam jacket and the atmospheric pressure are measured.

Activity 16.4

Using the above method a student obtained the following results for a liquid, Y, of empirical formula CH_2Cl_2:

volume of air in the syringe before injection of Y	= 12 cm^3
volume of air + vaporised Y after injection	= 95 cm^3
mass of injection syringe + liquid Y before injection	= 5.96 g
mass of syringe + remaining liquid Y after injection	= 5.72 g
temperature of steam jacket	= 100°C
atmospheric pressure	= 100 700 Pa

1. What volume of vapour did the sample of Y produce in this experiment?
2. Using the value for the gas constant given in Activity 16.3, express the volume of Y, the temperature and the pressure in consistent units.
3. Use the ideal gas equation to calculate the number of moles of Y vaporised in the experiment.
4. Calculate the relative molecular mass of Y and hence its molecular formula.

The methods based on measuring gas volume give approximate answers using relatively simple equipment and, given the empirical formula, are generally accurate enough to distinguish between the different possibilities to allow the molecular formula to be found. Mass spectrometry, outlined in section 1.10, provides a much more precise measurement of molecular mass and can even be used to establish molecular formula and structural information without the need to first determine empirical formula.

Low-resolution mass spectrometry

This refers to spectra where relative mass is recorded to the nearest whole number. The main lines in such a spectrum for an organic liquid, P, known to contain carbon, hydrogen and oxygen only, are shown in Figure 16.6.

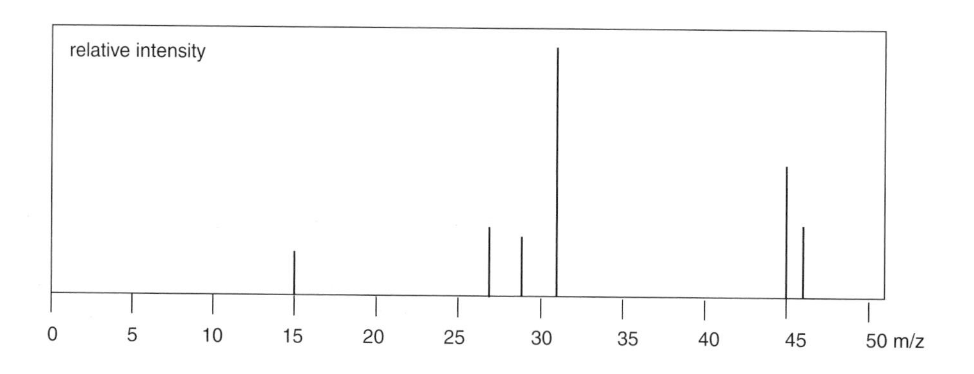

Figure 16.6
Low resolution mass spectrum of compound P, containing C, H and O

The peak at highest relative mass in this spectrum is due to the molecular ion, which is produced by removing a single electron from a molecule of P.

$$\underset{\text{parent molecule}}{P} \longrightarrow \underset{\text{molecular ion}}{P^+} + e^-$$

Although strictly speaking the horizontal scale is mass to charge ratio (m/z), if we assume that each molecule only has one electron knocked off it, then the charge is +1 and m/z is equal to the relative molecular mass. From this we can see that, to the nearest whole number, the relative molecular mass of P is 46. From the information given so far the molecular formula could be C_2H_6O or CH_2O_2. Of course, determination of the empirical formula by elemental analysis would distinguish between these but this is not necessary if high resolution mass spectra can be obtained.

The peaks at lower mass are due to the molecular ion splitting up into various fragments. If the molecular ion is very unstable it may not even produce a measurable peak. Another complication is that fragments may also recombine to give unexpected peaks:

$$\underset{\text{molecular ion}}{P^+} \longrightarrow \underset{\substack{\text{neutral fragment}\\\text{(not detected)}}}{X} + \underset{\substack{\text{ion fragment}\\\text{(may fragment further)}}}{Y^+}$$

The fragmentation pattern provides a molecular 'fingerprint' and, as we shall see later in this chapter, analysis of the individual peak masses can provide clues to parts of the structure.

If the sample contains an element that exists as different isotopes, then the molecular ion forms multiple peaks whose mass and relative intensity correspond to the various possible isotope combinations. Figure 16.7 shows the mass spectrum of dibromomethane, CH_2Br_2. Bromine exists in roughly equal amounts of isotopes with relative masses 79 and 81. This means that the molecular ion may have three different masses:

mass of 1st Br atom	mass of 2nd Br atom	mass of $CH_2Br_2^+$ ion
79	79	172
79	81	174
81	79	174
81	81	176

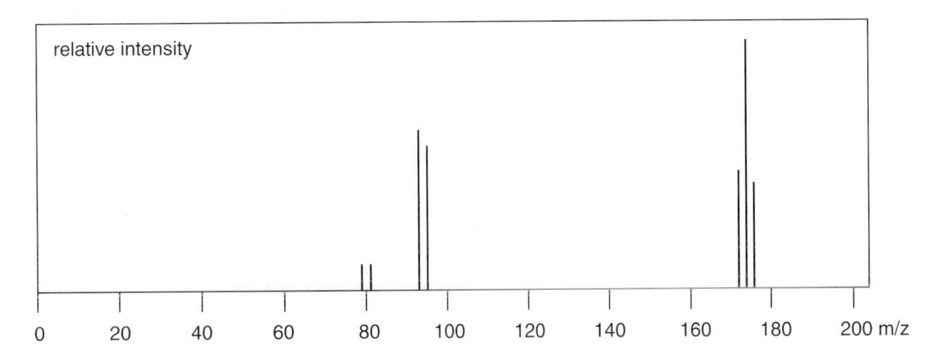

Figure 16.7
The mass spectrum of dibromomethane, CH_2Br_2

Since there is double the chance of obtaining a mass of 174, this peak is twice the size of those at 172 and 176. As you can see there is no other 'triplet' with this 1:2:1 pattern characteristic of fragments that contain two bromine atoms. There are 'doublet' 1:1 peaks at 93 and 95 which correspond to the CH_2Br^+ ion and at 79 and 81 due simply to Br^+.

Activity 16.5

This activity concerns tribromomethane, $CHBr_3$.

1. Deduce the masses and relative intensities of the peaks due to the molecular ion $CHBr_3^+$.
2. Suggest the formulae of the fragment ions responsible for each of the following sets of peaks present in the mass spectrum of tribromomethane:
   a) a triplet at masses 171, 173 and 175 in intensity ratio 1:2:1
   b) a doublet at masses 94 and 92 in intensity ratio 1:1
   c) a doublet at masses 93 and 91 in intensity ratio 1:1
   d) a doublet at masses 79 and 81 in intensity ratio 1:1.

High-resolution mass spectrometry

By definition, the relative mass of ^{12}C is exactly 12.000. However, at this precision the relative masses of other atoms are rarely exactly whole numbers, e.g.

$$^{1}H = 1.008 \quad ^{14}N = 14.007 \quad ^{16}O = 15.995$$

This means that molecules with very similar relative masses may be distinguished by more precise measurement. Thus, for example, carbon monoxide, nitrogen and ethene all have a relative molecular mass of approximately 28. Using the above figures we may calculate the following more accurate values.

$$CO = 12.000 + 15.995 = 27.995$$
$$N_2 = 2 \times 14.007 = 28.014$$
$$C_2H_4 = (2 \times 12.000) + (4 \times 1.008) = 28.032$$

High-resolution mass spectrometry measures relative masses sufficiently precisely to be able to distinguish between all possible molecular formulae without the need for elemental analysis.

Activity 16.6

The low-resolution mass spectrum of liquid P is shown in Figure 16.6. High-resolution mass spectrometry gives a value of 46.04 for the relative mass of the molecular ion.

1 What is the molecular formula of P?

2 Suggest two feasible structures consistent with this molecular formula.

16.4 Bits and pieces. Detecting groups of atoms

After completing Activity 16.6 we know the molecular formula of liquid P is C_2H_6O but to establish its structure we need to identify particular structural groups within the molecule. One way of doing this is to carry out specific chemical tests for functional groups, but the fragmentation pattern in the mass spectrum may also provide useful clues. Possible species that might account for various fragment peaks in the mass spectrum in Figure 16.6 include the following:

m/z value of fragment	15	29	31
possible species	CH_3^+	$CH_3CH_2^+$	CH_3O^+

Of the two structures drawn for P in Activity 16.6, only ethanol can produce all three of the above fragments simply by breaking one bond in each case, i.e.

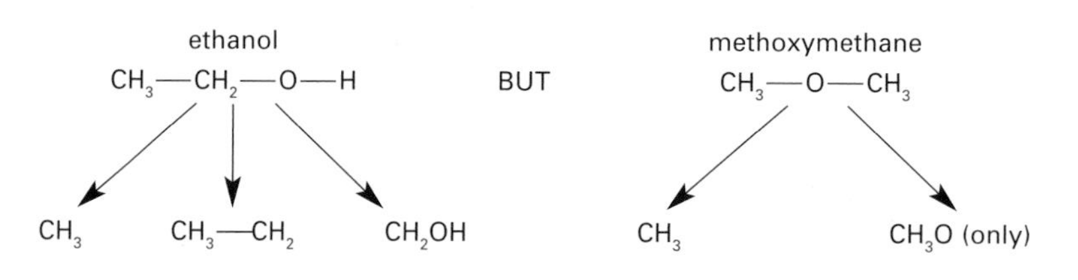

Activity 16.7

The mass spectrum fragmentation pattern of P strongly suggests, but does not prove, that it has the structure $CH_3—CH_2—O—H$ rather than $CH_3—O—CH_3$. Suggest chemical tests that would help to confirm this structure. (For help see Chapter 11).

Activity 16.8

Compound Q has the molecular formula $C_2H_4Br_2$.

1 Draw two possible structures for Q which are consistent with the following valencies of the elements: C(4), H(1), Br(1).

2 Suggest the formula of the species responsible for each of the following sets of peaks in the low-resolution mass spectrum of compound Q.
  a) a triplet at masses 186, 188, 190 in intensity ratio 1:2:1
  b) a doublet at masses 107 and 109 in intensity ratio 1:1
  c) a doublet at masses 93 and 95 in intensity ratio 1:1.

3 There are no peaks visible in the mass spectrum between 109 and 186. Explain which of the two structures you have drawn for Q is the more likely.

Focus 16c

1. The **molecular formula** can be deduced from the empirical formula and relative molecular mass.
2. If the compound is a gas or can easily be vaporised, the molecular mass may be found using the **ideal gas equation**; $pV = nRT$.
3. **Mass spectrometry** is the most widely used modern method of determining molecular mass. High-resolution mass spectrometry can give the molecular formula directly without the need to first establish one empirical formula.
4. The **fragmentation pattern** in a mass spectrum can give valuable clues to the structure of a compound.

One of the greatest weapons in the chemist's armoury for identifying structural features is molecular spectroscopy. Indeed this is so important that the whole of Chapter 17 is devoted to a consideration of the various types of molecular spectra and the information which may be obtained from them.

16.5 Shapes and sizes. Complete molecular geometry

Up to now all the methods we have examined have given only indirect evidence for structure. If a substance can be made into crystalline form, then X-ray diffraction studies may be used to establish its exact molecular geometry, including bond lengths and angles.

This technique relies upon the fact that the particles within a crystal lattice are arranged in a regular 3-D pattern that contains various sets of atom-bearing planes as illustrated in Figure 16.8.

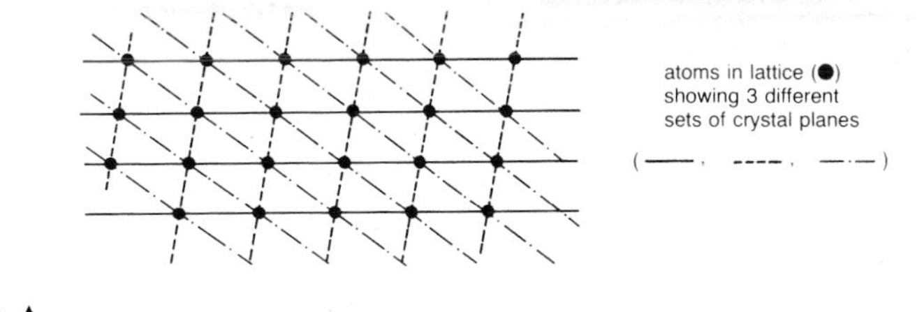

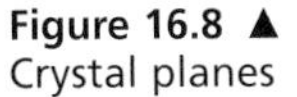

Figure 16.8 ▲
Crystal planes

The wavelength of X-rays is of the same order as the spacings between such planes and the crystal acts as a 3-D diffraction grating for X-rays. In effect, each set of crystal planes will 'reflect' X-rays only at certain angles which are characteristic of the interplanar spacing and the wavelength of the X-rays (if you are interested in this relationship then look at the extension box on the Bragg equation p.434). Figure 16.9 shows how the reflection angles for many sets of planes within a crystal can be recorded simultaneously using photographic film.

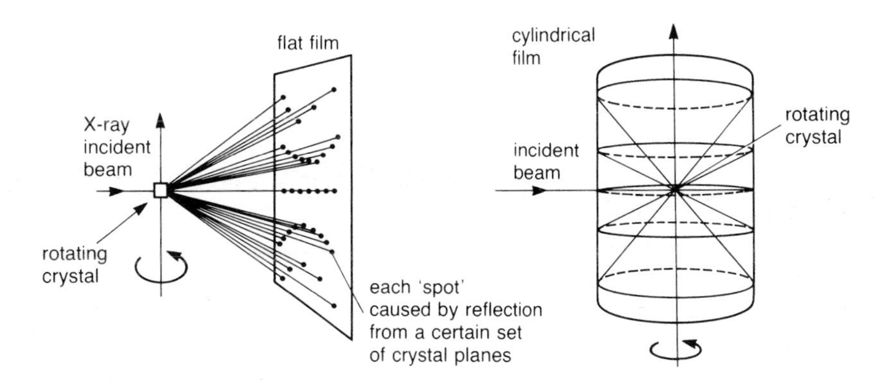

Figure 16.9 ►
The 'rotating crystal' method of X-ray diffraction which allows 'reflections' from many different crystal planes to be measured simultaneously. When the crystal is rotated about one of its axes, many sets of crystal planes will eventually come into strong reflection positions. All the reflected X-ray beams may be detected photographically using either a flat film placed behind the crystal or a cylindrical film placed around it. If the experiment is repeated, but rotating the crystal about its other two axes, then all possible reflections will be observed

The reflections are marked by 'spots' on the photographic film and each crystal structure will produce its own unique pattern of spots. Crystal structure analysis involves finding a 'model' structure which best matches the positions and relative intensities of the spots in this pattern and can be quite a tedious and time-consuming task. Fortunately, mathematical analysis of the reflection data can often be used to produce an electron density map of the structure in which the individual atoms may be identified. Such an electron density map for 4-methylbenzoic acid is shown with the structure superimposed in Figure 16.10.

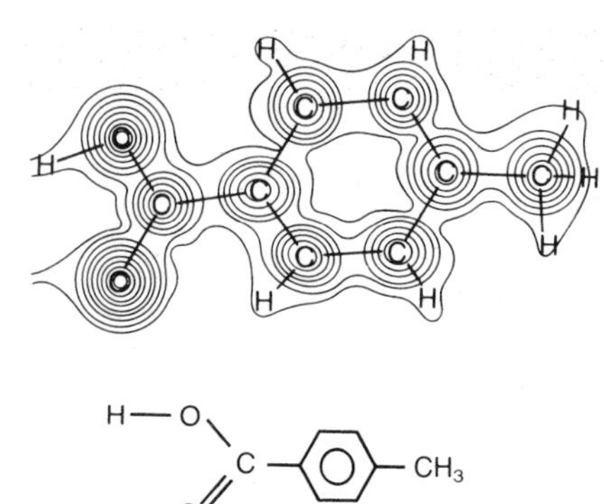

H—O C CH$_3$ O

4-methyl benzoic acid

Figure 16.10 ▲
Structural interpretation of the electron density map of a molecule of 4-methylbenzoic acid

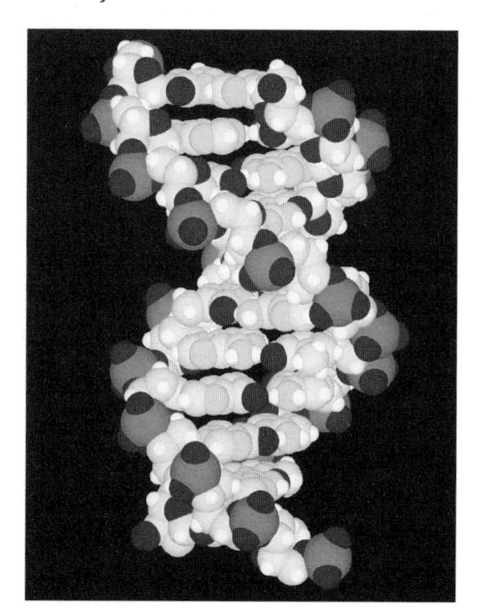

▲ Computer generated model illustrating the double-helix structure of DNA, confirmed by X-ray crystallography

A famous example of the use of X-ray crystallography was the confirmation by Crick and Watson in 1953 of the double-helix structure of deoxyribonucleic acid, DNA.

The Bragg equation

Monochromatic X-rays may be pictured as consisting of waves all of the same wavelength, λ. When a beam of this radiation strikes a crystal some waves are reflected from the surface plane, whilst others penetrate deeper and are reflected from successive planes within the same set. This means that the waves reflected from lower layers travel further than those reflected from the top layer. Figure 16.11 shows that for a set of planes with spacing, d, and a glancing angle for the X-rays beam of θ, the extra distance travelled by the beam as it penetrates each layer in the crystal plane is given by $2d \sin \theta$.

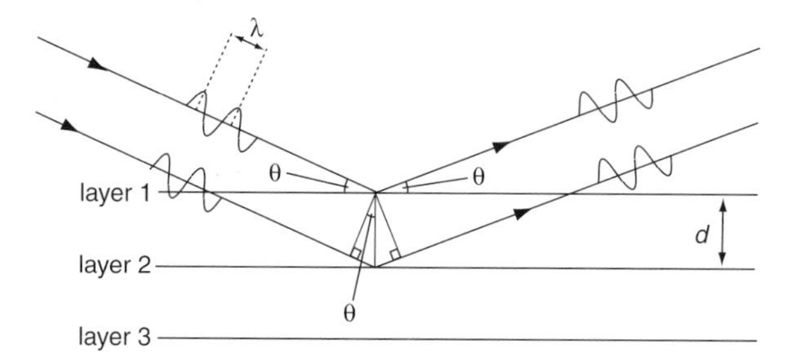

Figure 16.11 ▲
Conditions for 'strong' X-ray reflection from a set of crystal planes

Only if this distance is exactly equal to a whole number multiple, n, of the X-ray wavelength, λ, will the beams be 'in step' and therefore reinforce each other and produce a strong reflection. Thus for a strong reflection from any set of planes the following (Bragg) equation must be satisfied.

$$n\lambda = 2d \sin \theta$$

The lowest glancing angle for reflection from any set of planes corresponds to the case where $n = 1$. If the wavelength of the radiation is known, it is then a simple matter to calculate the interplanar spacing.

Questions on Chapter 16

1 During a study of air pollution, ice samples taken from drillings in Greenland were analysed for lead content.

a) Describe a method that might be used to measure the lead content of the ice sample. As new 'permafrost' is laid down in Greenland each year, the depth of the ice sample may be related to its age. The variation in lead content with age is shown in Figure 16.12.

b) The peak in lead content about 2500 years ago is thought to be due to open-air smelting of lead ore by the Greeks and Romans. What do you think has caused the large increase in lead pollution during the latter half of this century? Suggest why pollution from this source is likely to decline in the future.

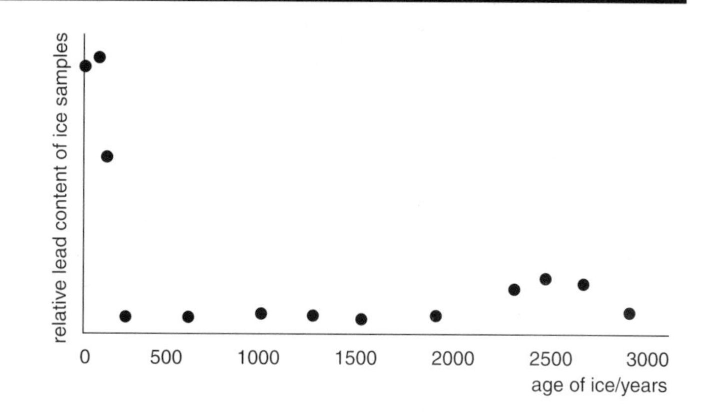

Figure 16.12 ▲
Lead content of Greenland permafrost

Questions on Chapter 16 *continued*

2 There are many possible structures that have the empirical formula CHCl.

a) Using the characteristic valencies C(4), H(1) and Cl(1), draw feasible structures which match the following:

i) a molecular formula of $C_2H_2Cl_2$

ii) a ring structure with relative molecular mass 194

iii) a chain structure which contains three carbon atoms.

b) Why is there no stable molecule CHCl?

3 Antabuse, a drug used to treat alcohol addiction, has the following structure:

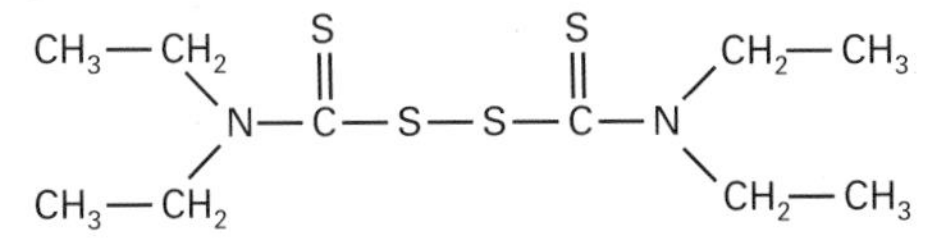

a) What is the molecular formula of antabuse?

b) What is its relative molecular mass?

c) What is its empirical formula?

4 Chlorofluorocarbons, CFCs, are volatile chemically inert compounds which, despite worries over damage to the ozone layer, have been used to form the bubbles in expanded polystyrene foam. One such compound, X, contains 58.7% by mass of chlorine and 31.4% by mass of fluorine and has the mass spectrum shown in Figure 16.13.

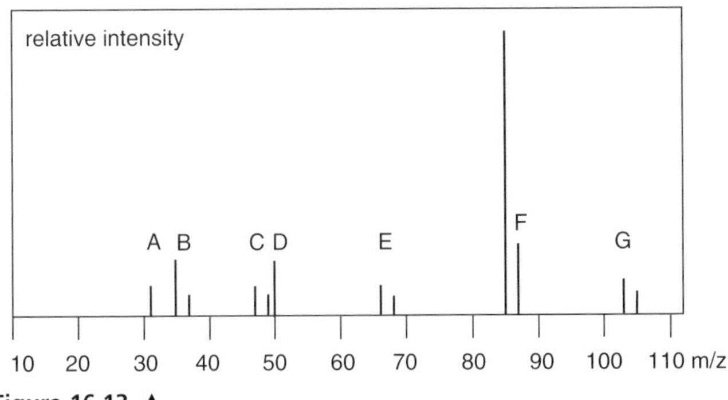

Figure 16.13 ▲
The mass spectrum of compound X

a) What is the percentage by mass of carbon in X?

b) Calculate the empirical formula of X?

c) What is the molecular formula of X?

d) Suggest species which may be responsible for each of the lettered peaks in the mass spectrum of X. (Chlorine exists as two natural isotopes ^{35}Cl and ^{37}Cl which occur naturally in a ratio of 3:1.)

e) There is no molecular ion peak in the mass spectrum of X. If this did exist, how many peaks would it contain and what would their relative masses be?

5 Another CFC, Y, was found to contain 76.9% by mass of chlorine. When 0.139 g of Y was vaporised in a gas syringe at 100°C and 101 kPa pressure it was found to have a volume of 30.7 cm^3.

a) Convert the above data where necessary into units consistent with the gas constant value of 8.31 J mol^{-1} K^{-1}.

b) Calculate the relative mass of Y.

c) Deduce the molecular formula of Y.

d) Explain the pattern of peaks you would expect to get for the molecular ion of Y in its mass spectrum.

6 The major peaks in the mass spectra of three hydrocarbons A, B and C with the same empirical formula are shown in Figures 16.14a, b and c respectively.

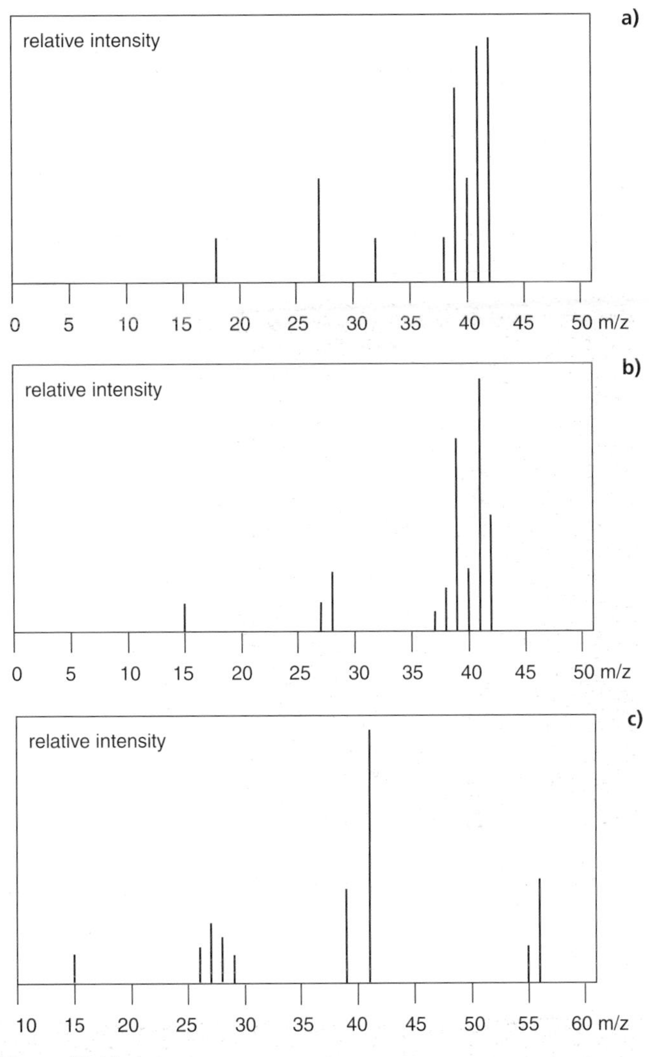

Figure 16.14 ▲
The mass spectra of hydrocarbons A, B and C

a) Give the relative molecular mass and molecular formula of each of the compounds.

b) What is their common empirical formula?

Iodine, I_2 undergoes a quantitative addition reaction with C=C double bonds as follows

$$C{=}C + I_2 \longrightarrow CI{-}CI$$

c) Compound A does not undergo an addition reaction with iodine at all. Suggest a likely structure for this compound.

d) Each of the other two compounds undergoes addition with an equal number of moles of iodine. Give the structure of B and two possible structures for C. Give the structures of the addition products that would be formed in each case with iodine.

Questions on Chapter 16 *continued*

7 A white crystalline solid, Z, obtained from rhubarb leaves, was found to contain 26.7% by mass of carbon, 71.1% of oxygen and 2.2% of hydrogen.

a) Calculate the empirical formula of Z. Could this also be the molecular formula of Z?

b) When added to sodium carbonate an aqueous solution of Z gives off carbon dioxide. What functional group must be present in Z?

c) Write at least two different feasible structures for Z which are consistent with the following typical valencies; C(4), O(2), H(1).

d) 1.125 g of *Z* was dissolved in water and the volume made up to exactly 250 cm^3. After thorough mixing, 25.0 cm^3 of this solution required exactly 25.0 cm^3 of 0.100 M NaOH solution for neutralisation. Calculate:

i) the number of moles of NaOH used in this experiment,

ii) the number of moles of the functional group you identified in **b)** above contained in the weighed sample of Z.

e) Use your previous answers to deduce the correct structure of Z.

f) When heated, Z decomposes before it vaporises. Consequently the mass spectrum of Z shown in Figure 16.15 does not show a peak corresponding to the molecular ion. Suggest possible decomposition products that might account for each of the peaks in the mass spectrum of Z.

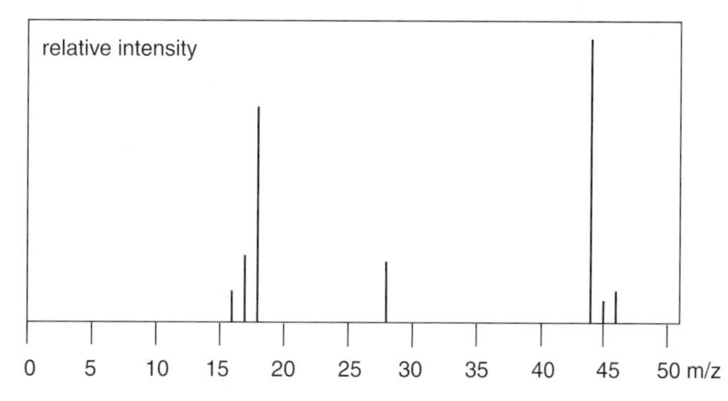

8 A gas, S, may contain any or all of the following elements; carbon, hydrogen, oxygen and nitrogen. The low-resolution mass spectrum of S contains a molecular ion peak at an *m/z* value of 30.

a) Give at least three different possible molecular formulae for S.

b) How could the actual formula be established using chemical methods?

c) High-resolution mass spectrometry gives a relative mass of 30.011 for the molecular ion peak. Use the following accurate relative masses for the most common isotopes of the elements concerned to identify the correct molecular formula for S.

^{1}H = 1.008 ^{12}C = 12.000 ^{14}N = 14.007
^{16}O = 15.995

d) Deduce the structure of S and show that your answer is consistent with the fragmentation pattern shown in Figure 16.16, which illustrates the low-resolution mass spectrum of S.

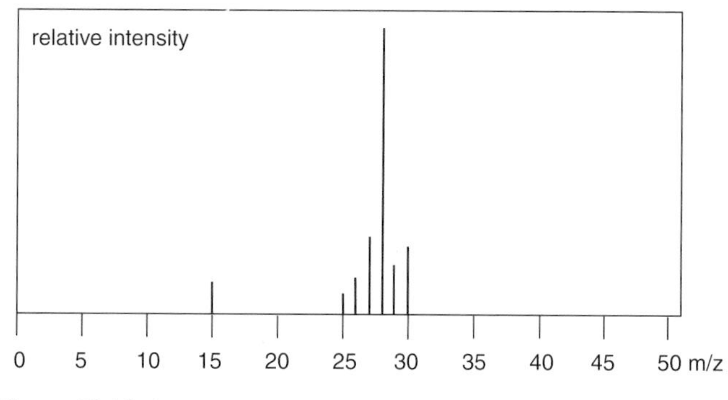

Figure 16.16 ▲
The mass spectrum of gas S

Figure 16.15 ◄
The mass spectrum of Z, a solid obtained from rhubarb leaves

Comments on the activities

Activity 16.1

1 The molecular formulae of ethanal is C_2H_4O and that of dioxan is $C_4H_8O_2$.

2 The empirical formula of both compounds is C_2H_4O.

3 There are several alternative answers to this question, some of which are shown in Figure 16.17. Make sure that each carbon atom forms four bonds, each oxygen two bonds and each hydrogen atom only a single bond. Compounds with the same molecular formula but different structures are referred to as structural isomers.

molecular formula C_2H_4O

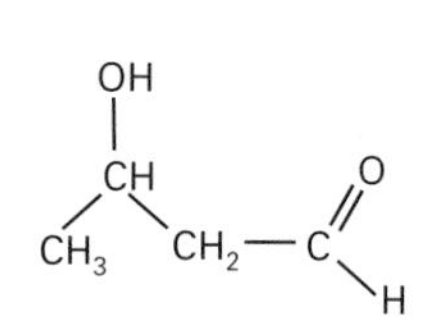

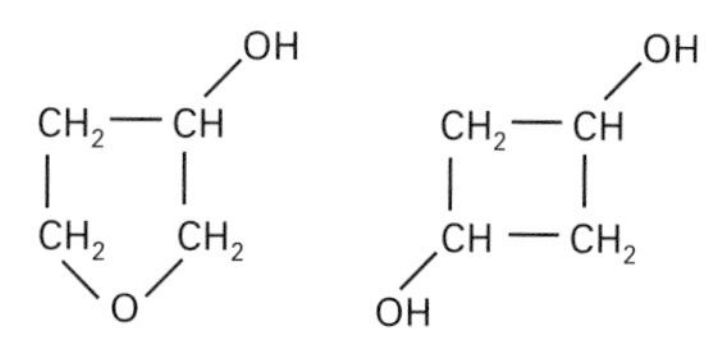

molecular formula $C_4H_8O_2$

Figure 16.17 ▲

Comments on the activities *continued*

4 Again there are several possible answers, including the following with molecular formula $C_6H_{12}O_3$:

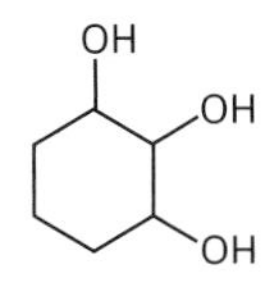

Activity 16.2

1 **a)** The cobalt chloride test indicates that water has been produced and hence that compound *A* must contain hydrogen. The calcium hydroxide solution test indicates that carbon dioxide is produced which means that A must also contain carbon.

b) 1 mole of H_2O contains 2 g of H and 16 g of O and weighs a total of 18 g, therefore:

0.900 g of H_2O contains $0.9 \times (2/18)$
= 0.100 g hydrogen

1 mole of CO_2 contains 12 g of C and 32 g of O and weighs a total of 44 g therefore:

2.200 g of CO_2 contains $2.200 \times 12/44$
= 0.600 g carbon

0.700 g of A therefore contains 0.100 g of H and 0.600 g of C (only).

The general equation for complete combustion of a hydrocarbon may be represented as:

$$C_xH_y + (x + 0.25y)O_2 \longrightarrow xCO_2 + 0.5yH_2O$$

c)

	C_x	H_y
0.700 g of compound A contains	0.600 g	0.100 g
i.e.	$\frac{0.600}{12}$	$\frac{0.100}{1}$
	= 0.050 mol C and	0.100 mol H
Dividing throughout by the smaller number gives us the simplest whole number mole ratio.	$\frac{0.05}{0.05}$ = 1C	$\frac{0.100}{0.050}$ = 2H

So the empirical formula of A is CH_2.

d) Decolourisation of bromine solution indicates that A is an unsaturated hydrocarbon, i.e. it contains a carbon–carbon multiple bond.

e) Since A must contain a carbon–carbon multiple bond, its molecular formula cannot be CH_2 and must, therefore, be a multiple of this. All simple alkenes share the same empirical formulae, e.g. ethene, $CH_2{=}CH_2$ and propene, $CH_2{=}CH{-}CH_3$

2 **a)**

	C_x	H_y	O_z
100 g of B contains	38.7 g	9.8 g	(100 – 38.7 – 9.8) = 51.5 g
i.e.	$\frac{38.7}{12}$ mol	$\frac{9.8}{1}$ mol	$\frac{51.5}{16}$ mol
	3.23 mol	9.8 mol	3.22 mol
Divide by smallest value	$\frac{3.23}{3.22}$	$\frac{9.8}{3.22}$	$\frac{3.22}{3.22}$
simplest whole number ratio	1	3	1

So the empirical formula of B is CH_3O

b) There is no feasible structure with the molecular formula CH_3O, but several possibilities exist for a molecule $(CH_3O)_2$ i.e. $C_2H_6O_2$, including: $HOCH_2{-}CH_2OH$ which would show the typical reactions of a primary alcohol, e.g. would turn acidified potassium dichromate(VI) from orange to green on warming.

3 **a)** Sodium fusion would convert any chlorine present in C into chloride ions. When dissolved in water this will give a white precipitate with silver nitrate solution that is insoluble in dilute nitric acid but soluble in dilute ammonia.

b)

	C_w	H_x	O_y	Cl_z
100 g of C contains	30.6 g	3.8 g	20.4 g	45.2 g
i.e.	$\frac{30.6}{12}$ mol	$\frac{3.8}{1}$ mol	$\frac{20.4}{16}$ mol	$\frac{45.2}{35.5}$ mol
=	2.55 mol	3.8 mol	1.28 mol	1.27 mol
Dividing by smallest number	$\frac{2.55}{1.27}$	$\frac{3.8}{1.27}$	$\frac{1.28}{1.27}$	$\frac{1.27}{1.27}$
Gives simplest ratio	2	3	1	1

So the empirical formula of C is C_2H_3OCl

c) Compound C contains, or produces in water, an acidic grouping. This maybe the carboxyl group —COOH. Alternatively it could contain the acid chloride group —COCl which hydrolyses in water to give the carboxyl group and hydrochloric acid.

d) We can draw an acid chloride with molecular formula C_2H_3OCl. Since a carboxylic acid molecule must contain two oxygen atoms, the simplest structure we can draw has the molecular formula $C_4H_6O_2Cl_2$.

$CH_3C(=O)Cl$ $\qquad$ $CHCl_2{-}CH_2{-}CH_2{-}C(=O)OH$

Comments on the activities *continued*

Activity 16.3

1 • 100 cm = 1 m, so $(100\ \text{cm})^3 = 1\,000\,000\ \text{cm}^3 = 1\ \text{m}^3$.

Thus volume of flask, $V = 500\ \text{cm}^3 = \dfrac{500}{1\,000\,000} = 0.000500\ \text{m}^3$

• Pressure, $p = 100.0\ \text{kPa} = 100.0 \times 1000\ \text{Pa} = 100\,000\ \text{Pa}$

• Temperature, $T = 23°\text{C} = (23 + 273)\text{K} = 296\text{K}$

2 $pV = nRT$, so $pV/RT = n = \dfrac{100\,000 \times 0.000500}{8.31 \times 296}$

$= 0.0203$ mol of X

3 Mass of gas X contained in the flask = 350.13 – 349.52 = 0.61 g

4 0.0203 mol of X weighs 0.61 g so 1 mol of *X* weighs 0.61/0.0203 = 30.0 g. The relative mass of the empirical formula, CH_2 is 14 and 30.0/14 = 2.15, i.e. 2 to the nearest whole number. The molecular formula of X is therefore $(CH_2)_2$ i.e. C_2H_4.

Activity 16.4

1 Volume of vapour produced by the sample of Y in this experiment = 95 – 12 = 83 cm^3

2 • Volume of vapour must be in $\text{m}^3 = \dfrac{83}{1\,000\,000} = 0.000083\ \text{m}^3$

• Pressure is already given in Pa = 100 700 Pa
• Temperature must be in K = 100 + 273 = 373 K

3 $pV = nRT$,

so $n = \dfrac{pV}{RT} = \dfrac{100\,700 \times 0.000083}{8.31 \times 373} = 0.0027$ mol

4 Mass of Y injected = 5.96 – 5.72 = 0.24 g.
Thus 0.0027 mol of Y weighs 0.24 g
so 1 mol of Y weighs $\dfrac{0.24}{0.0027} = 89$ g
Empirical formula mass of CH_2Cl_2 = 12 + 2 + 71 = 85 so this must also be the molecular formula.

mass of 1st Br atom	mass of 2nd Br atom	mass of 3rd Br atom	mass of $CHBr_3^+$ ion
79	79	79	250
79	79	81	252
79	81	79	252
81	79	79	252
79	81	81	254
81	79	81	254
81	81	79	254
81	81	81	256

Activity 16.5

1 For the molecular ion $CHBr_3^+$.
So there will be a 'quartet' (4 peaks) at masses 250, 252, 254 and 256 in intensity ratio 1:3:3:1

2 **a)** The triplet at masses 171, 173 and 175 in intensity ratio 1:2:1 must contain two Br atoms and be due to $CHBr_2^+$
Doublets in intensity ratio 1:1 correspond to fragments containing only one Br atom
b) the doublet at masses 94 and 92 is due to $CHBr^+$
c) the doublet at masses 93 and 91 is due to CBr^+
d) the doublet at masses 79 and 81 corresponds to the Br^+ ion

Activity 16.6

1 The following combinations of C, H and O atoms give relative masses of approximately 46:
• C_2H_6O accurate mass = (2 × 12.000) + (6 × 1.008) + 15.995 = 46.043
• CH_2O_2 accurate mass = 12.000 + (2 × 1.008) + (2 × 15.995) = 46.006
The molecular formula of P is therefore C_2H_6O.

2 Possible structures for P which are consistent with the valencies of the atoms are $CH_3—CH_2—OH$ (ethanol) and $CH_3—O—CH_3$ (methoxymethane).

Activity 16.7 (extension)

• Compounds containing the hydroxyl group, OH, including water and alcohols, react with sodium to give hydrogen gas which 'pops' when mixed with air and ignited.
• Ethanol is a primary alcohol and is readily oxidised by heating with acidic potassium dichromate(VI) solution. This reaction is accompanied by a colour change from orange to green.
• Ethanol reacts with hot alkaline iodine solution to give yellow crystals of triiodomethane (iodoform). This reaction is given only by compounds which contain the $CH_3—C{=}O$ or $CH_3—CH—OH$ groupings.

None of the above reactions are given by ethers such as methoxymethane.

Activity 16.8

1 Q could be 1,1-dibromoethane, $CHBr_2—CH_3$, or 1,2-dibromoethane, $CH_2Br—CH_2Br$.

2 **a)** The triplet at masses 186, 188 and 190 is due to the molecular ion, $C_2H_4Br_2^+$. The pattern confirms the presence of two Br atoms
b) The doublet at masses 107 and 109 is due to $C_2H_4Br^+$
c) The doublet at masses 93 and 95 is due to CH_2Br^+

3 1,1-dibromoethane might be expected to produce a fragment containing one C atom and two Br atoms, e.g. $CHBr_2^+$, giving another triplet signal at masses 171, 173 and 175. Since this is not observed Q is most likely to be 1,2-dibromoethane.

CHAPTER 17

Spectroscopy

Contents

Study Checklist

After studying this chapter you should be able to:

1. Describe electromagnetic (EM) radiation in terms of waves which travel at a fixed speed (in a vacuum).
2. State, and carry out calculations using, the equation $c = f\lambda$ which relates the speed of EM radiation, c, to its frequency, f, and wavelength, λ.
3. State that EM radiation is quantised, i.e. emitted only in fixed packets whose energy is given by Planck's equation $E = hf$, where h is Plank's constant.
4. List the main types of EM radiation in order of quantum energy, i.e. describe the EM spectrum.
5. Explain how information on electronic energy levels for the hydrogen atom may be obtained from the atomic spectrum (see also Chapter 16).
6. Explain the origin of the following types of energy which may be associated with molecules: electronic, vibrational and nuclear spin.
7. Recall that all types of molecular energy are quantised, i.e. can take only certain fixed values, and appreciate that a molecule can only change an energy state by absorbing (or emitting) a quantum of EM radiation of exactly the right size.
8. Recall that electronic transitions generally involve radiation in the ultraviolet (UV) and visible regions (covered in more detail in Chapter 18).
9. Recall that vibrational transitions involve infrared (IR) radiation.
10. Appreciate that particular covalent bonds within any molecule may often be identified by absorption of IR radiation at characteristic frequencies. Use tables of characteristic IR absorptions to identify such structural features from a given spectrum.
11. Recall that in certain regions IR absorption bands are characteristic of the molecule as a whole rather than individual bonds and that this 'fingerprint' may be used as an identifier.
12. State the conditions required for an atom to exhibit nuclear spin and appreciate that different spin states of a nucleus have different energies in an applied magnetic field.
13. Explain the origin of nuclear magnetic resonance (NMR) spectra in terms of absorption of radio waves 'flipping' the spin states of appropriate nuclei (including protons) in a magnetic field.
14. State that the position of an absorption in a proton magnetic resonance (PMR) spectrum depends upon molecular environment and that the area under the peak is proportional to the number of such equivalent protons.
15. Explain the use of tetramethylsilane as an internal standard in PMR spectroscopy and the δ scale of chemical shifts.
16. Use tables of characteristic chemical shifts for protons in specific molecular environments to interpret low resolution PMR spectra.

17 Recall that at high resolution the different nuclear spin states for a particular set of protons may split the signal from an adjacent set of protons and interpret simple cases of such spin–spin splitting in PMR spectra.

17.1 Electromagnetic radiation

Electromagnetic (EM) radiation may be pictured as waves. The wavelength, i.e. the distance between one wave peak and the next may vary from about one million millionth of a metre to many kilometres.

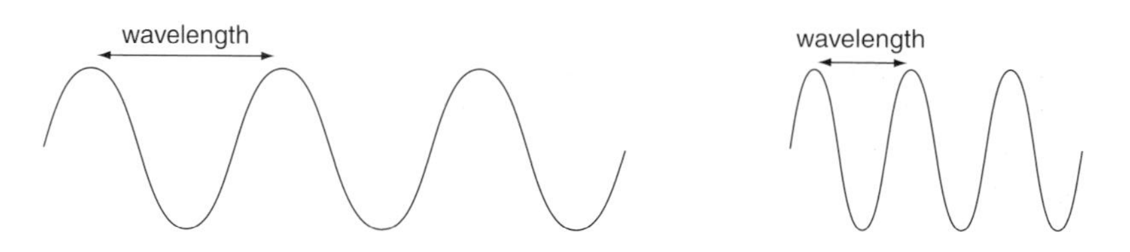

Many of the conveniences of modern day life depend upon EM radiation as an energy source and, for convenience, we divide up the full range, or EM spectrum, of radiation into various regions which are shown in Figure 17.1.

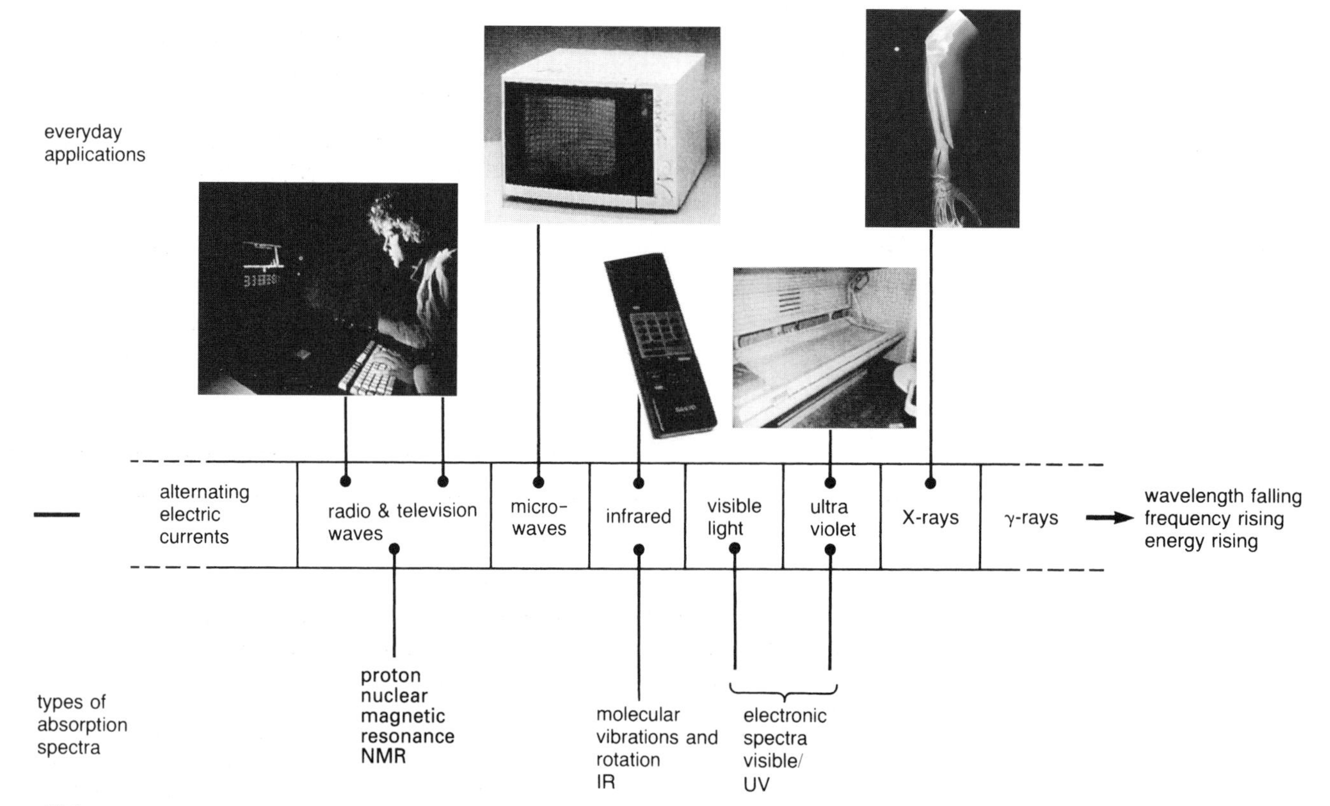

Figure 17.1
The electromagnetic spectrum – its applications in chemistry and everyday life

The frequency, measured in hertz (Hz), is the number of waves produced in one second. Since, regardless of wavelength, all EM radiation travels at the same speed through a vacuum we may write the equation:

$$c = f\lambda$$

where c is the speed of EM radiation, f is the frequency and λ is the wavelength.

The units of c and λ must be compatible, i.e. if the wavelength is given in metres then the speed must be given in units of metres second^{-1}. Another related quantity that is sometimes quoted is wavenumber, which refers to the number of waves of radiation contained in one centimetre.

$$\text{wavenumber} = \frac{\text{waves travelled in one second}}{\text{centimetres travelled in one second}}$$

$$= \frac{f}{100c},$$ where c is the speed of EM radiation in m s^{-1}.

Measuring the very small and the very big

The metre is a standard SI unit and may be used to express any distance. However, the EM spectrum covers such a huge wavelength range that the numbers involved can become very awkward and unwieldy. In order to simplify writing very small or very big numbers we may use the letters shown in Table 17.1 to prefix **any** standard unit.

Table 17.1

Prefix	relationship to standard unit		
p (pico)	10^{-12}	or	0.000 000 000 001
n (nano)	10^{-9}	or	0.000 000 001
μ (micro)	10^{-6}	or	0.000 001
m (milli)	10^{-3}	or	0.001
c (centi)	10^{-2}	or	0.01
d (deci)	10^{-1}	or	0.1
k (kilo)	10^{3}	or	1 000
M (mega)	10^{6}	or	1 000 000
G (giga)	10^{9}	or	1 000 000 000
T (tera)	10^{12}	or	1 000 000 000 000

Activity 17.1

EM radiation travels at a speed of about 300 000 000 m s^{-1} (or 186 000 miles per second.)

1. BBC Radio 4 is broadcast at a frequency of 94 MHz in the FM band. Calculate the wavelength at which Radio 4 FM is broadcast.
2. Radio 4 can also be found at a wavelength of 1515 m on the long-wave band. What is the frequency of Radio 4 long-wave?
3. Calculate the wavenumber values for each of the Radio 4 broadcasts.
4. Mains radios are driven by alternating electric current which is EM radiation operating at a frequency of 50 Hz. Calculate the wavelength of mains electricity.

The energy associated with EM radiation is said to be quantised, i.e. it is emitted or absorbed only in fixed packets known as quanta. You can perhaps picture this by imagining that the EM wave is cut up into equally sized 'wavicles' each of which has a fixed energy. Any object can absorb or emit only whole numbers of these 'wavicles'.

imaginary 'wavicles' that carry a fixed amount of energy; the shorter the wavelength the higher the energy of the quantum or packet

The quantum energy depends directly upon the frequency of the radiation and is given by Planck's equation.

$$E = hf$$

where h is Planck's constant (6.626×10^{-34} J s)

It is this quantum nature of EM radiation that makes it so useful in investigating atomic and molecular structure.

Focus 17a

1. Electromagnetic radiation may be pictured as waves that travel at a fixed speed, c, in a vacuum
2. The frequency, f, and wavelength, λ, of radiation are related to c by the equation $c = f\lambda$.
3. Radiation can only be absorbed or emitted in fixed energy packets called **quanta**. The energy of a quantum of radiation, E, is directly proportional to its frequency, $E = hf$, where h is Planck's constant.

Activity 17.2

1. Visible light forms a very small part of the EM spectrum, with colours ranging through the 'rainbow' from red, through yellow, green and blue to violet. Red light has a frequency of about 430 GHz whilst violet light has a wavelength of about 400 nm. Using any data provided in this section so far, calculate the quantum energies of red and violet light.
2. As well as visible light the sun also emits invisible radiation in both the infrared and ultraviolet regions (see Figure 17.1). Which of these types of EM radiation do you think is more likely to cause sunburn and why?

17.2 Atomic energy states

In the last chapter we explained how an element could be identified from the unique pattern of lines in its atomic spectrum. We shall now take a closer look at the atomic absorption spectrum of hydrogen. The full spectrum, shown in Figure 17.2, contains lines in the ultraviolet and infrared as well as in the visible region.

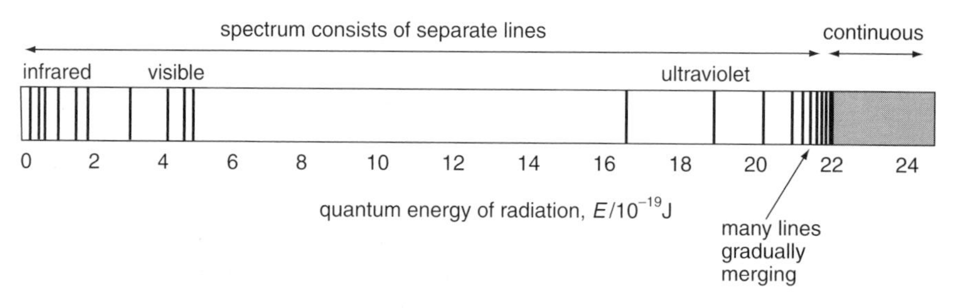

Figure 17.2
The atomic hydrogen spectrum

The quantum energies of the main lines shown are given ($E/10^{-19}$ J) in Table 17.2.

Table 17.2

ultraviolet	16.340		19.367		20.426		20.917		21.183
visible		3.027		4.086		4.577		4.843	
			1.059		1.550		1.816		
infrared				0.491		0.757			
					0.266				

Each line in this spectrum corresponds to the energy change involved in an electron jumping from a lower energy level to a higher one. We should, therefore, be able to work out a set of electronic energy levels that will account for all the lines in the spectrum.

A close look at the energy values shown in Figure 17.2 reveals an interesting pattern. The lines in the visible region of the spectrum have energies which correspond to *differences* in energy between lines in the higher energy ultraviolet region. Similarly, lines in the infrared correspond to differences between lines of higher energy. We can account for this by supposing that the series of lines in the ultraviolet is produced by electrons being promoted from the lowest possible energy level. The lines in the visible region are caused by electrons being promoted from the next lowest energy level, whilst infrared absorption results from promotion from yet higher energy levels. The relationship between the lines in the spectrum and the electronic jumps producing them is illustrated in Figure 17.3. It can be seen that the difference in electron jumps betwen B and C is equal to A in Figure 17.3.

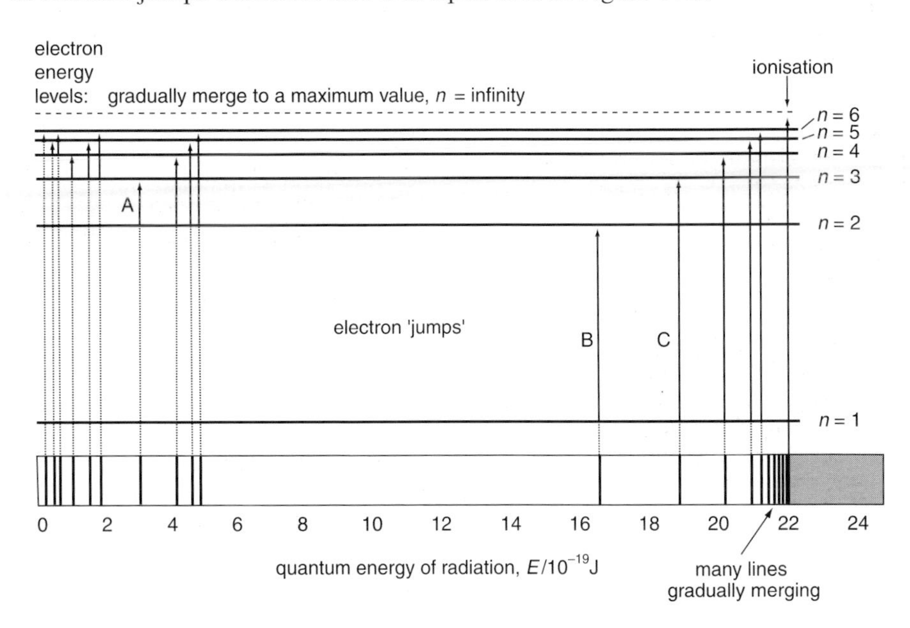

Figure 17.3
Origin of the lines in the atomic hydrogen absorption spectrum

As the electron is withdrawn further and further from the nucleus so the attraction decreases. This means that the difference in energy between successive levels gradually decreases. This causes the lines in the ultraviolet series to merge eventually at high energy into a region of continuous absorption. Since there are now no separate lines, at this convergence limit there are no distinct electronic energy levels. The electron no longer experiences attraction from the nucleus and has escaped from the atom. The energy required to promote the electron from the lowest possible energy level in the atom to the highest is the **ionisation energy** of the hydrogen atom.

Activity 17.3 Calculating the ionisation energy of hydrogen E

To determine the ionisation energy for the hydrogen atom we only need to find the exact point at which the lines merge to give continuous absorption in the ultraviolet region. However, it is difficult to judge this visually and the value is best determined graphically. As we rise through the energy levels, the energy gap between them gradually falls until at the convergence point eventually it becomes zero.

1. Draw and extrapolate the graph on the next page, using the data in Figure 17.2 in order to determine a value for the ionisation energy of the hydrogen atom.
2. Ionisation energies are normally quoted per mole of atoms. Avogadro's number (equivalent to the number of particles in one mole) is 6.02×10^{23}. Calculate the ionisation energy of hydrogen in kJ mol^{-1} and check your figure against that given in a data book.

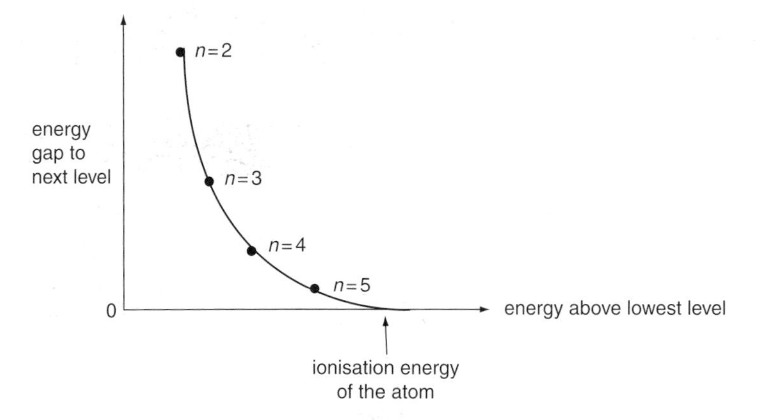

Focus 17b

1. Atomic line spectra can be interpreted in terms of electrons 'jumping' between certain fixed energy states.
2. In the case of hydrogen, where there is no inter-electronic repulsion, the atomic spectrum may be used to establish the pattern of electronic energy levels.
3. Each element has its own characteristic atomic spectrum, which may be used for identification and analysis (see also Chapter 15).

3 To simplify the calculations in this section we have given the 'energies'of the lines in the atomic hydrogen spectrum. In practice we would measure the wavelength, or frequency of the radiation. Calculate the wavelength, in nm, and the frequency of the radiation that corresponds to the point at which the lines in the ultraviolet region of the hydrogen spectrum just merge.

Whilst other elements have atomic spectra that are similar in appearance to the hydrogen spectrum, we cannot easily work out the exact pattern of their electron energy levels. The problem is that here the energy of an electron is not simply determined by nuclear attraction but also by repulsion from the other electrons whose positions are uncertain. However, the point at which the lines merge at high energy in the spectrum still gives a measure of the first ionisation energy of the atom concerned i.e. the energy needed to raise the most loosely bound electron from its lowest possible energy state to the highest in the atom.

17.3 Molecular energy states

In the last section we saw that the **electronic** energy of atoms was quantised, i.e. could only take certain fixed values. An electron in an isolated atom can only change its state by absorbing (or emitting) a quantum of EM radiation of exactly the right energy. We used this idea to establish the energy levels available to the electron in the hydrogen atom. Under normal conditions, electrons occupy the lower energy states so it is the transitions from these, which occur in the visible and ultraviolet regions of the EM spectrum, that are most important.

In a similar way the bonding electrons in molecules have various fixed energy states available to them and will, therefore, absorb visible and/or ultraviolet radiation at characteristic energies. The relationship between structure and bonding and visible/ultraviolet absorption is of great practical importance in many areas and will be considered in more detail in Chapter 18.

Unlike isolated atoms, molecules may also have a variety of other types of quantised energy associated with them. The bonds connecting the atoms behave like tiny springs and may vibrate or rock at characteristic frequencies as shown in Figure 17.4.

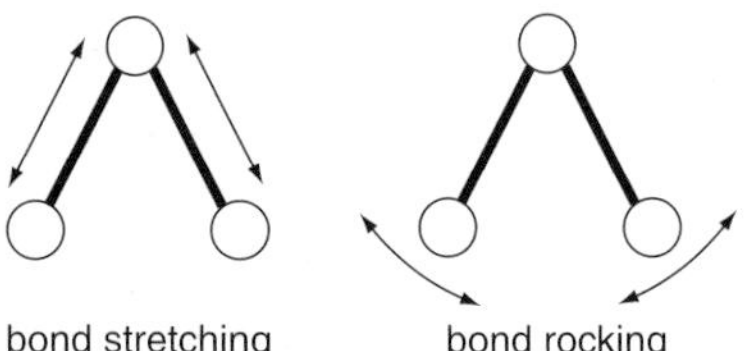

Figure 17.4
The quantum energies required for changes in such energy states fall into the infrared region of the spectrum

Predicting vibrational absorptions

A ball suspended on a stretched spring will vibrate up and down with a characteristic frequency when released. A heavier ball suspended from the same spring will vibrate more slowly.

In a similar way the stretching frequency of the bond between two atoms also depends upon their mass.

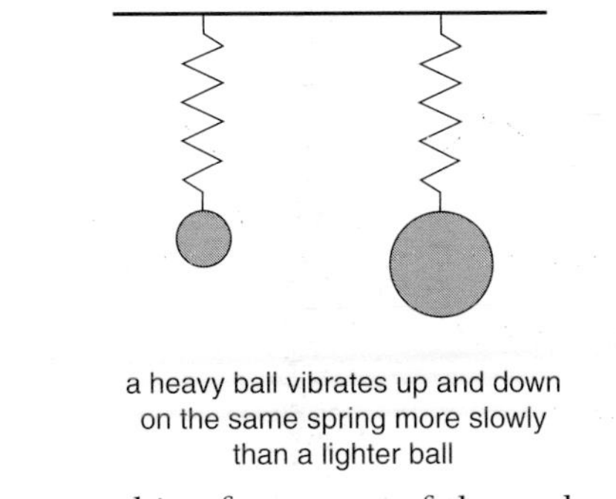

a heavy ball vibrates up and down on the same spring more slowly than a lighter ball

Thus the stretching frequency of the carbon–halogen bond drops as we pass down Group 7.

Bond	C—Cl	C—Br	C—I
Typical stretching frequency/THz	21	17	15

Taking our model a little further, the stretching frequency will also be affected by the strength of the spring or bond. The stronger the spring, or bond, the faster the vibration will be. This is nicely illustrated by the stretching frequencies of carbon–carbon bonds.

Bond	C—C	C═C	C≡C
Typical stretching frequency/THz	27	50	65

As you can see from the figures, increasing bond strength usually has a relatively greater effect on characteristic vibration frequency than changing the mass of an atom.

Activity 17.4

Place the following bonds in increasing order of their likely characteristic stretching frequencies and justify your answer:

C—N, C—O, C≡N and C═O

In practice not every molecular vibration gives rise to an infrared absorption. In order to interact with the oscillating electric field in electromagnetic radiation, the molecule must undergo a change in dipole moment, i.e. the distribution of electric charge must alter. Diatomic molecules of elements, such as hydrogen, oxygen, nitrogen and the halogens, have no permanent dipole and will not therefore absorb in the infrared region:

e.g. $\overleftarrow{\text{Cl}}$———$\overrightarrow{\text{Cl}}$ no permanent centres of charge and so zero dipole moment

Molecules composed of two different atoms will have a permanent dipole, owing to a difference in electronegativity. Vibration in this case will alter the positions of the centres of positive and negative charge:

e.g. $\overleftarrow{\text{H}}^{\delta+}$———$\overrightarrow{\text{Cl}}^{\delta-}$ charge separation increases as atoms move apart

In the case of triatomic molecules, there are various different modes of vibration, some of which may give rise to infrared absorption whilst others may be inactive. Again the test is, will the mode of vibration lead to a change in the overall dipole moment of the molecule i.e. do the centres of positive and negative charge move relative to each other? In the case of the linear carbon dioxide molecule we may identify three possible vibration modes:

$\overleftarrow{\text{O}}^{\delta-}=\text{C}^{\delta+}=\overrightarrow{\text{O}}^{\delta-}$	$\overleftarrow{\text{O}}^{\delta-}=\text{C}^{\delta+}=\overleftarrow{\text{O}}^{\delta-}$	$\text{O}^{\delta-}{\downarrow}=\text{C}^{\delta+}{\uparrow}=\text{O}^{\delta-}{\downarrow}$
symmetric stretch	asymmetric stretch	rocking

In symmetric stretching both bonds lengthen at the same time. Although the dipole moment of each bond increases, they act in opposite directions and so cancel out. You can tell this from the fact that the mid-point of the negative charge always coincides with the centre of positive charge, i.e. the $C^{\delta+}$. There is no change in the overall dipole moment of the CO_2 molecule and so symmetric stretching is infrared inactive. In both asymmetric stretching and rocking all the centres of charge move, resulting in a change in dipole moment and consequent infrared absorption.

Activity 17.5 Molecular shape and infrared absorption

Infrared absorption also depends upon molecular geometry. A water molecule has modes of vibration similar to those shown above for carbon dioxide but is an angular rather than a linear molecule.

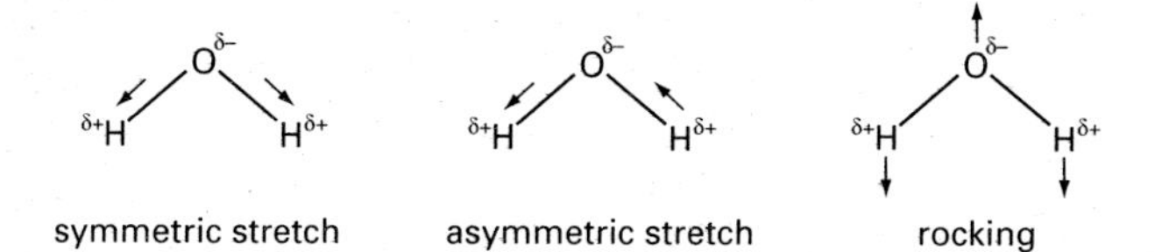

Which of these modes will be infrared active?

Nuclear magnetic resonance (NMR)

In section 2.12 we introduced the idea of electron spin. Two electrons can only occupy the same orbital if they have opposite spin, i.e. their spin values cancel out. In a similar way the protons and neutrons in the nucleus also possess spin. However, unless the mass number is odd, or the nucleus has an odd number of both protons and neutrons, the overall spin will be zero. If the nucleus does have a non-zero spin it behaves as a tiny magnet. The orientation of this nuclear magnet depends upon its spin state. Normally each nuclear spin state has the same energy but in a magnetic field a spin state which produces a nuclear magnet which is aligned with the applied field will be more stable. In effect the nucleus behaves like a compass needle that aligns itself with the Earth's magnetic field. The energy difference between the spin states increases as the applied field is increased. If the sample is exposed to radio waves of fixed frequency then at a certain field strength, the low energy spin state may be 'flipped' over into the high energy spin state by absorbing a quantum of the radiation. This effect is known as nuclear magnetic resonance (n.m.r).

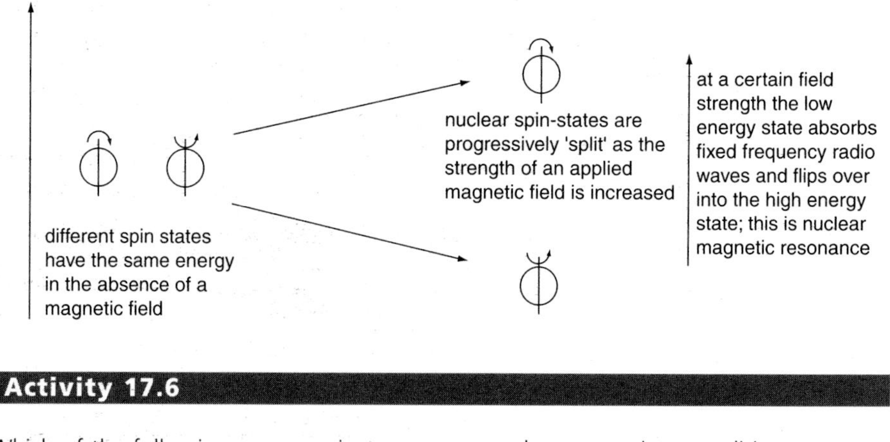

Activity 17.6

Which of the following common isotopes may, under appropriate conditions, undergo nuclear magnetic resonance?

^{1}H ^{12}C ^{16}O ^{19}F ^{35}Cl ^{37}Cl

Magnetic resonance imaging

Magnetic resonance imaging (MRI) uses NMR for medical diagnosis. The patient is placed inside a cylinder that contains a strong magnet. Radio waves then cause particular atoms of the body to resonate. Each type of body tissue emits characteristic signals from the nuclei of its atoms, and a computer translates these signals into a two-dimensional picture.

Unlike traditional X-rays or CAT scans, MRI does not use potentially damaging ionising radiation. The technique can also 'see through' bone and produce images of soft tissues such as blood vessels, cerebrospinal fluid, cartilage, bone marrow, muscles, and ligaments. MRI is particularly useful to detect brain tumours, damage caused by multiple sclerosis, joint injuries, and herniated disks. It is a harmless procedure to all apart from those with metallic implants, such as pacemakers, joint pins, or artificial heart valves. Such objects may be dislodged by the powerful magnetic field.

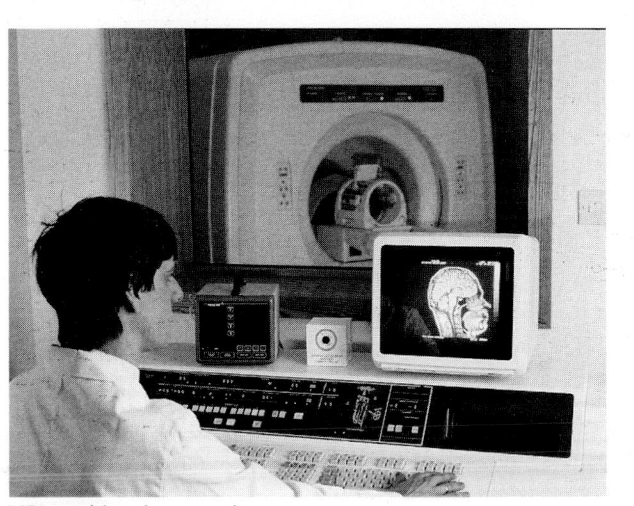

MRI machine in operation

Focus 17c

1. Molecules possess various kinds of energy, e.g. **electronic**, **vibrational** and **nuclear spin**, which are quantised, i.e. can only take certain fixed values.
2. When a molecule changes state it emits or absorbs a quantum of radiation energy equal to the difference between the initial and final states.
3. Analysis of molecular spectra can give much useful structural information.

17.4 Infrared spectra

An absorption spectrum simply measures the extent to which radiation is absorbed by a sample over a particular frequency range. The general principle of an absorption spectrometer is shown in Figure 17.5. The source produces radiation over the whole of the desired frequency range. This must be separated (generally by refraction through a prism or diffraction through a grating) so that each individual frequency is selected in turn. The beam is then equally split and passed through the sample and a blank. This is an identical container which contains everything but the sample under test. The relative intensities of the beam leaving the sample tube are then compared with those leaving the blank tube in the detector. Any difference can only have been caused by the sample absorbing the radiation. The results are then displayed using a screen or chart recorder.

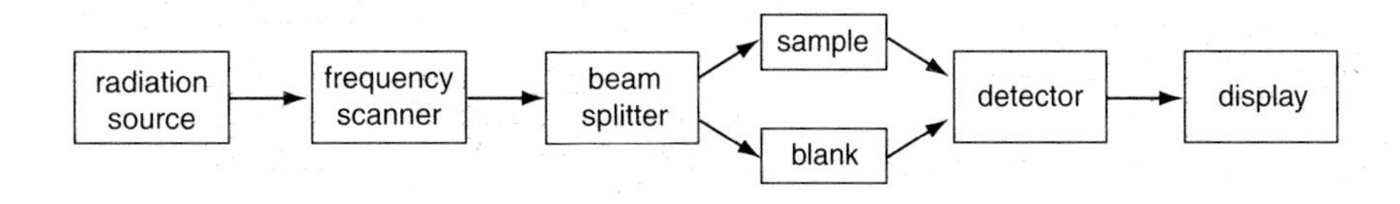

Figure 17.5
Principle of an absorption spectrometer

The IR spectra for a range of compounds are shown in Figure 17.6.

Preparation of samples for infrared spectroscopy

Sample preparation depends largely upon the properties of the material under test. The big problem is finding a container which is transparent to IR radiation since glass and many plastics, although transparent to visible light, absorb strongly in this region. The best materials for this purpose are alkali metal halides, e.g. KCl. A drop of a liquid sample may simply be sandwiched between two KCl plates to produce a thin film. This method can only be used for non-volatile liquids that do not vaporise quickly. Great care must be taken to ensure that the halide plates are never exposed to water or aqueous solutions.

A solid sample may be ground up with the metal halide. Compression of the mixture gives a transparent disposable sample disc. Gases may be enclosed in a sealed tube equipped with halide windows at each end. Because gases are more diffuse than solids or liquids a longer path length is required to produce measurable absorption. This method may also be used with samples in solution, although here you need an identical blank reference tube to compensate for any absorption by the solvent.

A quick, simple method that can be used for both solids and liquids is to mix the sample with a hydrocarbon oil called 'nujol'. This gives a mull that can be sandwiched between halide plates. The only disadvantage of this method is that the nujol itself shows absorption bands due to C—H and C—C bonds. Since these absorptions are, however, present in almost all organic compounds, this is not too critical.

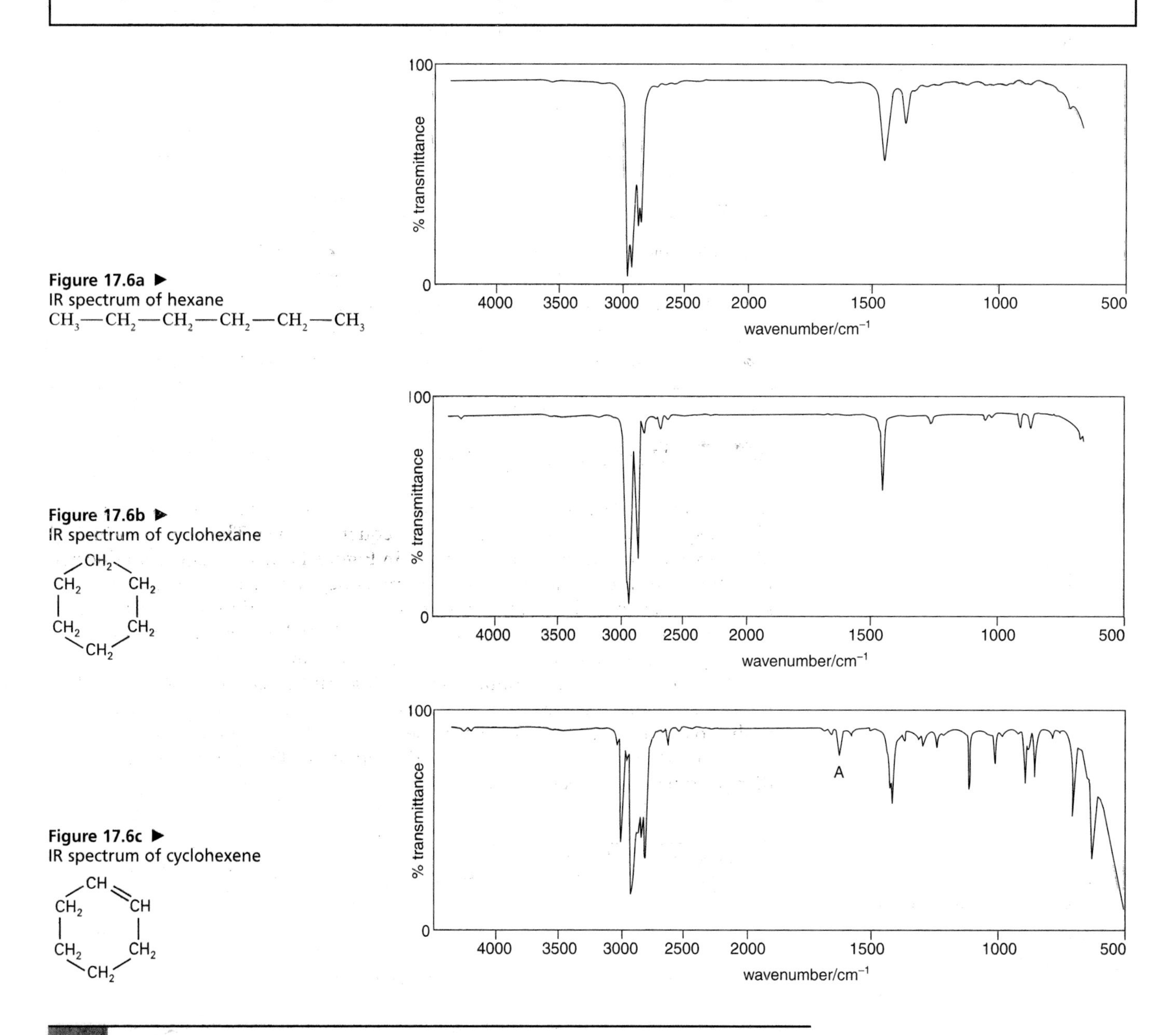

Figure 17.6a ▶ IR spectrum of hexane
$CH_3—CH_2—CH_2—CH_2—CH_2—CH_3$

Figure 17.6b ▶ IR spectrum of cyclohexane

Figure 17.6c ▶ IR spectrum of cyclohexene

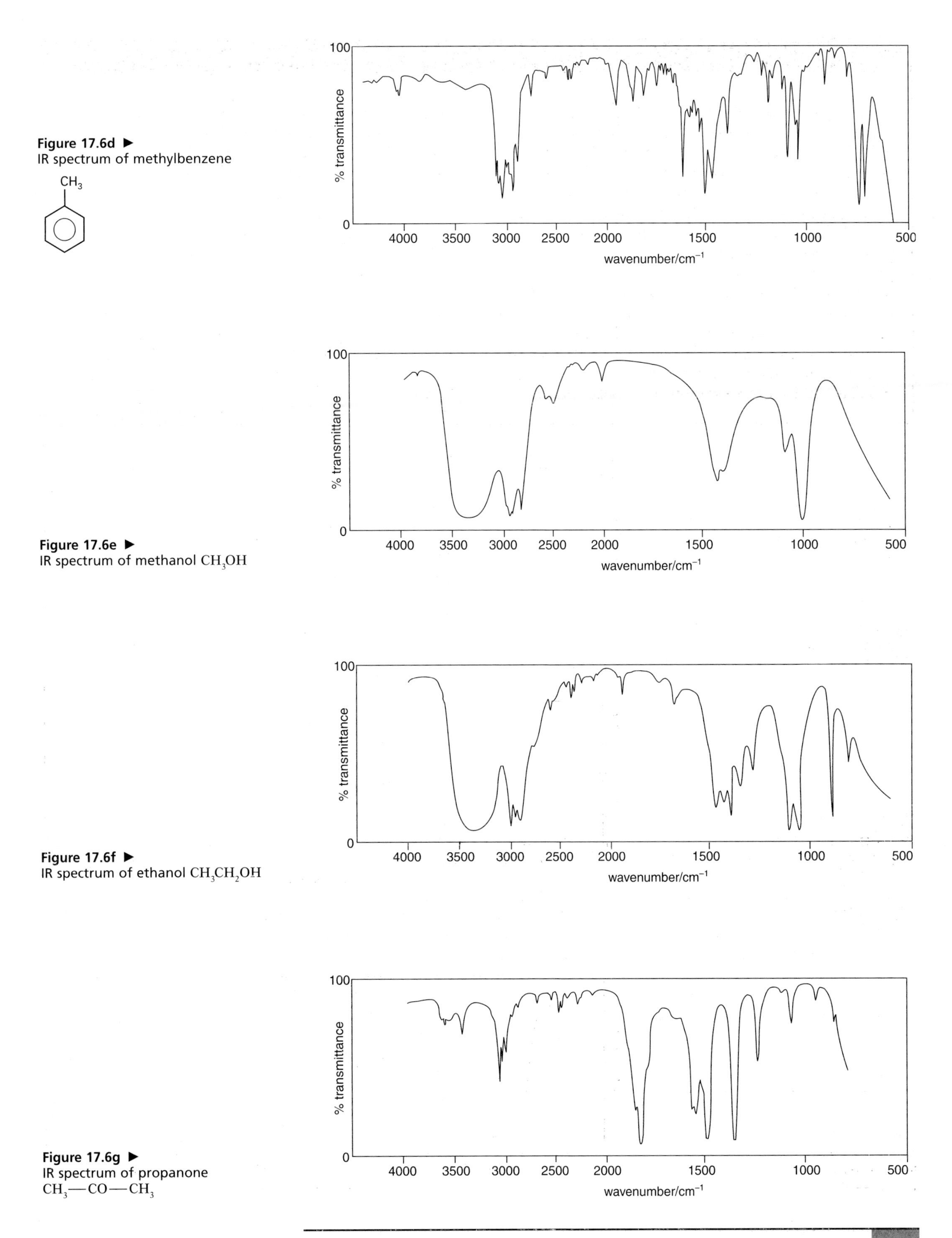

Figure 17.6d ▶
IR spectrum of methylbenzene

Figure 17.6e ▶
IR spectrum of methanol CH_3OH

Figure 17.6f ▶
IR spectrum of ethanol CH_3CH_2OH

Figure 17.6g ▶
IR spectrum of propanone
$CH_3—CO—CH_3$

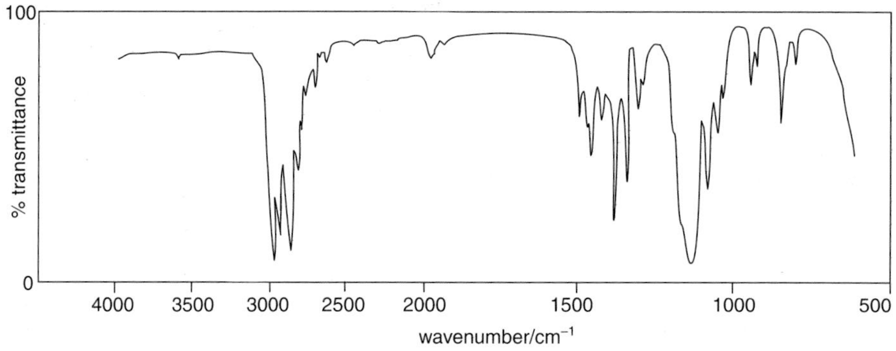

Figure 17.6h ▶
IR spectrum of ethoxyethane
$CH_3—CH_2—O—CH_2—CH_3$

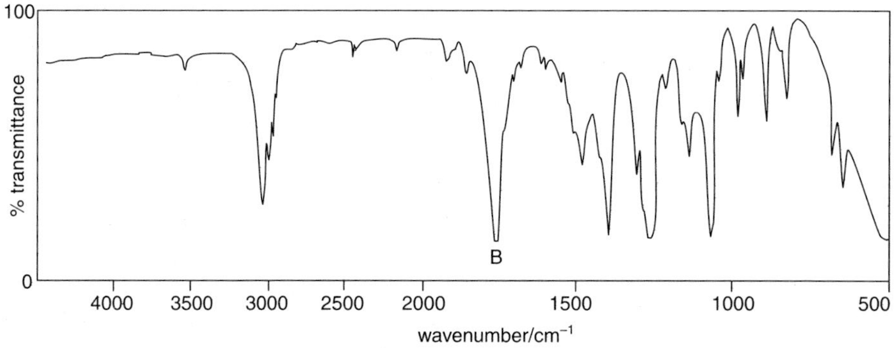

Figure 17.6i ▶
IR spectrum of ethyl ethanoate
$CH_3COOCH_2CH_3$

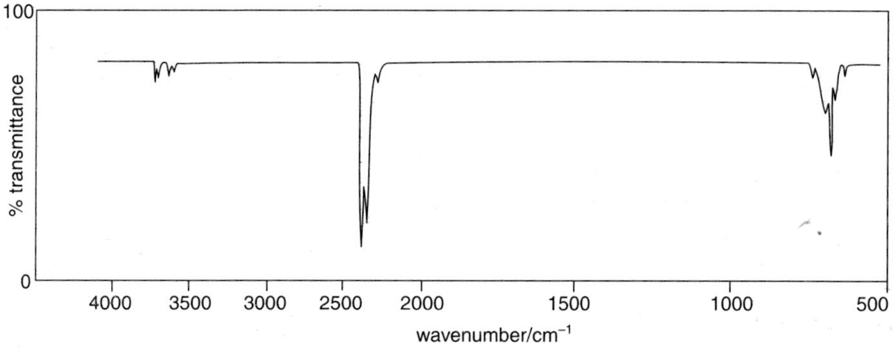

Figure 17.6j ▶
IR spectrum of carbon dioxide
$O{=}C{=}O$

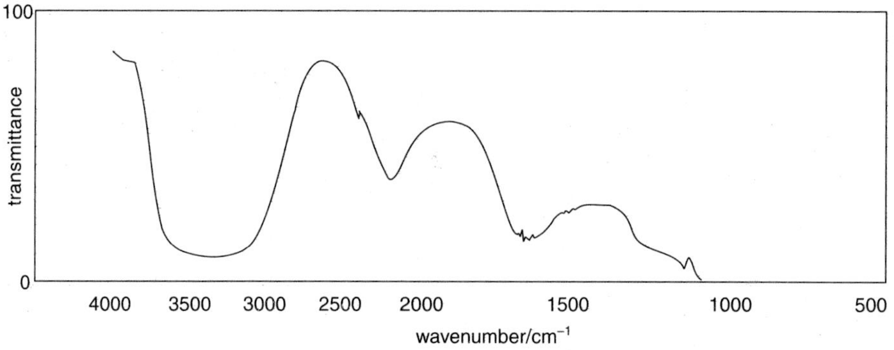

Figure 17.6k ▶
IR spectrum of water
H—O—H

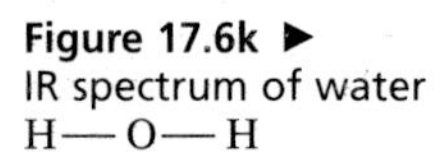

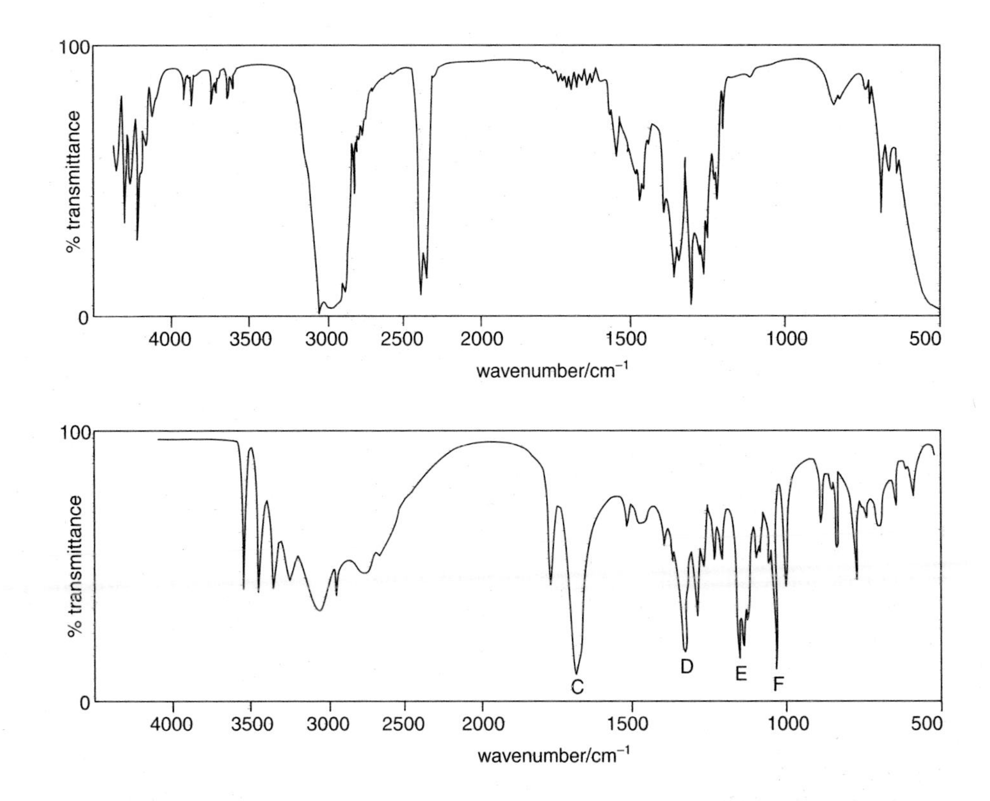

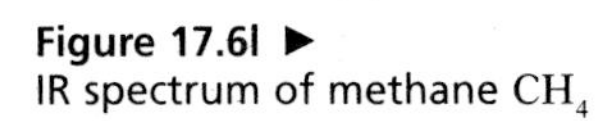

Figure 17.6l ▶
IR spectrum of methane CH_4

Figure 17.6m ▶
IR spectrum of vitamin C

You will notice that the spectra of all the organic compounds have a strong absorption at around 3000 cm^{-1} due to the C—H bond. Water and the alcohols methanol and ethanol have a broad absorption band between 3000 and 3500 cm^{-1} due to the O—H bond. In a similar way we can relate the strong absorption around 1700 cm^{-1} in carbon dioxide, propanone and ethyl ethanoate to the C=O bond. Almost all of the absorption bands above about 1500 cm^{-1} may be assigned to individual bond types and their position is only slightly affected by the remainder of the molecule. A reference table such as Table 17.3 may therefore be used to deduce likely structural features in an unknown sample.

Table 17.3 Characteristic absorption bands for some covalent bonds

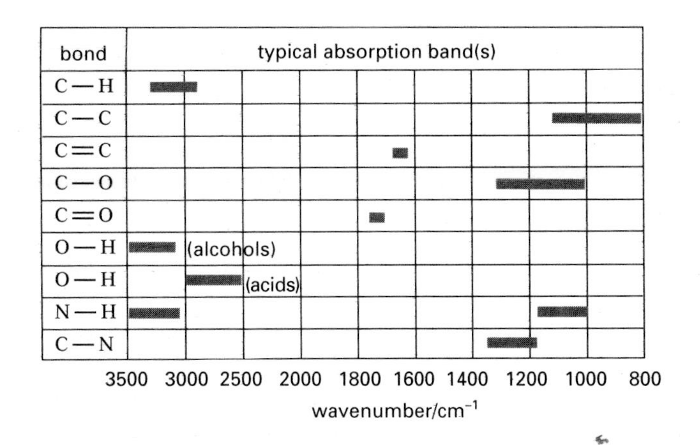

Activity 17.7

Use Table 17.3 (showing the characteristic absorption bands) where necessary to help you answer these questions.

1 What type of bond do you think is responsible for the absorption bands marked A in the spectrum of cyclohexene and B in the spectrum of ethyl ethanoate?

2 Vitamin C contains carbon, hydrogen and oxygen only. What structural information is suggested by the labelled absorption bands in the section of its IR spectrum shown in Figure 17.6m?

Whilst there are characteristic bond absorption bands below 1500 cm^{-1} these are less reliable indicators of specific bonds as this region also contains absorption bands which involve vibrations of larger parts of the structure. So, whilst the IR spectra of hexane and cyclohexane are very similar above 1500 cm^{-1} there are significant differences at lower wavenumbers. Since this part of the spectrum is determined by the structure as a whole rather than by individual bond types, it is referred to as the fingerprint region and acts as a unique identifier for any particular compound.

Cancer detection using infrared spectroscopy

Cervical cancer is traditionally diagnosed by expert microscopic examination of a smear test for the presence of abnormal cells. However, there are frequent reports of patients being recalled for re-testing owing to the poor reliability of the results which can be 50% or lower in some cases.

A new method based on infrared analysis of the smear is now being developed. Scientists in the USA and Canada have found significant differences in the infrared spectra of normal smears and those which are known to contain cancerous cells. The spectrum of a typical normal smear is compared with that of a cancerous smear in Figure 17.7. The peak at 1020 cm^{-1} is caused by glycogen. As you can see the amount of this compound is dramatically reduced in cancerous smears. Another difference occurs in the region between 1200 and 1350 cm^{-1} that contains two overlapping peaks due to phosphodiester groups in the cell's DNA. This grouping normally absorbs at about 1300 cm^{-1} but in cancerous cells the frequency is reduced to about 1250 cm^{-1} by hydrogen bonding to water molecules. Enlargement of this latter peak also indicates the presence of cancer cells. Smaller changes to the 'normal' spectrum may also indicate a pre-cancerous condition.

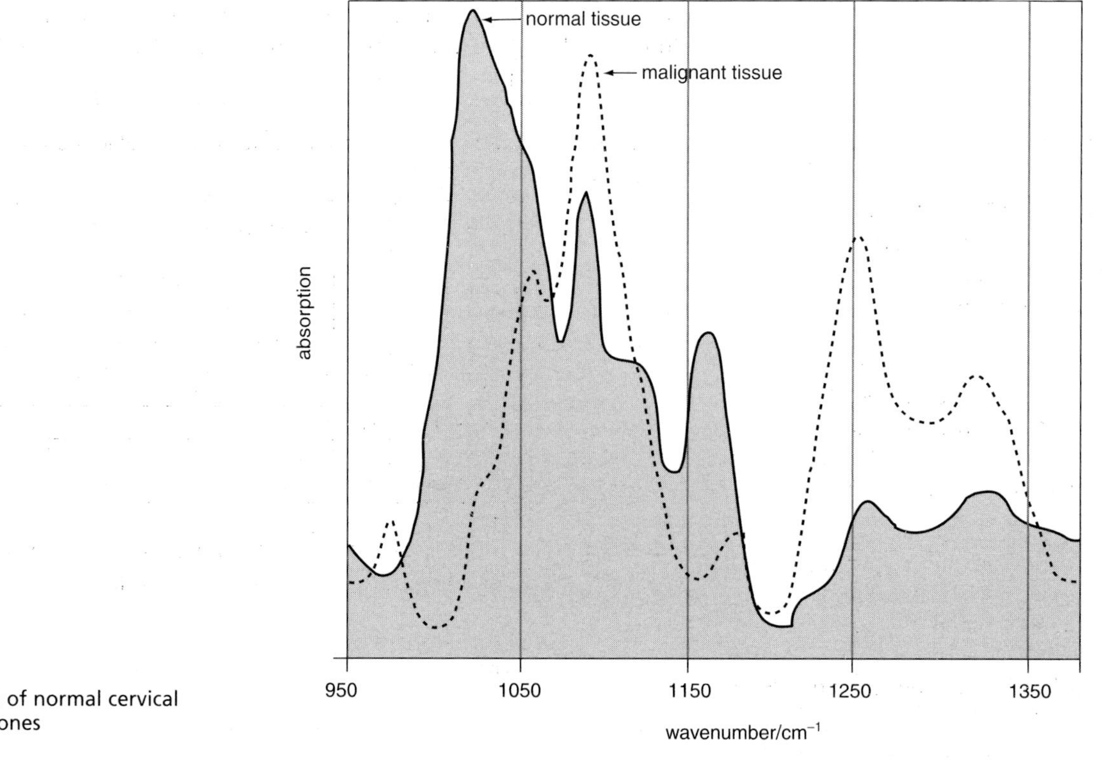

Figure 17.7
The infrared spectra of normal cervical cells and cancerous ones

Focus 17d

1. A molecule which can undergo a change of **dipole moment** by bond vibration will absorb infrared radiation.
2. Apart from the **fingerprint region**, which is characteristic of the molecule as a whole, most infrared absorption bands may be assigned to specific bond types and are therefore useful in structural analysis.

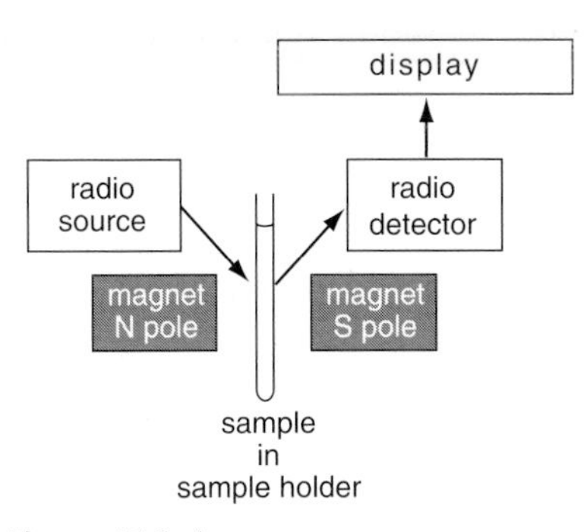

Figure 17.8 ▲
Principle of an NMR spectrometer

17.5 Nuclear magnetic resonance spectra

In this section we shall consider only proton magnetic resonance (PMR) spectra which is of universal application since ^1H is found in virtually all organic compounds. The general design of the spectrometer is outlined in Figure 17.8. The sample sits in a spinning tube between the poles of a strong electromagnet. Radio waves of a fixed frequency are passed into the sample chamber and the magnetic field strength is slowly varied. As particular nuclei resonate the strength of the radio signal received in the detector falls, producing a response on the display.

The positions of the PMR absorptions are measured relative to the signal produced by an internal standard added to the sample. Tetramethylsilane (TMS), $(CH_3)_4Si$, is commonly used for this as it is chemically unreactive and produces a large single peak well away from the proton signals normally found in organic compounds. In order that spectra from machines operating at different radio frequencies may be directly compared, signal positions are measured as **chemical shifts** on the δ scale in p.p.m. (parts per million) field strength difference from the TMS reference peak. The signal position of a proton depends upon its molecular environment and, as with IR spectra, reference tables of characteristic PMR bands, such as shown in Table 17.4, may be used to identify structural features.

Table 17.4 Characteristic PMR shifts (R = alkyl group, Ar = benzene ring)

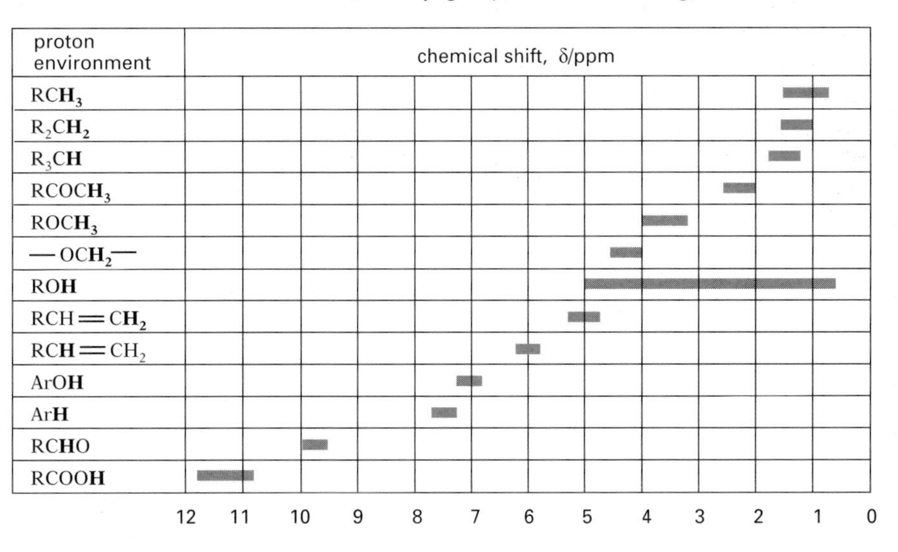

However, a PMR spectrum not only tells us how many different types of proton there are but also, by comparing the area under the peaks, their relative numbers. This is often indicated by an integration trace as shown in the low resolution PMR spectrum for ethanol in Figure 17.9.

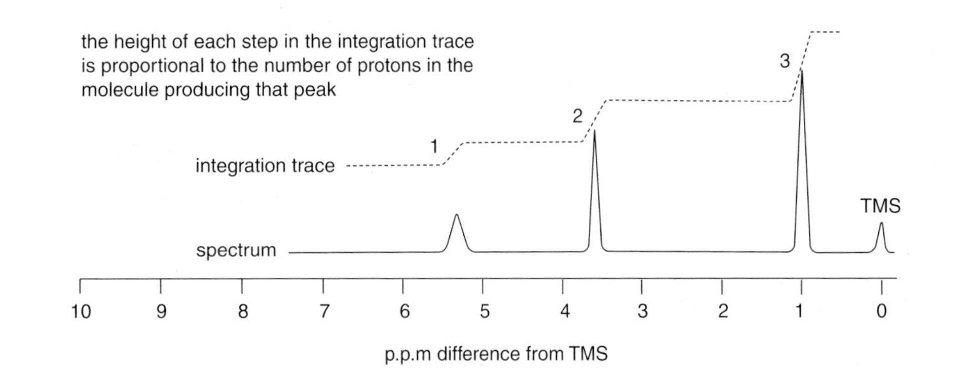

Figure 17.9 ▶
Integrated low resolution PMR spectrum of ethanol, CH_3CH_2OH. The height of each step in the integration trace is proportional to the number of protons in the molecule producing that peak

Confirmation that the small peak at $\delta = 5.4$ is due to the –OH proton is obtained by adding D_2O to the sample. D_2O is simply water in which the ^1H atoms are replaced by ^2H otherwise known as deuterium, D. The protons in –OH (and –OD) groups are **labile**, i.e. easily undergo exchange, so with ethanol and D_2O we get:

$$CH_3CH_2OH + D_2O \rightleftharpoons CH_3CH_3OD + HDO$$

Since deuterium nuclei do not undergo magnetic resonance, the area under the –OH peak in the spectrum is reduced on adding D_2O.

Activity 17.8

The low resolution PMR spectrum of a compound A with molecular formula $C_4H_8O_2$ is shown in Figure 17.10. Using the reference data in Table 17.4 and the integrated trace in the spectrum, suggest the structure of A.

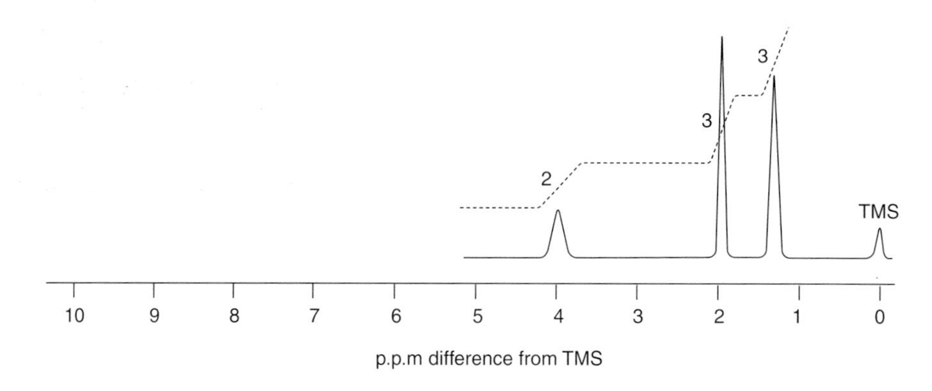

Figure 17.10
Low resolution PMR spectrum of compound A, $C_4H_8O_2$

At higher resolutions the single peak given by a group of equivalent protons is often seen to split into a characteristic pattern. An analysis of this **spin–spin splitting** or coupling can give further structural information. Spin–spin splitting arises from the fact that protons on **adjacent** carbon atoms may affect each others' signal.

Each proton may have either of two spins states which we shall call + and –. The number of possibilities for the overall spin for a set of protons may be worked out as shown in Table 17.5.

Table 17.5

Number of protons in the set	possible combinations of individual spins in this set	overall spin	number of combinations
1	+	+1	1
	–	–1	1
2	+ +	+2	1
	+ – or – +	0	2
	– –	–2	1
3	+ + +	+3	1
	+ + – or + – + or – + +	+1	3
	+ – – or – + – or – – +	–1	3
	– – –	–3	1

This means that a single proton may split the signal of adjacent protons into a 1:1 doublet, a pair of protons could cause a split into a 1:2:1 triplet and a set of three protons might produce a 1:3:3:1 quartet in adjacent protons. In other words, spin–spin coupling by an adjacent group of n protons will cause the signal to split into $n+1$ peaks. This is illustrated in Figure 17.11, which shows the high resolution PMR spectrum of ethanol.

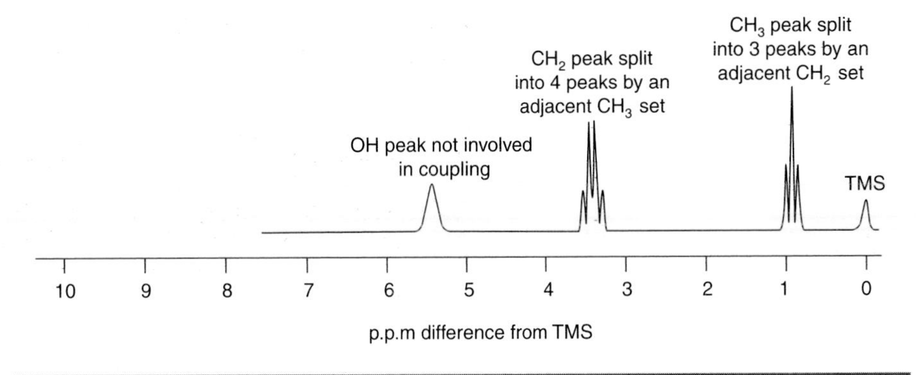

Figure 17.11 ► High resolution PMR spectrum of ethanol, CH_3–CH_2–OH

Activity 17.9

Deduce the structure of compound B which has the molecular formula C_4H_8O, from its high resolution PMR spectrum shown in Figure 17.12. Explain how you used the positions of the signals and their splitting patterns to arrive at your answer.

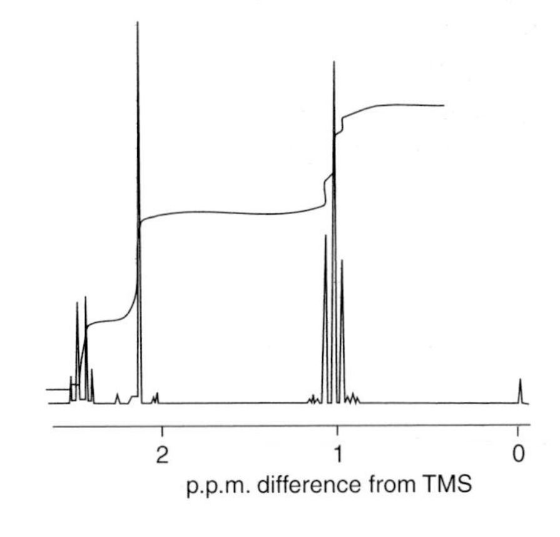

Figure 17.12 ► High resolution PMR spectrum of compound B, C_4H_8O

Focus 17e

1. Certain nuclei, including ^{1}H, have net spin and can exist in different energy states in a magnetic field.
2. In **proton magnetic resonance** (PMR) spectrometry, the nuclei are 'flipped' from one energy state to the other by absorption of radio waves.
3. The magnetic field strength required to make a particular proton 'flip' when exposed to fixed energy radio waves depends upon its environment within the molecule.
4. The relative size of the signal is directly proportional to the number of protons in that particular environment.
5. The signal for a particular proton type is split if there are hydrogen atoms attached to adjacent carbon atoms. Such **spin–spin splitting** provides useful structural information as ***n*** **neighbouring hydrogen atoms will split a signal into *n* +1 peaks**.

17.6 Structure analysis – putting it all together

Determination of structure usually involves a combination of some, or all, of the methods outlined in this chapter and Chapter 16. To illustrate this the next activity will lead you through a possible route for identifying an unknown organic compound. Other examples can be found in the study questions.

Activity 17.10

Compound C is known to contain carbon, hydrogen and oxygen only. Its mass spectrum and PMR spectrum are shown in Figure 17.13.

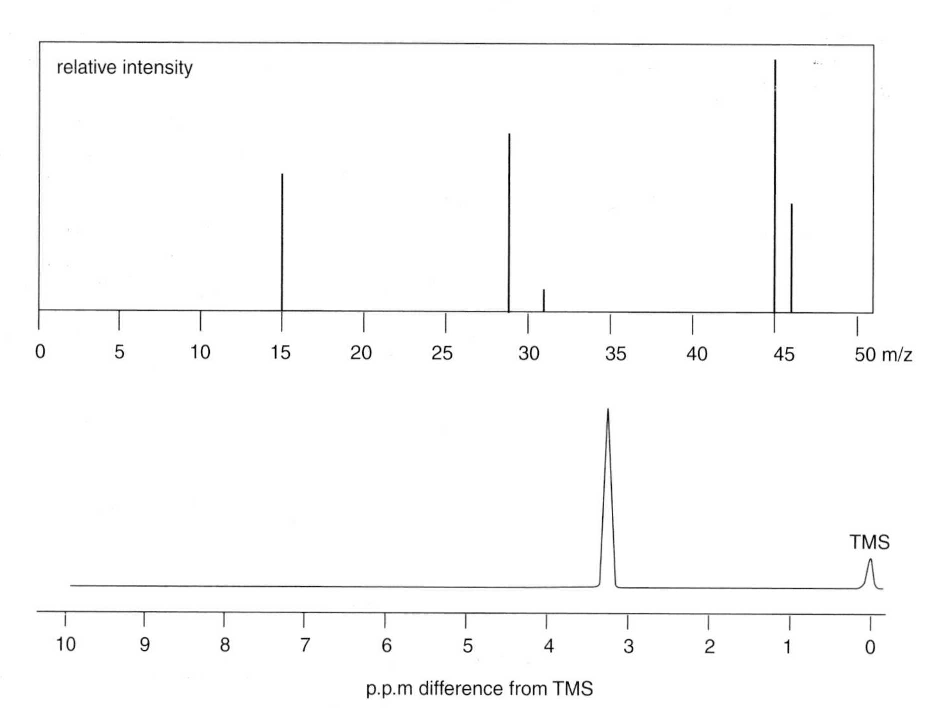

Figure 17.13
Low resolution mass spectrum of compound C

High resolution PMR spectrum of compound C

1. Assuming that the highest m/z peak in the mass spectrum is due to the molecular ion, what is the relative molecular mass of compound C?
2. Give two possible molecular formulae for C.
3. Elemental analysis shows that C contains 52.2% by mass of carbon. Which of the two possible molecular formulae in 2 above is correct?
4. Give one other way, apart from elemental analysis, that could have been used to establish the molecular formula of C.
5. Write two possible structures that match the molecular formula of C.
6. Explain how the PMR spectrum can be used to decide between the two possible structures you have given in 5 above.
7. Further evidence to support your answer to 6 above can be obtained from the IR spectrum of C. Explain how the IR spectrum of C would differ from that of the other structural isomer you identified in your answer to 5 above.

Questions on Chapter 17

1 BBC Radio 1 broadcasts at about 98 MHz on the FM band. Calculate the wavelength of this radio frequency.

2 BBC Radio 5 Live broadcasts on the medium wave band at wavelengths of 330 m and 431 m. What are the frequencies associated with each of these signals?

3 Chlorofluorocarbons, or CFCs, are chemically inert, volatile liquids that have been used for a variety of purposes. They are compounds containing carbon, fluorine and chlorine only. One of the simplest CFCs is dichlorodifluoromethane, CF_2Cl_2. The IR spectrum of this compound contains peaks corresponding to characteristic bond vibrations.

a) Explain which bond you would expect to vibrate faster, C—F or C—Cl.

b) What kind of atom in dichlorodifluoromethane would NOT be capable of undergoing nuclear magnetic resonance? Explain your answer.

4 Identify the different modes of vibration possible in the following molecules. In each case explain whether or not the particular vibrational mode will be infrared active.

a) O_2 **b)** CO **c)** H_2O **d)** HCN

5 The following polymers are commonly used as packaging in the food and drink industry.

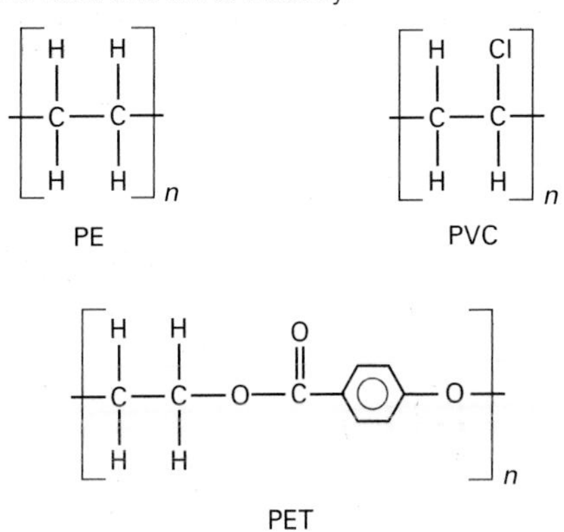

Match the IR spectra of the commercial plastics shown in Figure 17.14 with each of these polymers.

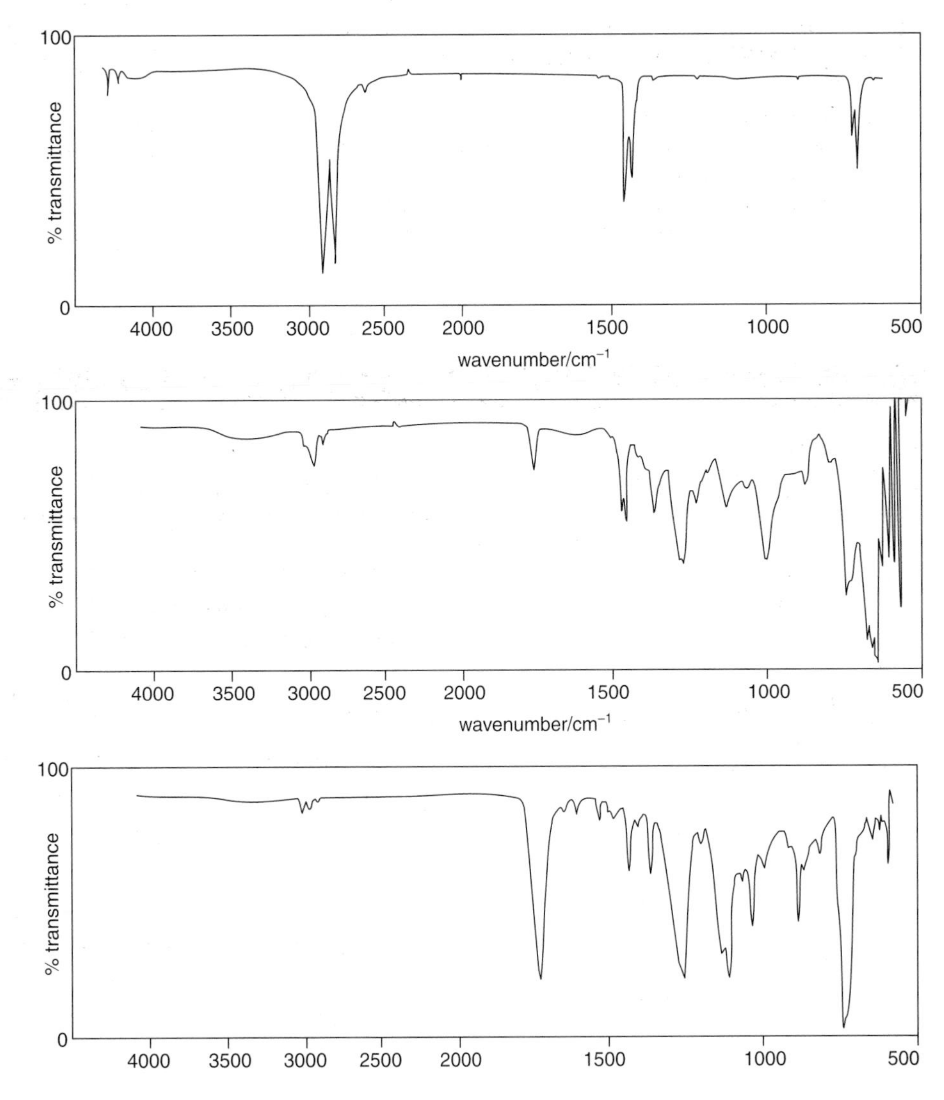

Figure 17.14a
IR spectrum of a plastic tomato ketchup bottle

Figure 17.14b
IR spectrum of cling film

Figure 17.14c
IR spectrum of a plastic lemonade bottle

Questions on Chapter 17 *continued*

6 When methanol, CH_3OH, is treated with an equal number of moles of ethene oxide, $(CH_2)_2O$, a product, X, is formed which has the PMR spectrum shown below.

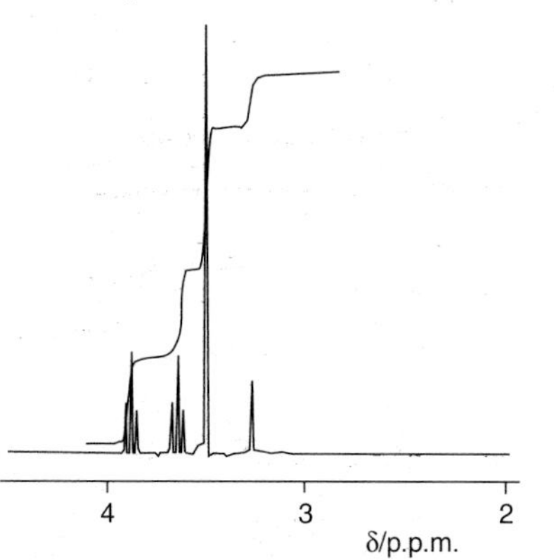

a) How many different types of proton are present in compound X?

b) What is the ratio of the numbers of each type of proton suggested by the integration trace?

c) Suggest a structure for X which would account for the following splitting pattern shown in the PMR spectrum of X: triplet at 3.9δ, triplet at 3.7δ, singlet at 3.5δ and singlet at 3.3δ.

d) What features would you look for in the mass spectrum and the IR spectrum of X as evidence to support your suggested structure?

e) An important detergent may be manufactured by treating X with a large excess of ethene oxide. The PMR spectrum of the product has peaks in the same position as X but the triplet signals are much stronger. What is the structure of this detergent?

f) A liquid used as antifreeze is obtained by treating ethene oxide with an excess of water. The PMR spectrum of this compound contains one single peak and a triplet which has double the area. What is the structure of the antifreeze molecule?

7 Infrared spectroscopy may used to estimate the concentration of ethanol in a driver's breath. However, care must be taken that other compounds that may be present do not affect the results. The IR spectrum of ethanol is shown in Figure 17.6f.

a) Ethanol shows a broad, strong IR absorption band at around 3300 cm^{-1}. What is this band due to and why would it be unsuitable for measuring the ethanol concentration in a driver's breath?

b) In fact the absorption band at about 2940 cm^{-1} is used to estimate ethanol concentration in breath. What causes this absorption band?

c) Other organic compounds also absorb in this region. For example, propanone may be found in the breath of people suffering from certain diseases or who have a poor diet. The amount of propanone present may be compensated for by measuring the intensity of a strong peak in the propanone spectrum that is absent from the ethanol spectrum. Suggest a peak in the IR spectrum of propanone, shown in Figure 17.6g, which could be used for this purpose. What bond is responsible for this absorption band?

Comments on the activities

Activity 17.1

1 94 MHz = 94 000 000 Hz.

$c = f\lambda$ so,

$300\,000\,000 \text{ m s}^{-1} = 94\,000\,000 \text{ waves s}^{-1} \times \lambda$

$$\text{so } \lambda = \frac{300\,000\,000}{94\,000\,000} = 3.19 \text{ m}$$

2 $c = f\lambda$ so,

$300\,000\,000 \text{ m s}^{-1} = f \times 1515 \text{ m}$

$$f = \frac{300\,000\,000}{1515} = 198\,000 \text{ s}^{-1},$$

i.e. approx. 198 000 Hz (or 198 kHz)

3 $$\text{wavenumber} = \frac{f}{100c}$$

for Radio 4 FM,

$$\text{wavenumber} = \frac{94\,000\,000 \text{ s}^{-1}}{30\,000\,000\,000 \text{ cm s}^{-1}} = 0.00313 \text{ cm}^{-1}$$

for Radio 4 LW,

$$\text{wavenumber} = \frac{198\,000 \text{ s}^{-1}}{30\,000\,000\,000 \text{ cm s}^{-1}} = 6.6 \times 10^{-6} \text{ cm}^{-1}$$

4 $c = f\lambda$ so,

$300\,000\,000 = 50\lambda$

$$\lambda = \frac{300\,000\,000 \text{ m s}^{-1}}{50 \text{ s}^{-1}} = 6\,000\,000 \text{ m (or } 6\,000 \text{ km)}$$

Activity 17.2

1 $E = hf$

So for red light,

$E = (6.626 \times 10^{-34} \text{ J s}) \times (430\,000\,000\,000 \text{ s}^{-1})$

Note: frequency must be in Hz

$= 2.85 \times 10^{-22}$ J

For violet light, $c = f\lambda$

$300\,000\,000 \text{ m s}^{-1} = f \times (400 \times 0.000001 \text{ m})$

Note: wavelength must be in metres

$$f = \frac{300\,000\,000}{0.000400} = 7.5 \times 10^{11} \text{ Hz}$$

$E = hf$ so for violet light:

$E = (6.626 \times 10^{-34} \text{ J s}) \times (7.5 \times 10^{11} \text{ s}^{-1})$

$= 4.97 \times 10^{-22}$ J

Comments on the activities *continued*

2 As you can see from above the quantum energy of radiation increases with frequency and decreases with wavelength. Figure 17.1 shows that ultraviolet radiation has a higher frequency (shorter wavelength) than infrared. It therefore carries more energy and is more likely to damage skin.

Activity 17.3 (extension)

1 You should have plotted the numbers in the second row of the following table against those in the bottom row. The value you obtain will depend upon exactly how you extrapolate the line in the graph to zero energy difference but it should be about 21.8×10^{-19} J

level number/*n*	2	3	4	5
energy of level (above lowest level) /10^{-19} J	16.340	19.367	20.426	20.917
energy difference (gap to next level) /10^{-19} J	3.027	1.059	0.491	0.266

2 Ionisation energy for 1 mole of atoms = ionisation energy for 1 atom × Avogadro's number
Taking the result from (1) above:
$= 21.8 \times 10^{-19}$ J atom^{-1} $\times 6.02 \times 10^{23}$ atoms mol^{-1}
$= 1\,312\,360$ J mol^{-1} , i.e. about 1310 kJ mol^{-1}
This figure is in good agreement with that given in chemical data books.

3 $E = hf$, so frequency of radiation $f = \frac{E}{h}$

$$= \frac{21.8 \times 10^{-19}\ \text{J}}{6.626 \times 10^{-34}\ \text{J s}} = 3.29 \times 10^{15}\ \text{Hz or } 3290\ \text{THz}$$

Activity 17.4

C—O, C—N, C=O, C≡N. For the single covalent bonds we would expect the C—N frequency to be higher than C—O as a nitrogen atom is lighter than an oxygen atom. The multiple bonds should vibrate more quickly with the stronger C≡N having a higher stretching frequency than C=O.

Activity 17.5

All of the vibration modes shown for the water molecule are infrared active. In a 'bent' molecule the changes in bond dipole for symmetric stretching are not in exactly opposite directions. They do not, therefore, cancel out and, since there is a change in the overall dipole moment of the molecule, infrared absorption occurs.

Activity 17.6

Only those nuclei with an odd mass number or an odd number of both protons and neutrons will have an overall non-zero nuclear spin and hence undergo magnetic resonance, i.e. ^{1}H, ^{19}F, ^{35}Cl and ^{37}Cl.

Activity 17.7

1 Absorption band A in the spectrum of cyclohexene is due to the C=C bond. (Note, this is absent in saturated hydrocarbons such as cyclohexane.) Band B in the spectrum of ethyl ethanoate is due to the carbonyl bond C=O in the ester group.

2 Absorption band C is characteristic of the carbonyl group C=O or C=C. Bands D, E and F could all be due to C—O vibrations although C—C bonds might be responsible for E or F. The structure of vitamin C is shown below.

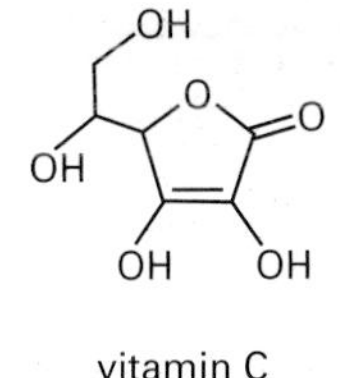

vitamin C
(ascorbic acid)

Activity 17.8

There are three sets of protons in the ratio 2:3:3. From the signal positions and the molecular formula these could be due to CH_2—O—C=O, CH_3—C=O and CH_3—C. The only structure consistent with this is ethyl ethanoate $CH_3COOCH_2CH_3$.

Activity 17.9

From the relative sizes of the steps in the integration trace there are the following sets of protons

set C: CH_2 at δ = 1.0
set B: CH_3 at δ = 2.1
set A: CH_3 at δ = 2.5

From the splitting patterns

- set B is a single peak unaffected by spin–spin splitting and, therefore, probably has no other protons immediately adjacent to it.
- set A is split into a triplet and so is adjacent to the CH_2 group.
- set C is split into a quartet and must be adjacent to the other CH_3 group.

This suggests the structure must be that of butanone:

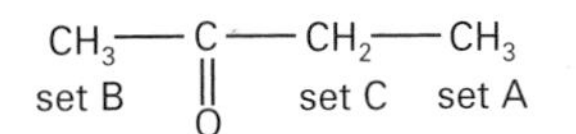

The positions of the peaks also fall within the ranges indicated by the typical absorption regions

Activity 17.10

1 The mass spectrum gives a value of 46 for the relative molecular mass of compound C.

2 Given that C contains carbon, hydrogen and oxygen only, possible combinations that produce a relative mass of 46 are: C_2H_6O and CH_2O_2.

Comments on the activities *continued*

3 The molecular formula C_2H_6O contains $\frac{24}{46} \times 100 = 52.2\%$ by mass of carbon so this must be the molecular formula of C.

4 Precise measurement of the mass of the molecular ion peak using high resolution mass spectrometry would also distinguish between similar molecular masses.

5 Possible structures which have the molecular formula C_2H_6O are: methoxymethane CH_3—O—CH_3 and ethanol CH_3—CH_2—OH.

6 C must be methoxymethane. All the protons in this molecule are in identical environments and so give rise to only a single peak in the PMR spectrum. There is no spin–spin splitting as the protons are not situated on adjacent carbon atoms. Ethanol has three distinct types of proton and gives the PMR spectra shown in Figures 17.9 and 17.11. The position of the peak in the PMR spectrum of compound C also falls within the region expected for protons in the CH—O—C environment.

7 The IR spectrum of methoxymethane would not contain a peak between 2500 and 3000 cm^{-1} characteristic of the O—H bond which would be present in the IR spectrum of ethanol.

CHAPTER

18

Colour chemistry

Contents

Study Checklist

After studying this chapter you should be able to:

1. State that white light is a mixture of all colours of the rainbow.
2. Explain that an object will appear 'coloured' if it reflects or transmits part of the visible spectrum and state that a 'black' object reflects no visible light, whilst a 'white' object reflects all visible wavelengths.
3. Explain how the Beer–Lambert law enables UV/visible spectrometry to be used for quantitative analysis.
4. Recall that in a free gaseous transition metal ion all five d orbitals are degenerate, i.e. have the same energy.
5. Explain how the approach of ligands during complex formation breaks this degeneracy.
6. State that radiation of quantum energy equal to the energy splitting of the d orbitals will be absorbed by a transition metal complex. If this falls in the visible region, the complex will appear coloured.
7. Realise that the observed colour of a complex will depend upon the degree to which the d orbitals are split which in turn depends upon the type(s) of the ligand, the co-ordination number and the shape of the complex.
8. Explain how transition metal complexes with a d^0 configuration may be coloured if the oxidation state of the metal is high enough to cause charge transfer, i.e. an electron jumping from the ligand to the metal.
9. Explain how σ and π bonding molecular orbitals are formed by the overlap of suitable atomic orbitals.
10. State that each bonding molecular orbital (σ or π) has a corresponding anti-bonding molecular orbital (σ^* or π^*) at higher energy and that electrons may be promoted from the bonding to the corresponding anti-bonding orbital by absorbing radiation of suitable quantum energy.
11. Recall that lone pairs are non-bonding electrons (n) that may also be promoted into anti-bonding orbitals.
12. State that transitions to σ^* orbitals generally involve UV absorption.
13. Give examples of coloured systems which involve the following types of transition:

    $n \longrightarrow \pi^*$ (e.g. azobenzene) and $\pi \longrightarrow \pi^*$ (e.g. carotenes)
14. State the meaning of the following terms: chromophore, auxochrome, bathochromic shift and hypsochromic shift.
15. Explain the difference between a pigment and a dye and state that the incorporation of a 'solubilising' group, such as $—SO_3Na$, into the structure of a pigment may convert it into a dye.
16. Recognise the structural features of the following types of dye: azo, anthraquinone, triarylmethane and indigoid.

17 Explain the mechanisms by which dyes may be attached to fibres and give examples of the following types: acid, basic, direct, disperse, vat and fibre-reactive.
18 Give examples of the following types of pigment: inorganic, azo, phthalocyanines and those derived from vat dyes.
19 Outline the use of pigments in the paint and printing industries.
20 Explain the origin of fluorescence and give an example of its use in 'optical brightening'.
21 Explain the origin of chemiluminescence and state a use for such 'cold' light.

18.1 What causes colour?

'White' light, such as sunlight, is a mixture of all visible wavelengths or colours as shown in Figure 18.1.

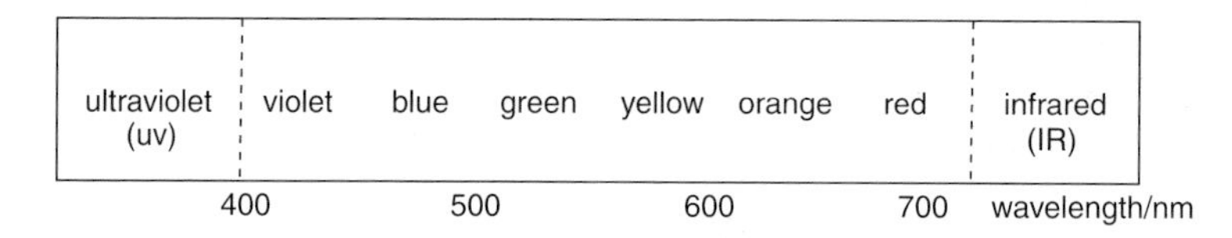

Figure 18.1 ▶
The visible spectrum

Any object that reflects, or transmits, all of these wavelengths therefore appears white or colourless. An object that absorbs all visible light and transmits none, looks black. Coloured objects are those which only absorb part of the visible spectrum as shown in the simplified diagram in Figure 18.2

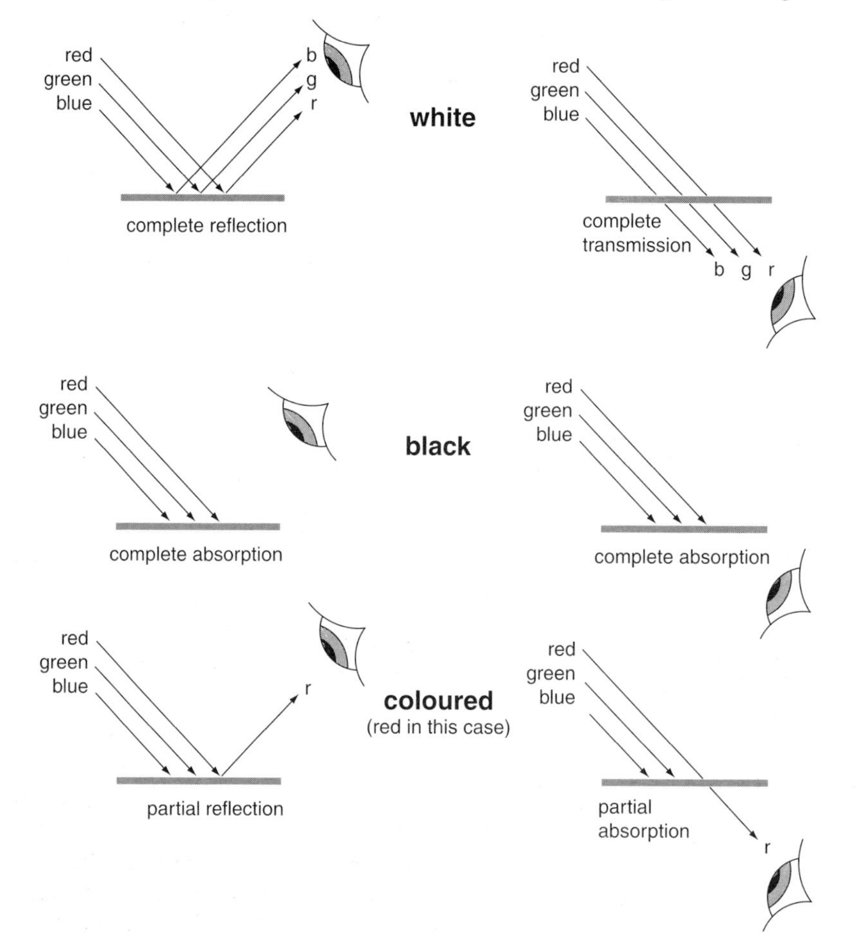

Figure 18.2 ▶
Colour and absorption of visible radiation

Activity 18.1

1. From the absorption spectra shown in Figure 18.3 and from Figure 18.1, identify those samples which, when viewed in white light, are likely to appear
   a) white or colourless
   b) black
   c) red
   d) blue.
2. If the samples were viewed in blue light only, which additional sample would appear black? (Note that visible and UV spectra are generally drawn in terms of ascending absorption rather than ascending transmittance, i.e. compared to IR spectra they appear 'upside down'.)

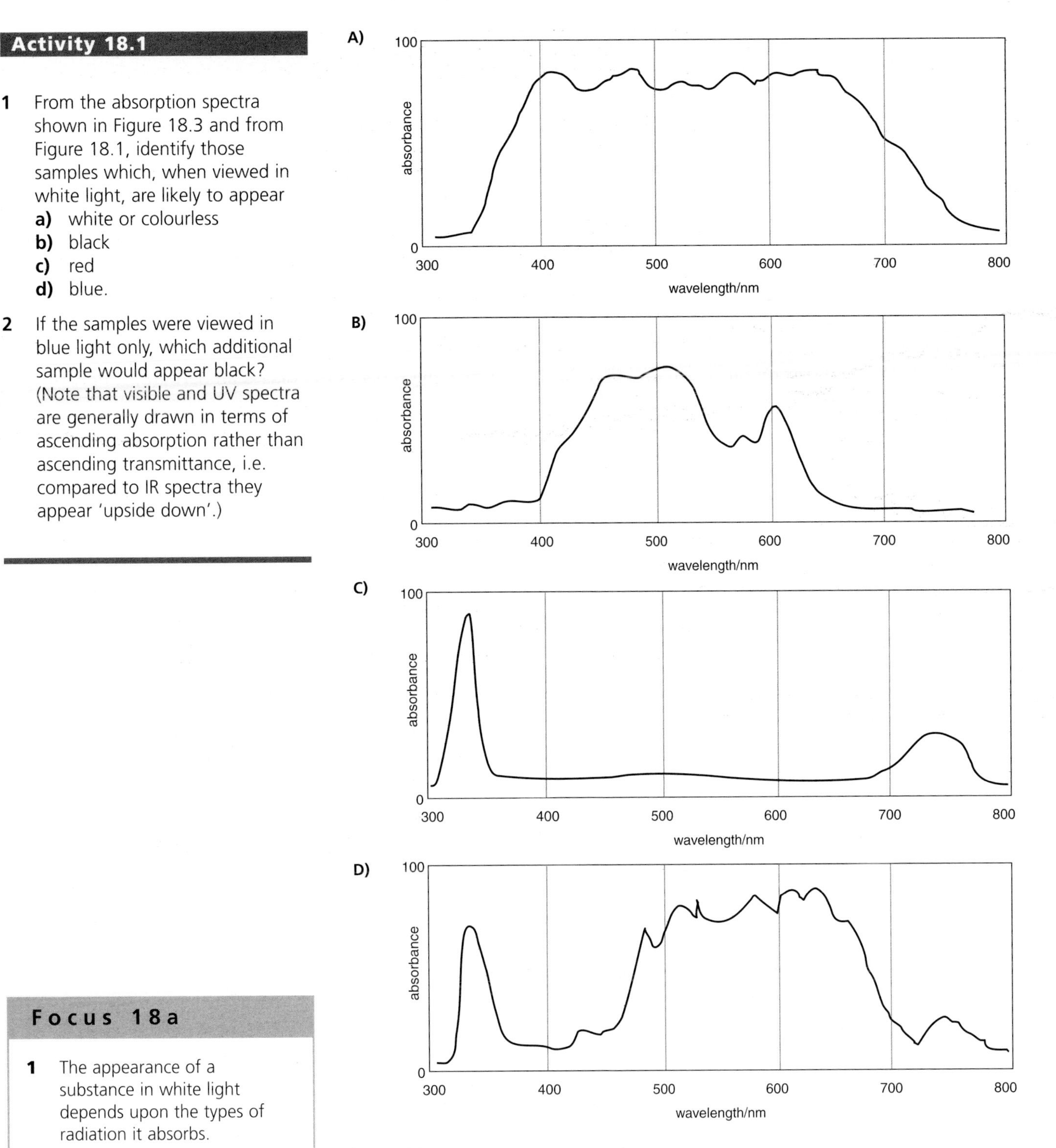

Figure 18.3
Absorption spectra for activity 18.1

Focus 18a

1. The appearance of a substance in white light depends upon the types of radiation it absorbs.
2. A black material absorbs all visible wavelengths.
3. A white material absorbs no visible wavelengths.
4. If a material absorbs some but not all visible wavelengths, its colour is a mixture of the visible radiation which is reflected or transmitted.

As with atomic spectra (covered in Chapter 17), a molecule can only absorb visible or ultraviolet radiation if an electron in its structure can be promoted from one level to another that differs in energy by an amount equal to the quantum energy of the radiation. The next two sections will look at examples of how this can arise in both inorganic and organic compounds.

Quantitative spectroscopy

The concentration of a sample in solution may be determined by measuring the transmittance of the sample at a wavelength, λ_{max}, corresponding to the position of maximum absorption. If the intensity of the incident light is I_0 and that of the light transmitted by the sample is I, then the percentage transmittance, T, is given by 100 I/I_0. This value will decrease as either the sample length or its concentration is increased. However, the relationship is not linear in either case.

However, if we define sample **absorbance**, A, as $\log(I_0/I)$ then, over a certain range, we find that this quantity is directly proportional to the sample length and its concentration. In this region the sample is said to obey the **Beer–Lambert law** and we may write the following expression:

$$A = \varepsilon lc$$

where, l is the sample length in cm, c is the molar concentration and ε is a constant known as the **molar absorption coefficient**, i.e. the absorbance produced by a solution of 1 M concentration and sample length 1 cm. If the molar absorption coefficient of a substance has been determined then it may be used to calculate the concentration of any sample, providing it obeys the Beer–Lambert law, simply by measuring its length and absorbance.

Activity 18.2

The variation in absorbance at a certain wavelength with concentration for solutions of compound X with a sample length of 2 cm is shown in the table below.

absorbance	0.27	0.54	0.82	1.08	1.35
molar concentration	0.010	0.020	0.030	0.040	0.050

1. Draw a graph to show that the samples obey the Beer–Lambert law.
2. From the graph, calculate a value for the molar absorption coefficient of X at this wavelength.
3. Estimate the molar concentration of a solution of X that has an absorbance of 0.55 at a sample length of 4 cm.
4. What is the percentage transmittance of the sample in 3 above?

18.2 Colour in transition metal complexes

Many, but not all, compounds of d block elements are coloured. Exceptions include scandium(III) compounds with an empty 3d sub-level and zinc(II) compounds which have a completely full 3d sub-level. Most d block compounds that contain a metal ion with a partially filled d sub-level are, however, coloured. Take, for example, copper(II) sulphate which dissolves in water to form a blue solution. We can explain the origin of this blue colour by considering the electronic structure of the Cu^{2+} ion shown in Figure 18.4.

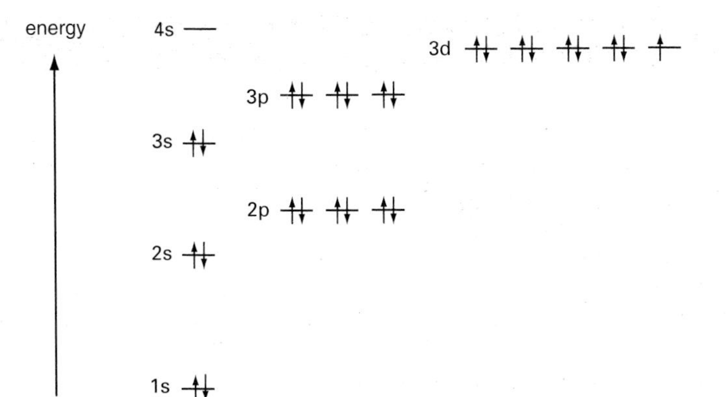

Figure 18.4
Electronic structure of Cu^{2+}

Here electrons may be promoted from the 3p into the 3d or from the 3d into the 4s. Neither of these transitions would, however explain why Sc(III) and Zn(II) compounds are white or colourless. With an empty 3d sub-level the 3p $\longrightarrow$ 3d transition is possible and with a full 3d sub-level we can still promote 3d $\longrightarrow$ 4s.

This leaves us only one possibility here for the origin of the blue colour; movement of an electron *between* the orbitals of the 3d sub-level. However, as the above diagram shows, in an isolated Cu^{2+} ion all five 3d orbitals are **degenerate**, i.e. have exactly the same energy. In this situation no energy is required to move one 3d electron into the half-filled orbital and so no absorption occurs. However, if we evaporate a solution of copper(II) sulphate to dryness and drive off all the water, the blue colour is lost and we are left with a white residue of anhydrous copper(II) sulphate. This suggests that the presence of water is needed to produce the colour, a fact confirmed by the fact that the blue colour is restored on adding water to anhydrous copper(II) sulphate. In Chapter 22 we explain what happens when a metal ion is dissolved in water. A lone pair from the oxygen on a water molecule may be donated to the metal ion to form a co-ordinate covalent (or dative) bond. In most cases, six water molecules may attach themselves to the metal ion.

The lone pair donor, water in this case, is called a **ligand**, the number of ligands attached to the metal is the **co-ordination number** (six in this case) and the resulting species, $[Cu(H_2O)_6]^{2+}$, is known as a **complex**:

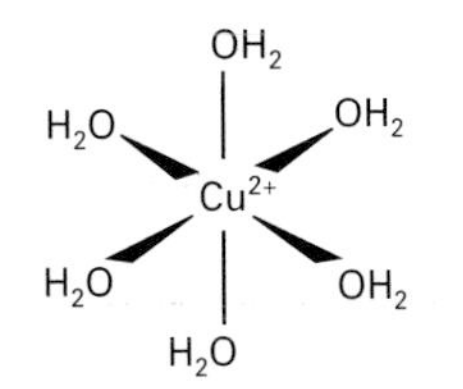

When the lone pairs on the ligands approach, they repel the electrons present in the metal d orbitals whose energy therefore rises. However, the octahedral shape of the complex ion means that the ligand lone pairs get much closer to two of the d orbitals than they do to the remaining three. The degeneracy of the d orbitals is broken and this energy level splits into two parts separated by the ligand field splitting energy, Δ, as shown in Figure 18.5. Promotion of an electron from the lower to the higher level within the 3d sub-level now involves absorption of energy.

If the ligand field splitting energy falls within the quantum energy range of visible radiation, then the complex will absorb light of corresponding wavelengths. The absorption spectrum for the $[Cu(H_2O)_6]^{2+}$ complex is shown in Figure 18.6a. Although maximum absorption occurs at about 800 nm, the band is very broad with measurable absorption down to a wavelength of about 600 nm. This broadening of the absorption band is caused by bond vibration. As the ligands move with respect to the metal ion, so the splitting energy will vary.

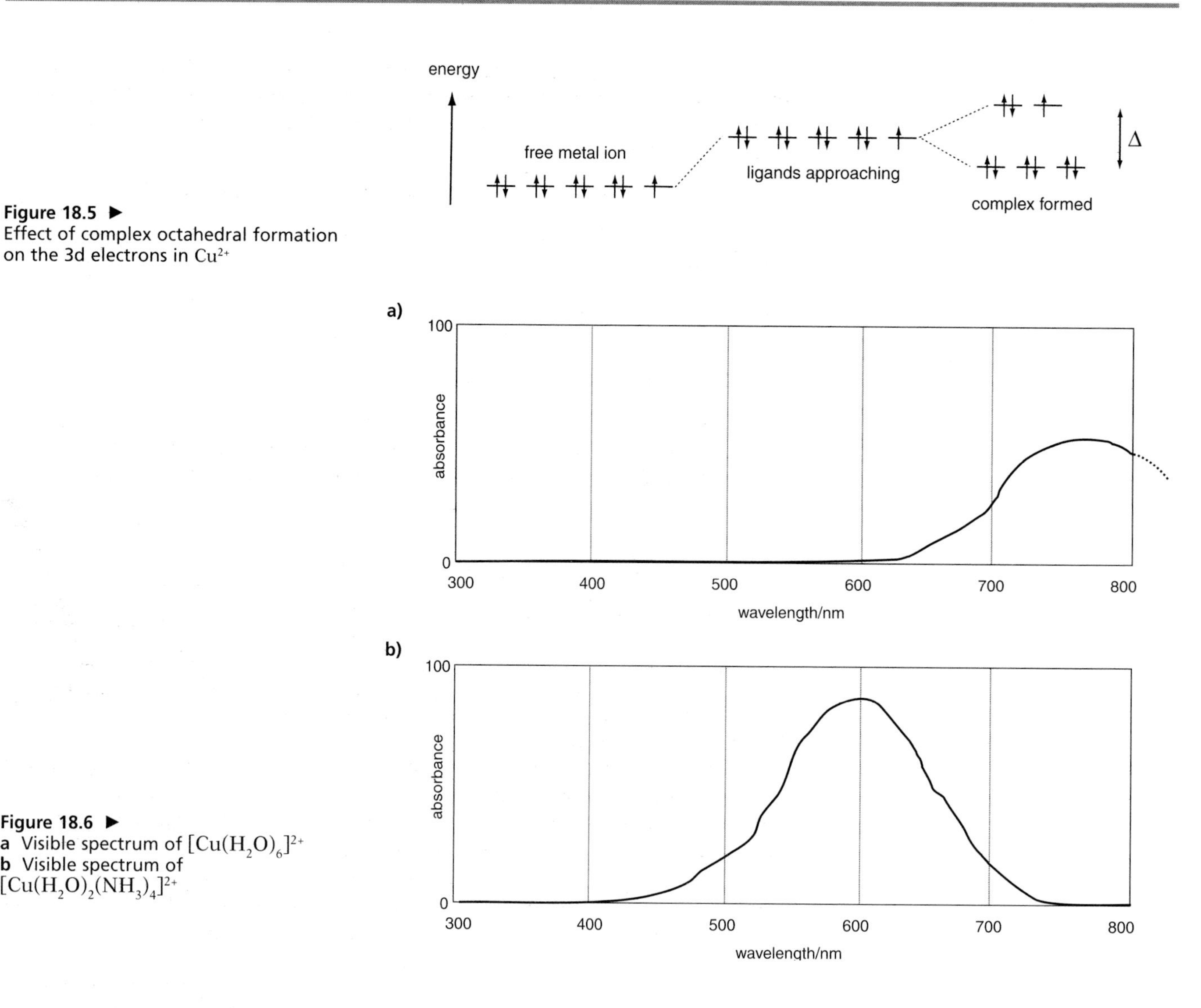

Figure 18.5 ▶
Effect of complex octahedral formation on the 3d electrons in Cu^{2+}

Figure 18.6 ▶
a Visible spectrum of $[Cu(H_2O)_6]^{2+}$
b Visible spectrum of $[Cu(H_2O)_2(NH_3)_4]^{2+}$

When an excess of concentrated ammonia solution is added to a solution containing $[Cu(H_2O)_6]^{2+}$, the colour changes from pale blue to a characteristic deep royal-blue owing to the formation of a new complex ion:

$$[Cu(H_2O)_6]^{2+}(aq) + 4NH_3(aq) \longrightarrow [Cu(H_2O)_2(NH_3)_4]^{2+}(aq) + 4H_2O(l)$$

The complex is still octahedral in shape but four of the water ligands have been replaced by ammonia:

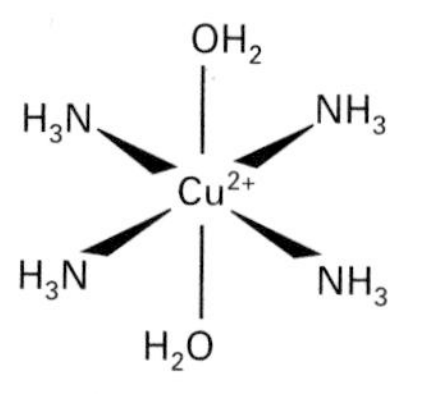

Maximum absorption now occurs at around 600 nm as shown in Figure 18.6b, i.e. at a shorter wavelength than for $[Cu(H_2O)_6]^{2+}(aq)$.

This means that the ligand field splitting energy has been increased by substituting ammonia ligands for water. The following **spectro-chemical series** lists some common ligands in order of their d orbital splitting ability.

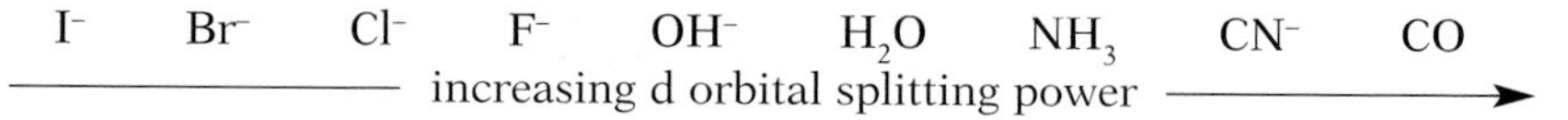

It is not only the type of ligand that affects splitting energy, but also the co-ordination number and the shape of the complex. Addition of concentrated hydrochloric acid to a solution containing $[Cu(H_2O)_6]^{2+}$ causes a colour change from blue to yellow-green:

$$[Cu(H_2O)_6]^{2+}(aq) + 4Cl^-(aq) \rightleftharpoons [CuCl_4]^{2-}(aq) + 6H_2O(l)$$

The new complex has only four ligands, arranged tetrahedrally around the metal ion:

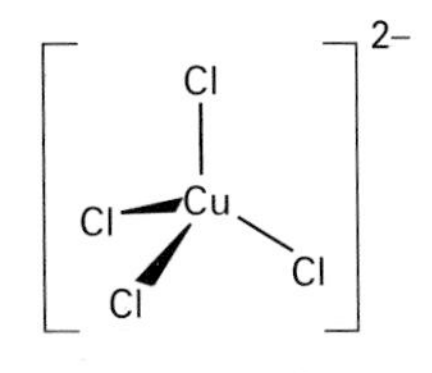

The d orbitals on the Cu^{2+} ion are again split into two sets but this time the pattern is reversed with three orbitals at a higher energy than the other two:

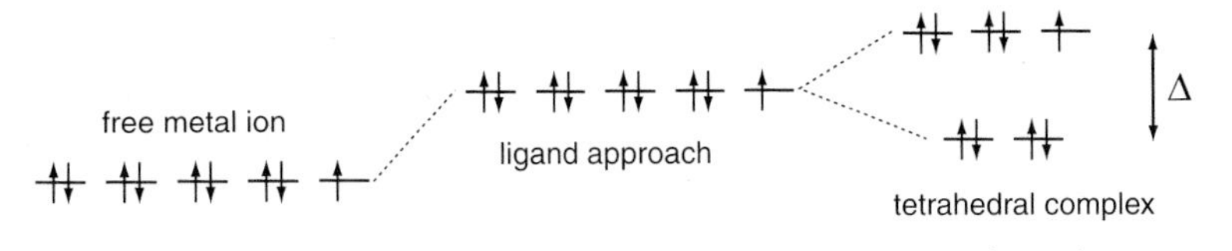

As there are only four ligands here, the splitting energy, Δ, is less than in a 6-co-ordinate octahedral complex. Combined with the weaker splitting caused by Cl^- ions compared to NH_3 and H_20 ligands, maximum absorption moves to lower energy, i.e. higher wavelength.

So far we have explained colour in transition metal compounds in terms of movement of electrons within a partially filled d sub-level which has been split by a ligand field. However there are coloured compounds in which the d sub-level is empty. This generally occurs when the oxidation state of the central metal ion is high, for instance manganate(VII), MnO_4^-, which is an intense purple. The complex is tetrahedral and may be considered as a central highly charged metal ion surrounded by four oxide ion ligands:

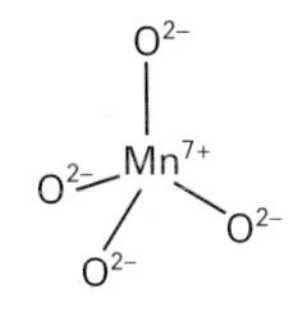

The attraction here is so strong that electrons on the ligands may jump across into the orbitals on the central metal ion. The energy difference between the ligand orbitals and the metal orbitals falls into the visible spectrum so the complex appears coloured. This effect is known as **charge transfer** and can give very intense absorption bands, as shown for $MnO_4^-(aq)$ in Figure 18.7.

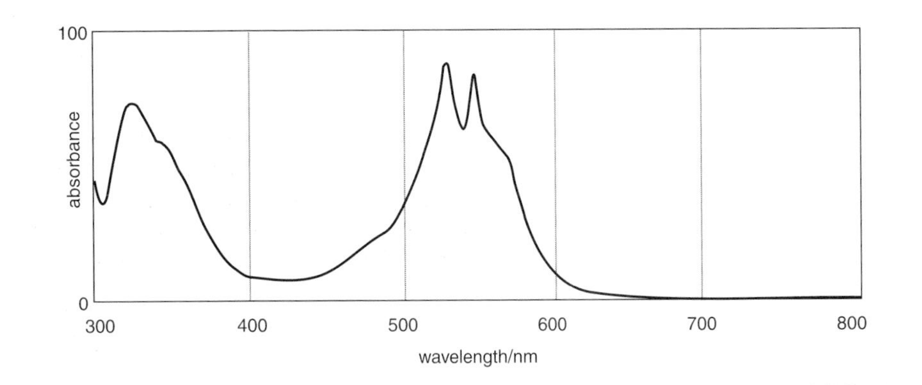

Figure 18.7
Visible spectrum of MnO_4^- due to charge transfer

Focus 18b

1. In transition metal complexes the d orbitals in the metal are not all at the same energy.
2. This d orbital splitting is caused by electrons in the metal being repelled by the ligands that approach some orbitals more closely than others.
3. The extent to which the energies of the d orbitals are split depends upon the geometry of the complex and the nature of the ligands.
4. If the **ligand field splitting energy** falls into the same region as that of visible light, the complex will absorb this radiation and hence appear coloured.
5. Complexes that contain a transition metal in a high oxidation state may be intensely coloured even if the d sub-level is empty. This is caused by **charge transfer**, i.e. electron transfer between the ligand and the central metal ion.

Activity 18.3

1. In dilute aqueous solution cobalt(II) chloride contains the complex ion $[Co(H_2O)_6]^{2+}(aq)$. On addition of concentrated hydrochloric acid, ligand exchange occurs to give $[CoCl_4]^{2-}(aq)$.

$$[Co(H_2O)_6]^{2+}(aq) + 4Cl^-(aq) \rightleftharpoons [CoCl_4]^{2-}(aq) + 6H_2O(l)$$

   a) What is the electron configuration of Co^{2+}?
   b) Both of the above complex ions are coloured. What causes this?
   c) Which of the above complexes has the higher ligand field splitting energy? Explain your answer.
   d) One of these cobalt complexes is pink, whilst the other is blue. Explain, with reasons, which is which.

2. The highest oxides of the metals at the start of the first row of the d block are as follows;

$$Sc_2O_3 \qquad TiO_2 \qquad V_2O_5 \qquad CrO_3 \qquad Mn_2O_7$$

   a) Deduce the oxidation state of the metal in each case and state the number of electrons in its 3d sub-level.
   b) How do you explain the fact that Sc_2O_3 and TiO_2 are white, whereas the other oxides in the list are highly coloured?

18.3 Colour in organic compounds

In Chapter 4, we described covalent bonding in terms of the overlap of suitable atomic orbitals to give σ or π bonding molecular orbitals. Here the maximum electron density is located between the two nuclei, so drawing them together by mutual attraction. A dot and cross diagram of the ethane molecule, C_2H_6, shows that each carbon atom shares one electron with the other to form a single bond.

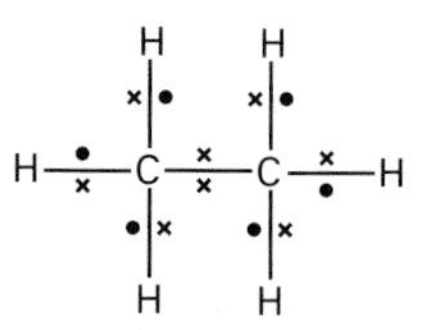

We may picture the overlap between atomic orbitals on the carbon atoms as follows:

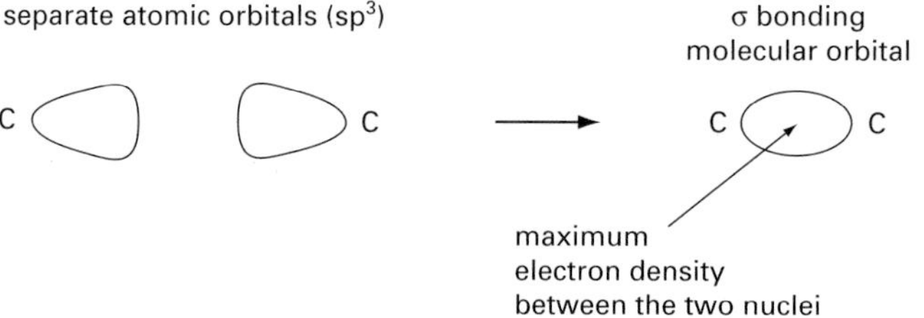

However, the two atomic orbitals may combine in a different way to give the molecular orbital shown below.

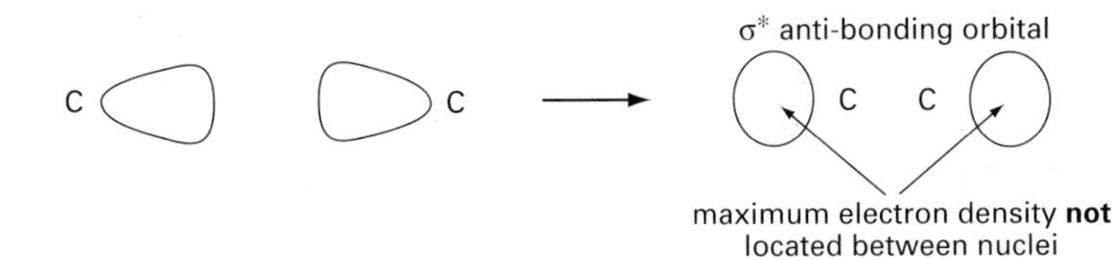

This is referred to as an anti-bonding orbital (denoted by *) since attraction between the two nuclei and the areas of maximum electron density now pull the atoms apart rather than bind them together. Whenever a covalent bond is formed, there is an alternative anti-bonding orbital available. Since energy is released when atoms bond, the bonding orbital is energetically more stable than the original atomic orbitals, whilst the anti-bonding orbital is less stable.

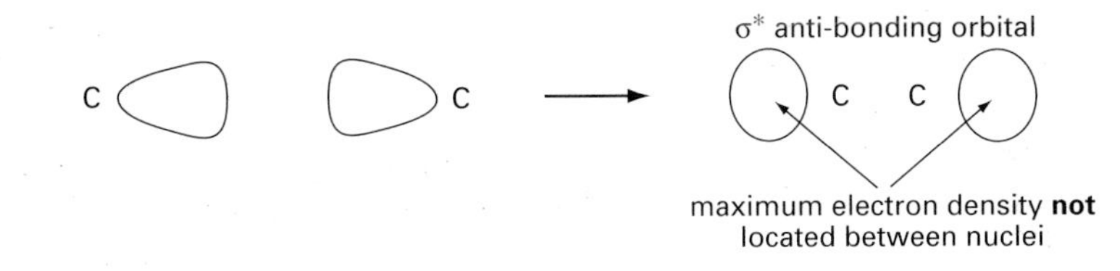

The two electrons occupy the lower energy σ (bonding) orbital.

σ* —

energy

σ ⇅

If the ethane molecule can absorb a quantum of radiation of just the right energy, electrons will be promoted from the σ bonding state to the σ^* anti-bonding state. The carbon atoms will then fly apart, destroying the molecule.

Exactly the same situation arises when atomic orbital overlap leads to the formation of a π bond, for example in ethene, C_2H_4:

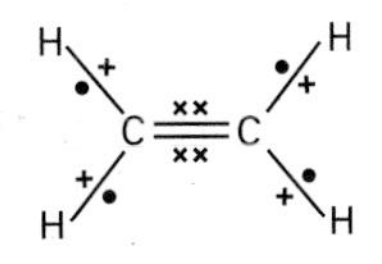

An anti-bonding π^* molecular orbital is available at higher energy. Since the attraction of the nuclei for the shared electron density in a π bond is generally rather weaker than that in a σ bond, the energy difference between a π–π^* pair of molecular orbitals is less than that between a σ–σ^* pair. Thus in the case of ethene, C_2H_4, we may represent the energy states available to the electrons bonding the two carbon atoms as follows:

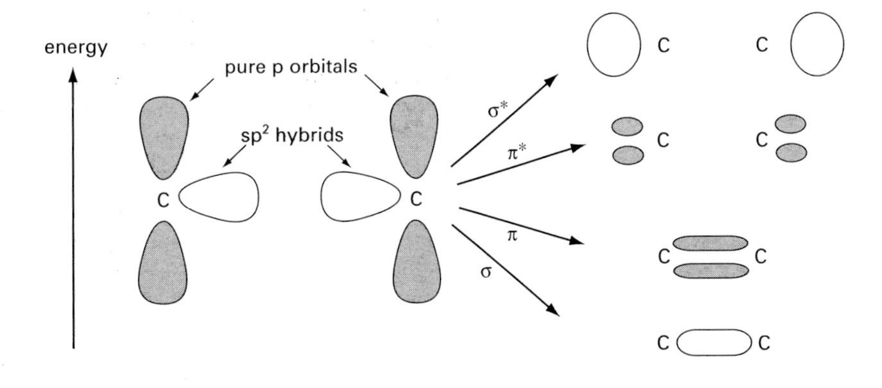

Electrons fill up the available molecular orbitals in order of increasing energy. Thus for the two carbon atoms in ethene we have,

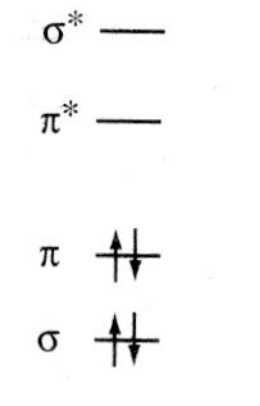

The following electron promotions are now possible; $\sigma \longrightarrow \sigma^*$, $\sigma \longrightarrow \pi^*$, $\pi \longrightarrow \pi^*$, $\pi^* \longrightarrow \sigma^*$. In the case of the $\pi \longrightarrow \pi^*$ transition, the strong σ bonding is unaffected and the molecule remains intact.

Similar σ and π overlap occurs in the carbonyl group (C=O) in, for example, methanal, H_2CO. Here, however, a dot and cross diagram shows that there are two lone pairs of electrons in the valency shell of the oxygen atom.

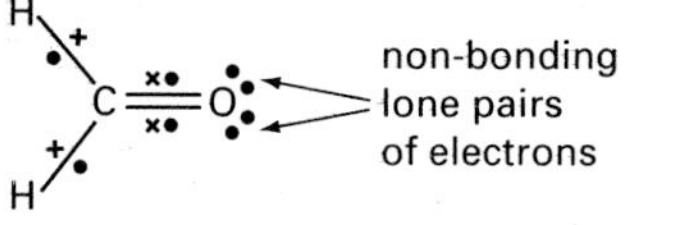

These electrons are referred to as **non-bonding** and, in an energy level diagram, such orbitals are labelled n.

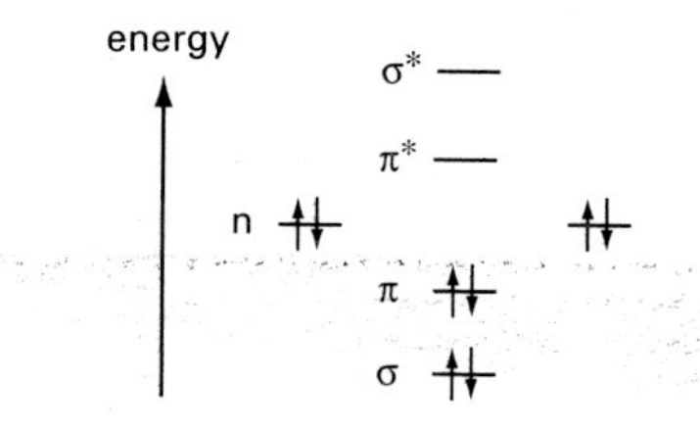

Extra transitions are now possible between the non-bonding n orbitals and the anti-bonding π^* and σ^* orbitals.

If an electron can be promoted from one state to a higher one by absorbing a quantum of visible radiation then the compound will appear coloured. Transitions to σ^* molecular orbitals generally require higher energies and correspond to absorptions in the ultraviolet region. However, in certain circumstances $n \longrightarrow \pi^*$ and $\pi \longrightarrow \pi^*$ transitions can give rise to colour. Although this means that organic compounds without multiple bonds are likely to be colourless, it does not follow that all compounds with such bonds will be coloured. Ethene, for example, is colourless although it contains a C=C double bond. The $\pi \longrightarrow \pi^*$ absorption band here is well into the ultraviolet region with λ_{max} at 165 nm. The structure of 1,3-butadiene may be drawn as containing two C=C bonds:

$$H_2C{=}CH{-}CH{=}CH_2$$

Maximum absorption in this compound occurs at about 215 nm, still in the ultraviolet but closer to the visible region. This shift is caused by the possibility of electron delocalisation in compounds containing alternate C=C and C—C bonds. In such a **conjugated system** the π molecular orbital in one C=C unit may interact with its neighbours to produce a new set of orbitals covering more than one pair of atoms.

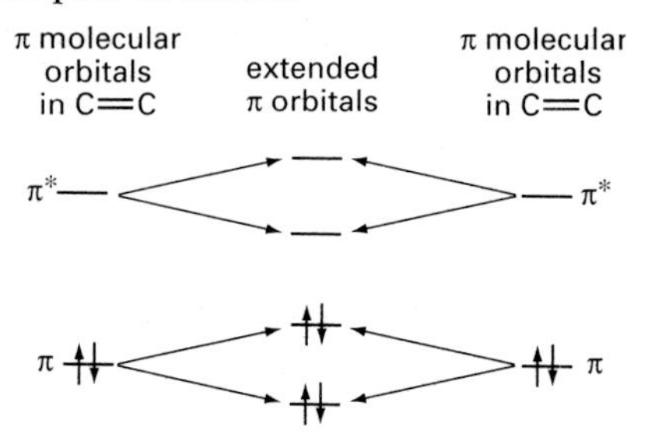

As you can see this produces more energy levels and therefore a greater number of possible transitions, including some which are of lower energy than that in the unconjugated ethene molecule. The number of possible extended molecular orbitals increases with the number of double bonds in the conjugated chain and eventually a number of the absorption bands will enter the blue end of the visible spectrum. Figure 18.8 shows the visible spectrum of β-carotene, which is responsible for the orange colour in carrots.

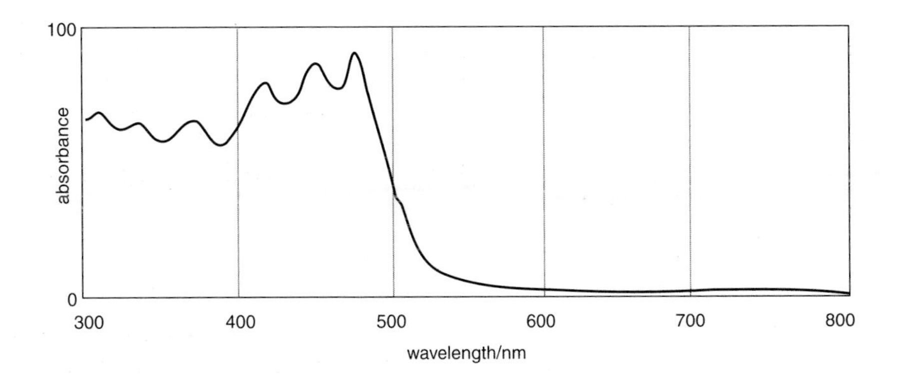

Figure 18.8
Visible spectrum of β-carotene

This molecule has a conjugated chain containing eleven C═C bonds in its structure.

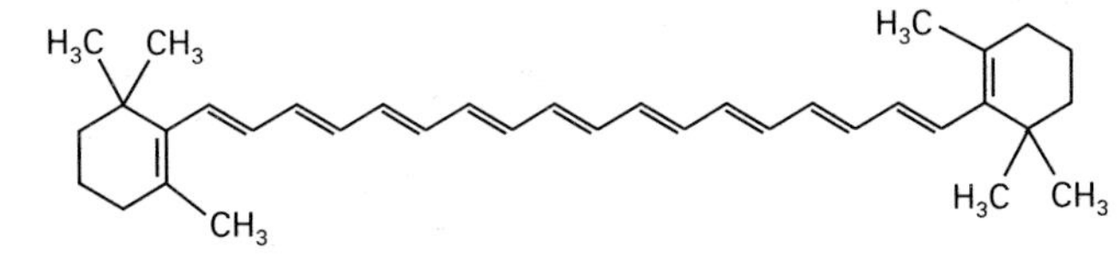

The colour of β-carotene is due to the presence of the C═C double bonds. Any group, such as this, which has the potential to produce colour is known as a **chromophore** and the substance which contains it is referred to as a **chromogen**. Other common chromophore groups containing multiple bonds include the following:

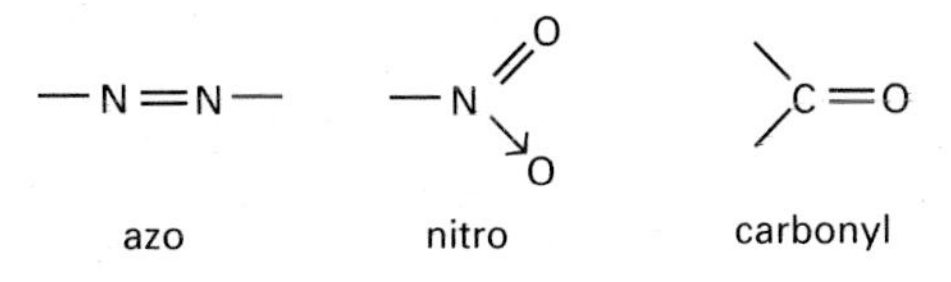

Chromogens that contain only a small number of chromophore groups may not be coloured since there may be insufficient conjugation to cause absorption in the visible region.

Activity 18.4 Colour in aromatic compounds

The structure of benzene is described in Chapter 9. The molecule contains a delocalised ring of six electrons in an extended π molecular orbital above and below the plane of the carbon atoms. In terms of absorption, this behaves like a conjugated system containing three C═C bonds.

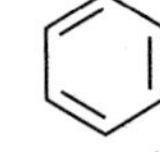

Benzene shows a series of absorption bands, the highest wavelength of which is at about 260 nm.

Focus 18c

1. Overlap of an atomic orbital from each of two adjacent atoms gives rise to a **bonding molecular orbital** and an **anti-bonding molecular orbital** at a higher energy.
2. In any system of alternating single and double bonds, such overlap extends over several atoms and leads to a more complex molecular orbital pattern.
3. If the degree of **conjugation** is sufficient then the energy required to promote an electron may fall into the visible region, leading to a coloured compound.
4. The colour of an organic compound may be modified by incorporating molecular groupings which alone do not generate colour.

Activity 18.4 Colour in aromatic compounds continued

1. Is benzene coloured? Explain your answer.
2. What kind of transition is responsible for the absorption bands in the electronic spectrum of benzene?

Azobenzene is a yellow solid. Its molecular structure consists of two benzene rings joined by an azo group.

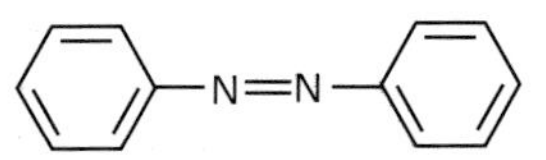

3. How many different types of chromophore does azobenzene contain?
4. How many alternate single and double bonds are present in an azobenzene molecule?
5. What extra type of electronic transition is possible in azobenzene that cannot occur in benzene?
6. Give two reasons that might explain why the absorption bands of azobenzene reach much higher wavelengths than those in benzene.

An **auxochrome** is a group which, although it may not contain a multiple bond, can still influence colour by shifting the absorption peak due to a chromophore to a higher or lower wavelength and may possibly also increase molar absorption. Many auxochromes contain atoms such as nitrogen or oxygen which possess lone pairs of electrons which may then be promoted into π^* anti-bonding molecular orbitals provide by the chromophore. Some common auxochromes are:

$$—OH, —NH_2 \text{ and } NR_2 \text{ (where R is a hydrocarbon group)}$$

In most cases inclusion of an auxochrome moves the absorption peak to a longer wavelength. This is known as a **bathochromic** effect and results from the provision of extra electronic transitions. A shift in absorption in the opposite direction, i.e. towards shorter wavelengths, is referred to as a **hypsochromic** shift (see Figure 18.9).

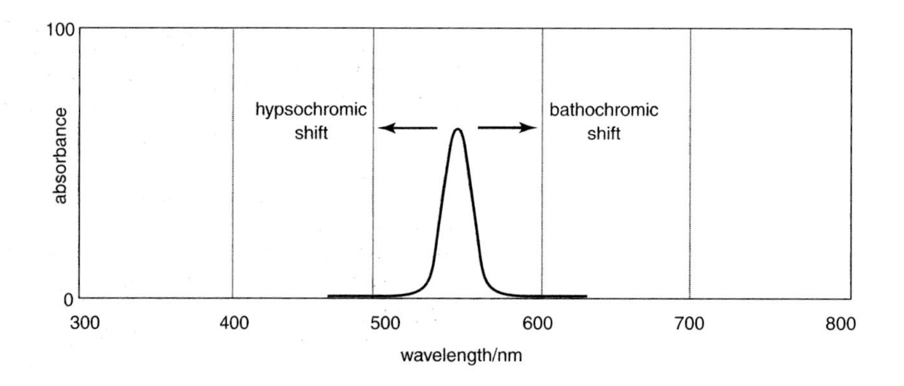

Figure 18.9
Absorption shifts caused by auxochromes

18.4 Dyes and pigments

Dyes and pigments are substances that are of practical use in the coloration of other materials. Not only must they have the desired colour but they should resist fading during washing and exposure to light or chemicals. Dyes and pigments may share similar colour-producing structures but, essentially, dyes are applied in solution whereas pigments are used as finely dispersed insoluble solids. Certain dyes may be converted into pigments by reducing their solubility and, conversely, the introduction of a 'solubilising' group, such as $—SO_3^-Na^+$, into a pigment molecule may produce a dye.

One way of classifying dyes is according to the type of colour-producing structure they contain. **Azo dyes** are based upon the azobenzene structure.

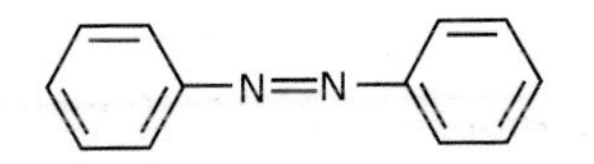

Two commercial azo dyes are shown in Figure 18.10.

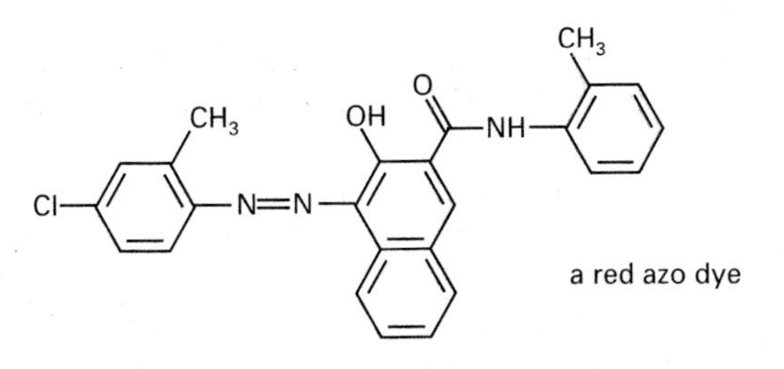

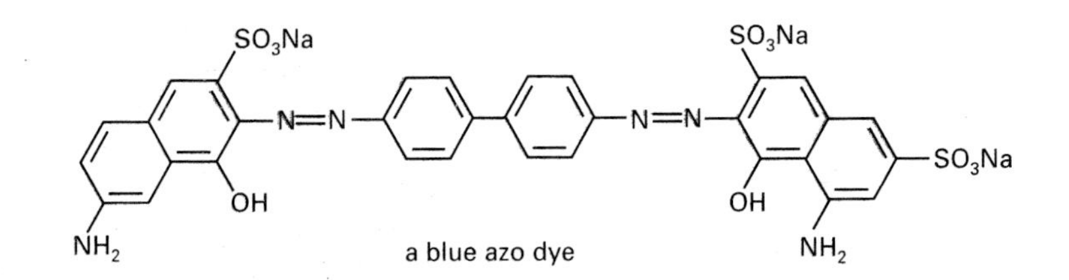

Figure 18.10
Structures of two commercial azo dyes. Can you identify the auxochromes present in each of these dyes? What do you think causes the difference in colour? Which dye is likely to be more soluble in water, and why?

Unless an azo dye contains polar or ionic groupings such as $—SO_3Na$, it is likely to be sparingly soluble in water. However, these dyes are generally formed in the material itself by the reaction of a solution containing a diazonium salt with a solution of a phenol or an amine (see Chapter 14) If the insoluble material is prepared beforehand it may be used as a pigment rather than a dye.

Triarylmethane dyes are produced from compounds based on triphenylmethane, for example:

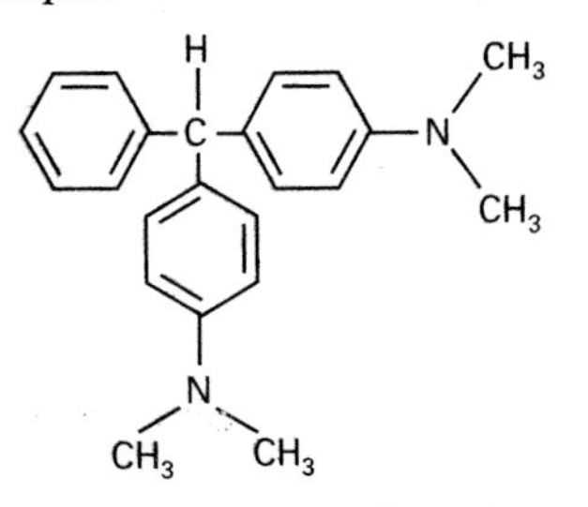

This compound is itself colourless and is referred to as a **leuco-base**. Colour is developed by oxidising this to a **colour-base**, followed by addition of acid to give the **dye**; in this case malachite green.

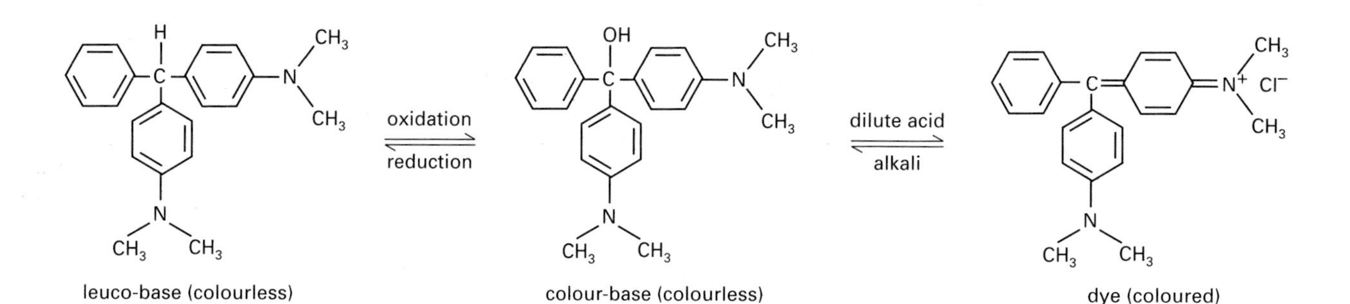

leuco-base (colourless) colour-base (colourless) dye (coloured)

Can you explain why the dye is coloured whilst both the leuco-base and colour-base are colourless?

Anthraquinone dyes are based on the following structure:

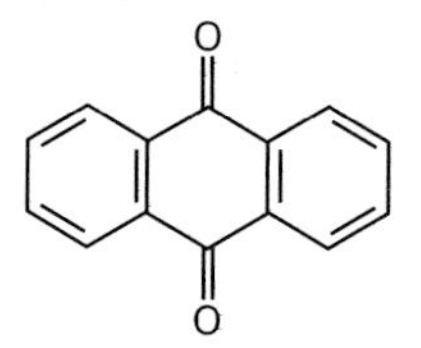

Unless polar or ionic solubilising groups are present, such compounds are rather insoluble. However, reduction under alkaline conditions converts the C=O groups into C—O^-Na^+, forming a soluble colourless leuco-base that, after application, may be re-oxidised to form the dye.

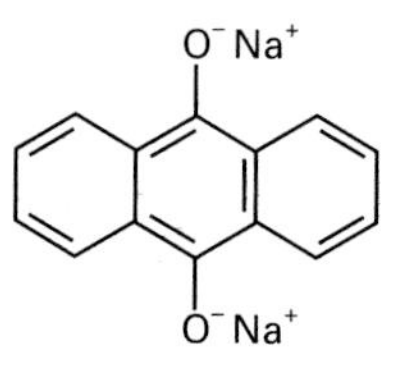

Indigoid dyes also contain carbonyl, C=O, groups and may similarly be reduced under alkaline conditions to give soluble leuco-bases.

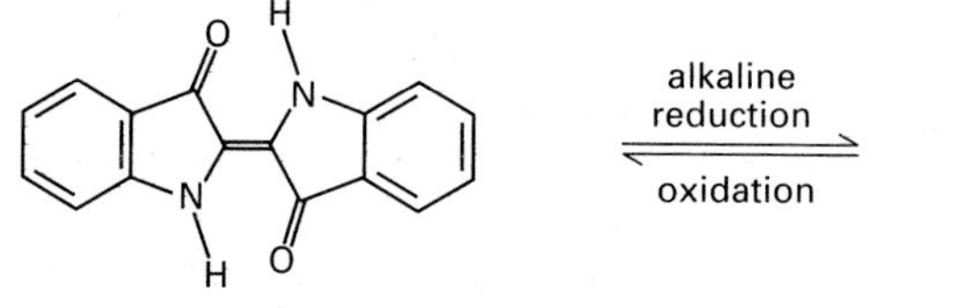

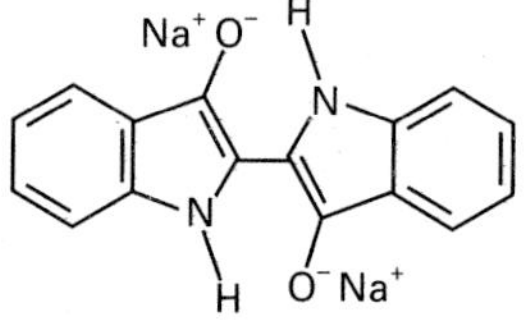

indigotin (coloured and insoluble) leuco-base (colourless and soluble)

Phthalocyanine pigments and dyes contain a metal ion co-ordinated to a conjugated chromophoric ligand. The colour may be varied by introducing different auxochromes into the ligand and by changing the metal ion.

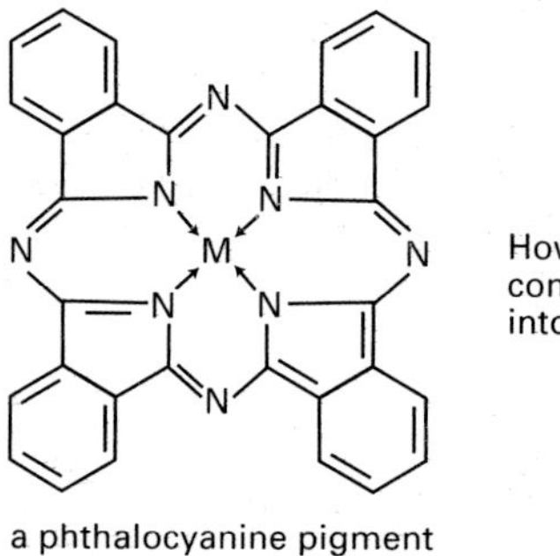

How might you convert this pigment into a soluble dye?

a phthalocyanine pigment
(M is a metal ion, e.g. Cu^{2+} or Mg^{2+})

Since dyes are generally more soluble than pigments, they may be washed out of fibres and fabrics unless they bind in some way to the polymer structure. The nature of this binding depends upon the chemical structure of the fibre, which may contain particular functional groups that can act as dye attachment sites.

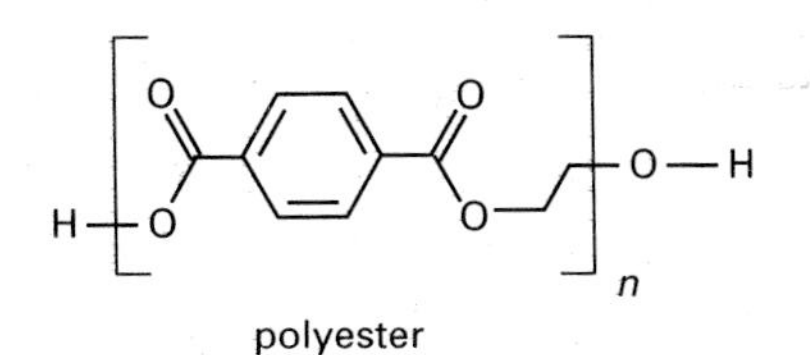

polyester

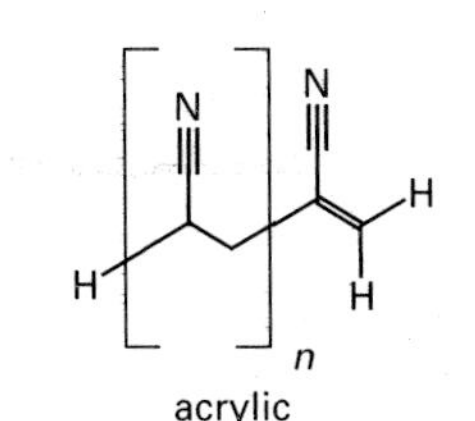

acrylic

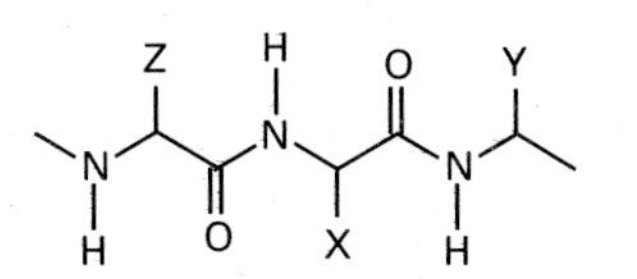

protein, e.g. wool and silk
(X, Y and Z are various side groups)

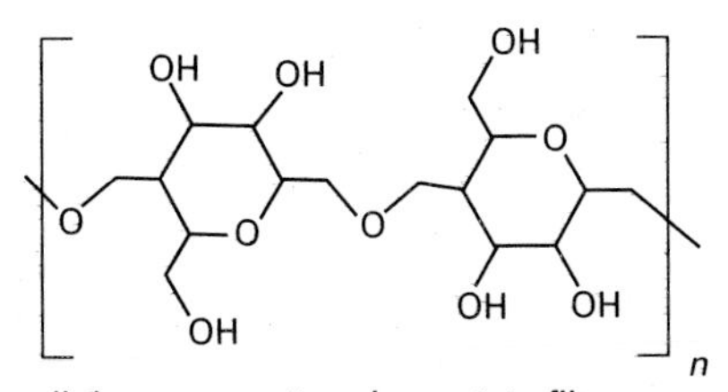

cellulose, e.g. cotton; in acetate fibres some or all of the — OH groups are converted into the ethanoate ester, — $COCH_3$

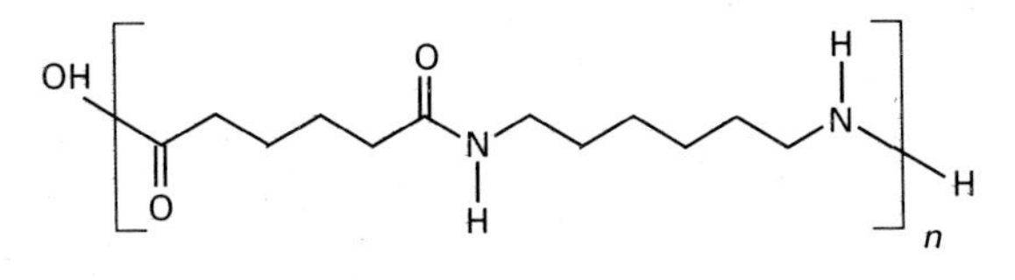

polyamide

Figure 18.11
The chemical structures of some natural and synthetic fibres

A variety of polymer structures are shown schematically in Figure 18.11. Cellulosic fibres, including cotton, linen and rayon, contain hydroxy groups, — OH. Wool and silk are proteins which contain the amide linkage, —CONH—, which is also present in synthetic polymers such as nylon-6 and nylon-6,6. Polyesters, such as terylene, contain the —COO— linkage, whereas polyalkene fibres have only a carbon backbone, with any functional groups present only as side chains.

Acid dyes contain the **sulphonic acid group**, $—OSO_3H$ (usually as the sodium salt). This forms a salt with basic groups such as —NH—, giving $—OSO_3^-H_2N^+—$ and will therefore bind strongly to protein fibres and synthetic polyamides.

Basic dyes contain **amino groups**, $—NH_2$, —NHR or NR_2 (or their hydrochloride salts) which attach readily to acid groups. Since these are not normally present in the fibre structure, they must be introduced using a suitable *co-monomer* during the polymerisation process. Thus the introduction of the sulphonic acid salt (shown below) during the manufacture of poly(ethyleneterephthalate) gives a basic dyeable polyester.

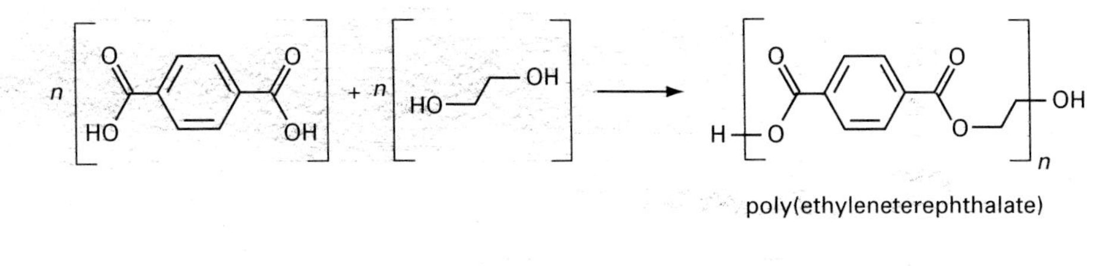

poly(ethyleneterephthalate)

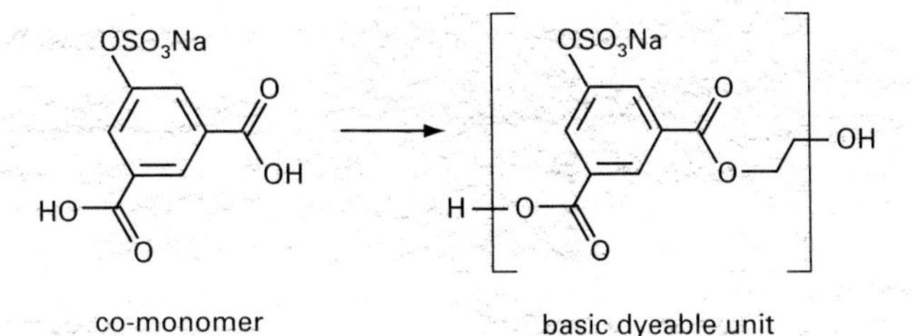

co-monomer

basic dyeable unit

Cellulose-based fibres, such as cotton, contain many hydroxyl groups capable of binding to electronegative atoms, such as nitrogen, on dye molecules by *hydrogen bonding*. In order to maximise this attraction, **direct dyes** usually consist of long molecules containing two or more azo, —N=N—, groups.

Disperse dyes are much less soluble than those mentioned above and so, once they penetrate the fibre structure, are less readily washed out. They are particularly useful for synthetic fibres with few acidic or basic groups, such as polyester, which can withstand the relatively high processing temperatures necessary to open up the polymer structure sufficiently to allow dye penetration. Simple **mono-azo** dyes i.e. those containing only one azo group are generally favoured in disperse dyeing as small molecules can penetrate polymer structures more easily than large ones.

Triarylmethane, anthraquinone and indigoid dyes, which are also usually relatively insoluble, may be applied by **vat dyeing**. In this process, the insoluble dye is reduced to the more soluble, colourless leuco-base which readily penetrates cellulose fibres. Oxidation by air, or a chemical oxidising agent such as potassium dichromate(VI), reconstitutes the insoluble dye.

Fibre-reactive dyes have excellent wash-fastness as they combine chemically with fibre molecules, becoming attached by strong covalent bonds. The hydroxyl groups in cellulose fibres condense with trichlorotriazine under alkaline conditions.

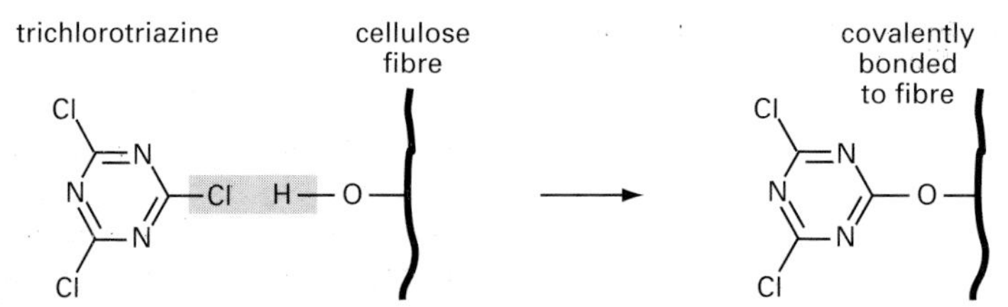

If one of the other chlorine atoms on the trichlorotriazine is replaced by a dye molecule, then this becomes firmly attached to the cellulose fibre. To aid fibre penetration the 'dye' portion generally contains several solubilising groups.

Activity 18.5

1 From the range of structures shown in Figure 18.12 select **one** example in each case, which:

Figure 18.12 ▼
Structures for use with Activity 18.5

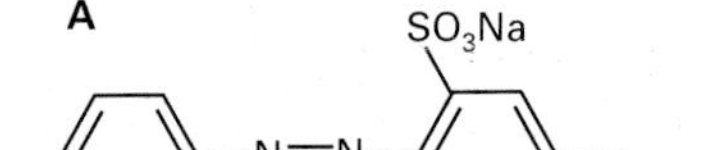

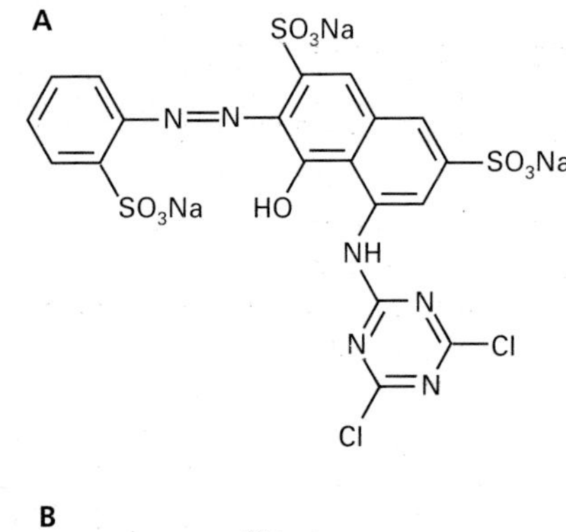

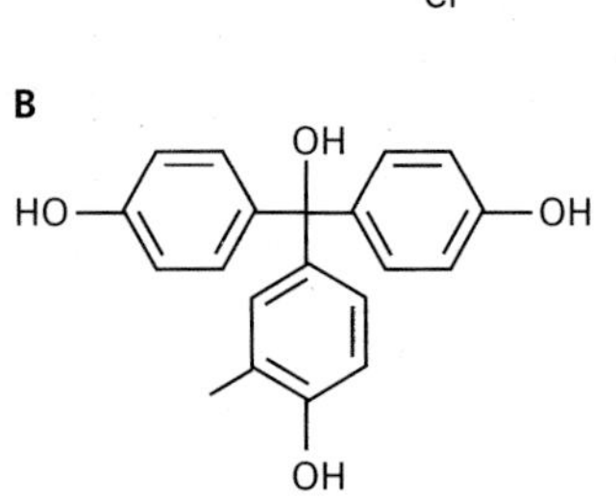

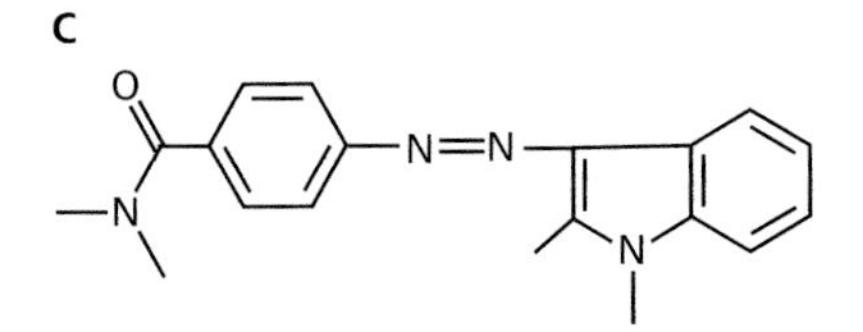

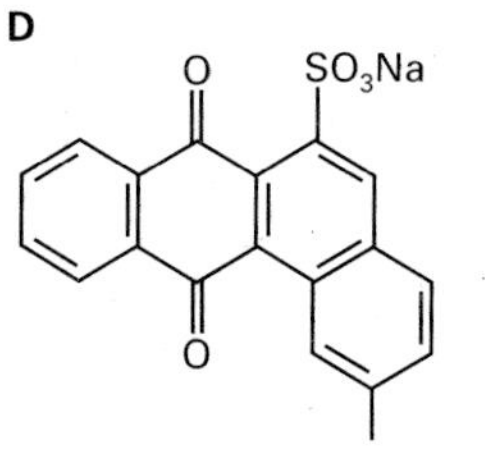

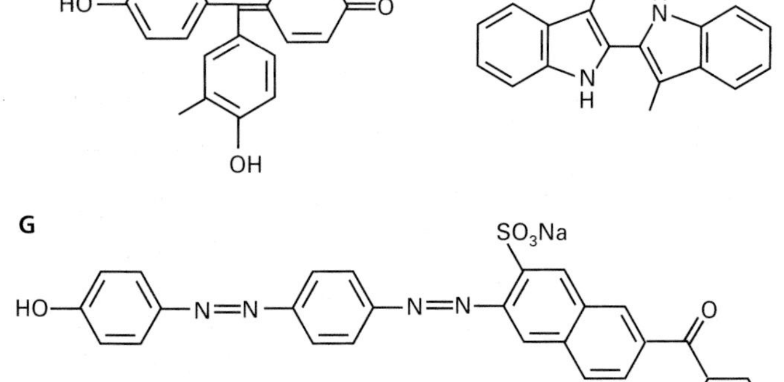

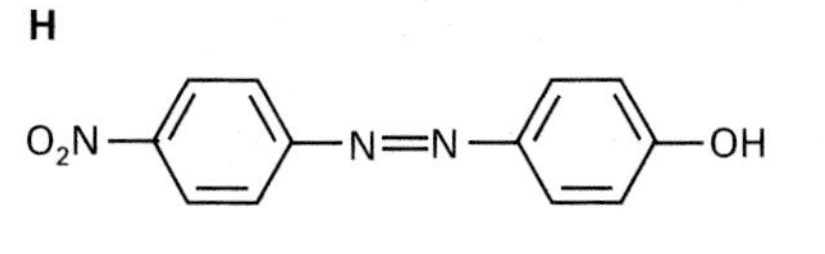

Activity 18.5 continued

a) contains one azo group
b) contains two azo groups
c) is an anthraquinone dye
d) is an insoluble triarylmethane dye
e) is colourless
f) is an indigoid dye
g) will bind covalently to cotton fibre
h) is likely to be used as a disperse dye
i) can be used as a vat-dye
j) is likely to be used as a direct dye
k) contains at least one 'acidic' grouping
l) contains at least one 'basic' grouping.

2 'Acrylic' fibre is made by the addition polymerisation of propenonitrile, $CH_2{=}CHCN$.
a) Draw the repeat unit in this polymer.
b) Why does the dyeing of acrylic fibre present a problem?
c) The dyeing properties of acrylic fibre may be improved by adding the following co-monomer:

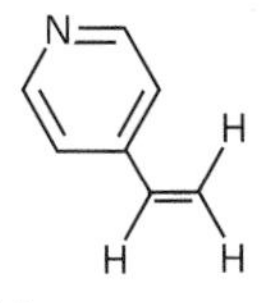

Why does the inclusion of this co-monomer improve the dyeing characteristics of this fibre?
d) Suggest one dye from Figure 18.12 that would be taken up much better by acrylic fibre modified in this way. Explain your answer.

Pigments in paints and inks

Natural pigments have been used by humans since pre-historic times. Cave paintings at Lascaux in France date from around 13 000 BC and depict the hunting of animals.

Pre-historic cave painting

Surprisingly, much of the basic technology used in such ancient times is mirrored in present-day paints and printing inks. Finely powdered carbon and iron(III) oxide are still used as black and brown pigments respectively, although nowadays zinc oxide or titanium dioxide is preferred to chalk for white.

In a **paint** the coloured pigment is dispersed as fine particles in a liquid medium known as the **vehicle**. After the paint has been applied, the vehicle must harden to form a protective film to protect the pigment and prevent it being rubbed off, or washed away. The vehicle in oil-based paints contains unsaturated linseed oil. This is known as a 'drying' oil as, on exposure to air, it forms a cross-linked solid polymeric structure which binds the pigment in place.

Pigments in paints and inks *continued*

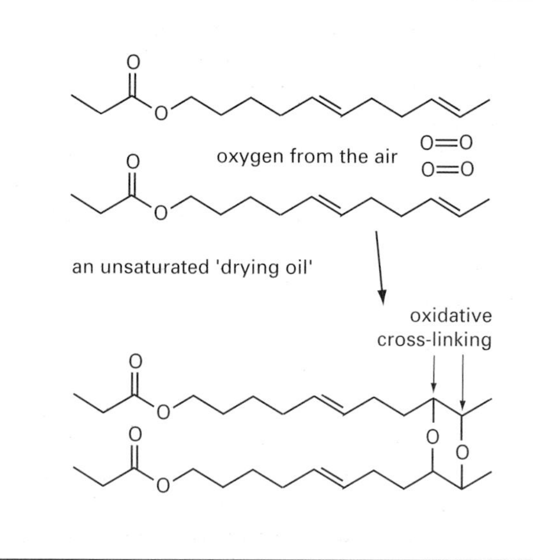

Similar vegetable oil residues together with egg white or milk-based protein binders have been found in cave paintings.

Water-based paints contain a solid binder, usually an acrylic-based resin, which, like the pigment, must be finely dispersed in the liquid carrier. No chemical reaction takes place when this dries; the water simply evaporates to leave the pigment trapped in the resin matrix.

An **ink** is essentially a 'paint-like' material used for writing or printing. Writing inks generally contain soluble dyes but most printing inks consist of pigment dispersed in a solution of a resin. On drying the resin binds the pigment to the printed article as a film that is, ideally, both wear and wash resistant.

The colour index

Most dyes and many pigments have complex structures and equally complex systematic names. The colour index, produced by the Society of Dyers and Colourists, lists all the colouring agents in current use. Each colouring agent is given its own 5-digit C.I. number that denotes its structural type as shown in Table 18.1. In additon to this a second reference system indicates how the dye or pigment is used. Thus Malachite Green has the 5-digit reference C.I. 42000 showing it has a triphenylmethane structure and is also referred to as C.I. Basic Green 4, i.e. it was the fourth basic dye to be included in the index.

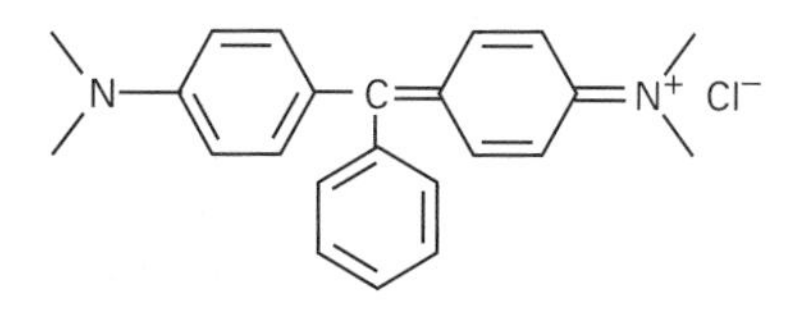

Table 18.1 Colour index 5-digit reference numbers and structural type

C.I. 5-digit number	structural type
10000 to 10299	nitroso
10300 to 10999	nitro
11000 to 19999	monoazo
20000 to 29999	disazo
30000 to 34999	trisazo
35000 to 36999	polyazo
37000 to 39999	azoic
40000 to 40799	stilbene
40800 to 40999	caroteinoid
41000 to 41999	diphenylmethane
42000 to 44900	triphenylmethane
45000 to 45999	xanthene
46000 to 46999	acridine
47000 to 47999	quinoline
48000 to 48999	methine
49000 to 49399	thiazole
49400 to 49699	indamine
49700 to 49999	indophenol
50000 to 50999	azine
51000 to 51999	oxazine
52000 to 52999	thiazine
53000 to 54999	sulphur
55000 to 55999	lactone
56000 to 56999	aminoketone
57000 to 57999	hydroxyketone
58000 to 72999	anthaquinone
73000 to 73999	indigo
74000 to 74999	phthalocyanine
75000 to 75999	natural
76000 to 76999	oxidation bases

18.5 Fluorescence

Any material that absorbs electromagnetic radiation and then re-emits radiation at a longer wavelength is said to be **fluorescent**. (Strictly speaking, a fluorescent material re-emits absorbed radiation very rapidly. If the process is slower it is referred to as phosphorescence.) Absorption of radiation promotes an electron and increases the vibrational energy of the structure. If the extra vibrational energy is quickly distributed to surrounding particles then, when the electron drops back to its original state, radiation of lower quantum energy is emitted.

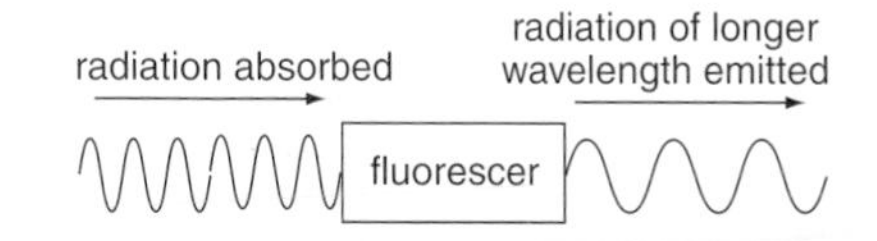

Fluorescers that absorb ultraviolet light and emit visible blue light are often added to washing powders as **optical brighteners**. Fabrics that are white when new often yellow slightly with age. This is caused by movement of the absorption band from the ultraviolet region into the blue end of the visible spectrum. If the fabric is treated with an optical brightener the missing blue component of the reflected light is replaced causing the material to regain its 'as new' white appearance. A white shirt or blouse washed in an optical brightener will glow blue-white under the ultraviolet lighting used in many night clubs.

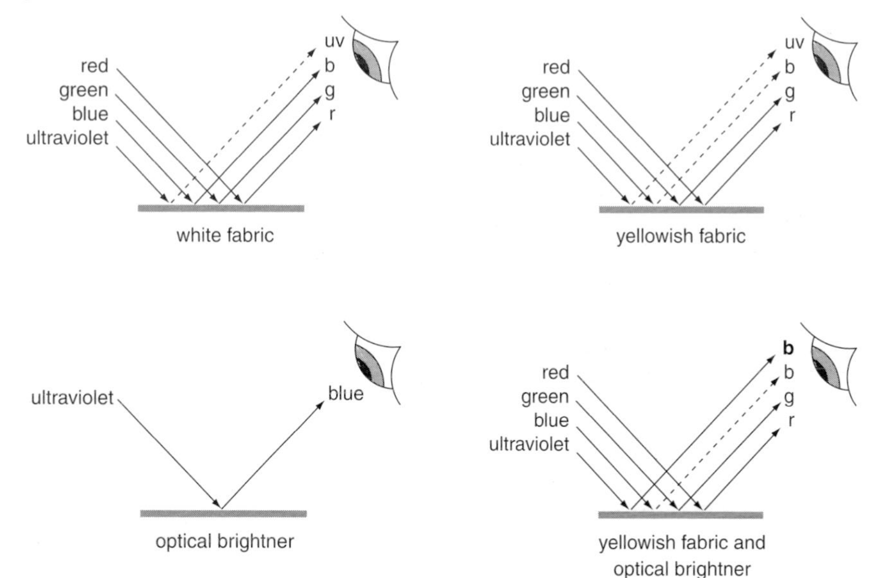

Fluorescent materials can only emit light if they can first absorb sufficient energy. In **chemiluminescence** the energy is provided by a chemical reaction rather than electromagnetic radiation. For example, the reaction between certain ethanedioate esters and hydrogen peroxide is thought to produce a highly unstable C_2O_4 molecule:

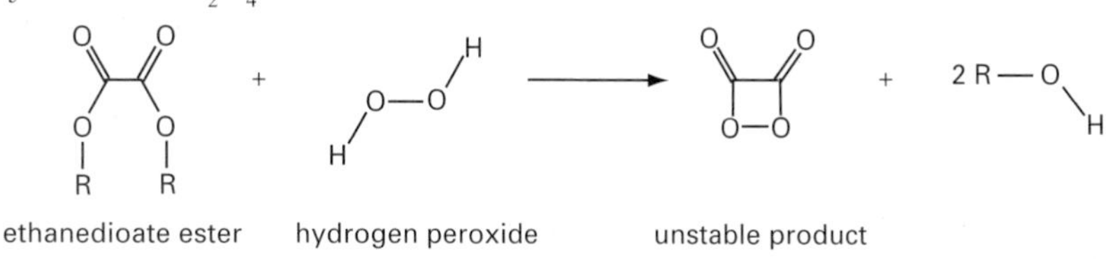

The C_2O_4 decomposes almost immediately into carbon dioxide releasing enough energy to excite a suitable fluorescer that is present in the reaction mixture. Light-sticks are flexible plastic tubes containing the separate

Focus 18d

1. Dyes and pigments are commercially important colouring materials.
2. Pigments are dispersed as insoluble fine powders, whereas most dyes are applied from solution.
3. Dyes may be classified according to their chemical structure, how they are applied, or how they bind to the material being coloured.

Activity 18.6

The structures of some optical brighteners are shown in Figure 18.13

1. In what way are the structures of these optical brighteners similar to those of dyes?
2. Why do you think they absorb in the ultraviolet region rather than the visible?
3. Which of the brighteners shown is most likely to dissolve during the fabric washing process? Explain your answer.

A

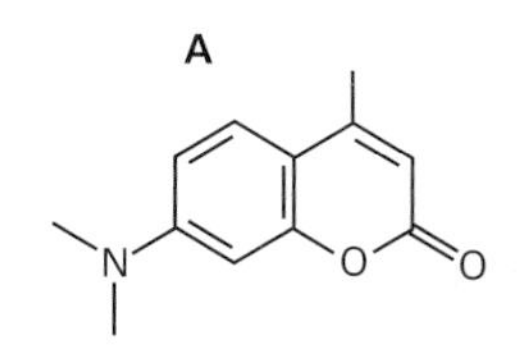

B

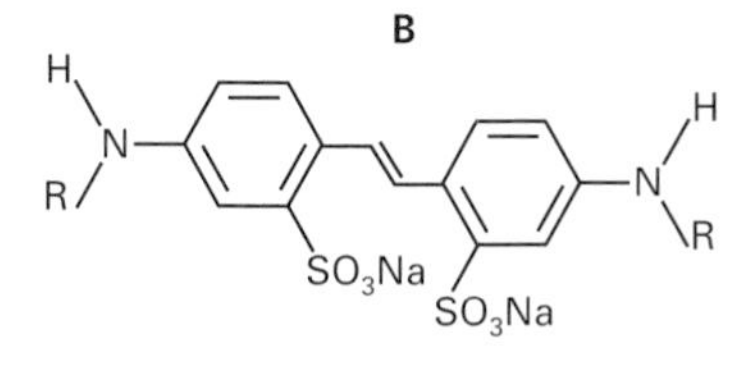

C

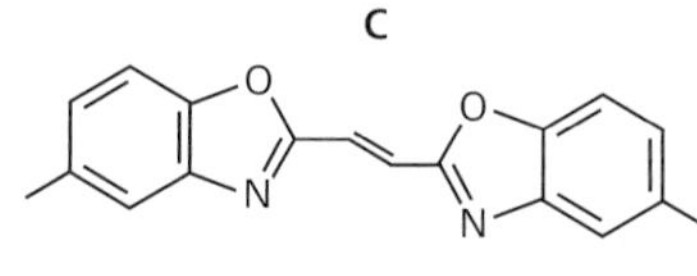

Figure 18.13
Structure of some optical brighteners

reactants in thin glass containers. Bending the stick breaks the containers, allows the reactants to mix, and produces 'cold' light for up to 12 hours. These sticks are supplied in survival packs used by sailors to enable them to be visible in the water in case of emergencies.

Fireflies and several other **bioluminescent** creatures signal by producing flashes of light in a similar way.

Focus 18e

1. A **fluorescent** material first absorbs energy and then emits it at a lower frequency.
2. Materials which absorb in the UV region and emit blue light may be used in washing powders as **optical brighteners**.
3. In **chemiluminescence** and **bioluminescence**, the original energy required to excite the molecule is provided not by EM radiation but by chemical or biochemical reactions respectively.

Questions on Chapter 18

1 a) Explain, in terms of the absorption and reflection of visible light, why grass is green, snow is white, blood is red and coal is black.

b) What colour would you expect each of the above to appear when viewed in blue light only?

2 What happens to a molecule when it absorbs a quantum of visible or ultraviolet radiation?

3 How do you explain the following observations?

a) Copper(I) chloride is white whereas copper(II) chloride is coloured.

b) Copper(II) chloride is blue in dilute aqueous solution but turns yellow-green on adding an excess of concentrated hydrochloric acid.

c) $[Co(NH_3)_6]^{2+}(aq)$ and $[Co(H_2O)_6]^{2+}(aq)$ have different colours.

d) Titanium(IV) oxide is white but chromium(VI) oxide is coloured.

4 a) Draw a dot and cross diagram showing the arrangement of outer electrons in the azo group —N=N—.

b) Draw a diagram illustrating the relative energies of the different types of molecular orbital (bonding, anti-bonding and non-bonding) in the azo group and show which are occupied by electrons in the ground state.

5 Organic compounds that contain multiple bonds may be coloured.

a) What kind of electronic transition between molecular orbitals might correspond to the absorption of visible radiation in such a compound?

b) What is meant by the term 'conjugation' and why does this increase the likelihood of colour in unsaturated systems?

6 Explain what is meant by each of the following terms, giving examples where appropriate:

a) chromophore

b) auxochrome

c) bathochromic shift

d) hypsochromic shift.

7 What is the essential difference between a dye and a pigment? How may the structure of a pigment sometimes be modified to convert it into a dye?

8 a) Classify each of the dyes in Figure 18.14 according to structural type.

b) Which of these can act as (i) acid dyes and (ii) basic dyes?

c) Which might be used in vat dyeing? Explain the principles on which this process is based.

d) What grouping could be attached to any of these structures to make a fibre-reactive cotton dye? Explain how this group binds to cotton fibre.

e) What feature of phthalocyanine dyes and pigments is not shown by any of the structures in Figure 18.14?

9 a) What is the function of linseed oil in oil-based paints?

b) Why are water-based emulsion paints generally much quicker drying than oil-based paints?

10 a) What is meant by fluorescence and how does it arise?

b) Explain the use of fluorescent materials in optical brighteners and light sticks.

Questions on Chapter 18 *continued*

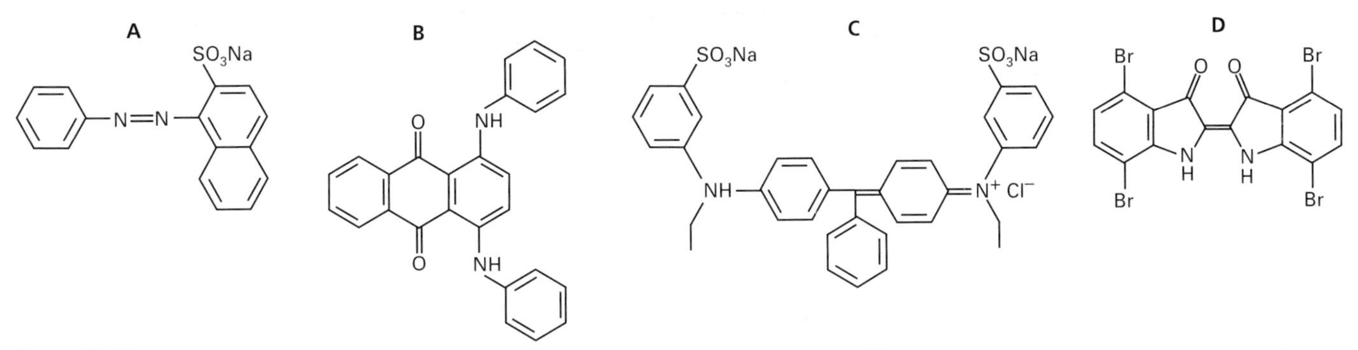

Figure 18.14

Comments on the activities

Activity 18.1

1 **a)** Sample C would be white or colourless, as it does not absorb at all in the visible region.
b) Sample A would appear black as it absorbs all visible wavelengths.
c) Sample B absorbs everything but red light and so appears red.
d) Sample D absorbs everything but violet/blue light as so appears blue.

2 Sample B only reflects red light so would appear black when viewed in blue light.

Activity 18.2

1 The graph below shows that the absorbance of the sample is directly proportional to its molar concentration at a fixed sample length of 2 cm, i.e. the Beer–Lambert law is obeyed.

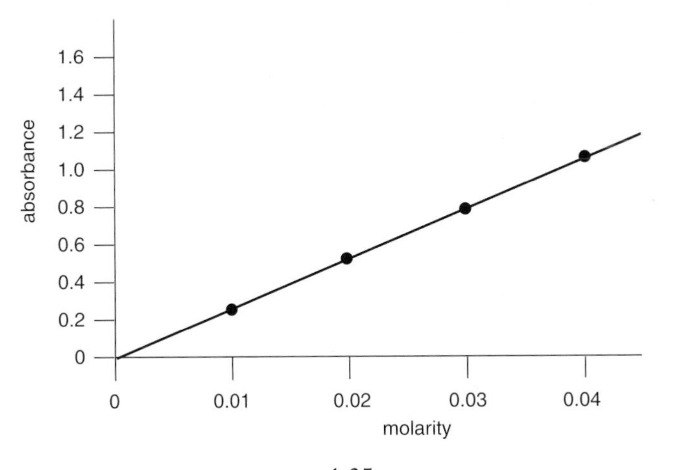

2 The slope of the graph is $\frac{1.35}{0.050} = 27.0$

This is the absorbance of a 1 M solution of path length 2 cm. The molar absorption coefficent of X is therefore $\frac{27.0}{2} = 13.5$

3 $A = \varepsilon lc$, so $0.55 = 13.5 \times 4 \times c$, i.e. $c = \frac{0.55}{13.5 \times 4} = \frac{0.55}{54}$ $= 0.0102$ M

4 $A = \log\left(\frac{I_0}{I}\right)$ so $\frac{I_0}{I}$ = antilog A = antilog (0.55) = 3.55

$$\frac{I}{I_0} = \frac{1}{3.55} = 0.282$$

so percentage transmittance $= 100\frac{I}{I_0} = 28.2$

Activity 18.3

1 **a)** The electron configuration of Co^{2+} is [Ar]$3d^7$.
b) The Co^{2+} ion has a partially filled d sub-level. Under the influence of a ligand field this energy level splits. This splitting energy must correspond to the quantum energy of radiation within the visible region of the spectrum in order to produce colour.
c) The splitting energy in $[Co(H_2O)_6]^{2+}$(aq) will be higher than in $[CoCl_4]^{2-}$(aq) as it contains a greater number of ligands. In addition the spectro-chemical series shows that the water molecule is more effective at d orbital splitting than the chloride ion.
d) The pink complex must absorb radiation towards the blue end of the visible spectrum. The blue complex must absorb at the red end of the spectrum. The quantum energy of blue light is greater than that of red light so the complex which appears pink must have the higher ligand field splitting energy, i.e. $[Co(H_2O)_6]^{2+}$(aq).

2 **a)**

oxide	Sc_2O_3	TiO_2	V_2O_5	CrO_3	Mn_2O_7
oxidation state	+3	+4	+5	+6	+7

In each of the above oxidation states the metal has no electrons in its 3d sub-level.

Comments on the activities *continued*

b) Since the d sub-level on the metal ion is empty, the only way in which colour can arise in these oxides is by charge transfer, i.e. electrons jumping from the oxide ions into orbitals on the central metal ion. The energy required for this transition will be too high to cause absorption in the visible region unless the charge on the metal ion, i.e. its oxidation state, is high enough, so Sc_2O_3 and TiO_2, which have the lowest charges on their metal ions, are white and the rest are coloured.

Activity 18.4

1 Since the highest wavelength at which benzene absorbs, 260 nm, is well into the ultraviolet region it is colourless.

2 The above absorption band in benzene is due to a $\pi \longrightarrow \pi^*$ transition.

3 Azobenzene contains two types of chromophore, C=C and N=N.

4 The azo group links the conjugated systems in each of the benzene rings. As a result the conjugated system is extended from three double bonds to seven.

5 Each of the nitrogen atoms in the azo group possesses a lone pair of electrons in its valency shell, leading to the possibility of $n \longrightarrow \pi^*$ transitions.

6 Greater conjugation and the presence of lone pairs of electrons both lead to a greater variety of transitions, some of which will require less energy and therefore involve absorption at longer wavelengths.

Activity 18.5

1 a) A, C and H are mono-azo dyes.

b) G contains two azo groups.

c) D is an anthaquinone dye.

d) E is an insoluble triarylmethane dye.

e) B is a colourless leuco-base (derived by reducing E).

f) F is an indigoid dye.

g) A will bind covalently to cotton fibre via its trichlorotriazine ring.

h) H (or C) is likely to be used as a disperse dye.

i) D, E/B may be used in vat-dyeing.

j) G is likely to be used as a direct dye.

k) A, D and G contain the —SO_3Na 'acidic' grouping.

l) A, C and F contain 'basic' amino groups.

2 a) The repeat unit of acrylic fibre is shown below.

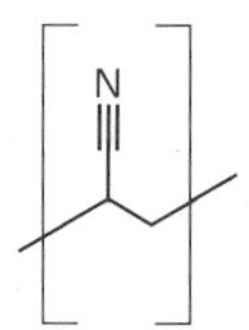

b) There are no convenient functional groups to attract dye molecules.

c) The co-monomer contains an amino group which is basic and will attract any acidic groups on the dye molecule.

d) Uptake of any of the acidic dyes, A, D or G, will be improved by incorporation of this basic co-monomer.

Activity 18.6

1 Like dyes, optical brighteners contain conjugated multiple bonds.

2 The degree of conjugation is less than in most dyes so brighteners absorb radiation of higher quantum energy, i.e. in the ultraviolet rather than the visible region.

3 Brightener B contains solubilising —SO_3Na groups and so is most likely to dissolve in the wash liquid.

THEME D

Inorganic chemistry: trends and patterns

CHAPTER 19

The s block

Contents

Study Checklist

After studying this chapter you should be able to:

1. Recall and explain the trends in the following properties for the elements of groups 1 and 2: atomic radius, ionisation energies, electronegativity, $E^{\ominus}$, melting point.
2. State and explain the common oxidation states shown by the s block metals.
3. Recall the reactions of the elements with oxygen, chlorine and water and relate to $E^{\ominus}$ values.
4. Recall the trends in thermal stability of the nitrates and carbonates and relate to the charge density of the cations.
5. Recall the trends in solubility of the sulphates and hydroxides and relate to lattice and hydration enthalpies.
6. Give examples of the covalent nature of lithium compounds and explain this in terms of the high charge density of the Li^+ ion.
7. Recall the main sources of sodium, potassium, magnesium and calcium and relate these to their solubility.

The elements of group 1 are known as the **alkali metals**, whereas those of group 2 are the **alkaline earth metals**. Despite their high reactivity the metals of the s block have several commercially important uses. Sodium and magnesium are the most widely used and are produced in bulk, though small quantities of the other metals are needed for special applications.

Over 100 000 tonnes of sodium are currently used annually, alloyed with lead in the manufacture of lead tetraethyl, an anti-knock additive for petrol.

$$Pb/4Na + 4C_2H_5Cl \longrightarrow Pb(C_2H_5)_4 + 4NaCl$$

However, given the increasing popularity of unleaded petrol, it seems likely that this use will decline.

The excellent thermal conductivities and low melting points of the group 1 metals make them very efficient coolants. 'Fast' nuclear reactors use circulating molten sodium as a primary coolant to absorb the heat produced in their cores.

Magnesium is widely used in low-density alloys, which have excellent strength and heat-resistant properties. These are used by the aircraft and aerospace industries and provide the fuel cans in Magnox nuclear reactors. Lithium is a constituent of specialist alloys and long-lasting lightweight batteries that are often found in products such as computer equipment and cameras.

Lightweight alloys containing magnesium are used in aircraft manufacture

Special methods for metal extraction that make use of the high reactivity of sodium and magnesium are covered in Chapter 23. Many compounds of s block elements are also of industrial importance and large-scale uses of salt (sodium chloride) and limestone (calcium carbonate) are also outlined in Chapter 23.

Poisonous metals

Although soluble lithium salts can be an effective treatment for manic depression, they are very poisonous. If the dose is not carefully controlled, severe side effects such as kidney damage, thyroid abnormalities and, in extreme cases, even death may result.

Beryllium is an effective neutron reflector, used in the casing for nuclear fission bombs. Without such a reflective casing, neutrons could escape from the imploding uranium, reducing the efficiency of the nuclear explosion. Also any beryllium nucleus that catches a fast neutron gives off two more, further increasing the 'efficiency' of the bomb. However, beryllium is even more toxic than lithium. The following article from the *New Scientist* describes the aftermath of an accident in which beryllium was released into the atmosphere.

Officials 'held back data' on Soviet blast

Secretive factory officials seriously delayed cleanup operations following last month's explosion and fire at a military plant in Utika, in the Soviet Union. The officials refused to hand over vital details to doctors and environmentalists, according to the Soviet newspaper *Izvestiya*.

As a result, local authorities responsible for the cleanup operation did not have a clear picture of what had happened until the Ministry of the Atomic Energy Industry released a statement two days after the accident.

The refusal to hand over information also made it far more difficult for officials to map the airborne spread of beryllium oxide in the first hour after the accident.

As many as 120 000 people may have been contaminated with beryllium as a result of the accident, according to Rishat Adamav, the chairman of Eastern Kazakhstan's regional environmental protection committee. Last week the area, which is close both to the border with China and to the nuclear weapons testing site at Semipalatinsk, was declared an ecological disaster zone by the Republic of Kazakhstan.

The republic's president has already asked Moscow for compensation for the health effects that the explosion is likely to have. He has also asked for independent experts to be sent to the area to see if the plant should be closed down.

The relationship between the emergency services and plants or laboratories engaged in classified work in the Soviet Union has never been a good one. Under Glasnost, however, it is no longer possible to conceal such accidents.

The city authorities of Ust-Kamenogorsk were told that beryllium contamination had occurred, and immediately launched a round-the-clock monitoring and cleanup operation. The contamination was high initially, with airborne beryllium at 60 times the permissible level in both the city centre and one of the suburbs.

By the second day, however, the contamination was reported by officials to have fallen below the permissible limit in all but one district of the city. Laboratories and medical centres found no evidence of beryllium contamination in members of the public; however scientists point out that the effects of beryllium poisoning can take up to 10 years to appear.

Complacency seems to be the order of the day at the Utika site. The explosion of hydrogen had destroyed part of the outer wall of one of the beryllium units, leaving it open to the air, and damaged the ventilation system which contained aerosols of beryllium. But the management of the plant was reluctant to divert workers from normal production, even for the time necessary to have the hole boarded up.

This week the Semipalatinsk regional council banned nuclear tests by the Soviet government at Semipalatinsk – the country's main testing site for nuclear weapons.

Source: *New Scientist*, 6 October 1990, Vol. 128 No. 1737 by Vera Rich

19.1 Atomic and physical properties

Although there are very close similarities in the chemistry of the elements within a particular group, Table 19.1 shows that there are also well-defined trends. As we shall see, many of these trends may be explained largely in terms of atomic and ionic size. *In particular the small sizes of the Li^+ and Be^{2+} ions lead to some unusual properties.*

Table 19.1 Data sheet for the elements of groups 1 and 2

Element	atomic number	electron configuration	atomic radius /nm	ionic radius /nm	ionisation energies /kJ mol⁻¹ 1st	2nd	3rd	electro-negativity	$E^{\ominus}$ /volts	density /g cm⁻³	m.p. /°C
Group 1											
lithium Li	3	[He] $2s^1$	0.152	0.060	519	7300	11800	1.0	−3.04	0.53	180
sodium Na	11	[Ne] $3s^1$	0.186	0.095	494	4560	6940	0.9	−2.71	0.97	98
potassium K	19	[Ar] $4s^1$	0.231	0.133	418	3070	4600	0.8	−2.92	0.86	64
rubidium Rb	37	[Kr] $5s^1$	0.244	0.148	402	2650	3850	0.8	−2.92	1.53	39
caesium Cs	55	[Xe] $6s^1$	0.262	0.169	376	2420	3300	0.7	−2.92	1.90	29
francium Fr	87	[Rn] $7s^1$	0.270	0.176	381	?	?	0.7	?	?	(27)
Group 2											
beryllium Be	4	[He] $2s^2$	0.112	(0.031)	900	1760	14800	1.5	−1.85	1.85	1280
magnesium Mg	12	[Ne] $3s^2$	0.160	0.065	736	1450	7740	1.2	−2.38	1.74	650
calcium Ca	20	[Ar] $4s^2$	0.197	0.099	590	1150	4940	1.0	−2.87	1.54	850
strontium Sr	38	[Kr] $5s^2$	0.215	0.113	548	1060	4120	1.0	−2.89	2.62	768
barium Ba	56	[Xe] $6s^2$	0.217	0.135	502	966	3390	0.9	−2.90	3.51	714
radium Ra	88	[Rn] $7s^2$	0.220	0.140	510	979	?	0.9	−2.92	5.0	700

What is the general trend in the value of standard electrode potential, $E^{\ominus}$, on passing (a) down a group, (b) from group 1 to group 2? The standard electrode potential of lithium does not fit this pattern. Can you explain why?

Atomic and ionic size

In any period, the group 2 atom is **smaller** than the group 1 atom since the outer electrons, which are in the same shell in both atoms, are attracted more strongly by the higher nuclear charge in the group 2 atom. On passing down either of the groups the effect of the increasing nuclear charge is outweighed by the addition of an extra shell and the consequent increase in the number of inner screening electrons, so the atomic size increases.

All the atoms form positive ions by losing their outer electrons. The remaining electrons are then more tightly held by the nucleus, making these ions smaller than the parent atoms.

Ionisation energies

The first **ionisation energy** of an element is the enthalpy change on converting one mole of its gaseous atoms into gaseous ions with a single positive charge, i.e.:

$$\text{1st ionisation energy: } M(g) \longrightarrow M^{+}(g) + e^{-}$$

First ionisation energy, therefore, depends upon the attraction of the nucleus for the outer electron in the atom. Table 19.1 shows that *first ionisation energy increases on passing from left to right across the periodic table* because the nuclear charge increases and the atomic radius falls. *On passing down either group, first ionisation energy falls* because the increase in atomic radius and the number of inner screening electrons outweigh the extra nuclear charge.

Successive ionisation energies involve the removal of further electrons from gaseous ions.

$$\text{2nd ionisation energy: } M^{+}(g) \longrightarrow M^{2+}(g) + e^{-}$$
$$\text{3rd ionisation energy: } M^{2+}(g) \longrightarrow M^{3+}(g) + e^{-} \text{ etc.}$$

Table 19.1 shows that there is a very large jump in ionisation energy when an inner electron shell is broken. This is because the electron concerned is closer to the nucleus and less well shielded by inner electrons. This effect is illustrated graphically for sodium and magnesium in Figure 19.1.

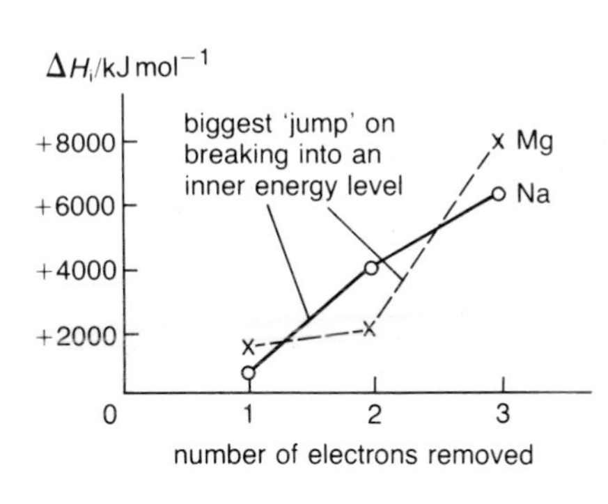

Figure 19.1 ▲
Increase in successive ionisation energies

Electronegativity

Electronegativity measures the ability of the atom to attract the electrons in a covalent bond. Figure 19.2 shows a very good correlation between electronegativity and first ionisation energy, which might be expected as both properties depend upon the attractive power of the nucleus.

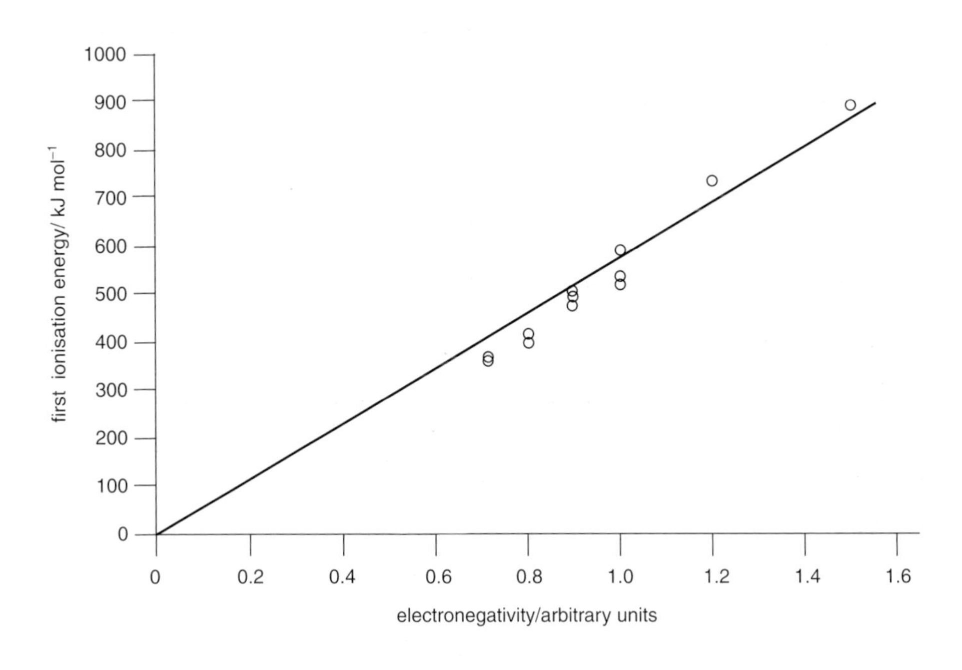

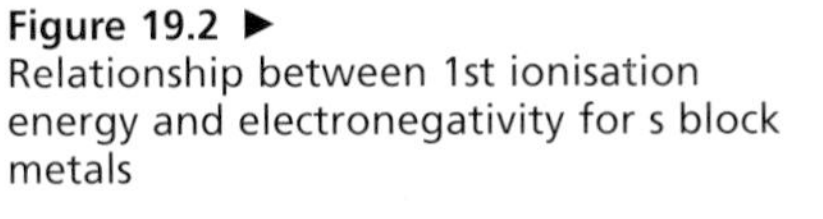
Figure 19.2 ►
Relationship between 1st ionisation energy and electronegativity for s block metals

Standard electrode potential

This property again relates to the tendency of an atom to lose electrons; in this case, however, forming ions in **aqueous solution**.

$$\text{for group 1: } M(s) \longrightarrow M^{+}(aq) + e^{-}$$
$$\text{for group 2: } M(s) \longrightarrow M^{2+}(aq) + 2e^{-}$$

The more negative the value of the standard electrode potential, $E^\ominus$, the more readily the element forms an aqueous ion. As we might expect, since the outer electrons are less strongly held by the nucleus, electrode potential generally becomes more negative on passing down a group. However, the variation is less clear cut than for first ionisation energy, and lithium is unusual in having the most negative $E^\ominus$ value of all. Although a complete explanation of this involves entropy change (Chapter 28), a consideration of the various enthalpy changes involved gives some insight into the reason for this.

$$\Delta H_a \text{ atomisation: } M(s) \longrightarrow M(g)$$
$$\Delta H_i(1) \text{ ionisation: } M(g) \longrightarrow M^+(g) + e^-$$
$$\Delta H_h \text{ hydration: } M^+(g) \longrightarrow M^+(aq)$$

The figures for group 1 are shown in Table 19.2.

Table 19.2

	ΔH_a	$\Delta H_i(1)$	ΔH_h	ΔH(total)	$E^\ominus$/volts
Li	161	519	–519	161	–3.04
Na	109	494	–406	197	–2.71
K	90	418	–322	186	–2.92
Rb	86	402	–301	187	–2.92
Cs	79	376	–276	179	–2.92

(all enthalpy changes given as kJ mol^{-1})

Although more energy is required to form the gaseous Li^+ ion, its small size and high charge density results in the release of more energy when it is hydrated.

Melting point

All the elements involved are metals, and melting point is a measure of the relative strength of bonding within the lattice. Treated simply, metallic bonding involves the delocalisation of outer electrons, which then hold the cations together by electrostatic attraction (see Figure 19.3).

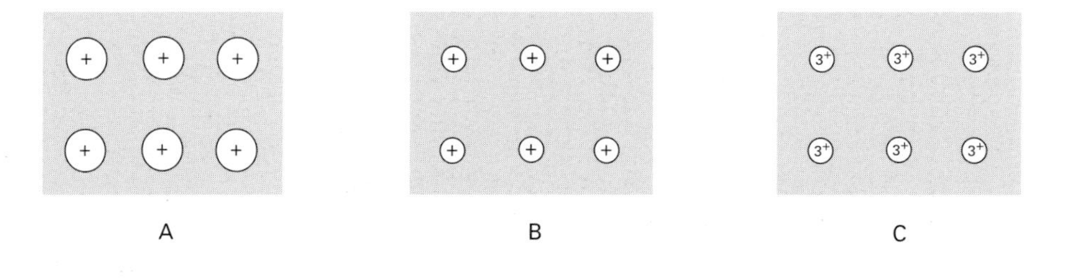

Figure 19.3
Metals with small highly charged ions form stronger metallic bonds. Here, metallic bond strength increases A < B < C

As we might expect, group 2 metals have higher melting points than group 1 metals since each atom contributes two electrons to the delocalised 'sea'.

Melting points fall on passing down group 1 since the charge density of the metal ion falls and so the attraction for the delocalised electrons weakens. However, differences in structural type on passing down group 2 spoil this simple pattern.

19.2 Chemical properties

Since the outer s electrons are relatively loosely held, the metals form predominantly ionic compounds in which they show an oxidation state equal to the group number. For example, all the elements burn vigorously on heating in chlorine as follows.

group 1: $2M(s) + Cl_2(g) \longrightarrow 2MCl(s)$
group 2: $M(s) + Cl_2(g) \longrightarrow MCl_2(s)$

The reasons behind this behaviour become clear if we construct a **Born–Haber cycle** showing all the various energy changes involved in forming one mole of an ionic solid.

Born–Haber cycles

All of the enthalpy changes given in this section are in kJ mol^{-1}.

For NaCl(s)

The **standard enthalpy of formation**, $\Delta H_f^{\ominus}$, is the enthalpy change on forming one mole of the compound from its elements with all substances in their standard (most stable) states at 1 atmosphere pressure and 25°C (298K):

$Na(s) + \frac{1}{2}Cl_2(g) \longrightarrow NaCl(s) \qquad \Delta H_f^{\ominus}[NaCl] = -411$

Theoretically, we can achieve the same conversion by a series of steps with named enthalpy changes.

The sodium metal may be converted into Na^+ in two stages. **Atomisation energy** is the enthalpy change on forming one mole of gaseous atoms from the element in its standard state:

$Na(s) \longrightarrow Na(g) \qquad \Delta H_a[Na] = +109$

1st ionisation energy is the enthalpy change on converting one mole of gaseous atoms into gaseous ions with a single positive charge:

$Na(g) \longrightarrow Na^+(g) \qquad \Delta H_i(1)[Na] = +494$

Chlorine gas is also converted into gaseous chloride ions in two stages. So the **atomisation energy** is needed,

$\frac{1}{2}Cl_2(g) \longrightarrow Cl(g) \qquad \Delta H_a[Cl] = +121$

followed by **1st electron affinity**, which is the enthalpy change on converting one mole of gaseous atoms into gaseous ions with a single negative charge:

$Cl(g) \longrightarrow Cl^-(g) \qquad \Delta H_e[Cl] = -364$

Although we now have Na^+ and Cl^- ions, they are separate and in the gaseous state. **Lattice energy** may be defined as the enthalpy change on forming one mole of an ionic solid from its isolated gaseous ions:

$Na^+(g) + Cl^-(g) \longrightarrow NaCl(s) \qquad \Delta H_l[NaCl] = -771$

Since **Hess's law** states that the overall enthalpy change for any process depends only upon the initial and final states and not upon the route taken, the sum of all these separate energy changes must equal the overall enthalpy of formation. This is illustrated graphically for sodium chloride, NaCl, in Figure 19.4a.

For $MgCl_2(s)$

The energy changes in the cycle are as follows:

overall formation:
$Mg(s) + Cl_2(g) \longrightarrow MgCl_2(s) \qquad \Delta H^{\ominus}_f[MgCl_2] \ -652$

atomisation:
$Mg(s) \longrightarrow Mg(g) \qquad \Delta H_a[Mg] \ +150$

ionisation:
$Mg(g) \longrightarrow Mg^+(g) \qquad \Delta H_i(1)[Mg] \ +736$
$Mg^+(g) \longrightarrow Mg^{2+}(g) \qquad \Delta H_i(2)[Mg] \ +1450$

atomisation:
$Cl_2(g) \longrightarrow 2Cl(g) \qquad \Delta H_a[Cl] \times 2 \ +242$

electron affinity:
$2Cl(g) \longrightarrow 2Cl^-(g) \qquad \Delta H_e[Cl] \times 2 \ -728$

lattice formation:
$Mg^{2+}(g) + Cl^-(g) \longrightarrow MCl_2(s) \qquad \Delta H_l[MgCl_2] \ -2502$

The Born–Haber cycle for $MgCl_2(s)$ is shown in Figure 19.4b.

Born–Haber cycles *continued*

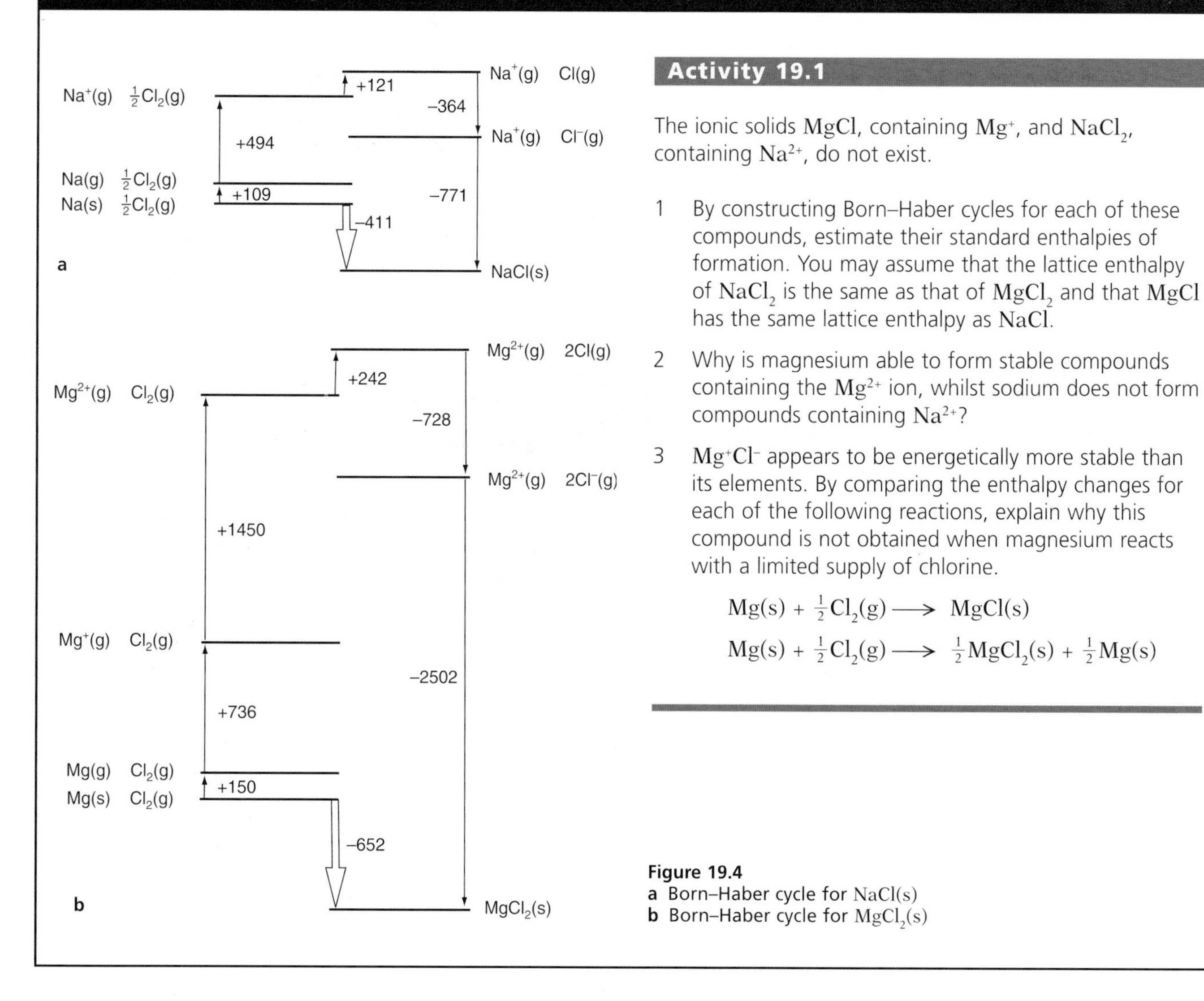

Activity 19.1

The ionic solids MgCl, containing Mg^+, and $NaCl_2$, containing Na^{2+}, do not exist.

1 By constructing Born–Haber cycles for each of these compounds, estimate their standard enthalpies of formation. You may assume that the lattice enthalpy of $NaCl_2$ is the same as that of $MgCl_2$ and that MgCl has the same lattice enthalpy as NaCl.

2 Why is magnesium able to form stable compounds containing the Mg^{2+} ion, whilst sodium does not form compounds containing Na^{2+}?

3 Mg^+Cl^- appears to be energetically more stable than its elements. By comparing the enthalpy changes for each of the following reactions, explain why this compound is not obtained when magnesium reacts with a limited supply of chlorine.

$$Mg(s) + \tfrac{1}{2}Cl_2(g) \longrightarrow MgCl(s)$$

$$Mg(s) + \tfrac{1}{2}Cl_2(g) \longrightarrow \tfrac{1}{2}MgCl_2(s) + \tfrac{1}{2}Mg(s)$$

Figure 19.4
a Born–Haber cycle for NaCl(s)
b Born–Haber cycle for $MgCl_2(s)$

Reaction with oxygen

Apart from beryllium and magnesium, the elements are oxidised by air at room temperature. They all burn on heating in oxygen and, though the oxidation state of each metal is fixed, different products are possible owing to the ability of oxygen to form different anions. Each of these is formed by 'persuading' an oxygen molecule to accept electrons.

normal oxides contain the O^{2-} ion: $O_2(g) + 4e^- \longrightarrow 2O^{2-}$
peroxides contain the O_2^{2-} ion: $O_2(g) + 2e^- \longrightarrow O_2^{2-}$
superoxides contain the O_2^- ion: $O_2(g) + e^- \longrightarrow O_2^-$

The type of compound formed by heating each element in excess oxygen is shown below.

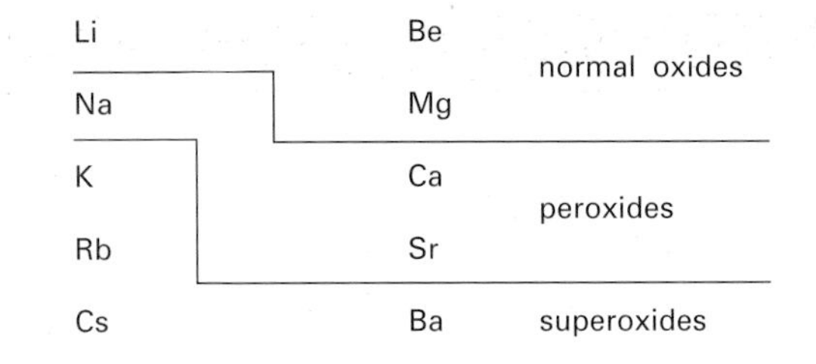

Lithium, beryllium and magnesium form normal oxides only, because their small ions attract electrons so strongly that they would break the O—O linkage in the peroxide or superoxide ion.

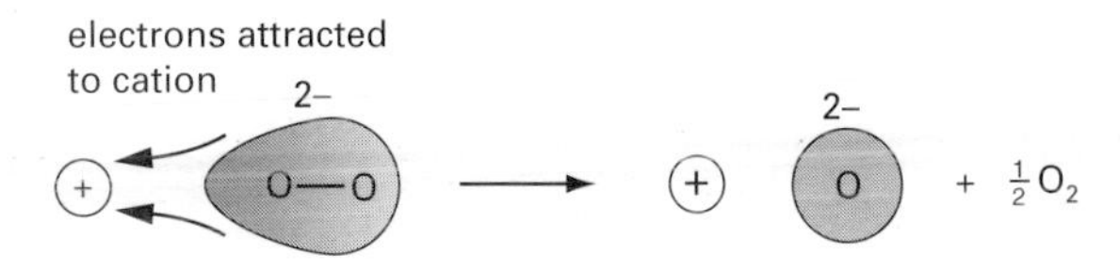

Reaction with water

Apart from beryllium and magnesium, the metals react readily with cold water to give hydrogen gas and a solution of the metal hydroxide.

$$\text{group 1: } M(s) + H_2O(l) \longrightarrow M^+(aq) + OH^-(aq) + \tfrac{1}{2}H_2(g)$$

$$\text{group 2: } M(s) + 2H_2O(l) \longrightarrow M^+(aq) + 2OH^-(aq) + H_2(g)$$

Again we can estimate the overall enthalpy change (kJ) by considering the individual stages, e.g. for sodium:

atomisation:	$Na(s) \longrightarrow Na(g)$	$\Delta H_a[Na]$	+109
ionisation:	$Na(g) \longrightarrow Na^+(g)$	$\Delta H_i(1)[Na]$	+494
hydration:	$Na^+(g) \longrightarrow Na^+(aq)$	$\Delta H_h[Na^+]$	–406
–(neutralisation):	$H_2O(l) \longrightarrow H^+(aq) + OH^-(aq)$	$-\Delta H_n$	+57
–(hydration):	$H^+(aq) \longrightarrow H^+(g)$	$-\Delta H_h[H^+]$	+1091
–(ionisation):	$H^+(g) \longrightarrow H(g)$	$-\Delta H_i[H]$	–1315
–(atomisation) :	$H(g) \longrightarrow \tfrac{1}{2}H_2(g)$	$-\Delta H_a[H]$	–218

overall reaction:

$$Na(s) + H_2O(l) \longrightarrow Na^+(aq) + OH^-(aq) + \tfrac{1}{2}H_2(g) \quad \Delta H^\ominus = -188$$

These changes are shown graphically in Figure 19.5.

Using the same method we can estimate the enthalpy change for the reaction of one mole of each of the alkali metals with water. These are shown in Table 19.3.

Table 19.3

	lithium	sodium	potassium	rubidium	caesium
ΔH/kJ mol^{-1}	–224	–188	–199	–198	–200

Since lithium releases the most energy we might expect this metal to react the most vigorously with water. However, the reactivity of the elements increases on passing down the group and so lithium is the least reactive alkali metal. The reason for this is that 'vigour' of a reaction depends upon its rate as well as the amount of energy released. If you look again at Figure 19.5 you can see that the metal must be ionised during the reaction. On passing down the group the ionisation energy, and hence the activation energy for the reaction, falls and the reaction with water becomes faster.

Activity 19.2

1 The hydration energy of the Mg^{2+} ion is –1920 kJ mol^{-1}. Use other enthalpy changes given in this chapter to estimate the enthalpy change for the following reaction:

$$Mg(s) + 2H_2O(l) \longrightarrow Mg^{2+}(aq) + 2OH^-(aq) + H_2(g)$$

2 How do you account for the fact that magnesium does not react readily with cold water but will react on heating strongly in steam?

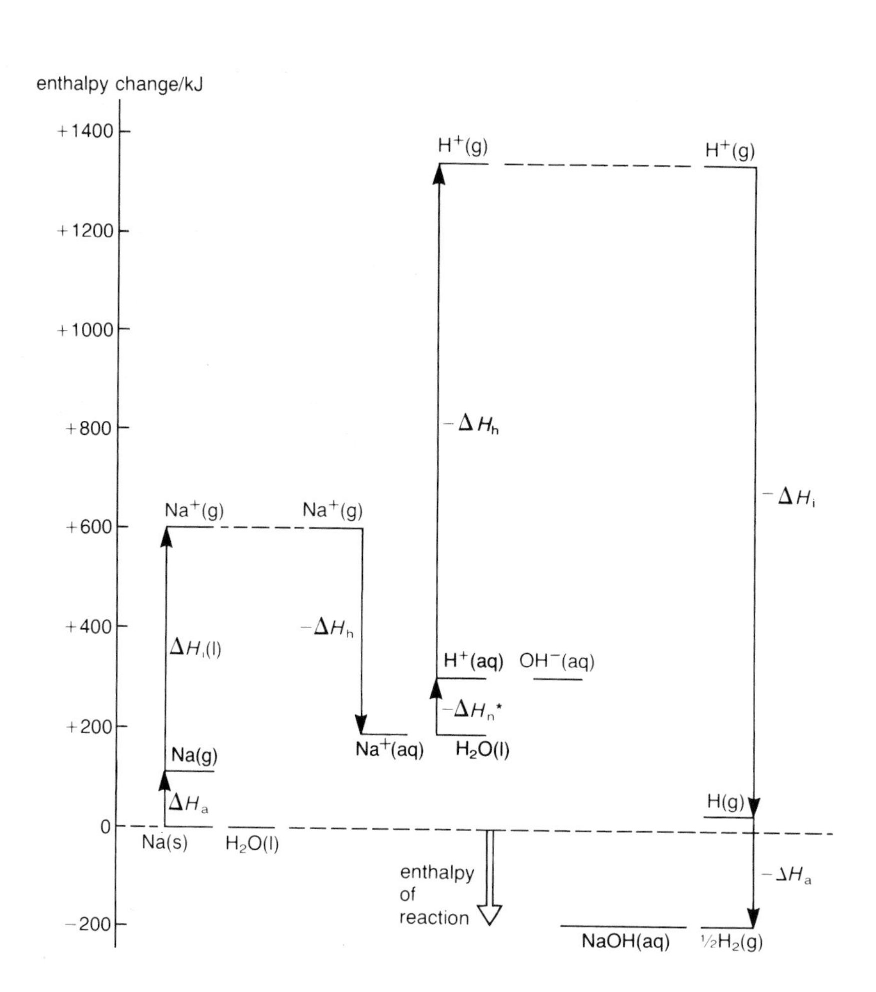

Figure 19.5
Enthalpy cycle for the reaction of sodium with water (approximately to scale)
ΔH_n is the enthalpy of neutralisation, i.e. for the reaction
$H^+(aq) + OH^-(aq) \longrightarrow H_2O(l)$

Focus 19a

1 The s block contains the most reactive metals in the periodic table; group 1, the alkali metals, and group 2, the alkaline earth metals.
2 The alkali metals are more reactive than the alkaline earth metals and reactivity increases on passing down each group.
3 All the elements form positive ions by losing their outer shell electron(s).

19.3 Properties of compounds

Most of the compounds of the alkali and alkaline earth metals have typical ionic properties, e.g. high melting point, crystalline solids that conduct electricity when molten or in solution. As with the elements, however, trends are apparent on passing down each group.

Water solubility

Although a full treatment of this topic involves free energy (see Chapter 28), we can explore the main trends more simply by considering only enthalpy changes. The overall process can be considered to take place in stages. Energy must be supplied to separate the ions in the lattice but is then released when the individual ions are hydrated. For sodium hydroxide the enthalpy changes are as follows:

vaporise lattice:	$Na^+OH^-(s) \longrightarrow Na^+(g) + OH^-(g)$	$-\Delta H_l[NaOH]$	+823
hydrate ions:	$Na^+(g) \longrightarrow Na^+(aq)$	$\Delta H_h[Na^+]$	–406
	$OH^-(g) \longrightarrow OH^-(aq)$	$\Delta H_h[OH^-]$	–460

The enthalpy change for the overall process, i.e. dissolving one mole of the solid in water, known as the enthalpy of solution, ΔH_{sol}, is obtained by adding the enthalpy changes for the individual steps.

overall process:	$Na^+OH^-(s) \longrightarrow Na^+(aq) + OH^-(aq)$	$\Delta H_{sol}[NaOH]$	–43

For magnesium hydroxide, the enthalpy changes are:

vaporise lattice:

$Mg^{2+}(OH^-)_2(s) \longrightarrow Mg^{2+}(g) + 2OH^-(g)$	$-\Delta H_l[NaOH]$	+2846

hydrate ions:

$Mg^{2+}(g) \longrightarrow Mg^{2+}(aq)$	$\Delta H_h[Mg^{2+}]$	–1920
$2OH^-(g) \longrightarrow 2OH^-(aq)$	$\Delta H_h[OH^-] \times 2$	–920

overall process:

$Mg^{2+}(OH^-)_2(s) \longrightarrow Mg^{2+}(aq) + 2OH^-(aq)$	$\Delta H_{sol}[Mg(OH)_2]$	+6

As you can see from the graphs in Figure 19.6, not only is the enthalpy of solution less favourable for magnesium hydroxide but also the higher lattice energy presents a bigger activation energy barrier to dissolving. Not surprisingly, sodium hydroxide is readily soluble in water, whereas magnesium hydroxide is almost insoluble.

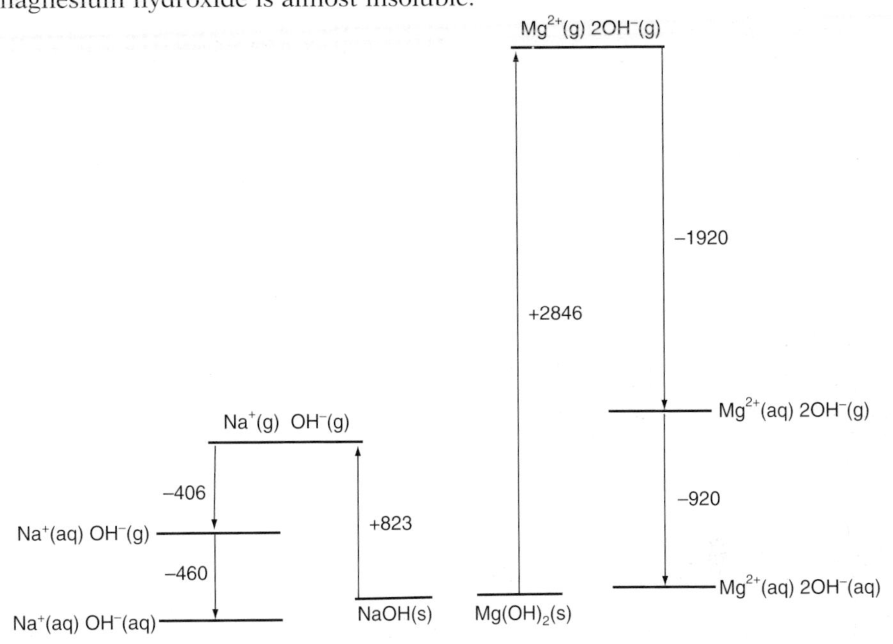

Figure 19.6
Energy changes on dissolving sodium and magnesium hydroxides in water. All enthalpy changes in kJ mol^{-1}

Activity 19.3

1 Copy and complete Table 19.4, which shows the estimated enthalpies of solution of the group 1 hydroxides (all values as kJ mol^{-1}).

Table 19.4

	Li^+	Na^+	K^+	Rb^+	Cs^+
$-\Delta H_l[MOH]$	+958	+823	+727	+698	+665
$\Delta H_h[M^+]$	–519	–406	–322	–301	–276
$\Delta H_h[OH^-]$	–460	–460	–460	–460	–460
$\Delta H_{sol}[MOH]$					

2 Explain why the lattice energies of the hydroxides fall on passing down group 1.

3 Explain why the hydration energies of the ions become less exothermic on passing down group 1.

4 How, and why, would you expect the water solubility of the hydroxides to vary on passing down group 1?

In other cases, prediction of solubility trends from enthalpy changes alone is not so successful. For example, the solubility of the sulphates tends to decrease on passing down group 2, even though less energy is needed to break up the lattice. Here we must take account of entropy change as explained in Chapter 28.

Table 19.5

	$BeSO_4$	$MgSO_4$	$CaSO_4$	$SrSO_4$	$BaSO_4$
ΔH_{sol}/kJ mol^{-1}		−91	−18	−9	+19
solubility/g dm^{-3}	390	330	2.1	0.013	0.0024

Solubility in organic solvents

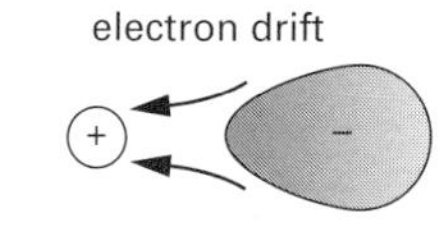

Lithium and beryllium chlorides are unusual in the s block, being appreciably soluble in polar organic solvents such as alcohols. This behaviour indicates considerable covalent character in the bonding. In all ionic compounds there is a tendency for the cation to attract electrons from the anion. If the effect is large enough, this leads to partial sharing of electrons from the anion by the cation, i.e. covalent character. The small sizes of Li^+ and Be^{2+}, and consequent high charge density, make them powerful electron attractors capable of deforming relatively large anions such as Cl^-, Br^- and I^-.

Lithium chloride and rust protection

The cost of protecting the internal steel box-girder decks that make up the northern span of the Humber bridge by painting is estimated at about £5 million. However, the Humber Bridge Board intends to stop corrosion using a desiccant dehumidifier costing only £10 000. This system is already used to control the atmosphere inside the concrete anchor chambers which house the suspension cables at either end of the bridge.

Damp air inside the chambers is drawn through a slowly rotating wheel impregnated with lithium chloride, which absorbs moisture from the air. This moisture is removed by blowing warm air through the wheel and then expelling it through a pipe to the outside of the bridge. The relative humidity inside the chambers is reduced to about 35%, which is low enough to stop the cables corroding.

Thermal stability of oxo-salts

The carbonates and nitrates of the s block elements usually decompose on heating. *In general, the alkali metal salts are more stable than those of the alkaline earth metals and thermal stability increases on passing down each group*. For example, lithium carbonate and the group 2 carbonates form the oxide and carbon dioxide on heating whilst the other group 1 carbonates are thermally stable.

$$\mathrm{Li_2CO_3(s) \longrightarrow Li_2O(s) + CO_2(g)}$$
$$\mathrm{MgCO_3(s) \longrightarrow MgO(s) + CO_2(g)}$$

In effect, such reactions simply involve the breakdown of the carbonate ion. This is promoted by the polarising effect of the cation that attracts electron density from the anion. If the cation is small enough, its charge density will cause sufficient polarisation to break the bonds within the carbonate ion.

A similar effect occurs with the s block nitrates. Only the nitrates of lithium and the alkaline earth metals decompose fully on heating to give the metal oxide.

$$\mathrm{4LiNO_3(s) \longrightarrow 2Li_2O(s) + 4NO_2(g) + O_2(g)}$$
$$\mathrm{2Ca(NO_3)_2(s) \longrightarrow 2CaO(s) + 4NO_2(g) + O_2(g)}$$

The other alkali metal nitrates do decompose on heating but only as far as the nitrite and oxygen.

$$\mathrm{2NaNO_3(s) \longrightarrow 2NaNO_2(s) + O_2(g)}$$

Figure 19.7 illustrates the effect of anion polarisation on the thermal decomposition of metal carbonates and nitrates together with hydrogencarbonates and hydroxides.

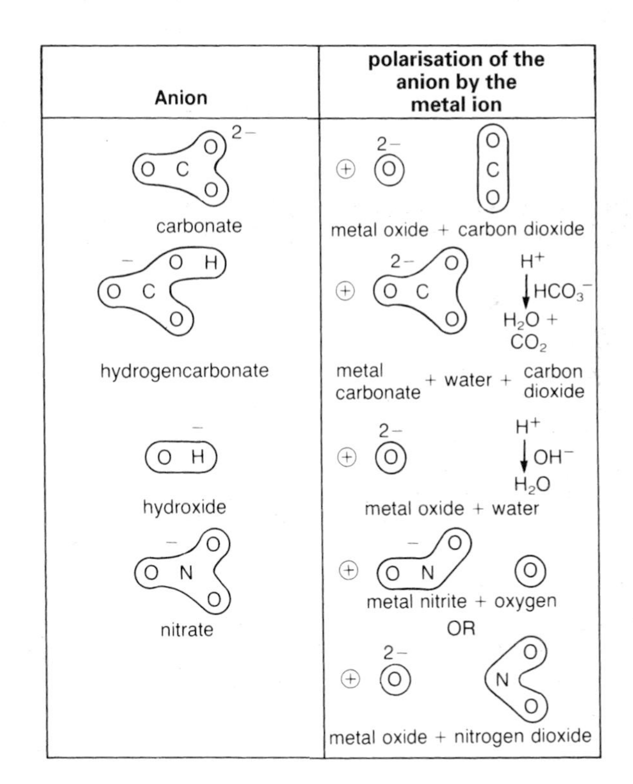

Figure 19.7
Thermal dissociation as a result of anion polarisation – which of the group 1 metal nitrates gives nitrogen dioxide on heating? Why is this?

Focus 19b

1. The thermal stability of the nitrates and carbonates decrease with increasing charge density of the metal ion.
2. Lithium and beryllium chlorides are appreciably soluble in ethanol, indicating considerable covalent character.
3. Virtually all the compounds of the alkali metals are water soluble, whereas the higher lattice energies of group 2 compounds tend to make them less so.

Table 19.6

metal	common sources
sodium	rocksalt, $NaCl$
potassium	carnallite, $KCl/MgCl_2$
magnesium	magnesite, $MgCO_3$ dolomite, $MgCO_3/CaCO_3$
calcium	limestone, $CaCO_3$

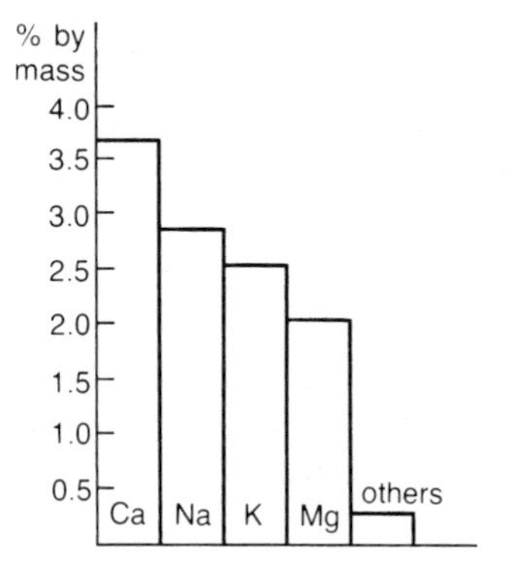

Figure 19.8
Relative abundance by mass of the alkali and alkaline earth metals

19.4 Occurrence and extraction

As shown in Figure 19.8 sodium, potassium, magnesium and calcium are quite abundant in nature but are too reactive to be found in the free state. Some of the main sources of these elements are shown in Table 19.6.

Since most sodium and potassium compounds are water soluble, they tend to be washed out into the sea, which contains on average about 3.5% by mass of dissolved salts.

Before the s block metals can be extracted, seawater must be evaporated to leave solid material. This is expensive in energy terms and it is often commercially preferable to use evaporite minerals such as rocksalt (NaCl) which have been formed by natural evaporation of ancient seas.

These highly reactive metals are extracted by electrolysis of their molten chlorides. Details of the manufacture of sodium using the Downs cell can be found in Chapter 23.

Activity 19.4

Table 19.7 shows the typical ionic content of seawater.

Table 19.7

cations	% by mass in seawater	anions	% by mass in seawater
Na^+	1.076	Cl^-	1.935
Mg^{2+}	0.130	SO_4^{2-}	0.271
Ca^{2+}	0.041	CO_3^{2-}	0.014
K^+	0.040	Br^-	0.007

1 How do you account for the fact that the most abundant s block metal in seawater is sodium, although in the earth's crust in general, calcium is most abundant?

2 This question concerns the production of 1 tonne (1000 kg) of sodium from seawater.
a) What mass of seawater contains this amount of sodium?
b) The molar enthalpy of evaporation of water is 41 kJ mol^{-1}. How much energy is required to evaporate the seawater?
c) The calorific value of natural gas is about 38 MJ m^{-3}. Calculate the volume of natural gas needed to evaporate the water. If the gas is charged at £0.15 m^{-3} what is the energy cost involved?

Questions on Chapter 19

1 Write balanced equations showing how each of the following metals reacts (if at all) with (i) chlorine, (ii) excess oxygen and (iii) cold water:

lithium, sodium, potassium, magnesium.

How do you account for any differences in behaviour?

2 All metal nitrates are soluble in water, whereas most metal carbonates are insoluble. How do you account for this?

3 How and why do the following properties vary on passing down group 2: atomic radius, 1st ionisation energy, electronegativity, standard electrode potential, melting point?

4 Write a balanced equation for the decomposition of the least thermally stable nitrate of an s block metal.

5 Which group 1 carbonate decomposes readily on heating? Why is this less stable than the other alkali metal carbonates?

6 Which group 1 chloride is appreciably soluble in ethanol? How do you account for this unusual behaviour?

7 a) Explain the trends in the lattice energies of the alkali metal fluorides and the hydration energies of the cations shown in Table 19.8.

Table 19.8

	lithium	sodium	potassium	rubidium	caesium
ΔH_l[MF] /kJ mol^{-1}	−1022	−902	−801	−767	−716
ΔH_h[M$^+$] /kJ mol^{-1}	−519	−406	−322	−301	−276

b) Given that the hydration energy of the fluoride ion is −506 kJ mol^{-1}, estimate the enthalpies of solution of the alkali metal fluorides and comment on the likely trend in water solubility on passing down the group.

8 Sketch Born–Haber cycles for the formation of NaBr(s) and $MgBr_2$(s). Use the information in the box on Born–Haber cycles, together with the following data, to estimate the lattice enthalpies of each of these compounds.

atomisation energy of bromine = +97 kJ mol^{-1}
1st electron affinity of bromine = −342 kJ mol^{-1}
standard enthalpies of formation:
NaBr(s) = −360 kJ mol^{-1}, $MgBr_2$(s) = −518 kJ mol^{-1}

How would you expect the lattice enthalpy of calcium bromide to compare with that of magnesium bromide? Explain your answer.

Comments on the activities

Activity 19.1

All enthalpy changes in kJ mol^{-1}

1 For MgCl(s)

atomisation:
$Mg(s) \longrightarrow Mg(g)$ $\quad \Delta H_a$[Mg] +150
ionisation:
$Mg(g) \longrightarrow Mg^+(g)$ $\quad \Delta H_i(1)$[Mg] +736
atomisation:
$\frac{1}{2}Cl_2(g) \longrightarrow Cl(g)$ $\quad \Delta H_a$[Cl] +121
electron affinity:
$Cl(g) \longrightarrow Cl^-(g)$ $\quad \Delta H_e$[Cl] −364
lattice formation:
$Mg^+(g) + Cl^-(g) \longrightarrow MgCl(s)$ $\quad \Delta H_l$[MgCl] −771
overall formation:
$Mg(s) + \frac{1}{2}Cl_2(g) \longrightarrow MgCl(s)$ $\quad \Delta H_f$[MgCl] −128

For NaCl$_2$(s)

atomisation:
$Na(s) \longrightarrow Na(g)$ $\quad \Delta H_a$, [Na] +109
ionisation:
$Na(g) \longrightarrow Na^+(g)$ $\quad \Delta H_i(1)$[Na] +494
$Na^+(g) \longrightarrow Na^{2+}(g)$ $\quad \Delta H_i(2)$[Na] +4560
atomisation:
$Cl_2(g) \longrightarrow 2Cl(g)$ $\quad \Delta H_a[Cl] \times 2$ +242
electron affinity:
$2Cl(g) \longrightarrow 2Cl^-(g)$ $\quad \Delta H_e[Cl] \times 2$ −728
lattice formation:
$Na^{2+}(g) + 2Cl^-(g) \longrightarrow NaCl_2(s)$ $\quad \Delta H_l[NaCl_2]$ −2502
overall formation:
$Na(s) + Cl_2(g) \longrightarrow NaCl_2(s)$ $\quad \Delta H_f[NaCl_2]$ +2175

The Born–Haber cycles for MgCl and $NaCl_2$ are shown graphically in Figures 19.9a and 19.9 b respectively.

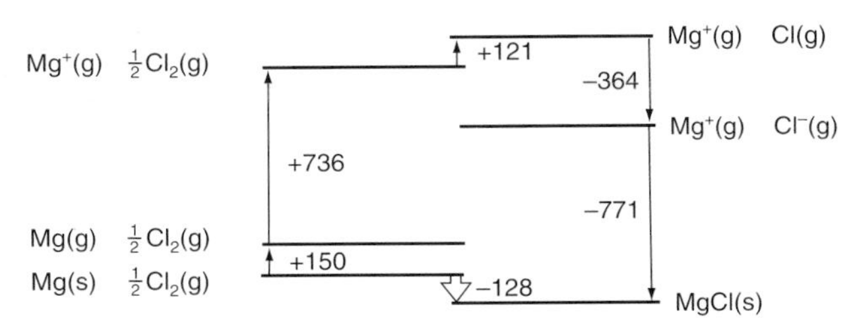

Figure 19.9a
Born–Haber cycle for MgCl(s). All enthalpy changes in kJ mol^{-1}

2 The very high second ionisation energy of sodium is mainly responsible for the instability of $NaCl_2$. The magnesium atom can lose its two outer electrons relatively easily before breaking into an inner shell.

3 The calculated enthalpy changes are

$Mg(s) + \frac{1}{2}Cl_2(g) \longrightarrow MgCl(s)$ $\quad \Delta H = -128$ kJ mol^{-1}

Comments on the activities *continued*

$Mg(s) + \frac{1}{2}Cl_2(g) \longrightarrow \frac{1}{2}MgCl_2(s) + \frac{1}{2}Mg(s)$

$\Delta H = -326\ kJ\ mol^{-1}$

Thus when magnesium is heated with a limited supply of chlorine it is still energetically more favourable to form $MgCl_2$ rather than MgCl.

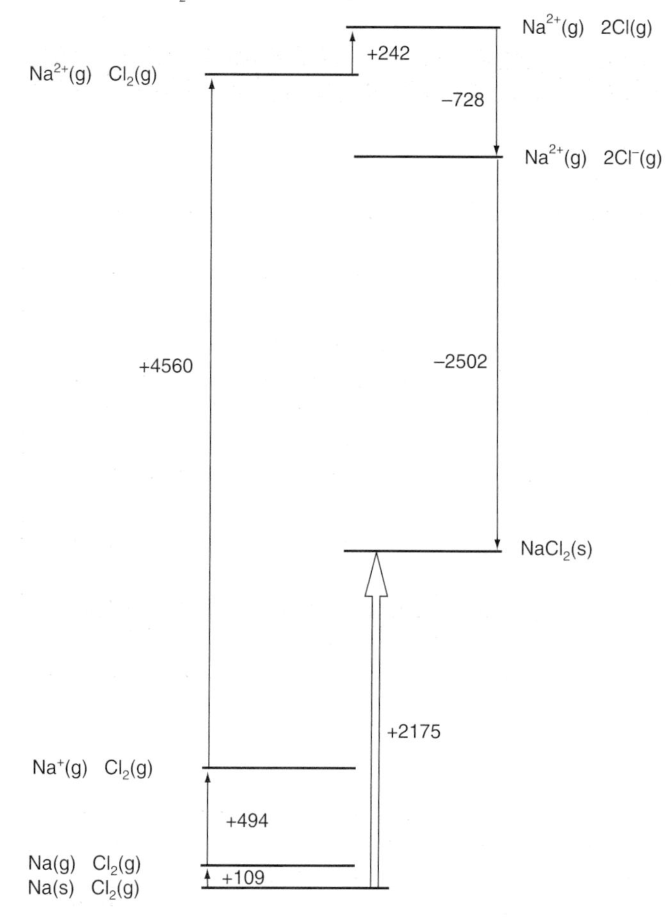

Figure 19.9b
Born–Haber cycle for $NaCl_2(s)$. All enthalpy changes in kJ mol^{-1}

Activity 19.2

1 For the reaction of magnesium with water

atomisation:

$Mg(s) \longrightarrow Mg(g)$	$\Delta H_a[Mg]$	+150

ionisation:

$Mg(g) \longrightarrow Mg^+(g)$	$\Delta H_i(1)[Mg]$	+736
$Mg^+(g) \longrightarrow Mg^{2+}(g)$	$\Delta H_i(2)[Mg]$	+1450

hydration:

$Mg^{2+}(g) \longrightarrow Mg^{2+}(aq)$	$\Delta H_h[Mg^{2+}]$	–1920

–(neutralisation):

$2H_2O(l) \longrightarrow 2H^+(aq) + 2OH^-(aq)$	$-\Delta H_n \times 2$	+114

–(hydration):

$2H^+(aq) \longrightarrow 2H^+(g)$	$-\Delta H_h[H^+] \times 2$	+2182

–(ionisation):

$2H^+(g) \longrightarrow 2H(g)$	$-\Delta H_i[H] \times 2$	–2630

–(atomisation):

$2H(g) \longrightarrow H_2(g)$	$-\Delta H_{at}[H] \times 2$	–436

overall process:

$Mg(s) + H_2O(l) \longrightarrow Mg^{2+}(aq) + 2OH^-(aq) + H_2(g)$

ΔH –354

2 The overall enthalpy change shows that the reaction of magnesium with water is energetically favourable but gives no information on the reaction rate. Magnesium does in fact react very slowly with cold water but, since the reaction proceeds readily on heating, it seems that the relatively high ionisation energies of magnesium must contribute to a high activation energy.

Activity 19.3

1

	Li^+	Na^+	K^+	Rb^+	Cs^+
ΔH_{sol} [MOH]	–21	–43	–55	–63	–71
solubility/ g dm^{-3} at 20°C	128	1090	1120	1770	3300

2 On passing down the group the metal ion gets bigger and its charge density falls. This means it is less strongly attracted to the anion and so the lattice enrgy falls.

3 Hydration energy is also related to the charge density of the ion. Larger ions are less attractive to the polar water molecules and so release less energy on hydration.

4 As more energy is released on solution and less energy is required to break the lattice, we would expect the solubility of the hydroxides to increase on passing down group 1. This is confirmed by the figures in the above table. Although the actual figures are much lower, a similar trend is found in the solubility of the alkaline earth metals' hydroxides.

Activity 19.4

1 All sodium compounds are water soluble but common calcium compounds, e.g. calcium sulphate and calcium carbonate, are only sparingly soluble in water and hence are not so readily leached out.

2 **a)** 100 t of seawater contains 1.076 t of sodium

$\frac{100}{1.076} = 92.9$ t of seawater contains 1 t of sodium

b) The mass of 1 mole of water is 18 g

92.9 t water must be evaporated,

i.e. $92.9 \times 1000 \times 1000 = 9.29 \times 10^7$ g

energy required $= \frac{41 \times 9.29 \times 10^7}{18} = 2.12 \times 10^8$ kJ,

i.e. 212 000 MJ

c) 1 m^3 of natural gas supplies 38 MJ of energy

hence $\frac{212000}{38} = 5579$ m^3 of gas is required

at £0.15 m^{-3} this will cost $0.15 \times 5579 = £837$

C H A P T E R

The p block

Contents

Study Checklist

After studying this chapter you should be able to:

1. Describe the bonding and structure of the chlorides of boron, aluminium, carbon and silicon and explain their behaviour towards water.
2. Recall the acid–base nature of the oxides of boron, aluminium, carbon, silicon and lead.
3. Recall that Sn(II) is reducing and that Pb(IV) is oxidising.
4. Use displacement reactions of the halogens as evidence for the decrease in oxidising power on passing down the group and show how this is applied in the extraction of bromine from sea water.
5. State and explain the effect of concentrated sulphuric acid on solid sodium halides.
6. Recall that HF differs from the other hydrogen halides in being a weak acid and having an unusually high boiling point.
7. Recall the disproportionation reactions of chlorine with aqueous alkali under different conditions to give either chlorate(I) or chlorate(V).

The s and d blocks of the periodic table each contain largely similar elements. In the p block, however, there is a much greater variation in properties: from typical metals such as aluminium, through semi-metals like germanium to the non-metals, which include all of the elements in groups 7 and 0. In this chapter we can only explore in detail selected examples of the chemistry of the p block but Tables 20.1 to 20.6 provide data for all the elements and are included for reference purposes.

Table 20.1 Data sheet for the elements of group 3

Element	atomic number	electron configuration	atomic radius* /nm	ionic radius† /nm	ionisation energies /kJ mol^{-1} 1st	2nd	3rd	4th	electro-negativity	$E^\ominus$ /volts‡	density /g cm^{-3}	m.p. /°C
boron B	5	$[He]2s^2 2p^1$	0.080	(0.020)	799	2420	3660	25000	2.0		2.34	2300
aluminium Al	13	$[Ne]3s^2 3p^1$	0.125	0.050	577	1820	2740	11600	1.5	−1.66	2.70	660
gallium Ga	31	$[Ar]3d^{10}4s^2 4p^1$	0.125	0.062	577	1980	2960	6190	1.6	−0.53	5.91	30
indium In	49	$[Kr]4d^{10}5s^2 5p^1$	0.150	0.081	556	1820	2700	5230	1.7	−0.34	7.30	157
thallium Tl	81	$[Xe]4f^{14}5d^{10}6s^2 6p^1$	0.155	0.095	590	1970	2870	4900	1.8	+0.72	11.8	304

*values are 'covalent' radii
† for the 3+ ions – the value for B^{3+} is theoretical only, since the ion does not exist. Why?
‡ for the process $M^{3+}(aq) + 3e^- \rightleftharpoons M(s)$

Why are the atomic radii of aluminium and gallium similar?
Which is the better oxidising agent $Al^{3+}(aq)$ or $Tl^{3+}(aq)$?

Table 20.2 Data sheet for the elements of group 4

Element	atomic number	electron configuration	atomic radius* /nm	ionic radius† /nm	ionisation energies /kJ mol^{-1} 1st	2nd	3rd	4th	5th	electro-negativity	density /g cm^{-3}	m.p. /°C
carbon C	6	$[He]2s^2 p^2$	0.077	–	1090	2350	4610	6220	37800	2.5	2.25$^{(1)}$ 3.51$^{(2)}$	3730 (sub)
silicon Si	14	$[Ne]3s^2 3p^2$	0.177	–	786	1580	3230	4360	16 000	1.8	2.33	1410
germanium Ge	32	$[Ar]3d^{10}4s^2 4p^2$	0.122	0.093	762	1540	3300	4390	8950	1.8	5.35	937
tin Sn	50	$[Kr]4d^{10}5s^2 5p^2$	0.140	0.112	707	1410	2940	3930	7780	1.8	7.28	232
lead Pb	82	$[Xe]4f^{14}5d^{10}6s^2 6p^2$	0.154	0.120	716	1450	3080	4080	6700	1.8	11.30	327

*values are 'covalent' radii
† for the 2+ ions
(1) graphite; (2) diamond

Why is diamond denser than graphite?
Why is there a 'big jump' between the values of the 4th and 5th ionisation energies for all these elements?

Table 20.3 Data sheet for the elements of group 5

Element	atomic number	electron configuration	atomic radius* /nm	ionisation energies/kJ mol^{-1} 1st	2nd	3rd	4th	5th	6th	electro-negativity	density /g cm^{-3}	m.p. /°C
nitrogen N	7	[He]$2s^2p^3$	0.074	1400	2860	4590	7480	9440	53200	3.0	0.81(1)	-210
phosphorus P	15	[Ne]$3s^23p^3$	0.110	1060	1900	2920	4960	6280	21200	2.1	1.82(2) 2.34(3)	44(2) 590(3)
arsenic As	33	[Ar]$3d^{10}4s^24p^3$	0.121	966	1950	2730	4850	6020	12300	2.0	5.72	613 (sub)
antimony Sb	51	[Kr]$4d^{10}5s^25p^3$	0.141	833	1590	2440	4270	5360	10400	1.9	6.62	630
bismuth Bi	83	[Xe]$4f^{14}5d^{10}6s^26p^3$	0.152	703	1610	2460	4350	5400	8500	1.9	9.80	271

*values are 'covalent' radii
(1) measured for the liquid at its boiling point (–196°C); (2) for 'white' phosphorus; (3) for 'red' phosphorus

How does electronegativity vary on passing down the group?
Can you explain this trend?

What do the comparative densities and melting points suggest about the structures of 'white' and 'red' phosphorus?

Table 20.4 Data sheet for the elements of group 6

Element	atomic number	electron configuration	atomic radius* /nm	ionisation energies/kJ mol^{-1} 1st	2nd	3rd	4th	5th	6th	7th	electro-negativity	density /g cm^{-3}	m.p. /°C
oxygen O	8	[He]$2s^22p^4$	0.074	1310	3390	5320	7450	11000	13300	71000	3.5	1.15†	–218
sulphur S	16	[Ne]$3s^23p^4$	0.104	1000	2260	3390	4540	6990	8490	27100	2.5	2.07 (rhombic)	113
selenium Se	34	[Ar]$3d^{10}4s^24p^4$	0.117	941	2080	3090	4140	7030	7870	16000	2.4	4.81	217
tellurium Te	52	[Kr]$4d^{10}5s^25p^4$	0.137	870	1800	3010	3680	5860	7000	13200	2.1	6.25	450
polonium Po	84	[Xe]$4f^{14}5d^{10}6s^26p^4$	0.140	812							2.0	9.32	254

*values are 'covalent' radii
† measured for the liquid at its boiling point (–183°C)
Why is the melting point of oxygen much lower than that of the other elements?
Why is there a 'big jump' between the values of the 6th and 7th ionisation energies for all these elements?

Table 20.5 Data sheet for the elements of group 7

Element	atomic number	electron configuration	atomic radius* /nm	ionisation energies /kJ mol⁻¹ 1st	2nd	3rd	4th	5th	6th	7th	8th	electron affinity /kJ mol⁻¹	electro-negativity	standard electrode potential $E^\ominus$/V†	density /g cm⁻³	m.p. /°C
fluorine F	9	[He]$2s^22p^5$	0.072	1680	3370	6040	8410	11000	15100	17900	91600	–348	4.0	+2.87	1.11‡	–220
chlorine Cl	17	[Ne]$3s^23p^5$	0.099	1260	2300	3850	5150	6540	9330	11000	33600	–364	3.0	+1.36	1.56‡	–101
bromine Br	35	[Ar]$3d^{10}4s^24p^5$	0.114	1140	2030	3460	4850	5770	8370	10000	20300	–342	2.8	+1.07	3.12	–7
iodine I	53	[Kr]$4d^{10}5s^25p^5$	0.133	1010	1840	3000	4030	5000	7400	8700	16400	–314	2.5	+0.54	4.93	114
astatine At	85	[Xe]$4f^{14}5d^{10}6s^26p^5$	0.140	920								?	2.2	?	?	302

*values are 'covalent' radii

†$E^\ominus$ is the standard electrode potential for the reaction $\frac{1}{2}X_2 + e \rightleftharpoons X^-$ in aqueous solution

‡ measured for the liquid at its boiling point

Are you surprised by the trend in the values of electron affinity?

What do the above standard electrode potentials show about the relative oxidising power of the halogens in aqueous solution?

Table 20.6 Data sheet for the elements of group 0

Element	atomic number	electron configuration	atomic radius* /nm	ionisation energies /kJ mol⁻¹ 1st	2nd	3rd	4th	5th	6th	7th	8th	density† /g cm⁻³	b.p. /°C	m.p. /°C
helium He	2	$1s^5$	0.120	2370	5250	–	–	–	–	–	–	0.147	–269	–270
neon Ne	10	[He]$2s^2p^6$	0.160	2080	3950	6150	9290	12100	15200	20000	23000	0.20	–246	–249
argon Ar	18	[Ne]$3s^23p^6$	0.192	1520	2660	3950	5770	7240	8790	12000	13800	1.40	–186	–189
krypton Kr	36	[Ar]$3d^{10}4s^24p^6$	0.197	1350	2370	3560	5020	6370	7570	10700	12200	2.16	–152	–157
xenon Xe	54	[Kr]$4d^{10}5s^25p^6$	0.217	1170	2050	3100	4300	5800	8000	9800	12200	3.52	–108	–112
radon Rn	86	[Xe]$4f^{14}5d^{10}6s^26p^6$	?	1040	1930	2890	4250	5310	?	?	?	4.4	–62	–71

*values are van der Waals' radii

† measured for the liquid at its boiling point

How and why do the boiling points of these elements vary on passing down the group?

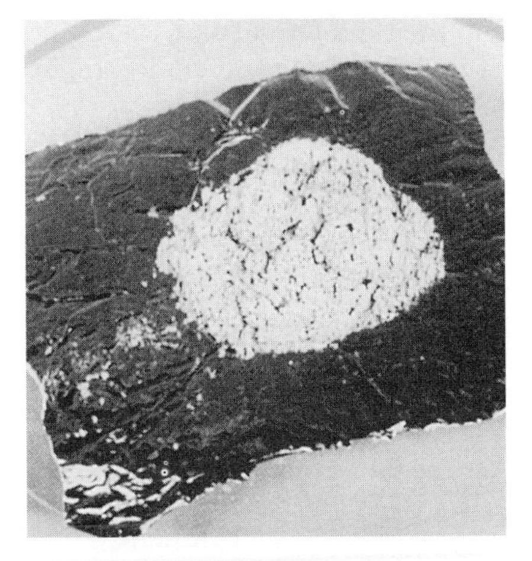

▲ The central area of this sheet of aluminium foil has reacted with moisture in the surrounding air following the removal of its oxide layer

20.1 Groups 3 and 4

Aluminium

Aluminium has a very wide range of uses, from cooking foil to aircraft and engine construction. Its low density and very good electrical conductivity, together with its excellent resistance to corrosion, makes it ideal for overhead electricity power cables. When alloyed with other metals, aluminium can be stronger than steel. Over a hundred different aluminium alloys are available, two of the commonest being:

duralumin	95% Al,	4% Cu,	0.5% Mg,	0.5% Mn
magnalium	95% Al,	5% Mg		

The corrosion resistance of aluminium results from the reaction of the metal with air, forming a thin, tough, impermeable **oxide layer** that protects the metal from further attack. If the oxide layer is chemically removed, e.g. by rubbing the surface of aluminium foil with mercury(II) chloride solution, the exposed metal reacts rapidly and exothermically with atmospheric moisture to give aluminium hydroxide.

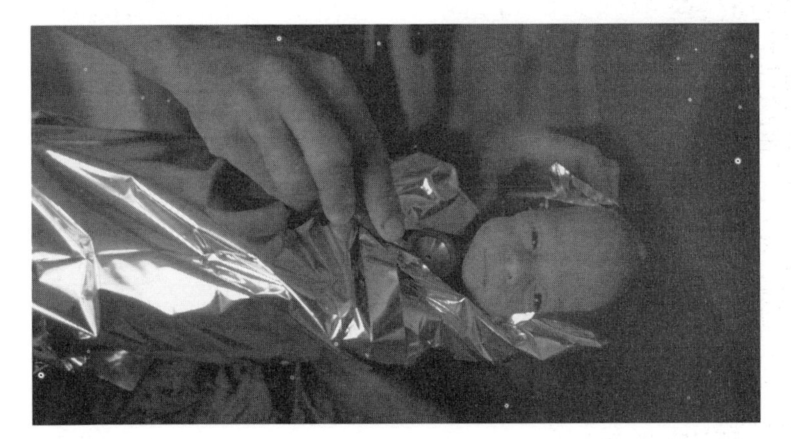

Aluminium is used to keep babies warm straight after being born ▶

The thickness of this protective oxide layer can be increased by **anodising**. The aluminium is used as the positive electrode or anode, in an electrolytic cell containing dilute sulphuric acid. Oxygen formed at the metal surface reacts to build up the oxide layer:

$$4OH^-(aq) \longrightarrow 2H_2O(l) + O_2(g) + 4e^-$$

Anodised aluminium can be coloured, as the oxide readily takes up a range of dyes. Aluminium sulphate solution may be used in a similar way as a **mordant** to fix dyes onto fabric. Treatment with an alkali precipitates aluminium hydroxide in the pores of the cloth.

Although aluminium is the most abundant metal in the Earth's crust, its low concentration in many minerals and its reactive nature, makes it quite difficult to extract. Details of the manufacture of aluminium are given in Chapter 23. Corundum is a naturally occurring form of aluminium oxide which, because of its extreme hardness, is used as an abrasive. Traces of iron and other transition metal impurities in corundum produce coloured gemstones such as ruby (containing chromium) and sapphire (containing titanium).

Focus 20a

1. Aluminium is the commonest metallic element in the Earth's crust.
2. Though quite reactive, it is generally protected from atmospheric attack by a thin impervious oxide layer that may be reinforced by anodising.
3. The main uses of aluminium depend upon its corrosion resistance and low density.

Carbon

Pure carbon can exist in several different forms or **allotropes** of which the best known are **diamond** and **graphite**. Though very different in many of their properties, both of these have regular macromolecular structures, shown in Figure 20.1.

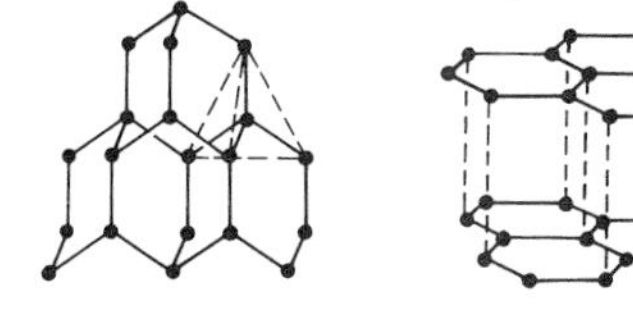

Figure 20.1 ▲
The structure of the allotropes of carbon

Diamond is the hardest natural material

The extreme hardness of diamond results from the rigid tetrahedral arrangement in which each carbon atom is connected to four others by single covalent bonds. Graphite is composed of flat hexagonal sheets of carbon atoms that can readily slide over each other, accounting for its 'slippery' feel. Since in this structure each atom has a free electron that is delocalised along the sheet, *graphite, unlike diamond, is a good conductor of electricity*.

Under normal circumstances graphite is energetically slightly more stable than diamond:

$$\text{C(graphite)} \longrightarrow \text{C(diamond)} \ \Delta H = +2 \text{ kJ}$$

At high temperature and pressure, however, the relative stabilities are reversed. This accounts for the fact that natural diamond is often found in places where molten magma from deep in the Earth has forced itself closer to the surface and solidified. Since the 1950s such conditions have been reproduced on a small scale in the manufacture of 'industrial' diamonds.

Although under normal conditions diamond is less stable than graphite, there is no chance that an expensive diamond ring will suddenly turn into graphite. The carbon–carbon bonds are so strong that the **activation energy** required for the conversion is far too high.

In 1985 **buckminsterfullerene**, a molecular allotrope of carbon, was first prepared by condensing carbon vapour, produced from graphite using a laser. In this structure 60 atoms are linked to form a more or less spherical molecule that resembles a soccer ball, leading to its nickname, '**buckyball**'. Unlike the macromolecular carbon allotropes, C_{60} is soluble in organic solvents such as benzene. A whole family of similar '**fullerenes**' with differing numbers of atoms has now been prepared together with cylindrical carbon molecules known as '**buckytubes**'. These novel forms of carbon have several unique and unexpected properties and are currently being investigated for applications as diverse as semiconductors, high temperature lubricants and photoluminescent displays.

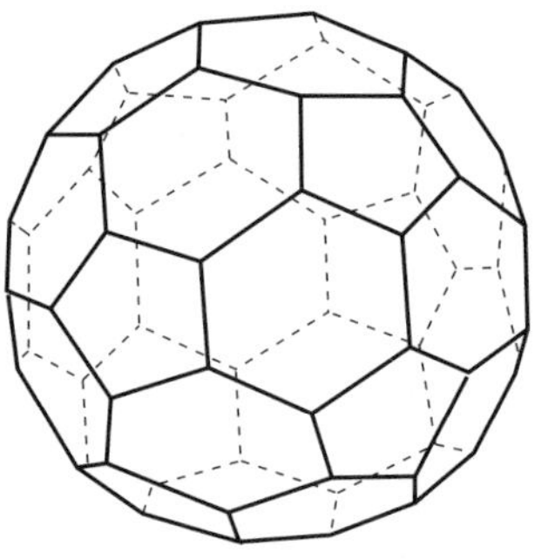

C_{60}: the carbon football

Focus 20b

1. Carbon exists as several different allotropes.
2. Diamond and graphite have quite different properties that depend upon their macromolecular structures.
3. Recently spherical molecular forms of carbon known as 'fullerenes' have been isolated.
4. Silicon adopts a diamond type macromolecular structure.
5. When doped with impurity atoms, the silicon lattice can produce different types of semiconductor.
6. Integration of large numbers of semiconductor devices on the same silicon chip has led to the miniaturisation revolution in electronics.

Silicon

Next to oxygen, silicon is the Earth's most common element. Although the element is not found in its free state, sand is largely silicon dioxide and silicate minerals constitute about 75% of the Earth's crust. Silicon is a component of glass and is used to make abrasives and waterproof polymers but it is in microelectronics, especially the computer revolution, that it has made the biggest impact.

Silicon chips

Silicon has a macromolecular structure very similar to diamond in which each atom is connected to four others by single covalent bonds (Figure 20.2a.) Since all four outer electrons are fixed in position, pure silicon does not conduct electricity under normal conditions. However, it is possible to replace some silicon atoms in the lattice with other elements in a process known as '**doping**'. To fit into the lattice, each of the impurity atoms 'tries' to bond in a similar way to silicon, i.e. by using four outer electrons to form single covalent bonds.

In the case of phosphorus or arsenic, from group 5, this leaves a single unpaired electron on each impurity atom which is free to move throughout the doped lattice (Figure 20.2b). Since there is a surplus of free electrons which are *negatively* charged this is known as **n type semiconductor**.

Group 3 elements, such as aluminium or gallium, can also replace silicon. Since these atoms have only three outer electrons, each leaves one bond incomplete (Figure 20.2c). Such *positive* **holes** can be filled by electrons moving in from elsewhere in the lattice but this, of course, produces new holes. Thus, as with the n type, electrons are free to move around within a **p type semiconductor**.

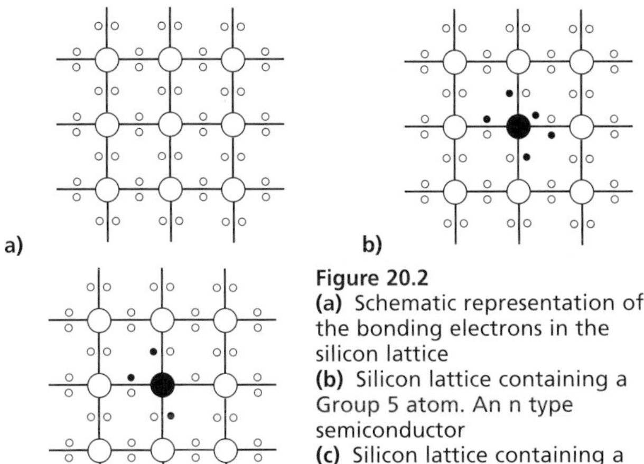

Figure 20.2
(a) Schematic representation of the bonding electrons in the silicon lattice
(b) Silicon lattice containing a Group 5 atom. An n type semiconductor
(c) Silicon lattice containing a Group 3 atom. A p type semiconductor

The basis of many miniaturised electronic components is the **junction diode** in which n and p regions are placed together. Figure 20.3 shows a circuit in which such a device acts as a rectifier, converting alternating current into pulsed direct current.

When electrons flow into the n side, they readily cross the junction to fill the holes in the p region, allowing a current to pass. However, the device cannot readily conduct electricity in the opposite direction as electrons would be trying to leave a positive region and enter a 'negative' one.

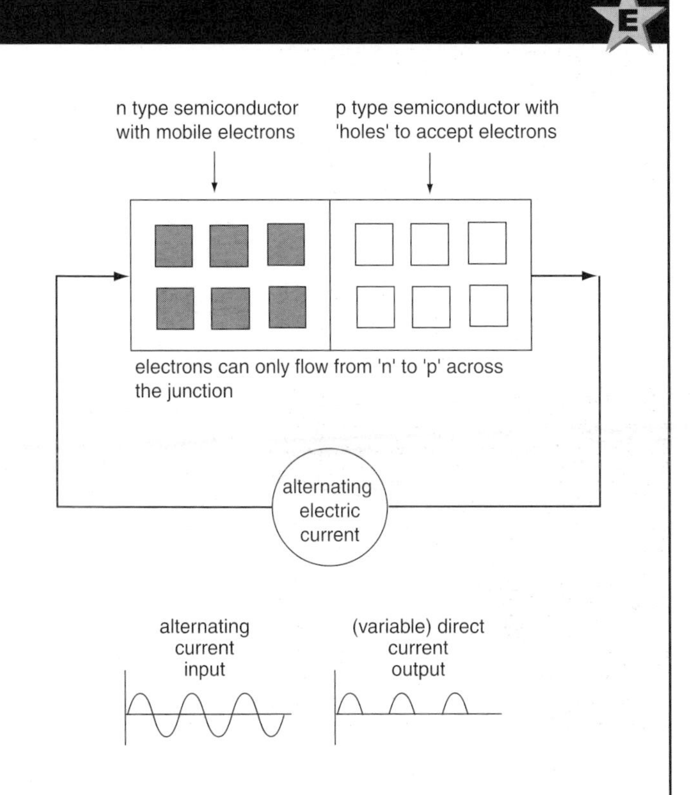

Figure 20.3 ▲
A junction diode rectifier

Literally millions of such similar components can now be integrated on a single silicon **chip** allowing a degree of miniaturisation in electronic systems that would have seemed incredible only a few years ago.

▲ An integrated circuit 'chip'. Gold wires connect the chip to copper tracks leading to other components

Compounds

Chlorides

Table 20.7 compares some of the properties of the chlorides of boron, aluminium, carbon and silicon. With the exception of aluminium chloride, they are liquids or gases at room temperature, which indicates a simple covalent molecular structure.

Table 20.7

	BCl_3	Al_2Cl_6	CCl_4	$SiCl_4$
m.p./°C	–107	192 sublimes	–23	–70
b.p./°C	13	–	77	58
cold water	violent reaction acidic	vigorous reaction acidic	no reaction	vigorous reaction acidic

Since aluminium chloride sublimes on heating it is conveniently prepared in the laboratory by direct combination of the elements in the apparatus shown in Figure 20.4.

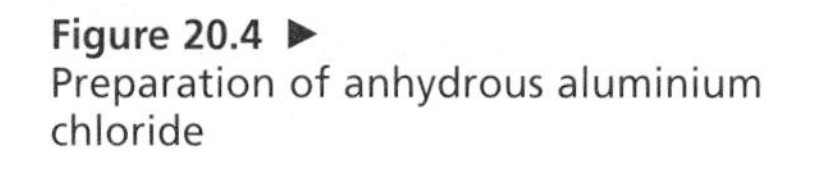

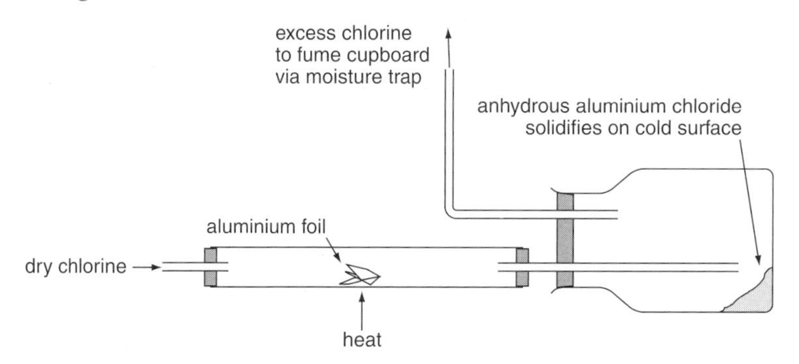

Figure 20.4 ▶ Preparation of anhydrous aluminium chloride

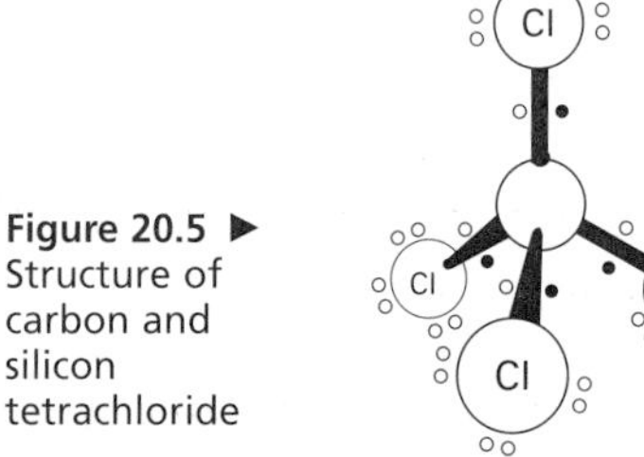

Figure 20.5 ▶ Structure of carbon and silicon tetrachloride

Figure 20.6 ▶ Structure of boron trichloride

Carbon and silicon chlorides have the tetrahedral structure shown in Figure 20.5 with each atom achieving a noble gas electron configuration. Boron forms only three covalent bonds in the triangular shaped molecule shown in Figure 20.6. As a result the boron outer shell is two electrons short of a noble gas configuration. This explains why boron trichloride reacts vigorously with electron pair donors such as water. At high temperatures aluminium chloride vapour contains similar triangular molecules but the solid consists of dimers in which two $AlCl_3$ units are joined by dative bonds between chlorine and aluminium atoms (Figure 20.7).

Even though the aluminium atoms have a noble gas electron configuration in this structure, aluminium chloride reacts vigorously with water to form aqueous ions:

$$Al_2Cl_6(s) \longrightarrow 2Al^{3+}(aq) + 6Cl^-(aq)$$

The high charge density of the Al^{3+} ion polarises co-ordinated water molecules giving a **strongly acidic** solution:

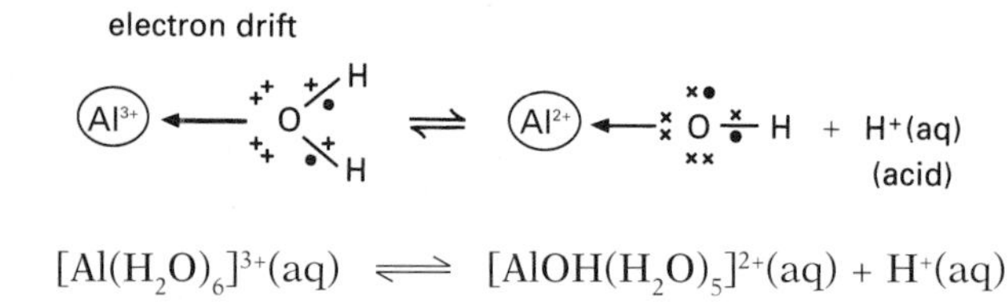

$$[Al(H_2O)_6]^{3+}(aq) \rightleftharpoons [AlOH(H_2O)_5]^{2+}(aq) + H^+(aq)$$

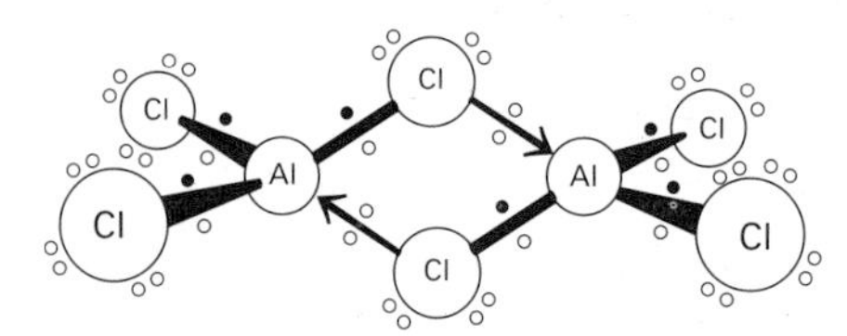

Figure 20.7 ▲ Structure of solid aluminium chloride

The main species in the dilute aqueous solution is still $[Al(H_2O)_6]^{3+}(aq)$ but the addition of alkali removes further protons, precipitating hydrated aluminium hydroxide, $[Al(OH)_3(H_2O)_3](s)$. This is an amphoteric (or amphiprotic) hydroxide, dissolving in both acid and excess strong alkali:

$$[Al(H_2O)_6]^{3+}(aq) \underset{\text{acid}}{\rightleftharpoons} [Al(OH)_3(H_2O)_3](s) \overset{\text{excess alkali}}{\rightleftharpoons} [Al(OH)_4(H_2O)_2]^-(aq)$$

A more detailed account of the acid–base reactions of aqueous metal ions is given in section 21.3.

Silicon tetrachloride reacts vigorously with cold water.

$$SiCl_4(l) + 4H_2O(l) \longrightarrow Si(OH)_4(s) + 4H^+(aq) + 4Cl^-(aq)$$

The initial attack takes place by the donation of a lone pair of electrons from the oxygen atom of the water molecule into a vacant orbital in the valency shell of the silicon atom. Such rapid attack cannot happen with CCl_4 since the valency shell of the central carbon atom is completely full.

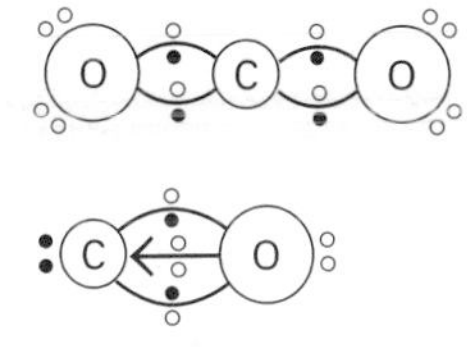

Figure 20.8 ▲
Multiple bonding in the oxides of carbon

Oxides

Table 20.8 compares the properties of the common oxides of boron, aluminium, carbon and silicon. Carbon monoxide and dioxide are the only **gases** and have simple covalent molecular structures in which the carbon atoms form **stable multiple bonds** (Figure 20.8).

Table 20.8

	B_2O_3	Al_2O_3	CO	CO_2	SiO_2
m.p./°C	460	2040	–205	–56	1710
b.p./°C	1860	2980	–191	–78	2230
cold water	soluble acidic	insoluble	insoluble	soluble weak acid	insoluble
acid	no reaction	salt formed	no reaction	no reaction	no reaction
alkali	salt formed	salt formed	salt formed	salt formed	salt formed

The high melting points of the remaining oxides suggest some sort of giant structure. Aluminium oxide conducts electricity when liquid indicating an ionic lattice composed of Al^{3+} and O^{2-} ions. Boron and silicon oxides are both non-conductors consisting of covalent macromolecules in which the oxygen atoms form two single covalent bonds (Figure 20.9).

All the oxides are **acidic** to some extent with both boron oxide and carbon dioxide dissolving in water:

$$B_2O_3(s) + 3H_2O(l) \longrightarrow 2H_3BO_3(aq)$$
$$CO_2(g) + H_2O(l) \rightleftharpoons H_2CO_3(aq)$$

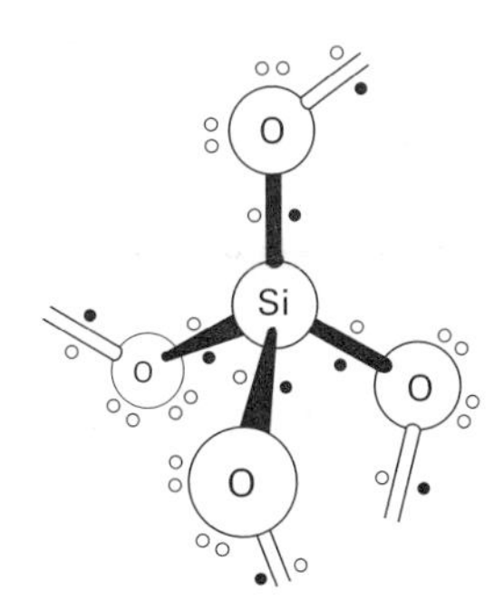

Figure 20.9 ▲
Macromolecular structure of silicon dioxide

Although the other oxides are insoluble in water, they show acidic properties by forming salts on heating with concentrated or molten alkalis:

$$Al_2O_3(s) + 2NaOH(aq) + 3H_2O(l) \longrightarrow 2NaAl(OH)_4(aq)$$
$$CO(g) + NaOH(aq) \longrightarrow HCOONa(aq)$$
$$SiO_2(s) + 2NaOH(l) \longrightarrow Na_2SiO_3(l) + H_2O(g)$$

Like most metal oxides, aluminium oxide also acts as a **base** by dissolving in acids to give salts:

$$Al_2O_3(s) + 6H^+(aq) \longrightarrow 2Al^{3+}(aq) + 3H_2O(l)$$

Oxides that show both acidic and basic properties are referred to as **amphoteric** or **amphiprotic**.

On passing down each group in the p block there is an increasing tendency for the elements not to use their outer s electrons in bonding. This is referred to as the **inert pair effect** and, in group 4, tin and lead show an oxidation state of +2 as well as +4. (Note that carbon at the top of group 4 also shows an oxidation state of +2 in carbon monoxide. In this case, however, it results from the ability of carbon atoms to form stable multiple bonds as shown in Figure 20.8.)

Sn	Sn^{2+}	Sn^{4+}
$[Kr]4d^{10}5s^{2}5p^{2}$	$[Kr]4d^{10}5s^{2}$	$[Kr]4d^{10}$
Pb	Pb^{2+}	Pb^{4+}
$[Xe]5d^{10}6s^{2}6p^{2}$	$[Xe]5d^{10}6s^{2}$	$[Xe]5d^{10}$

The relative stability of the two oxidation states can be judged from the standard electrode potential for the redox equilibrium between them:

$$Pb^{4+}(aq) + 2e^{-} \rightleftharpoons Pb^{2+}(aq) \qquad E^{\ominus} = +1.69\ V$$

$$Sn^{4+}(aq) + 2e^{-} \rightleftharpoons Sn^{2+}(aq) \qquad E^{\ominus} = +0.15\ V$$

The more positive the standard potential, the more readily the system accepts electrons and moves to the right to form the lower oxidation state. Thus lead(IV) is a powerful oxidising agent, whilst tin(II) acts as a reducing agent.

Focus 20c

1. The chlorides of boron, carbon, aluminium and silicon are essentially covalent in nature.
2. Boron trichloride exists as triangular molecules in which the boron atom has only six electrons in its valency shell. It therefore reacts as a lone-pair acceptor.
3. Carbon chlorides are unusual in being stable towards hydrolysis. Since the valency shell of the carbon atom is full it cannot readily accept a lone pair of electrons from the water molecule.
4. The oxides of boron, carbon, aluminium and silicon all show some acidic character. Aluminium oxide also reacts as a base and is therefore amphoteric.
5. Carbon oxides are unusual in being gaseous at room temperature. They exist as small covalent molecules in which the carbon atoms form stable multiple bonds.
6. On passing down the periodic table there is an increasing tendency for the elements to show an oxidation state of two less than the group number. Thus the most stable oxidation state of lead is +2, rather than +4. This is known as the 'inert pair effect'.

20.2 Groups 5 and 6

The most important elements here are the two non-metals at the top of each group.

Nitrogen and oxygen

Between them, these elements form almost 99% of pure dry air. Nitrogen boils at a lower temperature (77K) than oxygen (90K) and large quantities of each are separated by fractional distillation of liquid air. Although both elements exist as diatomic covalent molecules, oxygen is much more reactive than nitrogen. This difference in reactivity is reflected in the industrial uses of the elements. A major use of oxygen is in steel making where it burns off the excess carbon in molten iron. A typical use of nitrogen, on the other hand, is in the packaging of foods such as potato crisps to extend shelf-life by preventing spoilage by reaction with oxygen in the air. The N≡N bond is much stronger than other bonds involving nitrogen atoms, making most reactions involving the nitrogen molecule energetically unfavourable.

	N≡N	O=O	F—F
bond enthalpies/kJ mol^{-1}	944	496	158

Ozone ... friend or foe?

When exposed to high energy radiation, oxygen is converted into a less stable allotrope ozone, O_3:

$$O_2 \xrightarrow{\text{energy}} \bullet O \bullet + \bullet O \bullet$$

$$\bullet O \bullet + O_2 \longrightarrow O_3$$

In the upper atmosphere, about 20 to 25 km above the Earth's surface, the concentration of ozone rises to about 10 ppm. This **ozone layer** strongly absorbs ultraviolet radiation from the Sun, which if it reached the Earth's surface, would be dangerous to all forms of life. It would, for instance, cause an increase in human skin cancers and cataracts, as well as reducing the yield of food crops. The amount of ozone varies naturally in response to seasonal changes in solar radiation and temperature.

In recent years, however, the thickness of the ozone layer, particularly in the polar regions, has decreased. Such damage may be caused through the introduction of chemicals into the ozone layer, which produce **free radicals**. These are highly reactive species that contain an unpaired electron. An example is the chlorine atom, Cl•. If a chemical that can form free-radical chlorine reaches the ozone layer, the chlorine will react as follows:

$$O_3 + Cl\bullet \longrightarrow \bullet OCl + O_2 \text{ and } \bullet OCl + O_3 \longrightarrow Cl\bullet + 2O_2$$

The chemicals that have been linked most directly to this phenomenon are the **chlorofluorocarbons**, or **CFCs**, which were widely used as aerosol propellants, refrigerants, foaming agents for plastic packaging, and cleaning fluids. Concern over depletion of the ozone layer resulted in an international agreement, known as the Montreal Protocol, which, in 1987, called for a reduction in the use of CFCs. Two years later 93 developed countries agreed to phase out their own CFC production and to promote the development of safer substitutes. However, many of the suggested alternatives, such as hydrocarbons, are less versatile and more expensive than CFCs.

Although ozone in the upper atmosphere is essential in reducing the intensity of harmful UV radiation, it is quite poisonous and, even in small quantities, can cause breathing disorders such as asthma. Just as photochemical reactions high in the atmosphere produce ozone, similar reactions are also possible nearer the Earth's surface. One such mechanism responsible for forming **smog** (a mixture of smoke and other pollutants trapped at ground level) involves nitrogen monoxide, NO, which is emitted in traffic exhaust fumes. On contact with oxygen in the air, nitrogen monoxide is oxidised to the red-brown gas nitrogen dioxide, NO_2:

$$2NO + O_2 \longrightarrow 2NO_2$$

As well as being a dangerous pollutant in its own right, nitrogen dioxide can dissociate in sunlight to form nitrogen monoxide and atomic oxygen:

$$NO_2 \longrightarrow NO + \bullet O \bullet$$

Reaction of these oxygen atoms with oxygen molecules produces ozone.

Activity 20.1

1 Only chemicals that are sufficiently stable to persist long enough in the atmosphere can rise high enough to cause damage to the ozone layer. Read section 20.3 and then suggest why CFCs should be so stable.

2 Explain why it only takes relatively small amounts of CFCs to cause great damage to the ozone layer.

Ammonia, NH_3, is one of the most important nitrogen compounds as it forms the basis of the nitrogen fertiliser industry (see section 23.3). It is conveniently made in the laboratory by warming an ammonium salt with an alkali such as sodium hydroxide solution:

$$NH_4Cl(aq) + NaOH(aq) \longrightarrow NH_3(g) + NaCl(aq)$$

The relatively high boiling point of ammonia, −33°C, is due to the strong intermolecular hydrogen bonding which is a feature of the chemistry of the hydrides of oxygen and fluorine as well as nitrogen. Mutual hydrogen bonding also makes ammonia extremely soluble in water in which it forms a weakly alkaline solution,

$$NH_3(aq) + H_2O(l) \rightleftharpoons NH_4^+(aq) + OH^-(aq)$$

It forms ammonium salts with acids, e.g.

$$NH_3(aq) + HCl(aq) \longrightarrow NH_4Cl(aq)$$

and acts as a ligand with many metal ions, e.g.

$$[Cu(H_2O)_6]^{2+}(aq) + 4NH_3(aq) \longrightarrow [Cu(NH_3)_4(H_2O)_2]^{2+}(aq) + 4H_2O(l)$$

Phosphorus and sulphur

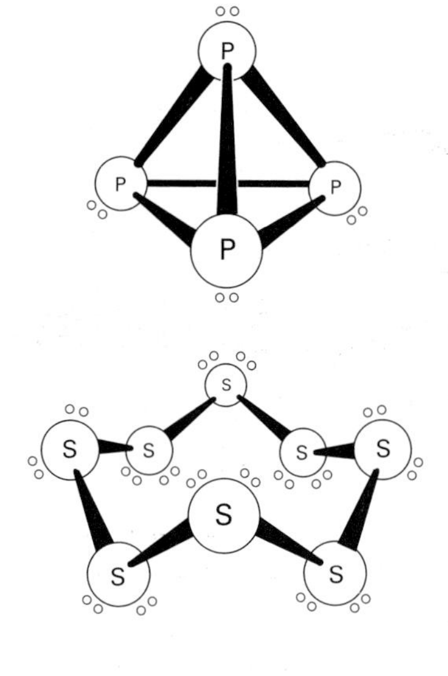

Figure 20.10
Molecular structures of white phosphorus and sulphur

Both phosphorus and sulphur are non-metals that exist in different forms, or allotropes. White phosphorus and the most common allotropes of sulphur consist of separate covalent molecules whose structures are shown in Figure 20.10

The bond angle in P_4 is only 60°, compared to about 108° in S_8, and the consequent strong repulsive forces make P_4 a much more **strained** molecule. This explains why white phosphorus is much more reactive than sulphur, being spontaneously inflammable in air at room temperature.

Nitric and sulphuric acids

The manufacture of these acids from nitrogen and sulphur respectively is outlined in Chapter 23. Large quantities of each are required by industry and some of their major uses are given in Table 20.9.

Table 20.9 Major uses of nitric and sulphuric acids

sulphuric acid	nitric acid
UK annual production – approx. 3×10^8 tonnes	UK annual production – approx. 8×10^5 tonnes
32% fertiliser production 16% paint manufacture 14% synthetic fibre manufacture 10% detergent production 9% plastics and polymers 19% other uses	90% fertiliser production 10% explosives, dyes, etc.

As well as behaving as typical strong acids, both nitric and sulphuric acids act as **oxidising agents**. Nitric acid, even when cold and dilute, does not generally form hydrogen when it reacts with metals (except with very electropositive metals such as magnesium) but instead forms nitrogen monoxide gas:

$$HNO_3 + 3H^+ + 3e^- \longrightarrow NO + 2H_2O$$

With concentrated nitric acid, the brown gas nitrogen dioxide is more likely to be formed than nitrogen monoxide:

$$HNO_3 + H^+ + e^- \longrightarrow NO_2 + H_2O$$

Hot, concentrated sulphuric acid oxidises both metallic and non-metallic elements, generally forming sulphur dioxide:

$$H_2SO_4 + 2H^+ + 2e^- \longrightarrow SO_2 + 2H_2O$$

We can derive the overall redox equations for such reactions by combining the reduction half-equation for the acid with the appropriate oxidation half-equation. For example, copper metal:

$$Cu \longrightarrow Cu^{2+} + 2e^-$$

1 With dilute nitric acid

$$(HNO_3 + 3H^+ + 3e^- \longrightarrow NO + 2H_2O) \times 2$$

$$(Cu \longrightarrow Cu^{2+} + 2e^-) \times 3$$

$$\Longrightarrow \quad 2HNO_3 + 6H^+ + 3Cu \longrightarrow 2NO + 4H_2O + 3Cu^{2+}$$

2 With concentrated nitric acid

$$(HNO_3 + H^+ + e^- \longrightarrow NO_2 + H_2O) \times 2$$

$$Cu \longrightarrow Cu^{2+} + 2e^-$$

$$\Longrightarrow \quad 2HNO_3 + 2H^+ + Cu \longrightarrow 2NO_2 + 2H_2O + Cu^{2+}$$

3 With hot, concentrated sulphuric acid

$$H_2SO_4 + 2H^+ + 2e^- \longrightarrow SO_2 + 2H_2O$$

$$Cu \longrightarrow Cu^{2+} + 2e^-$$

$$\Longrightarrow \quad H_2SO_4 + 2H^+ + Cu \longrightarrow SO_2 + 2H_2O + Cu^{2+}$$

Activity 20.2

1. Concentrated nitric or sulphuric acid will oxidise carbon to carbon dioxide according to the following half-equation:

   $$C + 2H_2O \longrightarrow CO_2 + 4H^+ + 4e^-$$

   Combine this with the reduction half-equations above to give the overall reaction for the oxidation of carbon by each acid.

2. Like carbon, sulphur is oxidised to the dioxide by concentrated nitric and sulphuric acid. Derive equations for these reactions.

Focus 20d

1. Nitrogen is much less reactive than oxygen largely as a result of the high strength of the $N \equiv N$ bond.
2. Ozone, O_3, is an allotrope of oxygen. Though poisonous, its presence in the upper atmosphere is essential to life as it absorbs much of the harmful UV radiation from the sun. Free radical producing substances, such as chlorofluorocarbons (CFCs), attack ozone and can damage this ozone layer.
3. White phosphorus consists of P_4 molecules. These are highly strained and account for the reactive nature of this allotrope. Common forms of sulphur contain S_8 crown-shaped molecules.
4. Nitric and sulphuric acids are industrially important compounds which can both act as oxidising agents.

20.3 Group 7

During this century world annual production of chlorine has risen from about 0.1 million tonnes to over 29 million tonnes. Table 20.10 shows that the petrochemical industry consumes well over half the chlorine produced in the UK. In carefully controlled amounts the poisonous nature of chlorine is used to kill bacteria in drinking water. Many chlorine compounds are toxic but, despite concern over environmental problems, continue to be used in bleaches, disinfectants, insecticides and weed-killers.

Table 20.10 Some major uses of chlorine in the UK

Approx. % of total chlorine production	intermediate product(s)	end uses(s)
Petrochemicals		
27	chloroethene (vinyl chloride) CH_2CHCl	poly(chloroethene, i.e. poly(vinyl chloride), PVC
17	chloromethanes $CHCl_3$, CCl_4, CH_3Cl	refrigerant liquids and aerosol 'propellants' petrol 'antiknock'additives
16	other chlorinated hydrocarbons, e.g. $CCl_2.CHCl$ $CCl_2.CCl_2$	metal 'degreasing' solvent 'dry-cleaning' solvent
6	propene oxide	car brake fluid polyurethane plastics many pharmaceutical products
Inorganics		
13		extraction of magnesium, titanium and bromine, manufacture of hydrochloric acid
1	sodium chlorate(I), NaClO	paper and pulp bleach
Miscellaneous		
20		sterilisation of water manufacture of disinfectants, anaesthetics, insecticides and dyestuffs, etc.

Fluorides are used in fluoride toothpaste and organic fluorine compounds are found in aerosol propellants and non-stick coatings on pans. Silver bromide and iodide are light-sensitive compounds used in photographic films.

The extraction of chlorine is outlined in Chapter 23.

The elements

The elements of group 7 are known collectively as the **halogens**. They are all non-metals and exist as simple diatomic covalent molecules, X_2:

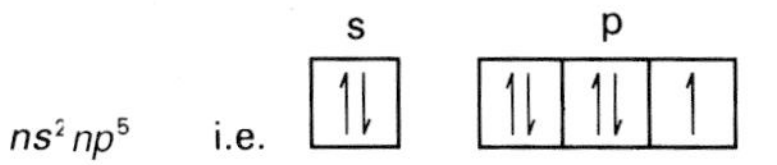

The properties of the halogens are compared in Table 20.5. *Fluorine, the first member of the group, is the most electronegative of all elements* and shows several unique features in its chemistry. All isotopes of astatine, the last member of the group, are intensely radioactive with short half-lives. All the halogen atoms have an outer electron configuration ns^2np^5:

ns^2np^5 i.e. s: ⇅ p: ⇅ ⇅ ↑

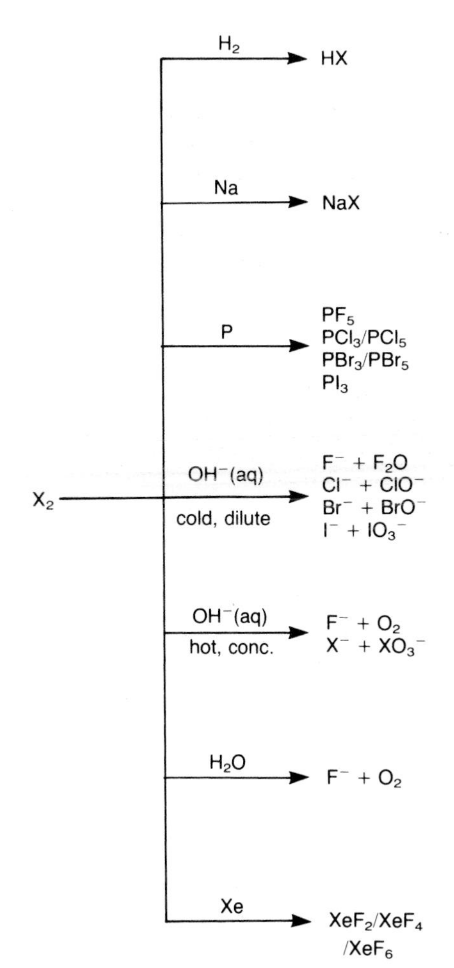

Figure 20.11
Some reactions of the halogens

On reacting with electropositive metals they can achieve a stable noble gas configuration by taking an electron from a metal, forming the halide ion, X^-:

$$\underset{ns^2np^5}{X} + e^- \longrightarrow \underset{ns^2np^6}{X^-}$$

With non-metals the halogen atom can complete its outer octet by sharing an electron in a single covalent bond, e.g. in the hydrogen halides, HX:

$$H \overset{\bullet}{\underset{\times}{-}} X$$

Except for fluorine, halogen atoms can also show covalencies of 3, 5 and 7 by promoting paired electrons into empty d orbitals in the valency shell. Iodine, for example, shows covalencies of 1, 3, 5 and 7 in the interhalogen compounds, i.e. compounds formed between two halogens:

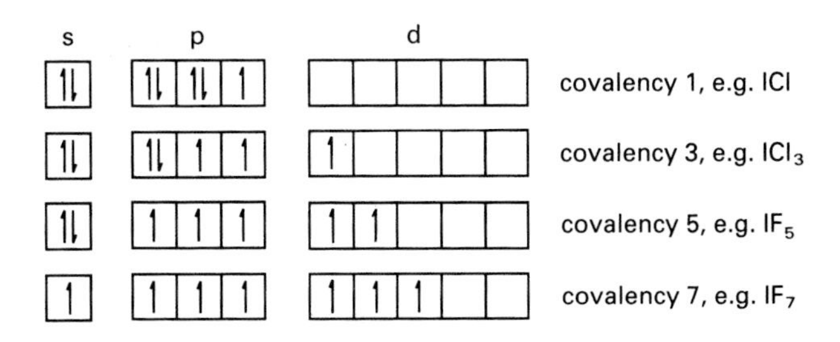

The halogens react with a wide range of metallic and non-metallic elements. Some of the main reactions are illustrated in Figure 20.11 which shows that whilst the chemistries of chlorine and bromine are virtually identical, fluorine, and to some extent iodine, show considerable differences.

The reactivity of the elements falls markedly on passing down the group. Fluorine, for instance, is the only halogen that combines directly with carbon:

$$C(s) + 2F_2(g) \longrightarrow CF_4(g)$$

Fluorine even combines directly with the noble gas xenon, e.g.

$$Xe(g) + 3F_2(g) \longrightarrow XeF_6(s)$$

All the elements react with hydrogen to form hydrogen halides but the conditions required vary considerably.

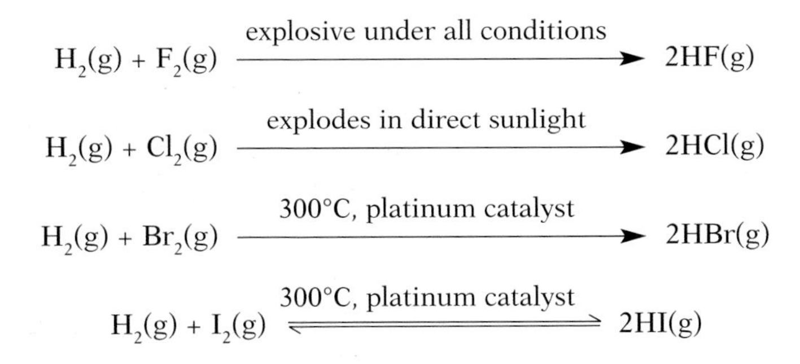

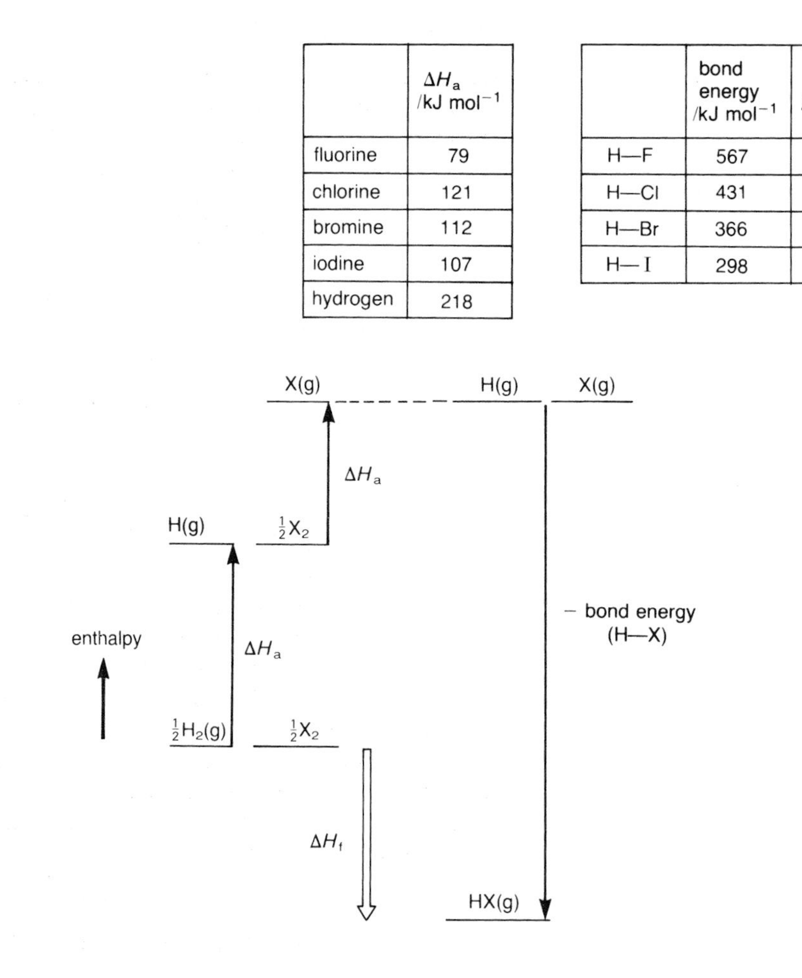

	ΔH_a /kJ mol^{-1}
fluorine	79
chlorine	121
bromine	112
iodine	107
hydrogen	218

	bond energy /kJ mol^{-1}	ΔH_f /kJ mol^{-1}
H—F	567	−269
H—Cl	431	−92
H—Br	366	−36
H—I	298	+27

Figure 20.12
Enthalpy cycle for the formation of a hydrogen halide, HX

We can explain this trend in reactivity by considering the enthalpy cycle and data shown in Figure 20.12.

The standard enthalpies of formation show that the hydrogen halides become progressively less stable with respect to the elements on passing down the group. The highly reactive nature of fluorine results from its comparatively low atomisation energy and the great strength of the H—F bond. At first sight it might seem strange that fluorine should form strong covalent bonds with other atoms, whilst the F—F bond is rather weak. In fact the strength of a covalent bond involving fluorine decreases with the number of lone pairs in the valency shell of the other atom:

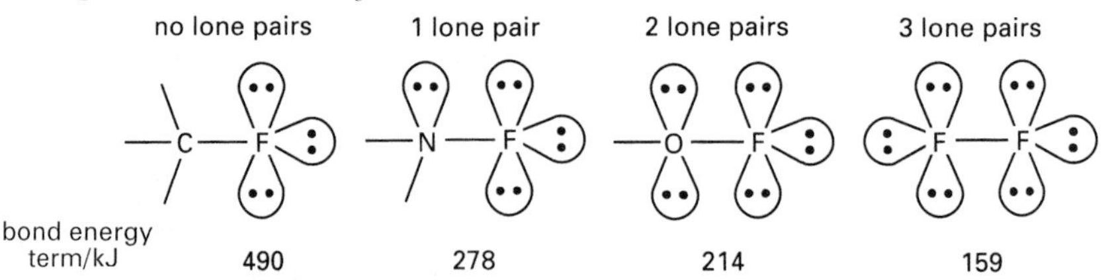

Increased repulsion across the bond between the lone pairs on the two atoms is responsible for the weakening of these short covalent bonds. With larger halogen atoms the bond is longer and so any lone pair repulsion is weaker.

The strength of the covalent bonds that fluorine forms with other atoms also helps to explain why it brings out their maximum covalency. Sulphur, for example, forms a hexafluoride, SF_6, whilst its highest chloride is only SCl_4. Of course, it is also easier to pack the smaller fluorine atoms around the central sulphur atom.

Oxidising power

A species is said to be oxidised when it has electrons totally or partially removed. All the halogens act as oxidising agents when they react with more electropositive elements. They will totally remove the outer electrons from reactive metals, being themselves reduced to halide ions, e.g.

	$2Na$	$\longrightarrow$	$2Na^+ + 2e^-$	oxidation
oxidation states	2(0)		2(+1)	
	$X_2 + 2e^-$	$\longrightarrow$	$2X^-$	reduction
oxidation states	2(0)		2(−1)	

Less electronegative non-metals can also be oxidised, although here the halogen atom only partially withdraws electrons, forming a polar covalent bond:

$$H{-}H + X{-}X \longrightarrow 2\,\overset{\delta+}{H}{-}\overset{\delta-}{X}$$

oxidation states: 2 (0)   2 (0)   2 (+1)   2 (−1)

We should be able to judge the relative oxidising power of the halogens by comparing their electron affinities, electronegativities and standard electrode potentials which are listed in Table 20.5.

Both electronegativity values and standard electrode potentials indicate a decrease in oxidising power on passing down the group. Except in the case of fluorine this is also confirmed by the electron affinity data. Despite having a less exothermic electron affinity, fluorine is a more powerful oxidising agent than chlorine. We must remember that, whilst we are concerned here with the reactions of the X_2 molecule, electron affinity refers to isolated atoms:

$$X(g) + e^- \longrightarrow X^-(g)$$

The total energy required to form the halide ion from the halogen molecule is the sum of the processes:

$$\tfrac{1}{2}X_2 \longrightarrow X(g) \quad \text{followed by} \quad X(g) + e^- \longrightarrow X^-(g)$$

If we add atomisation energy to electron affinity the figures confirm that fluorine should be the most powerful oxidising agent (Table 20.11).

Table 20.11

	F_2	Cl_2	Br_2	I_2
atomisation energy/kJ mol^{-1}	79	121	112	107
electron affinity/kJ mol^{-1}	−348	−364	−342	−314
atomisation energy + electron affinity/kJ mol^{-1}	−269	−243	−230	−207

Experimental confirmation of the prediction of decrease in oxidising power on passing down group 7 is provided by displacement reactions. *Any halogen will oxidise a halide ion from below it in the group*, e.g.

$$\underset{2(0)}{Cl_2(aq)} + \underset{2(-1)}{2Br^-(aq)} \longrightarrow \underset{2(-1)}{2Cl^-(aq)} + \underset{2(0)}{Br_2(aq)}$$

but

$$I_2(aq) + 2Br^-(aq) \longrightarrow \text{no reaction}$$

Activity 20.3

Use is made of halogen displacement reactions in the extraction of bromine from seawater which contains approximately 65 parts per million of bromide ions. The process involves several stages.

A Chlorine gas is bubbled through acidified seawater, displacing the bromine.

B Bromine vapour is removed by blowing air through the mixture.

C The bromine vapour is reduced by mixing with sulphur dioxide, SO_2, and steam. The hydrobromic acid, HBr, and sulphuric acid, H_2SO_4, that are formed remain dissolved when the steam condenses, giving a solution which is over 1000 times more concentrated in bromide than the original seawater.

D Bromine is reformed from this solution using a stream of steam and chlorine. The crude bromine, which separates as a lower layer beneath hydrochloric acid, is purified by distillation.

1 Copy, complete and balance the following ionic equations:
a) step A $Br^-(aq) + Cl_2(g) \longrightarrow$
b) step C $Br_2(g) + SO_2(g) + H_2O(g) \longrightarrow$

2 Assuming no losses at any stage, calculate the following for 1000 tonnes of seawater treated by the above process:
a) the mass of bromide ions present in the seawater,
b) the number of moles of bromide ions in the seawater,
c) the mass of chlorine needed to oxidise the bromide to bromine,
d) the mass of sulphur dioxide needed to reduce the bromine back to bromide in step C.
(NB 1 tonne = 1 000 kg or 1 000 000 g

3 In this process the bromide is oxidised by chlorine twice. Why do you think it is necessary to reduce the bromine back to bromide in step C?

We may explain the variation in oxidising power quite simply in terms of the relative sizes of the halogen atoms. A halide ion is oxidised by removing one of its outer electrons. In a large halide ion, the outer electrons are more easily removed as they are further away from the nucleus and better screened from its attraction by more inner electrons. Small halide ions, on the other hand, hold on to their electrons more strongly. Similarly, a small halogen atom can attract an extra electron more powerfully than a larger atom.

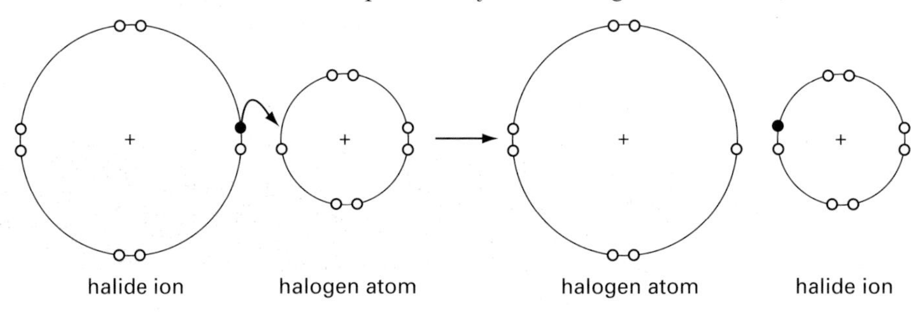

Hydrogen halides

Hydrogen fluoride and hydrogen chloride can both be prepared in the laboratory by the action of concentrated sulphuric acid on a metal halide, e.g.:

$$CaF_2(s) + 2H_2SO_4(l) \longrightarrow Ca(HSO_4)_2(s) + 2HF(g)$$
$$NaCl(s) + H_2SO_4(l) \longrightarrow NaHSO_4(s) + HCl(g)$$

However, the same method **cannot** be used to make hydrogen bromide or hydrogen iodide. With these compounds the sulphuric acid acts as an oxidising agent and converts the halide ion to the free halogen.

Concentrated sulphuric acid may be used to distinguish between solid chlorides, bromides and iodides as the products are easily identifiable using Table 20.12.

Table 20.12

Gaseous products	chloride	bromide	iodide
from the halide ion	HCl – white, steamy fumes with a pungent odour	(some) HBr – white, steamy pungent fumes	
		Br_2 – orange vapour	I_2 – purple vapour (and black solid)
from the reduction of sulphuric acid	none	SO_2 – white fumes smell of burning sulphur. Turn orange $K_2Cr_2O_7$ paper green	H_2S – 'bad egg' smell*

*care should be taken here as hydrogen sulphide is highly poisonous

Activity 20.4

1 Why does concentrated sulphuric acid oxidise Br^- and I^- but not Cl^- or F^-? (Hint: look at the previous section.)

2 Copy, complete and balance the following half-equations. In each case say whether the process represents oxidation or reduction.

$Br^- \longrightarrow Br_2$

$I^- \longrightarrow I_2$

$H_2SO_4 + H^+ \longrightarrow SO_2 + H_2O$

$H_2SO_4 + H^+ \longrightarrow H_2S + H_2O$

3 By combining the half-equations you wrote in question 2, complete and balance the following equations for redox reactions which take place when concentrated sulphuric acid is added to bromide or iodide.

$Br^- + H_2SO_4 + H^+ \longrightarrow$

$I^- + H_2SO_4 + H^+ \longrightarrow$

The use of silver nitrate to distinguish between halide ions in aqueous solution is described in Chapter 22.

Hydrogen bromide and hydrogen iodide can be prepared by forming the phosphorus(III) halide and then decomposing it with water, e.g.:

$$2P + 3Br_2 \longrightarrow 2PBr_3$$

followed by:

$$PBr_3 + 3H_2O \longrightarrow 3HBr + H_3PO_3$$

Some of the main properties of the hydrogen halides are compared in Table 20.13. You will notice that, in general, boiling point rises with relative molecular mass indicating an increase in intermolecular attraction. *Hydrogen fluoride, however, has by far the highest boiling point, owing to strong*

Table 20.13 Properties of the hydrogen halides

Hydrogen halide	boiling point/°C	ΔH_f /kJ mol–1	bond enthalpy, H—X /kJ mol^{-1}	acid dissociation constant, K_a, in aqueous solution/M
hydrogen fluoride, HF	20	–269	567	7×10^{-4}
hydrogen chloride, HCl	–85	–92	431	10^7
hydrogen bromide, HBr	–69	–36	366	$>10^7$
hydrogen iodide, HI	–35	+26	298	$>10^7$

How and why do (i) the boiling points and (ii) the strength of the H—X covalent bond vary on passing down the group?

intermolecular hydrogen bonding (see Chapter 4). In the liquid state its molecules are linked into long chains.

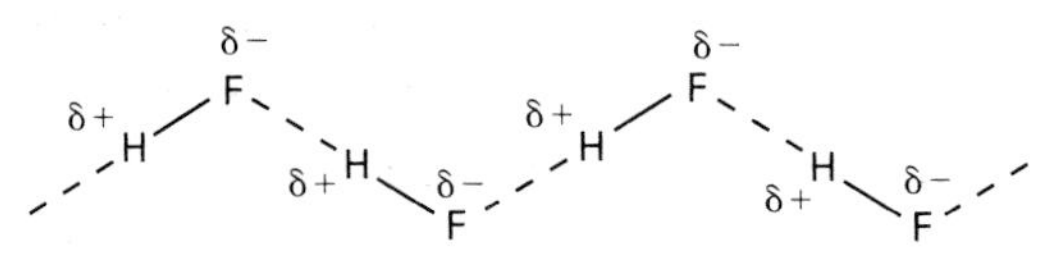

As we might expect from the weakening of the H—X bond and the trend in enthalpies of formation shown in Table 20.13, the hydrogen halides become thermally less stable on descending the group. This is especially noticeable with hydrogen iodide, which readily dissociates into the elements on heating:

$$2HI(g) \underset{\text{at } 100°C}{\overset{\text{about 30\% dissociation}}{\rightleftharpoons}} H_2(g) + I_2(g)$$

As we might expect the covalent hydrogen halide molecules are appreciably soluble without dissociation in organic solvents such as benzene. They are also very soluble in water but here ionisation occurs to give hydrohalic acid solutions:

$$HX(g) \longrightarrow H^+(aq) + X^-(aq)$$

The dissociation constants in Table 20.13 show that acid strength increases on passing down the group. We can relate this to the weakening of the H—X bond, which must be broken before ions can be formed.

Oxo-salts

All the halogens except fluorine react with alkalis to form oxo-salts. The most important reactions involve chlorine and sodium hydroxide solution where the product depends upon the conditions used.

When chlorine is bubbled through cold, dilute sodium hydroxide, a mixture of sodium chloride, NaCl, and sodium **chlorate(I)**, NaOCl is formed. We can represent the formation of these species by the following half-equations:

$$\tfrac{1}{2}Cl_2 + e^- \longrightarrow Cl^-$$

0   −1   i.e. the chlorine has been reduced

$$\tfrac{1}{2}Cl_2 + 2OH^- \longrightarrow OCl^- + H_2O + e^-$$

0   +1   i.e. the chlorine has been oxidised

Combining these, we get the overall redox equation:

$$\underset{2(0)}{Cl_2} + 2OH^- \longrightarrow \underset{-1}{Cl^-} + \underset{+1}{OCl^-} + H_2O$$

Here the chlorine has oxidised and reduced itself! Such *self oxidation–reduction reactions, referred to as disproportionation, are a feature of the chemistry of the halogen oxo-salts.*

If the alkali is hot and concentrated we get a different disproportionation reaction and **chlorate(V)** is formed together with chloride:

$$\tfrac{1}{2}Cl_2 + 6OH^- \longrightarrow ClO_3^- + 3H_2O + 5e^-$$

0   +5   i.e. the chlorine has been oxidised

Combining this with the above equation for chloride formation, we get the overall redox equation:

$$\underset{6(0)}{3Cl_2} + 6OH^- \longrightarrow \underset{5(-1)}{5Cl^-} + \underset{+5}{ClO_3^-} + 3H_2O$$

The manufacture and commercial uses of chlorates are covered in Chapter 23.

Activity 20.5

As might be expected, solutions of chlorate(I) breakdown on heating to give chlorate(V) and chloride.

1 Complete and balance the following half-equations, showing the changes in oxidation state of the chlorine and stating in each case whether oxidation or reduction has occurred.

$$ClO^- + H^+ \longrightarrow Cl^- + H_2O$$
$$ClO^- + H_2O \longrightarrow ClO_3^- + H^+$$

2 Combine these half-equations to give an overall equation for the thermal decomposition of chlorate(I) solutions. Is it correct to describe this as disproportionation? Explain your answer.

Focus 20e

1 The halogens are non-metallic elements that exist as diatomic molecules.

2 The reactivity and oxidising power of the halogens decrease on passing down the group.

3 Characteristic reactions with concentrated sulphuric acid may be used in the identification of halide ions.

4 The hydrogen halides all dissolve in water to give acidic solutions.

5 Hydrogen fluoride is a weaker acid than the other hydrogen halides. It also has a much higher boiling point resulting from appreciable intermolecular hydrogen bonding.

6 Chlorine forms chlorate(I) with cold, dilute sodium hydroxide and chlorate(V) if the alkali is hot or concentrated. In each case these disproportionation (self oxidation–reduction) reactions also give chloride ions.

Questions on Chapter 20

1 **a)** Sketch a dot and cross diagram showing the arrangement of outer electrons and the shape of the BF_3 molecule.

b) Boron trifluoride reacts with ammonia to form an addition compound $BF_3.NH_3$. Draw a dot and cross diagram showing the bonding in this molecule.

c) What happens to the F—B—F bond angle when boron trifluoride reacts with ammonia?

2 This question concerns the inert pair effect shown by elements towards the bottom of the p block. Find thallium in group 3 of the Periodic Table.

a) Write the electron configuration of the thallium atom.

b) Give the formula of **two** chlorides formed by thallium.

c) Which of the thallium chlorides is likely to be least stable and how would you expect it to react?

3 **a)** Why are the common oxides of carbon gaseous whereas the other group 4 elements form solid oxides?

b) An oxide of carbon contains 52.9% by mass of carbon. Calculate its empirical formula. (C = 12, O = 16)

c) The molecular shape of this oxide is linear. Draw a dot and cross diagram showing the arrangement of all the outer electrons.

4 **a)** Give the oxidation state of nitrogen in each of the following compounds:
N_2O, NH_3, N_2F_2, N_2H_4, HNO_3 and Mg_3N_2

b) Draw suitable diagrams to illustrate the bonding in each of the above.

5 Sulphur forms a variety of oxo-anions. On adding dilute acid, thiosulphate ions, $S_2O_3^{2-}$, decompose to give solid sulphur and sulphur dioxide.

a) Complete the following half-equations, labelling each as oxidation or reduction:

$$S_2O_3^{2-} + H^+ \longrightarrow S + H_2O$$
$$S_2O_3^{2-} + H_2O \longrightarrow SO_2 + H^+$$

b) Combine these half-equations to give an overall equation for the reaction of thiosulphate and acid. Is it correct to regard this as a disproportionation reaction?

Questions on Chapter 20 *continued*

6 An aqueous solution is thought to contain iodide ions. Describe how you could confirm this using an immiscible solvent, such as tetrachloromethane, and aqueous chlorine solution. How would your observations differ if the solution contained bromide instead of iodide?

7 When a mixture of xenon and fluorine is exposed to sunlight, white crystals are produced which contain 22.5% by mass of fluorine.

a) Calculate the empirical formula of this compound. (Xe = 131, F = 19).

b) Assuming the molecular formula is the same as the empirical formula, draw a dot and cross diagram showing all the outer electrons and state the likely shape of the molecule.

c) Estimate the enthalpy of formation of this compound given the following bond dissociation enthalpies: F—F 158 kJ mol^{-1}, Xe—F 130 kJ mol^{-1}.

Comments on the activities

Activity 20.1

1 Fluorine forms particularly strong covalent bonds with other elements. The C—F bond is harder to break than any other single covalent bond involving the carbon atom, making CFCs stable long-lived molecules.

Average bond enthalpies/kJ mol^{-1}:

C—F	C—Cl	C—Br	C—I	C—H	C—C
484	338	276	238	412	348

There is some concern amongst scientists that the concentration of CFCs already in the atmosphere may continue to cause damage to the ozone layer for some time after a successful ban is introduced.

2 If you look at the equations given in the extension box you can see that when •OCl reacts with ozone, the Cl• free radical is reformed. This can react with further oxygen molecules to give •OCl, which destroys more ozone. This is an example of a chain reaction. In a similar way, small quantities of nitrogen monoxide from exhaust fumes can produce significant amounts of ozone near ground level.

Activity 20.2

1
$$(HNO_3 + H^+ + e^- \longrightarrow NO_2 + H_2O) \times 4$$
$$C + 2H_2O \longrightarrow CO_2 + 4H^+ + 4e^-$$
$$\Longrightarrow 4HNO_3 + C \longrightarrow 4NO_2 + 2H_2O + CO_2$$
$$(H_2SO_4 + 2H^+ + 2e^- \longrightarrow SO_2 + 2H_2O) \times 2$$
$$C + 2H_2O \longrightarrow CO_2 + 4H^+ + 4e^-$$
$$\Longrightarrow 2H_2SO_4 + C \longrightarrow 2SO_2 + 2H_2O + CO_2$$

2
$$S + 2H_2O \longrightarrow SO_2 + 4H^+ + 4e^-$$
Sulphur therefore reacts in exactly the same way as carbon as shown in the equations in question 1.

Activity 20.3

1 step A $\quad 2Br^-(aq) + Cl_2(g) \longrightarrow Br_2(aq) + 2Cl^-(aq)$

step C:

$$\underset{2(0)}{Br_2(g)} + \underset{+4}{SO_2(g)} + 2H_2O(g) \longrightarrow \underset{2(-1)}{2Br^-(aq)} + \underset{+6}{H_2SO_4(aq)} + 2H^+(aq)$$

2 **a)** Seawater contains 65 ppm of bromide,
1 000 000 tonnes of sea water contains 65 tonnes of bromide
1 000 tonnes of sea water contains 0.065 tonnes of bromide
i.e. 0.065 × 1 000 000 = 65 000 g of Br^-

b) 1 mole of Br^- weighs 80 g so 65 000 g is
$$\frac{65\,000}{80} = 812.5 \text{ moles } Br^-$$

c) From the above equation for step A, each mole of Br^- needs 0.5 mole Cl_2 for oxidation, i.e.
$$\frac{812.5}{2} = 406.25 \text{ moles } Cl_2$$
Each mole of Cl_2 weighs 35.5 × 2 = 71 g, so 406.25 × 71 = 28 844 g Cl_2
i.e. 28.844 kg or 0.028844 tonnes of chlorine

d) From the above equation for step C, 1 mole of SO_2 is needed to reduce 1 mole of Br_2. Since 406.25 mole of Br_2 is formed in step A, 406.25 mole of SO_2 is required. Molecular mass of SO_2 is 64, so mass of SO_2 required is 64 × 406.25 = 26 000 g, or 26 kg or 0.026 tonnes

3 The reason the bromine is reduced in step C is to produce a more concentrated product. The bromine/air mixture from step A is too dilute to extract economically.

Activity 20.4

1 Larger halide ions are more easily oxidised than smaller ones since the electron to be removed is further from the nucleus and better screened from its attraction.

2 $2Br^- \longrightarrow Br_2 + 2e^-$ oxidation
$2I^- \longrightarrow I_2 + 2e^-$ oxidation
$H_2SO_4 + 2H^+ + 2e^- \longrightarrow SO_2 + 2H_2O$ reduction
$H_2SO_4 + 8H^+ + 8e^- \longrightarrow H_2S + 4H_2O$ reduction

3 $2Br^- + H_2SO_4 + 2H^+ \longrightarrow Br_2 + SO_2 + 2H_2O$
$8I^- + H_2SO_4 + 8H^+ \longrightarrow 4I_2 + H_2S + 4H_2O$

Comments on the activities *continued*

Activity 20.5

1 $ClO^- + 2H^+ + 2e^- \longrightarrow Cl^- + H_2O$

+1 → −1 the chlorine has been reduced

$ClO^- + 2H_2O \longrightarrow ClO_3^- + 4H^+ + 4e^-$

+1 → +5 the chlorine has been oxidised

2 $2ClO^- + 4H^+ + 4e^- \longrightarrow 2Cl^- + 2H_2O$

$ClO^- + 2H_2O \longrightarrow ClO_3^- + 4H^+ + 4e^-$

$\Longrightarrow \quad 3ClO^- \longrightarrow ClO_3^- + 2Cl^-$

3(+1) → +5, 2(−1)

Since this reaction involves self oxidation and reduction of chlorate(I), it is correctly described as disproportionation.

CHAPTER 21

The d block

Contents

Two ways in which 'd' block metals have been used from Roman times to the present day. Can you identify the metal in each case?

Study Checklist

After studying this chapter you should be able to:

1. Explain what is meant by a 'transition metal' and recall that these elements typically show variable oxidation states, form coloured compounds and are catalytically active.
2. Explain the formation of complexes in terms of co-ordinate bonding of ligands to a central metal ion.
3. Define co-ordination number and give examples of unidentate, bidentate and multidentate ligands.
4. Recall that most first row transition metal complexes are 6-co-ordinate octahedral but that the large size of the chloride ion leads to 4-co-ordinate tetrahedral complexes.
5. Give the systematic name for a complex from its formula (and vice versa.)
6. Write equations showing the acid reactions of metal ions in aqueous solution.
7. Explain the effect of charge density on the acid strength of metal–aqua complexes.
8. Give examples of the reactions of $M^{2+}(aq)$ and $M^{3+}(aq)$ ions with typical bases such as OH^-, NH_3 and CO_3^{2-}.
9. Explain and give examples of 'amphoteric character'.
10. Explain why $Fe_2(CO_3)_3$ does not exist whilst $FeCO_3$ does.
11. Give examples of ligand exchange reactions involving metal-aqua ions with Cl^- and NH_3.
12. Describe the reduction of each of the following by zinc in acid solution: V(V), Cr(VI) and Mn(VII).
13. Recall that the relative stabilities of oxidation states can be affected by conditions, giving the example of the preparation of a cobalt(III) ammine in alkaline conditions.
14. Outline the laboratory preparation of anhydrous iron(II) chloride and iron(III) chloride, explaining the choice of reagent in each case.
15. Give examples of the importance of complexes in the body (haemoglobin), in medicine (cisplatin) and in photography and electroplating (silver complexes).

21.1 Typical properties

Table 21.1 summarises some of the properties of selected elements from the d block of the periodic table. They are all fairly dense metals, usually with quite high melting points. (Note, however, that mercury is the only metal which is liquid at room temperature.) Such metals have played a vital part in our history. Copper and iron were amongst the first metals to be used for weapons and tools, whilst coins were minted from gold and silver. Although no longer generally used as coinage, reserves of such precious metals are still considered an important measure of a nation's wealth.

In more recent times some of the metals and their compounds have been found to act as catalysts in the manufacture of important chemicals. Many of the metals also have special uses that depend upon their individual properties. The extraction and commercial applications of metals are covered in section 23.4.

Table 21.1 Data sheet for the elements of the first row of the d block and silver

Element	atomic number	atomic radius /nm	ionisation energies /kJ mol^{-1} 1st	2nd	3rd	4th	5th	6th	maximum oxidation state	electro-negativity	$E^{\ominus}$* /volts	density /g cm^{-3}	m.p. /°C
Group 2													
calcium Ca	20	0.197	590	1150	4940	6480	8120	10700	+2	1.0	−2.87	1.54	850
scandium Sc	21	0.160	632	1240	2390	7110	8870	10700	+3	1.3		2.99	1540
titanium Ti	22	0.146	661	1310	2720	4170	9620	11600	+4	1.5	−1.63	4.54	1675
vanadium V	23	0.131	648	1370	2870	4600	6280	12400	+5	1.6	−1.2	5.96	1900
chromium Cr	24	0.125	653	1590	2990	4770	7070	8700	+6	1.6	−0.91	7.19	1890
manganese Mn	25	0.129	716	1510	3250	5190	7360	9750	+7	1.5	−1.18	7.20	1240
iron Fe	26	0.126	762	1560	2960	5400	7620	10100	+3 commonly	1.8	−0.44	7.86	1535
cobalt Co	27	0.125	757	1640	3230	5100	7910	10500	+3	1.8	−0.28	8.90	1492
nickel Ni	28	0.124	736	1750	3390	5400	7620	10900	+2	1.8	−0.25	8.90	1453
copper Cu	29	0.128	745	1960	3550	5690	7990	10500	+2	1.9	+0.34	8.92	1083
zinc Zn	30	0.133	908	1730	3828	5980	8280	11000	+2	1.6	−0.76	7.14	420
Group 3 gallium Ga	31	0.141	577	1980	2960	6190	8700	11400	+3	1.6		5.91	30
silver Ag	47	0.144	732	2070	3360	5000	6700	8400	+1	1.9	+0.80	10.5	961

Underlined figures show the LARGEST single jump between successive ionisation energies. What causes these?
Which of the above metals would you not expect to give hydrogen with dilute sulphuric acid and why? (section 26.1)
* for the process $M^{2+}(aq) + 2e^{-} \rightleftharpoons M(s)$

Most d block elements are **transition metals**. *These are generally defined as possessing an atom, or at least one common oxidation state, with a partially filled inner electron sub-level*. You can see from Table 21.2 that zinc is not a transition element and that scandium only fits this definition when it is in its elemental state.

Table 21.2 Electron configuration of common oxidation states in the first row of the d block

Oxidation state	Sc	Ti	V	Cr	Mn	Fe	Co	Ni	Cu	Zn
0	[Ar] $3d^1 4s^2$	$3d^2 4s^2$	$3d^3 4s^2$	$3d^5 \mathbf{4s}^1$	$3d^5 4s^2$	$3d^6 4s^2$	$3d^7 4s^2$	$3d^8 4s^2$	$3d^{10} \mathbf{4s}^1$	$3d^{10} 4s^2$
+1									$3d^{10} 4s^0$	
+2		$3d^2 4s^0$	$3d^3 4s^0$	$3d^4 4s^0$	$3d^5 4s^0$	$3d^6 4s^0$	$3d^7 4s^0$	$3d^8 4s^0$	$3d^9 4s^0$	$3d^{10} 4s^0$
+3	$3d^0 4s^0$	$3d^1 4s^0$	$3d^2 4s^0$	$3d^3 4s^0$	$3d^4 4s^0$	$3d^5 4s^0$	$3d^6 4s^0$			
+4		$3d^0 4s^0$	$3d^1 4s^0$		$3d^3 4s^0$					
+5			$3d^0 4s^0$							
+6				$3d^0 4s^0$	$3d^1 4s^0$					
+7					$3d^0 4s^0$					

Note that electrons are lost from the 4s sub-level before the 3d and that chromium and copper atoms are unusual in having only one 4s electron.

Why is this distinction between transition and non-transition metals so important? The answer to this question lies in the characteristic properties shown by species that possess a partially filled inner electron sub-level. Generally such elements show variable oxidation states, form coloured compounds and, as we have mentioned, are often good catalysts.

Activity 21.1

Silver and gold lie below copper in the second and third rows of the d block respectively. Both metals show a stable oxidation state of +1 and gold also forms some +3 compounds. Their atomic electron configurations are:

Ag [Kr] $4d^{10}5s^1$ Au [Xe] $5d^{10}6s^1$

1. Write the electron configurations of the common oxidation states of silver and gold.
2. According to the information above, which, if either, of the two metals is correctly regarded as a transition element?

Focus 21a

1. Transition metals may be defined as possessing at least one oxidation state that contains a partially filled inner electron sub-level. These elements generally have the following characteristic properties:
   - **variable oxidation states**
   - **coloured compounds**
   - **often good catalysts**
2. According to this definition not all d block elements are transition metals.

21.2 Complex formation

By and large d block compounds, especially in solution, contain **complexes** or **co-ordination compounds**. In such species the central metal ion is surrounded by electron pair donors called **ligands**. The number of ligands attached to the central metal ion by co-ordinate covalent bonds is known as the **co-ordination number**. For example, Co^{2+} ions in aqueous solution form the 6-co-ordinate complex $[Co(H_2O)_6]^{2+}$.

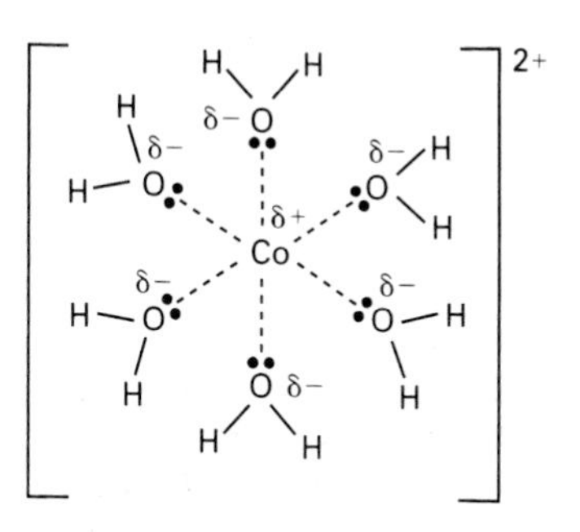

A ligand must have a lone pair of electrons that it can donate to the metal ion. It may be a negative ion, e.g. chloride or oxide, or a neutral molecule such as water or ammonia.

However, not all species that have a lone pair will act as ligands. Noble gas atoms such as neon, krypton and xenon have four pairs of electrons in their valency shells but do not donate these to metal ions.

All the ligands we have mentioned so far are **monodentate**, literally 'one-toothed', i.e. they can only form a single co-ordinate covalent bond with a particular metal ion. Figure 21.1 shows some **bidentate** and **polydentate** ligands, which have two or more atoms capable of donating a lone pair of electrons to the same metal atom. Such ligands form ring-type complexes known as **chelates**. The structures of some chelates are shown in Figure 21.2.

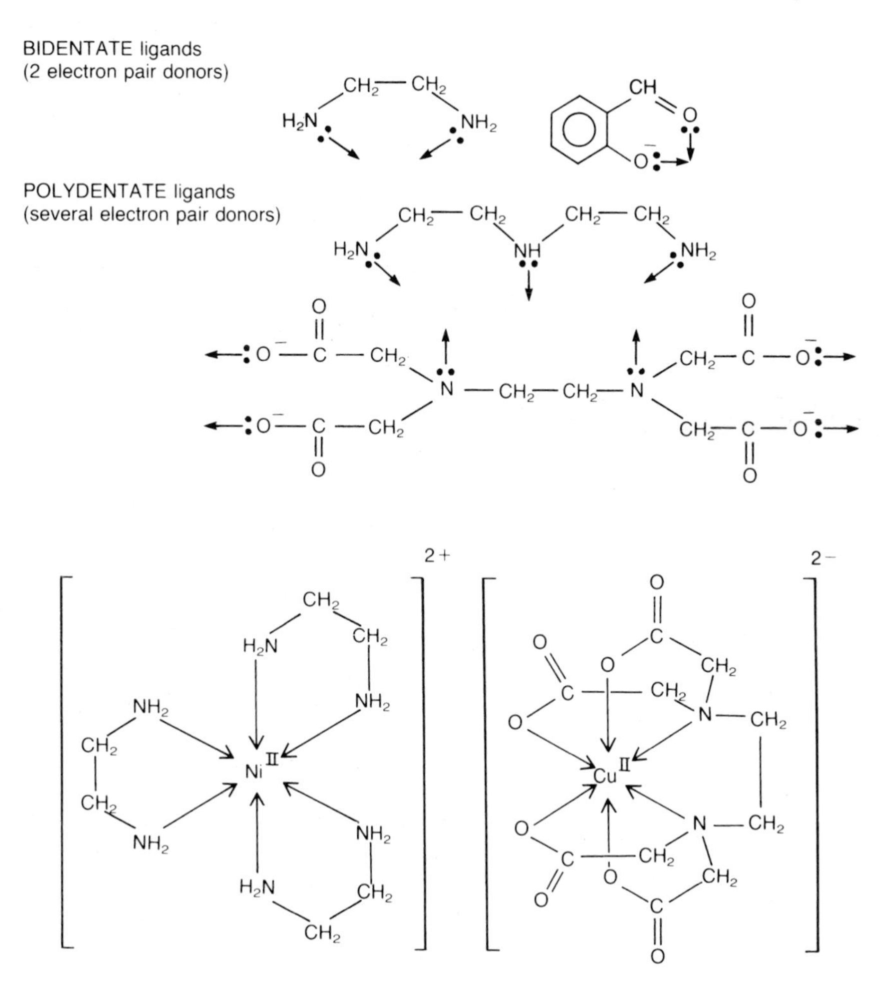

Figure 21.1 ▶ Examples of ligands which contain more than one electron pair donor atom

Figure 21.2 ▶ Examples of 'chelate' (ring-containing) complexes of bi- and polydentate ligands

6-co-ordinate complexes, such as the hexaaqua ions, are octahedral in shape. Most 4-co-ordinate complexes are tetrahedral, e.g. tetrachloro species, though some square planar species are known, e.g. $[Ni(CN)_4]^{2-}$. Copper(I) and silver form linear 2-co-ordinate complexes such as $[CuCl_2]^-$ and $[Ag(NH_3)_2]^+$.

Complexes that contain only neutral ligands are generally positively charged, e.g. $[Cu(H_2O)_6]^{2+}$, $[Cr(NH_3)_6]^{3+}$. Those which contain anions as ligands may be electrically neutral, e.g. $[Fe(OH)_3(H_2O)_3]$, or have an overall negative charge, e.g. $[CuCl_4]^{2-}$. As a 'rule of thumb': *electrically charged complexes attract polar molecules strongly and are therefore generally water soluble, whereas neutral complexes are insoluble and readily precipitate from solution.*

Activity 21.2

1 By considering their electronic structure, suggest which of the following molecules and ions might act as ligands. Which, if any, of these might act as bidentate ligands?

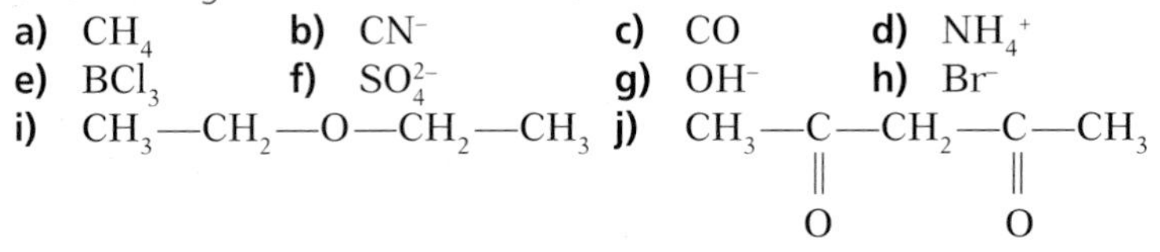

a) CH_4 **b)** CN^- **c)** CO **d)** NH_4^+
e) BCl_3 **f)** SO_4^{2-} **g)** OH^- **h)** Br^-
i) $CH_3—CH_2—O—CH_2—CH_3$ **j)** $CH_3—C(=O)—CH_2—C(=O)—CH_3$

2 Ammonia, NH_3, is a better ligand than nitrogen trichloride, NCl_3, but not so good as trimethylamine, $N(CH_3)_3$. Account for this in terms of the effect of the rest of the molecule upon the ease of donation of the lone pair of electrons on the nitrogen atoms of each of the species.

When writing formulae, a complex is normally enclosed within square brackets. Anything outside the square brackets is not part of the complex. Thus $[Fe(H_2O)_6]Cl_3$ contains free chloride ions, whereas in its isomer, $[FeCl_3(H_2O)_3](H_2O)_3$, the chloride ions are ligands.

Activity 21.3 Naming complexes

As with organic chemistry there are systematic rules for naming co-ordination compounds.

A All complexes have **single word** names in which the ligands precede the metal ion.

B i) Negatively charged ligands are named before neutral ligands, each group in alphabetical order.

ii) The names of negative ion ligands end in 'o', e.g. Cl^-, chloro. Neutral ligands generally keep their normal names. Exceptions include the common ligands H_2O, **aqua** and NH_3, **ammine**.

iii) The number of each type of ligand is indicated by the following prefixes: 1-mono, 2-di, 3-tri, 4-tetra, 5-penta, 6-hexa.

C If the complex has an overall negative charge the metal name is given the ending '**ate**'. The Latin form is sometimes used in this case, e.g. iron (ferrate), copper (cuprate) and silver (argentate). The oxidation state of the metal, written in roman numerals, is given in brackets after its name.

D In ionic compounds, the positive ion is named before the negative ion regardless of whether either, neither or both is a complex.

Though they may seem complicated, the following examples will show that, in practice, these rules are quite simple to apply.

Activity 21.3 Naming complexes continued

$[Fe(H_2O)_6]Cl_2$

This contains $[Fe(H_2O)_6]^{2+}$ and $2Cl^-$.

We name the cation first followed by the anion:
hexaaquairon(II) chloride.

Note the space between the two ions and that there is no need to state the number of chloride ions.

$[Cr(OH)_3(H_2O)_3]$

This is a neutral complex and so has a single word name:
trihydroxotriaquachromium(III).

Note that the negative ligand is named before the neutral ligand.

$K_2[CuCl_4]$

This contains $2K^+$ and $[CuCl_4]^{2-}$

We name the cation followed by the complex anion
potassium tetrachlorocuprate(II).

Note the 'ate' ending of negatively charged complex ions.

1 Can you name the following?

$[Cu(NH_3)_4(H_2O)_2]SO_4$ $[Ni(OH)_2(H_2O)_4]$ $Na[AgCl_2]$

2 Write down the formulae of these cobalt compounds.
hexaamminecobalt(III) chloride
trichlorotriamminecobalt(III)
sodium tetrachlorocobaltate(II)

3 Which of the complexes in questions 1 and 2 above would you expect to be in insoluble in water?

The next three sections give examples of typical reactions of metal complexes. More detail can be found in Tables 21.3 to 21.10 which outline the essential chemistry of some first row d block elements together with silver.

Table 21.3 Outline chemistry of chromium

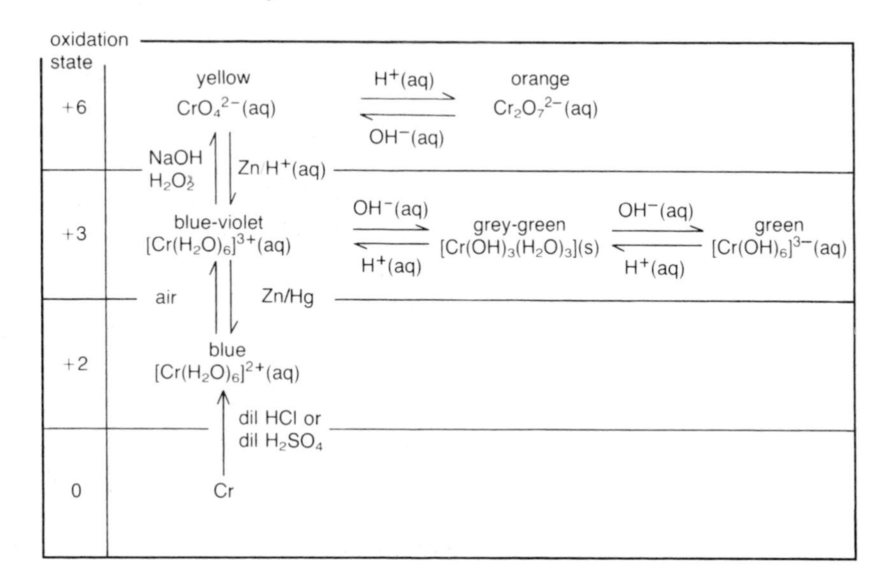

Table 21.4 Outline chemistry of manganese

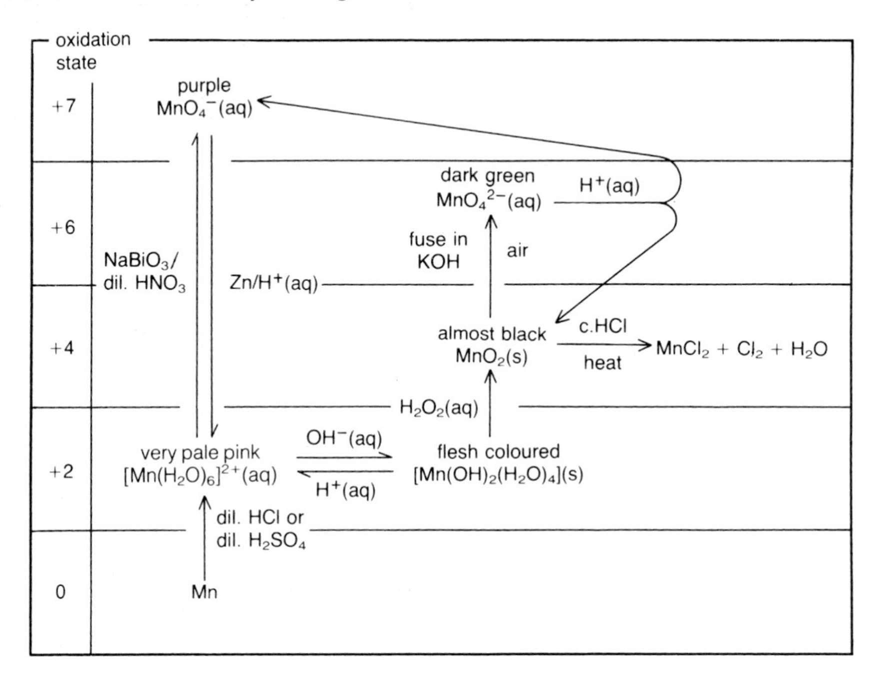

Table 21.5 Outline chemistry of iron

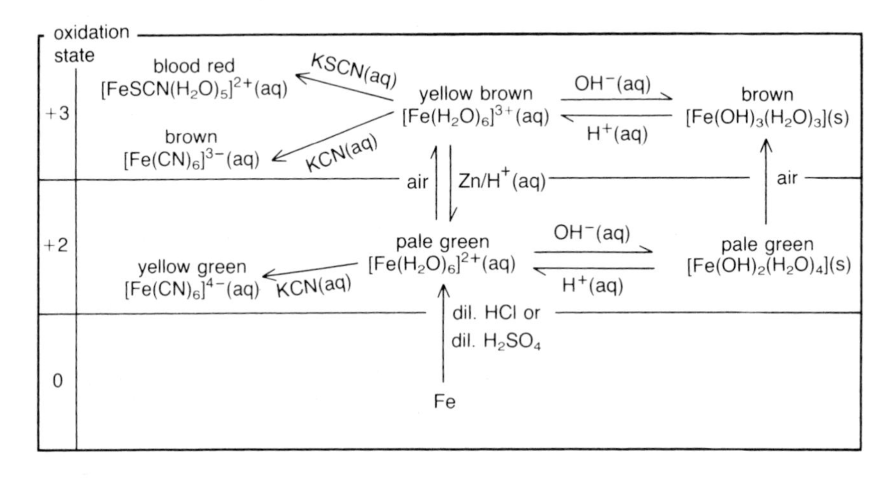

Table 21.6 Outline chemistry of cobalt

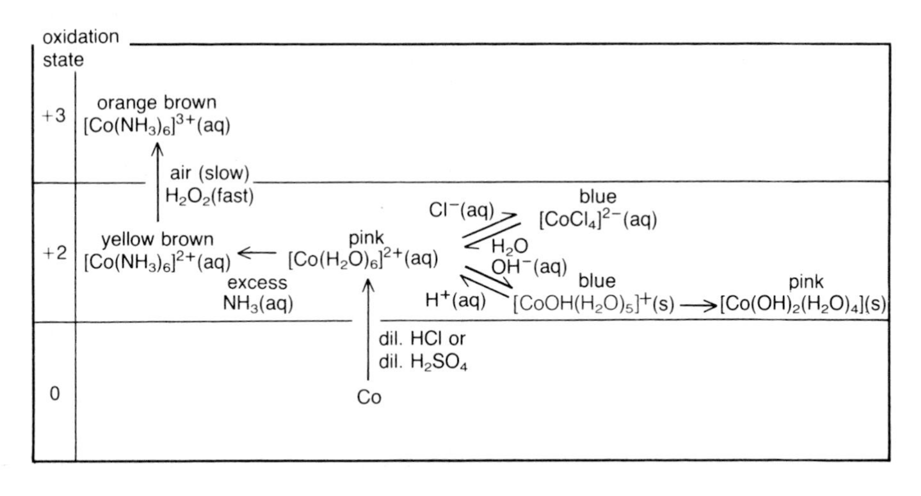

Table 21.7 Outline chemistry of nickel

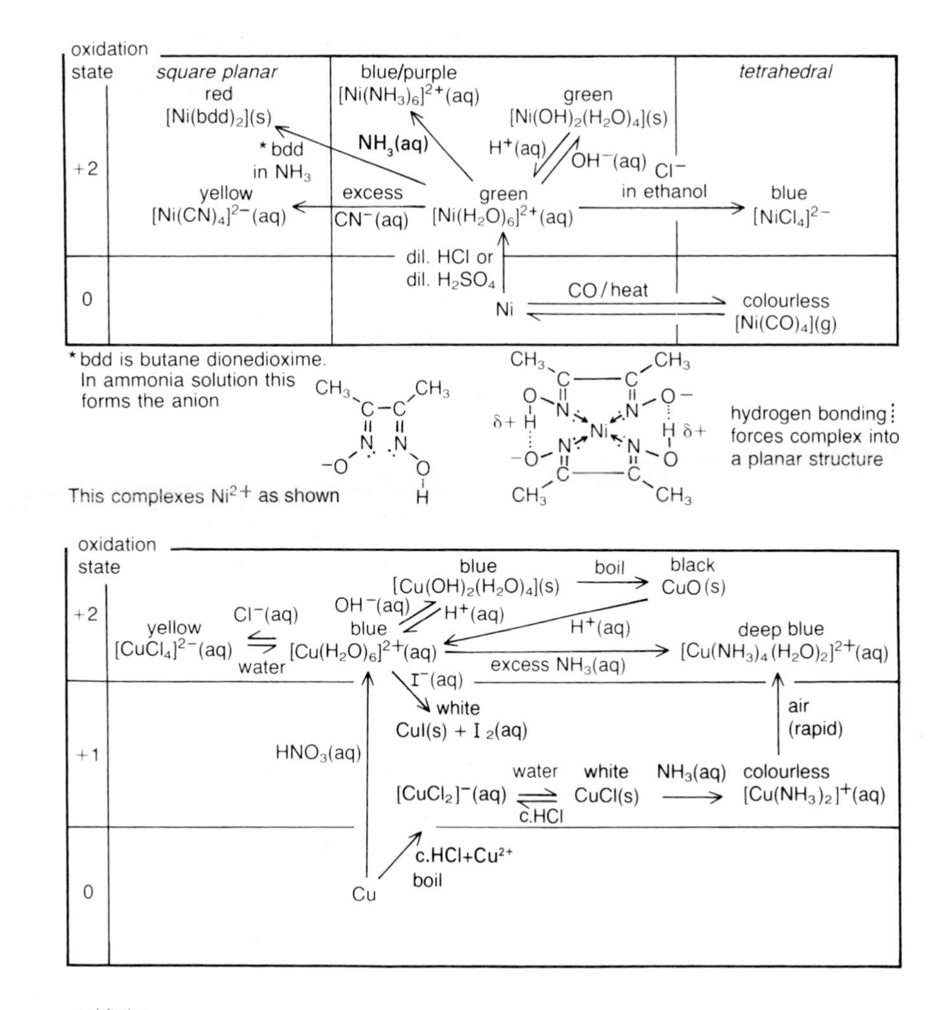

Table 21.8 Outline chemistry of copper

Table 21.9 Outline chemistry of zinc

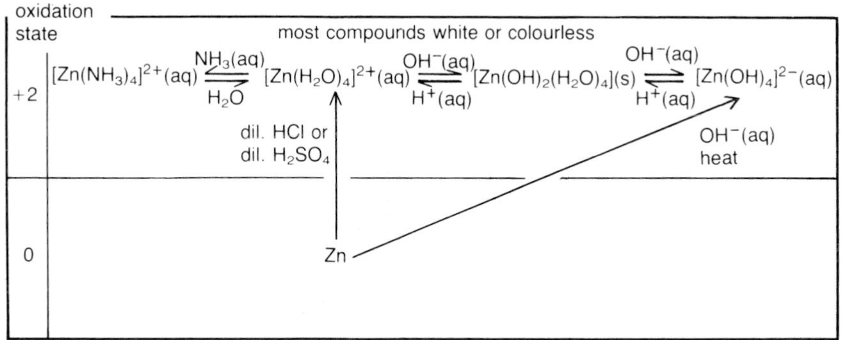

Table 21.10 Outline chemistry of silver

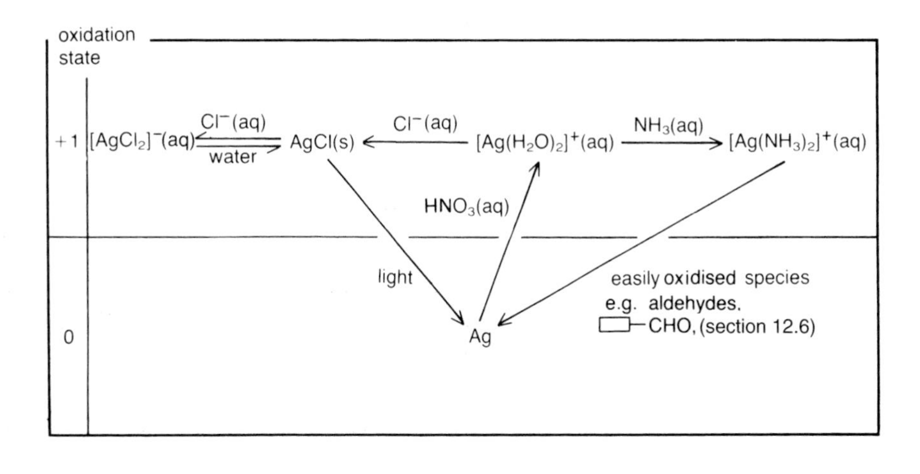

Focus 21b

1. Transition metals readily form **complexes** in which electron pair donors, known as **ligands**, surround a central metal ion
2. The number of ligands attached to a metal ion by **co-ordinate covalent bonds** is known as the **co-ordination number**. 6-co-ordinate complexes are **octahedral** in shape, most 4-co-ordinate complexes are **tetrahedral**, and the 2-co-ordinate complexes of copper(I) and silver are **linear**.
3. Ligands are normally neutral molecules or negative ions and may have two, **bidentate** or more, **polydentate** lone pairs of electrons capable of being donated to the same metal ion.
4. Complexes may be electrically neutral or have an overall positive or negative charge.
5. Complexes with an overall charge are generally soluble in water whilst electrically neutral complexes are insoluble.

21.3 Acid–base reactions of aqua complexes

In common with many other metals all the d block elements form **hexaaqua ions**, $[M(H_2O)_6]^{x+}$, in dilute aqueous solution. As the approximate pH values in Table 21.3 show, *M^{3+}(aq) ions are appreciably more acidic than M^{2+}(aq) species*. This acidity arises from the attraction of the metal ion for the electrons in the water ligands, releasing H^+(aq) ions into solution:

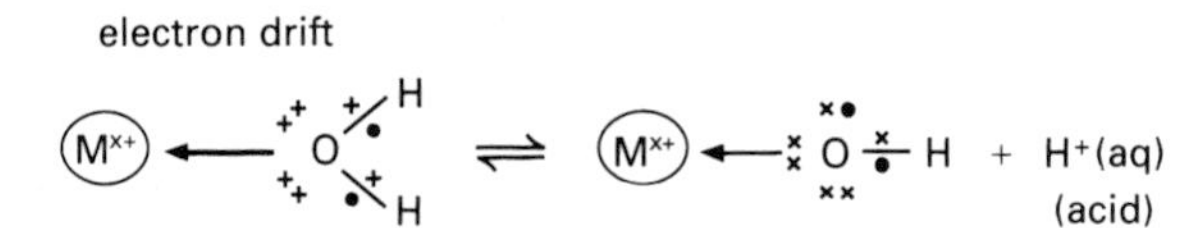

Table 21.11 The acidity of some d block cations in aqueous solution

Metal ion	approximate pH in dilute aqueous solution
Sc^{3+}(aq)	3
Ti^{3+}(aq)	2
V^{3+}(aq)	2
Cr^{3+}(aq)	3
Mn^{2+}(aq)	5
Fe^{2+}(aq)	5
Fe^{3+}(aq)	2
Co^{2+}(aq)	4
Ni^{2+}(aq)	5
Cu^{2+}(aq)	5
Zn^{2+}(aq)	5

Which is more acidic, M^{2+}(aq) or M^{3+}(aq)? How do you explain this?

Clearly, as the positive charge density of the metal ion increases, so does the polarisation of the water molecules and hence the acidity of the solution.

Theoretically, hexaaqua ions may act as polybasic acids by losing further H^+(aq) ions, e.g.

1 $[M(H_2O)_6]^{3+}(aq) \rightleftharpoons [MOH(H_2O)_5]^{2+}(aq) + H^+(aq)$

2 $[MOH(H_2O)_5]^{2+}(aq) \rightleftharpoons [M(OH)_2(H_2O)_4]^{+}(aq) + H^+(aq)$

3 $[M(OH)_2(H_2O)_4]^{+}(aq) \rightleftharpoons [M(OH)_3(H_2O)_3](s) + H^+(aq)$

Unless a base is added, only reaction 1 is important.

If a base is added to the hexaaqua ion solution, H^+(aq) ions are removed, pushing the above systems to the right. *The first stage is always the precipitation of the electrically neutral hydroxoaqua complex*, e.g. with NaOH(aq):

$$[Fe(H_2O)_6]^{2+}(aq) + 2OH^-(aq) \rightleftharpoons [Fe(OH)_2(H_2O)_4](s) + 2H_2O(l)$$

$$[Fe(H_2O)_6]^{3+}(aq) + 3OH^-(aq) \rightleftharpoons [Fe(OH)_3(H_2O)_3](s) + 3H_2O(l)$$

Note that these reactions are reversible. *The hydroxoaqua precipitate will act as a **base** by dissolving in dilute acid to reform the hexaaqua ion.*

In some cases, addition of excess strong alkali will also dissolve the precipitate giving a negatively charged complex, e.g.

$$[Cr(OH)_3](H_2O)_3](s) + 3OH^-(aq) \rightleftharpoons [Cr(OH)_6]^{3-}(aq) + 3H_2O(l)$$

In this reaction the precipitate acts as an ***acid*** *by donating further protons to the alkali*. Such species that can act as both bases and acids are referred to as **amphoteric** or **amphiprotic**.

Since most metal carbonates are insoluble, the addition of $Na_2CO_3(aq)$ to $M^{2+}(aq)$ generally results in precipitation of the metal carbonate, e.g.

$$[Fe(H_2O)_6]^{2+}(aq) + CO_3^{2-}(aq) \longrightarrow FeCO_3(s) + 6H_2O(l)$$

Although the addition of $Na_2CO_3(aq)$ to $M^{3+}(aq)$ also gives a precipitate, this is *not* the carbonate. The $M^{3+}(aq)$ is sufficiently acidic to decompose the CO_3^{2-} ion, releasing carbon dioxide gas and water:

$$CO_3^{2-}(aq) + 2H^+(aq) \longrightarrow CO_2(g) + H_2O(l)$$

Loss of protons from the hexaaqua ion precipitates the electrically neutral hydroxoaqua complex:

$$[Fe(H_2O)_6]^{3+}(aq) \rightleftharpoons [Fe(OH)_3(H_2O)_3](s) + 3H^+(aq)$$

Combining these two equations gives us the overall process:

$$2[Fe(H_2O)_6]^{3+}(aq) + 3CO_3^{2-}(aq) \rightleftharpoons 2[Fe(OH)_3(H_2O)_3](s) + 3CO_2(g) + 3H_2O(l)$$

Activity 21.4

Magnesium metal dissolves in acid, releasing hydrogen gas as follows:

$$Mg(s) + 2H^+(aq) \longrightarrow Mg^{2+}(aq) + H_2(g)$$

When a piece of magnesium ribbon is added to a solution of iron(III) chloride, bubbles of gas are evolved and a red-brown precipitate is formed. Write an overall equation for this reaction and account for the fact that iron(II) chloride solution does not give a similar reaction with magnesium.

21.4 Ligand exchange

If another ligand is added to a solution of a hexaaqua complex it may compete with the water molecules to form a complex with the metal ion. Such **ligand exchange** often causes a change in the colour of the solution, which can be useful for identification purposes. For example, addition of concentrated hydrochloric acid to pink $Co^{2+}(aq)$ increases the concentration of chloride ions and gives the deep blue tetrachloro complex:

$$\underset{\text{pink}}{[Co(H_2O)_6]^{2+}(aq)} + 4Cl^-(aq) \rightleftharpoons \underset{\text{blue}}{[CoCl_4]^{2-}(aq)} + 6H_2O(l)$$

This reversible reaction is the basis of the blue cobalt chloride paper test for water. Filter paper is soaked in cobalt chloride solution. On drying in an oven, the water evaporates leaving blue anhydrous cobalt(II) chloride. If this is dipped into a solution that contains water the colour changes to pink owing to the formation of the hexaaqua complex. Note that since Cl^- ions are larger than water molecules, the co-ordination number of the first row d block chloro complexes is 4 rather than 6.

Ammonia will also often displace water molecules either fully or partially from many hexaaqua complexes. Initially the ammonia acts as a base, precipitating the electrically neutral hydroxoaqua complex, e.g.

$$[Ni(H_2O)_6]^{2+}(aq) + 2NH_3(aq) \rightleftharpoons [Ni(OH)_2(H_2O)_4](s) + 2NH_4^+(aq)$$

Ligand exchange may then follow, e.g.

$$[Ni(OH)_2(H_2O)_4](s) + 6NH_3(aq) \longrightarrow [Ni(NH_3)_6]^{2+}(aq) + 2OH^-(aq)$$

green — blue-violet

The overall effect of adding excess ammonia to $Ni^{2+}(aq)$ is therefore given by the equation,

$$[Ni(H_2O)_6]^{2+}(aq) + 6NH_3(aq) \longrightarrow [Ni(NH_3)_6]^{2+}(aq) + 6H_2O(l)$$

With $Cu^{2+}(aq)$ aqueous ammonia only replaces four of the water ligands,

$$[Cu(H_2O)_6]^{2+}(aq) + 4NH_3(aq) \longrightarrow [Cu(H_2O)_2(NH_3)_4]^{2+} + 4H_2O(l)$$

pale blue — royal blue

Potassium thiocyanate solution, KSCN(aq), provides an extremely sensitive test for $Fe^{3+}(aq)$:

$$[Fe(H_2O)_6]^{3+}(aq) + SCN^-(aq) \longrightarrow [FeSCN(H_2O)_5]^{2+}(aq) + H_2O(l)$$

yellow-brown — deep blood-red

21.5 Redox reactions

Redox reactions are possible when a metal shows more than one oxidation state. Higher oxidation states may often be reduced by zinc in dilute sulphuric acid. Such reduction reactions, like ligand exchange, are generally accompanied by characteristic colour changes that are often of use in chemical analysis (Chapter 22).

$Fe^{3+}(aq)$	$\longrightarrow$	$Fe^{2+}(aq)$
+3		+2
yellow brown		pale green

If the metal shows more than one oxidation state then a series of colour changes indicates progressive reduction. For vanadate(V) the following sequence is observed:

$VO_3^-(aq)$	$\longrightarrow$	$VO^{2+}(aq)$	$\longrightarrow$	$V^{3+}(aq)$	$\longrightarrow$	$V^{2+}(aq)$
+5		+4		+3		+2
pale yellow		bright blue		green		violet

If the resulting solution is separated from the zinc and allowed to stand in air it reverts to a bright blue colour, showing than +4 is the most stable oxidation state of vanadium under these conditions.

Reduction of dichromate(VI) solution by zinc takes place as follows:

$Cr_2O_7^{2-}(aq)$	$\longrightarrow$	$Cr^{3+}(aq)$	$\longrightarrow$	$Cr^{2+}(aq)$
+6		+3		+2
orange		green		sky blue
		(most stable)		(very easily oxidised)

Intermediate oxidation states are not always observed and in a strongly acidic solution, manganate(VII) is reduced directly to manganese(II).

$MnO_4^-(aq)$	$\longrightarrow$	$Mn^{2+}(aq)$
+7		+2
deep purple		almost colourless (most stable)

Although we have indicated the most stable oxidation states in the above sequences, it should be noted that this can depend upon the conditions or the type of ligand. Thus $[Fe(H_2O)_6]^{2+}$ is oxidised by air to $[Fe(H_2O)_6]^{3+}$, slowly in acidic solution but rapidly on addition of alkali. If the water ligands are replaced by cyanide however, the iron(II) complex $[Fe(CN)_6]^{4-}$ is resistant to atmospheric oxidation.

Disproportionation involves self oxidation and reduction. For example, copper(I) oxide dissolves in dilute acid to give copper metal and $Cu^{2+}(aq)$:

	$Cu^+ + e^- \longrightarrow Cu(s)$	reduction
	$Cu^+ \longrightarrow Cu^{2+}(aq) + e^-$	oxidation
$\Longrightarrow$	$2Cu^+ \longrightarrow Cu(s) + Cu^{2+}(aq)$	redox

Activity 21.5

When cobalt(II) chloride is dissolved in water a pink solution containing hexaaquacobalt(II), (A) is formed which is unaffected by exposure to air or oxidising agents. When excess ammonia is added a pale brown solution containing hexaamminecobalt(II), (B) is formed which on standing in air slowly darkens at the surface. When an oxidising agent such as hydrogen peroxide is added the mixture now immediately turns dark brown. Addition of concentrated hydrochloric acid gives orange crystals (C) with the following composition by mass: Co = 22.1%, NH_3 = 38.1%, Cl = 39.8%. These crystals dissolve in water to give a solution that has neither oxidising nor reducing properties.

1 Write the formulae of complexes (A) and (B).

2 What kind of reaction is involved in the conversion of (A) into (B)?

3 How do you account for the surface discoloration of (B) on exposure to air?

4 Calculate the empirical formula of the orange crystals (C) (relative atomic masses: Co = 59, N = 14, H = 1, Cl = 35.5)

5 If (C) contains only ammonia as a ligand, give its full systematic name.

6 Comment on the relative stabilities of the two oxidation states of cobalt.

Redox chemistry and drug delivery

E

The stability of cobalt(III) complexes with nitrogen ligands may prove useful in treating certain types of cancer.

Solid tumours often have a restricted blood supply which makes them difficult to target by injected drugs. However, the low oxygen levels in such tumours that result from this may be used against them. Nitrogen mustard contains the chemical group bis(2-chloroethyl)amine, which causes cell damage similar to that of X-rays and is far too toxic to inject directly.

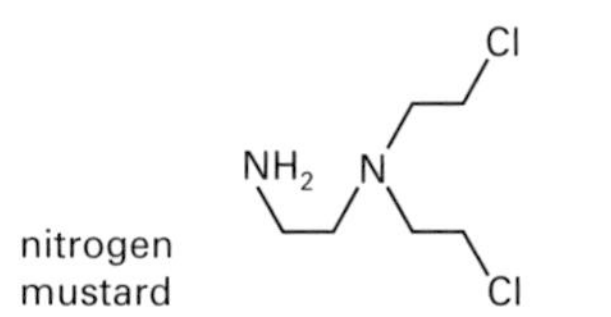

However, it binds tightly to cobalt(III) ions through its nitrogen atoms producing a complex that is much less toxic.

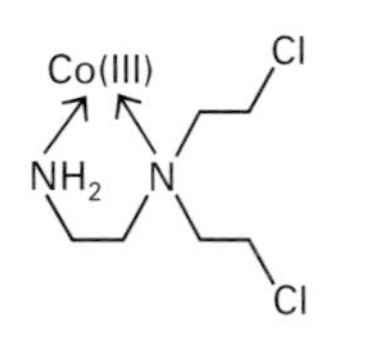

stable complex with Co(III)

Inside the body this complex is attacked by enzymes that quickly reduce cobalt(III), to cobalt(II) which does not bind to the nitrogen mustard so tightly. However, in the oxygen-rich blood supply of healthy cells, the cobalt(II) is rapidly re-oxidised to cobalt(III), allowing it to regain its tight grip on the ligand. The drug will, therefore, pass through the blood stream causing little damage to normal cells. When it reaches the tumour, however, there is insufficient oxygen to convert the cobalt(II) back into cobalt(III), allowing the nitrogen mustard to escape and kill the infected cells.

Before leaving the topic of oxidation and reduction let us consider the preparation from iron of the anhydrous chlorides, $FeCl_2$ and $FeCl_3$. Again the conditions are important in deciding which product is formed. If iron is heated in a stream of chlorine in the apparatus used to prepare aluminium chloride (Figure 20.4), the oxidising nature of the halogen means that the higher chloride is produced:

$$2Fe(s) + 3Cl_2(g) \longrightarrow 2FeCl_3(s)$$

This can be reduced to iron(II) chloride by heating in a stream of hydrogen:

$$2FeCl_3(s) + H_2(g) \longrightarrow 2FeCl_2(s) + 2HCl(g)$$

Iron(II) chloride can be prepared directly from the metal by heating in a stream of hydrogen chloride gas. The hydrogen that is also produced ensures that the lower oxidation state is formed:

$$Fe(s) + 2HCl(g) \longrightarrow FeCl_2(s) + H_2(g)$$

Focus 21c

Metal ions generally form hexaaqua complexes in dilute aqueous solution. These undergo the following characteristic types of reaction:

1. **Acid–base** by loss or recapture of protons from co-ordinated water molecules.
2. **Ligand exchange**, in which water molecules are substituted by other ligands such as chloride ions or ammonia molecules.
3. **Oxidation** or **reduction** of the central metal ion.

Many of these reactions are accompanied by characteristic colour changes.

Colour in transition metal complexes

Many, but not all, compounds of d block elements are coloured. Exceptions include scandium(III) compounds that have an empty 3d sub-level and zinc(II) compounds which have a completely full 3d sub-level. Most d block compounds that contain a metal ion with a partially filled d sub-level are, however, coloured. Take for example copper(II) sulphate that dissolves in water to form a blue solution. We can explain the origin of this blue colour by considering the electronic structure of the Cu^{2+} ion below.

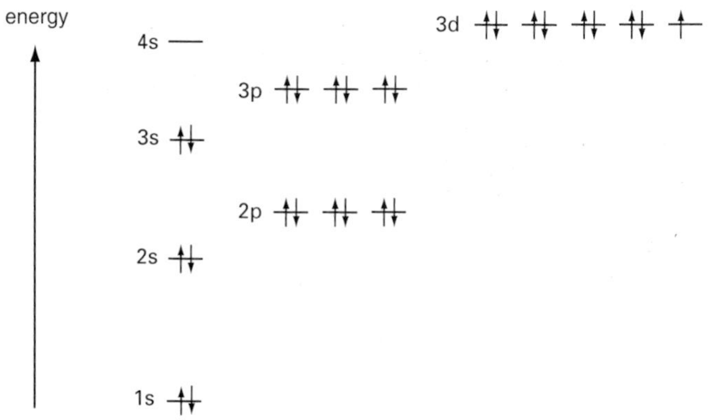

energy

3d ⥮ ⥮ ⥮ ⥮ ↿

4s —

3p ⥮ ⥮ ⥮

3s ⥮

2p ⥮ ⥮ ⥮

2s ⥮

1s ⥮

Although electrons could be promoted from the 3p into the 3d or from the 3d into the 4s, neither of these transitions would explain why Sc(III) and Zn(II) compounds are white or colourless. With an empty 3d sub-level the 3p $\longrightarrow$ 4s transition is possible and with a full 3d sub-level we can still promote 3d $\longrightarrow$ 4s.

This leaves us only one possibility here for the origin of the blue colour; movement of an electron *between* the orbitals of the 3d sub-level. However, as the above diagram shows, in an isolated Cu^{2+} ion all five 3d orbitals are **degenerate**, i.e. have exactly the same energy. In this situation no energy is required to move one 3d electron into the half-filled orbital and so no absorption occurs.

If we evaporate a solution of copper(II) sulphate to dryness and drive off all the water, the blue colour is lost and we are left with a white residue of anhydrous copper(II) sulphate. This suggests that the presence of water is needed to produce the colour, a fact confirmed by the fact that the blue colour is restored on adding water to anhydrous copper(II) sulphate. It is in fact the complex ion $[Cu(H_2O)_6]^{2+}$ that is responsible for the blue colour:

Colour in transition metal complexes *continued*

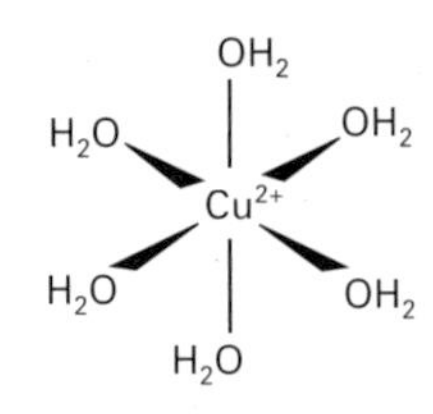

When the lone pairs on the ligands approach, they repel the electrons present in the metal d orbitals whose energy therefore rises. However, the octahedral shape of the complex ion means that the ligand lone pairs get much closer to two of the d orbitals than they do to the remaining three. The d orbitals therefore split into two sets separated by the ligand field splitting energy, Δ:

Cu^{2+} free metal ion

ligands approaching

octahedral complex formed

Δ

If the ligand field splitting energy falls within the quantum energy range of visible radiation then the complex will absorb light of a corresponding wavelength. Its observed colour will be a mixture of the visible radiation which is not absorbed.

Although maximum absorption for $[Cu(H_2O)_6]^{2+}$ occurs in the infrared at about 800 nm, the band is quite broad with measurable absorption of red light.

When an excess of concentrated ammonia solution is added to a solution containing $[Cu(H_2O)_6]^{2+}$, the colour changes from pale blue to a characteristic deep royal blue owing to the formation of a new complex ion:

$$[Cu(H_2O)_6]^{2+}(aq) + 4NH_3\,(aq) \longrightarrow [Cu(H_2O)_2(NH_3)_4]^{2+}(aq) + 4H_2O(l)$$

The complex is still octahedral in shape but four of the water ligands have been replaced by ammonia.

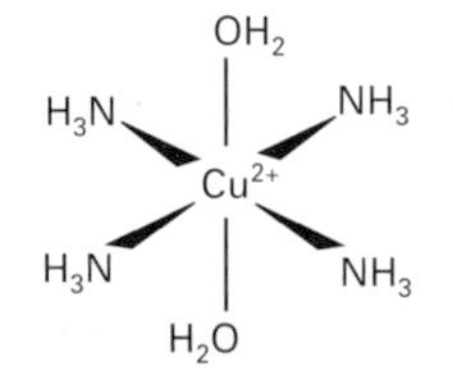

Maximum absorption now occurs at around 600 nm, i.e. at a shorter wavelength than that for $[Cu(H_2O)_6]^{2+}(aq)$. This means that the ligand field splitting energy has been increased by substituting ammonia ligands for water. The following **spectrochemical series** lists some common ligands in order of their d orbital splitting ability:

I^- Br^- Cl^- F^- OH^- H_2O NH_3 CN^- CO

increasing d orbital splitting power →

It is not only the type of ligand that affects splitting energy, but also the co-ordination number and the shape of the complex. Addition of concentrated hydrochloric acid to a solution containing $[Cu(H_2O)_6]^{2+}$ causes a colour change from blue to yellow-green:

$$[Cu(H_2O)_6]^{2+}(aq) + 4Cl^-(aq) \rightleftharpoons [CuCl_4]^{2-}(aq) + 6H_2O(l)$$

The new complex has only four ligands, arranged tetrahedrally around the metal ion:

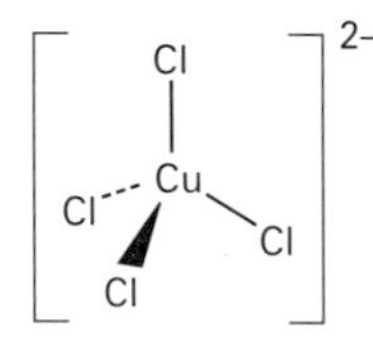

The d orbitals on the Cu^{2+} ion are again split into two sets but this time the pattern is reversed with three orbitals at higher energy than the other two.

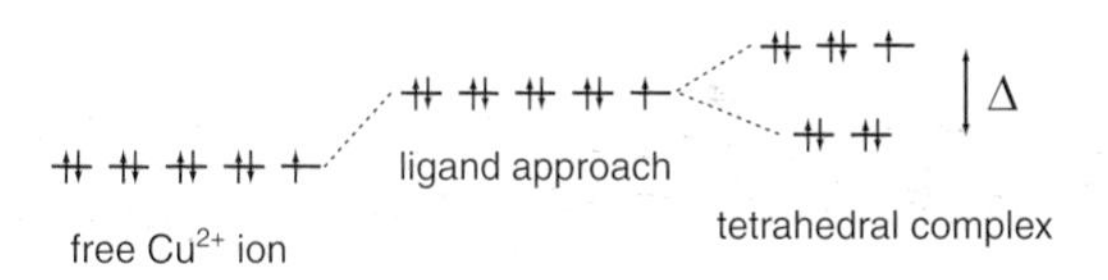

As there are only four ligands here, the splitting energy, Δ, is less than in a 6-co-ordinate octahedral complex. Combined with the weaker splitting caused by Cl^- ions compared to NH_3 and H_2O ligands, maximum absorption moves to lower energy, i.e. higher wavelength, causing the colour to change.

So far we have explained colour in transition metal compounds in terms of movement of electrons within a partially filled d sub-level that has been split by a ligand field. However there are coloured compounds in which the d sub-level is apparently empty. This generally occurs when the oxidation state of the central metal ion is high, for instance manganate(VII), MnO_4^-, which is an intense purple. The complex is tetrahedral and may be considered as a central highly charged metal ion surrounded by four oxide ion ligands:

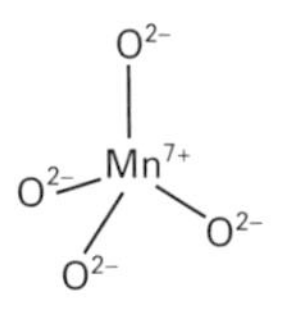

Colour in transition metal complexes *continued*

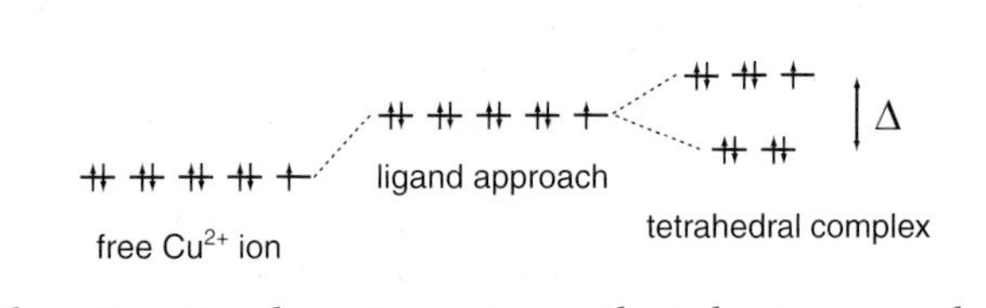

The attraction here is so strong that electrons on the ligands may jump across into the orbitals on the central metal ion. The energy difference between the ligand orbitals and the metal orbitals falls into the visible spectrum so the complex appears coloured. This effect is known as **charge transfer** and can give very intense absorption bands.

For more details on colour chemistry see Chapter 18.

21.6 Some useful complexes

Transition metal complexes are found in almost all areas of life. Indeed, such complexes are essential to life itself. Figure 21.3 illustrates the structure of **haemoglobin**, the substance in red blood cells that transports oxygen round our bodies.

This contains iron(II) complexed with a large protein. Anaemia caused by lack of iron in the diet may be treated simply by taking iron(II) sulphate tablets.

Another complex with important medical applications is the anti-cancer drug **cisplatin** whose structure is shown in Figure 21.4.

It works by binding to the DNA molecule which displaces the chloride ligands. This can only happen when chloride ligands are in the *cis* position and the *trans* isomer is ineffective in cancer treatment.

Other areas in which complexes have been used include the production of coloured pigments and in the preparation of electrolytes for electroplating with precious metals such as silver and gold. Thiosulphate complexes of silver are also involved in the processing of photographic film.

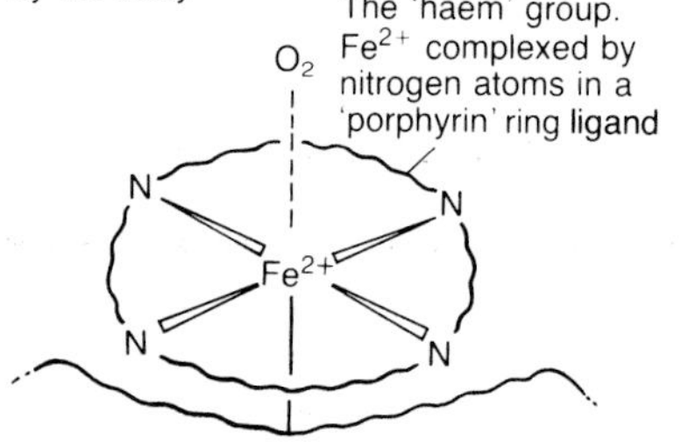

Figure 21.3 ▲
Schematic structure of haemoglobin and its function as an 'oxygen carrier' in the blood

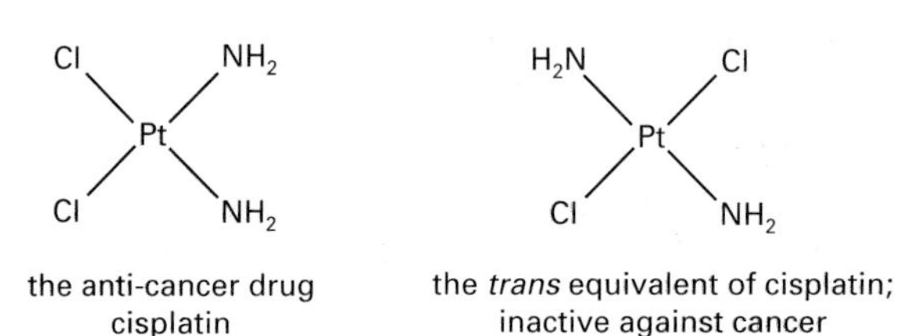

Figure 21.4 ▲
Geometrical isomers of the square planar complex dichlorodiammineplatinum(II)

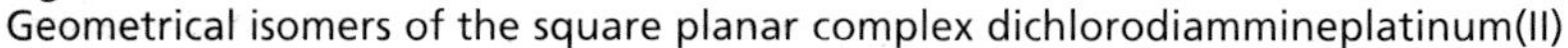

Activity 21.6

Black and white photographic film contains a gelatin emulsion of a silver halide coated onto a transparent, flexible plastic backing. On exposure to light the halide starts to decompose, forming metal atoms, e.g.

$$2AgBr \longrightarrow 2Ag + Br_2$$

At this stage the image is extremely faint and must be **developed** before it becomes visible.

Activity 21.6 continued

Hydroquinone, 1,4 dihydroxybenzene, in alkaline solution selectively reduces silver ions at a rate that is in proportion to previous exposure to light. The exposed parts of the film emulsion are thus converted into black metallic silver, forming a visible image. After developing, the film is still light sensitive as the unexposed parts still contain unchanged silver halide.

The **fixing** stage removes the remaining silver halide, thus stabilising the image to further light exposure. Photographers' 'hypo' solution contains sodium thiosulphate, $Na_2S_2O_3$. The thiosulphate ions dissolve away silver halide by forming the dithiosulphatoargentate(I) complex.

Of course this process gives a negative, i.e. those areas exposed to the greatest amount of light appear blackest on the film. For some purposes, e.g. medical X-rays, this is quite acceptable but a 'positive' image is normally produced by 'printing'. Here light passing through the transparent negative is allowed to fall onto opaque photographic paper that is processed as above.

1 Hydroquinone ionises as follows in alkaline solution:

$$(C_6H_4)(OH)_2(aq) + 2OH^-(aq) \longrightarrow (C_6H_4)(O^-)_2(aq) + 2H_2O(l)$$

This converts silver ions to silver metal, producing benzoquinone, $(C_6H_4)(O)_2(aq)$. Write a balanced equation for this redox reaction.

2 Give the formula and charge of the dithiosulphatoargentate(I) complex. Why is this complex water soluble and why is it important that it should be so?

Questions on Chapter 21

1 Using the species $[CrCl_3(H_2O)_3].3H_2O$, explain the terms: complex, ligand, and co-ordination number. How does the isomer $[Cr(H_2O)_6]Cl_3$ differ from the above species?

2 Use examples to illustrate the meaning of the following terms; anionic ligand, neutral ligand, cationic complex, neutral complex, anionic complex, bidentate ligand, polydentate ligand, chelate.

3 Reactions of complexes may be divided into three main types: acid–base, ligand exchange and redox. Identify the type(s) of reaction involved when a solution containing the complex $[Cr(H_2O)_6]^{3+}$ is converted in turn into each of the following.

a) $[Cr(H_2O)_6]^{2+}$ **b)** $[Cr(OH)_3(H_2O)_3]$ **c)** CrO_4^{2-}
d) $[Cr(NH_3)_6]^{3+}$ **e)** $[Cr(OH)_4(H_2O)_2]^-$

4 Give the full systematic name for each of the complexes given in question 3. Which one of these species is most likely to be insoluble in water and why?

5 When excess sodium hydroxide solution is added to a fresh solution of iron(II) sulphate, containing the pale green hexaaquairon(II) complex, a green precipitate (A) is formed.

a) Give the name and formula of this green precipitate and write an equation for its formation.

b) On standing in air, the green precipitate rapidly turns dark brown near the surface. Explain what causes this and give the name and formula of the dark brown precipitate (B).

c) The brown precipitate (B) dissolves in dilute acid to give a yellow brown solution containing another iron complex, (C). Give the name of this complex and write an equation for its formation.

d) When the original iron(II) sulphate solution is allowed to stand in air it is slowly converted into (C). Suggest a common laboratory reagent that might be added to slow down this conversion.

6 What is meant by the term amphoteric (or amphiprotic)? Give an example involving a transition metal complex to illustrate your answer.

7 Chromium(III) sulphate dissolves in water to give a solution containing hexaaquachromium(III). When excess sodium carbonate solution is added, a gas is evolved and a precipitate is formed. Name these products and write a balanced equation for the reaction. Explain why the metal carbonate is NOT produced in this case.

8 A dilute solution of cobalt(II) chloride is pale pink owing to complex (A). When filter paper soaked in this solution is heated in an oven, it turns dark blue owing to the formation of another cobalt complex (B). When a drop of water is placed on the blue paper it turns pink, reforming complex (A). Give the names and formulae of the two complexes and explain by a series of balanced equations the above sequence of reactions.

Questions on Chapter 21 *continued*

9 Manganate(VI) ions, MnO_4^{2-}, are stable only under very alkaline conditions. When a solution of this complex is acidified it disproportionates into manganate(VII) solution, MnO_4^-, and a precipitate of manganese(IV) oxide, MnO_2.

a) What is meant by 'disproportionation'?

b) Write half-equations for the conversion of manganate(VI) into manganate(VII) and manganese(IV) oxide.

c) Use your answers to b) above to derive an overall equation for the disproportionation of manganate(VI) in acid solution.

Comments on the activities

Activity 21.1

1 Ag(I) [Kr] $4d^{10}5s^0$ Au(I) [Xe] $5d^{10}6s^0$
Au(III) [Xe] $5d^8 6s^0$

2 Gold is a transition element by virtue of its +3 oxidation state. Silver has only one common oxidation state and is not generally regarded as a transition element.

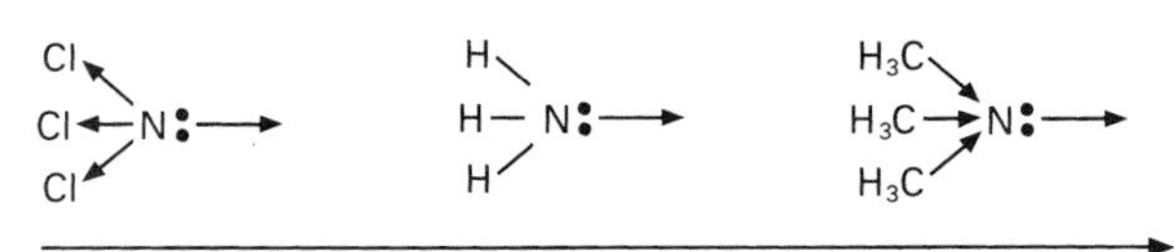

increasing electron density on the nitrogen atom improves complexing ability

Activity 21.2

1 Dot and cross electron diagrams reveal that the following possess atoms with 'lone' pairs of electrons and could, therefore, act as ligands:

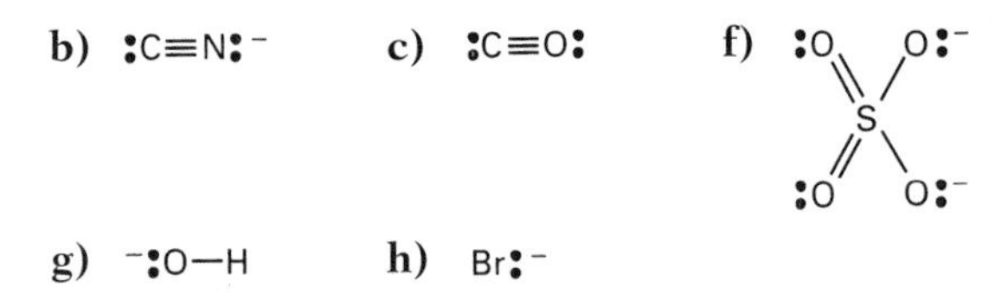

i) $CH_3—CH_2—\ddot{O}—CH_2—CH_3$

j) $CH_3—C(=\!:\!O)—CH_2—C(=O\!:)—CH_3$

(For clarity, any extra lone pairs on particular donor atoms have been omitted.) The remaining species cannot function as ligands as they possess no 'lone' pairs of electrons.

Since b), c), f) and j) above contain more than one atom with an available lone pair, we might expect them to behave as bi- (or in the case of SO_4^{2-} poly-) dentate ligands. In practice only species j) donates more than one pair of electrons to a particular metal ion, since only this molecule possesses the 'flexibility' required to bend itself into the necessary shape.

2 The key to complexing ability is the ease with which a pair of electrons is donated. In nitrogen trichloride the very electronegative chlorine atoms attract electron density away from the nitrogen atom, making it more *reluctant* to donate its lone pair. Methyl groups are, however, electron repelling, and in trimethylamine, *improve* the donating power of the nitrogen atom by *increasing* the electron density around it.

Activity 21.3

1 $[Cu(NH_3)_4(H_2O)_2]SO_4$ tetraamminediaquacopper(II) sulphate
$[Ni(OH)_2(H_2O)_4]$ dihydroxotetraaquanickel(II)*
$Na[AgCl_2]$ sodium dichloroargentate(I)

2 hexaamminecobalt(III) chloride $[Co(NH_3)_6]Cl_3$
trichlorotriamminecobalt(III) $[CoCl_3(NH_3)_3]$*
sodium tetrachlorocobaltate(II) $Na_2[CoCl_4]$

*electrically neutral complexes likely to be insoluble in water.

Activity 21.4

$Mg(s) + 2H^+(aq) \longrightarrow Mg^{2+}(aq) + H_2(g) \times 3$
$[Fe(H_2O)_6]^{3+}(aq) \longrightarrow [Fe(OH)_3(H_2O)_3](s) + 3H^+(aq) \times 2$
$\Longrightarrow 2[Fe(H_2O)_6]^{3+}(aq) + 3Mg(s)$
$\longrightarrow 2[Fe(OH)_3(H_2O)_3](s) + 3Mg^{2+}(aq) + 3H_2(g)$

$Fe^{2+}(aq)$ is not sufficiently acidic to give hydrogen with magnesium.

Activity 21.5

1 (A) $[Co(H_2O)_6]^{2+}$ (B) $[Co(NH_3)_6]^{2+}$

2 Conversion of (A) into (B) is ligand exchange.

3 Oxidation of the complex by oxygen in the air darkens the solution. This is confirmed by the effect of adding the oxidising agent hydrogen peroxide.

4

	Co	NH_3	Cl
in 100 g of (C)	22.1g	38.1g	39.8g
number of moles	$\frac{22.1}{59}$	$\frac{38.1}{17}$	$\frac{39.8}{35.5}$
	= 0.375	= 2.24	= 1.12 moles
simplest ratio	$\frac{0.375}{0.375}$	$\frac{2.24}{0.375}$	$\frac{1.12}{0.375}$
	= 1	= 6	= 3

Comments on the activities *continued*

The empirical formula of (C) is $Co(NH_3)_6Cl_3$

5 Since ammonia is the only ligand, (C) must be $[Co(NH_3)_6]Cl_3$ hexaamminecobalt(III) chloride. (As expected for an ionic complex this is water soluble.)

6 With water as ligand, +2 is by far the most stable oxidation state of cobalt. If the water is replaced by ammonia, however, cobalt(II) is readily oxidised, even by air, into cobalt(III).

Activity 21.6

1 $(C_6H_4)(O^-)_2(aq) \longrightarrow (C_6H_4)(O_2)(aq) + 2e^-$
$Ag^+(s) + e^- \longrightarrow Ag(s) \times 2$
$\Longrightarrow (C_6H_4)(O^-)_2(aq) + 2Ag^+(s)$
$\longrightarrow (C_6H_4)(O_2)(aq) + 2Ag(s)$

2 Dithiosulphatoargentate(I) is $[Ag(S_2O_3)_2]^{3-}$. As with most electrically charged species it is water soluble. This is important as all the excess insoluble silver halide must be washed off the film if it is to be stable to further exposure to light.

C H A P T E R

22

Inorganic analysis

Contents

Study Checklist

After studying this chapter you should be able to:

1. Describe the following tests for anions:
   silver nitrate and ammonia to identify halides
   barium chloride solution to identify sulphate
   dilute acid to identify carbonate and sulphite
2. Describe the use of flame tests and reactions involving precipitation and complex formation to identify common metal ions in solution.
3. Carry out calculations involving weights of precipitates (gravimetric analysis).
4. Describe the procedures used and carry out calculations involved in the following types of titration (volumetric analysis):
   acid–base
   acid–carbonate
   manganate(VII) oxidations
   iodine–thiosulphate

Safety first!

Many 'heavy metal' ions, such as lead, silver, mercury and barium, are toxic, especially in solution. Care should be taken when using barium chloride solution to test for sulphate ions and when testing for halides with silver nitrate solution.

22.1 What have we got? (qualitative analysis)

Anions in aqueous solution may often be detected by reactions that involve either the precipitation of an insoluble salt or the production of a gas.

Apart from fluoride, halide ions may be identified by precipitation with silver nitrate solution:

$$Ag^+(aq) + Cl^-(aq) \longrightarrow AgCl(s) \text{ white}$$
$$Ag^+(aq) + Br^-(aq) \longrightarrow AgBr(s) \text{ cream}$$
$$Ag^+(aq) + I^-(aq) \longrightarrow AgI(s) \text{ pale yellow}$$

Dilute nitric acid must be added to prevent the precipitation of other insoluble silver salts such as the carbonate. Confirmation of the suspected halide may be obtained by testing the solubility of the precipitate in ammonia solution. Silver chloride dissolves readily in dilute ammonia, whilst silver bromide only dissolves in concentrated ammonia and silver iodide is insoluble in both.

Sulphate ions give a white precipitate with barium chloride solution which is insoluble in dilute hydrochloric acid:

$$Ba^{2+}(aq) + SO_4^{2-}(aq) \longrightarrow BaSO_4(s)$$

Carbonate ions are precipitated by both silver nitrate and barium chloride solution:

$$2Ag^+(aq) + CO_3^{2-}(aq) \longrightarrow Ag_2CO_3(s) \text{ white}$$
$$Ba^{2+}(aq) + CO_3^{2-}(aq) \longrightarrow BaCO_3(s) \text{ white}$$

However, dilute acid decomposes carbonate ions, giving carbon dioxide gas and so sulphate and carbonate ions can be identified:

$$2H^+(aq) + CO_3^{2-}(aq) \longrightarrow H_2O(l) + CO_2(g)$$

Carbon dioxide is colourless and odourless but will turn lime-water (calcium hydroxide solution) cloudy white, owing to the precipitation of calcium carbonate:

$$\underset{\text{(lime-water)}}{Ca^{2+}(aq) + 2OH^-(aq)} + CO_2(g) \longrightarrow CaCO_3(s) + H_2O(l)$$

This test is conveniently carried out by suspending a drop of lime-water on the end of a glass rod above the solution.

Another anion that is decomposed by dilute acid releasing a gas is sulphite, SO_3^{2-}:

$$2H^+(aq) + SO_3^{2-}(aq) \longrightarrow H_2O(l) + SO_2(g)$$

Sulphur dioxide gas has the characteristic odour of burning sulphur and, since it is a reducing agent, it turns paper soaked in orange potassium dichromate solution bright green.

Activity 22.1 The nitrate problem

There is little doubt that the current world population could not be maintained without the help of artificial nitrogen fertilisers. However, their use is not without its problems as indicated by the following passage, taken from a report at the 1992 Earth Summit in Rio de Janeiro.

> 'We already have a full-occupancy planet,' says Noel Brown, North American director of the U.N. Environment Program. 'Today 80% of deforestation results from population growth. If the numbers keep rising until 2050, the U.N. estimates, an additional 5.9 million square kilometres (2.3 million square miles) of land will have to be turned over to farming, roads and urban uses. This is almost equivalent to the total size of protected natural areas on earth today. Most good agricultural land is already under the plough, and each year desertification, improper irrigation and overuse take millions of acres out of production. Farms may increase in productivity, but it will be much harder to match the gains of the past, and whether agricultural output can keep pace with population is an open question. The world has already overshot the saturation point in its ability to process many wastes. For instance, a doubling of human population would be likely to boost the concentrations of nitrates in rivers by 55%. Nitrates, which get into the water from air pollution and fertiliser run-off, are among the most difficult contaminants to remove. The chemicals cause human diseases and promote water conditions that kill fish and other aquatic life.'

High levels of nitrate in drinking water can cause a blood disorder in very young babies known as 'blue-baby syndrome' in which the infant's lips and body develop a noticeable blue hue. It is caused by the conversion of nitrate into nitrite, NO_2^-, by bacteria, either from within the child's gut or from an unsterilised feeding bottle. The haemoglobin in the baby's blood takes up the nitrite instead of oxygen and severe respiratory failure may result. Although the last reported case in Britain was in 1972, the World Health Organisation reported 2000 cases, including over 150 fatalities, between 1945 and 1986.

Activity 22.1 The nitrate problem continued

Nitrate, NO_3^-, is very difficult to remove from solution since it is stable to acid and alkali and forms no common insoluble salts. This also means that it cannot be detected as simply as other anions found in water supplies, such as chloride, sulphate and carbonate. However, in the laboratory nitrate can be converted into ammonia by reduction with aluminium powder in alkaline solution. Balance the following half-equations and combine them to give an overall equation for the process.

$$NO_3^-(aq) + H_2O(l) \longrightarrow NH_3(g) + OH^-(aq)$$
$$Al(s) \longrightarrow Al^{3+}(aq)$$

The alkaline gas ammonia may be detected by its pungent odour. If required, the amount of ammonia produced may be determined volumetrically by reaction with acid (Activity 22.5).

Metal ions

Several metal ions may be identified by the characteristic colour of the light they emit when heated in a bunsen flame. Traditionally, such flame tests are carried out using an inert metal wire, such as platinum, dipped into the sample. However, this is expensive and the wire is difficult to clean after the test. An equally effective method uses the bottom of a test-tube full of cold water. The glass surface is easily cleaned between tests by rinsing well with water. Table 22.1 lists the characteristic flame test colours of some common metal ions.

Table 22.1 Characteristic flame colorations of some metal ions

	metal	flame coloration
group 1	lithium	crimson
	sodium	golden yellow
	potassium	lilac
	rubidium	red
	caesium	blue
group 2	calcium	brick red
	strontium	crimson
	barium	apple green
	lead	blue
	copper	green

If the light emitted during a flame test is passed through a spectrometer it splits into sharp lines at particular wavelengths. Each element produces its own unique pattern of lines, known as an **atomic emission spectrum**. This technique is used in the analysis of metal alloys and can even give information about the composition of distant stars simply from their 'starlight'. Helium was, in fact, 'discovered' in the sun almost thirty years before it was isolated on the earth.

What is the origin of atomic spectra?

In Chapter 2 the atom was described as a central positively charged nucleus, surrounded by negatively charged electrons. We may imagine these electrons as tiny particles that may only occupy certain fixed orbits. Normally, electrons in an atom occupy the lowest possible energy levels nearest the nucleus but, if the sample is heated, they may absorb energy, overcome nuclear attraction, and be 'promoted' to higher levels. Such 'excited' atoms are unstable and the promoted electrons will drop back to lower levels, emitting energy as electromagnetic radiation, e.g. light. The wavelength or colour of the light depends upon the energy jump made by the electron. Since each atom has its own unique set of electronic energy levels, each element produces its own 'fingerprint' pattern of emission lines. This idea is developed in more detail in section 17.2.

when an atom absorbs energy its electrons may be 'excited', i.e. promoted to higher energy levels that are further from the nucleus

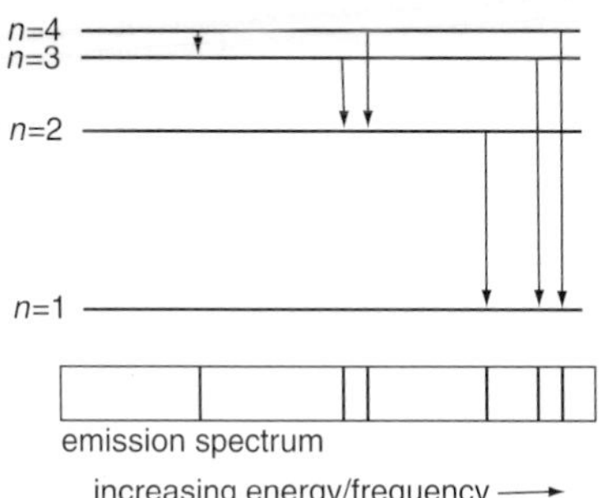

when an electron falls back to a lower energy level, the atom emits a quantum of electromagnetic radiation, corresponding to a line in the emission spectrum.

As with anions, many metal ions may be identified by simple test-tube reactions. The following sequence involves both precipitation and complex formation.

Divide the solution containing the unknown ions into 4 parts and test as follows:

1
- Add dilute hydrochloric acid
- White precipitates formed by Ag^+ and Pb^{2+}:
  $Ag^+(aq) + Cl^-(aq) \longrightarrow AgCl(s)$ $\quad Pb^{2+}(aq) + 2Cl^-(aq) \longrightarrow PbCl_2(s)$
- Lead chloride dissolves on heating the mixture

2 Add a few drops of sodium hydroxide solution. Insoluble hydroxides are formed:

white; Mg^{2+}, $*Al^{3+}$, $*Zn^{2+}$, $*Pb^{2+}$
brown; Ag^+ (oxide formed), Fe^{3+}
green; $*Cr^{3+}$, Fe^{2+} (turns brown in air)
blue; Cu^{2+}, Co^{2+} (turns pink on standing)

Precipitates marked * are amphoteric, dissolving on shaking in excess sodium hydroxide solution.

3 Add ammonia solution to precipitate the hydroxides as in test 2. With excess ammonia the following precipitates dissolve, forming soluble ammine complexes:

colourless; Ag^+, Zn^{2+}
purple; Cr^{3+}
yellow; Co^{2+}(darkens on exposure to air)
deep blue; Cu^{2+}

4 Add sodium carbonate solution. Ions with a high charge density, e.g. Al^{3+}, Cr^{3+}, Fe^{3+}, precipitate the hydroxide and give off carbon dioxide gas, whilst most others give a precipitate of the carbonate (see section 21.3).

Activity 22.2

Each of the following solutions contains one cation and one anion. Identify each from the test results given.

1. Golden yellow flame test, white precipitate on adding silver nitrate solution, insoluble in dilute nitric acid but soluble in dilute ammonia solution.
2. Green precipitate forms with sodium hydroxide solution, which dissolves in excess to give a green solution. White precipitate with barium chloride solution insoluble in dilute hydrochloric acid.
3. Blue precipitate with ammonia solution, which dissolves in excess ammonia to give a deep blue solution. Cream precipitate with silver nitrate solution soluble in concentrated, but not dilute, ammonia solution.
4. White precipitate with dilute hydrochloric acid, which dissolves on heating. Ammonia gas produced on warming with sodium hydroxide solution and aluminium powder.

Which, if any, of the above solutions would you expect to form a gas on adding sodium carbonate solution?

22.2 How much have we got? (quantitative analysis)

Precipitation reactions can again be useful here. Provided that the reaction is virtually complete, we can simply separate the precipitate, weigh it, and then calculate how much of the ion in question is present.

Activity 22.3

The iron deficiency disease anaemia may be treated by taking 'iron' tablets. The active ingredient is hydrated iron(II) sulphate, $FeSO_4.7H_2O$. The following experiment was carried out to find the percentage by mass of sulphate in iron tablets.

0.5010 g of powdered 'iron' tablets was stirred thoroughly with distilled water to dissolve all the active ingredient. After acidifying with hydrochloric acid and filtering, excess barium chloride solution was added to precipitate all the sulphate ions as barium sulphate, $BaSO_4$. After filtering off, washing and drying, the precipitate was found to weigh 0.256 g.

1. Write an ionic equation for the precipitation of barium sulphate.
2. Why is it necessary to acidify the solution before adding the barium chloride?
3. What is the relative formula mass of barium sulphate? (A_r: Ba 137, S 32, O 16)
4. Calculate the percentage by mass of sulphate in barium sulphate.
5. What mass of sulphate is present in the precipitate?
6. Assuming no other sulphates are present, calculate the mass of $FeSO_4.7H_2O$ in the 'iron' tablet sample.
7. Calculate the percentage by mass of $FeSO_4.7H_2O$ in the tablets.

In volumetric analysis we accurately measure the volumes of solutions which react with each other. If we know the **concentration** of a solution then its volume will tell us how many moles are present.

Note: Concentration is given as **molarity**, i.e. the number of **moles** of substance present in 1 dm^3 (i.e. 1 litre or 1000 cm^3) of solution. Thus 0.100 M sodium hydroxide solution contain 0.100 moles of NaOH per litre.

The following activities take you through examples of some common titrations that you may meet in practical work.

Apparatus used in volumetric analysis

Pipette This is used to transfer a FIXED volume of solution, usually 25 cm^3. It should be filled by suction using a special filler (NOT by mouth!). Always fill so that the BOTTOM of the meniscus is level with the mark.

Do Wash out the pipette with the solution you are going to use before filling it up (NOT distilled water, WHY?)

Do not Blow out the last drop of liquid (instead touch the tip on the side of the apparatus)

Do not Suck liquid up into the bulb of the pipette filler.

Do not Leave the pipette filler bulb squeezed in.

Burette This is used to add VARIABLE volumes of a solution and measures BY DIFFERENCE. It should be filled using a small funnel, which is removed before use. Again readings are normally taken from the BOTTOM of the meniscus.

Do Wash out the burette with the solution you are going to use before filling it up. Again, why not water?

Do not Forget to record BOTH starting and finishing volumes (to 0.05 cm^3 accuracy) in a titration.

Do not Leave the filling funnel in place during titrations.

Graduated flask Used to make up a solution of known concentration by dissolving a weighed amount of solid, or volume of liquid, in a FIXED volume, usually 250 cm^3. Again use the BOTTOM of the meniscus when making up the volume.

Do Wash out the graduated flask with DISTILLED WATER before use.

Do Dissolve solids in a beaker FIRST before transferring to the graduated flask.

Do Shake the solution THOROUGHLY for several minutes after making up to the final volume before using it.

Do not Try to weigh solids directly into a graduated flask.

Conical flask This is used to carry out the reaction, i.e. the location where one solution is added to the other. Normally, the first solution is transferred by pipette, an indicator is added, and the second solution is added from the burette until the indicator colour shows that the reaction is complete.

Do Wash the conical flask out with DISTILLED WATER before use. (Why not one of the solutions that you will use?)

Do not Try to use a conical flask as a graduated flask.

Some volumetric analysis apparatus in use

Activity 22.4 Acid–alkali titration

In this experiment you are going to find the exact concentration of some approximately 1 M hydrochloric acid using a solution of sodium hydroxide with a molarity of exactly 0.100 M. N.B. Any solution whose concentration is known exactly, is referred to as a **standard solution**.

Introduction

The equation for the reaction is as follows:

hydrochloric acid + sodium hydroxide ⟶ sodium chloride + water

$HCl(aq) + NaOH(aq) \longrightarrow NaCl(aq) + H_2O(l)$

1 mole   1 mole

Since the acid is much more concentrated than the sodium hydroxide we must first prepare a dilute solution for titration.

Method

Making a dilute acid solution

1 Pipette 25.0 cm^3 of the original hydrochloric acid solution into a clean 250 cm^3 graduated flask.

2 Dilute to the 250 cm^3 mark with distilled water and shake thoroughly for several minutes.

Titrating the sodium hydroxide with the diluted hydrochloric acid

3 Fill the burette with your diluted solution of hydrochloric acid.

4 Record the reading of the burette at the start (to 0.05 cm^3 accuracy)

5 Pipette 25.0 cm^3 of the exactly 0.100 M sodium hydroxide solution into a clean conical flask. Dilute to approximately 50 cm^3 with distilled water (volume not critical) and add not more than *two drops* of screened methyl orange indicator. The solution should turn pale green indicating alkaline conditions. In acid conditions the indicator turns red but when neutral it is almost colourless (pale grey)

6 Add the acid from the burette, swirling the flask, until the colour of the indicator JUST turns pink. Record the final burette reading and calculate the volume of acid added.

This first titration should be done fairly quickly to give you an approximate value. You should now repeat the experiment, adding acid very carefully near the end-point to get a grey/colourless appearance. You should aim to get at least two titrations, which are **concordant** (i.e. within 0.15 cm^3). Record ALL titration volumes but do not use non-concordant results in the calculation.

Observations (example)

Burette reading/cm^3		Titration/cm^3
start	finish	
0.00	24.70	(complete this column, noting any non-concordant results)
0.10	24.65	
0.15	24.65	

Activity 22.4 Acid–alkali titration continued

Questions (answer the following to get the final result)

1. How many moles of NaOH were used in each titration?
2. How many moles of HCl must have been added during the titration? (Hint: always check the balanced equation)
3. What was the average concordant titration volume of diluted acid?
4. How many moles of HCl are present in 1 cm^3 of diluted acid?
5. How many moles of HCl are present in 1 cm^3 of original acid?
6. How many moles of HCl are present in 1000 cm^3 of original acid? This is the molar concentration of the original acid.

Activity 22.5 Volumetric analysis of nitrate in water

Introduction

This analysis is based on the reduction of nitrate to ammonia by aluminium powder mentioned in Activity 22.1:

$$NO_3^-(aq) \longrightarrow NH_3(aq)$$

The ammonia may then be absorbed in a known excess of hydrochloric acid:

$$NH_3(aq) + HCl(aq) \longrightarrow NH_4Cl(aq)$$

The excess acid may then be determined by titration with standard sodium hydroxide solution as in Activity 22.4.

Method

1. A 100 cm^3 sample of water taken from a drain leading from a wheat field into a stream was pipetted into a distillation flask whose outlet dipped into 50.0 cm^3 of 0.1000 M hydrochloric acid.
2. The sample was made alkaline by the addition of about 10 cm^3 of 2 M NaOH solution and then about 3 g of aluminium powder was added.
3. After warming to start the reaction the mixture was allowed to stand for an hour to complete the reduction of nitrate to ammonia.
4. The mixture in the distillation flask was then boiled vigorously until its volume was reduced to about 20 cm^3. This ensured that all the ammonia passed into the hydrochloric acid.
5. The remaining hydrochloric acid was found to need 38.5 cm^3 of 0.1000 M sodium hydroxide solution to neutralise it when titrated using screened methyl orange as indicator.

Questions

1. How many moles of hydrochloric acid are present in 50.0 cm^3 of 0.1000 M HCl?
2. How many moles of hydrochloric acid were left after distillation of the ammonia?
3. How many moles of hydrochloric acid were used up by the ammonia?
4. What was the molarity of nitrate ions in the water sample?
5. Does the drainwater nitrate level exceed the target level of 0.5 g per litre?

Activity 22.6 Acid–carbonate titration

Introduction

Magnesite contains magnesium carbonate, which reacts with hydrochloric acid to form magnesium chloride solution, water and carbon dioxide gas:

$$MgCO_3(s) + 2HCl(aq) \longrightarrow MgCl_2(aq) + CO_2(aq) + H_2O(l)$$

The percentage of magnesium carbonate in a magnesite sample may be estimated by adding a weighed sample to a known excess of hydrochloric acid and then finding the amount of acid left by titration with sodium hydroxide solution of known molarity.

Method

1.000 g of powdered magnesite was dissolved in 50.0 cm^3 of 1.00 M hydrochloric acid. The solution was transferred to a graduated flask and made up to 250 cm^3 with distilled water. After shaking well, 25.0 cm^3 portions were titrated with 0.100 M sodium hydroxide solution, giving the following results:

Observations (sample)

Burette reading/cm^3		Titration/cm^3
start	finish	
0.00	30.00	(complete this column, noting any non-concordant results)
10.00	39.95	
5.50	35.55	

Questions

1. Why is it important to shake well after making up to the volume in the graduated flask? How can you tell from the results above that this has been done correctly?
2. Write balanced equations for the reaction of hydrochloric acid with:
   a) magnesium carbonate **b)** sodium hydroxide

Calculate an average value for the percentage of magnesium carbonate in the magnesite sample by calculating the following.

3. On average, how many moles of sodium hydroxide were needed to neutralise 25.0 cm^3 of the diluted solution?
4. How many moles of hydrochloric acid must have been present in the 250 cm^3 of solution in the graduated flask?
5. How many moles of hydrochloric acid were originally added to the magnesite sample?
6. How many moles of hydrochloric acid must have been used up by the magnesite sample?
7. How many moles of magnesium carbonate must have been present in the magnesite sample?
8. Calculate the percentage by mass of magnesium carbonate in the magnesite sample.
9. Magnesite also contains compounds other than magnesium carbonate. Explain how, if at all, the presence of each of the following impurities in the magnesite would affect the accuracy of the percentage of magnesium carbonate calculated from the above results:
   a) magnesium sulphate **b)** calcium carbonate

Activity 22.7 Determination of iron(II) in iron tablets by titration with potassium manganate(VII)

Introduction

Potassium manganate(VII), $KMnO_4$, or more exactly the manganate(VII) ion, MnO_4^-, is a powerful oxidising agent in acid solution, being converted into Mn^{2+} ions by the following reduction half-equation:

$$\underset{\text{deep purple}}{MnO_4^-} + 8H^+ + 5e^- \longrightarrow \underset{\text{almost colourless}}{Mn^{2+}} + 4H_2O$$

It can take the electrons it needs from a wide range of species, including Fe^{2+} ions, which react as shown by the following oxidation half-equation:

$$\underset{\text{pale green}}{Fe^{2+}} \longrightarrow \underset{\text{pale yellow}}{Fe^{3+} + e^-}$$

No indicator is needed when titrating Fe^{2+} with MnO_4^- since the latter is the only strongly coloured species. The first excess of manganate(VII) is signalled by the appearance of a faint but permanent pink tinge to the pale yellow solution.

Note: a definite brown coloration in the titration indicates lack of acid. If this occurs add more sulphuric acid before continuing.

Method

1 Weigh ***accurately*** between 10 and 12 g of powdered iron tablets into a small beaker. Stir with about 150 cm³ distilled water and transfer to a 250 cm³ graduated flask. Make up to the mark with dilute sulphuric acid and mix thoroughly.

2 Pipette 25.0 cm³ of this solution into a clean conical flask and add about an equal volume of dilute sulphuric acid. Titrate with 0.0200 M potassium manganate(VII) solution until the first permanent pink coloration.

   Note: because of its very intense colour it is difficult to see the bottom of the meniscus for potassium manganate(VII) solution in the burette. It is much easier to take *all* burette readings from the *top* of the meniscus in this case.

3 Repeat until at least two concordant results are obtained (within 0.15 cm³).

Observations (sample)

mass of sample bottle + iron tablets = 32.519 g

mass of empty sample bottle = 21.459 g

mass of iron tablets = 11.060 g

Burette reading/cm³		Titration/cm³
start	finish	
0.15	24.00	(complete this column, noting any non-concordant results)
24.00	47.95	
0.10	23.75	

Questions

1 Work out the balanced redox equation for the reaction between Fe^{2+} and MnO_4^- in acid solution.

2 Calculate the number of moles of MnO_4^- used in the average of the concordant titrations.

Activity 22.7 Determination of iron(II) in iron tablets by titration with potassium manganate(VII) continued

3 Use the balanced overall equation you worked out to find the number of moles of Fe^{2+} that react with this number of moles of MnO_4^-.

4 Calculate the mass of Fe^{2+} in the weighed sample and hence find the percentage by mass of $FeSO_4.7H_2O$ in the tablets.

5 Is your answer consistent with the percentage by mass of sulphate found in Activity 22.3?

Activity 22.8 Analysis of a copper(II) salt by titration with iodine-thiosulphate titration

Introduction

Cu^{2+} ions oxidise I^- in aqueous solution as follows:

$$2Cu^{2+} + 4I^- \longrightarrow 2CuI + I_2$$

The iodine formed may be reduced back to iodide by titrating with sodium thiosulphate solution, using starch as an indicator:

$$I_2 + 2S_2O_3^{2-} \longrightarrow 2I^- + S_4O_6^{2-}$$

Method

1 Weigh accurately between 5.0 and 5.5 g of the copper(II) salt. Dissolve this in distilled water and make up to 250 cm^3 in a graduated flask, shaking thoroughly to ensure mixing.

2 Pipette 25.0 cm^3 of the solution into a clean conical flask and add approximately 10 cm^3 of 10% potassium iodide solution.

3 Titrate the iodine formed with 0.1000 M sodium thiosulphate solution until the mixture is pale yellow. Add starch solution and continue the titration until the blue-black colour just disappears.

Note: do not worry if this colour reappears after standing for a while.

4 Repeat the titration until concordant results have been obtained.

Observations (sample)

mass of sample bottle + copper(II) salt = 25.906 g

mass of empty sample bottle = 20.462 g

mass of copper(II) salt = 5.444 g

Burette reading/cm^3		Titration/cm^3
start	finish	
0.00	21.80	(complete this column, noting any non-concordant results)
21.80	43.50	
0.25	22.05	

Activity 22.8 Analysis of a copper(II) salt by titration with iodine-thiosulphate titration continued

Questions

1 Calculate the number of moles of $S_2O_3^{2-}$ ions in the average concordant titration of 0.1000 M sodium thiosulphate solution.

2 How many moles of I_2 are liberated by 25.0 cm^3 of the copper salt solution?

3 How many moles of Cu^{2+} must have been present in the weighed sample of copper salt?

4 Calculate the percentage by mass of copper in the salt.

5 The copper salt is thought to be $CuSO_4.5H_2O$. Does the result you obtained above agree with this suggestion?

Questions on Chapter 22

1 It is suggested that a water-soluble green solid is chromium(III) chloride. Suggest a series of simple laboratory tests that could check this.

2 Identify the anion and cation present in a solution from the following test results:
a) flame test: green
b) dilute hydrochloric acid: no observable change
c) barium chloride solution: white precipitate insoluble in hydrochloric acid
d) ammonia solution: blue precipitate, which dissolves in excess to give a deep blue solution

3 A solution is thought to contain both Ag^+ and Pb^{2+} ions. Suggest a series of tests to find out if either, or both, of these ions are present.

4 Describe **one** test in each case which would distinguish between each of the following pairs of substances:
a) sodium chloride and potassium chloride
b) silver nitrate and lead(II) nitrate
c) potassium carbonate and potassium sulphate
d) iron(II) sulphate and iron(III) sulphate
e) sodium chloride and sodium bromide

5 The concentration of chloride ions in solution may be determined by adding excess silver nitrate solution and then weighing the amount of silver chloride precipitated:

$$Cl^-(aq) + AgNO_3(aq) \longrightarrow AgCl(s) + NO_3^-(aq)$$

a) What mass of silver chloride would be precipitated on adding excess silver nitrate to 10.0 cm^3 of 0.500 M magnesium chloride solution?
b) What volume of 0.100 M silver nitrate solution would be required to precipitate all the chloride from 25.0 cm^3 of 0.200 M sodium chloride solution?
c) If 30.0 cm^3 of 0.0500 M silver nitrate is needed to precipitate all the chloride from 12.5 cm^3 of a sample of hydrochloric acid, calculate the molarity of this acid.

6 The following experimental results were obtained in an experiment to find the number of moles of water of crystallisation in a sample of barium chloride crystals.

$$BaCl_2.xH_2O(s) \xrightarrow{\text{heat}} BaCl_2(s) + xH_2O(g)$$

mass of crucible + hydrated barium chloride sample	= 10.845 g
mass of crucible + sample after heating	= 10.779 g
mass of crucible + sample after re-heating	= 10.773 g
mass of crucible + sample after re-heating again	= 10.773 g
mass of empty crucible	= 10.357 g

a) Why was the sample heated until its mass remained constant?
b) How many moles of anhydrous barium chloride were left?
c) How many moles of water were lost during heating?
d) What is the value of *x* in the formula above?

7 Calculate the molarity of the following solutions:
a) 58.5 g of $NaCl$ in 2 litres
b) 19.6 g of H_2SO_4 in 1 litre
c) 4.0 g of $NaOH$ in 100 cm^3
d) 8.0 g of $CuSO_4$ in 250 cm^3
e) 20 g of NH_4NO_3 in 500 cm^3

8 Calculate the number of moles of:
a) H_2SO_4 in 2 litres of 1.5 M sulphuric acid
b) $NaOH$ in 50 cm^3 of 0.100 M sodium hydroxide
c) K^+ ions in 100 cm^3 of 0.20 M K_2SO_4 solution
d) HCl in 1 litre of hydrochloric acid which contains 73 g per litre
e) $NaOH$ needed to fully neutralise 1 litre of 1.0 M H_2SO_4 (CARE)

Questions on Chapter 22 *continued*

9 Calculate the mass (in grams) of each substance needed to prepare the following solutions:

a) 1 litre of 2.0 M HNO_3

b) 100 cm^3 of 0.100 M Na_2CO_3

c) 1 m^3 of 0.5 M HCl

d) 3 dm^3 of 1 M $(NH_4)_2SO_4$

e) 250 cm^3 of 1 M $CuSO_4.5H_2O$

10 Sodium carbonate reacts with sulphuric acid as follows:

$$Na_2CO_3 + H_2SO_4 \longrightarrow Na_2SO_4 + 2H_2O + CO_2$$

The following experiment was carried out to determine the concentration of a solution of sulphuric acid.
10.60 g of anhydrous sodium carbonate was dissolved in water and diluted to a total volume of 100 cm^3. 25.0 cm^3 of the original sulphuric acid was diluted to 250 cm^3 with distilled water. 10.0 cm^3 of the diluted acid needed 30.0 cm^3 of the sodium carbonate solution for complete neutralisation.

a) What was the molarity of the sodium carbonate solution?

b) How many moles of Na_2CO_3 were needed to neutralise the acid?

c) How many moles of H_2SO_4 were present in 10.0 cm^3 of the diluted acid?

d) What was the molarity of the diluted acid?

e) What was the molarity of the original sulphuric acid?

11 Vinegar is a dilute solution of ethanoic acid, CH_3COOH. The concentration of ethanoic acid in vinegar may be determined by neutralising it with sodium hydroxide solution. The equation for this reaction is:

$$CH_3COOH(aq) + NaOH(aq) \longrightarrow CH_3COONa(aq) + H_2O(l)$$

In an experiment, 25.0 cm^3 of a vinegar sample needed 20.8 cm^3 of 1.00 M sodium hydroxide for complete neutralisation.

a) Describe briefly, but including all essential details, how you would carry out this experiment.

b) Calculate the number of moles of sodium hydroxide needed to neutralise the vinegar sample.

c) Calculate the molarity of ethanoic acid in the vinegar.

d) Calculate the mass of ethanoic acid in a 500 cm^3 bottle of the vinegar.

12 Iron dissolves in dilute sulphuric acid to give iron(II):

$$Fe(s) + 2H^+(aq) \longrightarrow Fe^{2+}(aq) + H_2(g)$$

The Fe^{2+}(aq) concentration can be determined by titration with standard manganate(VII) solution,

$$5Fe^{2+}(aq) + MnO_4^-(aq) + 8H^+(aq) \longrightarrow 5Fe^{3+}(aq) + Mn^{2+}(aq) + 4H_2O(l)$$

2.544 g of steel wire was dissolved in dilute sulphuric acid and the volume diluted to 500 cm^3. 20.0 cm^3 of this solution needed 18.1 cm^3 of 0.0200 M $KMnO_4$(aq) for complete oxidation.

a) How many moles of MnO_4^-(aq) were used in the titration?

b) How many moles of Fe^{2+}(aq) were present in 20.0 cm^3 of solution?

c) How many moles of iron were present in the original sample?

d) What mass of iron did the original wire contain?

e) What was the percentage by mass of iron in the steel?

13 Iodine reacts with sodium thiosulphate solution as follows,

$$I_2(aq) + 2Na_2S_2O_3(aq) \longrightarrow 2NaI(aq) + Na_2S_4O_6(aq)$$

a) What mass of sodium thiosulphate is required to make up 100 cm^3 of 0.1000 M solution?

b) 25.0 cm^3 of an iodine solution needed 32.6 cm^3 of 0.1000 M sodium thiosulphate for complete reaction. What was the molarity of the iodine solution?

14 20.0 cm^3 of a solution of sulphuric acid needed 40.0 cm^3 of 0.0500 M sodium hydroxide for complete neutralisation.

a) Write an equation for the reaction.

b) Calculate the molarity of the sulphuric acid used.

c) What volume of the same sodium hydroxide solution would be needed to neutralise 20.0 cm^3 of nitric acid of the same molarity as the sulphuric acid?

15 1.050 g of a limestone sample (calcium carbonate) was dissolved in 100.0 cm^3 of a hydrochloric acid solution. It took 20.0 cm^3 of 2.00 M $NaOH$ to neutralise this solution and 30.0 cm^3 of 2.00 M $NaOH$ to neutralise 100 cm^3 of the original hydrochloric acid.

a) How many moles of hydrochloric acid did the limestone sample react with?

b) Write a balanced equation for the reaction of calcium carbonate, $CaCO_3$, with hydrochloric acid.

c) How many moles of calcium carbonate did the limestone sample contain?

d) What is the percentage by mass of calcium carbonate in the limestone sample?

Comments on the activities

Activity 22.1

$\{NO_3^-(aq) + 6H_2O(l) + 8e^- \longrightarrow NH_3(g) + 9OH^-(aq)\} \times 3$

$\{Al(s) \longrightarrow Al^{3+}(aq) + 3e^-\} \times 8$

$8Al(s) + 3NO_3^-(aq) + 18H_2O(l) \longrightarrow 8Al^{3+}(aq) + 3NH_3(g) + 27OH^-(aq)$

Activity 22.2

1 Golden yellow flame test: Na^+. White precipitate on adding silver nitrate solution, insoluble in dilute nitric acid but soluble in dilute ammonia solution: Cl^-.

2 Green precipitate formed with sodium hydroxide, which dissolves in excess to give a green solution: Cr^{3+}. White precipitate with barium chloride solution, insoluble in dilute hydrochloric acid: SO_4^{2-}.

3 Blue precipitate with ammonia solution, which dissolves in excess ammonia to give a deep blue solution, Cu^{2+}. Cream precipitate with silver nitrate solution, soluble in concentrated ammonia solution only: Br^-.

4 White precipitate with dilute hydrochloric acid, which dissolves on heating: Pb^{2+}. Ammonia gas produced on warming with sodium hydroxide solution and aluminium powder: NO_3^-.
Solution 2, which contains Cr^{3+}, would give carbon dioxide gas on adding sodium carbonate solution.

Activity 22.3

1 $Ba^{2+}(aq) + SO_4^{2-}(aq) \longrightarrow BaSO_4(s)$

2 Acid prevents the precipitation of other insoluble barium salts, e.g. carbonate.

3 M_r $BaSO_4$ = 137 + 32 + 64 = 223

4 % by mass of SO_4^{2-} in $BaSO_4$ $= 100 \times \frac{96}{233} = 41.2\%$

5 Mass of SO_4^{2-} present in the precipitate

$= 0.256 \times \frac{41.2}{100} = 0.1055$ g

6 % by mass of SO_4^{2-} in $FeSO_4.7H_2O$

$= 100 \times \frac{96}{278} = 34.5\%$

0.1055 g SO_4^{2-} is contained in

$0.1055 \times \frac{100}{34.5} = 0.3058$ g $FeSO_4.7H_2O$

7 % by mass of $FeSO_4.7H_2O$ in the tablets

$= 100 \times \frac{0.3058}{0.5010} = 61.0\%$

Activity 22.4

Titration results: 24.70 (non-concordant), 24.55, 24.50

1 25.0 cm³ 0.100 M NaOH contains

$25 \times \frac{0.100}{1000} = 0.0025$ moles NaOH

2 From the equation, this reacts with 0.0025 moles HCl

3 Average concordant titration

$= \frac{24.55 + 24.50}{2} = 24.53\ cm^3$

4 1 cm³ diluted acid contains

$\frac{0.0025}{24.53} = 0.0001019$ moles HCl

5 1 cm³ of original acid contains

$0.0001019 \times \frac{250}{25} = 0.001019$ moles HCl

6 1000 cm³ of original acid contains 1000 × 0.001019 = 1.019 moles HCl
thus concentration of original hydrochloric acid is 1.019 M

Activity 22.5

1 50.0 cm³ of 0.1000M HCl contains

$50.0 \times \frac{0.1000}{1000} = 0.00500$ moles HCl

2 Moles HCl left = moles NaOH used
= 38.5 × 0.1000/1000 = 0.00385 moles

3 0.00500 – 0.00385 = 0.00115 moles of HCl used up by the ammonia

4 Molarity of nitrate ions in the water sample
= 0.00115 × 10 = 0.0115 M

5 M_r NO_3^- = 14 + 3(16) = 62.
0.0115 moles NO_3^- weighs 62 × 0.0115 = 0.713 g.
Drainwater nitrate level therefore *does* exceed the target level of 0.5 per litre.

Activity 22.6

Titration results: 30.00, 29.95, 30.05. Average: 30.00

1 The results are quite concordant (within 0.1 cm³) which shows that the concentration of the acid solution in the volumetric flask must be uniform – hence well shaken!

2 $MgCO_3(s) + 2HCl(aq) \longrightarrow MgCl_2(aq) + CO_2(g) + H_2O(l)$
$NaOH(aq) + HCl(aq) \longrightarrow NaCl(aq) + H_2O(l)$

3 Average titration is 30.0 cm³ of 0.100 M NaOH which therefore contains 0.0030 moles NaOH

4 25 cm³ of solution must contain 0.0030 moles HCl
250 cm³ of solution must contain 0.0300 moles HCl

5 Originally 0.0500 moles HCl added to sample

6 1.000 g magnesite has reacted with 0.0200 moles HCl

7 1.000 g magnesite contains 0.0100 moles $MgCO_3$

8 1 mole of $MgCO_3$ weighs 24 + 12 + (3 × 16) = 84 g
1.000 g magnesite contains 0.84 g $MgCO_3$
The sample therefore contains 84% $MgCO_3$

9 Magnesium sulphate does not react with HCl and will not affect the accuracy of the result. Calcium carbonate reacts in the same way as magnesium carbonate and will make the answer HIGHER than it should be.

Comments on the activities *continued*

Activity 22.7

Titration results: 23.85, 23.95, 23.65 (non-concordant). Average concordant: 23.90

1 $MnO_4^- + 8H^+ + 5e^- \longrightarrow Mn^{2+} + 4H_2O$

$\{Fe^{2+} \longrightarrow Fe^{3+} + e^-\} \times 5$

$MnO_4^- + 8H^+ + 5Fe^{2+} \longrightarrow Mn^{2+} + 4H_2O + 5Fe^{3+}$

2 Moles of MnO_4^- in 23.90 cm^3 0.0200 M $KMnO_4$

$$= 23.9 \times \frac{0.0200}{1000} = 0.000478$$

3 From above equation this reacts with

$5 \times 0.000478 = 0.00239$ moles Fe^{2+}

4 Mass of Fe^{2+} in the weighed sample

$= 10 \times 56 \times 0.00239 = 1.3384$ g

1.3384 g of Fe^{2+} is contained in

$$1.3384 \times \frac{278}{56} = 6.6442 \text{ g } FeSO_4.7H_2O$$

Tablets contain $100 \times \frac{6.6442}{11.060} = 60.1\%$ $FeSO_4.7H_2O$

5 This value compares quite well with that obtained by gravimetric determination of sulphate in Activity 22.3.

Activity 22.8

Titration results: 21.80, 21.70, 21.80. Average: 21.77

1 21.77 cm^3 0.1000 M $Na_2S_2O_3$ contains

$$21.77 \times \frac{0.1000}{1000} = 0.002177 \text{ moles } S_2O_3^{2-}$$

2 This reacts with $\frac{0.002177}{2} = 0.001089$ moles I_2

3 0.001089 moles I_2 is formed by

0.001089×2 moles $Cu^{2+} = 0.002177$ moles.

25.0 cm^3 of copper salt solution contains

0.002177 moles Cu^{2+}

Weighed sample must contain

$0.002177 \times 10 = 0.02177$ moles Cu^{2+}

4 0.02177 moles Cu^{2+} weighs $64 \times 0.02177 = 1.393$ g

% by mass of Cu in the salt $= 100 \times \frac{1.393}{5.444} = 25.6\%$

5 $M_r(CuSO_4.5H_2O) = 64 + 32 + 4(16) + 5(2+16) = 250$

% of Cu $= 100 \times \frac{64}{250} = 25.6\%$

i.e. consistent with our result.

CHAPTER 23

Industrial inorganic chemistry

Contents

Study Checklist

After studying this chapter you should be able to:

1 Recall the uses of limestone in agriculture, as a construction material and in the manufacture of cement, glass and iron.

2 Describe the manufacture from salt and outline the main uses of the following: sodium, sodium hydroxide, chlorine, hydrogen and sodium chlorate(I).

3 State and explain the conditions used for the synthesis of ammonia from its elements by the Haber process.

4 Describe the conversion of ammonia into nitric acid.

5 Outline major uses of ammonia and nitric acid.

6 Explain how the method of extraction of a metal is related to its reactivity.

7 Describe the following methods of metal extraction:
aluminium by electrolysis of bauxite
iron from its oxide in the blast furnace
titanium from its oxide via the chloride using sodium (or magnesium)
tungsten by hydrogen reduction of its oxide
chromium by aluminium reduction of its oxide
copper by heating the sulphide in air

8 Analyse details of a given manufacturing process, both from a chemical and an economic viewpoint.

23.1 Limestone

Approximately 20% of all sedimentary rock is limestone, which is largely calcium carbonate, $CaCO_3$. Under the action of heat and pressure deep under the Earth's surface, limestone changes into marble. Most limestone is quite porous and often acts as a natural reservoir for groundwater, oil and gas. About 20% of the 60+ million tonnes of limestone quarried each year in the UK is used directly as building stone and aggregate, with most of the rest being used in the manufacture of calcium oxide (quicklime), cement, glass and iron.

Powdered limestone is used in some coal-fired power stations to remove sulphur dioxide from the flue gases. This produces gypsum (hydrated calcium sulphate) which is sold to plasterboard manufacturers and helps offset the running costs of this flue gas desulphurisation (FGD) process:

$$CaCO_3(s) + SO_2(aq) + \tfrac{1}{2}O_2(g) + 2H_2O(l) \longrightarrow \underset{\text{gypsum}}{CaSO_4.2H_2O(s)} + CO_2(g)$$

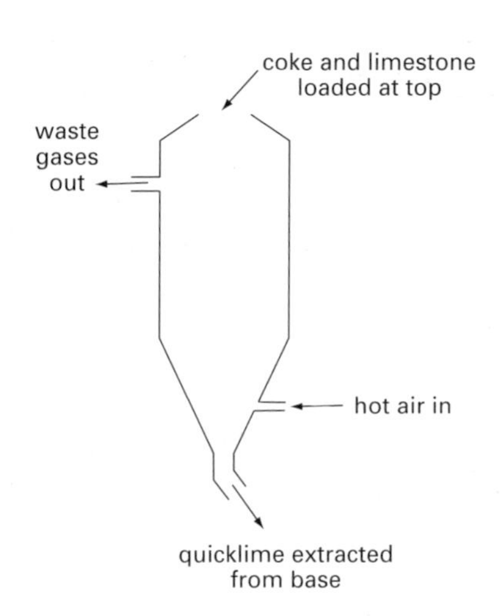

Figure 23.1
Calcium oxide production in a lime-kiln

Though limestone also provides a cheap way of removing excess acidity from soil, its insolubility makes it rather slow acting:

$$CaCO_3(s) + 2H^+(aq) \longrightarrow Ca^{2+}(aq) + H_2O(l) + CO_2(g)$$

Calcium oxide (quicklime) and calcium hydroxide (slaked lime) are much quicker acting soil conditioners made in large quantities from limestone. In section 19.3 we mentioned the thermal decomposition of group 2 carbonates. Strictly speaking this is a reversible process (see Chapter 25):

$$CaCO_3(s) \rightleftharpoons CaO(s) + CO_2(g) \qquad \Delta H = +180 \text{ kJ mol}^{-1}$$

Since the decomposition is endothermic, both the reaction rate and equilibrium position are favoured by high temperature. The equilibrium constant for this reaction is given by the equilibrium partial pressure of carbon dioxide:

$$K_p = p_{CO_2}$$

To maximise the yield of calcium oxide, it is important therefore to prevent the build-up of carbon dioxide. In a modern lime-kiln, such as that shown in Figure 23.1, a mixture of crushed limestone and coke is continuously fed in at the top. Hot air passes up from the bottom, burning the coke to provide temperatures of up to 1200°C and sweeping away the carbon dioxide. As the mixture passes down the kiln the coke fuel is used up and the calcium oxide cools before being removed at the bottom.

Calcium oxide reacts vigorously with water to give calcium hydroxide:

$$CaO(s) + H_2O(l) \longrightarrow Ca(OH)_2(s) \qquad \Delta H = -65 \text{ kJ mol}^{-1}$$

If excess water is added the calcium hydroxide dissolves to give a solution known as lime-water.

Cement and concrete

Cement is a material capable of bonding mineral fragments into a compact whole known as concrete. Portland cement is made from limestone and clay that contains silica and aluminium. The raw materials are mixed and heated in a large rotary kiln to a temperature of approximately 1350°C, producing 'clinker'. This is cooled and ground to a fine powder, and gypsum (calcium sulphate) is added to control the speed of setting when the cement is mixed with water. Portland cement has the following typical composition:

tricalcium silicate	$3CaO.SiO_3$	55%
dicalcium silicate	$2CaO.SiO_3$	25%
tricalcium aluminate	$3CaO.Al_2O_3$	10%

The chemical reaction of the two silicates with water produces calcium silicate hydrates and calcium hydroxide which make the largest contribution to the strength of the cement. Tricalcium aluminate aids the burning process in the kiln but contributes little to the strength of the cement.

Concrete is strong when compressed but relatively weak when stretched. In reinforced concrete, steel rods are placed where it is necessary for structural members to resist stretching forces. The steel reinforcement is bonded to the surrounding concrete so that stress is transferred between the two materials.

The steel reinforcing rods may be stretched before the development of bond between it and the surrounding concrete. After the concrete has set, the stretching force on the steel rods is released, so pre-compressing the concrete. When loads are applied to the structure the compressive force is reduced, but generally tensile cracking is avoided. Such concrete is known as pre-stressed concrete.

Concrete bridge supports

Glass

Glass is not a true solid but a super-cooled liquid. As molten glass is cooled the random molecular structure, typical of liquids, becomes 'frozen' into the lattice. Common glass, or soda-lime glass, is made by heating a mixture of sand, sodium carbonate (soda), and limestone to a temperature of about 1300°C. In the 'float process' a continuous strip of glass from the melting furnace floats onto the surface of a molten metal, usually tin, at a carefully controlled temperature. The flat surface of the molten metal gives the glass a smooth, undistorted surface as it cools. Glass containers such as bottles and jars are made by blowing hot glass into a mould on a continuous machine. Objects such as plates, tumblers, and vases can be made cheaply by pressing hot glass in a mould. Because soda-lime glass melts at a relatively low temperature, is easy to shape, has good chemical durability, and is inexpensive, it accounts for about 90% of all glass produced. A typical commercial soda-lime glass is composed of 73% silica, 15% soda, 5% lime, 4% magnesium oxide, 2% aluminium oxide, and 1% boron oxide.

Pyrex contains about 81% silica, 13% boron oxide, 4% soda, and 2% aluminium oxide. Although more expensive, it has better resistance to thermal shock and chemical attack than soda-lime glass and softens at a higher temperature. It is used in cooking and laboratory apparatus and for car headlights and optical instruments.

A glass of special interest is fused silica. Its very high melting point makes it expensive to produce, but its purity and excellent optical transparency make it ideal for high quality optical equipment such as telescope mirrors and optical fibre cables.

Although its lattice structure is extremely strong, glass itself is very brittle because of tiny surface flaws or cracks. When glass is subjected to a stretching force, these cracks spread and join up to create a break. This tendency may be reduced by a process called tempering. Here the surface of the glass is cooled more rapidly than the interior. The surface becomes rigid first, and when the interior cools and contracts it pulls on the surface, causing a residual compressive stress. This is commonly used for strengthening glass windows and doors.

Glass can be coloured by adding transition metal compounds: chromates for green, copper and cobalt for blue, and manganese for purple. The common green of wine bottles results from the addition of iron, and brown glass is produced by adding a combination of iron and sulphur. Very small metal particles in glass can colour it deeply; for instance, the addition of gold produces ruby coloured glass.

Optical fibre cables conduct light along their length with little or no losses

23.2 The chlor-alkali industry

Well over 1 000 000 tonnes each of chlorine and sodium hydroxide are currently produced annually in the UK by the electrolysis of brine (sodium chloride solution). Various types of cell are currently used but the overall process in each is the same:

$$2NaCl(aq) + 2H_2O(l) \longrightarrow 2NaOH(aq) + Cl_2(g) + H_2(g)$$

In the **diaphragm cell** shown in Figure 23.2, saturated brine is fed continuously into the anode compartment where chlorine gas is formed:

$$2Cl^-(aq) \longrightarrow Cl_2(g) + 2e^-$$

In the cathode compartment hydrogen gas is produced, leaving Na^+ and OH^- ions unchanged in solution:

$$2H^+(aq) + 2e^- \longrightarrow H_2(g)$$

The asbestos diaphragm keeps the chlorine and sodium hydroxide solution separate as they would otherwise react as follows:

$$2NaOH(aq) + Cl_2(g) \longrightarrow NaOCl(aq) + NaCl(aq)$$

In practice, the final solution leaving the cathode compartment contains roughly equal concentrations of Cl^- and OH^-. On partial evaporation, the less soluble sodium chloride crystallises out leaving a concentrated solution of sodium hydroxide. The recovered sodium chloride is, of course, recycled through the electrolysis cell.

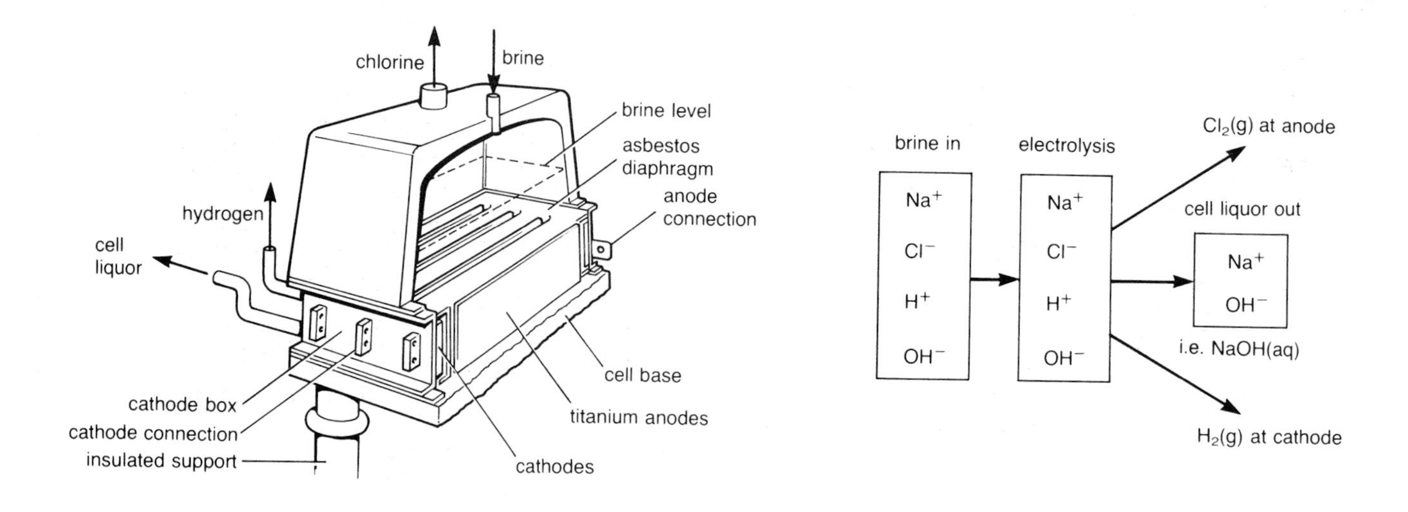

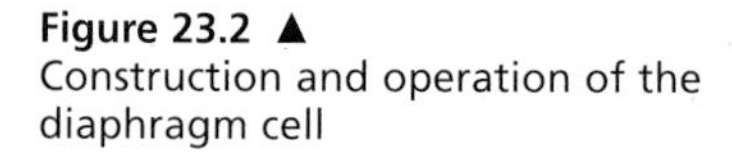

Figure 23.2 ▲
Construction and operation of the diaphragm cell

In the **mercury cell**, shown in Figure 23.3, saturated brine is passed between the titanium anodes and a flowing mercury cathode. Chlorine gas is formed as above at the anode but at the cathode, Na^+ ions are discharged rather than H^+ ions:

$$Na^+(aq) + e^- \longrightarrow Na$$

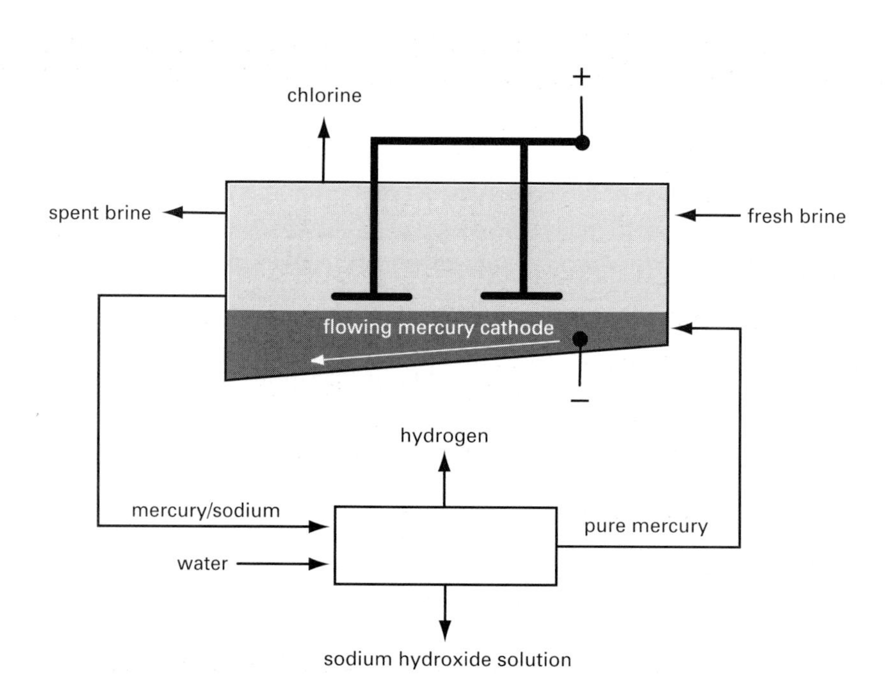

Figure 23.3 ▶
A flowing mercury cell

The sodium metal dissolves in the mercury to give an alloy, Na/Hg, known as an **amalgam**. On leaving the cell this is decomposed by water, giving hydrogen gas and sodium hydroxide solution:

$$2Na/Hg + 2H_2O(l) \longrightarrow 2Na^+(aq) + 2OH^-(aq) + H_2(g)$$

The mercury is recycled and returned to the electrolysis cell.

Which is the better process – diaphragm or mercury?

Since both processes are still used, each must have its advantages and disadvantages. The diaphragm cell is cheaper to construct and avoids the use of mercury which is very toxic. Against this, the asbestos diaphragms must be replaced frequently and the mercury cell produces sodium hydroxide solution of greater purity and higher concentration. Recently cells using an artificial membrane rather than an asbestos diaphragm have been introduced. Since the membrane allows positive ions but not negative ions to pass though it, there is no chance of the chlorine and sodium hydroxide mixing.

Electrical energy is a significant cost in all cells and a typical chlor-alkali plant can consume the total output of a large modern power station. The transport costs of the raw materials can also be high so the ICI installation at Runcorn on Merseyside was positioned so it can obtain its brine from the nearby Cheshire salt beds. Good road and rail links are also important for transport of the products, chlorine and sodium hydroxide, to customers. Both of these are hazardous chemicals that are usually transported as liquids in special tankers.

An alternative method of electrolysing brine to produce chlorine is to use a battery of mercury cells. This one is at a chemical plant in Cheshire

The major uses of chlorine are shown in Table 23.1.

Hydrogen is used in the Haber process to make ammonia, as fuel and in margarine production. Sodium hydroxide is used with chlorine to make sodium chlorate(I) bleach, in the paper and synthetic fibres industries, and in the production of soaps and detergents.

Chlorinated hydrocarbons

You can see from Table 23.1 that the use of these chemicals is very widespread. Indeed, decaffeinated coffee was originally made by treating the beans with chlorinated hydrocarbon solvents prior to roasting. (The caffeine is now extracted using carbon dioxide as the solvent.)

Many of the most widely used chlorinated hydrocarbon compounds, however, are extremely toxic and resistant to biological degradation. The so-called **dioxins** are unwanted by-products in the manufacture of certain organo-halogen compounds that are used in making herbicides and often remain as contaminants in the finished products. 2,3,7,8-tetrachlorodibenzo-*p*-dioxin (TCDD) is a trace contaminant in the herbicide 2,4,5-trichlorophenoxyethanoic acid (2,4,5-T).

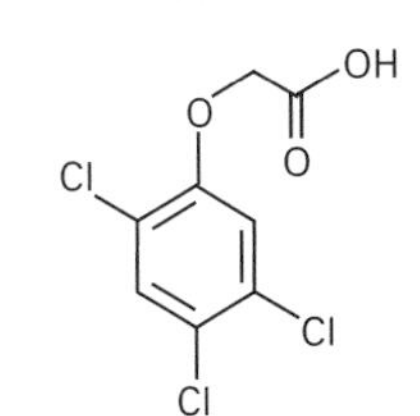

2,4,5-trichlorophenoxyethanoic acid (2,4,5-T)

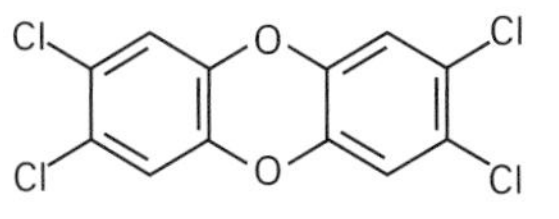

2,3,7,8-tetrachlorobenzo-*p*-dioxin (TCDD)

Chlorinated hydrocarbons, *continued*

Under its code name, Agent Orange, 2,4,5-T was used by the US military to defoliate large areas of land during the Vietnam war. The contaminant in this herbicide is believed to have caused innumerable animal deaths and birth defects in Vietnamese children. Some American soldiers who were exposed to Agent Orange also claim to have developed illnesses such as cancer as a result of exposure to TCDD. Similar dioxins have been found in many industrial wastes such as paper-mill effluents.

Herbicides are toxic, especially if they contain TCDD as an impurity

Table 23.1 Some major uses of chlorine in the UK

Approx. % of total chlorine production	intermediate product(s)	end use(s)
Petrochemicals		
27	chloroethene (vinyl chloride) CH_2CHCl	poly(chloroethene), i.e. poly(vinyl chloride), PVC
17	chloromethanes $CHCl_3$, CCl_4 CH_3Cl	refrigerant liquids and aerosol 'propellants' petrol 'antiknock' additives
16	other chlorinated hydrocarbons, e.g. $CCl_2.CHCl$ $CCl_2.CCl_2$	metal 'degreasing' solvent 'dry-cleaning' solvent
6	propene oxide	car brake fluid polyurethane plastics many pharmaceutical products
Inorganics		
13		extraction of magnesium, titanium and bromine manufacture of hydrochloric acid
1	sodium chlorate(I), NaClO	paper and pulp bleach
Miscellaneous		
20		sterilisation of water manufacture of disinfectants, anaesthetics, insecticides and dyestuffs, etc.

Activity 23.1 Analysis of commercial liquid bleach

Introduction

Liquid bleach contains chlorate(I), OCl^-, as its active ingredient. It is made by bubbling chlorine through sodium hydroxide solution:

$$2OH^-(aq) + Cl_2(g) \longrightarrow Cl^-(aq) + OCl^-(aq) + H_2O(l)$$

Under acid conditions this reaction is reversed, re-forming free chlorine. The mass of chlorine that is produced from 100 g of bleach is known as the percentage of 'available chlorine'.

1 Write half-equations for the conversion of chlorate(I) and chloride ions into chlorine gas and an overall equation for the action of acid on liquid bleach.

If an excess of potassium iodide solution is added, the chlorine oxidises the iodide ions to iodine.

2 Write half-equations showing what happens to the chlorine and the iodide ions in this step and so work out an overall balanced equation.

The iodine is reduced to iodide on titration with thiosulphate, $S_2O_3^{2-}$, which is itself converted to tetrathionate, $S_4O_6^{2-}$.

3 Write half-equations for the oxidation of thiosulphate and the reduction of iodine and derive the overall redox equation for this titration.

Method

- Weigh a clean, dry 100 cm^3 beaker to the nearest 0.01 g on a top-pan balance.
- Add between 12 g and 15 g of the liquid bleach and reweigh.
- Calculate the mass of bleach taken.
- Transfer all the bleach from the beaker to a 250 cm^3 graduated flask. Make up to the mark using distilled water and shake thoroughly. You may ignore any froth above the mark.
- Pipette 25 cm^3 of the diluted bleach into a conical flask and add an approximately equal volume of distilled water.
- Add approximately 2 g of solid potassium iodide and 10 cm^3 glacial ethanoic acid (so-called because it freezes in cold weather) from a small measuring cylinder.
- Titrate the liberated iodine with 0.100 M sodium thiosulphate solution until the mixture is pale yellow. Add starch solution and continue the titration until the blue colour just disappears.
- Repeat with fresh 25 cm^3 portions of the diluted bleach solution until you have obtained at least two concordant titrations (as explained in Chapter 22).

Observations (example)

mass of beaker + bleach = 73.98 g
mass of empty beaker = 62.54 g
mass of bleach used = ...

Burette reading/cm^3		Titration/cm^3
start	finish	
0.00	24.40	(complete this column, noting any non-concordant results)
24.40	48.75	
10.25	34.70	

Activity 23.1 Analysis of commercial liquid bleach continued

Calculation

Use the equations you derived above to help you answer the following questions.

4 How many moles of sodium thiosulphate are present in the average concordant titration?

5 How many moles of iodine did this react with?

6 How many moles of chlorine must have been produced on acidifying 25 cm^3 of the diluted bleach?

7 How many moles of chlorine would have been liberated on acidifying the whole of the weighed bleach sample?

8 What mass of chlorine would be produced from the whole of the weighed sample?

9 What is the percentage of 'available chlorine' in the bleach?

23.3 Ammonia and nitric acid

In the UK about 2 500 000 tonnes of ammonia are produced each year. It is used to make a variety of products, including nylon, but by far its largest use is in fertiliser manufacture. The method used to make ammonia is the **Haber process** which involves the direct combination of nitrogen and hydrogen.

The raw materials for the process are natural gas, water and air and the first stage involves the production of a mixture of nitrogen and hydrogen, known as **syngas** (or synthesis gas). Natural gas, mainly methane, is heated with steam to give a mixture of carbon monoxide and hydrogen:

$$CH_4(g) + H_2O(g) \rightleftharpoons CO(g) + 3H_2(g) \qquad \Delta H = +210 \text{ kJ}$$

This endothermic reaction is favoured by high temperature and at 750°C results in over 90% conversion of the methane.

Air is then added to give an overall nitrogen to hydrogen ratio of 1:3. At this stage some of the hydrogen reacts with the oxygen from the air to give steam:

$$2H_2(g) + O_2(g) \longrightarrow 2H_2O(g) \qquad \Delta H = -482 \text{ kJ}$$

The heat evolved by this exothermic reaction is used to preheat the reactants and drive steam turbines. Further reaction then takes place between the steam and the residual methane as above, reforming hydrogen gas.

The carbon monoxide is then converted into carbon dioxide by catalytic oxidation with steam:

$$CO(g) + H_2O(g) \rightleftharpoons CO_2(g) + H_2(g) \qquad \Delta H = -42 \text{ kJ}$$

'Scrubbing' the gas with potassium carbonate solution then removes the carbon dioxide giving potassium hydrogencarbonate:

$$K_2CO_3(aq) + CO_2(g) + H_2O(l) \rightleftharpoons 2KHCO_3(aq)$$

This reaction may be reversed by steam treatment allowing the potassium carbonate to be continually recycled.

The final composition of the syngas mixture is approximately 74.3% N_2, 24.6% H_2, 0.8% CH_4 and 0.3% Ar (from the air)

Ammonia is then formed by compressing the mixture to about 200 atmospheres and passing over an iron catalyst at about 450°C:

$$N_2(g) + 3H_2(g) \rightleftharpoons 2NH_3(g) \qquad \Delta H = -92 \text{ kJ}$$

A full discussion of the choice of conditions for this reversible reaction is given in Chapter 25. The theoretical conversion under these conditions should be about 60% but, in practice, equilibrium is never achieved and the mixture leaving the reaction chamber contains about 15% ammonia. Since the ammonia has a much higher boiling point than either of the reactants, it may be removed as a liquid by refrigeration. The unreacted syngas is recycled through the catalyst chamber but periodically some of the gas must be bled off to prevent a build up of methane and argon, which take no part in the reaction. The life expectancy of the catalyst is several years, enabling ammonia plants to operate continuously, apart from routine maintenance work or breakdowns. A simplified flowchart for an ammonia plant is shown in Figure 23.4.

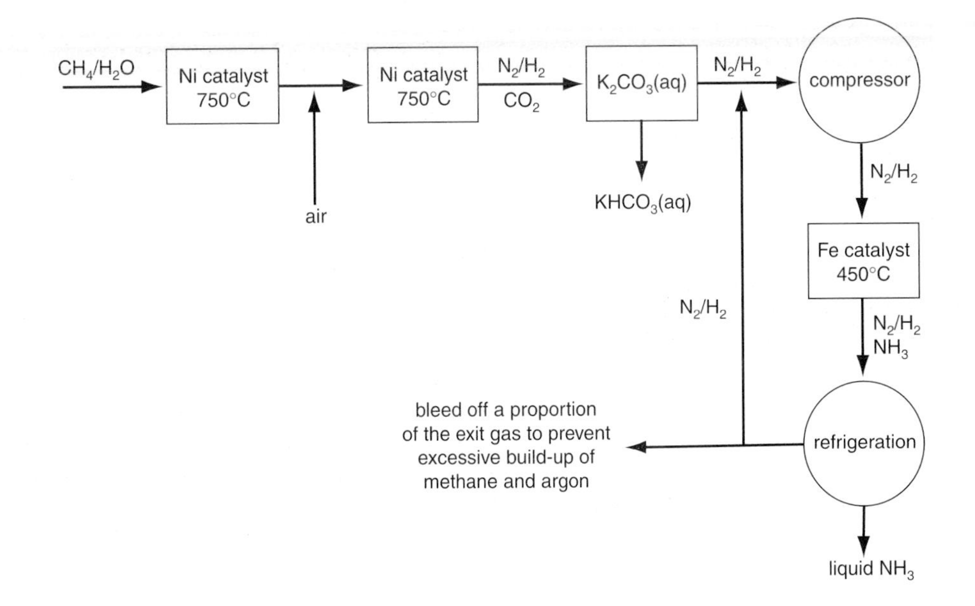

Figure 23.4
Simplified flow diagram for the Haber process

Although liquid ammonia may be applied directly by injecting it beneath the surface of the soil, it is generally more convenient to transport and apply a solid fertiliser. One of the most widely used is ammonium nitrate, NH_4NO_3, made by neutralising ammonia with nitric acid:

$$NH_3(aq) + HNO_3(aq) \longrightarrow NH_4NO_3(aq)$$

Over 700 000 tonnes of nitric acid per year is produced in the UK from ammonia by the **Ostwald process**, outlined in the flowchart in Figure 23.5.

In the first stage a mixture of air and ammonia is passed over a platinum catalyst at about 900°C and 10 atmospheres pressure:

$$4NH_3(g) + 5O_2(g) \longrightarrow 4NO(g) + 6H_2O(g) \qquad \Delta H = -1636 \text{ kJ}$$

The heat produced by this exothermic reaction is absorbed by steam turbines, cooling the mixture to around room temperature. Injection of more air then converts the nitrogen monoxide to nitrogen dioxide:

$$2NO(g) + O_2(g) \longrightarrow 2NO_2(g) \qquad \Delta H = -115 \text{ kJ}$$

This is dissolved in warm water where further oxidation takes place to give approximately 60% aqueous nitric acid:

$$4NO_2(g) + O_2(g) + 2H_2O(l) \longrightarrow 4HNO_3(aq)$$

Although about 90% of total production is used to make fertilisers, nitric acid is also important in the manufacture of a wide range of other products including dyestuffs and explosives.

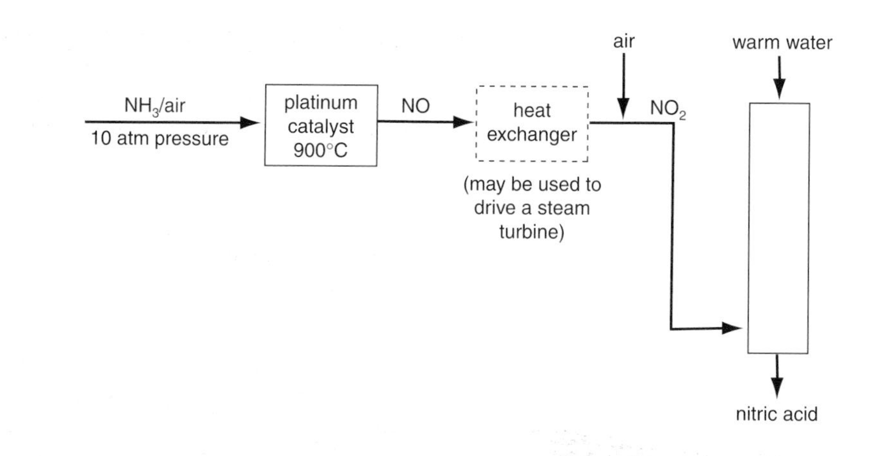

Figure 23.5
Simplified flow diagram for the manufacture of nitric acid by the Ostwald process

Fritz Haber (1868–1934)

This German chemist won the 1918 Nobel Prize for chemistry for his discovery of a process for making ammonia by the direct combination of nitrogen and hydrogen under high pressure. As well as opening the way for the manufacture of synthetic nitrogen fertilisers, his work made it possible for Germany to continue explosives production during World War I, after the Allied blockade had cut access to Chilean nitrate deposits. Haber also played a major role in the war effort by developing poison gases. Despite his scientific achievements it is perhaps ironic that, as a Jew, he was forced to leave Germany in 1933, when Hitler came to power, and died a year later in Switzerland.

Fritz Haber

Activity 23.2 Nitrogen fertilisers

As we mentioned in Chapter 22, overuse of nitrogen fertilisers can cause environmental problems. It is therefore important to apply the correct amount of nitrogen to each particular crop.

Suppose that the recommended nitrogen level for a certain grain is 50 kg per hectare and that analysis shows the soil already contains 15 kg per hectare. If 100 hectares of the grain are planted, calculate the following.

1 The total mass of nitrogen which should be applied to the land to reach the recommended level.
2 The mass of each of the following nitrogen fertilisers which would need to be applied.
   a) liquid ammonia, NH_3
   b) ammonium nitrate, NH_4NO_3
   c) sodium nitrate, $NaNO_3$
   d) carbamide, $CO(NH_2)_2$
   e) ammonium sulphate, $(NH_4)_2SO_4$
   (Relative atomic masses: N = 14, H = 1, C = 12, O = 16, Na = 23)
3 Why do you think that liquid ammonia is often used as a nitrogen fertiliser even though it is difficult to handle and apply?

The crop on the right is unfertilised whilst that on the left has had nitrogen fertiliser added to the soil

Activity 23.3 Siting a chemical plant

An ammonia producer presently buys in sulphuric acid to make ammonium sulphate fertiliser by the following reaction:

$$2NH_3(aq) + H_2SO_4(aq) \longrightarrow (NH_4)_2SO_4(aq)$$

However, the price of sulphuric acid has risen recently and the company is considering building its own sulphuric acid plant. The map in Figure 23.6 shows the locations of three sites under consideration.

Figure 23.6
Map showing possible sites for the sulphuric acid plant in Activity 23.3

Location A

This is where your ammonia plant is situated and where all of your sulphuric acid will be used. Although there is just enough land to build the new plant, this would mean having to find alternative storage space for the finished fertiliser before delivery to customers. The factory is close to London with good links to the motorway system.

Location B

This is a disused factory in Sheffield. There are several other chemical companies in the area that also use sulphuric acid in their manufacturing processes. The site has been unoccupied for some time and is for sale at a very reasonable price. It is near to the Yorkshire coalfields and a power station.

Activity 23.3 Siting a chemical plant continued

Location C

This is a completely undeveloped 'green field' site on the coast just north of Belfast in Northern Ireland. There is no traditional chemical industry in the region but it has good port facilities and, because of high local unemployment, government grants are available to developers. The proposed site is, however, close to an area of outstanding natural beauty and there is an active local conservation group.

Read through the account of the sulphuric acid manufacturing process and produce a report comparing the advantages and disadvantages of each site. You are not expected to make a final choice but should include a consideration of the following factors.

1. The cost of building the plant and supplying the main services, e.g. water and electricity.
2. The availability of fuel.
3. Transport costs: both to obtain raw materials and for moving the sulphuric acid (if necessary) after manufacture.
4. Possible extra sources of income: perhaps by selling surplus sulphuric acid or other products of the process.
5. The ease of recruiting the skilled and semi-skilled labour needed to operate and maintain the plant.
6. Environmental factors associated with the disposal of waste products or, more seriously, with a malfunction of the plant.

Sulphuric acid manufacture

Figure 23.7 shows a flow diagram for the **contact process** described below.

Raw materials

As well as air and water, the process uses sulphur, which is imported from the USA, Poland or Sicily.

Sulphur burning

The sulphur is burned in excess air to give sulphur dioxide,

$$S(s) + O_2(g) \longrightarrow SO_2(g) \qquad \Delta H = -297 \text{ kJ}$$

The mixture of sulphur dioxide and air leaves the burners at about 1000°C and must be cooled by passing it through a heat exchanger. This produces steam which may be sold for electricity production, thereby helping to reduce running costs.

Sulphur trioxide formation

The mixture of sulphur dioxide and excess air is passed over a vanadium(V) oxide catalyst at 450°C to give sulphur trioxide. Though this reaction is reversible, about 99.5% conversion is achieved in this exothermic step:

$$2SO_2(g) + O_2(g) \rightleftharpoons 2SO_3(g) \qquad \Delta H = -196 \text{ kJ}$$

Oleum production

The gases which contain about 10% of sulphur trioxide are cooled and dissolved in concentrated sulphuric acid to give 'fuming' sulphuric acid or oleum:

$$H_2SO_4(l) + SO_3(g) \longrightarrow H_2S_2O_7(l)$$

It is unsafe to dissolve the sulphur trioxide directly in water. The reaction is so vigorous that a corrosive mist of sulphuric acid droplets results.

Sulphuric acid manufacture *continued*

Sulphuric acid production

Oleum may be safely diluted with water to give 98% sulphuric acid. Part of this is recirculated to dissolve more sulphur trioxide and the waste gases are vented from the top of the vessel. Current regulations demand that the sulphur dioxide content of the emissions does not exceed 0.05%. At this level the contribution to acid rain is negligible compared with that from a conventional coal-fired power station.

Owing to the highly corrosive nature of sulphuric acid, it must be transported in tankers lined with an inert material such as polyethene.

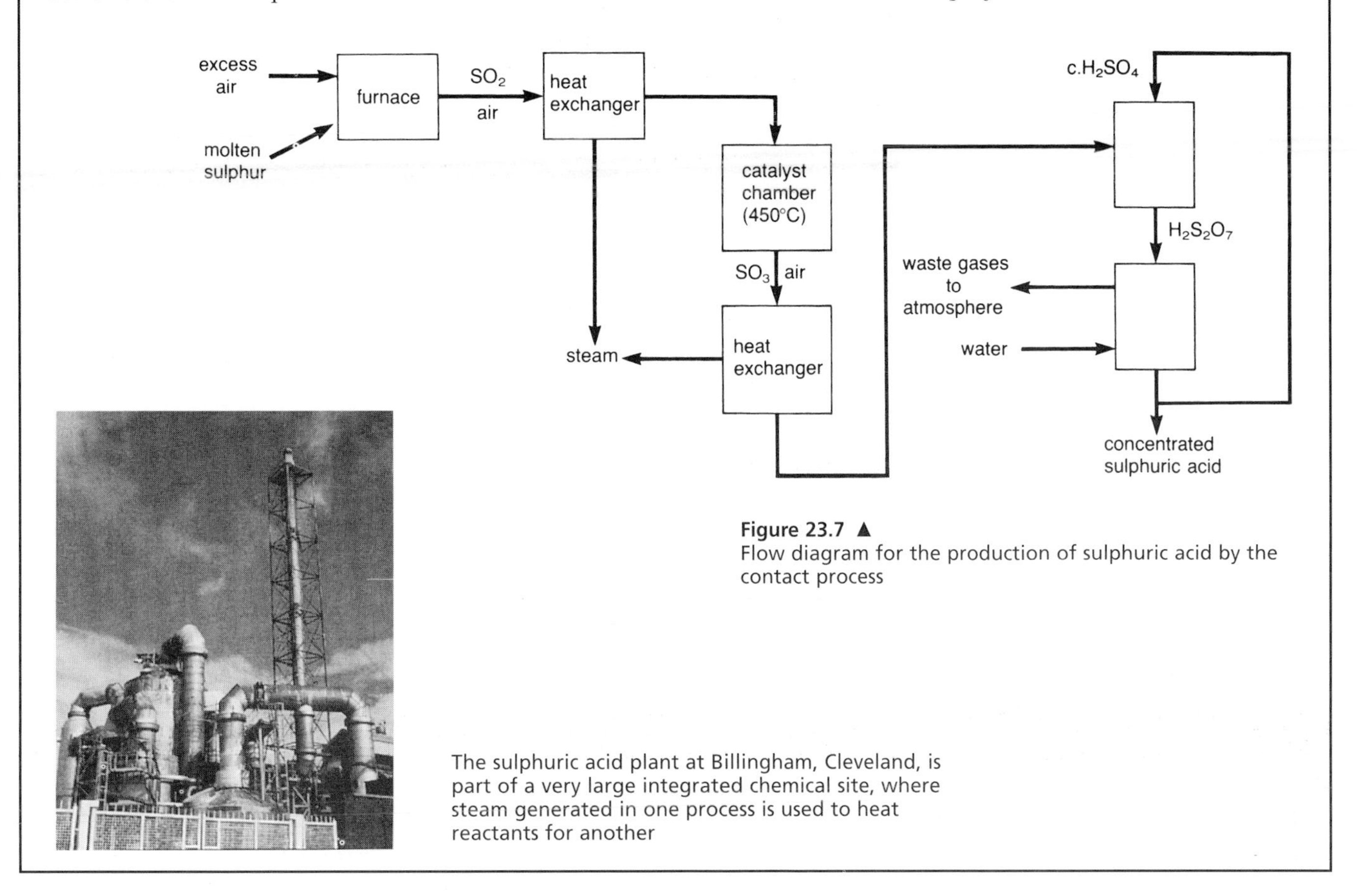

Figure 23.7 ▲
Flow diagram for the production of sulphuric acid by the contact process

The sulphuric acid plant at Billingham, Cleveland, is part of a very large integrated chemical site, where steam generated in one process is used to heat reactants for another

23.4 Metals

Metals have many unique and valuable properties. They are generally tough, flexible and readily conduct both heat and electricity. Apart from a few scarce unreactive metals, such as gold, they are not found naturally as the free elements but instead as chemical compounds. There are several stages involved in extracting a pure metal from a naturally occurring ore:

- mining the crude ore
- purifying the ore, i.e. removing non metal-bearing rock and waste
- extracting the metal from the purified ore
- if necessary, depending upon the end use, refining or purifying the metal.

In this section we shall deal mainly with the extraction of the metal from its purified ore.

In their chemical compounds, most metals exist as **positively charged** ions. Extraction of the element therefore involves a **reduction** process, i.e. the addition of electrons:

$$M^{x+} + xe^- \longrightarrow M$$

The reduction method chosen must give a product of acceptable quality as economically as possible. We are, in fact, reversing the natural tendency of most metals to form positive ions by losing electrons (see Chapters 26 and 28 for more detail). More reactive metals are therefore more difficult, and often much more expensive, to obtain. Table 23.2 outlines some general methods for metal extraction. In the rest of this section we shall look in more detail at the extraction and uses of certain metals.

Table 23.2 Methods used for the extraction of some metals from their purified ores

Metal	from	extraction method
lithium	LiCl	electrolysis of fused* compounds
sodium	NaCl	
potassium	KCl	
beryllium	BeF_2	
magnesium	$MgCl_2$	
calcium	$CaCl_2$	
strontium	$SrCl_2$	
barium	$BaCl_2$	
aluminium	Al_2O_3	
manganese	MnO_2	reduction with carbon
zinc	ZnO	
tin	SnO_2	
lead	PbO	
iron	Fe_2O_3	reduction with carbon monoxide
nickel	NiO	
titanium	$TiCl_4$	reduction with a more reactive metal
chromium	Cr_2O_3	
cobalt	Co_3O_4	
copper	Cu_2S	roasting the sulphide in air
mercury	HgS	

*Why can't these metals be extracted by electrolysis of aqueous solutions of their salts? (Hint: see section 26.1.)

Aluminium

Reactive metals such as aluminium are generally extracted by electrolysis of their molten oxide or chloride. This method is very expensive in terms of energy, since metal compounds usually have high melting points that are typical of ionic structures.

Aluminium is the most abundant metal comprising about 7.5% of the Earth's crust. Although it occurs in a variety of minerals, the only commercially important source is bauxite, which contains about 50% of aluminium oxide. Separation of pure alumina, Al_2O_3, from other material in the ore depends upon its amphoteric nature (it has acidic and basic properties). The bauxite is treated at high temperature and pressure with sodium hydroxide solution, which dissolves the aluminium oxide as sodium aluminate:

$$Al_2O_3(s) + 2NaOH(aq) + 7H_2O(l) \longrightarrow 2Na[Al(OH)_4(H_2O)_2]\ (aq)$$

Insoluble impurities, such as iron(III) oxide, are filtered off and discarded as 'red mud'. 'Seeding' the solution with a little of the pure solid causes hydrated alumina, $Al_2O_3.3H_2O$, to precipitate. The water of crystallisation is then driven off by heating, leaving pure anhydrous alumina:

$$Al_2O_3.3H_2O(s) \longrightarrow Al_2O_3(s) + 3H_2O(g)$$

Electrolysis of molten alumina gives aluminium at the cathode and oxygen at the anode:

cathode reaction:	$2[Al^{3+}(l) + 3e^- \longrightarrow Al(l)]$	reduction
anode reaction:	$3[O^{2-}(l) \longrightarrow \frac{1}{2}O_2(g) + 2e^-]$	oxidation
overall:	$Al_2O_3(l) \longrightarrow 2Al(l) + 1\frac{1}{2}O_2(g)$	redox

In practice, since the melting point of alumina is over 2000°C, it is dissolved in molten cryolite (sodium hexafluoroaluminate), $Na_3[AlF_6]$. This remains unchanged in the electrolysis cell but lowers the melting point of the mixture to about 1000°C with a consequent saving in energy costs. The process operates continuously in banks of cells, as shown diagramatically in Figure 23.8.

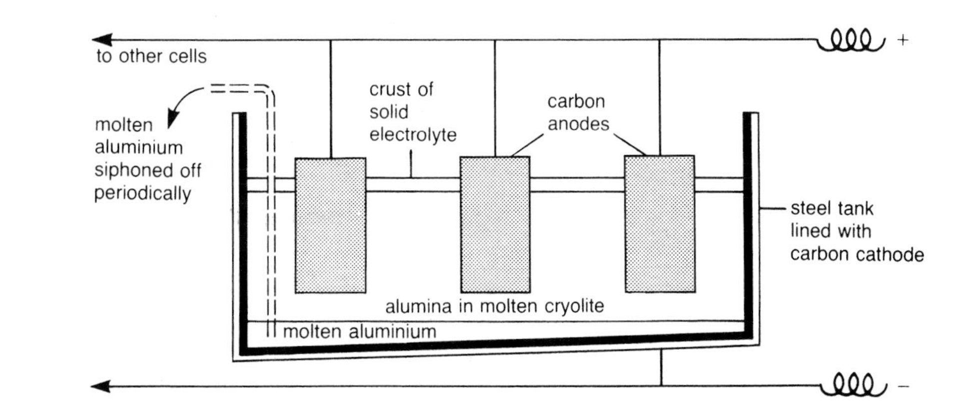

Figure 23.8 ▶ Production of aluminium by electrolysis of molten alumina

▲ Recycling aluminium from drink cans is becoming increasingly worthwhile

The surface crust of solid electrolyte is periodically broken to siphon off the molten aluminium and add further alumina to maintain its concentration at about 5%. At the operating temperature of the cell the carbon-based anodes gradually burn away in the oxygen formed and need to be replaced regularly.

Aluminium does not appear to be a very reactive metal, owing to the formation of a very thin but tough oxide layer which prevents further attack by air or water. Its density is less than a third of that of steel and many of its uses, outlined in section 20.1, depend upon its light weight and corrosion resistance. Because of the high energy costs involved in extracting aluminium from bauxite and as the richest, most accessible ores are being used up, it is becoming increasingly worthwhile to collect waste aluminium, e.g. drink cans, for recycling.

Iron and steel

Metals that are less reactive than aluminium are generally extracted more cheaply using chemical reducing agents such as carbon or carbon monoxide. Iron is extracted from oxide ores such as haematite, Fe_2O_3, in a blast furnace, as shown in Figure 23.9.

A mixture of the **iron ore**, **coke** to supply the carbon and **limestone** is fed in at the top of the furnace and meets a blast of hot air fed in from the base. The coke burns, producing heat and carbon dioxide:

$$C(s) + O_2(g) \longrightarrow CO_2(g) \qquad \Delta H = -394 \text{ kJ}$$

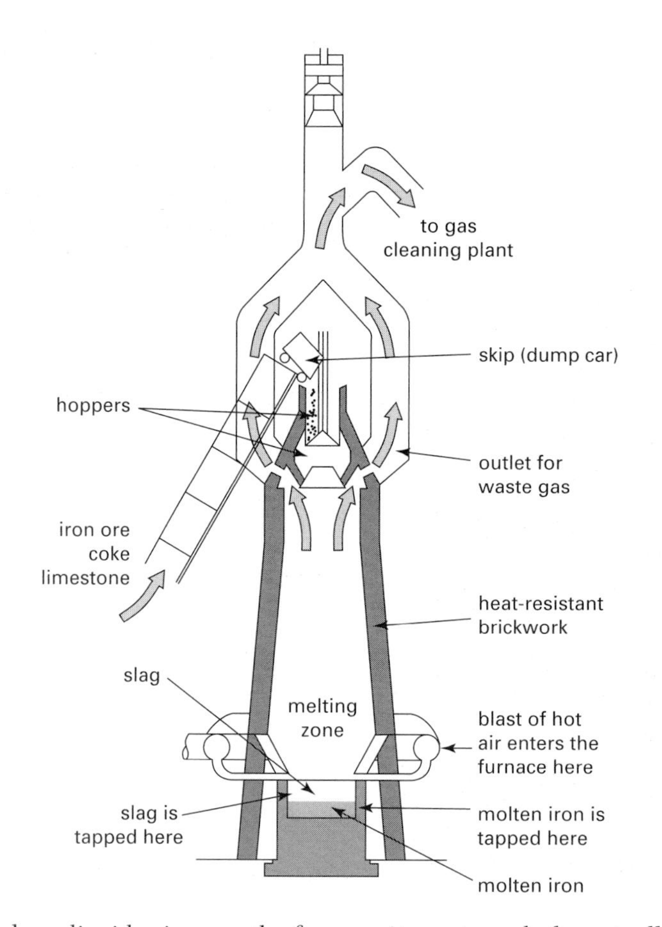

Figure 23.9
A blast furnace

As the carbon dioxide rises up the furnace it reacts endothermically with more of the coke to form carbon monoxide:

$$CO_2(g) + C(s) \longrightarrow 2CO(g) \qquad \Delta H = +111 \text{ kJ}$$

The carbon monoxide reduces the ore to **molten iron** that trickles down to the bottom of the furnace:

$$Fe_2O_3(s) + 3CO(g) \longrightarrow 2Fe(s) + 3CO_2(g)$$

The limestone is added to remove sand, silicon dioxide, which is a common impurity in the metal ore. The limestone first decomposes to give calcium oxide:

$$CaCO_3(s) \longrightarrow CaO(s) + CO_2(g)$$

This basic oxide reacts with the acidic silicon dioxide to give calcium silicate:

$$CaO(s) + SiO_2(s) \longrightarrow CaSiO_3(l)$$

Unlike silicon dioxide, the calcium silicate is liquid at the furnace temperature and forms a layer of **slag** floating on top of the molten iron. Periodically the slag and iron are tapped off separately from the bottom of the furnace. After cooling, the rock-like slag is often sold as a foundation material for road building.

The metal obtained from the blast furnace is known as **pig iron** and contains up to 5% carbon with smaller amounts of other impurities such as silicon and phosphorus oxides. The high carbon content makes the metal very hard but also brittle and most pig iron is converted into steel by removing most of the carbon. In the **basic oxygen process**, a high-pressure stream of pure oxygen is passed though the molten iron via a water-cooled lance. The oxygen burns off carbon and other unwanted impurities in a very exothermic and spectacular reaction.

For specialised uses, other metals may be added during steel-making to give alloy-steels. Examples of some common types of steel and their main uses are given in Table 23.3.

Table 23.3 The composition, properties and uses of some common steels

Type of steel	composition excluding iron	properties and uses
mild	0.1–0.2% C	soft but very malleable, readily shaped into sheet and rods
high carbon	0.7–1.5% C	very hard, used for tools such as hammers and punches
stainless	18% Cr, 8% Ni, some C	very corrosion resistant, used in cutlery
manganese	13% Mn, some C	very tough, used for drilling through rock
tungsten	5% W, some C	very hard, used in high-speed cutting tools
permalloy	78% Ni, some C	strongly magnetised in an electric field but loses magnetism when current is switched off, used in electromagnets

Steel was used extensively in the construction of the Lloyd's building in London

Titanium

Titanium is relatively abundant in nature being found in minerals such as ilmenite, $FeTiO_3$, and rutile, TiO_2. It is not practical to produce the pure metal by reducing the oxide with carbon as titanium reacts to form a stable carbide. Instead the ore is first heated with carbon and chlorine to give titanium(IV) chloride which may be separated from impurities by fractional distillation:

$$TiO_2(s) + C(s) + 2Cl_2(g) \longrightarrow TiCl_4(g) + CO_2(g)$$

The titanium(IV) chloride is then reduced by heating in an argon atmosphere with sodium:

$$TiCl_4(l) + 4Na(l) \longrightarrow Ti(s) + 4NaCl(s)$$

After washing out the sodium chloride with dilute hydrochloric acid, the titanium is melted and cast into ingots. Magnesium may be used as an alternative reducing agent in this process.

Like aluminium, titanium is lightweight and strong. Alloyed with aluminium and vanadium, it is used in aircraft construction and space capsules. Its relative inertness makes titanium useful in surgery as a replacement for bone and cartilage. Titanium(IV) oxide, TiO_2, is a brilliant white pigment used in paints, plastics, paper, synthetic fibres, and rubber.

Titanium and other similar metals are used to make artificial hip joints

Tungsten

Tungsten occurs together with other metals in such minerals as scheelite and wolframite. In the extraction process the ore is first fused with sodium carbonate to yield sodium tungstate, Na_2WO_4. After dissolving in hot water, addition of hydrochloric acid precipitates tungstic acid, H_2WO_4:

The tungsten filament in a 60W domestic light bulb, magnified 99 times

$$Na_2WO_4(aq) + 2H^+(aq) \longrightarrow H_2WO_4(s) + 2Na^+(aq)$$

Heating in air removes water, leaving tungsten(VI) oxide, WO_3:

$$H_2WO_4(s) \longrightarrow WO_3(s) + H_2O(g)$$

Like titanium, tungsten reacts with carbon to form a stable carbide. In this case hydrogen is used as the reducing agent to obtain the pure metal,

$$WO_3(s) + 3H_2(g) \longrightarrow W(s) + 3H_2O(g)$$

Tungsten has a very high melting point (3410°C) and is used for filaments in electric light bulbs, in electric furnace wiring, and in the production of hard, tough steel alloys. Tungsten carbide is an excellent abrasive that is used to tip drill bits.

Chromium

Chromium plated taps

The metal is extracted from chromium(III) oxide obtained from chromite ore, $FeCr_2O_4$. Although coke may be used as the reducing agent, the most important process uses powdered aluminium:

$$Cr_2O_3(s) + 2Al(s) \longrightarrow 2Cr(s) + Al_2O_3(s)$$

Most chromium is used to form alloys with iron, nickel, or cobalt. It imparts hardness, strength, and corrosion resistance to the alloys and stainless steels contain up to 18% chromium. An extremely hard alloy of chromium, cobalt, and tungsten is used for high-speed metal-cutting tools. Electroplated chromium provides an attractive shiny finish on articles such as car bumpers and bathroom fittings.

Copper

Plumbing in copper piping in a new building

This is one of the least reactive metals and is even found naturally as the free element. The most important ores are sulphides, e.g. chalcopyrite, $CuFeS_2$. The copper and iron may be separated by heating with sand and air, producing two immiscible liquid layers:

$$2CuFeS_2(s) + 5O_2(g) + 2SiO_2(s) \longrightarrow Cu_2S(l) + \underset{\text{slag}}{2FeSiO_3(l)} + 3SO_2(g)$$

No reducing agent need be added to extract the copper, since the sulphur is removed by oxygen from the air:

$$Cu_2S(l) + O_2(g) \longrightarrow 2Cu(l) + SO_2(g)$$

This produces up to 99.5% pure copper, which may be further refined by electrolysis. The impure copper is the anode in a cell that contains a small pure copper cathode and copper sulphate solution as electrolyte. The impure copper dissolves into solution at the anode:

$$Cu(s) \longrightarrow Cu^{2+}(aq) + 2e^-$$

and copper ions deposit from solution at the cathode:

$$Cu^{2+}(aq) + 2e^- \longrightarrow Cu(s)$$

Many valuable metal impurities, including silver, gold and platinum, do not pass into solution and may be recovered from the 'sludge' which collects beneath the anode. Other metals, such as iron and nickel, dissolve readily at the anode but remain in the electrolyte solution.

Copper is an excellent conductor of heat and electricity and is also resistant to corrosion. Its main uses include electrical wiring, central heating boilers and plumbing, 'copper' coinage, and the manufacture of alloys such as brass and bronze.

Activity 23.4 'Copper' coinage

If you test copper coins with a small magnet you may get a surprise. One penny and two pence coins minted before September 1992 are not attracted by a magnet, whilst newer coins are.

1 Can you suggest any explanations for this difference?

Here is a brief account of an investigation into copper coins carried out by A level students that you might like to repeat.

An old (minted before 1992) and new coin were taken and a little concentrated nitric acid added to the coins in separate, small beakers (CARE: Very corrosive!) Both coins reacted vigorously at first, giving off brown fumes of nitrogen dioxide (a fume cupboard is advisable). When the reaction was over the liquid surrounding each coin was diluted with water. On adding excess concentrated ammonia solution, both samples gave a deep blue solution but, in the case of the new coin, a thick brown precipitate was also formed. After the nitric acid treatment the old coin retained its normal 'copper' colour, whereas the new coin was silvery in appearance.

2 Which metal ion produces a deep blue solution with excess ammonia?

3 Can you suggest a metal ion that might cause the brown precipitate with ammonia solution?

4 What conclusions can you draw from the appearance of the coins after treatment with nitric acid?

On closer inspection, the new coins where found to be noticeably thicker than the old ones, even though their masses were identical.

5 Use a data book to check the densities of the metals you suspect may be in each type of coin. Do the values support your ideas?

Questions on Chapter 23

1 The main uses of phosphoric acid, H_3PO_4, are shown in the pie chart in Figure 23.10.

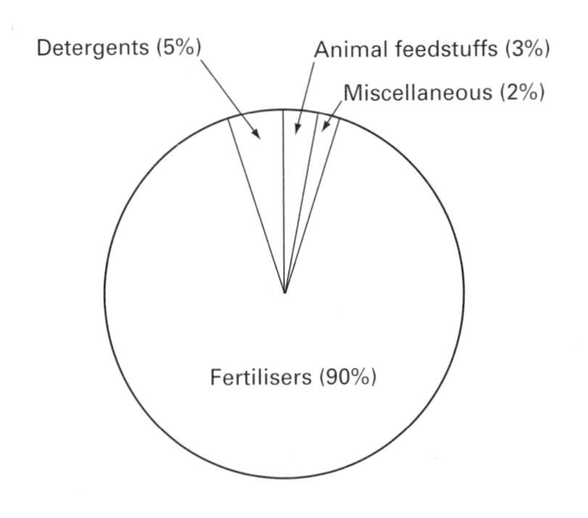

Figure 23.10
Main uses of phosphoric acid

If the annual UK production of phosphoric acid is 600 000 tonnes, estimate the amount used in the manufacture of synthetic detergents.

2 In the manufacture of phosphoric acid, rock containing calcium phosphate, $Ca_3(PO_4)_2$, is treated with sulphuric acid:

$$Ca_3(PO_4)_2(s) + 3H_2SO_4(aq) \longrightarrow 2H_3PO_4(aq) + 3CaSO_4(s)$$

The insoluble calcium sulphate is filtered off, leaving a solution of phosphoric acid.

Assuming 100% conversion, what masses of calcium phosphate and sulphuric acid are needed to produce one tonne of phosphoric acid?

3 To make fertiliser, phosphoric acid is partially neutralised by ammonia to give diammonium hydrogenphosphate, $(NH_4)_2HPO_4$.

a) Write a balanced equation for the production of diammonium hydrogenphosphate.

Questions on Chapter 23 *continued*

b) What is the maximum mass of diammonium hydrogenphosphate that could be made from one tonne (i.e.1000 kg) of phosphoric acid?

c) Calculate the percentage by mass of phosphorus in diammonium hydrogenphosphate.

d) An alternative method for making phosphoric acid involves burning liquid phosphorus, P_4, in air to give phosphorus(V) oxide, P_2O_5 which is then dissolved in water. Write balanced equations for this sequence of reactions.

4 Although sulphur occurs naturally as the free element, it is also extracted from crude petroleum, where it is present as hydrogen sulphide, H_2S.

After separation from the gas or oil, the hydrogen sulphide is burned in a limited supply of air to give sulphur vapour and steam.

When the hydrogen sulphide is burned, some sulphur dioxide is also produced. This is treated with more hydrogen sulphide in the presence of a bauxite catalyst to give sulphur and water only.

a) Why is it important that sulphur compounds are removed from gas and oil before they are used as fuels?
b) Write a balanced equation for the production of sulphur from hydrogen sulphide and calculate the maximum mass of sulphur that might be produced from one tonne of hydrogen sulphide.
c) Write a balanced equation for the reaction between hydrogen sulphide and sulphur dioxide.

Use the information in Tables 23.4a and b on the production and uses of selected metals to help you answer questions 5–9.

5 Many quite reactive metals are found in nature as compounds containing oxygen. These may be simple oxides or salts of acids that contain oxygen. Less reactive metals are often found as sulphides or even, sometimes, as the free metal.

a) Name one metal in the table which is obtained from an oxide ore.
b) Name one metal in the table that is extracted from its sulphide.
c) Which of the metals in the table is MOST likely to be found in the free state?

6 Ores must often be concentrated to remove waste material before the metal is extracted. In the blast furnace, for example, the ore used must contain at least 60% of iron.

a) What is the relative formula mass of iron(III) oxide, Fe_2O_3?
b) What is the percentage by mass of iron in this compound?
c) How many tonnes of iron could be made from 100 tonnes of Fe_2O_3?
d) How many tonnes of iron can be made from 100 tonnes of the main ore of iron?
e) Why can't the main ore of iron be used directly in the blast furnace?
f) What kind of pollution problem can the disposal of solid waste from metal ores cause?

7 Lead is obtained largely from a sulphide-containing ore. After concentration, lead(II) sulphide is roasted with air to give lead(II) oxide and sulphur dioxide. The sulphur dioxide is removed and used to produce sulphuric acid. The lead oxide is smelted with coke to give the metal and carbon monoxide.

a) Write a formula equation for the reaction of lead sulphide with air.
b) Write a formula equation for the reduction of lead oxide by coke.
c) Why can't the sulphur dioxide produced by roasting sulphide ores be allowed to escape directly into the air?

8 The price of a metal depends upon several factors, including:

- how rare the metal is,
- how easy it is to obtain a reasonably concentrated ore,
- how much energy is required to reduce the metal compound to the metal,
- how useful the metal is.

Suggest the MAIN reason for each of the following:

a) Silver is the most expensive metal in the table.
b) Aluminium is more expensive than iron.
c) Lead and zinc are both obtained from sulphide ores by the same process, but the price of zinc is almost twice that of lead.

9 Primary metal production involves extraction from a naturally occurring ore, whereas secondary production involves recycling waste metal. Whilst this saves the cost of reducing the ore, it can be quite expensive to recover and separate the metal from other waste.

a) Name a metal, NOT in the table, which would require more energy per tonne to produce than those shown.
b) Calculate the number of tonnes of lead recycled in a year.
c) From which of the uses of lead given in the table would it be most difficult to recycle the metal?
d) Why do you think that only 15% of tin is recycled, even though the metal is quite valuable?

Questions on Chapter 23 *continued*

Table 23.4a Statistics for the production of selected metals

Metal	aluminium	copper	iron	lead	silver	tin	zinc
abundance (% of Earth's crust)	8.1	0.006	5.0	0.0013	0.000007	0.0002	0.007
formula of metal-containing compound in the main ore	$Al_2O_3.3H_2O$	$CuFeS_2$	Fe_2O_3	PbS	Ag_2S	SnO_2	ZnS
% of metal in main ore	28	0.5	45	10	0.6	1.5	20
annual production (megatonnes*)	13.4	8.3	65	3.4	0.2	0.5	5.0
price (£/tonne)	900	1000	150	350	2 000 000	8000	600
% metal recycled	28	42	45	40	?	15	23
energy required to extract the metal (megajoules*/tonne)							
from ore	250	112	32	28	?	?	68
from scrap	12	20	12	12	?	?	20

*Mega means 'millions', so 1 megatonne equals 1 million tonnes.

Table 23.4b Major uses of selected metals

Metal	major uses (% shown in brackets)
aluminium	construction (50), light alloys (20), foil (10), transmission cables (10), other uses (10)
copper	electrical wiring (58), water pipes/roofing (19), boilers (17), coinage and other uses (6)
iron	steel (85), cast iron (10), other uses (5)
lead	car/lorry batteries (52), piping (14), petrol additives (8), other uses (26)
silver	photographic chemicals (50), jewellery (25), electrical contacts (10), other uses (15)
tin	tin-plated iron and steel (85), other uses (15)
zinc	galvanised iron (35), brass (20), chemicals (15), other uses (30)

Comments on the activities

Activity 23.1

1. $ClO^-(aq) + 2H^+(aq) + e^- \longrightarrow \frac{1}{2}Cl_2(g) + H_2O(l)$
   $Cl^-(aq) \longrightarrow \frac{1}{2}Cl_2(g) + e^-$

   $ClO^-(aq) + Cl^-(aq) + 2H^+(aq) \longrightarrow Cl_2(g) + H_2O(l)$

2. $2I^-(aq) \longrightarrow I_2(aq) + 2e^-$
   $Cl_2(g) + 2e^- \longrightarrow 2Cl^-(aq)$

   $2I^-(aq) + Cl_2(g) \rightleftharpoons I_2(aq) + 2Cl^-(aq)$

3. $I_2(aq) + 2e^- \longrightarrow 2I^-(aq)$
   $2S_2O_3^{2-}(aq) \longrightarrow S_4O_6^{2-}(aq) + 2e^-$

   $I_2(aq) + 2S_2O_3^{2-}(aq) \longrightarrow 2I^-(aq) + S_4O_6^{2-}(aq)$

   Mass of bleach used = 11.44 g
   Titration results: 24.40, 24.35, 24.45. Average: 24.40

4. Moles of sodium thiosulphate present in the average concordant titration
   $= 0.1000 \times \frac{24.40}{1000} = 0.00244$
5. Moles of iodine in titration $= \frac{0.00244}{2} = 0.00122$
6. Moles of chlorine produced from 25 cm^3 of the diluted bleach = 0.00122
7. Moles of chlorine liberated from whole of bleach sample = 0.0122
8. Mass of chlorine 'available' from whole of bleach sample = 71×0.0122 = 0.8662 g
9. Percentage of 'available chlorine' in the bleach
   $= \frac{0.8662}{11.44} \times 100 = 7.57\%$

Comments on the activities *continued*

Activity 23.2

1 Each hectare of land requires 50 – 15 = 35 kg of nitrogen
The total mass of nitrogen needed for 100 hectares = 3500 kg.

2 Perhaps the easiest way to calculate the mass of each fertiliser needed is to first work out the percentage by mass of nitrogen.

a) For liquid ammonia

	N	H_3	
	14	3	total 17

$$\% \text{ by mass of N} = 100 \times \frac{14}{17} = 82.4$$

i.e. 82.4 kg of nitrogen is provided by 100 kg of ammonia

$$1 \text{ kg of nitrogen is provided by } \frac{100}{82.4} \text{ kg of ammonia}$$

3500 kg of nitrogen is provided by

$$3500 \times \frac{100}{82.4} = 4248 \text{ kg ammonia}$$

Similar calculations show that

b) ammonium nitrate,
NH_4NO_3 (35% N) 10000 kg required

c) sodium nitrate,
$NaNO_3$ (16.5% N) 21212 kg required

d) carbamide,
$CO(NH_2)_2$ (46.7% N) 7495 kg required

e) ammonium sulphate,
$(NH_4)_2SO_4$ (21.2%) 16509 kg required

3 The advantage of liquid ammonia is its very high percentage of nitrogen. Less is therefore required which reduces transport and storage costs.

Activity 23.3

Here are some of the points that might be considered.

Building costs

The actual cost of building the plant should be lowest at site A, as the company already owns the land. Finding alternative storage space for the ammonium sulphate could be very expensive, however, given the high cost of land near London. Site C is likely to be more expensive to develop than site B, since mains water and electricity supplies will need to be provided. Government development grants may still make this an economically attractive proposition.

Fuel supplies

Since all the reactions involved are highly exothermic, the process is a net producer of energy. Fuel supply is therefore not a major consideration here.

Transport

Since sulphur must be imported, sites A and C have an advantage in being closer to major port facilities. Site A has the added advantage that the acid is used on site, avoiding the hazards of transporting the sulphuric acid by road or rail.

Extra income

Site B has two advantages here. There are potential customers for any surplus sulphuric acid production nearby and also a power station that might buy the steam generated during acid production.

Labour

Site A could use your existing management but extra labour might be more expensive here than in the more economically depressed areas around the other two sites. Skilled labour is more likely to be found at site B than at site C since the latter has no local chemical industry.

Environmental considerations

Although modern sulphuric acid plants are very safe and emit negligible quantities of sulphur dioxide and other pollutants, there is still likely to be stiff opposition from conservation groups if you decide to build at site C. If there were to be a serious accident, site C has the advantage that the area is least populated and that the prevailing westerly winds would probably blow the worst of any pollution out to sea.

Activity 23.4

1 The composition of the coins must be different. The new coins must contain a magnetic metal such as iron, nickel or cobalt.

2 Cu^{2+} ions give the soluble deep blue complex, $[Cu(NH_3)_4(H_2O)_2]^{2+}(aq)$ with excess ammonia (see section 21.4).

3 Fe^{3+} ions give the precipitate $[Fe(OH)_3(H_2O)_3](s)$ with ammonia (see section 21.3).

4 The old coin seems to have a uniform composition throughout, whereas the new coin appears to have a thin surface coating of copper over another metal, probably iron.

5 The density of copper is 8.92 g cm^{-3}, whilst that of iron is 7.86 g cm^{-3}. An iron-based coin would therefore need to be bigger than a copper coin of the same mass. Further experiments, including atomic emission spectroscopy carried out in the laboratories of a local company, supported the conclusion that old coins were largely pure copper whilst the newer coins were copper-plated steel. The Royal Mint confirmed that this change was carried out for economic reasons as the cost of producing traditional 'copper' coins had exceeded their face value.

THEME E

Physical chemistry: how far, how fast?

CHAPTER 24

Rates of reaction

Contents

Study Checklist

After studying this chapter you should be able to:

1. Appreciate that reactions may only occur when reactant particles collide with sufficient energy.
2. Sketch the Maxwell–Boltzmann distribution curve for molecular energies in a gas at two different temperatures and account qualitatively for the differences.
3. State and explain the effect of changes in gas pressure, solution concentration and particle size of solid on reaction rate.
4. Define activation energy as the minimum collision energy needed to result in a reaction and explain why most collisions do not give rise to a reaction.
5. Explain, in terms of activation energy and the Maxwell–Boltzmann energy distribution, why a small rise in temperature may give a dramatic increase in reaction rate.
6. Explain what is meant by a catalyst and know that it speeds up a reaction by providing an alternative reaction pathway with a lower activation energy.
7. Give examples of important large-scale use of catalysts, e.g. in the chemical industry and in the control of vehicle exhaust emissions by catalytic converters.
8. Explain the difference between homogeneous and heterogeneous catalysts.
9. Understand that a heterogeneous catalyst operates via surface adsorption of reactants and desorption of products and explain how 'poisoning' may occur.
10. Recall that a homogeneous catalyst works by forming an intermediate species and that when this involves a transition metal species a change in oxidation state is generally involved.
11. Appreciate the importance of enzymes as specific catalysts in particular biochemical reactions, including those taking place in living cells.
12. Name and define all the terms in a simple rate equation of the type rate = $k[A]^a[B]^b$, and be able to carry out simple calculations involving the rate equation.
13. Appreciate that many reactions occur as a series of steps and that the slowest step in the sequence will be rate determining.
14. Devise practical techniques for following the progress of a given reaction and explain how information on rate may be obtained from experimental measurements.
15. Recall that first order reactions have a constant half-life.

▲ Fresh food can be kept longer in this frozen-food store because bacteria breed much more slowly at low temperatures. Chickens, for example, can be stored for up to nine months in a freezer at −15°C

24.1 Why measure reaction rate?

Control of reaction rate may be of considerable importance for a variety of reasons.

- From a manufacturer's point of view, the economic success of a new product depends upon the margin between the price the customer is prepared to pay and manufacturing costs. In this case it is clearly important to be able to make the desired material as quickly and safely as possible. On the other hand, undesirable reactions, such as the rusting of iron or the spoilage of food, should be slowed down as far as possible to make the product more appealing to the customers.
- From a consumer viewpoint, it is important to have information on reaction rate when using certain products. For example, the cooking time and temperature for a food product, the setting time for an adhesive or the time that must elapse between spraying a pesticide and harvesting a crop.
- For the chemist, a study of reaction rate can give valuable information about the mechanism of a reaction. This in turn can lead to the development of new products or processes.

Controlled release of drugs

E

In the treatment of certain conditions a small dose of drugs needs to be delivered continuously over a prolonged period. For example, in hormone replacement therapy the drug may be contained within a biodegradable capsule inserted just under the skin. As the coating dissolves away the hormone is gradually released into the body, avoiding the need for regular injections or tablets.

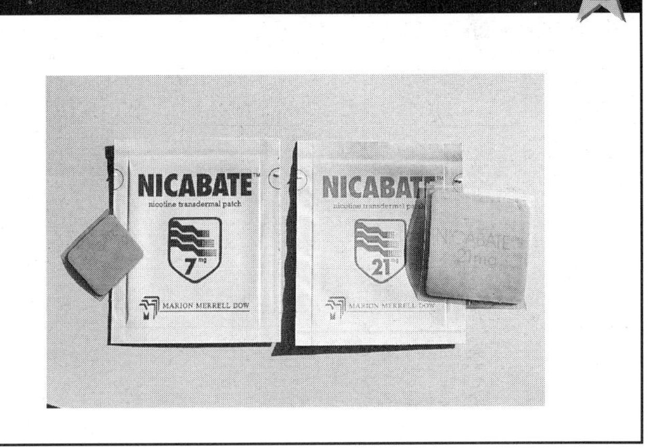

Patches that slowly release nicotine through the skin can help relieve the withdrawal symptoms of giving up smoking ▶

▲ Adhesive setting times must be predictable

24.2 What must happen to get a reaction?

It might seem self-evident, but a reaction can only occur if suitable particles **collide**. Even then, after most collisions the particles simply bounce away from each other unchanged. Only if the collision takes place with sufficient force will a reaction result. The minimum energy needed for a collision to result in a reaction is known as the **activation energy**, E_a. We can think of this as the energy needed to break bonds in the reactant particles in order for new bonds to form in the products. An interesting analogy is the traditional fairground 'coconut shy' shown in Figure 24.1.

To win a prize you have to knock a coconut off its stand by throwing a ball at it. To succeed, the following conditions must all be satisfied:

- there must be a collision, i.e. the ball must hit the coconut,
- the ball must have enough force to dislodge the coconut otherwise it will simply bounce off,

▲ Sufficient time must be allowed after crop spraying to ensure that pesticide levels have fallen to a 'safe' value

- an additional condition is that even if the ball is thrown hard enough it will only dislodge the coconut if it hits it at a suitable point. **Collision geometry** can also be important in chemistry, especially if the reactant molecules are large and complex.

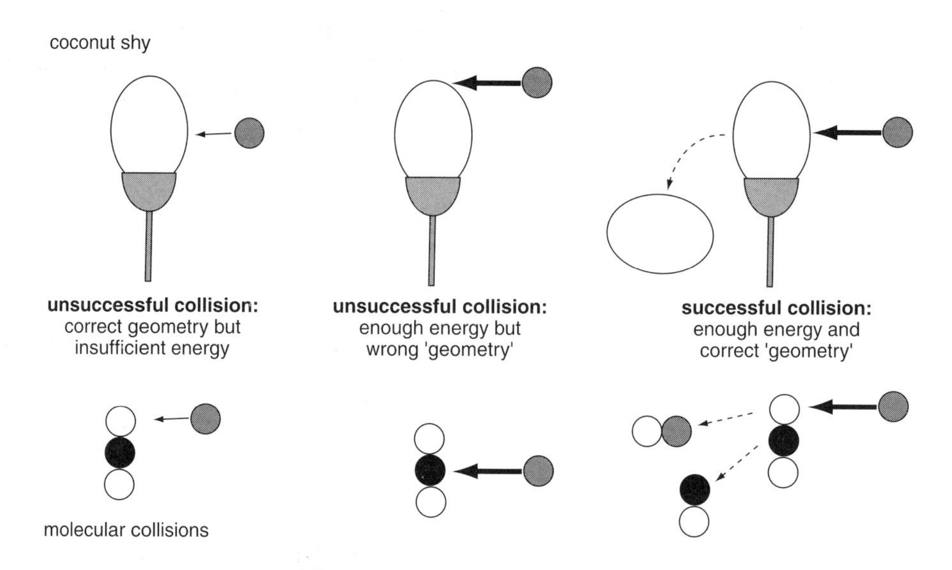

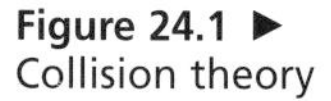
Figure 24.1 ▶
Collision theory

The rate of reaction depends upon the frequency of successful collisions. Ignoring the effect of collision geometry, this will depend upon the total collision frequency and the fraction of reactant molecules which possess enough energy to react on collision, i.e. the **activated fraction**:

$$\textbf{frequency of successful collisions} = \textbf{total collision frequency} \times \textbf{activated fraction}$$

Any change in conditions which increases either the total collision frequency or the activated fraction will, therefore, lead to a faster reaction rate.

24.3 What affects total collision frequency?

There are many ways in which the total collision frequency may be increased. As Figure 24.2 shows, increasing the concentration of a solution, or the pressure of a gas, provides more particles in the same space so they will collide more often. In the case of a solid its particles are not mobile but rely upon liquid or gas particles colliding with the solid's surface. If the amount of surface can be increased, total collision frequency will also rise. Figure 24.3 shows that separate small particles have a greater surface area than the same amount of a single large particle. Powdered solids therefore react more quickly than larger lumps.

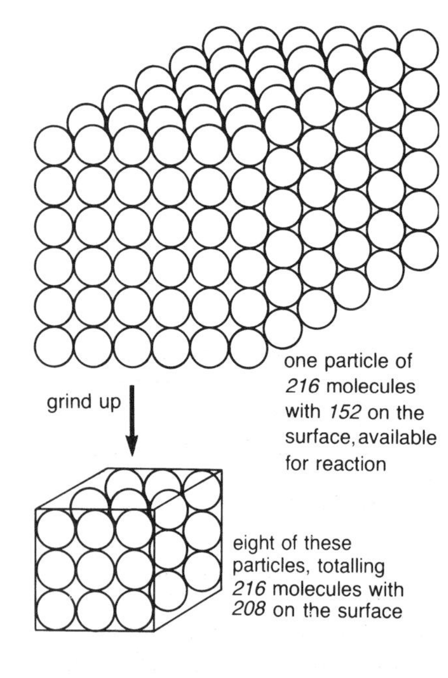

Figure 24.3 ▲
As the particles get smaller, their total surface area gets larger. Thus, more molecules are available for collision and the reaction rate increases

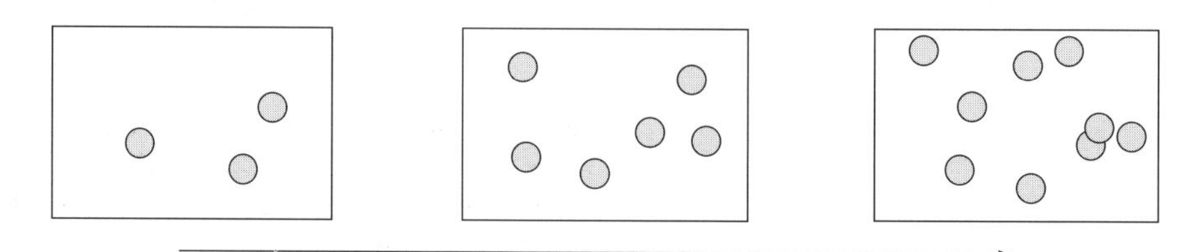

Figure 24.2 ▲

24.4 What affects the fraction of activated particles?

Remember that this is the fraction of the particles which have enough energy to react when they collide. The **total** kinetic energy of the reactant particles is proportional to the absolute temperature, T. The **distribution** of this energy between the individual molecules in a gas is given by the **Maxwell–Boltzmann** curve which has the form shown in Figure 24.4.

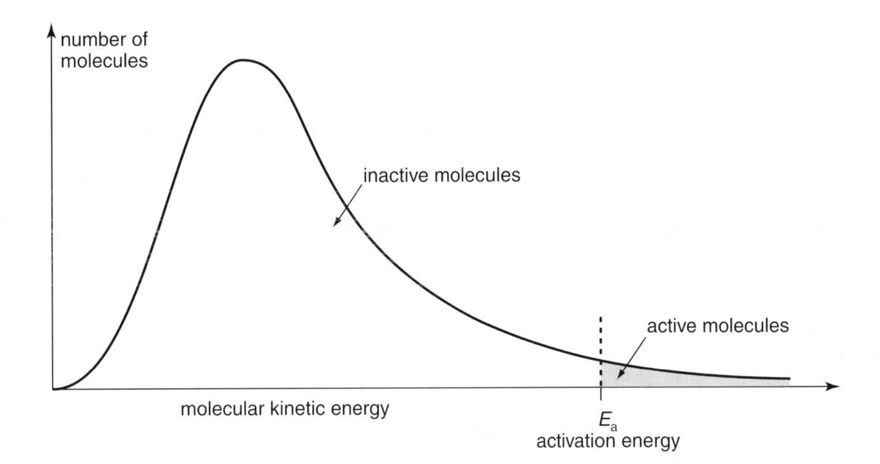

Figure 24.4 ▶ The Maxwell–Boltzmann molecular energy distribution curve for a gas

Even though energy is continually being transferred between particles when they collide, statistically the activated fraction will remain constant if the temperature is fixed. The total area under the energy distribution curve represents the total number of reactant particles, whilst that portion above the activation energy, E_a, represents the number of active particles.

The effect on the energy distribution curve of a small rise in temperature is shown in Figure 24.5.

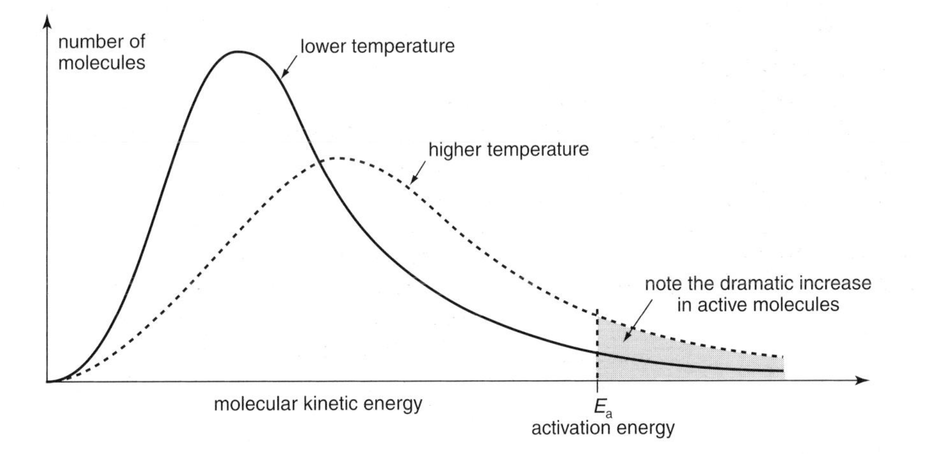

Figure 24.5 ▶ The effect on molecular energy distribution of a small temperature rise

The peak moves to the right indicating that the most common molecular energy has risen. Since the total area under the curve represents the total number of particles, this must remain unchanged. As a consequence the curve must flatten out somewhat. The important point in this case is the effect on the number of active molecules. As you can see from Figure 24.5, a small change in temperature can have a dramatic effect on the activated fraction and hence the rate of reaction. A temperature rise will, of course, also increase the velocities of the particles and so the total collision frequency will go up. However, since mean molecular velocity is proportional to the square root of the absolute temperature, a small temperature rise will increase total collision frequency only slightly and this effect may normally be ignored in comparison to the increase in the activated fraction.

Another way of increasing the activated fraction is to reduce the activation energy. To do this we must provide an alternative reaction pathway and this must involve adding an extra substance to the reaction mixture.

If the added material can be recovered chemically unchanged at the end of the reaction, it is known as a catalyst.

The effect of a catalyst on the energy changes involved in a reaction are illustrated in Figure 24.6.

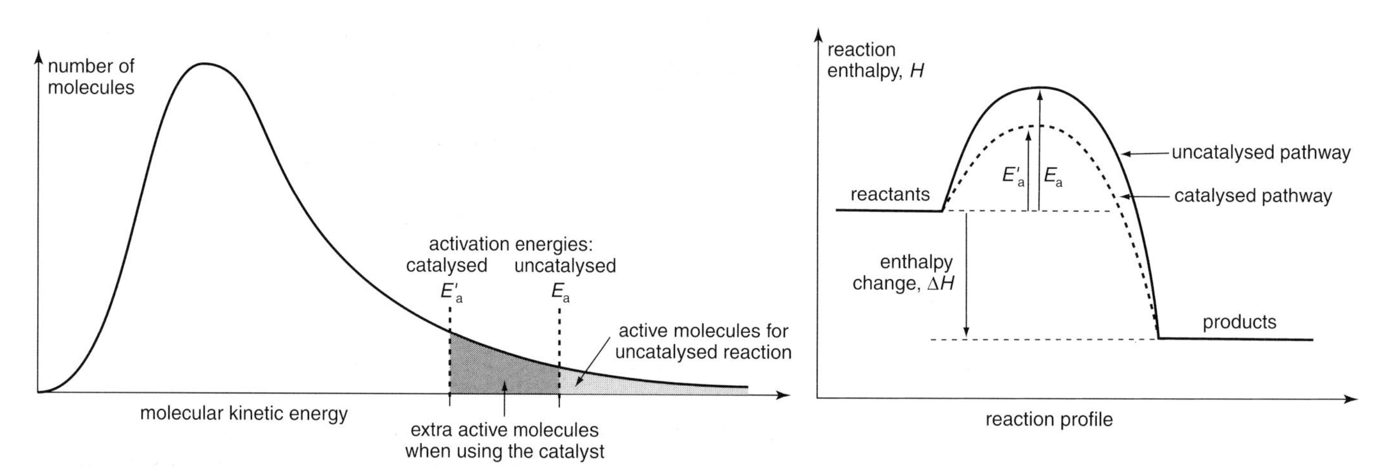

Figure 24.6
The effect of a catalyst on the energy changes involved in a reaction

There are two main ways in which a catalyst may interact with reactant particles to provide a pathway with a lower activation energy.

Heterogeneous catalysis

Here the catalyst and the reactants are in different phases, i.e. they are physically distinct. Many industrially important reactions in the gas phase use a heterogeneous solid catalyst (for example the Haber process and the contact process outlined in Chapter 23).

Only the catalyst **surface** is in contact with the reaction mixture and, as the catalyst increases the activated fraction, the collisions between reactant molecules here must require less energy for reaction than normal. We shall illustrate the general mechanism of heterogeneous catalysis using nickel in the hydrogenation of ethene into ethane:

$$H_2C{=}CH_2 + H_2 \longrightarrow H_3C{-}CH_3$$

In the first step reactant molecules are *adsorbed* onto the catalyst surface. (Note that the reactant molecules are *not absorbed*, i.e. taken inside the catalyst.) In this process reactant molecules stick to the catalyst by forming weak chemical bonds. Such **chemisorption** weakens the bonds within the reactant molecules.

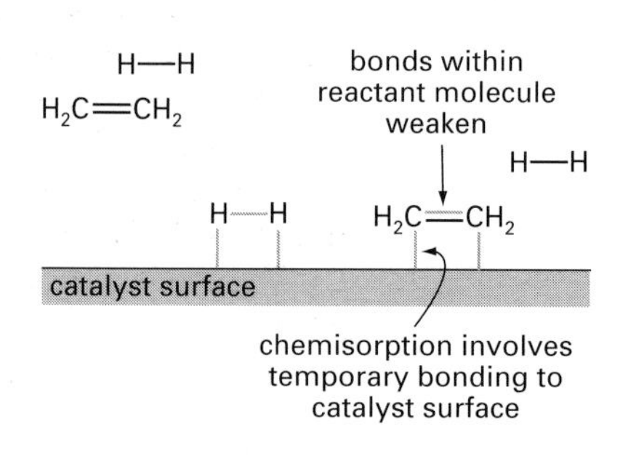

The adsorbed reactant molecules are now much more susceptible to reaction. They are still able to move over the catalyst surface and products are formed when they collide. In addition, the adsorption process effectively concentrates the reactant molecules on the catalyst surface, increasing the total collision frequency.

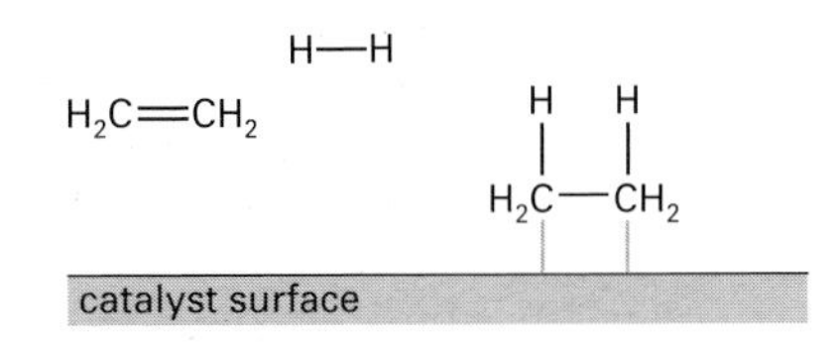

The final, and very important, step is **desorption** where the product molecule leaves the catalyst surface. If this does not happen the catalyst surface will not be available to adsorb fresh reactant molecules.

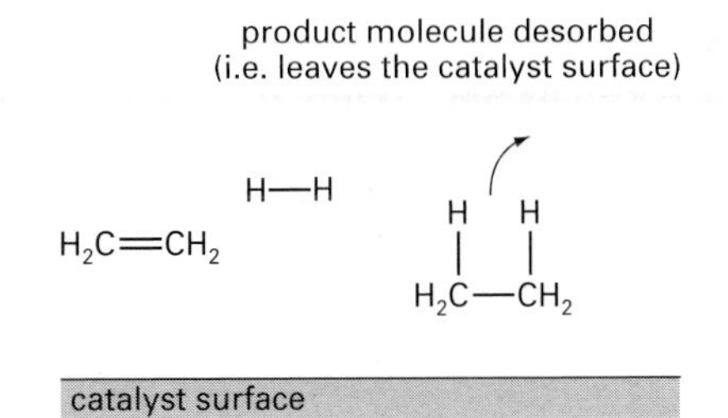

The efficiency of such a catalyst will depend upon how strongly it adsorbs reactant molecules. Up to a certain point catalytic activity increases as the strength of adsorption increases, since the bonds within the reactant molecules are further weakened. However, if the reactant molecules are adsorbed too strongly this prevents free movement across the surface and reduces the overall collision frequency.

Activity 24.1

1 Why are heterogeneous catalysts generally used in the form of fine powders or 'sponges'?

2 Heterogeneous catalysts are often very susceptible to 'poisoning'. Small amounts of certain types of impurity in a reaction mixture can drastically reduce their efficiency. How do you think such 'catalyst poisons' achieve this effect?

Homogeneous catalysis

Here the catalyst is in the same phase as the reactants, i.e. the whole reaction mixture has a uniform composition. Homogeneous catalysis may, therefore, involve reactions in the gas phase or in solution. Such catalysts operate by forming an **intermediate** with one of the reactants. The intermediate must undergo further reaction to give the product and reform the catalyst. There are few examples of industrial processes which use homogeneous catalysts although, before the development of the contact process, sulphuric acid was made by the lead chamber process.

The direct conversion of sulphur dioxide into sulphuric acid is normally impractical:

$$SO_2(g) + \tfrac{1}{2}O_2(g) + H_2O(l) \longrightarrow H_2SO_4(l)$$

Activity 24.2

1 Sulphuric acid is now made exclusively by the contact process which uses a heterogeneous catalyst rather than by the lead chamber process which operates a homogeneous catalyst. Can you suggest why heterogeneous catalysts seem to have an economic advantage over homogeneous catalysts?

2 An example of an undesirable homogeneous catalyst is offered by depletion of the ozone layer. Read through the information in the box on p.509 and identify the catalyst and intermediate involved in this process.

In the chamber process sulphur dioxide and air are mixed with nitrogen monoxide, NO, in the presence of water. A possible reaction sequence is as follows.

step 1:	$NO(g)$	+	$\frac{1}{2}O_2(g)$			$\longrightarrow$	$NO_2(g)$		
	catalyst		reactant				intermediate		
step 2:	$SO_2(g)$	+	$NO_2(g)$	+	$H_2O(l)$	$\longrightarrow$	$H_2SO_4(l)$	+	$NO(g)$
	reactant		intermediate		reactant		product		catalyst (reformed)
overall:	$SO_2(g)$	+	$\frac{1}{2}O_2(g)$	+	$H_2O(l)$	$\longrightarrow$	$H_2SO_4(l)$		

Incidentally, the action of the vanadium(V) oxide heterogeneous catalyst used in the contact process may also be explained in terms of the formation of an intermediate. The sulphur dioxide reduces vanadium(V) oxide to vanadium(IV) oxide which is then re-oxidised by the oxygen.

step 1:	$V_2O_5(s)$	+	$SO_2(g)$	$\longrightarrow$	$SO_3(g)$	+	$V_2O_4(s)$
	catalyst		reactant		product		intermediate
step 2:	$\frac{1}{2}O_2(g)$	+	$V_2O_4(s)$	$\longrightarrow$	$V_2O_5(s)$		
	reactant		intermediate		catalyst (reformed)		
overall:	$SO_2(g)$	+	$\frac{1}{2}O_2(g)$	$\longrightarrow$	$SO_3(g)$		

Enzymes

Many reactions essential to life are catalysed by complex protein molecules known as enzymes. For example, the breakdown of large indigestible starch molecules in food to give smaller sugar molecules which can be absorbed into the bloodstream is carried out rapidly at body temperature by an enzyme in saliva called amylase. Each enzyme is specific for a particular reaction and even very small changes in its structure may destroy its activity. For example, heating even to quite moderate temperatures often **denatures** the protein. This suggests that molecular shape plays an important part in enzyme action which may be likened to a **lock and key** mechanism. The surface folding of the protein is thought to provide an **active site** (the lock) which can only adsorb a **substrate** molecule of suitable shape and structure (its own particular key.) As in the case of a heterogeneous catalyst, this weakens the bonds within the substrate molecule, reducing the activation energy needed to form products. When the reaction has taken place, the products must leave the active site free for fresh incoming substrate.

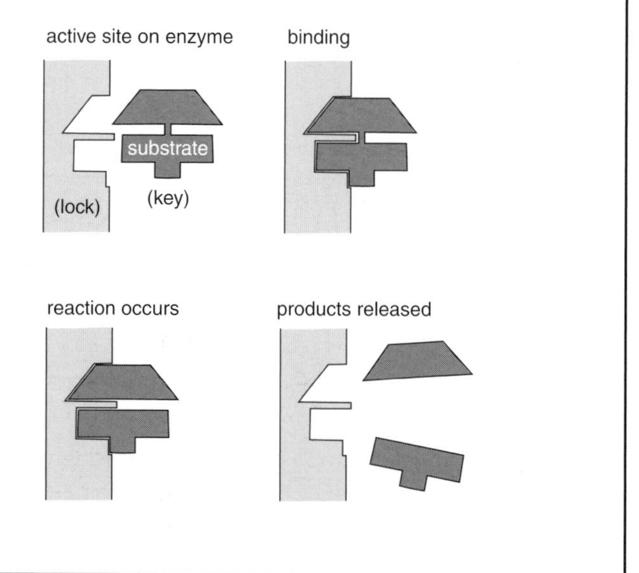

Catalytic converters

Vehicle exhaust fumes contain several dangerous pollutant gases. Under ideal conditions the fuel would undergo complete combustion to give carbon dioxide and water vapour only:

$$C_8H_{18}(g) + 12\tfrac{1}{2}O_2(g) \longrightarrow 8CO_2(g) + 9H_2O(l)$$

However, incomplete combustion leads to the production of the poisonous gas carbon monoxide, CO, and even unburnt hydrocarbons which are thought to be carcinogenic (cancer forming). At the high temperatures and pressures in an engine another problem is the combination of oxygen with nitrogen from the air to give toxic nitrogen oxides, e.g. NO and NO_2:

$$N_2(g) + O_2(g) \longrightarrow 2NO(g)$$

and

$$2NO(g) + O_2(g) \longrightarrow 2NO_2(g)$$

Many modern cars are fitted with catalytic converters which reduce both types of pollution. A typical converter consists of an expansion chamber which contains a mixture of catalysts supported on a fine honeycomb aluminium oxide mesh. Rhodium metal catalyses the **reduction** of nitrogen oxides to nitrogen, e.g.

$$2NO(g) \longrightarrow N_2(g) + O_2(g)$$

The other catalyst, usually, palladium metal, catalyses the **oxidation** of carbon monoxide and unburned hydrocarbons into carbon dioxide, e.g.

$$2CO(g) + O_2(g) \longrightarrow 2CO_2(g)$$

Note that the decomposition of nitrogen oxides by the first process provides more oxygen for this step.

Magnified section of a catalytic converter

Activity 24.3

1. Why are the catalysts supported on a fine honeycomb mesh in a converter?
2. The overall effect of a converter is to change poisonous pollutants in exhaust fumes into gases naturally present in the atmosphere. Write an overall equation for the conversion of carbon monoxide and nitrogen monoxide into 'safe' gases.
3. Use of even small amounts of leaded petrol will irreversibly destroy the efficiency of a catalytic converter. How do you think this might happen?
4. You may have noticed that the exhaust from a catalytic converter sometimes has a distinct smell of bad eggs, due to small amounts of hydrogen sulphide, H_2S. How do you account for the formation of this gas?

Focus 24a

1. Reactions result when particles collide with sufficient energy (and appropriate geometry).
2. An increase in concentration (for a solution), or pressure (for gases), increases reaction rate by increasing the total collision frequency.
3. An increase in temperature speeds up a reaction (largely) by increasing the proportion of particles that are sufficiently energetic to react on collision i.e. the **activated fraction**.
4. A catalyst speeds up a reaction by providing an alternative pathway in which the particles require less energy in order to react.
5. **Heterogeneous catalysts** are in a different phase to the reactants whilst **homogeneous** catalysts are in the same phase.
6. **Enzymes** are highly specific biological catalysts.

(a)

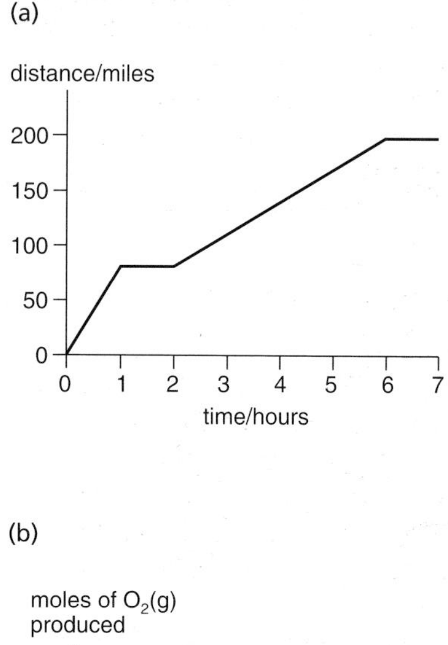

(b)

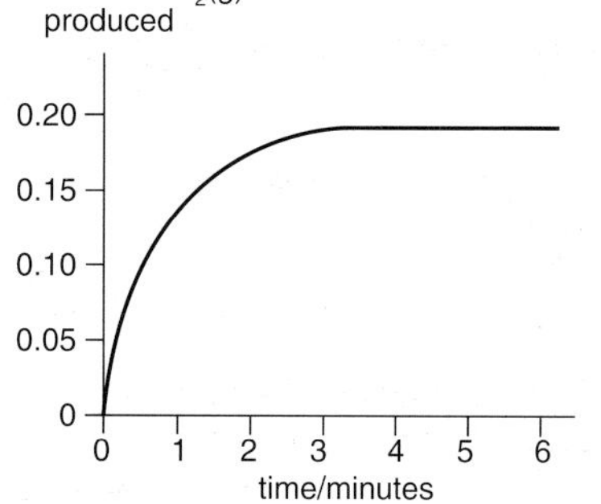

Figure 24.7 ▲
(a) Distance/time graph
(b) Amount/time graph
In both graphs the speed or rate at any given time is given by the slope of the line at that point; the steeper the slope, the faster the speed or rate

24.5 Following the progress of a reaction

Reaction rate is defined as the rate at which a specified product is formed or the rate at which a reactant is used up. It is important to specify which species is used to measure the rate as the rate at which a one substance is formed (or removed) may not be the same as that for another substance involved in the reaction. A useful analogy is the speed of a motor car during a journey.

For the car: rate of change of position (i.e. speed) $= \dfrac{\text{distance moved}}{\text{time taken}}$

For a reaction: rate of reaction $= \dfrac{\text{amount of product formed (or reactant used)}}{\text{time taken}}$

The speed of a car will be measured in units of: $\dfrac{\text{length}}{\text{time}}$, e.g. miles h^{-1}

The rate of reaction will be measured in units of: $\dfrac{\text{amount of substance}}{\text{time}}$, e.g. mol s^{-1}

If the speedometer on a car breaks, then it is possible to estimate speed by using road signs to judge the distances travelled after certain times. The graph for one such journey is shown in Figure 24.7a. Since there is no chemical 'speedometer' capable of directly measuring reaction rate, we may use a similar graphical method to follow the progress of a reaction. For the decomposition of hydrogen peroxide solution into water and oxygen gas we get a graph of the type shown in Figure 24.7b.

The overall balanced equation for this reaction is:

$$2H_2O_2(aq) \longrightarrow 2H_2O(l) + O_2(g)$$

Activity 24.4 A comparison of distance/time and amount/time graphs

Use the distance/time graph for the car journey shown in Figure 24.7a to answer the following questions.

1 What was the overall distance travelled, the total time taken and the average speed of the journey?

2 During which parts of the journey was the car:
a) stationary **b)** travelling a steady speed of about 70 mph
c) likely to have been travelling through a 30 mph speed limit zone?

To answer these questions you will have calculated the slope at various points on the graph. Use a similar method to answer the following questions about the amount/time graph for the decomposition of hydrogen peroxide.

3 What was the total number of moles of oxygen gas produced, the total time taken and the average reaction rate (in terms of oxygen formation)?

4 Calculate the average rate of reaction during each of the first three minutes of the reaction. What happens to the rate of the reaction as time goes on? Is this what you would expect? Explain your reasoning.

5 Suggest an experimental method for monitoring the number of moles of oxygen gas produced (i.e. the equivalent of 'mileposts' for the reaction).

6 How would the rate at which hydrogen peroxide is used up compare with the rate of oxygen formation?

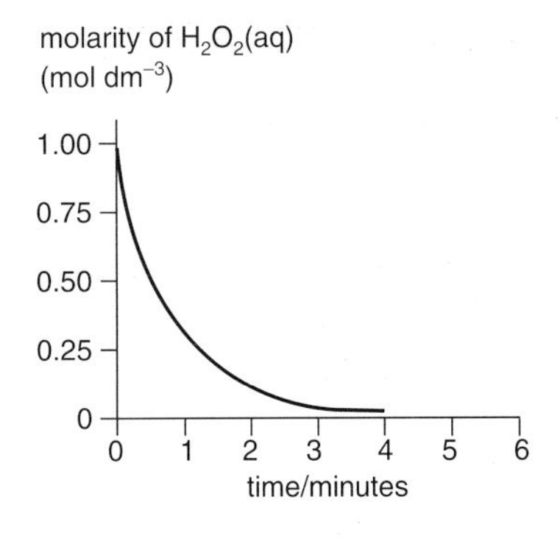

Figure 24.8
Concentration/time graph

For species in solution, any change in the number of moles present results in a proportional change in **molar concentration**. Instead of plotting moles of substance formed (or used up) we may therefore use a plot showing change in concentration with time, as shown in Figure 24.8 for the decomposition of hydrogen peroxide solution. We shall see in the next section that such graphs are particularly useful for correlating reaction rate with reactant concentration.

Other ways of monitoring reaction progress

We need to find some measurable 'property' of the system under observation which can be related to a concentration change. Ideally we should be able to monitor this property continuously without disturbing the reaction, for example measuring the volume of oxygen produced when hydrogen peroxide decomposes. Alternatively we might have chosen to carry out this reaction in a sealed vessel at constant volume. The increase in pressure could then be related to the number of moles of oxygen formed and hence to the change in concentration of the hydrogen peroxide ($pV = nRT$).

Other properties which may be monitored continuously are listed below.

Colour change

This is particularly useful when only one of the species in the reaction is coloured, e.g.

$$BrO_3^-(aq) + 5Br^-(aq) + 6H^+ \longrightarrow 3Br_2(aq) + 3H_2O(l)$$

Here the reactants are colourless and as the reaction proceeds the orange colour of bromine gradually deepens and this can be followed using a colorimeter or spectrometer.

Precipitation

If a precipitate is produced during the reaction, the solution will become cloudy, e.g.

$$S_2O_3^{2-}(aq) + 2H^+(aq) \longrightarrow S(s) + SO_2(aq) + H_2O(l)$$

Again a spectrometer may be used to follow the progress of the reaction. As the amount of suspended solid increases, the mixture will allow less light to pass through it.

Electrical conductivity

If the reactants and products have a different electrical conductivity this property will vary during the reaction. For example, in each of the two reactions above there will be a decreases in conductivity as ionic reactants are replaced by covalent products.

Optical activity

Organic compounds which contain a chiral carbon atom, i.e. one which is connected to four different groups, exist as optical isomers. If an optically active compound is produced, or removed, the change in the angle of rotation of the plane of polarised light may be followed in a polarimeter (see section 6.5).

In many cases, there is no suitable concentration-related property which may be continuously monitored. The progress of such reactions is generally followed by taking successive small **samples** and finding the concentration by chemical analysis. For example, in the following ester hydrolysis, the samples may be titrated with standard acid to give the concentration of alkali. Since the analysis may take a while, steps must be taken to stop the reaction continuing as soon as the sample is taken. This is often achieved by adding a large excess of iced water.

$$CH_3COOC_2H_5(aq) + NaOH(aq) \longrightarrow CH_3COONa(aq) + C_2H_5OH(aq)$$

Chapter 22 contains a selection of other procedures which might be useful in sample analysis.

Activity 24.5

1 Suggest ways in which the progress of each of the following reactions might be followed.

a) $CaCO_3(s) + 2HCl(aq) \longrightarrow CaCl_2(aq) + H_2O(l) + CO_2(g)$

b) $C_2H_4(g) + H_2(g) \longrightarrow C_2H_6(g)$

c) $C_6H_{12}(l) + Br_2(l) \longrightarrow C_6H_{12}Br_2(l)$

d) $2H_2S(aq) + SO_2(aq) \longrightarrow 3S(s) + 2H_2O(l)$

e) $H_2(g) + I_2(g) \longrightarrow 2HI(g)$

Note: $I_2(g)$ is purple and $HI(g)$ dissolves in water to give a strongly acidic solution.

2 Why does the addition of a large excess of iced water to a sample effectively 'stop' the reaction?

24.6 The rate equation

In this section we shall use some actual experimental data for the decomposition of hydrogen peroxide solution described above to illustrate what is meant by a **rate equation** and show how it may be established.

Experimental details

- 5 cm^3 of 2 mol dm^{-3} iron(III) chloride solution, which acts as a homogeneous catalyst for this reaction, was placed in a conical flask.
- 45 cm^3 of '2-volume' hydrogen peroxide solution (i.e. the hydrogen peroxide will produce twice its own volume of oxygen on complete decomposition) was quickly added from a measuring cylinder to the flask which was immediately connected to the gas syringe as shown in Figure 24.9. (**Note:** this makes the total volume of the mixture 50 cm^3)

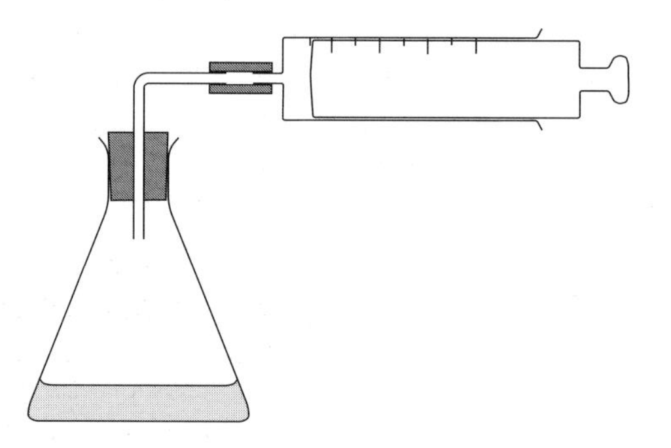

Figure 24.9
Apparatus for following the decomposition of hydrogen peroxide solution

- The volume of gas collected in the syringe was noted every half minute for 5 minutes. The apparatus was then left until the reaction had completely finished before the final volume of oxygen produced was measured. Room temperature was also noted. The results are shown in Table 24.1.

Results

Table 24.1 Data for the decomposition of 50 cm^3 of approx. 2-volume $H_2O_2(aq)$ using $FeCl_2$ as a homogeneous catalyst at room temperature

time /minutes	0	0.5	1.0	1.5	2.0	2.5	3.0	3.5	4.0	4.5	5.0
volume of gas/cm^3	0	18	35	48	58	66	72	78	81	84	87

- total volume of oxygen released when the reaction had finished = 96 cm^3
- room temperature = 21°C

Analysis of the results

The starting concentration of the hydrogen peroxide solution may be calculated from the total volume of oxygen released at the end of the reaction:

Assuming ideal gas behaviour, $pV = nRT$ so moles of gas, $n = pV/RT$.
Assuming normal atmosphere pressure i.e. about 101 KPa (check with a barometer for accurate work)
$n = pV/RT = (101\,000 \times 0.000096)/(8.31 \times 294) = 0.00397$
From the balanced equation for the reaction:

$$2H_2O_2(aq) \longrightarrow 2H_2O(l) + O_2(g)$$

$0.00397 \times 2 = 0.00794$ moles of H_2O_2 must have been used.
Since the volume of solution taken was 50 cm^3, the starting concentration of $H_2O_2(aq)$ in the reaction mixture is
$0.00794 \times 1000/50 = 0.159$ mol dm^{-3}.

As the reaction proceeds, hydrogen peroxide is used up and so its concentration falls. Since we have shown that 96 cm^3 oxygen corresponds to a fall in reactant concentration of 0.159 mol dm^{-3}, we may use simple ratio to calculate the concentration of $H_2O_2(aq)$ left in solution after any particular volume, v cm^3, of oxygen has been produced.

$$[H_2O_2(aq)] = 0.159 - \frac{0.159v}{96}$$

Using the experimental data in Table 24.1, we may produce a concentration–time table (Table 24.2).

Table 24.2 Data for the decomposition $H_2O_2(aq)$ showing the variation in reagent concentration

time/minutes	0	0.5	1.0	1.5	2.0	2.5	3.0	3.5	4.0	4.5	5.0
volume of gas/cm^3	0	18	35	48	58	66	72	78	81	84	87
$[H_2O_2(aq)]$/mol dm^{-3}	0.159	0.129	0.101	0.080	0.063	0.050	0.040	0.030	0.025	0.020	0.015

By measuring the slope of the line at various points on the graph of concentration against time shown in Figure 24.10, we can estimate the rate of decomposition of hydrogen peroxide at various concentrations.
At the start of the reaction $[H_2O_2(aq)] = 0.16$ mol dm^{-3}

$$\text{The slope of the graph at this point} = \frac{0.16 \text{ mol dm}^{-3}}{2.1 \text{ min}}$$

$$= 0.076 \text{ mol dm}^{-3} \text{ min}^{-1}$$

Similar calculations at other points on the graph give the following results

$[H_2O_2(aq)]$/mol dm^{-3}	0.16	0.12	0.08	0.04
rate/mol dm^{-3} min^{-1}	0.076	0.049	0.033	0.019

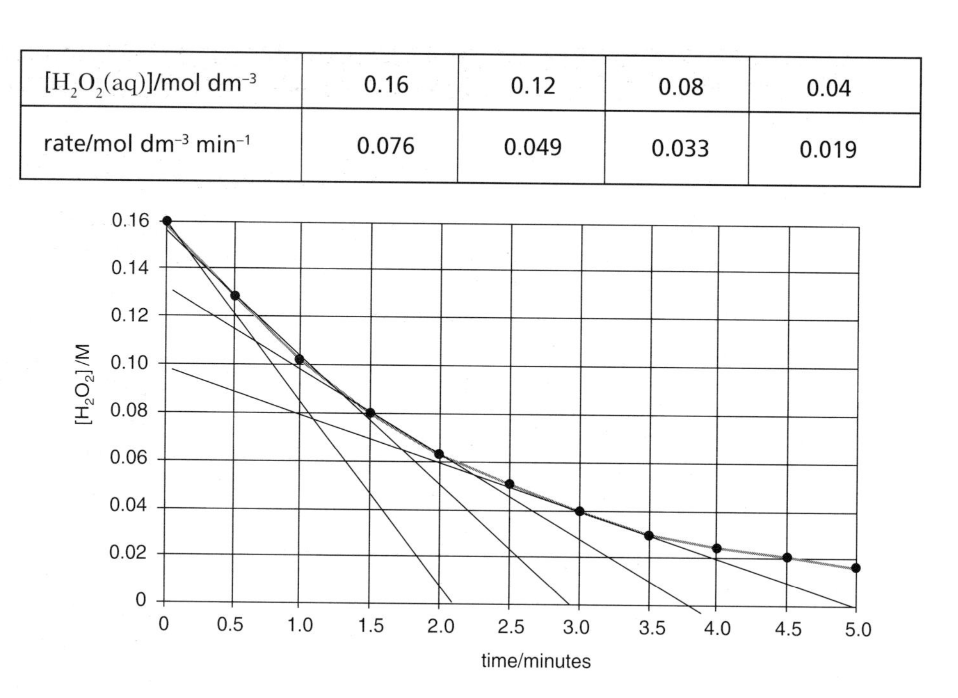

Figure 24.10 ▶ Concentration–time graph for hydrogen peroxide decomposition

The relationship between the concentration and rate of decomposition of hydrogen peroxide is shown by the rate–concentration graph in Figure 24.11. Within the limits of experimental error this gives a straight line, showing that the rate of decomposition of hydrogen peroxide is directly proportional to its concentration.

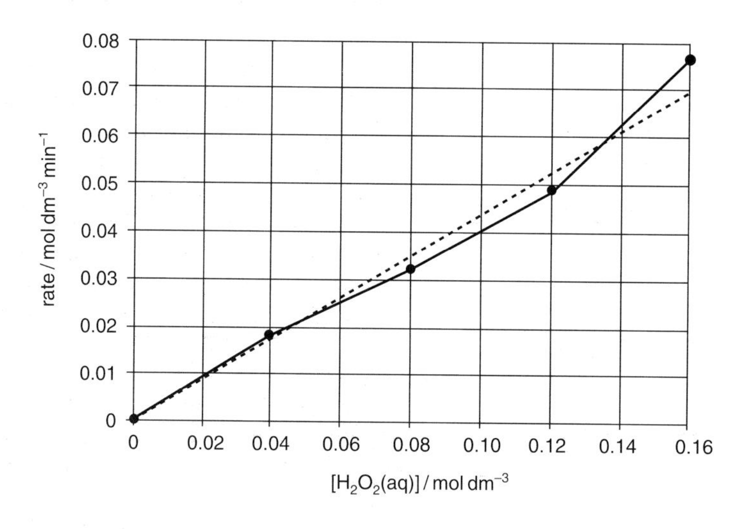

Figure 24.11 ▶ Rate–concentration graph for the decomposition of hydrogen peroxide

So we may write the **rate equation** as

$$\text{rate of decomposition of } H_2O_2(aq) = k[H_2O_2(aq)]$$

where k is a constant known as the **rate constant** for the reaction.

Since $k = \dfrac{\text{rate}}{\text{concentration}}$, its value is given by the slope of the graph, so:

$$k = \frac{0.070 \text{ mol dm}^{-3} \text{ min}^{-1}}{0.16 \text{ mol dm}^{-3}} = 0.44 \text{ min}^{-1}$$

Thus we have shown that under the conditions of the experiment the rate of decomposition of hydrogen peroxide at any concentration is given by,

$$\text{rate} = 0.44[H_2O_2(aq)]$$

Finding the rate equation by the initial rates method

Measuring rates from a concentration–time graph involves drawing a graph and measuring the slopes of tangents to the curve at various points. This can be quite a tricky operation, especially towards the end of the reaction when it is slowing down. This problem can be avoided by measuring the **initial rates** in a series of experiments where the starting concentration of the reactant is varied. Since it is only the initial rate we want, there is no need to plot concentration–time graphs, indeed all we need to do is measure the times taken to achieve the same (small) change in reactant concentration.

In the case of the hydrogen peroxide decomposition we could measure the time taken to produce, say, 10 cm^3 of oxygen using the original reactant concentration (this corresponds to a fall in $[H_2O_2(aq)]$ of 0.0166 mol dm^{-3}). Since the reactant concentration changes only slightly, the rate will be roughly constant over this period and the initial rate is found simply by dividing the change in concentration by the time taken. Solutions of known (lower) concentrations are prepared by dilution of the original with water and the experiment is repeated.

Activity 24.6

The following data was obtained by repeating the above experiment using various different starting concentrations of hydrogen peroxide and in each case measuring the time taken to collect 10 cm^3 of oxygen gas.

$[H_2O_2]$/mol dm^{-3}	0.16	0.12	0.08	0.04
Time taken to collect 10 cm^3 oxygen/s	14	19	28	57

1. Assuming that the production of 10 cm^3 of oxygen results in a fall in hydrogen peroxide concentration of 0.0166 mol dm^{-3}, calculate the initial rate of decomposition of hydrogen peroxide in each of these experiments in mol dm^{-3} min^{-1}.
2. Plot a graph of initial reaction rate against starting concentration of hydrogen peroxide and hence calculate a value for the rate constant under these conditions.

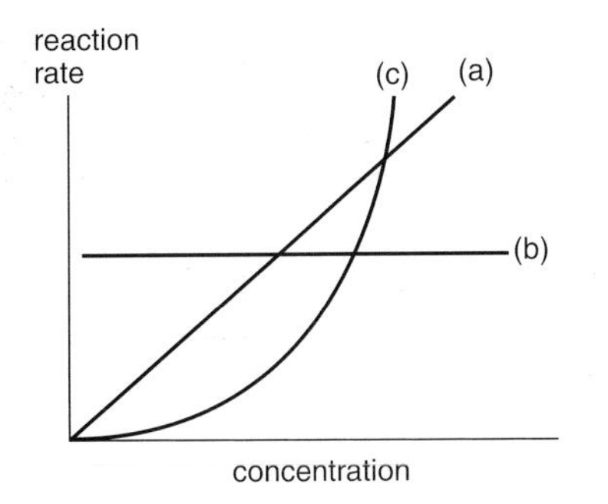

Figure 24.12
Examples of rate–concentration graphs

Not all reactions have such a simple rate equation as the decomposition of hydrogen peroxide. Some other possibilities for rate–concentration graphs are shown in Figure 24.12.

Curve (a) represents the kind of rate equation we have seen for the decomposition of hydrogen peroxide solution, i.e. rate = $k[\text{reactant}]^1$.

In the case of line (b), the reaction rate is independent of concentration (over a wide range – obviously there must be a lower concentration limit for this behaviour, i.e. no reaction is possible when the concentration is extremely low). The rate equation here takes the form rate = k, i.e. rate = $k[\text{reactant}]^0$.

This introduces the idea of **order of reaction**. *The order with respect to any reactant is simply the power to which its concentration is raised in the rate equation*. Thus for line (a) the order of reaction is 1, i.e. it is **first order** with respect to the reactant concerned. In the case of line (b), the rate is independent of concentration. The reaction is **zero order** with respect to the reactant.

Line (c) shows a reaction whose rate increases steeply with reactant concentration, i.e. whose order is greater than 1. One way of finding the actual order is to plot the rate against different powers of reactant concentration and see which gives a straight line. If the order is not a whole number, many trials may be necessary to discover it. Graphical methods for direct determination of order may be found in more advanced textbooks.

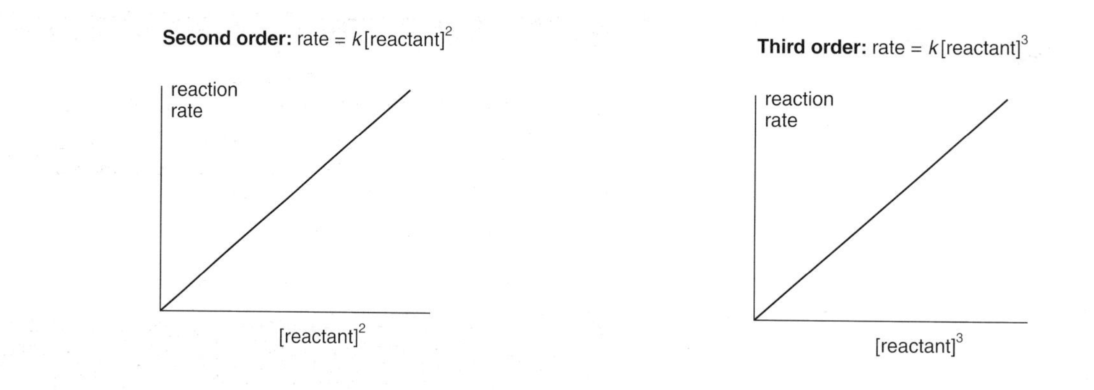

Activity 24.7 Finding order by comparison of initial rates

By careful choice of reactant concentrations we can find the order simply by comparing initial reaction rates. Tables 24.3(a)–(d) show how the initial rates of several reactions depend upon the concentration of the reactants.

Table 24.3
(a)

[A]/mol dm^{-3}	0.50	1.00	1.50	2.00
initial rate/mol dm^{-3} min^{-1}	0.21	0.42	0.63	0.84

(b)

[B]/mol dm^{-3}	0.50	1.00	1.50	2.00
initial rate/mol dm^{-3} min^{-1}	0.13	0.13	0.13	0.13

(c)

[C]/mol dm^{-3}	0.50	1.00	1.50	2.00
initial rate/mol dm^{-3} min^{-1}	1.00	4.00		

(d)

[D]/mol dm^{-3}	0.50	1.00	1.50	2.00
initial rate/mol dm^{-3} min^{-1}	1.00			

In the case of reaction (a), you can see that doubling the molar concentration of the reactant A from 0.50 mol dm^{-3} to 1.00 mol dm^{-3}, doubles the initial rate from 0.21 mol dm^{-3} min^{-1} to 0.42 mol dm^{-3} min^{-1}. Doubling the reactant concentration again to 2.00 mol dm^{-3} also doubles the rate. So, in this case:

rate α $[A]^1$, i.e. the reaction is 1st order with respect to reactant A.

1 Write the full rate equation for this reaction and calculate the rate constant, including appropriate units, from the information given.

In reaction (b), changing the molar concentration of reactant B has no effect on initial rate. Thus rate is independent of reactant concentration.

rate α $[B]^0$, i.e. the reaction is zero order with respect to reactant B.

2 What is the rate constant for this reaction, including units?

In reaction (c) doubling the concentration of reactant C, e.g. from 0.50 mol dm^{-3} to 1.00 mol dm^{-3} increases the initial rate from 1.00 mol dm^{-3} min^{-1} to 4.00 mol dm^{-3} min^{-1}, i.e. by a factor of 4 or 2^2. Here,

rate α $[C]^2$, i.e. the reaction is 2nd order with respect to reactant C.

Activity 24.7 Finding order by comparison of initial rates ***continued***

3. Complete the missing initial rates in Table 24.3c and calculate the rate constant, including units, under the conditions used.
4. Reaction (d) is 3rd order with respect to reactant D. Complete Table 24.3d to show the initial rates at all the concentrations shown and calculate a value for the rate constant, including units.

Half-lives – a quick way of identifying first order reactions

The **half-life** of a reaction is simply the time taken for the concentration of a reactant to fall to half its starting value. From Figure 24.10 you can see that it takes 1.5 minutes for the concentration of hydrogen peroxide to fall from its starting value of 0.16 mol dm^{-3} to 0.08 mol dm^{-3}. It also takes the same time for the concentration to halve again to 0.04 mol dm^{-3} and likewise to 0.02 mol dm^{-3}. This reaction therefore has a **constant** half-life of 1.5 minutes, regardless of the starting concentration. *This is only true for first order reactions and provides a convenient short-cut for identifying such cases.*

24.7 Rate equations for reactions with more than one reactant

So far we have assumed that the rate of reaction depends upon the concentration of only one reactant. In most cases however, there are two or more reactants involved, all of which can have a bearing on reaction rate. Thus for a reaction with several reactants:

$$A + B + C + \ldots \longrightarrow \text{products}$$

the rate equation is of the type:

$$\text{rate} = k[A]^a[B]^b[C]^c\ldots$$

where a, b, c... are the orders of the reaction with respect to the individual reactants A, B, C.... The **overall order** of such a reaction is given by the sum of all the individual orders, i.e.

$$\text{overall order of reaction} = a + b + c + \ldots$$

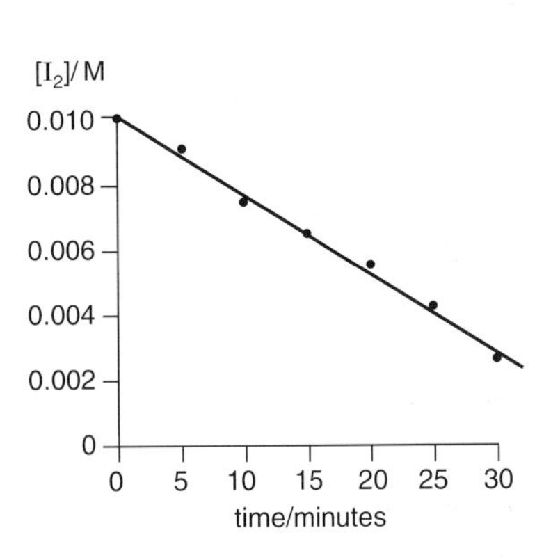

Figure 24.13
Experiment (a) excess of propanone

From a practical point of view it is more difficult to establish the rate equation if there are two or more reactants. In such a case the concentrations of **all** the reactants fall as the reaction proceeds and the problem is to identify the change in rate caused by the change in concentration of *each reactant separately*. To illustrate one way in which this may be done we shall look at some actual experimental results for the reaction between propanone and iodine in aqueous solution.

$$CH_3COCH_3(aq) + I_2(aq) \longrightarrow CH_3COCH_2I(aq) + HI(aq)$$

Iodine is the only strongly coloured species, so we may follow the progress of this reaction by measuring its concentration using a colorimeter. For the first experiment we start with a mixture containing a large excess of propanone and follow the change in concentration of iodine with time. The results of this are shown in Table 24.4a and graphically in Figure 24.13.

Table 24.4 Kinetic data for the reaction between propanone and iodine at room temperature

(a)

Time/min	0	5	10	15	20	25	30
$[I_2]$/mol dm^{-3}	0.0100	0.0091	0.0077	0.0066	0.0055	0.0047	0.0034
$[CH_3COCH_3]$ /mol dm^{-3}	0.250	(0.249)	(0.248)	(0.247)	(0.245)	(0.245)	(0.243)

(b)

Time/min	0	5	10	15	20	25	30
$[I_2]$/mol dm^{-3}	0.0100	0.0096	0.0091	0.0086	0.0079	0.0073	0.0067
$[CH_3COCH_3]$ /mol dm^{-3}	0.125						

(c)

Time/min	0	5	10	15	20	25	30
$[I_2]$/mol dm^{-3}	0.0100	0.0095	0.0091	0.0088	0.0084	0.0082	0.0080
$[CH_3COCH_3]$ /mol dm^{-3}	0.065						

As you can see from Table 24.4(a), because the initial concentration of propanone was much greater than that of iodine, in relative terms it hardly changes during the course of the reaction. This means that any change in the rate must be due only to change in the iodine concentration. Since the graph of $[I_2]$ against time shown in Figure 24.13 is a straight line, its gradient, i.e. the rate at which the iodine is used up, is constant. Since the rate does not change as the concentration of iodine falls it must be **zero order** with respect to iodine.

The concentration–time graphs for all three experiments are compared in Figure 24.14.

Activity 24.8

1 Calculate a value for the rate constant for the reaction between iodine and propanone at room temperature from the data above.

2 Predict the rate at which iodine would be used up if the experiment were repeated as above except with an initial propanone concentration of 0.65 mol dm^{-3}.

3 Sketch the shape of the graph of $[I_2]$ against time that you would expect if the reaction had been carried out with equal starting concentrations of iodine and propanone. Explain your answer.

4 What is the overall order of the reaction between iodine and propanone?

$[I_2]$/ M

0.010

0.008

0.006

0.004

0.002

0

0 5 10 15 20 25 30

time/minutes

(c) $[CH_3COCH_3]$ = 0.0625 M

(b) $[CH_3COCH_3]$ = 0.125 M

(a) $[CH_3COCH_3]$ = 0.250 M

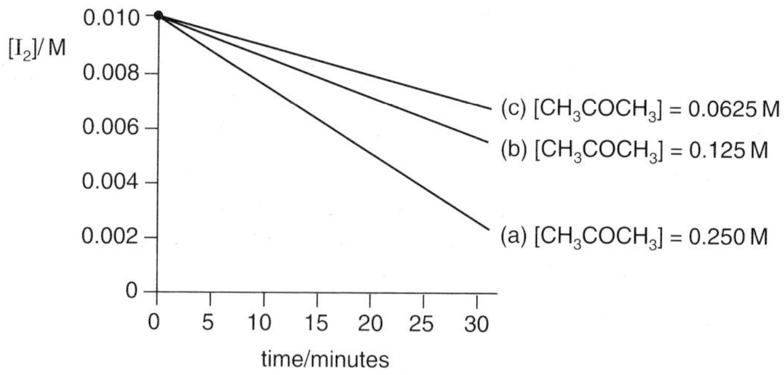

Figure 24.14

As the starting concentration of propanone is reduced, the slope of the graph, and hence the rate at which iodine is used up, also falls.

$[CH_3COCH_3]$/mol dm^{-3}	0.250	0.125	0.0625
rate of removal of I_2/mol dm^{-3} min^{-1}	0.022	0.011	0.006

You can see that doubling the initial concentration of propanone also doubles the rate at which iodine is used up, i.e. the reaction is **first order** with respect to propanone, i.e.

$$\text{rate at which iodine is used up} \propto [CH_3COCH_3]$$

Can we predict the order of a reaction from the chemical equation? E

The short answer to this question is no. Although, as we shall see in the following chapters, we can write the expression for an equilibrium constant simply by looking at the balanced equation for a reversible reaction, it is not possible to confidently predict orders in the same way.

The problem is that most reactions take place in a series of stages rather than in a single step. Again let us take a traffic analogy, this time using accidents involving cars and pedestrians in a city centre. An accident involving a single car and a single pedestrian might reasonably be assumed to occur as a result of a single collision.

$$1 \text{ car} + 1 \text{ pedestrian} \longrightarrow \text{accident (a)}$$

Doubling the number of cars on the road, or doubling the number of pedestrians might each be expected to double the chances of an accident. We would (correctly) expect this 'reaction' to be 'first order' with respect to both cars and pedestrians.

$$\text{accident (a) rate} = k[\text{cars}]^1[\text{pedestrians}]^1$$

However, some accidents involve two vehicles and a pedestrian.

$$2 \text{ cars} + 1 \text{ pedestrian} \longrightarrow \text{accident (b)}$$

It is tempting to write the 'rate expression' for this as follows:

$$\text{accident (b) rate} = k[\text{cars}]^2[\text{pedestrians}]^1$$

However, in this case we would almost certainly be wrong since the overall 'reaction' is most unlikely to proceed via a single step. The chances of a pedestrian being hit **simultaneously** by two cars is extremely remote. It is much more likely that one car will hit a pedestrian first and that a second car following too close behind will run into them shortly afterwards. In other words, we have a two-stage reaction.

step 1: $1 \text{ car} + 1 \text{ pedestrian} \longrightarrow \text{accident(a)}$
step 2: $\text{accident(a)} + 1 \text{ car} \longrightarrow \text{accident(b)}$

Whilst we can predict the orders of each of the individual steps, we cannot say what the overall order of the reaction will be. This is because the overall rate is governed by the slowest or **rate determining** step. Thus if **every** accident involved a second car the first of the above steps would be rate determining and the orders with respect to both cars and pedestrians would be one.

The **molecularity** of a reaction is the total number of particles involved in the rate determining step. In the above traffic example the molecularity of the accident (a) 'reaction' would be 2, i.e. 1 car and 1 pedestrian.

Once the orders of a reaction have been determined by experiment, we may use them to suggest possible reaction mechanisms. Look back to the reaction between propanone and iodine:

$$CH_3COCH_3(aq) + I_2(aq) \longrightarrow CH_3COCH_2I(aq) + HI(aq)$$

which has the rate equation:

$$\text{rate} = k[CH_3COCH_3]^1[I_2]^0$$

Since the iodine concentration has no effect on rate, I_2 cannot be involved in the rate determining step. The propanone is first slowly converted into an isomeric unsaturated alcohol by $H^+(aq)$ ions acting as a catalyst.

$$CH_3COCH_3 \longrightarrow CH_3C(OH){=}CH_2$$
(slow, therefore rate-determining)

Iodine then rapidly converts this intermediate into the final product:

$$CH_3C(OH){=}CH_2 + I_2 \longrightarrow CH_3COCH_2I + HI \text{ (fast).}$$

Focus 24b

1. The rate of any equation may be expressed in the form rate = $k[A]^x[B]^y$ ... etc. where the terms in square brackets stand for the molar concentrations of reactant, k is the **rate constant** and x and y are the **orders of reaction** with respect to the individual reactants.
2. The rate equation for any particular reation is dependent upon its mechanism and can only be established experimentally.
3. **Molecularity** is the number of particles involved in the rate determining any step of a reaction. In the case of a single-step reaction, molecularity equals the overall order.
4. A first order reaction may be recognised by its constant half-life, i.e. the time taken for the concentration of the reactant to fall to half its starting value is fixed at any given temperature.

Questions on Chapter 24

1 For a reaction involving two substances A and B forming one or more products, the rate is given by the following general expression:

$$\text{rate} = k[A]^a[B]^b$$

a) What name is given to such an expression?
b) What does [A] represent?
c) What name is given to k?
d) What is the term a known as?
e) What is the sum, $a + b$, known as?
f) Assuming A is nitrogen gas, B is hydrogen gas and the overall equation is $N_2(g) + 3H_2(g) \longrightarrow 2NH_3(g)$, can you predict the values of a and b? Explain your answer.

2 Thiosulphate ions decompose in aqueous acid solution to give sulphur and sulphur dioxide according to the equation:

$$S_2O_3^{2-}(aq) + 2H^+(aq) \longrightarrow S(s) + SO_2(aq) + H_2O(l)$$

The initial rate of this reaction may be estimated by timing how long it takes to produce a certain amount of sulphur.
a) Suggest an experimental method that would allow you to compare the times taken to produce the same amount of sulphur under different reaction conditions.

The following results were obtained in a series of experiments all carried out at a temperature of 20°C.

initial concentration of sodium thiosulphate/mol dm^{-3}	initial concentration of HCl(aq)/mol dm^{-3}	time taken to produce a set amount of sulphur/seconds
0.10	0.50	120
0.20	0.50	60
0.30	0.50	40
0.20	1.00	59
0.20	1.50	60

b) To the nearest whole number, deduce the order of reaction with respect to each of the reactants.
c) If possible, calculate the rate constant under these conditions from the above data. If this is not possible, explain why.
d) How, and why, would you expect the reaction times to vary if the experiments were repeated at 30°C rather than 20°C?
e) In the light of your answers so far explain why the reaction between thiosulphate and aqueous acid cannot occur via the single step shown in the overall balanced equation above. What can you deduce about the mechanism for this reaction?

3 Benzenediazonium chloride $C_6H_5N_2Cl$ decomposes in aqueous solution according to the following equation:

$$C_6H_5N_2Cl(aq) + H_2O(l) \longrightarrow C_6H_5OH(aq) + N_2(g) + HCl(aq)$$

a) Suggest an experimental method for following this reaction.

At 60°C the concentration of benzenediazonium chloride was found to change with time as shown in the table.

Questions on Chapter 24 *continued*

$[C_6H_5N_2Cl]$/mol dm^{-3}	0.080	0.058	0.043	0.032	0.024	0.017	0.008
time/minutes	0	2	4	6	8	10	15

b) Draw a graph of these results and use it to estimate the times taken for the concentration of benzenediazonium chloride to drop to ½ and ¼ of its initial value. What information does this give you about the kinetics of the reaction?

c) Write the rate equation for the reaction and use it to calculate a value for the rate constant at 60°C.

d) At temperatures below about 5°C, benzenediazonium chloride solution appears to be quite stable. How do you account for this?

4 Pure zinc reacts quite slowly with dilute sulphuric acid as follows:

$$Zn(s) + H_2SO_4(aq) \longrightarrow ZnSO_4(aq) + H_2(aq)$$

If a little copper sulphate solution is added, the reaction speeds up considerably but the blue colour of the copper sulphate solution slowly fades and a brownish solid is precipitated.

a) Is the copper sulphate solution acting as a catalyst in the reaction between zinc and sulphuric acid? Explain your answer.

b) Write an equation for a reaction which would account for the fading of the blue colour of the copper sulphate solution and the appearance of the brown precipitate.

5 Discuss the following statements as fully as you can in terms of the collision theory of reaction rates.

a) A reaction involving gases usually takes place more quickly if the pressure is increased.

b) The rate of a reaction usually falls as time goes on.

c) Small increases in temperature may produce dramatic increases in reaction rate.

d) Some substances may increase the rate of reaction yet be recovered chemically unchanged afterwards.

e) Solid reactants in industrial processes are often ground into a fine powder before use.

f) It is not possible to predict the orders of a reaction simply by looking at the overall balanced equation.

6 Sulphur dioxide is oxidised by air in the contact process at 450°C in the presence of vanadium(V) oxide:

$$2SO_2(g) + O_2(g) \rightleftharpoons 2SO_3(g)$$

a) The equilibrium yield of sulphur trioxide decreases as the temperature rises. Why is it necessary to use a temperature of 450°C in this industrial process?

b) Explain the function of the vanadium(V) oxide in the contact process.

c) Sulphur trioxide is also produced from sulphur dioxide during the following sequence of reactions in the atmosphere and is thought to contribute to the problem of acid rain:

$$2NO(g) + O_2(g) \longrightarrow 2NO_2(g)$$

$$SO_2(g) + NO_2(g) \longrightarrow SO_3(g) + NO(g)$$

In what ways is the function of the nitrogen monoxide, NO, similar to that of the vanadium(V) oxide in the contact process and in what ways is it different?

7 The following data was obtained from a series of experiments designed to find the rate expression for the oxidation of NO to NO_2 at a fixed temperature.

total initial pressure of NO and O_2/atm	3.00	2.00	1.50
initial partial pressure of O_2/atm	1.00	1.00	0.50
initial partial pressure of NO/atm	2.00		
initial reaction rate/atm s^{-1}	12.00	3.00	1.50

a) Copy and complete the missing partial pressures in the table above.

b) Deduce the orders of reaction with respect to each of the reactants

c) What is the overall order of this reaction?

d) Calculate a value for the rate constant at this temperature, clearly stating any units.

8 Ethyl ethanoate reacts with sodium hydroxide solution as follows:

$$CH_3COOC_2H_5(aq) + NaOH(aq) \longrightarrow CH_3COONa(aq) + C_2H_5OH(aq)$$

a) Suggest how samples taken from the mixture might be analysed to follow the progress of the reaction.

When a mixture containing an equal number of moles of ethyl ethanoate and sodium hydroxide was kept at 80°C, the concentration of ethyl ethanoate varied with time as follows:

$[CH_3COOC_2H_5]$ /mol dm^{-3}	0.80	0.45	0.33	0.27	0.22	0.16
time/minutes	0	10	20	30	40	60

b) Plot a graph of these results and use it to complete the following table

$[CH_3COOC_2H_5]$/mol dm^{-3}	0.80	0.60	0.40	0.20
rate at which $CH_3COOC_2H_5$ is used up/mol dm^{-3} min^{-1}				

c) By plotting a suitable graph deduce the overall order of the reaction.

d) How would you modify the starting mixture to determine the order of reaction with respect to sodium hydroxide alone?

Comments on the activities

Activity 24.1

1 Powders and sponges have a large surface area available for catalyst action, i.e. to adsorb reactant molecules.

2 Heterogeneous catalysts are poisoned by any substance which permanently affects its surface properties. This may occur when an impurity is strongly and irreversibly adsorbed onto its surface.

Activity 24.2

1 Since heterogeneous catalysts are not in the same phase as the reaction mixture, separation from the products does not present a problem. It is usually far more difficult to separate a homogeneous catalyst. Indeed one of the main drawbacks of sulphuric acid made by the chamber process is that it contains some nitric acid produced when oxides of nitrogen dissolve in the water.

2 The homogeneous catalyst is the chlorine radical, Cl•, which may be produced by the breakdown of CFCs. This reacts with ozone to form •OCl which acts as an intermediate.

$O_3(g)$	+	Cl•(g)	$\longrightarrow$	$O_2(g)$	+	•OCl(g)
reactant		catalyst		product		intermediate
$O_3(g)$	+	•OCl(g)	$\longrightarrow$	$2O_2(g)$	+	Cl•
reactant		intermediate		product		catalyst (reformed)

overall:

$2O_3(g) \longrightarrow 3O_2(g)$

Activity 24.3

1 The fine mesh provides a large surface area for contact between the catalysts and the exhaust gases.

2 $2NO(g) + 2CO(g) \longrightarrow N_2(g) + CO_2(g)$

3 Lead acts as a catalyst 'poison' by coating the surface and so disturbing the adsorption of reactant molecules.

4 Petrol may contain trace amounts of sulphur compounds. When the fuel is burned, this is converted into sulphur dioxide, SO_2. The rhodium catalyst in the converter can reduce this to hydrogen sulphide, H_2S.

Activity 24.4

Depending upon your reading of the graph, the answers you get may differ slightly from those below.

1 The overall journey time is ≈ 6 hours since after this time no further distance is covered. The overall distance travelled ≈ 190 miles so the average speed for the journey ≈ 190/6, i.e. 32 mph.

2 The car was:

a) stationary from 1 to 2 hours into the journey (distance travelled did not change),

b) travelling a steady speed of about 70 mph during the first hour,

c) likely to have been travelling through a 30 mph speed limit zone from 3 hours until the end of the journey as ≈ 120 miles covered in ≈ 4 hours so speed ≈ 120/4, i.e. about 30 mph.

3 Total number of moles of $O_2(g)$ formed ≈ 0.200 in 4 minutes (no further reaction occurs after this time) so overall rate ≈ 0.200/4, i.e. 0.050 mol min^{-1}.

4 Moles of $O_2(g)$ produced during:
1st minute ≈ 0.13 mol
2nd minute ≈ 0.04 mol
3rd minute ≈ 0.02 mol
We should expect the rate to fall as time passes. Reactant molecules are used up, reducing its concentration and lowering the overall collision frequency.

5 The volume of oxygen gas given off during this reaction may be measured using a gas syringe. The volume of gas is proportional to the number of moles of oxygen produced ($pV = nRT$).

6 The balanced equation for the reaction shows that 2 moles of hydrogen peroxide are used up for every mole of oxygen gas produced. Hydrogen peroxide is therefore used up twice as quickly as oxygen is formed.

Activity 24.5

1 Suggest ways in which the progress of each of the following reactions might be followed.

a) $CaCO_3(s) + 2HCl(aq) \longrightarrow CaCl_2(aq) + H_2O(l) + CO_2(g)$
Measure the volume of gas given off at constant pressure, or monitor the increase in pressure at constant volume. If the calcium carbonate is in the form of a fine suspension it would also be possible to follow the reduction in turbidity as it dissolves. Samples could be titrated with standard sodium hydroxide solution to determine the concentration of hydrochloric acid.

b) $C_2H_4(g) + H_2(g) \longrightarrow C_2H_6(g)$
Here two gas molecules react to produce only one. There will be a volume reduction at constant pressure, or a fall in pressure at constant volume.

c) $C_6H_{12}(l) + Br_2(l) \longrightarrow C_6H_{12}Br_2(l)$
Here only bromine is coloured so the reaction may be followed using a spectrometer or colorimeter.

d) $2H_2S(aq) + SO_2(aq) \longrightarrow 3S(s) + 2H_2O(l)$
A precipitate of solid sulphur will make the mixture progressively more cloudy.

e) $H_2(g) + I_2(g) \longrightarrow 2HI(g)$
There is no change in the number of moles of gas but the purple colour of the original mixture will fade as the iodine reacts. Samples could also be taken and:

- the concentration of iodine found by titration with sodium thiosulphate solution (see, for example, Activity 22.8), or,
- the concentration of HI found by dissolving in water and titrating with standard alkali.

Comments on the activities *continued*

2 The reaction rate is dramatically reduced for two reasons. A large excess of water reduces the reactant concentrations and the ice reduces the temperature. The amount of each substance present in the sample is therefore 'frozen' at the time of adding the iced water.

Activity 24.6

1

$[H_2O_2]$/mol dm^{-3}	0.16	0.12	0.08	0.04
time taken to collect 10 cm^3 oxygen/s	14	19	28	57
time taken to collect 10 cm^3 oxygen/min	$\frac{14}{60}$ = 0.23	$\frac{19}{60}$ = 0.32	$\frac{28}{60}$ = 0.47	$\frac{57}{60}$ = 0.95
initial rate of removal of H_2O_2/mol dm^{-3} min^{-1}	$\frac{0.0166}{0.23}$ = 0.072	$\frac{0.0166}{0.32}$ = 0.052	$\frac{0.0166}{0.47}$ = 0.035	$\frac{0.0166}{0.095}$ = 0.018

2 The rate constant is given by the slope of this graph = 0.072/0.16 = 0.45 mol dm^{-3} min^{-1}, i.e. in good agreement with the figure obtained by the previous method.

Activity 24.7

1 Rate ∝ $[A]^1$, so the full rate equation is rate = k[A]. The rate constant, k, may be calculated by substituting any of the data from table (a) into this equation. Thus when [A] = 0.5 mol dm^{-3},

$$0.21 \text{ mol dm}^{-3} \text{ min}^{-1} = k \times 0.50 \text{ mol dm}^{-3}$$
$$k = 0.21 \text{ mol dm}^{-3} \text{ min}^{-1}/0.50 \text{ mol dm}^{-3} = 0.42 \text{ min}^{-1}$$

2 For this zero order reaction rate = $[B]^0$, and the rate equation is simply rate = k. The rate constant is equal to the rate, i.e. 0.13 mol dm^{-3} min^{-1}.

[C]/mol dm^{-3}	0.50	1.00	1.50	2.00
initial rate/mol dm^{-3} min^{-1}	1.00	*4.00*	*9.00*	*16.00*

3

For this 2nd order reaction the rate is proportional to the square of the concentration. When [C] = 1.50 mol dm^{-3}, i.e. 3 times 0.50 mol dm^{-3}, the rate increases by a factor of 9 (i.e. 3^2) from 1.00 mol dm^{-3} min^{-1} to 9.00 mol dm^{-3} min^{-1}. Similarly the rate when [C] = 2.00 mol dm^{-3} is 16 times (i.e. 4^2) that when [C] = 0.50 mol dm^{-3}.

Substituting into the rate equation, rate = $k[C]^2$ so when [C] = 1.00 mol dm^{-3}:

$$4.00 \text{ mol dm}^{-3} \text{ min}^{-1} = k \times 1.00^2 \text{ mol}^2 \text{ dm}^{-6}$$
$$k = 4.00 \text{ mol dm}^{-3} \text{ min}^{-1}/1.00 \text{ mol}^2 \text{ dm}^{-6}$$
$$= 4.00 \text{ mol}^{-1} \text{ dm}^3 \text{ min}^{-1}$$

4

[D]/mol dm^{-3}	0.50	1.00	1.50	2.00
initial rate/mol dm^{-3} min^{-1}	1.00	*8.00*	*27.00*	*64.00*

For a 3rd order reaction the rate is proportional to the concentration cubed. When [D] is 1.00 mol dm^{-3}, i.e. 2 times 0.50 mol dm^{-3}, the rate goes up 8 times (i.e. 2^3) from 1.00 mol dm^{-3} min^{-1} to 8.00 mol dm^{-3} min^{-1}. The other values in the table above may be calculated in a similar way.
Substituting into the rate equation, rate = $k[D]^3$, so when [D] = 1.00 mol dm^{-3}:

$$8.00 \text{ mol dm}^{-3} \text{ min}^{-1} = k \times 1.00^3 \text{ mol}^3 \text{ dm}^{-9}$$
$$k = 8.00 \text{ mol dm}^{-3} \text{ min}^{-1}/1.00 \text{ mol}^3 \text{ dm}^{-9}$$
$$= 8.00 \text{ mol}^{-2} \text{ dm}^6 \text{ min}^{-1}$$

Activity 24.8

1 Rate = $k[CH_3COCH_3]$ so k = rate/$[CH_3COCH_3]$
From experiment a)
k = 0.022 mol dm^{-3} min^{-1}/0.25 mol dm^{-3} = 0.088 min^{-1}
(You may get slightly different answers using data from the other experiments.)

2 Rate = $k[CH_3COCH_3]$ and when $[CH_3COCH_3]$ = 0.65 mol dm^{-3},
rate = 0.088 min^{-1} × 0.65 mol dm^{-3}
= 0.057 mol dm^{-3} min^{-1}

3 The concentration of propanone will remain equal to that of iodine throughout the experiment (it is no longer in large excess). Though the change in iodine concentration has no effect on rate, the fall in $[CH_3COCH_3]$ does and the reaction will slow down. The graph of $[I_2]$ against time is, therefore, no longer a straight line but a curve of the type shown in Figure 24.8.

4 Order with respect to I_2 is 0, with respect to CH_3COCH_3 is 1 so overall order = 0 + 1 = 1

C H A P T E R

25

Reversible reactions

Contents

Study Checklist

After studying this chapter you should be able to:

1. Appreciate that a dynamic equilibrium results when the rates of the forward and backward reactions in a reversible system are equal.
2. Derive the expression for the equilibrium constant for any system in terms of molar concentration, K_c.
3. Derive the expression for the equilibrium constant for any reaction involving gases in terms of partial pressures, K_p.
4. Appreciate that the molar concentration (or partial pressure) of a pure solid (or liquid) is constant at any given temperature, and is therefore incorporated into the equilibrium constant.
5. Carry out calculations involving K_c and K_p.
6. For any system at equilibrium, predict the effect (if any) of changes in concentration, partial pressure, total pressure, and temperature.
7. Appreciate that a catalyst speeds up both forward and reverse reactions by the same factor and does not alter equilibrium position.
8. Recall that the equilibrium constant for any system changes only with temperature.

▲ Whisky distillation involves reversible processes

Many industrially important processes are reversible. For example, the Haber process for the production of ammonia:

$$N_2(g) + 3H_2(g) \rightleftharpoons 2NH_3(g)$$

In section 23.3 we gave details of the conditions under which this reaction is carried out, but how and why are such decisions made? Profitability is, of course, a major consideration and we must aim to produce the maximum amount of product in the shortest possible time. Already in this theme we have dealt with the **time** element by looking at reaction rates and in this chapter we shall investigate ways of maximising the **amount** of product we can get from a reversible reaction.

Though you may not suspect it, such diverse applications as whisky distillation, oil refining, electrical batteries and rust prevention also depend on an understanding of reversible systems.

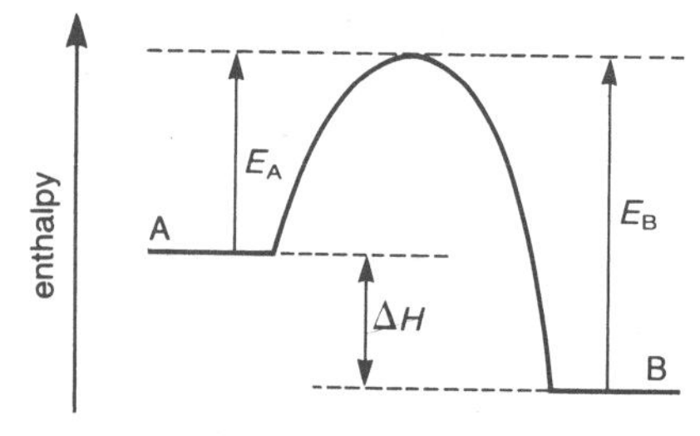

ΔH is the enthalpy change
−ve A → B or +ve A ← B

E_A is the activation energy for A → B

E_B is the activation energy for A ← B

Figure 25.1 ▲
Energy changes in a chemical reaction

25.1 Dynamic equilibria

Theoretically, all chemical reactions are reversible. For any general reaction:

$$A \rightleftharpoons B$$

we can draw an energy level diagram of the type shown in Figure 25.1.

However, a very high activation energy in one direction may act as an effective barrier, making the reaction one way for all practical purposes, i.e.

if $E_B >> E_A$, then only A $\longrightarrow$ B is likely to occur
if $E_A >> E_B$, then only A $\longleftarrow$ B is likely to occur

Provided that conditions allow both A and B molecules to reach their activation energy, then both forward and backward reactions will occur. *When the rates of the forward and backward reactions are equal, no further change will be observed*. Even though both reactions are still occurring, the concentrations of reactants and products remain constant and the system is said to be in **dynamic equilibrium**.

Activity 25.1 Simulated dynamic equilibrium

This simple experiment gives you considerable insight into the nature of a dynamic equilibrium and the theory of reversible reactions. We shall simulate the reaction,

$$A \rightleftharpoons B$$

by transferring water backwards and forwards between two troughs labelled A and B. The *depth* of water in each trough will be taken as a measure of the *amount* of each substance present.

You will need two glass troughs, labelled A and B, two beakers of equal size and one much smaller beaker.

a) Fill trough A with water so that one of the large beakers, lying on its side, is just submerged. If you wish, a little dye may be added to make it easier to see the level.
b) At the start of the experiment trough B is empty because no 'product' has been formed.
c) Now we may start the 'reaction'. Using the large beaker, scoop out as much liquid as you can from trough A and transfer it to trough B. You have now begun to form some 'product', B.
d) Since you now have some B, it will start to change back into A and so both reactions, A $\longrightarrow$ B and A $\longleftarrow$ B, will occur. To simulate this use the large beakers to scoop out liquid from each trough simultaneously and transfer it to the other (do not tilt the troughs when doing this).
e) Repeat step **d)** several times.

1 What happens to the relative liquid levels in A and B? Explain why this happens.

2 What eventually happens to the liquid levels in A and B? How do you account for this?

Now repeat the experiment as above but use the smaller beaker to transfer liquid from trough B to trough A.

3 What differences do you notice and how do you explain them?

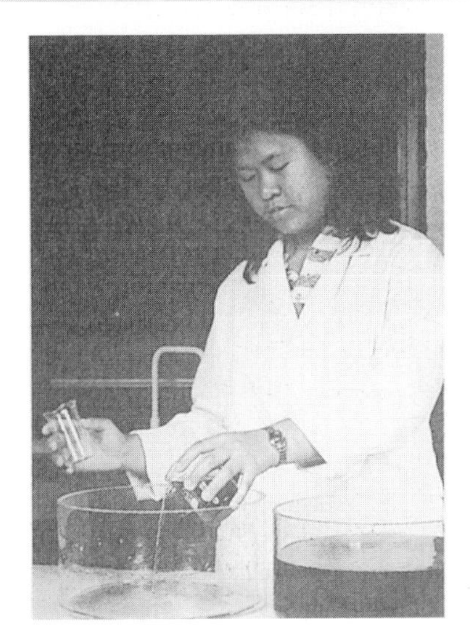

Simulating setting up of a dynamic equilibrium

The concept of phase is important in the study of equilibria. A phase is any physically distinct part of a system which itself has a uniform composition. For instance, when sand is shaken with water it does not dissolve and the system contains two phases, solid sand and liquid water. Similarly, a mixture of oil and water will separate into two distinct liquid phases. However, if salt is shaken with water it will dissolve to give a single phase of salt solution. Wine is an example of a single phase system which contains two miscible liquids, ethanol and water. Single phase systems are referred to as

homogeneous, whereas those consisting of two or more phases are known as **heterogeneous**.

Another way of subdividing reversible systems is by the type of change involved. A **redox** equilibrium involves the transfer of **electrons**, e.g.

$$Mg^{2+}(aq) + 2e^- \rightleftharpoons Mg(s)$$

Oxidation **i**nvolves **l**oss of electrons, whereas **r**eduction **i**nvolves **g**ain (use the words **oil rig** to remind you of this). In this example, Mg^{2+} ions are being reduced and the metal itself is being oxidised.

Transfer of **protons** takes place in **acid–base** equilibria. An **acid** donates protons to a **base**, e.g.

$$\underset{\text{acid}}{CH_3COOH(aq)} + \underset{\text{base}}{H_2O(l)} \rightleftharpoons \underset{\text{base}}{CH_3COO^-(aq)} + \underset{\text{acid}}{H_3O^+(aq)}$$

Some reversible systems involve physical rather than chemical change. **Phase** equilibria are largely concerned with changes of state, e.g.

$$H_2O(s) \rightleftharpoons H_2O(l) \quad \text{or} \quad C(\text{diamond}) \rightleftharpoons C(\text{graphite})$$

Activity 25.2 Classifying reversible systems

Classify each of the following systems as homogeneous or heterogeneous and identify those that are:

A redox equilibria
B acid–base equilibria
C physical equilibria

1. $CaCO_3(s) \rightleftharpoons CaO(s) + CO_2(g)$
2. $NH_3(g) + HCl(g) \rightleftharpoons NH_4^+Cl^-(s)$
3. $CO_2(s) \rightleftharpoons CO_2(g)$
4. $C_6H_{12}O_6(s) \rightleftharpoons C_6H_{12}O_6(aq)$
5. $Fe^{3+}(aq) + e^- \rightleftharpoons Fe^{2+}(aq)$
6. $HCl(g) + H_2O \rightleftharpoons H_3O^+(aq) + Cl^-(aq)$
7. $2SO_2(g) + O_2(g) \rightleftharpoons 2SO_3(g)$
8. $Zn(s) + Fe^{2+}(aq) \rightleftharpoons Zn^{2+}(aq) + Fe(s)$
9. $2H_2O(l) \rightleftharpoons H_3O^+(aq) + OH^-(aq)$
10. $3O_2(g) \rightleftharpoons 2O_3(g)$

Focus 25a

1. A **dynamic equilibrium** results when the rates of forward and backward reactions are equal.
2. Equilibria may be classified as **homogeneous** or **heterogeneous** or by the type of change involved, e.g. **acid–base**, **redox** or **physical**.

Later chapters in this theme will deal with reversible systems involving ions and those concerned with physical changes but here we shall concentrate on general chemical equilibria.

25.2 Homogeneous chemical equilibria

Remember, here all the reactants and products must be in the same phase. This includes reactions which take place **entirely in solution**, or which involve **only gases**. Consider the general reaction,

$$aA + bB \rightleftharpoons cC + dD$$

where a, b, c and d represent the number of particles of each substance in the balanced equation. Suppose that we start with a mixture of A and B only. If the reaction mechanism involves only a single step, then the rate of the forward reaction is given by,

$$\text{forward rate} = k_f[A]^a.[B]^b$$

where k_f is the rate constant for the forward reaction.

As the reaction proceeds the rate of this forward reaction will decrease as the concentrations of A and B fall. Once some C and D have been formed, the reverse reaction will start and this will have a rate given by,

$$\text{reverse rate} = k_r[C]^c.[D]^d$$

where k_r is the rate constant for the reverse direction.

As more C and D are formed this reverse rate will increase until eventually, as Figure 25.2 shows, it will equal the rate of the forward reaction.

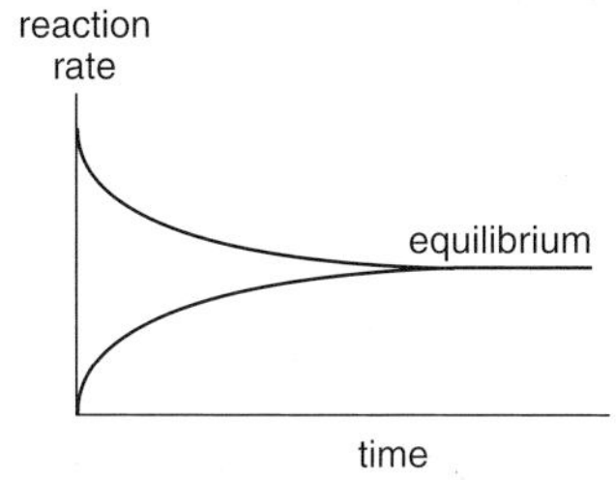

Figure 25.2
Changes in the forward and backward reaction rates in the system $A + B \rightleftharpoons C + D$, starting with A and B only

At this point, *the rate at which any substance is being used up equals its rate of production*. The concentration of each species will therefore remain constant and, though *both reactions still continue, there is no observable change in the system*. Unless conditions are altered, this dynamic equilibrium will remain unchanged. At equilibrium,

$$\text{forward rate} = \text{reverse rate}$$
$$k_f[A]^a.[B]^b = k_r[C]^c.[D]^d$$

using this expression we get:

$$K_c = \frac{k_f}{k_r} = \frac{[C]^c.[D]^d}{[A]^a.[B]^b}$$

Strictly speaking, this treatment applies only to reactions with simple one-step mechanisms. However this **equilibrium law**, as it is known, can be shown to apply generally to all reactions regardless of mechanism. $\boldsymbol{K_c}$ is known as the **equilibrium constant** for the reaction **in terms of molar concentration**. For any reaction its value is fixed only by temperature.

Although K_c may be calculated for any type of homogeneous system, it is often more convenient to use an alternative equilibrium constant for **gaseous** reactions, defined in terms of **partial pressures** rather than molar concentrations. For any mixture of gases, the total pressure P is the sum of the partial pressures of all the components, i.e. the pressure each alone would exert at the temperature and volume of the system.

$$\underset{\text{total pressure}}{P} = \underset{\text{sum of partial pressures}}{p_A + p_B + p_C + p_D}$$

Since the pressure exerted by a gas is proportional to the number of moles present, we may also write this as,

$$P = (f_A.P) + (f_B.P) + (f_C.P) + (f_D.P)$$

where f_A, f_B, f_C and f_D are the **mole fractions** of each component, i.e. number of moles of component/total number of moles in mixture, e.g.

$$f_A = \frac{\text{(moles of A)}}{\text{(moles of A + moles of B + moles of C + moles of D)}}$$

For the general case we have considered above,

$$K_p = \frac{(p_C)^c.(p_D)^d}{(p_A)^a.(p_B)^b}$$

or,

$$K_p = \frac{(f_C.P)^c.(f_D.P)^d}{(f_A.P)^a.(f_B.P)^b}$$

A term you will often hear is **equilibrium position**. This is often (mis)used to describe the relative amounts of reactants and products present at equilibrium. In fact the only objective measure of the equilibrium position is given by the value of the equilibrium constant. As we shall see later, *the* ***system*** *itself often shifts to preserve the equilibrium constant.*

Activity 25.3 Equilibrium constants and their units

1 For each of the systems below, write an expression for K_c.

a) $CH_3COOH + CH_3OH \rightleftharpoons CH_3COOCH_3 + H_2O$ (miscible liquid mixture)

b) $H_2(g) + I_2(g) \rightleftharpoons 2HI(g)$

c) $N_2O_4 \rightleftharpoons 2NO_2$ (in an organic solvent)

d) $2SO_2(g) + O_2(g) \rightleftharpoons 2SO_3(g)$

e) $3O_2(g) \rightleftharpoons 2O_3(g)$

Remembering that the units of concentration are mol dm^{-3} (or simply M), give the units of K_c for each system. For example, in the reaction,

$$N_2(g) + 3H_2(g) \rightleftharpoons 2NH_3(g)$$

$K_c = [NH_3]^2/[N_2].[H_2]^3$ and the units are $M^2/M.M^3 = M^2/M^4 = 1/M^2$ or M^{-2} converting back into mol dm^{-3}, M^{-2} = (mol $dm^{-3})^{-2}$ = $mol^{-2}\ dm^6$

2 For those reactions which involve only gases, write an expression for K_p and give the units in terms of atmospheres (pressure). Thus for the example above, $K_p = (pNH_3)^2/(pN_2).(pH_2)^3$ and the units are $atm^2/atm.atm^3 = atm^2/atm^4 = 1/atm^2$ or atm^{-2}.

The equilibrium constant for any system may be calculated from the balanced equation and the equilibrium concentrations, or partial pressures, of the components. For instance, after heating hydrogen and iodine vapour at 448°C in a sealed container, the equilibrium mixture was found to contain 7.83 moles H_2, 1.68 moles I_2 and 25.54 moles HI.

$$H_2(g) + I_2(g) \rightleftharpoons 2HI(g)$$

The equilibrium constant in terms of concentration is given by,

$$K_c = \frac{[HI]^2}{[H_2][I_2]}$$

Let the volume of the container be V dm^3.

$$[HI] = \frac{25.54}{V} \qquad [H_2] = \frac{7.83}{V} \qquad [I_2] = \frac{1.68}{V}$$

$$K_c = \frac{(25.54/V)^2}{(7.83/V)(1.68/V)} = \frac{25.54^2}{(7.83 \times 1.68)} = 49.6$$

Now we have the value of the equilibrium constant (K_c = 49.6), we may use it to predict the course of the reaction. Suppose 1 mole of hydrogen iodide is heated in a sealed container at 448°C. How many moles of each component will be present when equilibrium is reached?

Activity 25.4 Calculating K_p

For such a gaseous system we could have chosen to calculate K_p instead of K_c. In fact, when the equilibrium constant has no units, the two values are identical. Confirm this by calculating the mole fraction and partial pressure of each gas at a total pressure P.

	$H_2(g)$	+	$I_2(g)$	$\rightleftharpoons$	$2HI(g)$
moles at start	0		0		1

Let x = moles of H_2 formed at equilibrium,

moles at equilibrium	x	x	$(1 - 2x)$
concentration at equil.	$\frac{x}{V}$	$\frac{x}{V}$	$\frac{(1 - 2x)}{V}$

$$K_c = \frac{\{(1 - 2x)/V\}^2}{(x/V)^2} = \frac{(1 - 2x)^2}{x^2}$$

$$\sqrt{K_c} = \frac{(1 - 2x)}{x} = \sqrt{49.6} = 7.04$$

$$1 - 2x = 7.04x$$

$$9.04x = 1,\ x = \frac{1}{9.04} = 0.111$$

i.e. moles H_2 = moles I_2 = 0.111 and moles HI = 0.778
As a quick check, recalculate K_c with these figures:

$$K_c = \frac{(0.778)^2}{(0.111)^2} = 49.1$$

(allowing for mathematical 'rounding' errors, this is in agreement with the above figure).

If we now disturb the system by changing the concentration or partial pressure of one of the components then it will move in such a way as to re-establish equilibrium. For example, if we want more of the hydrogen iodide to decompose, we might remove some iodine vapour from the container. More hydrogen iodide will then decompose to increase the concentration of iodine vapour and so re-establish equilibrium. In this example changing the volume of the container, and hence the total pressure, has no effect on the system. Since there is an equal number of particles on each side of the equation, there is no total volume (or pressure) term in the expression for the equilibrium constant.

Any system which has unequal numbers of reactant and product particles will be affected by a change in total volume or pressure. Consider the manufacture of ammonia by the Haber process:

$$N_2(g) + 3H_2(g) \rightleftharpoons 2NH_3(g) \qquad \Delta H = -92\ \text{kJ}$$

$$K_p = \frac{(pNH_3)^2}{(pN_2)(pH_2)^3} = \frac{(fNH_3.P)^2}{(fN_2.P)(fH_2.P)^3}$$

$$= \frac{(fNH_3)^2.P^2}{(fN_2)(fH_2)^3.P^4} = \frac{(fNH_3)^2}{(fN_2)(fH_2)^3.P^2}$$

so, $$(fNH_3)^2 = K_p(fN_2)(fH_2)^3P^2$$

Since K_p is fixed by temperature, an **increase** in the total pressure, P, results in an increase in the mole fraction of ammonia at equilibrium. In fact, a pressure of several hundred atmospheres is used in industry to maximise the equilibrium yield of ammonia. Note that the formation of more ammonia has the effect of reducing the total number of particles in the system and so lowering the pressure.

Although changes in pressure or concentration often cause a shift in a system, only a temperature change will alter the equilibrium constant. The formation of ammonia from its elements in the Haber process is an exothermic reaction. An energy profile for this reaction is shown in Figure 25.3.

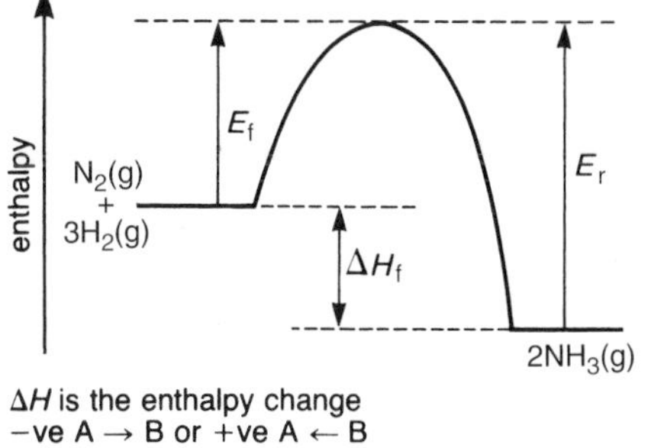

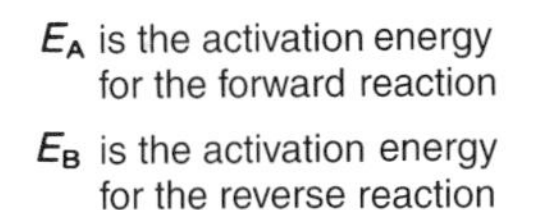

Figure 25.3
Energy profile for the Haber process: $N_2(g) + 3H_2(g) \rightleftharpoons 2NH_3(g)$

The activation energy for the reverse (endothermic) reaction is greater than that for the forward (exothermic) reaction. These activation energies are sketched on the Maxwell–Boltzmann distribution curve in Figure 25.4.

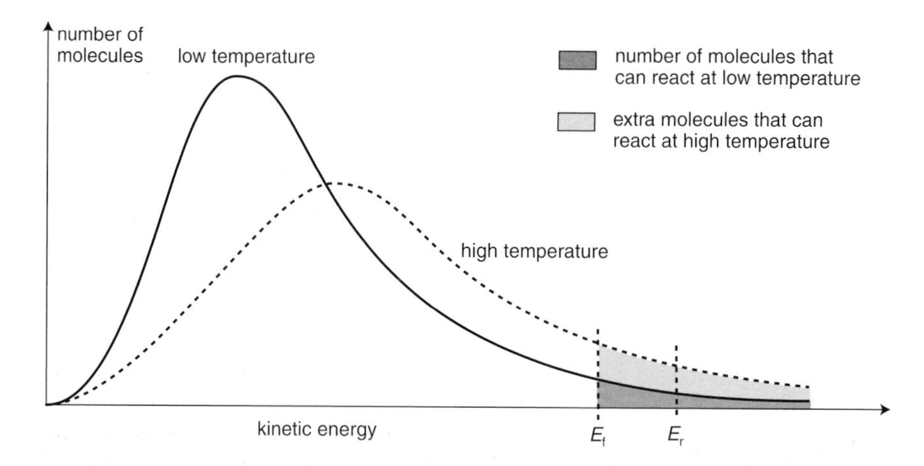

Figure 25.4
How activation energy influences the effect of temperature change on reaction rate

The number of molecules able to react on collision is represented by the area under the curve above the activation energy. As you can see, a temperature rise increases the number of molecules able to take part in both forward and reverse reactions but the **proportional** increase is greater for the reaction with the higher activation energy. This means that an increase in temperature always favours the endothermic process. In this case, an increase in temperature **reduces** the yield of ammonia. However, too low a temperature gives an unacceptably slow reaction rate and a compromise must be reached.

Temperature and equilibrium constant

The effect of a temperature change on an equilibrium constant is given by the van't Hoff equation:

$$\log_{10}\frac{K_1}{K_2} = -\Delta H^\ominus \frac{1/T_1 - 1/T_2}{2.303R}$$

where, K_1 and K_2 are the equilibrium constants at temperatures T_1 and T_2. If $T_2 > T_1$ and $\Delta H^\ominus$ is negative (i.e. an exothermic reaction) then $\log_{10}(K_1/K_2)$ is positive. This means that $K_1 > K_2$, i.e. the equilibrium constant falls if the temperature rises.

If we know the standard free energy change, $\Delta G^\ominus$, for the reaction then the equilibrium constant may be calculated as follows:

$$\log_{10}K = \frac{-\Delta G^\ominus}{2.303RT}$$

For spontaneous reactions, $\Delta G^\ominus$ is negative. This makes $\log_{10}K$ positive, i.e. $K > 1$ and the system lies to the right at equilibrium (see also Chapter 28.)

If you look carefully at the examples above you may notice an interesting connection, known as **Le Chatelier's principle**. *If we change the conditions of any system at equlibrium, then, if possible, the system will shift in such a direction as to reduce the imposed change in conditions*. For example in the Haber process, removing ammonia causes a shift to produce more ammonia. An increase in total pressure also produces more ammonia, reducing the total number of gas molecules and lowering the pressure again. As we have just seem, increasing the temperature favours the endothermic reaction, thus cooling the system down again. Although it gives no explanation for the effect, Le Chatelier's principle, used carefully, provides a quick way of predicting how an equilibrium system will respond to a change in conditions.

Activity 25.5 Using Le Chatelier's principle

The manufacture of sulphuric acid, described in section 23.3, involves the following reversible stage:

$$2SO_2(g) + O_2(g) \rightleftharpoons 2SO_3(g) \qquad \Delta H = -196 \text{ kJ}$$

1 Use Le Chatelier's principle to decide whether a high or low total pressure and a high or low temperature should be used to get the maximum equilibrium amount of sulphur trioxide.

2 Explain why the following conditions are used in practice:
a) an excess of air,
b) atmospheric pressure,
c) a catalyst at about 450°C.

Catalysts and equilibrium

A catalyst works by providing an alternative reaction route with a lower activation energy, as shown in Figure 25.5.

Since the reduction in activation energy is the identical for forward and reverse reactions, each is speeded up by the same factor. Although a catalyst enables equilibrium to be reached more quickly it does not affect the final position.

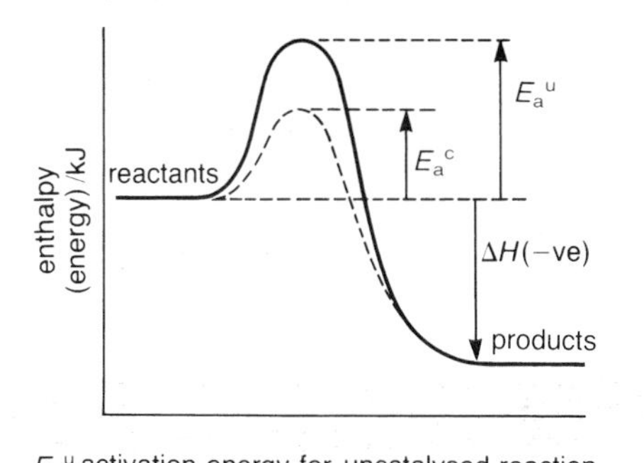

E_a^u activation energy for uncatalysed reaction
E_a^c activation energy for the catalysed reaction
$E_a^c < E_a^u$

Figure 25.5
The effect of a catalyst on the energetics of a reaction

25.3 Heterogeneous chemical equilibria

When copper metal is added to silver nitrate solution, the following heterogeneous equilibrium is established:

$$Cu(s) + 2Ag^+(aq) \rightleftharpoons Cu^{2+}(aq) + 2Ag(s)$$

Applying the equilibrium law,

$$\frac{[Cu^{2+}(aq)][Ag(s)]^2}{[Cu(s)][Ag^+(aq)]^2} = \text{constant}$$

However, *although the **amount** of a pure solid may change, its **concentration** is effectively fixed*. If you take a coin it contains a certain number of moles, m, in a fixed volume, v. The molar concentration is m/v. If we add another coin of the same type, we double the number of moles but also double the total volume, so molar concentration = $2m/2v = m/v$. The same is true if you compare, say, 10 cm^3 of water with 20 cm^3. This means that we may incorporate the concentrations of pure solids (and pure liquids) into the equilibrium constant expression.

$$\frac{[Cu^{2+}(aq)]\text{const}^2}{\text{const}[Ag^+(aq)]^2} = \text{constant}$$

$$\frac{[Cu^{2+}(aq)]}{[Ag^{+}(aq)]^2} = K_c$$

Activity 25.6

When excess copper metal is added to 2 M silver nitrate solution at 25°C, the concentration of aqueous silver ions falls to 1.8×10^{-8} M at equilibrium,

$$Cu(s) + 2Ag^{+}(aq) \rightleftharpoons Cu^{2+}(aq) + 2Ag(s)$$

1 What is the reduction in concentration of aqueous silver ions?

2 What is the concentration of the aqueous copper ions formed? (*Hint*: look carefully at the above equation.)

3 Calculate the equilibrium constant, K_c, for the above reaction.

4 What would you observe when a copper rod is left in a solution of silver nitrate?

The thermal decomposition of limestone is an industrially important heterogeneous equilibrium reaction involving a gas:

$$CaCO_3(s) \rightleftharpoons CaO(s) + CO_2(g) \qquad \Delta H = +180 \text{ kJ mol}^{-1}$$

Using the equilibrium law,

$$\frac{pCO_2(g).pCaO(s)}{pCaCO_3(s)} = \text{constant}$$

As with concentration, *the* **partial pressure** *exerted by any pure solid (or liquid) is fixed by temperature*. We may, therefore simplify the expression for K_p as follows.

$$\frac{pCO_2(g).\text{const}}{\text{const}} = \text{constant, i.e. } pCO_2(g) = K_p$$

As long as the partial pressure of carbon dioxide is kept below K_p, the system will move to the right, converting all the limestone into quicklime. The details of this process are given in section 23.1.

Thus, changing the amount of any pure solid (or liquid) has no effect on any equilibrium system and we can simply ignore these substances when writing the expression for the equilibrium constant.

Focus 25b

1 The **equilibrium constant** in terms of molar concentrations, K_c, may be used for any reversible system. K_p, the equilibrium constant in terms of **partial pressures**, may only be used for reactions involving gases.

2 The only factor that can alter the value of an equilibrium constant is temperature.

3 The concentration or partial pressure of any pure liquid or solid is virtually constant and so does not appear in the expression for K_c or K_p for a heterogeneous system.

4 The qualitative effect, if any, on a system at equilibrium of any change in conditions may be deduced using **Le Chatelier's principle**.

Activity 25.7

Hydrogen is formed by the action of steam on hot iron, according the the following equation:

$$3Fe(s) + 4H_2O(g) \rightleftharpoons Fe_3O_4(s) + 4H_2(g)$$

1 Write an expression for the equilibrium constant in terms of partial pressure.

2 Explain the effect, if any, of the following changes on this system at equilibrium in a closed container:

a) increasing the total pressure,
b) adding more steam to the system at constant volume,
c) adding more steam to the system but increasing the volume to keep the same partial pressure of steam
d) adding more iron,
e) removing some of the iron oxide.

Questions on Chapter 25

1 For the following reversible reactions, write expressions for K_c and, where appropriate, K_p including units:

a) $N_2O_4(g) \rightleftharpoons 2NO_2(g)$

b) $CH_3COOH(l) + CH_3OH(l) \rightleftharpoons CH_3COOCH_3(l) + H_2O(l)$

c) $CaCO_3(s) \rightleftharpoons CaO(s) + CO_2(g)$

d) $NH_3(g) + HCl(g) \rightleftharpoons NH_4Cl(s)$

How, if at all, will the amount of product change in each of the above systems if the total pressure on the system is doubled?

2 Ethyl methanoate can be made by reacting methanoic acid with ethene:

$$HCOOH + C_2H_4 \rightleftharpoons HCOOC_2H_5$$

The reaction may be carried out with all species dissolved in an inert solvent such as tetrachloromethane. In a series of experiments 1 mole of methanoic acid, 1 mole of ethene and a small amount of boron trifluoride were dissolved in 1 litre of solution and allowed to reach equilibrium at different temperatures. The boron trifluoride was unchanged at the end of each experiment and the following equilibrium yields of ethyl methanoate were obtained:

temperature/°C	50	60	70
moles product	0.5	0.45	0.40

a) Write an expression for K_c for this reaction.

b) Calculate the value of K_c at each of the above temperatures.

c) What can you deduce about the enthalpy change for the above reaction?

d) Why do you think boron trifluoride was added to the mixture?

e) If the above reactions had been carried out in a total solution volume of 2 litres, how (if at all) would this affect the yield of product? Explain your answer.

The reaction can also be carried out without the solvent at temperatures above 200°C in the gas phase.

f) Give **one** likely advantage of this method over the solution reaction and **one** disadvantage.

g) How, if at all, would an increase in pressure in the gas reaction affect the equilibrium yield of product? Explain your answer.

3 Gaseous hydrogen iodide dissociates as follows;

$$2HI(g) \rightleftharpoons H_2(g) + I_2(g)$$

a) Write an expression for the equilibrium constant, K_p.

b) At a certain temperature the fraction of HI molecules that have dissociated is 0.4. Calculate the value of K_p at this temperature.

c) If more hydrogen iodide dissociates at a higher temperature, what can you deduce about the enthalpy change for the above reaction?

d) How, and why, would you expect the value of K_p to change on doubling the total pressure?

4 2 moles of sulphur dioxide were mixed with 1 mole of oxygen at a constant temperature and a constant total pressure of 10 atmospheres in the presence of a platinum catalyst. At equilibrium, half of the sulphur dioxide had been converted into sulphur trioxide.

$$2SO_2(g) + O_2(g) \rightleftharpoons 2SO_3(g) \quad \Delta H = -196 \text{ kJ mol}^{-1}$$

a) Write an expression for an equilibrium constant for this reaction and calculate its value under these conditions.

b) State Le Chatelier's principle.
How would you expect the equilibrium ratio of SO_3:SO_2 and the value of the equilibrium constant to change (if at all) when:
i) the total pressure is increased,
ii) the temperature is decreased,
iii) the catalyst is changed to vanadium(V) oxide?

5 Ammonium hydrogensulphide dissociates on heating according to the following equation:

$$NH_4HS(s) \rightleftharpoons NH_3(g) + H_2S(g)$$

a) Write an expression for the equilibrium constant for this reaction in terms of molar concentration.
When solid ammonium hydrogensulphide was left in a 1 litre evacuated flask at 100°C, the pressure gradually rose to a constant value of 0.613 atmospheres i.e. 61.9 kPa.

b) Calculate the total number of moles of gas released by the decomposition of the solid.

c) Calculate a value for K_c under these conditions.

d) What deduction regarding the decomposition of ammonium hydrogensulphide can you make from the information that K_c at 200°C is 0.00072 M^2 ($mol^2\,dm^{-6}$)? Explain your answer.

6 Strontium carbonate decomposes endothermically on heating:

$$SrCO_3(s) \rightleftharpoons SrO(s) + CO_2(g)$$

Strontium carbonate is heated to 900°C in an evacuated sealed vessel until equilibrium is reached.

a) What would happen to the amount of strontium carbonate in the vessel if the following changes were made and the mixture left to regain equilibrium?
i) The temperature was raised to 1000°C.
ii) The volume of the container was reduced at 900°C.

b) What would happen to the value of the equilibrium constant for the reaction on making each of the above changes?

c) Imagine it was possible to remove 0.1 mole of the strontium carbonate without disturbing the equilibrium and replace it with 0.1 mole of radioactive strontium carbonate (containing the isotope ^{90}Sr). Would any of the strontium oxide in the system become radioactive? Explain your answer.

Questions on Chapter 25 *continued*

7 Pentyl ethanoate may be made from pentene and ethanoic acid:

$$C_5H_{10}(l) + CH_3COOH(l) \rightleftharpoons CH_3COOC_5H_{10}(l)$$

a) At 80°C the value of K_c is 540 M^{-1} and the equilibrium concentrations of pentene and ethanoic acid were found to be 0.0060 M and 0.0025 M respectively. What was the concentration of pentyl ethanoate in this system?

b) At 100°C, K_c for this reaction was found to be 440 M^{-1}. State, with reasons, whether the formation of pentyl ethanoate in this reaction is exothermic or endothermic.

8 When lead sulphate is added to aqueous sodium iodide, the following equilibrium is established:

$$PbSO_4(s) + 2I^-(aq) \rightleftharpoons PbI_2(s) + SO_4^{2-}(aq)$$

The equilibrium constant for this reaction may be determined by adding an excess of lead sulphate to a known volume of standard sodium iodide solution and allowing the mixture to reach equilibrium in a water bath maintained at the desired temperature. Excess iced water is then added and the mixture is titrated with standard silver nitrate solution.

a) Explain why it is not necessary to know the mass of lead sulphate added to the sodium iodide solution.

b) Why is iced water added to the mixture before the titration?

c) In an experiment using 50 cm^3 of 0.1 M sodium iodide, 31.0 cm^3 of 0.1 M silver nitrate was needed to precipitate all the iodide ions remaining in solution at equilibrium. Calculate the following:

i) the number of moles of I^- in the original sodium iodide solution,

ii) the number of moles of I^- ions in solution at equilibrium,

iii) the number of moles of SO_4^{2-} ions in solution at equilibrium,

iv) the concentrations of I^- and SO_4^{2-} ions in solution at equilibrium,

d) Write an expression for the equilibrium constant in this system and calculate its value from your answers to the above questions.

Comments on the activities

Activity 25.1

1 The level in A falls whilst the level in B *rises*. The *rate* at which liquid is transferred from a trough depends upon the *depth* of liquid in it. Since initially there is *more* water in trough A, the rate of transfer from A to B will be *greater* than from B to A,

i.e. forward rate > backward rate

2 As the 'reaction' proceeds the level of water in A, and hence the rate of transfer to B, *falls*, whereas since the level in B is *rising*, the rate of transfer to A *increases*. Eventually the two rates become equal,

i.e. forward rate = backward rate

An **equilibrium** has been established where both forward and backward 'reactions' are occurring at the *same* rate. No matter how long the process is continued, the water levels in each trough will remain *steady*, even though 'reactions' are still taking place. The equilibrium is, therefore, **dynamic**. Since the beakers used are of the same size, in this case we should expect the equilibrium levels in each trough to be identical.

3 When a *smaller* beaker is used to transfer water from B to A, a dynamic equilibrium is still reached, but here the final level of liquid in trough B is *higher* than that in A. The rate of transfer of water from B to A will be *lower* than before and the level of water in A will have to *fall* in order to make the rates of the forward and backward 'reactions' *equal* at equilibrium. This illustrates an important point in reversible reactions, that *though the forward and backward reaction rates must be equal at equilibrium, the concentrations of reactants and products need not be the same, and in fact very rarely are.*

Activity 25.2

1 heterogeneous
2 heterogeneous acid–base
3 heterogeneous physical
4 heterogeneous physical
5 homogeneous redox
6 homogeneous acid–base
7 homogeneous
8 heterogeneous redox
9 homogeneous acid–base
10 homogeneous

Activity 25.3

1 a) $K_c = \dfrac{[CH_3COOCH_3][H_2O]}{[CH_3COOH][CH_3OH]}$ $\dfrac{M.M}{M.M}$, no units

b) $K_c = \dfrac{[HI]^2}{[H_2][I_2]}$ $\dfrac{M^2}{M.M}$, no units

c) $K_c = \dfrac{[NO_2]^2}{[N_2O_4]}$ $\dfrac{M^2}{M}$, units are M (mol dm^{-3})

d) $K_c = \dfrac{[SO_3]^2}{[SO_2]^2[O_2]}$ $\dfrac{M^2}{M^2.M}$, units are M^{-1} (mol^{-1} dm^3)

e) $K_c = \dfrac{[O_3]^2}{[O_2]^3}$ $\dfrac{M^2}{M^3}$, units are M^{-1} (mol^{-1} dm^3)

Comments on the activities *continued*

2 **b)** $K_p = \frac{(pHI)^2}{(pH_2)(pI_2)}$ $atm^2/atm.atm$, no units

d) $K_p = \frac{(pSO_3)^2}{(pSO_2)^2(pO_2)}$ $atm^2/atm^2.atm$, units are atm^{-1}

e) $K_p = \frac{(pO_3)^2}{(pO_2)^3}$ atm^2/atm^3, units are atm^{-1}

Activity 25.4

total number of moles at equilibrium
$= 7.83 + 1.68 + 25.54 = 35.05$

mole fraction of HI $= 25.54/35.05 = 0.729$,
partial pressure $= 0.729P$

mole fraction of $H_2 = 7.83/35.05 = 0.223$,
partial pressure $= 0.223P$

mole fraction of $I_2 = 1.68/35.05 = 0.048$,
partial pressure $= 0.048P$

$K_p = (pHI)^2/(pH_2)(pI_2) = (0.729P)^2/(0.223P)(0.048P)$
$= 0.729^2/(0.223 \times 0.048)$
$= 49.6$, i.e. identical with the value for K_c

CARE: This does not apply unless the concentration (or pressure) terms in the equilibrium constant expression cancel out.

Activity 25.5

1 The formation of sulphur trioxide involves a reduction in the total number of gas molecules from 3 to 2 and a consequent reduction in total pressure. Le Chatelier's principle predicts that an increase in pressure will produce more sulphur trioxide. A high pressure will therefore give the best equilibrium yield.

The formation of sulphur trioxide is exothermic, i.e. involves a temperature rise. According to Le Chatelier's principle, lowering the temperature will favour this reaction. A low temperature will therefore give the best equilibrium yield of sulphur trioxide.

2 **a)** Air is much cheaper than sulphur dioxide. Adding excess of air will push the system to the right converting more sulphur dioxide into sulphur trioxide.

b) Although it is true that a high pressure will produce more sulphur trioxide, over 99% conversion is achieved at atmospheric pressure and it is simply not economically worthwhile to spend extra money on high pressure equipment.

c) Although lowering temperature favours the formation of sulphur trioxide, we must also bear in mind that it dramatically slows down the reaction. Economic considerations dictate a compromise temperature of about 450°C with a catalyst to boost the rate.

Activity 25.6

1 The decrease in $[Ag^+(aq)]$ is $2 - 1.8 \times 10^{-8}$ M, i.e. about 2 M.

2 Looking at the equation for the reaction, half as many moles of Cu^{2+} are formed, therefore at equilibrium $[Cu^{2+}(aq)] = 1$ M.

3
$$Cu(s) + 2Ag^+(aq) \rightleftharpoons Cu^{2+}(aq) + 2Ag(s)$$
at equilibrium 1.8×10^{-8} M   1 M

$$K_c = \frac{[Cu^{2+}(aq)]}{[Ag^+(aq)]^2} = \frac{1}{(1.8 \times 10^{-8})^2} = 3.09 \times 10^{15}\ M^{-1}$$

4 Silver metal would deposit on the surface of the copper and the solution would turn blue owing to the formation of $Cu^{2+}(aq)$ ions.

Activity 25.7

1 Ignoring the two solids,

$$K_p = \frac{(pH_2)^4}{p(H_2O)^4} = \left(\frac{pH_2}{pH_2O}\right)^4$$

2 **a)** Since there is no change in the total number of gas molecules, changing the overall pressure has no effect on the system.

b) pH_2O is increased whilst pH_2 remains the same. The system must move to the right to re-establish equilibrium.

c) Although pH_2O remains the same, the increase in volume reduces pH_2. Again the system must produce more hydrogen to re-establish equilibrium.

d) Changing the amount of pure solid present has no effect on the system (as long as some remains present).

e) As d).

CHAPTER 26

Ionic equilibria

Contents

Study Checklist

After studying this chapter you should be able to:

1 Define reduction as electron gain, oxidation as electron loss, and a redox equilibrium as electron transfer.

2 Use the IUPAC convention for writing:
- the half-equation for an electrode reaction (i.e. as a reduction process)
- a cell diagram (i.e. the negative electrode reaction followed by the positive electrode reaction)

3 Recall that an electrode potential may only be measured relative to another in an electrochemical cell, $E_{cell} = E_{rhs} - E_{lhs}$.

4 Define standard electrode potential as the potential measured relative to a hydrogen electrode at 25°C, 1 atmosphere pressure and with all solutions of concentration = 1 M.

5 Appreciate the practical difficulties in operating a standard hydrogen electrode and the consequent need for more convenient secondary standards, e.g. the calomel electrode.

6 Recognise the electrochemical series as a list of redox systems in order of $E^{\ominus}$ values, from the most negative at the top to the most positive at the bottom. Use this series to predict the feasibility of redox reactions.

7 Explain the electrochemical nature of the wet corrosion of iron and how this may be reduced by contact with a more reactive metal.

8 Outline other methods of rust prevention, including surface coatings.

9 Define acids as proton donors, bases as proton acceptors and acid–base reactions as proton transfer.

10 Recall that strong acids and bases ionise fully in solution, whereas weak acids and bases ionise only slightly.

11 Recall that water self-ionises to a slight extent and write an expression for the equilibrium constant K_w.

12 Write an expression for the dissociation constant K_a for a weak monobasic acid, HA.

13 Define pH, pK_w and pK_a as $-\log_{10}$ of the original quantity and explain the convenience of such a scale.

14 Carry out calculations involving K_w, K_a, $[H^+(aq)]$, pK_w, pK_a and pH.

15 Define buffers as solutions which resist change in pH on adding small amounts of acid or alkali, and give examples of their use.

16 Explain why the mixture of a weak acid (or base) and its salt with a strong base (or acid) acts as a buffer.

17 Sketch and explain the shape of pH curves during the titration of:
- strong (and weak) monoprotic acids with strong (and weak) monoprotic alkalis,

- carbonate with a strong monoprotic acid, e.g. HCl,
- ethanedioic acid with a strong monoprotic base, e.g. NaOH.

18 Appreciate that indicators are themselves weak acids (or bases) where the ionised and unionised forms have different colours.

19 Recall that (screened) methyl orange changes colour below pH 7, whilst phenolphthalein changes colour above pH 7, and select the appropriate indicator for any particular acid–base titration.

20 Define, and carry out calculations involving, solubility product, K_{sp}, as applied to sparingly soluble ionic compounds.

26.1 Redox equilibria

When a metal is dipped into a solution containing its own ions, a redox equilibrium is set up. By convention this is normally written as a reduction process, e.g.

$$M^{x+}(aq) + xe^- \rightleftharpoons M(s)$$

The ease with which a metal loses electrons depends upon its reactivity. Fairly reactive metals, such as zinc, ionise readily and the redox equilibrium lies well to the left:

$$Zn^{2+}(aq) + 2e^- \rightleftharpoons Zn(s)$$

Figure 26.1 shows that, as the metal electrode dissolves, it accumulates electrons and develops a negative charge, whereas the solution gains positive charge due to the release of metal ions.

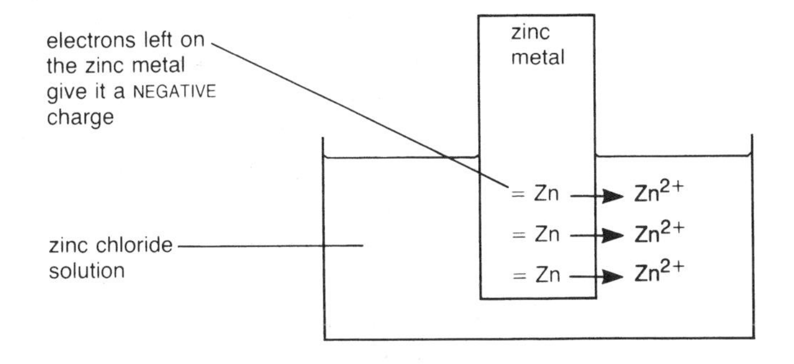

Figure 26.1 ▶
A metal in contact with a solution containing its own ions

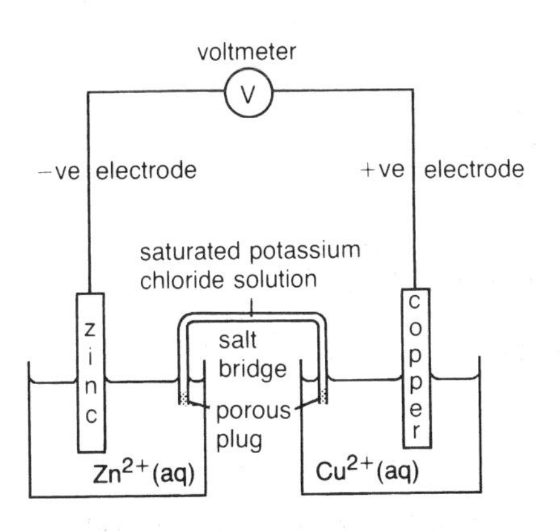

Figure 26.2 ▲
An electrochemical cell formed from a zinc electrode and a copper electrode

A less reactive metal, such as copper, does not ionise as readily and the equilibrium will lie further to the right:

$$Cu^{2+}(aq) + 2e^- \rightleftharpoons Cu(s)$$

As a result, fewer electrons collect on the metal and its charge will be less negative than for the zinc. There is no way of measuring the potential of an isolated electrode but the difference in potential between two electrodes may be measured by connecting them up in an electrochemical cell. Such a cell made from zinc and copper electrodes is shown in Figure 26.2.

Electrons can flow from the more negative zinc electrode to the copper via the voltmeter. The salt bridge is needed to allow the transfer of ions between the two solutions. When a current is taken from the cell, the equilibrium at each electrode is disturbed. Electrons are withdrawn from the zinc and added to the copper. By Le Chatelier's principle, this will cause each system to shift as follows:

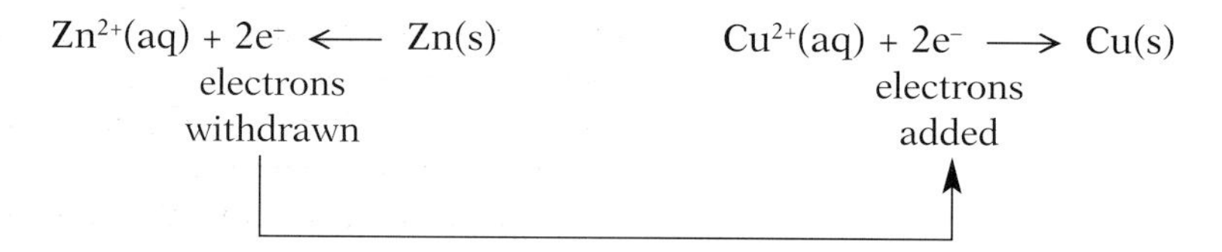

These processes may be represented using the following **cell diagram**:

[NEGATIVE electrode process]		[POSITIVE electrode process]
$Zn(s)\,|\,Zn^{2+}(aq)$	$||$	$Cu^{2+}(aq)\,|\,Cu(s)$

In this cell diagram, different phases in the same electrode are separated by the symbol | (a comma separates two species in the same phase at an electrode) whilst || represents the salt bridge linking the two solutions.

By convention the voltage of any working cell such as this is always positive, so,

$$\text{cell voltage} = E(\text{positive electrode}) - E(\text{negative electrode})$$

Since, by convention the negative electrode is always given first in the cell diagram, we may also write

$$\text{cell voltage} = E(\text{right hand electrode}) - E(\text{left hand electrode})$$

or

$$E_{\text{cell}} = E_{\text{rhs}} - E_{\text{lhs}}$$

Activity 26.1 Other types of electrode

Any redox system may be used as an electrode, even if it does not contain a metal, e.g.

gas/solution	$H^+(aq) + e^- \rightleftharpoons \frac{1}{2}H_2(g)$
	$\frac{1}{2}Cl_2(g) + e^- \rightleftharpoons Cl^-(aq)$
solution/solution	$Cr^{3+}(aq) + e^- \rightleftharpoons Cr^{2+}(aq)$

In order to connect such electrodes up into a cell, we must provide an inert conductor such as platinum metal. In the case of a redox system entirely in solution, we may simply dip the inert conductor into the liquid. If a gas is involved, we need the more complicated set-up shown in Figure 26.3.

metal wire

gas

aqueous solution containing ions produced from gas

platinum foil coated with 'platinum black' (finely divided platinum); the large surface area helps to establish equilibrium quickly

Figure 26.3
A gas/ion electrode

In the cell diagram, any species in the same phase, e.g. $Cr^{3+}(aq)$ and $Cr^{2+}(aq)$ above are separated by a comma, i.e. $Cr^{3+}(aq), Cr^{2+}(aq)$, and the inert conductor is also indicated, written as (Pt).

1. The hydrogen electrode above forms the negative electrode when combined in a cell with the chlorine electrode. Deduce the spontaneous reactions taking place at each electrode and draw the cell diagram.
2. Repeat question 1 for a cell formed between the aqueous Cr^{3+}/Cr^{2+} system and the hydrogen electrode, where the latter is the positive electrode.
3. Draw the diagram for the cell formed between the $Cr^{3+}(aq)/Cr^{2+}(aq)$ electrode and the chlorine electrode.

The voltage supplied by a cell is not constant but depends upon the current drawn from it. As electrons flow between the electrodes the equilibria are disturbed and the potential difference drops. The **electro-motive force (e.m.f.)**

of a cell is the potential difference when no current flows. This may be measured by using a voltmeter which has a very high resistance and so reduces the current flowing to a negligible value.

Any change in conditions which affects the equilibrium at one or both electrodes will also affect the cell e.m.f. For example, in the zinc electrode,

$$Zn^{2+}(aq) + 2e^- \rightleftharpoons Zn(s)$$

By Le Chatelier's principle, increasing the concentration of $Zn^{2+}(aq)$ will push the system to the right. This will use up electrons and make the electrode potential more positive. For an electrode which contains a gas, e.g.

$$H^+(aq) + e^- \rightleftharpoons \tfrac{1}{2}H_2(g)$$

Increasing the pressure of hydrogen will push the system to the left, releasing electrons and making the electrode potential more negative. In order to make a meaningful comparison between electrode potentials, we must define a set of standard conditions. ***Standard electrode potential, $E^\ominus$***, *refers to 25°C, 1 atmosphere pressure and with the concentrations of all aqueous species at 1 M.*

Up to now we have only considered potential differences. In order to give numerical values to individual electrode potentials, we need to pick one electrode to act as a reference point against which all others are measured. *The hydrogen electrode is taken to have a standard potential of zero volts.* In order to measure the standard potential of any other electrode, we therefore need to measure the e.m.f. of the cell shown in Figure 26.4.

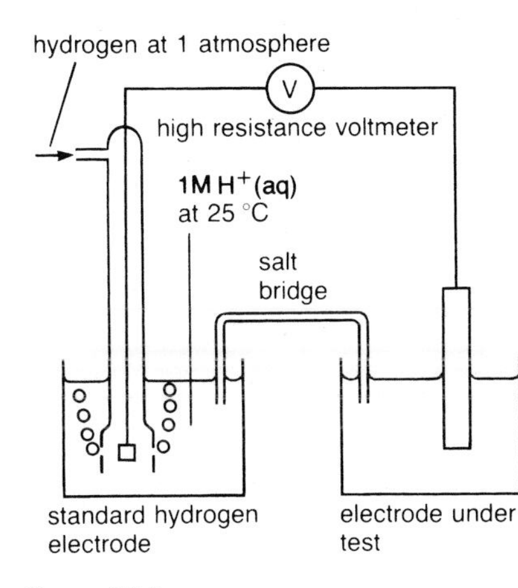

Figure 26.4
Experimental determination of electrode potential

The potential of the electrode under test is simply the e.m.f. of such a cell, with a negative value if it forms the negative pole in the cell. Of course, once the potential of any system has been measured relative to the standard hydrogen electrode, it may itself be used to find the potential of another system. In practice, such secondary standards are generally preferred, since hydrogen electrodes are quite tricky to set up and operate.

Focus 26a

1. Any reversible system that contains species related by the gain/loss of electrons can act as an electrode.
2. The **standard potential**, $E^\ominus$, of any electrode is measured relative to a hydrogen electrode at 25°C, 1 atmosphere pressure with all solutions at 1 M concentration.
3. The standard **e.m.f.** of a cell is given by the difference in standard potential of its electrodes.

Activity 26.2

The e.m.f. of a cell made up from a standard zinc electrode and a standard hydrogen electrode is 0.76 volts with the zinc as the negative pole.

1. Draw a cell diagram for this system and deduce the standard electrode potential of zinc.

When standard zinc and copper electrodes are connected into a cell, the e.m.f. is 1.10 volts with zinc as the negative pole.

2. Draw a cell diagram for this system and deduce the standard electrode potential of copper.
3. What is the e.m.f. of a cell made from standard hydrogen and copper electrodes? Draw a cell diagram for this system.

Table 26.1, which lists standard electrode potentials for redox systems in order from the most negative to the most positive, is known as the **electrochemical series (e.c.s.)**. As well as providing $E^\ominus$ values for calculating a cell e.m.f., it may be used to predict the direction of redox reactions. You will notice that the most reactive metals are at the top of the e.c.s. This is to be expected, since these elements are reducing agents which react by releasing electrons and so they have very negative $E^\ominus$ values. Non-metals, on the other hand, are generally oxidising agents which take electrons and have more positive $E^\ominus$ values. The e.c.s. therefore, compares systems in order of their reducing or oxidising power.

Table 26.1 Standard electrode potentials – an electrochemical series
Note: All potentials are quoted at 25°C, 1 atm pressure, with all aqueous concentrations being 1 M. Electrode reactions are written as reduction processes, i.e. with electrons added on the left hand side

Electrode process	$E^\ominus$/V
$Li^+(aq) + e^- \rightleftharpoons Li(s)$	−3.03
$Rb^+(aq) + e^- \rightleftharpoons Rb(s)$	−2.93
$K^+(aq) + e^- \rightleftharpoons K(s)$	−2.92
$Sr^{2+}(aq) + 2e^- \rightleftharpoons Sr(s)$	−2.89
$Ca^{2+}(aq) + 2e^- \rightleftharpoons Ca(s)$	−2.87
$Na^+(aq) + e^- \rightleftharpoons Na(s)$	−2.71
$Mg^{2+}(aq) + 2e^- \rightleftharpoons Mg(s)$	−2.37
$Be^{2+}(aq) + 2e^- \rightleftharpoons Be(s)$	−1.85
$Al^{3+}(aq) + 3e^- \rightleftharpoons Al(s)$	−1.66
$Mn^{2+}(aq) + 2e^- \rightleftharpoons Mn(s)$	−1.19
$Zn^{2+}(aq) + 2e^- \rightleftharpoons Zn(s)$	−0.76
$Cr^{3+}(aq) + 3e^- \rightleftharpoons Cr(s)$	−0.74
$2CO_2(g) + 2H^+(aq) + 2e^- \rightleftharpoons H_2C_2O_4(aq)$	−0.49
$Fe^{2+}(aq) + 2e^- \rightleftharpoons Fe(s)$	−0.44
$Cr^{3+}(aq) + e^- \rightleftharpoons Cr^{2+}(aq)$	−0.41
$Ti^{3+}(aq) + e^- \rightleftharpoons Ti^{2+}(aq)$	−0.37
$Ni^{2+}(aq) + 2e^- \rightleftharpoons Ni(s)$	−0.25
$Sn^{2+}(aq) + 2e^- \rightleftharpoons Sn(s)$	−0.14
$Pb^{2+}(aq) + 2e^- \rightleftharpoons Pb(s)$	−0.13
$H^+(aq) + e^- \rightleftharpoons \frac{1}{2}H_2(g)$	0.00
$Sn^{4+}(aq) + 2e^- \rightleftharpoons Sn^{2+}(aq)$	+0.15
$Cu^{2+}(aq) + e^- \rightleftharpoons Cu^+(aq)$	+0.15
$Cu^{2+}(aq) + 2e^- \rightleftharpoons Cu(s)$	+0.34
$\frac{1}{2}O_2(g) + H_2O(l) + 2e^- \rightleftharpoons 2OH^-(aq)$	+0.40
$Cu^+(aq) + e^- \rightleftharpoons Cu(s)$	+0.52
$\frac{1}{2}I_2(aq) + e^- \rightleftharpoons I^-(aq)$	+0.54
$MnO_4^-(aq) + e^- \rightleftharpoons MnO_4^{2-}(aq)$	+0.56
$Fe^{3+}(aq) + e^- \rightleftharpoons Fe^{2+}(aq)$	+0.77
$Ag^+(aq) + e^- \rightleftharpoons Ag(s)$	+0.80
$NO_3^-(aq) + 4H^+(aq) + 3e^- \rightleftharpoons NO(g) + 2H_2O(l)$	+0.96
$\frac{1}{2}Br_2(aq) + e^- \rightleftharpoons Br^-(aq)$	+1.09
$IO_3^-(aq) + 6H^+(aq) + 5e^- \rightleftharpoons \frac{1}{2}I_2(aq) + 3H_2O(l)$	+1.19
$MnO_2(s) + 4H^+(aq) + 2e^- \rightleftharpoons Mn^{2+}(aq) + 2H_2O(l)$	+1.23
$\frac{1}{2}Cr_2O_7^{2-}(aq) + 7H^+(aq) + 3e^- \rightleftharpoons Cr^{3+}(aq) + \frac{7}{2}H_2O(l)$	+1.33
$\frac{1}{2}Cl_2(aq) + e^- \rightleftharpoons Cl^-(aq)$	+1.36
$Mn^{3+}(aq) + e \rightleftharpoons Mn^{2+}(aq)$	+1.51
$MnO_4^-(aq) + 8H^+(aq) + 5e^- \rightleftharpoons Mn^{2+}(aq) + 4H_2O(l)$	+1.51
$Pb^{4+}(aq) + 2e^- \rightleftharpoons Pb^{2+}(aq)$	+1.69
$Co^{3+}(aq) + e^- \rightleftharpoons Co^{2+}(aq)$	+1.81
$S_2O_8^{2-}(aq) + 2e^- \rightleftharpoons 2SO_4^{2-}(aq)$	+2.01
$\frac{1}{2}F_2(g) + e^- \rightleftharpoons F^-(aq)$	+2.87

Are the species on the left hand side of these equations oxidising agents or reducing agents?

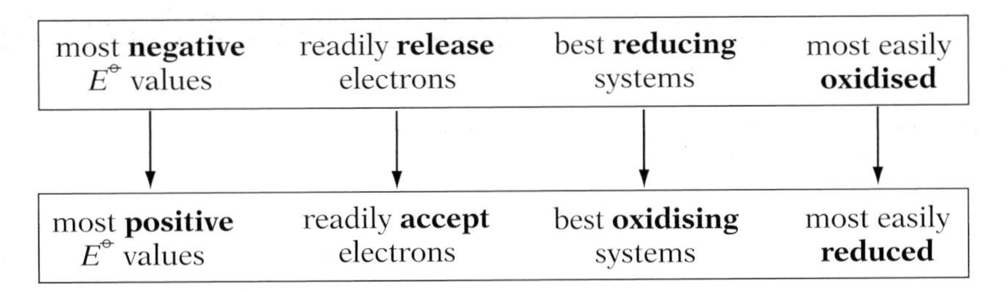

Take the standard zinc and copper systems mentioned in Activity 26.2. The zinc system will reduce the copper system that lies below it in the e.c.s. The spontaneous reactions will therefore be:

$$Zn^{2+}(aq) + 2e^- \longleftarrow Zn(s) \text{ and } Cu^{2+}(aq) + 2e^- \longrightarrow Cu(s)$$

If we add zinc metal to copper sulphate solution, we get a spontaneous redox reaction,

$$Zn(s) \longrightarrow Zn^{2+}(aq) + 2e^- \qquad \text{oxidation}$$

reducing agent

$$Cu^{2+}(aq) + 2e^- \longrightarrow Cu(s) \qquad \text{reduction}$$

oxidising agent

$$Zn(s) + Cu^{2+}(aq) \longrightarrow Zn^{2+}(aq) + Cu(s) \qquad \text{overall redox}$$

If, on the other hand, we add copper metal to a solution of zinc sulphate there will be no reaction.

Activity 26.3 Using the e.c.s. to predict redox reactions

Use the electrochemical series in Table 26.1 to predict the reaction, if any, when each of the following is added separately to hydrochloric acid:

1 magnesium ribbon,

2 copper turnings,

3 potassium manganate(VII) solution,

4 nitric acid.

We shall work through the first case to illustrate a strategy for tackling such problems.

First, identify all the species present and write down the redox systems which contain them:

	oxidising agents		reducing agents	
From the hydrochloric acid:	$\mathbf{H^+(aq)} + e^-$	$\rightleftharpoons$	$\frac{1}{2}H_2(g)$	$E^\ominus = 0.00$ V
	$\frac{1}{2}Cl_2(g) + e^-$	$\rightleftharpoons$	$\mathbf{Cl^-(aq)}$	$E^\ominus = +1.36$ V
From the magnesium:	$Mg^{2+}(aq) + 2e^-$	$\rightleftharpoons$	$\mathbf{Mg(s)}$	$E^\ominus = -2.37$ V

Now select the best oxidising agent present, i.e. the one in the system with the most positive $E^\ominus$ value. In this case,

$$\mathbf{H^+(aq)} + e^- \longrightarrow \tfrac{1}{2}H_2(g) \qquad E^\ominus = 0.00 \text{ V}$$

Now look at all the systems which have a **less positive** $E^\ominus$ value and identify any reducing agents present. If there is no suitable reducing agent there can be no reaction. In this case,

$$Mg^{2+}(aq) + 2e^- \longleftarrow \mathbf{Mg(s)} \qquad E^\ominus = -2.37 \text{ V}$$

Activity 26.3 Using the e.c.s. to predict redox reactions continued

Now combine the reduction and oxidation half-reactions so that the electrons cancel out to give the overall redox equation:

reduction	$\mathbf{2H^+(aq)} + 2e^- \longrightarrow H_2(g)$
oxidation	$\mathbf{Mg(s)} \longrightarrow Mg^{2+}(aq) + 2e^-$
redox	$\mathbf{2H^+(aq)} + \mathbf{Mg(s)} \longrightarrow Mg^{2+}(aq) + H_2(g)$

Thus we predict (correctly) that magnesium dissolves in hydrochloric acid to give magnesium ions and hydrogen gas. Now try the other questions in this activity.

Corrosion of iron

The rusting of iron involves an oxidation process,

$$Fe^{2+}(aq) + 2e^- \longleftarrow \mathbf{Fe(s)} \qquad E^\circ = -0.44\ V$$

Water and air contain systems capable of being reduced by iron,

$$\mathbf{H^+(aq)} + e^- \longrightarrow \tfrac{1}{2}H_2(g) \qquad E^\circ = 0.00\ V$$

$$\tfrac{1}{2}\mathbf{O_2(g)} + \mathbf{H_2O(l)} + 2e^- \longrightarrow 2OH^-(aq) \quad E^\circ = +0.40\ V$$

One way of preventing rusting is simply to cover the surface of the article with a coating that stops air and water reaching the surface of the iron, e.g. paint or grease. However the protection is lost if such a coating is damaged or scratched. Indeed, in the case of a tin-plated steel food can, rusting is actually speeded up. From Table 26.1,

$$Sn^{2+}(aq) + 2e^- \rightleftharpoons Sn(s) \qquad E^\circ = -0.14\ V$$

Any $Sn^{2+}(aq)$ will accelerate the oxidation of iron via the reaction,

$$Fe(s) + Sn^{2+}(aq) \longrightarrow Fe^{2+}(aq) + Sn(s)$$

However, this gives us a clue to another way of protecting iron. Since a less reactive metal, such as tin, accelerates rusting, perhaps a more reactive metal will reduce it? Galvanised iron has a thin surface coating of zinc, which has the following electrode potential:

$$Zn^{2+}(aq) + 2e^- \rightleftharpoons Zn(s) \quad E^\circ = -0.76\ V$$

If the surface layer is damaged, the zinc metal is the best reducing agent present and is **sacrificed**, converting Fe^{2+} back into metallic iron.

$$Zn(s) + Fe^{2+}(aq) \longrightarrow Zn^{2+}(aq) + Fe(s)$$

In the case of ships' hulls or bridge supports which are continually submerged in water, a piece of zinc is simply bolted onto the structure and is easily replaced as it dissolves away.

Rust can be an expensive problem

Cell e.m.f. and free energy change

The free energy change for a cell reaction may be calculated from its e.m.f. Under standard conditions,

$$\Delta G^\circ = -nFE^\circ$$

where n is the number of electrons transferred, ΔG° is the standard free energy change in joules and F is the charge on a mole of electrons, i.e. 96 500 coulombs.

Since the e.m.f. of a working cell is always taken as positive, the free energy change is negative, as required for a spontaneous reaction (see also Chapter 28).

Electrolysis reactions and rechargeable batteries

Spontaneous cell reactions may be reversed by applying an external voltage to drive the cell in reverse. We have already looked at some large-scale uses of such **electrolysis** reactions in the manufacture of aluminium (section 23.4) and the chlor-alkali industry (section 23.2).

In theory we should be able to recharge a chemical cell by simply reversing the spontaneous reaction. However, not all chemical cells are rechargeable. In an ordinary zinc–carbon battery the negative zinc electrode dissolves in use to give zinc ions:

$$Zn^{2+}(aq) + 2e^- \longleftarrow Zn(s) \qquad E^\ominus = -0.76\ V$$

If we try to reverse this process, then H^+ ions from the aqueous solution are more easily reduced:

$$H^+(aq) + e^- \longrightarrow \tfrac{1}{2}H_2(g) \qquad E^\ominus = 0.00\ V$$

Any attempt to recharge such a cell is likely to cause an explosion owing to the build up of hydrogen pressure.

In some cases, however, the original electrode reactions can be reversed by applying an external voltage. Ni–Cad batteries are used for small electrical appliances such as re-chargeable electric shavers, whilst car batteries use lead–acid cells in which plates of lead and lead coated with lead(IV) oxide, PbO_2, are immersed in fairly concentrated sulphuric acid. The redox equilibria involved are:

$$PbSO_4(s) + 2H^+(aq) + 2e^- \rightleftharpoons Pb(s) + H_2SO_4(aq) \qquad E^\ominus = -0.36\ V$$

$$PbO_2(s) + H_2SO_4(aq) + 2H^+(aq) + 2e^- \rightleftharpoons PbSO_4(s) + 2H_2O(l) \qquad E^\ominus = +1.68\ V$$

Activity 26.4

1. Write equations for the reactions occurring at each electrode in a lead–acid cell when it supplies a current.
2. Calculate the standard e.m.f. of a lead–acid cell.
3. When a lead–acid cell is completely discharged both electrodes are covered with a white solid. What is this?

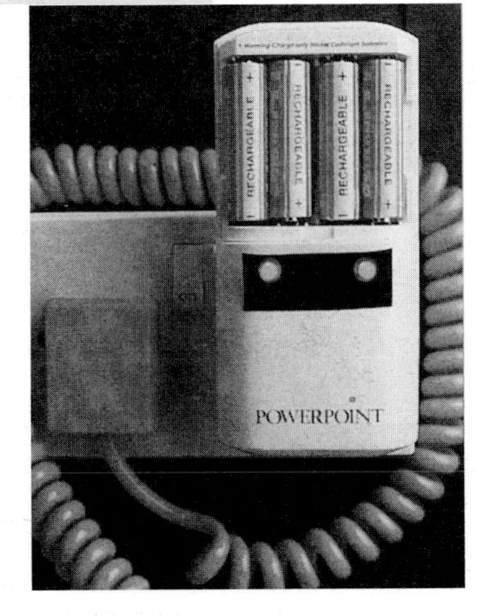

Spontaneous redox reactions are the source of the voltage supplied by all of these batteries. Only certain types of battery, such as the nickel–cadmium ones in the bottom photo, can readily be recharged by applying an external voltage

Focus 26b

1. The **electrochemical series** lists systems in order of standard potentials and may be used to predict the direction of redox reactions.
2. **Electrolysis** involves driving a redox reaction in the non-spontaneous direction by applying an external voltage.

26.2 Acid–base equilibria

A Bronsted–Lowry acid is defined as a proton donor and may only function as such in the presence of a proton acceptor, i.e. a base. Water is an **amphiprotic** solvent since it may either donate or accept protons. In fact, pure water self-ionises to a limited extent:

$$H_2O(l) + H_2O(l) \rightleftharpoons H_3O^+(aq) + OH^-(aq)$$

The equilibrium constant may be written,

$$K_c = \frac{[H_3O^+(aq)][OH^-(aq)]}{[H_2O(l)]^2}$$

Since the equilibrium lies well to the left, $[H_2O(l)]$ is virtually constant for pure water (and dilute solutions) and we may incorporate this into the equilibrium constant:

$$K_c = \frac{[H_3O^+(aq)][OH^-(aq)]}{\text{const}^2}$$

$$K_c.\text{const}^2 = K_w = [H_3O^+(aq)][OH^-(aq)]$$

Since the hydroxonium ion H_3O^+ is effectively a hydrated proton, we may simplify this to:

$$K_w = [H^+(aq)][OH^-(aq)]$$

K_w is known as the **ionic product of water** and has a value of 1×10^{-14} mol² dm⁻⁶ at 25°C. For pure water, $[H^+(aq)] = [OH^-(aq)]$ so:

$$K_w = [H^+(aq)]^2$$
$$\sqrt{K_w} = [H^+(aq)] = [OH^-(aq)] = 1 \times 10^{-7} \text{ mol dm}^{-3} \text{ at } 25°\text{C}$$

Any solution in which $[H^+(aq)] = [OH^-(aq)]$ is said to be **neutral**, i.e. neither acidic or basic. Addition of an acid increases $[H^+(aq)]$ and decreases $[OH^-(aq)]$ to maintain the value of K_w. A base has the opposite effect, increasing $[OH^-(aq)]$ at the expense of $[H^+(aq)]$.

Strong acids and alkalis ionise fully in aqueous solution, e.g.

1 M hydrochloric acid:	HCl(aq)	⟶	**H⁺(aq)**	+	Cl⁻(aq)
before ionisation	1 mol dm⁻³		0		(0)
after ionisation	0		1 mol dm⁻³		(1 mol dm⁻³)

$$K_w = [H^+(aq)][OH^-(aq)]$$
$$\text{at } 25°\text{C, } 10^{-14} \text{ mol}^2 \text{ dm}^{-6} = 1.[OH^-(aq)], \text{ i.e. } [OH^-(aq)] = 10^{-14} \text{ mol dm}^{-3}$$

Acids such as HCl are known as **monoprotic** since only one H^+ ion is produced from the formula unit. **Diprotic** acids, such as sulphuric acid, ionise to give two H^+ ions and **polyprotic** acids release even more H^+ ions per fomula unit. (You may also see the terms monobasic, dibasic and polybasic applied to such acids.)

1 M sodium hydroxide:	NaOH(aq)	⟶	Na⁺(aq)	+	**OH⁻(aq)**
before ionisation	1 mol dm⁻³		(0)		0
after ionisation	0		(1 mol dm⁻³)		1 mol dm⁻³

$$K_w = [H^+(aq)][OH^-(aq)]$$
$$\text{at } 25°\text{C, } 10^{-14} \text{ mol}^2 \text{ dm}^{-6} = [H^+(aq)].1, \text{ i.e. } [H^+(aq)] = 10^{-14} \text{ mol dm}^{-3}$$

Note that sodium hydroxide is **monoprotic** since each fomula unit can accept only one H^+ ion.

Activity 26.5 The pH scale and related values

Values of $[H^+(aq)]$ and $[OH^-(aq)]$ in dilute aqueous solutions cover a vast range. As we have seen, in 1 M HCl at 25°C, $[H^+(aq)] = 1$ mol dm⁻³ and $[OH^-(aq)] = 10^{-14}$ mol dm⁻³ whereas in 1 M NaOH these values are reversed. pH is a logarithmic scale which provides a more convenient, manageable range of numbers for measuring acidity and alkalinity.

$$[H^+(aq)] = 10^{-\text{pH}}$$

so,

$$\mathbf{pH = -log_{10}[H^+(aq)]}$$

Activity 26.5 The pH scale and related values continued

This means that the pH of 1 M hydrochloric acid is 1, whilst the pH of 1 M NaOH is 14. For pure water at 25°C, $[H^+(aq)] = 10^{-7}$ mol dm^{-3}, which corresponds to pH 7. A difference in pH of 1 corresponds to a tenfold change in $[H^+(aq)]$.

Note that log scales, such as pH, are simply numbers and have no units. Such scales are widely used in acid–base equilibria. For example;

$$pK_w = -\log_{10}K_w$$
$$pOH = -\log_{10}[OH^-(aq)]$$

1. Calculate the value of pK_w at 25°C.
2. Calculate the pOH value of: a) pure water, b) 1 M HCl, c) 1 M NaOH, at 25°C.
3. Calculate $[H^+(aq)]$, $[OH^-(aq)]$, pH and pOH for the following strong acid solutions at 25°C: a) 0.01 M HNO_3, b) 0.5 M HCl, c) 0.1 M H_2SO_4 (gives two protons per formula unit).
4. Calculate $[OH^-(aq)]$, $[H^+(aq)]$ pH and pOH for the following strong base solutions at 25°C: a) 0.1 M KOH, b) 0.5 M NaOH, c) 0.001 M $Ba(OH)_2$ (gives two hydroxide ions per formula unit).
5. The neutralisation of an acid by an alkali is exothermic and may be represented by the equation,

$$H^+(aq) + OH^-(aq) \longrightarrow H_2O(l)$$

How would you expect the values of a) K_w, b) pK_w, c) the pH of pure water, to change as the temperature is increased?

pH measurement

There are two main methods available to determine the hydrogen ion concentration of a solution and hence its pH.

An approximate value may be obtained quickly by the use of **universal indicator** paper or solution. The indicator develops a colour dependent upon the pH of the solution. This may be compared with a colour chart prepared by adding the indicator to solutions of known pH. The theory of acid–base indicators is developed more fully in this section.

Obviously the indicator method has its limitations. It is not suitable for highly coloured solutions, which would mask the indicator, and its accuracy depends on the skill and eyesight of the operator. A more accurate measure of pH may be found using a **pH meter**. This is an electrochemical cell which has one electrode whose potential changes with the hydrogen ion concentration of the solution. The obvious choice is the hydrogen electrode:

$$H^+(aq) + e^- \rightleftharpoons \tfrac{1}{2}H_2(g)$$

At any given temperature, increasing the concentration of hydrogen ions will push the system to the right thus making the electrode potential more positive. When the hydrogen electrode is connected to another electrode of known potential, e.g. a calomel electrode, the e.m.f. of the cell gives a measure of the pH of the solution into which the hydrogen electrode dips. In practice, since the hydrogen electrode is difficult to set up and maintain, other pH-sensitive electrodes are preferred.

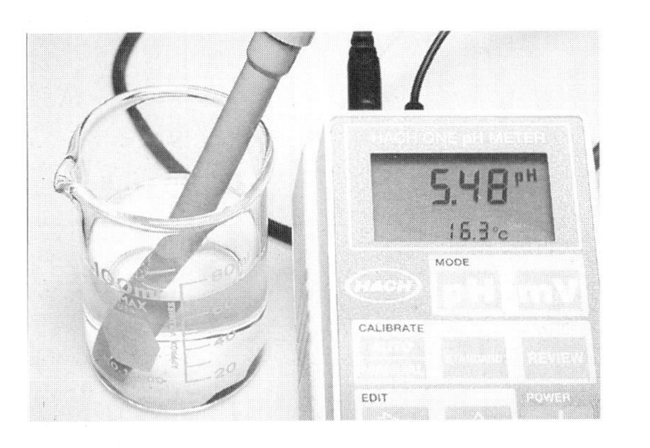

The pH meter reading shows that this solution is slightly acidic

Not all acids and bases ionise fully in solution. Whereas the pH of 0.1 M hydrochloric acid at room temperature is around 1, the pH of 0.1 M ethanoic acid, CH_3COOH, is about 3, i.e. $[H^+(aq)]$ is only about 0.001 mol dm^{-3}. This means that, like many organic acids, ethanoic acid is **weak**, i.e. it ionises only slightly in aqueous solution.

hydrochloric acid: STRONG – fully ionised in solution

$$HCl(aq) \longrightarrow H^+(aq) + Cl^-(aq)$$

ethanoic acid: WEAK – slightly ionised in solution

$$CH_3COOH(aq) \rightleftharpoons CH_3COO^-(aq) + H^+(aq)$$

The **strength** of such an acid, i.e. its readiness to ionise, depends upon its acid dissociation constant, K_a:

$$K_a = \frac{[CH_3COO^-(aq)].[H^+(aq)]}{[CH_3COOH(aq)]}$$

Using the approximate pH of 0.1 M ethanoic acid given above, we may estimate an approximate value for K_a at room temperature:

	$CH_3COOH(aq)$	$\rightleftharpoons$	$CH_3COO^-(aq)$	+	$H^+(aq)$
before ionisation:	0.1 mol dm^{-3}		0		0

If x mol dm^{-3} of the acid actually ionises, then

at equilibrium:	$(0.1 - x)$ mol dm^{-3}		x mol dm^{-3}		x mol dm^{-3}

If pH ~3 = $-\log_{10}[H^+(aq)]$ then $[H^+(aq)]$ = $-(\text{antilog}_{10}3)$ = 10^{-3} mol dm^{-3}, i.e. x = 0.001 mol dm^{-3}

$$K_a = \frac{[CH_3COO^-(aq)].[H^+(aq)]}{[CH_3COOH(aq)]}$$

$$= \frac{x^2}{(0.1 - x)} \text{ mol dm}^{-3} \simeq \frac{(0.001)^2}{(0.1 - 0.001)}$$

$$= \frac{0.000001}{0.099}$$

$$\simeq 0.00001 \text{ or } 10^{-5} \text{ mol dm}^{-3}$$

(Note that since the acid is weak there is very little error in taking the concentration of the unionised acid at equilibrium as equal to the total acid concentration.)

Now $pK_a = -\log_{10} K_a$, so for ethanoic acid at room temperature, $pK_a \simeq -\log_{10} 10^{-5}$, i.e. 5.

Similarly not all bases are strong. Ammonia, for example, ionises only slightly in aqueous solution:

$$NH_3(aq) + H_2O(l) \rightleftharpoons NH_4^+(aq) + OH^-(aq)$$

$$K_c = \frac{[NH_4^+(aq)].[OH^-(aq)]}{[NH_3(aq)].[H_2O(l)]}$$

In dilute solution, $[H_2O(l)]$ is virtually fixed so we may write a simplified expression for the base dissociation constant, K_b:

$$K_b = \frac{[NH_4^+(aq)].[OH^-(aq)]}{[NH_3(aq)]}$$

If we know the value of K_b (or pK_b) we may estimate the pH of an aqueous base of given concentration. So given that pK_b for ammonia at 25°C is 4.75, we can calculate the approximate pH of 0.01 M ammonia solution:

$$pK_b = 4.75 = -\log_{10}K_b, \text{ so } K_b = -\text{antilog}_{10}(4.75) = 0.0000178 \text{ mol dm}^{-3}$$

$$NH_3(aq) + H_2O(l) \rightleftharpoons NH_4^+(aq) + OH^-(aq)$$

before ionisation: $0.01\ mol\ dm^{-3}$ 0   0

Let x mol dm^{-3} of ammonia ionise, then

at equilibrium: $(0.01 - x)\ mol\ dm^{-3}$ $x\ mol\ dm^{-3}$ $x\ mol\ dm^{-3}$

$$K_b = \frac{[NH_4^+(aq)].[OH^-(aq)]}{[NH_3(aq)]} = \frac{x^2}{1 - x}$$

Since ammonia is weak, it ionises only slightly. This means that $0.01 >> x$ and, to a good approximation, $(0.01 - x) \simeq 0.001$.

$$K_b = 0.0000178\ mol\ dm^{-3} \simeq \frac{x^2}{0.01}$$

$$x^2 \approx 0.0000178 \times 0.01 = 0.000000178\ mol^2\ dm^{-6}$$

$$x = \sqrt{(0.000000178)}\ mol^2\ dm^{-6} = 0.000422\ mol\ dm^{-3} = [OH^-(aq)]$$

$$\text{At } 25°C,\ K_w = 10^{-14}\ mol^2\ dm^{-6} = [H^+(aq)].[OH^-(aq)]$$

$$= [H^+(aq)] \times 0.000422\ mol\ dm^{-3}$$

$$\text{Thus, } [H^+(aq)] = \frac{[10^{-14}]}{0.000422} = 2.37 \times 10^{-11}\ mol\ dm^{-3}$$

$$pH = -log^{10}[H^+(aq)] \simeq 10.6$$

Activity 26.6 Weak acids and bases

NOTE: All the following measurements refer to 25°C when $K_w = 1 \times 10^{-14}\ mol^2\ dm^{-6}$.

1 Given any two of the following pieces of information about a weak monoprotic acid solution it is possible to calculate the third: molar concentration, pH and K_a (or pK_a).

Fill in the spaces in the following table which shows the effect of various substituent groups, *X*, on the properties of ethanoic acid. Ethanoic acid itself has been done as an example.

Substituent	K_a/mol dm^{-3}	pK_a	pH of a 0.1 M solution
none: **H**—CH_2COOH	1.74×10^{-5}	4.76	2.9
hydroxy: **HO**—CH_2COOH	1.48×10^{-4}		
chloro: **Cl**—CH_2COOH			1.9
amino: $\mathbf{H_2N}$—CH_2COOH		9.87	

Arrange the acids in order of increasing strength.

2 In a similar way, complete the following table which shows the effect on base properties of successively substituting methyl groups for the hydrogen atoms in ammonia. Ammonia itself has been done as an example.

Base	K_b/mol dm^{-3}	pK_b	pH of a 0.1 M solution
ammonia: NH_3	1.78×10^{-5}	4.75	11.1
methylamine: CH_3NH_2		3.36	
dimethylamine: $(CH_3)_2NH$	5.25×10^{-4}		
trimethylamine: $(CH_3)_3N$			11.4

Arrange these bases in order of increasing strength.

A **salt** is produced when an acid is neutralised by a base, e.g.

BASE	+	ACID	$\longrightarrow$	SALT		
NaOH(aq)	+	HCl(aq)	$\longrightarrow$	NaCl(aq)	+	H_2O(l)
NH_3(aq)	+	HNO_3(aq)	$\longrightarrow$	NH_4NO_3(aq)		
KOH(aq)	+	CH_3COOH(aq)	$\longrightarrow$	CH_3COOK(aq)	+	H_2O(l)

However, as Table 26.2 shows, not all salts give neutral solutions. To explain this we must consider the possible reactions between the ions in the salt and water molecules.

Table 26.2

Salt	pH of solution	base	pK_b*	acid	pK_a*
sodium chloride NaCl	7.0	NaOH	s	HCl	s
sodium ethanoate CH_3COONa	9.4	NaOH	s	CH_3COOH	4.76
ammonium chloride NH_4Cl	4.6	NH_3	4.75	HCl	s
ammonium ethanoate CH_3COONH_4	7.0	NH_3	4.75	CH_3COOH	4.76

*s indicates a strong (fully ionised) acid or base

The cation in the salt, X^+, may combine with hydroxide ions;

$$X^+(aq) + OH^-(aq) \rightleftharpoons XOH(aq)$$

Such a reaction will not be important if XOH is a strong base. This fully ionises in solution and the above equilibrium will lie completely to the left, e.g. with Na^+:

$$Na^+(aq) + OH^-(aq) \longleftarrow \underset{\text{a strong base}}{NaOH(aq)}$$

However, if XOH is a weak base, OH^- ions are effectively removed from the solution tending to make it more acidic, e.g. with NH_4^+:

$$NH_4^+(aq) + OH^- \rightleftharpoons H_2O(l) + \underset{\text{a weak base}}{NH_3(aq)}$$

Similarly, the anion from the salt, Y^-, may also combine with ions from the water:

$$H^+(aq) + Y^-(aq) \rightleftharpoons HY(aq)$$

Again there will be virtually no effect if the acid formed, HY, is strong and therefore fully ionised in solution, e.g. with Cl^-:

$$H^+(aq) + Cl^-(aq) \longleftarrow \underset{\text{a strong acid}}{HCl(aq)}$$

If HY is weak, however, H^+(aq) are effectively removed from solution tending to make it more alkaline, e.g. with CH_3COO^-(aq),

$$H^+(aq) + CH_3COO^-(aq) \rightleftharpoons \underset{\text{a weak acid}}{CH_3COOH(aq)}$$

Focus 26c

1. Water self-ionises slightly. The extent of this ionisation is given by the **ionic product**, K_w, which increases with temperature.
2. An acid donates a proton, H^+, to a base.
3. A **monoprotic acid** (base) donates (accepts) one H^+ ion per formula unit, whereas **diprotic** and **polyprotic** acids (bases) donate (accept) two or more H^+ ions per formula unit respectively.
4. The **strength** of an acid (base) is a measure of its ability to donate (accept) H^+ ions. The **acid (base) dissociation constant**, K_a (K_b) is a measure of the strength of a weak acid (base).

Activity 26.7 pH changes during an acid–base titration

Consider adding 0.100 M NaOH to 25 cm^3 of 0.100 M HCl, both of which are fully ionised:

$$HCl(aq) \longrightarrow H^+(aq) + Cl^-(aq), \qquad NaOH(aq) \longrightarrow Na^+(aq) + OH^-(aq)$$

The neutralisation reaction may be represented simply as,

$$H^+(aq) + OH^-(aq) \longrightarrow H_2O(l)$$

Up to the equivalence point, we may calculate the number of moles of $H^+(aq)$ left. From the total volume of solution we can calculate $[H^+(aq)]$ and then pH.

At the start, when no NaOH has been added:

$$\text{moles of } H^+ \text{ at start} = 0.1 \times \frac{25}{1000} = 0.0025$$

$$\text{moles of } OH^- \text{ added} = 0.0000$$

$$\text{moles of } H^+ \text{ left} = 0.0025$$

$$\text{total volume of solution} = 25 \text{ cm}^3$$

$$[H^+(aq)] = 0.0025 \times \frac{1000}{25} = 0.100 \text{ mol dm}^{-3}$$

$$\text{pH} = -\log_{10}(0.100) = 1.00$$

After 10 cm^3 0.100 M NaOH has been added:

$$\text{moles of } H^+ \text{ at start} = 0.0025$$

$$\text{moles of } OH^- \text{ added} = 0.1 \times \frac{10}{1000} = 0.0010$$

$$\text{moles of } H^+ \text{ left} = 0.0025 - 0.0010 = 0.0015$$

$$\text{total volume of solution} = 25 + 10 = 35 \text{ cm}^3$$

$$[H^+(aq)] = 0.0015 \times \frac{1000}{35} = 0.04358 \text{ mol dm}^{-3}$$

$$\text{pH} = -\log_{10}[H^+(aq)] = 1.36$$

a) In a similar way complete the following table.

Volume of NaOH/cm^3	moles OH^- added	moles H^+ left	total volume/cm^3	[H+]/mol dm^{-3}	pH
0.0	0.0000	0.0025	25.0	0.10000	1.00
10.0	0.0010	0.0015	35.0	0.04358	1.36
15.0	0.0015	0.0010	40.0		
20.0	0.0020				
24.0					
24.5					
24.9					

After 25 cm^3 NaOH has been added there is no $H^+(aq)$ left in solution *from the acid*. We must, however, remember that pure water self-ionises and at 25°C the pH will be 7.

Addition of more NaOH will result in excess $OH^-(aq)$ ions in solution and we must calculate $[H^+(aq)]$ indirectly from $K_w = [H^+(aq)][OH^-(aq)] = 1 \times 10^{-14}$ mol^2 dm^{-6} (at 25°C), e.g. after 25.1 cm^3 NaOH has been added:

$$\text{moles of } H^+ \text{ at start} = 0.0025$$

$$\text{moles of } OH^- \text{ added} = 0.1 \times \frac{25.1}{1000} = 0.00251$$

Activity 26.7 pH changes during an acid–base titration continued

$$\text{moles of } OH^- \text{ in excess} = 0.00251 - 0.0025 = 0.00001$$

$$\text{total volume of solution} = 25 + 25.1 = 50.1 \text{ cm}^3$$

$$[OH^-(aq)] = 0.00001 \times \frac{1000}{50.1} = 0.00020 \text{ mol dm}^{-3}$$

$$[H^+(aq)] = \frac{1 \times 10^{-14}}{[OH^-]} = 5.01 \times 10^{-11} \text{ mol dm}^{-3}$$

$$\text{pH} = -\log_{10} [H^+(aq)] = 10.30$$

b) Complete the rest of this table.

volume of NaOH/cm³	moles OH^- in excess	total volume/cm³	$[OH^-]$ /mol dm⁻³	$[H^+]$ /mol dm⁻³	pH
25.1	0.00001	50.1	0.00020	5.01×10^{-11}	10.30
25.5	0.00005	50.5	0.00099	1.01×10^{-11}	11.00
26.0	0.0001				
30.0					
35.0					
40.0					
50.0					

The results may be plotted as a graph of pH against volume of NaOH added as shown in Figure 26.5a.

2 If either (or both) the acid or base is weak, this affects the pH curve. Account as fully as you can for the shapes of the titration curves shown in Figures 26.5b, c and d in particular for the numbered sections.

Figure 26.5 ▼
pH changes during acid–base titrations
(a) strong acid–strong base
(b) weak acid–strong base
(c) strong acid–weak base
(d) weak acid–weak base

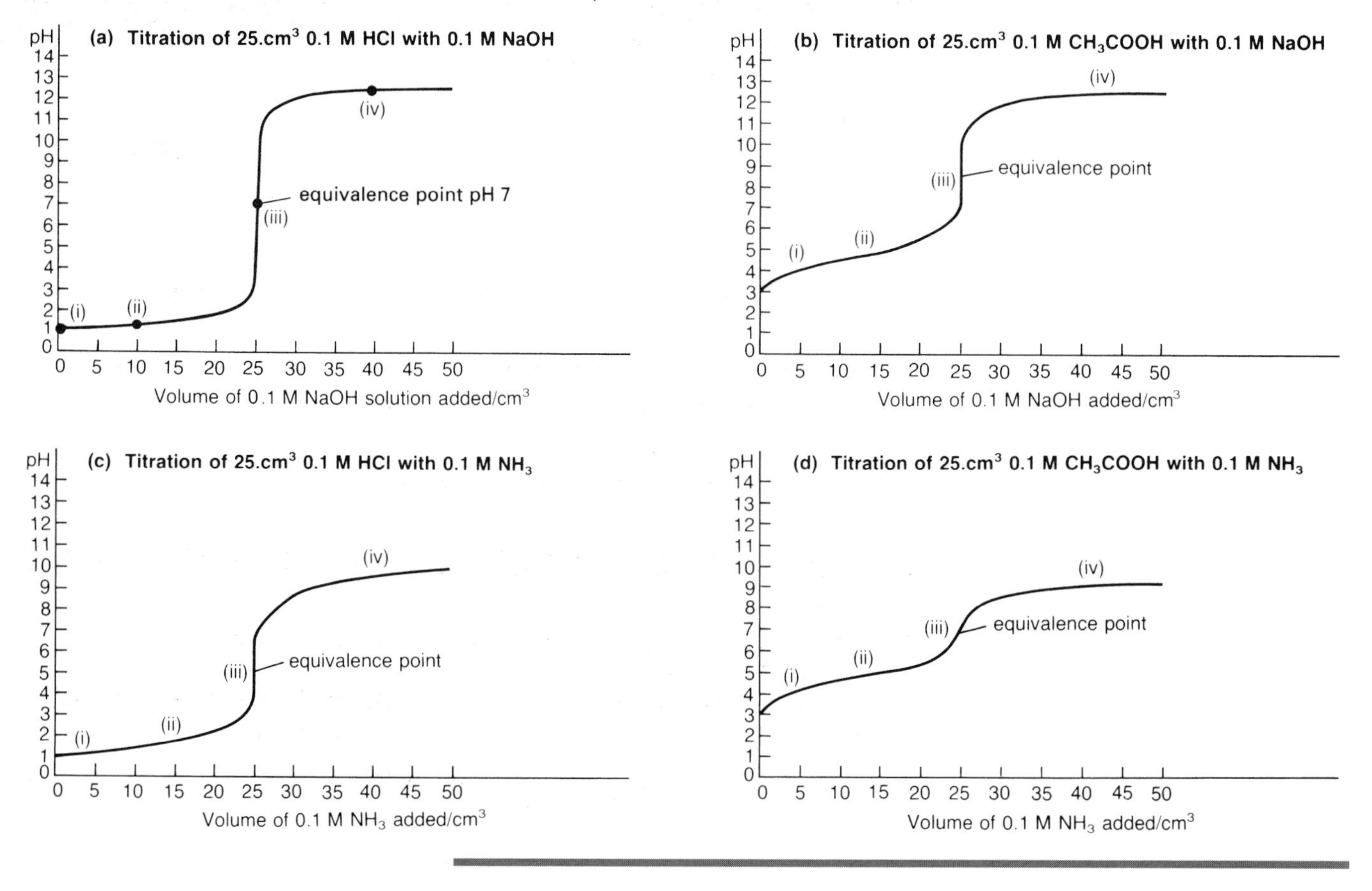

In order to find the equivalence point in a titration we need to detect the accompanying 'jump' in pH. This may be done by adding a suitable acid–base indicator. These are substances whose colour is pH dependent. Many indicators are themselves weak acids (or bases) whose unionised and ionised forms have different colours, i.e.

$$\underset{\text{colour 1 (unionised)}}{HIn(aq)} \rightleftharpoons H^+(aq) + \underset{\text{colour 2 (ionised)}}{In^-(aq)}$$

The acid dissociation constant in this case is known as the indicator constant, K_{in},

$$K_{in} = \frac{[H^+(aq)][In^-(aq)]}{[HIn(aq)]}$$

Rearranging we get:

$$\frac{K_{in}[HIn(aq)]}{[In^-(aq)]} = [H^+(aq)]$$

When $[H^+(aq)] = K_{in}$ (i.e. when $pH = pK_{in}$) the concentrations of ionised and unionised forms must be equal and so the indicator will appear as a equal mixture of the two coloured forms. Adding more acid means that [HIn(aq)] must predominate, whilst under more alkaline conditions the ionised form will be favoured. In practice the eye can only detect quite large changes in the ratios of the two forms, so the indicator appears to change from one colour to the other over a pH range of 2 or 3 units. If this change occurs within the pH jump in a titration, then the indicator can be used to detect the equivalence point. Table 26.3 shows the working pH range and colours of some common acid–base indicators.

Table 26.3 pH ranges over which some common acid–base indicators change colour

Indicator	pH 0	1	2	3	4	5	6	7	8	9	10	11	12	13
bromophenol blue		YELLOW		*	*	BLUE								
methyl orange[†]			RED	*	*		YELLOW							
methyl red				RED	*		*		YELLOW					
litmus[†]				RED			*		*	BLUE				
bromothymol blue					YELLOW		*	*		BLUE				
phenolphthalein[†]						COLOURLESS			*		*	RED		
thymolphthalein							COLOURLESS			*	*	BLUE		

[†]These indicators are amongst the commonest in general use.

What would be the approximate pH of a solution which gives a yellow colour with both methyl red and bromothymol blue?

Activity 26.8 Choosing and using acid-base indicators

1. Two indicators commonly used in acid–base titrations are methyl orange and phenolphthalein. Using the information in Table 26.3 explain whether either, both, or neither of these indicators could be used to determine the equivalence points for the titrations whose pH curves are shown in Figures 26.5a, b, c and d.
2. What method could you use to determine the equivalence point in a titration for which there is no suitable indicator?

Activity 26.8 Choosing and using acid–base indicators continued

3 Why is it important to use only a small amount of an indicator in a titration?

4 A **universal** indicator is a mixture that gives a variety of colours over a given pH range. Explain how the colour of a mixture of bromothymol blue and phenolphthalein would change as the pH is gradually increased from 5 to 10.

Activity 26.9 Acid–carbonate titrations

More complex pH curves are obtained if the acid can donate more than one H^+ ion or the base can accept more than one proton per formula unit. In the case of carbonate ions, for example,

$$CO_3^{2-}(aq) + H^+(aq) \rightleftharpoons HCO_3^-(aq)$$
$$HCO_3^-(aq) + H^+(aq) \rightleftharpoons H_2O(l) + CO_2(aq)$$

Each of these reactions has its own equivalence point. The pH curve for a titration of aqueous carbonate with a strong acid is shown in Figure 26.6. Careful choice of indicator allows us to detect either (or both) of these equivalence points.

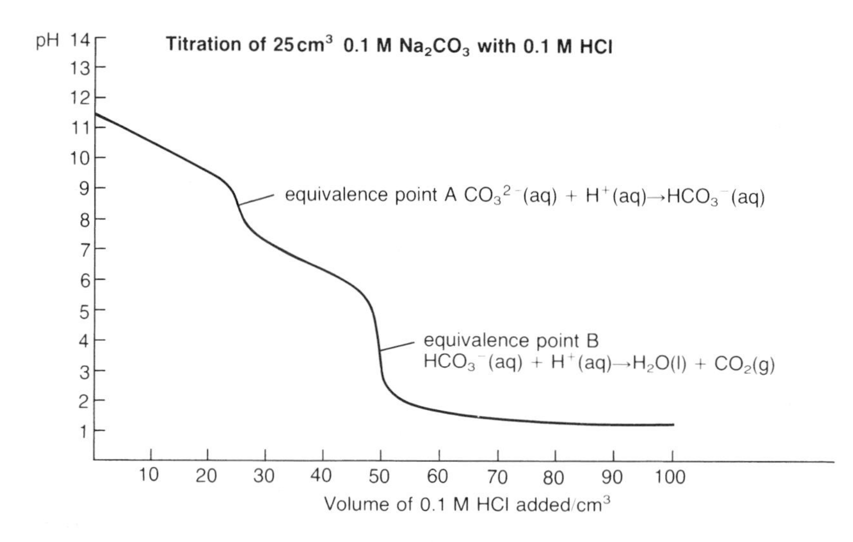

Figure 26.6
pH curve for a carbonate–strong acid titration

1 Which of the indicators mentioned in Activity 26.8 could be used to detect the equivalence point for the formation of:
   a) hydrogen carbonate, HCO_3^-
   b) carbon dioxide, CO_2?

The following experiment was carried out to find the ratio of carbonate ions to hydrogencarbonate ions in 25.0 cm³ of a solution. Phenolphthalein was first added and it took 10.0 cm³ of 0.100 0M hydrochloric acid for the initial colour to virtually disappear.

2 **a)** What was the initial colour on adding phenolphthalein?
   b) Write an ionic equation for the reaction taking place during the addition of 10.0 cm³ of acid.
   c) What is the main ion left in solution at this stage?

After adding methyl orange the titration was continued and it took a further 30.0 cm³ of 0.100 M HCl for this indicator to change colour.

Activity 26.9 Acid–carbonate titrations continued

3 **a)** What were the initial and final colours of methyl orange in this titration?
b) Write an ionic equation for the reaction occurring during this second addition of acid.
c) Why doesn't the phenolphthalein added earlier interfere in this second titration?

4 Calculate the number of moles of carbonate and hydrogencarbonate ions present in the original solution and so find the ratio of CO_3^{2-}:HCO_3^-.

Pure water at 25°C has a pH of exactly 7 but it takes only a minute amount of an acidic or basic impurity to affect this quite drastically. In fact the pH of distilled water is often as low as 5 owing to carbon dioxide dissolved from the air. A **buffer** solution has a known pH which is resistant to change on addition of moderate amounts of acid or base. Such a buffer must provide an acid–base equilibrium system of the following type:

$$\text{acid} \rightleftharpoons \text{base} + H^+(aq)$$

when extra H^+ is added the system moves $\longleftarrow$ (removing H^+)
when H^+ is removed the system moves $\longrightarrow$ (replacing H^+)

In an effective buffer there must be enough 'acid' present to replace any $H^+(aq)$ removed from the system and enough 'base' to mop up added $H^+(aq)$. One type of buffer is based upon a weak acid such as ethanoic acid,

$$CH_3COOH(aq) \rightleftharpoons CH_3COO^-(aq) + H^+(aq)$$

The acid is only slightly dissociated so there is plenty of the unionised form left which can split up to replace any $H^+(aq)$ removed from the solution. However, there are relatively few ethanoate ions, $CH_3COO^-(aq)$, available to absorb any added $H^+(aq)$. This can be remedied by adding a salt of the acid, e.g. sodium ethanoate, which fully dissociates in solution,

$$CH_3COONa(aq) \longrightarrow \underbrace{CH_3COO^-(aq)}_{\text{extra ethanoate ions}} + Na^+(aq)$$

Thus *a mixture of a weak acid and its salt of a strong base makes an effective 'acidic' buffer solution*. The actual pH of the solution depends upon the ratio of acid to salt used.

Buffer solutions are used to check the calibration of pH meters. Many important reactions, especially in biochemical systems, will only occur within quite narrow pH ranges which can be provided by a suitable buffer.

Activity 26.10 Buffering injections

Certain properties of blood, such as acidity and sugar level, must be kept kept within quite narrow limits. For example, human blood is very slightly alkaline and its pH must be maintained close to 7.4. If blood pH drops to 7.0 the individual may lapse into a fatal acidotic coma. On the other hand, a pH above 7.5 can cause an equally serious alkalosis tetany. Blood itself contains a natural buffer system consisting of carbonate and hydrogencarbonate ions:

$$HCO_3^-(aq) \rightleftharpoons CO_3^{2-}(aq) + H^+(aq) \qquad pK_a = 10.3$$

Solutions for injection are buffered to pH 7.4 often using the same system. Calculate the mole ratio of CO_3^{2-}:HCO_3^- which would give a buffer solution of the required pH.

Focus 26d

1. Acid–base indicators are themselves weak acids or bases where the ionised and unionised forms have different colours.
2. **Equivalence points** in acid–base titrations involve a jump in pH value. If large enough this may be detected using an appropriate indicator, otherwise a pH meter must be used.
3. A mixture of a weak acid with its salt (or a weak base with its salt) resists change in pH and is known as a **buffer solution**.

Can buffers help you slim?

A patent, filed by Pepsico in January 1992, describes a drink that is at least fifty times as acidic as lemonade. The tartness is disguised with additives and fruit flavouring. Citric and phosphoric acids are blended with still or carbonated water to give a very acidic solution. Citrate and phosphate salts are then added to buffer the mixture, keeping the pH above 2.5. The sourness is disguised by adding further fruit flavouring and artificial sweeteners such as saccharine or aspartame.

Volunteers who sampled the drink 20 minutes before they sat down to a meal of macaroni and beef thought it tasted like a conventional diet soft drink. However, they ate between 10 and 30% less of the meal that followed.

As yet there is no explanation of how the drink suppresses appetite.

26.3 Solubility equilibria

Addition of a solute with stirring to a fixed amount of solvent will eventually result in a dynamic equilibrium in which a **saturated** solution is in contact with the undissolved solid. At this point the rate at which the solid is dissolving equals the rate at which it is crystallising out of the solution.

The **solubility** of the solid is defined as the maximum amount which will dissolve in a specified amount of solvent at a given temperature. One very useful way of expressing solubility is to give the molar concentration of the solute, i.e. the number of moles present in 1 dm^3 of saturated solution.

In the case of an ionic compound which is sparingly soluble in water, the ions separate in solution. Thus for silver chloride we may write:

$$AgCl(s) \rightleftharpoons Ag^+(aq) + Cl^-(aq)$$

Since the concentration of any pure solid is fixed, we may incorporate this into the equilibrium constant for the system giving a constant known as the **solubility product**, K_{sp}:

$$K_{sp} = [Ag^+(aq)][Cl^-(aq)]$$

Note that the solubility product is NOT the same as solubility, although the two quantities are related.

Activity 26.11

At 25°C the maximum mass of silver chloride that can be dissolved in 1 dm^3 of water is 0.00193 g.

1. Calculate the solubility of silver chloride at this temperature in mol dm^{-3}.
2. Use your answer to the previous question to calculate a value of the solubility product of silver chloride at 25°C and give its units.
3. What will happen if a little concentrated hydrochloric acid is added to a saturated solution of silver chloride? Explain your answer.

Focus 26e

1. When an excess of a sparingly soluble salt is shaken with water, a heterogeneous dynamic equilibrium is set up between the solid and the saturated solution.
2. The maximum concentration of the aqueous ions which can co-exist in solution is given by the **solubility product**, K_{sp}. Increasing the concentration of one ion automatically reduces the concentration of the other which can remain in solution.

In many cases the ionic solid produces more than two ions per formula unit, e.g. for silver chromate(VI), Ag_2CrO_4:

$$Ag_2CrO_4(s) \rightleftharpoons 2Ag^+(aq) + CrO_4^{2-}(aq)$$

$$\text{so, } K_{sp} = [Ag^+(aq)]^2[CrO_4^{2-}(aq)]$$

Activity 26.12

The concentration of a saturated solution of silver chromate(VI) at room temperature was found to be 6.9×10^{-5} mol dm^{-3}.

1. What is the concentration of $Ag^+(aq)$ in this solution?
2. What is the concentration of $CrO_4^{2-}(aq)$ in this solution?
3. Calculate a value for the solubility product of silver chromate(VI) at this temperature and give its units.

Questions on Chapter 26

1 Describe, with the aid of a labelled diagram, how you would measure the standard electrode potential of a metal.

2 An electrochemical cell was set up using the following standard electrodes:

$$Pb^{2+}(aq) + 2e^- \rightleftharpoons Pb(s) \quad E^\ominus = -0.126 \text{ V}$$

$$Sn^{2+}(aq) + 2e^- \rightleftharpoons Sn(s) \quad E^\ominus = -0.136 \text{ V}$$

a) Under what conditions must standard electrode potentials be measured?
b) Which electrode system is used as the arbitrary zero against which all other electrode potentials are measured?
c) What is the e.m.f. of the above cell?
d) If the two metals are joined by a wire, in which direction will electrons flow? Explain your answer.
e) Draw the cell diagram showing the spontaneous chemical changes which take place when a current is taken from the cell.
f) What would happen to the magnitude of the cell e.m.f. if the solution of lead ions were diluted? Explain your answer.
g) In equilibria involving solid components, it is conventional to omit concentration terms for solids when writing the expression for K_c.
Write the expression for K_c for the following reaction:

$$Sn(s) + Pb^{2+}(aq) \rightleftharpoons Sn^{2+}(aq) + Pb(s)$$

h) An aqueous solution, initially 1.0 M in both tin and lead ions, is allowed to reach equilibrium with a mixture of both powdered metals. If the value of K_c under these conditions is 2.2, calculate the concentrations of each metal ion in solution at equilibrium.

3 What is meant by disproportionation? Use the following information to illustrate and explain this process.

$$MnO_4^-(aq) + e^- \rightleftharpoons MnO_4^{2-}(aq) \quad E^\ominus = +0.60 \text{ V}$$

$$4H^+(aq) + MnO_4^{2-}(aq) + 2e^- \rightleftharpoons MnO_2(s) + 2H_2O(l) \quad E^\ominus = +1.55 \text{ V}$$

4 Chlorine disproportionates in hot sodium hydroxide solution to give sodium chloride and sodium chlorate(V).
a) What is meant by disproportionation?
b) Deduce the oxidation state of chlorine in each of the following species: Cl_2, NaCl, $NaClO_3$.
c) Write half-equations for the conversion of chlorine, Cl_2, into:
i) chloride ions, Cl^-
ii) chlorate(V) ions, ClO_3^-.
d) Classify each of the changes in c) above as oxidation or reduction.
e) Write an ionic equation for the overall disproportionation reaction.

5 What is the change in oxidation state of each metal in the following conversions? By combining suitable half-equations deduce a balanced equation for the oxidation of Sn^{2+} by MnO_4^- in acid solution.

$$MnO_4^- \longrightarrow Mn^{2+} \qquad Sn^{2+} \longrightarrow Sn^{4+}$$

6 In aqueous solution, $HNO_3(l)$ acts as a proton donor, whereas in $H_2SO_4(l)$, it behaves as a proton acceptor. Write equations for both of these reactions and label all species as acid or base.

7 Hydrochloric acid ionises fully in solution:

$$HCl(aq) \longrightarrow H^+(aq) + Cl^-(aq)$$

Questions on Chapter 26 *continued*

What is the molarity of aqueous hydrochloric acid with a pH value of 3?

8 Outline two methods for determining the pH of an aqueous solution giving **one** advantage and disadvantage of each.

9 a) Define the terms pH, K_w and pK_w.
b) If the pH of pure water at 25°C is 7.0 what are the corresponding values of K_w and pK_w?
c) The pH of pure water at a different temperature, *T*, is 7.6. Is *T* above or below 25°C? Explain how you arrive at your answer.
d) Calculate the pH values of 0.05 M sulphuric acid and 0.05 M sodium hydroxide:
i) at 25°C,
ii) at temperature *T*.

10 This question concerns a weak monoprotic acid, HA.
a) Write an expression for its dissociation constant, K_a.
b) How is pK_a related to K_a?
c) Determine the value of K_a given that the pK_a of HA is 4.30.
d) Calculate the approximate pH of 2 M aqueous HA.
e) How could you convert the solution from d) above into a buffer solution?

11 Propanoic acid dissociates as follows in aqueous solution:

$$C_2H_5COOH(aq) \rightleftharpoons C_2H_5COO^-(aq) + H^+(aq)$$

a) Write an expression for the acid disocciation constant, K_a.
b) If the pH of 0.1M propanoic acid at 25°C is 4.6, calculate a value for K_a at this temperature.
c) Use your answer to part b) to calculate the approximate pH of 1.0 M propanoic acid at 25°C.
d) Describe an experiment you would carry out to find the molar concentration of a solution of propanoic acid.

12 Ammonia reacts as follows in aqueous solution:

$$NH_3(aq) + H_2O(l) \rightleftharpoons NH_4^+(aq) + OH^-(aq)$$

a) Write an expression for the base dissociation constant, K_b, for ammonia.
b) If you were given an aqueous solution of ammonia, describe the experiments you would carry out in order to determine the value of K_b at 25°C.

13 In an experiment, a solution which is 0.1 M with respect to chloride ions and 0.001 M with respect to chromate(VI) ions is titrated with silver nitrate solution. Answer the following questions using the following solubility product expressions for the sparingly soluble solids silver chloride, AgCl, and silver chromate(VI), Ag_2CrO_4.

AgCl, $K_{sp} = 1.8 \times 10^{-10}$ mol² dm⁻⁶
Ag_2CrO_4, $K_{sp} = 1.3 \times 10^{-12}$ mol³ dm⁻⁹

a) What concentration of $Ag^+(aq)$ is required to precipitate silver chloride from the original solution?
b) What concentration of $Ag^+(aq)$ is required to precipitate silver chromate(VI) from the original solution?
c) Given that silver chloride is white and that silver chromate(VI) is brick red, what would you observe during the above titration?
d) Do the chromate(VI) ions act as an indicator in the titration of chloride ions by silver nitrate solution?

Comments on the activities

Activity 26.1

1 Write the negative electrode on the left as an oxidation process, followed by the positive electrode written as a reduction reaction:

$\frac{1}{2}H_2(g) \longrightarrow H^+(aq) + e^-$, $\quad \frac{1}{2}Cl_2(g) + e^- \longrightarrow Cl^-(aq)$

This gives as the cell diagram,

$(Pt) \mid \frac{1}{2}H_2(g) \mid H^+(aq) \quad || \quad \frac{1}{2}Cl_2(g) \mid Cl^-(aq) \mid (Pt)$

2 $Cr^{2+}(aq) \longrightarrow Cr^{3+}(aq) + e^-$ $\quad H^+(aq) + e^- \longrightarrow \frac{1}{2}H_2(g)$

$(Pt) \mid Cr^{2+}(aq), Cr^{3+}(aq) \quad || \quad H^+(aq) \mid \frac{1}{2}H_2(g) \mid (Pt)$

3 Since the chlorine electrode is more positive than the hydrogen electrode and the Cr^{3+}/Cr^{2+} system is more negative,

$Cr^{2+}(aq) \longrightarrow Cr^{3+}(aq) + e^-$ $\quad \frac{1}{2}Cl_2(g) + e^- \longrightarrow Cl^-(aq)$

$(Pt) \mid Cr^{2+}(aq), Cr^{3+}(aq) \quad || \quad \frac{1}{2}Cl_2(g) \mid Cl^-(aq) \mid (Pt)$

Activity 26.2

1 $Zn(s) \longrightarrow Zn^{2+}(aq) + 2e^-$, $\quad 2H^+(aq) + 2e^- \longrightarrow H_2(g)$

$Zn(s) \mid Zn^{2+}(aq) \quad || \quad H^+(aq) \mid \frac{1}{2}H_2(g) \mid (Pt)$

$E_{cell} = E_{rhs} - E_{lhs}$
$0.76 = 0.00 - E_{lhs}$,
i.e $E^\ominus$ for the zinc electrode $= -0.76$ V

2 $Zn(s) \mid Zn^{2+}(aq) \quad || \quad Cu^{2+}(aq) \mid Cu(s)$

$E_{cell} = E_{rhs} - E_{lhs}$
$1.10 = E_{rhs} - (-0.76)$
$E_{rhs} = 1.10 - 0.76 = +0.34$ V

3 $E_{cell} = E_{rhs} - E_{lhs}$
Since the copper electrode has a more positive potential than the hydrogen electrode,
$E_{cell} = +0.34 - 0.00 = 0.34$ V

$(Pt) \mid \frac{1}{2}H_2(g) \mid H^+(aq) \quad || \quad Cu^{2+}(aq) \mid Cu(s)$

Comments on the activities *continued*

Activity 26.3

1 This was solved in the Activity.

2 Species present and redox systems which contain them:
From the hydrochloric acid:

oxidising agents		reducing agents	
$H^+(aq)$ + e^-	$\rightleftharpoons$	$\frac{1}{2}H_2(g)$	$E^\ominus = 0.00$ V
$\frac{1}{2}Cl_2(g) + e^-$	$\rightleftharpoons$	**$Cl^-(aq)$**	$E^\ominus = +1.36$ V

From the copper:

oxidising agents		reducing agents	
$Cu^{2+}(aq) + 2e^-$	$\rightleftharpoons$	**Cu(s)**	$E^\ominus = +0.34$ V
or, $Cu^+(aq) + e^-$	$\rightleftharpoons$	**Cu(s)**	$E^\ominus = +0.52$ V

The best oxidising agent present is $H^+(aq)$,

$$H+(aq) + e^- \longrightarrow \tfrac{1}{2}H_2(g) \qquad E^\ominus = 0.00\ V$$

Since all the reducing agents present are from systems with a more positive $E^\ominus$ value, there can be **no reaction** in this case.

3 Species present and redox systems which contain them:
From the hydrochloric acid:

oxidising agents		reducing agents	
$H^+(aq)$ + e^-	$\rightleftharpoons$	$\frac{1}{2}H_2(g)$	$E^\ominus = 0.00$ V
$\frac{1}{2}Cl_2(g) + e^-$	$\rightleftharpoons$	**$Cl^-(aq)$**	$E^\ominus = +1.36$ V

From the potassium manganate(VII):
oxidising agents **reducing agents**

$$MnO_4^-(aq) + 8H^+(aq) + 5e^- \rightleftharpoons Mn^{2+}(aq) + 4H_2O(l) \qquad E^\ominus = +1.51\ V$$

$$\text{or, } MnO_4^-(aq) + e^- \rightleftharpoons MnO_4^{2-}(aq) \qquad E^\ominus = +0.56\ V$$

The best oxidising agent present is MnO_4^-/H^+,

$$MnO_4^-(aq) + 8H^+(aq) + 5e^- \longrightarrow Mn^{2+}(aq) + 4H_2O(l) \qquad E^\ominus = +1.51\ V$$

The only reducing agent present from a system with a less positive $E^\ominus$ value is Cl^-,

$$\tfrac{1}{2}Cl_2(g) + e^- \rightleftharpoons Cl^-(aq) \qquad E^\ominus = +1.36\ V$$

Combining the half-equations,

$$MnO_4^-(aq) + 8H^+(aq) + 5e^- \longrightarrow Mn^{2+}(aq) + 4H_2O(l)$$

$$5Cl^-(aq) \longrightarrow 2\tfrac{1}{2}Cl_2(g) + 5e^-$$

overall,

$$MnO_4^-(aq) + 8H^+(aq) + 5Cl^-(aq) \longrightarrow Mn^{2+}(aq) + 4H_2O(l) + 2\tfrac{1}{2}Cl_2(g)$$

Thus we predict that potassium manganate(VII) will oxidise the chloride ions in hydrochloric acid to chlorine. This is why hydrochloric acid should never be used to acidify solutions in volumetric analysis involving manganate(VII).

4 Species present and redox systems which contain them.
From the hydrochloric acid:

oxidising agents		reducing agents	
$H^+(aq)$ + e^-	$\rightleftharpoons$	$\frac{1}{2}H_2(g)$	$E^\ominus = 0.00$ V
$\frac{1}{2}Cl_2(g) + e^-$	$\rightleftharpoons$	**$Cl^-(aq)$**	$E^\ominus = +1.36$ V

From the nitric acid:
H^+ system as above
oxidising agents **reducing agents**

$$NO_3^-(aq) + 4H^+(aq) + 3e^- \rightleftharpoons NO(g) + 2H_2O(l) \qquad E^\ominus = +0.96\ V$$

The best oxidising agent is NO_3^-,

$$NO_4^-(aq) + 4H^+(aq) + 3e^- \rightleftharpoons NO(g) + 2H_2O(l) \qquad E^\ominus = +0.96\ V$$

The only reducing agent present is Cl^- which belongs to a system with a more positive $E^\ominus$ value. We therefore predict that there will be no reaction.

Remember that these predictions use standard electrode potentials. Strictly speaking they refer only to 25°C, 1 atm pressure and 1 M aqueous concentrations and may not apply to very different conditions.

Activity 26.4

1 At –ve electrode

$$Pb(s) + H_2SO_4(aq) \longrightarrow PbSO_4(s) + 2H^+(aq) + 2e^-$$

At +ve electrode

$$PbO_2(s) + H_2SO_4(aq) + 2H^+(aq) + 2e^- \longrightarrow PbSO_4(s) + 2H_2O(l)$$

Overall

$$Pb(s) + PbO_2(s) + 2H_2SO_4(aq) \longrightarrow 2PbSO_4(a) + 2H_2O(l)$$

2 $E_{cell} = E_{rhs} - E_{lhs} = +1.68 - (-0.36) = 2.04$ V

3 The white solid coating the electrodes in a discharged lead–acid cell is lead(II) sulphate, $PbSO_4$.

Activity 26.5

1 At 25°C, $K_w = 10^{-14}$ mol² dm⁻⁶, $pK_w = -\log_{10}K_w = 14$

2 **a)** for pure water:
$[OH^-(aq)] = 10^{-7}$ mol dm^{-3}, $pOH = -\log_{10}[OH^-(aq)] = 7$

b) for 1 M HCl:
$[OH^-(aq)] = 10^{-14}$ mol dm^{-3}, $pOH = -\log_{10}[OH^-(aq)] = 14$

c) for 1 M NaOH:
$[OH^-(aq)] = 1$ mol dm^{-3}, $pOH = -\log_{10}[OH^-(aq)] = 1$

3 **a)**

	$HNO_3(aq)$ $\longrightarrow$	$H^+(aq) + NO_3^-(aq)$
before ionisation	0.01mol dm^{-3}	0
after ionisation	0	**0.01 mol dm^{-3}**

$$[OH^-(aq)] = \frac{K_w}{[H^+(aq)]} = \frac{10^{-14}}{0.01} = \mathbf{10^{-12}\ mol^2\ dm^{-6}}$$

$$pH = -\log_{10}[H^+(aq)] = -\log_{10}(0.01) = \mathbf{2}$$

$$pOH = -\log_{10}[OH^-(aq)] = -\log_{10}(10^{-12}) = \mathbf{12}$$

Comments on the activities *continued*

b) $HCl(aq) \longrightarrow H^+(aq) + Cl^-(aq)$

before ionisation 0.5 mol dm^{-3} 0

after ionisation 0 **0.5 mol dm^{-3}**

$$[OH^-(aq)] = \frac{K_w}{[H^+(aq)]} = \frac{10^{-14}}{0.5} = \mathbf{2 \times 10^{-14}\ mol\ dm^{-3}}$$

$pH = -\log_{10}[H^+(aq)] = -\log_{10}(0.5) = \mathbf{0.3}$

$pOH = -\log_{10}[OH^-(aq)] = -\log_{10}(2 \times 10^{-14}) = \mathbf{13.7}$

c) $H_2SO_4(aq) \longrightarrow 2H^+(aq) + SO_4^{2-}(aq)$

before ionisation 0.1 mol dm^{-3} 0

after ionisation 0 **0.2 mol dm^{-3}**

$$[OH^-(aq)] = \frac{K_w}{[H^+(aq)]} = \frac{10^{-14}}{0.2} = \mathbf{5 \times 10^{-14}\ mol\ dm^{-3}}$$

$pH = -\log_{10}[H^+(aq)] = -\log_{10}(0.2) = \mathbf{0.7}$

$pOH = -\log_{10}[OH^-(aq)] = -\log_{10}(5 \times 10^{-14}) = \mathbf{13.3}$

4 a) $KOH(aq) \longrightarrow K^+(aq) + OH^-(aq)$

before ionisation 0.1 mol dm^{-3} 0

after ionisation 0 **0.1 mol dm^{-3}**

$$[H^+(aq)] = \frac{K_w}{[OH^-(aq)]} = \frac{10^{-14}}{0.1} = \mathbf{10^{-13}\ mol\ dm^{-3}}$$

$pH = -\log_{10}[H^+(aq)] = -\log_{10}(10^{-13}) = \mathbf{13}$

$pOH = -\log_{10}[OH^-(aq)] = -\log_{10}(0.1) = \mathbf{1}$

b) $NaOH(aq) \longrightarrow Na^+(aq) + OH^-(aq)$

before ionisation 0.5 mol dm^{-3} 0

after ionisation 0 **0.5 mol dm^{-3}**

$$[H^+(aq)] = \frac{K_w}{[OH^-(aq)]} = \frac{10^{-14}}{0.5} = \mathbf{2 \times 10^{-14}\ mol\ dm^{-3}}$$

$pH = -\log_{10}[H^+(aq)] = -\log_{10}(2 \times 10^{-14}) = \mathbf{13.7}$

$pOH = -\log_{10}[OH^-(aq)] = -\log_{10}(0.5) = \mathbf{0.3}$

c) $Ba(OH)_2(aq) \longrightarrow Ba^{2+}(aq) + 2OH^-(aq)$

before ionisation 0.001mol dm^{-3} 0

after ionisation 0 **0.002 mol dm^{-3}**

$$[H^+(aq)] = \frac{K_w}{[OH^-(aq)]} = \frac{10^{-14}}{0.002} = \mathbf{5 \times 10^{-12}\ mol\ dm^{-3}}$$

$pH = -\log_{10}[H^+(aq)] = -\log_{10}(5 \times 10^{-12}) = \mathbf{11.3}$

$pOH = -\log_{10}[OH^-(aq)] = -\log_{10}(0.002) = \mathbf{2.7}$

5 The self-ionisation of water is the reverse of neutralisation and must therefore be an endothermic process. According to Le Chatelier's principle, an increase in temperature will favour the endothermic process and more water will ionise. This will increase the value of the ionic product of water, K_w, and also the concentration of hydrogen ions, $[H^+(aq)]$. This will mean that the values of pK_w and pH both decrease.

Activity 26.6

If you have difficulty with these calculations look at the examples in the text and then try to confirm the figures given for ethanoic acid and for ammonia.

1

Substituent	K_a/mol dm^{-3}	pK_a	pH of a 0.1 M solution
none: $H—CH_2COOH$	1.74×10^{-5}	4.76	2.9
hydroxy: $HO—CH_2COOH$	1.48×10^{-4}	3.83	2.4
chloro: $Cl—CH_2COOH$	1.58×10^{-3}	2.80	1.9
amino: $H_2N—CH_2COOH$	1.35×10^{-10}	9.87	5.4

Acid strength $H_2N– < H– < HO– < Cl–$

2

Base	K_a/mol dm^{-3}	pK_a	pH of a 0.1 M solution
ammonia: NH_3	1.78×10^{-5}	4.75	11.1
methylamine: CH_3NH_2	4.37×10^{-4}	3.36	11.8
dimethylamine: $(CH_3)_2NH$	5.25×10^{-4}	3.28	11.9
trimethylamine: $(CH_3)_3N$	6.31×10^{-5}	4.20	11.4

Base strength $NH_3 < (CH_3)_3N < CH_3NH_2 < (CH_3)_2NH$

Comments on the activities *continued*

Activity 26.7

1 a)

Volume of $NaOH/cm^3$	moles OH^- added	moles H^+ left	total volume/cm^3	$[H^+]$/mol dm^{-3}	pH
0.0	0.0000	0.0025	25.0	0.10000	1.00
10.0	0.0010	0.0015	35.0	0.04358	1.36
15.0	0.0015	0.0010	40.0	0.025	1.60
20.0	0.0020	0.0005	45.0	0.0111	1.95
24.0	0.0024	0.0001	49.0	0.00204	2.69
24.5	0.00245	0.00005	49.5	0.00101	3.00
24.9	0.00249	0.00001	49.9	0.00020	3.70

b)

Volume of $NaOH/cm^3$	moles OH^- in excess	total volume/cm^3	$[OH^-]$ /mol dm^{-3}	$[H^+]$ /mol dm^{-3}	pH
25.1	0.00001	50.1	0.00020	5.01×10^{-11}	10.30
25.5	0.00005	50.5	0.00099	1.01×10^{-11}	11.00
26.0	0.0001	50.9	0.00196	5.1×10^{-12}	11.29
30.0	0.0005	55.0	0.00909	1.10×10^{-12}	11.95
35.0	0.0010	60.0	0.01667	6.00×10^{-13}	12.22
40.0	0.0015	65.0	0.02307	4.33×10^{-13}	12.36
50.0	0.0025	75.0	0.03333	3.00×10^{-13}	12.52

2 All of the titration curves show a similar basic shape. The pH value changes only slowly whilst the acid or the base is in excess but there is a jump in pH at the equivalence point.

If the acid is weak, i.e. only slightly ionised, the starting pH is higher (curves b and d). Excess of a weak base gives a final plateau of lower pH (curves c and d). This means that the size of the jump in pH at the equivalence point is reduced if the acid and/or the base is weak.

Remember that the pH of the salt solution at the equivalence point depends upon the relative strengths of the acid and base involved in the titration (lowered by a weak base but increased by a weak acid).

Comments on the activities *continued*

Activity 26.8

1 Using Figure 26.5 and Table 26.2 we may compare the pH in each of the four titrations with the pH jumps in each of the four titrations with the pH range of each indicator.

Acid	base	steep pH jump		colour of indicator	
				methyl orange	phenolphthalein
Curve a)					
strong	strong	start	3	orange	colourless
		end	11	yellow	red
Curve b)					
weak	strong	start	7	yellow	colourless
		end	11	yellow	red
Curve c)					
strong	weak	start	3	orange	colourless
		end	7	yellow	colourless
Curve d)					
weak	weak	no steep jump		yellow	colourless
				yellow	colourless

The bigger the steep jump in pH at the equivalence point, the easier it is to detect using an indicator. Both of the indicators may be used in titration a) strong acid with a strong base.

If either the acid or base is weak then the pH jump is smaller and we must pick an indicator whose range corresponds to this. For titration b), weak acid–strong base, we need an indicator with a range above 7, i.e. phenolphthalein, whilst for titration c), strong acid–weak base, methyl orange should be used.

In titration d) where both the acid and base are weak there is no sudden jump in pH and the equivalence point cannot be detected accurately using either of the indicators.

2 The pH change in a titration may be followed continuously using a pH meter probe in the solution.

3 Indicators are themselves weak acids or bases so addition of excess could affect the accuracy of the results.

4 We can construct a table showing the colour changes with pH values for each indicator and the mixture (Y = yellow, R = red, B = blue, G = green, P = purple, – = colourless)

pH	5	6	7	8	9	10
bromothymol blue	Y	Y	Y/B	B	B	B
phenolphthalein	–	–	–	–	–/R	R
mixture	Y	Y	G	B	B/P	P

Activity 26.9

1 **a)** Phenolphthalein would detect conversion of CO_3^{2-} to HCO_3^- (pH jump about 9 to 7)

b) Methyl orange would detect conversion of HCO_3^- to CO_2 (pH jump about 6 to 2)

2 **a)** Phenolphthalein would be red at starting pH of around 11

b) $CO_3^{2-}(aq) + H^+(aq) \longrightarrow HCO_3^-(aq)$

c) The main ion left in solution at this stage is HCO_3^-

3 **a)** Methyl orange changes from yellow to orange which is not easy to see. In screened methyl orange a blue dye is added which gives a much more obvious colour change over the same pH range from red to green (grey at the end-point).

b) $HCO_3^-(aq) + H^+(aq) \longrightarrow H_2O(l) + CO_2(aq)$

c) The phenolphthalein added earlier does not interfere in this second titration because it is colourless below about pH 8.

4 Moles of HCl used in 1st titration

$$= 0.100 \times \frac{10}{1000} = 0.00100$$

this reacts with $\frac{0.00100}{2} = 0.00050$ moles CO_3^{2-}

moles of HCl used in 2nd titration $= 0.100 \times \frac{30}{100}$

$= 0.00300$

this reacts with $\frac{0.00300}{2} = 0.00150$ moles HCO_3^-

Comments on the activities *continued*

HOWEVER: 0.00100 moles HCO_3^- is formed in titration 1 (from CO_3^{2-}) so originally 0.00150 – 0.00050 = 0.00100 moles HCO_3^- present.
Ratio $CO_3^{2-}:HCO_3^-$ = 0.00050:0.00100 = 1:2

Activity 26.10

$K_a(HCO_3^-) = \text{antilog}_{10}(-10.3) = 5.01 \times 10^{-11}\ \text{mol dm}^{-3}$

$$= \frac{[CO_3^{2-}][H^+]}{[HCO_3^-]}$$

A blood pH of 7.4 corresponds to
$[H^+] = \text{antilog}_{10}(-7.4) = 3.9 \times 10^{-8}\ \text{mol dm}^{-3}$

$$\text{so, } 5.01 \times 10^{-11}\ \text{mol dm}^{-3} = \frac{[CO_3^{2-}]3.9 \times 10^{-8}\ \text{mol dm}^{-3}}{[HCO_3^-]}$$

$$\frac{5.01 \times 10^{-11}\ \text{mol dm}^{-3}}{3.9 \times 10^{-8}\ \text{mol dm}^{-3}} = \frac{[CO_3^{2-}]}{[HCO_3^-]}$$

$$0.00128 = \frac{[CO_3^{2-}]}{[HCO_3^-]}$$ i.e. ratio of about 1:778

Activity 26.11

1 Since the relative formula mass, M_r, of AgCl is 143.5, a solubility of 0.00193 g dm^{-3} of AgCl

$$= \frac{0.00193}{143.5} = 1.34 \times 10^{-5}\ \text{moles mol dm}^{-3}$$

2 $[Ag^+(aq)] = [Cl^-(aq)]$ = molarity of saturated solution
$= 1.34 \times 10^{-5}\ \text{mol dm}^{-3}$

$K_{sp} = [Ag^+(aq)][Cl^-(aq)] = (1.34 \times 10^{-5})^2$
$= 1.81 \times 10^{-10}\ \text{mol}^2\ \text{dm}^{-6}$

3 If a little concentrated hydrochloric acid is added to a saturated solution of silver chloride then solid silver chloride will precipitate out. If the temperature does not change then K_{sp} remains constant. Addition of hydrochloric acid increases $[Cl^-(aq)]$, so to maintain K_{sp}, $[Ag^+(aq)]$ must fall. The only way this can happen is if silver chloride precipitates out of solution:

$$AgCl(s) \rightleftharpoons Ag^+(aq) + Cl^-(aq)$$

increase in concentration of either EITHER aqueous ion shifts the system to the left, precipitating solid.

Activity 26.12

1,2 $Ag_2CrO_4(s) \rightleftharpoons 2Ag^+(aq) + CrO_4^{2-}(aq)$

$Ag_2CrO_4(s)$	$2Ag^+(aq)$	$CrO_4^{2-}(aq)$
6.9×10^{-5} mol dm^{-3}	13.8×10^{-5} mol dm^{-3}	6.9×10^{-5} mol dm^{-3}

3 $K_{sp} = [Ag^+(aq)]^2[CrO_4^{2-}(aq)]$
$= (1.38 \times 10^{-5})^2 \times (6.9 \times 10^{-5})$
$= 1.31 \times 10^{-12}\ \text{mol}^3\ \text{dm}^{-9}$

CHAPTER 27

Physical or phase equilibria

Contents

Study Checklist

After studying this chapter you should be able to:

1. Draw a vapour pressure–composition diagram for an ideal system of two miscible liquids and explain how deviations from this behaviour arise.
2. Sketch and interpret boiling–composition diagrams for miscible liquid mixtures, explaining how azeotropic mixtures may be formed during fractional distillation.
3. State that for an agitated mixture of two immiscible liquids each component exerts its full vapour pressure. Appreciate that such a mixture will therefore boil at a lower temperature than either of the pure components and that this is made use of in steam distillation.
4. Explain the conditions under which a mixture of solids may be separated by recrystallisation or fractional crystallisation.
5. Recall that when a solute is shaken with a mixture of two immiscible liquids it will distribute itself according to the partition law. Explain how this principle is used in the purification of a material by solvent extraction.
6. Appreciate that the melting point–composition diagram for a mixture of solids may have a minimum called a eutectic.
7. Recall that the eutectic mixture of tin and lead is used as electrical solder.

Here we shall deal with **physical** rather than chemical equilibria. Some of the most well-known physical changes involve a change of **phase**, where a phase is any physically distinct part of a system which has a uniform composition, e.g. melting ice and boiling water:

$$H_2O(s) \rightleftharpoons H_2O(l) \rightleftharpoons H_2O(g)$$

When glucose is dissolved in water, again its phase changes from solid to solution:

$$C_6H_{12}O_6(s) \rightleftharpoons C_6H_{12}O_6(aq)$$

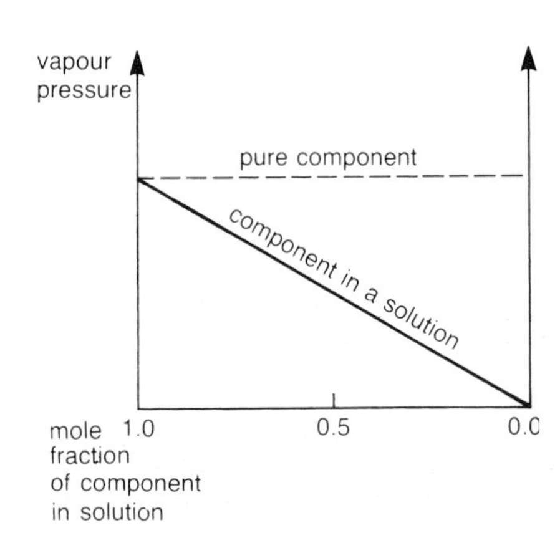

Figure 27.1 ▲
The relationship between vapour pressure and mole fraction for a component in an ideal mixture

27.1 Liquid–vapour equilibria

Any liquid left in a closed container at a fixed temperature will establish an equilibrium with its vapour:

$$\text{liquid} \rightleftharpoons \text{vapour}$$

The pressure exerted by the vapour at equilibrium is known as the **saturated vapour pressure (s.v.p.)** at that temperature.

If we take a mixture of two miscible liquids then, at equilibrium in an **ideal** system, **Raoult's law** states that each component will exert a vapour pressure proportional to its mole fraction. This means that a graph of mole fraction against vapour pressure for each of the liquids will be of the type shown in Figure 27.1, i.e. a straight line passing through the origin.

The total vapour pressure of the liquid mixture is found by adding the vapour pressures of the two components and it will vary with composition as shown in Figure 27.2.

The boiling point of a liquid is the temperature at which its vapour pressure reaches the external pressure. Since the component with the *lower* saturated vapour pressure will have a *higher* boiling point, we can draw a boiling point–composition diagram of the type shown in Figure 27.3.

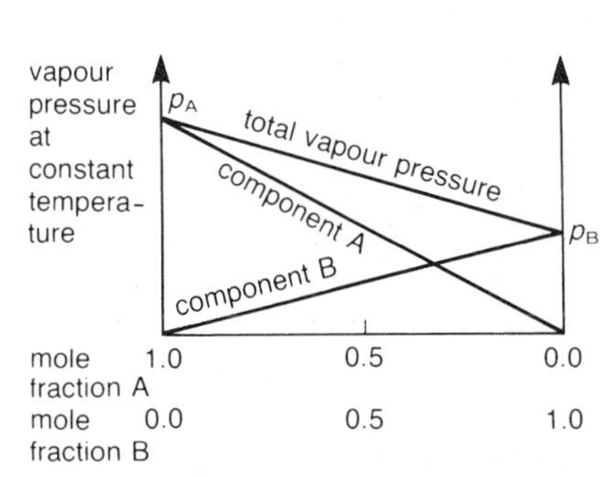

Figure 27.2 ▲
The relationship between mole fractions and partial and total vapour pressures for an ideal mixture of two liquids

Note that this graph has two curves since the liquid and vapour in equilibrium at any temperature will have different compositions. The vapour will always contain a higher proportion of the more volatile component, i.e. the one with the lower boiling point. In Figure 27.3, a liquid mixture of composition M when heated will boil at a temperature T_M. The vapour in equilibrium with liquid of composition L_1 will have the composition V_1, i.e. considerably richer in the lower boiling component A. This difference between the equilibrium liquid and vapour compositions enables a mixture to be separated by **distillation**. If the vapour V_1 is cooled it will condense to a liquid of the same composition, L_2. Boiling L_2 will produce a vapour V_2 which is even richer in A. If this process of condensing and boiling is continued, the final vapour will be virtually pure A. Meanwhile the proportion of A in the residual liquid will fall and its composition will follow the liquid curve up until it is virtually pure B.

Thus we may separate any mixture of liquids which obeys Raoult's law by repeated distillation, i.e. boiling and condensing. This is known a fractional distillation and is carried out using a fractionating column designed to allow many such steps to take place over its length. A typical laboratory column, shown in Figure 27.4, consists of a long glass tube packed with glass beads which provide a large surface area for the vapour to condense.

The vapour rising up the column condenses and is re-boiled by more hot vapours rising from the flask, thereby increasing the proportion of the lower boiling component. Each distillation step is known as a **theoretical plate**. If the column is efficient enough, i.e. contains enough theoretical plates, the thermometer at the top will indicate the boiling point of the pure more volatile (lower boiling point) component. Such fractional distillation processes are widely used in industry for separating liquid mixtures, e.g. pure oxygen, nitrogen and the noble gases from liquid air and useful **fractions** such as petrol and oil from crude petroleum.

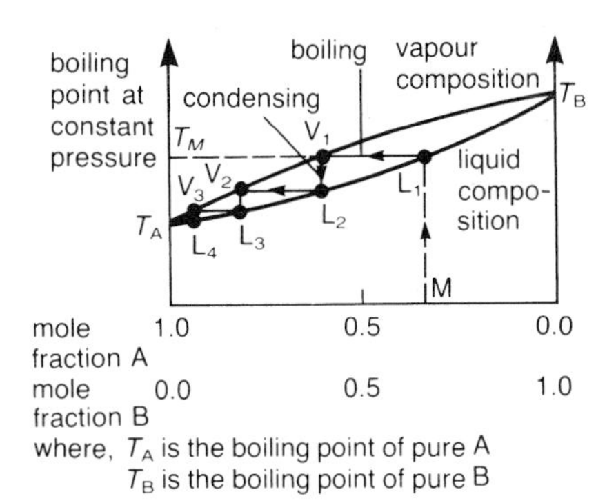

Figure 27.3 ▲
Boiling point–composition diagram for the ideal liquid mixture represented in Figure 27.2

Liquid mixtures which obey Raoult's law exactly are quite rare but often the deviations are so small that they may be successfully separated by fractional distillation. However, difficulties may arise with larger deviations from ideal behaviour.

When ethanol and water are mixed, there is a slight increase in the total volume. This happens because the very strong intermolecular hydrogen bonding in pure water is disrupted by ethanol molecules which can also form hydrogen bonds. This reduction in intermolecular attraction means that as

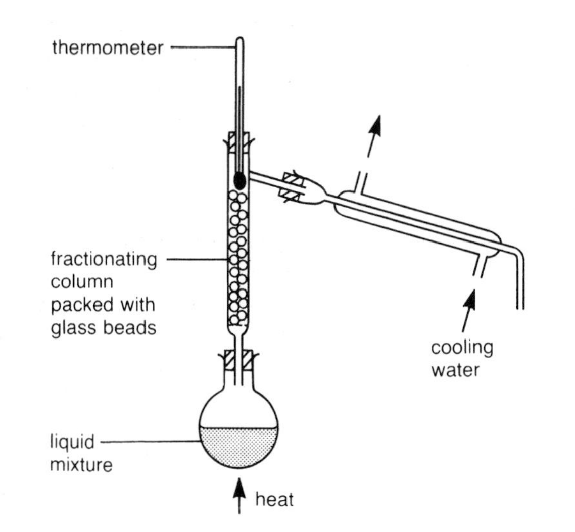

Figure 27.4 ►
Fractional distillation apparatus

well as an increase in volume, each component vaporises more easily and therefore contributes more than expected to the total vapour pressure. In this case, as Figure 27.5 shows, the **positive deviation** is so marked as to give a **maximum** in the vapour pressure–composition curve. In turn it follows that there must be a **minimum** in the boiling point–composition diagram as shown in Figure 27.6.

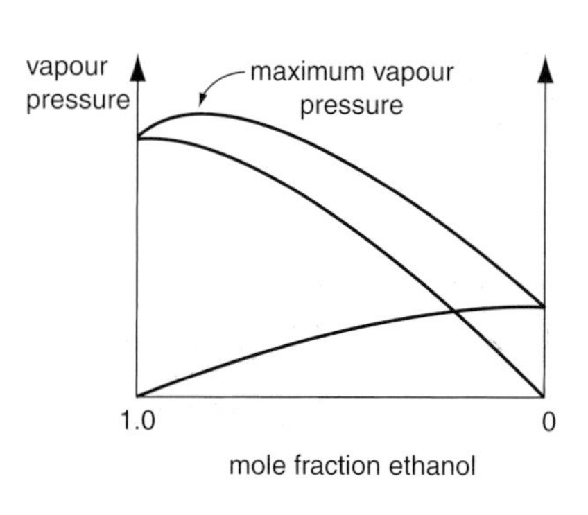

Figure 27.5 ▲
Vapour pressure–composition diagram for ethanol–water mixtures

Here, fractional distillation will not completely separate the mixture which effectively behaves as two independent systems joined at the minimum boiling point M. If we distil a mixture with composition L_1 then, applying the same arguments as for the ideal system above, the vapour will follow the composition curve downwards until it reaches the minimum boiling point. At this stage the liquid and vapour in equilibrium have the same composition. This means that the whole of the remaining mixture will distil over unchanged. Fortunately for whisky manufacturers and other 'distillers' this **minimum boiling azeotrope** contains over 95% by mass of ethanol.

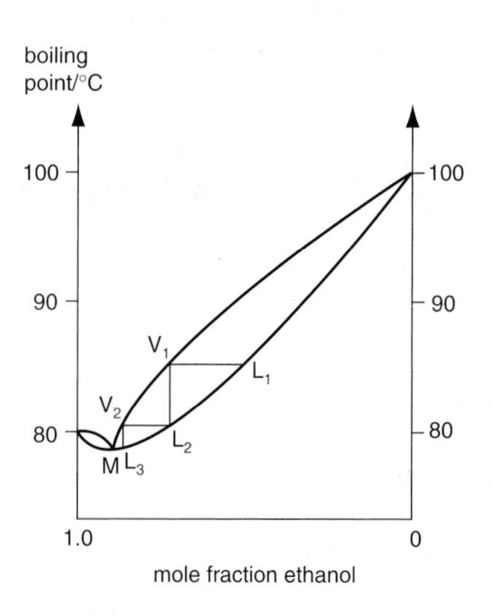

Figure 27.6 ▲
Boiling point–composition diagram for ethanol–water mixtures

Activity 27.1 Maximum boiling azeotropes

Propanone, $(CH_3)_2CO$, boils at 56°C and trichloromethane, $CHCl_3$, boils at 61°C. A mixture of these liquids shows a maximum in its boiling point curve at 65°C when the mole fraction of propanone is about 0.33.

1. Sketch the boiling point–composition curve for this mixture.
2. What will be the composition of the residue and of the distillate when an equimolar mixture of the two liquids is fractionally distilled?
3. This mixture is said to show marked **negative** deviation from Raoult's law. Explain this statement by sketching a vapour pressure–composition diagram for the system.
4. Explain in molecular terms how such deviation from ideal behaviour can occur.
5. When propanone and trichloromethane are mixed there is a change in temperature and total volume. Predict and explain the direction of each of these changes.

If the attraction between dissimilar molecules in a mixture of two liquids is very much less than the attraction between similar molecules then the two

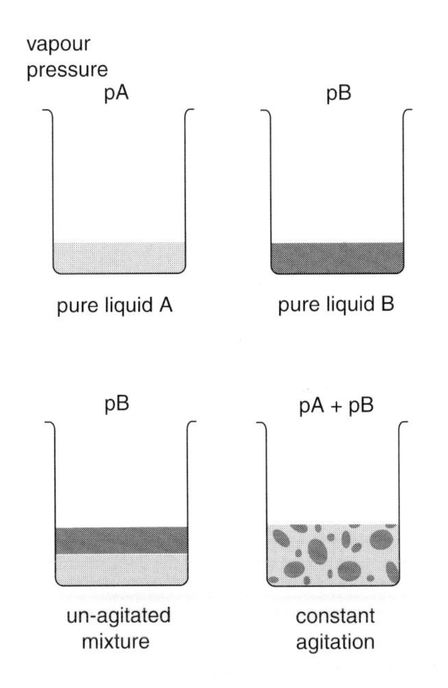

Figure 27.7 ▲
Vapour pressures of immiscible liquids

components may become immiscible, i.e. will separate into two layers. The pure less dense liquid will be in contact with the vapour and will therefore exert its full vapour pressure at any given temperature. Since the denser layer is not normally in contact with the surface it cannot exert a vapour pressure. However, as Figure 27.7 shows, if the mixture is constantly agitated the lower layer also reaches the surface and, as a result, will exert its full vapour pressure.

Since the boiling point is defined as the temperature at which the vapour pressure reaches the external pressure, the agitated mixture will boil at a lower temperature than either of the pure components.

Use is made of this behaviour in the purification of certain organic liquids by **steam distillation**. Phenylamine, $C_6H_5NH_2$, is a liquid which boils at about 184°C at atmospheric pressure but as it tends to decompose somewhat at this temperature it is difficult to purify by simple distillation. If it is mixed with water, with which it is immiscible, then the mixture boils at about 98°C and little decomposition occurs at this lower temperature. The apparatus for steam distillation is shown in Figure 27.8.

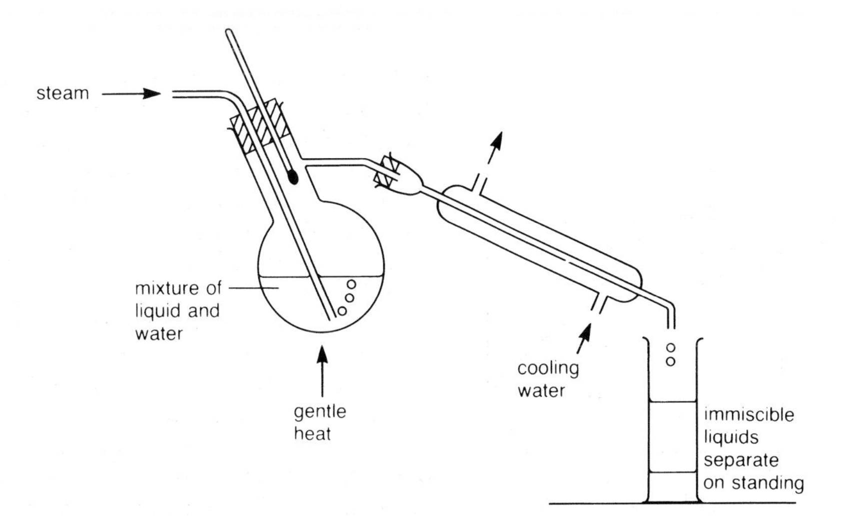

Figure 27.8 ►
Steam distillation apparatus

A mixture of impure phenylamine and water is both heated and agitated by bubbling steam through it. A mixture of steam and phenylamine vapour distils off and is condensed into a receiver. Gentle direct heat may also be applied to the mixture to maintain it at the boiling point. The molar amount of each component in the distillate is directly proportional to its vapour pressure.

Focus 27a

1. A liquid mixture in which the contribution to vapour pressure of each component is proportional to its mole fraction is said to be ideal and to obey **Raoult's law**.
2. Ideal liquid mixtures, and those that show only small deviations from Raoult's law, may be fully separated by **fractional distillation**.
3. Fractional distillation of a liquid mixture in which there is a maximum or minimum in the vapour pressure–composition diagram gives only one pure component and an **azeotropic mixture**.
4. In an agitated mixture of immiscible liquids each component exerts its full vapour pressure. The mixture will therefore boil at a temperature below the boiling points of each of the pure liquids. Use is made of this in the purification of liquids such as in the **steam distillation** of phenylamine.

27.2 Equilibria in solution

When an excess of a solute is shaken with a solvent, eventually a dynamic equilibrium is established when the rate at which the solid dissolves equals the rate at which it crystallises out of the solution. The solution is then said to be **saturated**. At any given temperature the maximum mass of solute which may be dissolved in a fixed mass of the solvent (usually 100 g) is known as the **solubility**. The solubility of a solute in any solvent depends upon temperature.

A mixture of solids may be separated using a solvent as long as their solubilities vary. In the simplest case, if a solvent is added which dissolves only one of the solids then filtration will isolate the insoluble component. If both solids dissolve appreciably in the solvent then a partial separation is still possible. The undissolved solid will be richer in the less soluble component than the original mixture. Further such treatments with fresh solvent will progressively increase the purity of this material. This technique, known as **fractional crystallisation**, is however very tedious and time consuming and is generally only used as a last resort.

Variation in solubility with temperature often makes the separation easier. Benzoic acid, for example, is much more soluble in hot water than in cold. Stirring with hot water will dissolve the benzoic acid and any water-soluble impurities, leaving any insoluble impurity to be filtered off. When the hot filtrate is allowed to cool, most of the benzoic acid will crystallise out of the solution, whilst (hopefully) any impurities which dissolved in the hot water will remain in solution on cooling. The purified benzoic acid may then be filtered off, washed with cold water and dried. This technique, known as **re-crystallisation**, may be used to purify any solid providing a solvent can be found where there is sufficient variation in solubility with temperature.

An alternative approach using variation in solubility is **solvent extraction**. If a solute is shaken with a mixture of two immiscible solvents at a fixed temperature then it will **distribute** itself according to the **partition law**. This states that, providing all the solute dissolves, its concentrations in the two liquid layers will be in a fixed ratio, given by:

$$K_d = \frac{\text{concentration of solute in solvent A}}{\text{concentration of solute in solvent B}}$$

where the constant K_d is known as the **distribution constant** or **partition constant** whose value will depend upon the relative solubilities of the solute in each solvent.

Activity 27.2

When an aqueous solution of butanedioic acid, $HOOC(CH_2)_2COOH$, is shaken with ethoxyethane at room temperature, some of the acid is transferred into the organic layer. When equilibrium was established the concentration of acid in the aqueous layer was found to be 2.60 g dm^{-3}, whereas in the organic layer the acid concentration was 0.50 g dm^{-3}.

1 Outline how you might determine the concentration of acid in each of the layers experimentally.

2 Calculate a value for the partition coefficient of butanedioic acid between water and ethoxyethane at this temperature.

3 If 100 cm^3 of an aqueous solution containing 0.500 g of butanedioic acid is shaken with 50 cm^3 of ethoxyethane, what mass of the acid will be extracted into the organic layer at room temperature?

Ethoxyethane is a very convenient liquid to use for the extraction of organic compounds from aqueous solution. Not only is it immiscible with water and an excellent solvent for most organic compounds, but it is also very volatile and therefore easily evaporated off afterwards. Care must be taken during this operation as ethoxyethane vapour is highly flammable and forms explosive mixtures with air. As the next (extension) activity will show, more solute is removed from an aqueous solution by several successive extractions using small portions of ethoxyethane than by using the same volume of solvent in a single extraction.

Activity 27.3 Efficiency of solvent extraction

The efficiency of a solvent extraction process may be defined as the percentage of solute which is extracted from the aqueous solution.

100 cm^3 of an aqueous solution contains 6.00 g of an organic compound A. If the partition coefficient of A between ethoxyethane and water is 2 then calculate the efficiency of extracting A under these conditions with 100 cm^3 of ethoxyethane in each of the following procedures.

1 A single extraction with 100 cm^3 of ethoxyethane.

2 Extraction with two separate, successive 50 cm^3 portions of ethoxyethane.

Did solvent extraction protect against BSE?

Extensive studies have demonstrated a link between cattle who contracted BSE and the presence in their feed of meat and bone meal derived from sheep who had suffered from a related disease known as scrapie. However, such sheep-based feed had been used for many years before the BSE outbreak in cattle so why did it suddenly cause a problem?

Since the 1920s, meat and bone meal had been obtained from sheep carcasses by a process which involved the use of solvent (generally trichloromethane, propanone, butanol and/or ethoxyethane) to extract the tallow. In the early 1980s this **solvent extraction** step was excluded from the process and it was this change which correlated with the outbreak of BSE. The solvent extraction was replaced by heat treatment to separate the tallow.

An article published in 1965 by Derek Mould and his co-workers at the Moredun Institute in Edinburgh showed that the infectivity of the scrapie agent is reduced by solvents such as trichloromethane and propanone, but that it can survive temperatures of up to 138°C.

New Scientist, 14 March 1998, by Fred Pearce

27.3 Solid–liquid equilibria

Just as a mixture of liquids may form an azeotrope with a minimum boiling point, so a mixture of solids may form a **eutectic** with a minimum melting point. The melting point–composition diagram for tin and lead mixtures is shown in Figure 27.9.

When lead is progressively added to tin, the melting point falls steadily from 232°C to a minimum of 183°C. After this **eutectic point** the melting point steadily rises again to that of pure lead at 327°C. The eutectic mixture in this system is used as electrical solder. When a liquid mixture of the two

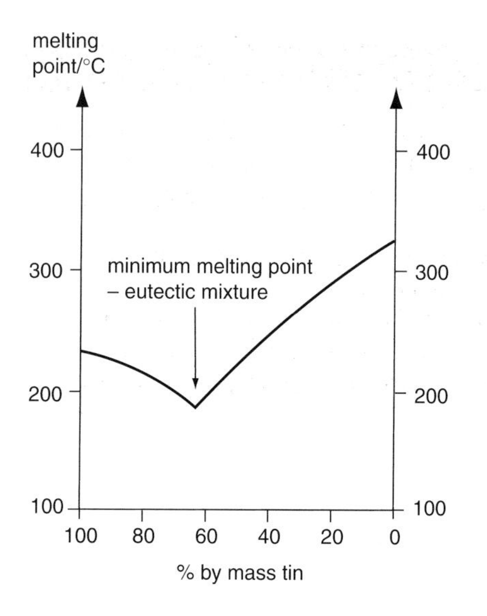

Figure 27.9
Melting point–composition diagram for tin–lead mixtures

metals is cooled, pure solid tin or lead first crystallises out, and the liquid composition follows the curve until the eutectic point is reached. On further cooling the liquid composition does not change and the solid crystallises as the eutectic mixture.

Eutectic mixtures and air conditioning

A simple air conditioning system uses off-peak electricity at night to either heat or freeze hollow plastic balls that then heat or cool the building during the day. It relies on the fact that as a solid melts it absorbs heat without rising in temperature. Similarly, the liquid releases the same amount of heat when it is refrozen back into a solid. The balls, filled with a so-called 'eutectic salt' that melts at 27°C, sit in a tank of water that is circulated through the air conditioning system.

In summer the solid eutectic salts melt during the day by absorbing heat from the warm water in the air conditioning system. In turn this absorbs heat from the air in the building so keeping it cool. The salts are refrozen at night using an electrical heat pump to cool the tank containing the balls. In winter the opposite occurs and the heat pump is used to warm up the balls during the night so that they can release heat during the day and help to keep the building warm. If there is a sufficiently large day–night temperature difference the system can even operate without the heat pump.

Focus 27b

1 Solids with different solubilities in a particular solvent may be separated by **fractional crystallisation**.
2 Solids may be **recrystallised** if they dissolve in a particular solvent when hot but not when cold.
3 A solid will **distribute** itself according to the **partition law** when it is shaken with a mixture of two immiscible solvents. This principle is made use of in **solvent extraction**.
4 The melting point–composition diagram for a mixture of two solids may show a minimum, known as a **eutectic**. Solder is a eutectic mixture of tin and lead.

Questions on Chapter 27

1 The boiling points of ethanol, C_2H_5OH, and benzene, C_6H_6, are 80°C and 79°C respectively. They form an azeotropic mixture of boiling point 68°C containing 68% by mass of benzene.
  - **a)** Sketch the boiling point–composition diagram for mixtures of these two liquids.
  - **b)** Sketch the likely appearance of the vapour pressure–composition diagram for this system.
  - **c)** This system is said to show a large positive deviation from Raoult's law. Explain this statement as fully as you can.
  - **d)** When benzene and ethanol are mixed, both the temperature and total volume change. Explain in molecular terms why this happens and predict the direction of each change.
  - **e)** What will be the composition of i) the distillate and ii) the residue when a mixture containing 50% by mass of both benzene and ethanol is fractionally distilled?
  - **f)** Benzene and water are immiscible. How would you expect the boiling point of an agitated mixture of these two liquids to compare with those of the separate pure liquids? Explain your answer.

Questions on Chapter 27 *continued*

2 The table below shows the relative solubilities of three solids, A, B and C, in three different solvents, water, ethanol and ethoxyethane, for both hot and cold conditions.

a) Which of the solids could be recrystallised using water?

b) Which of the solids could be recrystallised from ethanol?

c) Which of the solids could not be recrystallised using any of the solvents shown?

d) Outline how you would attempt to separate a mixture of all three solids using the solvents shown.

3 Benzene is an organic liquid which is immiscible with water. When 100 cm^3 of an aqueous solution containing 1.00 g of an organic solute A is shaken with 10 cm^3 of benzene, 0.40 g of A is extracted.

a) Calculate a value for the partition coefficient of A between benzene and water.

b) If the aqueous layer from a) above is shaken with a further 10 cm^3 portion of benzene, how much A will remain in the aqueous layer when equilibrium has been established?

	A	B	C
water	very soluble cold very soluble hot	insoluble cold soluble hot	insoluble cold insoluble hot
ethanol	insoluble cold insoluble hot	soluble cold soluble hot	insoluble cold soluble hot
ethoxyethane	insoluble	insoluble	soluble

4 Use Figure 27.9 to help you explain in detail what happens to a mixture of 50 g of tin and 50 g of lead as it is cooled from 350°C to room temperature.

Comments on the activities

Activity 27.1

1 The boiling point–composition curve for the propanone–trichloromethane system is sketched in Figure 27.10.

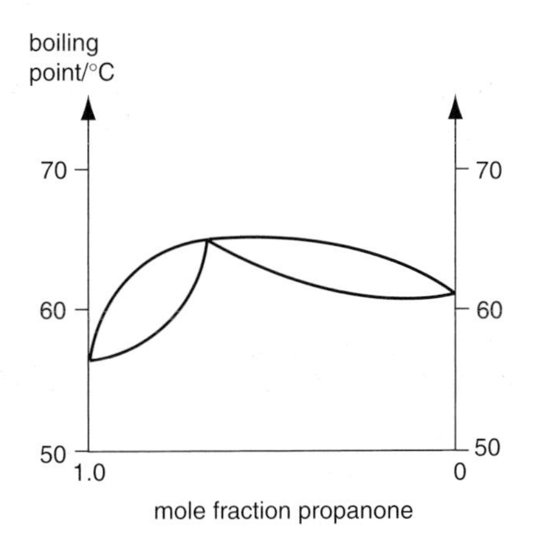

Figure 27.10 ▲
Boiling point–composition diagram for propanone–trichloromethane mixtures

2 The residue will be the maximum boiling point azeotrope and the distillate will be pure trichloromethane.

3 A maximum in the boiling point–composition curve means that there must be a corresponding minimum in the vapour pressure–composition graph, i.e. each component exerts **less** than its expected vapour pressure, as shown in Figure 27.11.

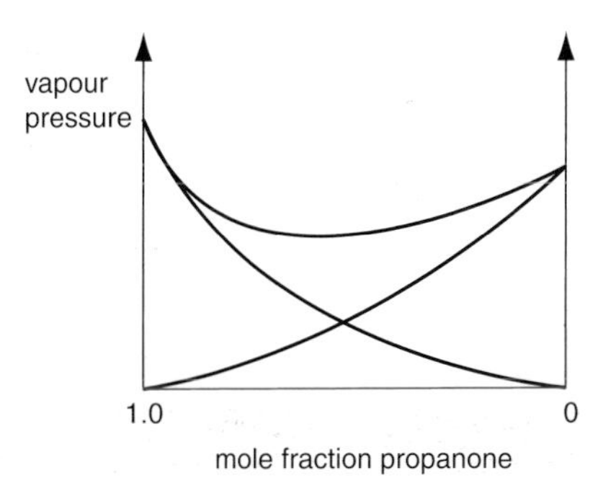

Figure 27.11 ▲
Vapour pressure–composition diagram for propanone–trichloromethane mixtures

4 Reduction in vapour pressure is caused by increased intermolecular attraction which makes it harder for the molecules to escape from the liquid. There must be a strong attraction between propanone and trichloromethane molecules.

5 The strong intermolecular attractions produced on mixing propanone and trichloromethane will cause a rise in temperature and a reduction in total volume.

Comments on the activities *continued*

Activity 27.2

1 A measured volume of each layer could be titrated with sodium hydroxide solution of known molarity using an indicator such as phenolphthalein. The sodium hydroxide will react with the acid as follows:

$HOOC(CH_2)_2COOH + 2NaOH(aq)$
$\longrightarrow 2Na^+(aq) + {}^-OOC(CH_2)_2COO^-(aq) + 2H_2O(l)$

2 The first named solvent is written on the top line of the expression for K_d. Since we are asked for the partition coefficient between water and ethoxyethane this gives:

$$K_d = \frac{\text{concentration in water}}{\text{concentration in ethoxyethane}} = \frac{2.60}{0.50} = 5.20$$

3 Let the mass of acid extracted into the organic layer be x grams.
Volume of ethoxyethane = 50 cm^3

$$\text{so,} \quad \text{concentration} = \frac{1000x}{50 \text{ g dm}^{-3}} = 20x \text{ g dm}^{-3}$$

The mass of acid remaining in the aqueous layer will be $(0.500 - x)$ grams
Volume of water = 100 cm^3

$$\text{so,} \quad \text{concentration} = \frac{1000(0.500 - x)}{100} = 5 - 10x$$

$$K_d = \frac{\text{concentration in water}}{\text{concentration in ethoxyethane}} = \frac{5 - 10x}{20x} = 5.2$$

Thus $5 - 10x = 5.2(20x) = 104x$, i.e. $5 = 114x$

$$\text{so,} \quad x = \frac{5}{114} = 0.0439 \text{ grams}$$

Activity 27.3

1 Let mass of A extracted by extraction with 100 cm^3 of ethoxyethane = x grams

$$\text{Concentration of A in ethoxyethane} = \frac{1000x}{100} = 10x$$

$$\text{Concentration of A in water} = \frac{1000(6.00 - x)}{100}$$
$$= 10(6.00 - x) = 60 - 10x$$

$$K_d = 2 = \frac{10x}{(60 - 10x)},$$

so $2(60 - 10x) = 10x$, $120 - 20x = 10x$, so $x = \frac{120}{30} = 4$ g

Efficiency of extraction = $100\left(\frac{4}{6}\right) \approx 67\%$

2 Let mass of A extracted by 1st portion of 50 cm^3 of ethoxyethane = y grams

$$\text{Concentration of A in ethoxyethane} = \frac{1000y}{50} = 20y$$

$$\text{Concentration of A in water} = \frac{1000(6.00 - y)}{100}$$
$$= 10(6.00 - y) = 60 - 10y$$

$$K_d = 2 = \frac{20y}{(60 - 10y)},$$

so $2(60 - 10y) = 20y$, $120 - 20y = 20y$, so $y = \frac{120}{40} = 3$ g

Let mass of A extracted by 2nd portion of 50 cm^3 of ethoxyethane = z grams

$$\text{Concentration of A in ethoxyethane} = \frac{1000z}{50} = 20z$$

$$\text{Concentration of A in water} = \frac{1000(3.00 - z)}{100}$$
$$= 10(3.00 - z) = 30 - 10z$$

$$K_d = 2 = \frac{20z}{(30 - 10z)},$$

so $2(30 - 10z) = 20z$, $60 - 20z = 20z$, so $z = \frac{60}{40} = 1.5$ g

Thus total mass of A extracted by two portions of ethoxyethane = 4.5 g
Efficiency of extraction = $100\left(\frac{4.5}{6}\right) = 75\%$

i.e. higher than with a single extraction using the same total volume of solvent.

C H A P T E R

28

Spontaneous reactions

Contents

Study Checklist

After studying this chapter you should be able to:

1. Appreciate the difficulty in explaining the existence of spontaneous endothermic reactions in terms of enthalpy change only.
2. Visualise entropy as a (statistical) measure of the 'degree of disorder' in a system.
3. Give examples of processes which involve positive and negative entropy changes and appreciate that 'the overall entropy of the universe is increasing'.
4. Express that free energy change as $\Delta G = \Delta H - T\Delta S$ and know that this must have a negative value for a spontaneous reaction.
5. Carry out calculations involving free energy, enthalpy and entropy changes and explain the effect of temperature change on the feasibility of a particular reaction.
6. Interpret Ellingham diagrams in which the free energy changes for processes involved in extraction of metals are shown graphically.
7. State that the e.m.f., *E*, of any working chemical cell is positive and that in this case free energy change is given by $\Delta G = -nFE$.

28.1 Spontaneous change

As we have seen earlier in this Theme all reactions are, at least in theory, reversible. However, everyday experience tells us that some changes seem to happen **naturally** in one direction rather than the reverse. For example, on ignition, methane gas burns **spontaneously** in air or oxygen, but the products of this reaction, carbon dioxide and water vapour, do not reform the original materials:

$$CH_4(g) + 2O_2(g) \longrightarrow CO_2(g) + 2H_2O(g)$$

Similarly, once milk has gone sour the process cannot be reversed.

A spontaneous change may occur when a system moves towards an equilibrium position. In the examples above the equilibrium position always lies well over to one particular side but sometimes the direction of spontaneous change may be reversed by changing the conditions. Thus ice melts to give liquid water above 0°C, but below this temperature water freezes.

The aim of this chapter is to investigate the factors which affect the equilibrium position of any system. We may then identify the direction of spontaneous change and so to predict the circumstances under which a particular process may be accomplished (or avoided.)

28.2 Energy changes

Most spontaneous changes are accompanied by a release of energy from the system involved. When a ball rolls down a slope, its potential energy is converted into kinetic energy which in turn is lost to the surroundings as frictional heat. Similarly, all combustion reactions are exothermic and involve loss of energy from the system.

However, not all spontaneous reactions do involve loss of energy from the system. Indeed many well-known spontaneous processes are endothermic, e.g. the melting of ice above 0°C and dissolving ammonium nitrate in water. We therefore need to find another factor which influences the spontaneity of a change. To do this we need to look at a spontaneous process in which there is no transfer of energy between the system and the surroundings.

28.3 The laws of chance

A gas spontaneously expands to fill any container in which it is placed. In the case of a (hypothetical) ideal gas, there are no intermolecular attractive forces and this process takes place without any exchange of energy. Consider eight molecules of an ideal gas in bulb A, connected by a tap to an identical empty container, bulb B:

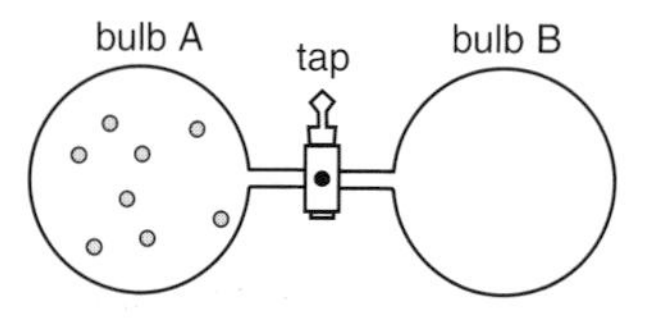

Each molecule in A is moving rapidly and, when the tap is open, is able to move into the empty bulb B. There is no reason why any molecule should prefer bulb B but since the motion is random, it will eventually find itself in B. At first, therefore, there will be a net transfer of gas molecules from A to B. Of course, once B contains gas molecules they can move back into A. Since molecular motion is completely random, we may use simple statistics to calculate the probability of any distribution of molecules between the two bulbs. The probability of any particular molecule being in bulb A at a particular time is 0.5. The probability of all eight molecules being in bulb A at the same time is $0.5^8 = 0.0039$, i.e. a 1 in 256 chance. Of course, this is exactly the same as the probability of any other particular distribution of the molecules but the difference is that there are many more ways in which the molecules can be distributed (i.e. more combinations of different molecules that give the same overall distribution) more evenly between the two bulbs. In fact the most likely distribution is an equal number of molecules in each bulb. Since there are 70 distinct ways of doing this the overall probability is:

$$\frac{70}{256} = 0.273.$$

Thus, although it is perfectly possible for all the molecules simultaneously to go back into bulb A, statistically the chances are against it. If this is true for our model with only eight molecules, think how much more unlikely it would be for all the trillions of molecules of air in a room suddenly to move spontaneously to one corner, leaving a vacuum elsewhere!

The spreading out, or diffusion, of a gas provides an example of a spontaneous process which is 'driven' by the laws of blind chance.

28.4 Order and chaos

When a gas diffuses, its molecules spread out and we can consider the system to have become **less ordered** or **more chaotic**. In an ordered system, each particle has less freedom of movement and therefore fewer positions which it may occupy. In our example above at the start each molecule must be in bulb A, whereas when the tap is opened it may be found in either bulb:

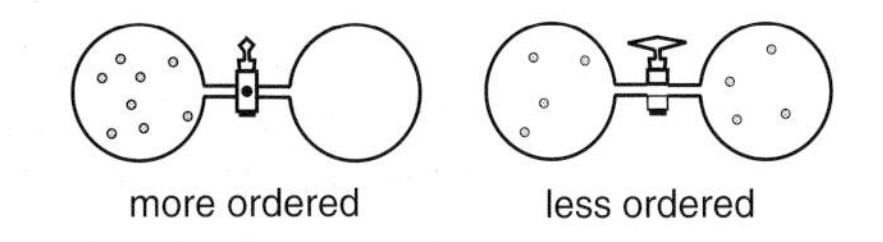

When a solid melts or dissolves, we get a similar increase in the freedom of particles to move around and, by the laws of probability, a more disordered arrangement:

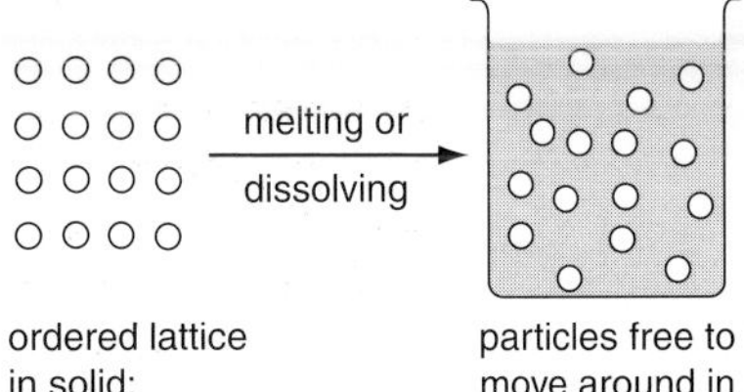

However, as with our consideration of energy changes, not all spontaneous processes are accompanied by an increase in the disorder of the system. For example when water freezes during cold weather, its molecules adopt a more ordered structure. However, it does appear that changes in both energy and degree of disorder are both implicated in determining the direction of spontaneous change.

One invariable rule is that a spontaneous process which is endothermic is always accompanied by an increase in disorder within the system, whilst a more ordered system can only result from an exothermic change.

Activity 28.1

For each of the following changes, explain with reasons whether the arrangement of particles will become more or less disordered.

a) A spoonful of sugar is stirred into a cup of coffee.
b) Raindrops turn to hailstones as they fall.
c) Puddles of water evaporate after a shower of rain.
d) A car tyre is punctured.
e) $NH_3(g) + HCl(g) \longrightarrow NH_4Cl(s)$
f) $2C(s) + O_2(g) \longrightarrow 2CO(g)$
g) $C_6H_{12}O_6(s) + 6O_2(g) \longrightarrow 6CO_2(g) + 6H_2O(l)$

28.5 Entropy and the second law of thermodynamics

So far we have used 'degree of disorder' in a qualitative or descriptive way but **entropy**, ***S***, gives a numerical value to this quantity. If the system becomes more disordered its entropy rises, i.e. the entropy change, ΔS, has a positive value. Only systems which are completely ordered, i.e. with no freedom at all for the particles to change position, have zero entropy. In practice this can never happen since molecular motion could only cease at a temperature of absolute zero. Clearly, however, when a substance is heated, its particles move faster and further and so its entropy rises with temperature.

The last sentence provides a big clue to the connection between energy change and entropy change. In an exothermic reaction, heat is released to the surroundings. When the particles in the surroundings absorb this energy they become more disordered, i.e. the entropy of the surroundings increases.

An antique entropy meter?

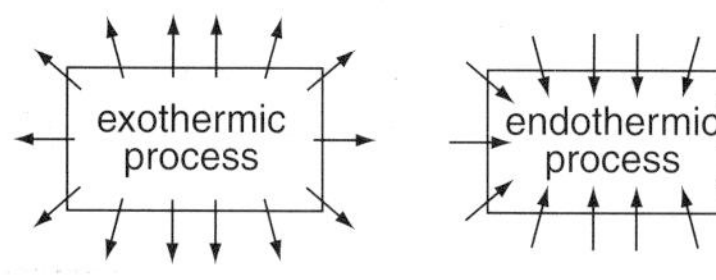

energy released to surroundings, the entropy of which therefore increases

energy absorbed from surroundings, the entropy of which therefore decreases

In order to account for spontaneous change therefore we must consider not only the **system** that undergoes the change but also its **surroundings**, i.e. the rest of the universe.

The second law of thermodynamics states that any spontaneous change leads to an overall increase in the entropy of the universe.

However, since it is rather difficult to have to consider the whole universe when studying a particular change we need some way of estimating the effect of any energy change on the surroundings.

The arrow of time?

According to the second law of thermodynamics, all spontaneous processes lead to an increase in the overall entropy of the universe. Since, by definition, such processes happen naturally, the passage of time is accompanied by an increase in the entropy or disorder of the universe. Indeed we can tell that time has passed by observing some process, for example the uncoiling of a 'clockwork' spring, which involves an increase in entropy.

Carried to its logical conclusion, the second law paints rather a depressing picture of the future. Eventually the universe should reach a completely disordered state with maximum entropy. Statistically speaking, from this most probable state it is very unlikely that, apart from minor local fluctuations, order could arise again. In this condition time would have no meaning because there would be no overall entropy change to observe.

The increase in entropy of the universe may be associated with the expansion caused by the 'big bang' at its birth. Some physicists argue that there is enough matter in the universe for gravitational forces eventually to overcome the expansion. If this is so, the universe may eventually start to contract, decreasing in entropy until we get the opposite of the 'big bang' … the 'big crunch'. Life during such contraction might be very different since the 'arrow of time' might be reversed. If this were true, natural processes would then result in an overall increase in order and 'effects' would precede their 'causes'. Can you imagine dying before you are born (or perhaps that should be undying before being unborn?).

Focus 28a

1. A **spontaneous** change is one that tends to occur naturally under specified conditions.
2. **Entropy** is a measure of the degree of disorder of a system.
3. A spontaneous change is always accompanied by an overall increase in entropy for the system and its surroundings (the universe).
4. An exothermic reaction releases heat energy and so increases the entropy of its surroundings, whereas an endothermic reaction reduces the entropy of its surroundings by absorbing heat energy.

28.6 Entropy change and free energy

Any process which involves an increase in entropy for the system and is exothermic must be spontaneous, since both factors increase the total entropy of the universe. Conversely, any endothermic process in which the entropy of the system decreases cannot possibly be spontaneous. However, when the enthalpy change for the system results in a change in entropy for the surroundings which is opposite in sign to the entropy change occurring within the system, we need to be able to find the sign of the combined entropy changes. This can be done by calculating a quantity known as **free energy change, ΔG**, which is related to the enthalpy and entropy changes for the system as follows:

$$\Delta G = \Delta H - T\Delta S$$

For a spontaneous change the value of ΔG must be negative.

This fits in with our previous findings since a negative ΔH and a positive ΔS (both factors favouring a spontaneous reaction) lead to a negative value for the free energy change. Note that the effect of entropy change becomes more important as the temperature rises.

Free energy change and the second law

E

The second law states that any spontaneous change must result in an increase in the overall entropy of the universe. This may be found by adding the entropy changes for the system and for the rest of the universe, referred to as the surroundings:

$$\Delta S_{universe} = \Delta S_{surroundings} + \Delta S_{system}$$

The entropy change for the surroundings results from the exchange of energy with the system, i.e. ΔH_{system}. For an exothermic reaction, energy is released into the surroundings causing an increase in disorder, i.e. a negative value for ΔH_{system} leads to a positive value for $\Delta S_{surroundings}$.

Now, the effect of the enthalpy change on $\Delta S_{surroundings}$ depends upon temperature. If the temperature is high the surroundings will be more disordered to begin with, so the effect of an energy change on its entropy will be less. Since the enthalpy change is effectively spread throughout the rest of the universe, the temperature of the surroundings does not change measurably and the entropy change of the surroundings is given by:

$$\Delta S_{surroundings} = \frac{-\Delta H_{system}}{T}$$

If we substitute this into the equation above we get,

$$\Delta S_{universe} = \frac{-\Delta H_{system}}{T} + \Delta S_{system}$$

It is really this expression that is the key to what is happening. The importance of enthalpy change becomes less important as the temperature rises because it has proportionately less effect on the entropy of the surroundings.

Taking the last equation and multiplying throughout by $-T$ we get,

$$-T\Delta S_{universe} = \Delta H_{system} - T\Delta S_{system}$$

Thus free energy change, ΔG, is simply $-T\Delta S_{universe}$. Since the second law states that for a spontaneous change the entropy of the universe must increase, the free energy change for the system must be negative.

28.7 Standard molar entropy

The absolute value of the entropy associated with any given amount of a particular substance under specified conditions may be calculated statistically in units of J mol^{-1} K^{-1}. Selected values for standard molar entropy values, $S^{\ominus}$, which refer to 1 mole at 25°C and 1 atmosphere pressure, are shown in Table 28.1, together with standard enthalpies of formation.

Table 28.1 Standard enthalpies of formation and standard molar entropies of selected substances

Substance	$\Delta H_f^{\ominus}$/kJ mol^{-1}	$S^{\ominus}$/J mol^{-1} K^{-1}
$H_2(g)$	0	130.6
$O_2(g)$	0	205.0
$H_2O(l)$	−285.8	69.9
$H_2O(g)$	−241.8	188.7
C(diamond)	+1.9	2.38
C(graphite)	0	5.74
$CH_4(g)$	−74.8	186.2
CO(g)	−110.5	197.6
$CO_2(g)$	−393.5	213.6
$C_2H_5OH(l)$	−277.7	160.7

We can use standard molar entropies to calculate $\Delta S^{\ominus}$ for any particular reaction simply by adding the standard entropies of the products and subtracting the standard entropies of the reactants. Thus for the formation of 1 mole of carbon dioxide from its elements:

$$\begin{array}{lcccccl} & \text{C(graphite)} & + & O_2(g) & \longrightarrow & CO_2(g) & \Delta H^{\ominus} = -393.5 \text{ kJ mol}^{-1} \\ S^{\ominus} & 5.74 & & 205.0 & & 213.6 & \Delta S^{\ominus} = +2.86 \text{ J mol}^{-1}\text{ K}^{-1} \end{array}$$

Since ΔH is negative and ΔS is positive, ΔG must be negative and the reaction is spontaneous under all conditions. However, although a process may be thermodynamically feasible this treatment gives no information as to the rate of reaction. In practice, carbon only combines with oxygen on heating.

Carrying out a similar calculation for the conversion of 1 mole of liquid water into vapour, we get:

$$\begin{array}{lcccl} & H_2O(l) & \longrightarrow & H_2O(g) & \Delta H^{\ominus} = +44.0 \text{ kJ mol}^{-1} \\ S^{\ominus} & 69.9 & & 188.7 & \Delta S^{\ominus} = +118.8 \text{ J mol}^{-1}\text{ K}^{-1} \end{array}$$

Check units!

Note that we have the enthalpy change in kJ but the entropy change in J. We must use consistent units, either kJ or J. In this case we have divided the entropy change by 1000 to convert J into kJ

Since the reaction is endothermic but has a positive entropy change, the sign of $\Delta G^{\ominus}$ will depend upon the temperature.

$$\Delta G^{\ominus} = \Delta H^{\ominus} - T\Delta S^{\ominus}$$

$$= +44.0 - T\left(+\frac{118.8}{1000}\right)$$

When $\Delta G^{\ominus} = 0$ then $+\dfrac{118.8T}{1000} = +44.0$, i.e. $T = \dfrac{44\,000}{118.8} = 370\text{K}$

This will be the temperature at which the sign of $\Delta G^{\ominus}$ changes. At higher temperatures $T\Delta S^{\ominus} > \Delta H^{\ominus}$, so $\Delta G^{\ominus}$ will be negative and the process becomes spontaneous. (Note: Because we have rounded off the values of $\Delta H^{\ominus}$ and $S^{\ominus}$, our calculated value for T, at 370K or 97°C, does not correspond exactly to the normal boiling point of water.)

Activity 28.2

Use the information in Table 28.1 to help you answer the following questions.

1 How do you account for the difference in molar entropy between liquid water and water vapour?

2 Calculate the standard molar enthalpy change and entropy change for the following reaction. At what temperatures, if any, is this process spontaneous?

$$\text{C(graphite)} + H_2O(g) \longrightarrow CO(g) + H_2(g)$$

28.8 Free energy change and temperature

In the last section we looked at examples where a change in temperature caused free energy change to become negative and so result in spontaneous reaction. We can simplify such calculations by plotting changes in free energy with temperature graphically. Consider the example in Activity 28.2, question 2, i.e. the formation of 'water gas' from carbon and steam.

$$\text{C(graphite)} + H_2O(g) \longrightarrow CO(g) + H_2(g) \qquad \Delta H^\ominus = +131.3\ \text{kJ}, \quad \Delta S^\ominus = +133.8\ \text{J mol}^{-1}\ \text{K}^{-1}$$

To calculate the standard enthalpy change for this reaction we used the standard enthalpies of formation of water vapour and carbon monoxide. The standard enthalpy and entropy changes for these reactions are as follows.

$$\text{C(graphite)} + \tfrac{1}{2}O_2(g) \longrightarrow CO(g) \qquad \Delta H^\ominus = -110.5\ \text{kJ}, \quad \Delta S^\ominus = +89.4\ \text{J mol}^{-1}\ \text{K}^{-1}$$

$$H_2(g) + \tfrac{1}{2}O_2(g) \longrightarrow H_2O(g) \qquad \Delta H^\ominus = -241.8\ \text{kJ}, \quad \Delta S^\ominus = -44.4\ \text{J mol}^{-1}\ \text{K}^{-1}$$

The variation of $\Delta G^\ominus$ with temperature for the overall process and for each of these steps is shown graphically in Figure 28.1.

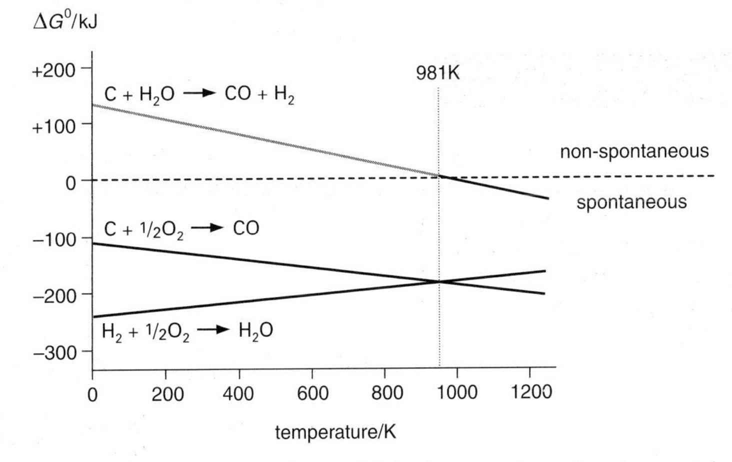

Figure 28.1
Variation in $\Delta G^\ominus$ with temperature

You can see that the temperature at which the reaction of carbon with steam becomes spontaneous, 981K, is also the point at which the lines for the formation of carbon monoxide and water intersect. In these steps, hydrogen and carbon are competing for oxygen. The reaction with the more negative free energy change will win the battle and so steam will not oxidise carbon spontaneously below the crossover temperature of 981K.

Ellingham diagrams extend this idea of showing graphically the competition between elements for oxygen upon the extraction of metals from their oxides.

Activity 28.3 Metal extraction

Figure 28.2 shows an Ellingham diagram in which the variation in free energy change on forming various oxides (using one mole of O_2 in each case) is plotted against temperature.

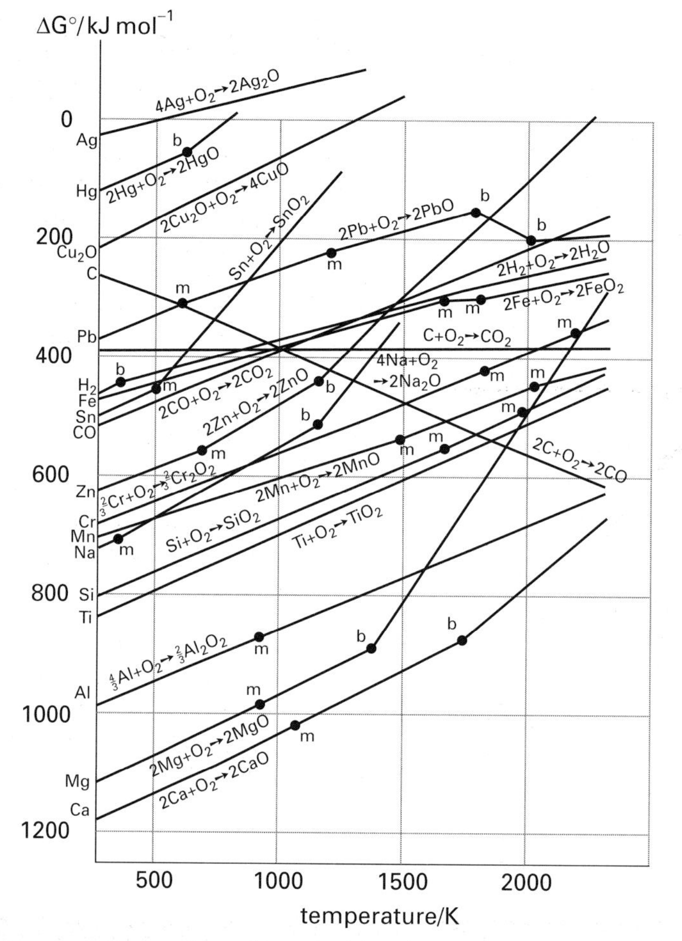

Figure 28.2
An Ellingham diagram

a) How does the stability of most oxides change on increasing the temperature?

b) Name one oxide which does not follow the above trend. How do you account for the difference?

c) Name one metal which may be obtained by simply heating its oxide in a bunsen flame (about 1000–1100 K).

Many metals may be obtained by heating their oxides with carbon but the temperatures needed for spontaneous reaction vary considerably.

d) What is the minimum temperature required for the spontaneous reduction of iron oxide by carbon? Give alternative equations for this process.

Sometimes, in order to make a metal of the desired purity, its oxide may be reduced by heating with another metal.

e) Which metal will reduce titanium(IV) oxide at temperatures similar to that used for carbon reduction?

Activity 28.3 Metal extraction continued

Many of the lines in this diagram change slope sharply at particular temperatures. How do you account for the following:

f) the decrease in slope for the H_2O line above 373K?
g) the increase in slope for the MgO line above 1400K?

Focus 28b

1. **Free energy change** relates the entropy change and enthalpy change for a given process to the overall enthalpy change for the universe.
2. Spontaneous reactions are always accompanied by a decrease in free energy.
3. Since free energy change is temperature dependent, the direction of spontaneous change may be reversed by heating or cooling.
4. Ellingham diagrams show the temperature dependence of free energy change graphically for the reaction of elements with oxygen.
5. Free energy change may be related to the equilibrium constant and to the e.m.f. of an electrochemical cell.

28.9 Free energy change, equilibrium constant and cell e.m.f.

Free energy change must be related to the equilibrium constant for a system. A large negative value for $\Delta G^\ominus$ indicates a spontaneous reaction, i.e. the products predominate at equilibrium and K_c will be large. Conversely, if $\Delta G^\ominus$ is positive the reaction is non-spontaneous, little product will be formed and K_c will be small.

The actual mathematical relationship between the free energy change and equilibrium constant is given by the following equation.

$$\Delta G^\ominus = -RT\log_e K$$

As expected, the bigger the value of K, the more negative is the free energy change.

In a working electrochemical cell, since spontaneous changes are occurring, $\Delta G^\ominus$ must be negative. The relationship between the free energy change and the cell e.m.f., $E^\ominus$ is given by:

$$\Delta G^\ominus = -nFE^\ominus$$

where n is the number of electrons transferred and F is the Faraday constant. As you can see, a negative free energy change will only result from a positive cell e.m.f. (If you are a mathematician, you may care to use the above equations to derive the relationship between the e.m.f. of a cell and the equilibrium constant of its redox system.)

Questions on Chapter 28

1 Explain in molecular terms why a hot cup of coffee cools down to the temperature of its surroundings but never warms up again spontaneously.

2 The following data refers to phosphorus(III) chloride, PCl_3.

	$\Delta H_f^\ominus$/kJ mol^{-1}	$S^\ominus$/J mol^{-1} K^{-1}
vapour state	–287.0	311.7
liquid state	–319.7	217.1

a) Calculate the standard enthalpy change and entropy change when one mole of this liquid vaporises.
b) Use your results to estimate the boiling point of phosphorus(III) chloride at 1 atmosphere pressure.

3 Using the information in Table 28.1, explain why the conversion of graphite into diamond can never be achieved at atmospheric pressure.

4 The following reaction takes place in a blast furnace:

$$C(graphite) + CO_2(g) \longrightarrow 2CO(g)$$

a) Use the information given in Table 28.1 to calculate the standard enthalpy and entropy changes for this process. From your results predict the minimum temperature required for this reaction to occur spontaneously. Compare your answer with that predicted from the Ellingham diagram shown in Figure 28.2.
b) Why is the free energy change on forming carbon dioxide from its elements much less temperature dependent than the formation of carbon monoxide?

Questions on Chapter 28 *continued*

5 **a)** Write an equation to show the relationship between the free energy change, enthalpy change and entropy change for any process at a temperature of T (K).

b) The standard free energy change and enthalpy change for the formation of magnesium oxide from its elements are –570 kJ mol^{-1} and –602 kJ mol^{-1} respectively. Calculate a value for the standard entropy change of this process and explain its sign in terms of the relative 'order' of the reactants and products.

6 Caesium chloride readily dissolves in water.

a) Use the following data to calculate the standard enthalpy change on dissolving 1 mole of this ionic solid in water:

	$\Delta H^\ominus$/kJ mol^{-1}
$Cs^+(g) + Cl^-(g) \longrightarrow CsCl(s)$	–645
$Cs^+(g) \longrightarrow Cs^+(aq)$	–276
$Cl^-(g) \longrightarrow Cl^-(aq)$	–364

b) Predict, with reasons, the sign of the entropy change on dissolving caesium chloride in water.

c) What is the sign of the free energy change for dissolving caesium chloride in water? Explain your answer.

d) What is likely to happen to the solubility of caesium chloride in water as the temperature falls? Again explain your reasoning.

7 Photosynthesis involves the formation of sugars from carbon dioxide and water, e.g.

$$6CO_2(g) + 6H_2O(l) \longrightarrow C_6H_{12}O_6(s) + 6O_2(g)$$

a) Predict, with reasons, the sign of the entropy change which occurs during photosynthesis.

b) The photosynthesis reaction is endothermic. Under what conditions, if any, can it be spontaneous?

c) How do you explain the fact that plants carry out photosynthesis only in sunlight?

Comments on the activities

Activity 28.1

a) As the sugar dissolves, its regular lattice structure is broken down and the structure will be more disordered.

b) Solid ice has a more ordered structure than liquid water.

c) Water vapour is far more disordered than liquid water.

d) The air molecules which were previously restricted to the space inside the car tyre can now move freely outside. There are, therefore, many more possible arrangements and the system therefore becomes more disordered.

e) $NH_3(g) + HCl(g) \longrightarrow NH_4Cl(s)$

Two moles of gas form an ordered solid crystal lattice so the system becomes more ordered.

f) $2C(s) + O_2(g) \longrightarrow 2CO(g)$

Gases are far more disordered than solids. Since two moles of gas are formed for each mole of gas used up, an increase in disorder is likely.

g) $C_6H_{12}O_6(s) + 6O_2(g) \longrightarrow 6CO_2(g) + 6H_2O(l)$

The total number of moles of gas in this reaction does not change but 1 mole of more ordered solid is used up whilst 6 moles of less ordered liquid is formed. The overall result of this respiration process should be an increase in disorder.

Activity 28.2

a) Water vapour has a higher molar entropy than liquid water because gases have a more 'disordered' structure than liquids.

b) To calculate the enthalpy change for

$C(graphite) + H_2O(g) \longrightarrow CO(g) + H_2(g)$

add the following

$C(graphite) + \frac{1}{2}O_2(g) \longrightarrow CO(g) \qquad \Delta H^\ominus = -110.5$ kJ

$H_2O(g) \longrightarrow H_2(g) + \frac{1}{2}O_2(g) \qquad \Delta H^\ominus = +241.8$ kJ

$C(graphite) + H_2O(g) \longrightarrow CO(g) + H_2(g) \qquad \Delta H^\ominus = +131.3$ kJ

To calculate the entropy change:

	C(graphite) +	$H_2O(g)$ ⟶	CO(g) +	$H_2(g)$
$S^\ominus$	5.74	188.7	197.6	130.6

$\Delta S^\ominus = +133.8$ J mol^{-1} K^{-1}

$$\Delta G^\ominus = \Delta H^\ominus - T\Delta S^\ominus$$
$$= +131.3 - T\left(+\frac{133.8}{1000}\right)$$
$$= +131.3 - 0.1338T$$

At low temperatures the positive value of $\Delta H^\ominus$ will dominate, leading to a positive free energy change and the reaction will not occur spontaneously. As the temperature rises, however, the positive entropy change will eventually outweigh the effect of the enthalpy change. The changeover to a spontaneous process will occur when the free energy change is zero.

$$\Delta G^\ominus = +131.3 - 0.1338T = 0$$
$$0 = +131.3 - 0.1338T$$
$$T = \frac{131.3}{0.1338} = 981\text{K}$$

Comments on the activities *continued*

Thus, at temperatures above about 708°C, carbon reacts spontaneously with steam to give a mixture of carbon monoxide and hydrogen. This is known as 'water gas' and is a useful fuel.

Activity 28.3

a) As their free energy change of formation becomes more positive, most oxides become less stable on heating. This is to be expected as oxide formation generally results in a decrease in the number of moles of gas in the system and hence a decrease in entropy.

b) Carbon monoxide, CO, becomes more stable on heating. Its formation involves an increase in the total number of moles of gas and hence a positive entropy change.

c) When the free energy change becomes positive the oxide will decompose spontaneously on heating. Thus silver and mercury may be obtained by simply heating their oxides to about 400 K and 1200 K respectively.

d) From the diagram the lines for CO and CO_2 both intersect the iron oxide line at about 1000K. Above this temperature, therefore, the following reactions are spontaneous:

$$\begin{array}{l} 2FeO \longrightarrow 2Fe + O_2 \\ 2C + O_2 \longrightarrow 2CO \\ \hline 2FeO + 2C \longrightarrow 2Fe + 2CO \end{array}$$

$$\begin{array}{l} 2FeO \longrightarrow 2Fe + O_2 \\ C + O_2 \longrightarrow CO_2 \\ \hline 2FeO + C \longrightarrow 2Fe + CO_2 \end{array}$$

In addition to these possibilities the line for the conversion of CO into CO_2 also intersects the FeO line at about the same temperature, so a third possibility is:

$$\begin{array}{l} 2FeO \longrightarrow 2Fe + O_2 \\ 2CO + O_2 \longrightarrow 2CO_2 \\ \hline 2FeO + 2CO \longrightarrow 2Fe + 2CO_2 \end{array}$$

e) Magnesium will reduce titanium(IV) oxide spontaneously at about the same temperature as carbon, i.e. just below 2000 K.

f) At 373K, water boils to give a gas which has a higher entropy than the liquid. The entropy change for the formation of H_2O from its gaseous elements is therefore less negative and hence $-T\Delta S$ has less effect on the value of ΔG.

g) Whereas the slope for water decreased with temperature, that for the MgO line increases sharply at about 1400 K. This must indicate a decrease in the entropy change on forming the oxide. In fact, it represents the increase in entropy of magnesium as it melts (1380 K).

Index